Handbook of Bioequivalence Testing

Second Edition

DRUGS AND THE PHARMACEUTICAL SCIENCES

A Series of Textbooks and Monographs

Series Executive Editor
James Swarbrick
PharmaceuTech, Inc.
Pinehurst, North Carolina

Recent Titles in Series

Handbook of Bioequivalence Testing, Second Edition, *Sarfaraz K. Niazi*

Generic Drug Product Development: Solid Oral Dosage Forms, Second Edition, *edited by Leon Shargel and Isadore Kanfer*

Drug Stereochemistry: Analytical Methods and Pharmacology, Third Edition, *edited by Krzysztof Jozwiak, W. J. Lough, and Irving W. Wainer*

Pharmaceutical Powder Compaction Technology, Second Edition, *edited by Metin Çelik*

Pharmaceutical Stress Testing: Predicting Drug Degradation, Second Edition, *edited by Steven W. Baertschi, Karen M. Alsante, and Robert A. Reed*

Pharmaceutical Process Scale-Up, Third Edition, *edited by Michael Levin*

Sterile Drug Products: Formulation, Packaging, Manufacturing and Quality, *Michael J. Akers*

Freeze-Drying/Lyophilization of Pharmaceutical and Biological Products, Third Edition, *edited by Louis Rey and Joan C. May*

Oral Drug Absorption: Prediction and Assessment, *edited by Jennifer B. Dressman and Christos Reppas*

Generic Drug Product Development: Specialty Dosage Forms, *edited by Leon Shargel and Isadore Kanfer*

Generic Drug Product Development: International Regulatory Requirements for Bioequivalence, *edited by Isadore Kanfer and Leon Shargel*

Active Pharmaceutical Ingredients: Development, Manufacturing, and Regulation, Second Edition, *edited by Stanley Nusim*

Pharmaceutical Statistics: Practical and Clinical Applications, Fifth Edition, *edited by Sanford Bolton and Charles Bon*

Biodrug Delivery Systems: Fundamentals, Applications and Clinical Development, *edited by Mariko Morishita and Kinam Park*

Biodrug Delivery Systems: Fundamentals, Applications and Clinical Development, *edited by Mariko Morishita and Kinam Park*

Handbook of Pharmaceutical Granulation Technology, Third Edition, *edited Dilip M. Parikh*

A complete listing of all volumes in this series can be found at ***www.crcpress.com***

Handbook of Bioequivalence Testing

Second Edition

Sarfaraz K. Niazi
Chairman and CEO, Therapeutic Proteins International, LLC
Chicago, Illinois, USA

CRC Press is an imprint of the
Taylor & Francis Group, an **informa** business

First published in paperback 2024

First published 2015
by CRC Press
2385 NW Executive Center Drive, Suite 320, Boca Raton FL 33431

and by CRC Press
4 Park Square, Milton Park, Abingdon, Oxon, OX14 4RN

CRC Press is an imprint of Taylor & Francis Group, LLC

Publisher's Note
The publisher has gone to great lengths to ensure the quality of this reprint but points out that some imperfections in the original copies may be apparent.

ISBN: 978-1-4822-2637-9 (hbk)
ISBN: 978-1-03-291735-1 (pbk)
ISBN: 978-0-429-15665-6 (ebk)

DOI: 10.1201/b17582

Visit the Taylor & Francis Web site at
http://www.taylorandfrancis.com

and the CRC Press Web site at
http://www.crcpress.com

To

My friend Barack Obama, who stood by me in helping the world by making drugs more affordable.

Other Books by the Author

Disposable Bioprocessing Systems, CRC Press, Boca Raton, FL, 2012
Filing Patents Online, CRC Press, Boca Raton, FL, 2003
Handbook of Bioequivalence Testing, Informa, New York, 2007
Handbook of Biogeneric Therapeutic Proteins: Manufacturing, Regulatory, Testing and Patent Issues, CRC Press, Boca Raton, FL, 2005
Handbook of Pharmaceutical Manufacturing Formulations, Volume 1, Second Edition: Compressed Solids, Informa, New York, 2009
Handbook of Pharmaceutical Manufacturing Formulations, Volume 2, Second Edition: Uncompressed Solids, Informa, New York, 2009
Handbook of Pharmaceutical Manufacturing Formulations, Volume 3, Second Edition: Liquid Products, Informa, New York, 2009
Handbook of Pharmaceutical Manufacturing Formulations, Volume 4, Second Edition: Semisolid Products, Informa, New York, 2009
Handbook of Pharmaceutical Manufacturing Formulations, Volume 5, Second Edition: Over the Counter Products, Informa, New York, 2009
Handbook of Pharmaceutical Manufacturing Formulations, Volume 6, Second Edition: Sterile Products, Informa, New York, 2009
Handbook of Preformulation: Chemical, Biological and Botanical Drugs, Informa Healthcare, New York, 2006
Love Sonnets of Ghalib: Translations, *Explication and Lexicon*, Ferozsons Publishers, Lahore, Pakistan, 2002 and Rupa Publications, New Delhi, India, 2002
Pharmacokinetic and pharmacodynamic modeling in early drug development, in Charles G. Smith and James T. O'Donnell (eds.), *The Process of New Drug Discovery and Development* (2nd edn.), CRC Press, New York, 2004
Textbook of Biopharmaceutics and Clinical Pharmacokinetics, John Wiley & Sons, New York, 1979
Textbook of Biopharmaceutics and Clinical Pharmacokinetics, The Book Syndicate, Hyderabad, India, 2010
The Omega Connection, Esquire Press, Springfield, IL, 1982
There is No Wisdom: Selected Love Poems of Bedil, Translations from Darri Farsi, Sarfaraz K. Niazi and Maryam Tawoosi, Ferozsons Private (Ltd), Lahore, Pakistan, 2013
Wine of Love: Complete Translations of Urdu Persian Love Poems of Ghalib, Sarfaraz K. Niazi, Ferozsons Private (Ltd), Lahore, Pakistan, 2013
Wine of Passion: Love Poems of Ghalib, Ferozsons (Pvt) Ltd., Lahore, Pakistan, 2010
Biosimilar and Interchangeable Products: From Cell Lines to Commercial Launch, CRC Press, Boca Raton, FL, 2015

Contents

List of Figures ... xxxv
List of Tables ... xxxvii
Preface ... xliii
Acknowledgments ... xlix
Author ... li

Chapter 1 Historical Perspective on Generic Pharmaceuticals ... 1

In the Beginning ... 1
History of Patent Medicine ... 2
Quackery ... 3
Regulations ... 5
Generic Era ... 6
Bioequivalence ... 7
Challenging Bioequivalence ... 8
The Big Future ... 8
Legal Challenges ... 9
Submarine Patenting ... 9
Conclusion ... 10

Chapter 2 Physicochemical Basis of Bioequivalence Testing ... 13

Background ... 13
Chemical Properties ... 14
Ionization ... 14
Henderson–Hasselbalch Equation ... 14
Partitioning ... 15
Distribution Coefficient ... 16
Chemical Structure and Form ... 19
Lipophilizing Modifications ... 19
Salt Forms ... 21
Physical Properties ... 22
Crystal Morphology ... 23
Polymorphism ... 24
Amorphous Forms ... 28
Solvates ... 29
Hydrates ... 29
Complexation ... 31
Surface Activity ... 31
Hygroscopicity ... 33
Particle Size ... 33
Solubility ... 35
Solubility ... 36
Molecular Size ... 38

Dissolution 40
Diffusion Model 40
Convection Model 40
Surface Reaction Model 41
Cube Root Model 41
Tablet Dissolution Model 42
Noyes–Whitney Model 42
Dissolution Factors 44
Concentration Gradient 44
Dissolution Constant 44
Dissolution Testing 45

Chapter 3 Drug Delivery Factors 47

Background 47
Solid Dosage Forms Considerations 48
Particle Size 48
Sieve Analysis 49
Particle Size Distribution 49
Surface Area 50
Porosity 50
Real, Tapped, and Bulk Density 51
True Density 51
Problems in Powder Handling 52
Mixing of Powders 52
Powder Flow Properties 53
Electrostaticity 53
Caking 54
Relative Humidity 54
Polymorphism 54
Milling 54
Powders 55
Oral Powders 55
Capsules 55
Inhalers and Lung Delivery 56
Modified-Release Products 56
Coated Particles 58
Tablets 58
Solutions 58
Solubility 59
Emulsion Formulations 60
Suspensions 60
Controlled-Release Dosage Forms 61
Therapeutic Systems 62
Evaluation of Drug Delivery Systems 63
Chemical Content 63
Content Uniformity 64
Presence of Contaminants 64
Disintegration Test 64
Dissolution Test 65

Absorption Principles 66
Passive Diffusion 66
Active Transport 67
Solvent Drag 67
Facilitated Transport 67
Ion-Pair Transport 68
Pinocytosis 68
Absorption Factors 68
Gastrointestinal Fluids 69
Gastric Emptying 70
Intestinal Transit 73
Blood Flow 73
Gastrointestinal Drug Biotransformation 73
Food Interactions 74
Pathophysiological Disorders 75
Age 77
First-Pass Biotransformation 77
Sublingual/Buccal Administration 78
Rectal Administration 78
Intravenous Administration 78
Intra-Arterial Administration 79
Intramuscular Administration 79
Subcutaneous Administration 80
Percutaneous Administration 81
Pulmonary Administration 82
Ophthalmic Administration 84
Nasal Administration 85
Miscellaneous Routes of Administration 86

Chapter 4 Pharmacokinetic/Pharmacodynamic (PK/PD) Modeling 87

Background 87
Pharmacokinetic Modeling Studies 87
Compartment Pharmacokinetic Modeling 87
Physiologically Based Pharmacokinetic Studies 90
Bioequivalence and Systemic Exposure Models 91
Deconvolution Techniques 92
Pharmacological Evaluation of Bioavailability 93

Chapter 5 Bioequivalence Testing Rationale and Principles 95

Background 95
Equivalence Documentation for Marketing Authorization 96
Overview of Bioequivalence Testing 97
How to Demonstrate Bioequivalence? 102
Rationale for Estimation 104
Evidence to Measure Bioequivalence 105
Measurement Indices 107
Dose Selection 107
Multiple Strengths of Solid Oral Dosage Forms 107
Manufacturing of Pilot Batch ("Biobatch") 107
Dosing by Labeled Concentration 108

Single-Dose versus Multiple-Dose Studies 108
Guidelines on the Design of a Single-Dose Study 108
Guidelines for Multiple-Dose Study 109
Fed versus Fasted State 109
Pharmacological Endpoint Studies 110
Clinical Endpoint Studies 110
Analytical Methods 111
Assay Consideration 111
Concentration Range and Linearity 111
Limit of Detection 112
Limit of Quantitation 112
Specificity 112
Accuracy: Recovery 112
Precision 112
Analyte Stability 112
Analytical System Stability 113
Quality Control Samples 113
Replicate and Repeat Analyses 113
Summary of Samples to Be Run with Each Analysis 113
Prior Review 113
Record Maintenance 114
Pharmacokinetic and Statistical Considerations in Study Design 115
Sampling Time Considerations 115
Protein Binding 116
Subject Number 116
Crossover and Parallel Design Considerations 116
Duration of Washout Time for Crossover Study 117
Pivotal Parameters for Blood Level Bioequivalence 117
Area under the Curve Estimates 117
AUC Measurements 118
Trapezoidal Rule 118
Integration Method 118
Computer Applications 119
Rate of Absorption 119
Determination of Product Bioequivalence 119
Errors in BE Studies 119
Failures in Bioequivalence Studies 120
Risk-Based Bioequivalence 122
Health Risk Categories 122
Definition 122
Critical Dose or Narrow Therapeutic Index Drugs 123
Clarification of Testing Limits 127
Clarification on Requirements 127
Typical Examples of Complex Bioequivalence 131
Digoxin 131
Levothyroxine 131
Warfarin Sodium 132
Albuterol Metered-Dose Inhalers 133
Bioequivalence Surrogates 133

In Vitro Systems 133
Disintegration of Dosage Form 133
Dissolution Tests 134
Everted Intestinal Sac 135
Isolated Perfused Liver 135
In Situ Methods 136
In Vivo Systems 136
LD50 Comparisons 136
Thiry–Vella Loop 136
Hepatobiliary Cannulation 136
Choice of Animal Species 136
Absorption Profiling 137
Statistical Analysis 138
Untransformed Data 139
Logarithmically Transformed Data 140
Animal Drug Bioequivalence Testing 141
Reference Product 142
Waiver of In Vivo Bioequivalence Study 142
Species Selection 142
Subject Characteristics 143
Human Food Safety Considerations 143
Locally Acting Gastrointestinal Drugs 144
Fed Bioequivalence Studies 145
Food Effects on Drug Products 145
Recommendations 146
For Immediate-Release Drugs 146
For Modified-Release Products 146
Study Design 147
General Design 147
Subject Selection 147
Dosage Strength 147
Test Meal 147
Administration 148
Sample Collection 148
Data Analysis and Labeling 148
Other Considerations 149
Sprinkles 149
Special Vehicles 149

Chapter 6 Bioequivalence Waivers 151

Background 151
IVIVC 152
Definitions 152
Purpose of IVIVC 152
Levels 153
Level A Correlation 153
Level B Correlation 154
Level C Correlation 154

Multiple Level C Correlation 154
Level D Correlation 154
Modeling 154
Relative Advantages of Modeling Methods 156
Deconvolution Model 157
Differential Equation–Based Approach 157
Predictability of Correlation 158
Internal Predictability 158
External Predictability 158
Limitation of Predictability Metrics 159
IVIVC Development 159
Step 1 159
Step 2 160
Step 3 161
BCS 161
Solubility 165
Permeability 165
Dissolution 165
Method of Classification 167
Determining Drug Substance Solubility Class 167
Determining Drug Substance Permeability Class 167
Pharmacokinetic Studies in Humans 168
Mass Balance Studies 168
Absolute Bioavailability Studies 168
Instability in the Gastrointestinal Tract 171
Determining Drug Product Dissolution Characteristics and Dissolution Profile Similarity 171
Additional Considerations 172
Excipients 172
Prodrugs 172
Stereochemistry 172
Exceptions 173
Narrow Therapeutic Range Drugs 173
Products Designed to Be Absorbed in the Oral Cavity 173
Reasons for Failure of Some IVIVC 173
Limitations in the IVIVC Arising from the In Vivo Data 173
Enteric-Coated Multiple Unit Dosage Form 174
Parenteral Controlled- or Sustained-Release Drug Delivery System 174
Buccal Tablet 174
Transdermal Drug Delivery System 174
Suppository 174
Nasal Drug Delivery System 174
Colonic Drug Delivery System 175
Biodegradable Parenteral Delivery System 175
Current Trends 175
Application in Drug Delivery System 175
Early Stage Drug Delivery 176
Formulation Assessment: In Vitro Dissolution 176
Dissolution Specifications 176
Mapping 176

Future Biowaivers 177
Parenteral Drug Delivery 177
Potent Drugs and Chronic Therapy 177
Regulatory Aspects 177
INDs/NDAs 177
ANDAs 178
Postapproval Changes 178
Manufacturing Changes 178
Biowaiver Risks 180
Data to Support Biowaiver 183
Data Supporting High Solubility 183
Data Supporting High Permeability 183
Data Supporting Rapid and Similar Dissolution 184
Surrogate Methods 184
Permeability Assays 185
Permeability Assay Protocol 185
PAMPA 186
Caco-2 Drug Transport Assays 187
Animal Model Testing 189
European Perspective on Biowaiver Criteria 190
Strength to Be Investigated 190
Linear Pharmacokinetics 190
Nonlinear Pharmacokinetics 191
Bracketing Approach 191
Fixed Combinations 191
BCS-Based Biowaiver 191
Introduction 191
Summary 192
Drug Substance 192
Drug Product 193

Chapter 7 Statistical Evaluation of Bioequivalence Data 195

Background 195
Experimental Designs 196
Comparisons of Two Formulations 196
Comparisons of More than Two Formulations 197
Number of Subjects in Study 197
Statistical Model 198
Statistical Approaches for Bioequivalence 199
Average Bioequivalence 199
Population Bioequivalence 199
Individual Bioequivalence 200
Standards 201
σ_{T0} and σ_{W0} 201
θ_P and θ_I 203
Study Design 203
Experimental Design 203
Nonreplicated Designs 203
Replicated Crossover Designs 203

Sample Size Determination 204
Computer Software for Power Analysis in Bioequivalence Trial with Interim Analysis 204
Power Analysis Using StudySize 3.0 in Bioequivalence Trial with Interim Analysis 204
Sample Size and Dropouts 210
Statistical Analysis 210
Logarithmic Transformation 210
General Procedures 210
Clinical Rationale 211
Pharmacokinetic Rationale 211
Presentation of Data 212
Data Analysis 212
Average Bioequivalence 212
Population Bioequivalence 214
Individual Bioequivalence 218
Methods for Obtaining Confidence Intervals for Individual and Population Bioequivalence Criteria 222
Individual Bioequivalence Method 1: Constrained REML 222
Individual Bioequivalence Method 2: Method of Moments 223
Population Bioequivalence 224
Miscellaneous Issues 225
Studies in Multiple Groups 225
Carryover Effects 225
Choice of Specific Replicated Crossover Designs 226
Reasons Unrelated to Carryover Effects 226
Reasons Related to Carryover Effects 227
Two-Period Replicated Crossover Designs 227
Outlier Considerations 227
Product Failure 228
Subject-by-Formulation Interaction 228
Discontinuity 228
Statistical Analysis of Bioequivalence of Highly Variable Drugs 228
Background 230
Change of Sample Size with Within-Subject Variation: Effects of Regulatory Requirements 233
Comparison of Sample Sizes Required by EMA and FDA 234
Designing Bioequivalence Studies for Highly Variable Drugs 236
Appendix 7A: SAS GLM Procedure 237
Overview 237
PROC GLM for Unbalanced ANOVA 238
Appendix 7B: Bioequivalence Testing Software 241

Chapter 8 Regulatory Inspection Process 245

Background 245
Protocols 245
Background 245
References 245
Policy 245
Procedures 246

Productivity Documentation 246
Background 246
Policy 246
Procedures 247
BE Studies 247
Dissolution Data (DIS) 247
Other (OTH) 247
Protocols 248
Controlled Correspondence 248
Processing of Work 248
Inspections 249
Background 249
Methods Validation for Abbreviated New Drug Applications 249
Background 249
References 250
Policy 250
Procedures 251
Regulatory Audit of Bioequivalence Studies Submitted 252
Part I: Background 252
Part II: Implementation 253
Objectives 253
Program Management Instructions 253
Coverage 253
Process 253
Part III: Inspectional Operations 254
Inspectional 254
Investigational 255
Refusals 255
Findings 255
Part IV: Regulatory/Administrative Strategy 256
Clinical Testing 256
Analytical Testing 256
Bioequivalence Inspection Report 256
Part I: Facilities and Procedures (Clinical and Analytical) 256
Facilities (Clinical and/or Analytical) 256
Personnel 257
Specimen Handling and Integrity 257
Electronic Records and Signatures 258
Clinical Data and Operations 259
General 259
Inspection Procedures 259
Study Responsibility and Administration 259
Protocol 259
Subjects' Records 260
Other Study Records 260
Consent of Human Subjects 261
Institutional Review Board 261
Sponsor 261
Test Article Accountability 261
Records Retention 262
Abbreviated Report Format 262

Analytical Data and Operations 263
Pre-Study Analysis 263
Protocol Acceptance 264
Equipment 264
Analytical Methods Validation 264
Sample Analyses 265
For Antibiotic Analyses 266
For Radiometric Analyses 267
Data Handling and Storage 267
Bioequivalence Testing Report Summary 267
Good Laboratory Practices 268
Inspections 269

Chapter 9 Fed Bioequivalence Studies 277

Introduction 277
Background 277
Potential Mechanisms of Food Effects on Bioavailability 277
Food Effects on Drug Products 277
Recommendations for Food-Effect Bioavailability and Fed Bioequivalence Studies 278
Immediate-Release Drug Products 278
INDs/NDAs 278
ANDAs 278
Modified-Release Drug Products 279
INDs/NDAs 279
ANDAs 279
Study Considerations 279
General Design 279
Subject Selection 280
Dosage Strength 280
Test Meal 280
Administration 280
Sample Collection 281
Data Analysis and Labeling 281
Other Considerations 282
Sprinkles 282
Special Vehicles 283

Chapter 10 Topical Drugs 285

Inactive Ingredients 285
Waiver of Bioequivalence 285
Bioequivalence Approaches 285
Dermatopharmacokinetic Approaches 286
Performance and Validation of the Skin-Stripping Technique 286
Sample Pilot Study 287
DPK Bioequivalence Study Protocol 288
Protocol and Subject Selection 288
Application and Removal of Test and Reference Products 289
Sites and Duration of Application 289

Collection of Sample 289
Procedure for Skin Stripping 290
Metrics and Statistical Analyses 291
Pharmacodynamic Approaches 291
In Vitro Release Approaches (Lower Strength) 291
In Vitro Release: Extension of the Methodology 292
Systemic Exposure Studies 292

Chapter 11 Bioequivalence of Nasal Products 293

Introduction 293
Background 293
Local Delivery Bioavailability/Bioequivalence Concepts 293
Systemic Exposure and Systemic Absorption Bioavailability/Bioequivalence Concepts 294
CMC Tests and In Vitro Bioavailability Tests (Noncomparative) versus Bioequivalence Tests (Comparative) 294
Formulation and Container and Closure System 295
Formulation 295
Container and Closure System 295
Documentation of Bioavailability and Bioequivalence 296
INDs/NDAs 296
ANDAs 296
Solution Formulations 296
Suspension Formulations with PK Systemic Exposure Data 297
Suspension Formulations without PK Systemic Exposure Data 297
Postapproval Change 297
Bioavailability and Bioequivalence: In Vitro Studies 297
Batches and Drug Product Sample Collection 297
INDs/NDAs 297
ANDAs 298
Tests and Metrics 298
Dose or Spray Content Uniformity through Container Life 299
Droplet and Drug Particle Size Distribution 299
Spray Pattern 301
Plume Geometry 302
Priming and Repriming 302
Tail-Off Profile 302
Bioavailability and Bioequivalence: Clinical Studies for Local Delivery 303
General Information 303
Bioequivalence Clinical Study Endpoints 303
Clinical Study Batches 303
Clinical Bioequivalence Study Designs and Subject Inclusion Criteria 303
Traditional Treatment Study 304
Day(s) in the Park (Outdoors) Study 304
Environmental Exposure Unit Study 304
Bioavailability and Bioequivalence: PK Systemic Exposure Studies 305
Bioavailability and Bioequivalence: PD or Clinical Studies for Systemic Absorption 305
General Information 305
Bioequivalence Study Endpoints for Corticosteroids 306

Clinical Study Batches 306
Clinical Study Designs and Subject Inclusion Criteria 306
Statistical Analyses 307
In Vitro Bioavailability Data 307
In Vitro Bioequivalence Data: Nonprofile Analyses Using a Confidence Interval Approach 307
Study Protocol 307
Criterion for Comparisons, Confidence Interval, and Bioequivalence Limit 308
In Vitro Bioequivalence Data: Supportive Nonprofile and Profile Analyses 309
In Vitro Bioequivalence Data: Profile Analyses Using a Confidence Interval Approach 310
Study Protocol 310
Criterion for Comparison 310
Determining a 95% Upper Bound 311
Specification of the Upper Limit 311
Multiple Strengths 311
Solution Formulation Nasal Sprays 311
Suspension Formulation Nasal Sprays 312
Smaller Container Sizes 312
Appendix 11A: In Vitro Profile Comparison Procedure Based on Chi-Square Differences 313
Appendix 11B: Determination of the 95% Upper Bound for In Vitro Profile Comparisons 314

Chapter 12 Bioequivalence of Complimentary and Alternate Medicines 315

Background 315
FDA Perspective on CAM 318

Chapter 13 Bioequivalence of Biosimilar Products 327

Background 327
U.S. Regulations 329
Biosimilarity 329
Basic Understanding 331
Scientific Basis 332
Manufacturing Process Considerations 332
Stepwise Approach 333
Mechanism of Action 333
Totality of Evidence 333
Product Specificity 334
Analytical Methodology 334
Functional Assays 335
Animal Data 335
Animal PK and PD Measures 336
Animal Immunogenicity Studies 336
Clinical Studies 336
Human Pharmacology Data 337
Clinical Immunogenicity Assessment 338
Clinical Safety and Effectiveness Data 339
Clinical Study Design Issues 340

Extrapolation of Clinical Data across Indications 342
Postmarketing Considerations 342
Summary Considerations 343
Extent of Similarity 343
Practical Issues 343
Quantitative Evaluation of Bioequivalence 344
Study Design 345
Statistical Methods 345
Special Considerations 346
Criteria, Design, and Statistical Methods for Biosimilarity 346
Criteria for biosimilarity 346
Study design 347
Statistical methods 347
Interchangeability 347
Definition and basic concepts 347
Switching and alternating 348
Study design 348
Biosimilarity Index 348
European Perspective 351
Background 351
Clinical Studies 353
Pharmacokinetic Studies 354
Pharmacodynamic Studies 354
Efficacy Trials 355
Study Designs 355
Efficacy Endpoints 355
Clinical Safety 356
Product Specific European Guidelines 357
Human Follicle-Stimulating Hormone 357
Toxicological Studies 358
Pharmacokinetic Studies 358
Pharmacodynamic Studies 359
Clinical Efficacy 359
Clinical Safety 359
Interferon Beta 360
Clinical Studies 361
Pharmacokinetics 361
Pharmacodynamics 361
Clinical Efficacy 361
Clinical Safety 363
Monoclonal Antibodies 363
In Vivo Studies 365
Clinical Studies 366
Pharmacokinetics 366
Study Design 366
Doses 367
Routes of Administration 368
Sampling Times 368
PK Parameters of Interest 368
Timing of the PK Evaluation 368
Pharmacodynamics 369

Clinical Efficacy ... 369
Clinical Studies ... 370
Erythropoietins ... 371
Pharmacodynamics Studies ... 372
Toxicological Studies ... 372
Clinical Studies ... 372
Demonstration of Efficacy for Both Routes of Administration ... 373
Demonstration of Efficacy for One Route of Administration ... 374
Clinical Safety ... 375
Low Molecular Weight Heparins ... 375
Pharmacodynamic Studies ... 376
In Vitro Studies ... 376
In Vivo Studies ... 376
Toxicological Studies ... 376
Clinical Studies ... 377
Pharmacokinetic/Pharmacodynamic Studies ... 377
Clinical Efficacy ... 377
Clinical Safety ... 378
Interferon Alpha 2a or 2b ... 378
Nonclinical Studies ... 379
Pharmacodynamics Studies ... 379
In Vitro Studies ... 379
In Vivo Studies ... 379
Toxicological Studies ... 379
Clinical Studies ... 380
Pharmacokinetic Studies ... 380
Pharmacodynamic Studies ... 380
Efficacy ... 380
Patient Population ... 380
Study Design and Duration ... 380
Endpoints ... 380
Safety ... 380
Immunogenicity ... 381
Extrapolation of Evidence ... 381
Human G-CSF ... 381
Nonclinical Studies ... 381
Pharmacodynamic Studies ... 382
In Vitro Studies ... 382
In Vivo Studies ... 382
Toxicological Studies ... 382
Clinical Studies ... 382
Pharmacokinetic Studies ... 382
Pharmacodynamic Studies ... 382
Clinical Efficacy Studies ... 382
Clinical Safety ... 383
Growth Hormone ... 383
Nonclinical Studies ... 383
Pharmacodynamics Studies ... 384
In Vitro Studies ... 384
In Vivo Studies ... 384
Toxicological Studies ... 384

Clinical Studies 384
Pharmacokinetic Studies 384
Pharmacodynamic Studies 384
Clinical Efficacy Studies 384
Clinical Safety 385
Human Insulin 385
Nonclinical Studies 386
Pharmacodynamic Studies 386
In Vitro Studies 386
In Vivo Studies 386
Toxicological Studies 386
Clinical Studies 386
Pharmacokinetic Studies 386
Pharmacodynamic Studies 387
Clinical Efficacy Studies 387
Clinical Safety 387
Immunogenicity 387
Local Reactions 387

Chapter 14 Bioequivalence Testing: The U.S. Perspective 389

FDA Guidance on Bioequivalent Testing 389
Background 389
Procedures for Determining the Bioavailability or Bioequivalence of Drug Products 389
Requirements for Submission of Bioavailability and Bioequivalence Data 389
Criteria for Waiver of Evidence of In Vivo Bioavailability or Bioequivalence 395
Basis for Measuring In Vivo Bioavailability or Demonstrating Bioequivalence 397
Types of Evidence to Measure Bioavailability or Establish Bioequivalence 398
Guidelines for the Conduct of an In Vivo Bioavailability Study 399
Guidelines on the Design of a Single-Dose In Vivo Bioavailability or Bioequivalence Study 401
Guidance on the Design of a Multiple-Dose In Vivo Bioavailability Study 402
Correlation of Bioavailability with an Acute Pharmacological Effect or Clinical Evidence 403
Analytical Methods for an In Vivo Bioavailability or Bioequivalence Study 403
Inquiries regarding Bioavailability and Bioequivalence Requirements and Review of Protocols by the FDA 404
Applicability of Requirements regarding an "Investigational New Drug Application" 404
Procedures for Establishing or Amending a Bioequivalence Requirement 405
Criteria and Evidence to Assess Actual or Potential Bioequivalence Problems 406
Requirements for Batch Testing and Certification by the FDA 407
Requirements for In Vitro Testing of Each Batch 407
Requirements for Maintenance of Records of Bioequivalence Testing 407
Retention of Bioavailability Samples 407
Bioequivalence Studies with Pharmacokinetic Endpoints for Drugs Submitted under an ANDA: New Guidance 409
Introduction 409
Background 409

Establishing Bioequivalence....409
Pharmacokinetic Studies....409
General Considerations....409
Pilot Study....410
Pivotal Bioequivalence Studies....410
Study Designs....410
Study Population....410
Single-Dose Studies....411
Steady-State Studies....411
Bioanalytical Methodology....411
Pharmacokinetic Measure of Rate and Extent of Exposure....411
Fed Bioequivalence Studies....412
Sprinkle Bioequivalence Studies....412
Bioequivalence Studies of Products Administered in Specific Beverages....412
General Considerations on Other Bioequivalence Studies....412
In Vitro Tests Predictive of Human In Vivo Bioavailability (In Vitro–In Vivo Correlation Studies)....413
Pharmacodynamic....413
Comparative Clinical Studies....413
In Vitro Studies....413
Establishing Bioequivalence for Different Dosage Forms....413
Oral Solutions....413
Immediate-Release Products: Capsules and Tablets....413
Preapproval....413
Postapproval....414
Suspensions....415
Modified-Release Products....415
Delayed-Release Products....415
Extended-Release Products....415
Bioequivalence Studies....415
Demonstration of Bioequivalence: Additional Strengths....415
Postapproval Changes....416
Chewable Tablets....416
Special Topics....416
Moieties to Be Measured....416
Parent Drug versus Metabolites....416
Enantiomers versus Racemates....416
Drug Products with Complex Mixtures as the Active Ingredients....417
Long Half-Life Drugs....417
First Point C_{max}....417
Alcoholic Beverage Effects on Modified-Release Drug Products....417
Endogenous Compounds....418
Orally Administered Drugs Intended for Local Action....418
In Vitro Dissolution Testing....418
Immediate-Release Products....418
Modified-Release Products....419
Attachment: General Design and Data Handling of Bioequivalence Studies with Pharmacokinetic Endpoints....419
Study Conduct....419

Fed Studies Test Meal Composition ... 420
Sample Collection and Sampling Times ... 420
Subjects with Predose Plasma Drug Concentrations ... 420
Data Deletion Because of Vomiting ... 421
FDA Recommends Applicants Provide the Following Pharmacokinetic Information in Their Submissions ... 421
Rounding Off CI Values ... 421

Chapter 15 Bioequivalence Testing: European Perspective ... 423

Background ... 423
European Legislation ... 423
Bioequivalence Metrics ... 427
Parent Compound versus Metabolites ... 427
Statistical Analysis and Acceptance Criteria ... 427
Highly Variable Drugs or Drug Products ... 428
Narrow Therapeutic Index Drugs ... 430
Dosage Strength(s) to Bioequivalence Investigated ... 432
BCS-Based Biowaivers ... 432
EMA Guideline ... 433
Executive Summary ... 433
Introduction ... 433
Background ... 433
Generic Medicinal Products ... 433
Other Types of Application ... 434
Scope ... 434
Legal Basis ... 434
Main Guideline Text ... 435
Design, Conduct, and Evaluation of Bioequivalence Studies ... 435
In Vitro Dissolution Tests ... 447
Study Report ... 448
Variation Applications ... 448
Definitions ... 449
Appendix 15A: Dissolution Testing and Similarity of Dissolution Profiles ... 449
General Aspects of Dissolution Testing as Related to BA ... 449
Similarity of Dissolution Profiles ... 450
Appendix 15B: Bioequivalence Study Requirements for Different Dosage Forms ... 451
Oral Immediate-Release Dosage Forms with Systemic Action ... 451
Orodispersible Tablets ... 451
Oral Solutions ... 452
Fixed Combination Dosage Forms ... 452
Non-Oral Immediate-Release Dosage Forms with Systemic Action ... 452
Parenteral Solutions ... 453
Liposomal, Micellar, and Emulsion Dosage Forms for Intravenous Use ... 453
Modified-Release Dosage Forms with Systemic Action ... 454
Locally Acting, Locally Applied Products ... 454
Gases ... 454

Appendix 15C: BCS-Based Biowaiver....454
Introduction....454
Summary Requirements....455
Drug Substance....455
Solubility....456
Absorption....456
Drug Product....456
In Vitro Dissolution....456
Excipients....457
Fixed Combinations....458

Chapter 16 Bioequivalence Testing: The ROW Perspective....459

Background....459
Australasia (Australia and New Zealand)....461
Brazil....461
Canada....462
European Union....462
India....462
Japan....463
South Africa....463
South America (Excluding Brazil): Pan American Health Organization....464
Brazil....464
Taiwan....465
Turkey....465
World Health Organization....465
Interpretation of Technical Terms....465
Comparator (Reference) Products....466
Generic Substitution (Interchangeability)....466
Global Diversity....468
Global Agencies....469
General Assessment of Bioequivalence....469
Pharmacokinetic Endpoint Studies....471
Pharmacodynamic Endpoint Studies....472
Clinical Endpoint Studies or Comparative Clinical Trials....473
In Vitro Endpoint Studies....473
Design and Analysis....473
Study Design....473
Bioanalysis....474
Selection of Appropriate Analyte(s)....478
Parent Drug versus Metabolite(s)....478
Enantiomers versus Racemates....485
Drug Products with Complex Mixtures....485
Bioequivalence Metrics and Data Treatment....488
Statistical Approaches....488
Study Power....489
75/75 Rule....489
90% Confidence Interval....489
Acceptance Criteria for Bioequivalence....490
General....490

For Highly Variable Drugs ... 490
For Narrow Therapeutic Index Drugs ... 491
World Health Organization Guidelines ... 492
Japanese Perspective on Bioequivalence Studies of Generic Products ... 493
Introduction ... 493
Terminology ... 493
Tests ... 494
Bioequivalence Studies ... 494
Test Methods ... 494
Assessment of Bioequivalence ... 499
Pharmacodynamic Studies ... 500
Clinical Studies ... 501
Dissolution Tests ... 501
Reporting of Test Results ... 506
Samples ... 506
Results ... 506
Oral Extended Release Products ... 507
Reference and Test Products ... 507
Bioequivalence Studies ... 508
Test Method ... 508
Assessment of Bioequivalence ... 508
Pharmacodynamic and Clinical Studies ... 508
Dissolution Tests ... 508
Reporting of Test Results ... 511
Non-Oral Dosage Forms ... 511
Reference and Test Products ... 511
Bioequivalence Studies ... 511
Pharmacodynamic and Clinical Studies ... 511
Dissolution (Release) Tests or Physicochemical Tests ... 512
Reporting of Test Results ... 512
Dosage Forms of Which Bioequivalence Studies Are Waived ... 512
Appendix 16A: f_2 (Similarity Factor) and Time Points for Comparisons ... 512
Definition of f_2 ... 512
Time Points for f_2 ... 512
Appendix 16B: Adjusting Dissolution Curves with Lag Times ... 512
Appendix 16C: List of Abbreviations of Parameters ... 513

Chapter 17 Bioequivalence Testing Protocols ... 515

Product-Specific Guidelines ... 516
Introduction ... 516
Background ... 522
What Are BE Studies? ... 522
How Did the Agency Make This Information Available in the Past? ... 522
Procedures for Making Recommendations Available ... 523

Chapter 18 Bioequivalence Documentation ... 587

Background ... 587

Submission of Summary BE Data for ANDAs for FDA 587
Introduction 587
Background 587
Submission of All BE Studies 591
What Types of ANDA Submissions Must Include All BE Studies? 591
What Format Should Be Used for a Summary Report? 593
Same Drug Product Formulation 593
Immediate-Release Drug Products 594
Extended-Release Drug Products: Nonrelease Controlling Excipients 595
Extended-Release Drug Products: Release Controlling Excipients 596
Semisolid Dosage Forms 597
Other Complex Dosage Forms 597
International Submissions 597
Format and Content of BE Study Report to be Submitted to the Central Administration of Pharmaceutical Affairs 598
Title Page 598
Report Contents 598
In Vitro Testing 599
Protocol, Approvals, and Details of In Vivo Study 599
Clinical Study 599
Assay Method and Validation 600
Pharmacokinetic Parameters 600
Statistical Analysis 600
Appendices 600
References 601
Screening Record 601
Formatting of BE Summary Tables 617

Chapter 19 Good Laboratory Practices 625

Background 625
Organization and Personnel 629
Personnel 629
Testing Facility Management 630
Study Director 630
Quality Assurance Unit 630
Facilities 631
General 631
Facilities for Handling Test and Control Articles 631
Laboratory Operation Areas 631
Specimen and Data Storage Facilities 632
Equipment 632
Equipment Design 632
Maintenance and Calibration of Equipment 632
Testing Facilities Operation 632
Standard Operating Procedures 632
Reagents and Solutions 633
Test and Control Articles 633
Test and Control Article Characterization 633
Test and Control Article Handling 633

Protocol for and Conduct of a Nonclinical Laboratory Study ... 633
Protocol ... 633
Conduct of a Nonclinical Laboratory Study ... 634
Records and Reports ... 634
Reporting of Nonclinical Laboratory Study Results ... 634
Storage and Retrieval of Records and Data ... 635
Retention of Records ... 635
Audit of Facilities for GLP Compliance ... 636
General Instructions to Investigators ... 636
Establishment Inspections ... 637
Data Audit ... 643
Computerized Systems ... 644
Personnel: Part III, C.1.c. (21 CFR 58.29) ... 644
Good Laboratory Practice Questions and Answers ... 645
Subpart A: General Provisions ... 645
Section 58.1: Scope ... 645
Section 58.3: Definitions ... 646
Section 58.10: Applicability to Studies Performed under Grants and Contracts ... 647
Section 58.15: Inspection of a Testing Facility ... 648
Subpart B: Organization and Personnel ... 650
Section 58.29: Personnel ... 650
Section 58.31: Testing Facility Management ... 650
Section 58.33: Study Director ... 650
Section 58.35: Quality Assurance Unit ... 650
Subpart C: Facilities ... 651
Section 58.41: General ... 651
Section 58.43: Animal Care Facilities ... 651
Section 58.45: Animal Supply Facilities ... 651
Section 58.47: Facilities for Handling Test and Control Articles ... 651
Section 58.49: Laboratory Operation Areas ... 652
Section 58.51: Specimen and Data Storage Facilities ... 652
Section 58.53: Administrative and Personnel Facilities ... 652
Subpart D: Equipment ... 652
Section 58.61: Equipment Design ... 652
Section 58.63 Maintenance and Calibration of Equipment ... 652
Subpart E: Testing Facilities Operation ... 652
Section 58.81: Standard Operating Procedures ... 652
Section 58.83: Reagents and Solutions ... 653
Section 58.90: Animal Care ... 653
Subpart F: Test and Control Articles ... 654
Section 58.105: Test and Control Article Characterization ... 654
Section 58.107: Test and Control Article Handling ... 655
Section 58.113: Mixtures of Articles with Carriers ... 655
Subpart G: Protocol for the Conduct of a Nonclinical Laboratory Study ... 656
Section 58.120: Protocol ... 656
Section 58.130: Conduct of a Nonclinical Laboratory Study ... 657
Subpart J: Records and Reports ... 657
Section 58.185: Reporting of Nonclinical Laboratory Study Results ... 657

Section 58.190: Storage and Retrieval of Records and Data 658
Section 58.195: Retention of Records 659
Conforming Amendments 659
General 659

Chapter 20 Bioanalytical Method Validation 661

Background 661
Method Classification Based on Data Types 661
Objective of an Analytical Method 663
Objective of the Pre-Study Validation Phase 663
Classical Design in Pre-Study Validation 663
Validation Criteria 664
Pre-Validation 665
Full Validation 666
Reference Standard 667
Method Development 667
Calibration 668
Computational Aspects 669
Linear and Polynomial Models 669
Nonlinear Models (PROC NLIN) 669
Nonlinear Models (PROC NLMIXED) 670
Precision Profile for Immunoassays 670
Back-Calculated Quantities or Inverse Predictions 671
Limits of Quantification and Range of the Assay 671
Limit of Detection 671
Specificity–Selectivity 672
Linearity 672
Accuracy and Precision 672
Accuracy (Trueness) 672
Precision 673
Total Error or Measurement Error 673
Decision Rule 674
Stability 675
Principles of Bioanalytical Method Validation and Establishment 676
Specific Recommendations for Method Validation 676
Microbiological and Ligand-Binding Assays 677
Application of Validated Method to Routine Analysis 679
Documentation 680
Conclusion 682
New Advisory from FDA 683
Introduction 683
Background 683
Chromatographic Methods 685
Ligand-Binding Assays 690
Incurred Sample Reanalysis 695
Additional Issues 695
Documentation 697

Chapter 21 Good Clinical Practice ... 703

Basic Principles ... 703
Nontherapeutic Biomedical Research Involving Human Subjects: Nonclinical Biomedical Research ... 704
Principles of ICH GCP ... 704
Institutional Review Board/Independent Ethics Committee ... 705
- Responsibilities ... 705
- Composition, Functions, and Operations ... 706
- Procedures ... 706
- Records ... 707

Investigator ... 707
- Investigator's Qualifications and Agreements ... 707
- Adequate Resources ... 707
- Medical Care of Trial Subjects ... 708
- Communication with Institutional Review Board/Independent Ethics Committee ... 708
- Compliance with Protocol ... 708
- Investigational Product(s) ... 709
- Randomization Procedures and Unblinding ... 709
- Informed Consent of Trial Subjects ... 709
- Records and Reports ... 712
- Progress Reports ... 713
- Safety Reporting ... 713
- Premature Termination or Suspension of a Trial ... 713
- Final Report(s) by Investigator ... 713

Sponsor ... 714
- Quality Assurance and Quality Control ... 714
- Contract Research Organization ... 714
- Medical Expertise ... 714
- Trial Design ... 714
- Trial Management, Data Handling, and Record Keeping ... 714
- Investigator Selection ... 716
- Allocation of Responsibilities ... 716
- Compensation to Subjects and Investigators ... 716
- Financing ... 716
- Notification/Submission to Regulatory Authority(ies) ... 716
- Confirmation of Review by Institutional Review Board/Independent Ethics Committee ... 717
- Information on Investigational Product(s) ... 717
- Manufacturing, Packaging, Labeling, and Coding Investigational Product(s) ... 717
- Supplying and Handling Investigational Product(s) ... 718
- Record Access ... 718
- Safety Information ... 718
- Adverse Drug Reaction Reporting ... 719
- Monitoring ... 719
- Audit ... 721

Noncompliance ... 722
Premature Termination or Suspension of a Trial ... 722
Clinical Trial/Study Reports ... 722
Multicenter Trials ... 722
Clinical Trial Protocol and Protocol Amendment(s) ... 722
General Information ... 723
Background Information ... 723
Trial Objectives and Purpose ... 723
Trial Design ... 723
Selection and Withdrawal of Subjects ... 724
Treatment of Subjects ... 724
Assessment of Efficacy ... 724
Assessment of Safety ... 724
Statistics ... 724
Direct Access to Source Data/Documents ... 725
Quality Control and Quality Assurance ... 725
Ethics ... 725
Data Handling and Record Keeping ... 725
Financing and Insurance ... 725
Publication Policy ... 725
Supplements ... 725
Investigator's Brochure ... 725
Introduction ... 725
General Considerations ... 726
Contents of the Investigator's Brochure ... 726
Appendix 21A ... 729
Appendix 21B ... 729
Essential Documents for the Conduct of a Clinical Trial ... 730
Introduction ... 730
Before the Clinical Phase of the Trial Commences ... 730
During the Clinical Conduct of the Trial ... 731
After Completion or Termination of the Trial ... 732

Chapter 22 Computer and Software Validation ... 735

Background ... 735
Data Handling and Storage Principles ... 737
Computer Access Controls ... 738
Audit Trails or Other Security Measures ... 738
Date/Time Stamps ... 739
Systems Features ... 739
Systems Used for Direct Entry of Data ... 739
Retrieval of Data and Record Retention ... 740
System Security ... 740
System Dependability ... 741
Legacy Systems ... 741
Off-the-Shelf Software ... 741
Change Control ... 742
Systems Control ... 742
Training of Personnel ... 743

Copies of Records 743
Electronic Signature Certification 743
General Principles of Software Validation 743
Quality System Regulations 744
Verification and Validation 744
IQ/OQ/PQ 745
Principles of Software Validation 746
Typical Tasks 748
Quality Planning 748
Requirements 749
Design 749
Construction or Coding 749
Testing by the Software Developer 749
User Site Testing 749
Validation of Automated Process Equipment and Quality System Software 750
User Requirements 751
Validation of OTS Software and Automated Equipment 752

Chapter 23 Outsourcing and Monitoring of Bioequivalence Studies 755

Background 755
Handling and Retention of Tested Samples 758
Introduction 758
Background 758
FDA Debarment 758
Termination 763
Sampling Techniques 764
Retention for Multiple Studies and Shipments 765
Quantity of Reserve Samples 765
Responsibilities in Various Study Settings 766
Studies Conducted at CROs, Universities, Hospitals, or Physicians' Offices 766
Studies Involving SMOs 766
Blinded Studies with Pharmacodynamic or Clinical Endpoints Involving an SMO 767
In-House Studies Conducted by a Study Sponsor and/or Drug Manufacturer 768
In Vitro Bioequivalence Studies 769
Exception for Inhalant Products 769
Risk-Based Monitoring of CROs 770
Introduction 770
Background 770
Current Monitoring Practices and FDA Guidance 771
FDA's Rationale for Risk-Based Monitoring 772
Overview of Monitoring Methods 773
On-Site and Centralized Monitoring 773
Examples of Alternative Monitoring Techniques 774
Risk-Based Monitoring 776
Identify Critical Data and Processes to Be Monitored 776
Risk Assessment 777

Factors to Consider When Developing a Monitoring Plan ... 778
Electronic Data Capture ... 779
Monitoring Plan ... 779
Documenting Monitoring Activities ... 781
Additional Strategies to Ensure Study Quality ... 781
Protocol and Case Report Form Design ... 781
Clinical Investigator Training and Communication ... 782
Delegation of Monitoring Responsibilities to a CRO ... 782
Clinical Investigator and Site Selection and Initiation ... 782
CRO Selection ... 782

Chapter 24 Epilog: Future of Bioequivalence Testing ... 785

Background ... 785
Pitfalls in Bioequivalence Testing ... 786
Site of Action Requirement ... 786
Statistical Modeling Errors ... 786
Waiver System ... 788
In Vitro–In Vivo Correlation ... 789
Biological Variability ... 789
Multiple Dosing ... 790
Dissolution Testing ... 790
Complex Systems ... 791
Onetime Testing ... 791
Time ... 791
Cost ... 791
Interchangeability ... 791
Global Harmonization ... 792
New Approach to Bioequivalence Demonstration ... 792
Thermodynamic Potential ... 793
Thermodynamic Equivalence Surrogate Test for Bioequivalence ... 794
Temperature ... 796
Dielectric Properties ... 796
Surfactants ... 797
pH ... 797
Osmolality ... 799
Lipophilicity ... 799
Bipolarity ... 799
Electrical Field ... 799
Physical Stress ... 800
Sink Condition ... 800
Duration of Testing ... 800
Proposed Protocols ... 800
Summary ... 800

Appendix A: Glossary of Terms ... 803

Appendix B: Dissolution Testing Requirements for U.S. FDA Submission ... 827

Bibliography ... 873

Index ... 937

List of Figures

FIGURE 1.1 Advertisement of a "patented" medicine. 3

FIGURE 2.1 Typical Bronsted acids and their conjugate bases. 14

FIGURE 2.2 The pH of common fluids. 16

FIGURE 2.3 Log *D* profile of an acid pK_a = 8. 17

FIGURE 2.4 Log *D* profile of a base pK_a = 8. 18

FIGURE 2.5 Log *D* profile of a zwitterion (base) pK_a (base) = 5.6 and pK_a (acid) = 7.0. 18

FIGURE 2.6 Optimal log *P* values for absorption from various sites of administration. 18

FIGURE 2.7 Crystal lattice. 23

FIGURE 2.8 Scalars of lattice structure. 23

FIGURE 2.9 Monotropic system as a function of temperature (*x*-axis). 25

FIGURE 2.10 Enantiotropic system as a function of temperature (*x*-axis). 25

FIGURE 2.11 Enantiotropic system with metastable phases as a function of temperature (*x*-axis). 26

FIGURE 2.12 Diffusion layer model of dissolution. 43

FIGURE 5.1 Examples of confidence-based failure of BE studies. 121

FIGURE 6.1 Convolution–deconvolution models. 160

FIGURE 6.2 Assay complexity versus correlation with human absorption. 184

FIGURE 6.3 96-Well permeability testing method. 185

FIGURE 6.4 MultiScreen Caco-2 assay system. 188

FIGURE 7.1 Comparison of sample sizes required by the U.S. FDA and EMA using true CVs but without the mixing effect and without the GMR constraint. 235

FIGURE 10.1 Schematic for drug application and removal sites for pilot study. 288

FIGURE 10.2 Schematic for drug uptake and drug elimination for BE study. 290

FIGURE 11.1 Decision tree for product quality for bioavailability and bioequivalence testing of nasal products. 295

FIGURE 13.1 Impact of variability on reproducibility. 350

FIGURE 16.1 BE study of oral dosage forms. 496

FIGURE 16.2 Judgment of dissolution similarity. 504

FIGURE 16.3 Judgment of dissolution equivalence. 510

FIGURE 24.1 Types of errors in hypothesis testing 787

List of Tables

TABLE 2.1 Examples of Solubility-Increasing Modifications to Drugs 20
TABLE 2.2 Partition Coefficient and Absorption of Barbiturates 20
TABLE 2.3 Examples of Enhanced Lipid Solubility 20
TABLE 2.4 Solubility of PAS as a Function of Its Salt Form 21
TABLE 2.5 Examples Where Salt Form Reduces Dissolution and BA 21
TABLE 2.6 pK_a of Common Weak Acids Used in Salt Formation 22
TABLE 2.7 The Seven Crystal Systems 24
TABLE 2.8 Thermodynamic Rules for Polymorphic Transitions 26
TABLE 2.9 Effect of Polymorphism on Dosage Form Characteristics 28
TABLE 2.10 Effects of Salvation on Drug Activity 29
TABLE 2.11 Some Drug Substance Hydrate Forms as Reported in the Pharmacopoeia 30
TABLE 2.12 Effect of Complexation on Release and BA 32
TABLE 2.13 Examples Where BA Has Been Increased as a Result of Addition of Surfactants 32
TABLE 2.14 The USP Solubility Classification 35
TABLE 2.15 Ratio of Total and Unionized Drug as a Function of Difference in pH and pK_a 37
TABLE 3.1 FDA Classifications of Powders 56
TABLE 3.2 FDA Classification of Capsule Types 57
TABLE 3.3 Selected Examples of Tablets Containing Small Amounts of Active Ingredients 64
TABLE 3.4 Influence of Food on the Absorption of Various Drugs in Man 71
TABLE 3.5 Examples of Relationships between Food Intake and Drug Regimens 71
TABLE 3.6 Influence of Various Factors on Gastric Emptying in Man 72
TABLE 3.7 Drug Interactant Influence on Drug Absorption 76
TABLE 3.8 Drugs for which First-Pass Hepatic Biotransformation Is Suspected, Possibly in Addition to Gastrointestinal Biotransformation 78
TABLE 3.9 Drugs that Can Undergo Biotransformation in the Lumen or during Absorption in the Mucosa 86
TABLE 4.1 Assumptions in Pharmacokinetic/Pharmacodynamic Modeling 88
TABLE 5.1 Data Requirement for Drug Approval in the United States 96
TABLE 5.2 Therapeutic Equivalence Code Classifications of the U.S. FDA 98
TABLE 5.3 Factors Determining the Establishment of BE Requirement by the FDA 102

TABLE 5.4 Drugs with Potential BE Problems 103

TABLE 5.5 Classification of Active Ingredients According to Their Health Risk 124

TABLE 6.1 Various Parameters Used in IVIVC Level Assignments 153

TABLE 6.2 Biopharmaceutical Drug Classification and Associated Likelihood of IVIVC for Immediate-Release Drug Products 162

TABLE 6.3 Biopharmaceutical Drug Classification and Likelihood of IVIVC for Extended-Release Drug Products 162

TABLE 6.4 BCS Class and Drug Delivery Technology 163

TABLE 6.5 BCS as Defined by the FDA and Modified by Recent Findings 164

TABLE 6.6 Model Drugs to Establish Permeability of Drugs 170

TABLE 6.7 Biowaiver Requirements Worldwide 181

TABLE 7.1 Randomized Balanced Crossover Design 196

TABLE 7.2 ANOVA for Crossover Design (in 10 Subjects) 197

TABLE 7.3 Latin Square Design 197

TABLE 7.4 Balanced Incomplete Block Design 197

TABLE 7.5 Four-Period, Two-Sequence, Two-Formulation Replicate BE Design 204

TABLE 7.6 Three-Period Replicate Design 204

TABLE 7.7 ABE: Estimated Numbers of Samples $\Delta = 0.05$ 205

TABLE 7.8 Population BE 205

TABLE 7.9 Individual BE, Estimated Numbers of Subjects: $\varepsilon_I = 0.05$, $\Delta = 0.05$ 206

TABLE 7.10 Individual BE, Estimated Numbers of Subjects: $\varepsilon_I = 0.05$, $\Delta = 0.10$ with Constraint on Δ ($0.8 \leq \exp(\Delta) \leq 1.25$) 206

TABLE 7.11 Some Examples for Individual BE, 95% Upper Confidence Bounds for Θ_S 224

TABLE 7.12 Some Examples for Individual BE, Parameter Estimates, and 90% CI Using Constrained REML 224

TABLE 7.13 Four-Sequence, Four-Period Design 226

TABLE 7.14 Two-Sequence, Three-Period Design 227

TABLE 7.15 Calculation of Number of Subjects in a Two-Way Crossover Design 229

TABLE 7.16 Sample Sizes for 90% Power Using EMA Criteria in Three-Period (TRR-RTR-RRT) Studies 231

TABLE 7.17 Sample Size for 90% Power Using EMA Criteria in Four-Period (TRTR-RTRT) Studies 231

TABLE 7.18 Sample Size Requirements to Meet the FDA Qualification in Three-Period Design (TRR-RTR-RRT) at 90% Power 232

TABLE 7.19 Sample Size Requirement to Meet the FDA Qualification in Four-Period (TRTR-RTRT) Design (TRR-RTR-RRT) at 90% Power 232

TABLE 7.20 Sample Sizes with EMA Conditions but without Mixed Procedure, GMR Constraint, and Cap on the Use of ABEL 233

TABLE 7.21 Sample Sizes with EMA Conditions Including Mixed Procedure but without GMR Constraint and Cap on the Use of ABEL 233

TABLE 7.22 Sample Sizes with EMA Conditions Including Mixed Procedure and GMR Constraint but without Cap on the Use of ABEL 234

TABLE 7.23 Sample Sizes with FDA Conditions Including Mixed Procedure and GMR Constraint 235

TABLE 7.A.1 Statements in the GLM Procedure 239

TABLE 7.A.2 A 2 × 2 ANOVA Model 239

TABLE 7.A.3 Class Level Information 240

TABLE 7.A.4 ANOVA Table and Tests of Effects 240

TABLE 11.1 In Vitro Bioavailability and Bioequivalence Studies for Nasal Aerosols and Nasal Sprays 311

TABLE 12.1 CAM System of Health-Care, Therapies, and Products 316

TABLE 14.1 Relevant Sections in the CFR Related to BA/BE Studies 390

TABLE 14.2 Final Biopharmaceutics Guidelines of the U.S. FDA 390

TABLE 14.3 Major BE-Related Legislative Events in the History of the United States 391

TABLE 16.1 Global Regulatory Agencies and Organizations 470

TABLE 16.2 Primary Pharmacokinetic Parameters Used in Bioavailability and Bioequivalence Testing 472

TABLE 16.3 Agencies Providing Specific Information on Drugs to Conduct Bioequivalence Studies 474

TABLE 16.4 Brief Description of Bioavailability and Bioequivalence Testing Designs 475

TABLE 16.5 Brief Description of the Criteria on Strength to be Investigated in Bioequivalence Studies 476

TABLE 16.6 Regulatory Criteria on Subject Demographics for Bioequivalence Studies 478

TABLE 16.7 Regulatory Criteria on Sample Size for Bioequivalence Studies 479

TABLE 16.8 Regulatory Criteria on Number of Studies Required for Conducting Bioequivalence Studies 481

TABLE 16.9 Regulatory Criteria for Conducting Fasting and Fed Bioequivalence Studies 483

TABLE 16.10 Regulatory Criteria on Fluid Intake, and Posture and Physical Activity for Bioequivalence Studies 484

TABLE 16.11 Regulatory "Add-On Criteria" for Conducting Bioequivalence Studies 485

TABLE 16.12 Regulatory Criteria on Sampling and Washout Period for Conducting Bioequivalence Studies 486

TABLE 16.13 Regulatory Acceptance Criteria for Bioequivalence 490

TABLE 16.14 Regulatory Bioequivalence Acceptance Criteria for Special Class Drugs 491

TABLE 16.15 Narrow Therapeutic Index Drugs (FDA) 492

TABLE 17.1 BE Reported on www.clinicaltrials.gov in October 2013 516

TABLE 17.2 Conditions and Frequency of Studies Reported 517

TABLE 18.1 Document Types and Management for CRO Monitoring 588

TABLE 18.2 List of Standard SOPs 592

TABLE 18.3 IR Formulations: Differences in Excipient Weights 594

TABLE 18.4 ER Formulations: Differences in Excipient Weights 596

TABLE 18.5 Submission Summary 617

TABLE 18.6 Bioanalytical Method Validation 618

TABLE 18.7 Demographic Profile of Subjects Completing the BE Study 618

TABLE 18.8 Incidence of Adverse Events in Individual Studies 619

TABLE 18.9 Study Information 619

TABLE 18.10 Product Information 620

TABLE 18.11 Dropout Information 620

TABLE 18.12 Protocol Deviations 620

TABLE 18.13 Summary of Standard Curve and QC Data for BE Sample Analyses 621

TABLE 18.14 SOP's Dealing with Bioanalytical Repeats of Study Samples 621

TABLE 18.15 Composition of Meal Used in Fed BE Study 621

TABLE 18.16 Summary of BA Studies 622

TABLE 18.17 Statistical Summary of the Comparative BA Data 623

TABLE 18.18 Summary of In Vitro Dissolution Studies 623

TABLE 18.19 Formulation Data 624

TABLE 18.20 Reanalysis of Study Samples 624

TABLE 20.1 EMA Bioanalytical Method Validation 662

TABLE 20.2 Minimal Sample Size for *r* Runs and *s* Replicates per Run for 10% Acceptance Limits 664

TABLE 20.3 Inverse Functions for Widely Used Response Functions 671

TABLE 20.4 Example of an Overall Summary Table for a Method Validation Report 700

TABLE 20.5 Example of Information for Reference Standards for Method Validation Conducted in Plasma Matrix 701

TABLE 23.1 Percentage of Companies and Extent of Sourcing 756

TABLE 23.2 FDA Debarment List 759

TABLE 23.3 List of a Few Potential CROs 783

TABLE 24.1 BCS and Correlation with IVIVC Distribution of Drugs in Various BCS Classes 788

TABLE 24.2 Dielectric Constant of Various Liquids 798

Preface

This second edition of the *Handbook of Bioequivalence Testing* brings significant and substantial updates to the first edition, which was a great success. Understanding of the scientific principles behind bioequivalence testing has evolved gradually over the past more than 50 years but the changes that have occurred over the past decade have been most significant in furthering the goal of making drugs more affordable. Unfortunately, the cost of generic drugs is rising much faster than before partly because of the increased cost of their approval, including the cost of bioequivalence testing. There is a dire need to reexamine the science behind bioequivalence testing to reduce the burden of development cost. In this revision of the book, I have made an effort to support this cause.

Spending on medicines will reach nearly a trillion dollars in 2015, despite a 50% decrease in the growth rate since 2005. The generic market share of all of these markets is fast expanding and in 2015 almost 40% of all spending will go to generic drugs, compared to only 20% in 2005 and all world regions will show significant increase in the generic share; the lowest share will continue to be held by the most generic-hostile market—Japan at about 9% while South Korea will lead the world reaching over 50%. However, given the absolute dollar value, the United States and Europe will remain the largest markets for generics and this will encourage companies to market their products in these regions. This second edition thus provides extensive details on the recent revisions in the bioequivalence testing requirements in the United States and European Union. However, given the past pace of how regulations were promulgated in the rest of the world (ROW) countries, I have included new chapters that cover these requirements also. This will be helpful to companies making global filing of their products.

Almost 30 years since the passage of the first generic drugs bill in the United States, we continue to question if the generics are as good as the original branded drugs. It has been clearly shown that, at least at a clinical level, generic medicines behave very similarly to their originator counterparts based on a study of over 2000 drugs approved by the FDA between 1996 and 2007. A few studies have shown that generic medicines may not work as well as the branded products and these include antidepressant drugs, bisphosphonates (due to local GI effects), and antiepileptic drugs such as carbamazepine. Some controversies refuse to abate, such as the use of generic levothyroxine, and many physicians continue to favor branded products despite the AB rating (therapeutically equivalent) of levothyroxine products.

While the active pharmaceutical ingredient (API) does not differ between the originator and the generic medicines, other (inactive) ingredients, known as excipients, may be different and a number of pharmaceutical excipients are known to have side effects or contraindications—a likely cause of differences in tolerance reported between the branded and the generic products. Note: while injectable product formulators are required to disclose full formulations details, other formulators are not required to do so. A few examples include the allergic reaction reported to croscarmellose sodium used as excipient in a generic furosemide preparation in a patient who had previously been taking branded furosemide without incident. Similarly, a lactose-intolerant patient with an arrhythmia who is switched from one formulation of antiarrhythmic drug to another that contains a lactose-based excipient may experience gastrointestinal disturbances, which could affect gut transport time and overall drug absorption, thereby affecting systemic levels of the drug. Allergies to excipients contained in topical steroids have also been well documented in both originator and generic preparations. Saccharose, an excipient with potential side effects, was seen in generic preparations of phenobarbitol used to treat epilepsy. Lactose and saccharose are contraindicated in people with lactase or saccharase deficiencies, and as the frequency of these enzyme disorders is high in African populations, it suggests the potential for negative clinical reaction to such medicines in African patients.

This book continues to be a practical treatise for all those who are involved in the planning and conducting of such studies. The roots to this book go back to the late 1970s when I began my teaching career at the University of Illinois College of Pharmacy in the field of biopharmaceutics and pharmacokinetics; the junior-level course that I taught had the routine difficulties of bringing mathematics to students as well as many emerging concepts relating to bioavailability and bioequivalence. To resolve these difficulties, I wrote a book that came mostly from my teaching notes and that became the first textbook on the subject, *Textbook of Biopharmaceutics and Clinical Pharmacokinetics*, which was published in 1977 by Appleton, then transferred to Prentice-Hall, to Wiley, and finally to ParmaMed Press. Surprisingly, the book remains in print today, 35 years later, and as a part of teaching curricula in pharmacy, nursing, and medicine.

Ten years after I began teaching in the field of biopharmaceutics, the U.S. FDA brought the most pivotal legislature, the Waxman–Hatch Act that created the category of generic drugs; however, this Act excluded biological drugs mainly because they were still in the primacy stage of development and poorly understood. Twenty-five years later, the FDA will bring another Act, The Biologics Price Competition and Innovation Act (BPCA) to include biological drugs, but these would be called biosimilar, not biogeneric; another category of substitutable biosimilars will be created, significantly complicating the science and the art of approving biosimilar drugs.

In 1998, I left my tenured position at the University of Illinois and began working for Abbott Laboratories in their international division. This was an era when the big pharmaceutical companies looked down upon generic drugs, Abbott being no exception; it was then that I made a strong case for Abbott to develop generic products, mainly for their overseas affiliates. I had good mentors at Abbott—Kamran Mirza, Rodney Horder, Romeo Bachand, Thomas Hodgson, and many more; we went ahead and developed products like diltiazam, diclofenac, and even OTC products including one of the still best-selling mosquito repellant sold by Abbott worldwide. In 2013, Abbott split into a generic company, Abbott, and a proprietary products company, Abbvie; prior to this, Abbott had already shed its hospital product division as Hospira.

In 1995, I left Abbott Laboratories after I was appointed to the fellowship of Volwiler Society at Abbott, to start my own consulting business, mainly to promote what I believed would be the new era in pharmaceutical products: generic biologics. I worked with leading scientists such as the late Arturto Falaschi, Francisco Baralle, and Sergio Tsminitsky to develop many cytokines and monoclonal antibodies and worked with Saul Mashaal to install manufacturing of insulin. I wrote the first book on the subject of developing and manufacturing bioequivalent therapeutic proteins (*Handbook of Recombinant Therapeutic Proteins*, CRC Press).

I was also fortunate to have the opportunity of meeting with Barack Obama, both when he was an Illinois senator and later the president, and I made many suggestions to him on creating a structure for biological generics in the United States. President Obama was of the opinion that even for the new products, the exclusivity should not be more than 7 years; that seems to have been tempered toward 12 years likely to appease the big pharma lobby, even though the BPCI that came out with the Affordable Healthcare Act had only 7 years of exclusivity. It is for this reason that I have dedicated this book to President Barack Obama. I am confident that the U.S. FDA will take a giant stride in allowing science-based approvals of biosimilars that will help reduce the cost of treatment by billions of dollars in the United States alone.

The 2009 BPCI that was part of the Affordable Healthcare Act did a total turn around in the scope of healthcare in the United States and allowed me to establish the first private company to manufacture recombinant drugs there, while every other competitor had already established plans to manufacture these outside of the United States.

To realize the full potential of generic drugs, the challenge remains on reducing the cost of their development. Whereas the regulatory agencies have come a long way in allowing biowaivers as well as limiting the dissolution testing that is needed to demonstrate equivalence of products, there remains a remarkable opportunity for further improvement to reduce the development costs. In this book, I propose a radical change in the process of biowaivers and hope that the developers of generic products and also all manufacturers when faced with conducting bioequivalence testing will challenge the regulatory authorities on the need for such studies. There is a dire need for both the developers as well as the regulatory agencies to take a fresh look at the current requirements of demonstrating bioequivalence. The FDA has made it abundantly clear that they will entertain any and all innovative suggestions to obviate bioequivalence studies. I have filed several petitions with the FDA to reconsider the entire methodology of requiring bioequivalence studies. My thesis starts with a broader view of this exercise and the last chapter of the book dwells on this subject in detail.

The theme of this book remains—a handbook, rather than an exhaustive treatise on the topics related to methods used to minimize bioequivalence variation, designing experiments to test bioequivalence and securing biowaiver, and preparing submission reports. The references quoted in the book should be of help to the reader but adequate material is enclosed here in sufficient depth to allow for a clear understanding of the difficult path that leads to bioequivalence demonstration.

A brief description of each of the chapters in the book follows; the appendices provide handy data and an extensive bibliography provides opportunities for further exploration of each subject.

Chapter 1: Historical Perspective on Generic Pharmaceuticals. This new chapter traces the history of drug use and development for millennia starting with the Egyptians until the common quackery in New York in the twentieth century—the era of patent medicines that led to regulations following multiple tragedies. The generic era that began in the 1980s changed the way we look at the quality of medicines and created the concept of demonstrating bioequivalence (BE) and later challenging the BE testing that led to waivers in BE studies for a class of drugs where BE is not considered to be a major issue. Whereas it has been more than 30 years since generic drugs came to life, the innovator companies continue to challenge the generic companies and a host of new legal challenges are brought into picture, which this chapter elaborates. The issues related to patenting are also discussed. Finally, a summary of the overall impact of generic drug industry on worldwide healthcare is presented.

Chapter 2: Physicochemical Properties Affecting Bioequivalence. Study of physicochemical properties remains the most important starting step in creating bioequivalent products. This chapter lists and expounds on all those factors that are responsible for introducing bioequivalence variability from the drug release point, particularly related to the API; sponsors need to understand why factors that may seem remote may have caused failure of bioequivalence. Details of dissolution models are also provided. This is the first step in establishing bioequivalence—the chemical equivalence.

Chapter 3: Drug Delivery Factors Affecting Bioequivalence. This chapter transitions from the properties of API to the properties of the drug delivery systems chosen; factors affecting drug release and assessment of these characteristics are provided in detail; this will be of great value to the formulation scientists both at the generic as well as innovator laboratories. Detailed discussions of mechanisms of absorption are also provided in light of release factors from dosage forms, but these principles also apply to several chapter discussions where drugs with potential bioequivalence problems are identified.

Chapter 4: Pharmacokinetic-Pharmacodynamic Modeling. This chapter fulfills the scientific inquiry needs addressed in subsequent chapters for a surrogate bioequivalence testing. A detailed mathematical treatment along with common assumptions is described. Bioequivalence and systemic exposure modeling equations are described to help create models. The deconvolution techniques, including computer software use, are described. Pharmacological evaluation of bioavailability is discussed.

Chapter 5: Bioequivalence Testing Rationale and Principles. This chapter has been updated with much new information to provide a historical perspective to the development of regulations that have lead to today's regulatory requirements and the scientific rationale behind the regulations. The newer guidelines that govern submissions related to BE studies along with therapeutic code classification are presented in detail. An overview of bioequivalence testing requirements and approaches establishes a scientific rationale for products that must demonstrate bioequivalence. This topic will be discussed in much greater depth in later chapters. Regulatory expectations are described in terms of what the FDA, EMA, and other agencies consider to be the significant factors of variability contributing to bioequivalence variation and various measurement indices and techniques described. Included in this chapter is the rationale for bioequivalence estimation along with a listing of drugs with historic bioequivalence problems. Broad guidelines to single dose, multiple-dose, and fed studies are described along with pharmacological and clinical end-point studies that can be substituted for traditional blood level studies. The use of precise and accurate analytical methods along with guidelines of validation is described. Pharmacokinetic and statistical considerations in experiment design are reviewed in this chapter as well. Errors in bioequivalence measurement are highlighted. Studies related to animals, locally administered drugs, and drugs given topically are also described.

Chapter 6: Bioequivalence Waivers. Waiver of bioequivalence studies is a pivotal chapter for both generic as well as innovator sponsors; changing trends at the regulatory level allow sponsors to make a strong case for reducing the scope of BA/BE studies; this is an extremely important factor for the industry. The chapter also includes a detailed description of the biopharmaceutics classification system with several recent modifications to the concept. Pharmacokinetic studies include discussion of absolute bioavailability, which, though of lesser importance to the generic sponsor, brings new possibilities where the drug delivery system is modified (improved). Details of all available surrogate methods to substitute bioavailability and bioequivalence data are given.

Chapter 7: Statistical Evaluation of Bioequivalence Data. This chapter is a detailed description of the statistical models, in theory and in practice, and the software available along with a full-length data analysis exercise. A list of available software and recommendations on choosing calculation support is provided. New additions include the models of noninferiority and demonstration of biosimilarity of large molecule drugs.

Chapter 8: Regulatory Inspection Process. Sponsors face pre-approval inspections from regulatory agencies and it is critical that they understand the type of inquiry coming their way, once the BE study has been filed. New companies inevitably fail these inspections not because of the rigor of manufacturing but the documentation required; advice is provided on how to succeed in this PAI process. Described in this chapter are the events that take place during regulatory audits; this is crucial for contract research organizations to understand what the regulatory authorities want to see when they land at the facility conducting the work.

Chapter 9: Fed Bioequivalence Studies. A large number of drugs are tested for their BE including the effect of food; this complicates and expands the study designs, and careful planning is required to make sure that the coefficient of variation of the data anticipated comports well with the experimental design. Detailed guidelines for such studies are also provided.

Chapter 10: Bioequivalence of Topical Drugs. While the toxicological implication of topical drugs is less important, the demonstration of BE becomes extremely complicated by the inherent variability in the absorption process through the skin. Many surrogate models available are discussed and detailed experimental designs specific to topical drugs are provided.

Chapter 11: Bioequivalence of Nasal Products. Delivery of drugs through the nose is an important and highly lucrative business opportunity. Details of various experimental designs and advice on how to make the studies less onerous and more productive are provided.

Chapter 12: Bioequivalence of Complementary and Alternate Medicines. This is a brand new topic of discussion now that the FDA allows registration of less defined drugs—the division of the FDA that deals with these drugs is still in the formative stages as far as the BE studies are concerned. This chapter therefore provides valuable advice on how to rationalize study designs. Almost inevitably, the sponsor will be challenging all existing concerns of the FDA and creating novel study models.

Chapter 13: Bioequivalence of Biosimilar Products. This new chapter in this edition provides the scientific discussion needed for companies planning to launch biosimilar therapeutic proteins and monoclonal antibodies—the hottest area of generic drug industry. Based on firsthand experience of the author, advice is provided on how to plan studies that will lead to FDA and EMA approvals as well as on understanding worldwide market potentials. Detailed guidelines on general advice to product-specific biosimilarity demonstration plans are provided.

Chapter 14: Bioequivalence Testing—The U.S. Perspective. The United States led the wave of legislation that created the generic pharmaceutical industry; dozens of guidelines that have are continuously evolving control the regulatory compliance requirements; this chapter provides a current view of how the U.S. filing of bioequivalence trials is organized.

Chapter 15: Bioequivalence Testing—The European Perspective. The European guidelines of bioequivalence along with a slightly different statistical modeling and conditions of waivers are described in detail to help sponsors to prepare filings that can be submitted globally, reducing the cost of approvals.

Chapter 16: Bioequivalence Testing—The ROW Perspective: While most of the countries either follow the U.S. or the EMA guidelines, subtle and sometimes substantial differences exist in the basic model of bioequivalence testing practiced worldwide. The guidelines also include the WHO perspective on BE testing.

Chapter 17: Bioequivalence Testing Protocols. Hundreds of drugs being approved as bioequivalents every year have produced a highly valuable database of what is construed as an approvable protocol; the FDA goes even further and provides a listing of suggested protocols that may be acceptable to the FDA. Sponsors must know that the FDA is never bound by any suggestions or guidelines provided by the FDA, so these protocols serve as a good starting point. The bibliography to this chapter includes an extensive list of the studies reported over the past two years.

Chapter 18: Bioequivalence Documentation. Bioequivalence Reports provides a practical guide to writing reports for regulatory submissions along with several examples of reports submitted to regulatory authorities. Also included is a list of standard SOPs that sponsors may want to consider adopting.

Chapter 19: Good Laboratory Practices. The practices of compliance for every study submitted to the FDA or the EMA must comply with certain documentation requirements. This chapter provides details that apply to all aspects related to the use of laboratory facilities and include data handling requirements. These are often the main areas of noncompliance during audit visits. Audit-related details are provided. Also added are typical questions and answers that sponsors of applications have frequently asked the FDA; this part of the chapter serves as a very good example of what may be forthcoming in a pre-approval inspection as the FDA inspectors are required to read these responses prior to their audits.

Chapter 20: Bioanalytical Method Validation. To be acceptable, the data must be collected using fully validated testing methods. Various guidelines from regulatory authorities along with real-time examples of validation are presented here. Statistical modeling of the robustness of methods is a key criterion of GMP compliance. This chapter now includes the additional guidance provided by the FDA in September 2013. Agencies are becoming particularly concerned about method validation and while some of these data will be generated in a contract research organizations (CRO) environment, the responsibility remains that of the sponsor to assure that statistical testing of all methods used is in place.

Chapter 21: Good Clinical Practice. When a drug is administered to humans, the studies are labeled as clinical and this includes all BE studies as volunteers are humans, even though they may not be tested for therapeutic response. This chapter begins with the Declaration of Helsinki and then provides the required compliance details and a long trail of documents needed to certify that the testing facility is GMP compliant.

Chapter 22: Computer and Software Validation. Regulatory agencies have raised the bar significantly over the past decade to ensure that the data collected and stored is fully protected from manipulation; this has caused a great concern for automated data handling. This chapter deals with a topic of greater importance as electronic submissions and the use of automated systems to collect and analyze the data are becoming norms; whereas the CRO is likely to use off-the-shelf and commercial hardware, it is important to understand what questions to ask when seeking validation proof. All phases of data collection, storage, and analysis are subject to strict regulations.

Chapter 23: Outsourcing and Monitoring of Bioequivalence Studies. This new chapter addresses the pitfalls in the common practice of outsourcing BE testing. This chapter provides a working model of arranging and managing CROs, from managing the day-to-day interaction with CROs to sample storage and data reproducibility.

Chapter 24: Epilogue: The Future of Bioequivalence Testing. This new chapter suggests a revolutionary approach to eliminating all testing of bioequivalence. This chapter discloses the pitfalls in the regulatory guidance available today and why they do not ensure a consistent quality of products. The basic definition of bioequivalence as currently used is challenged and developers are encouraged to develop surrogate tests that will obviate the need for any in vivo studies and save billions of dollars, eliminate unethical exposure of humans to drugs in testing, and allow developers to secure faster approval of drugs.

Appendix A: Glossary of Terms. While most terms used are fairly common practice, the regulatory agencies may assign specific meanings to various terms and these are listed in full detail for reference purposes.

Appendix B: Dissolution Testing Protocols. Just like the BE protocols, the U.S. FDA provides suggestions on dissolution testing conditions, sample sizes, and interpretation of dissolution data. While the FDA is not bound to accept the results of these protocols, generally, compliance with these protocols should form the first attempt to prove bioequivalence, particularly if the sponsor wishes to challenge the biowaiver status. The current listing of methods reported by the FDA was compiled in late 2013.

Bibliography: To conserve space and the volume in the book, all pertinent references for each of the chapters are presented at the end of the book instead of at the end of each chapter; though it may be slightly inconvenient, given that in books of the reference type, the reader does not frequently seek out the sources of information, this method of rearranging references should prove more practical.

Sarfaraz K. Niazi, PhD
Deerfield, Illinois

Acknowledgments

I highly appreciate the continuous support that I received from the publisher Taylor & Francis Group, who have also published several other books of mine. I am thankful to Barbara Ellen Norwitz, an executive editor at Taylor & Francis Group/CRC Press for suggesting that I revise my book and for giving me enough time to complete the task. The editorial staff at Taylor & Francis Group has always been very kind to me in correcting many inevitable, sometimes avoidable, errors in the manuscripts. The help of Naila Akiomv, Lillian Yanni, and Natalia Isaeva at my workplace was significant; Robert Salcedo provided me the direly needed encouragement and allowed me enough time to complete this work as scheduled; the continued patience of my wife Anjum in letting me have my time without interruption has served me well for over almost four decades. I cannot thank her enough.

As always, I acknowledge the contributions, direct or indirect, of all other scientists involved. It is quite impossible to mention all of their names here. Further, while I may have quoted works from other references, any errors that remain in this book are altogether mine and I would appreciate hearing from my readers; comments may be sent to niazi@niazi.com.

Author

Sarfaraz K. Niazi, PhD, is the chairman and CEO of Therapeutic Proteins International, a world-class developer and manufacturer of biosimilar recombinant drugs, located in the heart of Chicago. Dr. Niazi began his career teaching at the University of Illinois College of Pharmacy, where he held a tenured position before he joined the international division of Abbott Laboratories; he became a Volwiler fellow at Abbott Laboratories in 1992. In 1995, he left Abbott Laboratories to start his own consulting business in the field of biopharmaceutical manufacturing, an emerging field at that time. He helped establish a multitude of biopharmaceutical manufacturing operations worldwide that included all aspects from cell line creation to commercialization of a variety of recombinant drugs. In 2003, he established his own biotechnology company that now leads the world in using innovative methods of manufacturing biosimilar recombinant drugs coming off-patents. He employs hundreds of the world's top scientists working toward the goal of making life-saving biological drugs more affordable. He has published over 100 refereed research articles and abstracts; dozens of books, both technical and literary; and hundreds of literary writings that span the vast area of poetry, philosophy, rhetoric, irony, and modern dilemma. He also hosts a radio show on Voice of America that reaches billions of people around the world every week. He is the author of several first books including textbooks and handbooks in the field of biopharmaceutics, clinical pharmacokinetics, pharmaceutical formulations, recombinant manufacturing, and disposable bioprocessing; he has authored the first complete English translation of the most widely read book in the world, the 200-year-old Urdu and Persian poetry collection of the South Asian poet Ghalib; a book prominently placed in Barack Obama's bedroom. He has also authored the largest book on pharmaceutical manufacturing formulations ever published. Dr. Niazi continues to serve on the faculty of several academic institutions worldwide and has provided extensive support to several U.S. legislations on biological drugs, particularly the BPCI Act that opened up a pathway for more affordable drugs in the United States. He is also a prolific inventor with scores of patents relating to new drugs and delivery systems, bioprocessing, biosimilarity demonstration, and varied mechanical and energy designs. He is also a patent law practitioner and assists scientists in developing countries patent their inventions. He is a fellow of many learned academies and recipient of a number of research recognition awards including the Star of Distinction by the Government of Pakistan for his lifetime contributions toward promoting technology in developing countries.

1 Historical Perspective on Generic Pharmaceuticals

IN THE BEGINNING

Genetic measurements indicate that the ape lineage, which would lead to *Homo sapiens*, diverged from the lineage that would lead to chimpanzees (the closest living relative of modern humans) around five million years ago. It is assumed that the *Australopithecus* genus, which was likely the first ape to walk upright, eventually gave rise to genus *Homo*. Anatomically, modern humans arose in Africa about 200,000 years ago and reached behavioral modernity about 50,000 years ago. Modern humans spread rapidly from Africa into the frost-free zones of Europe and Asia around 60,000 years ago. The rapid expansion of humankind to North America and Oceania took place at the climax of the most recent Ice Age, when temperate regions of today were extremely inhospitable. Yet humans had colonized nearly all the ice-free parts of the globe by the end of the Ice Age, some 12,000 years ago. Other hominids such as *Homo erectus* had been using simple wood and stone tools for millennia, but as time progressed, tools became far more refined and complex. At some point, humans began using fire for heat and cooking. They also developed language in the Paleolithic period and a conceptual repertoire that included systematic burial of the dead and adornment of the living. Early artistic expression can be found in the form of cave paintings and sculptures made from wood and bone. During this period, all humans lived as hunter-gatherers and were generally nomadic.

The survival of mankind through this arduous history of time owes much to the ability of humans to fight of disease from the power of what is came from the pharmacy contained inside the human body. Later, this aspect of human survival will be discussed in terms of the biotechnology-driven drugs that hold the promise of improving our lives tremendously. Unbeknown to *Homo sapiens sapiens*, nature provided a pharmacy around where we lived, and surprisingly, the shapes of plant parts that yielded the active drugs often resembled the human organs that were ameliorated by the extracts of these plants. A digital leaf looks like the heart.

A critical analysis of life expectancy shows some dramatic findings; in 1850, the average life expectancy was about 40 years, but if we take out the infant mortality part, the average life expectancy goes up to 60 years. And in 2010, the average life expectancy was 67 years. So, despite spectacular changes in the field of medicine, we continue to live the same expanse of life as we did even in the classical Roman times when the average life expectancy was 28 years, but those who lived to be 15 lived 37 years longer. How did we survive so well as a species is a field of study with many surprises. The main pharmacy that protects us from illnesses resides inside our own body, and many have concluded that whatever more we need grows around us. What modern medicine did was to improve the quality of life and extend the productive years, and that led to Ronald Reagan's declaration that there should be no mandatory retirement age.

An interesting satirical history of medicine is very amusing:

4000 BC—Here, eat this root.
3000 BC—That root is heathen. Here, say this prayer.
2150 BC—That prayer is superstition. Here, drink this potion.

90 BC—That potion is snake oil. Here, swallow this pill.
70 BC—That pill is ineffective. Here, take this penicillin.
60 BC—Oops… bugs mutated. Here, take this tetracycline.
55–15 BC—39 more "oops." Here, take this more powerful antibiotic.
14 BC—The bugs have won! Here, eat this root.

—Anonymous

The ancient Egyptians had a system of medicine that was very advanced for its time and influenced later by medical traditions. The Egyptians and Babylonians both introduced the concepts of diagnosis, prognosis, and medical examination. The Greeks went even further and advanced as well medical ethics. The Hippocratic Oath, still taken by doctors today, was written in Greece in the fifth century BC. In the medieval era, surgical practices inherited from the ancient masters were improved and then systematized in Rogerius's The Practice of Surgery. Universities began systematic training of physicians in the 1220s in Italy. During the Renaissance, understanding of anatomy improved, and the microscope was invented. The germ theory of disease in the nineteenth century led to cures for many infectious diseases. Military doctors advanced the methods of trauma treatment and surgery. Public health measures were developed especially in the nineteenth century as the rapid growth of cities required systematic sanitary measures. Advanced research centers opened in the early twentieth century, often connected with major hospitals. The mid-twentieth century was characterized by new biological treatments, such as antibiotics. These advancements, along with developments in chemistry, genetics, and lab technology (such as the x-ray), led to modern medicine.

The history of using chemical drugs derived naturally or synthesized is less than 100 years old, but with making the potions stronger came many hazards that required monitoring of the safety of drugs.

HISTORY OF PATENT MEDICINE

The term *patent medicine* had become particularly associated with drug compounds in the eighteenth and nineteenth centuries, sold with colorful names and even more colorful claims. In ancient times, such medicines were called *nostrum remedium*, "our remedy," in Latin, hence the name "nostrum." Also known as proprietary medicines, these concoctions were, for the most part, trademarked medicines but *not* patented. Patent medicines should have referred to medications whose ingredients had been granted government protection for exclusivity. In actuality, the recipes of most nineteenth century patent medicines were not officially patented. Most producers (often small family operations) used ingredients quite similar to their competitors—vegetable extracts laced with ample doses of alcohol. These proprietary or "quack" medicines could be deadly, since there was no regulation on their ingredients. They were medicines with questionable effectiveness whose contents were usually kept secret.

In the modern context, a patented medicine is a product that has been granted a patent by a government authority to prevent others from practicing the invention for generally a period of 20 years from the date of filing of the patent. The purpose of awarding patents is to encourage innovation and to reward the inventors, but once the exclusivity period expires, the invention can be practiced freely to help mankind.

From quackery to submarine patenting (see more on this later), companies have tried to protect their formulations, rightly or wrongly. In 1984, the passage of law in the United States allowed generic drugs to enter the market, but not unlike the fierce protection of the "patent medicines" of the yore, the innovators fought back to protect their markets, and even today, about 30 years later, the courts are still sorting out the rights of the generic drug manufacturers vis-à-vis the claims of the innovators (Figure 1.1).

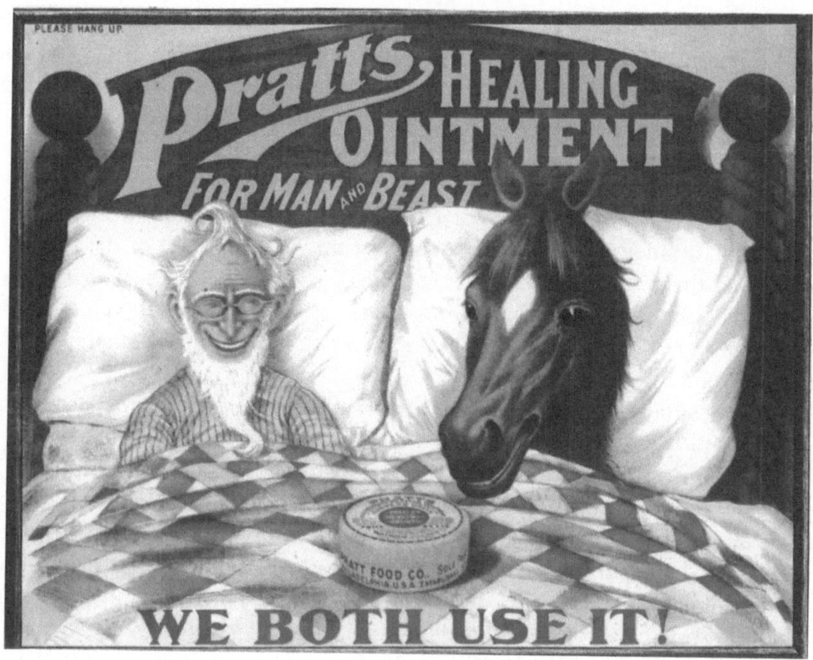

FIGURE 1.1 Advertisement of a "patented" medicine.

Quackery

With little understanding of the causes and mechanisms of illnesses, widely marketed "cures" (as opposed to locally produced and locally used remedies), often referred to as patent medicines, first came to prominence during the seventeenth and eighteenth centuries in Britain and the British colonies, including those in North America. Daffy's Elixir and Turlington's Balsam were among the first products that used branding (e.g., using highly distinctive containers) and mass marketing to create and maintain markets. A similar process occurred in other countries of Europe around the same time, for example, with the marketing of Eau de Cologne as a cure-all medicine by Johann Maria Farina and his imitators. Patent medicines often contained alcohol or opium.

The number of internationally marketed quack medicines increased in the later part of the eighteenth century; the majority of them originated in Britain and were exported throughout the British Empire. By 1830, British parliamentary records list over 1300 different "proprietary medicines," the majority of which were "quack" cures by modern standards. Examples are Dalby's Carminative, Daffy's Elixir, and Turlington's Balsam of Life dating back to the late eighteenth and early nineteenth centuries. These "typical" patent or quack medicines were marketed in very different and highly distinctive bottles. Each brand retained the same basic appearance for over 100 years.

In 1909, in an attempt to stop the sale of such medicines, the British Medical Association published *Secret Remedies, What They Cost and What They Contain*. The publication was composed of 20 chapters, organizing the work by sections according to the ailments the medicines claimed to treat. Each remedy was tested thoroughly, the preface stated: "Of the accuracy of the analytical data there can be no question; the investigation has been carried out with great care by a skilled analytical chemist." The book did lead to the end of some of the quack cures, but some survived the book by several decades. For example, Beecham's Pills (identified as containing only aloes, ginger, and soap but claiming to cure 31 medical conditions) were still on sale in 1997.

British patent medicines started to lose their dominance in the United States when they were denied access to the American market during the American Revolution and lost further ground for the same reason during the War of 1812. From the early nineteenth century, "home-grown" American brands started to fill the gap, reaching their peak in the years after the American Civil War. British medicines never regained their previous dominance in North America, and the subsequent era of mass marketing of American patent medicines is usually considered to have been a "golden age" of quackery in the United States. This was mirrored by similar growth in the marketing of quack medicines elsewhere in the world.

The Dutch Society Against Quackery was established in 1880. Within a short time, the society grew to more than 1100 members. Initially, quackery mainly consisted of the unauthorized practice of medicine and the peddling of "secret remedies." By the 1950s, their energy mostly shifted to magnetizers. Since the 1980s, the society has fought against the so-called alternative medicine. Their primary targets are Chinese acupuncture, homeopathy, manipulative therapy, complimentary medicine, and naturopathy.

In the United States, false medicines in this era were often denoted by the slang term snake oil, a reference to sales pitches for the false medicines that claimed exotic ingredients provided the supposed benefits. Those who sold them were called "snake oil salesmen" and usually sold their medicines with a fervent pitch similar to a fire and brimstone religious sermon. They often accompanied other theatrical and entertainment productions that traveled as a road show from town to town, leaving quickly before the falseness of their medicine was discovered. Not all quacks were restricted to such small-time businesses, however, and a number, especially in the United States, became enormously wealthy through national and international sales of their products.

One among many examples is that of William Radam, a German immigrant to the United States who, in the 1880s, started to sell his "Microbe Killer" throughout the United States and, soon afterward, in Britain and throughout the British colonies. His concoction was widely advertised as being able to "cure all diseases," and this phrase was even embossed on the glass bottles the medicine was sold in. In fact, Radam's medicine was a therapeutically useless (and in large quantities actively poisonous) dilute solution of sulfuric acid, colored with a little red wine. Radam's publicity materials, particularly his books, provide an insight into the role that pseudoscience played in the development and marketing of "quack" medicines toward the end of the nineteenth century.

Similar advertising claims to those of Radam can be found throughout the eighteenth, nineteenth, twentieth, and twenty-first centuries. "Dr. Sibley," an English patent medicine seller of the late eighteenth and early nineteenth centuries, even went so far as to claim that his Reanimating Solar Tincture would, as the name implies, "restore life in the event of sudden death." Another English quack, "Dr. Solomon," claimed that his Cordial Balm of Gilead cured almost anything but was particularly effective against all venereal complaints, from gonorrhea to masturbation. Although it was basically just brandy flavored with herbs, it retailed widely at 33 shillings a bottle in the period of the Napoleonic Wars, the equivalent of over $100 per bottle today.

Not all patent medicines were without merit. Turlington's Balsam of Life, first marketed in the mid-eighteenth century, did have genuinely beneficial properties. This medicine continued to be sold under the original name into the early twentieth century and can still be found in the British and American Pharmacopoeias as "compound tincture of benzoin." It can be argued that for some of these medicines, this is an example of the infinite monkey theorem in action.

The end of the road for the quack medicines now considered grossly fraudulent in the nations of North America and Europe came in the early twentieth century. February 21, 1906, saw the passage into law of the Pure Food and Drug Act in the United States. This was the result of decades of campaigning by both government departments and the medical establishment, supported by a number of publishers and journalists (one of the most effective of whom was Samuel Hopkins Adams, whose series *The Great American Fraud* was published in Colliers Weekly starting in late 1905). This American Act was followed 3 years later by similar legislation in Britain and in other European nations. Between them, these laws began to remove the more outrageously dangerous contents from patent and proprietary medicines and to force quack medicine proprietors to stop making some of their more blatantly dishonest claims.

Medical quackery and promotion of nostrums and worthless drugs were among the most prominent abuses that led to formal self-regulation in business and, in turn, to the creation of the Better Business Bureau.

Regulations

The history of legislation regulating drugs is barely a hundred years old in the United States, the regulations to assume bioequivalence (BE), less than 50 years. When the legislation to control drugs came into being, the main interest was safety; this was followed by proof of efficacy and finally the concept of similarity. The claims of similarity, however, date back to the 1930s when the proprietary medicines were copied by others and similarity claims and affidavits made. It was not until the 1980s that the term bioequivalence was coined, mainly for the purpose to meet the reimbursement requirements. The terminology of patented drugs also changed dramatically over the past century. A brief review of the history of the regulations that leads to BE testing is needed to fully appreciate how far we have come and also realize how far we still have to go now that the issue of similarity of biological drugs has sprung wide open (Chapter 13). The U.S. legislation led the world, so a review of the history in the United States is of prime importance and interest.

Notable regulations published relating to pharmaceutical regulation in the twentieth century began in 1906 with the Pure Drug and Cosmetic Act (PDCA) in the United States. In 1905, a book entitled *The Jungle* was published, in which Upton Sinclair wrote about the Chicago meat packing industry. This book is available free at the Gutenberg source (http://www.gutenberg.org/files/140/140-h/140-h.htm) as well as in print from many sources. The book described the unsanitary conditions in which animals were slaughtered and processed, including the practice of selling rotten or diseased meat to the public. This book had a major impact on the American people and led the U.S. Congress to pass the PDCA. With this new law, it became illegal to sell contaminated (officially interpreted as adulterated) food or meat, and for the first time, labeling of food and drugs had to be truthful—meaning that false or exaggerated claims could no longer be made on labels. The Act also required selected dangerous ingredients to be labeled on all drugs, and inaccurate or false labeling was called "misbranding" and also became illegal.

The U.S. Congress passed the Federal Food, Drug, and Cosmetic Act (FDCA) in 1938 to complement the PDCA. This was largely in response to a public health disaster with a medicine called Elixir Sulfanilamide in 1937. Elixir Sulfanilamide was a sulfa drug sold as an anti-infective. Over 100 people died, most of them children, following ingestion of this medicine due to the fact that it contained diethylene glycol (an antifreeze analog toxic to humans). The company that manufactured the medicine did not perform any toxicity testing prior to marketing the drug, as, at the time, there were no regulations requiring the premarketing safety testing of new medicines. The FDCA required, inter alia, that new drugs be demonstrated as safe to humans before marketing.

The Public Health Service (PHS) Act, which was passed in 1944, was the legal basis for the licensing and gaining of marketing approval of biological products. Biological products are medicinal products that include vaccines, blood and blood components, allergenics, somatic cells, gene therapy, tissues, and recombinant therapeutic proteins. Biologics can be composed of sugars, proteins, nucleic acids, or complex combinations of these substances or may be living entities such as cells and tissues. Biologics are isolated from a variety of natural sources—human, animal, or microorganism—and may be produced by biotechnology methods and other cutting-edge technologies. Gene-based and cellular biologics, for example, are often at the forefront of biomedical research and may be used to treat a variety of medical conditions for which no other treatments are available. Sixty-five years later, the legislation for producing biosimilar drugs will come into effect.

The 1962 Kefauver–Harris Drug Amendments (KHDA) added further protection to public health. The KHDA added the requirement that drugs be proven effective for their intended use. With both the 1938 and 1962 laws in place, U.S. regulators were now ensuring that drugs made available to the American public were relatively safe to consume, in addition to being proven effective in treating the disease or condition that they were being marketed in relation to.

In response to the emerging AIDS crisis in the 1980s, the Orphan Drug Act (ODA) was enacted in 1983 to encourage the development of medicines for conditions that affected small populations by providing monetary and marketing incentives to drug manufacturers. The following year, in 1984, the U.S. Congress also enacted the Hatch–Waxman Act (HWA), which provided for the marketing of generic medicines, the aim of which was to save Americans' money on their medicine bills.

Generic Era

In the 1960s, a government effort to prove the safety and effectiveness of pharmaceuticals manufactured prior to 1962 helped to launch the generic pharmaceutical industry. In 1962, the National Research Council of the National Academy of Sciences was instructed to evaluate all drugs that had been approved for use prior to that year. Under its Drug Efficacy Study Implementation (DESI) program, the National Research Council reviewed more than 3000 products. The list produced by this review described which products were effective for all claimed indications, which were probably or possibly effective for claimed indications, and which were ineffective for claimed indications.

As a result of the review of these products, generic manufacturers were able to file for approval to manufacture products that had been ruled effective without the need to conduct biostudies. Thus, a number of pre-1962 medications, if made to the prescribed chemical formula, were able to enter the market without additional study.

It was not until the passage of the Drug Price Competition and Patent Term Restoration Act of 1984, commonly known as Hatch–Waxman, that the generic industry truly blossomed. This landmark law created the regulatory mechanism under which the Food and Drug Administration (FDA) can approve affordable pharmaceuticals. As President Ronald Reagan said at the time, Hatch–Waxman provided "regulatory relief, increased competition, economy on government, and best of all, the American people will save money, and yet receive the best medicine that pharmaceutical science can provide."

Over the past two decades, President Reagan's prediction has proven to be true. The generic industry has grown dramatically, from $1 billion in annual revenues to $63 billion in the United States today. From a modest beginning, today, nearly 69% of all prescriptions are filled with generic medicines. And the value remains—roughly 16 cents of every dollar spent on prescriptions is spent on generic medicines.

The Generic Pharmaceutical Association (GPhA) itself was founded in 2001, following the merger of three industry trade organizations: the Generic Pharmaceutical Industry Association, the National Association of Pharmaceutical Manufacturers, and the National Pharmaceutical Alliance. The three groups represented similar members, with the former two placing a greater emphasis on scientific issues and the latter focusing on sales and marketing issues. With GPhA, the industry now speaks with a stronger, unified voice before federal and state governments, the courts, and the court of public opinion.

The Hatch–Waxman Act was an attempt to resolve two major issues: (1) regulatory delays in marketing of pharmaceutical products faced by innovator (also called pioneer or research) drug companies and (2) difficulties generic drug companies had at that time in marketing generic versions of pioneer products following expiry of pertinent patent(s). In practical terms, this Act made the following three important provisions: (1) it provided for the extension of the term of one existing patent for innovator drugs; (2) it made provisions for the marketing of generic versions of patented drugs on the day after patent expiry; and (3) it provided opportunities to challenge the validity of patents issued to innovator drug companies.

The U.S. FDA publishes a list of drug products and equivalents, *Approved Drug Products with Therapeutic Equivalence Evaluations*, commonly known as the "Orange Book." The FDA's designation of "therapeutic equivalence" indicates that the generic formulation is (among other things) bioequivalent to the innovator formulation and signifies the FDA's expectation that the formulations are likely "to have equivalent clinical effect and no difference in their potential for adverse effects." The assessment of "interchangeability" between the innovator and generic products is carried out by a study of "in vivo equivalence" or "BE." The pertinent situations in which BE studies are required include the following:

- When the proposed marketed dosage form is different from that used in pivotal clinical trials.
- When significant changes are made in the manufacture of the marketed formulation.
- When a new generic product is tested against the innovator's marketed product. Based on this background, BE information has been determined to have practical and public health value for pharmaceutical industries, regulatory agencies, patients, and practitioners.

The lower costs of generic drugs are possible through exemption of expensive preclinical, clinical, and bioavailability study requirements. BE is demonstrated through studies, which prove that the active ingredients work in the same way and in the same amount of time in the human body. Listed in the following are the FDA requirements for BE for a generic drug:

- Same active ingredients and strength as the brand-name product.
- Same dosage form and route of administration.
- BE—same amount of drug delivered in the same amount of time as the brand-name product.
- Same labeling except the name of the medication.
- Documented chemistry, manufacturing steps, and quality control measures. Raw materials and finished product specifications should meet the USP specifications.
- Potency and shelf-life/expiration date comparable to the brand-name product.
- Meet good manufacturing practices for the facilities used to manufacture, process, test, package, and label the generic product.

The 1984 legislation created a new class of drugs, generics or bioequivalent, which could be substituted by the pharmacist at the time of dispensing unless otherwise prohibited by the prescriber. The drug formularies at health-care institutions placed additional burden to switch to generics. The innovators at the time when the legislation came into effect strongly resisted the at. Drug developers now find it harder to ward off generics, which typically cost about 15% of the brand name's price and cause the innovator to quickly lose up to 90% of its market share. The stakes in the case are significant. Pharmaceutical sales in the United States totaled roughly $320 billion in 2011, according to IMS Health, and brand-name drugs accounted for 18% of the total prescriptions written by doctors in 2011 but 7% of consumer spending.

Bioequivalence

The concepts of BE have gained considerable importance during the last three decades and have become the cornerstones for the approval of brand-name and generic drugs globally. Consequently, regulatory authorities also started developing and formulating regulatory requirements for approval of generic drug products. It is encouraging to know that efforts by regulatory authorities and the scientific community at national as well as international levels are continuing, in order to understand and develop more efficient and scientifically valid approaches to assess BE of various dosage forms, including some of the complex special dosage forms.

Using BE as the basis for approving generic drugs was established by the Drug Price Competition and Patent Term Restoration Act of 1984 (Hatch–Waxman Act). Subsequently, various criteria and approaches for conducting and reporting BE studies for generic products from various regulatory authorities have been progressing.

The Biologics Price Competition and Innovation Act of 2009 (BPCI Act) was signed into law by President Barack Obama on March 23, 2010. The BPCI Act was an amendment to the PHS Act to create an abbreviated approval pathway for biological products that are demonstrated to be highly similar (biosimilar) to an FDA-approved biological product. This Act is similar, conceptually, to the Hatch–Waxman Act, and it aligns with the FDA's longstanding policy of permitting appropriate reliance on what is already known about a drug, thereby saving time and resources and avoiding unnecessary duplication of human or animal testing.

Other pieces of legislation have been, and continue to be, enacted to refine various aspects of pharmaceutical manufacturing and good manufacturing practices, in addition to ensuring that modern scientific practices and developments are incorporated into law.

CHALLENGING BIOEQUIVALENCE

Based on FDA analysis of hundreds of BE studies, FDA has determined that small differences in blood levels—less than 4%—may exist in some cases between a branded and its generic equivalent. But FDA has repeatedly pointed out that this minor difference is no greater than the difference that may exist between two different manufactured batches of the branded drug.

There is a frequent misstatement regarding the BE of generic drugs, which asserts that blood levels of the active ingredient in generic drugs may vary from –20% to +25% compared to the branded. This simply is not true, and the FDA would never approve any generic with a dramatic difference. The 20%–25% margin is one part of a complex statistical calculation used to help measure the BE. In no way does it represent the actual difference in the amount of active ingredient in a patient's bloodstream, which FDA has determined is typically less than 4% between generic and branded or between two different batches of a branded drug.

The Big Future

The U.S. FDA has initiated a risk-based regulatory compliance concept that has gradually reduced the burden on the generic industry, further reducing the cost of drug manufacturing. However, there remains a need to reevaluate the BE demonstration requirements in light of the success of the generic industry. The author is of opinion that most of the BE testing required by regulatory is redundant and serves little purpose to ensure quality products. This argument, filed under several petitions to the FDA, finds its basis from the following arguments:

1. Regulatory filing of BE data pertains to early stages of scale-up and does not necessarily represent the quality of the product over the period of its manufacturing; whereas limited BE may be required in some instances where change control is involved, it is relatively rare for the generic manufacturers to conduct additional BE studies.
2. Thirty years of history of approving generic drugs has taught us many lessons; first, rarely does a filing get rejected based on BE data; statistical models manipulate the observance of variability between the reference and test products. There is a need to examine in detail the drugs that should really be subject to BE studies. Chances are this list can be made much shorter than what is currently provided under the waiver to BE provision. The U.S. FDA must expand the list of waived products and instead add requirements or release requirements that will assure continued compliance.
3. Analytical advances over the past three decades now allow us to better test in vitro release patterns of the drug from the dosage form—the pivotal point in in vitro–in vivo correlation matrix. The onus of proving similarity between the reference product and the test product should move to the sponsors; if they can provide that under various conditions

of dissolution testing, where a differentiation in the release patterns of the reference drug product is achieved, a similarity of the test product should suffice without any need for BE testing.

4. If surrogate correlates to BE are developed, these should then become the part of the release requirement, and this will assure long-term compliance and improved standards of manufacture of generic drugs.

Any exposure to humans that is not necessary is unethical, and much of current thinking on BE testing needs to change dramatically; I predict that in the future, only limited drugs will be required for BE testing and even those requirements will go away once the regulatory agencies have clearly understood that the risk in these waivers is minimal.

Legal Challenges

Whereas the history of practice of generic approvals is well defined in the United States, yet many quirks and abuses remained. Since the passage of the Hatch–Waxman Act, there have been significant changes in the legal status of many drugs. For example, in a Supreme Court ruling on June 16, 2013 *(Federal Trade Commission [FTC] v. Actavis [No. 12-416])*, pharmaceutical companies that pay rivals to keep less-expensive generic versions of best-selling drugs off the market can expect greater federal scrutiny by FTC that can sue pharmaceutical companies for potential antitrust violations, a decision that is likely to increase the number of generic drugs in the marketplace and benefit consumers. The Supreme Court's decision adopted a different standard, known as the "rule of reason," which states that the agreements must be considered in the context of their possible benefits for consumers. In the case, a payment to Actavis by Solvay Pharmaceuticals, the holder of a patent on a testosterone gel known as AndroGel, represented an unlawful restraint of trade because it was intended to keep Actavis from producing its generic version of AndroGel for a certain number of years. This decision is likely to create considerable uncertainty in the drug business and shift an important balance of power to the generic companies.

In another landmark case, the Supreme Court in a ruling handed out in June 2013 (No. 12-142) protected generic drug companies from being sued for any harmful effects of drugs the FDA has approved. The Supreme Court ruled that generic drug makers cannot be sued under state law for adverse reactions to their products, and only the innovators are subject to this legal exposure. Obviously, it excludes any lapses in manufacturing that might bring a liability. Two years ago, the high court ruled that inadequate labeling of a generic drug was not a cause for a lawsuit against the maker. Branded drugs, in contrast, are liable if a product's safety label is inadequate. This causes a "regulatory gap" between generic and brand-name product safety. According to the FDA, the proposed new rule would make generic and brand-name producers equal with respect to updating safety labeling. Previous to this ruling, the innovators were allowed to change their labels without FDA approval, if they found additional adverse effects, while generic drug makers were restricted from updating safety labels to reflect new information unless it meets certain criteria. Generic manufacturers could only change product labeling if ordered by the FDA to do so or if its brand-name equivalent already has made a similar change. However, there were over 400 generic drugs for which no comparable brand-name product exists.

Submarine Patenting

Another impediment in the development of generic drugs that continues to haunt is called submarine patenting. A submarine patent is a patent whose issuance and publication are intentionally delayed by the applicant for a long time, such as several years. This strategy requires a patent system where, first, patent applications are not published and, second, patent term is measured from grant date, not from priority/filing date. In the United States, patent applications filed before November 2000 were not published and remained secret until they were granted. Analogous to a submarine, therefore, submarine patents could stay "under water" for long periods until they "emerged" and surprised the relevant market. Prior to requiring the publication of (U.S. patent) applications, the public would not learn of a patent until after it is issued, which is often several years after the application was filed. Some patentees took advantage of this practice to the extreme (with "submarine" patents) and intentionally delayed their patent issuance, and thus publication, of the patent for several years to allow potentially infringing industries to develop and expand, having no way to learn of the pending application."

Submarine patent practice was possible previously under the U.S. patent law but is no longer practical since the United States signed the TRIPS agreement of the WTO: since 1995, patent terms (20 years in the United States) are measured from the original filing or priority date, and not the date of issuance. A few potential submarine patents may result from pre-1995 filings that have yet to be granted and may remain unpublished until issuance. In the past, when the life of a U.S. patent was 17 years from the date it was granted, submarine patents could issue decades after the initial filing date. Therefore, an applicant for a U.S. patent could benefit by delaying the issuance, and thus expiration date, of a patent through the simple, but relatively costly, expedient of filing a succession of continuation applications. Some submarine patents emerged as much as 40 years after the date of filing of the corresponding application. During the extended prosecution period, the claims of the patent could be tweaked to more closely match whatever technology or products had become the industry standard.

Two examples show how the major pharmaceutical companies in the United States exploited submarine patenting. Interferon alpha used in the treatment of hepatitis C and etanercept, an antirheumatic monoclonal antibody, ended up getting patent coverage for decades as a result of submarine patenting in the United States.

CONCLUSION

Each year, more than 2.6 billion prescriptions are filled in the United States using generic versions of branded pharmaceuticals. This compares to approximately 1.2 billion brand-name prescriptions dispensed annually in the United States. However, there are no published studies or scientific evidence to show that the interchangeability of a generic for a brand-name drug presents difference in safety or effectiveness for the patient. FDA has confirmed on several occasions that BE requirements for generics and brands are rigorous and ensure that approved generics are therapeutically equivalent to their branded counterparts.

FDA's approval process for generic drugs is equally as stringent as the process followed to approve branded drugs. The rigorous chemistry, manufacturing, and control phase are applicable to both new branded drugs and generic drugs. Labeling and testing requirements also are the same for both branded and generics. The same FDA field inspectors evaluate the manufacturing facilities for generics and for branded products, using the same standards, to ensure compliance with all good manufacturing practices. The only meaningful difference between the generic and branded approval process is that human and animal clinical studies to show safety and efficacy are conducted for new branded drugs, whereas BE studies are used in place of clinical studies in approving generics. It should be noted that the branded drugs use BE studies during the approval process or after it is commercially available.

Generic drugs are less expensive than branded drugs in large part because generic manufacturers do not have to conduct costly clinical trials to test the safety and effectiveness of a generic version of a drug that has been safely and effectively used for several years. Rather, generic pharmaceutical companies perform smaller studies to confirm equivalence to the branded product. Moreover, generics do not spend hundreds of millions of dollars in advertising and promotion.

Some minor differences between branded and generic drugs are permitted. For example, a generic drug may have a different shape, size, or color compared to the branded product. However, BE testing assures that the generic drugs will have the same safety, efficacy, and therapeutic effect as the branded product.

We have come a long way in thwarting the belief inculcated by the innovator companies that generic drugs can only be inferior; later, we will examine the continued concerns of the prescribers. And why the past 30 years of use of generic drugs has subsided many concerns, and the history is repeated again as the biological drugs have begun to come of patents. The same hue and cry that was heard in the early 1980s is heard again, and such slogans as "product by process" have been raised to alert of the regulatory agencies in approved biotechnology-derived drugs. Fortunately, the greater cause of helping mankind has prevailed, and while Europe has approved dozens of biosimilar products (as the biological generics are called), the United States is just about getting ready with its own approval strategies.

2 Physicochemical Basis of Bioequivalence Testing

BACKGROUND

Chemical equivalence is relatively easy to establish; the gulf between the chemical equivalence of a multisource drug product and its bioequivalence (BE) is attributed mainly to the differences in the physical characteristics of the active drug, the choice and characteristics of excipients, the specification of the drug delivery system, and also how the drug is tested for BE. It is thus crucial for the formulators to be keenly aware of the subtleties in the physical and chemical properties that may lead to substantial differences in the BE of the tested dosage forms.

Drugs should be capable of existing either in a molecular dispersion such as solutions or in an aggregate state such as tablets, capsules, and suspensions that are readily rendered into finer state of dispersion and dissolution. Regardless of the stage of aggregation in the final formulation, the active pharmaceutical ingredient (API) must be released from the drug delivery system and, as the first step, dissolve in an aqueous environment; this will then be followed possibly by one or more transfers across nonaqueous barriers. Whereas the design of drug delivery systems can alter release characteristics to some extent, the basic permeation characteristics remain an innate property based on the physicochemical nature of the drug.

Drug absorption depends on the release of the drug substance from the drug product (dissolution), the solubility, and the permeability across the gastrointestinal tract. The release characteristics of a drug delivery system are often determined by the manufacture of the product and highly affected by drug solubility, which also affects dissolution rates. The release step is followed by dissolution of the active ingredient.

Absorption of drugs from the various cavities in human body follows certain general principles; for example, a drug must be present in a solution (monomolecular dispersion) or reasonably dispersed to be absorbed. The ionic forms of a drug are not readily absorbed, and similarly, the size of drug molecules is often critical. These properties become important since a drug molecule must traverse through several biologic barriers, both aqueous and nonaqueous; these barriers exist to protect our body from the noxious agents that can be toxic to our body. A compound highly soluble in water or highly insoluble in water would not be able to penetrate the deeper tissues and thus rendered ineffective. Neutral compounds without any polarizable centers often prove pharmacologically inert; take, for example, the fluorinated hydrocarbons like perfluorodecalin—a hexane structure with full fluorination. Fluorine is so highly electronegative that it pulls the electrons from the parent structure making it an inert compound. Interactions at the site of action are often electrically driven, and as a result, it is more likely that we will discover a compound that has weak acid or base properties as an active entity, which is more subject to variability in bioavailability (BA) and BE because of the physicochemical interactions with milieu interior.

CHEMICAL PROPERTIES

IONIZATION

Chemical moieties are known to attract to each other and, under appropriate conditions, disassociate; when this process is driven by the electrical charges on the components of the moiety, this phenomenon is known as ionization. The physicochemical properties of dissociated species differ significantly from the undissociated species and form a basis not only of the physicochemical stability but also of physiological activity of molecules and ions. Acids give rise to excess of H^+ in aqueous solution, whereas a base gives rise to excess of OH^- in solution (Bronsted–Lowry theory). A more general theory of acids and bases is the Lewis theory wherein an H^+ ion combines with an OH^- ion to form water. The pair of electrons that goes into the new covalent bond is donated by the OH^- ion, and thus, the Lewis theory argues that any substance that can act as an electron-pair donor is a Lewis base (such as the OH^- ion). Figure 2.1 shows the inverse relationship that exists between pK_a and pK_b values of typical acids and bases.

The strongest acids appear on the left side of the figure; the strongest bases on the right side of the figure. Any base can deprotonate any acid on the left side of it, a weaker base. Acetic acid, a weak acid, will ionize (or get deprotonated) water, methanol, or ammonia.

Henderson–Hasselbalch Equation

At a given temperature, the thermodynamic ionization constants are independent of concentration, and at a pH value equal to pK_a, the activity of ionized and neutral forms is equal. In many measurement techniques, we measure concentration rather than activity, such as in the use of spectroscopic methods. In such instance,

$$K_c a = [H^+ +][A^-]/[HA] \tag{2.1}$$

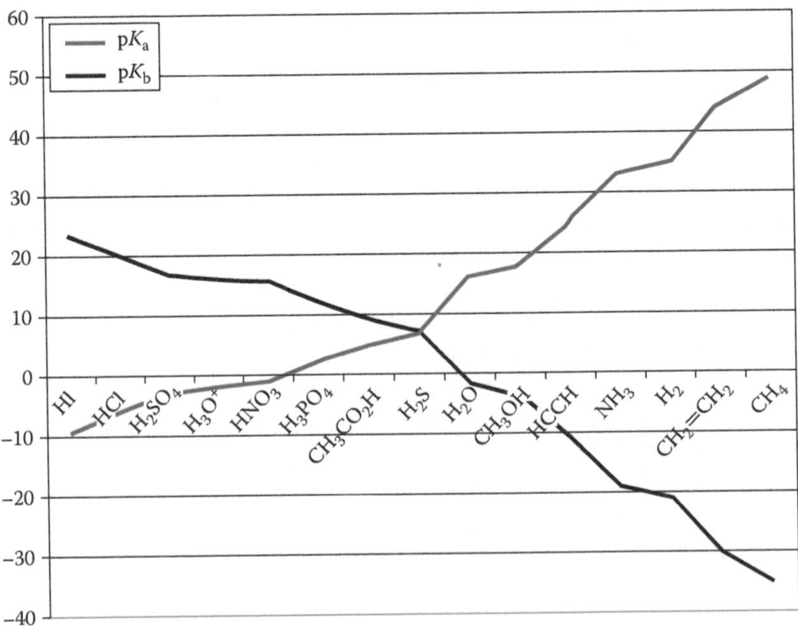

FIGURE 2.1 Typical Bronsted acids and their conjugate bases.

where values in brackets are observed concentration from spectroscopic measurements based on the Beer–Lambert law. The "thermodynamic" ionization coefficient is related to the "concentration" ionization coefficient by

$$K_a = K_c a \cdot \left(\frac{f_A f_H}{f_{HA}} \right) \tag{2.2}$$

where f is the activity coefficient.

The pK_a values are also temperature dependent, often in a nonlinear and unpredictable way. Samples measured by potentiometry are therefore held at a constant temperature bath, and therefore, pK_a value should be quoted at a specific temperature. Often a temperature of 25°C is chosen to reflect room temperature, whereas this may be quite different from the body temperature.

The Henderson–Hasselbalch equation defines the relationship between ionization and pH and is understood as follows (Equation 2.3). This equation relates the pK_a to the pH of the solution and the relative concentrations of the dissociated and undissociated parts of a weak acid:

$$\text{pH} = \text{p}K_a + \log[\text{A}^-]/[\text{HA}] \tag{2.3}$$

or

$$\text{pH} = \text{p}K_a + \log[\text{salt}]/[\text{acid}] \tag{2.4}$$

where

[A$^-$] is the concentration of the dissociated species
[HA] is the concentration of the undissociated species

This equation can be manipulated into the form given by Equation 2.4 to yield the percentage of a compound that will be ionized at any particular pH:

$$\%\text{ionized} = \frac{100}{1 + 10^{(\text{charge}(\text{pH} - \text{p}K_a))}} \tag{2.5}$$

One simple point to note about Equation 2.5 is the 50% dissociation (or ionization), pK_a = pH. It should also be noted that, usually, pK_a values are preferred for bases instead of pK_b values (pK_w = pK_a + pK_b). As a result, the extent of ionization of a compound will depend on the pH of medium. Figure 2.2 shows pH values of common fluids.

Partitioning

The partition coefficient is a measure of the extent a substance partitions between two phases, generally an oil phase and an aqueous phase. This ratio is often expressed as log P (logarithm of partition ratio). Both pK_a and log P measurements are useful parameters in understanding the dissolution and absorption behavior of drug molecules. The pK_a will determine the species of molecules, which is likely to be present at the site of absorption, and how quickly or completely would the species cross a large number of transport barriers in the body, regardless of the route of administration.

Partition coefficient is a ratio of the concentration in two immiscible solvents:

$$\text{Partition coefficient, } P = \frac{[\text{Organic}]}{[\text{Aqueous}]} \tag{2.6}$$

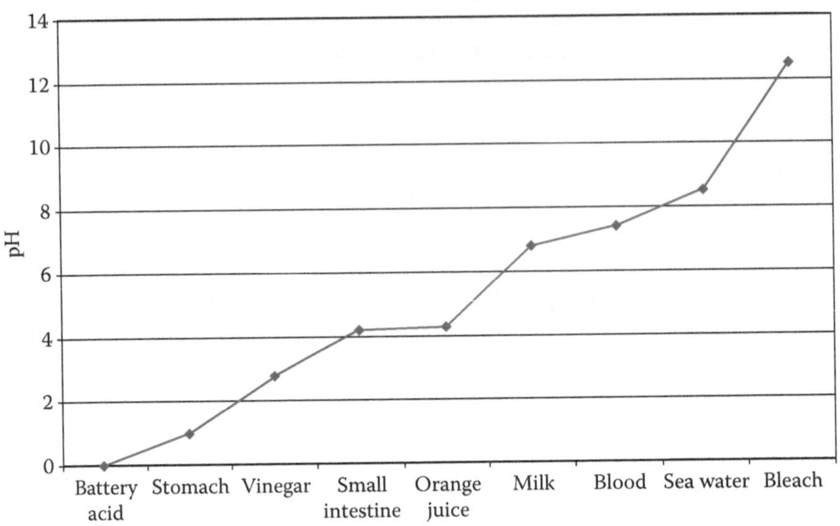

FIGURE 2.2 The pH of common fluids.

where the values in brackets describe measured concentrations.

$$\text{Log } P = \log_{10} \text{ (Partition coefficient)} \tag{2.7}$$

In practical terms, the uncharged or neutral molecule exists for bases >2pK_a units above the pK_a and for acids >2pK_a units below. In practice, the log P will vary according to the conditions under which it is measured and the choice of partitioning solvent.

It is worth noting that this is a logarithmic scale; therefore, a log $P = 0$ means that the compound is equally soluble in water and in the partitioning solvent. If the compound has a log $P = 5$, then the compound is 100,000 times more soluble in the partitioning solvent. A log $P = -2$ means that the compound is 100 times more soluble in water, that is, it is quite hydrophilic.

Log P values have been studied in approximately 100 organic liquid–water systems. Since it is virtually impossible to determine log P in a realistic biological medium, the octanol–water system has been widely adopted as a model of the lipid phase. While there has been much debate about the suitability of this system, it is the most widely used in pharmaceutical studies. Octanol and water are immiscible, but some water does dissolve in octanol in a hydrated state. This hydrated state contains 16 octanol aggregates, with the hydroxyl head groups surrounded by trapped aqueous solution. Lipophilic (unionized) species dissolve in the aliphatic regions.

Generally, compounds with log P values between 1 and 3 show good absorption, whereas those with log P greater than 6 or less than 3 often have poor transport characteristics. Highly nonpolar molecules have a preference to reside in the lipophilic regions of membranes, and very polar compounds show poor BA because of their inability to penetrate membrane barriers. Thus, there is a parabolic relationship between log P and transport, that is, candidate drugs that exhibit a balance between these two properties will probably show the best oral BA.

Distribution Coefficient

The partition coefficient refers to the intrinsic lipophilicity of the drug, in the context of the equilibrium of unionized drug between the aqueous and organic phases. If the drug has more than one ionization center, the distribution of species present will depend on the pH. The concentration of the ionized drug in the aqueous phase will therefore have an effect on the overall observed partition coefficient. This leads to the definition of the distribution coefficient (log D) of a compound, which takes into account the dissociation of weak acids and bases.

Since in the aqueous phase, the total concentration may comprise both ionized and unionized forms, the distribution is given as

$$\text{Distribution coefficient, } D = [\text{Unionized}]_{(o)}/[\text{Unionized}]_{(aq)} + [\text{Ionized}]_{(aq)} \quad (2.8)$$

$$\text{Log } D = \log_{10} (\text{Distribution coefficient}) \quad (2.9)$$

Log D is related to log P and the pK_a by the following equations:

$$\text{Log } D_{(\text{pH})} = \log P - \log[1+10^{(\text{pH}-\text{p}K_a)}] \text{ for acids} \quad (2.10)$$

$$\text{Log } D_{(\text{pH})} = \log P - \log[1+10^{(\text{p}K_a-\text{pH})}] \text{ for bases} \quad (2.11)$$

Log D is the log distribution coefficient at a particular pH. This is not constant and will vary according to the protogenic nature of the molecule. Log D at pH 7.4 is often quoted to give an indication of the lipophilicity of a drug at the pH of blood plasma. Figures 2.3 through 2.5 show the distribution profiles of various acids and bases.

It is important to understand that the species that partition are primarily the neutral molecules or molecules that appear neutral through such interaction as ion pairing, which allows transport of ionic species and thus complicating the calculations of log P and log D.

Besides projecting the solubility, the log P value has several important applications providing greater insight into how the molecule will cross various biological barriers and hence prove effective as a prospective new lead compound. In general, where passive absorption is assumed, the log P can be related to various fixed value ranges (Figure 2.6).

Generally, a low log P (below 0) is desirable for injectable products, whereas a medium (0–3) range is suitable for oral administration; transdermal administration requires a higher value (3–4), but once we reach in the range of 4–7, we risk accumulation of drug in the body fat that can be prove toxic due to accumulation of drug in multiple dosing situations. The renal clearance of drugs with log D (measured at pH 7.4) above zero will decrease renal clearance and increase metabolic clearance; the pK_a of drugs also plays an important role here as highly ionized drugs

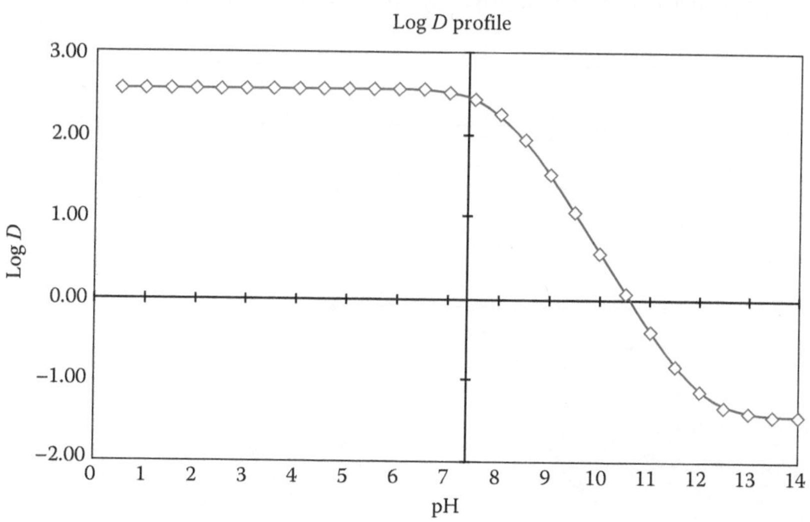

FIGURE 2.3 Log D profile of an acid pK_a = 8.

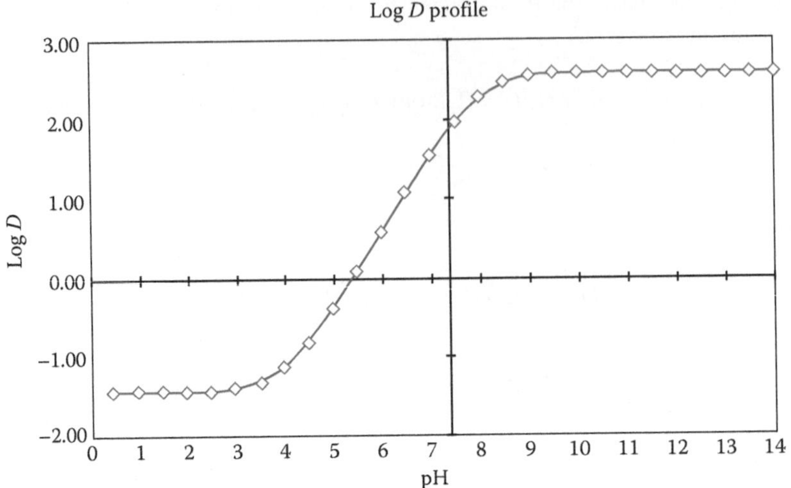

FIGURE 2.4 Log D profile of a base pK_a = 8.

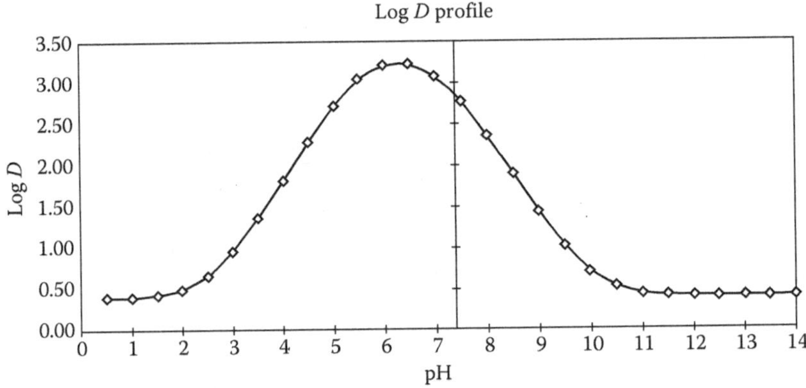

FIGURE 2.5 Log D profile of a zwitterion (base) pK_a (base) = 5.6 and pK_a (acid) = 7.0.

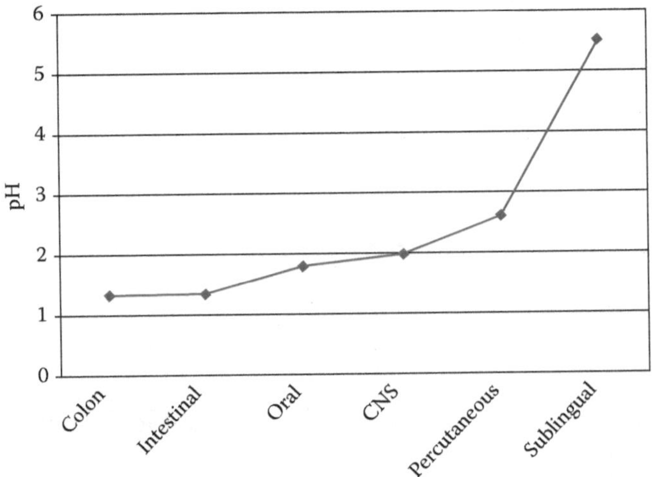

FIGURE 2.6 Optimal log P values for absorption from various sites of administration.

are kept out of cells and thus out of systemic toxicity; generally, a pK_a of 6–8 will be most optimal for transport across various biological membranes.

When making a choice, generally, a drug with lowest log P will be desirable; however, that may require making a choice between a high and a low molecular weight molecule; it is known that high molecular weight drugs are generally more allergenic. The goal should be to achieve a minimum hydrophobicity using a combination of log P, pK_a, and molecular size. The principle of minimum hydrophobicity keeps the drugs out of CNS that may produce side effects like depression, which means that most molecules should have a log P lower than 2.0; this technique was used in the design of the newer generation of nonsedative antihistamines. A very high lipophilicity should also be avoided because of adverse effects on protein binding and on drug absorption, including solubility.

Chemical Structure and Form

The chemical form significantly affects dissolution. Chemical modifications can involve changing the chemical structure of a drug to a form, which is significantly different from the active drug entity. This form can, however, provide a similar therapeutic response since within the body it breaks down into the active entity. Ideally, a drug molecule should have sufficient aqueous solubility for dissolution, an optimum oil/water partition coefficient to provide diffusion through several bilipid layers, and stable chemical groups that will interact with the receptor site. Such an ideal molecule does not usually exist in nature, however, and chemical modifications are generally directed toward that part of the molecule that is responsible for the hindrance in the overall absorption process. For example, it is desirable to restrict the absorption of a sulfonamide if it is to provide a local action in the gastrointestinal tract. This can be achieved by the synthesis of such chemical forms as succinylsulfathiazole, phthalylsulfathiazole, phthalyl sulfacetamide, and salazosulfapyridine, with a free carbonic acid structure that can ionize in the gut. These chemical modifications, which lead to an ionized species, decrease the lipid/water partition coefficient sufficiently to restrict the absorption of the sulfonamides. The antibacterial activity is unfolded when the amide links are broken down by hydrolysis, thus releasing the free and active sulfonamide structures. The aqueous solubility of drugs can be increased by such modifications as sulfacarboxychysoidine, a sulfonamide designed on the basis of insight gained with prontosil and prontosil rubrum. The introduction of dicarbonic acid and sugars into the chemical structure increases the aqueous solubility of tuberculostatic and thiosemicarbazone, and isonicotinic acid hydrazide, erythromycin, and chloramphenicol also provide increased aqueous solubility. Table 2.1 lists several examples of drugs whose water solubility has been increased as a result of chemical modifications. On the other hand, a decrease in the ionization will result in better absorption, as demonstrated for such ganglionic blocking agents of the onium type as hexamethonium. By switching to tertiary amines, such as mecamylamine and pempidine, one obtains drugs that are more steadily and completely absorbed.

Lipophilizing Modifications

Increasing lipid solubility through chemical modifications is exemplified by doxycycline, a derivative of tetracycline. This compound is more efficiently absorbed from the intestine than tetracycline partly because of better lipid solubility and partly because of a decreased tendency to form poorly soluble complexes with calcium and phosphate. This facilitated absorption decreases the risk of disturbances in the intestinal flora and of intestinal superinfection. Chemical changes related to lipid solubility and its effect on gastrointestinal absorption are best exemplified by barbiturates, in which an increase in lipid solubility is directly related to absorption from the colon (Tables 2.2 and 2.3).

TABLE 2.1
Examples of Solubility-Increasing Modifications to Drugs

Drug	High Water-Soluble Form
Tetracycline	Rolitetracycline, tetralysine
Theophylline	Diprophylline, soluphyllin
Theobromin	Isobromin
Prednisolone	Soludacortin, ultracorten, corticosol
Deoxycortone	Docaquosum, diethylstilbestrol, idroestril
Testosterone	Testosterone phosphate
Sulfanilamide	Rubiazole II, septosil, glucosyl sulfanilamide
Menadiol	Menadiol diphosphate, menadiol disulfate hemodol T
Tocopherol	Tocopheryl hemisuccinate
Chloramphenicol	Chloramphenicol hemisuccinate/hemiphthalate
Estriol	Estriol hemisuccinate
Phenetidine	Phenetidine hemisuccinate
Oxazepam	Oxazepam hemisuccinate
Hydroxydione	Hydroxydione dihemisuccinate
Griseofulvin	Succinate/oxime derivatives

TABLE 2.2
Partition Coefficient and Absorption of Barbiturates

Drug	Percent Absorption	Partition Coefficient
Barbital	12	0.7
Aprobarbital	17	4.9
Phenobarbital	20	4.8
Butalbital	23	10.5
Butethal	24	11.7
Cyclobarbital	24	13.9
Pentobarbital	30	28.0
Secobarbital	40	50.7
Hexethal	44	100.0

TABLE 2.3
Examples of Enhanced Lipid Solubility

Erythromycin	Erythromycin estolate
Tetracycline	Doxycycline
6-Azauridine	Triacetyluridine
Lincomycin	Propionate stearate and ethyl carbonate forms
Corticosteroids	Valerate forms for topical use
Nicotinic acid	Ester forms for topical use
Salicylic acid	Ester forms for topical use
Thiamine/other vitamins	Nonquaternary form

Salt Forms

Many important drugs are weak acids or bases. Salts of acidic or basic drugs have different solubility characteristics and show different BA. Sodium or potassium salts of weak acids dissolve much more rapidly than the corresponding free acid, regardless of the pH of the dissolution medium. The same is usually true of the hydrochloride salts or other strong acid salts of weak bases, such as tetracycline hydrochloride or atropine sulfate. The salt form of the drug is generally more soluble in an aqueous medium. However, the solubility of the salt depends on the strength and quantity of the counterions; the smaller the counterion, the more soluble is the salt. For example, *p*-amino salicylic acid exists in various salt forms, and this solubility is given in Table 2.4.

In comparing the absorption of para-aminosalicylic acid in humans, the salts provide clearly greater absorption than the acid form, and the rates of absorption are related to the solubility of the salt form. Novobiocin sodium salt is twice as bioavailable as the calcium salt and 50 times more available than the parent acid.

However, the use of salt forms is not always desirable such as demonstrated for several drugs as listed in Table 2.5.

One approach to the use of salt formation involves additives that provide an alkaline pH around the dissolving particles of weakly acidic drugs. This is best exemplified by the buffered aspirin formulation in which the sodium bicarbonate content provides the alkaline pH. Similarly, sodium phosphate also provides alkaline pH upon its hydrolysis in the gastrointestinal tract.

Combinatorial chemistry offers many advantages including synthesis of larger molecular weight drugs, which are mostly lipophilic; BA considerations require converting them to salt forms. This trend is apparent from recent regulatory approvals by FDA where more than 50% of new drugs approved have been in salt forms. There are fewer salt-forming species for weak acids than there are for weak bases, and the available information suggests that, in general, alkali metal salts exhibit greater solubility than the corresponding alkaline-earth salts. Among cations, the most frequently found ion is sodium (62%), followed by potassium and calcium (10%); this is followed by zinc and meglumine (3%), lithium, magnesium diethanolamine, benzathine,

TABLE 2.4
Solubility of PAS as a Function of Its Salt Form

Form	Solubility (mg/mL)
Unionized acid	1.7
Potassium salt	100
Calcium salt	143
Sodium salt	500

TABLE 2.5
Examples Where Salt Form Reduces Dissolution and BA

Example	Mechanism
Sodium phenobarbital	Swelling of table, retarded disintegration.
Aluminum aspirin	Water-insoluble aluminum coats the surface.
Chlortetracycline hydrochloride	Common ion suppression—excess chloride ions.
Sodium heptabarbital	Salt absorbed faster but incompletely due to large crystal precipitation.
Sodium warfarin/benzylamphetamine pamoate	Surface precipitation of free acid.

TABLE 2.6
pK_a of Common Weak Acids Used in Salt Formation

Acid	pK_a
Acetate	4.76
Ascorbate	4.21
Benzoate	4.20
Besylate	2.54
Citrate	3.13
Fumarate	3.0, 4.4
Gluconate	3.60
Hydrobromide	–8.0
Hydrochloride	–6.1
Malate	3.5, 5.1
Mesylate	1.92
Napsylate	0.17
Oleate	~4.0
Phosphate	2.15, 7.20, 12.38
Succinate	4.2, 5.6
Sulfate	–3.0
Tartrate	3.0, 4.3
Tosylate	–0.51

ethyldiamine, aluminum, chloroprocaine, and choline (in decreasing order of frequency). Among anions, the most frequently used counterion is hydrochloride (almost 50%), followed by sulfate (8%), bromide and chloride (5%), diphosphate, citrate, maleate (3%), iodine, mesylate, hydrobromide (2%), acetate, pamoate (1%), isothionate, methyl sulfate, salicylate, lactate, methyl bromide, nitrate, bitartrate, benzoate, dihydrochloride, gluconate, carbonate, edisylate, mandelate, methyl nitrate, subacetate, succinate, benzenesulfonate, calcium edentate, camsylate, edentate, fumarate, glutamate, hydrobromine, napsylate, pantothenate, stearate, gluceptate, bicarbonate, estolate, esylate, glycollylarsinate, hexylresorcinate, lactobionate, maleate, mucate, polygalacturonate, teoclate, and triethiodide (in decreasing order of frequency). The choice of counterions is a function of the pK_a of the weak acid involved in the formation of salt. Table 2.6 lists pK_a values of weak acids that are most frequently used in salt formation.

To form a salt of a basic compound, the pK_a of the salt-forming acid has to be less than or equal to the pK_a of the basic center of the compound—as a result, very weak basic compounds having a pK_a of around 2. Bases with higher pK_a have a greater range of possibilities for salt formation. Since most drugs are weak bases, it is not surprising that hydrochloride, sulfuric, and toluenesulfonic salts are very common.

Physical Properties

Physical properties as affected by solid-state properties can affect the activity of the drug as determined by the rate of delivery. Chemical stability as affected by physical properties can be significant. Whereas it is always desirable to enhance chemical stability (a pursuit of the synthetic chemist), modulation of physical properties like reducing hygroscopicity by increasing hydrophobicity of acid, by moving to carboxylic rather than sulfonic or mineral acid, or by using acid of higher pK_a to raise pH of solution often provides more stable compounds. Stability is also improved by decreasing solubility and an increase crystallinity by increase of melting. It is important to realize that factors that improve chemical stability often impact adversely on the physical properties. Therefore, a fine balance must be achieved when selecting between physical properties of a chemical property modulation.

Stability of the salt could also be an important issue, and depending on the pK_a, many properties can change including such indirectly related physical characteristics as volatility (e.g., hydrochloride salts are often more volatile than sulfate salts). Discoloration of salt form of drugs is also prominent for some specific forms as the oxidation reactions (often accompanied by hydrolysis) are a result of such factors as affinity for moisture and surface hydrophobicity. Hydrolysis of a salt back to the free base may also take place if the pK_a of the base is sufficiently weak.

Crystal Morphology

A crystalline species is defined as a solid that is composed of atoms, ions, or molecules arranged in a periodic, 3D pattern. A 3D array is called a lattice as shown in Figure 2.7. The requirement of a lattice is that each volume, which is called a unit cell, is surrounded by identical objects. Three vectors, *a*, *b*, and *c*, are defined in a right-handed sense for a unit cell. However, since three vectors are quite arbitrary, a unit cell is described by six scalars, *a*, *b*, *c*, α, β, and γ, without directions (Figure 2.8). Several kinds of unit cells are possible, for example, if $a = b = c$ and $\alpha = \beta = \gamma = 90°$, the unit cell is cubic. It turns out that only seven different kinds of unit cells are necessary to include all the possible lattices. These correspond to the seven crystal systems as shown in Table 2.7.

The seven different point lattices can be obtained simply by putting points at the corners of the unit cells of the seven crystal systems. However, there are more possible arrangements of points that do not violate the requirements of a lattice.

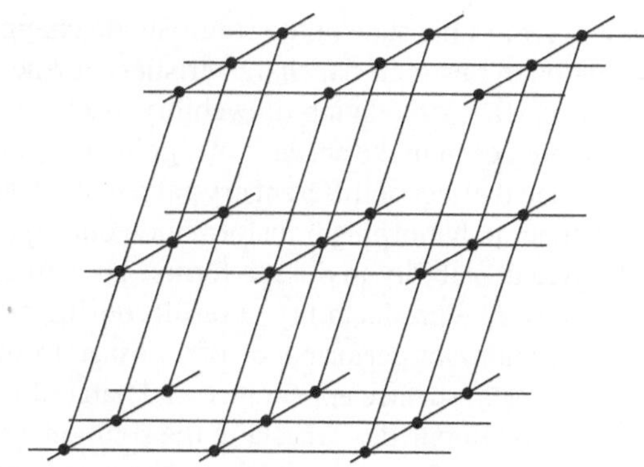

FIGURE 2.7 Crystal lattice.

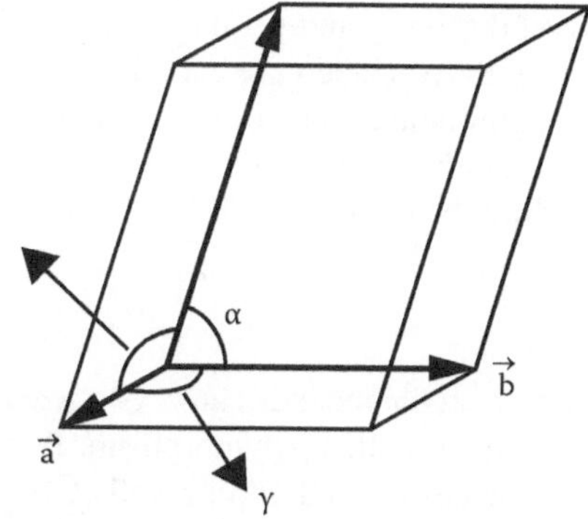

FIGURE 2.8 Scalars of lattice structure.

TABLE 2.7
The Seven Crystal Systems

Crystal System	Axial Lengths and Angles
Cubic	$a = b = c$ $\alpha = \beta = \gamma = 90°$
Tetragonal	$a = b \neq c$ $\alpha = \beta = \gamma = 90°$
Orthorhombic	$a \neq b \neq c$ $\alpha = \beta = \gamma = 90°$
Rhombohedral (Trigonal)	$a = b = c$ $\alpha = \beta = \gamma \neq 90°$
Hexagonal	$a = b \neq c$ $\alpha = \beta = 90°, \gamma = 120°$
Monoclinic	$a \neq b \neq c$ $\alpha = \gamma = 90° \neq \beta$
Triclinic	$a \neq b \neq c$ $\alpha \neq \beta \neq \gamma \neq 90°$

A crystalline particle is characterized by definite external and internal structures. Habit describes the external shape of a crystal, whereas polymorphic state refers to the definite arrangement of molecules inside the crystal lattice. Crystallization is invariably employed as the final step for purification of a solid. Use of different solvents and processing conditions may alter the habit of recrystallized particles, besides modifying the polymorphic state of the solid. Subtle changes in crystal habit at this stage can lead to significant variation in raw material characteristics. Furthermore, various indices of dosage form performance such as particle orientation, flowability, packing, compaction, suspension stability, and dissolution can be altered even in the absence of significantly altered polymorphic state. These effects are a result of the physical effect of different crystal habits. In addition, changes in crystal habit accompanied with or without polymorphic transformation during processing or storage can lead to serious implications of physical stability in dosage forms. Therefore, to minimize variations in raw material characteristics, to ensure reproducibility of results during preformulation, and to correctly judge the cause of instability and poor performance of a dosage form, it is essential to recognize the importance of changes in crystal surface appearance and habit of pharmaceutical powders.

The crystal habit is also affected by impurities present in the solution crystallization; often these impurities provide the earliest nucleation of crystal growth and become an integral part of the crystal. In some instances, the presence of impurities inhibits crystal growth as shown when certain dyes or heavy metals are mixed with solutions. If an impurity can adsorb at the growing face, it can significantly alter the course of crystal growth and geometry. The habits bound by plane faces are termed *euhedral*, and those with irregularly shaped are called *anhedral*. The symmetry of a crystal is generally studied by using optical goniometer that allows measurement of the angles between the crystal faces. This technique is of use only when good crystals of size >0.05 mm in each direction can be obtained, which is generally not the case.

Polymorphism

Both organic and inorganic pharmaceutical compounds can crystallize in two or more solid forms that have the same chemical composition; this is called polymorphism. Polymorphs have different relative intermolecular and/or interatomic distances as well as unit cells, resulting in different physical and chemical properties such as density, solubility, dissolution rate, and BA. When crystal structure contains solvents (or water), these are often called pseudopolymorphs with distinct physical and chemical properties. It is possible for each pseudopolymorph to have many polymorphs. In polymorphism, the crystal lattice formation can take place through two mechanisms: packing polymorphism and conformational polymorphism. Packing polymorphism represents formation of different crystal lattices of conformationally relatively rigid molecules that can be rearranged stably into different 3D structures through different intermolecular mechanisms. When a nonconformationally rigid molecule can be folded into alternative crystal structures, the polymorphism is categorized as conformational polymorphism.

Polymorphs and pseudopolymorphs can be also classified as either monotropes or enantiotropes, depending upon whether or not one form can transform reversibly to another. In a monotropic system, Form I does transform to Form II because the transition temperature cannot appear before the melting temperature (Figure 2.9, monotropy). In Figure 2.10 (enantiotropy), Form II is stable over a temperature range below the transition temperature at which two solubility curves meet and Form I is stable above the transition temperature. At the transition temperature, reversible transformation between two forms happens. Figure 2.11 (enantiotropy with metastable phases) shows the kinetic effects on thermodynamic property of solubility, which shows Ostwald ripening effect. An unstable system does not necessarily transform directly into the most stable state, but into one which most closely resembles its own, that is, into another transient state whose formation from the original is accompanied by the smallest loss of free energy.

When the decision on whether two polymorphs are enantiotropes or monotropes needs to be made, it is very useful to use the thermodynamic rules developed by Burger and Ramberger and tabulated in Table 2.8.

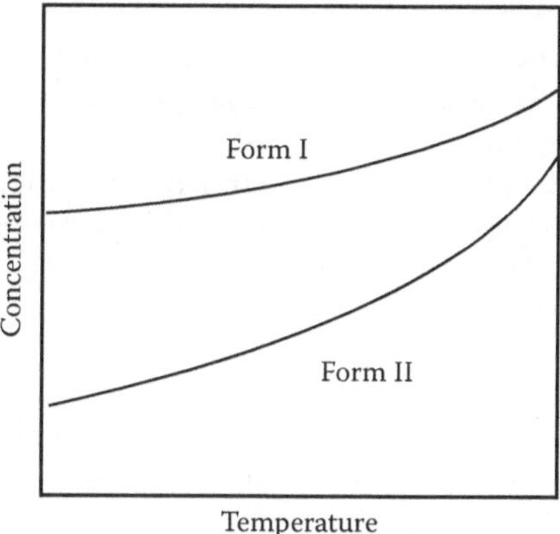

FIGURE 2.9 Monotropic system as a function of temperature (x-axis).

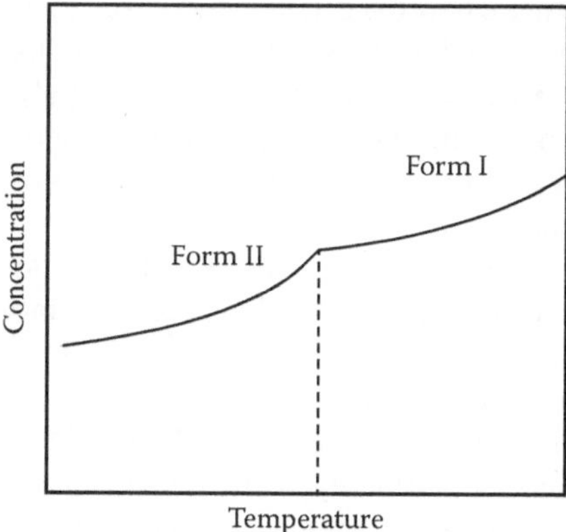

FIGURE 2.10 Enantiotropic system as a function of temperature (x-axis).

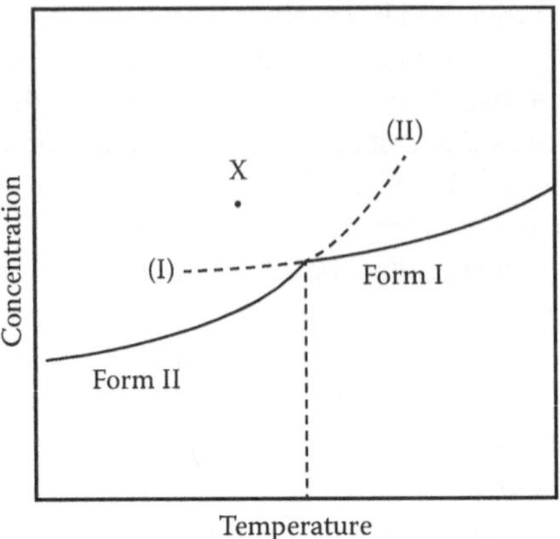

FIGURE 2.11 Enantiotropic system with metastable phases as a function of temperature (*x*-axis).

TABLE 2.8
Thermodynamic Rules for Polymorphic Transitions

Enantiotropy	Monotropy
Transition < melting I	Transition > melting I
I stable > transition	I always stable
II stable < transition	—
Transition reversible	Transition irreversible
Solubility I higher < transition	Transition I always lower
Solubility I lower > transition	—
Transition II → I endothermic	Transition II → I exothermic
$\Delta H_f^I < \Delta H_f^{II}$	$\Delta H_f^I > \Delta H_f^{II}$
IR peak I before II	IR peak I after II
Density I < II	Density I > density II

The stability of polymorphs is thermodynamically related to their free energy. The more stable polymorph has the lower free energy at a given temperature. The earlier classification of polymorphic substances into monotropic and enantiotropic classes from the view of the lattice theory is not always appropriate. There is a need to explore how the crystal lattice structures of polymorphs are related. At a transition point with the temperature and the pressure fixed, it is possible for interconversion to happen between two polymorphs only in the case that the structures of the polymorphs are related. If complete rearrangement is required by atoms or molecules during transformation, no point of contact for reversible interconversion exists. Therefore, the existence of enantiotropes or monotropes in thermodynamics and phase theory is corresponding to related or unrelated lattice structures in structural theory. Transformation between polymorphs that have completely different lattice structures exhibits the dramatic changes in properties. The difference in energy between polymorphs is not always considerable as shown with diamond/graphite. In most cases, polymorphs in this category are required to break bonds and rearrange atoms or molecules, and consequently, the polymorphs have a monotropic relation.

For the study of polymorphs that are structurally related, first, the structural relationships between the polymorphs should be established; secondly, it should be explained why a particular substance is able to arrange its structural units in two closely related lattices; and finally, there should be a description of the manner and conditions under which rearrangement of the units from one lattice type to another can happen. For drugs that undergo degradation in the solid state, the physical form of the drug influences degradation. Selection of a polymorph that is chemically more stable is a solution in many cases. Different polymorphs also lead to different morphology, tensile strength, and density of powder bed, which all contribute to compression characteristics of materials. Some investigation of polymorphism and crystal habit of a drug substance as it relates to pharmaceutical processing is desirable during its preformulation evaluation, especially when the active ingredient is expected to constitute the bulk of the tablet mass.

Various techniques are available for the investigation of the solid state. These include microscopy (including hot-stage microscopy), infrared spectrophotometry, single-crystal x-ray and x-ray powder diffraction (XRPD), thermal analysis, and dilatometry.

Most organic compounds are capable of exhibiting polymorphism because of their complex flexible structure; the window of physicochemical stress that a drug is generally subjected to during manufacturing is at times not able to adduce the differentiation of a drug into its possible polymorphic forms. For example, enantiotropic state is when one polymorph can be reversibly changed into another one by varying the temperature or pressure. One way of assessing whether the solid is a metastable form of the compound is to slurry the compound in a range of solvents. In this way, a solvent-mediated phase transformation may be detected using the usual techniques. The monotropic state exists when the change between the two forms is irreversible. Since all polymorphs are interchangeable, the lowest energy providing the lowest energy polymorph or the most able polymorph is often needed to assure consistency in the physicochemical properties; this is necessary for consistency in manufacturing procedures as well as in BA. The right polymorph at time is not necessarily the most stable polymorph; unstable forms like amorphous forms (that are most constrained) are often used because of their higher solubility and often a better BA profile.

The manufacturing factors that may be affected by the choice of a particular polymorphic form include granulation, milling and compression, stability (particularly for semisolid forms), amount of dose delivered in metered inhalers, crystallization from different solvents at different speeds and temperature, precipitation, concentration or evaporation, crystallization from the melt, grinding and compression, lyophilization, and spray drying. In the manufacturing processing, crystallization is a major problem, and it can be avoided by a careful study of polymorphic transition particularly in supercritical fluids.

Polymorphism is frequently a function of type of salt because the presence of counterions can make the crystals form differently leading to widely variable physicochemical properties as described earlier under the polymorphism description. Generally, salts exhibiting polymorphism should be avoided.

An interesting example of polymorphic structure differentiation is that of HIV protease inhibitors. The HIV protease inhibitors have serious problem in their BA. Invirase showed only modest market performance, and it was soon superseded by drugs such as ritonavir (Norvir) and indinavir sulfate (Crixivan®) that had better BA. Three years after initial approval, saquinavir was reintroduced in a formulation with sixfold higher oral BA relative to the original product. Ritonavir was originally launched as a semisolid dosage form, in which the waxy matrix contained dispersed drug in order to achieve acceptable oral BA. Two years after its introduction, ritonavir exhibited latent crystal polymorphism, which caused the semisolid capsule formulation of Norvir to be removed from the market.

Each polymorph has a certain thermodynamic energy associated with it as a result of strains in the bonds of the lattice structure, and therefore, one polymorph may be more stable than the others. At any given temperature and pressure, only one crystal form of a drug will be stable, and other forms will convert to this form. When the conversion is relatively slow, the polymorph is said to be metastable. The various polymorphic forms are chemically indistinguishable. However, they differ in physical properties, such as density, melting point, solubility, and dissolution rates. For example, riboflavin exists in several polymorphic forms with a 20-fold difference in their aqueous

TABLE 2.9
Effect of Polymorphism on Dosage Form Characteristics

Example	Explanation
Novobiocin	Increased BA from amorphous form, suspension stabilized by methyl cellulose
Sulfathiazole	Increased dissolution from amorphous form conversion stabilized by PEG 400
Lente insulin	Amorphous form for quick absorption, crystalline form giving sustained delivery
Theobroma oil	High m.p. form for room temp. stability
Penicillin G	Amorphous form less chemically stable
Chloramphenicol stearate	Amorphous form active
Aspirin, barbital, estrone, sulfonamides, chloramphenicol, chlordiazepoxide, adiphenine, erythromycin, methotrexate, cholesteryl palmitate	Altered BA

solubility. Amorphous forms in which no internal crystal structure exists have the highest solubilities, giving the order of dissolution rates for the crystal forms arranged as amorphous > metastable > stable forms.

It has been suggested that almost 40% of all organic compounds can exist in various polymorphic forms, sometimes in as many as five different forms, as in the case of cortisone acetate; almost 50% of all barbiturates and 70% of steroids exhibit polymorphism.

This premise, however, may not be applicable to all drugs, especially those that are absorbed by an active process, for example, various vitamins. Table 2.9 lists effects of polymorphism on drug and dosage form characteristics.

Amorphous Forms

Solid powders wherein there is no particular order of molecules are technically noncrystalline and called amorphous forms. The amorphous forms are formed by vapor condensation, supercooling of a melt, precipitation from solution, and milling and compaction of crystals. These are more like liquids where the molecular interaction has weakened; in most instances, there would be some crystalline forms among the amorphous forms as well. This two-state model is described in U.S. Pharmacopoeia (USP). The amorphous forms are thermodynamically unstable as they have high energy (that went into breaking intermolecular bonds), and as a result, they may turn into crystalline form, particularly in suspension dosage forms and even in solid dosage forms wherein atmosphere moisture may serve as nucleation points.

Discovery programs frequently yield amorphous compounds due to time pressures, the methods used to isolate them on small scales, and the increasing complexity of newly discovered molecules. Amorphous compounds carry inherent risks due to their physicochemical nature, and as a result, very few FDA-approved drugs appear in amorphous forms; examples include Accupril/Accuretic, itraconazole, Accolate (Zafirlukast), Viracept (Nelfinavir mesylate), and paroxetine. Other drugs that are available in amorphous form include celecoxib, amifostine, cefuroxime axetil, cefpodoxime proxetil, and novobiocin. In addition to being physically metastable physical form, amorphous forms are generally less stable chemically. They also tend to have very low bulk densities, making the materials difficult to isolate and handle. The irregular shape of powder of amorphous forms creates high surface area that attracts water molecules, making them inherently more hygroscopic.

Whereas all of these problems can be resolved, generally, the amorphous forms are to be avoided unless the differences in solubility make a significant impact on the BA.

Solvates

In addition to polymorphs, solvates (inclusion of the solvent of crystallization) are also often formed during the crystallization process. These forms are also called pseudopolymorphs. The solvent molecules fill spaces in the crystal lattice and generally reduce the solubility and dissolution rates. This phenomenon is thermodynamically driven. If the solvate contains an organic solvent, regulatory authorities would not admit this. According to the ICH guidelines, Class I solvents must be avoided as these are carcinogenic such as benzene, carbon tetrachloride, and 1,2-dichloromethane; Class II solvents should be limited and include nongenotoxic animal carcinogens such as cyclohexane and acetonitrile; and Class III solvents have low toxicity potential including acetic acid, alcohol, and acetone and are allowed as long as the daily permissible dose does not exceed 50 mg. Generally, an allowed solvate would likely be removed during the manufacturing process, but in some instances, the presence of solvate is desired like in beclomethasone dipropionate, product of Glaxo that includes trichlorofluoromethane solvate; this solvate prevents crystal growth in sprays containing trichlorofluoromethane as propellant.

During seeding, crystals may incorporate one or more of the molecules of the solvent into their structure, and the resultant forms are referred to as *solvates*. The solvates themselves may exist in various polymorphic forms and are referred to as *pseudopolymorphs*. Some examples of pseudopolymorphs include mercaptopurine, fluprednisolone, and succinylsulfathiazole. The number of drug solvates is well over 100, and some of the most common examples include steroids, antibiotics, sulfonamides, barbiturates, xanthines, and cardiac glycosides.

The use of a solvate in increasing BA is on the premise that some anhydrates dissolve faster than their corresponding hydrates in aqueous media. However, the relationship becomes much more complex when alcoholates or other nonaqueous solvate is dissolved in water. Table 2.10 lists several examples of effect of solvation on drug and dosage form.

Whereas the selection of an appropriate solvate or asolvate form can be advantageous in increasing the dissolution rates and BA in some instances, their use requires careful monitoring of manufacturing and storage conditions because of the possibility of inadvertent solvation or desolvation.

Hydrates

When solvate happens to be water, these are called hydrates wherein water is entrapped through hydrogen bonding inside the crystal and strengthens crystal structure and thereby invariably reduces the dissolution rate (Table 2.11). The water molecules can reside in the crystal either as isolate lattice where they are not in contact with each other or as lattice channel water where they fill space and metal-coordinated water in salts of weak acids where metal ion coordinates with water molecule. Metal ion coordinates may also fill channels such as in the case of nedocromil sodium trihydrate.

TABLE 2.10
Effects of Salvation on Drug Activity

Example	Explanation
Ampicillin	Anhydrate shows higher BA compared to trihydrate.
Hydrocortisone	Hemiacetone solvates show higher absorption compared to asolvates.
Caffeine, theophylline, glutethimide	Increased dissolution of anhydrous forms compared to hydrates.
Mercaptopurine, prednisolone	Higher dissolution and activity of asolvate from pellets implanted.
Griseofulvin	Chloroformate solvate gives higher BA compared to asolvate, also solvation–desolvation results in increased surface area.
Citric acid	Hydrate used to provide mole of water as granulating agent in effervescent preparations.

TABLE 2.11
Some Drug Substance Hydrate Forms as Reported in the Pharmacopoeia

Compound	Water of Hydration
Aminophylline	2
Ampicillin	3
Beclomethasone dipropionate	0 or 1
Caffeine	1
Calcium citrate	4
Calcium gluceptate	0 or variable; effloresces
Calcium gluconate	0 or 1
Dextrose	1
Diatrizoic acid	2
Dibasic sodium phosphate	0, 1, or 2
Ephedrine	½
Fluocinolone acetonide	2
Hydrocortisone hemisuccinate	1
Magnesium citrate oral solution	1
Magnesium gluconate	2
Magnesium sulfate	0, loses gradually
Monosodium sodium phosphate	0, 1, or 2
Naloxone hydrochloride	2
Nitrofurantoin	0 or 1
Potassium gluconate	0 or 1
Prednisolone	0 or 1
Saccharin sodium	1/3; effloresces
Sodium acetate	3
Sodium citrate	0 or 2
Sodium sulfate	0 or 1; effloresces
Succinyl chloride	2
Theophylline	0 or 1
Thioguanine	0 or ½
Thiothixene hydrochloride	0 or 2
Zinc sulfate	1 or 7

Source: United States Pharmacopoeia 24.

Crystalline hydrates have been classified by structural aspects into three classes: isolated lattice sites, lattice channels, and metal-ion-coordinated water. There are three classes that are discernible by the commonly available analytical techniques:

1. Class I includes isolated lattice sites that represent the structures with water molecules that are isolated and kept from contacting other water molecules directly in the lattice structure. Therefore, water molecules exposed to the surface of crystals may be easily lost. However, the creation of holes that were occupied by the water molecules on the surface of crystals does not provide access for water molecules inside the crystal lattice. The analyses by thermogravimetric analysis (TGA) and differential scanning calorimetry (DSC) for the hydrates in this class show sharp endotherms. Cephradine dihydrate is an example of this class of hydrates.
2. Class II includes hydrates that have water molecules in channels. The water molecules in this class lie continuously next to the other water molecules, forming channels through the

crystal. The TGA and DSC data show interesting characteristics of channel hydrate dehydration. Early-onset temperature of dehydration is expected, and broad dehydration is also characteristic for the channel hydrates. This is because the dehydration begins from the ends of channels that are open to the surface of crystals. Then dehydration keeps on happening until all water molecules are removed through the channels. Ampicillin trihydrate[17] belongs to this class. Some hydrates have water molecules in 2D space, and they are called planar hydrates.

3. Class III includes ion-associated hydrates. Hydrates contain metal-ion-coordinated water, and the interaction between the metal ions and water molecules is the major force in the structure of crystalline hydrates. The metal–water interactions may be quite strong relative to the other nonbonded interactions, and therefore, dehydration occurs at very high temperatures. In TGA and DSC thermograms, very sharp peaks corresponding to dehydration of water bonded with metal ions are expected at high temperatures.

Hydrates can also exist in various polymorphs such as in the case of amiloride hydrochloride. A myriad of methods are available to study hydrates and their polymorphs including differential thermal analysis (DTA), DSC, XRPD, and moisture uptake studies.

Complexation

A molecular complex consists of constituents held together by such weak forces as hydrogen bonds. The physical properties of drug complexes, such as solubility, molecular size, diffusivity, and lipid–water partition coefficient, can differ significantly from those of the drug itself, resulting in possible BA variations. Complexation will generally increase the total solubility of a poorly water-soluble drug if the complex itself is soluble in aqueous media. If the complexation process is reversible (drug + complexing agent complex), then the absorption rates and the extent of absorption will be increased for poorly soluble drugs. For drugs that are generally adequately absorbed, the complex formation may result in a slowing down of absorption, but the overall quantity of drug absorbed may not change.

The most frequently observed complex formation is between various drugs and macromolecules, such as gums, cellulose derivatives, high molecular weight polyols, and nonionic surfactants. Mostly, however, these complexations are reversible with little effect on the BA of drugs. But in those instances where the complex is insoluble in aqueous media, these interactions are clearly contraindicated. Table 2.12 lists several examples of complexation where BA has been altered.

Surface Activity

Surfactants have variable effects on the dissolution and absorption processes. The lowering of surface tension increases the dissolution rates by increasing the solubility of drugs if the concentration of the surfactant is above the critical micelle concentration. The lowering of surface tension also increases the diffusion of free molecules in the medium, increasing the contact between free drug and the absorption surface. The surfactants can also increase the membrane permeability, allowing greater absorption of most chemical structures.

The overall effect of surfactants on the BA of drugs is complex, since the molecules contained in the micelle are not available for absorption unless a quick equilibration between the free drug molecules and those inside the micelles can be established. A number of drugs have surface-active properties themselves and form their own micelles, thus facilitating the absorption. Examples of these drugs include potassium benzylpenicillin, mixtures of penicillin and streptomycin salts, amphetamine sulfate, cyclopentamine hydrochloride, ephedrine sulfate, propoxyphene hydrochloride, ionic derivatives of phenothiazines, dyes, quaternary ammonium salts of drugs, and liquorice.

TABLE 2.12
Effect of Complexation on Release and BA

Drug	Complexing Agent	Effect
Amphetamine	Na-CMC	Decreased BA
Phenobarbital	PEG 4000	Decreased BA
Tetracycline	Heavy	Decreased BA
Salicylic acid	Caffeine	Increased lipid solubility, no effect on BA due to rapid dissociation
Diphenhydramine	Methyl orange	
Acidic dyes	Basic nitrogen	
Atropine	Eosin B	
Prednisolone/prednisone	Propylamide/propionamides	Increased BA
Digoxin	Hydroquinone	Increased BA
Benzocaine, ergotamine	Caffeine	Increased dissolution
M-Benzoic acid	Tartaric acid	Increased dissolution
Maleic acid	Creatinine	Increased dissolution
Several antihistamines	Betacyclodextrin	Increased dissolution
Caffeine	Citric acid	Increased dissolution

TABLE 2.13
Examples Where BA Has Been Increased as a Result of Addition of Surfactants

Drug	Surfactant
A vitamin	Sodium lauryl sulfate
B-12 vitamin	Polysorbate 80/85 G-1096
O-Benzoylthiamine disulfide	Sodium lauryl sulfate
Cephaloridine	Various
G-Strophanthin	Sodium lauryl sulfate
Heparin	Sodium lauryl sulfate, dioctyl sodium sulfosuccinate
Iodoform	Polysorbate 80
Phenolsulfonphthalein	Dioctyl sodium sulfosuccinate
Riboflavin	Sodium deoxycholate
Salicylamide	Polysorbate 80
Salicylic acid	Various
Spironolactone	Polysorbate 80
Sulfasoxazole	Polysorbate 80
Thiourea	Alkylbenzene sulfonate, dodecyl trimethyl ammonium chloride

There are numerous examples in the literature where the use of surfactants in a formulation has resulted in increased absorption. Some of these examples are listed in Table 2.13.

The surface-active properties of gastrointestinal contents primarily involve the various fluids, for example, gastric juice and intestinal juice, and can be a major factor in the dissolution of several hydrophobic drugs. The most important component of gastrointestinal fluids is the high concentration of bile salts (100–200 mM) present, well above their critical micelle concentration (2–3 mM). The bile salt micelles solubilize polycyclic aromatic hydrocarbons, progesterone, cholesterol, estrone, griseofulvin, and reserpine and thus increase their absorption. Enhancement of drug absorption after meals is often related to the increased flow of bile that can solubilize the drug molecules, as demonstrated by griseofulvin. However, if micellization results in removing the otherwise available molecule for absorption, the BA may decrease.

Hygroscopicity

Water molecules have polar ends and readily form hydrogen bonding. As a result, several compounds interact with water molecules by surface adsorption, condensation in capillaries, bulk retention, and chemical interaction and are called hygroscopic. At times the interaction between compounds and water is so strong that interacting water vapors result in dissolving the compound, this process is called deliquescence, wherein there is formed a saturated layer of solution around particles. Most of these interactions are dependent on critical water vapor pressure or relative humidity. Moisture also induces hydrolysis and other degradation reactions; the presence of moisture also affects physical properties such as powder flow, dissolution, and even crystal structure. The impact of moisture on physical or chemical properties of compounds depends on the strength of bonding between water molecule and the surrounding space where water molecules are contained. In tightly bound state, the water molecules are generally not available to induce chemical reactions. Free water molecules can participate in the creation of a liquid environment around crystal lattice where the pH may be altered due to dissolution process. Similarly, water molecules held as crystal hydrates or trapped in an amorphous form are not available to modify the milieu interior of solid powders. It is noteworthy that some hydrates upon taking up moisture convert into hydrates; this transition can be useful in formulation studies, and this property should be tested for hygroscopic compounds.

The classification of compounds into different hygroscopic categories is based on two types of models: one where the relative humidity and temperature are kept constant and gain in the weight of compound is recorded such as the definitions of the European Pharmacopoeia and the compound tested is stored at 25°C for 24 h at 80% relative humidity. A slightly hygroscopic compound would show less than 2% m/m mass gain, hygroscopic compounds show less than 15%, and very hygroscopic compounds show more than 15% m/m mass gain; the deliquescent compounds simply liquefy. The dynamic model tests hygroscopic nature at various humidity; compounds showing no mass gain at 90% are called nonhygroscopic, while those not gaining at 90% are slightly hygroscopic, but those gaining 5% over a week period are called moderately hygroscopic. Where mass increases at 40%–50% humidity, these compounds are called very hygroscopic.

Generally, a compound that is very hygroscopic will be less desirable, but if studies show that, despite moisture uptake, the compound stays stable and workable in the formulation studies, this is an important consideration.

Particle Size

Of all the possible manipulations of the physical properties of drugs to yield better absorption, the reduction of particle size is the most widely exploited. Increased absorption due to reduction of particle size is a result of increased dissolution, which is in turn the result of a larger specific surface area being exposed to the fluids in the gastrointestinal tract or other sites of administration. For example, the breakdown of a 3 mm cu particle into 1 mm cu particles results in a 300% increase in the exposed surface area.

Reduction in particle size can be achieved by several methods, including milling, grinding, precipitating the drug on an absorbent, and dispersing the drug in an inert water-soluble carrier (referred to as a solid dispersion). The solid dispersion formulation techniques have received great attention in the recent past and provide an innovative method of particle size reduction. If a hydrophobic drug is dispersed in a hydrophilic medium in a solid state, a faster release of the drug can be expected from this system since the rate-limiting steps in the dissolution of the drug will be fewer. The state of drug dispersion can vary from microcrystalline to molecular, and thus, a wide range of dissolution rates are possible. For example, dispersion of sulfathiazole in urea results in a monomolecular dispersion (solid solution), and the dissolution rates are increased by almost 700 times. The dispersion of griseofulvin in polyethylene glycol 6000 results in an almost 100% increase in its BA as compared to the micronized form of griseofulvin. The solid dispersions are generally prepared by either fusing or dissolving the drug and the water-soluble carrier and then solidifying the melt or solution by cooling or evaporation. The drugs also often coprecipitate, as with the solid dispersion of reserpine and deoxycholic acid. Examples of drugs whose dissolution rates have been increased as a result of solid dispersion formulation include salicylic acid, reserpine, chloramphenicol, prednisone, salicylamide, and pentaerythritol.

Quite often solid dispersions can also be used to decrease the release of drugs, so as to provide sustained release as in the dispersion of chlorpheniramine in maleic anhydride copolymers.

The conventional methods of particle size reduction have long been employed to improve the BA of drugs; some of these examples include vitamin A, medroxyprogesterone acetate, 4-acetamidophenyl,2,2,2-trichlorethyl carbonate, nitrofurantoin, aspirin, phenobarbital, bishydroxycoumarin, phenacetin, chloramphenicol, procaine penicillin, cyheptamide, reserpine, digoxin, spironolactone, fluocinolone acetonide, sulfadiazine, griseofulvin, sulfisoxazole, *p*-hydroxypropiophenone, sulfur, and tolbutamide.

The reduction in the particle size is, however, not always desirable. For example, nitrofurantoin, when administered in its fine particle size, causes more gastrointestinal irritation than when administered in its coarser size. This is due to the higher plasma and gastrointestinal concentrations resulting from use of a fine particle size. The use of the coarser size is therefore preferred even though this results in retarded absorption. When chemical instability is a problem, the reduction of particle size is also contraindicated, as with penicillin G and erythromycin, which decompose in the gastrointestinal tract quickly. Even in the solid state, a small particle size means a greater surface area available for the absorption of moisture, which can result in an increased rate of decomposition. The reduction of the particle size of hydrophobic drugs also leads to increased surface charges (static), resulting in the agglomeration of the particles, especially in an aqueous media because of thermodynamic repulsion. This results in a significant decrease in the effective or exposed surface area available for dissolution. This problem can usually be resolved by adding appropriate surfactants that will reduce the interfacial tension and allow penetration of water molecules through the pocket of hydrophobic air surrounding these particles. It has been shown that the effect of particle size on dissolution of phenacetin with excellent correlation to the particle size. However, if the formulation adjuvants are removed, opposite effect is observed where decreasing particle size results in increased specific surface area that can absorb large quantities of air, reducing the effective surface area for dissolution. The addition of surfactants or other adjuvants that reduce this hydrophobic layer of air will increase the effective surface area and thus the dissolution rate. Gastric fluids have relatively lower surface tension, 43 dynes/cm compared to water, and may improve wetting effect of hydrophobic particles. The surface activity of gastric fluid is mainly due to the regurgitation of the intestinal fluids into the stomach. For example, phenacetin granules dissolve faster than the phenacetin powder in diluted gastric fluid.

The first step in the commencement of dissolution is the wettability of solid particles—there is a direct correlation between wettability and BA. Since the milieu of drug administration sites is mostly aqueous in nature, low wettability makes particles less hygroscopic.

Dissolution of salts leads to a change in the pH of the dissolution media because of the buffering effect; a base dissolved in acidic media increases the pH since the acidic counterions are trapped into salt forms. Similar as salts dissolve, the pH shift depends on whether it is acid or basic component that is weaker. The final balance is always dependent on the relative pK_a of the acidic and alkaline components. This is an important consideration as it explains the difference in the results obtained if the studies are conducted in water or buffer. When enteric protection is desired, the dissolution rates should be determined in 0.1 N HCl wherein many differences in the dissolution rates between water and buffer are obviated.

Dissolution of a solid usually takes place in two stages: salvation of the solute molecules by the solvent molecules followed by transport of these molecules from the interface into the bulk medium by convection or diffusion. The major factor that determines the dissolution rate is the aqueous solubility of the drug; however, other factors such as particle size, crystalline state (polymorphs, hydrates), pH, and buffer concentration can affect the rate. Moreover, physical properties such as viscosity and wettability can also influence the dissolution process.

Solubility

The discussions of pK_a, log P, and log D previously are relevant to understanding the factors that affect the solubility of a drug at the site of administration and thus determining the activity, toxicity, stability and dosage form, and route of administration. The USP classifies drugs based on their solubility (Table 2.14).

High solubility is defined as the highest dose strength that is soluble in 250 mL or less of aqueous media across the physiological pH range. Poorly soluble drugs can be defined as those with an aqueous solubility of less than 100 μg/mL. If a drug is poorly soluble, then it will only slowly dissolve, perhaps leading to incomplete absorption. Some general observations about the behavior of solutes in solution systems include the following:

1. Electrolytes dissolve in conducting solvents.
2. Solutes containing hydrogen capable of forming hydrogen bonds dissolve in solvents capable of accepting hydrogen bonds and vice versa.
3. Solutes having significant dipole moments dissolve in solvents having significant dipole moments.
4. Solutes with low or zero dipole moments dissolve in solvents with low or zero dipole moments.

There are always exceptions to these rules, but a good rule of thumb "like dissolves like" mostly applies. Therefore, solvents fall into three classes:

1. Protic solvents such as methanol and formamide that are hydrogen bond donors
2. Dipolar aprotic solvents (e.g., acetonitrile nitrobenzene) with dielectric constants greater than 15 but cannot form hydrogen bonds with the solute
3. Aprotic solvents in which the dielectric constant is weak and the solvent is nonpolar, for example, pentane or benzene

The solubility of ionizable compounds is pH dependent. For weak acids, as pH decreases, the solubility decreases. At equilibrium,

$$[\mathrm{HA}]\mathrm{solid} \rightarrow [\mathrm{HA}]\mathrm{solution} \tag{2.12}$$

TABLE 2.14
The USP Solubility Classification

Descriptive Term	Parts of Solvent Required for 1 Part of Solute
Very soluble	Less than 1
Freely soluble	From 1 to 10
Soluble	From 10 to 30
Sparingly soluble	From 30 to 100
Slightly soluble	From 100 to 1,000
Very slightly soluble	From 1,000 to 10,000
Practically insoluble or insoluble	10,000 and over

where S_o is the molar solubility assumed to be pH independent. The equilibrium dissociation constant is

$$K_a = [H_3O^+] * [A^-]/[HA] \quad (2.13)$$

or

$$[A^-] = K_a * [HA]/[H_3O^+] = K_a * S_o/[H_3O^+] \quad (2.14)$$

The total solubility (S) is expressed as

$$S = [A^-] + [HA] \quad (2.15)$$

or

$$S = K_a * S_o/[H_3O^+] + S_o \quad (2.16)$$

or

$$S - S_o = K_a * S_o/[H_3O^+] \quad (2.17)$$

$$\log (S - S_o) = \log K_a + \log S_o - \log [H_3O^+] \quad (2.18)$$

$$\log [(S - S_o)/S_o] = -\mathrm{p}K_a + \mathrm{pH} \quad (2.19)$$

Solubility

The solubility of a drug is the maximum amount of drug that dissolves at a specified temperature and solvent. It is a thermodynamic parameter given by

$$\ln N\,(\mathrm{sat}) = -\Delta H/R\,(1/T - 1/T_{mp}) \quad (2.20)$$

where

- N is the mole of solute dissolved in the solvent
- H is the enthalpy of dissolution
- T_{mp} is the melting point

Thus, the two characteristics that determine the solubility are the melting point and enthalpy of dissolution, both of which are dependent on such factors as chemical structure and physical state (polymorphism, etc.). These categories of factors will be discussed in detail.

Since the majority of drugs used are either weak acids or bases, their total solubility in any given media will depend on their ionization constant in the solution phase. The total solubility, S_t, of a drug is dependent on the relative contribution of the solubility of its unionized form, S_u, and the ionized form, S_i:

$$S_t = S_u + S_i \quad (2.21)$$

Whereas the solubility of the unionized form is a thermodynamic parameter, which is constant under given temperature, solvent, and pressure conditions, the fraction of the unionized form changes as a function of solution pH (Table 2.15).

In applying the equations given earlier to in vivo situations in man, one should remember that the stomach contents are usually in the pH range of 1–3 and the contents of the upper small intestine, where most drug absorption occurs, have a pH range of 5.5–7.0 and are not "alkaline," only less acidic. Weakly basic compounds will therefore generally dissolve faster in the gastric fluids,

TABLE 2.15
Ratio of Total and Unionized Drug as a Function of Difference in pH and pK_a

pH − pK_a	S_t/S_u (Acid)	S_t/S_u (Base)
−2.0	1.01	101.0
−1.5	1.03	32.6
−1.0	1.10	11.0
−0.5	1.32	4.16
0.0	2.0	2.0
+0.5	4.6	1.32
+1.0	11.0	1.10
+1.5	32.6	1.03
+2.0	101.0	1.01

and weakly acidic compounds will dissolve faster in the intestinal fluids. For example, salicylic acid with a pK_a of 3.0 shows an approximately 16-fold increase in its dissolution rate when the surrounding pH is changed from 1.5 to 6.8. However, for weak acids, a linear relationship between $1/H^+$ and its dissolution rate or, for weak bases, a linear relationship between the dissolution rate and H^+ may not always be possible, especially around neutral pH ranges. This is due to the mechanism of dissolution, which involves formation of a diffusion layer saturated with the dissolving compound, weak acid or base, which results in pH values that may not correspond to the pH values of the bulk medium. For example, dissolution of salicylic acid will result in a pH around the dissolving particles lower than the bulk pH of intestinal fluids. It is this pH of the diffusion layer that determines the actual rate of dissolution. As discussed earlier, inclusion of such agents as sodium bicarbonate in aspirin formulations results in a higher pH in the diffusion layer, increasing the dissolution rates.

In describing the dissolution rates by using the Noyes–Whitney equation, the term saturation solubility is replaced by the solubility in the boundary layer around the dissolving particle, and thus, for weak acids, the equation is transformed as follows:

$$\mathrm{d}C/\mathrm{d}t = K_s\,(C_s(1 + K_a/H^+) - C_t) \quad (2.22)$$

and for weak bases,

$$\mathrm{d}C/\mathrm{d}t = K_s\,(C_s(1 + H^+/K_a) - C_t) \quad (2.23)$$

Thus, for weak acids, dissolution rate increases at basic pH such as shown for tolbutamide, and for weak bases, the dissolution rate increases at acid pH, such as shown for tetracycline. Tolbutamide dissolution in gastric fluid is 15 times lower than in simulated intestinal fluid, whereas its salt has slightly higher dissolution in acid media. For tetracycline, the dissolution rate is decreased 2600 times in intestinal fluid, whereas its hydrochloride salt shows almost 100% increase in dissolution in the intestinal fluid compared to gastric fluid.

It should be noted that the dissolution of some dosage forms is dependent on pH such as used for enteric coated forms or other designs where the dissolution and disintegration of the dosage form may be highly pH dependent.

It is generally agreed that the unionized form of a drug is most suitable for gastrointestinal absorption. Thus, the efficiency of absorption of a weakly acid or weakly basic compound will change as the dosage form passes through various pH conditions in the gastrointestinal tract. This theory is also referred to as the pH-partition theory and holds true for a variety of drugs. However, if one takes into account the large differences in the absorption surface areas of the stomach, the small intestine, and the colon, it seems logical to assume that most of the drugs will show sufficient absorption from the upper part of the small intestine if equilibrium is always established between the unionized absorbable species and the ionized form of the drug. For example, in situ studies show that at pH 6.8, where salicylic acid is almost 100% ionized, the absorption is very fast from the rat intestine (50% absorbed in 7 min).

Molecular Size

Large organic molecules have a smaller aqueous solubility than smaller molecules, this being due to interactions between the nonpolar groups and water, that is, solubility is dependent on the number of solvent molecules that can pack around the solute molecule.

Poorly soluble compounds represent an estimated 60% of compounds in development and many major marketed drugs. It is important to measure and predict solubility and permeability accurately at an early stage and interpret these data to help assess the potential for the development of candidates. This requires developing an effective strategy to select the most appropriate tools to examine and improve solubility in each phase of development and optimization of solid-state approaches to enhance solubility including the use of polymorphs, cocrystals, and amorphous solids. All of these would affect the dissolution rates and BA that can be studied with nanocrystal technology.

With this trend of increasingly insoluble drugs stretching resources, many companies are now reevaluating their strategy. They know that there are many available technologies to measure and predict and finally improve solubility and several new techniques emerging. Studies that encompass this scope would include how membrane permeation of drugs can be enhanced by means of solubilizing agents, how the solid state is characterized and modified to improve solubility and drug performance, and how salt screening and selection can impact on dissolution rate and oral absorption, apply nanocrystal technology to increase dissolution rate, and analyze the use of pharmaceutical cocrystals in enhancing drug properties.

Many different approaches have been developed to overcome the solubility problem of poorly soluble drugs, for example, solubilization, inclusion compounds, and complexation. A basic disadvantage in these formulation approaches is that these can only be applied to a certain number of drugs exhibiting special features required for implementing the formulation principle (e.g., molecule fits into the cavity of the cyclodextrin ring). The use of solvent mixtures is also very limited due to toxicological considerations. In addition, more and more newly developed drugs are poorly soluble in aqueous media and simultaneously in organic media, thus excluding the use of solvent mixtures. Ideally, the formulation principle should be able to be applied to all or at least most of the poorly soluble drugs.

Solubilizers (e.g., organic solvents, detergents, pluronics) are often used to solubilize drugs in aqueous solution without considering their effects on biological systems such as lipid membranes and multidrug resistance efflux transporters (e.g., *P*-Glycoprotein or MDR1). Liposomal solubilization is an effective approach for the delivery of potent, insoluble drug candidates.

An alternative to other methods developed is the production of drug nanoparticles by high-pressure homogenization either as pearl milling or the continuous high-pressure homogenization. Of importance is the consideration of metallic contamination during fast-speed milling processes to keep it less than 1 ppm. Drug nanoparticles are produced by dispersing the drug powder in an aqueous surfactant solution; the obtained presuspension is passed through a high-pressure piston-gap homogenizer, for example, 5–20 homogenization cycles at typically 1000–1500 bars, and works on the principle that cavitation occurs in the aqueous phase. The particle suspension has a very high flow velocity when passing the tiny gap of the homogenizer, the static pressure on the water decreases below the vapor pressure of water, the water starts boiling at room temperature leading to the formation of gas bubbles, and at the exit of the gap, the gas bubbles implode. The implosion shock waves disintegrate the drug particles to drug nanoparticles. Further improvement on nanoparticle production includes homogenization in nonaqueous phases or with reduced water content to produce more pronounced cavitation at higher temperatures. The chemical stability of drugs is less impaired when homogenizing in nonaqueous or water-reduced media at low temperatures. The drug powder is dispersed in a nonaqueous medium (e.g., PEG 600, Miglyol 812) or a water-reduced mixture (e.g., water–ethanol) and the presuspension homogenized in a piston-gap homogenizer. A suitable machine for lab scale is the Micron Lab 40 (APV Deutschland GmbH, Lübeck, Germany). Ostwald ripening occurs due to different saturation solubilities in the vicinity of very small and of larger particles. The particles produced are relatively homogeneous. The differences in the size in combination with the generally poor solubility of the drug nanoparticles are sufficiently low to avoid Ostwald ripening. Aqueous drug nanoparticle suspensions generally prove to be physically stable for several years.

The application of micronization and nanonization is increasing surface area leading to an increased dissolution rate according to the Noyes–Whitney equation. However, this is only one aspect. The dissolution pressure is a function of the curvature of the surface that is much stronger for a curved surface of nanoparticles. Below a size of approximately 1–2 µm, the dissolution pressure increases distinctly leading to an increase in saturation solubility. In addition, the diffusion distance h on the surface of drug nanoparticles is decreased, thus leading to an increased concentration gradient $(C_s - C_x)/h$. The increase in surface area and increase in concentration gradient lead to a greater increase in the dissolution velocity compared to a micronized product. In addition, the saturation solubility is increased as well, even though it is a thermodynamic parameter; the increase in solubility occurs as the supersaturation stage is reached. Saturation solubility and dissolution velocity are important parameters affecting the BA of orally administered drugs. From this, nanoparticles have the potential to overcome these limiting steps.

Nanoparticle-based products are likely to have some unique characteristics: general adhesiveness of nanoparticles to the gut wall; adhesion to the gut wall being a reproducible process, thus minimizing variation in drug absorption; increase in dissolution velocity overcoming this rate-limiting step; and additionally increase in saturation solubility leading to an increased concentration gradient between gut and blood. Orally administered drug nanoparticles can increase the BA and can be the only tool available to achieve a sufficient BA with poorly soluble drugs. However, the possibility of faster absorption may have its own drawbacks, both from pharmacology and stability in the gut. For intravenous administration, the drug nanoparticles should possess a bulk population in the nanometer range by simultaneously having a low microparticle content, that is, especially particles larger than 5 µm that can cause capillary blockade. The homogenization process yields a product with minimized content of particles larger 1 µm. Intravenous administration of drug nanoparticles allows achievement of sufficient blood levels and finds good application in the evaluation of new compounds. In addition, toxicologically critical excipients such as Cremophor EL used in Taxol formulations can be avoided when stabilizing the drug nanoparticles with accepted emulsifiers, for example, lecithin or Tween 80. It is interested to note that when Taxol is administered with a Cremophor EL, the pharmacokinetics of drug turn out to be nonlinear. For intravenous administration, a small particle size below 150 nm is only desirable in case one wants to pass fenestrated endothelia (e.g., treatment of tumors); however, this is a very limited case. More realistic and short-term achievable goals are passive targeting of drugs to treat mucopolysaccharidosis (MPS) infections (i.e., targeting to the macrophages, e.g., treatment of *Mycobacterium tuberculosis* and *M. avium* infections, especially in HIV patients). Here, it is more desirable to have larger particles to ensure fast and efficient removal from the blood streams by the macrophages. Another therapeutic goal is the creation of stealth drug nanoparticles circulating in the blood, minimizing free drug concentration, but simultaneously prolonging the drug release by slow dissolution. For this purpose, very small particles are not suitable because they will dissolve too fast. Another therapeutic goal is targeting to non-MPS targets, for example, the brain and the bone marrow.

The particle size should be customized depending on the therapeutic requirements and purpose. The nanoparticle suspensions are physically stable on long term in case they are stabilized by emulsifiers/polymers in optimized composition. However, aqueous suspensions might not be the most convenient dosage form for the patient. The nanoparticle suspension can be used as granulation fluid to produce tablets or as wetting liquid for pellet production. The dispersions can also be spray dried to be filled into hard gelatin capsules or sachets. Drug nanoparticles produced in Peg 600 or Miglyol can directly be filled into soft gelatin capsules. Lyophilization of drug nanoparticles produced in water-reduced media can be used to produce fast-dissolving delivery systems. For parenteral application, nanoparticles can be lyophilized and reconstituted prior to injection with isotonic media (e.g., water with glycerol). There are also other areas of application, for example, ocular delivery (prolonged retention time) or topical application (increased saturation solubility leading to increased diffusion pressure into the skin).

DISSOLUTION

Dissolution is the conversion of solid state (highly aggregated state) to a solution state (highly dispersed state). Some of the key factors that affect this transition are the solubility of the drug, the diffusion process, hydrodynamic processes, and possible reactivity of the solute to solvents. The dissolution rates of dosage forms are affected by additional factors that almost invariably affect BE of the drug products. In order to understand the basic factors that affect dissolution rates, it is necessary to examine various mathematical models that describe the kinetic phenomenon of dissolution. The dissolution models are derived from the known principles of physics and chemistry such as Fick's laws of diffusion, concentration or chemical potential gradients, and the hydrodynamic principles. Since all of these models require simplification, sometimes oversimplification of the actual mechanism of dissolution, more often than not, deviations are observed between theoretical and actual rates of dissolution, which are often corrected by introducing a variety of empirical constants.

Diffusion Model

This is the simplest model where the solvent phase in contact with the solid surface becomes saturated with the solute, and if there is no turbulence in the system, the liquid at the surface remains motionless, and the dissolution across the liquid is primarily the function of the diffusion of molecules across and is given by

$$M = 2C_sADt \tag{2.24}$$

where

- M is the amount dissolved at time t
- C_s is the solubility in the medium
- A is the surface area
- D is the diffusion coefficient in the medium

This is entirely a diffusion-dependent model that does not take into consideration buildup of drug concentration in the solution (sink condition) and movement of liquid past the dissolving surface. Since the concentration buildup in the bulk solvent decreases the concentration gradient, dissolution process slows down with time.

Convection Model

If the liquid is moving past the dissolving surface, convection process is set up along with the diffusion process. It should be noted that the diffusion process is taking place normal to the dissolving surface, whereas the fluid flow is parallel to it as shown in Figure 4.3. The following equation describes the dissolution process in convection:

$$M = 0.81D^{2/3}\,C_s x^{1/3} b L^{2/3} t \tag{2.25}$$

where
- D is the rate of shear
- b is the width of the dissolving surface
- L is the length in the direction of flow

Unlike the previous model where a buildup of concentration in the medium slows down the dissolution, in this model, a sink condition exists, and no change in the dissolution is noted. The sink condition is defined as the state where C is less than 10% of the C_s. However, as the fluid flows past the surface, the dissolution rate due to diffusion slows down toward the ends of the surface due to concentration buildup in the diffusion layer as shown in Figure 4.3.

Surface Reaction Model

The models described previously assume that a saturation concentration of the drug is maintained in the diffusion layer. However, in some instances, the surface reaction leading to that state may be rate-limiting step, and thus, the dissolution rate is given by the following equation:

$$M = AC_s t/(1/K_s + h/D) \tag{2.26}$$

where
- K is the surface rate constant
- h is the thickness of the hypothetical diffusion layer

The concept of having an unstirred diffusion layer in this model is often questioned on the basis of hydrodynamic theory.

Cube Root Model

In most instances, dissolution occurs from a suspended particle wherein the total surface area changes with time; the dissolution rate for such a model is given by

$$M_o^{1/3} - \left(M_o - M\right)^{1/3} = Kt \tag{2.27}$$

where
- M is the mass of the powder dissolved at time t
- M_o is the initial mass

The constant K is directly proportional to the diffusion coefficient, the solubility, and the cube root of the number of particles, the particle size, and the diffusion layer thickness and is thus referred to as model-dependent constant. This model provides excellent fit for dissolution of single particles. In an actual system, a large number of particles exist with different (often log normal/distribution) diameters, which decrease with time. These situations are also well characterized by the previous model.

Tablet Dissolution Model

The dissolution of dosage forms, for example, a tablet, is preceded by its disintegration, which is given by

$$q = \mathrm{d}m/K\mathrm{d}t + m \tag{2.28}$$

where
q is the fraction of the tablet disintegrated
m is the fraction that has dissolved

As the disintegration rate becomes faster, the shape of the dissolution curve becomes more exponential and less sigmoidal. The fitting of dissolution data to equations is an empirical process such as used in "sigma-minus" plots where log of the amount of drug remaining in the dosage form versus time is plotted:

$$\ln (M_0 - M) = -Kt \tag{2.29}$$

where
M_0 is the dose (either actual or the amount dissolved at infinity time, a very important factor)
K is the first-order rate constant

Since there is often a lag time before the dissolution becomes appreciable, this function can also be introduced into the dissolution equation:

$$\ln (1 - M/M_0) = -K(t - T) \tag{2.30}$$

where T is the lag time. This lag time may also be associated with dosage form characteristics such as breaking down of a coating. This lag function can also be introduced into the cube root model described earlier (Equation 2.27).

Another common equation used to fit dissolution data is called Rosin–Rammler–Sperling–Weibull equation where T is lag time and a and b are adjustable parameters, wherein a is the scale parameter and b is the shape parameter:

$$-\ln (1 - M) = (t - T)b/a \tag{2.31}$$

The value of b determines whether the curve has sigmoidal or exponential shape. This flexibility in the use of equation allows its use in most common types of dissolution curves. The use of log-normal probability graph paper has also been made to linearize dissolution rate data.

Noyes–Whitney Model

The classic model for describing dissolution rates is given by Noyes–Whitney equation:

$$\mathrm{d}C/\mathrm{d}t = KS(C_s - C) \tag{2.32}$$

where
$\mathrm{d}C/\mathrm{d}t$ is the rate of dissolution
C_s is the saturation concentration of drug in the diffusion layer
C_t is the concentration of drug in dissolution media (or the bulk)
S is the surface area of the dissolving solid

K is the dissolution rate constant and is given by

$$K = \frac{D}{h} \tag{2.33}$$

where
D is the diffusion coefficient
h is the thickness of diffusion layer

This equation is of great value in the formulation studies wherein increase in the surface area of aggregates is the most powerful tool to optimize dissolution. In dissolution theory, it is assumed that an aqueous diffusion layer or stagnant liquid film of thickness h exists at the surface of a solid undergoing dissolution, as observed in Figure 2.12. This thickness h represents a stationary layer of solvent in which the solute molecules exist in concentrations from C_s to C. Beyond the static diffusion layer, at x greater than h, mixing occurs in the solution, and the drug is found at a uniform concentration, C, throughout the bulk phase (Figure 2.12).

The diffusion layer model of dissolution assumes that the dissolution of drug at the solid/liquid interface into a concentrated layer surrounding the solid particle is more rapid than the diffusion of dissolved drug from that layer into the bulk solution. This diffusion is therefore rate limiting in observed dissolution. Since diffusion involves kinetic energy, it is highly dependent on the temperature. For an ideal solution, no heat is absorbed or given off upon dissolution; however, for a real solution, the heat of solution (ΔH) can be either negative (heat is given off) or positive (heat is absorbed). The mathematical relationship of solubility (C_s) to temperature is

$$\log C_s = (-\Delta H/2.303RT) + \text{Constant} \tag{2.34}$$

where
R is the gas constant
T is the absolute temperature

A plot of log C_s versus $1/T$ gives the value of the constant. A heat effect depends on whether the material absorbs heat (an endothermic process) or gives off heat (an exothermic process) when it dissolves. Most materials absorb heat as they dissolve. According to Le Chatelier's principle, a system at equilibrium will adjust in such a manner as to reduce external stress. Therefore, if a substance absorbs heat when it dissolves and heat is added to the system, equilibrium can be restored, that is, the external stress can be reduced, by the absorption of heat. This can only be done in such a system by the dissolution of more of the substance, that is, an increase in solubility at the higher temperature until the equilibrium is restored.

The thermodynamic driving force for dissolution is therefore the heat of solution of the substance. For a crystalline solid, this represents the difference between the heat of sublimation of the

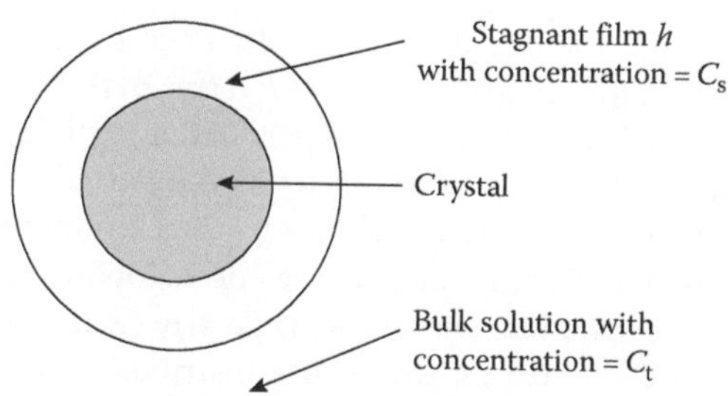

FIGURE 2.12 Diffusion layer model of dissolution.

compound and the heat of hydration of the ions. The heat of sublimation is the heat required to bring ions from the solid state to the gaseous state and is a measure of the energy required to pull apart the crystalline lattice. The heat of hydration is the heat given off by the hydration of those ions. For dissolution to be an endothermic process, the heat of sublimation is greater than the heat of hydration and ΔH is positive, that is, the heat is absorbed upon dissolution; therefore, solubility increases with an increase in temperature. If heat of sublimation is equal to the heat of hydration, solubility is independent of temperature.

DISSOLUTION FACTORS

Regardless of the dissolution model chosen, fundamental considerations of the physicochemical nature of the drug significantly affect dissolution.

CONCENTRATION GRADIENT

The saturation solubility of the drug in the diffusion layer determines the dissolution rate by providing the driving force for dissolution, the difference between C_s, the saturation C, and the concentration of dissolved drug in the bulk fluids, for example, in the gastrointestinal tract. The C is generally much smaller, meaning that dissolution occurs under sink conditions (defined accurately as condition where C_s 0.1°C). However, if a drug is absorbed slowly, the concentration in the gastrointestinal fluids may increase, thus decreasing the concentration gradient and the dissolution rate.

Additionally, the bulk fluids may not be identical to the dissolving fluids of the diffusion layer. A drug may be more soluble in the diffusion layer and then precipitate in the bulk fluids, especially if the pH differs. However, these precipitated particles are generally quite small and redissolve rapidly. If the drug is more soluble in the bulk fluids than in the diffusion layer because of a difference in pH or due to the complexation with other components in the bulk fluid, the concentration may increase and dissolution may increase, but if the solubility of the drug is lower in the bulk medium, the dissolution rate will slow down or even stop. Furthermore, the volume of the bulk fluids is much larger than the volume of the diffusion layer, resulting in smaller C despite dissolution of large amounts. In most instances, an increase in C that would affect dissolution rate would only occur when other process such as membrane transport or stomach emptying becomes the rate-limiting step in drug absorption. However, in general, it is advised that patients take their oral medications with a full glass of water in order to ensure that the drug dissolves adequately. Another advantage in taking drugs with large volumes of fluid is that it results in greater contact with the absorption surface, resulting in faster and higher absorption.

The buffering agent provides higher pH in the diffusion layer increasing the dissolution of weakly acidic drug, which precipitates and redissolves in the gastrointestinal tract. In the absence of buffering agent, the pH of the diffusion layer will be acidic as a result of the dissolution of weak acid.

DISSOLUTION CONSTANT

Although it is possible to control the dissolution rate of a drug by controlling its particle size and solubility, there is very little, if any, control over the D/h term. In this equation, it is assumed that h, the thickness of the stationary diffusion layer, is independent of particle size. In fact, this may not be true. The diffusion layer thickness generally increases as the particle size increases. Furthermore, h decreases as the "stirring rate" increases. In vivo, as gastrointestinal motility increases or decreases, h would be expected to decrease or increase correspondingly. Another assumption made here is that all the particles are spherical and of the same size, whereas in reality, the particles are polydisperse and of multiparticulate nature whose size distribution in terms of the number of particles tends to be skewed toward the smaller particles. Furthermore, as dissolution proceeds, the particles become smaller, and hence, h is more a variable than a constant.

Another uncontrollable term is D, the diffusion coefficient of the drug. For a spherical molecule in solution, it is given by the following:

$$D = kT/6nr \tag{2.35}$$

where

k is the Boltzmann's constant
T is the absolute temperature
r is the radius of molecule in solution
n is the viscosity of the solution

The two variables in this equation are the viscosity and temperature. Increasing the viscosity of the gastrointestinal fluid will decrease dissolution and will slow gastric emptying, thus delaying delivery of the drug to the absorption site. Increasing the temperature of the gastrointestinal fluids increases diffusion, and thus, taking oral dosage forms with warm liquids may be advised. However, extremely hot liquids generally delay stomach emptying. Thus, whereas D and h are regarded as constants, these may be variable under vivo conditions.

DISSOLUTION TESTING

Ideally, dissolution should simulate in vivo conditions. To do this, it should be carried out in a large volume of dissolution medium, or there must be some mechanism whereby the dissolution medium is constantly replenished by fresh solvent. Provided this condition, if met, the dissolution testing is defined as taking place under sink conditions. Conversely, if there is a concentration increase during dissolution testing, such that the dissolution is retarded by a concentration gradient, the dissolution is said to be nonsink. While the use of the USP paddle dissolution apparatus is mandatory when developing a tablet, the rotating disk method has great utility with regard to preformulation studies. The intrinsic dissolution rate is the dissolution rate of the compound under the condition of constant surface area. The rationale for the use of a compressed disk of pure material is that the intrinsic tendency of the test material to dissolve can be evaluated without formulation excipients.

The dissolution testing is performed not only on the finished products but also on the pure drug and in combination with various excipients to ascertain individual contributions of the components to overall dissolution. Basically, the dissolution test systems are of two types: the stirred-vessel type and the flow-through column. In the stirred type, agitation is provided by some kind of paddle, whereas in the column type, the solvent flows over the drug. A large number of variations of these systems are currently used. However, the USP apparatus is used for official certification of batches. The monographs describe the specific temperature, the dissolution medium (distilled water, simulated gastric fluid, or simulated intestinal fluid), the rotation speed of the basket (60–150 rpm), and the percentage of drug to be dissolved as an endpoint. These conditions are determined by the intrinsic properties of the drug and its dissolution behavior. The list of drugs included in official compendium where dissolution test must be conducted as requirement of dosage form release is extensive and likely to grow; some examples include acetohexamide, nitrofurantoin, digoxin, phenylbutazone, ergotamine tartrate and caffeine tablets, prednisolone, hydrochlorothiazide, prednisone, lithium carbonate, sulfamethoxazole, meprobamate, sulfisoxazole, methaqualone, theophylline, ephedrine, methylprednisolone, hydrochloride and phenobarbital tablets, and tolbutamide.

In the official dissolution tests, 6 or 12 tablets or capsules are tested individually for their dissolution properties. In the first stage, 6 units are tested and each unit must fall within less than 5% of the specified limit (e.g., 60% dissolved in 30 min). If one or more units fail, then another 6 units are tested, and the average of 12 units (6 from first test) should be equal to or greater than the specified percentage, and no unit should be less than 15% of the specified limit. If this stage also fails, then additional 12 units are tested, and the average of all (now 24) should be equal to or greater than the specified limit, and not more than 2 units can be less than 15% off the limit.

An inherent problem in this type of testing is that it requires the use of labeled amount of drug for calculation purposes, and any content variability is not considered. Conceivably, large variations in the dissolution rates are possible due to these differences, for example, tablets containing 80%–120% of the labeled amount will require 75%–50% dissolution if the requirement is 60%. In addition, the statistical design of the dissolution testing allows a batch with 20% defective tablets to pass 58% of the time. There are also serious problems in the reproducibility of dissolution data since the dissolution is dependent on human errors and such subtle factors as the vibrations in the room. Despite these drawbacks, FDA considers dissolution testing to be the most discriminating in vitro test with which to establish in vivo correlations.

The dissolution media comprise the fluids at the site of administration. Composition of these fluids may increase or decrease the solubility of a drug. In the case of salt, those that increase the solubility are said to "salt in" the solute, and those that decrease the solubility to "salt out" the solute. The effect of the additive depends very much on the influence it has on the structure of the water or its ability to compete with solvent–water molecules. Both effects are described by the empirically derived Setschenow equation:

$$\text{Log } [S_o]/[S] = kM \tag{2.36}$$

The aforementioned equation describes the relationship between the aqueous solubility of sparingly soluble salts (S_o) and the empirical Setschenow salting-out constant $k = 0.217/S_o$. This relationship and the Setschenow equation are valid only at low concentrations of added salt. As the concentration of added salt increases, the apparent k value is not constant but is dependent on solubility and the rate of change of solubility with added salt concentration. It was concluded that the Setschenow treatment is generally inappropriate for description and analysis of common ion equilibrium.

Therefore, since S_o is assumed to remain constant and pK_a is a constant, if pH decreases, the value of S must also decrease. In a similar manner, the solubility of a weak base decreases as pH increases.

Another aspect of the effect of electrolytes on the solubility of a salt is the concept of the solubility product for poorly soluble substances. The experimental consequences of this phenomenon are that if the concentration of a common ion is high, then the other ion must be come low in a saturated solution of the substance, that is, precipitation will occur. Conversely, the effect of foreign ions on the solubility of sparingly soluble slats is just the opposite, and the solubility increases. This is called salt effect.

Since dissolution is usually an endothermic process, increasing solubility of solids with a rise in temperature is the general rule. Therefore, most graphs of solubility plotted against temperature show a continuous rise, but there are exceptions, for example, the solubility of sodium chloride is almost invariant, while that for calcium hydroxide falls slightly from a solubility of 0.185 g/mL at 0°C to 0.077 g/mL at 100°C.

3 Drug Delivery Factors

BACKGROUND

Drugs must be present in a solution form to traverse the biological barriers such as gastrointestinal (GI) mucosa. Thus, the process of dissolution becomes an integral part of the various rate-limiting steps leading to a clinical response. However, dissolution alone is not sufficient to provide the absorption of drugs. The drug molecules must have the characteristics required for crossing the various lipoid layers or membranes in order to reach the general circulation. Lack of sufficient aqueous solubility is usually the rate-limiting step in the dissolution process, and lack of sufficient lipophilic properties is the usual rate-limiting step in the penetration of the lipoid barriers. Attempts to rectify dissolution problems can therefore lead to problems in membrane transport, and vice versa. A fine balance between the hydrophilic and lipophilic properties is needed to provide optimum delivery of drugs to the site of action.

The variations in bioavailability extend to almost all classes of drugs, and as a result, the chemical modifications required for optimum bioavailability are difficult to summarize. For example, bioavailability of several antibiotics varies broadly even though they may be in the same structural or pharmacological class. Besides chemical modifications, formulation manipulations also significantly affect bioavailability. For example, almost 60-fold differences have been reported in the rates of absorption of different formulations of spironolactone. The regulatory agencies have well recognized the drugs with potential bioequivalency problems. Most of these drugs are highly potent, with log-dose–response curves and some exhibiting "all or none" effects. It is important to recognize that bioavailability variations can cause significant pharmacological response variability. For those drugs where a minimum therapeutic level (e.g., minimum inhibitor concentration of antibiotics) must be achieved, lower bioavailability may mean totally ineffective dose. It should be noted that similar variations in bioavailability result in different changes in pharmacological response depending on the potency of drugs and the segment of log-dose–response curve at which the dose is administered. For instance, doses at less steep ranges of very low or very high doses result in proportionally smaller changes in the pharmacological responses as a result of bioavailability variations.

An identical variation in the bioavailability results in a significantly higher variation in the pharmacological response for high-potency drugs compared with low-potency drugs.

The purpose of dosage formulation is to design a dosage form with a suitable combination of the following attributes:

- Contains the labeled amount of drug in an active form
- Is free from extraneous materials
- Consistently delivers the drug to the general circulation at an optimum rate and to an optimum extent
- Is suitable for administration through an appropriate route
- Is acceptable to patients

The dosage form characteristics, such as particle size, salt form, solvent type, and dissolution rate, as well as the various additives, all contribute to the dosage form design. The additives may be pharmacologically inert, as with tablet binders and lubricants, or they may have the function of modifying the absorption, the biotransformation, or the excretion of the primary therapeutic agents.

A large number of formulation factors are common to many dosage forms and some of these are summarized here to make the reader aware of their complexity:

- The vehicle must be either miscible or spreadable throughout the biologic tissue before partitioning and absorption can take place.
- Sugars in the formulation increase viscosity, delay gastric emptying, and also alter passive drug diffusion by fluid uptake and other mechanisms.
- Various buffer systems affect surface tension, pH, and fluid uptake and this causes altered drug absorption.
- Surfactants affect solubility, dissolution, diffusion across lumen, and GI membrane permeability.
- Complexing agents affect solubility; partition coefficient can form nonabsorbable complexes.
- Chelating agents added to retard oxidation affect intestinal membrane permeability.
- Dyes adsorb on crystal surfaces and often retard dissolution.
- Absorbents such as kaolin, attapulgite, talc, and activated charcoal reduce the rate and extent of drug absorption.

In many instances, several different dosage forms are available for a given drug, and an appropriate selection must be made based on the attributes listed earlier. In general, the dissolution and hence absorption of drugs from the dosage form depends on the degree of dispersion. The following discussion attempts to characterize various dosage forms and provides a rational basis for their selection.

SOLID DOSAGE FORMS CONSIDERATIONS

Most pharmaceutical companies would rather have their new molecule enter the market as a tablet or capsule for a variety of safety, cost, and marketing considerations. As a result, almost 70% of all drugs administered today are in solid dosage forms. When so intended, the default form should be solid dosage form (unless it is predetermined in the case of therapeutic proteins or other drugs that must be administered by parenteral route or other specific routes for specificity of activity desired). The typical parameter studies for solid dosage forms relate to the ability of a powder mix, to flow well in manufacturing machines, as well as to the intrinsic characteristics that make it compressible. Some examples of properties studied include crystal structures (polymorphs); external shapes (habits); compression properties; cohesion; powder flow; micromeritics; crystallization; yield strengths and effects of moisture and hygroscopicity; particle size; true, bulk, and tapped density; and surface area.

Particle Size

The particle size of new drug substance is a critical parameter as it affects every phase of formulation and its effectiveness. Appropriate particle size is required to achieve optimal dissolution rate in solid dosage forms, control sedimentation, and flocculation in suspensions; small particle size (2–5 μm) is required for inhalation therapy; and content uniformity and compressibility is governed by particle size. As a result, the preformulation studies must develop a specification of particle size as early as possible in the course of studies and develop specifications that need to be adhered to throughout the studies.

Conventional methods of grinding in mortar or ball milling (where sample quantity is sufficient; generally, it is not limited to about 25–100 mg) or micronization techniques are used to reduce the particle size. The method used can have significant effect on the crystallinity, polymorphic structures (often to amorphous forms), and drug substance stability that can range from discoloration to significant chemical degradation. Changes in polymorphic forms can be determined by performing XRPD before and after milling.

Micronization where possible allows increase in the surface area to the maximum, which can impact on the solubility, dissolution, and, as a result, bioavailability. Since the aim of most preformulation studies is to determine if a solid dosage form can be administered, knowing that reduction of particle size where it changes dissolution rates can be pivotal in decision making for the selection of dosage forms. In the process of micronization, the drug substance is fed into a confined circular chamber where it is suspended in a high-velocity stream of air. Interparticulate collisions result in a size reduction. Smaller particles are removed from the chamber by the escaping air stream toward the center of the mill where they are discharged and collected. Larger particles recirculate until their particle size is reduced. Micronized particles are typically less than 10 μm in diameter. In some instances, micronization can prove counterproductive, where it results in increased aggregation (leading to reduced surface area) or alteration of crystallinity, which must be studied using such methods as microcalorimetry, dynamic vapor sorption (DVS), or inverse gas chromatography.

The introduction of DVS in 1994 revolutionized the world of gravimetric moisture sorption measurement, bringing outdated, time- and labor-intensive desiccator used into the modern world of cutting-edge instrumentation and overnight vapor sorption isotherms. With a resolution down to 0.1 μg, a 1% change in mass of a 10 mg sample on exposure to the humidity-controlled gas flow is both easily discernable and reproducible. DVS is a valued tool for studies related to polymorphism, compound stability, and bulk and surface adsorption effects of water and organic vapors. The DVS studies would typically show percent mass increases, but often, a hysteresis loop relationship is observed where there is crystallization of compound that results in the expelling of excess moisture. This effect can be important in some formulations, such as dry powder inhaler (DPI) devices, since it can cause agglomeration of the powders and variable flow properties. The DVS is a useful study when amorphous forms are involved upon size reduction; in many cases, a low level of amorphous character cannot be detected by techniques such as XRPD; microcalorimetry can detect <10% amorphous content (the limit of detection is 1% or less). The amorphous content of a micronized drug can be determined by measuring the heat output caused by the water vapor inducing crystallization of the amorphous regions.

Sieve Analysis

Dry sieving allows the fractionation of relatively coarse powders and granules. Sieves are stacked (*nested*) with the largest apertures at the top and the smallest at the bottom. A sample of powder is placed on the top sieve and shaken for a fixed time period at a given amplitude and pulse frequency. Variation in sieve sizes for powders is known to produce bioequivalency problems.

Particle Size Distribution

Sieving is a common method for establishing the distribution of particle size in a powder sample. It is a simple method that works well for powders in the size ranges used most often in the pharmaceutical industry. Sieves are limited in that they cannot be made with very small openings. The current lower limit is 43 mm, which corresponds to a No. 325 sieve. The sieve number or mesh number refers to the number of openings per linear inch. You can easily calculate the opening size in millimeters. For example, a No. 2 sieve has an opening of 9.52 mm, while a No. 200 sieve has an opening of 0.074 mm.

A frequency histogram is a useful tool in understanding the nature of a sample of powder. It is a bar graph with the size range on the *x*-axis and the number or weight of each segment of the powder on the *y*-axis.

Surface Area

The large surface area of powders provides greater opportunity for the production of static electricity during the friction of flow and handling. Make sure all equipment is well grounded or else significant segregation and impeded flow of powder can result. Monodisperse systems of particles of regular shape, such as perfect cubes or spheres, can be described completely by a single parameter; however, when either nonuniform size distribution or anisometric shapes exist, any single parameter is incapable of totally defining the powder. In addition to a value for the average particle size, often, we use frequency histograms to help describe the powder. We also use other measures of powder characteristics such as angle of repose and bulk or tap density. Lastly, we use compressibility and the powder's ability to undergo plastic deformation.

Since the surface area exposed to the site of administration determines how fast a particle dissolves in accordance with the Noyes–Whitney equation, these determinations are important. Also in those instances where the particle size is difficult to measure, a gross estimation of surface area is the second-best parameter to have to characterize the drug. The most common methods of surface area measurement include gas adsorption (nitrogen or krypton) based on what is most commonly described as the Braunauer, Emmet, and Teller, or BET, method applied either as a multipoint or single-point determination.

Porosity

Most solid powders contain a certain void volume of empty space. This is distributed within the solid mass in the form of pores, cavities, and cracks of various shapes and sizes. The total sum of the void volume is called the porosity. Porosity strongly determines important physical properties of materials such as durability, mechanical strength, permeability, and adsorption properties. The knowledge of pore structure is an important step in characterizing materials, predicting their behavior.

There are two main and important typologies of pores: closed and open pores. Closed pores are completely isolated from the external surface, not allowing the access of external fluids in neither liquid nor gaseous phase. Closed pores influence parameters like density and mechanical and thermal properties. Open pores are connected to the external surface and are therefore accessible to fluids, depending on the pore nature/size and the nature of fluid. Open pores can be further divided in dead-end or interconnected pores. Further classification is related to the pore shape, whenever it is possible to determine it. The characterization of solids in terms of porosity consists in determining the following parameters:

- *Pore size*: Pore dimensions cover a very wide range. Pores are classified according to three main groups depending on the access size:
    - Micropores: less than 2 nm diameter
    - Mesopores: between 2 and 50 nm diameter
    - Macropores: larger than 50 nm diameter
- *Specific pore volume and porosity*: The internal void space in a porous material can be measured. It is generally expressed as a void volume (in cc or mL) divided by a mass unit (g).
- *Pore size distribution*: It is generally represented as the relative abundance of the pore volume (as a percentage or a derivative) as a function of the pore size.
- *Bulk density*: Bulk density (or envelope density) is calculated by the ratio between the dry sample mass and the external sample volume.
- *Percentage porosity*: The percentage porosity is represented by ratio between the total pore volume and the external (envelope) sample volume multiplied by 100.
- *Surface area*: See above for discussion

Real, Tapped, and Bulk Density

Bulk or *tapped density* is a measure of the degree of packing or, conversely, the amount of space between the particles in the powder. Bulk density is determined by placing a sample of powder of known weight in a graduated cylinder. Tap density is determined by tapping the powder in the graduate until it no longer settles.

Many methods are also used to determine the true density of the powder (e.g., helium pycnometer or gas adsorption). Dividing the true density by the bulk or tap density yields a number that is related to the amount of space in the powder. If the particles are a sphere, the value is about 0.53, while irregular-shaped particles can have values of 0.74 or more.

The *real density* of a powder sample is the weight per unit volume of the material with no air spaces between particles. Therefore, if a material has a true density of 1 g/cm^3, 100 g of material will occupy 100 mL, assuming individual particles fit together exactly. In practice, most powders do not fit together very well. Therefore, if one fills a graduated cylinder to 100 mL with a powder, the weight of powder required may only be 70 g. This apparent density is known as the *bulk* or *expanded density* (0.7 g/cm^3). If the 100 mL cylinder is subsequently tapped, the particles slide past each other and become consolidated. The 70 g of particles that once occupied 100 mL may now only occupy 80 mL. They have an apparent *packed* or *tapped* density (g/cm^3) of 0.875 g/cm^3. Carr's index is a measure of interparticulate forces. If the interparticulate forces are high, powders will have a low bulk density because bridging will occur between particles. This results in a large Carr's index and a large change in volume caused by tapping. If the interparticulate forces are low, particles will have little affinity for one another and will compact spontaneously. Under these circumstances, Carr's index is small and little change in apparent density is induced by tapping. Porosity is the volume ratio occupied by air spaces (voids) between particles of a powder sample.

True Density

Density is the ratio of the mass of an object to its volume and for solids, this term describes the arrangement of molecules. The study of compaction of powders is described by the Heckel equation:

$$\ln\left[\frac{1}{1-D}\right] = KP + A \tag{3.1}$$

where

- D is the relative density, which is the ratio of the apparent density to the true density
- K is determined from the linear portion of the Heckel plot
- P is the pressure

The densities of molecular crystals can be increased by compression. Information about the true density of a powder can be used to predict whether a compound will cream or sediment in a suspension such as metered dose inhaler (MDI) formulation. Therefore, suspensions of compounds that have a true density less than these figures will cream (rise to the surface), and those that are denser will sediment. It should be noted, however, that the physical stability of a suspension is not merely a function of the true density of the material. The true density is thus a property of the material and is independent of the method of determination. In this respect, the determination of the true density can be determined using three methods: displacement of a liquid, displacement of a gas (pycnometry), or floatation in a liquid. The liquid displacement is tedious and tends to underestimate the true density; displacement of a gas is more accurate but needs relatively expensive instrumentation. As an alternative, the floatation method is simple to use and inexpensive.

Problems in Powder Handling

A sample of powder is the most complex physical system. No two particles are identical. The properties of the powder are dependent on both the chemical and physical nature of the component and the nature of the interactions between the particles in the powder.

The ability of a powder to pack is dependent on the shape, size, and porosity of the particle.

Powder materials exhibit a number of technological challenges with their manufacture, storage, transportation, mixing, dusting, characterization, packing, crushing, and milling.

Symptoms of a nonoptimized product system utilizing a powder include unacceptable rehydration, dissolution, and solubility rate/reproducibility of the powder mixture; degradation, loss of drug activity, and reduction of product shelf life; drug mixture heterogeneity both before and during use; clogging of spray nozzle; and loss of delivered drug. The following can have a significant impact on the performance of a product using a powder:

- Utilization of the appropriate binders and adhesives
- Disintegrating agents
- Fillers
- Lubricants
- Wetting agents/surfactants
- Glidants
- Flavoring and sweetening agents

Typical powder dispersion problems include

- Chemical and morphological heterogeneity of the surface
- Dissolution or isomorphous substitution of constituent components (metals)
- Dependency of the surface and solution (dissolved or added) ion species

A number of interrelated physicochemical properties, such as pH (acidity), pI (ionic strength), pe (redox), and pc (concentration), influence the properties of the dispersion besides pressure and temperature.

Mixing of Powders

Three primary mechanisms are responsible for mixing:

1. Convective movement of relatively large portions of the powder
2. Shear failure, which primarily reduces the scale of segregation
3. Diffusive movement of individual particles

Mixing solids involves a combination of one or more mechanisms of convection, shear, and diffusive mixing. Convection mixing is achieved by the transport of solids such as by blades or screws. Shear mixing results from the forces within the particulate mass; slip planes are set up. This can take place singularly or as a laminar flow. When shear occurs between regions of different composition and parallel to their interface, it reduces the scale of segregation by thinning the dissimilar layers. Shear occurring in a direction normal to the interface of such layers is also effective because it reduces segregation. The diffusive mixing is the random motion of particles.

Particulate solids tend to segregate by virtue of differences in the size, density, shape, and other properties; it can happen during mixing or subsequent storage handling as well. It is important to note that powders that are difficult to flow do not segregate easily because of high interparticulate adhesion; however, because powders must be rendered flowable for the purpose of filling capsules or in bottles or sachet, the segregation phenomenon becomes very important. Note that often, after the addition of magnesium stearate, it is advisable to mix the product only for a limited time because electrical charges on the particles may cause segregation. Often, additives are included in formulations to reduce the tendency of segregation; these components have polarity similar to the components of the formulation. A variety of mixers are designed to counter the segregation during mixing. Regardless of the formulation or equipment used, however, the formulator must conduct a validation study to assure that the product before filling is not segregated and that detailed manufacturing directions consequently include such conditions as humidity, mixing speeds, mixing times, and grounding of equipment. It is often said that longer mixing causes unmixing; this occurs because of segregation as well as abrasion of particles, which alters the particle size distribution profile.

Powder Flow Properties

The flow properties of a powder will determine the nature and quantity of excipients needed to prepare a compressed or powder dosage form. This refers mainly to factors such as ability to process the powder through machines. To make a quick evaluation, the compound is compressed using an infrared (IR) press and die under 10 tons of pressure with variable dwell times, and the resulting tablets are tested with regard to their crushing strength after storing the tablets for about 24 h. If longer dwell times result in higher crushing strength, then the material is likely plastic; elastic material will show capping at low dwell times; the brittle material will not show any effect of dwell times. It is recommended that the compressed tablets be subject to XPRD to record any changes in the polymorphic forms.

During many pharmaceutical production processes, it is necessary to transfer large quantities of powder from one location to another in a controlled manner, for example, in powder blending, powder filling into containers (e.g., dusting powders), powder flow into capsules, and powder filling into the dies of a tablet press.

One method of assessing flow properties is the *angle of repose*, which is another measure of the nature of the powder. It estimates the adhesive force between the particles. Uniform glass beads, which will show good flow properties, have an angle of repose of 23°. As the adhesive force between the particles increases, the angle increases. In rare cases, it can exceed 90°.

Powder is allowed to flow freely through a funnel onto the center of an upturned petri dish of known radius. When the powder reaches the side of the petri dish, the height of the cylindrical cone is determined. From the petri dish radius (r, cm) and cone height (h, cm), the angle of repose (between the petri dish and base of the powder cone) can be calculated. *Flow rate* can also be determined by measuring how fast a powder flows through an aperture. Free-flowing powders exhibit a high flow rate and a smaller angle of repose. Angle of repose and flow rate depend on particle size, shape, and surface roughness. Flow properties are frequently enhanced by the use of *glidants*.

Electrostaticity

When subjected to attrition, powders can acquire an electrostatic charge, the intensity of which is often proportional to physical force applied as static electrification of two dissimilar materials occurs by the making and breaking of surface contacts (triboelectrification or friction electrification). Electrostatic charges are often used to induce adhesive character to bind drugs to carrier systems, for example, glass beads coated with HPMC-containing drugs. The net charge on a powder may be either electropositive or electronegative depending on the direction of electron transfer. The mass charge density can vary from 10^{-a} to 100 μC/kg depending on the stress, ranging from gentle sieving to micronization process. This can be determined using electric detectors to determine polarity as well as the electrostatic field. The electrostaticity results in significant changes in the powder flow properties.

Caking

Powders cake due to agglomeration as a result of such factors as static electricity, hygroscopicity, particle size, impurities of the powder, storage conditions, stress, temperature, relative humidity (RH), and storage time. The mechanisms involved in caking are based on the formation of five types of interparticle bonds such as bonds resulting from mechanical tangling, bonds resulting from steric effects, bonds via static electricity, bonds due to free liquid, and bonds due to solid bridges. During the process of micronization, the formation of localized amorphous zones can lead to caking as these zones are more reactive to factors described earlier especially when exposed to moisture; the mechanisms involve moisture sorption due to surface sintering and recrystallization at well below the critical RH. In most instances, increase in RH begin to show some impact at values above 20% resulting in most dramatic effects above 75%–80% RH for powders that are subject to humidity effects.

Relative Humidity

Relative humidity in the filling and storage areas is more important for powders than for other dosage forms because of the large specific surface area (area/weight), which can result in significant moisture uptake. The gelatin capsule shells are also susceptible to moisture and degradation at high moisture. In addition, at very low moisture, gelatin in capsules can become very brittle; therefore, an appropriate humidity level must be maintained.

Polymorphism

Because polymorphism can have an effect on so many aspects of drug development, it is important to fix the polymorph (usually the stable form) as early as possible in the development cycle. Whereas it is not necessary to create additional solid-state forms by techniques or conditions unrelated to the synthetic process for the purpose of clinical trials, regulatory submission of a thorough study of the effects of solvent, temperature, and possibly pressure on the stability of the solid-state forms is advised. A conclusion that polymorphism does not occur with a compound must be substantiated by crystallization experiments from a range of solvents. This should also include solvents that may be involved in the manufacture of the drug product, for example, during granulation.

MILLING

Mixing of powders is easier if all components are of the same dimension in particle size. Granulation of powders is done to provide a more uniform particle size; this is a common practice in tablet, capsule, and powder suspension formulations. Milling of granulated mass produces uniform particle size; where dyes are used, milling provides a more uniform mixing and spread of dyes. Lubricants act by coating the particles and require the presence of a certain amount of fines. Size distribution profiles are routinely prepared as part of the development pharmaceutics process, especially where high-speed filling machines are used. Frequency and cumulative plots are made to validate the process. Probability function values found in statistics books should be consulted when designing a robust evaluation program. Particles are measured either microscopically or by weight fractions through a stack of sieves. A sedimentation method is also used for particles in the range of 1–200 mm to obtain a size–weight distribution. Other methods include adsorption, electrical conductivity, light and x-ray scattering, permeametry, and particle trajectory.

During the process of milling or comminution, the particles undergo transformation based on the strain applied, which produces stress, and size reduction begins with the opening of new cracks. If the force applied is not sufficient, then the particle returns to its original state from a stressed state and does not yield. The type of mill used is important, such as a cutter, fluid energy, hammer, or roller because each provides a special pattern of comminution. For example, it is useful for fibrous material, but not for friable material; it produces a product size of 20–80 mesh. The fluid energy mill can produce 1–30 mm particles and is more suitable for soft and sticky materials. The most common mill is the hammer mill, which is useful for abrasive materials and produces 4–325 mesh particles. In a hammer mill, it matters whether the blades are forward or reversed.

POWDERS

The formulation and bioavailability problems associated with suspensions are also characteristic of powders, whereby the active ingredient is mixed with inert diluents and administered either directly or in a capsulated form. An additional problem therefore arises due to possible adsorption of drugs onto diluents, from which the drug may not be released quickly enough for adequate absorption. For example, only 40% of thiamine and 79% of riboflavin are available for absorption from capsules containing Fuller's earth, which adsorbs these drugs. Similarly, calcium phosphate used as a diluent in tetracycline capsules reduces absorption by the formation of insoluble complexes.

The particle size of powders is significant in their dissolution and bioavailability, as demonstrated by spironolactone and griseofulvin, the micronization of which leads to significantly higher absorption in humans. However, smaller particle powders have a greater tendency to adsorb moisture from the atmosphere, which results in possibly unstable preparations. Smaller particle size also means increased electrostatic charges on the particle surface, especially with hydrophobic drugs. This might result in aggregation and the consequent loss of an effective or exposed surface area for dissolution.

An example in which smaller particle size is not always desirable, even though it does increase bioavailability, is the use of nitrofurantoin. The use of a larger particle size is recommended to avoid the GI irritation and accompanying nausea that occurs very frequently with the use of fine particles in oral dosage forms.

When powders are administered in a gelatin capsule, the capsule shell itself may affect the absorption process. Hard gelatin capsules dissolve more readily in the GI fluids than soft gelatin capsules. For example, the slow absorption of vitamin B from soft capsules may be attributed to the slow dissolution of capsules themselves. However, a recent study showed that soft capsules might produce an unexpected increase in the absorption of digoxin, with which absorption rates even higher than the solution dosage forms were obtained. This finding was attributed to the possible interaction of digoxin with the soft elastic capsule walls and also to the protection of digoxin against possible chemical decomposition from GI fluids.

A large number of drugs are administered in powder form, such as iodochlorhydroxyquin and methylbenzethonium chloride, or contained in a capsule, as are phenytoin, chloramphenicol, erythromycin, tetracycline, lithium carbonate, quinine sulfate, chlordiazepoxide hydrochloride, cephalexin, and propoxyphene.

ORAL POWDERS

Oral powders include headache powders, dusting powders (such as antifungal powders), powders to be reconstituted (such as antibiotics), and insufflations, which are powders intended to be blown into a body cavity such as in the ear or nose. Powder mixtures as a means of measuring small quantities of powders are called trituration. FDA classification of powders is given in Table 3.1.

CAPSULES

Capsules are solid dosage forms in which one or more medicinal ingredients and/or inert substances are enclosed within a small shell or container generally prepared from a suitable form of

TABLE 3.1
FDA Classifications of Powders

Powder (PWD 110)	An intimate mixture of dry, finely divided drugs or chemicals that may be intended for internal or external use.
Powder, dentifrice (PWD DENT 115)	A powder formulation intended to clean and polish the teeth and may contain certain additional agents.
Powder, for solution (PWD F/SOL 833)	An intimate mixture of dry, finely divided drugs or chemicals that, upon the addition of suitable vehicles, yields a solution.
Powder, for suspension (PWD F/SUSP 834)	An intimate mixture of dry, finely divided drugs or chemicals that, upon the addition of suitable vehicles, yields a suspension (a liquid preparation containing the solid particles dispersed in the liquid vehicle).
Powder, metered (PWD MET 841)	A powder dosage form that is situated inside a container, which has a mechanism to deliver a specified quantity.

gelatin. The FDA provides a classification of different types of capsules (Table 3.2). The amount of active ingredient per dose has a direct bearing on the proper size capsule to use. Because capsules usually require less excipients and additives, it is easier to get a more potent dosage without having to use a large-size capsule.

Very often, strict governmental regulations are placed on products that are being consumed by the public for health reasons. In most cases, pharmaceutical applications face different regulatory constraints than do dietary supplements.

INHALERS AND LUNG DELIVERY

Key factors that contribute to the aerodynamic properties of aerosol particles are found in Stokes' law. These factors may be monitored or controlled to optimize drug delivery to the lungs. Predictions of the aerodynamic behavior of therapeutic aerosols can be derived in terms of the physical implications of particle slip, shape, and density. The manner in which each of these properties has been used or studied by pharmaceutical scientists to improve lung delivery of drugs is readily understood in the context of aerosol physics. Additional improvement upon current aerosol delivery of particulates may be predicted by further theoretical scrutiny.

The history of inhaler development in modern times can be traced to the metering valve and propellants (MDIs, pressurized [pMDI]) used in the treatment of asthma in the 1950s. This was followed closely by somewhat primitive DPIs in the 1970s. Throughout this period, nebulizers were employed to deliver drugs in aqueous solution. In the past decade, research and development in the field has broadened.

The field of inhalation science is expanding rapidly as scientists are designing delivery systems for proteins and peptides using nanoparticle inhalation systems; the quick absorption through lung surface offers an excellent administration route.

MODIFIED-RELEASE PRODUCTS

The capsulation process offers many advantages for designing modified-release products. The simple process of loading the drug onto nonpareil sugar beads and then coating them with a variety of release profiles offers the opportunity of not only separating the incompatible components but also mixing granules that provide different release profiles, from instant release to step release to prolonged release. Equipment is available to fill several beads simultaneously into capsules, thus assuring dosing accuracies. (If granules with different coatings are mixed, segregation is likely because of the differences in their density.) Coated granules, if compressed, lose their release profiles.

TABLE 3.2
FDA Classification of Capsule Types

Capsule 600	A solid dosage form in which the drug is enclosed within either a hard or soft soluble container or "shell" made from a suitable form of gelatin.
Capsule, coated 602	A solid dosage form in which the drug is enclosed within either a hard or soft soluble container or "shell" made from a suitable form of gelatin; additionally, the capsule is covered in a designated coating.
Capsule, coated, extended release 611	A solid dosage form in which the drug is enclosed within either a hard or soft soluble container or "shell" made from a suitable form of gelatin; in addition, the capsule is covered in a designated coating, which releases a drug (or drugs) in such a manner to allow at least a reduction in dosing frequency as compared with the same drug (or drugs) presented as a conventional dosage form.
Capsule, coated pellets 603	A solid dosage form in which the drug is enclosed within either a hard or soft soluble container or "shell" made from a suitable form of gelatin; the drug itself is in the form of granules to which varying amounts of coating have been applied.
Capsule, delayed release 620	A solid dosage form in which the drug is enclosed within either a hard or soft soluble container made from a suitable form of gelatin, which releases a drug (or drugs) at a time other than promptly after administration. Enteric-coated articles are delayed-release dosage forms.
Capsule, delayed-release pellets 621	A solid dosage form in which the drug is enclosed within either a hard or soft soluble container or "shell" made from a suitable form of gelatin; the drug itself is in the form of granules to which enteric coating has been applied, thus delaying the release of the drug until its passage into the intestines.
Capsule, extended release 610	A solid dosage form in which the drug is enclosed within either a hard or soft soluble container made from a suitable form of gelatin, which releases a drug (or drugs) in such a manner to allow a reduction in dosing frequency as compared with the same drug (or drugs) presented as a conventional dosage form.
Capsule, film coated, extended release 612	A solid dosage form in which the drug is enclosed within either a hard or soft soluble container or "shell" made from a suitable form of gelatin; in addition, the capsule is covered in a designated film coating, which releases a drug (or drugs) in such a manner to allow at least a reduction in dosing frequency as compared with the same drug (or drugs) presented as a conventional dosage form.
Capsule, gelatin coated 605	A solid dosage form in which the drug is enclosed within either a hard or soft soluble container made from a suitable form of gelatin; through a banding process, the capsule is coated with additional layers of gelatin so as to form a complete seal.
Capsule, liquid filled 606	A solid dosage form in which the drug is enclosed within a soluble, gelatin shell that is plasticized by the addition of a polyol, such as sorbitol or glycerin, and is therefore of a somewhat thicker consistency than that of a hard shell capsule; typically, the active ingredients are dissolved or suspended in a liquid vehicle.

COATED PARTICLES

Use of hard gelatin capsules allows for the preparation of coated particles to provide modified release or stability; however, the possibilities of creating innovative dosage forms using different size of particles makes this dosage form highly desirable for many unstable drugs.

TABLETS

Whereas solutions represent a state of maximum dispersion, compressed tablets have the closest proximity of particles. Complexities in dissolution and bioavailability are generally inversely proportional to the degree of dispersion—compressed tablets are thus most prone to bioavailability problems. This is primarily due to the smaller surface area exposed for dissolution until the tablets break down into smaller particles (Scheme 5.1). Factors responsible for the primary breakdown of tablets into granules and their subsequent breakdown into finer particles include such parameters as the concentrations of binder, disintegrant, and lubricant; the hydrophobicity of the drug; and the adjuvants; therefore, it can be expected that a significant difference is always possible in the dissolution and bioavailability of various tablets.

The problem of tablet disintegration is well demonstrated by such drugs as dipyridamole, thioridazine, and digoxin, which exhibit higher blood levels if the tablets are crushed before administration.

The disintegration test for tablets has long been used to detect ineffective products, as determined by a lack of disintegration into large particles within a given period of time. This test allows monitoring of batch-to-batch variations in the manufacturing process. However, adequate disintegration alone does not assure ultimate dissolution, which may be retarded by the absorption of drug on hydrophobic lubricants in the formulation, the recrystallization of drugs, the presence of large primary granules, and the failure of these granules to break down further into finer particles. The importance of using smaller particles in tablet formulations is well demonstrated in the use of griseofulvin, with which the reduction of particle size has been consistently related to bioavailability. Recently, a solid dispersion of griseofulvin was formulated, which contained ultramicrosize particles of the drug, resulting in an almost 100% improvement in its bioavailability compared to the micronized forms.

The coatings of tablets, which are applied for a variety of reasons, add another rate-limiting factor, since a coating must dissolve or disrupt before the tablet can disintegrate and the dissolution process begins. The sugar coating used to mask unpleasant taste, appearance, and odor, or to protect a tablet ingredient from decomposition during storage, consists of an application of poorly soluble polymers that can interfere with the disintegration of tablets.

Film coatings are generally less problematic, but enteric coatings used to protect both the gastric mucosa from the drugs and the drugs from the gastric fluids give the most variable bioavailability, since their disintegration is often dependent on GI pH and other highly variable physiological and physicochemical factors.

SOLUTIONS

Solutions are thermodynamically stable monomolecular dispersions of drug molecules in a liquid or solid phase. Absorption from aqueous solutions is generally very fast and complete from all sites of administration, provided that penetration through the absorption barrier (such as the GI membrane) is not a rate-limiting factor. The rate-limiting steps as disintegration and dissolution are minimal in the use of solutions. For example, potassium penicillin V gives higher blood levels than benzathine penicillin V when both are administered orally in tablet form, but solutions of the two drugs yield essentially equal blood levels of penicillin.

Besides providing the highest bioavailability, solutions are also convenient for administration to pediatric and geriatric patients. In some instances, the use of solutions is a crucial part of the drug delivery. For example, calcium must be administered as a solution in its citrate form to achlorhydric patients, since the solid carbonate form will not dissolve sufficiently in the GI tract without the presence of hydrochloric acid. An analogous problem exists in the administration of sodium salts of weakly acidic drugs, which precipitate in the stomach in crystalline form. These crystals are usually very fine and redissolve quickly, but there is always a possibility of retarded absorption due either to precipitation as large particles or to coating of the particles with hydrophobic acid, as demonstrated with such poorly water-soluble drugs as warfarin and phenytoin. These drugs can therefore be absorbed better from a suspension dosage form than from a solution of their sodium salts.

Quite often, solutions of poorly water-soluble drugs are affected by adding cosolvents, such as alcohol or propylene glycol, and by adding complexing agents, which form a water-soluble complex with the drug or with surfactants that solubilize the drugs. In all of these instances, the drugs will precipitate because of the dilution effect and are subject to possible difficulty in redissolution.

Sometimes, nonaqueous solutions provide better absorption than aqueous solutions, as demonstrated by indoxole. A solution of indoxole in oil, administered as an oil-in-water emulsion, shows three times better absorption than the aqueous solution.

The use of solid solutions is a novel application of dispersion techniques, whereby the drug is dispersed in a solid water-soluble vehicle, such as urea, succinic acid, or polyethylene glycols, which dissolves rapidly in water, releasing the macrocrystalline or monomolecular form of the drug. Although there is a large volume of data about the applications of the principles of solid dispersions and solutions, only one product is currently available that utilizes this concept, that is, Gris-PEG, a dispersion of griseofulvin in polyethylene glycol.

Solution dosage forms offer several advantages, particularly the resolution of bioavailability problems and instant administration as injectable forms (though nonsolution forms are also given parenterally). At the preformulation stage, more important factors are the solubility (and any pH dependence) and stability of the new compound.

SOLUBILITY

Where a solution form is desired and the compound has low solubility, there are several techniques, some very simple to some very complex, to achieve the desirable property of the lead drug including pH manipulation, use of cosolvents, surfactants, emulsion formation, and adding complexing agents. On a more complex stage, the liposomes or similar drug delivery systems can be used.

Since many compounds are weak acids or bases, their solubility will then be a function of pH. However, the ionic strength of medium plays a significant role, and as a result, most parenteral formulations are buffered to prevent the crystallization of drugs.

The use of cosolvents improves solubility as a result of the polarity of the cosolvent mixture being closer to the drug than it is in water:

$$\text{Log } S_m = f \log S_c + (1 - f) \log S_w \tag{3.2}$$

where

S_m is the solubility of the compound in the solvent mix
S_w is the solubility in water
S_c is the solubility of the compound in pure cosolvent
f is the volume fraction of cosolvent
σ is the slope of the plot of log (S_m/S_w) versus f

There is a definite correlation between the s value to indices of cosolvent polarity such as the dielectric constant, solubility parameter, surface tension, interfacial tension, and octanol–water partition coefficient. The aprotic cosolvents give a much higher degree of solubility than the amphiprotic cosolvents. This means that if a cosolvent can donate a hydrogen bond, it may be an important factor in determining whether it is a good cosolvent. Use of cosolvents with polar drugs can reduce the solubility.

EMULSION FORMULATIONS

For drugs with poor water solubility, an emulsion formulation such as oil-in-water (o/w), where the drug has good partitioning in the oil phase chosen, offers often an excellent choice. The particle size of the emulsion and its stability (physical and chemical) then become significant factors since larger globule sizes may lead to phlebitis. To achieve smaller particle size, the technique of microfluidization is often used among other such homogenization available methods. Phospholipids added stabilize emulsions through surface charge changes as well as provision of a good mechanical barrier.

Many drugs show surface-active behavior because they have the correct mix of chemical groups that are typical of surfactants. The surface activity of drugs can be important if they show a tendency to, for example, adhere a surface or if solutions foam. Not all surface-active drugs form micelle because of steric hindrances.

SUSPENSIONS

Where the drug has limitations in its solubility and efforts to enhance fail, where there is a tendency for fast crystallization from solutions, or even where chemical stability is a problem, often, formulating suspension dosage forms obviates some of these drawbacks. However, suspensions, by nature, must have higher viscosity to prevent settling of particles and thus create problems in pourability, syringability, etc. Appropriate selection of a vehicle that provides an ideal compromise among all characteristics thus becomes a critical factor because the intent is to have as little solubility in the vehicle as possible to prevent crystallization from the solution that surrounds the suspended particles; as a result, weak acids and bases appear as poor choice for suspension formulation. In some instances, it may be possible to prepare a derivative with larger hydrophobic groups or salt formation that would have lower solubility if preparing a suspension dosage form is particularly desired. Compounds that can form hydrates while in suspension state can create stability problem. A significant thermodynamic problem in suspension formulation comes from Ostwald ripening, crystal growth, not due to phase change but as a result of differences in the solubility as a function of crystal size:

$$\frac{RT}{M}\ln\left(\frac{S_2}{S_1}\right)=\frac{2\sigma}{\rho}\left(\frac{1}{r_1}-\frac{1}{r_2}\right) \tag{3.3}$$

where

R is the gas constant
T is the absolute temperature
S_1 and S_2 are the solubilities of crystals of radii r_1 and r_2, respectively
σ is the specific surface energy
ρ is the density
M is the molecular weight of the solute molecules

Temperature fluctuations are obviously one factor that promotes Ostwald ripening. Whereas phase changes can be studied using such standard techniques as DSC, hot stage microscopy, or XRPD, Ostwald ripening is best studied using microscopic methods. The art of suspension formulation is complex as a large number of factors including additives can have a significant influence on crystal growth; for example, dye molecules often attach to high-energy points on crystals, affecting their growth; similarly, it is reported that PVP, a common ingredient of many suspension formulations, inhibits crystal growth. Albumin is also known to have similar impact. The choice of additives is also governed by the final form of suspension; if it has to be sterilized, the additives must be able to sustain autoclave temperatures; besides, autoclaving itself can affect both physical and chemical stability of the drug. Zeta potential measurements of suspensions often prove useful.

Suspensions require the dissolution of particles before they can be absorbed. The dissolution process can be rate limiting, depending on the aqueous solubility of the drug and the formulation additives involved. Thus, there are many more factors that can affect drug absorption in the use of suspensions than are possible in the use of solutions. Generally, however, suspensions will provide better absorption than such other dosage forms as capsules and tablets such as shown by trimethoprim and sulfamethoxazole combinations and sulfadimethoxine. Suspensions are also used when a slow release of the drug is desired, as with intramuscular administration of triamcinolone acetonide or with tetracycline ophthalmic suspensions. Since suspensions provide a large surface area, various antacid products are most effectively administered as suspensions, since the mode of action involves both the chemical neutralization of hydrochloric acid and its physical adsorption onto the suspended particles. It is interesting to note that a majority of official oral suspensions in current use involve antiinfective agents, for example, pyrantel pamoate, pyrvinium pamoate, thiabendazole, chloramphenicol palmitate, demeclocycline, methacycline, oxytetracycline, penicillin, tetracycline, methenamine mandelate, nitrofurantoin, sulfonamides, trisulfapyrimidines, and nystatin. Most of the antiinfective agents are chemically unstable, can cause GI irritation, and are often erratically absorbed from such solid dosage forms as tablets and capsules. The use of suspensions for these drugs provides an ideal mechanism for solving formulation problems related to these attributes. Consider a drug with a decomposition constant of 0.21 h and aqueous solubility of 4 g/L. A 500 mg dose in aqueous solution will have a shelf life (10% decomposition) of 0.53 h, whereas 500 mg suspended in 10 mL of saturated solution will have a shelf life of 6.25 h, indicating a more than 1000% increase in drug stability.

The use of suspensions is also advantageous in pediatric or geriatric practice, where they can be accurately and conveniently administered using droppers or oral syringes. Suspension dosage forms are utilized for all routes of administration except intravascular.

CONTROLLED-RELEASE DOSAGE FORMS

Unless specific formulation efforts are made to control the release of drugs, the rates of drug absorption are generally proportional to the amount of drug at the site of absorption. In many instances, it is necessary to prolong the action of drugs by sustaining their absorption over a longer period of time.

The design of oral prolonged-action dosage forms includes such modifications as

- Barrier coating, whereby the drug diffuses out through a membrane within which it may be dissolved by the penetrating GI fluids.
- Fat embedment, which involves suspending the drug in a fatty medium in a solid dosage form from which the drug is released by erosion, hydrolysis of fat, and direct dissolution.
- Plastic matrices, which allow leaching and diffusion of drugs from a solid plastic matrix, which is left intact after the drug has been released.
- Repeat action tablets, utilizing a double coating that releases an initial dose followed by another dose released either instantaneously or by slow diffusion.
- Ion-exchange resins, which provide prolonged dissolution by the formation of drug salts with resins, which then react with either hydrochloric acid in the stomach or sodium chloride in the intestine to exchange the drug.
- Hydrophilic matrices, utilizing hydrophilic gums for compression into tablets, which undergo gelation and release the drug by diffusion.
- Polymer resin beads, in which the drug is first dissolved or suspended in plastic monomers and then polymerized. The beads are then either filled into a capsule or compressed into tablets. Drug release is controlled by the dissolution and swelling of the resin and the diffusion of drug from the beads.

- Soft gelatin depot capsules involve the dissolution or suspension of drugs in sponge-forming solutions and consequent filling into capsules, which leave a solid skeleton upon diffusion of the drugs.
- Drug complexes utilizing macromolecules provide prolonged release upon the hydrolysis of the complex.

The release of drugs administered parenterally can also be controlled by the following methods:

- *Pharmacological methods*: Intramuscular or subcutaneous administration instead of intravenous. Simultaneous administration of vasoconstrictors (adrenalin in local anesthetics, ephedrine in heparin solution), blocking elimination of drugs through the kidney by simultaneous administration of a blocking agent such as probenecid with penicillin or p-Aminosalicylic acid.
- *Chemical methods*: Use of salts, esters, ethers, and complexes of the active ingredient with low solubility.
- *Physical methods*: Selection of a proper vehicle giving prolonged release, as with the use of oleaginous solutions instead of aqueous solution; the addition of macromolecules, which increase the viscosity, such as carboxymethylcellulose and tragacanth; the use of swelling material to increase the viscosity of oleaginous solutions, as with aluminum monostearate; the addition of absorbents; the use of a solution from which the drug is precipitated upon contact with body fluids; the use of aqueous and oleaginous suspensions; and the use of implants.

THERAPEUTIC SYSTEMS

Several dosage forms, termed Therapeutic Systems, have recently been marketed in this county. The Therapeutic System is a dosage form that provides preprogrammed, unattended delivery of drugs at a rate and, for a given time period, designed to meet a specific therapeutic need. These systems have been developed for introducing drug substances both via the systemic circulation and directly to specific target organs. Many new drug delivery techniques have been developed, including

- Diffusion of drugs through rate-controlling membranes
- Osmotic pumping
- Biodegradable polymer matrices
- Polymer-bound active species
- Nanosystem

The Therapeutic Systems are composed of an active drug in a delivery module, which consists of a drug reservoir, which may be a single- or multicompartment element; a rate controller; and an energy source to effect the release of the drug molecules through a delivery portal. The drug delivery module is housed in a "platform" that is compatible with the tissues and couples the system to the body site in which it is deployed. The platform may be either fixed or mobile within a defined area. Some examples include the ocular platform, which is designed so that it can float comfortably and inconspicuously in the tear film on the eye beneath the eyelid for controlled delivery (Occusert), and the T-shaped progesterone-impregnated polymer unit for intrauterine deployment for fertility control (Progestasert).

The osmotic drug delivery system resembles an ordinary tablet in appearance and is composed of a solid core of drug surrounded by a semipermeable membrane with a single minute orifice. The membrane allows steady entry of water at a predetermined rate to dissolve the drug. Drug solution is then continuously pumped through the orifice, providing a constant rate of release.

Other novel ideas include a transdermal therapeutic system consisting of a disk 0.2 mm thick and 2 cm in diameter, which is worn behind the ear like a tiny adhesive bandage and releases scopolamine for its antiemetic properties and the use of nitroglycerin patches for angina pectoris. The use of biodegradable polymers has also been suggested for implant systems for the controlled release of drugs.

The foregoing innovations are cited here to make the reader aware of the possibilities of bioavailability variation as a result of a large number of physicochemical and technological implementations in the design of dosage forms. The complexities in the design of dosage forms necessitate the development of an elaborate system to evaluate dosage forms and systems on the basis of the attributes listed at the beginning of this chapter.

EVALUATION OF DRUG DELIVERY SYSTEMS

It is not possible to predict if the administered will result in a consistent desirable therapeutic response. However, several tests can be conducted to assure some measure of reliability in dosage form functions. These include the following.

Chemical Content

It is essential that dosage forms contain the labeled amount of the active drug. Chemicals that are biologically active are also highly chemically reactive and can therefore undergo chemical decomposition reactions that result in a loss of content. For example, aspirin decomposes to salicylic acid and acetic acid. Salicylic acid is undesirable because it causes more GI irritation than aspirin and also because it may not possess a therapeutic activity equivalent to aspirin. Para-aminosalicylic acid decomposes via decarboxylation to meta-amino phenol, resulting in discoloration and enhanced toxicity. Tetracycline converts to epianhydrotetracycline, which is highly toxic to the kidneys.

Although not all chemical decomposition reactions result in a toxic product, a change in the color or the consistency of a preparation will quite often make it unacceptable to the patient. It is therefore necessary to provide a shelf life or expiration date on the products. A 3–5-year expiration date is rather common for a relatively stable product, and sometimes, a shelf life of only a few days or weeks is assigned to highly reactive drugs or radiolabeled compounds.

In order to account for the loss of drug during shelf life, overage additions are often made at the time of manufacture. This overage addition is necessary to compensate for the high reactivity of the active components. However, large overage additions cannot be allowed for drugs used internally or for those for which a narrow plasma concentration fluctuation has to be maintained, as with cardiac glycosides and anticoagulants, which have narrow therapeutic indices.

Chemical decomposition is not the only way in which active ingredients are lost from dosage forms. For some compounds with a low boiling point, the active principle can be lost by evaporation or volatilization. For instance, nitroglycerin tablets, if dispensed in a plastic container, have been reported to lose up to 80% of their active components in 2 years. Nitroglycerin tablets are therefore required by law to be dispensed in the original glass container and even then there can be a significant drug loss within the unopened container.

The container also plays an important role in determining dosage form effectiveness. For example, the increasing use of plastic large-volume parenteral containers has created an unanticipated problem of loss of drugs due to absorption onto the plastic and absorption through it, resulting in a significant loss of drugs such as vitamin A.

Chemical content evaluations are therefore fundamental in determining dosage form effectiveness. Hundreds of drugs have been recalled by the FDA due to subpotent or in some instances superpotent products, making potency one of the primary criteria in the evaluation of dosage forms.

TABLE 3.3
Selected Examples of Tablets Containing Small Amounts of Active Ingredients

Drug Component	Drug Available Tablet Strength (mg)
Atropine sulfate	0.3
Colchicine	0.5
Dexamethasone	0.25
Diethylstilbestrol	0.1
Ethinyl estradiol	0.05
Digitoxin	0.05
Digoxin	0.125
Reserpine	0.1

Content Uniformity

The chemical equivalence testing described earlier is generally performed on a large number of dosage form units (e.g., 20 tablets) at one time. This testing determines the average amount of active ingredients(s). It will not, however, reveal variations in drug content among the units. For example, the oral contraceptive Ortho-Novum 1/50 contains 1 mg of norethindrone and 0.05 mg of mestranol per tablet. What if one tablet contains 0.1 mg of mestranol while another tablet contains none? Although the two tablets combined will pass the chemical equivalence test, a course of therapy with tablets of this quality might result in an unanticipated pregnancy. The problem of content uniformity, therefore, exists for all products containing minute amounts of active ingredients, as is shown in Table 3.3.

The problems of content uniformity arise mainly from the mixing of small amounts of drugs into large batches where a uniform distribution must be assured. Again, the FDA has recalled many products in the last few years due to noncompliance with the *United States Pharmacopeia* (USP) content uniformity requirements of 5%.

Presence of Contaminants

Contaminants are defined as any undesirable substances contained in a formulation. Contamination of the drug product may occur during processing from impurities in raw materials, heavy metal ions from manufacturing equipment, microorganisms, or chemical decomposition products, which may be toxic as noted earlier or inactive, as the product of reaction between isoproterenol and bisulfite preservatives. Another source of contamination is dust spreading during the manufacturing process, when several products are handled simultaneously in a manufacturing facility. Although the presence of contaminants may not always be deleterious, it is always desirable to have as few as possible to prevent changes in the physical or aesthetic appearance of a product as well as unanticipated adverse reactions.

Disintegration Test

The disintegration test ascertains the time required for a compressed tablet to break up into granules. The first official disintegration test was included in Pharmacopoeia Helvetica in 1934. Since then, most official pharmacopeias have included these tests to formulate a basis for prediction of the availability of drugs from dosage forms. Up until the 1950s, disintegration was the key word and any dosage form that disintegrated within a prescribed time was assumed to provide adequate bioavailability.

A large number of formulation factors can affect the rate of tablet or capsule disintegration, including

Diluents or fillers
Manufacturing methods, such as dry or wet granulation
Compression pressure in capsulation
Hardness
Concentration of disintegrant and the method of its addition
Types and concentrations of lubricants, surfactants, and binders
Drug properties such as particle size, surface characteristics, solubility, and crystallinity
Composition and properties of capsule shell
Type and composition of coating
Age of finished product and storage conditions

The USP disintegration method involves a basket-rack assembly, which is moved up and down 30 times a minute. At specific times, the number of tablets or capsules disintegrated is determined. The disintegration time allowed varies from 5 min to 1 h. For example, aspirin tablets have a time limit of 5 min.

The present USP and National Formulary (NF) disintegration tests measure only the physical breakup of the tablet or capsule, which may not necessarily correlate with drug bioavailability. In order for a drug to be absorbed, it must be present in a solution form. It is possible that the particles from disintegrated tablets might not further disintegrate or dissolve and thus no bioavailability assurance can be obtained from formulations meeting only the official disintegration tests.

Dissolution Test

A dissolution is much more discriminating than the disintegration test. It is a better estimate of bioavailability, though it is still not fool proof. Dissolution rate test can be used to predict bioavailability if these two conditions are met:

1. The dissolved drug remains free and intact in the GI tract. If the dissolved drug complexes with a component of the GI tract, and if drug decomposition occurs in the GI tract, then the dissolution test cannot be a very good index of bioavailability.
2. Absorption is not the rate-limiting step. If the solution formed is quickly absorbed, then the amount absorbed can be correlated with the in vitro dissolution rate. However, when absorption is slow or limited, bioavailability may not be proportional to the dissolution rate.

The formulation factors listed as affecting the disintegration rates also affect the dissolution rates. A large volume of data has been reported, which correlates various formulation factors and the dissolution rates. For example, the particle size of a drug is most clearly related to the dissolution rate. Addition of surfactants quite often substantially increases the dissolution rates of hydrophobic drugs by the removal of air pockets around the particles, thus facilitating the contact of the dissolution medium with the drug. An important source of surface activity is the gastric fluid, where the surface tension varies between 38 and 52 dynes/cm. This lower surface tension allows better wetting of particles and promotes dissolution. The primary cause of surface activity in the gastric fluids is the reflux of intestinal contents into the stomach. The intestinal fluids have significant surface activity, as may be expected because of the lecithins, bile salts, etc.

The fillers and diluents used in a formulation have a significant effect on its dissolution. If the drug is hydrophobic, a hydrophilic filler will tend to enhance dissolution, especially if this filler is at the same time a disintegrant. Starch has hydrophilic properties and is an effective disintegrant and thus proves to be an excellent filler.

The lubricants used may have varying effects. If the granule particles are hydrophilic and disintegrate quickly, a surface-active lubricant will have little effect. If the granule particles are less hydrophilic and do not disintegrate as quickly, a surface-active lubricant may enhance dissolution. The use of such hydrophobic lubricants as stearates decreases dissolution rates, but this effect is minimal if their concentration is less than 1%.

The effect of compression pressure on dissolution rates is the most difficult to predict. Dissolution rates will generally decrease with increasing compression pressure due to a closer binding of the granules to each other. At higher pressure, a crushing of the granules and perhaps even of the drug crystals would occur, resulting in an increased surface area and an increased dissolution rate. A further increase in pressure may make the bonding more important than the crushing, resulting in a decrease in the dissolution rates. Where the bonding is not significant, a direct increase in the dissolution rates can be expected with increasing compression pressure at higher pressures.

The effect of tablet storage on the dissolution rate can also be important and reports have been made suggesting both increasing and decreasing dissolution rates.

In view of the importance of dissolution tests in predicting drug bioavailability, the official compendia continue to require dissolution test as part of the regulatory requirements, such as for acetohexamide, nitrofurantoin, digoxin, phenylbutazone, ergotamine tartrate and caffeine tablets, prednisolone, hydrochlorothiazide, prednisone, lithium carbonate, sulfamethoxazole, meprobamate, sulfisoxazole, methaqualone, theophylline, ephedrine hydrochloride and phenobarbital tablets, methylprednisolone, and tolbutamide. Appendix B lists the dissolution conditions for various approved drugs.

In those instances where a relatively insoluble drug is given orally, the role of dissolution rates can be ascertained from the blood levels achieved as a function of dose.

Absorption Principles

When a drug is introduced into the GI tract and is present in a form that can be absorbed, the process of absorption may be categorized as either "passive diffusion" or "active transport."

Passive Diffusion

This process describes the movement of drug molecules from a region of high relative concentration to a region of lower relative concentration. It also includes the movement of ions from a region of high ionic charge of one type to a region of lower charge of the same type or of opposite charge:

$$\frac{dX_a}{dt} = -DA(C_{gut} - C) \tag{3.4}$$

where

X_a is the amount of drug at the absorption site
D is the diffusion coefficient
A is the area of absorption surface
C_{gut} is the concentration of drug in the GI tract
C is the concentration of drug in the plasma

The driving force for passive diffusion is the concentration or the electrical gradient across the membrane that separates the GI lumen from the circulating blood. The concentration gradient is, however, more appropriately viewed as the chemical potential, represented by the number of molecules or ions that are free to move across a membrane and not by the total concentration in the lumen or plasma. In many instances, the plasma concentration is much lower than the concentration in the GI tract due to the rapid removal of the absorbed drug by the circulating blood, making the rate of transport across the membrane proportional to the chemical potential only in the GI tract.

Apart from the concentration gradient, diffusion rates depend also on the permeability characteristics of the membrane. The GI membrane acts like lipid barrier that permits the passage of lipid-soluble drugs, but across which lipid-insoluble but water-soluble molecules pass with difficulty; some of them may pass across the membrane through numerous pores that are too small to be seen even with the aid of an electron microscope, but for which strong evidence exists.

Active Transport

"Active transport" is a specialized process, which requires the expenditure of energy. The various active transport processes found in the GI tract are relatively structure specific and serve primarily in the absorption of natural substances, such as monosaccharides, 1-amino acids, pyrimidines, bile salts, and certain vitamins. However, there is evidence that certain drugs may also be absorbed by one of these active processes, if their chemical structures are sufficiently similar to that of the natural substrate. The anticancer drug 5-fluorouracil is an example of an actively transported drug. It is similar in structure to the natural substance, uracil, which is absorbed by means of the pyrimidine transport system.

Active transport is specific not only in terms of chemical structure but also with respect to direction, transporting molecules mainly from the mucosal side to the serosal side of the GI tract. The transport can also take place against the concentration gradient, that is, from a region of lower concentration or activity to the region of higher concentration or activity. Since active transport involves enzymes, these can be saturated at higher concentrations of the drug and may be subject to competitive inhibition in the presence of other drugs. Since active transport processes consume energy, they can be inhibited by various metabolic poisons such as fluoride or dinitrophenol, as well as by lack of oxygen.

Quite often, an active transport of drug molecules occurs concomitantly with passive diffusion. Faster absorption rates can generally be expected at lower concentrations due to the contributions of the active process but the passive diffusion becomes more important due to possible saturation of the active transport process at higher concentrations. Poor absorption or permeation is more likely when there are more than 5 H-bond donors and 10 H-bond acceptors, the molecular weight is greater than 500, and the calculated log P is greater than 5. This is also often referred to as Rule 5 of Lipinski. However, Lipinski specifically states that the Rule of 5 only holds for compounds that are *not* substrates for active transporters. Since almost all drugs are substrates for some transporter, much remains to be studied about Lipinski's rule. In addition, unless a drug molecule can passively gain intracellular access, it is not possible to simply investigate whether the molecule is a substrate for efflux transporters.

Solvent Drag

There are some variants of the two major types of transport processes described earlier. Water flux, in the same direction as drug movement, can increase the diffusion rate of a substance across the GI membrane. This is known as "solvent drag."

Facilitated Transport

Some substances are transported by a process that does not take place against a concentration gradient, but which involves a carrier that is subject to competition by other substances of similar structure and is affected by the metabolic inhibitors. This absorption process appears to be an active one and is referred to as "facilitated transport." The classical example of facilitated transport is the absorption of vitamin B_{12}. Vitamin B_{12} forms a complex with the intrinsic factor produced by the stomach wall and is transported in the form of this complex.

The observed low permeability of some drug substances in humans could be caused by efflux of drugs via membrane transporters such as P-glycoprotein (P-gp). When the efflux transporters are absent in these models, or their degree of expression is low compared to that in humans, there may be a greater likelihood of misclassification of permeability class for a drug subject to efflux compared to a drug transported passively. Expression of known transporters in selected study systems should be characterized. Functional expression of efflux systems (e.g., P-gp) can be demonstrated with techniques such as bidirectional transport studies, demonstrating a higher rate of transport in the basolateral-to-apical direction as compared to apical-to-basolateral direction using selected model drugs or chemicals at concentrations that do not saturate the efflux system (e.g., cyclosporine A, vinblastine, rhodamine 123).

An acceptance criterion for intestinal efflux that should be present in a test system cannot be set at this time. Instead, this guidance recommends limiting the use of nonhuman permeability test methods for drug substances that are transported by passive mechanisms. Pharmacokinetic studies on dose linearity or proportionality may provide useful information for evaluating the relevance of observed in vitro efflux of a drug. For example, there may be fewer concerns associated with the use of in vitro methods for a drug that has a higher rate of transport in the basolateral-to-apical direction at low drug concentrations but exhibits linear pharmacokinetics in humans.

Ion-Pair Transport

The absorption of highly ionized compounds at GI pH cannot be explained by passive diffusion or other mechanisms. A hypothesis has been suggested whereby highly ionized compounds (such as quaternary structures and sulfonic acids) form neutral complexes with other ions in the GI tract (such as mucin) and these ion-pair complexes are then absorbed by passive diffusion, since the complex has both the required lipid and aqueous solubility (Figure 6.3). This mechanism is referred to as "ion-pair transport."

Pinocytosis

Another mechanism of absorption is that of "pinocytosis," a process of physical absorption whereby an invagination of the cell membrane engulfs the particulate or droplet material. It is the only transport mechanism whereby a drug does not have to be in aqueous solution in order to be absorbed. Only a few compounds are absorbed by this mechanism, including vitamins A, D, E, and K. Pinocytosis is of significant importance in the uptake of nutrients.

Absorption Factors

The GI tract is composed of heterogeneous anatomic regions. As drug molecules descend through the GI tract, they encounter different environments that vary in pH, nature and concentration of enzymes, and fluidity of contents, as well as in the area available for absorption.

While the differences between the pH of the gastric and the intestinal fluids can account to some extent for the different rates of absorption of certain drugs from these two zones, the main reason is the difference in the absorption surface areas. Anatomically, the small intestine is much better designed for absorption than the stomach. The intestinal mucosa is covered by numerous villi and microvilli, providing a large surface area of approximately 120 m^2 (the intestine without the villi and microvilli would have a surface area of only 4 m^2). The large intestine has no villi and little drug absorption takes place from this region.

As the drug passes through the small intestine, the consistency of GI contents changes from fluid to paste due to the absorption of water. Thus, the drug particles that have not been dissolved in the stomach or upper small intestine will encounter difficulty in their dissolution in the lower intestine. Even if the drug is dissolved, it may not be absorbed quickly from the lower part of the intestine due to the retarded diffusion of molecules through pasty contents. Thus, in addition to the differences in absorption rates in different regions of the GI tract due to pH differences and absorption surface area, the consistency of the contents is also an important factor. In general, therefore, the upper part of the small intestine is the most important zone for the absorption of drugs, whether acids or bases.

Except for the colon, all other regions of the GI tract have areas for the specific transport of compounds. For example, iron absorption occurs mainly in the proximal part of the small intestine and decreases progressively in the intestine, thiamine absorption occurs mainly in the proximal region, and vitamin 12 is absorbed from the ileum. Therefore, if a drug is absorbed primarily through a specific GI area, it should not be administered by the rectal route.

For most drugs in general, and especially for those that are absorbed from a specific part of the GI tract, the extent and rates of absorption are dependent on the rate of passage of contents through the GI tract. Depending on the rate of passage, there may be only a limited time available for the dissolution of a solid particle and for the modification of its molecules into absorbable forms. This is exceptionally critical if the optimum absorption site is the proximal section of the small intestine. The rate of passage of intestinal contents through the upper small intestine is higher than it is through the lower part. Thus, if a drug is not present in an absorbable form within indicated time limits, it may be propelled past its absorption site and excreted totally or in part in the feces.

Gastrointestinal Fluids

Drugs must dissolve in the GI fluids before they can be absorbed; poorly water-soluble drugs have therefore inherent problems in their bioavailability. Any changes in the composition of GI fluids such as increased viscosity due to meal ingestion can reduce the dissolution of drugs. A moderate volume of fluid is also essential for optimal absorption since in addition to providing dissolution it also helps spread the drug over a larger area for absorption. It should be noted that larger fluid volumes also decrease the concentration gradient, yet the effect of increased contact area with the intestine overcomes this loss of driving force (concentration gradient). It is therefore advisable to take drugs with moderate volumes of fluids.

The pH of the GI fluids varies from about 1 to 3 in the stomach to about 8 in the large intestine. The factors that affect the pH include

- Type of diet
- Use of soft drinks
- Stress
- GI disease
- General health

Since the unionized forms of the drug is generally more lipid soluble, higher rates of absorption are observed at pH where the drug molecules are present predominantly in a unionized form. Several studies have confirmed this theory referred to as pH partition theory. In some instances, lack of conformation with this theory has been explained on the basis of a virtual membrane pH that may be different from the pH of the lumen.

Whereas the pH partition theory holds in principle, the majority of drugs whether acids or bases are primarily absorbed from the small intestine where much larger surface area is provided for absorption compared to the stomach and large intestine. It is interesting to note that the pH factors, which make the drug molecules more absorbable, can reduce the dissolution of drug molecules; for example, if tetracycline hydrochloride is administered with sodium bicarbonate (Figure 6.4) in a capsule form, the total absorption is significantly decreased due to decreased dissolutions of tetracycline at the alkaline pH due to sodium bicarbonate. If a solution of tetracycline hydrochloride is administered with sodium bicarbonate, no such effects are noted.

It should be noted that several formulation requires exposures to specific pH for their disintegration and changes in their gastric residence time can significantly alter the absorption. Exposure to intestinal pH may result in reduced absorption for some drugs due to the formation of insoluble hydroxides such as demonstrated when aluminum aspirin was used in chewable formulations and in the absorption of various iron preparations.

Drugs that are unstable at acidic pH show reduced absorption if the gastric residence times are prolonged such as shown for penicillin and erythromycin. In some instances, the use of a chemical modification such as erythromycin in an ester form helps to overcome this problem.

Intestinal fluids also contain a variety of components that may interact with drug molecules such as bile salts that may increase drug absorption by solubilization of drugs, increasing the diffusion of drug molecules across the lumen of the tract, and modifying the permeability of the intestinal membrane. Besides bile salts, there are several other naturally occurring surfactants of the intestinal fluid that regurgitate into the stomach and are primarily responsible for the lower surface tension (ca. 35–50 dynes/cm) of gastric fluid. Drugs generally dissolve faster in gastric fluid than in 0.1 N HCl.

Since the secretion of bile is not a continuous phenomenon, a variety of factors can affect the total content of bile and thus absorption of various drugs. For example, food causes increased secretion of bile and thus increased absorption of drugs like griseofulvin is observed. In addition to their solubilization effects, bile salts and synthetic derivatives such as dehydrocholic acid increase intestinal membrane permeability by a mucolytic action that reduces the barrier effect of intestinal mucins and by increasing biliary secretion due to their hydrocholerectic effects. A good example is the increased absorption of certain quaternary hypotensive agents, which bind to mucin. Another drug that seems to bind with mucin is tetracycline. It has been suggested to use pharmacologically inert quaternary compounds that can competitively bind to mucin and thus increase absorption of the active quaternary compounds.

Gastric Emptying

The gastric emptying rate affects the absorption rate primarily because of the pH differences between stomach and intestine. For example, weakly basic drug such as amphetamine and codeine will be absorbed primarily from the small intestine rather than from the stomach, and any delay in gastric emptying will tend to delay the absorption and thus the therapeutic response. Slow gastric emptying can also affect the bioavailability of drugs that are unstable in gastric fluids, for example, L-dopa, since the extent of degradation is proportional to the time for which the drug is exposed to the low pH and the enzymes of the stomach.

Some of the factors that affect the gastric emptying rates are as follows:

a. *Type of food*: The type of food will affect the stomach emptying rate significantly. For example, fats decrease the rate, proteins effect a lesser decrease, and carbohydrates retard gastric emptying the least. A fatty meal can therefore retard absorption rates of the drug and delay the onset of action. However, with such water-insoluble drugs as griseofulvin, the absorption can be increased as a result of retarded gastric emptying. The reason is that griseofulvin passes slowly to the small intestine, and therefore the longer duration of contact of griseofulvin with the intestine results in a greater chance for it to be dissolved and absorbed through a specific region. Table 3.4 highlights some examples where food affects absorption of various drugs.

   A faster gastric emptying rate is desirable for drugs that are not absorbed in the stomach. These should be taken either on an empty stomach or an hour before or 2 h after meals (Table 3.5).
b. *Volume of fluid or food*: The volume of fluid or food has a definite influence on the gastric emptying rate. The rate with which gastric contents leave the stomach is proportional to their volume. With small volumes, there is an initial lag time before gastric emptying begins, while with higher volumes, there is an initial phase of more rapid emptying. The fluid intake also affects the dissolution rate and forms the integral part of certain drug actions, for example, the use of bulk laxatives.
c. *Osmotic pressure*: The emptying rates are also dependent on the osmotic pressure of the liquids. For example, water leaves the stomach with a half life of about 5 min (a glass of water leaves the stomach in about 5–20 min). Hypertonic or hypotonic solutions generally leave the stomach at a slower rate than isotonic solutions.
d. *Acidity*: Gastric emptying is also retarded by increased acidity of gastric fluids. The use of antacid compounds increases gastric emptying rates. An interesting application of this property is the administration of L-dopa with sodium bicarbonate. Since L-dopa decomposes in the stomach, its administration with sodium bicarbonate increases its bioavailability due to decreased decomposition in the stomach—the result of both decreased acidity and increased emptying of the contents into the intestine.
e. *Food temperatures*: Hot or cold foods or fluids prolong the gastric emptying. For example, water taken at 25°C leaves the stomach at one-third the rate water taken at 37°C.

TABLE 3.4
Influence of Food on the Absorption of Various Drugs in Man

Drug	Influence on Absorption
Acetaminophen	Reduction in rate but not extent of absorption
Aspirin	Reduction in rate but not extent of absorption
Bretylium tosylate	Reduction in rate and extent of absorption
Capuride	Reduction in rate but not extent of absorption
Cephradine and cephalexin	Reduction in rate but not extent of absorption
Clindamycin	Reduction in rate but not extent of absorption
Digoxin	Reduction in rate but not extent of absorption
L-Dopa	Any factor reducing emptying rate will reduce rate and extent of absorption
Ethanol	Milk reduces the rate of absorption
Ethanol	Reduction in rate of absorption
Fenoprofen	Reduction in rate of absorption
Lincomycin	Reduction in rate and extent of absorption
Nitrofurantoin	Reduction in rate but increase in extent of absorption
Propantheline	Reduction in magnitude of pharmacological response
Rifampicin	Reduction in rate and extent of absorption
Theophylline	No noticeable influence on absorption

TABLE 3.5
Examples of Relationships between Food Intake and Drug Regimens

On Empty Stomach	1 h before or 2 h after Meals	1 h before Meals
Piperazine citrate	Tetracycline	Anticholinergic agents
Bephenium hydroxynaphthoate	Ampicillin	Methantheline bromide
Castor oil	Cefazolin sodium	Mepenzolate bromide
Pentaerythritol tetranitrate	Sulfisoxazole	Pancrelipase
Lincomycin	Trimethoprim	Sitosterols
Isosorbide dinitrate	Demeclocycline hydrochloride	Chlordiazepoxide hydrochloride
Dicloxacillin sodium	Fenfluramine hydrochloride	Anisotropine methylbromide
	Erythromycin	Diethylpropion hydrochloride
	Penicillin	Phenmetrazine hydrochloride
	Cholestyramine	Hexocyclium methylsulfate
	Rifampin	Propantheline bromide
	Methacycline	Glycopyrrolate
	Troleandomycin	Mazindol
	Nafcillin	
	Oxytetracycline	
	Hetacillin	

f. *Viscosity*: Liquids of low "viscosity" are emptied faster than liquids of higher viscosity. Solutions or suspensions of fine particles leave the stomach at a higher rate than lumpy substances.

g. *Psychological state*: The psychological state of an individual also affects gastric emptying. Depression, injury, and trauma lead to prolonged emptying. Agitation and excitement increase the peristaltic movement, thus increasing the rate of gastric emptying.

h. *Body posture*: Body posture can also significantly affect gastric emptying. Lying on the right side and standing may facilitate emptying, whereas the supine position may retard emptying.
i. *Drugs*: A number of drugs are capable of affecting the gastric emptying rate, usually through some central mechanism, such as anticholinergic drugs atropine, antihistamines, tranquilizers, aspirin, and morphine derivatives. Table 3.6 summarizes the various factors affecting gastric emptying.

TABLE 3.6
Influence of Various Factors on Gastric Emptying in Man

Factor	Influence on Gastric Emptying
1. Volume	The larger the starting volume, the greater the initial rate of emptying. After this initial period, the larger the original volume, the slower the rate of emptying.
2. Type of meal	
a. Fatty acids	Reduction in rate of emptying in direct proportion to their concentration and carbon chain length. Little difference from acetic to octanoic acids. Major inhibitory influence seen in chain length greater than 10 carbons (decanoic to steric acids).
b. Triglycerides	Reduction in rate of emptying. Unsaturated triglycerides are more effective than saturated ones. The most effective in reducing emptying rate were linseed and olive oils.
c. Carbohydrates	Reduction in rate of emptying primarily as a result of osmotic pressure. Inhibition of emptying increases as concentration increases.
d. Amino acids	Reduction in rate of emptying to an extent directly dependent upon concentration. Probably as a result of osmotic pressure.
3. Osmotic pressure	Reduction in rate emptying to an extent dependent upon concentration for salts and nonelectrolytes. Rate of emptying may increase at lower concentrations and then decrease at higher concentrations.
4. Physical state of gastric contents	Solutions or suspensions of small particles empty more rapidly than chunks of material that must first be reduced in size prior to emptying.
5. Chemicals	
a. Acids	Reduction in rate of emptying dependent upon concentration and molecular weight of the acid. Lower molecular weight acids are more effective than those of higher molecular weight. (In order of decreasing effectiveness: HCl, acetic, lactic, tartaric, citric acids.)
b. Alkali ($NaHCO_3$)	Increased rate of emptying at low concentrations (1%) and decreased rate at higher concentrations (5%).
6. Drugs	
a. Anticholinergics	Reduction in rate of emptying.
b. Narcotic analgesics	Reduction in rate of emptying.
c. Metoclopramide	Increase in rate of emptying.
d. Ethanol	Reduction in rate of emptying.
7. Miscellaneous	
a. Body position	Rate of emptying is reduced in a patient lying on the left side.
b. Viscosity	Rate of emptying is reduced with viscous solutions.
c. Emotional state	Aggressive or stressful emotional states increase stomach contractions and emptying rate. Depression reduces stomach contraction and emptying.
d. Bile salts	Rate of emptying is reduced.
e. Disease states	Rate of emptying is reduced in some diabetics, local pyloric lesions (pyloric ulcers, pyloric stenosis), and hypothyroidism. Gastric emptying rate is increased in hyperthyroidism and in the presence of duodenal ulcers.
f. Exercise	Vigorous exercise reduces emptying rate.
g. Gastric surgery	Gastric emptying difficulties often encountered after gastric surgery.

Intestinal Transit

The residence time in the intestine has a direct bearing on the amount of drug absorbed, however, absorption may be reached if the drug is unstable in intestinal fluids or binds irreversibly to the intestinal contents. For dosage forms where the drug is released only in the small intestine, for example, enteric-coated forms, intestinal transit times are of utmost importance.

The peristaltic and mixing movements of the intestine are also important in affecting the dissolution of the drug. Even though food greatly increases intestinal movements, administration of drug with food is generally not recommended because of other interactions with food.

Once the drug passes through to the colon, very little absorption can take place since the main function of this part of the intestine is to absorb water.

Blood Flow

The splanchnic circulation receives about 28% of the cardiac output that passes through the liver via a portal vein to the general circulation; thus, a significant metabolism of drug can take place in the liver before reaching the general circulation. The high perfusion of the GI tract creates a "sink" for the diffusion of drug molecules across the membrane. For most drugs, blood flow does not affect absorption rates unless

- The drug is actively absorbed where blood flow provides the energy for absorption process
- The absorption is very fast where it is more dependent on blood flow rate than on the transit across the membrane

Ingestion of meal increases flow rates, whereas extraneous exercise reduces blood flow rates to the GI tract.

Gastrointestinal Drug Biotransformation

The bioavailability of orally administered drugs can be affected due to biotransformation in the GI tract and the various organs (e.g., the liver) through which the drug molecules pass before reaching the general circulation. For example, the chromotropic activity of isoproterenol is about 1000 times greater when administered intravenously than through oral administration, largely due to the biotransformation of isoproterenol into an inactive sulfate during the transfer across the gut wall and passage through the liver. Similarly, some of the steroids are also extensively biotransformed during absorption. Since biotransformation reactions require the presence of enzymes, the saturation of these enzymes at higher drug concentrations results in dose-dependent effects, which have been noted for L-dopa and para-amino hippuric acid.

The intestinal microflora also plays an important role, causing biotransformation of such drugs as methotrexate, succinylsulfathiazole, and certain coumarin derivatives. Generally, the microflora has little effect on drug absorption except for drugs dissolving slowly or contained in slow-release dosage forms since these forms reach the distal end where most of the biotransformation takes place. The conjugates of many drugs can be cleaved by microorganisms that may cause recycling of drug molecules, a phenomenon that is altered in antibiotic therapy. However, for drugs that may generally be biotransformed in the gut, antibiotic therapy can increase their bioavailability. The following are some interesting examples:

- Antibiotic therapy alters the pattern of L-dopa metabolites excreted in urine.
- Conjugates of isonicotinic acid are hydrolyzed by intestinal bacteria with subsequent reabsorption of isonicotinic acid.
- Lanatoside C is converted to digoxin by intestinal bacteria.

- Cyclamate is converted to toxic metabolites cyclohexylamine in the intestine.
- Gut wall metabolism of salicylic acid and various steroids results in glucuronidation.
- Nonspecific esterase hydrolyze aspirin in the gut wall.
- L-Dopa is metabolized by decarboxylase enzyme in gastric mucosa.

The overall impact of gut metabolism on drug bioavailability and toxicity has not been fully evaluated due to lack of awareness of these mechanisms and analytic methods to monitor the metabolites.

FOOD INTERACTIONS

Food affects drug bioavailability by several mechanism including

- Changes in gastric and intestinal transit times
- Increased GI secretions
- Adsorption of drug onto food
- Competition of food components with drug for absorption
- Physicochemical interactions between food and drug
- Increased viscosity of GI fluids

Drugs that are actively absorbed from the GI tract may show competitive absorption with food components such as amino acids. Examples include L-dopa and several anticancer drugs.

It is often taken for granted that food impairs the absorption of drugs and that the drugs should be taken on an empty stomach. Some misleading assumptions include the following: (1) drugs should be administered with food only if they are irritant, (2) a reduction in drug absorption with food intake occurs due to decreased gastric emptying, and (3) most drugs are absorbed by a passive diffusion process. Recent findings dispute these assumptions and show that food can improve the bioavailability of several drugs. It has also been observed that food may influence the rate of drug absorption without affecting the extent of absorption. Furthermore, active intestinal transport mechanisms may be more important than hitherto recognized. Food may also affect the first-pass biotransformation of drugs in the gut and in the liver. The effect of food composition such as carbohydrate/protein ratios may change the elimination rates of some drugs.

Several aspects must be considered in studying the effect of food on the bioavailability of drugs:

- Food induces changes in the gastric emptying rate, intestinal transit time, and/or in gastroenterohepatic secretion of hydrochloric acid, bicarbonate, enzymes, and bile.
- Specific food components and contaminants can alter metabolic transformation of drugs in the gut and in the liver.
- Food refers to different kinds of meals and that one type of meal or food component may have both qualitatively and quantitatively different effects on drug bioavailability than other.
- Different preparations of the same drug may interact differently with food.
- Findings based on single-meal, single-dose studies in healthy volunteers may not necessarily be relevant as to food effects on the steady plasma level of drug during its long-term use in patients.

Rifampicin absorption is also reduced when given with food. Since this drug is mostly given qd, generally an hour before breakfast, food interaction does not present a problem in therapeutic management. Thus, for all those drugs that are generally given once a day, food interactions are of less significance unless an increased absorption is possible, in which case the dosing must be carefully monitored.

The absorption interactions of tetracyclines are well known. The absorption of first-generation tetracyclines such as oxytetracycline or tetracycline is drastically reduced by intake of antacids, or of calcium-containing food items, such as milk and cheese. It is a well-established fact that nonabsorbable chelates are formed between the metals (Al, Mg, Ca, Fe) and the tetracyclines. In addition, the pH raising influence of food and antacids is important, as the solubility of tetracycline is reduced with increasing pH. Absorption of newer tetracycline analogs, doxycycline and minocycline, is not significantly affected by food but is inhibited by antacids and iron preparations.

Absorption of penicillin, ampicillin, oxacillin, dicloxacillin, lincomycin, and some erythromycin preparations is reduced when taken with food. However, bioavailability of amoxicillin or that of ampicillin when given in esterified form is not affected by food, and recent studies indicate increased absorption of erythromycin stearate when given with food. In view of the irritant properties of ampicillin and erythromycin, these should be administered with food.

The bioavailability of nitrofurantoin is increased from both macro- and microcrystalline forms when given with food due to reduced gastric emptying, which allows greater time for the dissolution of the drug.

PATHOPHYSIOLOGICAL DISORDERS

Drug bioavailability is significantly altered in the presence of various pathophysiological disorders. The following are some specific observations:

Alterations in gastric pH have the following implications:

- pH partitioning and dissolution of poorly soluble drugs can be significantly affected, for example, aspirin is better absorbed in achlorhydric patients.
- Changes secondary to pH change may include epithelium integrity and blood flow rates that can directly affect rate and extent of drug absorption.
- Stability of acid labile drugs can be significantly altered.
- Several disease states including gastric cancer have been identified when the gastric pH is elevated.

Gastric emptying is hampered after gastric surgery and gastrectomy increases gastric emptying. The various effects observed are as follows:

- Drugs requiring exposure to gastric environment for dissolution show reduced bioavailability in gastrectomy.
- Enteric-coated tablets will show specific absorption problem in pyloric stenosis.
- Gastric emptying is delayed in labor and further exacerbated due to narcotic analgesics; serious consequences may result due to regurgitation of gastric fluid into respiratory tract.

The effect of intestinal transit on drug bioavailability is also very pronounced. Intestinal transit rate is decreased when

- Digestive juice secretion is reduced
- Thyroxine secretion is reduced
- Hypothyroidism exists
- Insulin hypoglycemia exists
- Chronic diarrhea exists

A variety of malabsorption syndromes affect drug absorption as well:

- In steatorrhea, absorption of phenoxymethyl penicillin is reduced.
- Ampicillin and nalidixic acid are less absorbed in shigellosis.
- Propranolol availability is increased in coeliac diseases due to reduced intestinal metabolism.
- Riboflavin absorption is impaired in biliary atresia.

A variety of drugs administered to treat pathophysiological orders show interactions resulting in alteration of absorption. Table 3.7 lists some examples of these interactions.

TABLE 3.7
Drug Interactant Influence on Drug Absorption

1. pH		
Folic acid	NaHCO	Reduced rate and possibly extent of absorption. Mechanism unknown but may be related to drug ionization.
	Diphenylhydantoin	Reduced absorption. Possibly due to alkalinization of gut fluids by the anticonvulsant.
Tetracycline	NaHCO	Reduced rate and extent of absorption from capsules. No influence on absorption from solution. Effect due to reduced drug dissolution rate from capsules.
2. Gastric emptying and intestinal motility		
Acetaminophen	Diacetylmorphine (IM)	Reduced rate but not extent of absorption.
	Meperidine (IM)	Reduced rate but not extent of absorption.
	Metoclopramide (IV)	Increased rate but not extent of absorption.
	Propantheline (IV)	Reduced rate but not extent of absorption.
Bishydroxycoumarin	Heptabarbital	Reduced absorption possibly due to increased intestinal transit rate.
Chlordiazepoxide	Antacid	Reduced rate but not extent of absorption. Possibly due to decreased gastric emptying rate.
Diazepam	Metoclopramide (IV)	Increased rate of absorption.
Digoxin	Propantheline	Increased rate and extent of absorption from a slowly dissolving tablet. No influence on absorption from a rapidly dissolving product.
	Metoclopramide	Reduced rate and extent of absorption.
L-Dopa	Antacid	Increased rate and extent of absorption possibly due to increased gastric emptying rate. Effect seems variable.
	Imipramine	Reduced rate of absorption.
	Metoclopramide	Increased rate and extent of absorption.
Ethanol	Metoclopramide (IV and p.o.)	Increased rate and possibly extent of absorption.
	Propantheline (IV and p.o.)	Decreased rate and possibly extent of absorption after i.v. propantheline. Oral propantheline has no influence on ethanol absorption.
Griseofulvin	Phenobarbital	Reduced absorption possibly due to increased intestinal transit rate.
Isoniazid	Antacid	Reduced rate and possibly extent of absorption possibly due to decreased gastric emptying rate.
Nitrofurantoin	Propantheline	Increased extent of absorption.
Phenolsulfonphthalein	Propantheline	Reduced rate but increased extent of absorption.
Phenylbutazone	Desmethylimipramine	Reduced rate of absorption.
Pivampicillin	Metoclopramide (IM)	Increased rate of absorption.
	Atropine (s.c.)	Reduced rate of absorption.
Riboflavin	Propantheline	Reduced rate but increased extent of absorption.
Sulfamethoxazole	Propantheline	Reduced rate of absorption.
Tetracycline	Atropine (s.c.)	Reduced rate of absorption.
	Metoclopramide (IM)	Increased rate of absorption.
3. Adsorption		
Acetaminophen	Charcoal	Reduced extent of absorption.
Aspirin	Charcoal	Reduced rate and extent of absorption.
	Cholestyramine	Reduced rate and possibly extent of absorption.
Chlorpromazine	Antacid	Reduced rate of absorption.

(Continued)

TABLE 3.7 (*Continued*)
Drug Interactant Influence on Drug Absorption

Chlorothiazide	Colestipol	Reduced extent of absorption.
Lincomycin	Antidiarrheal absorption	Reduced extent of absorption.
Nortriptyline	Charcoal	Reduced extent of absorption.
Phenylpropanolamine	Charcoal	Reduced extent of absorption.
Promazine	Charcoal	Reduced rate and extent of absorption.
	Antidiarrheal preparation	Reduced rate and extent of absorption.
Propantheline	Charcoal	Reduced response.
Propoxyphene	Charcoal	Reduced extent of absorption.
Pseudoephedrine	Kaolin	Reduced rate of absorption.
Salicylamide	Charcoal	Reduced extent of absorption.
Thyroxine	Cholestyramine	Reduced extent of absorption.
Vitamin B_{12}	Cholestyramine	Reduced extent of absorption.
Warfarin	Cholestyramine	Reduced rate and extent of absorption.
4. Complexation		
Bishydroxycoumarin	Milk of magnesia	Increased rate and extent of absorption.
Chlortetracycline	Antacids	Reduced extent of absorption.
Tetracycline	Antacids	Reduced extent of absorption.
5. Miscellaneous		
Ergotamine	Caffeine	Increased rate and extent of absorption possibly due to complexation.
Folic acid	Salicylazosulfapyridine	Reduced extent of absorption. Mechanism not known.
Iron	Hexocyclium methosulfate	Reduced absorption possibly due to decreased secretion of a gastric factor needed for absorption.
Pseudoephedrine	Antacid	Increased rate of absorption. Effect may be due to changes in gut pH or gastric emptying.
Rifampicin	Para-aminosalicylic acid	Reduced rate and extent of absorption. Mechanism not known.
Vitamin B_{12}	Para-aminosalicylic acid	Reduced absorption by an unknown mechanism.

AGE

Several GI functions mature with age including specialized absorption mechanisms. For example, sugar absorption is very inefficient in younger children. Whereas significant changes in the structural and functional properties of GI tract and blood flow occur in the elderly, no studies have demonstrated changes in the bioavailability of drugs in the elderly.

FIRST-PASS BIOTRANSFORMATION

A distinction can often be made between the biotransformation in the intestine and that in the liver during the first pass by administering the drugs either intraperitoneally or directly into portal vein. Some mathematical approaches have also been used and will be discussed later. Table 3.8 lists the drugs that are suspected of first-pass or gastric hepatic biotransformation. Whereas 100% of the orally administered dose goes through the liver, only about 25%–30% of the intravenously or intramuscularly administered dose passes through the liver, which may partly explain the differences in responses observed as a function of route of administration. The hepatic clearance of drugs depends on two factors:

1. Blood flow to liver
2. Capacity of liver to remove drug

TABLE 3.8
Drugs for which First-Pass Hepatic Biotransformation Is Suspected, Possibly in Addition to Gastrointestinal Biotransformation

Alprenolol	Pheniprazine
Desmethylimipramine	Propranolol
Dopamine	Reserpine
Lidocaine	Serotonin
Nortriptyline	Tryptophan
Oxyphenbutazone	

SUBLINGUAL/BUCCAL ADMINISTRATION

Some drugs are administered by placing them beneath the tongue or in the cheek pouch. A rapid absorption of drugs is thereby generally expected due to the high vascularity of this region. The pH of saliva is about 6, and drugs are absorbed by passive diffusion with a slightly higher requirement for lipid solubility than is needed for intestinal absorption.

A significant advantage of this route is that GI degradation and biotransformation are bypassed along with hepatic first-pass biotransformation. A variety of drugs can be administered by this route, including nitrates and such hormones as methyltestosterone, testosterone propionate, and oxytocin. Few studies have reported on the effective use of this route of administration. One such study reports significantly higher blood levels of methyltestosterone from sublingual tablets than are obtained from other routes. Absorption properties of sympathomimetic amines, methadone, meperidine, lidocaine, chlorpheniramine, imipramine, desipramine, and barbiturates have also been studied. Recently, chewing-gum-based drug delivery systems have been developed for several drugs.

RECTAL ADMINISTRATION

Some drugs are administered rectally either in suppository or in solution form, for example, retention enema. The solution yield better absorption provided that they are retained for a sufficient length of time in the rectum. The suppositories are the most commonly used dosage forms for both local and systemic effect. Examples of drugs administered rectally for systemic action include aspirin, acetaminophen, indomethacin, diazepam, theophylline, prochlorperazine, cyclizine, promethazine, and barbiturates.

The absorption mechanism mainly involves passive diffusion with no sites for active transport. The absorption rate and bioavailability are more erratic than observed with oral administration, due to such added factors as the presence of feces retarding absorption or irritant suppository bases such as carbowaxes causing early evacuation. The use of an enema before drug administration generally increase the absorption significantly.

The rectal route of administration is not suitable for irritant drugs such as tetracycline or penicillin. A large number of studies have attempted to develop an "ideal" base for suppositories or formulation for a microenema, but little has been reported regarding their comparative bioavailability in humans. Thus, a conclusion cannot be drawn regarding the relative merits of this route of administration compared with other routes.

INTRAVENOUS ADMINISTRATION

The direct administration of drugs into veins is the only route where bioavailability considerations are not relevant. This route provides an almost instantaneous response with controllability of the rate of drug input into the body. This route is especially suitable for those drugs that cannot be absorbed adequately from the GI tract or tissue depots (e.g., intramuscular administration) or where there is a significant first-pass effect upon oral administration. The drugs that would be intolerably painful in the subcutaneous or muscle tissues by virtue of their irritant properties may be injected slowly into a vein without much difficulty, for example, nitrogen mustard in cancer chemotherapy.

There are, however, several disadvantages with the use of the intravenous route. A drug administered intravenously cannot be recalled, whereas some such measures can be taken with other routes. Rapid intravenous injection may evoke catastrophic effects in the circulatory and respiratory systems due to the transient wave of concentrated solute suddenly reaching the myocardium and the chemoreceptors in the aortic arch and carotid sinus. Intravenous injections should, therefore, be administered slowly, preferably over a period of 1 min or more, during which time the blood completes its circulation. The possibility of anaphylactoid reactions is much greater than with any other route of administration.

The tonicity of solution is also important since hypotonic or hypertonic solution can cause hemolysis or agglutination of erythrocytes. The damage of the vascular wall also leads to local reactions, especially after prolonged infusions.

The possibility of microbiological contamination and pyrexia due to pyrogens is a serious concern in the use of intravenous administration.

The intravenous route is especially suitable when a rapid response is required, as in the treatment of epileptic seizures, acute asthmatic attacks, cardiac arrhythmias, etc. The fluctuation of plasma concentration is generally very small if a drug is administered by slow intravenous infusion, as is employed for lidocaine, theophylline, and many antibiotics. A caution is needed for drugs with poor water solubility that can precipitate resulting in thrombosis and removal of drug from circulation and deposition of the precipitate in various tissues resulting in reduced apparent bioavailability. Also drugs that bind to plasma proteins extensively may show altered response depending on the rate of injection since the initial binding and concentration at site of action can vary significantly.

INTRA-ARTERIAL ADMINISTRATION

This route is used for the injection of substances used in diagnosis. A typical example is the injection of a radiopaque compound into the carotid artery to trace the circulation of the brain by roentgenography. In addition, certain specialized techniques in cancer chemotherapy call for regional infusion of drugs by arterial routes, which may provide a significant advantage over other routes.

INTRAMUSCULAR ADMINISTRATION

More than 50% of hospitalized patients receive intramuscular drug administration. The popularity of this route is due to the decreased hazard of administration when compared with the intravenous route. Large volumes of solution can be injected (2–10 mL) by this route, generally with less pain and irritation than is encountered with the subcutaneous route.

Aqueous solution of drugs are usually absorbed from intramuscular administration sites within 10–30 min, but faster or slower absorption rates are possible depending on the vascularity of the site (blood flow rates range from 0.02 to 0.07 mL/min), the ionization and lipid solubility of the drug, the volume of injection, the osmolality of the solution, and other variables, including coadministered drugs and adjuvants in the formulation.

The small molecules are absorbed directly into the capillaries from the intramuscular administration sites, whereas large molecules gain access to the circulation by way of the lymphatic channels.

The drugs that are poorly water soluble, such as digoxin and diazepam, or those drugs that dissolve at pH values far above the physiological range, are often administered in nonaqueous media such as propylene glycol or in strongly acid or alkaline aqueous solutions. However, after intramuscular administration, these drugs may not stay in solution, resulting in slow or incomplete absorption. In some instances, the total bioavailability may be less than that from oral administration, as is demonstrated with phenytoin, diazepam, and cefamandole.

The high lipid-soluble molecules are quickly absorbed from intramuscular administration sites, whereas lipid-insoluble molecules diffuse between interstitial fluid and plasma only through the pores in the capillary membrane; this is generally not the rate-limiting step in the absorption. Only very large lipid-insoluble molecules that must be absorbed through the lymphatic system have a rate limitation in their absorption, due to the slow rate of lymph flow (0.1% of the plasma flow).

The concentration of the injected solution can also affect the rate of absorption. For example, atropine is absorbed more rapidly when administered in a smaller volume of more concentrated solution. Absorption rates can be accelerated by spreading the solution over large tissue areas, for example, by massaging or using high-pressure injection devices.

The blood flow to the administration site is often the rate-limiting step in the absorption of drugs. Absorption is more rapid after injection into the deltoid than into the vastus lateralis and is slowest after gluteal muscle injection. The drugs can be absorbed faster after administration into the buttock in males compared with females due to greater adipose tissue in females. Absorption rates increase during exercise regardless of the site of intramuscular administration, since this results in increased blood flow to skeletal muscles. Conversely, absorption rates decrease in circulatory shocks, hypotension, CHF, myxedema, and other disturbances of the circulatory system.

Absorption rates can often be quite erratic upon intramuscular administration of drugs. This is due to increased membrane contact as the solution spreads, change in drug concentration as a result of absorption, a possible hypertonic effect drawing water to the site, or to the precipitation of the drugs. The precipitation can lead to incomplete absorption due to extremely slow redissolution or to phagocytosis of the drug particles. Examples of these incompletely absorbed drugs are ampicillin, cephaloridine, cephradine, phenytoin, and quinidine. Conversely, the slow absorption of drugs can itself be exploited to produce prolonged administration. The slow absorption can be accomplished by the use of injection vehicles of high viscosity, such as glycerin, cottonseed oil, sesame oil, or polyethylene glycols. Another technique involves preparation of fatty acid ester derivatives, such as decanoate derivative of fluphenazine, which hydrolyzes slowly and provides gradual release. Benzathine penicillin and procaine penicillin are injected as water-insoluble suspensions for the same purpose. Slowly released preparations of antipsychotic agents have been useful in the maintenance therapy of schizophrenia.

The side effects of intramuscular administration include pain, elevation of serum creatine phosphokinase as a result of trauma, and often sciatic nerve damage following gluteal injections. Other complications include skin pigmentation, hemorrhage, septic or sterile abscesses, cellulitis, muscular fibrosis, tissue necrosis, and gangrene.

SUBCUTANEOUS ADMINISTRATION

The factors affecting intramuscular drug absorption also determine subcutaneous drug availability. The blood flow rates are poorer than in muscles and so are the rates of absorption. Yet some drugs are absorbed as rapidly from a subcutaneous site as from intramuscular administration, for example, anionic dye, phenolsulfonphthalein, and insulin.

A prime determinant of the absorption rate of a subcutaneous depot is the total surface area over which the absorption can occur. Although the subcutaneous tissues are somewhat loose, and moderate amounts of fluids can be administered, the normal connective tissue prevents indefinite lateral spread of the injected solutions. These barriers can be bypassed with the aid of hyaluronidase, an enzyme that breaks down mucopolysaccharides of the connective tissue matrix and results in wider spreading of solutions and faster absorption rates. The absorption rates can also be increased by massage or by application of heat to increase blood flow. Quite frequently, drugs affect their own rates of absorption if they alter the blood supply or capillary permeability. For example, methacholine, a cholinergic drug, causes vasodilation, which results in an immediate systemic response following subcutaneous administration.

The absorption of drugs from the depots formed following subcutaneous administration can be retarded to provide prolonged effect by such techniques as immobilization of the limb, local cooling to cause vasoconstriction, and the application of a tourniquet proximal to the injection site to block the superficial venous drainage and lymphatic flow. Inclusion of minute amounts of epinephrine (1:100,000 or 1:2,000,000) in the subcutaneous injection may retard absorption by constricting the veins to elicit local rather than systemic effects that are desired, for example, in the administration of local anesthetics.

The subcutaneous route of administration has frequently been used to provide prolonged release of drugs by incorporating the drugs into compressed pellets that can be implanted under the skin. The drug must be present in a relatively insoluble form and the pellet must resist disintegration by the subcutaneous fluid environment and mechanical stress. These conditions have been achieved with certain steroid hormones. For example, cylindrical pellets of testosterone, about 5 mm in thickness and diameter and weighing about 100 mg, are implanted subcutaneously in humans. The absorption of about 1% per day is generally obtained during the steady state for up to 2 months.

An ideal shape for achieving constant rates of absorption is a flat disc. A change in the weight of the disk due to absorption results in very little change in the total surface area exposed since the release of the drug takes place from the flat surfaces.

For spherical pellets, the ratio of surface area to volume increases with decreasing diameter. Thus, when drugs are prepared as spheres of known diameter, the rate of absorption can be predicted: the larger the sphere, the slower the rate of absorption. This principle has been used in the design of long-acting insulin preparations. Prompt insulin consists of small particles, and extended insulin is made up of relatively large particles.

Two examples of pellets used for subcutaneous implantation are Oreton pellets (75 mg testosterone) and Progynon pellets (24 mg estradiol).

Some drugs produce severe pain when injected subcutaneously. Local necrosis and sterile abscesses may also occur. Such drugs may have to be administered intravenously because no solution concentrated enough to be useful can be given subcutaneously or intramuscularly.

PERCUTANEOUS ADMINISTRATION

The absorption of drugs through the skin should be a difficult matter since the function of the skin is to act as a barrier between the outside environment and the vulnerable tissues under the skin. Yet drugs are absorbed, sometimes quite efficiently, from the skin.

A major function of skin is to retard the diffusion and evaporation of water from within the body, except at the sweat glands. The stratum corneum, also known as the horny layer, which is densely packed with keratin, is responsible for this retardation. Beneath the horny layer, separating it from the underlying granular layer of epithelial cells, is the so-called barrier area, a clear dense region that is quite different from the horny layer both in microscopic appearance and in chemical properties. If the horny layer is stripped but the barrier area is left intact, little change in permeability occurs although water loss increases. However, removal of the barrier area leads to an abrupt increase in permeability for all kinds of molecules, large or small, lipid or water soluble. The dermis is generally freely permeable to all types of molecules.

The penetration rates of drugs through the skin are determined largely by their lipid/water partition coefficients excluding significant absorption of ions or water-soluble structures, except for very small molecules. Highly lipid-soluble molecules also penetrate the skin slowly compared to their penetration through other membranes.

Drugs may be applied to the skin for a local effect, especially on the superficial layers of the epidermis. The drugs are incorporated into vehicles that adhere to the skin, allowing diffusion of drug molecules out of the vehicle and into the epidermis. If a pathological condition exists in the deeper layers of the skin, the systemic administration may be more desirable, especially if the drug is water soluble. For example, antifungal and antibacterial agents are often much more effective in skin infections when given orally or by injection than when applied to the skin. Highly lipid-soluble drugs, such as griseofulvin, are also effective in systemic administration for local skin infections.

Some recent studies suggest the use of pharmacologically inactive solvents, such as dimethyl sulfoxide, to facilitate the absorption of drugs through the skin. Examples of drugs whose absorption has been increased are corticosteroids, antineoplastics, antibiotics, carcinogens, and insulin. There is, however, great controversy on the toxicity of these solvents in topical formulations and it is difficult to justify their use at the present time.

A recent approach to utilizing Therapeutic Systems consists of a multilaminate structure of small size, which is worn in the postauricular region, providing optimum drug permeability. Scopolamine is the first drug applied in this way for prevention or treatment of motion-induced nausea with reduced parasympatholytic effects. A large number of "patches" are currently available for delivery of nitroglycerin through skin. These dosage forms are designed such that the rate-limiting factor in the absorption is the release form and dosage forms and not the skin permeability. This is necessary to reduce variability in absorption due to biological factors pertaining to skin permeability.

The ionic drugs can often be administered through the skin by applying electrical gradients to the skin. This method of iontophoresis involves applying galvanic current to electrodes placed at the absorption site and at other parts of the body.

The fast absorption of lipid-soluble molecules through the skin indicates an environmental hazard that continues to grow with increasing pollution in the atmosphere. For example, carbon tetrachloride and other organic solvents prevents the body through the skin and cause serious toxic effects.

Organic phosphates (DFP, parathion, malathion) and nicotine insecticides have caused deaths in agricultural workers as a result of percutaneous absorption upon in-field contact. Chlorovinyl arsine dichloride (lewisite), a mustard gas, is readily absorbed through the skin and has been used in chemical warfare. Most of the carcinogens in the atmosphere can be efficiently absorbed through the skin and it is no wonder that there is a higher cancer incidence rate in people living around the industrial centers, even though these people may not be directly exposed to these chemicals.

Topical delivery of drugs using semisolid, controlled release patches and many other delivery systems dosage forms offers advantages including reduced blood level fluctuation, obviating the first-pass effect, and protection from GI pH. Where localized action is desired, this dosage form offers remarkable opportunity for drug action. However, skin is a poor medium to deliver drugs because by its very design, it is supposed to prevent the entry of chemicals (though it fails miserably as we know from chemical warfare agents). Generally, large polar molecules do not penetrate the stratum corneum well. The intrinsic physicochemical properties of candidate drugs important in expediting delivery across skin include molecular weight and volume, aqueous solubility, melting point, and log P. For weakly acidic or basic drugs, the skin pH will play a strong role in their transport. Drugs that form zwitterions can be made more penetrable by using appropriate salt forms.

The formulation additives strongly impact on transdermal delivery as the variety of dosage forms such as creams, ointments, lotions, gels, patches offer a wide variety of formulation additives. The problems related to crystallization of drugs as discussed under suspension dosage forms also apply here just as do considerations that optimize physical and chemical stability. Entire textbooks have been dedicated to formulating semisolid and topical delivery dosage forms that describe in detail how the choice of basic drugs structure and additives affects stability. Where salt forms are available, it is often difficult to predict the stability profile including such factors as photostability, a test that must be conducted for all dosage forms. It is known that different salt forms can show differences in their photostability profile.

PULMONARY ADMINISTRATION

Drugs can be introduced into the pulmonary system as gases or in aerosol forms. An almost instantaneous absorption can be expected due to the extremely large surface area available for absorption.

The primary mechanism of absorption is passive diffusion but the lipid solubility tends to play a smaller role than in GI absorption. The main limiting step in the utilization of this route has been the need to design dosage forms that accurately deliver the drugs. Most of these drugs are administered as aerosols, and their delivery to a great extent is dependent on the particle size distribution. Particles greater than 10 μm are almost completely removed by impaction in the nasal passages. "Impaction" refers to the deposition of particles in the respiratory tract. The precipitation of particles arises from the tendency of a particle moving in a stream of air to continue in its original direction when the air current changes direction at bronchial branch points and at curves in the bronchial tree. Impaction due to diffusion is negligible except for very small particles. Particles below 10 μm in diameter are of great significance since these include bacteria, viruses, smoke, industrial fumes, dust laden with fission product, pollens, insecticide dusts and sprays, and inhalant sprays used in the therapy of pulmonary diseases.

In order for a drug to be absorbed from an aerosol, its particles must impact, preferably in the alveolar sacs, and dissolve in the available fluids. Larger particles are retained in the upper respiratory tract and smaller particles penetrate deeper into the pulmonary tree. Particles larger than 2 μm in diameter probably do not reach the alveolar sacs. Particle sizes approximating 1 μm are most desirable, but there is a greater tendency for these particles to be exhaled without being impacted. Thus, many formulations include hygroscopic substances in the formulation to increase the size of particles deeper into the trachea. The tidal volume is also an important consideration. At a given respiratory rate, the air stream velocity is greater at high tidal volumes and thus particles of all sizes tend to be driven deeper into the pulmonary tree before impaction.

Pulmonary administration has been used mainly for local therapy. For example, aerosols of epinephrine, isoproterenol, and dexamethasone are commonly used for acute asthmatic attacks, and antibiotics are sometimes incorporated for the treatment of complicated bronchopulmonary infections. In some instances, the systemic absorption of drugs administered for local action may be appreciable. For example, isoproterenol in a 0.5% aerosol is an effective bronchodilator, but a 1% aerosol is apt to cause undesirable cardioaccelerator and hypertensive actions after only a few inhalations. The quick responses can, however, be beneficial in the treatment of anaphylactic episodes, as in the use of epinephrine.

Although the pulmonary route is used mainly for local effects, several drugs have been successfully administered in this way for systemic effect, including penicillin, glycosides, diuretics, and tranquilizers. More recently, an inhalation form of insulin has been marketed where the primary mechanism is impaction of very fine particles.

The problem of accurate dosing in pulmonary dosing in pulmonary administration remains a serious obstacle to greater use of this route. The use of metered dose devices is certainly an improvement and some products use the drug as a powder aerosol. The powder particle sizes range primarily between 2 and 6 μm. This device, currently used for disodium cromoglycate (Aarane [cromolyn sodium] inhaler), provides a greater and more consistent absorption than can be obtained from other metered dose devices.

The pMDIs the use of environmentally friendly propellants means choice of hydrofluoroalkanes wherein the dosage form can be a suspension of solution form. The problems of formulating suspensions as discussed earlier apply here as well but particularly with respect to interactions with the formulation components specific to pressurized inhaler systems. Solution dosage forms require the selection of propellants wherein the drug can dissolve without crystallizing and may require the addition of surfactants and cosolvents. However, there are toxicological issues with the use of surfactants. The solubility of drugs in solvents is determined by filtering the suspension in pressurized can into another can and then evaporating the clear solution (bringing to room temperature) and determining the amount of drug in it. High solubility in propellants can lead to crystal growth as propellants evaporate. Ostwald ripening common to suspensions applies to inhalation suspensions; the changes in the property of suspension can be studied using microscopy and observing changes in the axial ratio of crystal.

Drugs for inhalation therapy in a powder form required particular particle size that is achieved by the process of micronization between 1 and 6 μm to allow deep penetration through the lung alveoli system. There are a number of devices that can deliver drugs to the lungs as dry powders, for example, Turbuhaler™ or Diskhaler™. These dosage forms rely on a larger carrier particle, such as α-lactose monohydrate, to which the drug is attached. The lactose is usually fractionated such that it lies in the size range 63–90 μm. Upon delivery, the drug detaches from the lactose and, because the drug is micronized, it is delivered to the lung, whereas the lactose is eventually swallowed. It should be realized that the polymorphic form of the lactose used could affect the aerosolization properties of the formulation. The β-forms were easily entrained but held onto the drug particles most strongly when flow properties are studied. The anhydrous α-form shows an opposite behavior and the monohydrate α-form demonstrates intermediate behavior. Interactions with packaging materials can also alter powder characteristics; for example, long contact times with PVC, polyethylene, or aluminum should be avoided since the adhesion force between the drug and these surfaces is much higher than between it and the lactose carrier. Thus, detachment and loss of drug in the formulation could occur. Because lactose is widely used as a carrier, its compatibility with the new drugs should be studied in detail especially if there are any amino groups in the structure. The surface property of lactose is also important. With increasing specific surface area and roughness, the effective index of inhalation decreases due to the drug being held more tightly in the inhaled airstreams. Therefore, characterization of the carrier particles by, for example, surface area measurements, SEM, and other solid-state techniques are recommended preformulation activities.

The recent approval by the U.S. FDA of Exubera, an inhalation form of insulin, is a classical example where the dosage form is an integral part of drug action. Using the Nektar company's delivery system to create a fine powder mist, insulin in Exubera is absorbed as the mist of fine reaches into the deep portions of lung structure without getting impacted. Whereas reduction in particle size is pivotal to pulmonary delivery of drugs, micronization makes powders difficult to flow, and these changes should be studied using such techniques as DVS, microcalorimetry, and IGC. The high energy at the surface of micronized powders can often be relieved by exposing it to higher-humidity air that can crystallize the amorphous high-energy regions. As a result, the common preformulate stage evaluations include measurements of the micromeritic, RH, and electrostatic properties of the powder. Different salt forms show variant flow properties; for example, stearate salts generally are better for aerosol formulation.

Nebulizer formulations are normally solutions but suspensions (particle size of less than 2 μm) are also used. Important preformulation considerations include stability, solubility, viscosity, and surface tension of the solution of suspension.

OPHTHALMIC ADMINISTRATION

As with permeability in most other routes of administration, the permeability of drugs into and through the cornea is a function of their lipoid and aqueous solubility. The cornea is composed of three distinct layers: the outer epithelium, an inner stroma, and the endothelium. The epithelium and endothelium are much more lipoidal than stroma. Therefore, drugs must possess biphasic solubility characteristics in order to be absorbed through this route.

Weakly basic drugs, such as tropicanade, epinephrine, pilocarpine, atropine, homatropine, or cyclopentolate, freely penetrate the cornea because of rapid equilibration between their lipid-soluble unionized forms and their water-soluble ionized forms. The penetration of quaternary ammonium compounds, such as carbachol, echothiophate iodide, and demecarium bromide, which are charged and water soluble at all pH values, is postulated on a binding mechanism that permits a small but sufficient quantity of these potent antiglaucomic agents to reach aqueous humor and evoke a therapeutic response. Tetracycline, gentamicin, carbenicillin, and methicillin do not penetrate the cornea because of their low lipid solubility, but chloramphenicol shows good penetration.

Fluorescein is used for diagnostic purposes because of its high lipid solubility, which prevents its entry into the stroma unless there is abrasion. If there is an abrasion, fluorescein enters the stroma and possibly the aqueous humor, giving a brilliant green color due to its alkaline pH. In the precorneal film, fluorescein exists in a yellow or orange form.

A variety of physiological factors influence corneal drug absorption. Lacrimal drainage of an instilled drug solution competes for drug with corneal penetration and can account for a considerable loss of drug. When a drop of solution is applied to the eye, two processes occur simultaneously: the solution is diluted by reflex tearing and the added volume in excess of the normal lacrimal volume is drained from the eye, which is partly facilitated by reflex blinking. In humans, administration of 25 μL of solution to the eye at 3 min intervals will minimize volume buildup, dilution, excess drainage, and overflow. Shorter intervals of administration would reduce ophthalmic bioavailability. The normal lacrimal volume in humans is about 7 μL, and if blinking does not occur, the human eye can hold approximately 10 μL. Since the size of commercial ophthalmic drops is between 50 and 75 μL, the loss of drug due to spillage out of the eye can be considered a significant factor in the reduction of bioavailability.

Ophthalmic dosage forms include solutions, ointments, suspensions, lyophilized powders, and oily solutions. Several new dosage forms have recently been introduced to the market. One is an ophthalmic insert, an elliptical device consisting of a drug-containing core surrounded by a flexible copolymer membrane through which pilocarpine diffuses while the ocular delivery system remains in contact with the conjunctiva (Occusert). A spray device has also been designed for accurate delivery of drugs.

Polymers such as methylcellulose, hydroxypropyl methylcellulose, and polyvinyl alcohol decrease the surface tension and increase the viscosity of solutions, thus enhancing bioavailability. Soft contact lenses soaked in pilocarpine have also been used. Biodegradable polymers have been employed for the controlled delivery of hydrocortisone and tetracycline.

NASAL ADMINISTRATION

The nasal cavity provides an ideal opportunity for the delivery of drugs. The nasal mucosa has high vascularity and offers very little formability of local biotransformation. The pH of the surface is shown 7.2 and drugs are generally absorbed by passive diffusion based on their lipid solubility. A number of drugs are administered intranasally for their local affects such as antibiotics, decongestants, and antihistamines. The systemic delivery has generally been limited to only a few preparations such as extracts of the posterior lobe of pituitary gland to treat diabetes insipidus. Drug developers and researchers are discovering that the accessibility and vascular structure of the nose make nasal drug delivery an attractive method for delivering both small molecule drugs and biologics, systemically as well as across the blood–brain barrier to the CNS. Nasal delivery offers the potential for faster onset of action and less frequent dosing relative to oral drugs. Nasal delivery of systemic drugs will grow at the expense of the predominant drug delivery methods (oral and parenteral), which cannot be readily optimized for the delivery and dosing of a significant portion of biologically derived drug substances. Recent developments have suggested that insulin, contraceptives, promabotol, lorazepam, midazolam, butorphanol, hydromorphone, several steroid hormones, and vaccines can be administered intranasally for their systemic effects. Significant problems in nasal delivery include the use of aerosol particles, the mucociliary clearance, and clearance to the lung. Generally, particles larger than 4 μm do not pass into the lung when given nasally. It is expected that in the near future, several drugs may be administered intranasally, especially those that undergo first-pass metabolism or show poor stability in the GI tract. However, the recent developments in nanoparticle research are likely to make intranasal drug delivery one of the most prominent areas of pharmaceutical research (Table 3.9).

TABLE 3.9
Drugs that Can Undergo Biotransformation in the Lumen or during Absorption in the Mucosa

Acetylsalicylic acid	Meperidine
Aldosterone	Methadone
Aminobenzoic acid	α-Methyl dopa
Aminohippuric acid	Nitrates, organic
Chlorpromazine	Pentazocine
Cortisone	Progesterone
Dexamethasone	Propoxyphene
L-Dopa	Salicylamide
Estrogens	Stilbestrol
Hippuric acid	Sulfonamides
Hydrocortisone	Testosterone
Isoproterenol terbutaline	

MISCELLANEOUS ROUTES OF ADMINISTRATION

Drugs are also administered through such routes as the urethra, vagina, and spinal cord. For example, urethral suppositories are frequently used for treatment of localized infections. Anesthetics are often administered in the spinal fluid, as are other drugs on occasion for localized effect.

Recent studies suggest that vaginal administration of drugs for systemic effect may be a valid alternative to rectal or even oral routes of administration because of fast and complete absorption from this site. Direct controlled delivery of fertility-controlling hormones has also been successfully made.

Smart dosage forms embedded with electronic sensors are likely to make drug delivery systems more controllable from outside of the body opening up an entirely new area of bioequivalence testing.

4 Pharmacokinetic/ Pharmacodynamic (PK/PD) Modeling

BACKGROUND

In the process of pharmacokinetic/pharmacodynamic (PK/PD) modeling, it is important to describe, prospectively, the objectives of the modeling, the study design, and the available PK and PD data. The assumptions of the model can be related to dose–response, PK, PD, or one or more of the assumptions listed in Table 4.1. A PK/PD model can be related to dose–response, to PK, to PD, or to one or more of the following assumptions.

The assumptions can be based on previous data or based on the results of any available current analysis. What constitutes an appropriate model depends on the mechanism of the drug's action, the assumptions made, and the intended use of the model in decision making. If the assumptions do not lead to a mechanistic model, an empirical model can be selected, in which case, validating the model's predictability becomes especially important. (Note that nonmechanistic models do not get good reviews from the Food and Drug Administration [FDA].) The model selection process comprises a series of trial-and-error steps in which different model structures or newly added or dropped components to an existing model can be assessed by visual inspection and can be tested using one of several objective criteria. New assumptions can be added when emerging data justifies it.

PHARMACOKINETIC MODELING STUDIES

Compartment Pharmacokinetic Modeling

Pharmacokinetics is the study of the movement of drug molecules in the body, requiring appropriate differential calculus equations to study various rates and processes. The rate of elimination of a drug is described as being dependent on, or proportional to, the amount of drug remaining to be eliminated, a process that obeys first-order kinetics. The rate of elimination can, therefore, be described as

$$\frac{dX}{dt} = -k \cdot X \tag{4.1}$$

where

k is a mere proportionality constant or a rate constant

X is the amount remaining to be eliminated (and therefore X_0 is the initial amount or the dose administered)

Integration allows converting Equation 4.1 to

$$X = X_0 \cdot e^{k_{el} \cdot t} \tag{4.2}$$

TABLE 4.1
Assumptions in Pharmacokinetic/Pharmacodynamic Modeling

The mechanism of the drug actions for efficacy and adverse effects	Presence or absence of active metabolites and their contribution to clinical effects
Development of tolerance or absence of tolerance	Immediate or cumulative clinical effects
Disease state progression	Drug-induced inhibition or induction of PK processes
Circadian variations in basal conditions	Response in a placebo group
Absence or presence of an effect compartment	Influential covariates
The PK model of absorption and disposition and the parameters to be estimated	The PD model of effect and the parameters to be estimated
Inclusion or exclusion of specific patient data	Distributions of intra- and interindividual variability in parameters
Distribution of PK and PD measures and parameters	

Because the amount X is proportional to the concentration, a similar equation describes the time-decay profile of the drug concentration instead of the amount

$$Cp^t = Cp^0 \cdot e^{-k_{el} \cdot t} \tag{4.3}$$

This simple, first-order relationship allows a linear association between the log (more appropriately, the natural logarithm) of concentration and time. It is noteworthy that this concentration is the "effective" concentration, not necessarily the measured concentration. *Effective* refers to a thermodynamic activity rather than the physical concentration. Drugs decay in proportion to the concentration of "free" drug molecules, and whatever is bound to proteins may not be available for disposition. This extrapolation becomes more complex when we take into account other factors that might alter the "activity" (in a thermodynamic sense) of the drug in a biological fluid. For example, structuring of water inside protoplasm imparts lipophilic characteristics, which create significant differences in available concentration gradients. This is a primary reason why it is not always possible to correlate measured concentrations with pharmacological responses because the level of drug at the site of action or at the receptors depends highly on the thermodynamic activity of the drug, which is difficult to assess.

The relationship between the amount of drug and its concentration is classically represented by the following equation, which functions as if there were a physical space (called distribution volume) throughout, which the drug distributes evenly:

$$V = \frac{\text{Amount of drug in the body}}{\text{Concentration measured in plasma}} \tag{4.4}$$

That relationship is an oversimplification of the distribution characteristics of drug molecules in the body and can provide results in volumes often much larger than the body weight. For example, if a drug were selectively stored in different parts of the body, like digoxin or diazepam are, the apparent distribution volumes, using Equation 4.4, would be several multiples of the body's weight. Because the distribution of a drug is a time-dependent process, even within the same "compartment," I suggested that this parameter be treated as a time-dependent variable; treating a "bolus" dose as a short-term infusion improves the results of the deconvolution of integrated equations. That assumption allows a more accurate physical representation of the PK models because an "instantaneous" intravenous (IV) injection is treated as a very short duration, zero-order, input function. As we shall see, this consideration is more important as we integrate

PD models where the action and effect of the drug is delayed for several reasons, including the input and distribution variables.

The area under the plasma concentration–time curve, the *AUC*, is a useful parameter in defining the overall body exposure to a drug; that parameter integrates the concentration-over-time function:

$$\text{AUC} = \int_{t=0}^{t=\infty} Cp^t \cdot dt \tag{4.5}$$

Because the time function of drug concentration is dependent on the rate at which the drug is cleared from the hypothetical "volume," the *AUC* function is dependent on total body clearance, *CL*:

$$V = \frac{\text{Dose}}{\text{AUC} \cdot k_{\text{el}}} \tag{4.6}$$

Clearance (*CL*) is a product of volume, *V*, and the elimination-rate constant, k_{el}, such as when the drug is removed in the urine or metabolized or removed from the sampled compartment by another means. This description of clearance often confuses students of PK. Clearance is an inherent phenomenon, in which distribution volumes are high and rate constants are small to compensate for the distribution. Both volume and the rate constant are derived phenomenon and do not determine clearance. Note that total body clearance is a composite of all pathways that clear or remove the drug from the sampled compartment or the compartment from which the drug is cleared; this is based on the mathematical relationship between the observed elimination-rate constant and its components: each of the pathways involved in the turnover of the drug within the body. Using the parameters described earlier, it is possible to "simulate" a sampled compartment (of fluid) concentration as a function of time in a single- or multiple-dose application using simple, iterative programs. Numerous computer programs are now available that are drug and model specific, which allow simulations of steady-state blood levels that depend on various body functions and body characteristics that affect the clearance of drug. Mixed-models, involving a zero-order infusion, a bolus, or other similar combinations, can be made to estimate blood concentrations under different circumstances related to drug administration.

When drugs are received by routes other than IV injection, input is not "instantaneous" or a short-order zero, and the function must often be represented as a mixed-order, primarily a first-order, process, which must then be taken into account in simulating drug concentration. Drug clearance, however, is not always a constant parameter, especially when an organ like the liver is involved in the removal of the drug from the body:

$$\text{Organ clearance} = \frac{Q\left(C_{\text{a}} - C_{\text{v}}\right)}{C_{\text{a}}} = Q \cdot E \tag{4.7}$$

where

- Q is the blood-flow rate to the organ
- C_a is the concentration of the drug in the blood when entering the organ (in the arterial blood)
- C_v is the concentration of drug in the blood when leaving the organ (in the venous blood)
- The term E is the steady-state extraction ratio

High E values mean high clearance by the liver and, thus, extensive metabolism. The liver blood-flow rate is a physiological parameter that can be altered in disease states. The extraction ratio depends not only on the function of liver but also on the nature of the drug. Both the hepatic clearance and the extraction ratio are empirical parameters and depend on the total hepatic blood flow, the unbound fraction of the drug, and the intrinsic clearance rate. Intrinsic clearance is differentiated from total clearance; the former is the ability to transform when other factors are not present. In other words, the intrinsic clearance is the property of a body organ that clears the drug such as the liver or kidney; for example, the maximum clearance in kidneys cannot exceed the total blood rate to the kidneys and the hepatic clearance cannot exceed the total blood flow to the liver. The actual clearance of a drug from the body depends on the intrinsic clearance as well as its concentration in the fluid that is being cleared; a lower concentration resulting from distribution to body tissues will reduce the total clearance but will have no effect on the intrinsic clearance:

$$\mathrm{CL} = Q \cdot \frac{fu \cdot \mathrm{CL}_{\mathrm{int}}}{Q + (fu \cdot \mathrm{CL}_{\mathrm{int}})} = \frac{Q \cdot \mathrm{CL}_{\mathrm{int}}^{\mathrm{total}}}{Q + \mathrm{CL}_{\mathrm{int}}^{\mathrm{total}}} \tag{4.8}$$

which makes the extraction ratio

$$\text{with } E = \frac{fu \cdot \mathrm{CL}_{\mathrm{int}}}{Q + \left(fu \cdot \mathrm{CL}_{\mathrm{int}}\right)} \tag{4.9}$$

High-clearance drugs are those for which there is no saturation of the reaction that converts the drug, and therefore, the clearance rate approaches the blood-flow rate. For capacity-limited drugs, flow rate is irrelevant, and clearance is a simple product of the unbound fraction and the intrinsic clearance.

The traditional method of PK data analysis uses a two-stage approach: estimation of PK parameters through nonlinear regression using an individual's extensive concentration–time data and using these data parameters as input data for the second-stage calculation of descriptive summary statistics on the sample. Those statistics typically include the mean parameter estimates, the variance, and the covariance of the individual parameter estimates. Analysis of dependencies between parameters and covariates using classical, statistical approaches (linear stepwise regression, covariance analysis, cluster analysis) can be included in the second stage. The two-stage approach yields adequate estimates of population characteristics. Mean estimates of parameters are usually unbiased, but the random effects (variance and covariance) are likely to be overestimated in all realistic situations. Refinements have been proposed (such as the global, two-stage approach) to improve the traditional approach through bias correction for the random effects of covariance and differential weighting of individual data according to the data's quality and quantity.

PHYSIOLOGICALLY BASED PHARMACOKINETIC STUDIES

Physiologically based PK studies take a different perspective in modeling drug disposition in the human body—a mechanistic physiological distribution model. This approach had been in use in other disciplines long before the compartment kinetic modeling was applied to studying drugs. In 1937, the mathematical basis for physiological PK modeling was established by Torsten Teorell, but the solution to the equations was too difficult to obtain before the invention of the digital computer. An automatic solution of a physiologically realistic, mathematical description of the uptake, distribution, and clearance of a chemical agent was proposed by Kenneth Bischoff in the early 1960s. At that time, computation limitations forced several simplifications to the models, including the assumption that the distribution of the drug between tissues and blood is instantaneously at equilibrium, which led to physiological models with blood flow–limited delivery of chemicals to tissues. The inhalation PK models using instantaneous distribution are well known. Physiological PK studies progressed no further until the early 1970s, when the physiological parameters of human organ system became better known and digital computers became more widely available. Today, physiological PK modeling is critical to understanding the behavior of a drug at the site of action.

Exposure modeling studies are often based on the physiological functions that determine uptake, distribution, and elimination of drugs from the body. This approach was pioneered using anesthetics in which physical distribution determines both the onset and termination of action. Similar results have been reported for other compounds like D_2O and ethanol, propranolol, and inulin and protein-bound antibiotics. The modeling is based on a quantitative description of distribution process using standardized organ weights and blood-flow rates. A simpler model assumes no solute binding and a tissue/plasma equilibrium coefficient of 10 for all tissues, except for muscles where this value is 3.62 and for fat where the value is 2.42 as used for propranolol. Also, in the simple model, there is no first-pass effect, and kidney excretion is the only mechanism of drug removal from the body; thus, the input function is equal to systemic availability. In more complex models, tissue binding and other factors that produce nonequilibrium of the tissue/plasma ratios are introduced. The simple model, when used to determine bolus-response function, is well described by a simple two-exponential function; in the more complicated models, three exponents generally provide good fit, and often, going to higher exponents does not improve the predictability. More important is the timing of the first data point obtained in the bolus-response function. This should, ideally, be obtained at or before the end of the constant infusion. (Note that better estimates are obtained from infusion studies than from single-bolus doses because there is always an inevitable delay in the dispersion of drug in the bolus dosing, but the model assumes no delay.) When a deconvolution method is used (see below), the robustness of analysis depends on the accuracy of venous-concentration data because the response function $r(t)$ is established from these data; therefore, any errors in this function reduces the reliability of the analysis, particularly when a later time sample, such as 10 min, is used as the first data point.

BIOEQUIVALENCE AND SYSTEMIC EXPOSURE MODELS

Screening drug molecules for suitability for use in humans is often subjected to certain basic toxicity or workability solutions to reduce the cost of screening. The human body must be able to remove the drug in a reasonable time. Drug clearance is an intrinsic parameter; however, body clearance (extent of drug removal) is dependent on cardiac output and the overall extraction ratio:

$$\text{Body clearance}\left(\text{plasma, blood}\right) = \dot{Q} \times \text{ER} \tag{4.10}$$

The *ER* is the extraction ratio that ranges from 0 to 1, and the cardiac output is proportional to body size:

$$\dot{Q}\left(\text{mL/kg/min}\right) = 180\ \text{BW}\left(\text{kg}\right)^{-0.19} \tag{4.11}$$

Cross-species comparisons can be made for crude estimates and, generally, drugs that have clearance of less than 4 mL/min/kg would be evaluated only if there are special reasons that the mechanisms of actions need to be evaluated.

In addition to the removal potential of a drug, the entry potential is also a good screening parameter; for drugs that are poorly bioavailable, further development should proceed only if proper modification to the molecular structure or to the drug delivery system is made to provide a reasonable possibility of entry. When evaluating bioavailability, it is important to first establish a PK basis because of the large variation in bioavailability as a result of the differences in population pharmacokinetics. Population models are most appropriate for this type of evaluation. Obviously, the consideration of bioequivalence in establishing compliance of generic products is important, and the guidelines for these measurements are defined in the *United States Pharmacopeia* and other guidelines provided by the FDA. It should be noted that the purpose of these studies is to compare the systemic exposure of the body to the drug molecules; this requires measurement of both the extent of absorption and the rate of absorption. Traditionally, parameters like *AUC*, t_{max}, and C_{max}

are studied using specified statistical models. For drugs given orally, these studies cannot be substituted with PD studies, which may be required for some drugs in which the plasma or sample tissue concentration is not available.

DECONVOLUTION TECHNIQUES

The bolus-response function $r(t)$ is generally described using a multiexponential function:

$$r(t) = \sum_{i=1}^{\rho} a_1(e^{-t/T_i}) \tag{4.12}$$

The optimized values for a_i and T_i are determined by using a mathematical approach without any significance attached to it for physiological reasons. Generally, the resorting required to use a three-exponential term takes the estimates out of the population parameters or global minimum.

The three parameters in γ-distribution are chosen by minimizing the error function:

$$\text{Error function} = \sum_i \frac{\left|(ygam)_i - (ydat)_i\right|}{(ydat)_i + noise} \tag{4.13}$$

in which $(ydat)_i$ is the sum of overall data points for the experimental venous concentration and $(ygam)_i$ is the venous concentration–determined y convoluting the γ-distribution input using a polyexponential equation as described earlier for $r(t)$. The *noise* factor in Equation 4.13 determines the weighting of each data point used. When there is no error of *noise*, the error is simplified for each point. When the error is large, the term $(ydat)_i$ drops out in the denominator, and the error is proportional to the numerator of the error term. Because the γ-distribution function is a highly nonlinear process, it is important to use a global annealing procedure such as that used in PKQuest (Minneapolis, MN; www.pkquest.com) requiring Maple software (Maplesoft, Ontario, Canada, www.maplesoft.com) and then follow it with nonlinear minimization. The venous concentration is fitted by using interpolation, meaning that it goes through each data point or uses a smoothing cubic-spline function and then performs the deconvolution. The B-spline function defines the number and position of "breakpoints" and the order of the spline function. Highly sophisticated models have been used for this purpose.

The course of systemic exposure to a drug is studied by comparing IV administration studies using deconvolution approach in which the systemic concentration, $r(t)$, produced from IV administration (also called bolus function) and $I(t)$ is the systemic input rate (in units such as g/min) from the nonintravenous (non-IV) route:

$$c(t) = \int_0^t r(t-\tau)\left[I(\tau)\right]d\tau \tag{4.14}$$

If there is no first-pass effect involved, then $I(t)$ is equal to the rate of intestinal absorption upon administration of equal doses (in IV and non-IV form). In first-pass metabolism, $I(t)$ is the systemic availability of the drug upon oral (or sublingual, rectal, or buccal, etc.) dosing. The function $r(t)$ is obtained by fitting the data upon IV administration to a variety of exponential equations and selecting the best fit through residual mean error of fit. The duration of infusion can be instantaneous (a few seconds for bolus input) but, more realistically, is usually a few minutes. Whereas it is desirable to obtain the sample as early as possible, sampling earlier than 2 min after injection is not advised to allow time for venous mixing. Longer-term IV infusions are also used to obtain the $r(t)$ function. Mathematical solutions of the deconvolution are easily obtained by using such validated software as PKQuest requiring Maple software. Several methods are used for deconvolution; γ-distribution input is a parametric-fitting technique. Whereas polyexponential fitting techniques are widely used, better fits are obtained by using a parametric approach for simulating $I(t)$ where A is the amount of drug reaching circulation, Γ is the γ-function, a is the γ-number that ranges from one to six, and b has inverse time units:

$$I(t)=\left[(Ab)^{a}\times t^{a-1}\times e^{-bt}\right]\Big/\left[\Gamma(a)\right] \quad \text{or} \quad I(t)=\frac{(Ab)^{a}\times t^{a-1}\times e^{-bt}}{\Gamma(a)} \tag{4.15}$$

This approach offers a superior simulation, particularly in situations in which there is a delay in the input function such as in intestinal absorption and gastric emptying variations. The three parameters given earlier are estimated by global (also called *simulated annealing*) and local (also called *Powell*) nonlinear optimization. The fitting of data using γ-deconvolution method smoothes data noise, and with no user adjustable parameters, the bias is removed. If the input is not possible to define using a single γ-distribution, then other deconvolution approaches, such as analytical, spline, or uniform approaches, which remove the *roughness* of the input rate, are used; thc choice of parameters is additionally improved by experimental Akaike criterion and the *generalized cross validation.*

The analytical deconvolution involves approximation of $C(t)$ by an interpolating or smoothed spline function and the deconvolution. The analytical deconvolution method is most commonly used for the advantage of being fast and where data are exact, excellent results are obtained; however, the robustness of this approach depends on the value chosen for the smoothing parameter, which is poorly estimated even when standard deviation is available (very rare). Where there is noisy data, it adds more error in analytical deconvolution compared to spline and uniform methods. Also, analytical deconvolution does not allow the use of negative values for input. In spline function input consideration, the input $I(t)$ is parameterized using a general B-spline function and then obtaining deconvolution by a constrained regression. In using uniform input, $I(t)$ is estimated on dense uniform sequence of time points and then using stochastic regularization procedure for deconvolution.

PHARMACOLOGICAL EVALUATION OF BIOAVAILABILITY

The estimations of bioavailability discussed earlier are based on plasma and/or urine levels of the drug and/or its biotransformation products. It is understood in these calculations that these concentrations relate in some manner to the pharmacological or clinical response of the drug. Ideally, therefore, it is desirable to measure bioavailability as a function of pharmacological or clinical effect. In order to do so, a specific and discriminating test is needed. Some quantitative endpoint must be available, which measures efficacy or quantitates the drug effect. For example, lowering of blood sugar by an antidiabetic agent, lowering of blood pressure by a hypotensive agent, weight loss produced by an anorexic agent, etc., would be appropriate measures. Less reliable measures, such as psychological rating score and a physician's opinion of efficacy, cannot be of great value in these studies.

Before any comparative bioavailability testing is performed using pharmacological or clinical response, a satisfactory dose–response curve should be obtained on one of the formulations to be included in the study. The success of the application of the dose–response curve should be established on one of the formulations to be included in the study. The success of the application of this dose–response curve depends on two factors. First, the curve should be steep, indicating that significant changes in pharmacological response occur with a small change in the dose, and second, the dose contained in the formulations should be such that the response lies between 20% and 80% of maximum response to assure linear measurements. Responses falling beyond these ranges are more difficult to quantitate.

Few studies have reported the use of dose–response curves in bioavailability measurements, but the idea is certainly attractive and relevant to drug therapy.

Studies in healthy volunteers or patients using PD measurements may be used for establishing equivalence between two pharmaceutical products. These studies may become necessary if quantitative analysis of the drug and/or metabolite(s) in plasma or urine cannot be made with sufficient accuracy and sensitivity. Furthermore, PD studies in humans are required if measurements of drug concentrations cannot be used as surrogate endpoints for the demonstration of efficacy and safety of the particular pharmaceutical product, for example, for topical products without an intended absorption of the drug into the systemic circulation.

If PD studies are to be used, they must be performed as rigorously as bioequivalence studies, and the principles of good clinical practice (GCP) (see WHO Guideline for GCP for Trials on Pharmaceutical Products) must be followed.

The following requirements must be recognized when planning, conducting, and assessing the results of a study intended to demonstrate equivalence by means of measuring PD drug responses:

1. The response that is being measured should be a pharmacological or therapeutic effect that is relevant to the claims of efficacy and/or safety.
2. The methodology must be validated for precision, accuracy, reproducibility, specificity, and ruggedness.
3. Neither the test nor the reference product should produce a maximal response in the course of the study, since it may be impossible to distinguish differences between formulations given in doses that give maximum or near-maximum effects. Investigation of dose–response relationships may be a necessary part of the design.
4. The response should be measured quantitatively under double-blind conditions and be recorded in an instrument-produced or instrument-recorded fashion on a repetitive basis to provide a record of the PD events that are substitutes for plasma concentrations. In those instances where such measurements are not possible, recordings on visual analogue scales may be used. In other instances where the data are limited to qualitative (categorized) measurements, appropriate special statistical analysis will be required.
5. Nonresponders should be excluded from the study through prior screening. The criteria by which responders versus nonresponders are identified must be stated in the protocol.
6. In instances where an important placebo effect can occur, comparison between pharmaceutical products can only be made by a priori consideration of the placebo effect in the study design. This may be achieved by adding a third phase with placebo treatment in the design of the study.
7. The underlying pathology and natural history of the condition must be considered in the study design. There should be knowledge of the reproducibility of baseline conditions.
8. A crossover design may be used. Where this is not appropriate, a parallel group study design should be chosen.

In studies in which continuous variables could be recorded, the time course of the intensity of the drug action can be described in the same way as in a study in which plasma concentrations were measured, and parameters can be derived that describe the area under the effect–time curve, the maximum response, and the time when maximum response occurred.

The statistical considerations for the assessment of the outcome of the study are, in principle, the same as outlined for the bioequivalence studies. However, a correction for the potential nonlinearity of the relationship between the dose and the area under the effect–time curve should be performed on the basis of the outcome of the dose-ranging study as mentioned earlier. However, it should be noted that the conventional acceptance range as applied for bioequivalence assessment is not appropriate (too large) in most of the cases but should be defined on a case-by-case basis and described in the protocol.

5 Bioequivalence Testing Rationale and Principles

BACKGROUND

The regulation of drug quality involves three arrangements in the United States that has led the world in creating and promulgating ordinances to regulate the drug industry. First, the U.S. Congress gave the U.S. Pharmacopeia (USP) and the National Formulary (NF) revision committees the authority to set standards of strength, quality, and purity of drugs and their finished preparations. Nevertheless, the USP and NF remain as private entity with no authority over the Food and Drug Administration (FDA). The FDA, also authorized by the U.S. Congress, establishes regulations for the development and manufacture of safe and effective drugs. Finally, in-house good manufacturing practices of the manufacturer, mostly dictated by the FDA regulations, assure quality of drug products. The FDA has also decreed on the bioavailability (BA) and bioequivalence (BE) of drug products. All new drug applications (NDAs) and amended new drug applications must demonstrate in vivo BA of the drug product that is followed by an in vitro test, usually a dissolution test, of individual batches to assure the quality. Table 5.1 shows a comparison of regulatory filing requirements under various applications.

Submitting an NDA or new animal drug application (NADA) under the provisions of section 505(b) in the Federal Food, Drug, and Cosmetic (FD&C) Act (the Act) are required to document BA (21 CFR 320.21(a)). If approved, an NDA drug product may subsequently become a reference-listed drug (RLD). Under section 505(j) of the Act, a sponsor of an abbreviated new drug application (ANDA) or abbreviated new animal drug application (ANADA) must document first pharmaceutical equivalence and then BE to be deemed therapeutically equivalent to a RLD. Defined as relative BA, BE is documented by comparing the performance of the generic (test) and listed (reference) products. (Pharmaceutical equivalents are drugs that have the same active ingredient, in the same strength, and the same dosage form and route of administration and have comparable labeling and meet compendia or other standards of identity, strength, quality, purity, and potency.)

In addition to the standard chemistry, manufacturing, and control (CMC) tests, the active bulk drug substance for an NDA should be studied and controlled via appropriate specifications for polymorphic form, particle size distribution, and other attributes important to the quality of the resulting drug product. To the extent possible and using compendial monographs where appropriate, sponsors of ANDAs should attempt to duplicate the specifications considered important for the RLD. Where the necessary information is not available, applicants may wish to rely on in vitro release to ensure batch-to-batch consistency. CMC guidances available from FDA are generally applicable to ensure the identity, strength, quality, purity, and potency of the drug substance and drug product for a topical dermatological drug product.

As stated at 21 CFR 320.24, approaches to document BE in order of preference are (1) pharmacokinetic (PK) measurements based on measurement of an active drug and/or metabolite in blood, plasma, and/or urine, (2) pharmacodynamic (PD) measurements, (3) comparative clinical trials, and (4) in vitro studies.

BE is defined in 21 CFR 320.1 as "the absence of a significant difference in the rate and extent to which the active ingredient or active moiety in pharmaceutical equivalents or pharmaceutical

TABLE 5.1
Data Requirement for Drug Approval in the United States

Application	FD&C 505(b)(1) NDA	FD&C 505(b)(2) NDA	FD&C 505(j) ANDA	PHS 351(a) BLA	PHS 351(k) ABLA
Preclinical	Yes	Yes/no	No	Yes	Yes
Clinical	Yes	Yes/no	No	Yes	To be mutually agreed
CMC	Yes	Yes	Yes (PE)	Yes	Yes
PK and BE	Yes	Yes		Yes	Yes
Labeling	Yes	Yes	Yes	Yes	Yes

alternatives becomes available at the site of drug action when administered at the same molar dose under similar conditions in an appropriately designed study." FDA usually considers that the plasma concentration of a drug is a surrogate for the concentration at the site of action for a systemically acting drug. 21 CFR 320.24 outlines options for BE testing. Proving equivalence therefore requires integration of several studies such as PK, PD, controlled-clinical (CC), in vitro studies and any other specific model or study that may prove useful in proving equivalence.

EQUIVALENCE DOCUMENTATION FOR MARKETING AUTHORIZATION

Pharmaceutically equivalent multisource pharmaceutical products must be verified to be therapeutically equivalent to one another in order to be considered interchangeable. Several test methods are available to assess equivalence, including

- Comparative BA (BE) studies, in which the active drug substance or one or more metabolites are measured in an accessible biological fluid such as plasma, blood, or urine
- Comparative PD studies in humans
- Comparative clinical trials
- In vitro dissolution tests in combination with the Biopharmaceutics Classification System (BCS, see later text)

Acceptance of any test procedure in the equivalence documentation between two pharmaceutical products by a drug regulatory authority depends on many factors, including characteristics of the active drug substance and the drug product and the availability of resources to carry out a specific type of study. Wherever a drug produces meaningful concentrations in an accessible biological fluid, such as plasma, BE studies are preferred. Wherever a drug does not produce measurable concentrations in an accessible biological fluid, comparative clinical trials or PD studies may be necessary to document equivalence. In vitro testing, preferably based on a documented in vitro/in vivo correlation or on consideration based on the BCS, may sometimes provide an indication of equivalence between two pharmaceutical products.

Oral drugs/drug products for which in vivo equivalence documentation is important: Regulatory authorities require equivalence documentation for multisource pharmaceutical products in which the product is compared to the reference pharmaceutical product. Studies must be carried out using the formulation proposed for marketing. For certain drugs and dosage forms, in vivo equivalence documentation, through either a BE study, a comparative clinical PD study, or a comparative clinical trial, is considered especially important. The following are the factors for oral drug products that should be considered when requiring in vivo equivalence documentation.

Immediate-release oral pharmaceutical products with systemic action when one or more of the following criteria apply:

1. Indicated for serious conditions requiring definite therapeutic response
2. Narrow therapeutic window/safety margin, steep dose–response curve
3. PKs complicated by variable or incomplete absorption or absorption window, nonlinear PKs, and presystemic elimination/high first-pass metabolism >70%
4. Unfavorable physicochemical properties, for example, low solubility, instability, metastable modifications, and poor permeability
5. Documented evidence of BA problems related to the drug or drugs of similar chemical structure or formulations
6. Where there is a high ratio of excipients to active ingredients

Non-oral and nonparenteral pharmaceutical products designed to act through systemic absorption (such as transdermal patches, suppositories): Plasma concentration measurements over time (BE) are normally sufficient proof for efficacy and safety.

Sustained or otherwise modified-release pharmaceutical products designed to act through systemic absorption: Plasma concentration measurements over time (BE) are normally sufficient proof for efficacy and safety.

Fixed combination products (see WHO Technical Report Series No. 825, 1992) with systemic action: Plasma concentration measurements over time (BE) are normally sufficient proof for efficacy and safety.

Nonsolution pharmaceutical products for nonsystemic use (oral, nasal, ocular, dermal, rectal, vaginal, etc., application) and intended to act without systemic absorption: In these cases, the BE concept is not suitable, and comparative clinical or PD studies are required to prove equivalence. This does not, however, exclude the potential need for drug concentration measurements in order to assess unintended partial absorption.

FDA has also provided a therapeutic classification of drugs and dosage forms for the purpose of BE testing (Table 5.2).

OVERVIEW OF BIOEQUIVALENCE TESTING

The submission of an NDA, ANDA, or supplemental application requires that it contains in vivo BA and BE data either by direct measurement of in vivo BA of the drug product that is the subject of the application or information to permit FDA to waive the submission of evidence measuring in vivo BA. The supplemental application involves a change in the manufacturing site or a change in the manufacturing process, including a change in product formulation or dosage strength, beyond the variations provided for in the approved application, or a change in the labeling to provide for a new indication for use of the drug product, for which a new clinical trial may be required.

FDA may approve a full NDA or a supplemental application proposing any of the changes set forth previously that does not contain evidence of in vivo BA or information to permit waiver of the requirement for in vivo BA data:

- For certain drug products, the in vivo BA or BE of the drug product may be self-evident. FDA shall waive the requirement for the submission of evidence obtained in vivo measuring the BA or demonstrating the BE of these drug products. A drug product's in vivo BA or BE may be considered self-evident based on other data in the application.
- If the drug product is a parenteral solution intended solely for administration by injection, or an ophthalmic or otic solution, and contains the same active and inactive ingredients in the same concentration as a drug product that is the subject of an approved full NDA or ANDA.

TABLE 5.2
Therapeutic Equivalence Code Classifications of the U.S. FDA

Name	Definition	FDA Code
Products in conventional dosage forms not presenting BE problems	Products coded as AA contain active ingredients and dosage forms that are not regarded as presenting either actual or potential BE problems or drug quality or standards issues. However, all oral dosage forms must, nonetheless, meet an appropriate in vitro test(s) for approval.	AA
Products meeting necessary BE requirements	Products generally will be coded AB if a study is submitted demonstrating BE. Even though drug products of distributors and/or repackagers are not included in the list, they are considered therapeutically equivalent to the application holder's drug product if the application holder's drug product is rated AB or is single source in the list. The only instance in which a multisource product will be rated AB on the basis of BA rather than BE is where the innovator product is the only one listed under that drug ingredient heading and has completed an acceptable BA study. However, it does not signify that this product is therapeutically equivalent to the other drugs under the same heading. Drugs coded AB under an ingredient heading are considered therapeutically equivalent only to other drugs coded AB under that heading.	AB
Solutions and powders for aerosolization	Uncertainty regarding the therapeutic equivalence of aerosolized products arises primarily because of differences in the drug delivery system. Solutions and powders intended for aerosolization that are marketed for use in any of several delivery systems are considered to be pharmaceutically and therapeutically equivalent and are coded AN. Those products that are compatible only with a specific delivery system or those products that are packaged in and with a specific delivery system are coded BN, unless they have met an appropriate BE standard because drug products in their respective delivery systems are not necessarily pharmaceutically equivalent to each other and, therefore, are not therapeutically equivalent.	AN
Injectable oil solutions	The absorption of drugs in injectable (parenteral) oil solutions may vary substantially with the type of oil employed as a vehicle and the concentration of the active ingredient. Injectable oil solutions are therefore considered to be pharmaceutically and therapeutically equivalent only when the active ingredient, its concentration, and the type of oil used as a vehicle are all identical.	AO
Injectable aqueous solutions	It should be noted that even though injectable (parenteral) products under a specific listing may be evaluated as therapeutically equivalent, there may be important differences among the products in the general category, injectable, injection. For example, some injectable products that are rated therapeutically equivalent are labeled for different routes of administration. In addition, some products evaluated as therapeutically equivalent may have different preservatives or no preservatives at all. Injectable products available as dry powders for reconstitution, concentrated sterile solutions for dilution, or sterile solutions ready for injection are all considered to be pharmaceutically and therapeutically equivalent provided they are designed to produce the same concentration prior to injection and are similarly labeled. Consistent with accepted professional practice, it is the responsibility of the prescriber, dispenser, or individual administering the product to be familiar with a product's labeling to assure that it is given only by the route(s) of administration stated in the labeling.	AP

(Continued)

TABLE 5.2 (*Continued*)
Therapeutic Equivalence Code Classifications of the U.S. FDA

Name	Definition	FDA Code
	Certain commonly used large volume intravenous products in glass containers are not included on the list (e.g., dextrose injection 5%, dextrose injection 10%, sodium chloride injection 0.9%) since these products are on the market without FDA approval and the FDA has not published conditions for marketing such parenteral products under approved NDAs. When packaged in plastic containers, however, FDA regulations require approved applications prior to marketing. Approval then depends on, among other things, the extent of the available safety data involving the specific plastic component of the product. All large volume parenteral products are manufactured under similar standards, regardless of whether they are packaged in glass or plastic. Thus, FDA has no reason to believe that the packaging container of large volume parenteral drug products that are pharmaceutically equivalent would have any effect on their therapeutic equivalence.	
Topical products	There are a variety of topical dosage forms available for dermatologic, ophthalmic, otic, rectal, and vaginal administration, including solutions, creams, ointments, gels, lotions, pastes, sprays, and suppositories. Even though different topical dosage forms may contain the same active ingredient and potency, these dosage forms are not considered pharmaceutically equivalent. Therefore, they are not considered therapeutically equivalent. All solutions and DESI drug products containing the same active ingredient in the same topical dosage form for which a waiver of in vivo BE has been granted and for which chemistry and manufacturing processes are adequate are considered therapeutically equivalent and coded AT. Pharmaceutically equivalent topical products that raise questions of BE including all post1962 topical drug products are coded AB when supported by adequate BE data and BT in the absence of such data.	AT
Extended-release dosage forms (capsules, injectables, and tablets)	An extended-release dosage form is defined by the official compendia as one that allows at least a twofold reduction in dosing frequency as compared to that drug presented as a conventional dosage form (e.g., as a solution or a prompt drug-releasing, conventional solid dosage form). Although BA studies have been conducted on these dosage forms, they are subject to BA differences, primarily because firms developing extended-release products for the same active ingredient rarely employ the same formulation approach. FDA, therefore, does not consider different extended-release dosage forms containing the same active ingredient in equal strength to be therapeutically equivalent unless equivalence between individual products in both rate and extent has been specifically demonstrated through appropriate BE studies. Extended-release products for which such BE data have not been submitted are coded BC, while those for which such data are available have been coded AB.	BC
Active ingredients and dosage forms with documented BE problems	The BD code denotes products containing active ingredients with known BE problems and for which adequate studies have not been submitted to FDA demonstrating BE. Where studies showing BE have been submitted, the product has been coded AB.	BD

(*Continued*)

TABLE 5.2 (*Continued*)
Therapeutic Equivalence Code Classifications of the U.S. FDA

Name	Definition	FDA Code
Delayed-release oral dosage forms	A delayed-release dosage form is defined by the official compendia as one that releases a drug (or drugs) at a time other than promptly after administration. Enteric-coated articles are delayed-release dosage forms. Drug products in delayed-release dosage forms containing the same active ingredients are subject to significant differences in absorption. Unless otherwise specifically noted, the agency considers different delayed-release products containing the same active ingredients as presenting a potential BE problem and codes these products BE in the absence of in vivo studies showing BE. If adequate in vivo studies have demonstrated the BE of specific delayed-release products, such products are coded AB.	BE
Products in aerosol-nebulizer drug delivery systems	This code applies to drug solutions or powders that are marketed only as a component of, or as compatible with, a specific drug delivery system. There may, for example, be significant differences in the dose of drug and particle size delivered by different products of this type. Therefore, the agency does not consider different metered aerosol dosage forms containing the same active ingredient(s) in equal strengths to be therapeutically equivalent unless the drug products meet an appropriate BE standard.	BN
Active ingredients and dosage forms with potential BE problems	FDA's BE regulations (21 CFR 320.33) contain criteria and procedures for determining whether a specific active ingredient in a specific dosage form has a potential for causing a BE problem. It is FDA's policy to consider an ingredient meeting these criteria as having a potential BE problem even in the absence of positive data demonstrating inequivalence. Pharmaceutically equivalent products containing these ingredients in oral dosage forms are coded BP until adequate in vivo BE data are submitted. Injectable suspensions containing an active ingredient suspended in an aqueous or oleaginous vehicle have also been coded BP. Injectable suspensions are subject to BE problems because differences in particle size, polymorphic structure of the suspended active ingredient, or the suspension formulation can significantly affect the rate of release and absorption. FDA does not consider pharmaceutical equivalents of these products bioequivalent without adequate evidence of BE.	BP
Suppositories or enemas that deliver drugs for systemic absorption	The absorption of active ingredients from suppositories or enemas that are intended to have a systemic effect (as distinct from suppositories administered for local effect) can vary significantly from product to product. Therefore, FDA considers pharmaceutically equivalent systemic suppositories or enemas bioequivalent only if in vivo evidence of BE is available. In those cases where in vivo evidence is available, the product is coded AB. If such evidence is not available, the products are coded BR.	BR
Products having drug standard deficiencies	If the drug standards for an active ingredient in a particular dosage form are found by FDA to be deficient so as to prevent an FDA evaluation of either pharmaceutical or therapeutic equivalence, all drug products containing that active ingredient in that dosage form are coded BS. For example, if the standards permit a wide variation in pharmacologically active components of the active ingredient such that pharmaceutical equivalence is in question, all products containing that active ingredient in that dosage form are coded BS.	BS

(*Continued*)

TABLE 5.2 (*Continued*)
Therapeutic Equivalence Code Classifications of the U.S. FDA

Name	Definition	FDA Code
Topical products with BE issues	This code applies mainly to post1962 dermatologic, ophthalmic, otic, rectal, and vaginal products for topical administration, including creams, ointments, gels, lotions, pastes, and sprays, as well as suppositories not intended for systemic drug absorption. Topical products evaluated as having acceptable clinical performance, but that are not bioequivalent to other pharmaceutically equivalent products or that lack sufficient evidence of BE will be coded BT.	BT
Drug products for which the data are insufficient to determine therapeutic equivalence	The code BX is assigned to specific drug products for which the data that have been reviewed by the agency are insufficient to determine therapeutic equivalence under the policies stated in this document. In these situations, the drug products are presumed to be therapeutically inequivalent until the agency has determined that there is adequate information to make a full evaluation of therapeutic equivalence.	BX

- If the drug product is administered by inhalation as a gas, for example, a medicinal or an inhalation anesthetic, and contains an active ingredient in the same dosage form as a drug product that is the subject of an approved full NDA or ANDA.
- If the drug product is a solution for application to the skin, an oral solution, elixir, syrup, tincture, a solution for aerosolization or nebulization, a nasal solution, or similar other solubilized form and contains an active drug ingredient in the same concentration and dosage form as a drug product that is the subject of an approved full NDA or ANDA and contains no inactive ingredient or other change in formulation from the drug product that is the subject of the approved full NDA or ANDA that may significantly affect absorption of the active drug ingredient or active moiety for products that are systemically absorbed or that may significantly affect systemic or local availability for products intended to act locally.

FDA also waives the requirement for the submission of evidence measuring the in vivo BA or demonstrating the in vivo BE of a solid oral dosage form (other than a delayed-release or extended-release dosage form) of a drug product determined to be effective for at least one indication in a Drug Efficacy Study Implementation (DESI) notice or which is identical, related, or similar (IRS) to such a drug product unless FDA has evaluated the drug product, included the drug product in the Approved Drug Products with Therapeutic Equivalence Evaluations List, and rated the drug product as having a known or potential BE problem. A drug product so rated reflects a determination by FDA that an in vivo BE study is required. (A *DESI* drug is any drug that lacks substantial evidence of effectiveness [less than effective, LTE] and is subject by the *FDA* to a Notice of Opportunity for Hearing [NOOH]. This includes drugs that are IRS to *DESI* drugs. Valid values: 2, safe and effective or non-*DESI* drug; 3, drug under review [no NOOH issued]; 4, LTE/IRS drug for *some* indications; 5, LTE/IRS drug for *all* indications; 6, LTE/IRS drug withdrawn from market.)

For certain drug products, BA may be measured or BE may be demonstrated by evidence obtained in vitro in lieu of in vivo data. FDA shall waive the requirement for the submission of evidence obtained in vivo measuring the BA or demonstrating the BE of the drug product if the drug product meets one of the following criteria:

- The drug product is in the same dosage form, but in a different strength, and is proportionally similar in its active and inactive ingredients to another drug product for which the same manufacturer has obtained approval and the following conditions are met that the BA of this other drug product has been measured and both drug products meet an appropriate in vitro test approved by FDA, and the applicant submits evidence showing that both drug products are proportionally similar in their active and inactive ingredients (except for the delayed-release or extended-release products).

- The drug product is, on the basis of scientific evidence submitted in the application, shown to meet an in vitro test that has been correlated with in vivo data.
- The drug product is a reformulated product that is identical, except for a different color, flavor, or preservative that could not affect the BA of the reformulated product, to another drug product for which the same manufacturer has obtained approval and the following conditions are met: The BA of the other product has been measured and both drug products meet an appropriate in vitro test approved by FDA.

FDA, for good cause, may waive a requirement for the submission of evidence of in vivo BA or BE if waiver is compatible with the protection of the public health. For full NDAs, FDA may defer a requirement for the submission of evidence of in vivo BA if deferral is compatible with the protection of the public health.

FDA, for good cause, may require evidence of in vivo BA or BE for any drug product if the agency determines that any difference between the drug product and a listed drug may affect the BA or BE of the drug product.

How to Demonstrate Bioequivalence?

A list of therapeutic, PK, and physicochemical factors has been compiled to classify which product needs demonstration of BE by in vivo testing (Table 5.3). A large number of drugs have been classified in this category (Table 5.4). All enteric-coated and controlled-release dosage forms of any solid oral dosage form require in vivo BA testing. It is generally suggested that if more than 25% intra-batch or batch-to-batch variability in BA is observed, in vivo tests will be required

TABLE 5.3
Factors Determining the Establishment of BE Requirement by the FDA

Therapeutic factors evidence from
- Clinical trials
- Controlled observations on patients
- Well-controlled BE studies that
  - The drug exhibits a low therapeutic ratio
  - The drug requires careful dosage titration
  - Bioinequivalence would produce adverse prophylactic or therapeutic effects

PK factor evidence that the drug entity
- Is absorbed from localized sites in the GI tract
- Is subject to poor absorption
- Is subject to first-pass metabolism
- Requires rapid dissolution and absorption for effectiveness
- Is unstable in specific portions of the GI tract
- Is subject to dose-dependent kinetics in or near the therapeutic range

Physiochemical factor evidence that the drug
- Possesses low solubility in water or gastric fluids
- Is dissolved slowly from one or more of its dosage forms
- Particle size and/or surface area affects BA

Exhibits certain physical structural characteristics, e.g., polymorphism and solvates, which modify its BA
- Has a high ratio of excipients to active ingredients as formulated
- Has a BA that may be affected by the presence or absence of hydrophilic or hydrophobic excipients and lubricant

TABLE 5.4
Drugs with Potential BE Problems

Acetazolamide	Acetyldigitoxin	Alseroxylon	Aminophylline
Aminosalicylic acid	Bendroflumethiazide	Benzthiazide	Betamethasone
Bishydroxycoumarin	Chlorambucil	Chlordiazepoxide	Chlorpromazine
Chlorothiazide	Cortisone acetate	Deserpidine	Dexamethasone
Dichlorphenamide	Dienestrol	Diethylstilbestrol	Dyphylline
Ethinyl estradiol	Ethosuximide	Ethotoin	Ethoxzolamide
Fludrocortisone	Fluphenazine	Fluprednisolone	Hydralazine
Hydrochlorothiazide	Hydroflumethiazide	Imipramine	Isoproterenol
Liothyronine	Menadione	Mephenytoin	Methazolamide
Methyclothizide	Methylprednisolone	Methyltestosterone	Nitrofurantoin
Oxtriphylline	Para-aminosalicylic acid	Para-methadione	Perphenazine
Phenacemide	Phensuximide	Phenylaminosalicylate	Phenytoin
Phytonadione	Polythiazide	Prednisolone	Primidone
Probenecid	Procainamide	Prochlorperazine	Promazine
Promethazine	Propylthiouracil	Pyrimethamine	Quinethiazide
Quinidine	Rauwolfia serpentina	Rescinnamine	Reserpine
Salicylazosulfapyridine	Sodium sulfoxone	Spironolactone	Sulfadiazine
Sulfadimethoxine	Sulfamerazine	Sulfaphenazole	Sulfasomidine
Sulfasoxazole	Theophylline	Thioridazine	Tolbutamide
Triamcinolone	Trichlormethiazide	Triethyl melamine	Trifluoperazine
Triflupromazine	Trimeprazine	Trimethadione	Uracil mustard
Warfarin			

for batch certification. Any changes in the manufacturing process, including product formulation or dosage strength change, beyond that suggested in the NDA or ANDA and changes in labeling for a new indication or new dosage regimen also require in vivo BA testing.

The pharmacotherapeutic nature of the drug plays an important role in the regulations regarding its BA. Drugs that exhibit narrow therapeutic index (NTI), that is, less than a twofold difference in median lethal dose (LD50) and median effective dose (ED50) values (or less than a twofold difference in the minimum effective concentration [MEC] and minimum toxic concentration [MTC] in the blood), require careful demonstration of BA and the consistency with which this requirement is met. Further consideration is needed in the type of side effects occurring if a toxic level is reached. For example, the therapeutic index (the U.S. FDA prefers to call this therapeutic range) for salicylates is smaller than cardiac glycosides; it does not mean that cardiac glycosides are less toxic. It merely signifies that the concentration of salicylates for therapeutic response is closer to the concentration where undesirable side effects start to appear. Another consideration along the same line is the potency of drug in question. Generally, highly potent drugs will require greater control of BA than the one with lesser potency. Because of the logarithmic nature of the response, the curves flatten out at low and high doses. Thus, a highly potent drug used in large doses will show lesser variability in response due to BA factor than a low-potency drug used at a dose level where the response is log-linear. Any such comparison, however, should take into account the relative nature of the slope of the response to dose.

The physicochemical evidence needed to establish a BE includes low water solubility, for example, less than 5 mg/mL, or if dissolution in the stomach is critical to absorption, the volume of gastric fluids required to dissolve the recommended dose (gastric fluid content is assumed to be 100 mL for adults and is prorated for infants and children). The dissolution rates are also taken into consideration if less than 50% of the drug dissolves in 30 min using official methods. Also included under physicochemical evidence are particle size and surface area of the active drug ingredient. Certain physical structural characteristics of the active drug ingredient, for example, polymorphism and solvation, are also considered. Drug products that have a high ratio of excipients to active ingredients (e.g., greater than 5:1) may also be subjected to BE demonstration. Other evidence includes specific absorption sites or where the available dose is less than 50% of an administered dose. Drugs that are rapidly biotransformed in the intestinal wall or liver during absorption and drugs that are unstable in specific portions of the gastrointestinal (GI) tract requiring special coating or formulations are also subjected to BE requirements, as are drugs that show dose-dependent absorption, distribution, biotransformation, or elimination.

For some dosage forms, BE requirements can be waived such as with topical products, oral dosage forms not intended for absorption, inhalations, and solutions if there is sufficient evidence that the inactive ingredients do not affect the release and delivery of drugs from the dosage form.

Rationale for Estimation

The BA of a drug is controlled by three factors, namely,

1. The rate and extent of release of the drug from the dosage form
2. Its subsequent absorption from the solution state
3. The biotransformation during the process of absorption

In all quantitative determinations of BA, concentration is measured in blood, plasma, and urine. Plasma concentrations following the oral administration of a drug assume four sequential phases depending on the magnitude of absorption and elimination:

1. Absorption > elimination
2. Absorption = elimination
3. Absorption < elimination
4. Absorption = elimination = 0

The shape of the plasma concentration profile depends on the relative rates of absorption and elimination, and thus, the plasma concentration profiles may be quite different with different routes of administration. Intravenous and sometimes intramuscular routes yield an early peak due to fast or almost instantaneous absorption, whereas oral, subcutaneous, rectal, and other routes may show delayed peaks due to slower rates of absorption. It should be noted that the rate of elimination is considered constant since it depends primarily in the specific nature of the active drug ingredient.

The purpose of BA studies is to demonstrate therapeutic equivalence. However, depending on the mechanism of action, more meaningful comparisons can be made from such parameters as peak plasma concentration or the time to reach peak plasma concentration. For example, in the case of antibiotics, it is important to know how soon the minimum inhibitory concentration (MIC) is reached and maintained. The choice of single-dose versus multiple-dose study depends on the mechanism of drug action. For example, antidepressants like imipramine show delayed action, a characteristic of many psychotropic and antihypertensive agents. In these instances, a new product should be judged for its quality from repeated administration because in these example, the peak concentration or time for peak concentration is relatively unimportant. It is therefore important to isolate the clinically important parameter, but in all instances, the area under the curve (AUC) must be monitored since it represents the proportionality to the total amount of drug eliminated from the body and hence absorbed.

The estimation of BA from plasma concentration profiles requires a thorough understanding of the nature of plasma level profiles. For example, a higher or earlier peak does not necessarily mean greater overall absorption than from a product giving a smaller or delayed peak. The total absorption of drugs is, therefore, proportional not only to the plasma concentrations achieved but also to the length of time these concentrations persist in the blood. One parameter that characterizes this aspect is the area under the plasma concentration versus time profile.

The major contribution to the AUC for a fast-absorbed formulation is due to the high peak concentration, whereas for a slowly absorbed formulation, the area is mainly due to sustained or prolonged plasma concentration. It should be noted that the area under the plasma concentration versus time profile (AUC) is only proportional to the total amount of drug absorbed and cannot be used to determine the actual amount of drug administered unless it is compared with a known standard, whereby the extent of absorption is either measured by other methods or assumed to be 100%, as in the case of intravenous administration.

The in vivo BA of a drug product is measured if the product's rate and extent of absorption, as determined by comparison of measured parameters, for example, concentration of the active drug ingredient in the blood, urinary excretion rates, or pharmacological effects, do not indicate a significant difference from the reference material's rate and extent of absorption. For drug products that are not intended to be absorbed into the bloodstream, BA may be assessed by measurements intended to reflect the rate and extent to which the active ingredient or active moiety becomes available at the site of action.

Statistical techniques used in establishing BE shall be of sufficient sensitivity to detect differences in rate and extent of absorption that are not attributable to subject variability.

A drug product that differs from the reference material in its rate of absorption, but not in its extent of absorption, may be considered to be bioavailable if the difference in the rate of absorption is intentional, is appropriately reflected in the labeling, is not essential to the attainment of effective body drug concentrations on chronic use, and is considered medically insignificant for the drug product.

Two drug products will be considered bioequivalent drug products if they are pharmaceutical equivalents or pharmaceutical alternatives whose rate and extent of absorption do not show a significant difference when administered at the same molar dose of the active moiety under similar experimental conditions, either single dose or multiple dose. Some pharmaceutical equivalents or pharmaceutical alternatives may be equivalent in the extent of their absorption but not in their rate of absorption and yet may be considered bioequivalent because such differences in the rate of absorption are intentional and are reflected in the labeling, are not essential to the attainment of effective body drug concentrations on chronic use, and are considered medically insignificant for the particular drug product studied.

Evidence to Measure Bioequivalence

In vivo BE may be determined by one of several direct or indirect methods. Selection of the method depends upon the purpose of the study, the analytical method available, and the nature of the drug product. BE testing should be conducted using the most appropriate method available for the specific use of the product.

The preferred hierarchy of BE studies (in descending order of sensitivity) is the blood level study, pharmacological endpoint study, and clinical endpoint study. When absorption of the drug is sufficient to measure drug concentration directly in the blood (or other appropriate biological fluids or tissues) and systemic absorption is relevant to the drug action, then a blood (or other biological fluid or tissue) level BE study should be conducted. The blood level study is generally preferred above all others as the most sensitive measure of BE. The sponsor should provide justification for choosing either a pharmacological or clinical endpoint study over a blood level (or other biological fluids or tissues) study.

When the measurement of the rate and extent of absorption of the drug in biological fluids cannot be achieved or is unrelated to drug action, a pharmacological endpoint (i.e., drug-induced physiological change that is related to the approved indications for use) study may be conducted. Lastly, in order of preference, if drug concentrations in blood (or fluids or tissues) are not measurable or are inappropriate and there are no appropriate pharmacological effects that can be monitored, then a clinical endpoint study may be conducted, comparing the test (generic) product to the reference (pioneer) product and a placebo (or negative) control.

BA may be measured or BE may be demonstrated by several in vivo and in vitro methods. FDA may require in vivo or in vitro testing, or both, to measure the BA of a drug product or establish the BE of specific drug products. Information on BE requirements for specific products is included in the current edition of FDA's publication *Approved Drug Products with Therapeutic Equivalence Evaluations* and any current supplement to the publication. The selection of the method used to meet an in vivo or in vitro testing requirement depends upon the purpose of the study, the analytical methods available, and the nature of the drug product. The following in vivo and in vitro approaches, in descending order of accuracy, sensitivity, and reproducibility, are acceptable for determining the BA or BE of a drug product:

- An in vivo test in humans in which the concentration of the active ingredient or active moiety, and, when appropriate, its active metabolite(s), in whole blood, plasma, serum, or other appropriate biological fluid, is measured as a function of time. This approach is particularly applicable to dosage forms intended to deliver the active moiety to the bloodstream for systemic distribution within the body.
- An in vitro test that has been correlated with and is predictive of human in vivo BA data.
- An in vivo test in humans in which the urinary excretion of the active moiety, and, when appropriate, its active metabolite(s), is measured as a function of time. The intervals at which measurements are taken should ordinarily be as short as possible so that the measure of the rate of elimination is as accurate as possible. Depending on the nature of the drug product, this approach may be applicable to the category of dosage forms described in paragraph (b)(1)(i) of this section. This method is not appropriate where urinary excretion is not a significant mechanism of elimination.
- An in vivo test in humans in which an appropriate acute pharmacological effect of the active moiety, and, when appropriate, its active metabolite(s), is measured as a function of time if such effect can be measured with sufficient accuracy, sensitivity, and reproducibility. This approach is applicable only when appropriate methods are not available for measurement of the concentration of the moiety, and, when appropriate, its active metabolite(s), in biological fluids or excretory products, but a method is available for the measurement of an appropriate acute pharmacological effect. This approach may be particularly applicable to dosage forms that are not intended to deliver the active moiety to the bloodstream for systemic distribution.
- Well-controlled clinical trials that establish the safety and effectiveness of the drug product, for purposes of measuring BA, or appropriately designed comparative clinical trials, for purposes of demonstrating BE. This approach is the least accurate, sensitive, and reproducible of the general approaches for measuring BA or demonstrating BE. For dosage forms intended to deliver the active moiety to the bloodstream for systemic distribution, this approach may be considered acceptable only when analytical methods cannot be developed to permit use of one of the approaches outlined earlier is not available. This approach may also be considered sufficiently accurate for measuring BA or demonstrating BE of dosage forms intended to deliver the active moiety locally, for example, topical preparations for the skin, eye, and mucous membranes; oral dosage forms not intended to be absorbed, for example, an antacid or radiopaque medium; and bronchodilators administered by inhalation if the onset and duration of pharmacological activity are defined.
- A currently available in vitro test acceptable to FDA (usually a dissolution rate test) that ensures human in vivo BA.
- Any other approach deemed adequate by FDA to measure BA or establish BE.

FDA may require in vivo testing in humans of a product at any time if the agency has evidence that the product

- May not produce therapeutic effects comparable to a pharmaceutical equivalent or alternative with which it is intended to be used interchangeably
- May not be bioequivalent to a pharmaceutical equivalent or alternative with which it is intended to be used interchangeably
- Has greater than anticipated potential toxicity related to PK or other characteristics

MEASUREMENT INDICES

Whenever comparison of the test product and the reference material is to be based on blood concentration–time curves or cumulative urinary excretion–time curves at steady state, appropriate dosage administration and sampling should be carried out to document attainment of steady state. A more complete characterization of the blood concentration or urinary excretion rate during the absorption and elimination phases of a single-dose administered at steady state is encouraged to permit estimation of the total area under concentration–time curves or cumulative urinary excretion–time curves and to obtain PK information, for example, half-life or blood clearance, that is essential in preparing adequate labeling for the drug product.

When comparison of the test product and the reference material is to be based on acute pharmacological effect–time curves, measurements of this effect should be made with sufficient frequency to demonstrate a maximum effect and a lack of significant difference between the test product and the reference material.

Dose Selection

Dose selection will depend upon the label claims, consideration of assay sensitivity, and relevance to the practical use conditions of the reference product. A blood level BE study should generally be conducted at the highest dose approved for the pioneer product.

However, the FDA will consider a BE study conducted at a higher than approved dose in certain cases. Such a study may be appropriate when a multiple of the highest approved dose achieves measurable blood levels, but the highest approved dose does not. In general, the study would be limited to 2–3× the highest dose approved for the pioneer product. The pioneer product should have an adequate margin of safety at the higher than approved dose level. The generic sponsor should also confirm (e.g., through literature) that the drug follows linear kinetics. A higher than approved dose BE study in food animal species would be accompanied by a tissue residue withdrawal study conducted at the highest approved dose for the pioneer product.

For products labeled for multiple claims involving different pharmacological actions at a broad dose range (e.g., therapeutic and production claims), a single BE study at the highest approved dose will usually be adequate. However, multiple BE studies at different doses may be needed if the drug is known to follow nonlinear kinetics. The sponsor should consult with FDA to discuss the BE study or studies appropriate to a particular drug.

Multiple Strengths of Solid Oral Dosage Forms

The generic sponsor should discuss with FDA the appropriate in vivo BE testing and in vitro dissolution testing to obtain approval for multiple strengths (or concentrations) of solid oral dosage forms. FDA will consider the ratio of active to inactive ingredients and the in vitro dissolution profiles of the different strengths, the water solubility of the drug, and the range of strengths for which approval is sought. One in vivo BE study with highest strength product may suffice if the multiple strength products have the same ratio of active to inactive ingredients and are otherwise identical in formulation. In vitro dissolution testing should be conducted using an FDA-approved method, to compare each strength of the generic product to the corresponding strength of the reference product.

Manufacturing of Pilot Batch ("Biobatch")

A pilot batch or "biobatch" should be the source of the finished drug product used in the pivotal studies (i.e., BE studies and tissue residue studies), stability studies, and the validation studies for the proposed analytical and stability-indicating methods. Individual batch testing is necessary to assure that all batches of the same drug product meet an appropriate in vitro test. The commissioner will ordinarily terminate a requirement for a manufacturer to submit samples for batch testing on a finding that the manufacturer has produced four consecutive batches that were tested by the FDA and found to meet the BE requirement, unless the public health requires that batch testing be extended to additional batches.

If a BE requirement specifies a currently available in vitro test or an in vitro BE standard comparing the drug product to a reference standard, the manufacturer shall conduct the test on a sample of each batch of the drug product to assure batch-to-batch uniformity.

Dosing by Labeled Concentration

The potency of the pioneer and generic products should be assayed prior to conducting the BE study to ensure that FDA or compendia specifications are met. The center recommends that the potency of the pioneer and generic lots should differ by no more than ±5% for dosage form products.

The animals should be dosed according to the labeled concentration or strength of the product, rather than the assayed potency of the individual batch (i.e., the dose should not be corrected for the assayed potency of the product). The BE data or derived parameters should not be normalized to account for any potency differences between the pioneer and generic product lots.

Single-Dose versus Multiple-Dose Studies

A single-dose study at the highest approved dose will generally be adequate for the demonstration of BE. A single-dose study at a higher than approved dose may be appropriate for certain drugs.

A multiple-dose study may be appropriate when there are concerns regarding poorly predictable drug accumulation (e.g., a drug with nonlinear kinetics) or a drug with a narrow therapeutic window. A multiple-dose study may also be needed when assay sensitivity is inadequate to permit drug quantification out to three terminal elimination half-lives beyond the time when maximum blood concentrations (C_{max}) are achieved or in cases where prolonged or delayed absorption exist. The determination of prolonged or delayed absorption (i.e., flip-flop kinetics) may be made from pilot data, from the literature, or from information contained with freedom of information (FOI) summaries pertaining to the particular drug or family of drugs.

Guidelines on the Design of a Single-Dose Study

A BE study should be a single-dose comparison of the drug product to be tested and the appropriate reference material conducted in normal adults. The test product and the reference material should be administered to subjects in the fasting state, unless some other approach is more appropriate for valid scientific reasons. A single-dose study should be crossover in design, unless a parallel design or other design is more appropriate for valid scientific reasons, and should provide for a drug elimination period. Unless some other approach is appropriate for valid scientific reasons, the drug elimination period should be either at least three times the half-life of the active drug ingredient or therapeutic moiety, or its metabolite(s), measured in the blood or urine or at least three times the half-life of decay of the acute pharmacological effect.

When comparison of the test product and the reference material is to be based on blood concentration–time curves, unless some other approach is more appropriate for valid scientific reasons, blood samples should be taken with sufficient frequency to permit an estimate of both the peak concentration in the blood of the active drug ingredient or therapeutic moiety, or its metabolite(s), measured and the total AUC for a time period at least three times the half-life of the active drug ingredient or therapeutic moiety, or its metabolite(s), measured.

In a study comparing oral dosage forms, the sampling times should be identical. In a study comparing an intravenous dosage form and an oral dosage form, the sampling times should be those needed to describe both the distribution and elimination phase of the intravenous dosage form and the absorption and elimination phase of the oral dosage form.

In a study comparing drug delivery systems other than oral or intravenous dosage forms with an appropriate reference standard, the sampling times should be based on valid scientific reasons.

When comparison of the test product and the reference material is to be based on cumulative urinary excretion–time curves, unless some other approach is more appropriate for valid scientific reasons, samples of the urine should be collected with sufficient frequency to permit an estimate of the rate and extent of urinary excretion of the active drug ingredient or therapeutic moiety, or its metabolite(s), measured.

When comparison of the test product and the reference material is to be based on acute pharmacological effect–time curves, measurements of this effect should be made with sufficient frequency to permit a reasonable estimate of the total AUC for a time period at least three times the half-life of decay of the pharmacological effect, unless some other approach is more appropriate for valid scientific reasons.

The use of an acute pharmacological effect to determine BA may further require demonstration of dose-related response. In such a case, BA may be determined by comparison of the dose–response curves as well as the total area under the acute pharmacological effect–time curves for any given dose.

Guidelines for Multiple-Dose Study

In selected circumstances, it may be necessary for the test product and the reference material to be compared after repeated administration to determine steady-state levels of the active drug ingredient or therapeutic moiety in the body. The test product and the reference material should be administered to subjects in the fasting or nonfasting state, depending upon the conditions reflected in the proposed labeling of the test product.

A multiple-dose study may be required to determine the BA of a drug product in the following circumstances that there is a difference in the rate of absorption but not in the extent of absorption; there is excessive variability in BA from subject to subject; the concentration of the active drug ingredient or therapeutic moiety, or its metabolite(s), in the blood resulting from a single dose is too low for accurate determination by the analytical method; and the drug product is an extended-release dosage form.

A multiple-dose study should be crossover in design, unless a parallel design or other design is more appropriate for valid scientific reasons, and should provide for a drug elimination period if steady-state conditions are not achieved. A multiple-dose study is not required to be of crossover design if the study is to establish dose proportionality under a multiple-dose regimen or to establish the PK profile of a new drug product, a new drug delivery system, or an extended-release dosage form.

If a drug elimination period is required, unless some other approach is more appropriate for valid scientific reasons, the drug elimination period should be either at least five times the half-life of the active drug ingredient or therapeutic moiety, or its active metabolite(s), measured in the blood or urine or at least five times the half-life of decay of the acute pharmacological effect.

Whenever a multiple-dose study is conducted, unless some other approach is more appropriate for valid scientific reasons, sufficient doses of the test product and reference material should be administered in accordance with the labeling to achieve steady-state conditions.

Fed versus Fasted State

Feeding may either enhance or interfere with drug absorption, depending upon the characteristics of the drug and the formulation. Feeding may also increase the inter- and intrasubject variability in the rate and extent of drug absorption. The rationale for conducting each BE study under fasting or fed conditions should be provided in the protocol. Fasting conditions, if used, should be fully described, giving careful consideration to the PKs of the drug and the humane treatment of the test animals. The protocol should describe the diet and feeding regime that will be used in the study.

If a pioneer product label indicates that the product is limited to administration either in the fed or fasted state, then the BE study should be conducted accordingly. If the BE study parameters pass the agreed-upon confidence intervals (CIs), then the single study is acceptable as the basis for approval of the generic drug.

However, for certain product classifications or drug entities, such as enteric-coated and oral sustained-release products, demonstration of BE in both the fasted and the fed states may be necessary, if drug BA is highly variable under feeding conditions, as determined from the literature or from pilot data. A BE study conducted under fasted conditions may be necessary to pass the CIs. A second smaller study may be necessary to examine meal effects. FDA will evaluate the smaller study with respect to the means of the pivotal parameters (AUC, C_{max}). The sponsors should consult with FDA prior to conducting the studies.

Pharmacological Endpoint Studies

Where the direct measurement of the rate and extent of absorption of the new animal drug in biological fluids is inappropriate or impractical, the evaluation of a pharmacological endpoint related to the labeled indications for use will be acceptable.

Typically, the design of a pharmacological endpoint study should follow the same general considerations as the blood level studies. However, specifics such as the number of subjects or sampling times will depend on the pharmacological endpoint monitored. The parameters to be measured will also depend upon the pharmacological endpoints and may differ from those used in blood level studies. As with blood level studies, when pharmacological endpoint studies are used to demonstrate BE, a tissue residue study will also be required in food-producing animals.

For parameters that can be measured over time, a time versus effect profile is generated, and equivalence is determined with the method of statistical analysis essentially the same as for the blood level BE study.

For pharmacological effects for which effect versus time curves cannot be generated, alternative procedures for statistical analysis should be discussed with FDA prior to conducting the study.

Clinical Endpoint Studies

If measurement of the drug or its metabolites in blood, biological fluids, or tissues is inappropriate or impractical and there are no appropriate pharmacological endpoints to monitor (e.g., most production drugs and some coccidiostats and anthelmintics), then well-controlled clinical endpoint studies are acceptable for the demonstration of BE.

Generally, a parallel group design with three treatment groups should be used. The groups should be a placebo (or negative) control, a positive control (reference/pioneer product), and the test (generic) product. The purpose of the placebo (or negative) control is to confirm the sensitivity or validity of the study. Dosage(s) approved for the pioneer product should be used in the study. Dosage(s) should be selected following consultation with FDA and should reflect consideration for experimental sensitivity and relevance to the common use of the pioneer product.

Studies should generally be conducted using the target animal species, with consideration for the sex, class, body weight, age, health status, and feeding and husbandry conditions, as described on the pioneer product labeling. In general, the length of time that the study is conducted should be consistent with the duration of use on the pioneer product labeling.

In general, the response(s) to be measured in a clinical endpoint study should be based upon the labeling claims of the pioneer product and selected in consultation with Center for Veterinary Medicine (CVM). It may not be necessary to collect data on some overlapping claims (e.g., for a production drug that is added at the same amount per ton of feed for both growth rate and feed efficiency, data from only one of the two responses need to be collected).

When considering sample size, it is important to note that the pen, not the individual animal, is often the experimental unit. As with blood level BE studies, FDA is advocating the use of 90% CIs as the best method for evaluating clinical endpoint studies. The bounds for confidence limits (e.g., ±20% of the improvement over placebo [or negative] control) for the particular drug should be agreed upon with FDA prior to initiation of the study.

The analysis should be used to compare the test product and the reference product. In addition, a traditional hypothesis test should be performed comparing both the test and reference products separately to the placebo (or negative) control. The hypothesis test is conducted to ensure that the study has adequate sensitivity to detect differences when they actually occur. If no significant improvement ($\alpha = 0.05$) is seen in the parameter (i.e., the mean of the test and the mean of the reference products are each not significantly better than the mean of the placebo [or negative] control), generally, the study will be considered inadequate to evaluate BE.

Assuming that the test and reference products have been shown to be superior to the placebo (or negative) control, the determination of BE is based upon the CI of the difference between the two products.

Some clinical endpoint studies may not include a placebo (or negative) control for ethical and/or practical considerations. If the placebo is omitted, then the response(s) to the test and reference products should each provide a statistically significant improvement over baseline.

If the results are ordered categorical data (e.g., excellent, good, fair, or poor), a nonparametric hypothesis test of no difference between test product and placebo (or negative) control and between the reference product and placebo (or negative) control should be performed. As previously, if these tests result in significant differences between the test product and control and the reference product and control, then a nonparametric CI on the difference between the test and reference products is calculated.

Another acceptable approach for categorical data is to calculate the CI on the odds ratio between the test and reference products after showing that the test and reference products are significantly better than the control.

ANALYTICAL METHODS

The analytical method used in an in vivo BA or BE study to measure the concentration of the active drug ingredient or therapeutic moiety, or its active metabolite(s), in body fluids or excretory products, or the method used to measure an acute pharmacological effect shall be demonstrated to be accurate and of sufficient sensitivity to measure, with appropriate precision, the actual concentration of the active drug ingredient or therapeutic moiety, or its active metabolite(s), achieved in the body. When the analytical method is not sensitive enough to measure accurately the concentration of the active drug ingredient or therapeutic moiety, or its active metabolite(s), in body fluids or excretory products produced by a single dose of the test product, two or more single doses may be given together to produce higher concentration.

Assay Consideration

A properly validated assay method is pivotal to the acceptability of any PK study. Sponsors should discuss any questions or problems concerning the analytical methodology with CVM before undertaking the BE studies. The ANADA submission should contain adequate information necessary for the CVM reviewer to determine the validity of the analytical method used to quantitate the level of drug in the biological matrix (e.g., blood).

The following aspects should be addressed in assessing method performance.

CONCENTRATION RANGE AND LINEARITY

The quantitative relationship between concentration and response should be adequately characterized over the entire range of expected sample concentrations. For linear relationships, a standard curve should be defined by at least five concentrations. If the concentration response function is nonlinear, additional points would be necessary to define the nonlinear portions of the curve. Extrapolation beyond a standard curve is not acceptable.

LIMIT OF DETECTION

The standard deviation of the background signal and limit of detection (LOD) should be determined. The LOD is estimated as the response value calculated by adding three times the standard deviation of the background response to the average background response.

LIMIT OF QUANTITATION

The initial determination of limit of quantitation (LOQ) should involve the addition of 10 times the standard deviation of the background response to the average background response. The second step in determining LOQ is assessing the precision (reproducibility) and accuracy (recovery) of the method at the LOQ. The LOQ will generally be the lowest concentration on the standard curve that can be quantified with acceptable accuracy and precision.

SPECIFICITY

The absence of matrix interferences should be demonstrated by the analysis of six independent sources of control matrix. The effect of environmental, physiological, or procedural variables on the matrix should be assessed. Each independent control matrix will be used to produce a standard curve, which will be compared to a standard curve produced under chemically defined conditions. The comparison of curves should exhibit parallelism and superimposability within the limits of analytical variation established for the chemically defined standard curve.

ACCURACY: RECOVERY

This parameter should be evaluated using at least three known concentrations of analyte freshly spiked in control matrix, one being at a point two standard deviations above the LOQ, one in the middle of the range of the standard curve ("midrange"), and one at a point two standard deviations below the upper quantitative limit of the standard curve. The accuracy of the method, based upon the mean value of six replicate injections, at each concentration level, should be within 80%–120% of the nominal concentration at each level (high, midrange, and LOQ).

PRECISION

This parameter should be evaluated using at least three known concentrations of analyte freshly spiked in control matrix, at the same points used for determination of accuracy. The coefficient of variation (CV) of six replicates should be ±10% for concentrations at or above 0.1 ppm (0.1 μg/mL). A CV of ±20% is acceptable for concentrations below 0.1 ppm.

ANALYTE STABILITY

Stability of the analyte in the biological matrix under the conditions of the experiment (including any period for which samples are stored before analyses) should be established. It is recommended that the stability be determined with incurred analyte in the matrix of dosed animals in addition to, or instead of, control matrix spiked with pure analyte. Also, the influence of three freeze–thaw cycles at two concentrations should be determined.

Stability samples at three concentrations should be stored with the study samples and analyzed through the period of time in which study samples are analyzed. These analyses will establish whether or not analyte levels have decreased during the time of analysis.

ANALYTICAL SYSTEM STABILITY

To assure that the analytical system remains stable over the time course of the assay, the reproducibility of the standard curve should be monitored during the assay. A minimal design would be to run analytical standards at the beginning and at the end of the analytical run.

QUALITY CONTROL SAMPLES

The purpose of quality control (QC) samples is to assure that the complete analytical method, sample preparation, extraction, cleanup, and instrumental analysis perform according to acceptable criteria. The stability of the drug in the text matrix for the QC samples should be known, and any tendency for the drug to bind to tissue or serum components over time should also be known.

Drug-free control matrix (e.g., tissue and serum) that is freshly spiked with known quantities of test drug should be analyzed contemporaneously with test samples, evenly dispersed throughout each analytical run. This can be met by the determination of accuracy and precision of each analytical run.

REPLICATE AND REPEAT ANALYSES

Single rather than replicate analyses are recommended, unless the reproducibility and/or accuracy of the method is borderline. Criteria for repeat analyses should be determined prior to running the study and recorded in the method SOP.

SUMMARY OF SAMPLES TO BE RUN WITH EACH ANALYSIS

a. Accuracy estimate
b. Precision estimate
c. Analytical system stability
d. Analyte stability samples

PRIOR REVIEW

The Commissioner of Food and Drugs strongly recommends that, to avoid the conduct of an improper study and unnecessary human research, any person planning to conduct a BA or BE study submits the proposed protocol for the study to FDA for review prior to the initiation of the study. FDA may review a proposed protocol for a BE study and will offer advice with respect to whether the conditions an appropriate design, the choice of reference product, and the proposed chemical and statistical analysis methods are met.

The Commissioner of Food and Drugs shall consider the following factors, when supported by well-documented evidence, to identify specific pharmaceutical equivalents and pharmaceutical alternatives that are not or may not be bioequivalent drug products:

- Evidence from well-controlled clinical trials or controlled observations in patients that such drug products do not give comparable therapeutic effects.
- Evidence from well-controlled BE studies that such products are not bioequivalent drug products.
- Evidence that the drug products exhibit a narrow therapeutic ratio, for example, there is less than a twofold difference in LD50 and ED50 values, or have less than a twofold difference in the MTCs and MECs in the blood and safe and effective use of the drug products requires careful dosage titration and patient monitoring.
- Competent medical determination that a lack of BE would have a serious adverse effect in the treatment or prevention of a serious disease or condition.

- The physicochemical evidences that the active drug ingredient has a low solubility in water, for example, less than 5 µg/1 mL; or, if dissolution in the stomach is critical to absorption, the volume of gastric fluids required to dissolve the recommended dose far exceeds the volume of fluids present in the stomach (taken to be 100 mL for adults and prorated for infants and children); or the dissolution rate of one or more such products is slow, for example, less than 50% in 30 min when tested using either a general method specified in an official compendium or a paddle method at 50 rpm in 900 mL of distilled or deionized water at 37°C, or differs significantly from that of an appropriate reference material such as an identical drug product that is the subject of an approved full DNA; or the particle size and/or surface area of the active drug ingredient is critical in determining its BA; or certain physical structural characteristics of the active drug ingredient, for example, polymorphic forms, conforms, solvates, complexes, and crystal modifications, dissolve poorly and this poor dissolution may affect absorption; or such drug products have a high ratio of excipients to active ingredients, for example, greater than 5–1; or specific inactive ingredients, for example, hydrophilic or hydrophobic excipients and lubricants, either may be required for absorption of the active drug ingredient or therapeutic moiety or, alternatively, if present, may interfere with such absorption.
- The PK evidence that the active drug ingredient, therapeutic moiety, or its precursor is absorbed in large part in a particular segment of the GI tract or is absorbed from a localized site; or the degree of absorption of the active drug ingredient, therapeutic moiety, or its precursor is poor, for example, less than 50%, ordinarily in comparison to an intravenous dose, even when it is administered in pure form, for example, in solution; or there is rapid metabolism of the therapeutic moiety in the intestinal wall or liver during the process of absorption (first-class metabolism) so the therapeutic effect and/or toxicity of such drug product is determined by the rate as well as the degree of absorption; or the therapeutic moiety is rapidly metabolized or excreted so that rapid dissolution and absorption are required for effectiveness; or the active drug ingredient or therapeutic moiety is unstable in specific portions of the GI tract and requires special coatings or formulations, for example, buffers, enteric coatings, and film coatings, to assure adequate absorption; or the drug product is subject to dose-dependent kinetics in or near the therapeutic range and the rate and extent of absorption are important to BE.

RECORD MAINTENANCE

All records of in vivo or in vitro tests conducted on any marketed batch of a drug product to assure that the product meets a BE requirement shall be maintained by the manufacturer for at least 2 years after the approval of the application submitted and would be available to the FDA on request:

- If the formulation of the test article is the same as the formulation(s) used in the clinical studies demonstrating substantial evidence of safety and effectiveness for the test article's claimed indications, a reserve sample of the test article is used to conduct an in vivo BA study comparing the test article to a reference oral solution, suspension, or injection.
- If the formulation of the test article differs from the formulation(s) used in the clinical studies demonstrating substantial evidence of safety and effectiveness for the test article's claimed indications, a reserve sample of the test article and of the reference standard is used to conduct an in vivo BE study comparing the test article to the formulation(s) (reference standard) used in the clinical studies.
- For a new formulation, new dosage form, or a new salt or ester of an active drug ingredient or therapeutic moiety that has been approved for marketing, a reserve sample of the test article and of the reference standard is used to conduct an in vivo BE study comparing the test article to a marketed product (reference standard) that contains the same active drug ingredient or therapeutic moiety.

Each reserve sample shall consist of a sufficient quantity to permit FDA to perform five times all of the release tests required in the application or supplemental application. Each reserve sample shall be adequately identified so that the reserve sample can be positively identified as having come from the same sample as used in the specific BA study. Each reserve sample shall be stored under conditions consistent with product labeling and in an area segregated from the area where testing is conducted and with access limited to authorized personnel. Each reserve sample shall be retained for a period of at least 5 years following the date on which the application or supplemental application is approved or, if such application or supplemental application is not approved, at least 5 years following the date of completion of the BA study in which the sample from which the reserve sample was obtained was used.

Authorized FDA personnel will ordinarily collect reserve samples directly from the applicant or contract research organization at the storage site during a preapproval inspection. If authorized FDA personnel are unable to collect samples, FDA may require the applicant or contract research organization to submit the reserve samples to the place identified in the agency's request. If FDA has not collected or requested delivery of a reserve sample or if FDA has not collected or requested delivery of any portion of a reserve sample, the applicant or contract research organization shall retain the sample or remaining sample for the 5-year period.

Upon release of the reserve samples to FDA, the applicant or contract research organization shall provide a written assurance that, to the best knowledge and belief of the individual executing the assurance, the reserve samples came from the same samples as used in the specific BA or BE study identified by the agency. The assurance shall be executed by an individual authorized to act for the applicant or contract research organization in releasing the reserve samples to FDA.

A contract research organization may contract with an appropriate, independent third party to provide storage of reserve samples provided that the sponsor of the study has been notified in writing of the name and address of the facility at which the reserve samples will be stored. If a contract research organization conducting a BA or BE study that requires reserve sample retention goes out of business, it shall transfer its reserve samples to an appropriate, independent third party and shall notify in writing the sponsor of the study of the transfer and provide the study sponsor with the name and address of the facility to which the reserve samples have been transferred.

The applicant of an abbreviated application or a supplemental application submitted under section 505 of the FD&C Act or, if BE testing was performed under contract, the contract research organization shall retain reserve samples of any test article and reference standard used in conducting an in vivo or in vitro BE study required for approval of the abbreviated application or supplemental application and beyond as required.

Pharmacokinetic and Statistical Considerations in Study Design

Sampling Time Considerations

The total number of sampling times necessary to characterize the blood level profiles will depend upon the curvature of the profiles and the magnitude of variability associated with the BA data (including PK variability, assay error, and interproduct differences in absorption kinetics).

The sampling times should adequately define peak concentration(s) and the extent of absorption. The sampling times should extend to at least three terminal elimination half-lives beyond t_{max}. The sponsor should consult with FDA prior to conducting the pivotal BE study if the assay is unable to quantify samples to three half-lives.

Maximum sampling time efficiency may be achieved by conducting a pilot investigation. The pilot study should identify the general shapes of the test and reference curves, the magnitude of the difference in product profiles, and the noise associated with each blood sampling time (e.g., variability attributable to assay error and the variability between subjects, for parallel study designs, or within subjects, for crossover study designs). This information should be applied to the determination of an optimum blood sampling schedule. Depending upon these variability estimates, it may be more efficient to cluster several blood samples rather than to have samples that are periodically dispersed throughout the duration of blood sampling.

Protein Binding

In general, product BE should be based upon total (free plus protein bound) concentrations of the parent drug (or metabolite, when applicable). However, if nonlinear protein binding is known to occur within the therapeutic dosing range (as determined from literature or pilot data), then sponsors may need to submit data on both the free and total drug concentrations for the generic and pioneer products.

Similarly, if the drug is known to enter blood erythrocytes, the protocol should address the issue of potential nonlinearity in erythrocyte uptake of the drug administered within the labeled therapeutic dosing range.

The BE protocol or completed study report should provide any information available from the literature regarding erythrocyte uptake and protein binding characteristics of the drug or drug class, including the magnitude of protein binding and the type of blood protein to which it binds.

Subject Number

Pilot studies are recommended as a means of estimating the appropriate sample size for the pivotal BE study. Estimated sample size will vary depending upon whether the data are analyzed on a log or linear scale. Useful references for sample size estimates include Westlake, Hauschke, and Steinijans.

Crossover and Parallel Design Considerations

A two-period crossover design is commonly used in blood level studies. The use of crossover designs eliminates a major source of study variability, between subject differences in the rates of drug absorption, drug clearance, and the volume of drug distribution.

In a typical two-period crossover design, subjects are randomly assigned to either sequence A or sequence B with the restriction that equal numbers of subjects are initially assigned to each sequence. The design is as follows:

	Sequence A	Sequence B
Period 1	Test	Reference
Period 2	Reference	Test

A crucial assumption in the two-period crossover design is that of equal residual effects. Unequal residual effects may result, for example, from an inadequate washout period. Another assumption of the crossover (or extended period) design is that there is no subject by formulation interaction. In other words, the assumption is that all subjects are from a relatively homogeneous population and will exhibit similar relative BA of the test and reference products. If there are subpopulations of subjects, such that the relationship between product BAs is a function of the subpopulation within which they are being tested, then a subject by formulation interaction is said to exist.

A one-period parallel design may be preferable in the following situations:

- The drug induces physiological changes in the animal (e.g., liver microsomal enzyme induction) that persist after total drug clearance and alter the BA of the product administered in the second period.
- The drug has a very long terminal elimination half-life, creating a risk of residual drug present in the animal at the time of the second-period dosing.

- The duration of the washout time for the two-period crossover study is so long as to result in significant maturational changes in the study subjects.
- The drug follows delayed or prolonged absorption (flip-flop kinetics[2]), where the slope of the beta-elimination phase is dictated by the rate of drug absorption rather than the rate of drug elimination from one or both products.

Other designs, such as the two-period design with four treatment sequences (test/test, reference/reference, test/reference, and reference/test) or the extended period design, may be appropriate depending on the circumstances. The use of alternative study designs should be discussed with FDA prior to conducting the BE study. Pilot data or literature may be used in support of alternative study designs.

Duration of Washout Time for Crossover Study

For drugs that follow a one- or two-compartment open-body model, the duration of the washout time should be approximately 10× the plasma apparent terminal elimination half-life, to provide for 99.9% of the administered dose to be eliminated from the body. If more highly complex kinetic models are anticipated (e.g., drugs for which long withdrawal times have been assigned due to prolonged tissue binding) or for drugs with the potential for physiological carryover effects, the washout time should be adjusted accordingly. The washout period should be sufficiently long to allow the second period of the crossover study to be applicable in the statistical analysis. However, if sequence effects are noted, the data from the first period may be evaluated as a parallel design study.

Pivotal Parameters for Blood Level Bioequivalence

The sponsor is encouraged to calculate parameters using formulas that involve only the raw data (i.e., the so-called model independent methods).

AREA UNDER THE CURVE ESTIMATES

The extent of product BA is estimated by the area under the blood concentration versus time curve (AUC). AUC is most frequently estimated using the linear trapezoidal rule. Other methods for AUC estimation may be proposed by the sponsor and should be accompanied by appropriate literature references during protocol development. For a single-dose BE study, AUC should be calculated from time 0 (predose) to the last sampling time associated with quantifiable drug concentration AUC(0–LOQ). The comparison of the test and reference product value for this noninfinity estimate provides the closest approximation of the measure of uncertainty (variance) and the relative BA estimate associated with AUC(0–INF), the full extent of product BA. The relative AUC values generally change very little once the absorption of both products has been completed. However, because of the possibility of multifunctional absorption kinetics, it cannot always be determined when the available drug has been completely absorbed. Therefore, FDA recommends extending the duration of sampling until such time that AUC(0–LOQ)/AUC(0–INF) ≥ 0.80. Generally, the sampling times should extend to at least three multiples of the drug's apparent terminal elimination half-life, beyond the time when maximum blood concentrations are achieved.

AUC(0–INF) should be used to demonstrate that the concentration–time curve can be quantitated such that AUC(0–LOQ)/AUC(0–INF) ≥ 0.80. The method for estimating the terminal elimination phase should be described in the protocol and the final study report. The AUC(0–LOQ)/AUC(0–INF) is calculated to determine whether AUC(0–LOQ) adequately reflects the extent of absorption.

The sponsor should consult with FDA if AUC(0–LOQ)/AUC(0–INF) is determined to be <0.80. If AUC(0–LOQ)/AUC(0–INF) is <0.80, then a multiple-dose study to steady state may be needed to allow an accurate assessment of AUC(0–INF) (where AUC(0–INF) = AUC(0–*t*) at steady state and *t* is the dosing interval).

In a multiple-dose study, the AUC should be calculated over one complete dosing interval AUC(0–*t*). Under steady-state conditions, AUC(0–*t*) equals the full extent of BA of the individual dose

AUC(0–INF) assuming linear kinetics. For drugs that are known to follow nonlinear kinetics, the sponsor should consult with FDA to determine the appropriate parameters for the BE determination.

AUC Measurements

Trapezoidal Rule

This is the simplest of all the methods and involves the breaking up of the plasma concentration versus time profile into several trapezoids, calculating the areas of individual trapezoids, and then adding up these areas to arrive at a cumulative AUC:

$$\left(\frac{(C_0+C_1)}{2}\right)(t_1-t_0)+\left(\frac{(C_1+C_2)}{2}\right)(t_2-t_1)+\cdots+\left(\frac{(C_{n-1}+C_n)}{2}\right)(t_n-t_{n-1}) \tag{5.1}$$

The units for the AUC are concentration × time, for example, μg·h/mL or mg·min/L. As a general rule, the larger the number of segments or trapezoids formed, the greater is the accuracy achieved. In other words, the closer the interval between each plasma concentration reading taken, the more accurate will be the results. If the plasma concentration values are quite far apart, a smooth curve may be drawn, which then can be broken up into a large number of trapezoids.

The AUC is proportional to the dose absorbed only when the calculations are extended to the point where the plasma concentration approaches zero. This may not be possible in some instances, in which case the comparisons can either be made up to a given time or the plasma concentrations can be extended to follow the shape of the curve, both of these approaches adding to the errors in the bioavailability estimations.

Integration Method

The rate of change of plasma concentration (C) is described as

$$\frac{\mathrm{d}C}{\mathrm{d}t} = \text{Rate of absorption} - \text{Rate of elimination} \tag{5.2}$$

$$= K_\mathrm{a}X_\mathrm{a} - KX \tag{5.3}$$

where

K_a and K are absorption and elimination rate constants

X_a and X are the amounts of drug in the gatrointestinal GI tract and the body, respectively

An integration of this equation between limits of time for which the drug remains in the body, as reflected by the plasma concentration, gives

$$C = A\left(e^{-kt} - e^{-K_\mathrm{a}t}\right) \tag{5.4}$$

and the total AUC, for which the total integral between time zero and infinity, is given by

$$\mathrm{AUC} = A\left(\frac{1}{K} - \frac{1}{K_\mathrm{a}}\right) \tag{5.5}$$

Thus, if C can be fitted to an equation that will allow calculation of the absorption and elimination rate constants, the exact calculation of AUC can be made very easily. This approach is identical to the trapezoidal rule method described earlier, except that it uses trapezoids whose time differential is approaching zero.

Computer Applications

Whereas in the past a variety of physical methods were used for area comparison, most of these including those described earlier have now been replaced with sophisticated computer programs that often combine the statistical evaluation of differences using a variety of linear and nonlinear approaches. These will be described in detail in another chapter.

RATE OF ABSORPTION

The rate of absorption will be estimated by the maximum observed drug concentration (C_{max}) and the corresponding time to reach this maximum concentration (t_{max}). When conducting a steady-state investigation, data on the minimum drug concentrations (trough values) observed during a single-dosing interval (C_{min}) should also be collected. Generally, three successive C_{min} values should be provided to verify that steady-state conditions have been achieved. Although C_{min} most frequently occurs immediately prior to the next successive dose, situations do occur with C_{min} observed subsequent to dosing. To determine a steady-state concentration, the C_{min} values should be regressed over time, and the resultant slope should be tested for its difference from zero.

DETERMINATION OF PRODUCT BIOEQUIVALENCE

Unless otherwise indicated by FDA during the protocol development for a given application, the pivotal BE parameters will be C_{max} and AUC(0–LOQ) (for a single-dose study) or AUC(0–t) (for a multiple-dose study). To be indicative of product BE, the pivotal metrics should be associated with CIs that fall within a set of acceptability limits.

The sponsor and FDA should agree to the acceptable bounds for the confidence limits for the particular drug and formulation during protocol development. If studies or literature demonstrates that the pioneer drug product exhibits highly variable kinetics, then the generic drug sponsor may propose alternatives to the generally acceptable bounds for the confidence limits. t_{max} in single-dose studies and C_{min} in multiple-dose studies will be assessed by clinical judgment.

Errors in BE Studies

Erroneous conclusions can easily be made if the logic behind BA studies is not clearly understood. The following are the important highlights of the most common errors:

1. When concentrations are monitored in the biological fluids, the specificity of the assay methods is of utmost importance. This is especially applicable to single-dose studies in which small concentrations should be monitored in order to allow study of the complete elimination of the drug from the body.
2. It is generally assumed that the absorption rates of drugs are higher than the rates of elimination, but there can be exceptions, in which case the terminal plasma concentration profiles would represent both the absorption and elimination processes and the mathematical/statistical models used should take this into account.
3. The extrapolation of plasma or urinary concentration data to compensate for missing experimental points always introduces some error in the calculations; it is desirable to extend the study to at least three elimination half-lives when plasma concentration is monitored and for at least seven half lives when monitoring urinary excretion of drugs to estimate their BA.
4. There is often a lack of sufficient data points to characterize the plasma concentration profiles. Significant area can be lost if sufficient points are not collected during the peaking of the concentration. In general, there should be at least three data points before the peak occurs and at least four or five values after the peak, if possible.

5. The variation among individuals in the elimination rates of a drug should be considered. The proportionality between AUC and BA is based on the assumption that the elimination rates are invariant; any deviation from the norm will result in significant error. Correction of this error can be made if the elimination rate constants are calculated for each subject and the AUC is corrected as follows:

$$\mathrm{AUC}_{\mathrm{corrected}} = \mathrm{AUC}_{\mathrm{apparent}}\left(K\right) \tag{5.6}$$

   If a drug is eliminated fast, K will be large, accounting for possible underestimation of the AUC.
6. Comparison of data for different studies that may not be well matched in terms of the characteristics of the subject population, study conditions, or routes of drug administration should be made with due consideration to these factors. It is ironic that such cross-study comparisons are both very common and very misleading.
7. When identical drug concentrations are obtained in the plasma following administration of equimolar doses from different formulations, these formulations are considered bioequivalent, and the principle is referred to as the superimposition principle. In using this principle, one must choose a number of subjects in accordance with the statistical criteria that will demonstrate at least 20% differences in the means of values in order to make them clinically significant. This criterion can be applied to the concentration at each sampling time, to the peak concentration, and to the time of the peak concentrations and the AUCs.
8. It should be noted that just because a drug product meets compendial standards of purity and other criteria, its BA is not assured. In fact, compendial requirements fall far short of assuring the efficiency of dosage forms in releasing drugs. The latest edition of USP and NF requires demonstration of sufficient dissolution for many drugs where evidence of dissolution affecting BA has been suggested. A large number of drugs remain to be included in this list, and it is hoped that eventually the demonstration of BA will become a compendial requirement. The costs of performing BA studies make such requirements impractical for some drugs. However, without such requirements, it is difficult to justify the rejection of a product on the grounds that its chemical equivalence varies by more than 10%, when its biological equivalent is allowed to vary to any degree.

Failures in Bioequivalence Studies

Almost one out of every eight ANDA BE studies submitted to FDA fail. From July 2009 to September 2011, FDA reviewed 199, of which 50 were surveyed and 1 out of 8 ANDA BE data failed. A typical BE study is

- A two-way crossover, single-dose study
- Normal volunteers
- Fasting
  - BE criteria: 90% CI between 80% and 125%
    - AUC
    - C_{max}
- Fed (point estimate between 80 and 125)

The sponsors only submit the studies that are not "failing." For NDAs, all human investigations made to show whether or not such drug is safe for use and whether such drug is effective must be submitted (FD&C 505[b] [1] [A]). For ANDAs (FD&C 505 [j]), similar language was not included in the Act. Sponsors have interpreted this to mean that "failed" BE studies need not be submitted in ANDAs.

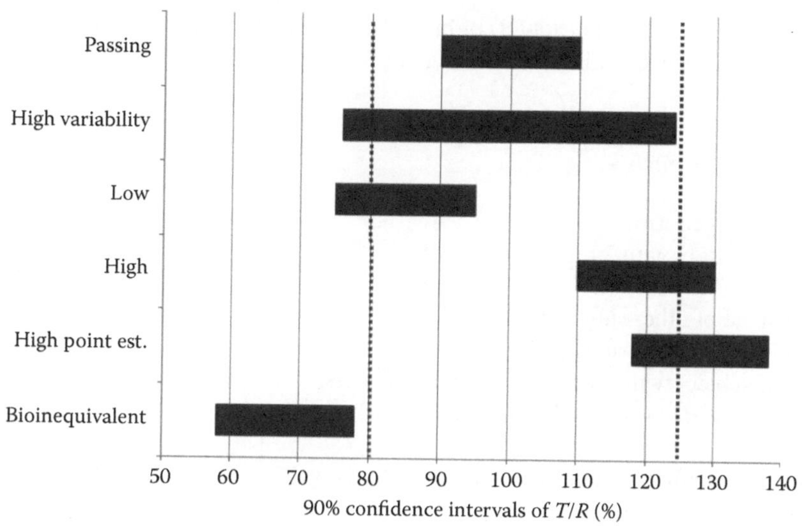

FIGURE 5.1 Examples of confidence-based failure of BE studies.

Other important considerations for "failed" submissions:

- Application may not contain untrue statements of material fact (FD&C 505[j][3][K], 21 CFR 314.127(a)[13]).
- Selective reporting of data may constitute untrue statements of material fact (e.g., FR 32982 June 26, 1995).
- Failure to report failed studies may be considered selective reporting.

Reasons for "failure" of BE studies (Figure 5.1):

- Underpowered study
- Unusual study designs
- "Outlier" response from one or more subjects
- Assay issues
- Wrong reference
- Baseline corrected versus not corrected
- Incorrect statistical analysis
- Compliance issues
- Formulation that is not truly bioequivalent to the reference

Example 1—drug *X* oral liquid:

- Liquid dosage form mixed with beverage prior to administration.
- ANDA studies performed with one beverage.
- Additional study in another beverage completed before ANDA approval.
- Product was not bioequivalent under this labeled administration condition.
- Study was not submitted before ANDA was approved.

Example 2—drug *Y* solid oral modified release:

- Discovered by compliance on inspection of another study
- First study against a lot (A) of brand product failed—C_{max} (105, 130)

- Second study against another lot (B) of brand passed (submitted in ANDA)
- Third study performed testing lot A versus lot B—failed: C_{max} (111, 131)
- Problem with RLD

Example 3—"outlier" response:

- Solid oral dosage form.
- Standard BE study with 24 normal subjects.
- One subject's *T*/*R* was ~4.
- Point estimate of other subjects' *T*/*R* ~1.1.
- Study did not pass CI criteria with all subjects.
- Restudy of subject with four other original subjects.
- Subject's new *T*/*R* = 1.05—subject data dropped from original study.

The questions that remained unanswered are the following:

- Should sponsors submit the results of all BE studies performed on the to-be-marketed ANDA formulation?
- Full reports or complete summaries?
- What should FDA do with this information?
    - Complete review
    - Brief, but careful, examination

RISK-BASED BIOEQUIVALENCE

Health Risk Categories

The selection of active ingredients for which BE studies should be required is a public health decision and as such should take into account the benefit–risk ratio of the same. This situation leads to the health risk concept, that is, which active ingredients require rigorous handling to prevent public health problems. One way of doing this is to take into account which active ingredients, because of their pharmacological characteristics, should be controlled through blood determinations.

Definition

As operational definition, the health risk concept should be established in the context of the problems of BE. For this purpose, it would be reasonable to establish what are the health consequences when the drug is outside (under or above) the therapeutic window (the margin determined by the nontoxic maximum concentration and the effective minimum concentration). Thus, in relating the therapeutic window (the margin whose limits are the nontoxic maximum and effective minimum concentrations) and adverse effects of the drugs, three risk levels can be established, as described in the following:

High health risk: This is the probability of the appearance of threatening complications of the disease for the life or the psychophysical integrity of the person and/or serious adverse reactions (death, patient hospitalization, extension of the hospitalization, significant or persistent disability, or threat of death), when the blood concentration of the active ingredient is not within the therapeutic window. For purposes of the selection, this risk level was assigned a score of 3 (three).

Intermediate health risk: This is the probability of the appearance of nonthreatening complications of the disease for the life or the psychophysical integrity of the person and/or adverse reactions, not necessarily serious, when the blood concentration of the active ingredient is not found within the therapeutic window. For purposes of the selection, this risk level was assigned a score of 2 (two).

Low health risk: This is the probability of the appearance of a minor complication of the disease and/or mild adverse reactions, when the blood concentration of the active ingredient is not within the therapeutic window. For purposes of the selection, this risk level was assigned a score of 1 (one).

While there are other factors to be considered such as the physicochemical and PK parameters, from the standpoint of public health, the most important element to take into account is the health risk. Table 5.5 lists the active ingredients classified in accordance with their health risk and the established scores.

CRITICAL DOSE OR NARROW THERAPEUTIC INDEX DRUGS

These are drugs in which comparatively small differences in dose or concentration may lead to serious therapeutic failures and/or serious adverse drug reactions. A variety of terms are used to describe them:

- NTI—narrow therapeutic window
- Narrow therapeutic range—critical dose (CD) drugs
- Narrow therapeutic ratio

Narrow therapeutic ratio is defined where less than a twofold difference in LD50 and ED50 values—or—less than twofold difference in the MTCs and MECs in the blood. These are drug products that are subject to therapeutic drug concentration or PD monitoring. Examples include digoxin, lithium, phenytoin, and warfarin. The traditional BE limit of 80%–125% remains unchanged for these products in the United States. A comparison of acceptance criteria across international agencies shows the following:

- Australia, Canada, EU, Japan, and South Africa: All had stricter acceptance criteria for NTI/CD drugs.
- Health Canada and the Japanese National Institute of Health Sciences (NIHS) publish lists of NTI/CD drugs.
- South African Medicine Control Council publishes a list of "bioproblem" drugs that include NTI/CD drugs.
- European Agency for the Evaluation of Medicinal Products (EMA) and Australian Therapeutic Goods Administration to date have not published a list of NTI/CD drugs.
- Canada—Health Canada, usual BE acceptance criteria:
  - AUC—90% CI of *T*/*R* ratio should fall within 80%–125%.
  - C_{max}—*T*/*R* point estimate should fall within 80%–125%.
  - Recommended BE acceptance criteria for generic CD drugs
    - Both AUC and C_{max}—90% CI of *T*/*R* ratios should meet acceptance criteria.
    - AUC—90%–112%.
    - C_{max}—80%–125%.
- Drugs considered NTI are cyclosporine, digoxin, flecainide, lithium, phenytoin, sirolimus, theophylline, and warfarin.

TABLE 5.5
Classification of Active Ingredients According to Their Health Risk

Active Ingredient	Health Risk
Acetazolamide	1
Allopurinol	1
Calcium folinate	1
Captopril	1
Clomifene	1
Cloxacillin	1
Dexamethasone	1
Diazepam	1
Folic acid + ferrous sulfate	1
Ibuprofen	1
Isosorbide dinitrate	1
Levamisole	1
Mebendazole	1
Mefloquine	1
Nalidixic acid	1
Niclosamide	1
Nifedipine	1
Nystatin	1
Phenoxymethylpenicillin	1
Phytomenadione	1
Pirantelo	1
Praziquantel	1
Pyrazinamide	1
Sulfasalazine	1
Amiloride	2
Amitriptyline	2
Amoxicillin	2
Atenolol	2
Azathioprine	2
Biperiden	2
Chloramphenicol	2
Cimetidine	2
Ciprofloxacin	2
Clofazimine	2
Clomipramine	2
Chlorpromazine	2
Co-trimoxazole	2
Cyclophosphamide	2
Dapsone	2
Diethylcarbamazine	2
Doxycycline	2
Erythromycin	2
Ethinyl estradiol	2
Etoposide	2
Flucytosine	2
Fludrocortisone	2
Furosemide	2

(Continued)

TABLE 5.5 (*Continued*)
Classification of Active Ingredients According to Their Health Risk

Active Ingredient	Health Risk
Haloperidol	2
Hydrochlorothiazide	2
Indometacin	2
Isoniazid	2
Ketoconazole	2
Levodopa + inhib. DDC	2
Levonorgestrel	2
Levotiroxina	2
6-Mercaptopurine	2
Methotrexate	2
Methyldopa	2
Metoclopramide	2
Metronidazole	2
Nitrofurantoin	2
Noretisterona	2
Oxamniquine	2
Paracetamol	2
Penicillamine	2
Piperazine	2
Piridostigmina	2
Procarbazine	2
Promethazine	2
Propranolol	2
Propylthiouracil	2
Pyrimethamine	2
Quinine	2
Rifampicin	2
Salbutamol, sulfate	2
Spironolactone	2
Tamoxifen	2
Tetracycline	2
Carbamazepine	3
Cyclosporine	3
Digoxin	3
Ethambutol	3
Ethosuximide	3
Griseofulvin	3
Lithium carbonate	3
Oxcarbazepine	3
Phenytoin	3
Procainamide	3
Quinidine	3
Theophylline	3
Tolbutamide	3
Valproic acid	3
Verapamil	3
Warfarin	3

European Union—EMA, usual BE acceptance criteria:

- Both AUC and C_{max}—90% CI of *T*/*R* ratios should fall within 80%–125%.
- Recommended BE acceptance criteria for generic NTI drugs.
- Both AUC and C_{max}—the usual 80%–125% acceptance interval "may need to be tightened."
- Has no listing of NTI drugs.

Japan—NIHS, usual BE acceptance criteria:

- Both AUC and C_{max}—90% CI of *T*/*R* ratios should fall—125%.
- Recommended BE acceptance criteria for generic NTI drugs.
- No change in acceptance criteria for AUC and C_{max}; however, if dissolution profiles of lower strengths of modified-release NTI drugs are not "equivalent" (f_2 analysis) to corresponding reference product profiles, then in vivo studies must be done (no biowaivers).
- A list of 26 NTI drugs includes digoxin, lithium, phenytoin, tacrolimus, theophylline, and warfarin and adds others such as carbamazepine, ethinyl estradiol, and quinidine.

South Africa—Medicines Control Council (MCC), usual BE acceptance criteria:

- AUC—90% CI of *T*/*R* ratio should fall within 80%–125%.
- C_{max}—90% CI of *T*/*R* ratio should fall within 70%–133%.
- Recommended BE acceptance criteria for generic NTI drugs
    - Both AUC and C_{max}—90% CI of *T*/*R* ratios should fall within 80%–125%.
- A list of "bioproblem" drugs (37) includes NTI drugs; substitution guideline states "unless adequate provision is made for monitoring the patient during the transition period, substitution should not occur when prescribing and dispensing generic mediations that:
    - Have a narrow therapeutic range;
    - Have been known to show erratic intra- and interpatient responses;
    - Are contained in dosage forms likely to give rise to clinically significant bioavailability problems (i.e., superbioavailability); or
    - Are intend for the critically ill and/or geriatric or pediatric patient"

Australia—Therapeutic Goods Administration (TGA):

- Follows EMA guidelines for usual BE acceptance criteria and recommended BE criteria for generic NTI drugs
- Has no list of NTI drugs

All five regulatory agencies discussed request log transformation of AUC and C_{max} data for BE statistical analysis. There is an ongoing difference of opinion among health-care providers, scientists, regulatory agencies, pharmaceutical companies, and consumer advocates whether the current BE criteria are appropriate for all drugs, specifically whether CD drugs require special consideration. Some have expressed concern that bioequivalent generic and brand-name CD drugs may not be equivalent in their effects on various clinical parameters. A retrospective analysis of the PK data from over 2000 studies showed that the average percent differences between generic and innovator geometric means was 4.35%, AUC average difference 3.56%; 98% of BE studies AUC differed by <10%. However, there are potential risks of applying current BE standards to CD drugs since they are based on the assumption that 20% deviation of plasma concentration is not clinically significant. For CD drugs, a 20% fluctuation in plasma concentration may be significant. The CD drugs often have low variability, 90% CI could actually be 85%–90% CI or 115%–120% CI, and the CI close to the boundary is highly associated with uncertainty of therapeutic equivalence. Switching between 85%–90% and 115%–120% CI generic products allowed. Potential future changes at TGA may include

- Point estimate on AUC within 90%–111%
- Point estimate on AUC within 95%–105%
- 90% CI on AUC within 90%–111%
- 90% CI on AUC within 95%–105%
- 90% CI must include 100%
- Replicate design study with limits set by variability of the reference product

Tighter BE limits (reducing the range) ensure smaller differences in mean BA; however, the differences in variability between products are not addressed by tightening the limits as the sources of product variability include formulation design and manufacturing quality (dose uniformity). A replicate design study provides variability quantification of test and reference products by comparing *T/R* distribution to *R/R*. The generic product should not be more variable than the reference product. Reference scaling includes models such as mixed scaling for highly variable drugs where the minimum three-period (2 *R*, 1 *T*) study with CV > 30% are scaled.

Clarification of Testing Limits

Clarification on Requirements

After the revision of the Note for Guidance (NfG) on the Investigation on Bioavailability and Bioequivalence in 2002 (http://www.emea.eu.int/pdfs/human/ewp/140198en.pdf), it appears that some harmonization in the interpretation of critical parts of the guideline is needed.

In which cases is it allowed to use a wider acceptance range for the ratio of C_{max}?

The NfG states under 3.6.2 that "With respect to the ratio of C_{max} the 90% confidence interval for this measure of relative bioavailability should lie within an acceptance range of 0.80–1.25. In specific cases, such as a narrow therapeutic range, the acceptance interval may need to be tightened."

The NfG also states that "In certain cases a wider interval may be acceptable. The interval must be *prospectively defined*, e.g. 0.75–1.33, *and justified* addressing in particular any safety or efficacy concerns for patients switched between formulations."

The possibility offered here by the guideline to widen the acceptance range of 0.80–1.25 for the ratio of C_{max} (not for AUC) should be considered exceptional and limited to a small widening (0.75–1.33). Furthermore, this possibility is restricted to those products for which at least one of the following criteria applies:

1. Data regarding *PK–PD relationships* for safety and efficacy are adequate to demonstrate that the proposed wider acceptance range for C_{max} does not affect PDs in a clinically significant way.
2. If PK–PD data are either inconclusive or not available, clinical safety and efficacy data may still be used for the same purpose, but these data should be specific for the compound to be studied and persuasive.
3. The reference product has a highly variable within-subject BA. Please refer to the question on highly variable drug or drug products for guidance on how to address this issue at the planning stage of the BE trial.

A post hoc justification of an acceptance range wider than defined in the protocol cannot be accepted. Information that would be required to justify results lying outside the conventional acceptance range at the post hoc stage should be utilized at the planning stage, either for a scientific justification of a wider acceptance range for C_{max} or for selecting an experimental approach that allows the assessment of different sources of variability.

When can subjects classified as outliers be excluded from the analysis in BE studies?

Under 3.6.3, the NfG states that "Post-hoc exclusion of outliers is generally not accepted" but at the same time acknowledges that "the protocol should also specify methods for identifying biologically implausible outliers."

Unbiased assessment of results from randomized studies requires that all subjects are observed and treated according to the same rules that should be independent from treatment or outcome. In consequence, PK data can only be excluded based on nonstatistical reasons that have been either defined previously in the protocol or, at the very least, established before reviewing the data. Acceptable explanations to exclude PK data or to exclude a subject would be protocol violations like vomiting, diarrhea, and analytical failure. The search for such explanations must apply to all subjects in all groups independently of the size of the observed PK parameters or its outlying position. Exclusion of data can never be accepted on the basis of statistical analysis or for PK reasons alone, because it is impossible to distinguish between formulation effects and PK effects.

Exceptional reasons may justify post hoc data exclusion, but this should be considered with utmost care. In such a case, the applicant must demonstrate that the condition stated to cause the deviation is present in the outlier(s) only and absence of this condition has been investigated using the same criteria for all other subjects.

Results of statistical analyses with and without the group of excluded subjects should be provided.

If one side of the 90% CI of a PK variable for testing BE lies on 0.80 or 1.25, can we conclude that the products are bioequivalent?

For establishing BE, the 90% CI should lie *within* the acceptance interval (in most cases, 0.80–1.25), the borders being included. The conclusion that products are bioequivalent is based on the overall scientific assessment of the PK studies, not only on meeting the acceptance range.

In which cases may a nonparametric statistical model be used?

The NfG states under 3.6.1—statistical analysis: "AUC and C_{max} should be analysed using ANOVA after log transformation."

The reasons for this request are the following:

a. The AUC and C_{max} values as biological parameters are usually not normally distributed.
b. A multiplicative model may be plausible.
c. After log transformation, the distribution may allow a parametric analysis.

However, the true distribution in a PK data set usually cannot be characterized due to the small sample size, so it is *not recommended* to have the analysis strategy depend on a pretest for normality. Parametric testing using analysis of variance (ANOVA) on log-transformed data should be the rule. Results from nonparametric statistical methods or other statistical approaches are nevertheless welcome as sensitivity analyses. Such analyses can provide reassurance that conclusions from the experiment are robust against violations of the assumptions underlying the analysis strategy.

For t_{max}, the use of nonparametric methods on the original data set is recommended.

When should metabolite data be used to establish BE?

According to the guideline, the only situations where metabolite data *can be used* to establish BE are the following:

- "If the concentration of the active substance is too low to be accurately measured in the biological matrix, thus giving rise to significant variability." *Comments:* Metabolite data can only be used if the applicant presents convincing, state-of-the-art arguments that measurements of the parent compound are unreliable. Even so, it is important to point out that C_{max} of the metabolite is less sensitive to differences in the rate of absorption than C_{max} of the parent drug. Therefore, when the rate of absorption is considered of clinical importance, BE should, if possible, be determined for C_{max} of the parent compound, if necessary at a higher dose. Furthermore, when using metabolite data as a substitute for parent drug concentrations, the applicant should present data supporting the view that the parent drug exposure will be reflected by metabolite exposure.

- "If metabolites significantly contribute to the net activity of an active substance *and* the pharmacokinetic system is nonlinear." *Comments*: To evaluate the significance of the contribution of metabolites, relative AUCs and nonclinical or clinical PD activities should be compared with those of the parent drug. PK–PD modeling may be useful. If criteria for significant contribution to activity and PK nonlinearity are met, then "it is necessary to measure both parent drug and active metabolite plasma concentrations and evaluate them separately." Any discrepancy between the results obtained with the parent compound and the metabolites should be discussed based on relative activities and AUCs. If the discrepancy lies in C_{max}, the results of the parent compound should usually prevail. Pooling of the plasma concentrations or PK parameters of the parent drug and its metabolite for calculation of BE is not acceptable.

When using metabolite data to establish BE, may one use the same justification for widening the C_{max} acceptance criteria as in the case of the parent compound?

In principle, the same criteria apply as for the parent drug (see question on widening the acceptance range for C_{max}). However, as stated earlier (see question regarding when metabolite data can be used), C_{max} of the metabolite is less sensitive to differences in the rate of absorption than C_{max} of the parent drug. Therefore, widening the C_{max} acceptance range when using metabolites instead of the parent compound is generally not accepted. When the metabolite has a major contribution to, or is completely responsible for, the therapeutic effect and if it can be demonstrated that a widened acceptance range would not lead to any safety or efficacy concerns, which will usually prove more difficult than for the parent compound (see question on widening the acceptance range for C_{max}), then a widened acceptance range for C_{max} of metabolite may be accepted.

What is a "highly variable drug or drug product"?

The standard approach to the analysis of a two-treatment, two-sequence, two-period crossover trial is an ANOVA for the log-transformed PK parameters, where the factor formulation, period, sequence, and subject nested within sequence are used to explain overall variability in the observations. The residual CV is a measure of the variability that is unexplained by the aforementioned factors. Among others, within-subject variability, formulation variability, analytical errors, and subject by formulation interaction can contribute to this residual variance.

A drug product is called highly variable if its intraindividual (i.e., within subject) variability is greater than 30%. A high CV as estimated from the ANOVA model is thus an indicator for high within-subject variability. However, a replicate design is needed to assess within-subject variability.

When testing for BE of a product with a nonlinear PK, how should one select the strengths with the largest sensitivity to detect differences in the two products?

Section 5.4 of the guideline states "If a new application concerns several strengths of the active substance a bioequivalence study investigating only one strength may be acceptable" provided five conditions are fulfilled, among which, when PK is not linear over the therapeutic dose range: "the strengths where the sensitivity is largest to identify differences in the two products should be used." Nonlinear PK, in this case, should reflect a nonlinear drug input rate as stated in the guideline.

Generally, it is the studied dose and not the studied formulation strength that is of importance when considering BE for drugs with nonlinear PK characteristics. An exception is when BA is governed by the solubility of the active ingredient. Then BE studies should include the highest formulation strength.

When studies are warranted at the high-dose range, they should be performed at the highest commonly recommended dose. If this dose cannot be administered to volunteers, the study may need to be performed in patients. If the study is conducted at the highest acceptable dose in volunteers, the applicant should justify this and discuss how BE determined at this dose can be extrapolated to the highest commonly recommended dose.

When proof of linear absorption or elimination kinetics is lacking or if evidence of nonlinearity is available, BE between test and reference formulations should be established with both the lowest and the highest doses unless adequately justified by the applicant. This approach is the most sensitive for detecting differences in rate and extent of absorption for substances with dose-dependent PKs. On the other hand, if only one dose is chosen in the BE studies, which dose to choose depends on the cause of nonlinearity. For instance, single-strength studies may be conducted:

- On the highest dose for drugs with a demonstrated greater than proportional increase in AUC or C_{max} with increasing dose during single- or multiple-dose studies. In this case, an additional steady-state study may be needed if the drug accumulates (steady-state concentrations are higher than those reached after single-dose administration).
- On the lowest dose (or a dose in the linear range) for drugs with a demonstrated less than proportional increase in AUC or C_{max} with increasing dose, for example, if this phenomenon is due to saturable absorption.

When BA of a substance with nonlinear PK is governed by the solubility of the active substance, resulting in a less than proportional increase in AUC with increasing dose, BE should be established with both the lowest and the highest dose (which may exceed the recommended initial dose) and should include the highest formulation strength.

It is worth mentioning that in case of linear kinetics but low or critical solubility, there is a similar need to test the highest strength and dose.

What are the conditions for using urinary PK data for BE assessment?

Section 3.3 of the guideline states "The use of urinary excretion data may be advantageous in determining the extent of drug input in case of products predominantly excreted renally, but has to be justified when used to estimate the rate of absorption."

The extent of drug input may be determined by the use of urinary excretion data provided elimination is dose linear and is predominantly renal as intact drug. However, the use of urinary data has to be carefully justified when used to estimate the *rate of absorption*. If a reliable plasma C_{max} can be determined, this should be combined with urinary data on the extent of absorption for assessing BE.

Standardization of BE studies with regard to food intake. How strictly should the guideline be interpreted?

Section 3.2.2 of the guideline states "If the Summary of Product Characteristics (SPC) of the reference product contains specific recommendations in relation to food intake related to food interaction the study should be designed accordingly."

The recommendations concerning food intake in the SPC are not sufficient for regulatory decisions on the adequacy of BE studies. Preferably, the following conditions should be considered separately when the SPC recommends administration of the substance together with food intake:

- If the recommendation of food intake in the SPC is based on PK properties such as higher BA, then a BE study under fed conditions is generally required.
- If the recommendation of food intake is intended to decrease adverse events or to improve tolerability, a BE study under fasting conditions is considered acceptable although it would be advisable to perform the study under fed conditions.
- If the SPC leaves a choice between fasting and fed conditions, then BE should preferably be tested under fasting conditions as this situation will be more sensitive to differences in PKs.

The composition of the meal should be described and taken into account, since a light meal might sometimes be preferable to mimic clinical conditions, especially when the fed state is expected to be less sensitive to differences in PKs. However, for modified-release products, a high-fat meal is required.

For products with release characteristics differing from conventional immediate release (e.g., improved release, dissolution, or absorption), even if they cannot be classified as modified-release products with prolonged or delayed release, BE studies may be necessary in both the fasted and fed states.

Typical Examples of Complex Bioequivalence

Digoxin

Digoxin in tablet form is not listed in the Orange Book, since this is a "grandfathered" dosage form of digoxin. Since the tablet formulation of digoxin was established in clinical use before the passage of the FD&C Act of 1938, generic versions of digoxin tablets may be marketed without an approved ANDA. Data showing BE of generic digoxin tablet products to the innovator product Lanoxin are generally not available or forthcoming, so that comparable rate and extent of absorption between generic products and Lanoxin brand tablets, or between different generic products, are not ensured. Seventeen generic digoxin tablets (0.25 mg) have been listed as currently marketed, though some of these may be marketed by suppliers or distributors of another manufacturer's product. Without PK data to verify the BE of these products to Lanoxin, the clinical responses (both therapeutic and toxic) from these generic products compared with Lanoxin are unpredictable. This inability to guarantee therapeutic equivalence to a reference product opposes the entire premise of generic substitution: the practitioner should expect the same responses (no more, no less) from a therapeutically equivalent generic product. Consequently, generic substitution is not advised. Use of a generic digoxin product as initial therapy may result in lower or higher than expected BA, requiring additional monitoring and dosage adjustment and ultimately increasing costs of therapy far above the cost savings from a less expensive generic product.

Levothyroxine

Levothyroxine sodium tablets are also currently not listed in the Orange Book. In the words of the FDA, levothyroxine sodium was first introduced into the market before 1962 without an approved NDA, apparently in the belief that it was not a new drug. The lack of BE data of generic preparations to the two major brand-name products Synthroid and Levothroid has been noted, along with the adoption in 1984 of *USP* guidelines for potency of levothyroxine sodium tablets. However, between 1987 and 1994, a total of 58 adverse drug experience reports with levothyroxine sodium tablets were received by the FDA, with 47 of the incidences apparently related to subpotency and 9 incidences related to superpotency. These adverse events were caused not only by switching product brands but also by inconsistencies in BA between different lots from the same source. BE issues regarding levothyroxine sodium tablets were highlighted when the results of a BE study comparing the innovator product Synthroid with several generic brands finally appeared in the literature. The study sponsor (the marketer of Synthroid) attempted to prevent publication of these results, which claimed BE of Synthroid to three other levothyroxine sodium products. After publication of these study results, advertisements appeared in journals and trade magazines advocating the substitution of other brand-name levothyroxine sodium products (e.g., Levothroid, Levoxyl) for Synthroid. In addition, statements were made such as "Feel comfortable using Levothroid, Levoxyl, or Synthroid in hypothyroid patients. These three are bioequivalent…even though they're not AB-rated."

Several points should be considered before routinely switching marketed brands of levothyroxine sodium tablets (at least 24 products for the 0.1 mg tablet are listed). First, although the conclusions stated in the peer-reviewed BE study cited appear to be generally accepted, the results of this study were not subjected to the scrutiny of the FDA review process. In view of significant stability and potency problems, the FDA has issued a Federal Register notice stating that (1) orally administered levothyroxine sodium products are now considered new drugs and (2) manufacturers who intend to continue marketing these products must submit an NDA within 3 years to obtain approval. Recently, the FDA extended this deadline for an additional year. Second, the impression that all levothyroxine sodium tablet formulations are likely to be bioequivalent is not currently supported with FDA-substantiated BE data; routine substitution of these products for refills of existing prescriptions is not advisable until FDA review is complete. Third, practitioners must always comply with the substitution laws in their individual states. If a statute mandates substitution of a therapeutically equivalent or bioequivalent product, reliance upon data reported in the scientific literature may not always guarantee these requirements will be satisfied.

Warfarin Sodium

Three approved generic versions of warfarin sodium tablets (seven strengths) are currently listed in the Orange Book. Before approval of these generic warfarin sodium products, several states either enacted or were considering legislation to require pharmacists to obtain prescriber and patient approval for generic substitution of drugs with an NTI. In response, the FDA issued a position statement. The FDA's position is clear with regard to the issue of tightening CIs and changing study designs for BE determinations of NTI drugs: the present requirements to prove BE, at least in the United States and Canada, are already so difficult and constrained that there is no possibility, even for NTI drugs, that dosage forms meeting the criteria could lead to therapeutic problems. Drugs approved through the NDA process with NTIs, by definition, must have low intrasubject variability. Otherwise, patients would have cycles of toxicity and lack of efficacy, and therapeutic drug monitoring would be useless. The low intrasubject variability associated with NTI drugs ensures that patient response to a specific drug should be consistent and the statistical criteria required by the FDA for BE appear more than adequate for confidence in generic substitution. This is especially true in light of the notable absence of data that prove otherwise. For the most part, the arguments against generic substitution of NTI drugs appear to be based on economic considerations. Commentaries debating the suitability of generic warfarin products have focused on the results from reports of clinical studies with generic warfarin and the content uniformity requirements for warfarin sodium tablets. As indicated in a letter addressing these issues, no convincing and substantiated scientific data have been published showing bioinequivalence of generic warfarin products or product failure of these products in clinical studies. Recently, an evidence-based medicine approach was used to compare the results reported with Coumadin and a generic warfarin product in clinical studies. No significant differences were found in the international normalized ratio (INR), number of dosage changes to adjust INR in range, or number of hospitalizations or incidences of bleeding between the reference and generic warfarin products. Physicians may sometimes encounter difficulties in maintaining stabilized INR in patients anticoagulated with warfarin, since multiple drug interactions and patient variables affect warfarin levels and create difficulty in achieving consistently therapeutic INR values. However, factors such as diet, concurrent illnesses, interacting drugs, and noncompliance are *intersubject* variables that are unrelated to the BE issue. For crossover studies using log-transformed data, it is largely the within-subject distribution of values (*intrasubject* variability) that determines the validity and efficiency of the standard parametric methods of analysis. For NTI drugs such as warfarin, intrasubject variability, by definition, is low, and the available clinical data indicate that lack of BE does not appear to be the explanation for problems experienced during warfarin therapy. Another article introduces the concept of "switchability," that is, the substitution of one approved generic product for another generic product. BE studies submitted to the FDA through an ANDA are conducted by comparing data from the proposed generic product and a reference product. The reference product is selected by the FDA and is typically the innovator or pioneer product that was originally introduced into the market. Suppose approved generic product A differed from the reference product in at least one parameter (e.g., mean AUC values) by +4% and that approved generic product B differed from the reference product by −4%. The net difference of generic products A and B would then be 8%; could this magnitude of difference result in bioinequivalence and lack of equivalent therapeutic response for an NTI drug? No data were presented from any clinical studies that could support the contention that switchability for NTI drugs is problematic. Rather, phrases such as "… with NTI drugs, small variations in BA can potentially pose problems" and conceptual arguments are used to suggest the need for special BE criteria to be applied to NTI drugs. Reference is made to the FDA's draft guidance for population and individual BE studies, which proposes the use of reference scaling (essentially, modifying the BE criteria to account for the variability of the reference product) for NTI drugs, regardless of the intrasubject variability of the reference product. Since NTI drugs have low intrasubject variability as discussed, this approach would likely result in narrower CI requirements. Finally, a recent report further confirms the BE of generic warfarin to the innovator product. More than 100 subjects anticoagulated with Coumadin were switched to a generic warfarin product for 8 weeks in a nonrandomized comparative clinical observational study. The overall conclusion was that the variability in INR in patients receiving generic warfarin was not statistically significant from that seen in the control group receiving Coumadin. These investigators identified associated factors not related to the product change in subjects whose INR varied by >1.0 from baseline. This further emphasizes the critical role of interpatient factors (physical activity, dietary vitamin K, noncompliance, drug interactions, congestive heart failure, diarrhea, alcohol consumption) affecting the anticoagulant response with warfarin.

Albuterol Metered-Dose Inhalers

Four approved generic versions of albuterol metered-dose inhalers are currently listed in the Orange Book as therapeutically equivalent (AB-rated) to the reference product Ventolin. The Proventil product is rated BN, or not therapeutically equivalent to Ventolin or the four generic products. For products administered by metered-dose inhalation and intended for local therapeutic effects, the typical PK methods for evaluating BE cannot be used. Rather, an approach based on acute PD response (forced expiratory volume in 1 s [FEV_1]) was proposed, with asthmatic patients as subjects. The statistical criteria and appropriate CIs for BE determination are not as rigidly defined for PD methods as for PK methods. Consequently, variability in patient response may be of slightly greater concern, since albuterol metered-dose inhalers are used as "rescue inhalers" for nocturnal asthma attacks (even though they are not considered NTI drugs). However, the FDA is satisfied that these products will produce equivalent therapeutic responses.

BIOEQUIVALENCE SURROGATES

It is not always possible or necessary to utilize in vivo human data in the evaluation of drug BA. The large number of physicochemical, physiological, and pharmaceutical factors that affect absorption of drugs need to be studied in detail, and thus, models have to be designed to study the effect of many of these factors with aim to optimize absorption. These models are primarily classified as in vitro, in situ, and in vivo.

In Vitro Systems

Several systems have been developed to simulate absorption of drugs across biological membranes, especially GI membranes in vitro. These models pertain either to the membrane permeability aspect or the release of drug from dosage forms.

Disintegration of Dosage Form

It is one of the most widely used test for the release of drugs from dosage forms. In most situations, poor correlations are found between the disintegration and absorption of drugs. Scheme 1.1 shows the various rate-limiting steps in the absorption of drugs, and it is evident that this test alone does not assure proper release and absorption characteristics. The official disintegration test is performed by placing one tablet each in the six tubes placed in a basket that is moved up and down in the immersion fluid at frequency rate between 28 and 32 cycles/min through a distance of not less than 5 cm and not more than 6 cm. For uncoated, buccal, and sublingual tablets, the immersion fluid is water at 37°C. However, for the latter two types of tablets, the disk placed on top of the tablets is omitted. For coated tablets, the immersion fluid is gastric fluid for 30 min followed by intestinal fluid. If one or two tablets fail to disintegrate within the specified time, then the test is repeated on 12 additional tablets; not less than 16 of the total of 18 tablets must disintegrate completely. The most common disintegration requirement is 30 min for uncoated tablets.

Dissolution Tests

Dissolution testing is performed not only on the finished products but also on the pure drug and in combination with various excipients to ascertain individual contributions of the components to overall dissolution. Basically, the dissolution test systems are of two types: the stirred-vessel type and the flow-through column. In the stirred type, agitation is provided by some kind of paddle, whereas in the column type, the solvent flows over the drug. A large number of variations of these systems are currently used. However, the USP apparatus is used for official certification of batches. The monographs describe the specific temperature, the dissolution medium (distilled water, simulated gastric fluid, or simulated intestinal fluid), the rotation speed of the basket (60–150 rpm), and the percentage of drug to be dissolved as an endpoint. These conditions are determined by the intrinsic properties of the drug and its dissolution behavior. Appendix B lists the drugs for which a dissolution test is required along with dissolution conditions as accepted by the FDA. Some key examples include acetohexamide, nitrofurantoin, digoxin, phenylbutazone, ergotamine tartrate and caffeine tablets, prednisolone, hydrochlorothiazide, prednisone, lithium carbonate, sulfamethoxazole, meprobamate, sulfisoxazole, methaqualone, theophylline, ephedrine hydrochloride and phenobarbital, methylprednisolone tablets, and tolbutamide.

In the official dissolution tests, 6 or 12 tablets or capsules are tested individually for their dissolution properties. In the first stage, 6 units are tested, and each unit must fall to within less than 5% of the specified limit (e.g., 60% dissolved in 30 min). If one or more units fail, then another 6 units are tested, and the average of 12 units (6 from first test) should be equal to or greater than the specified percentage, and no unit should be less than 15% of the specified limit. If this stage also fails, then additional 12 units are tested and the average of all (now 24) should be equal to or greater than the specified limit, and not more than 2 units can be less than 15% off the limit.

An inherent problem in this type of testing is that it requires the use of labeled amount of drug for calculation purposes, and any content variability is not considered. Conceivably, large variations in the dissolution rates are possible due to these differences, for example, tablets containing 80%–120% of the labeled amount will require 75%–50% dissolution if the requirement is 60%. In addition, the statistical design of the dissolution testing allows a batch with 20% defective tablets to pass 58% of the time. There are also serious problems in the reproducibility of dissolution data since the dissolution is dependent on human errors and such subtle factors as the vibrations in the room. Despite these drawbacks, FDA considers dissolution testing to be the most discriminating in vitro test with which to establish in vivo correlations.

In vitro dissolution profile comparison: model-independent approach using similarity factor. Dissolution profiles may be compared using the following equation that defines a similarity factor (f_2): $f_2 = 50$ LOG {[1 + 1/n 3n ($R - T$)^2]−0.5 × 100} 2 t = 1 t t, where LOG = logarithm to base 10, n = number of sampling time points, 3 = summation over all time points, Rt = dissolution at time point t of the reference (unchanged drug product, i.e., prechange batch), and Tt = dissolution at time point t of the test (changed drug product, i.e., postchange batch). For comparison of multipoint dissolution profiles obtained in multiple media, similarity testing should be performed using pairwise dissolution profiles (i.e., for the unchanged and changed product) obtained in each individual medium.

It is recommended that only one point past the plateau of the profiles be used in calculating the f_2 value. A correction for a lag time prior to similarity testing should not be performed unless justified.

An f_2 value between 50 and 100 suggests the two dissolution profiles are similar. Also, the average difference at any dissolution sampling time point should not be greater than 15% between the changed drug product and the biobatch or marketed batch (unchanged drug product) dissolution profiles. An appropriate reference for this comparison should represent an average dissolution profile derived from at least three consecutive recent batches of the unchanged drug product (biobatch or marketed batch). Finally, the dissolution data obtained under the application/compendial dissolution testing conditions (media, agitation, etc.) on both the changed drug product and the biobatch or marketed batch (unchanged drug product) should be within the application/compendial specifications.

An f_2 value less than 50 does not necessarily indicate lack of similarity. If the sponsor is of the opinion that the differences observed related to this calculation of f_2 are typical for the drug product involved in this scale-up and post approval change (SUPAC) situation, an appropriate justification can be submitted, but only as part of a prior approval supplement. This justification should include additional data to support the claim of similarity, as well as supporting statistical analysis (e.g., 90% CI analysis). If this justification is not found acceptable, the potential effect of the proposed change on the differences in dissolution on BA should be determined.

Dissolution profiles can also be compared using other model-independent or model-dependent methods.

Everted Intestinal Sac

Whereas disintegration and dissolution tests characterize the release characteristics of the products, the transport across biological membrane is studied by everted intestine method. The procedure involves isolating a small segment of the intestine of a laboratory animal such as hamster, guinea pig, and rat, everting the segment, filling the sac, and ligating it at both ends after filling it with a small volume of buffer. The ligated intestinal sac is incubated in oxygenated buffer solution containing drug at 37°C. The eversion of intestine helps expose the mucosal side to oxygenation, whereas the small volume on the serosal side allows analysis of low concentration of drugs. Several modifications of this concept in use include the one that allows multiple sample of the serosal side or replacement of membrane during experiment. The permeability characteristic is expressed in terms of the lag time before any drug appears on the serosal side and the cumulative amount of drug transferred in 60 min/unit concentration of the drug in mucosal solution. A large lag time or small transport characterizes drugs where absorption may be limited by the transport process itself. When these data are coupled with release characteristics such as dissolution test, more useful information can be obtained regarding possible hindrance in the absorption of drugs.

The everted sac method can also be utilized to study the effect of various formulation additives on absorption of the active drug. Several such examples have been reported including reduced absorption of chloramphenicol by adjuvants, increase absorption rates of drugs due to *N*-methyl glucamin, effect of surfactants on the permeability of soluble corticosteroids, effect of physiological surfactants on the permeability of salicylates, and the effects of complex formation on the permeability of salicylamide. Other factors studied include differentiation between active and passive transport, potential binding of drugs to the intestinal wall, GI metabolism, and the effects of electrolytes and sugars on membrane permeability of drugs. The GI metabolism of drugs has also been studied by using everted intestinal ring and slices.

Isolated Perfused Liver

Many drugs undergo metabolism in the liver before reaching the general circulation, and thus the BA is reduced. In order to estimate the extent of the loss of drug through this first-pass effect, drug solution is perfused through the liver that is maintained in physiologically active state by oxygenation and providing nutrients. Excellent in vitro–in vivo correlation for drug metabolism has been obtained using this technique. Some examples of drugs where such experiments have been of great value include aspirin, salicylamide, propranolol, acetaminophen, and phenacetin.

In Situ Methods

The in situ system better represents the in vivo systems since the blood supply to the absorption organs is maintained. It involves ligating segments of the GI tracts and perfusing drug solutions through the segment and recording the amount of drug lost as a function of time. Several procedures including that described by Levine et al. are widely used. These models are useful in characterizing the transport characteristics.

In Vivo Systems

LD50 Comparisons

Comparison of LD50 with intravenous route of administration or within a group of formulations provides a useful tool in determining BA. The advantages include quick results and no analysis of blood samples. This technique can be used to monitor both the rate and extent of absorption.

Thiry–Vella Loop

For chronic studies, a loop of small intestine is isolated and exteriorized with intact blood, nerve, and lymph supplies. With access to the proximal and distal ends of the intestinal segment, repeated use of the loop can be made for BA studies. This procedure, however, requires larger species such as dogs and offers a distinct advantage of running intravenous and GI absorption studies in the same animals.

Hepatobiliary Cannulation

Catheterization of hepatoportal vein to the outside allows direct administration of drug into the hepatic system, assuring 100% passage through the liver prior to entrance into the general circulation. This allows quick estimation whether the BA problem is due to GI factors or to postabsorption factors. Significantly lower levels following portal administration indicate a high degree of metabolism in the first pass through the liver. Cannulation of bile duct in vivo is of value in determining the extent of drug excreted in the GI tract and thus allows evaluation of degree of recirculation of the drug. A large number of polar compounds and compounds of high molecular weight are excreted in bile including digoxin, diazepam, pivampicillin, ampicillin, nitrofurantoin, dioctyl sodium sulfosuccinate, erythromycin, tetracyclines, and fluphenazine.

Choice of Animal Species

In the in vivo models suggested earlier, the choice of animal species depends on the factors including similarity of GI anatomy and physiology to humans. For example, cattle have a very different system of food and drug digestion and transport, and thus, it will not be a good choice. A good model is a dog wherein, like humans, no continuous secretion of hydrochloric acid and bile is recorded. It should be possible to administer the given dosage form. Many small species such as rats, mice, and hamsters cannot be used since it will not be possible to administer a full-size capsule or tablet to them. It should also be possible to obtain period biological samples such as blood or urine. Again, small species offer significant problem in this respect. Miniature pig seems to be an ideal species for this purpose.

Disposition kinetic characteristics should be as close to humans as possible. This is probably the most difficult factor to control. However, some species can be totally ruled out depending on the specific drug example if they metabolize, excrete, or distribute the drug differently. Also, if a species shows selective absorption of a drug in contrast to humans, it should also be ruled out. Monkey and other primates seem to meet many of these requirements. However, it is generally difficult to obtain primates for studies and cost a lot more too, a factor of great importance in the initial screening of the drugs and dosage forms.

Absorption Profiling

The following are factors and oral drugs/drug products that should be considered when requesting a waiver of evidence of in vivo BA or BE documentation. Generally, both in vivo and in vitro testing are necessary for orally administered drug products. In vivo testing is required for all generic drug products with certain exceptions. Based on scientific information, regulatory authorities may waive the requirement for BA or BE.

1. For certain formulations and under certain circumstances, equivalence between two pharmaceutical products may be considered self-evident, and no further documentation is required. For example:
   a. When multisource pharmaceutical or generic products are to be administered parenterally (e.g., intravenous, intramuscular, subcutaneous, intrathecal administration) as aqueous solutions and contain the same active substance(s) in the same concentration and the same excipients in comparable concentrations.
   b. When multisource pharmaceutical or generic products are solutions for oral use, contain the active substance in the same concentration, and do not contain an excipient that is known or suspected to affect GI transit or absorption of the active substance.
   c. Gas-based multisource pharmaceutical or generic products
   d. When the multisource pharmaceutical or generic products are powders for reconstitution as a solution and the solution meets either criterion (a) or criterion (b) previously.
   e. When multisource pharmaceutical or generic products are otic or ophthalmic products prepared as aqueous solutions, containing the same active substance(s) in the same concentration and essentially the same excipients in comparable concentrations.
   f. When multisource pharmaceutical or generic products are topical products prepared as aqueous solutions, containing the same active substance(s) in the same concentration and essentially the same excipients in comparable concentrations.
   g. When multisource pharmaceutical or generic products are inhalation or nasal spray products, tested to be administered with or without essentially the same device, prepared as aqueous solutions, and containing the same active substance(s) in the same concentration and essentially the same excipients in comparable concentrations. Special in vitro testing should be required to document comparable device performance of the multisource inhalation product.

   For elements (e), (f), and (g) previously, it is incumbent upon the applicant to demonstrate that the excipients in the multisource product are essentially the same and in comparable concentrations as those in the reference product.
2. In the event the applicant cannot provide this information about the reference product and the drug regulatory authority does not have access to these data or the data are protected under data exclusivity rights according to local regulations, in vivo studies should be performed.
3. For certain drug products, BA or BE may be demonstrated by evidence obtained in vitro in lieu of in vivo data. Regulatory authorities should waive the requirement for the submission of evidence obtained in vivo demonstrating the BA of the drug product if the drug product meets one of the following criteria:
   a. The drug product is in the same dosage form, but in a different strength, and is proportionally similar in its active and inactive ingredients to another drug product manufactured at the same site for which the same manufacturer has obtained approval and the following conditions are met:
   b. The BA of this other drug product has been demonstrated.

c. Both drug products meet an appropriate in vitro test approved by a drug regulatory authority and/or accepted reference pharmacopeias or have demonstrated in vivo–in vitro correlation (e.g., correlation level A).
d. The applicant submits evidence showing that both drug products are proportionally similar in their active and inactive ingredients. That is, the ratio of active ingredients and excipients between strengths is essentially the same.
e. The drug product is a reformulated product that is identical, except for a different color, flavor, or preservative that could not affect the BA of the reformulated product, to another drug product for which the same manufacturer has obtained approval and the following conditions are met:
f. The BA of the other product has been demonstrated.
g. Both drug products meet an appropriate in vitro test approved by the regulatory authority.
h. Regulatory authorities, for good cause, may require evidence of in vivo BA or BE for any drug product if the agency determines that any difference between the drug product and a listed drug may affect the BA or BE of the drug product. The Bioavailability and Bioequivalence Working Group strongly recommends that in the case of antiretroviral drug products, proof of pharmaceutical equivalence and BE be required to infer therapeutic equivalence.

STATISTICAL ANALYSIS

The U.S. FDA describes BE as the absence of a significant difference in the rate and extent to which the active ingredient or active moiety in pharmaceutical equivalents or pharmaceutical alternatives becomes available at the site of drug action when administered at the same molar dose under similar conditions in an appropriately designed study (21 CFR § 320.1). Evaluation of BE has gone through many significant changes over the past 40 years, and for drugs considered critical drugs, there has been controversy and scientific challenges to the position taken by the regulatory agencies.

The first rule for testing BE was the 75/75 (or 75/75–125) rule that was originally proposed in the late 1970s as an alternative means of testing the BE of two formulations of a pharmaceutical agent. The rule specified that the ratio of test-to-reference formulation of a BA measure arising in a BE study must be between 75% and 125% of unity in at least 75% of subjects to declare two formulations bioequivalent. This rule faced remarkable criticism in the literature. In the early 1980s, a "power approach" was applied to AUC and C_{max}. This approach consisted of two tests—test of null hypothesis of no difference between formulations and evaluation of the power of the test to detect a 20% mean difference in treatment. It was used in conjunction with the 75/75 rule at times, but the use of these methods discontinued by the agency in 1986. This rule was heavily criticized, and as a result, the U.S. FDA conducted a public hearing in 1986 and a Bioequivalence Task Force was formed to investigate the scientific issues raised at the hearing and the task force report was released in 1989, and subsequently, FDA issued guidance on statistical procedures for BE studies in July of 1992.

Current practice is to use a two one-sided tests procedure (also called 90% CI approach); this recognizes there will be a difference in mean values between treatments and provides reasonable assurance that mean treatment differences are acceptable. Before July 1992, 90% CIs for AUC and C_{max} had to be within the range of 80%–120% around the RLD mean value; as of July 1992, the statistical procedure guidance recommends CIs of 80%–125% for AUC and C_{max} after log transformation.

- BE criteria: Two one-sided tests procedure—test (T) is not significantly less than reference.
- Reference (R) is not significantly less than test.
- Significant difference is 20% ($\alpha = 0.05$ significance level)$T/R = 80/100 = 80\% R/T = 80\%$ (all data expressed as T/R so this becomes 100/80 = 125%).

The statistical models used in the evaluation of BE data have been evolving over the past few decades. The standard statistical method of null hypothesis was the first to be used where no difference is proved and rejection of null indicates statistically significant different ($p < .05$). A problem arises since small differences with $p < .05$ may be unimportant and large differences with $p > .05$ may be important. This prompted FDA to solve the problem by requesting power analysis CI test of Schuirmann where two one-sided comparisons are made; this also evolved in the use of the famous 75–125 rule to deal with individual effects.

FDA advocates the use of 90% CIs, as the best available method for evaluating BE study data. The CI approach should be applied to the individual parameters of interest (e.g., AUC and C_{max}). The sponsor may use untransformed or log-transformed data. However, the choice of untransformed or log-transformed data should be made by the sponsor with concurrence by FDA prior to conducting the study.

UNTRANSFORMED DATA

If we let $\bar{X}_{T1}$ be the mean for the test drug in period 1, $\bar{X}_{T2}$ the mean for the test drug in period 2, and $\bar{X}_{R1}$ and $\bar{X}_{R2}$ the respective means for the reference drug, then the estimates for the drugs averaged over both periods are $\bar{X}_T = (1/2)(\bar{X}_{T1} + \bar{X}_{T2})$ for the test drug and $\bar{X}_R = (1/2)(\bar{X}_{R1} + \bar{X}_{R2})$ for the reference drug. Although both sequence groups usually start with the same number of animals, the number of animals in each sequence group (n_A and n_B) that successfully finish the study may not be equal. The formulas earlier utilize the marginal or least squares estimates of μT and μR, the corresponding means in the target population. These means are not a function of the sample size in each sequence.

An ANOVA is needed to obtain the estimate of σ^2 the error variance. The estimator, s^2, which will be used in the calculation of the 90% CI, should be obtained from the "error" mean square term found in the following ANOVA table:

Source	Degrees of Freedom
Sequence	1
Animal (sequence)	$n_A + n_B - 2$
Period	1
Formulation	1
Error	$n_A + n_B - 2$
Total	$2n_A + 2n_B - 1$

Lower and upper 90% CIs are then found by formulas based on student's *t*-distribution:

$$L = \left(\bar{X}_T - \bar{X}_R\right) - t_{n_A} + n_B{}^{-2;0.05} s\sqrt{\frac{1}{2}\left(\frac{1}{n_A} + \frac{1}{n_B}\right)} \tag{5.7}$$

$$U = \left(\bar{X}_T - \bar{X}_R\right) + t_n + n^{-2;0.05} s\sqrt{\frac{1}{2}\left(\frac{1}{n_A} + \frac{1}{n_B}\right)} \tag{5.8}$$

The procedure of declaring two formulations bioequivalent if the 90% CI is completely contained in some fixed interval is statistically equivalent to performing two one-sided statistical tests ($\alpha = 0.05$) at the endpoints of the interval.

Consider the following example with $L = 3$, $U = 17$, $\bar{X}_T = 100$, and $\bar{X}_R = 100$. By the traditional hypothesis testing approach, the result would be considered statistically significant since the CI does not include 0. Using the CI approach, the entire CI lies within 17% of $\bar{X}_R$. (The lower end of the CI lies within $L/\bar{X}_R = 3/100 = 3\%$ of $\bar{X}_R$, while the upper end of the CI lies within $U/\bar{X}_R = 17/100 = 17\%$ of $\bar{X}_R$.) If it were determined by FDA that only differences larger than 20% were biomedically important, then using the CI approach, the results of this study would be considered adequate to demonstrate BE.

Now, consider an example with $L = -4$, $U = 24$, $\bar{X}_T = 110$, and $\bar{X}_R = 100$. In this case, by the traditional hypothesis testing approach, the result would not be considered statistically significant since the CI includes 0. However, the CI extends as far as 24% from $\bar{X}_R$. (The lower end of the CI lies within $L/\bar{X}_R = -4/100 = -4\%$ of $\bar{X}_R$, while the upper end of the CI extends to $U/\bar{X}_R = 24/100 = 24\%$ of $\bar{X}_R$.) If it were determined by FDA that only differences larger than 20% were biomedically important, then the results of this study would be considered inadequate to demonstrate BE, since the entire CI is not within 20% of $\bar{X}_R$.

LOGARITHMICALLY TRANSFORMED DATA

This section discusses how the 90% CI approach should be applied to log-transformed data. In this situation, the individual animal AUC and C_{max} values are log transformed, and the analysis is done on the transformed data. For a two-period crossover study, as described in D.1, the ANOVA model used to calculate estimates of the error variance and the least square means are identical for both transformed and untransformed data. The procedural difference that comes after the lower and upper 90% CIs are found by formulas based on student's *t*-distribution.

The lower and upper confidence bounds of the log-transformed data will then need to be back-transformed in order to be expressed on the original scale of the measurement. One thing to keep in mind when moving between the logarithm scale and the original scale is that the back-transformed mean of a set of data that has been transformed to the logarithm scale is not strictly equivalent to the mean that would be calculated from the data on the original scale of measurement. This back-transformed mean is known instead as the geometric mean.

It may help to see the calculations involved. If the AUC from each animal has been transformed to the logarithm scale, we can express the transformed AUC as LnAUC. Then, the mean on the logarithm scale is as follows:

$$\overline{Ln}\mathrm{AUC}_t = \sum_{i=1}^{n} \frac{Ln\mathrm{AUC}_t}{n} \tag{5.9}$$

where the subscript *t* represents the AUC determinations for the test article, *i* is the AUC of the *i*th animal, and *n* is the total number of animals receiving the text article. When this mean is back-transformed, it becomes the geometric mean $e^{(\overline{Ln}\mathrm{AUC}_t)}$. This geometric mean will be on the original scale of the measurement. It will be close to but not exactly equal to the mean obtained on the original scale of the measurement. The back-transformation of the confidence bounds is accomplished in the following way:

$$\text{Lower bound (expressed as a percentage)} = (e^{\mathrm{L}} - 1) \times 100$$

$$\text{Upper bound (expressed as a percentage)} = (e^{\mathrm{U}} - 1) \times 100$$

where

L is the lower 90% CI and calculated on the log-transformed data
U is the upper 90% CI and calculated on the log-transformed data

As an example, consider the data for AUC from a hypothetical crossover study in the following table:

		Reference Article		Test Article	
Animal	**Crossover Sequence**	**AUC**	**LogAUC**	**AUC**	**LogAUC**
1	1	518.0	6.25	317.8	5.76
2	1	454.9	6.12	465.0	6.14
3	1	232.8	5.45	548.4	6.31
4	1	311.1	5.74	334.8	5.81
5	2	340.4	5.83	224.7	5.41
6	2	497.7	6.21	249.2	5.52
7	2	652.0	6.48	625.4	6.44
8	2	464.1	6.14	848.7	6.74
	Mean	433.8	6.03	451.7	8602
	Standard deviation	133.3	0.33	214.3	047
	Geometric mean		414.7		

The statistics for AUC will be calculated from the log-transformed data. In this example, *L*, the lower 90% CI calculated on the log scale, is −0.395. *U*, the upper 90% CI calculated on the log scale, is 0.372. To back-transform these intervals and express them as percentages, we do the following:

Back-transformed lower bound:

$$(e - 0.395 - 1) \times 100 = (0.674 - 1) \times 100 = (-0.326) \times 100 = -32.6\%$$

Back-transformed upper bound:

$$e(0.372 - 1)^{s/s} \times 100 = (1.451 - 1) \times 100 = (0.451) \times 100 = 45.1\%$$

Therefore, the lower end of the confidence bound lies within −32.6% of the geometric mean of the reference article, while the upper end of the CI lies within 45.1% of the geometric mean of the reference article. If it were determined by FDA that the acceptable confidence bound was 80%–125% of the geometric mean of the reference article in order to demonstrate BE, then the back-transformed lower bound can be as low as −20%, and the back-transformed upper bound can be as high as 25%. In this example, we would determine that the study had not demonstrated an acceptable level of BE between the test article and the reference article.

The width of the CI is determined by the within-subject variance (between subject variance for parallel group studies) and the number of subjects in the study. In general, the CI for untransformed data should be 80%–120% (the CI should lie within ±20% of the mean of the reference product). For logarithmically transformed data, the CI is generally 80%–125% (the CI should lie within −20% to +25% of the mean of the reference product). The sponsor and FDA should determine the acceptable bounds for confidence limits for the particular drug and formulation during protocol development.

ANIMAL DRUG BIOEQUIVALENCE TESTING

A BE study may also be part of a NADA or supplemental NADA for approval of an alternative dosage form, new route of administration, or a significant manufacturing change that may affect drug BA. Many requirements described earlier for human studies also apply to animal studies; various descriptions of experimental design and data handling are common to both. FDA has concluded that the tissue residue depletion of the generic product is not adequately addressed through BE studies.

Therefore, sponsors of ANADAs for drug products for food-producing animals will generally be asked to include BE and tissue residue studies (21 U.S.C. 360b[n][1][E]). A tissue residue study should generally accompany clinical endpoint and pharmacological endpoint BE studies and blood level BE studies that cannot quantify the concentration of the drug in blood throughout the established withdrawal period (21 U.S.C. 360b[n][1][A][ii]). BE studies (i.e., blood level, pharmacological endpoint, and clinical endpoint studies) and tissue residue depletion studies should be conducted in accordance with good laboratory practice (GLP) regulations (21 CFR part 58). Whereas the focus of the guidance is BE testing for ANADA approval, the general principles also apply to relative BA studies conducted for NADAs.

REFERENCE PRODUCT

As a general rule, the proposed generic product should be tested against the original pioneer product. If the original pioneer product is no longer marketed, but remains eligible to be copied, then, the first approved and available generic copy of the pioneer should be used as the reference product for BE testing against the proposed new generic product.

If several approved NADAs exist for the same drug product and each approved product is labeled differently (i.e., different species and/or claims), then the generic sponsor must clearly identify which product label is the intended pioneer. BE testing should be conducted against the single approved product that bears the labeling that the generic sponsor intends to copy. The generic sponsor should consult with CVM (FDA) regarding selection of the appropriate reference product before conducting the BE study.

WAIVER OF IN VIVO BIOEQUIVALENCE STUDY

The requirement for the in vivo BE study may be waived for certain generic products (21 U.S.C. 360b[n][1][E]). Categories of products that may be eligible for waivers include, but are not limited to, the following:

- Parenteral solutions intended for injection by the intravenous, subcutaneous, or intramuscular routes of administration
- Oral solutions or other solubilized forms
- Topically applied solutions intended for local therapeutic effects. Other topically applied dosage forms intended for local therapeutic effects for nonfood animals only
- Inhalant volatile anesthetic solutions

In general, the generic product being considered for a waiver contains the same active and inactive ingredients in the same dosage form and concentration and has the same pH and physicochemical characteristics as an approved pioneer product.

However, the CVM will consider BE waivers for nonfood animal topical products with certain differences in the inactive ingredients of the pioneer and generic products.

If a waiver of the in vivo BE and/or the tissue residue study/studies is granted for a food animal drug product, then the withdrawal period established for the pioneer product will be assigned to the generic product. Sponsors may apply for waivers of in vivo BE studies prior to submission of the ANADAs.

SPECIES SELECTION

A BE study generally should be conducted for each species for which the pioneer product is approved on the label, with the exception of "minor" species (as defined in section 514.1 [d] [1] of title 21 of the Code of Federal Regulations) on the label.

SUBJECT CHARACTERISTICS

Ordinarily, studies should be conducted with healthy animals representative of the species, class, gender, and physiological maturity for which the drug is approved. The BE study may be conducted with a single gender for which the pioneer product is approved, unless there is a known interaction of formulation with gender. An attempt should be made to restrict the weight of the test animals to a narrow range in order to maintain the same total dose across study subjects. The animals should not receive any medication prior to testing for a period of two weeks or more, depending upon the biological half-life of the ancillary drug.

HUMAN FOOD SAFETY CONSIDERATIONS

The toxicology and tolerance developed for the pioneer animal drug are applied to generic copies of the drug. The CVM has concluded that in addition to a BE study, a tissue residue depletion study should be conducted for approval of a generic animal drug product in a food-producing species. Two drug products may have the same plasma disposition profile at the concentrations used to assess product BE, but may have very different tissue disposition kinetics when followed out to the withdrawal time for the pioneer product. Therefore, to show the withdrawal period at which residues of the generic product will be consistent with the tolerance for the pioneer product, a tissue residue depletion study is necessary.

The results of a BE study or tissue residue depletion study in one animal species cannot generally be extrapolated to another species. Possible species differences in drug partitioning or binding in tissues could magnify a small difference in the rate or extent of drug absorbed into a large difference in marker residue concentrations in the target tissue. Therefore, for a pioneer product labeled for more than one food-producing species, a BE study and a tissue residue depletion study will generally be requested for each major food-producing species on the label.

A traditional withdrawal study, as described in CVM's guidance number 3, *General Principles for Evaluating the Safety of Compounds Used in Food-Producing Animals*, is considered the best design for collecting data useful for the calculation of a preslaughter withdrawal period for drugs used in food-producing animals. In the traditional withdrawal study, 20 animals are divided into four or five groups of 4–5 animals each. Groups of animals are slaughtered at carefully preselected time points following the last administration of the test product, and the edible tissues are collected for residue analysis. A statistical tolerance limit approach is used to determine when, with 95% confidence, 99% of treated animals would have tissue residues below the codified limits.

For purposes of calculating a withdrawal period for a generic animal drug, only the generic product would be tested (i.e., not the pioneer product), and only the marker residue in the target tissue would be analyzed. Other study designs will be considered on a case-by-case basis. Sponsors are encouraged to submit the proposed tissue residue depletion protocol for CVM concurrence before proceeding with the withdrawal study.

The generic animal drug will be assigned the withdrawal time supported by the residue depletion data or the withdrawal time currently assigned to the pioneer product, whichever is the longer.

The generic animal drug sponsor may request a shorter withdrawal period for the generic product by supplementing the ANADA and providing tissue residue data necessary to support the shorter withdrawal period request. Such a supplement will be reviewed under the agency's policy for Category II supplements. For a Category II supplement, a reevaluation of the safety (or effectiveness) data in the parent application (i.e., the pioneer NADA) may be required (21 CFR 514.106(b)(2)). The CVM will ordinarily approve a request for a shorter withdrawal period when the residue data are adequate and when no other human food safety concerns for the drug are evident.

Under 21 CFR 514.1(b)(7), applications are required to include a description of practicable methods for determining the quantity, if any, of the new animal drug in or on food, and any substance formed in or on food because of its use, and the proposed tolerance or withdrawal period or other use restrictions to ensure that the proposed use of the drug will be safe. For certain drug products, a tissue residue depletion study is not needed to ensure that residues of the test product will be consistent with the codified drug tolerance at the withdrawal time assigned to the reference product. These include but may not be limited to products for which a waiver of in vivo BE testing is granted and products for which the assay method used in the blood level BE study is sensitive enough to measure blood levels of the drug for the entire withdrawal period assigned to the reference product. Other requests for waiver of the tissue residue study will be considered on a case-by-case basis.

CVM will not request that the assay methodology used to determine the withdrawal period for the generic product be more rigorous than the approved methodology used to determine the existing withdrawal period for the pioneer product. If an analytical method other than the approved method of analysis is used, the generic sponsor should provide data comparing the alternate method to the approved method.

LOCALLY ACTING GASTROINTESTINAL DRUGS

For drugs whose site of action is the GI tract, determination of BE is more complicated as local drug concentrations cannot be measured directly requiring evaluation of PKs, its relationship in vitro tests including dissolution and binding assays, and correlation with clinical studies.

The PK studies for locally acting drugs provide safety data, and whereas PK studies may not correlate with therapeutic effectiveness, the relationship with BE is not so straightforward. If a drug is acting locally and also absorbed in the systemic circulation, the PK studies would still reflect the dosage form factors even though the site of action is also local. The premise here remains the same, any differences noted in the C_{max} of AUC are due to differences in absorption rates and extent attributable to dosage form differences such as release of drug. However, when plasma levels can be connected to product effectiveness, then we can determine the significance of differences in product performance. When the connection to efficacy is broken, we do not have a simple way to say what difference in PK is significant. In this sense, downstream PK is similar to a PD endpoint for which a dose–response curve needs to be established. Another concern about PK studies on locally acting drugs is that the drug may be able to reach the plasma without passing the site of action. An example is an inhalation product for which some of the dose is swallowed and potentially absorbed orally. An important distinction is between parallel and sequential absorption paths. In the inhalation example, the drug either goes to the lung or to the stomach or could appear in plasma at the same time by either path. In a locally acting GI drug, the absorption process is sequential, so drug absorbed from the intestine appears before drug absorbed in the colon and thus can be distinguished.

The PK studies often fail for locally acting drugs because of the very low concentration observed in plasma and even at the site of local action. For example, mesalamine must reach the mucosal surface lining the GI tract in order to exert its pharmacological effect, which is dependent on the dissolution rate; for other dosage forms that dissolve instantly, the rate-limiting factors would be transit rate in the GI tract. The use of dissolution thus becomes an important tool to demonstrate BE. Some GI-acting drugs are formulated to target different regions of the GI tract, often via coatings that lead to pH-dependent dissolution. Comparative dissolution testing at different pH could demonstrate that test and reference products are targeting the same region of the GI tract. Biowaivers for BCS class I drugs formulated in rapidly dissolving immediate-release solid oral dosage forms are well established. Since a GI-acting drug does not need to be absorbed, application of the scientific basis of the BCS would suggest that a high solubility drug in a rapidly dissolving formulation with no excipients that affect product performance may be eligible for a biowaiver.

Generally, studies that measure the concentration of drug in the small intestinal mucosa could provide more direct evidence of equivalent tissue concentration at the site of action. But those studies are difficult to conduct, and interspecies correlations often add a lot of variability; as a result, there is a consensus developing that comparative clinical trials be conducted to demonstrate BE but only in those situations where other methods fail since not only are these expensive to conduct but also these can often be insensitive to formulation differences—the purpose of the study.

FED BIOEQUIVALENCE STUDIES

Food-effect BA studies are usually conducted for new drugs and drug products during the IND period to assess the effects of food on the rate and extent of absorption of a drug when the drug product is administered shortly after a meal (fed conditions), as compared to administration under fasting conditions. Fed BE studies, on the other hand, are conducted for ANDAs to demonstrate their BE to the RLD under fed conditions. Food can influence the BE between test and reference products. Food effects on BA can have clinically significant consequences. Food can alter BA by various means, including

- Delaying gastric emptying
- Stimulating bile flow
- Changing GI pH
- Increasing splanchnic blood flow
- Changing luminal metabolism of a drug substance
- Physically or chemically interacting with a dosage form or a drug substance

Food effects on BA are generally greatest when the drug product is administered shortly after a meal is ingested. The nutrient and caloric contents of the meal, the meal volume, and the meal temperature can cause physiological changes in the GI tract in a way that affects drug product transit time, luminal dissolution, drug permeability, and systemic availability. In general, meals that are high in total calories and fat content are more likely to affect the GI physiology and thereby result in a larger effect on the BA of a drug substance or drug product. It is recommended to use high-calorie and high-fat meals during food-effect fed BE studies.

FOOD EFFECTS ON DRUG PRODUCTS

Administration of a drug product with food may change the BA by affecting either the drug substance or the drug product. In practice, it is difficult to determine the exact mechanism by which food changes the BA of a drug product without performing specific mechanistic studies. Important food effects on BA are least likely to occur with many rapidly dissolving, immediate-release drug products containing highly soluble and highly permeable drug substances (BCS class I) because absorption of the drug substances in class I is usually pH- and site-independent and thus insensitive to differences in dissolution (http://www.fda.gov/cder/guidance/5194fnl.htm-_ftn3#_ftn3). However, for some drugs in this class, food can influence BA when there is a high first-pass effect, extensive adsorption, complexation, or instability of the drug substance in the GI tract. In some cases, excipients or interactions between excipients and the food-induced changes in gut physiology can contribute to these food effects and influence the demonstration of BE. For rapidly dissolving formulations of BCS class I drug substances, food can affect C_{max} and the time at which this occurs (t_{max}) by delaying gastric emptying and prolonging intestinal transit time. However, we expect the food effect on these measures to be similar for test and reference products in fed BE studies.

For other immediate-release drug products (BCS class II, III, and IV) and for all modified-release drug products, food effects are most likely to result from a more complex combination of factors that influence the in vivo dissolution of the drug product and/or the absorption of the drug substance. In these cases, the relative direction and magnitude of food effects on formulation BA and the effects on the demonstration of BE are difficult, if not impossible, to predict without conducting a fed BE study.

RECOMMENDATIONS

For Immediate-Release Drugs

- For uncomplicated drugs in immediate-release dosage forms, BE must be demonstrated under fasted conditions. In addition to a BE study under fasting conditions, we recommend a BE study under fed conditions for all orally administered immediate-release drug products, with the following exceptions:
  - When both test product and RLD are rapidly dissolving, have similar dissolution profiles, and contain a drug substance with high solubility and high permeability (BCS class I) (see footnote 3)
  - When the Dosage and Administration section of the RLD label states that the product should be taken only on an empty stomach
  - When the RLD label does not make any statements about the effect of food on absorption or administration
  - When the reference-listed product label does not make any statements about the effect of food on absorption or administration
- For complicated drugs in immediate-release dosage forms, for example, narrow therapeutic range drugs (drugs with a steep dose–response curve, critical drugs), highly toxic drugs, and drugs known to have nonlinear PKs, BE must be demonstrated under both fasted and fed conditions.
- Nonlinear drugs. BE must be demonstrated under both fasted and fed conditions unless the nonlinearity occurs after the drug enters the systemic circulation and there is no evidence that the product exhibits a food effect.
- Drugs in modified-release dosage forms. BE must be demonstrated under both fasted and fed conditions.

For Modified-Release Products

In addition to a BE study under fasting conditions, a BE study under fed conditions should be conducted for all orally administered modified-release drug products. It is recommended that food-effect BA and fed NE studies be conducted using meal conditions that are expected to provide the greatest effects on GI physiology so that systemic drug availability is maximally affected. A high-fat (approximately 50% of total caloric content of the meal) and high-calorie (approximately 800–1000 cal) meal is recommended as a test meal for food-effect BA and fed BE studies. This test meal should derive approximately 150, 250, and 500–600 cal from protein, carbohydrate, and fat, respectively. The caloric breakdown of the test meal should be provided in the study report.

For fasting administration, following an overnight fast of at least 10 h, subjects should be administered the drug product with 240 mL (8 fluid ounces) of water. No food should be allowed for at least 4 h post dose. Water may be allowed as desired, except 1 h before and after drug administration. Subjects should receive standardized meals scheduled at the same time in each period of the study.

For fed administration, following an overnight fast of at least 10 h, subjects should start the recommended meal 30 min prior to the administration of the drug product. Study subjects should eat this meal in 30 min or less; however, the drug product should be administered 30 min after start of the meal. The drug product should be administered with 240 mL (8 fluid ounces) of water. No food should be allowed for at least 4 h post dose. Water may be allowed as desired, except 1 h before and after drug administration. Subjects should receive standardized meals scheduled at the same time in each period of the study.

STUDY DESIGN

A sponsor may propose any study designs and data analyses. The scientific rationale and justification for these study designs and analyses should be provided in the study protocol. Sponsors may choose to conduct additional studies for a better understanding of the drug product and to provide optimal labeling statements for dosage and administration (e.g., different meals and different times of drug intake in relation to meals). In studying modified-release dosage forms, consideration should be given to the possibility that coadministration with food can result in *dose dumping*, in which the complete dose may be more rapidly released from the dosage form than intended, creating a potential safety risk for the study subjects.

GENERAL DESIGN

A randomized, balanced, single-dose, two-treatment (fed vs. fasting), two-period, two-sequence crossover design is recommended for studying the effects of food on the BE of either an immediate-release or a modified-release drug product. The formulation to be tested should be administered following a test meal (fed condition). The treatments should consist of both test and reference formulations administered following a test meal (fed condition). An adequate washout period should separate the two treatments in food-effect BE studies.

SUBJECT SELECTION

Fed BE studies can be carried out in healthy volunteers drawn from the general population. Studies in the patient population are also appropriate if safety concerns preclude the enrollment of healthy subjects. A sufficient number of subjects should complete the study to achieve adequate power for a statistical assessment of food effects. A minimum of 12 subjects should complete the fed BE studies.

DOSAGE STRENGTH

In general, the highest strength of a drug product intended to be marketed should be tested in fed BE studies. In some cases, clinical safety concerns can prevent the use of the highest strength and warrant the use of lower strengths of the dosage form. For ANDAs, the same lot and strength used in the fasting BE study should be tested in the fed BE study. For products with multiple strengths in ANDAs, if a fed BE study has been performed on the highest strength, BE determination of one or more lower strengths can be waived based on dissolution profile comparisons.

TEST MEAL

The fed BE studies be conducted using meal conditions that are expected to provide the greatest effects on GI physiology so that systemic drug availability is maximally affected. A high-fat (approximately 50% of total caloric content of the meal) and high-calorie (approximately 800–1000 cal) meal is recommended as a test meal for food-effect BA and fed BE studies. This test meal should derive approximately 150, 250, and 500–600 cal from protein, carbohydrate, and fat, respectively. (An example test meal would be two eggs fried in butter, two strips of bacon, two slices of toast with butter, 4 ounces of hash brown potatoes, and 8 ounces of whole milk.) Substitutions in this test meal can be made as long as the meal provides a similar amount of calories from protein, carbohydrate, and fat and has comparable meal volume and viscosity. The caloric breakdown of the test meal should be provided in the study report. If the caloric breakdown of the meal is significantly different from the one described earlier, the sponsor should provide a scientific rationale for this difference.

ADMINISTRATION

Fed Treatments: Following an overnight fast of at least 10 h, subjects should start the recommended meal 30 min prior to administration of the drug product. Study subjects should eat this meal in 30 min or less; however, the drug product should be administered 30 min after start of the meal. The drug product should be administered with 240 mL (8 fluid ounces) of water. No food should be allowed for at least 4 h post dose. Water can be allowed as desired except for 1 h before and after drug administration. Subjects should receive standardized meals scheduled at the same time in each period of the study.

SAMPLE COLLECTION

Timed samples in biological fluid, usually plasma, should be collected from the subjects to permit characterization of the complete shape of the plasma concentration–time profile for the parent drug. It may be advisable to measure other moieties in the plasma, such as active metabolites. Consideration should be given to the possibility that coadministration of a dosage form with food can alter the time course of plasma drug concentrations so that fasted and fed treatments can have different sample collection times.

DATA ANALYSIS AND LABELING

The following exposure measures and PK parameters should be obtained from the resulting concentration–time curves for the test and reference products:

- Total exposure or area under the concentration–time curve (AUC_{0-inf}, AUC_{0-t})
- Peak exposure (C_{max})
- Time to peak exposure (t_{max})
- Lag time (t_{lag}) for modified-release products, if present
- Terminal elimination half-life
- Other relevant PK parameters

Individual subject measurements, as well as summary statistics (e.g., group averages, standard deviations, coefficients of variation), should be reported. An equivalence approach is recommended analyzing data using an average criterion. Log transformation of exposure measurements (AUC and C_{max}) prior to analysis is recommended. The 90% CI for the ratio of population geometric means between test and reference products should be provided for AUC_{0-inf}, AUC_{0-t}, and C_{max}. For ANDA fed BE studies, the RLD administered under fed condition serves as the reference treatment.

For an ANDA, BE of a test product to the RLD product under fed conditions is concluded when the 90% CI for the ratio of population geometric means between the test and RLD product, based on log-transformed data, is contained in the BE limits of 80%–125% for AUC and C_{max}. Although no criterion applies to t_{max}, the t_{max} values for the test and reference products are expected to be comparable based on clinical relevance. The conclusion of BE under fed conditions indicates that with regard to food, the language in the package insert of the test product can be the same as the reference product.

OTHER CONSIDERATIONS

Sprinkles

In ANDAs, BE of the test to the RLD is demonstrated in a single-dose crossover study. Both treatments should be sprinkled on one of the soft foods mentioned in the labeling, usually applesauce. The BE data should be analyzed using average BE, and the 90% CI criteria should be used to declare BE. If there are questions about other foods, the design, or the analysis of such BE studies, the sponsors and/or applicants should contact the Office of Generic Drugs.

Special Vehicles

In ANDAs, BE of the test to the RLD is demonstrated in a single-dose crossover study. Both treatments should be mixed with one of the beverages mentioned in the labeling. Sponsors should provide evidence that BE differences would not be expected from the use of other listed vehicles. The BE data should be analyzed using average BE, and the 90% CI criteria should be used to declare BE.

6 Bioequivalence Waivers

BACKGROUND

This chapter covers the most significant aspect of the scope of this book—the relevance of bioequivalence (BE) testing. Establishing BE between two products extends to several types and situations:

- Prototype formulations during early development and pivotal clinical trial formulations
- Innovator formulations that differ from their new drug application (NDA) formulations as a result of scale-up and postapproval changes (SUPAC)
- Implementation of improved manufacturing technologies
- Scale-up changes in manufacturing locations
- Multisource products filed for approval under abbreviated new drug application (ANDA)

The demonstration of BE is assumed to represent product quality throughout its life cycle forming the basis for waiver of conducting expensive clinical studies to demonstrate safety and efficacy.

In the past few years, the development of in vitro model systems has led to prediction of the pharmacokinetics (PK) and the application of in vitro–in vivo correlation (IVIVC) for pharmaceutical dosage forms; this has been a main focus of attention of pharmaceutical industry, academia, and regulatory sectors. The main reason for adopting the IVIVC is to avoid any unnecessary exposure to human beings. 21 CFR 320.25(a) codifies the universal belief that "No unnecessary human testing should be performed" and goes on to suggest: "The basic principle in an in vivo bioavailability (BA) study is that no unnecessary human research should be done." Somewhere along the line, the basic tenet of this code was lost as the developers assumed that human studies are essential to approvals by the regulatory agencies. In the 1980s when the ANDA route was established, the BE testing was done indiscriminately merely to check off the filing requirements. In most instances, an ANDA would have required multiple studies including studies conducting to first establish the powering strategies. However, as we began seeing the development of highly active and therefore potentially toxic drugs, the indiscriminate exposure of these drugs to healthy subjects was unethical, and therefore there is a greater need now to reexamine the basic paradigm of BE testing.

Another driving force for IVIVC was to reduce the cost of drug development. The generic products coming into markets reduce the cost of treatment substantially, yet one of the most expensive and time-consuming development component of the filing remains the BE studies.

As a result of the IVIVC, a large number of drugs were waived from demonstrating BE as allowed by several regulatory agencies, except Japan. This is achieved by an elaborate biopharmaceutical classification system (BCS) that differentiates the drugs based on their dissolution rates and permeability across biological membranes.

Another waiver in BE comes in selecting only a specific dose when filing for approval of multiple strengths, to reduce the cost of studies.

If a drug does not fall in the classification where a waiver can be granted, the sponsors have the right to challenge and request waivers based on additional surrogate tests developed.

In the United States, the waivers are intended to apply to the following:

- Subsequent in vivo BA or BE studies of formulations after the initial establishment of the in vivo BA of immediate-release dosage forms during the investigational new drug application (IND) period.
- In vivo BE studies of instant release dosage forms in ANDAs. Regulations at 21 CFR part 320 address the requirements for BA and BE data for approval of drug applications and supplemental applications.

Provision for waivers of in vivo BA/BE studies (biowaivers) under certain conditions is provided at 21 CFR 320.22.

IVIVC

Definitions

United States Pharmacopoeia (USP): The establishment of a rational relationship between a biological property, or a parameter derived from a biological property produced by a dosage form, and a physicochemical property or characteristic of the same dosage form.

Food and Drug Administration (FDA): An IVIVC has been defined by the FDA as "a predictive mathematical model describing the relationship between an in-vitro property of a dosage form and an in-vivo response."

The IVIVC is a mathematical relationship between in vitro properties of a dosage form and its in vivo performance. The in vitro release data of a dosage form containing the active substance serve as characteristic in vitro property, while the in vivo performance is generally represented by the time course of the plasma concentration of the active substance. These in vitro and in vivo data are then treated scientifically to determine correlations. The term correlation is frequently employed within the pharmaceutical and related sciences to describe the relationship that exists between variables. From a biopharmaceutical standpoint, correlation could be referred to as the relationship between appropriate in vitro release characteristics and in vivo BA parameters. For oral dosage forms, the in vitro release is usually measured and considered as dissolution rate. The relationship between the in vitro and in vivo characteristics can be expressed mathematically by a linear or nonlinear correlation. However, the plasma concentration cannot be directly correlated to the in vitro release rate; it has to be converted to the in vivo release or absorption data, either by PK compartment model analysis or by linear system analysis.

Purpose of IVIVC

The purpose of IVIVC is to use drug dissolution results from two or more products to predict similarity or dissimilarity of expected plasma drug concentration (profiles):

- *Reduction of regulatory burden*: IVIVC can be used as substitute for additional in vivo experiments, under certain conditions.
- *Optimization of formulation*: The optimization of formulations may require changes in the composition, manufacturing process, equipment, and batch sizes. In order to prove the validity of a new formulation, which is bioequivalent with a target formulation, a considerable amount of effort is required to study BE/BA.
- *Justification for "therapeutic" product quality*: IVIVC is often adequate for justification of therapeutically meaningful release specifications of the formulation.

- *SUPAC (time and cost saving during the product development)*: Validated IVIVC also serves as justification for biowaivers in filings of a Level 3 (or type II in Europe) variation, either during scale-up or postapproval, as well as for line extensions (e.g., different dosage strengths).
- *IVIVC as surrogate for in vivo BE and to support biowaivers (time and cost saving)*: The main purpose of an IVIVC model is to utilize in vitro dissolution profiles as a surrogate for in vivo BE and to support biowaivers.

Levels

There are five correlation levels defined in the IVIVC by FDA (Table 6.1).

Level A Correlation

This level of correlation is the highest category of correlation and represents a point-to-point relationship between in vitro dissolution rate and in vivo input rate of the drug from the dosage form. A level A correlation is generally linear and the in vitro dissolution and in vivo input curves may be directly superimposable or may be made to be superimposable by the use of a scaling factor. Nonlinear correlations, while uncommon, may also be appropriate. When the in vitro and in vivo curves are superimposable, it is said to be a 1:1 relationship, while if a scaling factor is required to make the curve superimposable, then the relationship is called a point-to-point relationship. The percent of drug absorbed may be calculated by means of model-dependent techniques such as the Wagner–Nelson procedure or the Loo–Riegelman method or by model-independent numerical deconvolution method. These techniques represent a major advance over the single point approach in that these methodologies utilize all of the dissolution and plasma level data available to develop the correlation.

The purpose of a level A correlation is to define a direct relationship with in vivo data such that the measurement of the in vitro dissolution rate alone is sufficient to determine the biopharmaceutical rate of the dosage form. In the case of a level A correlation, an in vitro dissolution curve can serve as a surrogate for in vivo performance. Therefore, a change in manufacturing site, method of manufacture, raw material supplier, minor formulation modification, and even change in product strength of the same formulation can be justified without the need for additional human studies. It is an excellent quality control procedure since it is predictive of the in vivo performance of dosage form.

TABLE 6.1
Various Parameters Used in IVIVC Level Assignments

Level	In Vitro	In Vivo
A	Dissolution curve	Input (absorption) curves
B	Statistical moments: MDT	Statistical moments: MRT, mean absorption time (MAT), etc.
C	Disintegration time; time to have 10%, 50%, 90% dissolved; dissolution rate; dissolution efficiency (DE)	C_{max}; observed at time (t_{max}); absorption constant (K_a); time to have 10%, 50%, 90% absorbed; AUC (total or cumulative)
D	Rank order and qualitative analysis	Not useful for regulatory consideration but for formulation and process development

Notes: A, One-to-one relationship between in vitro and in vivo data, for example, in vitro dissolution versus in vivo absorption; B, Correlation based on statistical moments, for example, in vitro MDT versus in vivo MRT or MAT; C, Point-to-point relationship between a dissolution and a PK parameter, for example, in vitro T50% versus in vivo t_{max}; Multiple C, Relationship between one or several PK parameters and amount dissolved at several time points.

Level B Correlation

A level B IVIVC utilizes the principles of statistical moment analysis. In this level of correlation, the mean in vitro dissolution time (MDT_{vitro}) of the product is compared to either mean in vivo residence time or the mean in vivo dissolution time. Although a level B correlation uses all of the in vitro and in vivo data, it is not considered to be a point-to-point correlation, since there are a number of different in vivo curves that will produce similar mean residence time (MRT) values. A level B correlation does not uniquely reflect the actual in vivo plasma level curve. Therefore, one cannot rely upon a level B correlation alone to predict the effects of formulation modification, manufacturing site change, excipient source change, etc. In addition, in vitro data from such a correlation cannot be used to justify the extremes of quality control standards and it is the least useful for regulatory purposes.

Level C Correlation

In this level of correlation, one dissolution time point (*t*50%, *t*90%, etc.) is compared to one mean PK parameter such as area under the curve (AUC), t_{max}, or maximum observed concentration (C_{max}). Therefore, it represents a single point correlation and does not reflect the entire shape of the plasma drug concentration–time curve, which is the crucial indicator of the performance of a modified-release product. This is the weakest level of correlation as only a partial relationship between absorption and dissolution is established. Due to its obvious limitations, a level C correlation is limited in predicting in vivo drug performance. The usefulness of a level C correlation is subject to the same limitations as a level B correlation in its ability to support product and manufacturing site changes as well as justify quality control standard extremes. Level C correlations can be useful in the early stages of formulation development when pilot formulations are being selected. While a level C correlation may be useful in formulation development, waiver of an in vivo BE study (biowaiver) is generally not possible.

Multiple Level C Correlation

A multiple level C correlation relates one or several PK parameters of interest (C_{max}, AUC, or any other suitable parameters) to the amount of drug dissolved at several time points of the dissolution profile. A multiple point level C correlation may be used to justify a biowaiver provided that the correlation has been established over the entire dissolution profile for one or more PK parameters of interest. A relationship should be demonstrated at each time point for the same parameter such that the effect on the in vivo performance of any change in dissolution can be assessed. If such a multiple level C correlation is achieved, then the development of a level A correlation is also likely. A multiple level C correlation should be based on at least three dissolution time points covering the early, middle, and late stages of the dissolution profile.

Level D Correlation

A level D correlation is a rank order and qualitative analysis and is not considered useful for regulatory purposes. It is not a formal correlation but serves as an aid in the development of a formulation or processing procedure.

On the whole, the FDA ranking can be summarized as a level A IVIVC being most informative and recommended, if possible followed by a multiple level C, level C, level B, and level D correlation.

Modeling

The selection of a drug candidate marks the most crucial stage in the life cycle of drug development. Such selection is primarily based on the drug "developability" criteria, which include physicochemical properties of the drug and the results obtained from preliminary studies involving several in vitro systems and in vivo animal models, which address efficacy and toxicity issues. During this stage, exploring the relationship between in vitro and in vivo properties of the drug in animal models provides an idea about the feasibility of the drug delivery system for a given drug. In such correlations, study designs including study of more than one formulation of the modified-release dosage forms and a rank order of release (fast/slow) of the formulations should be incorporated. Even though the formulations and methods used at this stage are not optimal, they prompt better design and development efforts in the future.

The process of obtaining a drug profile from dissolution results is known as convolution. The opposite of this, that is, obtaining or extracting a dissolution profile from a blood profile, is known as deconvolution. In the development of convolution model, the drug concentration–time profiles obtained from dissolution results may be evaluated using criteria for in vivo BA/BE assessment, based on C_{max} and AUC parameters.

In mathematical terminology, dissolution results become an input function, and plasma concentrations become a weighting factor or function resulting in an output function representing plasma concentrations for the solid oral product.

Implementation of convolution-based method involves the production of a user-written subroutine for the NONMEM software package, has shown that a convolution-based method. Using the NONMEM package, a nonlinear mixed effects model can be fitted to the data with a timescale model linking the in vitro and in vivo components.

It has been demonstrated that the convolution-based and differential equation–based models can be mathematically equivalent. A software has been developed that implements a differential equation–based approach. This method utilizes existing NONMEM libraries and is an accurate method of modeling, which is far more straightforward for users to implement. This research shows that, when the system being modeled is linear, the use of differential equations will produce results that are practically identical to those obtained from the convolution method. But this is a task that can be time consuming and complex, and as a result, this methodology, despite its advantages over the deconvolution-based approach, is not in widespread use.

The most basic IVIVC models are expressed as a simple linear equation between the in vivo drug absorption and in vitro drug dissolved (released):

$$Y \text{ (In vivo absorbed)} = mX \text{ (In vitro drug dissolved)} + C \tag{6.1}$$

In this equation, m is the slope of the relationship and C is the intercept. Ideally, $m = 1$ and $C = 0$, indicating a linear relationship. However, depending on the nature of the modified-release system, some data are better fitted using nonlinear models, such as Sigmoid, Weibull, Higuchi, or Hixson–Crowell.

In vivo release rate (X'_{vivo}) can also be expressed as a function of in vitro release rate ($X'_{rel,vitro}$) with parameters (a, b), which may be empirically selected and refined using appropriate mathematical processes:

$$X'_{vivo}(t) = X'_{rel,vitro}(a + bt) \tag{6.2}$$

An iterative process may be used to compute the time-scaling and time-shifting parameters. Integral to the model development exercise is model validation, which can be accomplished using data from the formulations used to build the model (internal validation) or using data obtained from a different (new) formulation (external validation). While internal validation serves the purpose of providing basis for the acceptability of the model, external validation is superior and affords greater "confidence" in the model.

Generally, a plot of the fraction of drug absorbed (F_a) against the fraction of drug dissolved (F_d) is made wherein the fraction absorption absorbed is obtained by deconvoluting the plasma profile.

Often, the goal is to develop a profile that need not a priori be a linear or even a predefined function. For example,

$$F_a = \frac{1}{f_a}\left(1 - \frac{\alpha}{\alpha - 1}(1 - F_d) + \frac{1}{\alpha - 1}(1 - F_d)^{\alpha}\right) \tag{6.3}$$

where

F_a is the fraction of the total amount of drug absorbed at time t

f_a is the fraction of the dose absorbed at $t = \#$

α is the ratio of the apparent first-order permeation rate constant (k_{paap}) to the first-order dissolution rate constant (k_d)

F_d is the fraction of drug dose dissolved at time t

For establishing external predictability, the exposure parameters for a new formulation are predicted using its in vitro dissolution profile and the IVIVC model, and the predicted parameters are compared to the observed parameters. The prediction errors are computed as for the internal validation. For C_{max} and AUC, the prediction error for the external validation formulation should not exceed 10%. A prediction error of 10%–20% indicates inconclusive predictability and illustrates the need for further study using additional data sets. For drugs with narrow therapeutic index (NTI), external validation is required despite acceptable internal validation, whereas internal validation is usually sufficient with non-NTI drugs.

Several commercial software programs are available to study IVIVC, for example, PDx-IVIVC (http://www.globomaxservice.com/pdxivivc.htm), which is a comprehensive IVIVC software program that performs deconvolution, calculating the fraction or percentage of drug absorbed and correlating it with in vitro fraction or percentage dissolved data. It also allows level C correlations (single or multiple) wherein a single point relationship between a dissolution parameter, for example, percent dissolved in 4 h, and a PK parameter (e.g., AUC, C_{max}, t_{max}) is determined. A successful IVIVC model can be developed if in vitro dissolution is the rate-limiting step in the sequence of events leading to appearance of the drug in the systemic circulation following oral or other routes of administration. Thus, the dissolution test can be utilized as a surrogate for BE studies (involving human subjects) if the developed IVIVC is predictive of in vivo performance of the product.

Relative Advantages of Modeling Methods

A convolution model is represented as

$$C(t) = C_{\delta}(t)F(t) = \int_0^t C_{\delta}(\tau)F(t-\tau)\mathrm{d}\tau \tag{6.4}$$

where

$C(t)$ is plasma concentration after oral dose

$C_{\delta}(t)$ is plasma concentration after an intravenous (IV) dose or a dose of oral solution

Upon taking the derivative of $C(t)$ with respect to time,

$$C(t) = C_{\delta}(t)F(t) + C_{\delta}(0)\int_0^t F(\tau)\mathrm{d}\tau \tag{6.5}$$

When $C_{\delta}(0) = 0$,

$$C(t) = C_{\delta}(t) * F(t) \tag{6.6}$$

The advantages of this approach relative to deconvolution-based IVIVC approaches include the following:

- The relationship between measured quantities (in vitro release and plasma drug concentrations) is modeled directly in a single stage rather than via an indirect two-stage approach.
- The model directly predicts the plasma concentration–time course. As a result, the modeling focuses on the ability to predict measured quantities (not indirectly calculated quantities such as the cumulative amount absorbed). The results are more readily interpreted in terms of the effect of in vitro release on conventional BE metrics.

Deconvolution Model

Deconvolution is a numerical method used to estimate the time course of drug input using a mathematical model based on the convolution integral. The deconvolution technique requires the comparison of in vivo dissolution curves that are obtained from the blood profiles with in vitro dissolution profiles. The observed fraction of the drug absorbed is estimated based on the Wagner–Nelson method. IV, IM, or oral solution is attempted as the reference. Then, the PK parameters are estimated using a nonlinear regression tool or obtained from literatures reported previously. Based on the IVIVC model, the predicted fraction of the drug absorbed is calculated from the observed fraction of the drug dissolved. It is the most commonly cited and used method in the literature. However, this approach is conceptually difficult to use. For example,

- Extracting in vivo dissolution data from a blood profile often requires elaborate mathematical and computing expertise.
- Fitting mathematical models is usually subjective in nature and thus does not provide an unbiased approach in evaluating in vivo dissolution results/profiles. Even when in vivo dissolution curves are obtained, there is no parameter available with associated statistical confidence and physiological relevance, which would be used to establish the similarity or dissimilarity of the curves.
- A more serious limitation of this approach is that it often requires multiple products having potentially different in vivo release characteristics (slow, medium, fast). These products are then used to define experimental conditions (medium, apparatus, etc.) for an appropriate dissolution test to reflect their in vivo behavior.
- This approach is more suited for method/apparatus development as release characteristics of test products are to be known (slow, medium, fast) rather than product evaluation.

Differential Equation–Based Approach

Another approach that has been proposed is based on systems of differential equations. The use of a differential equation–based model could also allow for the possibility of accurately modeling nonlinear systems, and further investigation is being carried out into the case where the drug is eliminated by a nonlinear, saturable process. The convolution and deconvolution methods assume that the system being modeled is linear but, in practice, this is not always the case. Work to date has shown that the convolution-based method is superior, but when presented with nonlinear data, even this approach will fail. It is expected that, in the nonlinear case, the use of a differential equation–based method would lead to more accurate predictions of plasma concentration.

The incorporation of time scaling in the PDx-IVIVC equation allows this parameter to be estimated directly from the in vivo and in vitro release data. As a result, the predictability of an IVIVC model can be evaluated over the entire in vivo time course. Internal predictability of the IVIVC model was assessed using convolution. The PDx-IVIVC model equation is as follows:

$$x_{\text{vivo}}(t) = \begin{cases} 0, & t \geq 0 \\ a_1 + a_2 x_{\text{vitro}}(-b_1 + b_2 t), & t \geq 0 \end{cases} \tag{6.7}$$

For orally administered drugs, IVIVC is expected for highly permeable drugs, or drugs under dissolution rate-limiting conditions, which is supported by the BCS. For extended-release formulations following oral administration, modified BCS containing the three classes (high aqueous solubility, low aqueous solubility, and variable solubility) is proposed.

Predictability of Correlation

The objective of IVIVC evaluation is to estimate the magnitude of the error in predicting the in vivo BA results from in vitro dissolution data. This objective should guide the choice and interpretation of evaluation methods. Any appropriate approach related to this objective may be used for the evaluation of predictability. It can be calculated by prediction error, which is the error in prediction of the in vivo property from the in vitro property of the drug.

Internal Predictability

Evaluation of internal predictability is based on the initial data used to define the IVIVC model. Internal predictability is applied to IVIVC established using formulations with three or more release rates for non-NTI drugs exhibiting conclusive prediction error. If two formulations with different release rates are used to develop IVIVC, then the application of IVIVC would be limited to specified categories. The BA (C_{max}, t_{max}/AUC) of formulation that is used in the development of IVIVC is predicted from its in vitro property using IVIVC. Comparison between predicted BA and observed BA is done and prediction error (%PE) is calculated. According to FDA guidelines, the average absolute %PE should be below 10% and %PE for individual formulation should be below 15% for establishment of IVIVC.

Under these circumstances, for complete evaluation and subsequent full application of the IVIVC, prediction of error externally is recommended. According to FDA guidance, the acceptance criteria is

1. ≤15% for absolute %PE of each formulation
2. ≤10% for mean absolute %PE

External Predictability

Most important when using an IVIVC as a surrogate for BE is confidence that the IVIVC can predict in vivo performance of subsequent lots of the drug product. Therefore, it may be important to establish the external predictability of the IVIVC.

Evaluation of external predictability is based on additional test data sets. External predictability evaluation is not necessary unless the drug has an NTI, or only two release rates were used to develop the IVIVC, or if the internal predictability criteria are not met, that is, prediction error internally is inconclusive. The predicted BA is compared with known BA and %PE is calculated. The prediction error for external validation should be below 10%, whereas prediction error between 10% and 20% indicates inconclusive predictability and need of further study using additional data sets.

Using the IVIVC model, for each formulation, the relevant exposure parameters (C_{max} and AUC) are predicted and compared to the actual (observed) values. The prediction errors are calculated using the following:

$$\text{Prediction error}\left(\%\text{PE}\right) = \left[\frac{\left(C_{\text{max, observed}} - C_{\text{max, predicted}}\right)}{C_{\text{max, observed}}}\right] \times 100 \tag{6.8}$$

The C_{max} can be replaced with corresponding AUC. The criteria set in the FDA guidance on IVIVC are as follows: for C_{max} and AUC, the mean absolute percent prediction error (%PE) should not exceed 10%, and the prediction error for individual formulations should not exceed 15%.

Limitation of Predictability Metrics

Metrics used to evaluate the predictability described simply the prediction error (%PE) for only two PK parameters, that is, C_{max} and AUC. E_{max} predicted with the IVIVC model represents the maximum of the mean plasma profiles but is compared with the mean C_{max} (observed) calculated as the average of individual profile at different T. But t_{max} is not included in predictability metrics.

Depending on the intended application of an IVIVC and the therapeutic index of the drug, evaluation of prediction error internally and/or externally may be appropriate.

IVIVC Development

Any well-designed and scientifically sound approach would be acceptable for the establishment of an IVIVC. For the development and validation of an IVIVC model, two or three different formulations with different release rates, such as slow, medium, fast, should be studied in vitro and in vivo. Figure 6.1 shows the various steps and types of convolution models in use.

A number of products with different release rates are usually manufactured by varying the primary rate controlling variable (e.g., the amount of excipient or a property of the drug substance such as particle size) but within the same qualitative formulation. To develop a discriminative in vitro dissolution method, several method variables together with formulation variables are studied, for example, different pH values, dissolution apparatus, and agitation speeds. Essentially at this stage, a level A correlation is assumed and the formulation strategy is initiated with the objective of achieving the target in vitro profile. Development of a level A IVIVC model includes several steps.

In the context of understanding the applications of IVIVR throughout the product development cycle, it is useful to become familiar with the following terms as they relate to a typical product development cycle for oral extended-release product.

An assumed IVIVC is the one that provides the initial guidance and direction for the early formulation development activity. Thus, during step 1 and with a particular desired product, appropriate in vitro targets are established to meet the desired in vivo profile specification. This assumed model can be the subject of revision as prototype formulations are developed and characterized in vivo, with the results often leading to a further cycle of prototype formulation and in vivo characterization.

Out of this product development cycle and in vivo characterization and, of course, extensive in vitro testing often developed what can be referred to as retrospective IVIVC.

The defined formulation that meets the in vivo specification is employed for stage 2. At this stage, based on a greater understanding and appreciation of defined formulation and its characteristics, a prospective IVIVC is established through a well-defined prospective IVIVC study.

Step 1

In the first step, the in vivo input profile of the drug from different formulations is calculated from drug concentrations in plasma (Figure 6.1). The target in vivo profile needs to be first established, based on, if possible, PK/pharmacodynamic models. Certainly, step 1 activity should culminate in a pilot PK study. This is typically a four- or five-arm crossover study. The size of this pilot PK study will vary depending on the inherent variability of the drug itself but typically range from 6 to 10 subjects. The results of this pilot PK study provide the basis for establishing what has been referred to as a retrospective IVIVC. To separate drug input from drug distribution and elimination, model-dependent approaches, such as Wagner–Nelson and Loo–Riegelman, or model-independent procedures, based on numerical deconvolution, may be utilized.

In step 1, the parameters that describe drug input rate, drug distribution, and/or elimination are determined. In the model-dependent approaches, the distribution and elimination rate constants describe PK after absorption. In the numerical deconvolution approach, the drug unit impulse response function describes distribution and elimination phases, respectively. The physicochemical

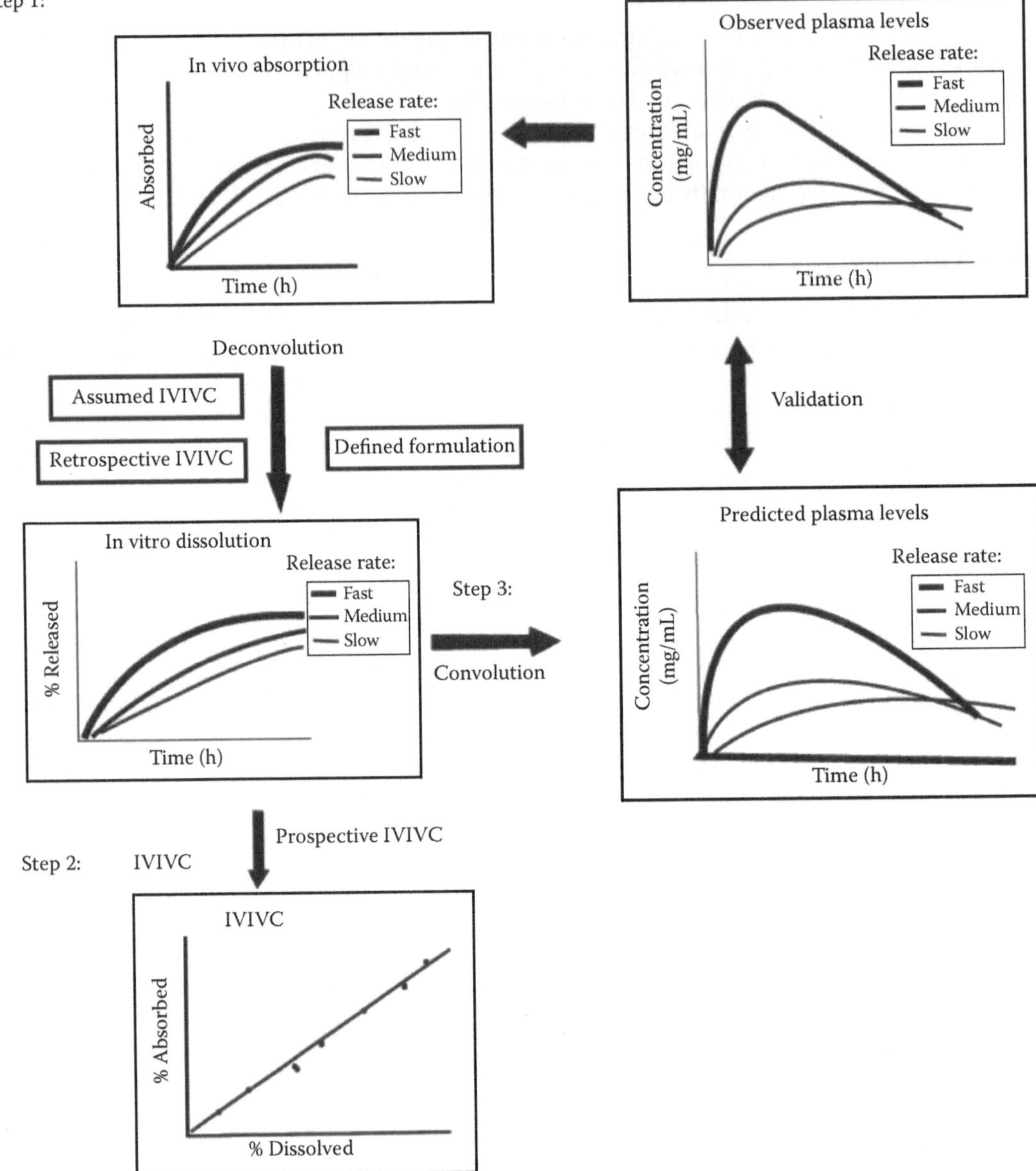

FIGURE 6.1 Convolution–deconvolution models.

characteristics of the drug substance itself, in relevance to formulation approach and dissolution at distal sites in the gastrointestinal tract, need to be taken into account. Based on this information, a priori in vitro methods are usually then developed and a theoretical in vitro target is established, which should achieve the desired absorption profile.

Step 2

By this phase of the development process, a defined formulation that meets the in vivo targets has been achieved. Extensive in vitro characterization is again performed across pH, media, and apparatus, along with the consideration of the results of stage 1. This leads to execution of a prospective

IVIVC study. The IVIVC is developed and defined after an analysis of the result of that prospective in vivo study. It can often involve further in vitro method development in the context of the observed results, but clearly with the objective of establishing a definitive IVIVC. In this step, the relationship between in vitro dissolution and the in vivo drug input profile is determined (Figure 6.1). Either a linear or nonlinear relationship may be found. In some cases, time scaling of in vitro data must be used, because in vitro dissolution and in vivo input may follow the same kinetics but still have different timescales. The time-scaling factor should be the same for all formulations if an IVIVC at level A is sought. During the early stages of correlation development, dissolution conditions may be altered to attempt to develop a one-to-one correlation between the in vitro dissolution profile and the in vivo dissolution profile. This work should also result in the definitive in vitro method that has been shown to be correlated with in vivo performance and sensitive to the specific formulation variables.

Step 3

In this phase, plasma drug concentration profiles are predicted and compared to the observed time courses for different formulations (Figure 6.1). To generate predicted time courses, the drug input profile is predicted based on in vitro dissolution data and the in vitro–in vivo relationship generated in step 2. In the convolution process, the predicted drug input and parameters describing drug distribution and/or elimination phases are combined in order to get predicted time courses. This procedure, which includes steps 1–3, is called two-stage deconvolution. Alternatively, a drug input profile based on in vitro dissolution data can be solved together with parameters describing systemic PK, that is, distribution and elimination. This approach is called direct convolution.

Different IVIVC models are used as a tool for formulation development and evaluation of immediate- and extended-release dosage forms for setting a dissolution specification and as a surrogate for BE testing.

As a result, considerable effort goes into their development and the main outcome is "the ability to predict, accurately and precisely, expected bioavailability characteristics for an extended release (ER) drug product from dissolution profile characteristics."

Once the IVIVC is established and defined, it can be then used to guide the final cycle of formulation and process optimization program of statistically based experimental design studies looking at critical formulation and process variables. This information can also be used into the activities of scale-up, pivotal batch manufacture and process validation culminating in registration, approval, and subsequent postapproval scale-up and other changes. Thus, rather than viewing the IVIVC as a single exercise at a given point in a development program, one should view it as a parallel development in itself starting at the initial assumed level and being built on and modified through experience and leading ultimately to a prospective IVIVC.

BCS

The BCS is a scientific framework for classifying drug substances based on their aqueous solubility and intestinal permeability. When combined with the dissolution of the drug product, the BCS takes into account three major factors that govern the rate and extent of drug absorption from instant release solid oral dosage forms: dissolution, solubility, and intestinal permeability. According to the BCS, drug substances are classified as follows:

Class 1: High solubility–high permeability
Class 2: Low solubility–high permeability
Class 3: High solubility–low permeability
Class 4: Low solubility–low permeability

In addition, instant release solid oral dosage forms are categorized as having rapid or slow dissolution. Within this framework, when certain criteria are met, the BCS can be used as a drug development tool to help sponsors justify requests for biowaivers. There are several factors that affect the classification of drugs in different classes.

The IVIVC is normally expected for highly permeable drugs under dissolution rate-limiting conditions. This statement is supported by the BCS, which anticipates a successful IVIVC for highly permeable drugs (Class I). BCS is a fundamental guideline for determining the conditions under which an IVIVC is expected. It is also used as a tool for developing the in vitro dissolution specification.

The BCS defines three dimensionless numbers, dose number (D_0), dissolution number (D_n), and absorption number (A_n), to characterize a drug substance. These numbers are combinations of physicochemical and physiological parameters and represent the most fundamental view of gastrointestinal drug absorption. The A_n is the ratio of the MRT to the absorption time. D_n is a ratio of MRT to mean dissolution time (MDT). D_0 is the mass divided by an uptake volume of 250 mL and the drug's solubility. The MRT here is the average of the residence time in the stomach, small intestine, and colon. The fraction of dose absorbed can then be predicted based on these three parameters. For example, an A_n of 10 means that the permeation across the intestinal membrane is 10 times faster than the transit through the small intestine indicating 100% drug absorbed. In the BCS, a drug is classified in one of four classes based solely on its solubility and intestinal permeability.

Tables 6.2 through 6.4 illustrate the BCS and the expected IVIVC for immediate- and extended-release formulations.

Class I drugs such as metoprolol exhibit high A_n and D_n values. The rate-limiting step to drug absorption is drug dissolution or gastric emptying rate if dissolution is very rapid. Class II drugs such as phenytoin have a high A_n but low D_n. The in vivo drug dissolution is then the rate-limiting

TABLE 6.2
Biopharmaceutical Drug Classification and Associated Likelihood of IVIVC for Immediate-Release Drug Products

Class	Solubility	Permeability	Likelihood of IVIVC
I	High	High	IVIVC expected (if dissolution is rate-limiting step)
II	Low	High	IVIVC expected
III	High	Low	Little or no IVIVC
IV	Low	Low	Little or no IVIVC

TABLE 6.3
Biopharmaceutical Drug Classification and Likelihood of IVIVC for Extended-Release Drug Products

Class	Solubility	Permeability	Likelihood of IVIVC
IA	High and site independent	High and site independent	IVIVC level A expected
IB	High and site independent	Dependent on site and narrow absorption window	IVIVC level C expected
IIa	Low and site independent	High and site independent	IVIVC level A expected
IIb	Low and site independent	Dependent on site and narrow absorption window	Little or no IVIVC
Va acidic	Variable	Variable	Little or no IVIVC
Vb basic	Variable	Variable	IVIVC level A expected

TABLE 6.4
BCS Class and Drug Delivery Technology

BCS Class	Examples	Drug Delivery Technology
I	Metoprolol, diltiazem, verapamil, propranolol, acyclovir, atropine	MacroCap, Micropump, multiporous oral drug absorption system (MODAS), single composition osmotic tablet system (SCOT), and stabilized pellet delivery system (SPDS).
II	Phenytoin, danazol, ketoconazole, mefenamic acid, tacrolimus, piroxicam, griseofulvin, warfarin	Micronization, stabilization of high-energy states (including lyophilized fast-melt systems), use of surfactants, emulsion or microemulsion systems, solid dispersion and use of complexing agent such as cyclodextrins; e.g., nanosuspension and nanocrystals are treated as hopeful means of increasing solubility and BA of poorly water-soluble active ingredients.
III	Cimetidine, neomycin, ranitidine, amoxicillin	Oral vaccine system, gastric retention system, high-frequency capsule and telemetric capsule.
IV	Cyclosporin A, furosemide, ritonavir, saquinavir, and taxol	The Class IV drugs present a major challenge for the development of drug delivery systems due to their poor solubility and permeability characteristics. These are administered by parenteral route with the formulation containing solubility enhancers.

step for absorption. For Class III drugs, permeability is the rate controlling drug absorption. Class IV drugs are low-solubility and low-permeability drugs. The absorption for Class II drugs is usually slower than for Class I and occurs over a longer period of time. IVIVC is usually expected for Class I and Class II drugs.

In Table 6.3, a new subclassification for drugs that have high and site-independent solubility or low and site-independent solubility has been suggested. Drugs that have high permeability and site-independent permeability are classified as Class Ia or Class IIa, and drugs that have site-dependent permeability and narrow absorption window are classified as Class Ib or Class IIb. Because little or no IVIVC is expected for Class III or Class IV drugs, there is no other subclassification for these groups. Drugs that have variable solubility and variable permeability are classified as Class V. Class Va includes acidic drugs and Class Vb includes basic drugs. By the addition of these properties of drugs (site-dependent or site-independent solubility, site-dependent or site-independent permeability, and narrow absorption window), the number of drugs classified as Class I, II, III, or IV decreases and the prediction of IVIVC becomes safer.

Possible new criteria and class boundaries are proposed for additional biowaivers based on the underlying physiology of the gastrointestinal tract. The proposed changes in new class boundaries for solubility and permeability are narrowing down the required solubility pH range from 1.0–7.5 to 1.0–6.8 and reducing the high-permeability requirement from 90% to 85%. The proposed new criterion for potential biowaiver extension is to define a new intermediate-permeability class boundary and allow biowaivers for highly soluble and intermediately permeable drugs in IR solid oral dosage forms with no less than 85% dissolved in 15 min in all physiologically relevant dissolution media, provided these IR products contain only known excipients that do not affect the oral drug absorption. The areas that require further research are increase in the dose volume for solubility classification to 500 mL, inclusion of bile salt (BS) in the solubility measurement, use of intrinsic dissolution method for solubility classification, definition of an intermediate-solubility class for BCS Class II drugs, and inclusion of surfactants in in vitro dissolution testing.

Table 6.5 expands this classification to include a more detailed description including the effect of transporter efflux factors.

Observed in vivo differences in the rate and extent of absorption of a drug from two pharmaceutically equivalent solid oral products may be due to differences in drug dissolution in vivo. However, when the in vivo dissolution of an instant release solid oral dosage form is rapid in relation to gastric

TABLE 6.5
BCS as Defined by the FDA and Modified by Recent Findings

	High solubility (e.g., when the highest dose strength is soluble in 250 mL or less of aqueous media over a pH range of 1–7. 5 at 37°C)	Low solubility
High permeability (e.g., absorption >90% compared to IV dose [drug + metabolite])	Class 1 (generally about 8% of new leads) • High solubility • High permeability • Rapid dissolution for biowaiver • Route of elimination: metabolism, extensive • Transporter effects: minimal Examples: abacavir, acetaminophen, *acyclovir*^{b}, *amiloride*^{S,I}, amitriptyline^{S,I}, antipyrine, *atropine*, **buspirone**^{c}, caffeine, *captopril*, chloroquine^{S,I}, **chlorpheniramine**, cyclophosphamide, desipramine, **diazepam**; **diltiazem**^{S,I}, **diphenhydramine**, disopyramide, **doxepin**, oxycycline, enalapril, ephedrine, ergonovine, ethambutol, ethinyl estradiol, fluoxetine^{I}, glucose, imipramine^{I}, ketoprofen, **ketorolac**, labetalol, levodopa^{S}, levofloxacin^{S}, **lidocaine**^{I}, lomefloxacin, **meperidine**, metoprolol, metronidazole, **midazolam**^{S,I}, **minocycline**, misoprostol, **nifedipine**^{S}, phenobarbital, phenylalanine, prednisolone. **primaquine**^{S}, promazine, propranolol^{I}, **quinidine**^{S,I}, **rosiglitazone**, salicylic acid, theophylline, valproic acid, **verapamil**^{I}, zidovudine	Class 2 • Low solubility • High permeability • Route of elimination: metabolism, extensive • Transporter: efflux transporter effects predominant Examples: **amiodarone**^{I}, **atorvastatin**^{S,I}, **azithromycin**^{S,I}, **carbamazepine**^{S,I}, **carvedilol**, chlorpromazine^{I}, *ciprofloxacin*^{S}, **cisapride**^{S}, **cyclosporine**^{S,I}, **danazol**, **dapsone**, diclofenac, diflunisal, digoxin^{S}, *erythromycin*^{S,I}. flurbiprofen, **glipizide**, glyburide^{S,I}, griseofulvin, ibuprofen, **indinavir**^{S}, **indomethacin**, **itraconazole**^{S,I}, **ketoconazole**^{I}, **lansoprazole**^{I}, **lovastatin**^{S,I}, *mebendazole*, naproxen, nelfinavir^{S,I}, ofloxacin, oxaprozin, phenazopyridine, phenytoin^{S}, piroxicam, Raloxifene^{S}, **ritonavir**^{S,I}, **saquinavir**^{S,I}, **sirolimus**^{S}, spironolactone^{I}, **tacrolimus**^{S,I}, talinolol^{S}, **tamoxifen**^{I}; **terfenadine**^{I}; warfarin
Low permeability	Class 3 • High solubility • Low permeability • Route of elimination: renal and/or biliary elimination of unchanged drug; metabolism poor • Transporter: absorptive effects predominant Examples: *acyclovir*, *amiloride*^{S,I}, amoxicillin^{S,I}, atenolol, *atropine*, bidisomide, bisphosphonates; *captopril*, cefazolin, cetirizine, cimetidine^{S}, *ciprofloxacin*^{S}, cloxacillin, dicloxacillin^{S}, *erythromycin*^{S,I}, famotidine, fexofenadine^{S}, folinic acid, *furosemide*, ganciclovir, *hydrochlorothiazide*, lisinopril, metformin, *methotrexate*, nadolol, penicillins, pravastatin^{S}, ranitidine^{S}, tetracycline, trimethoprim^{S}, valsartan, zalcitabine	Class 4 • Low solubility • Low permeability • Route of elimination: renal and/or biliary elimination of unchanged drug; metabolism poor • Transporter: absorptive and efflux transporters can be predominant Examples: amphotericin B, chlorothiazide, chlorthalidone, *ciprofloxacin*^{S}, colistin, *furosemide*, *hydrochlorothiazide*, *mebendazole*, *methotrexate*, neomycin

Notes: The compounds listed in *italic* are those falling in more than one category by different authors, which could be a result of the definition of the experimental conditions. The compounds listed in bold are primarily CYP3A substrates where metabolism accounts for more than 70% of the elimination; superscript I and/or S indicate P-gp inhibitors and/or substrate, respectively. The Class 1 and Class 2 compounds are eliminated primarily via metabolism, whereas Class 3 and Class 4 compounds are primarily eliminated unchanged into the urine and bile.

emptying and the drug has high permeability, the rate and extent of drug absorption is unlikely to be dependent on drug dissolution and/or gastrointestinal transit time. Under such circumstances, demonstration of in vivo BA or BE may not be necessary for drug products containing Class 1 drug substances, as long as the inactive ingredients used in the dosage form do not significantly affect absorption of the active ingredients. The BCS approach outlined in this guidance can be used to justify biowaivers for *highly soluble* and *highly permeable* drug substances (i.e., Class 1) in instant release solid oral dosage forms that exhibit *rapid* in vitro *dissolution* using the recommended test methods (21 CFR 320.22(e)). The recommended methods for determining solubility, permeability, and in vitro dissolution are discussed in the following.

Solubility

The solubility class boundary is based on the highest dose strength of an instant release product that is the subject of a biowaiver request. A drug substance is considered *highly soluble* when the highest dose strength is soluble in 250 mL or less of aqueous media over the pH range of 1–7.5. The volume estimate of 250 mL is derived from typical BE study protocols that prescribe administration of a drug product to fasting human volunteers with a glass (about 8 ounces) of water.

Permeability

The permeability class boundary is based indirectly on the extent of absorption (fraction of dose absorbed, not systemic BA) of a drug substance in humans and directly on measurements of the rate of mass transfer across human intestinal membrane. Alternatively, nonhuman systems capable of predicting the extent of drug absorption in humans can be used (e.g., in vitro epithelial cell culture methods). In the absence of evidence suggesting instability in the gastrointestinal tract, a drug substance is considered to be *highly permeable* when the extent of absorption in humans is determined to be 90% or more of an administered dose based on a mass balance determination or in comparison to an IV reference dose.

Dissolution

An instant release drug product is considered *rapidly dissolving* when no less than 85% of the labeled amount of the drug substance dissolves within 30 min, using *USP* Apparatus I at 100 rpm (or Apparatus II at 50 rpm) in a volume of 900 mL or less in each of the following media: (1) 0.1 N HCl or simulated gastric fluid USP without enzymes, (2) a pH 4.5 buffer, and (3) a pH 6.8 buffer or simulated intestinal fluid USP without enzymes. Appendix B provides the conditions of dissolution provided by the U.S. FDA if it is not listed in the USP or where the FDA has considered a different testing protocol to be more suitable. Whereas the FDA lists these conditions, the onus of proving that these conditions apply to the product remains with the sponsor.

The purpose of in vitro dissolution studies in the early stage of drug development is to select the optimum formulation, evaluate the active pharmaceutical ingredient (API) and excipients, and assess any minor changes in the drug product. However, from the IVIVC perspective, dissolution is proposed to be a surrogate of drug BA. Thus, a more rigorous dissolution standard may be necessary for the in vivo waiver. Generally, a dissolution methodology, which is able to discriminate between study formulations and which best reflects the in vivo behavior, is selected.

The USP and National Formulary (NF) recognize seven types of dissolution apparatus and describe the allowable modifications in detail. The choice should be considered during the development of the dissolution method since it can affect the results and duration of the test. The type of dosage form under investigation is the primary consideration in apparatus selection. The compendia apparatus for dissolution as per the USP are Apparatus 1 (rotating basket), Apparatus 2 (paddle assembly), Apparatus 3 (reciprocating cylinder), Apparatus 4 (flow-through cell), Apparatus 5 (paddle over disk), Apparatus 6 (cylinder), and Apparatus 7 (reciprocating holder). The European Pharmacopoeia (EP) has adopted some of the designs described in the USP with some minor modifications in the specifications.

The most important parameters to consider in simulating in vivo conditions are pH, buffer composition, buffer capacity, temperature, volume, and hydrodynamics. Noncompendia media have shown better IVIVC than compendia media listed in official monographs. In fact, noncompendia media have discriminating power and are widely used. Basically, pH increases from the small intestine to the large intestine (pH 6.7–8) requiring dissolution testing of extended-release drug products to be carried out through the entire physiological pH range (6.7–8). The ionic strength of the dissolution media also plays a vital role in dissolution testing. Ions present in food and food-induced secretions in the GIT cause changes in ionic strength of the gastrointestinal fluid. Buffer capacity is also important in dissolution testing of formulations that contain acidic or basic excipients. Studies have shown that the buffer capacity of a medium is an important criterion in designing dissolution media for investigating IVIVC.

Due to the nature of the test method, "quality by design" is an important qualification aspect for in vitro dissolution test equipment. The suitability of apparatus for dissolution/drug release testing depends on both the physical and chemical calibrations, which qualify the equipment for further analysis. Besides geometrical and dimensional accuracy and precision, as described in USP 27 and the EP, any irregularities such as vibration or undesired agitation due to mechanical imperfections are to be avoided. Temperature of the test medium, rotation speed/flow rate, volume sampling probes, and procedures need to be monitored periodically. Another vital aspect of qualification and validation is the "apparatus suitability test." The use of USP calibrator tablets (for Apparatus 1 and 2 disintegrating as well as nondisintegrating calibrator tablets) is the only standardized approach to establish apparatus suitability for conducting dissolution tests and is able to identify operator failures. Suitability tests using specific calibrators have also been developed for Apparatus 3; the aim is to generate a set of calibrators for each and every compendia dissolution test apparatus.

Comparison between dissolution profiles could be achieved using a difference factor (f_1) and a similarity factor (f_2), which originates from simple model-independent approach. The difference factor calculates the percent difference between the two curves at each time point and is a measurement of the relative error between the two curves:

$$f_1 = \left\{ \frac{\left[\sum_{t=1}^{n} |R_t - T_t| \right]}{\left[\sum_{t=1}^{n} |R_t| \right]} * 100 \right\} \tag{6.9}$$

where

n is the number of time points
R_t is the dissolution value of the reference batch at time t
T_t is the dissolution value of the test batch at time t

The similarity factor is a logarithmic reciprocal square root transformation of the sum squared error and is a measurement of the similarity in the percent dissolution between the two curves:

$$f_2 = 50 * \log \left\{ \left[1 + \left(\frac{1}{n} \right) \sum_{t=1}^{n} (R_t - T_t)^2 \right]^{-0.5} * 100 \right\} \tag{6.10}$$

Generally, f_1 values up to 15 (0–15) and f_2 values greater than 50 (50–100) ensure sameness or equivalence of the two curves. The MDT_{vitro} is the mean time for the drug to dissolve under in vitro dissolution medium conditions. This is calculated using the following:

$$MDT_{vitro} = \int_0^{\infty} \left(M_\infty - M(t)\right) \frac{dt}{M_\infty} \tag{6.11}$$

For the IVIVC development, the dissolution profiles of at least 12 individual dosage units from each lot should be determined. The coefficient of variation (CV) for mean dissolution profiles of a single batch should be less than 10%. Since dissolution apparatus tend to become less discriminative when operated at faster speeds, lower stirring speeds should be evaluated and an appropriate speed chosen in accordance with the test data. Using the basket method, the common agitation is 50–100 rpm; with the paddle method, it is 50–75 and 25 rpm for suspension.

Method of Classification

The following approaches are recommended for classifying a drug substance and determining the dissolution characteristics of an instant release drug product according to the BCS.

Determining Drug Substance Solubility Class

An objective of the BCS approach is to determine the equilibrium solubility of a drug substance under physiological pH conditions. The pH-solubility profile of the test drug substance should be determined at 37°C ± 1°C in aqueous media with a pH in the range of 1–7.5. A sufficient number of pH conditions should be evaluated to accurately define the pH-solubility profile. The number of pH conditions for a solubility determination can be based on the ionization characteristics of the test drug substance. For example, when the pK_a of a drug is in the range of 3–5, solubility should be determined at pH = pK_a, pH = $pK_a + 1$, pH = $pK_a - 1$, and pH = 1 and 7.5. A minimum of three replicate determinations of solubility in each pH condition is recommended. Depending on study variability, additional replication may be necessary to provide a reliable estimate of solubility. Standard buffer solutions described in the USP are considered appropriate for use in solubility studies. If these buffers are not suitable for physical or chemical reasons, other buffer solutions can be used. Solution pH should be verified after addition of the drug substance to a buffer. Methods other than the traditional shake-flask method, such as acid or base titration methods, can also be used with justification to support the ability of such methods to predict equilibrium solubility of the test drug substance. Concentration of the drug substance in selected buffers (or pH conditions) should be determined using a validated stability-indicating assay that can distinguish the drug substance from its degradation products. If degradation of the drug substance is observed as a function of buffer composition and/or pH, it should be reported along with other stability data.

The solubility class should be determined by calculating the volume of an aqueous medium sufficient to dissolve the highest dose strength in the pH range of 1–7.5. A drug substance should be classified as highly soluble when the highest dose strength is soluble in ≤250 mL of aqueous media over the pH range of 1–7.5.

Determining Drug Substance Permeability Class

The permeability class of a drug substance can be determined in human subjects using mass balance, absolute BA, or intestinal perfusion approaches. Recommended methods not involving human subjects include in vivo or in situ intestinal perfusion in a suitable animal model (e.g., rats), and/or in vitro permeability methods using excised intestinal tissues, or monolayers of suitable epithelial cells. In many cases, a single method may be sufficient (e.g., when the absolute BA is 90% or more or when 90% or more of the administered drug is recovered in urine). When a single method fails to conclusively demonstrate a permeability classification, two different methods may be advisable. Chemical structure and/or certain physicochemical attributes of a drug substance (e.g., partition coefficient in suitable systems) can provide useful information about its permeability characteristics. Sponsors may wish to consider use of such information to further support a classification.

Pharmacokinetic Studies in Humans

Mass Balance Studies

PK mass balance studies using unlabeled, stable isotopes or a radiolabeled drug substance can be used to document the extent of absorption of a drug. Depending on the variability of the studies, a sufficient number of subjects should be enrolled to provide a reliable estimate of extent of absorption. Because this method can provide highly variable estimates of drug absorption for many drugs, other methods described in the following may be preferable.

Absolute Bioavailability Studies

Oral BA determination can utilize the IV administration where possible as a reference can be used. Depending on the variability of the studies, a sufficient number of subjects should be enrolled in a study to provide a reliable estimate of the extent of absorption. When the absolute BA of a drug is shown to be 90% or more, additional data to document drug stability in the gastrointestinal fluid are not necessary.

The following methods can be used to determine the permeability of a drug substance from the gastrointestinal tract: (1) in vivo intestinal perfusion studies in humans, (2) in vivo or in situ intestinal perfusion studies using suitable animal models, (3) in vitro permeation studies using excised human or animal intestinal tissues, or (4) in vitro permeation studies across a monolayer of cultured epithelial cells.

In vivo or in situ animal models and in vitro methods, such as those using cultured monolayers of animal or human epithelial cells, are considered appropriate for passively transported drugs. The observed low permeability of some drug substances in humans could be caused by efflux of drugs via membrane transporters such as P-glycoprotein (P-gp). When the efflux transporters are absent in these models, or their degree of expression is low compared to that in humans, there may be a greater likelihood of misclassification of permeability class for a drug subject to efflux compared to a drug transported passively. Expression of known transporters in selected study systems should be characterized. Functional expression of efflux systems (e.g., P-gp) can be demonstrated with techniques such as bidirectional transport studies, demonstrating a higher rate of transport in the basolateral-to-apical direction as compared to apical-to-basolateral direction using selected model drugs or chemicals at concentrations that do not saturate the efflux system (e.g., cyclosporine A, vinblastine, rhodamine 123). An acceptance criterion for intestinal efflux that should be present in a test system cannot be set at this time. Instead, this guidance recommends limiting the use of nonhuman permeability test methods for drug substances that are transported by passive mechanisms. PK studies on dose linearity or proportionality may provide useful information for evaluating the relevance of observed in vitro efflux of a drug. For example, there may be fewer concerns associated with the use of in vitro methods for a drug that has a higher rate of transport in the basolateral-to-apical direction at low drug concentrations but exhibits linear PK in humans.

Poor absorption or permeation is more likely when there are more than 5 H-bond donors, there are 10 H-bond acceptors, the molecular weight is greater than 500, and the calculated log P is greater than 5. This is also often referred to as Rule of 5 of Lipinski. However, Lipinski specifically states that the Rule of 5 only holds for compounds that are *not* substrates for active transporters. Since almost all drugs are substrates for some transporter, much remains to be studied about the Lipinski's rule. In addition, unless a drug molecule can passively gain intracellular access, it is not possible to simply investigate whether the molecule is a substrate for efflux transporters.

Several generalizations can be made about the interplay of transporters and the BDCS classification:

a. *Transporter effects are minimal for Class 1 compounds.* The high permeability/high solubility of such compounds allows high concentrations in the gut to saturate any transporter, both efflux and absorptive. Class 1 compounds may be substrates for both uptake and efflux transporters in vitro in cellular systems under the right conditions (e.g., midazolam and nifedipine are substrates for P-gp), but transporter effects will not be important clinically. It is therefore possible that some compounds that should be considered Class 1 in terms of drug absorption and disposition are not Class 1 in BDCS due to the requirement of good solubility and rapid dissolution at low pH values. Such pH effects would not be limiting in vivo where absorption takes place from the intestine. Examples of this include the NSAIDs diclofenac, diflunisal, flurbiprofen, indomethacin, naproxen, and piroxicam; warfarin is almost completely bioavailable. In contrast, ofloxacin is listed as Class 2 because of its low solubility at pH 7.5.
b. *Efflux transporter effects will predominate for Class 2 compounds.* The high permeability of these compounds will allow ready access into the gut membranes, and uptake transporters will have no effect on absorption, but the low solubility will limit the concentrations coming into the enterocytes, thereby preventing saturation of the efflux transporters. Consequently, efflux transporters will affect the extent of oral BA and the rate of absorption of Class 2 compounds.
c. *Transporter–enzyme interplay in the intestines will be important primarily for Class 2 compounds that are substrates for cytochrome P450 3A4 (CYP3A) and Phase 2 conjugation enzymes.* For such compounds, intestinal uptake transporters will generally be unimportant due to the rapid permeation of the drug molecule into the enterocytes as a function of their high lipid solubility. That is, absorption of Class 2 compounds is primarily passive and a function of lipophilicity. However, due to the low solubility of these compounds, there will be little opportunity to saturate apical efflux transporters and intestinal enzymes such as CYP 3A4 and UDP-glucuronosyltransferases (UGTs). Thus, changes in transporter expression and inhibition or induction of efflux transporters will cause changes in intestinal metabolism of drugs that are substrates for the intestinal metabolic enzymes. Note the large number of Class 2 compounds in Table I that are primarily substrates for CYP3A (compounds listed in bold) as well as substrates or inhibitors of the efflux transporter P-gp (indicated by superscripts S and I, respectively). Work in our laboratory has characterized this interplay in the absorptive process for the investigational cysteine protease inhibitor K77 and sirolimus, substrates for CYP3A and P-gp, and more recently for raloxifene (33), a substrate for UGTs and P-gp.
d. *Absorptive transporter effects will predominate for Class 3 compounds.* For Class 3 compounds, sufficient drug will be available in the gut lumen due to good solubility, but an absorptive transporter will be necessary to overcome the poor permeability characteristics of these compounds. However, intestinal apical efflux transporters may also be important for the absorption of such compounds when sufficient enterocyte penetration is achieved via an uptake transporter.

Table 6.6 lists model drugs suggested for use in establishing the suitability of a permeability method. The permeability of these compounds was determined based on data available to the FDA. Potential *internal standards* (ISs) and *efflux pump substrates* (ESs) are also identified.

For application of the BCS, an apparent passive transport mechanism can be assumed when one of the following conditions is satisfied:

- A linear (PK) relationship between the dose (e.g., relevant clinical dose range) and measures of BA (area under the concentration–time curve) of a drug is demonstrated in humans.

TABLE 6.6
Model Drugs to Establish Permeability of Drugs

Drug	Permeability Class
Antipyrine	High (potential IS candidate)
Caffeine	High
Carbamazepine	High
Fluvastatin	High
Ketoprofen	High
Metoprolol	High (potential IS candidate)
Naproxen	High
Propranolol	High
Theophylline	High
Verapamil	High (potential ES candidate)
Amoxicillin	Low
Atenolol	Low
Furosemide	Low
Hydrochlorothiazide	Low
Mannitol	Low (potential IS candidate)
Methyldopa	Low
Polyethylene glycol (400)	Low
Polyethylene glycol (1000)	Low
Polyethylene glycol (4000)	Low (zero permeability marker)
Ranitidine	Low

- Lack of dependence of the measured in vivo or in situ permeability is demonstrated in an animal model on initial drug concentration (e.g., 0.01, 0.1, and 1 times the highest dose strength dissolved in 250 mL) in the perfusion fluid.
- Lack of dependence of the measured in vitro permeability on initial drug concentration (e.g., 0.01, 0.1, and 1 times the highest dose strength dissolved in 250 mL) is demonstrated in donor fluid and transport direction (e.g., no statistically significant difference in the rate of transport between the apical-to-basolateral and basolateral-to-apical direction for the drug concentrations selected) using a suitable in vitro cell culture method that has been shown to express known efflux transporters (e.g., P-gp).

To demonstrate the suitability of a permeability method intended for application of the BCS, a rank order relationship between test permeability values and the extent of drug absorption data in human subjects should be established using a sufficient number of model drugs. For in vivo intestinal perfusion studies in humans, six model drugs are recommended. For in vivo or in situ intestinal perfusion studies in animals and for in vitro cell culture methods, 20 model drugs are recommended. Depending on study variability, a sufficient number of subjects, animals, excised tissue samples, or cell monolayers should be used in a study to provide a reliable estimate of drug permeability. This relationship should allow precise differentiation between drug substances of low and high intestinal permeability attributes.

For demonstration of suitability of a method, model drugs should represent a range of low (e.g., <50%), moderate (e.g., 50%–89%), and high (≥90%) absorption. Sponsors may select compounds from the list of drugs and/or chemicals provided in Attachment A or they may choose to select other drugs for which there is information available on the mechanism of absorption and reliable estimates of the extent of drug absorption in humans.

After demonstrating the suitability of a method and maintaining the same study protocol, it is not necessary to retest all selected model drugs for subsequent studies intended to classify a drug substance. Instead, a low- and a high-permeability model drug should be used as ISs (i.e., included in the perfusion fluid or donor fluid along with the test drug substance). These two ISs are in addition to the fluid volume marker (or a zero permeability compound such as polyethylene glycol (PEG) 4000) that is included in certain types of perfusion techniques (e.g., closed loop techniques). The choice of ISs should be based on compatibility with the test drug substance (i.e., they should not exhibit any significant physical, chemical, or permeation interactions). When it is not feasible to follow this protocol, the permeability of ISs should be determined in the same subjects, animals, tissues, or monolayers, following evaluation of the test drug substance. The permeability values of the two ISs should not differ significantly between different tests, including those conducted to demonstrate suitability of the method. At the end of an in situ or in vitro test, the amount of drug in the membrane should be determined.

For a given test method with set conditions, the selection of a high-permeability IS with permeability in close proximity to the low-/high-permeability class boundary may facilitate classification of a test drug substance. For instance, a test drug substance may be determined to be highly permeable when its permeability value is equal to or greater than that of the selected IS with high permeability.

INSTABILITY IN THE GASTROINTESTINAL TRACT

Determining the extent of absorption in humans based on mass balance studies using total radioactivity in urine does not take into consideration the extent of degradation of a drug in the gastrointestinal fluid prior to intestinal membrane permeation. In addition, some methods for determining permeability could be based on loss or clearance of a drug from fluids perfused into the human and/or animal gastrointestinal tract either in vivo or in situ. Documenting the fact that drug loss from the gastrointestinal tract arises from intestinal membrane permeation, rather than a degradation process, will help establish permeability. Stability in the gastrointestinal tract may be documented using gastric and intestinal fluids obtained from human subjects. Drug solutions in these fluids should be incubated at 37°C for a period that is representative of in vivo drug contact with these fluids, for example, 1 h in gastric fluid and 3 h in intestinal fluid. Drug concentrations should then be determined using a validated stability-indicating assay method. Significant degradation (>5%) of a drug in this protocol could suggest potential instability. Obtaining gastrointestinal fluids from human subjects requires intubation and may be difficult in some cases. Use of gastrointestinal fluids from suitable animal models and/or simulated fluids such as gastric and intestinal fluids USP can be substituted when properly justified.

DETERMINING DRUG PRODUCT DISSOLUTION CHARACTERISTICS AND DISSOLUTION PROFILE SIMILARITY

Dissolution testing should be carried out in USP Apparatus I at 100 rpm or Apparatus II at 50 rpm using 900 mL of the following dissolution media: (1) 0.1 N HCl or simulated gastric fluid USP without enzymes, (2) a pH 4.5 buffer, and (3) a pH 6.8 buffer or simulated intestinal fluid USP without enzymes. For capsules and tablets with gelatin coating, simulated gastric and intestinal fluids USP (with enzymes) can be used.

Dissolution testing apparatus used in this evaluation should conform to the requirements in USP (<711> dissolution). Selection of the dissolution testing apparatus (USP Apparatus I or II) during drug development should be based on a comparison of in vitro dissolution and in vivo PK data available for the product. The USP Apparatus I (*basket method*) is generally preferred for capsules and products that tend to float, and USP Apparatus II (*paddle method*) is generally preferred for tablets. For some tablet dosage forms, in vitro (but not in vivo) dissolution may be slow due to the manner in which the disintegrated product settles at the bottom of a dissolution vessel. In such situations, USP Apparatus I may be preferred over Apparatus II. If the testing conditions need to be modified to better reflect rapid in vivo dissolution (e.g., use of a different rotating speed), such modifications can be justified by comparing in vitro dissolution with in vivo absorption data (e.g., a relative BA study using a simple aqueous solution as the reference product).

A minimum of 12 dosage units of a drug product should be evaluated to support a biowaiver request. Samples should be collected at a sufficient number of intervals to characterize the dissolution profile of the drug product (e.g., 10, 15, 20, and 30 min).

When comparing the test and reference products, dissolution profiles should be compared using a similarity factor (f_2). The similarity factor is a logarithmic reciprocal square root transformation of the sum of squared error and is a measurement of the similarity in the percent (%) of dissolution between the two curves:

$$f_2 = 50 * \log\left\{\left[1+\left(\frac{1}{n}\right)\sum(R_t - T_t)^2\right]^{-0.5} * 100\right\} \tag{6.12}$$

Two dissolution profiles are considered similar when the f_2 value is ≥50. To allow the use of mean data, the CV should not be more than 20% at the earlier time points (e.g., 10 min) and should not be more than 10% at other time points. Note that when both test and reference products dissolve 85% or more of the label amount of the drug in 15 min using all three dissolution media recommended earlier, the profile comparison with an f_2 test is unnecessary.

Additional Considerations

When requesting a BCS-based waiver for in vivo BA/BE studies for instant release solid oral dosage forms, applicants should note that the following factors can affect their request or the documentation of their request.

Excipients

Excipients can sometimes affect the rate and extent of drug absorption. In general, using excipients that are currently in FDA-approved instant release solid oral dosage forms will not affect the rate or extent of absorption of a highly soluble and highly permeable drug substance that is formulated in a rapidly dissolving instant release product. To support a biowaiver request, the quantity of excipients in the instant release drug product should be consistent with the intended function (e.g., lubricant). When new excipients or atypically large amounts of commonly used excipients are included in an instant release solid dosage form, additional information documenting the absence of an impact on BA of the drug may be requested by the agency. Such information can be provided with a relative BA study using a simple aqueous solution as the reference product. Large quantities of certain excipients such as surfactants (e.g., polysorbate 80) and sweeteners (e.g., mannitol or sorbitol) may be problematic, and sponsors are encouraged to contact the review division when this is a factor.

Prodrugs

Permeability of prodrugs will depend on the mechanism and (anatomical) site of conversion to the drug substance. When the prodrug-to-drug conversion is shown to occur predominantly after intestinal membrane permeation, the permeability of the prodrug should be measured. When this conversion occurs prior to intestinal permeation, the permeability of the drug should be determined. Dissolution and pH-solubility data on both prodrug and drug can be relevant. Sponsors may wish to consult with appropriate review staff before applying the BCS approach to instant release products containing prodrugs.

Stereochemistry

When one enantiomer has higher affinity toward receptors than its antipode, it gives rise to stereoselectivity in PK and/or pharmacodynamics. If such stereoisomers are administered orally in the form of a racemate, one form may have higher BA than the other. Obviously, in vitro dissolution data for the racemate will not be useful in the development of IVIVC and subsequent prediction of the in vivo availability of the active enantiomer. Thus, consideration of stereoisomerism is necessary in the development of IVIVC to provide a more meaningful relationship.

Exceptions

BCS-based biowaivers are not applicable for the following.

Narrow Therapeutic Range Drugs

The narrow therapeutic range drug products are defined as those containing certain drug substances that are subject to therapeutic drug concentration or pharmacodynamic monitoring and/or where product labeling indicates a narrow therapeutic range designation. Examples include digoxin, lithium, phenytoin, theophylline, and warfarin. Because not all drugs subject to therapeutic drug concentration or pharmacodynamic monitoring are narrow therapeutic range drugs, sponsors should contact the appropriate review division to determine whether a drug should be considered to have a narrow therapeutic range.

Products Designed to Be Absorbed in the Oral Cavity

A request for a waiver of in vivo BA/BE studies based on the BCS is not appropriate for dosage forms intended for absorption in the oral cavity (e.g., sublingual or buccal tablets).

REASONS FOR FAILURE OF SOME IVIVC

The best dissolution method for IVIVC is, obviously, the method that describes what happens in vivo. Although there are many published examples of drugs with dissolution data that correlate well with drug absorption in the body, there are also many examples of a poor correlation between dissolution and drug absorption. The problem of no correlation between BA and dissolution may be due to the complexity of drug absorption and the weakness of the drug dissolution design. For example, a product that involves fatty components may be subject to longer retention in the gastrointestinal tract. Ionic strength and pH change of the gastrointestinal fluid, peristaltic movement, digestive enzymes, bile, and presence of surfactants and other excipients in the formulation are the factors that can influence drug release in vivo. All of these factors cannot be easily reproduced in vitro by a simple dissolution method. In addition, the relative importance of these factors is not similar if the subject is in fed or fasted state; in the fed state, food also has a direct influence on the API or formulation behavior.

Limitations in the IVIVC Arising from the In Vivo Data

It is easily understood why the following applies:

- More than one dosage form is needed and if possible IV or solution is essential to calculate deconvolution.
- PK and absorption of the drug should be "linear." If the PK processes are dependent on the fraction of dose reaching the systemic blood flow (or of the dose administered) or on the rate of absorption, comparison between formulation and simulation cannot be made. This nonlinearity may be owing to saturable absorption processes (active absorption); induction or inhibition of the metabolism; the first past effect, which is rate/absorption dependent; etc. Those points must be studied before any attempt to establish an IVIVC.
- Absorption should not be the limiting factor; if the solubility is not the limiting factor in comparison to the drug release, an IVIVC may be attempted. The release must depend on the formulation and must be the slowest phenomenon versus dissolution and absorption.

Enteric-Coated Multiple Unit Dosage Form

In this dosage form, an individual unit empties gradually and separately from the stomach to the duodenum. Simulation of these conditions in vitro is troublesome and may be impossible. Direct prediction of the in vivo absorption profile from in vitro dissolution data for a multiple unit system was difficult but the convolution method overcame this problem. A good correlation (level A) was obtained for multiple-unit enteric-coated granules using the convolution method.

Parenteral Controlled- or Sustained-Release Drug Delivery System

A few methods for an in vitro drug release study of microparticles for parenteral administration have been established so far. These include flow-through cell and dialysis technique among others.

Buccal Tablet

Interesting correlations have been developed between the in vivo residence time of mucoadhesive tablets in the mouth and their in vitro bending point. Linear regression models permit optimization of buccal tablets to enhance the adhesion time using in vitro bending point as a selection criterion.

Transdermal Drug Delivery System

The variability of human skin thickness and resulting differences in the permeation of drugs in the case of transdermal drug delivery (TDD) is a major problem that still needs to be addressed. The USP 29 gives methods for in vitro drug release testing of transdermal patches like the paddle over disk, cylinder method, and reciprocating disk method. The Franz diffusion cell has been extensively used. In most instances, steady-state flux has been extrapolated to determine the in vivo plasma concentrations using PK modeling. Current in vitro methods are not sufficient for product development purposes but are good for candidate selection purposes. More comprehensive research should be done in this area to come up with better techniques to accurately predict the permeation characteristics of drugs from different formulation.

Suppository

Modified basket or paddle methods are recommended for lipophilic suppositories, while conventional basket, paddle, or flow-through cells are recommended for hydrophilic suppositories. The in vitro dissolution kinetics of the suppository formulations of indomethacin using two different apparatus, namely, a flow-through cell (Dissotest®) and a dialysis rotating cell (Pharmatest®), show that the distinction between dissolution profiles was more evident using the flow-through cell and it is possible to establish a linear relationship between the in vitro results and the in vivo percentages absorbed than using the Wagner–Nelson method.

Nasal Drug Delivery System

A variety of methods on in vitro testing of nasal drug delivery systems are available such as emitted dose, droplet or particle size distribution, spray pattern, and plume geometry. FDA guidance recommends these methods as a means of predicting BA and bioinequivalence for topically acting solution formulations because they can be performed reproducibly and are more discriminating between products.

Colonic Drug Delivery System

In the last decades, there has been continual interest in site-specific delivery to the colon. Recently, new types of site-specific delivery formulations have been developed for the treatment of ulcerative colitis and other colon-related diseases. USP Apparatus 3 is used for in vitro study and the passage through the gastrointestinal tract is simulated with a physiologically based pH gradient. Subsequently, the fraction of drug release in vitro is compared with the fraction of drug release in vivo.

Biodegradable Parenteral Delivery System

Various apparatus and methods have been developed to establish IVIVC for biodegradable parenteral delivery systems. However, only few examples can be found in the literature where an in vitro drug method accurately predicts the in vivo release profile for parenteral biodegradable depot systems. This demonstrates the difficulties in establishing IVIVC for this class of formulations due to the large number of parameters potentially affecting drug release in vivo and in vitro. As, in most cases, diffusion, dissolution, and erosion govern drug release, a simple kinetic model is unlikely to explain the overall in vivo release behavior. A level A correlation was established for the formulation with predominant diffusion-controlled release. A level B correlation, however, was achieved even when drug release occurred by a combination of diffusion and erosion processes.

CURRENT TRENDS

Current research on IVIVC deployment includes microparticles of tramadol hydrochloride, low-molecular-weight model compound, [18F]-2-fluoro-2-deoxy-D-glucose ([18F] FDG) from liposomes, amiodarone, and sodium diatrizoate (DTZ) disappearance profile obtained from the donor compartment of the rotating dialysis cell model and the joint disappearance profile following intra-articular administration, sustained-release formulations elaborated by a one-step melt granulation method using theophylline as model drug, clarithromycin granular suspensions, a novel two-phase modified-release formulation for diltiazem hydrochloride, metoprolol tartrate sustained-release capsule, gliclazide extended-release formulations, and amoxicillin dispersible tablets.

Dissolution profiles of solid formulations of a poorly soluble model compound have been compared in biorelevant dissolution medium (BDM) simulating fasted and two levels of fed state. A nonphysiologically relevant medium containing the cationic surfactant cetrimide has also been investigated. The IVIVC models developed indicated that fed state media (BDM 3) containing high levels of BSs and lipolysis products (LPs) were best able to predict in vivo PK parameters (C_{max} and AUC) with PE < 10%. It was inferred overall that design and use of appropriate media for in vitro dissolution are extremely important. This study demonstrated the potential value of physiologically relevant media containing both BS and LP in formulation and early drug development.

The predictive accuracy or "predictability" of two IVIVC modeling techniques, convolution and deconvolution, shows that by means of a simulation study, the convolution produces far more accurate results, accurately predicting the observed plasma concentration–time profile and, therefore, comfortably meeting the FDA validation criteria. The fact that the model using the deconvolution-based techniques often fails to describe the simulated data and thus fails the FDA validation test is of great concern.

Application in Drug Delivery System

Various rate controlling technologies are used as the basis for modified-release dosage forms, for example, diffusion–dissolution, matrix retardation, and osmosis, to control and prolong the release of drugs, for the administration by oral or parenteral route.

The novel drug delivery systems have been developed such as osmotic-controlled release and drug delivery system, liposomes, niosomes, pharmacosomes, microspheres, nanoparticles, implants, in situ gelling system, organogels, TDD systems, and parenteral depots as a substitute for conventional dosage forms. The obvious objective of these dosage forms is to achieve zero-order, long-term, pulsatile, or "on demand" delivery. Major applications of IVIVC related to oral drug delivery and a few issues related to the development of IVIVC models for parenteral drug delivery are addressed herewith.

Early Stage Drug Delivery

The most crucial stage in drug development is drug candidate selection. Such selection is primarily based on the drug "developability" criteria, which include physicochemical properties of the drug and the results obtained from preformulation, preliminary studies involving several in vitro systems and in vivo animal models, which address efficacy and toxicity issues. During this stage, IVIVC (exploring the relationship between in vitro and in vivo properties) of the drug in animal models provides an idea about the feasibility of the drug delivery system for a given drug candidate. In such correlations, study designs including study of more than one formulation of the modified-release dosage forms and a rank order of release (fast/slow) of the formulations should be incorporated. Even though the formulations and methods used at this stage are not optimal, they promise better design and development efforts in the future.

Formulation Assessment: In Vitro Dissolution

A suitable dissolution method that is capable of distinguishing the performance of formulations with different release rates in vitro and in vivo is an important tool in product development. Depending on the nature of the correlation, further changes to the dissolution method can be made. When the discriminatory in vitro method is validated, further formulation development can be relied on the in vitro dissolution only.

Dissolution Specifications

Modified-release dosage forms typically require dissolution testing over multiple time points, and IVIVC plays an important role in setting these specifications. Specification time points are usually chosen in the early, middle, and late stages of the dissolution profiles. In the absence of an IVIVC, the range of the dissolution specification rarely exceeds 10% of the dissolution of the pivotal clinical batch. However, in the presence of IVIVC, wider specifications may be applicable based on the predicted concentration–time profiles of test batches being bioequivalent to the reference batch.

The process of setting dissolution specifications in the presence of an IVIVC starts by obtaining the reference (pivotal clinical batch) dissolution profile. The dissolution of batches with different dissolution properties (slowest and fastest batches included) should be used along with the IVIVC model, and prediction of the concentration–time profiles should be made using an appropriate convolution method. Specifications should optimally be established such that all batches with dissolution profiles between the fastest and slowest batches are bioequivalent and less optimally bioequivalent to the reference batch. The earlier exercise in achieving the widest possible dissolution specification allows the majority of batches to pass and is possible only if a valid level A model is available.

Mapping

Mapping is a process that relates critical manufacturing variables (CMVs), including formulation, processes, and equipment variables that can significantly affect drug release from the product. The mapping process defines boundaries of in vitro dissolution profiles on the basis of acceptable

BE criteria. The optimum goal is to develop product specifications that will ensure BE of future batches prepared within the limits of acceptable dissolution specifications. Dissolution specifications based on mapping would increase the credibility of dissolution as a bioequivalence surrogate marker and will provide continuous assurance and predictability of the product performance.

Future Biowaivers

Frequently, drug development requires changes in formulations due to a variety of reasons, such as unexpected problems in stability and development, availability of better materials, and better processing results. Having an established IVIVC can help avoid BE studies by using the dissolution profile from the changed formulation and subsequently predicting the in vivo concentration–time profile.

This predicted profile could act as a surrogate of the in vivo BE study. This has enormous cost-saving benefit in the form of reduced drug development spending and speedy implementation of postapproval changes. The nature of postapproval changes could range from minor (such as a change in non–release-controlling excipient) to major (such as site change, equipment change, or change in method of manufacture).

Parenteral Drug Delivery

IVIVC can be developed and applied to parenteral dosage forms, such as controlled-release particulate systems, depot system, and implants, which are either injected or implanted. However, there are relatively fewer successes in the development of IVIVC for such dosage forms, which could be due to several reasons, a few of which are discussed further. Sophisticated modeling techniques are needed to correlate the in vitro and in vivo data, in case of burst release, which is unpredictable and unavoidable.

POTENT DRUGS AND CHRONIC THERAPY

In general, several parenteral drug delivery systems are developed for potent drugs (e.g., hormones, growth factors, and antibiotics) and for long-term delivery (anywhere from a day to a few weeks to months). In such instances, to establish a good IVIVC model, the drug concentrations should be monitored in the tissue fluids at the site of administration by techniques such as microdialysis, and then the correlation should be established to the in vitro release.

Regulatory Aspects

INDs/NDAs

Evidence demonstrating in vivo BA or information to permit FDA to waive this evidence must be included in NDAs (21 CFR 320.21(a)). A specific objective is to establish in vivo performance of the dosage form used in the clinical studies that provided primary evidence of efficacy and safety. The sponsor may wish to determine the relative BA of an instant release solid oral dosage form by comparison with an oral solution, suspension, or IV injection (21 CFR 320.25 (d)(2) and 320.25 (d (3)). The BA of the clinical trial dosage form should be optimized during the IND period.

Once the in vivo BA of a formulation is established during the IND period, waivers of subsequent in vivo BE studies, following major changes in components, composition, and/or method of manufacture (e.g., similar to SUPAC instant release Level 3 changes), may be possible using the BCS. BCS-based biowaivers are applicable to the to-be-marketed formulation when changes in components, composition, and/or method of manufacture occur to the clinical trial formulation, as long as the dosage forms have rapid and similar in vitro dissolution profiles. This approach is useful only when the drug substance is highly soluble and highly permeable (BCS Class 1) and the formulations pre- and postchange are *pharmaceutical equivalents* (under the definition at 21 CFR 320.1 (c)). BCS-based biowaivers are intended only for BE studies. They do not apply to food-effect BA studies or other PK studies.

ANDAs

BCS-based biowaivers can be requested for rapidly dissolving instant release test products containing highly soluble and highly permeable drug substances, provided that the reference listed drug product is also rapidly dissolving and the test product exhibits similar dissolution profiles to the reference listed drug product. This approach is useful when the test and reference dosage forms are pharmaceutical equivalents. The choice of dissolution apparatus (USP Apparatus I or II) should be the same as that established for the reference listed drug product.

Postapproval Changes

BCS-based biowaivers can be requested for significant postapproval changes (e.g., Level 3 changes in components and composition) to a rapidly dissolving instant release product containing a highly soluble, highly permeable drug substance, provided that dissolution remains rapid for the postchange product and both pre- and postchange products exhibit similar dissolution profiles. This approach is useful only when the drug products pre- and postchange are pharmaceutical equivalents.

Manufacturing Changes

Manufacturing changes may involve the manufacturing process itself (critical manufacturing variable). If a manufacturer wishes to use a manufacturing process that is not identical in every respect to the original manufacturing process used in the approved application, appropriate validation studies should be conducted to demonstrate that the new process is similar to the original process. For modified-release solid oral dosage forms, consideration should be given as to whether or not the change in manufacturing process is critical to drug release (critical processing variable). For purposes of categorizing the level of changes, process change may be considered only to affect a release controlling excipient when both types of excipients (i.e., nonrelease and release controlling) are present during the unit operation undergoing a change.

Level 1 Change

1. *Definition of Level*
   Process changes involving adjustment of equipment operating conditions such as mixing times and operating speeds within original approved application ranges affecting the nonrelease controlling and/or release controlling excipient(s). The sponsor should provide appropriate justifications for claiming any excipient(s) as a nonrelease controlling or a release controlling excipient in the formulation of the modified-release solid oral dosage form.
2. *Test Documentation*
   a. Chemistry Documentation
      None beyond application/compendial product release requirements. Notification of the change and submission of the updated executed batch records.
   b. Dissolution Documentation
      None beyond application/compendial release requirements.
   c. BE documentation
      None.
3. *Filing Documentation*
   Annual report.

Level 2 Change

1. *Definition of Level*

   This category includes process changes involving adjustment of equipment operating conditions such as mixing times and operating speeds outside of the originally approved application ranges.

2. *Test Documentation*

   a. *Chemistry Documentation*

      Application/compendial product release requirements. Notification of change and submission of updated executed batch records.

      Stability: One batch with 3 months' accelerated stability data reported in changes being effected supplement and long-term stability data of first production batch reported in annual report.

   b. *Dissolution Documentation*

      Extended release: In addition to application/compendial release requirements, multipoint dissolution profiles should be obtained in three other media, for example, in water, 0.1 N HCl, and USP buffer media at pH 4.5 and 6.8 for the changed drug product and the biobatch or marketed batch (unchanged drug product). Adequate sampling should be performed, for example, at 1, 2, and 4 h and every 2 h thereafter until either 80% of the drug from the drug product is released or an asymptote is reached. A surfactant may be used with appropriate justification.

      Delayed release: In addition to application/compendial release requirements, dissolution tests should be performed in 0.1 N HCl for 2 h (acid stage) followed by testing in USP buffer media, in the range of pH 4.5–7.5 (buffer stage) under standard (application/compendial) test conditions and two additional agitation speeds using the application/compendial test apparatus (three additional test conditions). If the application/compendial test apparatus is the rotating basket method (Apparatus 1), a rotation speed of 50, 100, and 150 rpm may be used, and if the application/compendial test apparatus is the rotating paddle method (Apparatus 2), a rotation speed of 50, 75, and 100 rpm may be used. Multipoint dissolution profiles should be obtained during the buffer stage of testing. Adequate sampling should be performed, for example, at 15, 30, 45, 60, and 120 min (following the time from which the dosage form is placed in the buffer) until either 80% of the drug from the drug product is released or an asymptote is reached. The earlier dissolution testing should be performed using the changed drug product and the biobatch or marketed batch (unchanged drug product).

      All modified-release solid oral dosage forms: In the presence of an established in vitro/in vivo correlation, only application/compendial dissolution testing should be performed (i.e., only in vitro release data by the correlating method should be submitted). The dissolution profiles of the changed drug product and the biobatch or marketed batch (unchanged drug product) should be similar. The sponsor should apply appropriate statistical testing with justifications (e.g., the f_2 equation) for comparing dissolution profiles. Similarity testing for the two dissolution profiles (i.e., for the unchanged drug product and the changed drug product) obtained in each individual medium is appropriate.

   c. *BE Documentation*

      None.

3. *Filing Documentation*

   Changes being effected supplement (all information including accelerated stability data) and annual report (long-term stability data).

Level 3 Change

1. *Definition of Level*
   This category includes change in the type of process used in the manufacture of the product, such as a change from wet granulation to direct compression of dry powder.
2. *Test Documentation*
   a. *Chemistry Documentation*
      Application/compendial product release requirements. Notification of change and submission of updated executed batch records.
      Stability: Three batches with 3 months' accelerated stability data reported in prior approval supplement and long-term stability data of first three production batches reported in annual report.
   b. *Dissolution Documentation*
      Extended release: In addition to application/compendial release requirements, a multipoint dissolution profile should be obtained using application/compendial test conditions for the changed drug product and the biobatch or marketed batch (unchanged drug product). Adequate sampling should be performed, for example, at 1, 2, and 4 h and every 2 h thereafter until either 80% of the drug from the drug product is released or an asymptote is reached.
      Delayed release: In addition to application/compendial release requirements, a multipoint dissolution profile should be obtained during the buffer stage of testing using the application/compendial test conditions for the changed drug product and the biobatch or marketed batch (unchanged drug product). Adequate sampling should be performed, for example, at 15, 30, 45, 60, and 120 min (following the time from which the dosage form is placed in the buffer) until either 80% of the drug from the drug product is released or an asymptote is reached.
   c. *BE Documentation*
      A single-dose BE study. The BE study may be waived in the presence of an established in vitro/in vivo correlation.
3. *Filing Documentation*
   Prior approval supplement (all information including accelerated stability data) and annual report (long-term stability data).

Biowaiver Risks

The risk of applying the practice of biowaivers in resulting a nonequivalent product approved is related to three things: the severity or the impact of such BE, the probability that it is not detected prior to batch release, and the probability of incidence that the product is produced bioinequivalent:

$$\text{Risk} = \text{Probability of incidence} \times \text{Probability of detection} \times \text{Severity}$$

This equation is used to calculate a risk priority number (RPN) in failure mode and effect analysis (FMEA) (ICH Guidelines).

Table 6.7 summarizes the conditions laid down by the current (updated) guidance documents as published by the FDA, EMA, and WHO; Japan is excluded since it does not allow biowaivers. The current guidance documents describe only dissolution as the comparative in vitro tool: dissolution testing at three pH levels is to be applied when the conditions listed in Table 6.7 are fulfilled. Comparative dissolution testing can be used when differences in the extent or rate of dissolution are expected to determine a potential difference in BA. The API should be highly soluble and show rapid dissolution from the finished product. Rapid dissolution is considered as taking place within the time frame of gastric emptying. In addition, the excipients present should not be expected to have an effect on the BA of the API via mechanisms other than dissolution or disintegration.

TABLE 6.7
Biowaiver Requirements Worldwide

	EMA		FDA	WHO		
Formulation						
API	**BCS Class I**	**BCS Class III**	**BCS Class I**	**BCS Class I**	**BCS Class II**	**BCS Class III**
Excipients	Excipients that might affect BA are qualitatively the same	Excipients that might affect BA are qualitatively and quantitatively the same	Excipients that are currently in FDA-approved instant release solid oral dosage forms; not large quantities of certain excipients that might affect BA	It should be demonstrated that the excipients are well established for use in products containing the API and will not lead to differences with respect to processes affecting absorption or which might lead to interactions that alter the PK of the API.		
Drug type	Not for "NTI" drugs		Not for NTI drugs and products designed to be absorbed in the oral cavity	Both the indication and therapeutic index are important considerations in determining whether the biowaiver based on BCS can be applied.		
Dissolution formulation	Very rapid (>85% within 15 min) or rapid dissolution (85% within 30 min)	Very rapid dissolution (>85% within 15 min)	Rapid dissolution (NLT 85% in 30 min)	Rapid dissolution (NLT 85% in 30 min)	Dose: solubility ratio of 250 mL at pH 6.8 and rapid dissolution (NLT 85% in 30 min) in pH 6.8	Very rapid dissolution (>85% within 15 min)
Comparative in vitro test						
In vitro dissolution testing	pH 1–6.8 (at least pH 1.2, 4.5, and 6.8); no surfactant; enzymes for gelatin only		pH 1.2, 4.5, and 6.8 or simulated gastric resp. intestinal fluid; no surfactant; enzymes for gelatin only	pH 1.2, 4.5, and 6.8		
Equivalence acceptance criteria						
	Similarity (f_2 calculation 50–100) or other appropriate statistical methods		Similarity (f_2 calculation)	Similarity (f_2 calculation 50–100) or other appropriate statistical methods, provided that the same criterion is used for acceptance (maximum 10% difference between the profiles)		

All guidelines agree that BCS Class I APIs are open for biowaivers; BCS Class III substances are considered by EMA and WHO guidance as well. Products containing APIs with low solubility are considered to have a higher risk of being bioinequivalent and dissolution testing may not be sufficiently discriminating. BCS Class II and IV compounds are therefore not open for applying for a biowaiver according to the EMA and FDA. Only the WHO considers biowaivers for BCS Class II compounds with specific characteristics.

The safety of biowaiving is evaluated based on the therapeutic consequences of a potential difference in BA. Acceptance criteria define the accepted difference between two dissolution profiles and thus indirectly limit the severity of a potential difference. For example, the severity of any occurring and undetected bioinequivalence of two NTI drugs is ranked higher than that of other drugs, and the dissolution test and limits are not to be applied for such products.

In calculating the risk, let us first evaluate the situation for highly soluble drugs. When the API is of BCS Class I, say propranolol, a biowaiver could be acceptable based on the guidelines of the EMA, FDA, and WHO, if appropriately justified. Assuming that mannitol is present in a quantity from which no effect on the BA of propranolol is expected, the risk calculation would then result in low incidence (BCS Class I, rapid dissolution, acceptable excipients) × high detection (dissolution testing possible) × low severity (not NTI). The overview of available Federation International Pharmaceutique (FIP) biowaiver monographs confirms that, for the APIs categorized as BCS Class I, indeed a positive biowaiver recommendation was given.

If the API is of BCS Class III, say acyclovir, the biowaiver is less straightforward. From the FDA viewpoint, the biowaiver will not be accepted due to the BCS classification. From the WHO and EMA perspective, the discussion would probably focus on the presence of mannitol. Is it expected to affect BA? If so, a biowaiver would only be acceptable, in the case that all excipients were qualitatively and quantitatively the same. Alternatively, if it is argued that mannitol is present in a quantity from which no effect on the BA of acyclovir is expected (as assumed earlier), a biowaiver may be considered acceptable. The risk calculation would then be as presented earlier and the discussion would concern the probability of incidence: Is it acceptably low? The overview of the available FIP biowaiver monographs includes several APIs classified as BCS Class III.

Survey studies to determine if the BCS may help to predict in vivo BE outcome show that the BE failure rate was generally low and similar (~11%) for BCS Class I and III compounds. In addition, solubility appeared to be the most discriminating factor with regard to BE outcome. These data confirm the approach taken by the EMA and WHO, allowing biowaivers for highly soluble BCS Class I and III compounds, while requiring comparative dissolution studies as a surrogate in vitro test.

The situation is different for low-solubility drugs. If the API is of BCS Class II, a biowaiver could be considered based on the WHO guidance only. The highly variable Class II compounds show the highest BE failure rate (54%). If a biowaiver for a BCS Class II compound is considered, additional factors such as physical characteristics of the API, excipients, and formulation aspects should thus be evaluated critically, and the biopredictive power of the applied dissolution media should be reconsidered. In line with these findings, the EMA and FDA consider the potential effects of excipients and formulation on the in vivo PK as being insufficiently covered by a dissolution test. Translating these aspects to the FMEA calculation, the probability of detection is the risk factor of importance for these APIs.

A biowaiver for an API of BCS Class IV would not be accepted by any of the guidelines. The potential effects of excipients on the in vivo solubility and dissolution of the API from the formulation as well as potential effects on the absorption, accessibility, and transit time are considered too complex to allow for a biowaiver approach. The risk of bioinequivalence, when applying the dissolution test only, is considered too high. A low probability of detection thus leads to an unacceptably high outcome of the risk calculation. Surprisingly, the failure rate for Class IV compounds is relatively low (10%) when tested in in vivo BE studies. This does not necessarily mean that the assumptions of the guideline are not correct, as their conclusion was not discussed in relation to dissolution data. Furthermore, adequate and perhaps relatively intensive product development studies could reduce the risk for a bioinequivalence of a product containing a BCS Class IV API.

Regardless of the BCS classification, all guidelines exclude NTI drugs from biowaiving. The severity of undetected BE would be higher. In addition, in vitro testing is not considered sufficiently sensitive to detect bioinequivalence for NTI drugs and consequentially, NTI drugs are excluded from the option of a biowaiver. Translating this to the risk calculation means that the probability of detection of bioinequivalence of NTI drugs is considered too low with the currently available dissolution test and the severity of undetected BE is considered too high, both factors leading to an unacceptably high risk.

Data to Support Biowaiver

The drug substance for which a waiver is being requested should be highly soluble and highly permeable. Sponsors requesting biowaivers based on the BCS should submit the following information to the agency for review by the Office of Clinical Pharmacology and Biopharmaceutics (for NDAs) or Office of Generic Drugs, Division of Bioequivalence (for ANDAs).

Data Supporting High Solubility

Data supporting high solubility of the test drug substance should be developed. The following information should be included in the application:

- A description of test methods, including information on analytical method and composition of the buffer solutions
- Information on chemical structure, molecular weight, nature of the drug substance (acid, base, amphoteric, or neutral), and dissociation constants (pK_a[s])
- Test results (mean, standard deviation, and CV) summarized in a table under solution pH, drug solubility (e.g., mg/mL), and volume of media required to dissolve the highest dose strength
- A graphic representation of mean pH-solubility profile

Data Supporting High Permeability

Data supporting high permeability of the test drug substance should be developed. The following information should be included in the application:

- For human PK studies, information on study design and methods used along with the PK data
- For direct permeability methods, information supporting the suitability of a selected method that encompasses a description of the study method; criteria for selection of human subjects, animals, or epithelial cell line; drug concentrations in the donor fluid; description of the analytical method; method used to calculate extent of absorption or permeability; and, where appropriate, information on efflux potential (e.g., bidirectional transport data)

A list of selected model drugs along with data on extent of absorption in humans (mean, standard deviation, CV) is used to establish suitability of a method, permeability values for each model drug (mean, standard deviation, CV), permeability class of each model drug, and a plot of the extent of absorption as a function of permeability (mean, standard deviation, or 95% confidence interval) with identification of the low-/high-permeability class boundary and selected IS. Information to support high permeability of a test drug substance should include permeability data on the test drug substance, the ISs (mean, standard deviation, CV), stability information, data supporting passive transport mechanism where appropriate, and methods used to establish high permeability of the test drug substance.

Data Supporting Rapid and Similar Dissolution

For submission of a biowaiver request, an instant release product should be rapidly dissolving. Data supporting rapid dissolution attributes of the test and reference products should be developed. The following information should be included in the application:

- A brief description of the instant release products used for dissolution testing, including information on batch or lot number, expiry date, dimensions, strength, and weight.
- Dissolution data obtained with 12 individual units of the test and reference products using recommended test methods. The percentage of labeled claim dissolved at each specified testing interval should be reported for each individual dosage unit. The mean percent dissolved, range (highest and lowest) of dissolution, and CV (relative standard deviation) should be tabulated. A graphic representation of the mean dissolution profiles for the test and reference products in the three media should also be included.
- Data supporting similarity in dissolution profiles between the test and reference products in each of the three media, using the f_2 metric.

Additional Information

The manufacturing process used to make the test product should be described briefly to provide information on the method of manufacture (e.g., wet granulation vs. direct compression). A list of excipients used, the amount used, and their intended functions should be provided. Excipients used in the test product should have been used previously in FDA-approved instant release solid oral dosage forms.

SURROGATE METHODS

As the U.S. FDA has begun accepting recommendations for waiver of BE requirement, protocols that prove extremely expensive in the drug development cycle, there is a greater need to develop surrogate models that one day may prove useful in securing waivers for all classes of drugs. Generally, the methods available currently show that the complexity of assay is directly proportional to its correlation with absorption of drugs in humans (Figure 6.2). Studies that correlated log *P* with human absorption profile and the suitability or lead candidates were elaborated in Chapter 2. In this chapter, we will examine more complex assay systems.

Drug transport across epithelial cell barriers, especially the human small intestine, is difficult to predict. The intestinal epithelial cell barrier is a sophisticated organ that has evolved over hundreds of millions of years to become a "smart," effective, and selective xenobiotic screen.

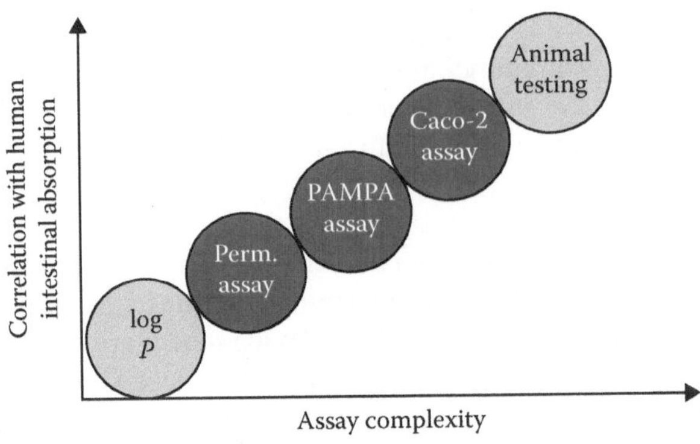

FIGURE 6.2 Assay complexity versus correlation with human absorption. Data from both complex biological and artificial permeation assays can provide valuable information regarding the absorption of a drug. (Courtesy of Millipore Corporation, EMD Millipore, Billerica, MA.)

Nevertheless, there is large interindividual variability in the intestinal transport of drugs. Genetic variability in key proteins is believed to be causal. There is a pressing need to better understand the key processes and how the system components interact at the molecular, cellular, and tissue level to control drug transport and determine drug absorption in small intestine.

Is it feasible to construct an in silico framework to represent the drug absorption in the small intestine at the cellular level with internal dynamic property and concert with the update molecular biochemical mechanism? This new generation of models and computational tools might integrate the available and emerging information at different levels to better account for and predict observed experimental results. Predicting aqueous solubility with in silico tools is a key drug property. It is, however, difficult to measure accurately, especially for poorly soluble compounds, and thus numerous in silico models have been developed for its prediction. Some in silico models can predict aqueous solubility of simple, uncharged organic chemicals reasonably well; however, solubility prediction for charged species and drug-like chemicals is not very accurate. However, extrapolating solubility data to intestinal absorption from PK and physicochemical data and elucidating crucial parameters for absorption and the potential for improvement of BA are important at the preformulation stages.

The poor oral BA of drugs is generally assumed to be due to physiochemical problems, which result in poor solubility in the gastrointestinal tract (GI tract) or difficulty in diffusion through small intestine epithelial membrane. Furthermore, the biochemical process also contributes to oral BA. The in vitro cell culture models of the intestinal epithelial cell barrier have evolved to become widely used experimental devices.

In the previous chapter, the log *P* factor was discussed in detail; in this chapter, we will examine other methods of testing transport across membranes.

Permeability Assays

The permeability assay uses an artificial membrane composed of hexadecane. The automated systems comprise multiwell systems as shown in Figure 6.3.

Permeability Assay Protocol

a. Into each well, add 15 μL of a 5% solution of hexadecane in hexane.
b. Dry for 45 min–1 h to ensure complete evaporation of hexane.
c. Add 300 μL of buffer with 5% dimethylsulfoxide (DMSO) at desired pH to acceptor plate.
d. Place donor plate into the acceptor plate making sure underside of membrane is in contact with buffer.
e. Dissolve drugs of interest to the desired concentration. Add 150 μL of the drug at the desired concentration in 5% DMSO/PBS (phosphate buffer saline) at the desired pH to each well in the donor plate.
f. Cover and incubate at room temperature for 4–6 h.

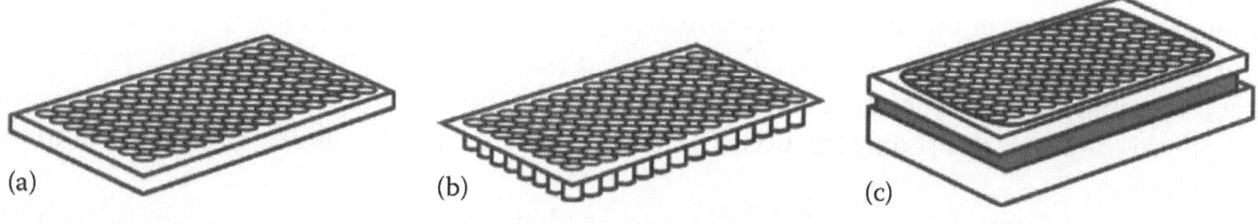

FIGURE 6.3 96-Well permeability testing method. The support membrane is a 3 μm track etched polycarbonate, 10 μm thick; the artificial membrane is hexadecane and the recommended incubation time is 4–6 h. (a) Lid, (b) donor plate, 96-well filter plate, and (c) acceptor plate, 96-well plate insert in single-well tray (supplied with plate) or PTFE acceptor plate. (Courtesy of Millipore Corporation, EMD Millipore, Billerica, MA.)

g. Transfer 100 μL/well from the donor plate and 250 μL/well from the acceptor plate to separate UV/Vis compatible plates, and measure the UV/Vis absorption from 250 to 500 nm (SPECTRAmax® plate reader, molecular devices) for both plates.
h. Prepare drug solutions at the theoretical equilibrium (i.e., the resulting concentration if the donor and acceptor solutions were simply combined), and measure UV/Vis absorption from 250 to 500 nm for 250 μL/well of each.
i. Calculate log P_e and membrane retention using the following equations:

$$\log\ P_e = \log\left\{C\cdot -\ln\left(1-\frac{[\text{Drug}]_{\text{Acceptor}}}{[\text{Drug}]_{\text{equilibrium}}}\right)\right\} \quad \text{where } C = \left(\frac{V_D \cdot V_A}{(V_D+V_A)\text{Area}\cdot\text{Time}}\right) \tag{6.13}$$

PAMPA

Early drug discovery absorption, distribution, metabolism, and excretion (ADME) assays, such as fast Caco-2 screens, can help in rejecting test compounds that lack good pharmaceutical profiles. A cost-effective, high-throughput method—parallel artificial membrane permeability analysis (PAMPA)—which uses a phospholipid artificial membrane that models passive transport of epithelial cells, is becoming increasingly popular. The PAMPA assay utilizes a range of lipid components that model a variety of different plasma membranes. The support membrane is 0.45 μm hydrophobic polyvinylidene fluoride, 130 μm thick, and the artificial membrane is lecithin in dodecane; recommended incubation time is 16–24 h. The protocol involves the following:

a. Dissolve drugs of interest to the desired concentration.
b. Add 300 μL of buffer with 5% DMSO at desired pH to acceptor plate.
c. Into each well, add 5 μL of lipids in organic solvent (e.g., 2% lecithin in dodecane).
d. Add 150 μL of the drug at the desired pH and concentration in 5% DMSO/PBS to each well in the donor plate.
e. Place donor plate into the acceptor plate making sure underside of membrane is in contact with buffer. Steps 3–5 should be completed quickly, within 10 min.
f. Cover and incubate at room temperature for 16–24 h.
g. Transfer 100 μL/well from the donor plate and 250 μL/well from the acceptor plate to separate UV/Vis compatible plates, and measure the UV/Vis absorption from 250 to 500 nm (SPECTRAmax® plate reader, molecular devices) for both plates.
h. Prepare drug solutions at the theoretical equilibrium (i.e., the resulting concentration if the donor and acceptor solutions were simply combined), and measure UV/Vis absorption from 250 to 500 nm for 250 μL/well of each.
i. Calculate log P_e and membrane retention using Equation 6.13.

The permeability and PAMPA assays as described earlier are robust and reproducible assays for determining passive, transcellular compound permeability. Permeability and PAMPA are automation compatible, relatively fast (4–16 h), inexpensive, and straightforward and their results correlate with human drug absorption values from published methods. The PAMPA assay provides the benefits of a more biologically relevant system. It is also possible to tailor the lipophilic constituents so that they mimic specific membranes such as the blood–brain barrier. Optimization of incubation time, lipid mixture, and lipid concentration will also enhance the assay's ability to predict compound permeability.

Modifications of permeability and PAMPA systems have been reported, for example, using the pION PAMPA Evolution 96 System with Double-Sink and Gut-Box (http://www.pion-inc.com/products.htm) as a new surrogate assay that predicts the gastrointestinal tract absorption of candidate drug molecules at different pH conditions. Using Beckman Coulter's Biomek FX Single-Bridge

Laboratory Automation Workstation PAMPA Assay System that features a 30 min incubation time and an on-deck integrated Gut-Box and a SPECTRAMax microplate spectrophotometer, the permeability coefficients of drug standards with diverse physiochemical properties can be compared from both PAMPA and Caco-2 assays automated using the Biomek FX Workstation.

These automated assays can be used for high-throughput ADME screening in early drug discovery. The Double-Sink PAMPA permeability assay mimics in vivo conditions by the use of a chemical sink in the acceptor wells and pH gradient in the donor wells. The use of the pION Gut-Box integrated on the deck has shortened the PAMPA assay incubation time to 30 min. The permeability coefficient and rank order correlate well with data obtained using the in vitro Caco-2 assay and in vivo permeability properties measured in rat intestinal perfusions.

Caco-2 Drug Transport Assays

Drug absorption generally occurs through either passive transcellular or paracellular diffusion, active carrier transport, or active efflux mechanisms. Several methods have been developed to aid in the understanding of the absorption of new lead compounds. The most common ones use an immortalized cell line (e.g., Caco-2 and maldin darby canine kidney (MDCK)) to mimic the intestinal epithelium. These in vitro models provide more predictive permeability information than artificial membrane systems (i.e., PAMPA and permeability assays, described in the preceding text) based on the cells' ability to promote (active transport) or resist (efflux) transport. Various in vitro methods that are listed in the U.S. FDA guidelines, acceptable to evaluate the permeability of a drug substance, include monolayer of suitable epithelial cells, and one such epithelial cell line that has been widely used as a model system of intestinal permeability is the Caco-2 cell line.

The kinetics of intestinal drug absorption, permeation enhancement, chemical moiety structure–permeability relationships, dissolution testing, in vitro/in vivo correlation, BE, and the development of novel polymeric materials are closely associated with the concept of Caco-2. Since most drugs are known to be absorbed via the intestines without using cellular pumps, passive permeability models came in the limelight. In a typical Caco-2 experiment, a monolayer of cells is grown on a filter separating two stacked microwell plates. The permeability of drugs through the cells is determined after the introduction of a drug on one side of the filter. The entire process is automated, and when used in conjunction with chromatography and/or mass spectroscopy detection, it enables any drug's permeability to be determined. The method requires careful sample analysis to calculate permeability correctly. Limitations of Caco-2 experiments are as follows: 21 days for preparing a stable monolayer and stringent storage conditions; however, tight-junction formation prior to use is the better choice. The villus in the small intestine contains more than one cell type, the Caco-2 cell line does not produce the mucus as observed in the small intestine, and no P450 metabolizing enzyme activity has been found in the Caco-2 cell line. Test compound solubility may pose a problem in Caco-2 assays because of the assay conditions. Finally, Caco-2 cells also contain endogenous transporter and efflux systems, the latter of which work against the permeability process and can complicate data interpretation for some drugs.

The Caco-2 cell line is heterogeneous and was derived from a human colorectal adenocarcinoma. Caco-2 cells are used as in vitro permeability models to predict human intestinal absorption because they exhibit many features of absorptive intestinal cells. This includes their ability to spontaneously differentiate into polarized enterocytes that express high levels of brush border hydrolases and form well-developed junctional complexes. Consequently, it becomes possible to determine whether passage is transcellular or paracellular based on a compound's transport rate. Caco-2 cells also express a variety of transport systems including dipeptide transporters and P-gp's. Due to these features, drug permeability in Caco-2 cells correlates well with human oral absorption, making Caco-2 an ideal in vitro permeability model. Additional information can be gained on metabolism and potential drug–drug interactions as the drug undergoes transcellular diffusion through the Caco-2 transport model.

Although accurate and well researched, the Caco-2 cell model requires a high investment of time and resources. Depending on a number of factors, including initial seeding density, culturing conditions, and passage number, Caco-2 cells can take as much as 20 days to reach confluence and achieve full differentiation. During this 20-day period, they require manual or automated exchange of media as frequently as every other day. The transport assays consume valuable drug compounds and normally require expensive, posttransport sample analyses (e.g., LC/MS). Therefore, the use of the Caco-2 transport model in a high-throughput laboratory setting is only possible if the platform is robust, automation compatible, reproducible, and provides high-quality data that correlate well with established methodologies.

The Millipore MultiScreen Caco-2 assay system is a reliable 96-well platform for predicting human oral absorption of drug compounds (using Caco-2 cells or other cell lines whose drug transport properties have been well characterized). The MultiScreen system format is automation compatible and is designed to offer more cost-effective, higher-throughput screening of drugs than a 24-well system. The MultiScreen Caco-2 assay system exhibits good uniformity of cell growth and drug permeability across all 96 wells and low variability between production lots. The plate design supports the use of lower volumes of expensive media and reduced amounts of test compounds. Using the MultiScreen Caco-2 assay system, standard drug compounds are successfully categorized as either "high" or "low" permeable, as defined by the FDA, and the permeability data correlate well with established human absorption values. The components of Caco-2 assay system are shown in Figure 6.4.

The apparent permeability (P_{app}), in units of centimeters per second, can be calculated for Caco-2 drug transport assays using the following equation:

$$P_{app} = \left(\frac{V_A}{\text{Area} \cdot \text{Time}} \right) \cdot \left(\frac{[\text{Drug}]_{\text{acceptor}}}{[\text{Drug}]_{\text{Initial, donor}}} \right) \tag{6.14}$$

where

V_A is the volume (in mL) in the acceptor well

Area is the surface area of the membrane (0.11 cm^2 for MultiScreen Caco-2 plate and 0.3 cm^2 for the 24-well plate)

Time is the total transport time in seconds

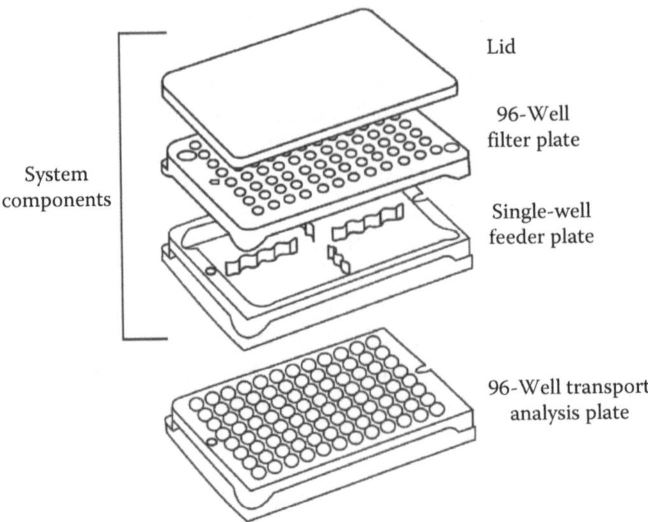

FIGURE 6.4 MultiScreen Caco-2 assay system. Components with single-well feeder plate and 96-well transport analysis plate. (Courtesy of Millipore Corporation, EMD Millipore, Billerica, MA.)

For radiolabeled drug transport experiments, the CPM units obtained from the Trilux Multiwell Plate Scintillation Counter are used directly for the drug acceptor and initial concentrations such that the formula becomes

$$P_{\text{app}} = \left(\frac{V_{\text{A}}}{\text{Area} \cdot \text{Time}}\right) \cdot \left(\frac{\text{CPM}_{\text{acceptor}}}{\text{CPM}_{\text{Initial, donor}}}\right) \tag{6.15}$$

Historically, it has been shown that a sigmoidal relationship exists between drug absorption rates as measured with the in vitro Caco-2 model and human absorption:

$$\frac{\left(\%\ \text{Human absorption} = 100 \times \exp\left(a + b \times P_{\text{app}}\right)\right)}{\left(1 + \exp\left(a + b \times P_{\text{app}}\right)\right)} \tag{6.16}$$

Caco-2 cells are heterogeneous and their properties in final culture may differ based on the selection pressures of a particular laboratory. Direct comparison of compound permeability rates between laboratories is not possible unless the same Caco-2 cells and conditions are used. Therefore, transport rates and permeability classification ranges of specific drugs are expected to vary between reported studies. Most important is the ability to successfully classify compounds as low-, medium-, or high-permeability drugs and produce transport results that correlate to established human absorption values.

Several modifications of Caco-2 cell model have been tested; for example, CYP3A4-transfected Caco-2 cells are also used to define the biochemical absorption barriers. Oral BA and intestinal drug absorption can be significantly limited by metabolizing enzymes and efflux transporters in the gut. The most prevalent oxidative drug-metabolizing enzyme present in the intestine is CYP3A4. Currently, more than 50% of the drugs on the market metabolized by P450 enzymes are metabolized by CYP3A4. Oral absorption of CYP3A4 substrates can also be limited by the multidrug resistance transporter P-gp, because there is extensive substrate overlap between these two proteins. P-gp is an ATP-dependent transporter on the apical plasma membrane of enterocytes that functions to limit the entry of drugs into the cell. There is significant interaction between CYP3A4 and P-gp in the intestine. Although Caco-2 cells express a variety of uptake and efflux transporters found in the human intestine, a major drawback to the use of Caco-2 cells is that they lack CYP3A4. As such, no data regarding the importance of intestinal metabolism on limiting drug absorption can be obtained from normal Caco-2 cells. Caco-2 cells pretreated with 1, 25-dihydroxyvitamin-D_3 (vitamin D_3) express higher levels of CYP3A4 compared with Caco-2 but still underestimate the amount of CYP3A4 in the human intestine. CYP3A4-transfected Caco-2 cells show that P-gp can enhance the metabolism of orally dosed drugs by repeated cycling of the drug at the apical membrane.

Animal Model Testing

Whereas the quantity of substance available at the preformulation stages is generally small, in some instances, early animal testing for absorption potential is needed, particularly if the solid form of the new drug offers many options such as amorphous forms and solvates. The absorption models used in animals are well described and would not be discussed here. Establishing good IVIVC at this stage proves useful because of limited access to sufficient compound to run the entire absorption profiles. The "IVIVC" analysis can be made extensive or general conclusions drawn from limited studies; the choice depends on the amount of compound available and the nature or robustness of correlation observed.

EUROPEAN PERSPECTIVE ON BIOWAIVER CRITERIA

Strength to Be Investigated

If several strengths of a test product are applied for, it may be sufficient to establish BE at only one or two strengths, depending on the proportionality in composition between the different strengths and other product-related issues described in the succeeding text. The strength(s) to evaluate depends on the linearity in PK of the active substance. In case of nonlinear PK (i.e., not proportional increase in AUC with increased dose), there may be a difference between different strengths in the sensitivity to detect potential differences between formulations. In the context of this guideline, PK is considered to be linear if the difference in dose-adjusted mean AUCs is no more than 25% when comparing the studied strength 12/27 (or strength in the planned BE study) and the strength(s) for which a waiver is considered. In order to assess linearity, the applicant should consider all data available in the public domain with regard to the dose proportionality and review the data critically. Assessment of linearity will consider whether differences in dose-adjusted AUC meet a criterion of ±25%. If BE has been demonstrated at the strength(s) that are most sensitive to detect a potential difference between products, in vivo BE studies for the other strength(s) can be waived.

The following general requirements must be met where a waiver for additional strength(s) is claimed:

a. The pharmaceutical products are manufactured by the same manufacturing process.
b. The qualitative composition of the different strengths is the same.
c. The composition of the strengths is quantitatively proportional, that is, the ratio between the amount of each excipient and the amount of active substance(s) is the same for all strengths (for immediate-release products, coating components, capsule shell, color agents, and flavors are not required to follow this rule). If there is some deviation from quantitatively proportional composition, condition (c) is still considered fulfilled if conditions (i) and (ii) or (i) and (iii) in the following apply to the strength used in the BE study and the strength(s) for which a waiver is considered:
    i. If the amount of the active substance(s) is less than 5% of the tablet core weight, the weight of the capsule content.
    ii. The amounts of the different core excipients or capsule content are the same for the concerned strengths and only the amount of active substance is changed.
    iii. The amount of a filler is changed to account for the change in amount of active substance. The amounts of other core excipients or capsule content should be the same for the concerned strengths.
d. Appropriate in vitro dissolution data should confirm the adequacy of waiving additional in vivo BE testing.

Linear Pharmacokinetics

For products where all the conditions given earlier, (a) to (d), are fulfilled, it is sufficient to establish BE with only one strength. The BE study should in general be conducted at the highest strength. For products with linear PK and where the drug substance is highly soluble (see Appendix B), selection of a lower strength than the highest is also acceptable. Selection of a lower strength may also be justified if the highest strength cannot be administered to healthy volunteers for safety/tolerability reasons. Further, if problems of sensitivity of the analytical method preclude sufficiently precise plasma concentration measurements after single-dose administration of the highest strength, a higher dose may be selected (preferably using multiple tablets of the highest strength). The selected dose may be higher than the highest therapeutic dose provided that this single dose is well tolerated in healthy volunteers and that there are no absorption or solubility limitations at this dose.

Nonlinear Pharmacokinetics

For drugs with nonlinear PK characterized by a more than proportional increase in AUC with increasing dose over the therapeutic dose range, the BE study should in general be conducted at the highest strength. As for drugs with linear PK, a lower strength may be justified if the highest strength cannot be administered to healthy volunteers for safety/tolerability reasons. Likewise, a higher dose may be used in case of sensitivity problems of the analytical method in line with the recommendations given for products with linear PK earlier.

For drugs with a less than proportional increase in AUC with increasing dose over the therapeutic dose range, BE should in most cases be established both at the highest strength and at the lowest strength (or a strength in the linear range), that is, in this situation, two BE studies are needed.

If the nonlinearity is not caused by limited solubility but is due to, for example, saturation of uptake transporters and provided that conditions (a) to (d) earlier are fulfilled and the test and reference products do not contain any excipients that may affect gastrointestinal motility or transport proteins, it is sufficient to demonstrate BE at the lowest strength (or a strength in the linear range).

Selection of other strengths may be justified if there are analytical sensitivity problems preventing a study at the lowest strength or if the highest strength cannot be administered to healthy volunteers for safety/tolerability reasons.

Bracketing Approach

Where BE assessment at more than two strengths is needed, for example, because of deviation from proportional composition, a bracketing approach may be used. In this situation, it can be acceptable to conduct two BE studies, if the strengths selected represent the extremes, for example, the highest and the lowest strength or the two strengths differing most in composition, so that any differences in composition in the remaining strengths is covered by the two conducted studies. Where BE assessment is needed both in fasting and in fed state and at two strengths due to nonlinear absorption or deviation from proportional composition, it may be sufficient to assess BE in both fasting and fed state at only one of the strengths.

Waiver of either the fasting or the fed study at the other strength(s) may be justified based on previous knowledge and/or PK data from the study conducted at the strength tested in both fasted and fed state. The condition selected (fasting or fed) to test the other strength(s) should be the one that is most sensitive to detect a difference between products.

Fixed Combinations

The conditions regarding proportional composition should be fulfilled for all active substances of fixed combinations (FCs). When considering the amount of each active substance in an FC, the other active substance(s) can be considered as excipients. In the case of bilayer tablets, each layer may be considered independently.

BCS-BASED BIOWAIVER

Introduction

The BCS-based biowaiver approach is meant to reduce in vivo BE studies, that is, it may represent a surrogate for in vivo BE. In vivo BE studies may be exempted if an assumption of equivalence in in vivo performance can be justified by satisfactory in vitro data. Applying for a BCS-based biowaiver is restricted to highly soluble drug substances with known human absorption and considered not to have an NTI. The concept is applicable to immediate-release, solid pharmaceutical products for oral administration and systemic action having the same pharmaceutical form. However, it is not applicable for sublingual, buccal, and modified-release formulations. For orodispersible formulations, the BCS-based biowaiver approach may only be applicable when absorption in the oral cavity can be excluded.

BCS-based biowaivers are intended to address the question of BE between specific test and reference products. The principles may be used to establish BE in applications for generic medicinal products, extensions of innovator products, variations that require BE testing, and between early clinical trial products and to-be-marketed products.

Summary

Requirements for BCS-based biowaiver are applicable for an immediate-release drug product if the following applies:

a. The drug substance has been proven to exhibit high solubility and complete absorption (BCS Class I).
b. Either very rapid (>85% within 15 min) or similarly rapid (85% within 30 min) in vitro dissolution characteristics of the test and reference product have been demonstrated considering specific requirements.
c. Excipients that might affect BA are qualitatively and quantitatively the same. In general, the use of the same excipients in similar amounts is preferred.

BCS-based biowaiver are also applicable for an immediate-release drug product if

a. The drug substance has been proven to exhibit high solubility and limited absorption (BCS Class III)
b. Very rapid (>85% within 15 min) in vitro dissolution of the test and reference product has been demonstrated considering specific requirements
c. Excipients that might affect BA are qualitatively and quantitatively the same and other excipients are qualitatively the same and quantitatively very similar

Generally, the risks of an inappropriate biowaiver decision should be more critically reviewed (e.g., site-specific absorption, risk for transport protein interactions at the absorption site, excipient composition, and therapeutic risks) for products containing BCS Class III than for BCS Class I drug substances.

Drug Substance

Generally, sound peer-reviewed literature may be acceptable for known compounds to describe the drug substance characteristics of importance for the biowaiver concept. Biowaiver may be applicable when the active substance(s) in test and reference products are identical. Biowaiver may also be applicable if test and reference contain different salts provided that both belong to BCS Class I (high solubility and complete absorption).

Biowaiver is not applicable when the test product contains a different ester, ether, isomer, mixture of isomers, and complex or derivative of an active substance from that of the reference product, since these differences may lead to different bioavailabilities not deducible by means of experiments used in the BCS-based biowaiver concept. The drug substance should not belong to the group of "NTI" drugs.

Solubility

The pH-solubility profile of the drug substance should be determined and discussed. The drug substance is considered highly soluble if the highest single dose administered as immediate-release formulation(s) is completely dissolved in 250 mL of buffers within the range of pH 1–6.8 at 37°C ± 1°C. This demonstration requires the investigation in at least three buffers within this range (preferably at pH 1.2, 4.5, and 6.8) and in addition at the pK_a, if it is within the specified pH range.

Replicate determinations at each pH condition may be necessary to achieve an unequivocal solubility classification (e.g., shake-flask method or other justified method). Solution pH should be verified prior and after addition of the drug substance to a buffer.

Absorption

The demonstration of complete absorption in humans is preferred for BCS-based biowaiver applications. For this purpose, complete absorption is considered to be established where the measured extent of absorption is 85%. Complete absorption is generally related to high permeability. Complete drug absorption should be justified based on reliable investigations in human. Data from absolute BA or mass balance studies could be used to support this claim.

When data from mass balance studies are used to support complete absorption, it must be ensured that the metabolites taken into account in determination of fraction absorbed are formed after absorption. Hence, when referring to total radioactivity excreted in urine, it should be ensured that there is no degradation or metabolism of the unchanged drug substance in the gastric or intestinal fluid. Phase 1 oxidative and Phase 2 conjugative metabolism can only occur after absorption (i.e., cannot occur in the gastric or intestinal fluid). Hence, data from mass balance studies support complete absorption if the sum of urinary recovery of parent compound and urinary and fecal recovery of Phase 1 oxidative and Phase 2 conjugative drug metabolites accounts for 85% of the dose.

In addition, highly soluble drug substances with incomplete absorption, that is, BCS Class III compounds, could be eligible for a biowaiver provided certain prerequisites are fulfilled regarding product composition and in vitro dissolution. The more restrictive requirements will also apply for compounds proposed to be BCS Class I but where complete absorption could not convincingly be demonstrated. Reported BE between aqueous and solid formulations of a particular compound administered via the oral route may be supportive as it indicates that absorption limitations due to (immediate release) formulation characteristics may be considered negligible.

Well-performed in vitro permeability investigations including reference standards may also be considered supportive to in vivo data.

Drug Product

General aspect investigations related to the medicinal product should ensure immediate-release properties and prove similarity between the investigative products, that is, test and reference show similar in vitro dissolution under physiologically relevant experimental pH conditions. However, this does not establish an in vitro/in vivo correlation. In vitro dissolution should be investigated within the range of pH 1–6.8 (at least pH 1.2, 4.5, and 6.8). Additional investigations may be required at pH values in which the drug substance has minimum solubility. The use of any surfactant is not acceptable. Test and reference products should meet requirements as outlined in Section 4.1.2 of the main guideline text. In line with these requirements, it is advisable to investigate more than one single batch of the test and reference products.

Comparative in vitro dissolution experiments should follow current compendial standards. Hence, thorough description of experimental settings and analytical methods including validation data should be provided. It is recommended to use 12 units of the product for each experiment to enable statistical evaluation. Usual experimental conditions are as follows:

- Apparatus: paddle or basket.
- Volume of dissolution medium: 900 mL or less.
- Temperature of the dissolution medium: 37°C ± 1°C.
- Agitation: paddle apparatus, usually 50 rpm; basket apparatus, usually 100 rpm.
- Sampling schedule: for example, 10, 15, 20, 30, and 45 min.

- Buffer: pH 1.0–1.2 (usually 0.1 N HCl or SGF without enzymes), pH 4.5, and pH 6.8 (or SIF without enzymes) (pH should be ensured throughout the experiment; Ph.Eur. buffers recommended).
- Other conditions: no surfactant; in case of gelatin capsules or tablets with gelatin coatings, the use of enzymes may be acceptable.

Complete documentation of in vitro dissolution experiments is required including a study protocol, batch information on test and reference batches, detailed experimental conditions, validation of experimental methods, individual and mean results, and respective summary statistics.

Drug products are considered "very rapidly" dissolving when more than 85% of the labeled amount is dissolved within 15 min. In cases where this is ensured for the test and reference product, the similarity of dissolution profiles may be accepted as demonstrated without any mathematical calculation. Absence of relevant differences (similarity) should be demonstrated in cases where it takes more than 15 min but not more than 30 min to achieve almost complete (at least 85% of labeled amount) dissolution. F_2-testing or other suitable tests should be used to demonstrate profile similarity of test and reference. However, discussion of dissolution profile differences in terms of their clinical/therapeutic relevance is considered inappropriate since the investigations do not reflect any in vitro/in vivo correlation.

Although the impact of excipients in immediate-release dosage forms on BA of highly soluble and completely absorbable drug substances (i.e., BCS Class I) is considered rather unlikely, it cannot be completely excluded. Therefore, even in the case of Class I drugs, it is advisable to use similar amounts of the same excipients in the composition of test like in the reference product.

If a biowaiver is applied for a BCS Class III drug substance, excipients have to be qualitatively the same and quantitatively very similar in order to exclude different effects on membrane transporters. As a general rule, for both BCS Class I and III drug substances, well-established excipients in usual amounts should be employed and possible interactions affecting drug BA and/or solubility characteristics should be considered and discussed. A description of the function of the excipients is required with a justification whether the amount of each excipient is within the normal range. Excipients that might affect BA, for example, sorbitol, mannitol, sodium lauryl sulfate, or other surfactants, should be identified as well as their possible impact on the following:

- Gastrointestinal motility.
- Susceptibility of interactions with the drug substance (e.g., complexation).
- Drug permeability.
- Interaction with membrane transporters. Excipients that might affect BA should be qualitatively and quantitatively the same in the test product and the reference product.

FC BCS-based biowaiver is applicable for immediate-release FC products if all active substances in the FC belong to BCS Class I or III and the excipients fulfill the requirements outlined earlier. Otherwise, in vivo BE testing is required.

7 Statistical Evaluation of Bioequivalence Data

BACKGROUND

Defined as *relative bioavailability* (BA), bioequivalence (BE) involves comparison between a test (*T*) and reference (*R*) drug product, where *T* and *R* can vary, depending on the comparison to be performed (e.g., to-be-marketed dosage form vs. clinical trial material, generic drug vs. reference listed drug, drug product changed after approval vs. drug product before the change). Although BA and BE are closely related, BE comparisons normally rely on (1) a criterion, (2) a confidence interval (CI) for the criterion, and (3) a predetermined BE limit. BE comparisons could also be used in certain pharmaceutical product line extensions, such as additional strengths, new dosage forms (e.g., changes from immediate release to extended release), and new routes of administration. In these settings, the approaches described in this guidance can be used to determine BE. The general approaches discussed in this guidance may also be useful when assessing pharmaceutical equivalence or performing equivalence comparisons in clinical pharmacology studies and other areas.

In the July 1992 guidance on *Statistical Procedures for Bioequivalence Studies Using a Standard Two-Treatment Crossover Design* (the 1992 guidance), Center for Drug Evaluation and Research (CDER) recommended that a standard in vivo BE study design be based on the administration of either single or multiple doses of the *T* and *R* products to healthy subjects on separate occasions, with random assignment to the two possible sequences of drug product administration. The 1992 guidance further recommended that statistical analysis for pharmacokinetic (PK) measures, such as area under the curve (AUC) and peak concentration (C_{max}), be based on the *two one-sided test* (*TOST*) *procedure* to determine whether the average values for the PK measures determined after administration of the *T* and *R* products were comparable. This approach is termed *average bioequivalence* (ABE) and involves the calculation of a 90% CI for the ratio of the averages (population geometric means) of the measures for the *T* and *R* products. To establish BE, the calculated CI should fall within a BE limit, usually 80%–125% for the ratio of the product averages. (For a broad range of drugs, BE limits of 80%–125% for the ratio of the product averages have been adopted for use of an ABE criterion. Generally, the BE limit of 80%–125% is based on a clinical judgment that a *T* product with BA measures outside this range should be denied market access.) In addition to this general approach, the 1992 guidance provided specific recommendations for (1) logarithmic transformation of PK data, (2) methods to evaluate sequence effects, and (3) methods to evaluate outlier data.

Although ABE is recommended for a comparison of BA measures in most BE studies, the current method of testing requires evaluating both the *population* and *individual BE*. This is useful, in some instances, for analyzing in vitro and in vivo BE studies. The ABE approach focuses only on the comparison of population averages of a BE measure of interest and not on the variances of the measure for the *T* and *R* products. The ABE method does not assess a subject-by-formulation interaction (SFI) variance, that is, the variation in the average *T* and *R* difference among individuals. In contrast, population and individual BE approaches include comparisons of both averages and variances of the measure. The population BE approach assesses total variability of the measure in the population. The individual BE approach assesses within-subject variability for the T and R products, as well as the SFI.

EXPERIMENTAL DESIGNS

A good knowledge of statistics is essential in designing meaningful BA studies. There are too many factors that, if not properly considered, can render BA data less meaningful such as healthy subjects versus patients; effects of disease, age, diet, and environmental conditions; the number of subjects, their sex, and age; choice of blood levels versus urinary excretion; monitoring metabolites versus parent drug; frequency and number of blood/urine samples obtained; relative importance of the various phases of plasma concentration to therapeutic or toxic response; single-dose studies versus multiple-dose steady-state estimations; and sensitivity of analytic methodology.

In comparing formulations, the basic statistical designs depend on whether it is a comparison between two formulations or more than two formulations:

Comparisons of Two Formulations

When two formulations are compared to each other, one of the formulations serves as the standard. The standard is generally the innovator's product whose BA has been accepted as a standard by the Food and Drug Administration (FDA). Two basic designs are possible for this type of comparison, that is, parallel groups or crossover. The parallel group design takes two groups of subjects in equal number and matched as much as possible for age, sex, height, weight, etc. An assumption is made that the overall biological variability is equal in both groups. One group is given the standard formulation, and the other group receives the formulation being evaluated. However, it can be understood why there may be significant variability between groups, a factor that can be significantly reduced by giving both formulation to all subjects in the study and thus each subject serving its own control. There must, however, be sufficient time between trials to wash out all the drug from the body that requires at least 10 half-lives (to reduce levels to below 0.1% level), and it is generally about 1 week for most drugs. However, for drugs like digoxin, phenobarbital, or reserpine that have long half-lives, more than 1 week may be needed to wash out all the drug. Further care must be exercised to take into account disposition of metabolites. It will be necessary to rid the body of all as well since there may be an interaction in disposition between metabolite and the drug. The crossover designs are called randomized balanced crossover designs as given in Table 7.1 and analysis in Table 7.2.

TABLE 7.1
Randomized Balanced Crossover Design

Subject No.	Week 1	Week 2
1	A	B
2	B	A
3	B	A
4	B	A
5	A	B
6	B	A
7	A	B
8	A	B
9	B	A
10	A	B

TABLE 7.2
ANOVA for Crossover Design (in 10 Subjects)

Source of Variation	Degrees of Freedom
Days	1
Subjects	9
Formulations	1
Error	8

TABLE 7.3
Latin Square Design

Subject No.	Week 1	Week 2	Week 3
1	A	B	
2	C	A	
3	B	C	

TABLE 7.4
Balanced Incomplete Block Design

Subject No.	Week 1	Week 2	Subject No.	Week 1	Week 2
1	A	B	7	C	A
2	B	C	8	C	D
3	D	B	9	A	D
4	B	A	10	B	D
5	A	C	11	C	B
6	D	C	12	D	A

Comparisons of More than Two Formulations

If a crossover design is used, the study becomes too large to handle. Also the order in which the formulations are given becomes important. To overcome these disadvantages, for a small number of formulations, a Latin square design can be used (Table 7.3).

For large number of formulations, a balanced incomplete block design (Table 7.4) is used where each formulation occurs in the same number of times and every pair of formulation occurs together in the same number of subjects.

Number of Subjects in Study

The number of subjects used in the study depends on two factors: the level of statistical significance (generally taken as 95%) and the degree of difference within formulations that makes them different (generally taken as 20%). The number of subjects is calculated by

$$N = 20 \times \frac{E}{M_d} \tag{7.1}$$

This equation assumes that there be at least 10 subjects in the study, where E is the error of variance per observation and M_d is the minimum difference between formulations. Thus, depending on the degree of variability of the AUC or another parameter monitored, the number of subjects can increase, but it is not less than 10.

STATISTICAL MODEL

The suggested regulatory model for testing BE is expressed as follows:

$$\frac{[(\mu_T - \mu_R)^2 + (\sigma_{WT}^2 - \sigma_{WR}^2) + \sigma_D^2]}{\sigma_W^2} \leq \theta_I \quad (7.2)$$

The model has three components in its numerator. The first term is the squared difference between the means ($[\mu_T - \mu_R]^2$) of the T and R formulations and a measure of ABE. The second component compares the within-subject variances of the two drug products ($\sigma_{WT}^2 - \sigma_{WR}^2$). The third term is the variance component for the SFI (σ_D^2). The model requires that the sum of the three terms, normalized by a variance (σ_W^2) and with an associated CI, should not exceed a preset regulatory limit (θ_I). The *trade-off* of the first and second terms supposedly provides a reward for a better formulation. SFI is thought to have potential clinical importance. Also the model claims to evaluate the "switchability" rather than the "prescribability" of drug formulations.

The normalizing variance term is constant ($\sigma_W^2 = \sigma_0^2$) if the intrasubject variance of the R formulation does not exceed a preset value ($\sigma_{WR}^2 \leq \sigma_0^2$) but takes the value of $\sigma_W^2 = \sigma_{WR}^2$ at larger variances ($\sigma_{WR}^2 > \sigma_0^2$). $\sigma_0^2 = 0.04$ has been suggested to separate the "constant-scaled" and "reference-scaled" calculations.

Several concerns arise in the use of this model:

- There is no clinical evidence of the inadequacy of the ABE approach.
- There are only tenuous indications of the prevalence of SFIs.
- The original assumptions for the properties of the interaction term had to be modified in this model.
- The mean-variability trade-off is asymmetric and can result not only in rewards but also in penalties, which, in the presence of random variations, can be large.
- The usefulness of the proposed aggregate criterion is questioned since it intends to accomplish various goals simultaneously, whereas stepwise procedures could be less problematic and more effective.

Statistical analyses of BE data are typically based on a statistical model for the logarithm of the BA measures (e.g., AUC and C_{max}). The model is a mixed-effects or two-stage linear model. Each subject, j, theoretically provides a mean for the log-transformed BA measure for each formulation, μ_{Tj} and μ_{Rj} for the T and R formulations, respectively. The model assumes that these subject-specific means come from a distribution with population means μ_T and μ_R and between-subject variances σ_{BT}^2 and σ_{BR}^2, respectively. The model allows for a correlation between μ_{Tj} and μ_{Rj}. The SFI variance component, σ_D^2, is related to these parameters as follows:

$$\sigma_D^2 = \text{Variance of } (\mu_{Tj} - \mu_{Rj}) = (\sigma_{BT} - \sigma_{BR})^2 + 2(1-\rho)\sigma_{BT}\sigma_{BR} \quad (7.3)$$

For a given subject, the observed data for the log-transformed BA measure are assumed to be independent observations from distributions with means μ_{Tj} and μ_{Rj} and within-subject variances σ_{WT}^2 and σ_{WR}^2. The total variances for each formulation are defined as the sum of the within- and between-subject components (i.e., $\mu_{TT}^2 = \mu_{WT}^2 + \mu_{BT}^2$ and $\mu_{TR}^2 = \mu_{WR}^2 + \mu_{BR}^2$). For analysis of crossover studies, the means are given additional structure by the inclusion of period and sequence effect terms.

STATISTICAL APPROACHES FOR BIOEQUIVALENCE

The general structure of a BE criterion is that a function (Θ) of population measures should be demonstrated to be no greater than a specified value (θ). Using the terminology of statistical hypothesis testing, this is accomplished by testing the hypothesis H_0: $\Theta > \theta$ versus H_A: $\Theta \cdot \theta$ at a desired level of significance, often 5%. Rejection of the null hypothesis H_0 (i.e., demonstrating that the estimate of Θ is statistically significantly less than θ) results in a conclusion of BE. The choice of Θ and θ differs in average, population, and individual BE approaches.

A general objective in assessing BE is to compare the log-transformed BA measure after administration of the *T* and *R* products. The population and individual approaches are based on the comparison of an expected squared distance between the *T* and *R* formulations to the expected squared distance between two administrations of the *R* formulation. An acceptable *T* formulation is one where the *T*–*R* distance is not substantially greater than the *R*–*R* distance. In both population and individual BE approaches, this comparison appears as a comparison to the reference variance, which is referred to as *scaling to the reference variability.*

Population and individual BE approaches, but not the ABE approach, allow two types of scaling: reference scaling and constant scaling. Reference scaling means that the criterion used is scaled to the variability of the *R* product, which effectively widens the BE limit for more variable *R* products. Although generally sufficient, the use of reference scaling alone could unnecessarily narrow the BE limit for drugs and/or drug products that have low variability but a wide therapeutic range. This guidance, therefore, recommends mixed scaling for the population and individual BE approaches. With mixed scaling, the reference-scaled form of the criterion should be used if the *R* product is highly variable; otherwise, the constant-scaled form should be used.

AVERAGE BIOEQUIVALENCE

The following criterion is recommended for ABE:

$$(\mu_T - \mu_R)^2 \leq \theta_A^2 \tag{7.4}$$

where

μ_T is the population average response of the log-transformed measure for the *T* formulation
μ_R is the population average response of the log-transformed measure for the *R* formulation

This criterion is equivalent to

$$-\theta_A \leq (\mu_T - \mu_R) \leq \theta_A \tag{7.5}$$

and, usually, $\theta_A = \ln(1.25)$.

POPULATION BIOEQUIVALENCE

The following mixed-scaling approach is recommended for population BE (i.e., use the reference-scaled method if the estimate of $\sigma_{TR} > \sigma_{T0}$ and the constant-scaled method if the estimate of $\sigma_{TR} \cdot \sigma_{T0}$).

The recommended criteria are as follows:
Reference scaled

$$\frac{(\mu_T - \mu_R)^2 + (\sigma_{TT}^2 - \sigma_{TR}^2)}{\sigma_{TR}^2} \leq \theta_P \tag{7.6}$$

or

Constant scaled

$$\frac{(\mu_T - \mu_R)^2 + (\sigma_{TT}^2 - \sigma_{TR}^2)}{\sigma_{T0}^2} \leq \theta_P \tag{7.7}$$

where

- μ_T is the population average response of the log-transformed measure for the *T* formulation
- μ_R is the population average response of the log-transformed measure for the *R* formulation
- σ_{TT}^2 is the total variance (i.e., sum of within- and between-subject variances) of the *T* formulation
- σ_{TR}^2 is the total variance (i.e., sum of within- and between-subject variances) of the *R* formulation
- σ_{T0}^2 is the specified constant total variance
- θ_P is the BE limit

Equations 7.4 and 7.5 represent an aggregate approach where a single criterion on the left-hand side of the equation encompasses two major components: (1) the difference between the *T* and *R* population averages ($\mu_T - \mu_R$) and (2) the difference between the *T* and *R* total variances ($\sigma_{TT}^2 - \sigma_{TR}^2$). This aggregate measure is scaled to the total variance of the *R* product or to a constant value (σ_{T0}^2, a standard that relates to a limit for the total variance), whichever is greater.

The specification of both σ_{T0} and σ_P relies on the establishment of standards. The generation of these standards is discussed in Appendix 7A. When the population BE approach is used, in addition to meeting the BE limit based on confidence bounds, the point estimate of the geometric test/reference mean should fall within 80%–125%.

Individual Bioequivalence

The following mixed-scaling approach is one approach for individual BE (i.e., use the reference-scaled method if the estimate of $\sigma_{WR} > \sigma_{W0}$ and the constant-scaled method if the estimate of $\sigma_{WR} \cdot \sigma_{W0}$) (discontinuity, for further discussion).

The recommended criteria are as follows:
Reference scaled

$$\frac{(\mu_T - \mu_R)^2 + \sigma_D^2 (\sigma_{WT}^2 - \sigma_{WR}^2)}{\sigma_{WR}^2} \leq \theta_I \tag{7.8}$$

or

Constant scaled

$$\frac{(\mu_T - \mu_R)^2 + \sigma_D^2 + (\sigma_{WT}^2 - \sigma_{WR}^2)}{\sigma_{W0}^2} \le \theta_I \tag{7.9}$$

where

μ_T is the population average response of the log-transformed measure for the *T* formulation
μ_R is the population average response of the log-transformed measure for the *R* formulation
σ_D^2 is the SFI variance component
σ_{WT}^2 is the within-subject variance of the *T* formulation
σ_{WR}^2 is the within-subject variance of the *R* formulation
σ_{W0}^2 is the specified constant within-subject variance
θ_I is the BE limit

Equations 7.6 and 7.7 represent an aggregate approach where a single criterion on the left-hand side of the equation encompasses three major components: (1) the difference between the *T* and *R* population averages ($\mu_T - \mu_R$), (2) SFI (σ_D^2), and (3) the difference between the *T* and *R* within-subject variances ($\sigma_{WT}^2 - \sigma_{WR}^2$). This aggregate measure is scaled to the within-subject variance of the *R* product or to a constant value (σ_{W0}^2, a standard that relates to a limit for the within-subject variance), whichever is greater.

The specification of both σ_{W0} and θ_I relies on the establishment of standards. The generation of these standards is described in the following. When the individual BE approach is used, in addition to meeting the BE limit based on confidence bounds, the point estimate of the geometric test/reference mean ratio should fall within 80%–125%.

Standards

The equations for standards to be established (i.e., σ_{T0} and θ_P for assessment of population BE and σ_{W0} and θ_I for individual BE). The recommended approach to establishing these standards is described in the following.

σ_{T0} and σ_{W0}

A general objective in assessing BE should be to compare the difference in the BA log measure of interest after the administration of the *T* and *R* formulations, *T–R*, with the difference in the same log metric after two administrations of the *R* formulation, *R–R*.

Population Bioequivalence

For population BE, the comparisons of interest should be expressed in terms of the ratio of the expected squared difference between *T* and *R* (administered to different individuals) and the expected squared difference between *R* and *R* (administered to different individuals), as shown in the following:

$$E(T-R)^2 = (\mu_T - \mu_R)^2 + \sigma_{TT}^2 + \sigma_{TR}^2 \tag{7.10}$$

$$E(R-R')^2 = 2\sigma_{TR}^2 \tag{7.11}$$

$$\frac{E(T-R)^2}{E(R-R')^2} = \frac{\{(\mu_T - \mu_R)^2 + \sigma_{TT}^2 + \sigma_{TR}^2\}}{2\sigma_{TR}^2} \tag{7.12}$$

The population bioequivalence criterion (PBC) in Equation 7.4 is derived from Equation 7.10, such that the criterion equals zero for two identical formulations. The square root of Equation 7.10 yields the "population difference ratio" (PDR):

$$\text{PDR} = \frac{[(\mu_T - \mu_R)^2 + \sigma_{TT}^2 + \sigma_{TR}^2]}{2\sigma_{TR}^2} \tag{7.13}$$

The PDR is the square root of the ratio of the expected squared *T–R* difference compared to the expected squared *R–R′* difference in the population. It should be noted that the PDR is monotonically related to the PBC described in Equation 7.4 as follows:

$$\text{PDR} = (\text{PBC}/2 + 1) \tag{7.14}$$

Individual Bioequivalence

For individual BE, the comparisons of interest should be expressed in terms of the ratio of the expected squared difference between *T* and *R* (administered to the same individual) and the expected squared difference between *R* and *R* (two administrations of *R* to the same individual), as shown in the following:

$$E(T - R)^2 = (\mu_T - \mu_R)^2 + \sigma_D^2 + \sigma_{WT}^2 + \sigma_{WR}^2 \tag{7.15}$$

$$E(R - R')^2 = 2\sigma_{WR}^2 \tag{7.16}$$

$$\frac{E(T - R)^2}{E(R - R')^2} = \frac{\{(\mu_T - \mu_R)^2 + \sigma_D^2 + \sigma_{WT}^2 + \sigma_{WR}^2\}}{2\sigma_{WR}^2} \tag{7.17}$$

The individual bioequivalence criterion (IBC) in Equation 7.6 is derived from Equation 7.15, such that the criterion equals zero for two identical formulations. The square root of Equation 7.15 is the *individual difference ratio* (IDR):

$$\text{IDR} = \frac{[(\mu_T - \mu_R)^2 + \sigma_D^2 + \sigma_{WT}^2 + \sigma_{WR}^2]}{2\sigma_{WR}^2} \tag{7.18}$$

The IDR is the square root of the ratio of the expected squared *T–R* difference compared to the expected squared *R–R′* difference within an individual. The IDR is monotonically related to the IBC described in Equation 7.6 as follows:

$$\text{IDR} = (\text{IBC}/2 + 1) \tag{7.19}$$

This guidance recommends that $\sigma_{W0} = 0.2$, based on the consideration of the maximum allowable IDR of 1.25. (The IDR upper bound of 1.25 is drawn from the currently used BE limit of 1.25 for the ABE criterion.)

θ_P and θ_I

The determination of θ_P and θ_I should be based on the consideration of ABE criterion and the addition of variance terms to the PBC and IBC, as expressed by the following formula:

$$\theta = \frac{(\text{ABE limit} + \text{Variance factor})}{\text{Variance}}$$

Population Bioequivalence

$$\theta_P = \frac{-(\ln 1.25)^2 + \varepsilon_P}{\sigma_{T0}^2} \tag{7.20}$$

The value of ε_P for population BE is guided by the consideration of the variance term $(\sigma_{TT}^2 - \sigma_{TR}^2)$ added to the ABE criterion. Sponsors or applicants wishing to use the population BE approach should contact the agency for further information on ε_P and θ_P.

Individual Bioequivalence

$$\theta_P = \frac{-(\ln 1.25)^2 + \varepsilon_I}{\sigma_{W0}^2} \tag{7.21}$$

The value of ε_I for individual BE is guided by the consideration of the estimate of SFI (σ_D) as well as the difference in within-subject variability $(\sigma_{WT}^2 - \sigma_{WR}^2)$ added to the ABE criterion. The recommended allowance for the variance term $(\sigma_{WT}^2 - \sigma_{WR}^2)$ is 0.02. In addition, this guidance recommends a σ_D^2 allowance of 0.03. The σ magnitude of σ_D is associated with the percentage of individuals whose average *T*/*R* ratios lie outside 0.8–1.25. It is estimated that if $\sigma_D = 0.1356$, ~10% of the individuals would have their average ratios outside 0.8–1.25, even if $\mu_T - \mu_R = 0$. When $\sigma_D = 0.1741$, the probability is ~20%.

Accordingly, on the basis of consideration for both σ_D and variability $(\sigma_{WT}^2 - \sigma_{WR}^2)$ in the criterion, this guidance recommends that $\varepsilon_I = 0.05$.

STUDY DESIGN

Experimental Design

Nonreplicated Designs

A conventional nonreplicated design, such as the standard two-formulation, two-period, two-sequence crossover design, can be used to generate data where an average or population approach is chosen for BE comparisons. Under certain circumstances, parallel designs can also be used.

Replicated Crossover Designs

Replicated crossover designs can be used irrespective of which approach is selected to establish BE, although they are not necessary when an average or population approach is used. Replicated crossover designs are critical when an individual BE approach is used to allow estimation of within-subject variances for the *T* and *R* measures and the SFI variance component. The following four-period, two-sequence, two-formulation design is recommended for replicated BE studies (Table 7.5).

For this design, the same lots of the *T* and *R* formulations should be used for the replicated administration. Each period should be separated by an adequate washout period.

Other replicated crossover designs are possible, for example, a three-period design, as shown in the following (Table 7.6).

A greater number of subjects would be encouraged for the three-period design compared to the recommended four-period design to achieve the same statistical power to conclude BE.

TABLE 7.5
Four-Period, Two-Sequence, Two-Formulation Replicate BE Design

	Period			
Sequence	**1**	**2**	**3**	**4**
1	*T*	*R*	*T*	*R*
2	*R*	*T*	*R*	*T*

TABLE 7.6
Three-Period Replicate Design

	Period		
Sequence	**1**	**2**	**3**
1	*T*	*R*	*T*
2	*R*	*T*	*R*

Sample Size Determination

Sample sizes for ABE should be obtained using published formulas. Sample sizes for population and individual BE should be based on simulated data. The simulations should be conducted using a default situation allowing the two formulations to vary as much as 5% in average BA with equal variances and certain magnitude of SFI. The study should have 80% or 90% power to conclude BE between these two formulations. Sample size also depends on the magnitude of variability and the design of the study. Variance estimates to determine the number of subjects for a specific drug can be obtained from the biomedical literature and/or pilot studies.

Tables 7.7 through 7.10 give sample sizes for 80% and 90% power using the specified study design, given a selection of within-subject standard deviations (natural log scale), between-subject standard deviations (natural log scale), and SFI, as appropriate.

While the aforementioned sample sizes assume equal within-subject standard deviations, simulation studies for three-period and four-period designs reveal that if $\Delta = 0$ and $\sigma_{WT}^2 - \sigma_{WR}^2 = 0.05$, the sample sizes given will provide either 80% or 90% power for these studies.

To maintain consistency with Section V.C, which suggests a minimum of 12 subjects in all BE studies, the one case where $n = 10$ provides 80% power should be increased to $n = 12$.

Computer Software for Power Analysis in Bioequivalence Trial with Interim Analysis

Whereas a large volume of literature data reports methods to calculate an appropriate sample size, a handy statistical analysis software, StudySize 3.0 (http://www.studysize.com/), works well for most types of analyses encountered. Given in the following is an exercise showing how to perform power analysis in sample size analyses. (Courtesy of Studysize.com.)

Power Analysis Using StudySize 3.0 in Bioequivalence Trial with Interim Analysis

A new formulation of a drug has been developed. A two-way crossover study is planned to compare this new formulation with the existing formulation. The concentration of the active

TABLE 7.7
ABE: Estimated Numbers of Samples $\Delta = 0.05$

		80%	Power	90%	Power
$\sigma_{WT} = \sigma_{WR}$	σ_D	2P	4P	2P	4P
0.15	0.01	12	6	16	8
	0.10	14	10	18	12
	0.15	16	12	22	16
0.23	0.01	24	12	32	16
	0.10	26	16	36	20
	0.15	30	18	38	24
0.30	0.01	40	20	54	28
	0.10	42	24	56	30
	0.15	44	26	60	34
0.50	0.01	108	54	144	72
	0.10	110	58	148	76
	0.15	112	60	150	80

Notes:
1. Results for two-period designs use method of Diletti et al. (1991).
2. Results for four-period designs use relative efficiency data of Liu (1995).

TABLE 7.8
Population BE

$\sigma_{WR} = \sigma_{WT}$	$\sigma_{BR} = \sigma_{BT}$	80% Power	90% Power
0.15	0.15	18	22
	0.30	24	32
0.23	0.23	22	28
	0.46	24	32
0.30	0.30	22	28
	0.60	26	34
0.50	0.50	22	28
	1.00	26	34

Four-period design (RTRT/TRTR), estimated numbers of subjects: $\varepsilon = 0.02$; $\Delta = 0.05$.

Notes: Results for population BE are approximates from simulation studies (1540 simulations for each parameter combination), assuming two-sequence, four-period trials with a balanced design across sequences.

substance is measured over a 24 h time interval, and the AUC is calculated. The new formulation is considered bioequivalent to the old one if the ratio of the true mean AUC can be concluded to be within the interval 0.80–1.25. The null hypothesis is that the true ratio is outside the interval. BE is concluded if the null hypothesis is rejected. The null hypothesis is rejected at an upper significance level of 0.05 if the TOST for testing the ratio is less than 0.8 and greater than 1.25,

TABLE 7.9
Individual BE, Estimated Numbers of Subjects: $\varepsilon_I = 0.05$, $\Delta = 0.05$

		80%	Power	90%	Power
$\sigma_{WT} = \sigma_{WR}$	σ_D	3P 3P2S [3]	4P	3P 3P2S	4P
0.15	0.01	14	10	18	12
	0.10	18	14	24	16
	0.15	28	22	36	26
0.23	0.01	42	22	54	30
	0.10	56	30	74	40
	0.15	76	42	100	56
0.30	0.01	52	28	70	36
	0.10	60	32	82	42
	0.15	76	42	100	56
0.50	0.01	52	28	70	36
	0.10	60	32	82	42
	0.15	76	42	100	56

Notes: Results for individual BE are approximates using simulations (5000 simulations for each parameter combination). The designs used in simulations are RTR/TRT (3P) and RTRT/TRTR (4P) assuming two-sequence trials with a balanced design across sequences.

TABLE 7.10
Individual BE, Estimated Numbers of Subjects: $\varepsilon_I = 0.05$, $\Delta = 0.10$ with Constraint on Δ ($0.8 \leq \exp(\Delta) \leq 1.25$)

		80% Power	90% Power
$\sigma_{WT} = \sigma_{WR}$	σ_D	4P	4P
0.30	0.01	30	40
	0.10	36	48
	0.15	42	56
0.50	0.01	34	46
	0.10	36	48
	0.15	42	

Notes: Results for individual BE are approximates using simulations (5000 simulations Δ for each parameter combination). The designs used in simulations are RTRT/TRTR (4P), assuming two-sequence trials with a balanced design across sequences. When = 0.05, sample sizes remain the same as given in Table 6.3. This is because the studies are already powered for variance estimation and inference, and therefore, a constraint on the point estimate of Δ has little influence on the sample size for small values of Δ.

respectively; both are rejected at the significance level 0.05 (TOST situation). It can be shown that this is equivalent to a CI for the true ratio, with confidence level 0.90, which is entirely within the interval 0.8–1.25.

The analysis for the two-way crossover design is performed using an analysis of variance (ANOVA) on the log-transformed AUC values. The study is planned to have a power of 0.80 to conclude BE if the true ratio is approximately 1.05 at the significance level 0.05. The expected residual standard deviation in the ANOVA (the within-subject standard deviation) for the log-transformed AUC values is assumed to be in the range 0.15–0.25.

Open the File menu and choose New Table.
Open the Test Procedure menu and choose Bioequivalence test.
Set the following options and press OK.

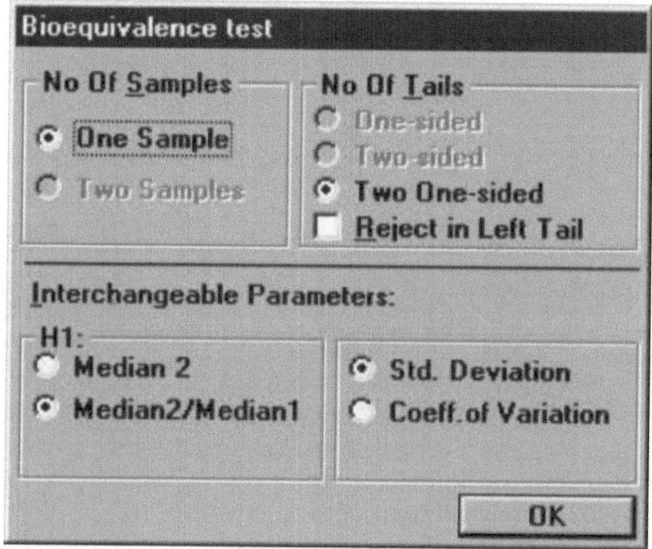

Set the following values and press OK.

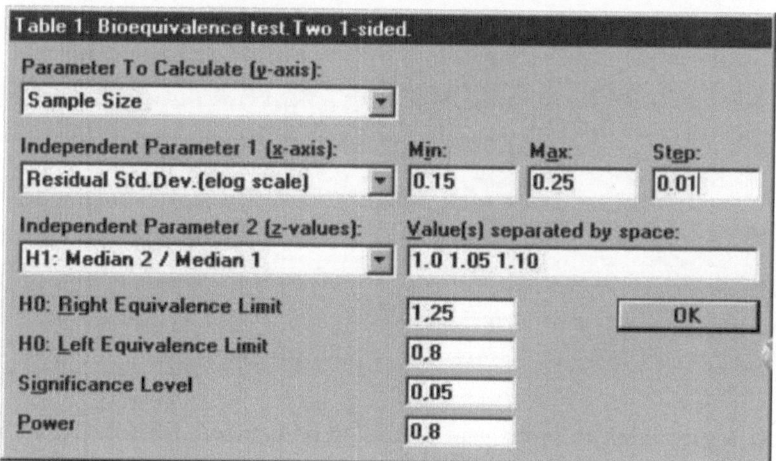

This will create a table where the sample size is calculated for a range of values for the residual standard deviation and for three distinct values of the true AUC ratio, 1.0, 1.05, and 1.10. Note that the sample size is treated as a continuous parameter.

Table 1

Bioequivalence test.Two 1-sided.
Sample Size as a function of Residual Std.Dev.(elog scale) and H1: Median 2 / Median 1.
H0: Right Equivalence Limit=1,25 H0: Left Equivalence Limit=0,8 Significance Level=0,05
Power=0,8

Residual Std.Dev.(elog scale)	H1: Median 2 / Median 1		
	1,00	1,05	1,10
0,15	9,5	11,1	18,6
0,16	10,5	12,4	20,9
0,17	11,6	13,7	23,4
0,18	12,8	15,2	26,0
0,19	14,0	16,7	28,8
0,20	15,3	18,3	31,7
0,21	16,7	20,0	34,8
0,22	18,2	21,8	38,1
0,23	19,7	23,7	41,5
0,24	21,3	25,7	45,0
0,25	23,0	27,7	48,7

Edit Table

The table shows the sample sizes needed if the true AUC ratio is 1.05 and the residual standard deviation is in the interval 0.15–0.25. It also shows how much smaller and larger the sample size has to be if the assumption about the true AUC ratio instead is set to 1.0 and 1.10.

Since the study is planned to be a two-period crossover study and we usually want the same number of patients in the two possible formulation sequences, the sample sizes should be rounded upward to the nearest even integer. For example, with a ratio of 1.05 and a residual standard deviation of 0.20, the sample size 18.3 should be rounded upward to 20.

To calculate the power for 20 subject,
Open the File menu and choose New Calculation.
Open the Test Procedure menu and choose Bioequivalence test.
The dialog box with the retained values will show up. Press the OK button.
Set the following values except for power and then press the Power button.

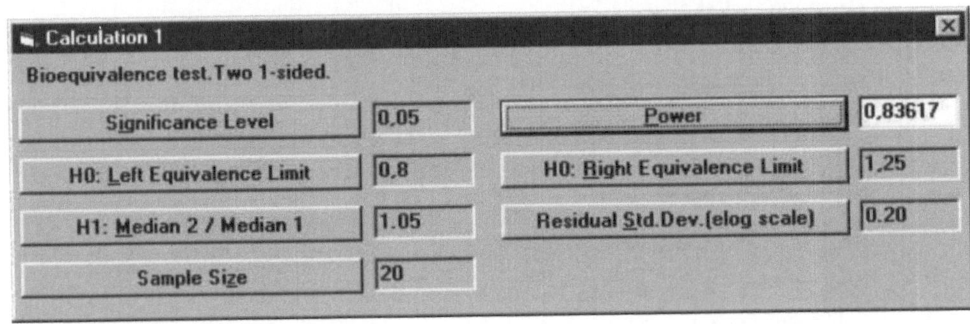

There is now a suggestion to investigate if BE can be concluded before all 20 subjects have entered the study.

Open the File Menu and choose New Monte Carlo Simulation.
Open the Test Procedure menu and choose Bioequivalence test.
Set the number of interim analyses to 1 in the dialog box. The old parameter settings are retained.
Press the OK button.

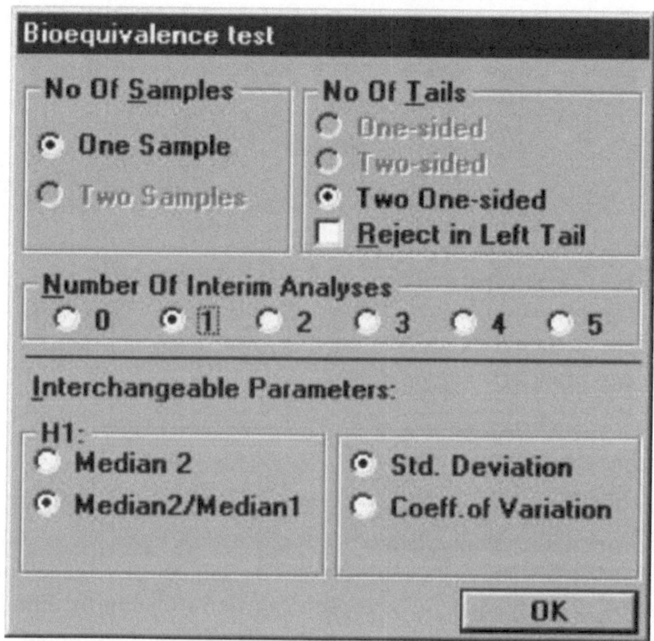

A new dialog box will be shown.

To compensate for two analyses, one has to choose the significance level at the interim analysis and the final analysis in such a way that the overall significance level will be 0.05. There are many possibilities for such a design. Set, for example, the values as shown in the following and press OK.

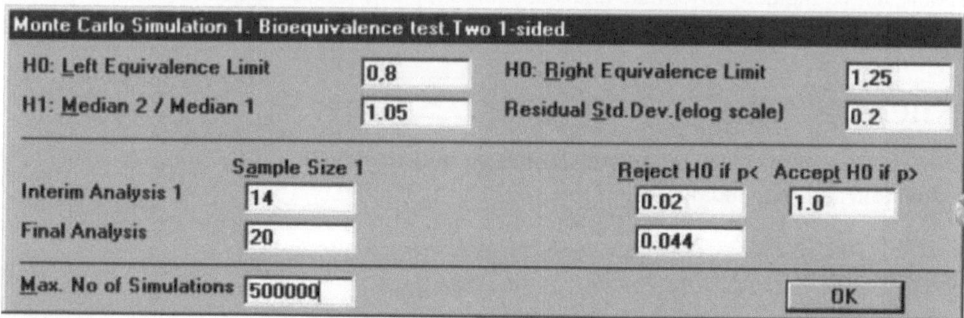

If one prefers to use CIs instead of the p-values, the corresponding confidence levels are 96% at the interim analysis and 91.2% at the final analysis. BE is concluded if the respective CI is within the interval 0.80–1.25.

The results are presented in the following after 500.000 Monte Carlo simulations.

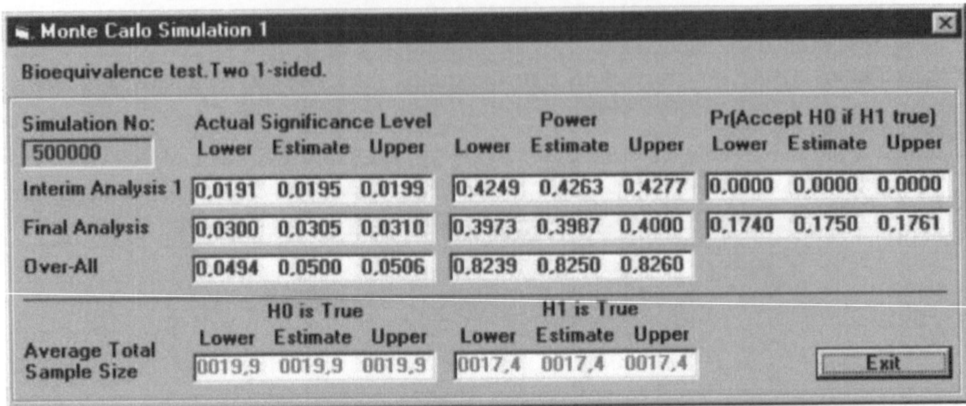
Monte Carlo Simulation 1

Bioequivalence test.Two 1-sided.

Simulation No: 500000	Actual Significance Level			Power			Pr(Accept H0 if H1 true)		
	Lower	Estimate	Upper	Lower	Estimate	Upper	Lower	Estimate	Upper
Interim Analysis 1	0.0191	0.0195	0.0199	0.4249	0.4263	0.4277	0.0000	0.0000	0.0000
Final Analysis	0.0300	0.0305	0.0310	0.3973	0.3987	0.4000	0.1740	0.1750	0.1761
Over-All	0.0494	0.0500	0.0506	0.8239	0.8250	0.8260			

	H0 is True			H1 is True		
	Lower	Estimate	Upper	Lower	Estimate	Upper
Average Total Sample Size	0019.9	0019.9	0019.9	0017.4	0017.4	0017.4

Exit

The significance level is controlled and approximately 0.05 and the power is 0.825. The average number of subjects when H_1 is true is 0.4263 * 14 + (1 – 0.4263) * 20 = 17.4, a gain of 2 subjects compared to 20 but with somewhat lower power. However, there is a large chance (0.4263) that 14 subjects will suffice to show BE.

Sample Size and Dropouts

A minimum number of 12 evaluable subjects should be included in any BE study. When an ABE approach is selected using either nonreplicated or replicated designs, methods appropriate to the study design should be used to estimate sample sizes. The number of subjects for BE studies based on either the population or individual BE approach can be estimated by simulation if analytical approaches for estimation are not available. Further information on sample size is provided in Appendix 7C.

Sponsors should enter a sufficient number of subjects in the study to allow for dropouts. Because replacement of subjects during the study could complicate the statistical model and analysis, dropouts generally should not be replaced. Sponsors who wish to replace dropouts during the study should indicate this intention in the protocol. The protocol should also state whether samples from replacement subjects, if not used, will be assayed. If the dropout rate is high and sponsors wish to add more subjects, a modification of the statistical analysis may be recommended. Additional subjects should not be included after data analysis unless the trial was designed from the beginning as a sequential or group sequential design.

STATISTICAL ANALYSIS

The following sections provide recommendations on statistical methodology for assessment of average, population, and individual BE.

Logarithmic Transformation

General Procedures

This guidance recommends that BE measures (e.g., AUC and C_{max}) be log-transformed using either common logarithms to the base 10 or natural logarithms. The choice of common or natural logs should be consistent and should be stated in the study report. The limited sample size in a typical BE study precludes a reliable determination of the distribution of the data set. Sponsors and/or applicants are not encouraged to test for normality of error distribution after log transformation, nor should they use normality of error distribution as a reason for carrying out the statistical analysis on the original scale. Justification should be provided if sponsors or applicants believe that their BE study data should be statistically analyzed on the original rather than on the log scale.

Clinical Rationale

The FDA Generic Drugs Advisory Committee recommended in 1991 that the primary comparison of interest in a BE study is the ratio, rather than the difference, between average parameter data from the T and R formulations. Using logarithmic transformation, the general linear statistical model employed in the analysis of BE data allows inferences about the difference between the two means on the log scale, which can then be retransformed into inferences about the ratio of the two averages (means or medians) on the original scale. Logarithmic transformation thus achieves a general comparison based on the ratio rather than the differences.

Pharmacokinetic Rationale

Westlake observed that a multiplicative model is postulated for PK measures in BA/BE studies (i.e., AUC and $C_{\max}$, but not $t_{\max}$). Assuming that elimination of the drug is first order and only occurs from the central compartment, the following equation holds after an extravascular route of administration:

$$\mathrm{AUC}_{0-\infty} = \frac{FD}{\mathrm{CL}} \tag{7.22}$$

$$= \frac{FD}{(VK_e)} \tag{7.23}$$

where

F is the fraction absorbed
D is the administered dose
FD is the amount of drug absorbed

CL is the clearance of a given subject that is the product of the apparent volume of distribution (V) and the elimination rate constant (K_e). (Note that a more general equation can be written for any multicompartmental model as

$$\mathrm{AUC}_{0-\infty} = \frac{FD}{V_{d\beta}\lambda_n} \tag{7.24}$$

where

$V_{d\beta}$ is the volume of distribution relating drug concentration in plasma or blood to the amount of drug in the body during the terminal exponential phase
λ_n is the terminal slope of the concentration–time curve.)

The use of AUC as a measure of the amount of drug absorbed involves a multiplicative term (CL) that might be regarded as a function of the subject. For this reason, Westlake contends that the subject effect is not additive if the data are analyzed on the original scale of measurement.

Logarithmic transformation of the AUC data will bring the CL (VK_e) term into the following equation in an additive fashion:

$$\ln \mathrm{AUC}_{0-\infty} = \ln F + \ln D - \ln V - \ln K_e \quad (7.25)$$

Similar arguments were given for C_{max}. The following equation applies for a drug exhibiting one compartmental characteristic:

$$C_{max} = \left(\frac{FD}{V}\right) \times e^{-k_e * T_{max}} \quad (7.26)$$

where again *F*, *D*, and *V* are introduced into the model in a multiplicative manner. However, after logarithmic transformation, the equation becomes

$$\ln C_{max} = \ln F + \ln D - \ln V - K_e t_{max} \quad (7.27)$$

Thus, log transformation of the C_{max} data also results in the additive treatment of the *V* term.

Presentation of Data

The drug concentration in biological fluid determined at each sampling time point should be furnished on the original scale for each subject participating in the study. The PK measures of systemic exposure should also be furnished on the original scale. The mean, standard deviation, and coefficient of variation (CV) for each variable should be computed and tabulated in the final report.

In addition to the arithmetic mean and associated standard deviation (or CV) for the *T* and *R* products, geometric means (antilog of the means of the logs) should be calculated for selected BE measures. To facilitate BE comparisons, the measures for each individual should be displayed in parallel for the formulations tested. In particular, for each BE measure, the ratio of the individual geometric mean of the *T* product to the individual geometric mean of the *R* product should be tabulated side by side for each subject. The summary tables should indicate in which sequence each subject received the product.

Data Analysis

Average Bioequivalence

Overview

Parametric (normal-theory) methods are recommended for the analysis of log-transformed BE measures. For ABE using the criterion stated in Equations 7.2 or 7.3, the general approach is to construct a 90% CI for the quantity $\mu_T - \mu_R$ and to reach a conclusion of ABE if this CI is contained in the interval $[-\theta_A, \theta_A]$. Due to the nature of normal-theory CIs, this is equivalent to carrying out TOST of hypothesis at the 5% level of significance.

The 90% CI for the difference in the means of the log-transformed data should be calculated using methods appropriate to the experimental design. The antilogs of the confidence limits obtained constitute the 90% CI for the ratio of the geometric means between the *T* and *R* products.

Nonreplicated Crossover Designs

For nonreplicated crossover designs, this guidance recommends parametric (normal-theory) procedures to analyze log-transformed BA measures. General linear model procedures available in PROC GLM in SAS or equivalent software are preferred, although linear mixed-effects model procedures can also be indicated for analysis of nonreplicated crossover studies.

For example, for a conventional two-treatment, two-period, two-sequence (2 × 2) randomized crossover design, the statistical model typically includes factors accounting for the following sources of variation: sequence, subjects nested in sequences, period, and treatment. The ESTIMATE statement in SAS PROC GLM, or equivalent statement in other software, should be used to obtain estimates for the adjusted differences between treatment means and the standard error associated with these differences.

Replicated Crossover Designs

Linear mixed-effects model procedures, available in PROC MIXED in SAS or equivalent software, should be used for the analysis of replicated crossover studies for ABE. The following illustrates an example of program statements to run the ABE analysis using PROC MIXED in SAS version 6.12, with SEQ, SUBJ, PER, and TRT identifying sequence, subject, period, and treatment variables, respectively, and Y denoting the response measure (e.g., log(AUC), log(C_{max})) being analyzed:

```
PROC MIXED;
CLASSES SEQ SUBJ PER TRT;
MODEL Y = SEQ PER TRT/DDFM = SATTERTH;
RANDOM TRT/TYPE = FA0(2) SUB = SUBJ G;
REPEATED/GRP = TRT SUB = SUBJ;
ESTIMATE 'T vs. R' TRT 1 -1/CL ALPHA = 0.1;
```

The ESTIMATE statement assumes that the code for the *T* formulation precedes the code for the *R* formulation in sort order (this would be the case, e.g., if *T* were coded as 1 and *R* were coded as 2). If the *R* code precedes the *T* code in sort order, the coefficients in the ESTIMATE statement would be changed to −11.

In the RANDOM statement, TYPE = FA0(2) could possibly be replaced by TYPE = CSH. This guidance recommends that TYPE = UN not be used, as it could result in an invalid (i.e., not non-negative definite) estimated covariance matrix.

Additions and modifications to these statements can be made if the study is carried out in more than one group of subjects.

(A new SAS software-based statistical application "EquivEasy" has been designed and developed with the aim of making the actual execution of equivalence testing procedures easier for a typical final user (e.g., a researcher in a pharmaceutical company) while at the same time ensuring both a high level of statistical competency and access to other powerful features [e.g., import formats, other statistical procedures] of the SAS software.) Although SAS GLM and MIXED procedures can be used for "standard" BE testing (e.g., 90% CIs and TOST), their use is not straightforward because the test results have to be calculated (using appropriate formulas) from SAS procedure outputs. Also it is usually necessary to first examine the results from the model with carryover effects (in case of crossover design) and then (in the absence of significant carryover effects) from the model without carryover effects. Furthermore, reporting on BE studies requires that some standard tables, figures, and listings (TFLs) (e.g., means, CV, ratios, estimates of inter- and intra-subject variability) be supplied in addition to the BE test results. These usually require that appropriate manipulations and transformations be applied to the data before TFLs are made. The purpose of EquivEasy application is

- To raise the likelihood of proper reporting on BE studies (for data from two-treatment, two-period crossover design and 3 × 3 crossover [Williams] design)
- To minimize the errors in report preparation (increased quality)
- To minimize the maximum time required for studies (increased efficiency)
- To reduce the need for in-house SAS expertise (i.e., so as to simplify use and to reduce training costs)

- To maximize the uniformity of reporting (standardization)
- To minimize additional validation costs by using prevalidated SAS Institute software procedures wherever possible

Parallel Designs

For parallel designs, the CI for the difference of means in the log scale can be computed using the total between-subject variance. As in the analysis for replicated designs, equal variances should not be assumed.

Population Bioequivalence

Overview

Analysis of BE data using the population approach should focus first on estimation of the mean difference between the *T* and *R* for the log-transformed BA measure and estimation of the total variance for each of the two formulations. This can be done using relatively simple unbiased estimators such as the method of moments (MM). After the estimation of the mean difference and the variances has been completed, a 95% upper confidence bound for the PBC can be obtained, or equivalently a 95% upper confidence bound for a linearized form of the PBC can be obtained. Population BE should be considered to be established for a particular log-transformed BA measure if the 95% upper confidence bound for the criterion is less than or equal to the BE limit, θ_P, or equivalently if the 95% upper confidence bound for the linearized criterion is less than or equal to 0.

To obtain the 95% upper confidence bound of the criterion, intervals based on validated approaches can be used. The procedure involves the computation of a test statistic that is either positive (does not conclude population BE) or negative (concludes population BE).

Consider the following statistical model that assumes a four-period design with equal replication of *T* and *R* in each of *s* sequences with an assumption of no (or equal) carryover effects (equal carryovers go into the period effects):

$$Y_{ijkl} = \mu_k + \gamma_{ikl} + \delta_{ijk} + \varepsilon_{ijkl} \quad (7.28)$$

where

$i = 1, \ldots s$ indicates sequence
$j = 1, \ldots n_i$ indicates subject within sequence i
$k = R, T$ indicates treatment
$l = 1, 2$ indicates replicate on treatment k for subjects within sequence
Y_{ijkl} is the response of replicate l on treatment k for subject j in sequence i
Y_{ikl} represents the fixed effect of replicate l on treatment k in sequence i
δ_{ijk} is the random subject effect for subject j in sequence i on treatment k
ε_{ijkl} is the random error for subject j within sequence i on replicate l of treatment k

The ε_{ijkl}'s are assumed to be mutually independent and identically distributed as $\varepsilon_{ijkl} \sim N(0, \sigma^2_{Wk})$ for $i = 1, \ldots s, j = 1, \ldots n_i, k = R, T,$ and $l = 1, 2$. Also the random subject effects

$$\delta_{ij} = \left(\mu_R + \delta_{ijR}, \mu_T + \delta_{ijT}\right)' \quad (7.29)$$

are assumed to be mutually independent and distributed as

$$\delta_{ij} \sim N_2\left[\begin{pmatrix} \mu_R \\ \mu_T \end{pmatrix}, \begin{pmatrix} \sigma^2_{BR} & \rho\sigma_{BT}\sigma_{BR} \\ \rho\sigma_{BT}\sigma_{BR} & \sigma^2_{BT} \end{pmatrix}\right]. \quad (7.30)$$

The following constraint is applied to the nuisance parameters to avoid overparameterization of the model for $k = R, T$:

$$\sum_{i=1}^{s}\sum_{l=1}^{2}\gamma_{ikl} = 0 \tag{7.31}$$

This statistical model assumes $s*p$ location parameters (where p is the number of periods) that can be partitioned into t treatment parameters and $sp - t$ nuisance parameters. This produces a saturated model. The various *nuisance* parameters are estimated in this model, but the focus is on the parameters needed for population BE. In some designs, the sequence and period effects can be estimated through a reparameterization of the nuisance effects. This model definition can be extended to other crossover designs.

Linearized Criteria

$$\text{Reference scaled}: \eta_1 = \left(\mu_T - \mu_R\right)^2 + \left(\sigma_{TT}^2 - \sigma_{TR}^2\right) - \theta_P \cdot \sigma_{TR}^2 < 0 \tag{7.32}$$

$$\text{Constant scaled}: \eta_2 = \left(\mu_T - \mu_R\right)^2 + \left(\sigma_{TT}^2 - \sigma_{TR}^2\right) - \theta_P \cdot \sigma_{T0}^2 < 0 \tag{7.33}$$

Estimating the Linearized Criteria

The estimation of the linearized criteria depends on study designs. The remaining estimation and CI procedures assume a four-period design with equal replication of T and R in each of s sequences. The reparameterizations are defined as

$$U_{Tij} = \frac{1}{2} * \left(Y_{ijT1} + Y_{ijT2}\right) \tag{7.34}$$

$$U_{Rij} = \frac{1}{2} * \left(Y_{ijR1} + Y_{ijR2}\right) \tag{7.35}$$

$$V_{Tij} = \frac{1}{\sqrt{2}} * \left(Y_{ijT1} - Y_{ijT2}\right) \tag{7.36}$$

$$V_{Rij} = \frac{1}{\sqrt{2}} * \left(Y_{ijR1} - Y_{ijR2}\right) \tag{7.37}$$

$$I_{ij} = Y_{ijT\cdot} - Y_{ijR\cdot}, \tag{6.38}$$

for $i = 1,\ldots,s$ and $j = 1,\ldots,n_i$, where

$$Y_{ijT\cdot} = \frac{1}{2}\left(Y_{ijT1} + Y_{ijT2}\right) \quad \text{and} \quad Y_{ijR\cdot} = \frac{1}{2}\left(Y_{ijR1} + Y_{ijR2}\right).$$

Compute the formulation means pooling across sequences:

$$\hat{\mu}_k = \frac{1}{s}\sum_{i=1}^{s} \bar{Y}_{i\cdot k\cdot}, \quad k = R,T \text{ and } \hat{\Delta} = \hat{\mu}_T - \hat{\mu}_R$$

where

$$\bar{Y}_{i\cdot k\cdot} = \frac{1}{n_i}\sum_{j=1}^{n_i} \frac{1}{2}\sum_{l=1}^{2} Y_{ijkl}.$$

Compute the variances of $U_{\mathrm{T}ij}$, $U_{\mathrm{R}ij}$, $V_{\mathrm{T}ij}$, $V_{\mathrm{R}ij}$, pooling across sequences, and denote these variance estimates by $MU_{\mathrm{T}}, MU_{\mathrm{R}}, MV_{\mathrm{T}}, MV_{\mathrm{R}}$, respectively. Specifically,

$$MU_{\mathrm{T}} = \frac{1}{n_{U_{\mathrm{T}}}}\sum_{i=1}^{s}\sum_{j=1}^{n_i}\left(U_{\mathrm{T}ij} - \bar{U}_{\mathrm{T}i}\right)^2 \tag{7.38}$$

$$MV_{\mathrm{T}} = \frac{1}{n_{V_{\mathrm{T}}}}\sum_{i=1}^{s}\sum_{j=1}^{n_i}\left(V_{\mathrm{T}ij} - \bar{V}_{\mathrm{T}i}\right)^2 \tag{7.39}$$

$$MU_{\mathrm{R}} = \frac{1}{n_{U_{\mathrm{R}}}}\sum_{i=1}^{s}\sum_{j=1}^{n_i}\left(U_{\mathrm{R}ij} - \bar{U}_{\mathrm{R}i}\right)^2 \tag{7.40}$$

$$MV_{\mathrm{R}} = \frac{1}{n_{V_{\mathrm{R}}}}\sum_{i=1}^{s}\sum_{j=1}^{n_i}\left(V_{\mathrm{R}ij} - \bar{V}_{\mathrm{R}i}\right)^2$$

$$n_{\mathrm{I}} = n_{U_{\mathrm{T}}} = n_{U_{\mathrm{R}}} = n_{V_{\mathrm{T}}} = n_{V_{\mathrm{R}}} = \left(\sum_{i=1}^{s} n_i\right) - s \tag{7.41}$$

Then, the linearized criteria are estimated by the following:

Reference scaled

$$\hat{\eta}_1 = \hat{\Delta}^2 + MU_{\mathrm{T}} + 0.5\cdot MV_{\mathrm{T}} - (1+\theta_{\mathrm{P}})\cdot[MU_{\mathrm{R}} + 0.5\cdot MV_{\mathrm{R}}] \tag{7.42}$$

Constant scaled

$$\hat{\eta}_2 = \hat{\Delta}^2 + MU_{\mathrm{T}} + 0.5\cdot MV_{\mathrm{T}} - (1)\cdot[MU_{\mathrm{R}} + 0.5\cdot MV_{\mathrm{R}}] - \theta_{\mathrm{P}}\cdot\sigma_{\mathrm{T0}} \tag{7.43}$$

95% Upper Confidence Bounds for Criteria

The following table illustrates the construction of a $(1 - \alpha)$ level upper confidence bound based on the two-sequence, four-period design, for the reference-scaled criterion, $\hat{\eta}_1$. Use $\alpha = 0.05$ for a 95% upper confidence bound.

H_q = Confidence Bound	E_q = Point Estimate	$U_q = (H_q - E_q)^2$
$H_D = \left(\left\|\hat{\Delta}\right\| + t_{1-\alpha,n-s}\left(\frac{1}{s^2}\sum_{i=1}^{s} n_i^{-1} M_I\right)^{1/2}\right)^2$	$E_D = \hat{\Delta}^2$	U_D
$H_1 = \frac{(n-s)\cdot E_1}{\chi^2_{n-s,\alpha}}$	$MU_T = E_1$	U_1
$H_2 = \frac{(n-s)\cdot E_2}{\chi^2_{n-s,\alpha}}$	$0.5\cdot MV_T = E_2$	U_2
$H_3rs = \frac{(n-s)\cdot E_3rs}{\chi^2_{n-s,1-\alpha}}$	$-(1+\theta_P)MU_R = E_3rs$	U_3rs
$H_4rs = \frac{(n-s)\cdot E_4rs}{\chi^2_{n-s,1-\alpha}}$	$-(1+\theta_P)\cdot 0.5\cdot MV_R = E_4rs$	U_4rs
$H_{\eta_1} = \sum E_q + \left(\sum U_q\right)^{1/2}$		

$H_{\eta_1} = \sum E_q + \left(\sum U_q\right)^{1/2}$ is the upper 95% confidence bound for $\hat{\eta}_1$. Note $n = \sum_{i=1}^{s} n_i$, where s is the number of sequences, n_i is the number of subjects per sequence, and $\chi^2_{\alpha,n-s}$ is from the cumulative distribution function of the chi-square distribution with $n-s$ degrees of freedom, that is, $\Pr\left(\chi^2_{n-s} \le \chi^2_{\alpha,n-s}\right) = \alpha$. The confidence bound for $\hat{\eta}_2$ is computed similarly, adjusting the constants associated with the variance components where appropriate (in particular, the constant associated with MU_R and MV_R).

H_q = Confidence Bound	E_q = Point Estimate	$U_q = (H_q - E_q)^2$
$H_D = \left(\left\|\hat{\Delta}\right\| + t_{1-\alpha,n-s}\left(\frac{1}{s^2}\sum_{i=1}^{s} n_i^{-1} M_I\right)^{1/2}\right)^2$	$E_D = \hat{\Delta}^2$	U_D
$H_1 = \frac{(n-s)\cdot E_1}{\chi^2_{n-s,\alpha}}$	$MU_T = E_1$	U_1
$H_2 = \frac{(n-s)\cdot E_2}{\chi^2_{n-s,\alpha}}$	$0.5\cdot MV_T = E_2$	U_2
$H_3cs = \frac{(n-s)E_3cs}{\chi^2_{n-s,1-\alpha}}$	$-1\cdot MU_R = E_3cs$	U_3cs
$H_4cs = \frac{(n-s)E_4cs}{\chi^2_{n-s,1-\alpha}}$	$-0.5\cdot MV_R = E_4cs$	U_4cs
$H_{\eta_2} = \sum E_q - \theta_P\cdot\sigma^2_{T0} + \left(\sum U_q\right)^{1/2}$		

Using the mixed-scaling approach, to test for population BE, compute the 95% upper confidence bound of either the reference-scaled or constant-scaled linearized criterion. The selection of either reference-scaled or constant-scaled approach depends on the study estimate of total standard deviation of the R product (estimated by $[MU_R + 0.5 \cdot MV_R]^{1/2}$ in the four-period design). If the study estimate of standard deviation is $\leq \sigma_{T0}$, the constant-scaled criterion and its associated CI should be computed. Otherwise, the reference-scaled criterion and its CI should be computed. The procedure for computing each of the confidence bounds is described earlier. If the upper confidence bound for the appropriate criterion is negative or zero, conclude population BE. If the upper bound is positive, do not conclude population BE.

Nonreplicated Crossover Designs

For nonreplicated crossover studies, any available method (e.g., SAS PROC GLM or equivalent software) can be used to obtain an unbiased estimate of the mean difference in log-transformed BA measures between the T and R products. The total variance for each formulation should be estimated by the usual sample variance, computed separately in each sequence, and then pooled across sequences.

Replicated Crossover Designs

For replicated crossover studies, the approach should be the same as for nonreplicated crossover designs, but care should be taken to obtain proper estimates of the total variances. One approach is to estimate the within- and between-subject components separately, as for individual BE, and then sum them to obtain the total variance. The method for the upper confidence bound should be consistent with the method used for estimating the variances.

Parallel Designs

The estimate of the means and variances from parallel designs should be the same as for nonreplicated crossover designs. The method for the upper confidence bound should be modified to reflect independent rather than paired samples and to allow for unequal variances.

Individual Bioequivalence

Analysis of BE data using an individual BE approach should focus on estimation of the mean difference between T and R for the log-transformed BA measure, the SFI variance, and the within-subject variance for each of the two formulations. For this purpose, we recommend the MM approach.

To obtain the 95% upper confidence bound of a linearized form of the IBC, intervals based on validated approaches can be used. The procedure involves the computation of a test statistic that is either positive (does not conclude individual BE) or negative (concludes individual BE).

Consider the following statistical model that assumes a four-period design with equal replication of T and R in each of s sequences with an assumption of no (or equal) carryover effects (equal carryovers go into the period effects):

$$Y_{ijkl} = \mu_k + \gamma_{ijk} + \delta_{ijk} + \varepsilon_{ijkl} \tag{7.44}$$

where

$i = 1,\ldots s$ indicates sequence
$j = 1,\ldots n_i$ indicates subject within sequence i
$k = R$
T indicates treatment
$l = 1, 2$ indicates replicate on treatment k for subjects within sequence i
Y_{ijkl} is the response of replicate l on treatment k for subject j in sequence i
γ_{ikl} represents the fixed effect of replicate l on treatment k in sequence i
δ_{ijk} is the random subject effect for subject j in sequence i on treatment k
ε_{ijkl} is the random error for subject j within sequence i on replicate l of treatment k

The ε_{ijkl}'s are assumed to be mutually independent and identically distributed as

$$\varepsilon_{ijkl} \sim N(0, \sigma_{Wk}{}^2) \tag{7.45}$$

for $i = 1,\ldots s, j = 1,\ldots n_i, k = R, T$, and $l = 1, 2$. Also the random subject effects $\delta_{ij} = (\mu_R + \delta_{ijR}, \mu_T + \delta_{ijT})'$ are assumed to be mutually independent and distributed as

$$\delta_{ij} \sim N_2\left[\begin{pmatrix}\mu_R\\ \mu_T\end{pmatrix}, \begin{pmatrix}\sigma_{BR}^2 & \rho\sigma_{BT}\sigma_{BR}\\ \rho\sigma_{BT}\sigma_{BR} & \sigma_{BT}^2\end{pmatrix}\right] \tag{7.46}$$

The following constraint is applied to the nuisance parameters to avoid overparameterization of the model for $k = R, T$:

$$\sum_{i=1}^{s}\sum_{l=1}^{2}\gamma_{ikl} = 0 \tag{7.47}$$

This statistical model proposed by Chinchilli and Esinhart assumes $s*p$ location parameters (where p is the number of periods) that can be partitioned into t treatment parameters and $sp - t$ nuisance parameters. This produces a saturated model. The various *nuisance* parameters are estimated in this model, but the focus is on the parameters needed for individual BE. In some designs, the sequence and period effects can be estimated through a reparameterization of the nuisance effects.

This model definition can be extended to other crossover designs.

Linearized Criteria

Reference scaled

$$\eta_1 = (\mu_T - \mu_R)^2 + (\sigma_{TT}^2 - \sigma_{TR}^2) - \theta_p\sigma_{TR}^2 < 0 \tag{7.48}$$

Constant scaled

$$\eta_2 = (\mu_T - \mu_R)^2 + \sigma_D^2 + (\sigma_{WT}^2 - \sigma_{WR}^2) - \theta_I \cdot \sigma_{WO}^2 < 0 \tag{7.49}$$

Estimating the Linearized Criteria

The estimation of the linearized criteria depends on study designs. The remaining estimation and CI procedures assume a four-period design with equal replication of T and R in each of s sequences. The reparameterizations are defined as

$$I_{ij} = Y_{ijT\cdot} - Y_{ijR} \tag{7.50}$$

$$T_{ij} = Y_{ijT1} - Y_{ijT2} \tag{7.51}$$

$$R_{ij} = Y_{ijR1} - Y_{ijR2} \tag{7.52}$$

for $i = 1,\ldots s$ and $j = 1,\ldots n_i$, where

$$Y_{ijT0} = \frac{1}{2}(Y_{ijT1} + Y_{ijT2}) \quad \text{and} \quad Y_{ijR\cdot} = \frac{1}{2}(Y_{ijR1} + Y_{ijR2})$$

Compute the formulation means; the variances of I_{ij}, T_{ij}, and R_{ij}; and pooling across sequences and denote these variance estimates by M_{I}, M_{T}, and M_{R}, respectively, where

$$\hat{\mu}_k = \frac{1}{s}\sum_{i=1}^{s} \bar{Y}_{i\cdot k\cdot}, \quad k = R, T \text{ and } \hat{\Delta} = \hat{\mu}_{\text{T}} - \hat{\mu}_{\text{R}}$$

$$\bar{Y}_{i\cdot k\cdot} = \frac{1}{n_i}\sum_{j=1}^{n_i} \frac{1}{2}\sum_{l=1}^{2} Y_{ijkl}$$

$$M_{\text{I}} = \hat{\sigma}_{\text{I}}^2 = \frac{1}{n_{\text{I}}}\sum_{i=1}^{s}\sum_{j=1}^{n_i}\left(I_{ij} - \bar{I}_i\right)^2$$

$$n_{\text{I}} = n_{\text{T}} = n_{\text{R}} = \left(\sum_{i=1}^{s} n_i\right) - s$$

$$M_{\text{T}} = \hat{\sigma}_{\text{WT}}^2 = \frac{1}{2n_{\text{T}}}\sum_{i=1}^{s}\sum_{j=1}^{n_i}\left(T_{ij} - \bar{T}_i\right)^2$$

$$M_{\text{R}} = \hat{\sigma}_{\text{WR}}^2 = \frac{1}{2n_{\text{R}}}\sum_{i=1}^{s}\sum_{j=1}^{n_i}\left(R_{ij} - \bar{R}_i\right)^2.$$

Then the linearized criteria are estimated by the following:

Reference scaled

$$\hat{\eta}_1 = \hat{\Delta}^2 + M_{\text{I}} + 0.5 \cdot M_{\text{T}} - (1.5 + \theta_{\text{I}}) \cdot M_{\text{R}} \tag{7.53}$$

Constant scaled

$$\hat{\eta}_2 = \hat{\Delta}^2 + M_{\text{I}} + 0.5 \cdot M_{\text{T}} - 1.5 \cdot M_{\text{R}} - \theta_{\text{I}} \cdot \sigma_{\text{WO}}^2 \tag{7.54}$$

and the SFI variance component can be estimated by the following:

$$\hat{\sigma}_{\text{D}}^2 = \hat{\sigma}_{\text{I}}^2 - \frac{1}{2}\left(\hat{\sigma}_{\text{WT}}^2 + \hat{\sigma}_{\text{WR}}^2\right) \tag{7.55}$$

95% Upper Confidence Bounds for Criteria

The following table illustrates the construction of a $(1 - \alpha)$ level upper confidence bound based on the two-sequence, four-period design, for the reference-scaled criterion, $\hat{\eta}_1$. Use $\alpha = 0.05$ for a 95% upper confidence bound.

H_q = Confidence Bound	E_q = Point Estimate	$U_q = (H_q - E_q)^2$
$H_D = \left(\lvert \hat{\Delta} \rvert + t_{1-\alpha,n-s} \left(\frac{1}{s^2} \sum_{i=1}^{s} n_i^{-1} M_I \right)^{1/2} \right)^2$	$E_D = \hat{\Delta}^2$	U_D
$H_I = \frac{(n-s) \cdot M_I}{\chi^2_{\alpha,n-s}}$	$E_I = M_I$	U_I
$H_T = \frac{0.5 \cdot (n-s) \cdot M_T}{\chi^2_{\alpha,n-s}}$	$E_T = 0.5 \cdot M_T$	U_T
$H_R = \frac{-(1.5+\theta_I) \cdot (n-s) \cdot M_R}{\chi^2_{1-\alpha,n-s}}$	$E_R = -(1.5+\theta_I) \cdot M_R$	U_R
$H_{\eta_1} = \sum E_q + \left(\sum U_q \right)^{1/2}$		

where

$$n = \sum_{i=1}^{s} n_i,$$

s is the number of sequences, and $\chi^2_{\alpha,n-s}$ is from the cumulative distribution function of the chi-square distribution with $n-s$ degrees of freedom, that is, $\Pr(\chi^2_{n-s} \le \chi^2_{\alpha,n-s}) = \alpha$. Then

$$H_{\eta_1} = \sum E_q + \left(\sum U_q \right)^{1/2}$$

is the upper 95% confidence bound for $\hat{\eta}_1$. The confidence bound for $\hat{\eta}_2$ is computed similarly, adjusting the constants associated with the variance components where appropriate (in particular, the constant associated with M_R).

H_q = Confidence Bound	E_q = Point Estimate	$U_q = (H_q - E_q)^2$
$H_D = \left(\lvert \hat{\Delta} \rvert + t_{1-\alpha,n-s} \left(\frac{1}{s^2} \sum_{i=1}^{s} n_i^{-1} M_I \right)^{1/2} \right)^2$	$E_D = \hat{\Delta}^2$	U_D
$H_I = \frac{(n-s) \cdot M_I}{\chi^2_{\alpha,n-s}}$	$E_I = M_I$	U_I
$H_T = \frac{0.5 \cdot (n-s) \cdot M_T}{\chi^2_{\alpha,n-s}}$	$E_T = 0.5 \cdot M_T$	U_T
$H_R = \frac{-(1.5) \cdot (n-s) \cdot M_R}{\chi^2_{1-\alpha,n-s}}$	$E_R = -(1.5) \cdot M_R$	U_R
$H_{\eta_2} = \sum E_q - \theta_I \cdot \sigma^2_{W0} + \left(\sum U_q \right)^{1/2}$		

Using the mixed-scaling approach, to test for individual BE, compute the 95% upper confidence bound of either the reference-scaled or constant-scaled linearized criterion. The selection of either reference-scaled or constant-scaled criterion depends on the study estimate of within-subject standard deviation of the R product. If the study estimate of standard deviation is $\le \sigma_{W0}$, the constant-scaled criterion and its associated CI should be computed. Otherwise, the reference-scaled criterion and its CI should be computed. The procedure for computing each of the confidence bounds is described earlier. If the upper confidence bound for the appropriate criterion is negative or zero, conclude individual BE. If the upper bound is positive, do not conclude individual BE.

This guidance recommends that sponsors use either reference scaling or constant scaling at the changeover point. To test for individual BE, compute the 95% upper confidence bounds of both reference-scaled and constant-scaled linearized criteria. The procedure for computing these confidence bounds is described earlier. If the upper bound of either criterion is negative or zero (either H_{η_1} or H_{η_2}), conclude individual BE. If the upper bounds of both criteria are positive, do not conclude individual BE.

After the estimation of the mean difference and the variances has been completed, a 95% upper confidence bound for the IBC can be obtained, or equivalently a 95% upper confidence bound for a linearized form of the IBC can be obtained. Individual BE should be considered to be established for a particular log-transformed BA measure if the 95% upper confidence bound for the criterion is less than or equal to the BE limit, U_I, or equivalently if the 95% upper confidence bound for the linearized criterion is less than or equal to 0.

The restricted maximum likelihood (REML) method may be useful to estimate mean differences and variances when subjects with some missing data are included in the statistical analysis. A key distinction between the REML and MM methods relates to differences in estimating variance terms. If alternative methods to REML or MM are envisioned, the sponsor is encouraged to discuss it with appropriate CDER review staff prior to submitting their applications.

Variance Estimation

Relatively simple unbiased estimators, the MM or the REML method, can be used to estimate the mean and variance parameters in the individual BE approach. A key distinction between the REML and MM methods relates to differences in estimating variance terms. The REML method estimates each of the three variances, σ_D^2, σ_{WR}^2, σ_{WT}^2, separately and then combines them in the IBC. The REML estimate of σ_D^2 is found from estimates of σ_{BR}^2, σ_{BT}^2, and the correlation. The MM approach is to estimate the sum of the variance terms in the numerator of the criterion, $\sigma_D^2 + \sigma_{WT}^2 - \sigma_{WR}^2$, and does not necessarily estimate each component separately. One consequence of this difference is that the MM estimator of σ_D^2 is unbiased but could be negative. The REML approach can also lead to negative estimates, but if the covariance matrix of the random effects is forced to be a proper covariance matrix, the estimate of σ_D^2 can be made to be nonnegative. This forced nonnegativity has the effect of making the estimate positively biased and introduces a small amount of conservatism to the confidence bound. The REML method can be used in special cases (e.g., when substantial missing data are present). In addition, the MM approaches have not yet been adapted to models that allow assessment of carryover effects.

Methods for Obtaining Confidence Intervals for Individual and Population Bioequivalence Criteria

Individual Bioequivalence Method 1: Constrained REML

Statistical Model

- Mixed-effects ANOVA model in natural log scale.
- Subjects within sequence as random effects.
- Within- and between-subject variances allowed to differ by formulation.
- Fixed effects are formulation, period, sequence, and period–sequence interaction (nested within formulation).

Parameter Estimation

- REML estimates of random effects.
- Choose estimation procedure so that the between-subject covariance is nonnegative definite, that is, so that the correlation does not exceed 1.0. (This is the constraint in

"constrained REML." REML without the constraint, similar to Method 2, is possible but has not been evaluated.)

- Generalized least-squares estimates of fixed effects (type III coefficients).

Confidence Intervals

- Ninety-five percent upper confidence bounds using nonparametric percentile bootstrap CI procedure (upper bound of the 90% two-sided CI).
- Use minimum of 1500 (2000 recommended) bootstrap samples that preserve the number of subjects per sequence.

Example of an Implementation: SAS

- SAS PROC MIXED version 6.10 for Windows™ 6.09 maintenance release for UNIX, or equivalent or later release (needed for CSH covariance structure).
- The following is some SAS code for the aforementioned model and four-period designs:
    - proc mixed method = reml maxiter = 200;
    - class formulat subj_id period sequence;
    - model lnmetric = formulat period sequence period*sequence(formulat);
    - random formulat/subject = subj_id type = csh;
    - repeated/group = formulat;
    - estimate 'T – R' formulat -1 1;
- For three-period designs, the simple estimate statement is usually not enough. The coefficients for the estimable function need to be specified.
- Note that the type = un covariance structure is nominally the same as csh for this model (with a 2 × 2 covariance matrix). However, there are some differences. Csh forces the covariance matrix to be nonnegative definite and is the approach that has been evaluated. Un may yield estimates of correlation greater than 1.0 and, hence, estimates of the SFI (σ_D) that are negative. However, the un structure sometimes finds better estimates than csh (in terms of likelihood) when it does find a positive-definite covariance matrix.
- Bootstrapping uses SAS macros. (Do not use a BY statement to bootstrap PROC MIXED.)

Individual Bioequivalence Method 2: Method of Moments

Parameter Estimation (Complete Data)

- Use MM to obtain unbiased estimates of the components of the criterion, difference of means, sum of variance terms in numerator, and the within-subject variance of the reference and then bootstrap to obtain CIs. The implied estimate of the subject-by-formulation term may then be negative ("unconstrained"). The between-subject components of variance are not estimated.
- Chinchilli provides estimates of the means.
- Variance estimates are standard unbiased estimates pooled across sequences.
- Formulas depend on design and so are not given here.
- Much simpler to implement than Method 1. For example, implementation in SAS requires only PROCs MEANS, SUMMARY, and TRANSPOSE.
- Compared to Method 1, Method 2 tends to yield larger estimates of the within-subject variances and smaller estimates of the SFI.
- This approach is possible with three-period designs that replicate only the reference. However, evaluation has considered only four-period designs.

Parameter Estimation (Incomplete Data)

- One approach to missing data with Method 2 is to use the methods described earlier for complete data with the following additions:
    - Any subject missing a formulation metric will not be used in the calculation of intra-subject variance component for that formulation.

- Any subject with any missing T or R will not be used in the estimate of subject by formulation interaction.
- In all cases, the denominator of the variance estimate will be corrected to reflect the number of subjects actually used in calculations for the particular variance component.

- There may be alternate approaches for handling missing data with Method 2. If there is more than minimal missing data, Method 1 is likely preferable to the approach described here for Method 2.

Population Bioequivalence

Parameter Estimation

- As for individual BE Method 2, obtain unbiased estimates of the components of the criterion, difference of means, sum of variances in numerator, and the total variance of the reference and then bootstrap to obtain CIs.
- Use any available method, such as SAS PROC GLM, to obtain unbiased estimates of the difference of means.
- For standard, two-period, two-sequence crossover designs, the variance estimates are the standard unbiased estimates pooled across sequences (Tables 7.11 and 7.12).

TABLE 7.11

Some Examples for Individual BE, 95% Upper Confidence Bounds for Θ_S

Allowable Upper Limit 2.495 (ε = .05)	Constrained REML		Method of Moments	
Scaled to	Reference	Constant	Reference	Constant
Furosemide	11.293	4.660	10.188	4.638
Verapamil	**2.179**	**1.898**	**1.644**	**1.505**
ac-5-ASA	**1.720**	**1.600**	2.649	**1.079**

Note: Bold text shows criterion selected by mixed scaling. Figures in bold satisfy the IBC at the 5% level.

TABLE 7.12

Some Examples for Individual BE, Parameter Estimates, and 90% CI Using Constrained REML

		Intrasubject		SD (σ_W)	Intersubject	SD (σ_B)	
Data Set	N	Ratio of Means (Original Scale)	Reference	T/R Ratio	Reference	T/R Ratio	SFI (σ_D)
Furosemide, all	8	**98.1** **(81.2–116.9)**	.242 (.108–.263)	.986 (.543–1.813)	.059 (.000–.157)	5.487 (.665–6.725)	.274 (.033–.373)
Subj. 8, per. 1 removed	8	**103.7** **(91.6–120.5)**	.225 (.111–.263)	.756 (.432–1.581)	.106 (.000–.167)	2.669 (.000, 3.563)	.177 (.012–.238)
Verapamil	23	**98.3** **(91.1–106.0)**	.257 (.172–.278)	1.248 (.924–1.700)	.533 (.348–.594)	.904 (.705–1.070)	.051 (.005–.126)
ac-5-ASA	10	115.9 (106.0–127.3)	.327 (.137–.392)	.827 (.493–1.463)	.517 (.000–.634)	.853 (.635–1.058)	.076 (.008–.147)

Note: Mean ratios in bold satisfy the 80/125 ABE criterion at the 5% level.

MISCELLANEOUS ISSUES

Studies in Multiple Groups

If a crossover study is carried out in two or more groups of subjects (e.g., if, for logistical reasons, only a limited number of subjects can be studied at one time), the statistical model should be modified to reflect the multigroup nature of the study. In particular, the model should reflect the fact that the periods for the first group are different from the periods for the second group. This applies to all of the approaches (average, population, and individual BE) described in this guidance.

If the study is carried out in two or more groups and those groups are studied at different clinical sites or at the same site but greatly separated in time (e.g., months apart), questions may arise as to whether the results from the several groups should be combined in a single analysis. Such cases should be discussed with the appropriate CDER review division.

A *sequential* design, in which the decision to study a second group of subjects is based on the results from the first group, calls for different statistical methods and is outside the scope of this guidance. Those wishing to use a sequential design should consult the appropriate CDER review division.

Carryover Effects

The use of crossover designs for BE studies allows each subject to serve as his or her own control to improve the precision of the comparison. One of the assumptions underlying this principle is that *carryover effects* (also called *residual effects*) are either absent (the response to a formulation administered in a particular period of the design is unaffected by formulations administered in earlier periods) or equal for each formulation and preceding formulation. If carryover effects are present in a crossover study and are not equal, the usual crossover estimate of μ_{T-R} could be biased. One limitation of a conventional two-formulation, two-period, two-sequence crossover design is that the only statistical test available for the presence of unequal carryover effects is the sequence test in the ANOVA for the crossover design. This is a between-subject test, which would be expected to have poor discriminating power in a typical BE study. Furthermore, if the possibility of unequal carryover effects cannot be ruled out, no unbiased estimate of μ_{T-R} based on within-subject comparisons can be obtained with this design.

For replicated crossover studies, a within-subject test for unequal carryover effects can be obtained under certain assumptions. Typically, only first-order carryover effects are considered of concern (i.e., the carryover effects, if they occur, only affect the response to the formulation administered in the next period of the design). Under this assumption, consideration of carryover effects could be more complicated for replicated crossover studies than for nonreplicated studies. The carryover effect could depend not only on the formulation that preceded the current period but also on the formulation that is administered in the current period. This is called a *direct-by-carryover* interaction. The need to consider more than just *simple* first-order carryover effects has been emphasized (Fleiss 1989). With a replicated crossover design, a within-subject estimate of μ_{T-R} unbiased by general first-order carryover effects can be obtained, but such an estimate could be imprecise, reducing the power of the study to conclude BE.

In most cases, for both replicated and nonreplicated crossover designs, the possibility of unequal carryover effects is considered unlikely in a BE study under the following circumstances:

- It is a single-dose study.
- The drug is not an endogenous entity.
- More than an adequate washout period has been allowed between periods of the study, and in the subsequent periods, the predose biological matrix samples do not exhibit a detectable drug level in any of the subjects.
- The study meets all scientific criteria (e.g., it is based on an acceptable study protocol, and it contains sufficient validated assay methodology).

The possibility of unequal carryover effects can also be discounted for multiple-dose studies and/or studies in patients, provided that the drug is not an endogenous entity and the studies meet all scientific criteria as described earlier. Under all other circumstances, the sponsor or applicant could be asked to consider the possibility of unequal carryover effects, including a direct-by-carryover interaction. If there is evidence of carryover effects, sponsors should describe their proposed approach in the study protocol, including statistical tests for the presence of such effects and procedures to be followed. Sponsors who suspect that carryover effects might be an issue may wish to conduct a BE study with parallel designs.

Choice of Specific Replicated Crossover Designs

Reasons Unrelated to Carryover Effects

Each unique combination of sequence and period in a replicated crossover design can be called a *cell* of the design. For example, the two-sequence, four-period design recommended has eight cells. The four-sequence, four-period design in the following has 16 cells (Table 7.13).

The total number of degrees of freedom attributable to comparisons among the cells is just the number of cells minus one (unless there are cells with no observations).

The fixed effects that are usually included in the statistical analysis are sequence, period, and treatment (i.e., formulation). The number of degrees of freedom attributable to each fixed effect is generally equal to the number of levels of the effect minus one. Thus, in the case of the two-sequence, four-period design recommended in Section V.A.1, there would be 2 – 1 = 1 degree of freedom due to sequence, 4 – 1 = 3 degrees of freedom due to period, and 2 – 1 = 1 degree of freedom due to treatment, for a total of 1 + 3 + 1 = 5 degrees of freedom due to the three fixed effects. Because these 5 degrees of freedom do not account for all 7 degrees of freedom attributable to the eight cells of the design, the fixed-effects model is not *saturated*. There could be some controversy as to whether a fixed-effects model that accounts for more or all of the degrees of freedom due to cells (i.e., a more saturated fixed-effects model) should be used. For example, an effect for sequence-by-treatment interaction might be included in addition to the three *main effects*—sequence, period, and treatment. Alternatively, a sequence-by-period interaction effect might be included, which would fully saturate the fixed-effects model.

If the replicated crossover design has only two sequences, the use of only the three main effects (sequence, period, and treatment) in the fixed-effects model or the use of a more saturated model makes little difference to the results of the analysis, provided there are no missing observations and the study is carried out in one group of subjects. The least-squares estimate of $\sigma_T - \sigma_R$ will be the same for the main-effects model and for the saturated model. Also the MM estimators of the variance terms in the model used in some approaches to assessment of population and individual BE, which represent within-sequence comparisons, are generally fully efficient regardless of whether the main-effects model or the saturated model is used.

TABLE 7.13
Four-Sequence, Four-Period Design

	Period			
Sequence	**1**	**2**	**3**	**4**
1	*T*	*R*	*R*	*T*
2	*R*	*T*	*T*	*R*
3	*T*	*T*	*R*	*R*
4	*R*	*R*	*T*	*T*

TABLE 7.14
Two-Sequence, Three-Period Design

	Period		
Sequence	**1**	**2**	**3**
1	*T*	*R*	*R*
2	*R*	*T*	*T*

If the replicated crossover design has more than two sequences, these advantages are no longer present. Main-effects models will generally produce different estimates of μ_{T-R} than saturated models (unless the number of subjects in each sequence is equal), and there is no well-accepted basis for choosing between these different estimates. Also MM estimators of variance terms will be fully efficient only for saturated models, while for main-effects models, fully efficient estimators would have to include some between-sequence components, complicating the analysis. Thus, use of designs with only two sequences minimizes or avoids certain ambiguities due to the method of estimating variances or due to specific choices of fixed effects to be included in the statistical model.

Reasons Related to Carryover Effects

One of the reasons to use the four-sequence, four-period design described earlier is that it is thought to be optimal if carryover effects are included in the model. Similarly, the two-sequence, three-period design is thought to be optimal among three-period replicated crossover designs. Both of these designs are *strongly balanced for carryover effects*, meaning that each treatment is preceded by each other treatment *and itself* an equal number of times (Table 7.14).

With these designs, no efficiency is lost by including *simple* first-order carryover effects in the statistical model. However, if the possibility of carryover effects is to be considered in the statistical analysis of BE studies, the possibility of direct-by-carryover interaction should also be considered. If direct-by-carryover interaction is present in the statistical model, these favored designs are no longer optimal. Indeed, the TRR/RTT design does not permit an unbiased within-subject estimate of $\mu_T - \sigma_R$ in the presence of general direct-by-carryover interaction.

The issue of whether a purely main-effects model or a more saturated model should be specified, as described in the previous section, also is affected by possible carryover effects. If carryover effects, including direct-by-carryover interaction, are included in the statistical model, these effects will be partially confounded with sequence-by-treatment interaction in four-sequence or six-sequence replicated crossover designs, but not in two-sequence designs.

In the case of the four-period and three-period designs, the estimate of $\mu_T - \mu_R$, adjusted for first-order carryover effects including direct-by-carryover interaction, is as efficient or more efficient than for any other two-treatment replicated crossover designs.

Two-Period Replicated Crossover Designs

For the majority of drug products, two-period replicated crossover designs such as the Balaam design (which uses the sequences TR, RT, TT, and RR) should be avoided for individual BE because subjects in the TT or RR sequence do not provide any information on SFI. However, the Balaam design may be useful for particular drug products (e.g., a long half-life drug for which a two-period study would be feasible but a three- or more period study would not).

Outlier Considerations

Outlier data in BE studies are defined as subject data for one or more BA measures that are discordant with corresponding data for that subject and/or for the rest of the subjects in a study. Because BE studies are usually carried out as crossover studies, the most important type of subject outlier is the within-subject outlier, where one subject or a few subjects differ notably from the rest of the subjects with respect to a within-subject *T–R* comparison. The existence of a subject outlier with no protocol violations could indicate one of the following situations:

Product Failure

Product failure could occur, for example, when a subject exhibits an unusually high or low response to one or the other of the products because of a problem with the specific dosage unit administered. This could occur, for example, with a sustained and/or delayed-release dosage form exhibiting dose dumping or a dosage unit with a coating that inhibits dissolution.

Subject-by-Formulation Interaction

An SFI could occur when an individual is a representative of subjects present in the general population in low numbers, for whom the relative BA of the two products is markedly different than for the majority of the population and for whom the two products are not bioequivalent, even though they might be bioequivalent in the majority of the population.

In the case of product failure, the unusual response could be present for either the *T* or *R* product. However, in the case of a subpopulation, even if the unusual response is observed on the *R* product, there could still be concern for lack of interchangeability of the two products. For these reasons, deletion of outlier values is generally discouraged, particularly for nonreplicated designs. With replicated crossover designs, the *retest* character of these designs should indicate whether to delete an outlier value or not. Sponsors or applicants with these types of data sets may wish to review how to handle outliers with appropriate review staff.

Discontinuity

The mixed-scaling approach has a discontinuity at the changeover point, σ_{W0} (IBC) or σ_{T0} (PBC), from constant to reference scaling. For example, if the estimate of the within-subject standard deviation of the reference is just above the changeover point, the CI will be wider than just below. In this context, the CI could pass the predetermined BE limit if the estimate is just below the boundary and could fail if just above. This guidance recommends that sponsors applying the individual BE approach may use either reference scaling or constant scaling at either side of the changeover point. With this approach, the multiple testing inflates the type I error rate slightly, to approximately 6.5%, but only over a small interval of σ_{WR} (about 0.18–0.20).

STATISTICAL ANALYSIS OF BIOEQUIVALENCE OF HIGHLY VARIABLE DRUGS

In a certain class of drugs, the variability in blood levels is very large, for a variety of reasons; some associated with their absorption patterns and the other with their disposition patterns. For these drugs, if standard methods of statistical modeling are used, the number of subjects required to establish BE is very large. Given in the following is a tabulation of the number of subjects required within the normal variation from 20% to 80% using the standard two-way crossover design, TOST, and other conditions such as 90% power, same number of subjects.

As seen in Table 7.15, the number of subjects that need to be included in each of the *T*/*R* group can get very large, and this negates the very philosophy of the generic drug development, where the purpose is to keep the cost of development as low as possible. As a result, the regulatory agencies have come up with statistical models that allow using a reduced number of subjects to conduct these studies yet arrive at the same robustness of conclusions about their BE.

The Committee for Medicinal Products for Human Use (CHMP) of the European Medicines Agency (EMA) issued in 2010 a substantially revised BE guideline, which included new approaches

TABLE 7.15
Calculation of Number of Subjects in a Two-Way Crossover Design[a]

CV	Significance Level			
	0.001	0.025	0.050	0.100
0.20	20.1	11.3	9.3	7.2
0.22	23.7	13.4	11.0	8.6
0.24	27.7	15.6	12.9	10.1
0.26	31.9	18.1	15.0	11.7
0.28	36.4	20.7	17.1	13.4
0.30	41.2	23.5	19.5	15.3
0.32	46.3	26.4	21.9	17.2
0.34	51.7	29.6	24.5	19.3
0.36	57.3	32.8	27.2	21.4
0.38	63.2	36.2	30.0	23.6
0.40	69.3	39.7	33.0	26.0
0.42	75.6	43.4	36.0	28.4
0.44	82.1	47.2	39.2	30.9
0.46	88.8	51.1	42.4	33.4
0.48	95.8	55.1	45.8	36.1
0.50	102.9	59.2	49.2	38.8
0.52	110.2	63.5	52.7	41.6
0.54	117.6	67.8	56.3	44.4
0.56	125.2	72.2	60.0	47.3
0.58	133.0	76.7	63.7	50.3
0.60	140.9	81.2	67.5	53.3
0.62	148.9	85.9	71.4	56.4
0.64	157.0	90.6	75.3	59.5
0.66	165.2	95.4	79.3	62.6
0.68	173.6	100.2	83.3	65.8
0.70	182.0	105.1	87.4	69.0
0.72	190.5	110.0	91.5	72.3
0.74	199.1	115.0	95.6	75.5
0.76	207.7	120.0	99.8	78.8
0.78	216.4	125.0	104.0	82.2
0.80	225.2	130.1	108.2	85.5

Note: The number of subjects listed in the table is for each group, so the total number of subjects will be twice the number listed; contrast this with the number of subjects given in the following for other statistical designs.

[a] Test approach, using a two-way one-sided and using the sample size sequence 1 as a function of CV and significance level at power = 0.9, H_0: left equivalence limit = 0.8; H_0: right equivalence limit = 1.25; H_1: median 2/median 1 = 1; sample size seq. 2/seq. 1 = 1.

for the determination of BE for HV drugs. For the same purpose, members of a working group of the FDA in the United States published a procedure. The approaches of both EMA and FDA are based on the method of scaled average bioequivalence (SABE) or its close variant with a few differentiating elements, yet the main goal remains to lower the required sample sizes in comparison with the expectations of the customarily applied unscaled ABE.

Background

The usually applied criterion for the determination of BE requires the TOST procedure. Accordingly, the average logarithmic kinetic responses of the *T* and *R* formulations (μ_T and μ_R, respectively) are contrasted. The acceptance of BE is stated if the 90% CI for the difference between the estimated logarithmic means is between preset regulatory limits. The limits (θ_A) are generally symmetrical on the logarithmic scale and usually equal to ln(1.25). Consequently, the criterion for the determination of ABE is schematically

$$-\theta_A \leq \mu_T - \mu_R \leq \theta_A \tag{7.56}$$

In a BE study, the individual kinetic responses are evaluated from the measured concentrations. The means of the logarithmic responses of the two formulations (m_T and m_R) are calculated; these sample averages estimate the true population means (μ_T and μ_R). The true values of the means are not known, and therefore, their estimates have to be used:

$$-\theta_A \leq m_T - m_R \leq \theta_A \tag{7.57}$$

Also the within-subject variance is estimated (s^2w), in replicate-design studies with three or four periods, for each BE metric.

Equation 7.2 implies that, for the declaration of BE, the difference between the estimated logarithmic means, together with their 90% CI, should be within the regulatory limits of $\pm\theta_A$. The succeeding expressions have the corresponding interpretation.

The method of EMA substitutes the fixed θ_A regulatory constant with limits, which depend on s^2w:

$$-ksW \leq m_T - m_R \leq ksW \tag{7.58}$$

This is still ABE but with expanding limits (ABEL). Therefore, the TOST procedure of Schuirmann can be directly applied. Based on statistical recommendations, EMA proposed 0.76 as the value of the regulatory constant *k*. An alternative, related definition of the regulatory constant, preferred by some, is

$$\sigma_0 = \frac{\ln(1.25)}{k} \tag{7.59}$$

The expectation of EMA for the value of σ_0 is 0.294.

The European guideline prescribes additional conditions and requirements:

1. Subject variance must be evaluated using the data only of the reference drug. Accordingly, the within-subject variation of the *R* product should be used as subject variance in Equation 7.3 to calculate the limits. The determination of subject variance for reference requires three- or four-period crossover designs.
2. Apply the so-called mixed procedure for ABE; when the estimated within-subject CV does not exceed 30%, apply unscaled ABE as usual. If the variation is higher than 30% (corresponding to subject variance for reference > 0.294), then apply ABEL.
3. Geometric mean ratio (GMR) constraint: the point estimate for the GMR, exp ($m_T - m_R$), must be between 0.80 and 1.25.
4. A cap is placed on the expansion of the limits. If the CV exceeds 50%, then the limits may not expand anymore, but ABE should be applied again with limits of ±ln(1.43).

TABLE 7.16
Sample Sizes for 90% Power Using EMA Criteria in Three-Period (TRR-RTR-RRT) Studies

CV (%)	GMR	0.85	0.90	0.95	1.00	1.05	1.10	1.15	1.20
30		>201	74	36	28	36	62	147	>201
35		181	70	39	32	39	63	117	>201
40		130	61	38	33	39	57	94	>201
45		132	55	37	33	38	51	85	>201
50		158	55	39	34	38	51	84	>201
55		178	59	41	37	41	53	97	>201
60		199	64	45	41	46	60	112	>201
65		>201	72	51	46	51	67	125	>201
70		>201	82	57	52	57	76	141	>201
75		>201	93	66	58	64	85	161	>201
80		>201	100	70	63	71	93	176	>201

TABLE 7.17
Sample Size for 90% Power Using EMA Criteria in Four-Period (TRTR-RTRT) Studies

CV (%)	GMR	0.85	0.90	0.95	1.00	1.05	1.10	1.15	1.20
30		180	49	25	19	24	42	95	>201
35		123	48	27	22	27	43	80	>201
40		93	42	26	23	26	39	66	165
45		90	40	27	24	27	37	59	181
50		102	39	27	25	27	36	60	>201
55		123	41	29	26	29	38	63	>201
60		139	45	32	29	31	41	71	>201
65		159	51	36	32	35	46	81	>201
70		172	55	40	36	40	52	97	>201
75		195	62	43	39	44	58	106	>201
80		>201	69	49	45	49	62	113	>201

Table 7.16 shows sample sizes with the EMA regulatory criteria for three-period design, and Table 7.17 shows sample sizes with the EMA regulatory criteria using a four-period design, both at 90% power.

The U.S. FDA uses SABE, which utilizes a rearranged form of Equation 7.3:

$$k \leq (m_T - m_R)/sW \leq k \tag{7.60}$$

The FDA proposes that the value of the regulatory constant be $k = 0.893$ or $\sigma_0 = 0.25$ (3).

The additional conditions and requirements of the FDA procedure are as follows:

1. Substitute subject variance in Equation 7.59 by the estimated within-subject variation of the *R* product. Evaluate subject variance of reference either in three-period crossover studies in which the *R* formulation is administered twice (RRT-RTR-TRR) or in four-period investigations.
2. Apply the mixed procedure for ABE.

3. The CI for SABE cannot be calculated analytically. The FDA recommends a numerical approach based on the approximate linearization of Equation 7.60.
4. Apply the GMR constraint.

The FDA does not impose further restrictions such as a cap on the application of SABE. Overall, the statistical properties of the methods proposed by EMA and FDA are rather complex as a result of the additional conditions and requirements (mixed procedure, GMR constraint, and, for EMA, a cap on the limits). Furthermore, the tests required by both EMA and FDA are dependent on each other, which makes the theoretical treatment very complicated.

Sample sizes for the FDA design studies are given in Tables 7.18 and 7.19.

The GMRs are considered from 0.85 to 1.20. More than 200 subjects are always required when the GMR is outside this range. Within-subject variations of the *R* product are shown between CVs of 30% and 80%.

All entries in these tables refer to total sample sizes in an investigation. Consequently, the number of subjects within a study sequence is either one-third or one-half of these figures. Two general conclusions can be drawn. First, the four-period design requires about 30% fewer subjects

TABLE 7.18
Sample Size Requirements to Meet the FDA Qualification in Three-Period Design (TRR-RTR-RRT) at 90% Power

CV (%)	GMR	0.85	0.90	0.95	1.00	1.05	1.10	1.15	1.20
30		>201	65	33	26	32	55	122	>201
35		106	51	32	28	32	47	77	186
40		99	45	31	28	31	43	68	>201
45		128	43	30	28	30	40	69	>201
50		158	45	31	28	30	40	79	>201
55		178	50	31	28	31	42	96	>201
60		199	54	33	30	34	50	112	>201
65		>201	61	35	32	36	53	125	>201
70		>201	68	39	34	37	61	141	>201
75		>201	80	43	37	41	68	161	>201
80		>201	83	48	41	47	75	176	>201

TABLE 7.19
Sample Size Requirement to Meet the FDA Qualification in Four-Period (TRTR-RTRT) Design (TRR-RTR-RRT) at 90% Power

CV (%)	GMR	0.85	0.90	0.95	1.00	1.05	1.10	1.15	1.20
30		152	44	23	18	22	38	81	>201
35		80	38	23	20	23	34	55	128
40		70	32	22	20	22	30	48	158
45		84	32	22	20	22	30	49	181
50		102	32	23	20	22	30	54	>201
55		123	34	23	21	22	31	61	>201
60		139	38	24	22	24	33	71	>201
65		159	44	26	23	25	35	81	>201
70		172	46	26	24	27	43	97	>201
75		195	53	29	26	29	48	106	>201
80		>201	60	33	28	31	51	113	>201

in comparison with the three-period investigations. Second, the sample size requirement is lower with the FDA than with the EMA approach.

Change of Sample Size with Within-Subject Variation: Effects of Regulatory Requirements

As noted earlier, the regulatory conditions and requirements of both EMA and FDA are complicated and contain various stipulations. As a result, the change of sample sizes with increasing variation is also complicated. For instance, when GMR is removed from 1.00, the required sample size, typically, initially decreases with rising variation and later increases.

Table 7.20 shows the required sample sizes without the additional regulatory conditions, that is, without the mixed procedure, GMR constraint, and the cap on the regulatory limits. Table 7.21 shows that the required sample size does not change with increasing CV when GMR = 1.00 and decreases with rising CV when GMR deviates from unity.

The explanation lies in the features of the scaled difference of the means, which is shown in Equation 7.59. The scaled difference follows the noncentral t-distribution. When GMR = 1.00, the noncentrality parameter is zero. In this case, the scaled difference has a standard t-distribution. This means that the width of the CI is independent of CV and depends only on the sample size. Consequently, with GMR = 1.00, the number of volunteers is also independent of CV. This is shown in Table 7.21 (apart of small, random fluctuations due to the simulation error).

TABLE 7.20
Sample Sizes with EMA Conditions but without Mixed Procedure, GMR Constraint, and Cap on the Use of ABEL

			80%			90%	
CV (%)	GMR	1.0	Power 1.1	1.2	1.0	Power 1.1	1.2
30		26	54	>201	33	74	>201
35		26	45	198	33	63	>201
40		27	41	123	33	56	172
50		26	34	73	33	47	104
60		26	32	55	33	43	77
70		27	30	46	33	40	64
80		27	30	43	33	39	58

TABLE 7.21
Sample Sizes with EMA Conditions Including Mixed Procedure but without GMR Constraint and Cap on the Use of ABEL

			80%			90%	
CV (%)	GMR	1.0	Power 1.1	1.2	1.0	Power 1.1	1.2
30		23	46	>201	28	64	>201
35		25	45	>201	32	63	>201
40		26	40	124	33	55	175
50		27	35	73	33	47	103
60		26	32	56	33	43	78
70		26	31	48	33	41	64
80		27	30	42	33	39	59

TABLE 7.22
Sample Sizes with EMA Conditions Including Mixed Procedure and GMR Constraint but without Cap on the Use of ABEL

		80%			90%		
CV (%) GMR	1.0	Power 1.1	1.2	1.0	Power 1.1	1.2
30	23	46	>201	28	63	>201
35	26	45	>201	32	62	>201
40	26	41	138	33	56	>201
50	27	36	137	33	51	>201
60	27	34	199	35	52	>201
70	28	37	>201	37	63	>201
80	30	38	>201	42	72	>201

When GMR deviates from 1.00, then the noncentrality parameter raises the upper limit of the CI. The noncentrality parameter is proportional to log(GMR) and reciprocally proportional to sW. Thus, at a given sample size and log(GMR), the rise in the CI gets smaller as sW (i.e., CV) increases. Correspondingly, the required sample size declines as sW (i.e., CV) increases.

Similar considerations apply when the approach of ABEL is used since the power of ABE can be characterized by a bivariate noncentral *t*-distribution.

Table 7.21 shows *the effect of including the mixed strategy.*

With the mixed strategy, the method of evaluation depends on the estimated CV of the *R* product. When the true CV is 30%, then half of the simulated trials are evaluated by the unscaled and the other half by the scaled approach. When the true CV is higher than 30%, say 34%, then the estimated CV is still below 30% in many studies, which are then evaluated by unscaled ABE. Thus, instead of using ABEL (or SABE), which would have narrower limits ("goalposts") in this region, unscaled ABE is applied with its wider, more relaxed 0.80–1.25 limits/goalposts. Consequently, on applying the mixed strategy, the required sample size becomes lower at and near CV = 30%. Table 7.22 presents sample sizes *in the additional presence of the GMR constraint.* (These results do not assume the cap on the use of ABEL, which is an expectation of EMA.)

Consequently, in comparison with Table 7.3, the effect of including the GMR constraint can be observed. The constraint raises the required sample size when the CV is high. This is understandable since at large CVs, the chances of observing very high (or very low) GMR are high. The effect of the GMR constraint is particularly conspicuous when the true GMR deviates from 1.00.

Table 7.23 shows sample sizes with similar conditions as Table 7.22 except that the value of the regulatory constant is $k = 0.893$ or $\sigma_0 = 0.25$ and thereby *corresponds to the requirements of FDA.*

The number of subjects is lower with the FDA than the EMA requirements. The deviation is meaningful at moderately high variations but diminishes at still higher CVs.

The results in Table 7.23 can be compared also with those given in Table 7.16. The sample sizes in Table 7.22 were obtained by using ABEL (in order to be able to compare them directly with those in Table 7.22), whereas those in Table 7.17 were computed with SABE as required by FDA. Therefore, small differences between entries in the two tables are due to the two algorithms.

Comparison of Sample Sizes Required by EMA and FDA

The regulatory requirements of FDA call for fewer subjects than those of EMA. At first sight, the requirements of FDA are more favorable to sponsors than those of EMA. However, statistical reasoning supports the recommendations of EMA. Figure 7.1 illustrates the sample sizes required by

TABLE 7.23
Sample Sizes with FDA Conditions Including Mixed Procedure and GMR Constraint

		80%			90%		
CV (%)	GMR	1.0	Power 1.1	1.2	1.0	Power 1.1	1.2
30		21	38	>201	26	55	>201
35		21	31	103	26	44	178
40		21	28	96	26	39	>201
50		21	27	137	27	40	>201
60		22	28	194	29	48	>201
70		24	31	>201	34	59	>201
80		27	38	>201	41	76	>201

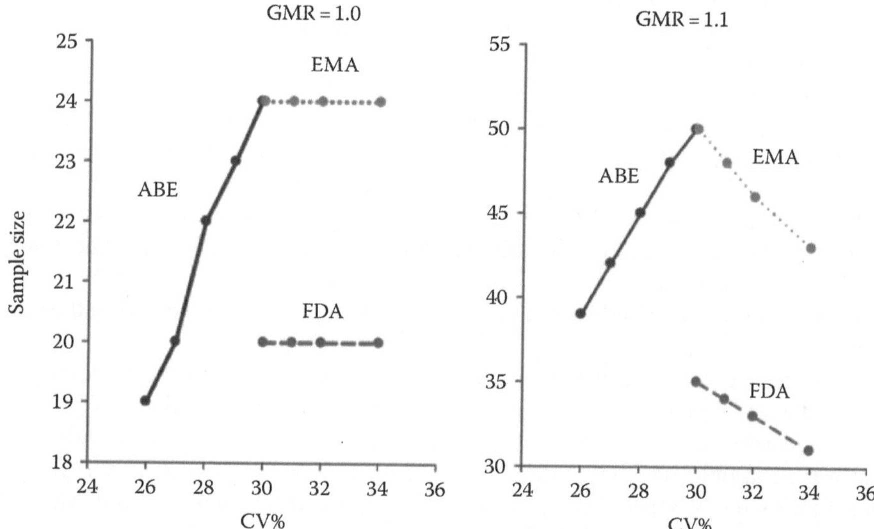

FIGURE 7.1 Comparison of sample sizes required by the U.S. FDA and EMA using true CVs but without the mixing effect and without the GMR constraint.

the two agencies at and just above CV = 30%. The figure shows also the sample sizes needed by unscaled ABE just below and up to this variation.

As expected, the required sample size increases with rising variation when the results are evaluated by unscaled ABE. Above CV = 30% and with the application of ABEL, the sample size is independent of the variation when the true GMR = 1.0. However, when the true GMR deviates from 1.0 (e.g., when GMR = 1.1), the required sample size initially decreases with rising CV. Importantly, the required sample size changes *continuously* around CV = 30% when the requirements of EMA are followed (i.e., with the regulatory constant of $k = 0.76$ or $\sigma_0 = 0.294$). In other words, at CV = 30%, the same sample size is obtained regardless whether ABE or ABEL is applied.

In contrast, there is a *discontinuity* of sample sizes when the regulatory conditions of FDA are used (i.e., with the regulatory constant of $k = 0.893$ or $\sigma_0 = 0.25$). In other words, at CV = 30%, larger sample size is required by applying ABE than ABEL. In fact, using ABEL within a range above CV = 30% requires a smaller sample than applying ABE within a range below CV = 30% (Figure 7.1).

By requiring smaller samples above than below CV = 30%, the FDA regulatory condition could tempt some sponsors to prefer higher variations.

Other disadvantageous consequences of the proposed FDA regulation include the higher consumer risk for non–highly variable drugs. Further features of sample sizes arising from the EMA and FDA requirements can be noted. The difference between the two sample sizes decreases as the variation gets higher, toward 40%–50%. The reason is that the GMR constraint influences the outcome of the BE decision sooner with the FDA than with the EMA condition.

At still higher variations, at CV > 50%, the required sample size increases more rapidly with rising CV with the EMA than with the FDA requirements. The reason is the cap on the use of ABEL at CV = 50%, which EMA imposes. FDA does not apply a similar cap. Consequently, with the EMA requirements, unscaled ABE is used again at CVs exceeding 50%, thereby leading to stricter study requirements.

Designing Bioequivalence Studies for Highly Variable Drugs

Sample sizes for designing BE studies that involve non–highly variable drugs are typically estimated by assuming a within-subject (or a residual) variation and using a sample size table such as that of Hauschke et al. The sample size is usually selected at a 5% deviation between the means, that is, at a true GMR = 1.05.

Larger absolute differences between the two logarithmic means can be noted in the various BE studies when the within-subject variation is higher. Therefore, it is recommended that a 10% deviation between the means, that is, a true GMR = 1.10, be considered when the sample size tables in Appendix 7A are used.

With the approach of FDA, the minimum number of subjects with the three-period partially replicating design is 28 and 40 subjects for 80% and 90% power, respectively. With a four-period replicated design, these numbers are 21 and 30 for 80% and 90% power, respectively. The suggestion of the inclusion of at least 24 subjects may be considered as an absolute minimum.

With the procedure of EMA, the minimum number of subjects with the three-period partially replicating design is 37 and 51 subjects for 80% and 90% power, respectively. With a four-period replicated design, these numbers are 27 and 36 for 80% and 90% power, respectively.

The estimated sample sizes depend on the within-subject variation of the T product. If it is lower than that of the R formulation, then fewer volunteers are needed to achieve the stated power. Conversely, if the variability of the T formulation is higher than that of the R product, then more subjects are needed than shown in the tables in Appendix 7A. In practice, however, the samples are too small to make judgments with adequate power about the relative variances of the two products, and consequently, the assumption of identical variabilities is generally reasonable.

In view of the consequences of the mixed approach, it could be judicious to consider larger numbers of subjects at variations fairly close to 30%.

Both EMA and FDA developed the approaches for highly variable drugs in order to reduce the regulatory burden, that is, to lower the required number of subjects in BE studies. The sample size tables in Appendix 7A demonstrate that both authorities achieve this goal.

In conclusion, when the two drug products have truly the same kinetic metrics (GMR = 1.00) and without the additional regulatory conditions, the required sample size is independent of the within-subject variation of the R formulation. In other words, the producer risk is independent of the variation.

Each of the additional regulatory conditions and requirements yields complications in the relationship between sample size and variation when the true GMR deviates from 1.00. Use of the mixed strategy lowers the sample size near CV = 30%. A constraint on the point estimate of GMR increases the required sample size at higher variations. A cap on applying ABEL at a variation of 50% raises the sample size.

The sample sizes are lower with the requirements suggested by FDA than by those of EMA. However, as illustrated in Figure 7.1, the regulatory constant of FDA leads to a discontinuity of the required sample size and to statistically inconsistent behavior around the switching variation of 30%.

APPENDIX 7A: SAS GLM PROCEDURE

Overview

PROC GLM is for complete data (subjects completed all treatment periods), whereas PROC MIXED is for incomplete (some data are missing), irrespective of the number of treatments. GLM uses an exact solution, whereas MIXED is iterative; when complete data are available, both can be used but GLM is faster.

The GLM procedure uses the method of least squares to fit general linear models. Among the statistical methods available in PROC GLM are regression, ANOVA, analysis of covariance, multivariate analysis of variance (MANOVA), and partial correlation. PROC GLM analyzes data within the framework of general linear models. PROC GLM handles models relating one or several continuous dependent variables to one or several independent variables. The independent variables may be either *classification* variables, which divide the observations into discrete groups, or *continuous* variables. Thus, the GLM procedure can be used for many different analyses, including

- Simple regression
- Multiple regression
- ANOVA, especially for unbalanced data
- Analysis of covariance
- Response-surface models
- Weighted regression
- Polynomial regression
- Partial correlation
- MANOVA
- Repeated-measures ANOVA

As described previously, PROC GLM can be used for many different analyses and has many special features not available in other SAS procedures. The following procedures perform some of the same analyses as PROC GLM:

ANOVA
Performs ANOVA for balanced designs. The ANOVA procedure is generally more efficient than PROC GLM for these designs.

MIXED
Fits mixed linear models by incorporating covariance structures in the model fitting process. Its RANDOM and REPEATED statements are similar to those in PROC GLM but offer different functionalities.

NESTED
Performs ANOVA and estimates variance components for nested random models. The NESTED procedure is generally more efficient than PROC GLM for these models.

NPAR1WAY
Performs nonparametric one-way analysis of rank scores. This can also be done using the RANK procedure and PROC GLM.

REG
Performs simple linear regression. The REG procedure allows several MODEL statements and gives additional regression diagnostics, especially for detection of collinearity. PROC REG also creates plots of model summary statistics and regression diagnostics.

RSREG
Performs quadratic response-surface regression and canonical and ridge analysis. The RSREG procedure is generally recommended for data from a response-surface experiment.

TTEST
Compares the means of two groups of observations. Also tests for equality of variances for the two groups are available. The TTEST procedure is usually more efficient than PROC GLM for this type of data.

VARCOMP
Estimates variance components for a general linear model.

The following statements are available in PROC GLM:

```
PROCGLM <options>;
   CLASS variables;
   MODEL dependents = independents </options>;
   ABSORB variables;
   BY variables;
   FREQ variable;
   ID variables;
   WEIGHT variable;
   CONTRAST 'label' effect values <… effect values> </options>;
   ESTIMATE 'label' effect values <… effect values> </options>;
   LSMEANS effects </options>;
   MANOVA <test-options> </detail-options>;
   MEANS effects </options>;
   OUTPUT <OUT = SAS-data-set>
   keyword = names <… keyword = names> </option>;
   RANDOM effects </options>;
   REPEATED factor-specification </options>;
   TEST < H = effects > E = effect </options>;
```

Although there are numerous statements and options available in PROC GLM, many applications use only a few of them. To use PROC GLM, the PROC GLM and MODEL statements are required. You can specify only one MODEL statement (in contrast to the REG procedure, e.g., which allows several MODEL statements in the same PROC REG run). If your model contains classification effects, the classification variables must be listed in a CLASS statement, and the CLASS statement must appear before the MODEL statement. In addition, if you use a CONTRAST statement in combination with a MANOVA, RANDOM, REPEATED, or TEST statement, the CONTRAST statement must be entered first in order for the contrast to be included in the MANOVA, RANDOM, REPEATED, or TEST analysis (Table 7.A.1).

PROC GLM FOR UNBALANCED ANOVA

ANOVA typically refers to partitioning the variation in a variable's values into variation between and within several groups or classes of observations. The GLM procedure can perform simple or complicated ANOVA for balanced or unbalanced data.

TABLE 7.A.1
Statements in the GLM Procedure

Statement	Description
ABSORB	Absorbs classification effects in a model
BY	Specifies variables to define subgroups for the analysis
CLASS	Declares classification variables
CONTRAST	Constructs and tests linear functions of the parameters
ESTIMATE	Estimates linear functions of the parameters
FREQ	Specifies a frequency variable
ID	Identifies observations on output
LSMEANS	Computes least-squares (marginal) means
MANOVA	Performs a MANOVA
MEANS	Computes and optionally compares arithmetic means
MODEL	Defines the model to be fit
OUTPUT	Requests an output data set containing diagnostics for each observation
RANDOM	Declares certain effects to be random and computes expected mean squares
REPEATED	Performs multivariate and univariate repeated-measures ANOVA
TEST	Constructs tests using the sums of squares for effects and the error term you specify
WEIGHT	Specifies a variable for weighting observations

TABLE 7.A.2
A 2 × 2 ANOVA Model

		A	
		1	2
B	1	12	20
		14	18
	2	11	17
		9	

This example discusses a 2 × 2 ANOVA model. The experimental design is a full factorial, in which each level of one treatment factor occurs at each level of the other treatment factor. The data are shown in a table and then read into a SAS data set (Table 7.A.2).

```
title 'Analysis of Unbalanced 2-by-2 Factorial';
data exp;
input A $ B $ Y @@;
datalines;
A1 B1 12 A1 B1 14 A1 B2 11 A1 B2 9
A2 B1 20 A2 B1 18 A2 B2 17
;
```

Note that there is only one value for the cell with A = 'A2' and B = 'B2'. Since one cell contains a different number of values from the other cells in the table, this is an unbalanced design.

The following PROC GLM invocation produces the analysis:

```
proc glm;
class A B;
model Y = A B A*B;
run;
```

Both treatments are listed in the CLASS statement because they are classification variables. A*B denotes the interaction of the A effect and the B effect:

Table 7.A.3 displays information about the classes as well as the number of observations in the data set. Table 7.A.4 shows the ANOVA table, simple statistics, and tests of effects.

The degrees of freedom may be used to check your data. The model degrees of freedom for a 2×2 factorial design with interaction are $(ab - 1)$, where a is the number of levels of A and b is the

TABLE 7.A.3
Class Level Information

Analysis of Unbalanced 2-by-2 Factorial
The GLM Procedure

Class Level Information

Class	Levels	Values
A	2	A1 A2
B	2	B1 B2
Number of observations		7

TABLE 7.A.4
ANOVA Table and Tests of Effects

Analysis of Unbalanced 2-by-2 Factorial
The GLM Procedure
Dependent Variable: Y

Source	DF	Sum of Squares	Mean Square	F Value	Pr > F
Model	3	91.71428571	30.57142857	15.29	0.0253
Error	3	6.00000000	2.00000000		
Corrected Total	6	97.71428571			

R-Square	Coeff. Var.	Root MSE	Y Mean
0.938596	9.801480	1.414214	14.42857

Source	DF	Type I SS	Mean Square	F Value	Pr > F
A	1	80.04761905	80.04761905	40.02	0.0080
B	1	11.26666667	11.26666667	5.63	0.0982
A*B	1	0.40000000	0.40000000	0.20	0.6850

Source	DF	Type III SS	Mean Square	F Value	Pr > F
A	1	67.60000000	67.60000000	33.80	0.0101
B	1	10.00000000	10.00000000	5.00	0.1114
A*B	1	0.40000000	0.40000000	0.20	0.6850

number of levels of B; in this case, (2 × 2 − 1) = 3. The corrected total degrees of freedom are always one less than the number of observations used in the analysis; in this case, 7 − 1 = 6.

The overall F test is significant ($F = 15.29$, $p = 0.0253$), indicating strong evidence that the means for the four different A × B cells are different. You can further analyze this difference by examining the individual tests for each effect.

Four types of estimable functions of parameters are available for testing hypotheses in PROC GLM. For data with no missing cells, type III and type IV estimable functions are the same and test the same hypotheses that would be tested if the data were balanced. Type I and type III sums of squares are typically not equal when the data are unbalanced; type III sums of squares are preferred in testing effects in unbalanced cases because they test a function of the underlying parameters that is independent of the number of observations per treatment combination.

According to a significance level of 5% ($\alpha = 0.05$), the A*B interaction is not significant ($F = 0.20$, $p = 0.6850$). This indicates that the effect of A does not depend on the level of B and vice versa. Therefore, the tests for the individual effects are valid, showing a significant A effect ($F = 33.80$, $p = 0.0101$) but no significant B effect ($F = 5.00$, $p = 0.1114$).

APPENDIX 7B: BIOEQUIVALENCE TESTING SOFTWARE

ABSPLOTS	Lotus 123 spreadsheet for Wagner–Nelson calculations.
acslXtreme	Physiologically based pharmacokinetic (PBPK) and pharmacodynamic (PD) simulation software.
acslXtreme Pharmacokinetic Toolkit	PBPK and PD effects with the acslXtreme Pharmacokinetic Toolkit.
ADAPT II	Supplied as FORTRAN code for VAX VMS, MS DOS, and SUN UNIX system. This program performs simulations, nonlinear regression, and optimal sampling and includes extended least squares and Bayesian optimization. Models can be expressed as integrated or differential equations using FORTRAN statements.
ATIS	Nonlinear least squares.
AUC-RPP	Noncompartmental evaluation of PK parameters.
BIOEQV52, BIOPAR40, and BIOEQNEW	BE calculations including statistical power. (Reference: Wijnand H.P. 1994 Updates of bioequivalence programs [including statistical power approximated by Student's t], Computer Methods and Programs in Biomedicine 42, 275–281)
Biokmod	A Mathematica toolbox for solving systems of differential equations, fitting coefficients, convolution, and more, with applications for modeling linear and nonlinear systems.
BIOPAK	Statistical analysis package for BA/BE studies.
BOOMER/MULTI-FORTE	Supplied as compiled programs for Macintosh (including PowerMac), MS DOS, and VAX VMS systems. This program performs simulations and nonlinear regression and includes Bayesian optimization. Models, integrated or differential equations, can be expressed as a sequence of parameters (BOOMER) or using FORTRAN statements (MULTI-FORTE).
CSTRIP	Polyexponential stripping.
CompleX Tools (CXT)	CXT for linear dynamic system analysis from BIO-LAB Bratislava uses the frequency response method to model PK and/or PD data.
Cyber Patient	Multimedia PK simulation program that can be used for development and presentation of problem-solving case studies.
EASYFIT	Analysis of compartmental models.
EDFAST	Fitting and simulating linear PK models.
EquivEasy	Modeling and BE testing program—an interface for SAS PROC GLM and other modules and an excellent choice for a CFR compliant software.
GastroPlus™	Simulates absorption and PBPK for orally dosed drugs.
INTELLIPHARM PK	Simulates drug dissolution, absorption, and pharmacokinetics.

(Continued)

JavaPK for Desktop	Bayesian individualized PK/PD parameters estimation (UDBM) for analyzing batch input data. Users can define their own model with population PK/PD parameters using a single-dose, integral equation (for the multiple dosed) or a steady-state integral equation.
JGuiB	Includes three most commonly used functions of Boomer in PK/PD modeling: normal fitting, simulation and Bayesian estimation.
KINBES	BA and rate of drug absorption by various methods such as numerical deconvolution and WLS reconvolution; also featuring a number of statistical tests on BE (e.g., ANOVA, FDA 75/75 rule).
Kinetica	Thermo Electron's PK analysis tool offers fast high-throughput data analysis for clinical, preclinical, discovery, drug metabolism, and drug delivery settings. This tool standardizes analyses across the organization and minimizes variability between PK analysts and analyses. Together with EP, it becomes a fully FDA CFR 21, part 11 compliant PK/PD DB, enabling full audit trail from protocol inception to final report.
MKMODEL	The program, for MS DOS systems, performs nonlinear least-squares regression with extended least squares. Models can be represented by integrated or differential equations.
ModKine	Modeling program with custom features for pharmacokinetics and pharmacodynamics (Windows) from Biosoft.
NCOMP	Noncompartmental analysis of PK data.
NLMEM SAS/IML macro	The macro is designed for hierarchical nonlinear mixed-effects models. The program invokes part of the code contained in the SAS/NLINMIX macro developed by SAS technical support and can be considered as an interface for the NLINMIX macro to SAS/IML. The macro runs under SAS system and is an attractive alternative to NONMEM software.
NONMEM	The program is provided as FORTRAN source code for UNIX, IBM, and other computers. The program performs nonlinear regression analysis of individual or population data.
Nonparametric expectation maximization (NPEM)	This is part of the USC PACK collection.
Nonparametric maximum likelihood (NPML)	NPML estimation procedure by A. Mallet. (Reference: Mallet, A. 1986 A maximum likelihood estimation method for random coefficient regression models. *Biometrika*, 73: 654–656)
PCDCON	By W.R. Gillespie. (gillespiew@donald.cder.fda.gov). It performs deconvolution analysis. This program is available as a compiled program for the IBM PC. (Reference: Karol, M., Gillespie, W.R., and Veng-Pederson, P. 1991 AAPS Short Course: Convolution, Deconvolution and Linear Systems, AAPS, Washington, DC, Nov. 17)
PDx-IVIVC	Comprehensive toolset for in vitro–in vivo correlation.
PDx-Pop	Integrates with NONMEM and other existing software to expedite population modeling and analysis.
PH\EDSIM	A universal PK–PD modeling tool that enables the user to create custom-made PK, PD, or PK–PD models in a graphical way without the need for programming. Created models may be used for simulation and fitting purposes.
Physiological parameters for PBPK modeling version 1.0 (P3M)	P3M provides a convenient tool for parameterization of PBPK models of interindividual variation.
PK functions for Microsoft Excel	By Joel Usansky, Atul Desai, and Diane Tang-Liu. Download the word document first for a description and installation instructions.
PK simulations	By Guenther Hochhaus.
PK Solutions	It is an Excel-based noncompartmental pharmacokinetics software program.
PKAnalyst for Windows	It provides the capability of simulation and parameter estimation for PK models.
PKBugs	It is an efficient and user-friendly interface for specifying complex population PK/PD models within the widely used WinBUGS software.
POP3CM	A Free Visual Compartmental Population Analysis Program. The program POP3CM provides a graphical user interface for the analysis of a three-compartment model.

(Continued)

PopKinetics	PopKinetics is a population analysis program. It is a companion application to SAAM II that uses parametric algorithms, standard two-stage and iterated two-stage, to compute population parameters and their CIs. PopKinetics operates directly on SAAM II study files.
SAAM II	It is a compartmental (differential equations) and numerical (algebraic equations) modeling program that can be used in the analysis of PK, PD, and enzyme kinetic studies. It is designed to help researchers easily create models, perform simulations, and fit experimental kinetic data resulting in parameter estimates and their associated errors. SAAM II has a user-friendly graphical user interface that is fully menu driven. Development of SAAM II, at the University of Washington, Seattle, was supported by a research resource grant from the National Institutes of Health (NIH). SAAM II is available for PC Windows (Win95/98, NT). The Macintosh version (68030 or higher [with FPU], PowerMac) is still available but no longer supported. Available from the SAAM Institute, info@saam.com, phone (206)729-1315, fax (206)729-7854. A demo version is available on the website.
SAAM/CONSAM	It is available from L.A. Zech and P.C. Greif, Laboratory of Mathematical Biology, Building 10, Room 4B-56, NIH/NCI, Bethesda, MD, 20892, Internet: zech@ncifcrf.gov. The program is provided as compiled programs for VAX VMS and MS DOS computers. The program performs nonlinear regression in batch (SAAM) or conversational mode (CONSAM). The SAAM/CONSAM programs are kindly provided by the USPHS/NIH/DRR-NHLBI-NCI joint development project.
SAS	Most important yet difficult to use software; PROC GLM is the preferred tool by the FDA.
Simcyp	It includes a fully automated whole-body PBPK model that incorporates enzyme kinetic data from routine in vitro studies.
TopFit	This MS DOS program performs noncompartmental and model-based analyses.
WinNonlin	It provides an easy-to-use Windows application for PK, PK/PD, and noncompartmental analysis. WinNonlin includes extensive libraries of PK and PK/PD models and provides tools for table generation, scripting, and data management.
WinNonMix	It is a program for nonlinear mixed-effects modeling provided in an interactive and easy-to-use Windows application.
WinSAAM	It is a Windows version of the original interactive biological modeling program, CONSAAM, developed in 1980 at NIH.
Xpose	It is an *R*-based model building aid for population analysis using NONMEM. It facilitates data set checkout, exploration and visualization, model diagnostics, candidate covariate identification, and model comparison.

8 Regulatory Inspection Process

BACKGROUND

The bioequivalence (BE) review process establishes that the proposed generic drug is bioequivalent to the reference listed drug, based upon a demonstration that both the rate and extent of absorption of the active ingredient of the generic drug fall within established parameters when compared to that of the reference listed drug.

The Food and Drug Administration (FDA) requires an applicant to provide detailed information to establish bioequivalency. Applicants may request a waiver from performing in vivo (testing done in humans) BE studies for certain drug products where bioavailability (the rate and extent to which the active ingredient or active moiety is absorbed from the drug product and becomes available at the site of action) may be demonstrated by submitting data such as (1) a formulation comparison for products whose bioavailability is self-evident, for example, oral solutions, injectables, or ophthalmic solutions where the formulations are identical or (2) comparative dissolution.

Alternatively, in vivo BE testing comparing the rate and extent of absorption of the generic versus the reference product is required for most tablet and capsule dosage forms. For certain products, a head-to-head evaluation of comparative efficacy based upon clinical endpoints may be required.

The Manual of Policies and Procedures of the Center for Drug Evaluation and Research (CDER) (Generic Drugs) (MAPP 5210.7) describes the following procedures for review of BE study protocols.

PROTOCOLS

Background

BE studies are frequently needed to support the filing and approval of abbreviated new drug applications (ANDAs). To conduct an adequate study and avoid unnecessary human research, any sponsor planning to conduct a bioavailability or BE study should submit the proposed study protocol to the Office of Generic Drugs (OGD) for review prior to the initiation of the study. OGD reviews the protocol and provides advice on appropriate study design, reference material, and the proposed analytical and statistical methods to be used. Sponsors or contract research organizations (CROs) can submit protocols.

References

21 CFR 320.30, Inquiries regarding bioavailability and BE requirements and review of protocols by the Food and Drug Administration; 21 CFR 10.90, Food and Drug Administration regulations, recommendations, and agreements.

Policy

The division of bioequivalence (DBE) will review submitted BE protocols. The protocols will be randomly assigned to BE reviewers, unless a protocol requires the expertise of a particular reviewer.

The reviewers will perform a search of the literature and the agency's databases and prepare the review. After the protocol review, comments will be provided in a letter to the generic firm.

Procedures

When a protocol is received in the DBE, the project manager (PM) assigns it randomly to the next available reviewer. All protocols received are entered in the protocol tracking system and assigned a control number. The protocol receipt date, firm name, drug name, reviewer assigned, and date of assignment are recorded. The reviewer searches the literature and the agency's databases (e.g., Excalibur, WinBio, drug files [hard copy and electronic]). If a protocol has been previously submitted and found acceptable by the division, this should be used as a model in the preparation of responses to subsequent protocols for the same drug. The reviewer should state in the review whether other protocols for the same drug have been previously reviewed. If no other protocols have been reviewed for the product, a statement to that effect should be included in the review. The reviewer prepares a review with recommendations to the requestor. The review must have the concurrence of the team leader and division director. If the reviewer discovers discrepancies in BE criteria or appropriate study design in recommendations provided to industry in previous protocols or correspondence for the same drug product, the reviewer prepares a memorandum to the team leaders and division director. The memo should specify the name of the sponsor or CRO that received conflicting information/guidance in protocol responses. ANDAs affected by this information should also be noted. Once the review is finalized and has the concurrence of the division director, it is forwarded to the PM. The PM or TIA drafts a letter based on the reviewer's recommendation. The PM ensures that all recommendations are provided to the firm. The letter will be routed through the team leader for corrections and endorsement and to the division director for signature. Once the letter is signed by the division director, the PM or TIA enters into the protocol tracking system the date the review was finalized and the date the letter was issued. The protocol is then forwarded to the document room. Document room personnel mail the letter and store the protocol in the designated area. The PM drafts letters to sponsors or CROs that have received outdated information to ensure that consistent information is provided to industry.

PRODUCTIVITY DOCUMENTATION

Background

The COMIS database was created, in part, to keep track of the workload of all divisions. Information on all submissions received in OGD on ANDAs is entered into this system, including the applicant's name, ANDA number, drug name, dosage form, strengths, letter date, and receipt date. The BE section of an ANDA contains data on the demonstration of BE, such as BE studies, studies with clinical endpoints, dissolution data, and waiver requests. The BE data entry screen in COMIS keeps a record of (1) the reviewer assigned to the submission, (2) the type of studies submitted in the BE section, and (3) the dates when the review was initiated and satisfactorily completed by the reviewer. Other work, such as controlled correspondence and protocols, is tracked in separate databases. The overall productivity of the division and the reviewers is monitored using the information in COMIS and the other databases.

Policy

Information entered into the COMIS database on the study types in the BE section of an ANDA documents the overall productivity of the reviewers and the division. Consistent and fair classification of these study types ensures objective evaluation of reviewers. Non-ANDA-related work is tracked in separate databases. That information includes a control number, name of sponsor, drug name, name of assigned reviewer, date of assignment, date of completion of the review, and dates when letters are issued.

Procedures

When the document room assigns an ANDA to the DBE, a description of the BE section is entered into the BE data entry screen in COMIS, using the following study types:

BE Studies

1. *Fasting study* (STF). This includes replicate study designs and combined studies (e.g., combined fasting and multiple-dose studies where the same subjects are used).
2. *Food study* (STP).
3. *Multiple-dose study* (STM).
4. *Study* (STU). This category is generally used for a BE study with clinical endpoints, in vitro studies for metered-dose inhalers (MDIs) and nasal sprays, pilot and pivotal studies for vasoconstrictors, or any pharmacokinetic/pharmacodynamic study other than a standard BE study (such as a–c previously).

Dissolution Data (DIS)

This code is usually used when dissolution data are the only basis for approval. Examples are AA drugs and supplements for which changes in formulation or manufacturing require dissolution data only. In vitro release data for topical products may also be coded under DIS. *Note*: Dissolution data submitted for the same strength of drug that was the subject of a BE study are not separately coded. The dissolution information is considered part of the study.

Other (OTH)

Study amendment (STA). This category is for responses to deficiency comments. Whether the amendment contains dissolution data or addresses a deficiency such as incomplete information on analytical methods or a study, the submission should be coded as STA unless a new study is submitted for review. In that case, the appropriate code under BE studies should be selected. If an amendment to a previously submitted BE study is included with a new, not previously submitted, BE study required to establish BE, then STA should be coded for the amendment, and the new study should be coded separately. Retesting of subjects classified as outliers in the original submission should not be classified as a separate study, but as part of the original study. Frequently, the division telephones sponsors to request information needed to finalize the review. These requests should be made for information the sponsor can respond to within 10 working days and should be coded as STA. If the sponsor submits incorrect information or partial data, the submission should be coded as new correspondence (NC). Once the correct information is received, the submission should be coded as STA.

Waiver (WAI). This category is used for injectable, ophthalmic, otic, oral, and topical solutions. A formulation in the same concentration packaged in different sizes is not coded separately, but different concentrations of the same product are coded separately.

Dissolution waiver (DIW). This code is used for lower strengths that can be approved based on proportionality of the formulation and an acceptable study on the highest strength or the strength of the reference listed drug. A dissolution waiver should be coded for each strength for which dissolution data are submitted, except the strength for which BE studies have been conducted.

Other (OTH). This category is used for correspondence or addenda revising the original review. The Division of Scientific Investigations (DSI) inspection reports may generate an addendum to the review. If a significant statistical analysis is needed based on the recommendation of the DSI or if the issuance of a Form 483 (Inspectional Observations) indicates serious violations by the laboratory, then the review of the DSI report may be coded as OTHER. If the DSI report is acceptable, the DSI report should be filed in the ANDA, and no addendum to the review is necessary. Addenda to the reviews are entered as U.S. documents (FDA generated), because these reviews are not prompted by industry submissions, but are due to internal policy changes or inspection reports. Diskettes containing the data already coded in a previous submission will not be coded separately.

Protocols

1. Protocol (PRO). This is used for protocols submitted as part of an investigational new drug (IND) application or an ANDA. An example of a protocol submitted as part of an ANDA would be a skin irritation study protocol.
2. Protocol amendment (PRA). Amendment to a protocol.
3. Other protocols. There are also protocols sent to the DBE for review to obtain comments on the proposed study design prior to the submission of ANDAs. Pilot studies submitted with a protocol to justify a particular study design are not coded separately. A review is generated and comments are provided to the firm by letter. This is not recorded in COMIS. It is tracked in a separate database and is counted as part of the overall productivity of individual reviewers. Occasionally, sponsors submit protocols for studies that are not necessary (i.e., a waiver request for in vivo testing). In this case, the additional protocol does not have to be reviewed, and credit will not be given.

Controlled Correspondence

BE information requests sent as correspondence are also randomly assigned to DBE reviewers for evaluation and generation of a review. These reviews are not recorded in COMIS, but are tracked in a separate database and counted as part of the overall productivity of individual reviewers. A citizen petition is counted as controlled correspondence. If additional information is submitted for pending correspondence and/or citizen petitions prior to the completion of the response to the original piece, the issues raised by the additional supplement to the submission should be addressed in the review underway. If a review has been finalized and an additional supplement is submitted raising new issues, another review can be generated.

PROCESSING OF WORK

The reviewers sign their names in the assignment logbook. When an assignment is available, the BE PM assigns it to the next reviewer and enters the reviewer's name and date of assignment in the appropriate database (COMIS, protocols, controlled correspondence) and the assignment logbook. The PM also verifies study codes at this time. The reviewer obtains the submission from the document room. When the review is completed, the reviewer states on the last page of the review the study types reviewed in the submission and comments on the acceptability of the data provided by the firm. The following decision codes should be used when determining the acceptability of each study type.

AC—Acceptable. The submission was complete and all data were found acceptable.
UN—Unacceptable. A study failed to meet standard criteria for BE (e.g., 90% clinical investigator for fasting study, incorrect dissolution methods).
IC—Incomplete. Information was missing from the submission.
NC—No Action. No action or review was necessary.

The team leaders verify that study codes and decision codes are accurate. Once the review is finalized and has the division director's concurrence, it is forwarded to the BE PM, who forwards acceptable comments to the chemistry PM or prepares fax cover sheets for deficiencies to be transmitted to the firm. The BE PMs then deliver acceptable completed reviews to the document room. Reviews containing deficiencies to be transmitted to the firm are delivered to the review support branch chief, who gathers any comments from other disciplines (chemistry, labeling, microbiology) and faxes all deficiencies and comments together. The document room staff enters data into the BE data entry screen in COMIS, including the completion date (the date when the director of BE signed the review). The document room staff also verifies study codes and enters decision codes. This closes the submission, indicating that the review has been completed. Once the submission is closed, reviewers are credited for their work.

INSPECTIONS

Background

This MAPP outlines policies and procedures to use in (1) identifying when to request inspections of clinical facilities or analytical laboratories associated with BE studies and (2) applying inspection information to the review of ANDAs.

In vivo BE studies are used to support the approval of many ANDAs. To help ensure that these studies are reliable, the OGD needs information on the inspection status of clinical facilities and analytical laboratories where the studies are conducted.

OGD requests information on the compliance status of relevant clinical facilities and laboratories from the Good Laboratory Practice (GLP)/Bioequivalence Investigations Branch (GBIB), DSI, Office of Medical Policy.

OGD requests that GBIB initiate a *routine inspection* of clinical facilities or analytical laboratories conducting BE studies included in an unapproved ANDA if

A clinical facility or analytical testing site is identified in the ANDA that has no inspection history, was classified OAI on its last inspection, or has not been inspected within the past 3 years

A clinical facility and/or analytical laboratory is performing a nonconventional BE study for which it has never been inspected by DSI (e.g., a study using pharmacodynamic endpoints to assess BE)

OGD requests a *directed inspection* of a facility if there is a question about the quality or integrity of the data submitted in an ANDA. Instances of suspect data may include missing data points, errors in calculation, or inadequate documentation.

METHODS VALIDATION FOR ABBREVIATED NEW DRUG APPLICATIONS

Background

Since 1981, methods validation has not been an approval criterion for new drug applications (NDAs). Until 1997, however, OGD's policy was to require satisfactory methods validation before approval of ANDAs for noncompendial drug products. In some cases, ANDA approvals were delayed pending completion of methods validation. Validation of the analytical methods and testing procedures was considered an important component when ensuring application approvability. However, there were circumstances when a delay in completion of the method validation process was beyond the control of the applicant. In those instances, OGD wanted to ensure that an application that was otherwise eligible for approval was approved without undue delay. Therefore, in November 1998, OGD revised its policy regarding method validation for applications that have been recommended for approval to allow approval of an ANDA if (1) there was no undue delay in sample submission by the applicant, (2) there is no apparent problem with the validation in progress or the validation has not been initiated by the servicing laboratory, and (3) there is a commitment from the applicant to resolve any problems with method validation. Now, to better use the limited resources of the program to ensure adequacy of critical and/or complex methods, OGD has determined that there are other situations in which method validation is not needed to support approval of ANDAs. Consequently, OGD is revising its policy regarding method validation consistent with this determination.

References

21 CFR 314.50(e), Samples and labeling; 21 CFR 314.70, Supplements and other changes to an approved application; Compliance Program on Preapproval Inspections CP7346.832.

Policy

Method validation requests will be limited to noncompendial drug products and, with team leader and division director (or deputy) concurrence, will be further subject to reviewer discretion because of specific concerns (i.e., for cause) relating to a drug product or an analytical method. Representative for cause examples include (but are not limited to)

- New emerging analytical technologies
- Analytical methods for novel/complex drug delivery systems (e.g., transdermal delivery system [TDS], MDI, nasal spray)
- Chromatographic methods for quantitation of low-dose drugs
- Chromatographic methods for resolving multiple drug components with concomitant impurities/degradants

OGD does not require or request method verification by an FDA laboratory of a product for which a USP monograph exists. However, FDA laboratories may conduct method verification analyses of compendial products at their option. Application approval is not dependent on receipt of these test results. Proposals for alternative analytical methods for products that are the subject of a USP monograph will be evaluated during the review process. There is no need for FDA laboratories to validate the alternative methods since the official methods for regulatory purposes are those of the USP and, therefore, OGD does not request method validation for alternative methods for compendial products.

If there is no USP monograph for a drug substance or drug product, the applicant's proposed regulatory analytical methods may be validated by an FDA laboratory.

Under certain other circumstances, method validation for an ANDA for a noncompendial drug product may clearly be waived. The final decision should be documented in the application. Circumstances that support a waiver include, but are not limited to, the following:

- The proposed analytical methods have been validated previously in an FDA laboratory under another of the same applicant's ANDAs for a similar drug product (e.g., different strength, different packaging configuration).
- There exists in the compendium a monograph for a similar dosage form (e.g., for injection versus injection) containing the applicant's proposed regulatory methods, and the reviewer has verified that the change in dosage form will cause no analytical interferences in the compendial procedures. That is, the reviewer has verified the suitability of the compendial methods under actual use conditions.
- The division director will sign off on an approval package if all aspects of the ANDA are complete and satisfactory, excluding method validation and establishment evaluation report (EER) results.
- OGD will not wait for completion of method validation to begin the administrative review process.

Upon completion of the administrative review process, the application will be approved if all other aspects of the ANDA, including the EER and office-level BE review, are satisfactory and the following criteria are met:

- There is no undue delay in sample submission by the applicant.
- There is no apparent problem encountered with the validation in progress, or the validation has not been initiated by the servicing laboratory.
- There is a commitment from the applicant in the ANDA to resolve any problems with method validation.

OGD expects the applicant to provide samples to the servicing laboratory within 10 working days of the request and will consider longer time frames to be *undue delay*. If it is determined that there were delays in the provision of samples to the laboratory or if significant problems are identified in the course of method validation, OGD will not approve the application before the *completion* of the method validation and the resolution of the deficiencies. Whether pre- or postapproval, the chemistry review branch will evaluate negative laboratory findings and determine their impact on the applicable submission.

Procedures

A request for validation of the applicant's proposed regulatory analytical methods is sent by the review chemist to the Office of Regulatory Affairs (ORA) coordinator in the Division of Field Science (DFS) using form FDA 2871a. This action should be taken as soon as the need is identified and the test methods are determined to be adequate by the review chemist.

- A copy of the methods, testing specifications, and composition statement is to be included with the request. The package is sent to DFS by current procedures.
- Requests are processed and carried out as detailed in the Supplement to the Compliance Program on Preapproval Inspections CP7346.832.

The chemistry/microbiology review is included in the approval package, along with the BE and labeling reviews. Upon concurrence by the chemistry team leader, the package proceeds through the final administrative review channels. If, after administrative review, the application remains approvable (including an acceptable EER and office-level BE endorsement), the project manager determines the status of the method validation process. The application can be approved with or without results of the method validation, except under the following circumstances:

- There was an undue delay in sample submission by the applicant.
- There are problems identified in the course of methods validation by the servicing laboratory.
- There is no commitment from the applicant to resolve any problems subsequently found by the FDA laboratory.

Any problem identified with the method or the product is evaluated by the review chemist for its significance. Any problem that potentially affects the quality of the drug product must be resolved before application approval. When approval is granted in the absence of a completed method validation, the approval letter is revised to include the following statement as the last paragraph. *Validation of the regulatory methods has not been completed. It is the general policy of the OGD not to withhold approval until the validation is complete.*

The approval letter is endorsed by the chemistry reviewer and team leader as well as the division director. If the laboratory results are received during the administrative review process for approval and they reveal problems with the methods or the product, the approval of the application is delayed and the results transmitted to the applicant. The applicant is asked to address these issues as soon as possible in an amendment to the application. This amendment is given priority review in consultation, if necessary, with the servicing laboratory. If the amended methods are satisfactory to OGD and they address the concerns of the laboratory, the application can then be approved, provided all other aspects of the application are acceptable. Out-of-specification results on products already expired at the time of testing are evaluated for their significance and relevance. Any product failures must be satisfactorily resolved before application approval. Routine revalidation can be done after approval of the application. The review chemist can request testing at a second FDA laboratory to resolve conflicting results obtained by an applicant and by the FDA servicing laboratory. The team leader and the division director must concur with the request. For method validation completed after an application is approved, any deficiencies identified are communicated promptly to the applicant. Generally, the response addressing the deficiencies can be submitted as a changes-being-effected supplement. If the method validation is waived, this fact must be documented and filed in the ANDA.

Regulatory Audit of Bioequivalence Studies Submitted

When BE studies are submitted as part of an aNDA, the U.S. FDA inspections include an audit of the studies submitted under the Compliance Program 7348.001. It is important to review these directives since it allows firms to prepare studies and have them ready for presentation in a format that is readily accessible and comprehensible. This applies to both domestic and international inspections. When the clinical and analytical portions of a study have been performed at separate locations, separate reports should be prepared and submitted for each site.

PART I: BACKGROUND

The BE regulations (21 CFR 320) of January 7, 1977 and its amendments stated the requirements for submission of in vivo bioavailability and BE data as a condition of marketing a new (i.e., new chemical compound, new formulation, new dosage form, or new route of administration of a marketed drug) or generic drug. 21 CFR 320 also provided general guidance concerning the design and conduct of bioavailability/BE studies. However, it should be noted that BE studies conducted to support ANDAs involve testing of already approved drug entities and, therefore, generally do not require an IND application. However, sponsors of generic drugs need to file INDs when studies involve a route of administration or dosage level or use in a patient population or other factor that significantly increases the risks (or decreases the acceptability of the risks) associated with the use of the drug product (21 CFR 312.2(b)(iii)).

The FDA does not require BE studies on pre-1938 drug products. It is however the responsibility of the firm to assure that the studies are submitted in accordance with the most current guidelines as amended.

BE studies involve both a *clinical component* and an *analytical component*. The objective of a typical BE study is to demonstrate that the test and reference products achieve a similar pharmacokinetic profile in plasma, serum, and/or urine. BE studies usually involve administration of test and reference drug formulations to 18–36 normal healthy subjects, but patients with a target disease may also be used. Formulations to be tested are administered either as a single dose or as multiple doses. Sometimes, formulations can be labeled with a radioactive component to facilitate subsequent analysis. In a BE study, serial samples of biological fluid (plasma, serum, or urine) are collected just before and at various times after dose administration. These samples are later analyzed for drug and/or metabolite concentrations. The study data are used in subsequent pharmacokinetic analyses to establish BE.

In some situations, the clinical and analytical facilities for a study may be part of the same organization and therefore may be covered by one district. In other situations, the two facilities may be located in different districts. For the purpose of this program, the district where the clinical facility is located will be referred to as the clinical component district, and the district where the analytical facility is located will be referred to as the analytical component district.

PART II: IMPLEMENTATION

Objectives

1. To verify the quality and integrity of scientific data from BE studies submitted to the CDER
2. To ensure that the rights and welfare of human subjects participating in drug testing are protected
3. To ensure compliance with the regulations (21 CFR 312, 320, 50, and 56) and promptly follow up on significant problems, such as research misconduct or fraud

Program Management Instructions

Coverage

It is important to draw distinctions between a clinical laboratory, a clinical facility, and an analytical facility. A clinical laboratory generally uses blood and/or urine to conduct medical screening or diagnostic tests such as blood counts (CBC), liver function tests (ALT, AST), or kidney function (BUN, creatinine clearance, etc.) tests Clinical laboratories are usually certified under programs based on the Clinical Laboratories Improvement Act (42 USC 263a) and are not routinely inspected by the FDA. A clinical laboratory may be visited during a BE study audit to confirm that reported screening or diagnostic laboratory work was indeed performed. The clinical facility and the analytical facility as described earlier are the laboratories that will be routinely inspected under this program.

1. *Clinical Facilities*
   Clinical facilities conduct BE studies (including screening, dosing, monitoring of subjects' safety) in order to obtain biological specimens (e.g., plasma, serum, urine) for analysis of drug and/or drug metabolite concentrations. Facilities that conduct BE studies in human research subjects for pharmacodynamic measurements (i.e., clinical or pharmacological effects) are also included.
2. *Analytical Facilities*
   Analytical facilities analyze biological specimens collected in BE studies and other human clinical studies for drug and/or metabolite concentrations to measure the absorption and disposition of the drug.
3. *Clinical and Analytical Investigators*
   The clinical investigator in a BE study is involved in the screening and dosing of human subjects and will ordinarily be a physician. PhD clinical pharmacologists and Pharm. D.'s are acceptable if a physician is available to cover medical emergencies. The clinical investigator may also perform pharmacodynamic measurement(s) and evaluation activities of clinical or pharmacological endpoints. The analytical investigator in a BE study is the scientist in the analytical facility responsible for assay development and validation and analyses of biological specimens, for example, scientific director or laboratory director.

Process

Facilities where BE studies are conducted are to include a review of the clinical and analytical testing procedures plus an audit of source data from one or more specified studies.

Assignments under this program are of two basic categories:

1. *Directed data audit*—Covers studies and/or facilities in which gross problems/inadequacies are suspected (including, but not limited to, research misconduct or fraud). Such assignments require rapid evaluation and resolution.
2. *Routine data audit*—Covers (1) pivotal studies under current review in the Divisions of Pharmaceutical Evaluation I (HFD-860), II (HFD-870), or III (HFD-880) in the Office of Clinical Pharmacology and Biopharmaceutics (HFD-850) and (2) BE studies supporting the approval of a generic product.

Assignments will be issued by the GLP and Bioequivalence Investigations Branch (GBIB, HFD-48) to the field. For each assignment, a scientific reviewer in GBIB with expertise in chemical assays, bioavailability/BE, biopharmaceutics, pharmacokinetics, or pharmacodynamics will (1) assist the field in coordinating and as necessary conducting the inspection, (2) provide technical guidance and on-site support to the field as necessary, and (3) serve as the liaison between the field investigator(s) and the review divisions in CDER.

GBIB will generate assignments under this program based on information provided by the review divisions in CDER. GBIB will send assignment memos to the director of the investigations branch in the appropriate district office(s) (for domestic inspections) or the Division of Emergency and Investigational Operations, International Operations Group (for foreign inspections).

The assignment memo will include the following information:

a. NDA/ANDA number
b. Name of the drug
c. Name of the sponsor
d. Study/protocol number(s)
e. Title of each study identified for inspection
f. Address(es) of the clinical and analytical facilities
g. Instructions on inspectional areas
h. Deadline(s) such as preferred date for completion of inspection, review division action goal date, or the user fee goal date
i. The name of the GBIB contact

After a field investigator has been assigned, background material (including source data from the specific study[ies]) will be forwarded to the field investigator. In the event that a clinical or analytical facility designated for inspection is found to be located elsewhere, the district should contact GBIB immediately in order to redirect the assignment.

For all inspections in which a FDA 483 is issued, a copy of the FDA 483 should be forwarded by facsimile to the GBIB contact or the branch chief of GBIB.

PART III: INSPECTIONAL OPERATIONS

Inspectional

A complete inspection report under this compliance program consists of inspectional findings covering:

1. *Clinical testing*, which includes the adequacy of facilities and procedures utilized by the clinical investigator along with a data audit of the specific study(ies) identified by GBIB

2. *Analytical testing*, which includes the adequacy of the facilities, equipment, personnel, and methods and procedures utilized at the analytical facility including an audit of the method validation and analytical data for the study(ies) identified by GBIB

A full narrative report of any deviations from existing regulations is required. Deviation(s) must be documented sufficiently to support legal or administrative action. For example, any records containing data that are inconsistent with data submitted to FDA should be copied, and the investigator should identify the discrepancy. Generally, serious violations will require more extensive documentation of the discussion between the inspector and his supervisor and the appropriate center contact prior to embarking on this type of coverage.

Investigational

If inspections of institutional review boards (IRBs) and/or clinical laboratories are indicated, the inspector is required to contact his supervisor and GBIB for guidance prior to initiating the inspection.

Refusals

If access to or copying of records is refused for any reason, the inspector promptly contacts his supervisor so that the GBIB contact can be advised of the refusal. Send follow-up information via EMS to GBIB and ORO contacts. The same procedure is followed when it becomes evident that delays by the firm constitute a de facto refusal.

If actions by the firm take the form of a partial refusal for inspection of documents or areas to which FDA is entitled under the law, inspector calls attention to 301(e) and (f) and 505(k)(2) of the FD&C Act; if the refusal persists, his telephones his supervisor and the GBIB contact for instructions.

If the proper course of action to deal with a refusal cannot be resolved expeditiously by GBIB or ORO, GBIB will notify the bioresearch program coordinator (HFC-230).

Findings

1. If the inspector encounters serious problems with the data, methodology, quality control (QC) practices, etc., his will continue with the originally assigned inspection, but contact GBIB for advice on possibly expanding the inspection. GBIB will determine if an in-depth inspection, involving additional BE studies, should be initiated.
2. If the inspector encounters questionable or suspicious records and is unable to review or copy them immediately and have reason to preserve their integrity by officially sealing them, the inspector contacts his supervisor immediately for instructions. Procedures exist for the inspector district to clear this type of action by telephone with the ORA/bioresearch program coordinator (HFC-230). See *Inspection Operations Manual*, section 453.5.
3. Issuance of a FDA 483, Inspectional Observations, is appropriate when (1) practice at the clinical site deviates from the standards for conduct of a clinical study as set forth in 21 CFR 312 and 320 and 361, (2) practice at the analytical site deviates from the standards of laboratory practices as set forth in 21 CFR 320, and (3) discrepancies have occurred between source data and reported data in the case report forms. Items that need to be checked for compliance to study standards are provided in Attachment A. Examples of noncompliance to study standards at the clinical and analytical sites are listed in part V of this guidance. Observed deficient practices should be discussed with the responsible officials.

PART IV: REGULATORY/ADMINISTRATIVE STRATEGY

Clinical Testing

Examples of noncompliance:

1. Subjects not receiving the test or reference drug formulation according to the study randomization codes
2. Biological samples compromised by improper identification, handling, or storage
3. Failure to report adverse experiences, such as vomiting and diarrhea, which may affect absorption and elimination of drugs
4. Inadequate drug accountability records
5. Inadequate medical supervision and coverage
6. Significant problems/protocol deviations/adverse events not reported to the sponsor
7. Failure to adhere to the inclusion/exclusion criteria of the approved protocol
8. Inadequate or missing informed consent for participating subjects
9. Any other situation in which the health and welfare of the subjects are compromised

Analytical Testing

Examples of noncompliance:

1. Inconsistencies between data reported to FDA and at the site.
2. Inadequate or missing validation of assay methodology with respect to specificity (related chemicals, degradation products, metabolites), linearity, sensitivity, precision, and reproducibility.
3. Failure to employ standard, scientifically sound QC techniques, such as use of appropriate standard curves and/or analyte controls that span the range of subjects' analyte levels.
4. Failure to include all data points, not otherwise documented as rejected for a scientifically sound reason, in determination of assay method precision, sensitivity, accuracy, etc.
5. Samples are allowed to remain for prolonged periods of time without proper storage.
6. Failure to maintain source data, for example, source data written on scrap paper and/or discarded in trash after transferring to analytical documents.
7. Lack of objective standard for data acceptance of calibration standards, QCs, etc.
8. Unskilled personnel conducting analytical procedures.
9. No documentation of analytical findings.
10. Inadequate or no written procedures for drug sample receipt and handling.
11. Inadequate or missing standard operating procedures (SOPs).

Note: These lists are not all-inclusive lists of examples of clinical and analytical noncompliance.

BIOEQUIVALENCE INSPECTION REPORT

Part I: Facilities and Procedures (Clinical and Analytical)

Facilities (Clinical and/or Analytical)

1. Evaluate the general facilities for adequate space, work flow patterns, separation of operations, etc.
2. Comment on potential or actual problems, such as the following:
   a. Adjacent clinic room housing concurrent studies.
   b. Open windows allowing ingress of unauthorized food, drugs, etc., into clinic rooms.

c. Are dropped ceilings sealed or monitored to prevent storage of nonpermitted materials?
d. Other conditions that may compromise study security and contribute to the potential for sample mix-up, sample contamination/degradation, etc.

3. Comment if the facilities do not appear adequate to support their normal workload.
4. Are there written, dated, and approved SOPs readily available to all personnel in their work areas? Are working copies kept current?
5. Are outdated procedures archived for future reference?
6. Are visitors to the clinical facility permitted? How are visitors monitored to prevent passage of nonpermitted materials to the study subjects?
7. Are off-site trips for smoking or other reasons monitored to prevent consumption of nonpermitted materials or passage of such materials to or from unauthorized persons?

Personnel

1. Check the relevant qualifications, training, and experience of personnel. Assess staff's ability to perform assigned functions. Document any deficiencies that relate to the audited study(ies).

Specimen Handling and Integrity

In the Clinic. Check and describe

1. Procedures for positive subject and sample identification so that study, drug, subject, sampling time, etc., are linked
2. Procedures for adherence to processing time, temperature, and light conditions as specified by analytical method
3. Storage conditions before and after processing as well as during transit to the laboratory
4. Precautions against sample loss and mix-up during storage, processing, and transit to the laboratory

In the Analytical Laboratory

1. Determine if the analytical facility receives BE samples from other locations. If yes,
   a. Are there freight receipts for sending/receiving samples?
   b. Is a documented history of sample integrity available (e.g., the sample storage time and conditions prior to shipment)?
   c. Is the length of time in shipment recorded?
   d. Evaluate the type of transportation employed and type of protection provided (e.g., shipped by air in insulated containers of dry ice). Report any questionable practices.
   e. What arrangement(s) can be made for receiving shipments outside of normal working hours?
   f. Are the conditions of the samples noted upon arrival at the analytical laboratory, along with the identity of the person(s) receiving the samples?
   g. Are there procedures and documentation to assure that the samples remained at the proper temperature during shipment and holding?
2. Describe the storage equipment for BE samples until analysis (e.g., GE freezer, chest type, Model #417, etc.).
3. Evaluate the equipment and procedures (e.g., ultraviolet light protection) for storing and maintaining BE samples, prior to and during analysis.
   a. Compare storage capacity versus number of samples in storage.
   b. Examine set points for alarms and temperature controlling/recording devices.

   c. Review procedures for calibration and maintenance of alarms and controllers/recorders.
   d. Determine practices for monitoring, review, and storage of temperature records.
   e. Report any evidence of sample thawing.
   f. Check integrity of study samples.
   g. Determine if action plans are in place in case of power loss leading to abnormal storage conditions, that is, emergency procedures.
4. Determine if samples are labeled and separated in storage and during analysis to prevent sample loss or mix-up between studies, subjects, and test/reference drug.
5. Examine how sample identification is maintained through transfer steps during analysis.
6. Is there accurate documentation to show how many freeze and thaw cycles the samples have been subjected to, including accidental thawing due to equipment failure(s)?

ELECTRONIC RECORDS AND SIGNATURES

FDA published the Electronic Records; Electronic Signatures; Final Rule (21 CFR 11) on March 20, 1997. The rule became effective on August 20, 1997. Records in electronic form that are created, modified, maintained, archived, retrieved, or transmitted under any record requirement set forth in agency regulations must comply with 21 CFR 11. The following questions are provided to aid evaluation of electronic records and electronic signatures:

1. Are electronic data systems used to gather clinical (e.g., adverse experiences, concomitant medications) and analytical data (e.g., peak heights, peak areas of chromatograms)? Are such systems used to store, analyze, and/or calculate pharmacokinetic/pharmacodynamic modeling or to transmit clinical and analytical data to the sponsor? If so, identify the system(s), and summarize the system(s)' capabilities. If electronic data systems are not used, omit coverage of the remainder of this section.
2. Determine the source(s) of data entered into the computer for accuracy, security, and traceability.
   a. Direct electronic transfer of online instrument data
   b. Case report forms, analytical worksheets, or similar records requiring manual data entry
   c. Chromatograms requiring evaluation prior to manual extraction of data
   d. Other
3. Determine the following:
   a. Who enters data and when?
   b. Who verifies data entry and when?
   c. Who has access to computer and security codes?
   d. How are data in computers changed? By whom? Audit trail?
4. Determine if the sponsor gets source data or tabulated, evaluated data.
5. Determine how data are transmitted to sponsor (hard copy, computer disk, fax, modem, etc.).
6. If the *sponsor* discovers errors, omissions, etc., in the final report, what contacts are made with the investigator, how are corrections effected, and how are they documented?
7. Determine how data are retained by the investigator (hard copy, electronic, etc.).
8. Determine if the firm has SOPs for validation of computer systems involved in storing, analyzing, calculating, modeling, and/or transmitting clinical and analytical data. Have the computer systems been validated according to the SOPs? Are results of the validations documented and available for audit? Summarize the validated capabilities of the computer systems with respect to their effect on the validity of the study data.

CLINICAL DATA AND OPERATIONS

General

Inspections of clinical facilities should include a comparison of the practices and procedures of the clinical investigator with the requirements of 21 CFR 312, 320.

Inspections should also include a comparison of the source data in the clinical investigator's files with the data submitted to the FDA. Original records should be reviewed, including medical records, dosing records, clinical laboratory test reports, adverse reaction reports, concomitant medications records, and nurses' notes.

Inspection Procedures

This part identifies the minimum information that must be obtained during an inspection to determine if the clinical investigator is complying with the regulations. Each FDA investigator should expand the inspection as facts emerge. The inspections should be sufficient in scope to determine the clinical investigator's general practices for each point identified as well as the particular practices employed for the study(ies) under audit.

Study Responsibility and Administration

1. Determine if the clinical investigator was aware of the status of the test article(s), nature of the protocol, and the obligations of the clinical investigator.
2. Determine whether authority for the conduct of various aspects of the study was delegated properly so that the investigator retained control and knowledge of the study.
3. Determine if the investigator discontinued the study before completion. If so, provide reason.
4. Determine the name and address of any clinical laboratory performing clinical laboratory tests for qualifying and/or safety monitoring of study subjects.
   a. If any clinical laboratory testing was performed in the investigator's own facility, determine whether that facility is equipped to perform each test specified.
   b. Determine if individuals performing the clinical tests are adequately qualified.

Protocol

Obtain a copy of the written protocol. Unavailability should be reported and documented. If a copy of the protocol is sent with the assignment background material, it should be compared to the protocol on site. If the protocols are identical, a duplicate copy does not need to be obtained. The narrative should note that the protocols were identical. If the protocol has been accepted by a review division in CDER, a copy of the acceptance letter should be attached to the establishment inspection report (EIR). If the agency has recommended the incorporation of additional material, method, or information into the protocol, verify that appropriate modifications were made.

1. Compare the written protocol and all IRB-approved modifications against the protocol provided with the assignment package. Report and document any differences.
2. Determine if the approved protocol was followed with respect to
   a. Subject selection (inclusion/exclusion criteria)
   b. Number of subjects
   c. Drug dose form, strength, and route of administration
   d. Frequency of subject dosing, monitoring, and sampling
   e. Washout period between study arms (test versus reference drug)
   f. Other (specify)

3. Determine whether all significant changes to the protocol were
   a. Documented by an approved amendment that is maintained with the protocol
   b. Dated by the investigator
   c. Approved by the IRB and reported to the sponsor before implementation except where necessary to eliminate apparent immediate hazard to human subjects
   d. Implemented after IRB approval

Note: Changes in protocol are not violations of protocol.

Subjects' Records

1. Describe the investigator's source data files in terms of their organization, condition, accessibility, completeness, and legibility.
2. Determine whether there is adequate documentation to assure that all audited subjects did exist and were alive and available for the duration of their stated participation in the study.
3. Compare the source data in the clinical investigator's records with the case reports completed for the sponsor. Determine whether clinical laboratory testing (including blood work, ECGs, x-rays, eye exams), as noted in the case report forms, was documented by the presence of completed laboratory records among the source data.
4. Determine whether all adverse experiences were reported in the case report forms. Determine whether they were regarded as caused by or associated with the test article and if they were previously anticipated (specificity, severity) in any written information regarding the test article.
5. Concomitant therapy and/or intercurrent illnesses might interfere with the evaluation of the effect of the test article. Check whether concomitant therapy or illness occurred. If so, was such information included in the case report forms?
6. Determine whether the number and type of subjects entered into the study were confined to the protocol limitations and whether each record contains
   a. Observations, information, and data on the condition of each subject at the time the subject entered into the clinical study
   b. Records of exposure of each subject to the test article
   c. Observations and data on the condition of each subject throughout participation in the investigation including time(s) of drug administration; dosing according to preestablished, randomization schedules; results of lab tests; development of unrelated illness; bleeding times and any other specimen collections; washout periods for subjects; and other factors that might alter the effects of the test article
   d. The identity of all persons and locations obtaining source data or involved in the collection or analysis of such data

Other Study Records

Review information in the clinical investigator's records that would be helpful in assessing any underreporting of adverse experiences by the sponsor to the agency. The following information will ordinarily be obtained from the sponsor and sent with the assignment:

a. The total number of subjects entered into the study
b. The total number of dropouts from the study (identified by subject number)
c. The number of evaluable subjects and the number of nonevaluable subjects (the latter identified by subject number)
d. The adverse experiences identified by subject number and a description of the adverse experience

Compare the information submitted to the sponsor according to the clinical investigator's files with the information obtained from the sponsor, and document any discrepancies found.

Consent of Human Subjects

1. Obtain a copy of the consent form actually used.
2. Determine whether proper informed consent was obtained from *all* subjects *prior* to their entry into the study. Identify the staff who obtain and witness the signing of informed consent for study subjects.

Institutional Review Board

1. Identify the name, address, and chairperson of the IRB for this study.
2. Determine whether the investigator maintains copies of all reports submitted to the IRB and reports of all actions by the IRB. Determine the nature and frequency of periodic reports submitted to the IRB.
3. Determine whether the investigator submitted reports to the IRB of all deaths and serious adverse experiences and unanticipated problems involving risk to human subjects (21 CFR 312.66).
4. Determine if the investigator submitted to and obtained IRB approval of the following *before* subjects were allowed to participate in the investigation:
   a. Protocol
   b. Modifications to the protocol
   c. Materials to obtain human subject consent
   d. Media advertisements for subject recruitment
5. Determine if the investigator disseminated any promotional material or otherwise represented that the test article was safe and effective for the purpose for which it was under investigation. Were the promotional material(s) submitted to the IRB for review and approval before use?

Sponsor

1. Did the investigator provide a copy of the IRB-approved consent form to the sponsor?
2. Determine whether the investigator maintains copies of all reports submitted to the sponsor.
3. Determine if and how the investigator submitted any report(s) of deaths and adverse experiences to the sponsor.
4. Determine whether all intercurrent illnesses and/or concomitant therapy(ies) were reported to the sponsor.
5. Determine whether all case report forms on subjects were submitted to the sponsor shortly (within 6 months) after completion.
6. Determine whether all dropouts and the reasons therefore were reported to the sponsor.
7. Did the sponsor monitor the progress of the study to assure that investigator obligations were fulfilled? Briefly describe the method (on-site visit, telephone, contract research organization, etc.) and *frequency* of monitoring. Do the study records include a log of on-site monitoring visits and telephone contacts?

Test Article Accountability

1. Determine whether unqualified or unauthorized persons administered or dispensed the test article(s).
2. What names are listed on the FDA-1571 (for sponsor–investigator) and FDA-1572 (for studies conducted under an IND)? Obtain a copy of all FDA-1572s.

3. Determine accounting procedures for test articles:
   a. Receipt date(s) and quantities
   b. Dates and quantities dispensed
   c. Quantities of BE testing samples retained (see Sample Collection section under part III)
4. Inspect storage area.
   a. Reconcile amounts of test article used with amounts received, returned, and retained. Report any discrepancy.
   b. If not previously sampled under CP 7346.832, collect samples of both the test and reference products for FDA analysis.
5. If test articles are controlled substances, determine if proper security is provided.

Records Retention

1. Determine who maintains custody of the required records and the means by which prompt access can be assured.
2. Determine whether the investigator notified the sponsor in writing regarding alternate custody of required records, if the investigator does not maintain them.
3. Be aware that records should be retained at the study site for the specified time as follows:
   a. Two years following the date on which the test article is approved by FDA for marketing for the purposes that were the subject of the clinical investigation.
   b. Two years following the date on which the entire clinical investigation (not just the investigator's part in it) is terminated or discontinued *by the sponsor.* If the investigator was terminated or discontinued, was FDA notified?

ABBREVIATED REPORT FORMAT

For inspection of a clinical facility, abbreviated report is allowed (1) if there are no significant violations and no FDA 483 is issued and (2) in cases where there are objectionable findings but the findings are not serious and clearly do not have any impact on data integrity and study outcomes. The following is a guideline for preparation of the abbreviated report:

1. Reason for inspection:
   a. Identify the headquarters unit that initiated and/or issued the assignment.
   b. State the purpose of the inspection.
2. What was covered:
   a. Identify the clinical study, protocol number, sponsor, NDA, ANDA, etc.
   b. Location of study.
3. Administrative procedures:
   a. Report the name, title, and authority of the person to whom credentials were shown and FDA-482 Notice of Inspection was issued.
   b. Persons interviewed.
   c. Who accompanied the inspector during establishment inspection.
   d. Who provided relevant information.
   e. Identify the IRB.
   f. Prior inspectional history.
4. Individual responsibilities:
   a. Identify study personnel and summarize their responsibilities relative to the clinical study (e.g., who screened the subjects, who administered the drugs, who supervised collection, identification, and processing of samples).
   b. A statement about (i) who obtained informed consent, (ii) how it was obtained, and (iii) was informed consent signed by each subject.
   c. Identify by whom the clinical study was monitored, when, etc.

5. Inspectional findings:
   a. A statement regarding the comparison of data on the case report forms to the source data at the investigator's site. Indicate the number of records compared and what was compared (patient charts, hospital records, lab slips, etc.) and specific information about any discrepancies.
   b. A statement indicating if the drug accountability records were sufficient to reconcile the amount of drug received, dispensed, returned, and retained.
   c. A statement about protocol adherence. Describe in detail any nonadherence.
   d. A statement concerning doses in accordance with preestablished, randomization schedules.
   e. The EIR should identify the IRB and state if it approved the study and was kept informed of the progress of the study.
   f. A statement on (i) follow-up activities in response to reports of adverse experiences (including death) if any occurred and (ii) whether there was evidence of underreporting of adverse experiences/events.
   g. Discussion of 483 observations; reference the exhibits/documentation collected.
6. Discussion with management:
   a. Discussion of 483 observations and non-483 observations
   b. Clinical investigator's response to observations

Remember that the aforementioned list deals with abbreviated reports, not abbreviated inspections. All assignments issued for cause must have full reporting. The assignment EMS or memo will indicate the need for full reporting for any special inspection.

ANALYTICAL DATA AND OPERATIONS

Information required by this section must be obtained with the assistance of a qualified analyst from the field and/or a reviewer in GBIB with expertise in the type of analysis used in the BE study under review.

At random, compare the analytical source data with data provided in the inspection assignment for accuracy of transference and for scientific soundness/bearing on the validity of the study. Analytical source data are as follows: codes used to blind samples; data establishing the sensitivity, linearity, specificity, and precision of the analytical assay; data determining the stability of the drug in the biological specimen; all standard curves; blinded and unblinded spiked control samples; blanks; data on reagent preparation; instrumental readings; calculations; etc. The data comparison and the testing procedural review should include an evaluation of any discrepancies found.

PRE-STUDY ANALYSIS

If the analytical laboratory is involved in analysis of drug standards and products employed in the BE studies, determine if the following applies:

1. Appropriate samples were analyzed by the laboratory to determine potency and content uniformity for tablets and capsules. Include a description of procedures used to prepare the sample(s) used in the study.
2. If testing of the samples described earlier was not performed by the analytical laboratory, did the sponsor provide test results to the laboratory?
3. For both the test and reference drug products studied, were the products' appearance, potency, dosage form (capsule, tablet, suspension, controlled release, etc.), lot numbers, and expiration dates the same as that reported to FDA?

Protocol Acceptance

If the review division reviewed the protocol and recommended protocol modifications, verify that the modifications were incorporated into the protocol.

Equipment

Check on the following with respect to both current equipment and practices and those in place at the time of the study:

1. Does the laboratory have the same type, brand, and model (not serial) numbers of all major pieces of analytical equipment and instrumentation used in their testing procedures, as reported in the ANDA or NDA (e.g., gas chromatographs, high-performance liquid chromatographs, ultraviolet spectrophotometers, colorimeter, fluorescence or atomic absorption spectrophotometer, pH meter)? If not, describe the discrepancy and include its effect on the validity of the study data.
2. Assess the general condition of the major pieces of equipment (e.g., gross mistreatment) that may render them inaccurate or unreliable. Examples are damaged gas chromatograph inlet port and dry pH meter electrodes. Review maintenance and repair logs for indications of past problems.
3. Are there written operating instructions for these major pieces of equipment, and are they available to the laboratory personnel?
4. Are there written and scheduled calibration/standardization procedures and preventative maintenance procedure for all analytical instruments employed in the study? Determine whether these calibration/standardization procedures are actually employed and documented. If not, describe the deficiencies, and determine whether the instruments have been calibrated during the time of the study.
5. Were specific instrument operating parameters documented during the study? If so, where?

Analytical Methods Validation

Determine through data and procedural review if the following applies:

1. The analytical laboratory has scientifically sound data to support claims for the specificity of the assay employed in this study. Ascertain the laboratory's justification for noninterferences, both endogenous and exogenous (e.g., metabolites, solvent contamination) in measuring the analytes (drug, metabolites, etc.) studied.
2. The analytical laboratory has data to support the claims for the linearity of the assay employed in this study.
3. The laboratory analyst who analyzed the biological samples has generated data demonstrating the sensitivity of the assay using the same instrumentation as that employed in the BE study. The sensitivity of the assay (or limit of detection) may be defined as the lowest quantifiable limit that can be *reproducibly determined* for the measured analyte(s) being carried through the method.
4. The laboratory analyst who analyzed the biological specimen has generated data demonstrating the precision of the assay using the instrumentation employed in the BE study. The data should be available for both standard and QC samples and should include the consistency of precision of the standard and control samples carried through the assay procedure. Ascertain the laboratory's justification for the precision based on the separation procedure, instrumentation, and analyte concentration levels in the biological fluids.

5. The laboratory has data to demonstrate drug recoveries (percent recovery) for the measured analyte(s). This should include both analyte extraction efficiency from the biological fluid *and* recovery of the analyte(s) carried through the analytical testing procedure.
6. The analytical laboratory determined the stability of the drug both in the biological specimen and in the sample preparation medium under the same condition as in actual analysis of subject samples.
7. The analytical laboratory showed that the storage procedures (e.g., freezing and number of freeze–thaw cycles) have no adverse effect on drug stability for the period of time the samples were stored, from subject dosing until last sample analysis.
8. The water quality specified for sample and reagent preparation is consistently and readily available in the lab.

Sample Analyses

Determine if the following applies:

1. The analytical assay employed was the same as that specified in the ANDA or NDA.
2. The assay parameters observed for the study's sample analysis are similar to those (e.g., specificity, precision) obtained during method validation. Review study subjects' source analytical data to check this; pay particular attention to analytical runs determined toward the end of analytical testing.
3. Coding techniques were used to blind the analytical laboratory to the sample. Was the code available to the analytical chemist?
4. The samples were analyzed in a randomized fashion or in some specific order. Were samples of test and reference products for the same subject analyzed at the same time under identical conditions with the same standard curve, same control, and same instrument?
5. Standard curves are prepared each time a batch of unknown samples is assayed. If not, how often are standards run? Have all the standard curves run during the study been reported? How many standards are used to define each standard curve? (Should be 5–8, excluding blank.) Does the laboratory have scientifically sound procedures for acceptance or rejection of a standard point and/or a standard curve?
6. The standard curve encompasses the concentration values reported. Were any values reported that were derived from points extrapolated on the standard curve?
7. The laboratory has a scientifically sound SOP in place to guide the acceptance/rejection of data. Did the laboratory adhere to the SOPs in the reporting of repeated determinations, or was supervisory discretion used to accept/reject data points?
8. Blinded or nonblinded spiked control samples have been included and reported with each run. Who prepared these samples? Were the controls made from a standard weight different from the standard weight used to prepare standards for the standard curve (i.e., two separate independent weighings for calibration standards and QC stock solutions)? Do the controls span the expected analyte concentration range (low, midrange, and high) found in the subjects' samples? Have all control values been reported individually, as opposed to averages?
9. The control samples were processed and analyzed exactly the same as the unknown samples. Were the controls interspersed throughout the entire analytical run?
10. The source of blank biological fluids. (Was each subject's zero hour serum used as the blank, pooled plasma, etc.?) Were interferences noted in the analytical source data for these samples? Specifications should be established to assure that blank biological fluids are as similar as possible to the biological matrix for the subject samples.
11. The source of the drug standards used for the in vivo sample analysis. If not compendial standards, how was the quality and purity of the standard assured?

12. All sample values were recorded and reported. If not, were reasons for rejection documented and justified? Were any samples rerun? When repeated determinations were made, were new standard curves and control samples run concurrently?
13. The procedure employed for determining which value of a rerun sample is reported. Was this procedure scientifically sound and consistently followed? Was an established written procedure followed?
14. The submitted chromatograms are representative of the quality of the chromatograms generated throughout the study.
15. There are written procedures for preparing reagents used in these assays. Are reagents properly labeled with date of preparation, storage requirements, and chemist who prepared them? Were the original weighings for calibration standard and QC stock solutions checked and countersigned by a second party?
16. Copies of the following chromatograms are available. (If not submitted by the applicant, the field investigator or chemist should obtain copies.)
    a. Reagent blank
    b. Sample blank
    c. Internal standard
    d. A standard run
    e. A QC run
    f. A set of chromatograms for one subject over the entire span of the study

For Antibiotic Analyses

Determine the following:

1. Are incubators available? Specify dimensions and type.
2. Whether
    a. The bench tops are level
    b. The room temperature is controlled and, if so, what are the temperature tolerances
    c. Agar, propagation cultures, and other necessary resources are available and properly monitored
    d. Zone readers are available; if so, specify type
    e. Autoclaves are available, and, if so, specify type and determine if the autoclave sterilization process has been validated
3. The room where these studies are conducted is "environmentally sterile" and what monitoring is done to determine the degree of "environmental sterility."
4. Whether the samples were run properly through the incubator, that is, times and temperatures are controlled to desired specifications and properly documented.
5. Whether the standards, controls, and samples are incubated at the same time, in the same incubator.
6. Whether the microorganisms used in the media are the same as described in the AADA.
7. Whether a burner is used to heat the wire for transfer purposes.
8. Whether calibrated zone readers were used for zone size determinations.
9. Whether turbidimetric methodology was employed. Also, determine the type of spectrophotometry used.
10. Whether the turbidimetric standardization procedure was the same as that specified in the AADA. If not, describe differences.
11. Whether all samples were read in duplicate. Were all samples read by the same person? Did zone diameters or turbidimetric readings correlate with drug concentration levels?
12. Are SOPs in place to calibrate the incubator, autoclave, etc., used in antibiotic analysis? Are the SOPs readily available to laboratory personnel?

For Radiometric Analyses

In addition to the general guidance earlier, determine the following:

1. How the specific activity of the radiochemical standards employed was determined.
2. Whether all counts specified in records submitted to the agency were actually counted for the time interval specified.
3. Whether an inventory of all radiolabeled compounds is maintained by the laboratory.
4. If the background level has been determined. If yes, by what method?
5. For RIA methodology, determine if a commercial kit was used in the analysis. If so, report the type of kit, the expiration date, and whether the laboratory validated the accuracy, specificity, precision, sensitivity, and linearity of the kit assay in relation to the reported study assay procedure.

Data Handling and Storage

Determine the following:

1. Whether bound notebooks and/or source data worksheets are used by the laboratory.
2. If bound notebooks are used, are the pages filled in sequentially on a chronological basis? Does the analyst sign the notebook/worksheet daily? Does a supervisor initial the notebook/worksheet after checking it for accuracy?
3. Whether the laboratory retains all source data, such as notebooks, worksheets, chromatograms, and standard curves. Is there justification for source data excluded from the study report, such as rejected runs and missing samples?
4. Whether the analyst(s) sign and date all source data records.
5. How long the source data is retained.
6. Describe the maintenance and accessibility of laboratory source data (e.g., repeated determinations, rejected analytical runs). Document problems with data recording and verification, such as lack of dates and signatures, erasures, and whiteout.

BIOEQUIVALENCE TESTING REPORT SUMMARY

1. District:
2. Date(s) of inspection:
3. Application no. (if applicable):
4. Application sponsor (if any):
   a. Name:
   b. Address:
   c. City: State: Zip:
5. Location where testing performed:
   a. Clinical facility name: Address:
   b. City: State: Zip:
   c. Central file no.:
   d. Analytical facility name:
   e. Address:
   f. City: State: Zip:
   g. Central file no.:
6. Responsible official (recipient of *notice of inspection*):
   a. Name and title:

7. Person receiving FDA 483 (if issued):
   a. Name and title:
8. Drug under study:
   a. Generic name:
   b. Trade name:
   c. Dosage form:
   d. Strength(s):
9. Number of subjects in clinical test:
10. Status of clinical testing:
    a. Date started:
    b. Completion date:
11. Sample collection sample lot # _______________
12. FDA investigator(s):
13. Remarks:

GOOD LABORATORY PRACTICES

In the 1970s, FDA inspections of nonclinical laboratories revealed that some studies submitted in support of the safety of regulated products had not been conducted in accord with acceptable practice and that accordingly data from such studies were not always of the quality and integrity to assure product safety. As a result of these findings, FDA promulgated the GLP Regulations, 21 CFR part 58, on December 22, 1978 (43 FR 59986). The regulations became effective June 1979. The regulations establish standards for the conduct and reporting of nonclinical laboratory studies and are intended to assure the quality and integrity of safety data submitted to FDA.

FDA relies on documented adherence to GLP requirements by nonclinical laboratories in judging the acceptability of safety data submitted in support of research and/or marketing permits. FDA has implemented this program of regular inspections and data audits to monitor laboratory compliance with the GLP requirements.

The objective of this program is

1. To verify the quality and integrity of data submitted in a research or marketing application
2. To inspect (approximately every 2 years) nonclinical laboratories conducting safety studies that are intended to support applications for research or marketing of regulated products
3. To audit safety studies and determine the degree of compliance with GLP regulations
   a. Types of inspections
      i. Surveillance inspections
         Surveillance inspections are periodic, routine determinations of a laboratory's compliance with GLP regulations. These inspections include a facility inspection and audits of ongoing and/or recently completed studies.
      ii. Directed inspections
         Directed inspections are assigned to achieve a specific purpose, such as
         - Verifying the reliability, integrity, and compliance of critical safety studies being reviewed in support of pending applications
         - Investigating issues involving potentially unreliable safety data and/or violative conditions brought to FDA's attention
         - Reinspecting laboratories previously classified Office of Analytical Inspection (OAI) (usually within 6 months after the firm responds to a warning letter)
         - Verifying the results from third-party audits or sponsor audits submitted to FDA for consideration in determining whether to accept or reject questionable or suspect studies

INSPECTIONS

1. The investigator will *determine* the current state of GLP compliance by evaluating the laboratory facilities, operations, and study performance.
2. Organization chart—If the facility maintains an organization chart, *obtain* a current version of the chart for use during the inspection and *submit* it in the EIR.
3. Facility floor-plan diagram—*Obtain* a diagram of the facility. The diagram may identify areas that are not used for GLP activities. If it does not, request that appropriate facility personnel identify any areas that are not used for GLP activities. Use during the inspection and *submit* it in the EIR.
4. Master schedule sheet—*Obtain* a copy of the firm's master schedule sheet for all studies listed since the last GLP inspection or last 2 years and select studies as defined in 21 CFR 58.3(d). If the inspection is the first inspection of the facility, review the entire master schedule. If studies are identified as non-GLP, *determine* the nature of several studies to verify the accuracy of this designation. See 21 CFR 58.1 and 58.3(d). In contract laboratories, *determine* who decides if a study is a GLP study.
5. Identification of studies
    a. Directed inspections—Inspection assignments will identify studies to be audited.
    b. Surveillance inspections—Inspection assignments may identify one or more studies to be audited. If the assignment does not identify a study for coverage or if the referenced study is not suitable to assess all portions of current GLP compliance, the investigator will select studies as necessary to evaluate all areas of laboratory operations. When additional studies are selected, first priority should be given to FDA studies for submission to the assigning center.
6. Ongoing studies—*Obtain* a copy of the study protocol and *determine* the schedule of activities that will be underway during the inspection. This information should be used to schedule inspections of ongoing laboratory operations, as well as equipment and facilities associated with the study. If there are no activities underway in a given area for the study selected, evaluate the area based on ongoing activities.
7. Completed studies—The data audit should be carried out as outlined in part III, D. If possible, accompany laboratory personnel when they retrieve the study data to assess the adequacy of data retention, storage, and retrieval as described in part III, C 10.

The facility inspection should be guided by the GLP regulations. The following areas should be evaluated and described as appropriate:

1. Organization and personnel (21 CFR 58.29, 58.31, 58.33)
    a. Purpose: To determine whether the organizational structure is appropriate to ensure that studies are conducted in compliance with GLP regulations and to determine whether management, study directors, and laboratory personnel are fulfilling their responsibilities under the GLPs.
    b. Management responsibilities (21 CFR 58.31): Identify the various organizational units, their role in carrying out GLP study activities, and the management responsible for these organizational units. This includes identifying personnel who are performing duties at locations other than the test facility and identifying their line of authority. If the facility has an organization chart, much of this information can be determined from the chart.
2. *Determine* if management has procedures for assuring that the responsibilities in 58.31 can be carried out. Look for evidence of management involvement, or lack thereof, in the following areas:
    a. Assigning and replacing study directors.
    b. Control of study director workload (use the master schedule to assess workload).

c. Establishment and support of the quality assurance unit (QAU), including assuring that deficiencies reported by the QAU are communicated to the study directors and acted upon.
d. Assuring that test and control articles or mixtures are appropriately tested for identity, strength, purity, stability, and uniformity.
e. Assuring that all study personnel are informed of and follow any special test and control article handling and storage procedures.
f. Providing required study personnel, resources, facilities, equipment, and materials.
g. Reviewing and approving protocols and SOPs.
h. Providing GLP or appropriate technical training.

3. Personnel (21 CFR 58.29)—Identify key laboratory and management personnel, including any consultants or contractors used, and review personnel records, policies, and operations to *determine* if
   a. Summaries of training and position descriptions are maintained and are current for selected employees
   b. Personnel have been adequately trained to carry out the study functions that they perform
   c. Personnel have been trained in GLPs
   d. Practices are in place to ensure that employees take necessary health precautions, wear appropriate clothing, and report illnesses to avoid contamination of the test and control articles and test systems
4. If the firm has computerized operations, *determine* the following:
   a. Who was involved in the design, development, and validation of the computer system?
   b. Who is responsible for the operation of the computer system, including inputs, processing, and output of data?
   c. Whether computer system personnel have training commensurate with their responsibilities, including professional training and training in GLPs?
   d. Whether some computer system personnel are contractors who are present on-site full time or nearly full time. The investigation should include these contractors as though they were employees of the firm. Specific inquiry may be needed to identify these contractors, as they may not appear on organization charts.
   e. Interview and observe personnel using the computerized systems to assess their training and performance of assigned duties.
5. Study director (21 CFR 58.33)
   a. Assess the extent of the study director's actual involvement and participation in the study. In those instances when the study director is located off-site, review any correspondence/records between the testing facility management and QAU and the off-site study director. *Determine* that the study director is being kept immediately apprised of any problems that may affect the quality and integrity of the study.
   b. Assess the procedures by which the study director.
   c. Assures the protocol and any amendments have been properly approved and are followed.
   d. Assures that all data are accurately recorded and verified.
   e. Assures that data are collected according to the protocol and SOPs.
   f. Documents unforeseen circumstances that may affect the quality and integrity of the study and implements corrective action.
   g. Assures that study personnel are familiar with and adhere to the study protocol and SOPs.
   h. Assures that study data are transferred to the archives at the close of the study.

6. EIR documentation and reporting—Collect exhibits to document deficiencies. This may include SOPs, organizational charts, position descriptions, and curriculum vitae (CVs), as well as study-related memos, records, and reports for the studies selected for review. *The use of outside or contract facilities must be noted in the EIR. The assigning center should be contacted for guidance on inspection of these facilities.*
7. QAU (21 CFR 58.35)
   a. Purpose: To determine if the test facility has an effective, independent QAU that monitors significant study events and facility operations, reviews records and reports, and assures management of GLP compliance.
8. QAU operations (21 CFR 58.35(b-d)): Review QAU SOPs to assure that they cover all methods and procedures for carrying out the required QAU functions, and confirm that they are being followed. *Verify* that SOPs exist and are being followed for QAU activities including, but not limited to, the following:
   a. Maintenance of a master schedule sheet.
   b. Maintenance of copies of all protocols and amendments.
   c. Scheduling of its in-process inspections and audits.
   d. Inspection of each nonclinical laboratory study at intervals adequate to assure the integrity of the study and maintenance of records of each inspection.
   e. Immediately notify the study director and management of any problems that are likely to affect the integrity of the study.
   f. Submission of periodic status reports on each study to the study director and management.
   g. Review of the final study report.
   h. Preparation of a statement to be included in the final report that specifies the dates inspections were made and findings reported to management and to the study director.
9. Inspection of computer operations
   a. *Verify* that, for any given study, the QAU is entirely separate from and independent of the personnel engaged in the conduct and direction of that study. Evaluate the time QAU personnel spend in performing in-process inspection and final report audits. *Determine* if the time spent is sufficient to detect problems in critical study phases and if there are adequate personnel to perform the required functions.
   b. *Note*: The investigator may request the firm's management to certify in writing that inspections are being implemented, performed, documented, and followed up in accordance with this section (see 58.35[d]).
10. EIR documentation and reporting—*Obtain* a copy of the master schedule sheet dating from the last routine GLP inspection or covering the past 2 years. If the master schedule is too voluminous, *obtain* representative pages to permit headquarters review. When master schedule entries are coded, *obtain* the code key. Deficiencies should be fully reported and documented in the EIR. Documentation to support deviations may include copies of QAU SOPs, list of QAU personnel, their CVs or position descriptions, study-related records, protocols, and final reports.
11. Facilities (21 CFR 58.41-51)
    a. Purpose: Assess whether the facilities are of adequate size and design.
    b. Facility Inspection
       i. Review environmental controls and monitoring procedures for critical areas (i.e., animal rooms, test article storage areas, laboratory areas, handling of biohazardous material), and *determine* if they appear adequate and are being followed.
       ii. Review the SOPs that identify materials used for cleaning critical areas and equipment, and assess the facility's current cleanliness.

   iii. *Determine* whether there are appropriate areas for the receipt, storage, mixing, and handling of the test and control articles.
   iv. *Determine* whether separation is maintained in rooms where two or more functions requiring separation are performed.
   v. *Determine* that computerized operations and archived computer data are housed under appropriate environmental conditions (e.g., protected from heat, water, and electromagnetic forces).

12. EIR documentation and reporting—Identify which facilities, operations, SOPs, etc., were inspected. Only significant changes in the facility from previous inspections need to be described. Facility floor plans may be collected to illustrate problems or changes. Document any conditions that would lead to contamination of test articles or to unusual stress of test systems.
13. Equipment (21 CFR 58.61-63)
   a. Purpose: To assess whether equipment is appropriately designed and of adequate capacity and is maintained and operated in a manner that ensures valid results
   b. Equipment inspection—Assess the following:
      i. The general condition, cleanliness, and ease of maintenance of equipment in various parts of the facility.
      ii. The heating, ventilation, and air-conditioning system design and maintenance, including documentation of filter changes and temperature/humidity monitoring in critical areas.
      iii. Whether equipment is located where it is used and that it is located in a controlled environment, when required.
      iv. Nondedicated equipment for preparation of test and control article carrier mixtures is cleaned and decontaminated to prevent cross contamination.
      v. For representative pieces of equipment, check the availability of the following:
         - SOPS and/or operating manuals
         - Maintenance schedule and log
         - Standardization/calibration procedure, schedule, and log
         - Standards used for calibration and standardization
      vi. For computer systems, assess that the following procedures exist and are documented:
         - Validation study, including validation plan and documentation of the plan's completion
         - Maintenance of equipment, including storage capacity and backup procedures.
         - Control measures over changes made to the computer system, which include the evaluation of the change, necessary test design, test data, and final acceptance of the change.
         - Evaluation of test data to assure that data are accurately transmitted and handled properly when analytical equipment is directly interfaced to the computer.
         - Procedures for emergency backup of the computer system (e.g., backup battery system and data forms for recording data in the event of a computer failure or power outage).
14. EIR documentation and reporting—The EIR should list which equipment, records, and procedures were inspected and the studies to which they are related. Detail any deficiencies that might result in contamination of test articles, uncontrolled stress to test systems, and/or erroneous test results.
15. Testing facility operations (21 CFR 58.81)
   a. Purpose: To determine if the facility has established and follows written SOPs necessary to carry out study operations in a manner designed to ensure the quality and integrity of the data.
   b. SOP Evaluation

i. Review the SOP index and representative samples of SOPs to ensure that written procedures exist to cover at least all of the areas identified in 58.81(b).
ii. *Verify* that only current SOPs are available at the personnel workstations.
iii. Review key SOPs in detail and check for proper authorization signatures and dates and general adequacy with respect to the content (i.e., SOPs are clear, complete, and can be followed by a trained individual).
iv. *Verify* that changes to SOPs are properly authorized and dated and that a historical file of SOPs is maintained.
v. Ensure that there are procedures for familiarizing employees with SOPs.
vi. *Determine* that there are SOPs to ensure the quality and integrity of data, including input (data checking and verification), output (data control), and an audit trail covering all data changes.
vii. *Verify* that a historical file of outdated or modified computer programs is maintained. If the firm does not maintain old programs in digital form, ensure that a hard copy of all programs has been made and stored.
viii. *Verify* that SOPs are periodically reviewed for current applicability and that they are representative of the actual procedures in use.
ix. Review selected SOPs and observe employees performing the operation to evaluate SOP adherence and familiarity.EIR documentation and reporting—*Submit* SOPs, data collection forms, and raw data records as exhibits that are necessary to support and illustrate deficiencies.

16. Reagents and solutions (21 CFR 58.83)
    a. Purpose: To determine that the facility ensures the quality of reagents at the time of receipt and subsequent use.
        i. Review the procedures used to purchase, receive, label, and *determine* the acceptability of reagents and solutions for use in the studies.
        ii. *Verify* that reagents and solutions are labeled to indicate identity, titer or concentration, storage requirements, and expiration date.
        iii. *Verify* that for automated analytical equipment, the profile data accompanying each batch of control reagents are used.
        iv. Check that storage requirements are being followed.
17. Test and control articles (21 CFR 58.105–113)
    a. Purpose: To determine that procedures exist to assure that test and control articles and mixtures of articles with carriers meet protocol specifications throughout the course of the study and that accountability is maintained
    b. Characterization and stability of test articles (21 CFR 58.105)—The responsibility for carrying out appropriate characterization and stability testing may be assumed by the facility performing the study or by the study sponsor. *When test article characterization and stability testing is performed by the sponsor, verify that the test facility has received documentation that this testing has been conducted.*
    c. *Verify* that procedures are in place to ensure that
        i. The acquisition, receipt and storage of test articles, and means used to prevent deterioration and contamination are as specified
        ii. The identity, strength, purity, and composition (i.e., characterization) to define the test and control articles are determined for each batch and are documented
        iii. The stability of test and control articles is documented
        iv. The transfer of samples from the point of collection to the analytical laboratory is documented
        v. Storage containers are appropriately labeled and assigned for the duration of the study
        vi. Reserve samples of test and control articles for each batch are retained for studies lasting more than four weeks

   d. Test and control article handling (21 CFR 58.107). *Determine* that there are adequate procedures for
      i. Do*cumentati*on for receipt and distribution
      ii. Proper identification and storage
      iii. Precluding contamination, deterioration, or damage during distribution
      iv. Inspect test and control article storage areas to *verify* that environmental controls, container labeling, and storage are adequate.
      v. Observe test and control article handling and identification during the distribution and administration to the test system.
      vi. Review a representative sample of accountability records, and, if possible, *verify* their accuracy by comparing actual amounts in the inventory. For completed studies, *verify* documentation of final test and control article reconciliation.
18. Protocol and conduct of nonclinical laboratory study (21 CFR 58.120 130)
   a. Purpose: To determine if study protocols are properly written and authorized and that studies are conducted in accordance with the protocol and SOPs
   b. Study protocol (21 CFR 58.120)
      i. Review SOPs for protocol preparation and approval and *verify* they are followed.
      ii. Review the protocol to *determine* if it contains required elements.
      iii. Review all changes, revisions, or amendments to the protocol to ensure that they are authorized, signed, and dated by the study director.
      iv. *Verify* that all copies of the approved protocol contain all changes, revisions, or amendments.
19. Conduct of the nonclinical laboratory study (21 CFR 58.130). Evaluate the following laboratory operations, facilities, and equipment to *verify* conformity with protocol and SOP requirements for
   a. Test system monitoring
   b. Recording of raw data (manual and automated)
   c. Corrections to raw data (corrections must not obscure the original entry and must be dated, initialed, and explained)
   d. Randomization of test systems
   e. Collection and identification of specimens
   f. Authorized access to data and computerized systems
20. Records and reports (21 CFR 58.185–195)
   a. Purpose: To assess how the test facility stores and retrieves raw data, documentation, protocols, final reports, and specimens
   b. Reporting of study results (21 CFR 58.185)—*Determine* if the facility prepares a final report for each study conducted. For selected studies, *obtain* the final report, and *verify* that it contains the following:
21. Storage and retrieval of records and data (21 CFR 58.190)
   a. *Verify* that raw data, documentation, protocols, final reports, and specimens have been retained.
   b. *Identify* the individual responsible for the archives. Determine if delegation of duties to other individuals in maintaining the archives has occurred.
   c. *Verify* that archived material retained or referred to in the archives is indexed to permit expedient retrieval. It is not necessary that all data and specimens be in the same archive location. For raw data and specimens retained elsewhere, the archives index must make specific reference to those other locations.
   d. *Verify* that access to the archives is controlled and *determine* that environmental controls minimize deterioration.

e. Ensure that there are controlled procedures for adding or removing material. Review archive records for the removal and return of data and specimens. Check for unexplained or prolonged removals.
f. *Determine* how and where computer data and backup copies are stored, that records are indexed in a way to allow access to data stored on electronic media, and that environmental conditions minimize deterioration.
g. *Determine* to what electronic media such as tape cassettes or ultrahigh capacity portable disks the test facility has the capacity of copying records in electronic form. *Report* names and identifying numbers of both copying equipment type and electronic medium type to enable agency personnel to bring electronic media to future inspections for collecting exhibits.

22. Data audit. In addition to the procedures outlined earlier for evaluating the overall GLP compliance of a firm, the inspection should include the audit of at least one completed study. Studies for audit may be assigned by the center or selected by the investigator as described in part III, A. The audit will include a comparison of the protocol (including amendments to the protocol), raw data, records, and specimens against the final report to substantiate that protocol requirements were met and that findings were fully and accurately reported. For each study audited, the study records should be reviewed for quality to ensure that data are the following:
    a. Attributable—the raw data can be traced, by signature or initials and date to the individual observing and recording the data. Should more than one individual observe or record the data, that fact should be reflected in the data.
    b. Legible—the raw data are readable and recorded in a permanent medium. If changes are made to original entries, the changes
        i. Must not obscure the original entry
        ii. Indicate the reason for change
        iii. Must be signed or initialed and dated by the person making the change
    c. Contemporaneous—the raw data are recorded at the time of the observation.
    d. Original—the first recording of the data.
    e. Accurate—the raw data are true and complete observations. For data entry forms that require the same data to be entered repeatedly, all fields should be completed, or a written explanation for any empty fields should be retained with the study records.
23. General
    a. *Determine* if there were any significant changes in the facilities, operations, and QAU functions other than those previously reported.
    b. *Determine* whether the equipment used was inspected, standardized, and calibrated prior to, during, and after use in the study. If equipment malfunctioned, review the remedial action, and ensure that the final report addresses whether the malfunction affected the study.
    c. Determine if approved SOPs existed during the conduct of the study. Compare the content of the protocol with the requirements in 21 CFR. Review the final report for the study director's dated signature and the QAU statement as required in 21 CFR 58.35(b)(7).
24. Protocol versus final report. Study methods described in the final report should be compared against the protocol and the SOPs to confirm those requirements were met.
25. Final report versus raw data. The audit should include a detailed review of records, memorandum, and other raw data to confirm that the findings in the final report completely and accurately reflect the raw data. Representative samples of raw data should be audited against the final report.

26. Samples. Collection of samples should be considered when the situation under audit or surveillance suggests that the facility had, or is having, problems in the area of characterization, stability, storage, contamination, or dosage preparation.
27. Inspectional observations. An FDA 483 listing inspectional observations will be issued under this program. Findings should not be listed on the FDA 483 if in the opinion of the field investigator,
    a. The findings are problems that have been observed and corrected by the firm through its internal procedures
    b. The findings are minor and are one-time occurrences that have no impact on the firm's operations, study conduct, or data integrity
    c. Findings that are not considered significant enough to be listed on the FDA 483 may be discussed with the firm's management. Such discussions must be reported in the EIR.

9 Fed Bioequivalence Studies

INTRODUCTION

There are specific recommendations on studying the effect of food on the bioavailability of drugs; these considerations are important in conducting bioavailability and bioequivalence trials, particularly the oral solid dosage forms.

BACKGROUND

Food-effect bioavailability studies are usually conducted for new drugs and drug products during the investigational new drug (IND) period to assess the effects of food on the rate and extent of absorption of a drug when the drug product is administered shortly after a meal (fed conditions), as compared to administration under fasting conditions. Fed bioequivalence studies, on the other hand, are conducted for abbreviated new drug applications (ANDAs) to demonstrate their bioequivalence to the reference listed drug (RLD) under fed conditions.

Potential Mechanisms of Food Effects on Bioavailability

Food can change the bioavailability of a drug and can influence the bioequivalence between test and reference products. Food effects on bioavailability can have clinically significant consequences. Food can alter bioavailability by various means, including

- Delaying gastric emptying
- Stimulating bile flow
- Changing gastrointestinal (GI) pH
- Increasing splanchnic blood flow
- Changing luminal metabolism of a drug substance
- Physically or chemically interacting with a dosage form or a drug substance

Food effects on bioavailability are generally greatest when the drug product is administered shortly after a meal is ingested. The nutrient and caloric contents of the meal, the meal volume, and the meal temperature can cause physiological changes in the GI tract in a way that affects drug product transit time, luminal dissolution, drug permeability, and systemic availability. In general, meals that are high in total calories and fat content are more likely to affect the GI physiology and thereby result in a larger effect on the bioavailability of a drug substance or drug product. The FDA recommends use of high-calorie and high-fat meals during food-effect bioavailability and fed bioequivalence studies.

Food Effects on Drug Products

Administration of a drug product with food may change the bioavailability by affecting either the drug substance or the drug product. In practice, it is difficult to determine the exact mechanism by which food changes the bioavailability of a drug product without performing specific mechanistic studies. Important food effects on bioavailability are least likely to occur with many rapidly dissolving, immediate-release drug products containing highly soluble and highly permeable drug substances (biopharmaceutics classification system [BCS] Class I) because absorption of the drug substances in Class I is usually pH and site independent and thus insensitive to differences in dissolution. However, for some drugs in this class, food can influence bioavailability when there is a high first-pass effect, extensive adsorption, complexation, or instability of the drug substance in the GI tract. In some cases, excipients or interactions between excipients and the food-induced changes in gut physiology can contribute to these food effects and influence the demonstration of bioequivalence. For rapidly dissolving formulations of BCS Class I drug substances, food can affect C_{max} and the time at which this occurs (t_{max}) by delaying gastric emptying and prolonging intestinal transit time. However, we expect the food effect on these measures to be similar for test and reference products in fed bioequivalence studies.

For other immediate-release drug products (BCS Class II, III, and IV) and for all modified-release drug products, food effects are most likely to result from a more complex combination of factors that influence the in vivo dissolution of the drug product and/or the absorption of the drug substance. In these cases, the relative direction and magnitude of food effects on formulation bioavailability and the effects on the demonstration of bioequivalence are difficult, if not impossible, to predict without conducting a fed bioequivalence study.

RECOMMENDATIONS FOR FOOD-EFFECT BIOAVAILABILITY AND FED BIOEQUIVALENCE STUDIES

Given later are the recommendations on when food-effect bioavailability studies should be conducted as part of INDs and new drug applications (NDAs) and when fed bioequivalence studies should be conducted as part of ANDAs. For postapproval changes in an approved immediate- or modified-release drug product that requires in vivo redocumentation of bioequivalence under fasting conditions, fed bioequivalence studies are generally unnecessary.

Immediate-Release Drug Products

INDs/NDAs

The U.S. FDA recommends that a food-effect bioavailability study be conducted for all new chemical entities (NCEs) during the IND period.

Food-effect bioavailability studies should be conducted early in the drug development process to guide and select formulations for further development. Food-effect bioavailability information should be available to design clinical safety and efficacy studies and to provide information for the Clinical Pharmacology and/or Dosage and Administration sections of product labels. If a sponsor makes changes in components, composition, and/or method of manufacture in the clinical trial formulation prior to approval, bioequivalence should be demonstrated between the to-be-marketed formulation and the clinical trial formulation.

Sponsors may wish to use relevant principles described in the guidance for industry on *SUPAC-IR: Immediate Release Solid Oral Dosage Forms*, subtitle *Scale-Up and Postapproval Changes: Chemistry, Manufacturing, and Controls, In Vitro Dissolution Testing, and In Vivo Bioequivalence Documentation* (SUPAC-IR guidance), to determine if in vivo bioequivalence studies are recommended. These bioequivalence studies, if indicated, should generally be conducted under fasting conditions.

ANDAs

In addition to a bioequivalence study under fasting conditions, we recommend a bioequivalence study under fed conditions for all orally administered immediate-release drug products, with the following exceptions:

- When both test product and RLD are rapidly dissolving, have similar dissolution profiles, and contain a drug substance with high solubility and high permeability (BCS Class I)

- When the Dosage and Administration section of the RLD label states that the product should be taken only on an empty stomach
- When the RLD label does not make any statements about the effect of food on absorption or administration

Modified-Release Drug Products

The FDA recommends that food-effect bioavailability and fed bioequivalence studies be performed for all modified-release dosage forms.

INDs/NDAs

The FDA recommends a study comparing the bioavailability under fasting and fed conditions for all orally administered modified-release drug products.

When changes occur in components, composition, and/or method of manufacture between the to-be-marketed formulation and the primary clinical trial material, the sponsor may wish to use relevant principles described in the guidance for industry on *SUPAC-MR: Modified Release Solid Oral Dosage Forms, subtitle Scale-Up and Postapproval Changes: Chemistry, Manufacturing, and Controls, In Vitro Dissolution Testing and In Vivo Bioequivalence Documentation* (SUPAC-MR guidance), to determine if documentation of in vivo bioequivalence is recommended. These bioequivalence studies, if indicated, should generally be conducted under fasting conditions.

ANDAs

In addition to a bioequivalence study under fasting conditions, a bioequivalence study under fed conditions should be conducted for all orally administered modified-release drug products.

STUDY CONSIDERATIONS

Given in the following are general considerations for designing food-effect bioavailability and fed bioequivalence studies. A sponsor may propose alternative study designs and data analyses. The scientific rationale and justification for these study designs and analyses should be provided in the study protocol. Sponsors may choose to conduct additional studies for a better understanding of the drug product and to provide optimal labeling statements for dosage and administration (e.g., different meals and different times of drug intake in relation to meals). In studying modified-release dosage forms, consideration should be given to the possibility that coadministration with food can result in *dose dumping*, in which the complete dose may be more rapidly released from the dosage form than intended, creating a potential safety risk for the study subjects.

General Design

The FDA recommends a randomized, balanced, single-dose, two-treatment (fed vs. fasting), two-period, two-sequence crossover design for studying the effects of food on the bioavailability of either an immediate-release or a modified-release drug product. The formulation to be tested should be administered on an empty stomach (fasting condition) in one period and following a test meal (fed condition) in the other period. The FDA recommends a similar, two-treatment, two-period, two-sequence crossover design for a fed bioequivalence study except that the treatments should consist of both test and reference formulations administered following a test meal (fed condition). An adequate washout period should separate the two treatments in food-effect bioavailability and fed bioequivalence studies.

Subject Selection

Both food-effect bioavailability and fed bioequivalence studies can be carried out in healthy volunteers drawn from the general population. Studies in the patient population are also appropriate if safety concerns preclude the enrollment of healthy subjects. A sufficient number of subjects should complete the study to achieve adequate power for a statistical assessment of food effects on bioavailability to claim an absence of food effects or to claim bioequivalence in a fed bioequivalence study. A minimum of 12 subjects should complete the food-effect bioavailability and fed bioequivalence studies.

Dosage Strength

In general, the highest strength of a drug product intended to be marketed should be tested in food-effect bioavailability and fed bioequivalence studies. In some cases, clinical safety concerns can prevent the use of the highest strength and warrant the use of lower strengths of the dosage form. For ANDAs, the same lot and strength used in the fasting bioequivalence study should be tested in the fed bioequivalence study. For products with multiple strengths in ANDAs, if a fed bioequivalence study has been performed on the highest strength, bioequivalence determination of one or more lower strengths can be waived based on dissolution profile comparisons.

Test Meal

The FDA recommends that food-effect bioavailability and fed bioequivalence studies be conducted using meal conditions that are expected to provide the greatest effects on GI physiology so that systemic drug availability is maximally affected. A high-fat (approximately 50% of total caloric content of the meal) and high-calorie (approximately 800–1000 cal) meal is recommended as a test meal for food-effect bioavailability and fed bioequivalence studies. This test meal should derive approximately 150, 250, and 500–600 cal from protein, carbohydrate, and fat, respectively. The caloric breakdown of the test meal should be provided in the study report. If the caloric breakdown of the meal is significantly different from the one described earlier, the sponsor should provide a scientific rationale for this difference. In NDAs, it is recognized that a sponsor can choose to conduct food-effect bioavailability studies using meals with different combinations of fats, carbohydrates, and proteins for exploratory or label purposes. However, one of the meals for the food-effect bioavailability studies should be the high-fat, high-calorie test meal described earlier.

An example test meal would be two eggs fried in butter, two strips of bacon, two slices of toast with butter, four ounces of hash brown potatoes, and eight ounces of whole milk. Substitutions in this test meal can be made as long as the meal provides a similar amount of calories from protein, carbohydrate, and fat and has comparable meal volume and viscosity.

Administration

Fasted treatments: Following an overnight fast of at least 10 h, subjects should be administered the drug product with 240 mL (8 fluid ounces) of water. No food should be allowed for at least 4 h post-dose. Water can be allowed as desired except for 1 h before and after drug administration. Subjects should receive standardized meals scheduled at the same time in each period of the study.

Fed treatments: Following an overnight fast of at least 10 h, subjects should start the recommended meal 30 min prior to administration of the drug product. Study subjects should eat this meal in 30 min or less; however, the drug product should be administered 30 min after start of the meal.

The drug product should be administered with 240 mL (8 fluid ounces) of water. No food should be allowed for at least 4 h post-dose. Water can be allowed as desired except for 1 h before and after drug administration. Subjects should receive standardized meals scheduled at the same time in each period of the study.

Sample Collection

For both fasted and fed treatment periods, timed samples in biological fluid, usually plasma, should be collected from the subjects to permit characterization of the complete shape of the plasma concentration-time profile for the parent drug. It may be advisable to measure other moieties in the plasma, such as active metabolites, and sponsors should refer to the guidance on *Bioavailability and Bioequivalence Studies for Orally Administered Drug Products—General Considerations* for recommendations on these issues. Consideration should be given to the possibility that coadministration of a dosage form with food can alter the time course of plasma drug concentrations so that fasted and fed treatments can have different sample collection times.

DATA ANALYSIS AND LABELING

Food-effect bioavailability studies may be exploratory and descriptive, or a sponsor may want to use a food-effect bioavailability study to make a label claim. The following exposure measures and pharmacokinetic parameters should be obtained from the resulting concentration-time curves for the test and reference products in food-effect bioavailability and fed bioequivalence studies:

- Total exposure or area under the concentration-time curve ($AUC_{0\text{-inf}}$, AUC_{0-t})
- Peak exposure (C_{max})
- Time to peak exposure (t_{max})
- Lag time (t_{lag}) for modified-release products, if present
- Terminal elimination half-life
- Other relevant pharmacokinetic parameters

Individual subject measurements, as well as summary statistics (e.g., group averages, standard deviations, coefficients of variation), should be reported. An equivalence approach is recommended for food-effect bioavailability (to make a claim of no food effects) and fed bioequivalence studies, analyzing data using an average criterion. Log transformation of exposure measurements (AUC and C_{max}) prior to analysis is recommended. The 90% confidence interval (CI) for the ratio of population geometric means between test and reference products should be provided for $AUC_{0\text{-inf}}$, AUC_{0-t}, and C_{max}. For IND or NDA food-effect bioavailability studies, the fasted treatment serves as the reference. For ANDA fed bioequivalence studies, the RLD administered under fed condition serves as the reference treatment.

The effect of food on the absorption and bioavailability of a drug product should be described in the Clinical Pharmacology section of the labeling. In addition, the Dosage and Administration section of the labeling should provide instructions for drug administration in relation to food based on clinical relevance (i.e., whether or not the changes in systemic exposure caused by coadministration with food results in safety or efficacy concerns or when there is no important change in systemic exposure but there is a possibility that the drug substance causes GI irritation when taken without food).

For an NDA, an absence of food effect on bioavailability is not established if the 90% CI for the ratio of population geometric means between fed and fasted treatments, based on log-transformed data, is not contained in the equivalence limits of 80%–125% for either $AUC_{0\text{-inf}}$ (AUC_{0-t} when appropriate) or C_{max}. When the 90% CI fails to meet the limits of 80%–125%, the sponsor should provide specific recommendations on the clinical significance of the food effect based on what is known from the total clinical database about dose–response (exposure–response) and/or pharmacokinetic–pharmacodynamic relationships of the drug under study. The clinical relevance of any difference in t_{max} and t_{lag} should also be indicated by the sponsor. The results of the food-effect bioavailability study should be reported factually in the Clinical Pharmacology section of the labeling and should form the basis for making label recommendations (e.g., *take only on an empty stomach*) in the Dosage and Administration section of the labeling. The following are examples of language for the package insert:

> A food-effect study involving administration of (the drug product) to healthy volunteers under fasting conditions and with a high-fat meal indicated that the C_{max} and AUC were increased 57% and 45%, respectively, under fed conditions. This increase in exposure can be clinically significant, and therefore (the drug) should be taken only on an empty stomach (1 h before or 2 h after a meal).
>
> A food-effect study involving administration of (the drug product) to healthy volunteers under fasting conditions and with a high-fat meal indicated that the C_{max} was decreased 15% while the AUC remained unchanged. This decrease in exposure is not clinically significant, and therefore (the drug) could be taken without regard to meals.

An absence of food effect on bioavailability is indicated when the 90% CI for the ratio of population geometric means between fed and fasted treatments, based on log-transformed data, is contained in the equivalence limits of 80%–125% for AUC_{0-inf} (AUC_{0-t} when appropriate) and C_{max}. In this case, a sponsor can make a specific claim in the Clinical Pharmacology or Dosage and Administration section of the label that no food effect on bioavailability is expected provided that the t_{max} differences between the fasted and fed treatments are not clinically relevant. The following is an example of language for the package insert:

> The C_{max} and AUC data from a food-effect study involving administration of (the drug product) to healthy volunteers under fasting conditions and with a high-fat meal indicated that exposure to the drug is not affected by food. Therefore, (the drug product) may be taken without regard to meals.

For an ANDA, bioequivalence of a test product to the RLD product under fed conditions is concluded when the 90% CI for the ratio of population geometric means between the test and RLD product, based on log-transformed data, is contained in the bioequivalence limits of 80%–125% for AUC and C_{max}. Although no criterion applies to t_{max}, the t_{max} values for the test and reference products are expected to be comparable based on clinical relevance. The conclusion of bioequivalence under fed conditions indicates that with regard to food, the language in the package insert of the test product can be the same as the reference product.

OTHER CONSIDERATIONS

Sprinkles

In NDAs, the labeling of certain drug products (e.g., controlled-release capsules containing beads) can recommend that the product be sprinkled on soft foods, such as applesauce, and swallowed without chewing. For the labeling to indicate that the drug product can be sprinkled on soft foods, additional in vivo relative bioavailability studies should be performed by sprinkling the product on the soft foods to be listed in the labeling (test treatment) and comparing it to the product administered in the intact form (reference treatment), then administering both on an empty stomach.

In ANDAs, bioequivalence of the test to the RLD is demonstrated in a single-dose crossover study. Both treatments should be sprinkled on one of the soft foods mentioned in the labeling, usually applesauce. The bioequivalence data should be analyzed using average bioequivalence, and the 90% CI criteria should be used to declare bioequivalence.

Special Vehicles

For NDAs, the labeling for certain oral solution products (e.g., cyclosporine oral solution, modified) recommends that the solution be mixed with a beverage prior to administration. The bioavailability of these products can change when mixed with different beverages due to the formation of complex mixtures and other physical–chemical and/or physiological factors.

In ANDAs, bioequivalence of the test to the RLD is demonstrated in a single-dose crossover study. Both treatments should be mixed with one of the beverages mentioned in the labeling. Sponsors should provide evidence that bioequivalence differences would not be expected from the use of other listed vehicles. The bioequivalence data should be analyzed using average BE, and the 90% CI criteria should be used to declare bioequivalence.

10 Topical Drugs

For topical dermatological drug products, pharmacokinetic (PK) measurements in blood, plasma, and/or urine are usually not feasible to document bioequivalence (BE) because topical dermatological products generally do not produce measurable concentrations in extracutaneous biological fluids. The BE determination for these products is thus often based on pharmacodynamic or clinical studies. An additional approach is to document BE through reliance on measurement of the active moiety(ies) in the stratum corneum. This approach is termed dermatopharmacokinetics (DPK). Although measurement of the active moiety(ies) in blood or urine is not regarded as an acceptable measurement of BE for dermatological drug products, it may be used to measure systemic exposure.

INACTIVE INGREDIENTS

During the investigational new drug (IND) process for a new drug application (NDA), the safety of inactive ingredients in a topical drug product should be documented by specific studies or may be based on a prior history of successful use in the same amount administered via the same route of administration in an approved product. The requisite safety studies to establish the safety of a new excipient during the IND process should be discussed with appropriate review staff at the Food and Drug Administration (FDA). For an abbreviated NDA (ANDA), the safety of inactive ingredients in an ANDA can be based on a prior history of successful use in an NDA or ANDA. If the inactive ingredients in an ANDA are not the same as the reference listed drug (RLD), the applicant should demonstrate to the agency that the change(s) does not affect the safety and/or efficacy of the proposed drug product. In some instances, a comparative bioavailability (BA) study will satisfy this recommendation. If preclinical or clinical studies are needed to demonstrate the safety of inactive ingredients(s) in the generic drug product, the ANDA may not be approved. In this circumstance, the applicant may wish to resubmit their application as an NDA under the provisions of 505(b)(1) or (b)(2) of the Act.

WAIVER OF BIOEQUIVALENCE

In accordance with 21 CFR 314.94(a)(9)(v), generally, the test (generic) product intended for topical use must contain the same inactive ingredients as the RLD. For all topical drug products intended for marketing under an abbreviated application, documentation of in vivo BE is required under 21 CFR 320.21(b). For a topical solution drug product, in vivo BE may be waived if the inactive ingredients in the product are qualitatively identical and quantitatively essentially the same compared to the listed drug. In this setting, quantitatively *essentially the same* means that the amount/concentration of the inactive ingredient(s) in the test product cannot differ by more than ±5% of the amount/concentration of the listed drug. Where a test solution differs qualitatively or quantitatively from the listed drug, in vivo BE may be waived, provided the sponsor submits evidence that the difference does not affect safety and/or efficacy of the product at the time a waiver is requested.

BIOEQUIVALENCE APPROACHES

Comparative clinical trials are generally difficult to perform, highly variable, and insensitive. For these reasons, other approaches, such as DPK or pharmacodynamic, may be used for BE determination.

Dermatopharmacokinetic Approaches

The DPK approach is comparable to a blood, plasma, and urine PK approach applied to the stratum corneum. DPK encompasses drug concentration measurements with respect to time and provides information on drug uptake, apparent steady-state levels, and drug elimination from the stratum corneum based on a stratum corneum concentration–time curve.

When applied to diseased skin, topical drug products induce one or more therapeutic responses, where onset, duration, and magnitude depend on the relative efficiency of three sequential processes, namely, (1) the release of the drug from the dosage form, (2) penetration of the drug through the skin barrier, and (3) generation of the desired pharmacological effect. Because topical products deliver the drug directly to or near the intended site of action, measurement of the drug uptake into and drug elimination from the stratum corneum can provide a DPK means of assessing the BE of two topical drug products. Presumably, two formulations that produce comparable stratum corneum concentration–time curves may be BE, just as two oral formulations are judged BE if they produce comparable plasma concentration–time curves. Even though the target site for topical dermatological drug products in some instances may not be the stratum corneum, the topical drug must still pass through the stratum corneum, except in instances of damage, to reach deeper sites of action. In certain instances, the stratum corneum itself is the site of action. For example, in fungal infections of the skin, fungi reside in the stratum corneum, and therefore DPK measurement of an antifungal drug in the stratum corneum represents direct measurement of drug concentration at the site of action. In instances where the stratum corneum is disrupted or damaged, in vitro drug release may provide additional information toward the BE assessment. In this context, the drug release rate may reflect drug delivery directly to the dermal skin site without passage through the stratum corneum. For antiacne drug products, target sites are the hair follicles and sebaceous glands. In this setting, the drug diffuses through the stratum corneum, epidermis, and dermis to reach the site of action. The drug may also follow follicular pathways to reach the sites of action. The extent of follicular penetration depends on the particle size of the active ingredient if it is in the form of a suspension. Under these circumstances, the DPK approach is still expected to be applicable because studies indicate a positive correlation between the stratum corneum and follicular concentrations. Although the exact mechanism of action for some dermatological drugs is unclear, the DPK approach may still be useful as a measure of BE because it has been demonstrated that the stratum corneum functions as a reservoir, and stratum corneum concentration is a predictor of the amount of drug absorbed.

For reasons thus cited, DPK principles should be generally applicable to all topical dermatological drug products including antifungal, antiviral, antiacne, antibiotic, corticosteroid, and vaginally applied drug products. The DPK approach can thus be the primary means to document BA/BE. Additional information, such as comparative in vitro release data and particle size distribution of the active ingredient between the RLD and the test product, may provide additional supportive information. Generally, BE determinations using DPK studies are performed in healthy subjects because the skin where disease is present demonstrates high variability and changes over time. The use of healthy subjects is consistent with similar use in BE studies for oral drug products.

A DPK approach is not generally applicable (1) when a single application of the dermatological preparation damages the stratum corneum, (2) for otic preparations except when the product is intended for otic inflammation of the skin, and (3) for ophthalmic preparations because the cornea is structurally different from the stratum corneum. The following three sections of the guidance provide general procedures for conducting a BA/BE study using DPK methodology.

Performance and Validation of the Skin-Stripping Technique

DPK studies should include validation of both analytical methods and the technique of skin stripping. Since the DPK approach involves two components of validation (sampling and analytical method), overall DPK variability may be greater than with other methodologies. For analytical methods, levels of accuracy, precision, sensitivity, specificity, and reproducibility should be documented according to established procedures. Although the forearm, back, thigh, or other parts of the body can be used for skin-stripping studies, most studies are conducted on the forearm, for reasons of convenience. Care should be taken to avoid any damage with physical, mechanical, or chemical irritants (e.g., soaps, detergents, agents). Usual hydration and environmental conditions should be maintained. After washing prior to treatment, sufficient time, preferably 2 h, should be allowed to normalize the skin surface. Detailed and workable standard operating procedures (SOPs) for area and amount of drug application, excess drug removal, and skin-stripping methodology should be developed. The product's stability during the course of the study should be established. If the product is unstable, the rate and extent of degradation in situ over the period should be determined accurately so that a correction factor may be applied. Skin on both left and right arms of healthy subjects may be used to provide eight or more sites per arm. The size of the skin-stripping area is important to allow collection of a sufficient drug in a sample to achieve adequate analytical detectability. Inter- and intra-arm variability should be assessed, and the treatment sites should be randomized appropriately. If a sponsor or applicant is using multiple investigators to conduct a single study, the reproducibility of skin-stripping data between the investigators should be established. Either of the following approaches is recommended:

- A dose–response relationship between the drug concentration in the applied dosage form and the drug concentration in the stratum corneum should be established using the skin-stripping method. A DPK dose–response relationship is analogous to a dose proportionality study performed with solid oral dosage forms. This type of study can be readily performed using three different strengths of the formulations. These can be marketed or specially manufactured products. Alternatively, a solution of the active drug representing three concentrations can be prepared for this purpose. The amount of drug in the stratum corneum at the end of a specified time interval, such as 3 h, can provide a dose–response relationship.
- The skin-stripping method should be capable of detecting differences of ±25% in the strength of a product. This can be determined by applying different concentrations (e.g., 75%, 100%, 125%) of a test dosage form such as a simple solution to the skin surface for a specified exposure time such as 3 h, executing the skin-stripping method, and performing the appropriate statistical tests comparing the strength applied to the measured drug concentration in the stratum corneum.

Using the reference product, the approximate minimum time required for drug to reach saturation level in the stratum corneum should be determined. This study establishes the time point at which the elimination phase of the study may be initiated.

The drug concentration–time profile may vary with the drug, the drug potency class, formulation, subject, sites of application, circadian rhythm, ambient temperature, and humidity. These factors should be considered and controlled as necessary.

Circadian rhythms may be present and may affect the measurement of skin-stripping drug concentration if the drug is also an endogenous chemical (e.g., corticosteroid or retinoic acid). In such circumstances, the baseline concentration of the endogenous compound should be measured over time from sites where no drug product has been applied.

Sample Pilot Study

The reference drug product is randomly applied to eight sites on one forearm, with skin stripping performed at incremental times after application (e.g., 15, 30, 60, and 180 min) (Figure 10.1).

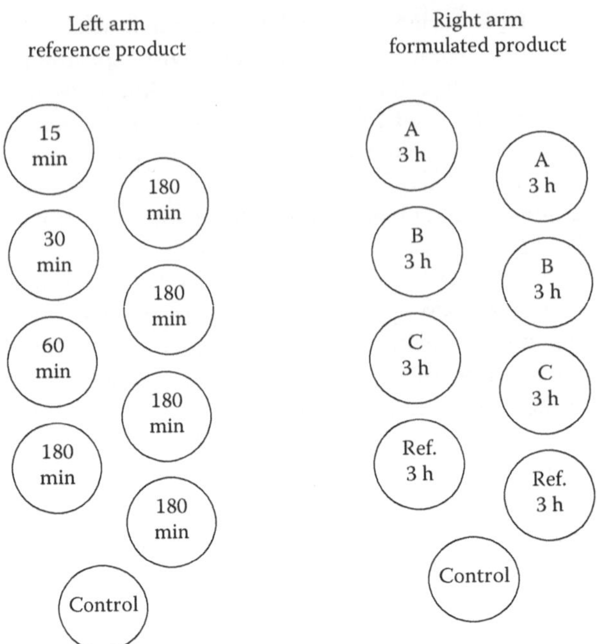

FIGURE 10.1 Schematic for drug application and removal sites for pilot study. A, B, and C represent three concentrations of the drug product or drug solution.

One site is used for each time point. Four additional sites at 180 min on the same arm should be assessed to provide a total of five replicates for the same time point. An additional site with no application of a drug product should be sampled as a control, yielding a total of nine sampling sites. The contralateral forearm may be used to assess dose–response and sensitivity relationships by applying at least three concentrations of the drug product or simple drug solution for 180 min in duplicates. Two additional applications of the reference drug product on the same arm should be tested for 180 min as well to provide additional information about inter- and intra-arm variability and reproducibility. A control site with no drug application should also be included for a total of nine sites on the contralateral arm. The pilot study should be carried out in at least six subjects. Stratum corneum samples are removed according to procedures described below and analyzed for drug concentration. Standard procedures should be followed in all elements of the study and should be carried through all subsequent studies.

DPK Bioequivalence Study Protocol

Protocol and Subject Selection

Healthy volunteers with no history of previous skin disease or atopic dermatitis and with a healthy, homogeneous forearm (or other) skin areas sufficient to accommodate at least eight (8) treatment and measurement sites (time points) should be recruited. The number of subjects to be entered may be obtained from power calculations using intra- and intersubject variability from the pilot study. Because skin stripping is highly sensitive to specific study site factors, care should be taken to perfecting the technique and enrolling a sufficient number of subjects. The following study design is based on a crossover study design, where the crossover occurs at the same time using both arms of a single subject. A crossover design in which subjects are studied on two different occasions may also be employed. If this design is employed, at least 28 days should be allowed to rejuvenate the harvested stratum corneum.

Application and Removal of Test and Reference Products

The treatment areas are marked using a template without disturbing or injuring the stratum corneum/skin. The size of the treatment area will depend on multiple factors including drug strength, analytical sensitivity, the extent of drug diffusion, and exposure time. The stratum corneum is highly sensitive to certain environmental factors. To avoid bias and to remain within the limits of experimental convenience and accuracy, the treatment sites and arms should be randomized. Uptake, steady-state, and elimination phases, as described in more detail later, may be randomized between the right and left arms in a subject. Exposure time points in each phase may be randomized among various sites on each arm. The test and reference products for a particular exposure time point may be applied on adjacent sites to minimize differences. Test and reference products should be applied concurrently on the same subjects according to a SOP that has been previously developed and validated. The premarked sites are treated with predetermined amounts of the products (e.g., 5 mg/cm^2) and covered with a nonocclusive guard. Occlusion is used only if recommended in product labeling. Removal of the drug product is performed according to SOPs at the designated time points, using multiple cotton swabs or Q-tips with care to avoid stratum corneum damage. In case of certain oily preparations such as ointments, washing the area with a mild soap may be needed before skin stripping. If washing is carried out, it should be part of an SOP.

Sites and Duration of Application

The BE study should include measurements of drug uptake into the stratum corneum and drug elimination from the skin. Each of these elements is important to establish bioavailability and/or BE of two products, and each may be affected by the excipients present in the product. A minimum of eight sites should be employed to assess uptake/elimination from each product. The time to reach steady state in the stratum corneum should be used to determine timing of samples. For example, if the drug reaches steady state in 3 h, 0.25, 0.5, 1, and 3 h posttreatment may be selected to determine uptake and 4, 6, 8, and 24 h may be used to assess elimination. A *zero* time point (control site away from test sites) on each subject should be selected to provide baseline data. If the test/reference drug products are studied on both forearms, randomly selected sites on one arm may be designated to measure drug uptake/steady state. Sites on the contralateral arm may then be designated to measure drug elimination. During drug uptake, both the excess drug removal and stratum corneum stripping times are the same so that the stratum corneum stripping immediately follows the removal of the excess drug. In the elimination phase, the excess drug is removed from the sites at the steady-state time point, and the stratum corneum is harvested at succeeding times over 24 h to provide an estimate of an elimination phase (Figure 10.2).

Collection of Sample

Skin stripping proceeds first with the removal of the first 1–2 layers of stratum corneum with two adhesive tapes strip/disk applications, using a commercially available product (e.g., D-Squame, Transpore). These first two tape strips contain the generally unabsorbed, as opposed to penetrated or absorbed, drug and therefore should be analyzed separately from the rest of the tape strips. The remaining stratum corneum layers from each site are stripped at the designated time intervals. This is achieved by stripping the site with an additional 10 adhesive tape strips. All 10 tape strips obtained from a given time point are combined and extracted, with drug content determined using a validated analytical method. The values are generally expressed as amounts/area (e.g., ng/cm) to maintain uniformity in reported values. Data may be computed to obtain full drug concentration–time profiles, C_{max-ss}, t_{max-ss}, and AUCs for the test and reference products.

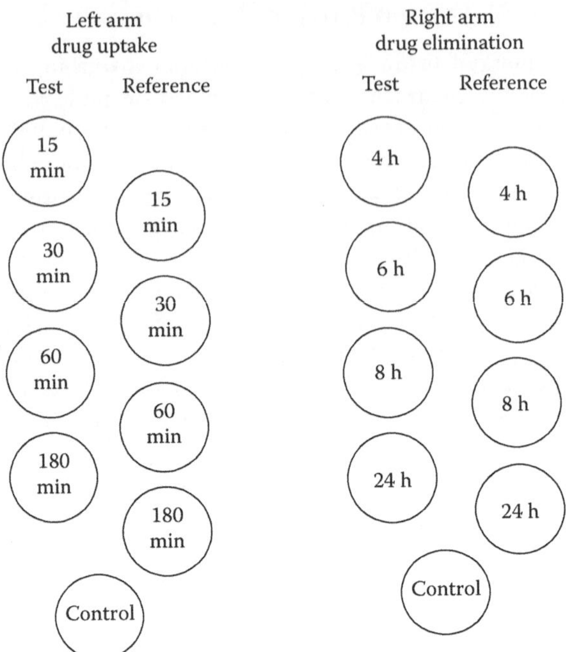

FIGURE 10.2 Schematic for drug uptake and drug elimination for BE study.

Procedure for Skin Stripping

The general test procedures in either the pilot study or the pivotal BA/BE study are summarized in the following:

To assess drug uptake:

- Apply the test and/or reference drug products concurrently at multiple sites.
- After an appropriate interval, remove the excess drug from a specific site by wiping three times lightly with a tissue or cotton swab.
- Using information from the pilot study, determine the appropriate times of sample collection to assess drug uptake.
- Repeat the application of adhesive tape two times, using uniform pressure, discarding these first two tape strips.
- Continue stripping at the same site to collect ten more stratum corneum samples.
- Care should be taken to avoid contamination with other sites.
- Repeat the procedure for each site at other designated time points.
- Extract the drug from the combined ten skin stripping and determine the concentration using a validated analytical method.
- Express the results as amount of drug per square cm treatment area of the adhesive tape.

To assess drug elimination:

- Apply the test and reference drug product concurrently at multiple sites chosen based on the results of the pilot study. Allow sufficient exposure period to reach apparent steady-state level.
- Remove any excess drug from the skin surface as described previously, including the first two skin strippings.
- Collect skin-stripping samples using ten successive tape strips at time intervals based on the pilot study and analyze them for drug content.

Metrics and Statistical Analyses

A plot of stratum corneum drug concentration versus a time profile should be constructed to yield stratum corneum metrics of C_{max}, t_{max}, and AUC. The two one-sided hypotheses at the $p = 0.05$ level of significance should be tested for AUC and C_{max} by constructing the 90% confidence interval (CI) for the ratio between the test and reference averages. Individual subject parameters, as well as summary statistics (average, standard deviation, coefficient of variation, 90% CI), should be reported. For the test product to be bioequivalent, the 90% CI for the ratio of means (population geometric means based on log-transformed data) of test and reference treatments should fall within 80%–125% for AUC and 70%–143% for C_{max}. Alternate approaches in the calculation of metrics and statistics are acceptable with justification.

PHARMACODYNAMIC APPROACHES

Sometimes, topically applied dermatological drug products produce direct/indirect pharmacodynamic responses that may be useful to measure BE. For example, topically applied corticosteroids produce a vasoconstrictor effect that results in skin blanching. This pharmacodynamic response has been correlated with corticosteroid potency and efficacy. Based on this pharmacodynamic response, FDA issued a guidance entitled *Topical Dermatological Corticosteroids: In Vivo Bioequivalence* (June 1995). The guidance recommends that a pilot study be conducted to assess the dose–response characteristics of the corticosteroid followed by a formal study to assess BE. Topically applied retinoid produces transepidermal water loss that may be used as a pharmacodynamic measure to assess BE.

In Vitro Release Approaches (Lower Strength)

Usually only one strength of a topical dermatological drug product is available although sometimes two or, rarely, three strengths may be marketed. When multiple strengths are available, a standard practice is to create lower strengths by altering the percentage of active ingredients without otherwise changing the formulation or its manufacturing process. Topical dermatological drug products usually contain relatively small amounts of the active drug substance, usually ≤5% and frequently ≤1%. In this setting, changes in the active ingredient may have little impact on the overall formulation.

Safety and efficacy should be documented for all strengths of topical drug products in the NDA submissions. Using some of the approaches suggested in this guidance, BA may also be documented for the highest strength. For lower strengths, where documentation of BA is considered important, this guidance suggests that in vitro release may be performed. Similarly, for an ANDA, when BE has been documented for the highest strength, in vitro release may also be used to waive in vivo studies to assess BE between these lower strengths and the corresponding strengths of the RLD. If this approach suggests bioinequivalence, further studies may be important.

To support the BE of lower strengths in an ANDA, the following conditions are important:

- Formulations of the two strengths should differ only in the concentration of the active ingredient and equivalent amount of the diluent.
- No differences should exist in manufacturing process and equipment between the two strengths.
- For an ANDA, the RLD should be marketed at both higher and lower strengths.
- For an ANDA, the higher strength of the test product should be BE to the higher strength of RLD.

In vitro drug release rate studies should be measured under the same test conditions for all strengths of both the test and RLD products. The in vitro release rate should be compared between

(1) the RLD at both the higher (RHS, reference high strength) and lower strengths (RLS, reference low strength) and (2) the test (generic) products at both higher (THS, test high strength) and lower strengths (TLS, test low strength). Using the in vitro release rate, the following ratios and comparisons should be made:

Release rate of RHS/Release rate of RLS ≈ Release rate of THS/Release rate of TLS

The ratio of the release rates of the two strengths of the test products should be about the same as the ratio of the release rate of reference products, that is,

$$\frac{(\text{Release rate of RHS} \times \text{Release rate of TLS})}{(\text{Release rate of RLS} \times \text{Release rate of THS})} \approx 1$$

Using appropriate statistical methods, the standard BE interval (80–120) for a lower strength comparison of test and reference products should be used.

After approval, a sponsor may wish to develop an intermediate strength of a topical dermatological drug product when two strengths have been approved and are in the marketplace. In this case, the in vitro release rate of the intermediate strength should fall between the in vitro release rates of the upper and lower strengths. Modifications of the approach described in this section of the guidance can thus be applied, provided all strengths differ only in the amount of active ingredient and do not differ in manufacturing processes and equipment.

In Vitro Release: Extension of the Methodology

Drug release from semisolid formulations is a property of the dosage form. Current scientific consensus is that in vitro release is an acceptable regulatory measure to signal inequivalence in the presence of certain formulation and manufacturing changes. With suitable validation, in vitro release may be used to assess batch-to-batch quality, replacing a series of tests that in the aggregate assess product quality and drug release (e.g., particle size determination, viscosity, and rheology). Because topical dosage forms are complex dosage forms, manufacturers should optimize the in vitro release test procedure for their product in a manner analogous to the use of in vitro dissolution to assess the quality of extended release products from batch to batch. In addition, in vitro release might be used in a sponsor-specific comparability protocol to allow more extensive postapproval changes in formulation and/or manufacturing, provided that BE between two products representing the extremes of the formulation and manufacturing changes has been shown to be bioequivalent, using approaches recommended earlier in this document.

SYSTEMIC EXPOSURE STUDIES

To ensure safety, and, when appropriate, comparable safety, information on systemic exposure is important for certain types of topical dermatological drug products, such as retinoid and high-potency corticosteroids. The degree of systemic exposure for the majority of topical dermatological drug products may be determined via standard in vivo blood, plasma, or urine PK techniques. For corticosteroids, an in vivo assessment of the hypothalamic–pituitary–adrenal (HPA) axis suppression test may provide the information. For other topical dermatological drug products, such tests may not be needed.

11 Bioequivalence of Nasal Products

INTRODUCTION

Given here are recommendations for planning product quality studies to measure bioavailability and/or establish bioequivalence in support of new drug applications (NDAs) or abbreviated new drug applications (ANDAs) for locally acting drugs in nasal aerosols (metered-dose inhalers [MDIs]) and nasal sprays (metered-dose spray pumps). Product quality includes chemistry, manufacturing, and controls (CMCs), microbiology, certain bioavailability information, and bioequivalence information (i.e., information that pertains to the identity, strength, quality, purity, and potency of a drug product). Product quality bioavailability and bioequivalence are reflective of potency, in that release of the drug substance from the drug product should be assessed and controlled to achieve a reproducibly potent product. The bioavailability studies can address many questions, but the studies that focus on product performance (i.e., release of drug substance from drug product) are important to consider. A bioequivalence study is normally used to compare a test product (*T*) to a precursor product (*R*)—the to-be-marketed product is compared to a pivotal clinical trial material; a generic product is compared to a reference listed drug.

Product quality approaches should be similar for all nasal aerosols and nasal sprays where the active ingredient/active moiety is intended for local action, regardless of drug or drug class. Product quality information is different from, yet complementary to, the clinical safety and efficacy information that supports approval of an NDA.

BACKGROUND

Bioavailability and bioequivalence may be established by in vivo (pharmacokinetic [PK], pharmacodynamic [PD], or clinical) and in vitro studies or, with suitable justification, by in vitro studies alone. The bioavailability and bioequivalence assessments for locally acting nasal aerosols and sprays are complicated because the delivery to the sites of action does not occur primarily after systemic absorption. Droplets and/or drug particles are deposited topically, which are then absorbed and become available at local sites of action. Systemic exposure following nasal administration can occur either from drug absorbed into the systemic circulation from the nasal mucosa or after ingestion and absorption from the gastrointestinal (GI) tract. A drug administered nasally and intended for local action is therefore likely to produce systemic activity, although plasma levels of the drug do not reflect the amount of the drug reaching nasal sites of action. For these reasons, bioavailability and bioequivalence studies should consider both local delivery and systemic exposure or systemic absorption.

Local Delivery Bioavailability/Bioequivalence Concepts

For local delivery, bioavailability is determined by several factors, including release of drug substance from the drug product and availability to local sites of action. Release of drug from the drug product is characterized by distribution patterns and droplet or drug particle size within the nose that are dependent upon drug substance, formulation, and device characteristics. Availability to local sites of action is a function of the aforementioned release factors, as well as drug dissolution in the case of suspension products, absorption across mucosal barriers to nasal receptors, and rate of removal from the nose. From a product quality perspective, the critical issues are release of drug substance from drug product and delivery to the mucosa. Other factors are of lesser importance. A critical question in assessing product quality bioavailability and bioequivalence is the extent to which one can rely on in vitro methods alone or upon in vitro methods plus clinical endpoints, to measure (benchmark) bioavailability and/or establish bioequivalence. In vitro methods are less variable, easier to control, and more likely to detect differences between products if they exist, but the clinical relevance of these tests, or the magnitude of the differences in the tests, is not always clearly established. Clinical endpoints may be highly variable and relatively insensitive in detecting differences between products but can unequivocally establish effectiveness.

The recommended approach for solution formulations of locally acting nasal drug products is to rely on in vitro methods to assess bioavailability and bioequivalence. This approach is based on the assumption that in vitro studies would be more sensitive indicators of drug delivery to nasal sites of action than would be clinical studies. Drug particle size distribution (PSD) in suspension formulations has the potential to influence the rate and extent of availability to nasal sites of action and to the systemic circulation. For suspension formulation products, however, due to the inability to adequately characterize drug PSD, in vivo studies should be conducted as part of the studies establishing product quality bioavailability and bioequivalence. In vitro studies should be coupled with a clinical study for bioavailability, or a bioequivalence study with a clinical endpoint for bioequivalence, to determine the delivery of drug substance to local nasal sites of action. An in vivo systemic exposure or systemic absorption study should also be conducted for suspensions.

Systemic Exposure and Systemic Absorption Bioavailability/Bioequivalence Concepts

Locally acting drugs are intended to produce their effects upon delivery to nasal sites of action without relying upon systemic absorption. Although systemic absorption may contribute to clinical efficacy for certain corticosteroids and antihistamines, the consequences of systemic absorption (e.g., HPA suppression by corticosteroids) are generally undesirable. In the absence of validated in vitro methodology for characterization of drug PSD for suspension products, and when measurable plasma levels can be obtained, it is recommended that PK studies be conducted to measure systemic exposure bioavailability or establish systemic exposure bioequivalence. For suspension products that do not produce sufficient concentrations to assess systemic exposure, clinical studies or bioequivalence studies with a clinical endpoint should be used to measure systemic absorption bioavailability and establish systemic absorption bioequivalence, respectively. For a schematic representation of recommended studies, see the decision tree for in vivo product quality bioavailability and bioequivalence studies for nasal aerosols and nasal sprays (Figure 11.1).

The investigational new drugs (INDs) and NDAs, not only should product quality bioavailability be provided but bioavailability/PK studies should also be included in the human PK section of the NDA for nasal aerosols and nasal sprays for local action, whether formulated as solutions or suspensions and whether or not validated methods of determining drug PSD are available. These PK data provide biopharmaceutic and clinical pharmacology information beyond product quality bioavailability characterization.

CMC Tests and In Vitro Bioavailability Tests (Noncomparative) versus Bioequivalence Tests (Comparative)

Generally, CMC tests help characterize the identity, strength, quality, purity, and potency of the drug product and assist in setting specifications (tests, methods, acceptance criteria) to allow batch release. These tests have a different purpose than do bioavailability/bioequivalence tests, which focus on release of drug substance from drug product. Some of the in vitro bioavailability/bioequivalence tests nasal aerosols and sprays may be the same as CMC tests for characterization and/or batch release. A specification (test, method, acceptance criterion) for a CMC test for

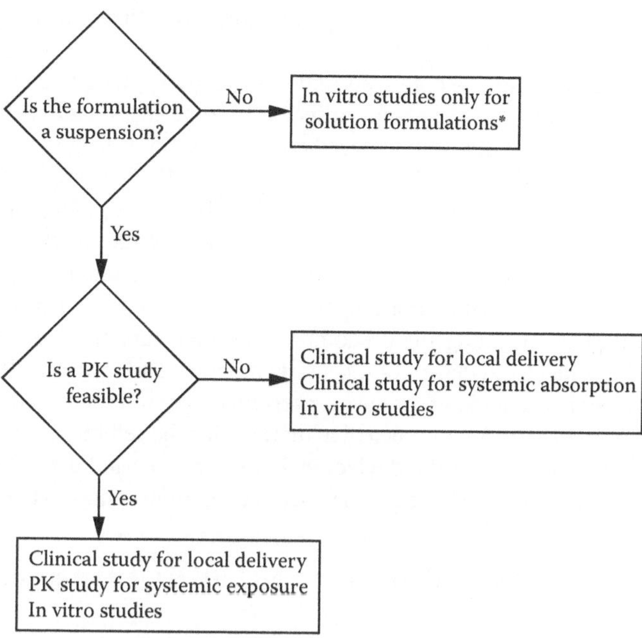

FIGURE 11.1 Decision tree for product quality for bioavailability and bioequivalence testing of nasal products. *Note*: * as defined by FDA.

batch release is usually based on general or specific manufacturing experience. For example, a CMC test such as dose content uniformity has acceptance criteria based on repeated manufacturing of batches. Bioequivalence limits for bioequivalence studies are not usually based on manufacturing experience but are part of equivalence comparisons between test and reference products. Equivalence comparisons normally include (1) a criterion to allow the comparison, (2) a confidence interval for the criterion, and (3) a bioequivalence limit for the criterion. Bioequivalence limits may be based on a priori judgments and may be scaled to variability of the reference product. When conducting premarket for an NDA, some of the in vitro bioavailability tests can be noncomparative and serve primarily to document (benchmark) the product quality bioavailability of a pioneer product.

FORMULATION AND CONTAINER AND CLOSURE SYSTEM

Formulation

Particle size, morphic form, and state of solvation of the active ingredient have the potential to affect the bioavailability of the drug product as a result of different solubilities and/or rates of dissolution. For an ANDA of a suspension formulation, the PSD of the active drug in the dosage form should be the same as that of the reference listed drug. Comparative information on the morphic form of the drug particles, and size and number of drug aggregates in the dosage form should be provided. In addition, documentation of the same anhydrous or solvate form should be provided. For suspension formulations marketed in more than one strength, the drug substance in each strength product should be micronized under identical parameters, and the PSD of the resultant bulk drug should be identical in each strength product.

Container and Closure System

Nasal aerosols consist of the formulation, container, valve, actuator, dust cap, associated accessories (e.g., spacers), and protective packaging, which together constitute the drug product. Similarly, nasal sprays consist of the formulation, container, pump, actuator, protection cap, and protective packaging, which together constitute the drug product.

For nasal aerosols and nasal sprays approved under an ANDA, bioequivalence should be documented on the basis of validated in vivo and vitro tests, or, in some cases, validated in vitro tests alone may be appropriate. Assurance of equivalence on the basis of in vitro tests is greatest when the test product uses the same brand and model of devices (particularly the metering valve or pump and the actuator) as used in the reference product. If this is not feasible, valve, pump, and actuator designs should be as close as possible in all critical dimensions to those of the reference product. Metering chamber volumes should be the same. For nasal aerosols, overall actuator design, including actuator orifice diameter, should be the same. For a nasal spray, spray characteristics may be affected by features of the pump design, including the precompression mechanism; actuator design, including specific geometry of the orifice; and design of the swirl chamber. The external dimensions of the test actuator should ensure comparable depth of nasal insertion to the reference actuator. A test product should attain prime within the labeled number of actuations for the reference product. Consideration should be given to the *dead volume* of the device, including the internal diameter and length of the dip tube, because this volume can influence the number of actuations required to prime a spray pump.

DOCUMENTATION OF BIOAVAILABILITY AND BIOEQUIVALENCE

INDs/NDAs

For INDs/NDAs, in vitro bioavailability studies for solutions and suspensions, and in vivo studies for suspensions, should be provided. These data are useful as a benchmark to characterize the in vitro performance and, for suspensions, the in vivo performance of the product based on the clinical efficacy and either systemic exposure for a PK study or systemic absorption for a clinical safety study. Where the formulation and/or method of manufacture of the pivotal clinical trial product changes in terms of physicochemical characteristics of the drug substance, the excipients, or the device characteristics, bioequivalence data using in vitro tests (for solutions and suspensions) and in vivo tests (for suspensions) may be useful in certain circumstances during the preapproval period to ensure that the to-be-marketed product (*T*) is comparable to very similar clinical trial batches and/or to batches used for stability testing (*R*).

ANDAs

Solution Formulations

In vivo studies, such as seasonal allergic rhinitis (SAR) studies to establish equivalent delivery to nasal sites or HPA suppression studies for corticosteroids to establish equivalent systemic absorption, are not considered necessary for nasally administered solution drug products intended for local action. Thus, reliance on in vitro tests alone to document bioequivalence is suitable for nasal solution formulation products intended for local action. This approach is based on an understanding that for solution products, equivalent in vitro performance, inactive ingredients that are qualitatively the same and quantitatively essentially the same as the inactive ingredients in the reference listed drug, and adherence to container and closure recommendations of "Formulation and container and closure system" section will ensure comparable delivery to the nasal mucosa and to the GI tract. Quantitatively *essentially the same* has been determined by Center for Drug Evaluation and Research (CDER) to mean that the concentration or amount of the inactive ingredient(s) in the test product should not differ by more than ±5% of the concentration or amount in the reference listed drug. When in vitro data fail to meet acceptance criteria, the applicant is encouraged to modify the test product to attain equivalent in vitro performance. Because of insensitivity to potential differences between *T* and *R*, in vivo studies will not be sufficient in the face of in vitro studies that fail to document bioequivalence.

Suspension Formulations with PK Systemic Exposure Data

To document bioequivalence for suspension nasal formulation products intended for local action, both in vitro and in vivo data should be used. Inactive ingredients also should be qualitatively the same and quantitatively essentially the same as the inactive ingredients in the reference listed drug, and the container and closure recommendations of "Formulation and container and closure system" section should be followed. In vivo studies should include both a PK study (systemic exposure) and a bioequivalence study with a clinical endpoint (local delivery). This approach is only applicable for those suspension formulation products that produce sufficiently high drug concentrations in blood or plasma after nasal administration to obtain meaningful AUC and C_{max} data.

As with solutions, in vivo studies will not be sufficient in the face of in vitro studies that fail to establish bioequivalence (i.e., in vitro bioequivalence studies that fail to meet the statistical test result in a failed bioequivalence study) even though the bioequivalence study with a clinical endpoint or the PK study meets the statistical test.

Suspension Formulations without PK Systemic Exposure Data

For suspension nasal formulation products, inactive ingredients should be qualitatively the same and quantitatively essentially the same as the inactive ingredients in the reference listed drug, and the container and closure recommendations of "Formulation and container and closure system" section should be followed. In addition, for those products intended for local action that produce blood or plasma levels that are too low for adequate measurement, given current assay constraints, a bioequivalence study with a clinical endpoint to establish equivalent local delivery to nasal sites and a study with a PD or clinical endpoint to establish equivalent systemic absorption are recommended. In vivo studies that meet the statistical test will not be sufficient in the face of in vitro studies that fail to document bioequivalence.

Postapproval Change

For an NDA submitted under 505(b)(1) of the Food, Drug, and Cosmetic Act, the primary need for bioequivalence documentation would be between the reference product before and the reference product after very limited changes. For an ANDA and for an NDA submitted in accordance with section 505(b)(2) of the Food, Drug, and Cosmetic Act, the primary documentation of bioequivalence for the changed product is the reference or pioneer product. At this time, no guidance from FDA as to when bioequivalence should be redocumented in the presence of any postapproval changes, either for an NDA or ANDA. Sponsors planning such changes should contact the appropriate review division prior to instituting the change.

BIOAVAILABILITY AND BIOEQUIVALENCE: IN VITRO STUDIES

Batches and Drug Product Sample Collection

INDs/NDAs

In vitro product quality bioavailability studies for nasal aerosols and sprays should generally be performed on samples from three batches. The batches should include a pivotal clinical trial batch, a primary stability batch, and, if feasible, a production scale batch, to provide linkage of in vitro performance to in vivo data. If a production scale batch is not available, a second pivotal clinical trial batch can be substituted.

The aforementioned bioavailability batches should be equivalent to the to-be-marketed product. The manufacturing process of these batches should simulate that of large-scale production batches for marketing (additional information on large-scale batches is provided in the International Conference on Harmonisation [ICH] guidance for industry Q1A *Stability Testing of New Drug Substances and Products* [September 1994], see "Tests and metrics" section). Complete batch records, including batch numbers of device components used in the batches, should accompany the bioavailability submission.

In vitro bioavailability studies are intended to characterize the means and variances of measures of interest for canisters (nasal aerosols) or bottles (nasal sprays) within a batch and between batches, where applicable. However, under 21 CFR 320.1 and 320.21, the studies may be noncomparative to other formulations or products. The in vitro tests and metrics are described in "Tests and metrics" section. The test method or standard operating procedure (SOP) for each test should accompany the data in the submission.

ANDAs

In vitro bioequivalence studies for nasal aerosols and sprays should generally be performed on samples from each of three batches of the test product and three batches of the reference listed drug. Test product samples should be from the primary stability batches used to establish the expiration dating period. Test product should preferably be manufactured from three different batches of the drug substance, different batches of critical excipients, and container and closure components. For nasal sprays formulated as solutions, in vitro bioequivalence tests can alternatively be performed on three sublots of product prepared from one batch of the solution.

The aforementioned bioequivalence batches should be equivalent to the to-be-marketed product. The manufacturing process of these batches should simulate that of large-scale production batches for marketing (ICH Q1A *Stability Testing of New Drug Substances and Products* (September 1994), see "Tests and metrics" section). Complete batch records, including batch numbers of device components used in the batches or sublots (for solution nasal sprays), should accompany the bioequivalence submission.

Reference product samples should be from three different batches available in the marketplace. The recommended in vitro tests and metrics are described in "Tests and metrics" section. The recommended number of canisters or bottles of each product and batch to be used in the aforementioned studies, and recommended statistical approaches, including suggested boundaries for each of the studies, are described in "Statistical analyses" section.

Tests and Metrics

In vitro bioavailability and bioequivalence for locally acting drugs delivered by nasal aerosol or nasal spray are characterized by six tests:

1. Dose or spray content uniformity through container life
2. Droplet and drug PSD
3. Spray pattern
4. Plume geometry
5. Priming and repriming
6. Tail-off profile

For solution formulation nasal sprays, variability in in vitro bioequivalence study data between batches is expected to be due primarily to variability in the device components of the product rather than in the solution. Therefore, a single batch of solution may be split filled into three equal size sublots of product. The sublots should be prepared from three different batches of the same device (pump and actuator) components.

All in vitro tests should be conducted on test canisters or bottles selected in a randomized manner from the test batch, including units from the beginning, middle, and end of the production run. Bioequivalence tests should be conducted in a blinded manner or should use another approach that removes potential analyst bias, without interfering with product performance. Automated actuation stations are recommended for all comparative in vitro bioequivalence tests to decrease variability in drug delivery due to operator factors (including removal of potential analyst bias in actuation) and increase the sensitivity for detecting potential differences between products in any of the aforementioned tests. The blinding procedure should also be extended to postactuation evaluations. The randomization procedure and the test method or SOP for each test should accompany the data in the submission.

Dose or Spray Content Uniformity through Container Life

Sampling apparatus for collection of dosage units from aerosols is described in *U.S. Pharmacopeia 23/National Formulary 18* (Tenth Suppl., May 15, 1998). A suitable apparatus should be used for collection of dosage units from nasal sprays. For both solution and suspension formulations of nasal aerosols and nasal sprays, the mass of drug delivered per single (unit) dose should be determined based on a stability-indicating chemical assay. A single dose represents the minimum number of sprays per nostril specified in the product labeling. For a nasal product for which the minimum single usual dose is one actuation in each nostril, the single dose should be based on one actuation. For a nasal product for which the minimum usual dose is two actuations in each nostril, the single dose should not exceed two actuations. For bioavailability and bioequivalence studies, dose or spray content uniformity data should be determined on primed units at the beginning of unit life, at the middle of unit life, and at the end of unit life for nasal aerosols and at beginning and end of unit life for nasal sprays. Mean dose or spray content uniformity and variability in content uniformity is to be determined based on within and between canister or bottle data and, for nasal aerosols and suspension formulation nasal sprays, between batch data.

Automated actuation stations may be stand-alone systems or accessories for laser diffraction instruments. Stations may include settings for actuation force, actuation velocity, hold time, return time, delay time between actuations, length of stroke, and number of actuations. Selection of appropriate settings should be relevant to proper usage of the nasal aerosol or nasal spray by the trained patient and should be documented based on exploratory studies in which actuation force, actuation time, and other relevant parameters are varied. These studies should accompany the validation data. Selected settings used for the comparative in vitro study should be specified in the SOP for each test for which the automatic device is employed.

Based on the labeled number of full medication doses, the FDA uses the terms *beginning life stage*, *middle life stage*, and *end life stage* interchangeably with the terms *beginning of unit life* (the first actuation(s) following the labeled number of priming actuations), *middle of unit life* (the actuation(s) corresponding to 50% of the labeled number of full medication doses), and *end of unit life* (the actuation(s) corresponding to the label claim number of full medication doses).

Analytical data should be validated, and the analytical validation report should accompany the content uniformity report.

Droplet and Drug Particle Size Distribution

To increase nasal deposition and minimize deposition in the lungs and GI tract, aerosol droplets should generally have a mass median aerodynamic diameter (MMAD) greater than 10–20 μm. As MMAD decreases over the 5–20 μm range, it indicates that reduced nasopharyngeal deposition and increased pulmonary deposition may occur. Droplet size distribution measurements are thus critical to delivery of drug to the nose. For bioavailability and bioequivalence, studies of droplet size distribution and PSD by validated methods should be performed. For suspension products, drug particle size may be important to rate of dissolution and availability to sites of action within the nose. Therefore, drug or drug and aggregate PSD should be characterized in the formulation both within the can or bottle and within the aerosolized droplets. Present agency experience suggests that drug and drug aggregate PSD characterization cannot be acceptably validated for nasal aerosols and nasal sprays. In this circumstance, drug and drug aggregate PSD studies should be performed, and these supportive characterization data, along with available validation information, should be submitted.

Particle Size Distributions

Droplet Size Distribution For all nasal aerosols and nasal sprays, whether formulated as solution or suspension products, droplet size distribution should be determined utilizing a method suitable for fully characterizing the droplet size. Laser diffraction methodology, or appropriately validated alternate methodology, is recommended.

Particle Size Distribution For all nasal aerosols and nasal sprays, whether formulated as solution or suspension products, PSD should be determined using a suitable aerodynamic method (e.g., multistage cascade impactor [CI], multistage liquid impinger [MSLI]).

Drug and Aggregate PSDs. Nasal spray suspension formulations typically contain micronized drug within an aqueous vehicle with partially undissolved suspending agents and other ingredients. Nasal aerosol suspension formulations contain micronized drug suspended within propellants and may contain a surfactant and/or cosolvent. Light microscopy may be considered for estimating drug and drug aggregate PSD of these products.

Instrumental Methods

Laser Diffraction Laser diffraction is a nonaerodynamic optical method of droplet or particle sizing that measures the geometric size of droplets or particles in flight. To characterize the beginning, middle, and end of the plume, measurements should be made at three distances from the delivery orifice. Multiple actuations may be performed at each life stage to assess precision. The droplet size distributions due to each actuation and the means, standard deviations (SDs), and percent coefficients variation (CVs) should be reported. At each distance, measurements should be made at different delay times in order to characterize the size distribution of droplets or particles within the plume upon formation, as the plume has started to dissipate, and at some intermediate time (Sciarra and Cutie, 1989). Selected delay times may be based on obscuration levels or other suitable means.

Droplet size distribution data (D_{10}, D_{50}, D_{90}) and span (($D_{90} - D_{10}$)/D_{50}) should be reported based on volume (mass). Droplet size distribution data by count (number of droplets) are not requested. All instrument/computer printouts should be submitted, including cumulative percent undersize tables, histograms of PSD, obscuration values, and other details and statistics. The manufacturer's recommended obscuration ranges for the laser diffraction instrument should be submitted.

Obscuration refers to the percentage of laser light obscured or scattered out of the beam by the sample and is influenced by sample concentration and width of the plume. Following actuation, obscuration levels are initially low, increase as the plume develops, and then decrease as the plume dissipates.

Comparative laser diffraction data are requested at beginning, middle, and end of unit life. For bioequivalence, statistical comparisons should be based on D_{50} and span.

Multistage Cascade Impaction or Multistage Liquid Impinger Sizing of droplets or particles by CI or MSLI measures aerodynamic diameter based on inertial impaction, an important factor in the deposition of drug in the nasal passages. CI or MSLI data should be provided for all nasal sprays and nasal aerosols to characterize the size distribution of drug based on aerodynamic mass diameters. The greatest percentage of the emitted dose is deposited prior to or on the first stage of the CI for both nasal aerosols and nasal sprays. Thus, the equivalence of aerodynamic drug PSD of test and reference products, although conducted by validated procedures, does not ensure equivalent PSD of drug within the aerosolized droplets. Characterization of drug PSD by CI or MSLI, along with the other recommended in vitro tests, does not allow waiver of in vivo bioequivalence studies for suspension formulation products.

For bioavailability and bioequivalence, CI or MSLI drug deposition profile data should be based on three size range groups. Group 1 includes summation of drug deposition in or on the valve stem, actuator, inlet port, and upper stage, which should have a nominal effective cutoff diameter (ECD) (e.g., greater than or equal to 9.0, 10.0, 13.0, or 16.0 μm). Group 2 includes drug deposition on the stage immediately below the upper stage (e.g., greater than or equal to 5.0 μm). Group 3 includes summation of drug deposition below the Group 2 stage, including the filter. For Group 1 only, deposition should also be reported for each of the individual accessories and the upper stage. Deposition should be reported in mass units. Mass balance accountability (sum of all drug deposited from the valve stem to the filter) should be documented.

Selection of the most suitable CI may be influenced by the ECDs of stages of various brands of CIs, the geometry of the induction port, and other factors. Studies should use the fewest number of actuations justified by the sensitivity of the analytical method (generally not exceeding 10), in order to be more reflective of the PSD of individual doses. Analytical data should be based on a validated chemical assay. The analytical validation report should accompany the CI data report. The SOP or validation report should indicate the minimum quantifiable amount of drug deposited on each of the three groups of deposition sites and on each accessory or stage of the Group 1 data.

For bioavailability and bioequivalence, CI data are requested at the beginning and end of unit life. The middle of unit life data is not requested. For bioequivalence, statistical comparisons of drug deposition on the three groups should be based on profile analysis.

Light Microscopy Light microscopy may provide drug and aggregate PSD data. However, the method is limited in its ability to fully characterize PSD by the resolution limit of light microscopy (about 0.5 μm or higher), which may not be adequate for sizing micronized drug. A second limitation is potential difficulty in distinguishing drug from undissolved excipient in suspension formulation nasal sprays. Due to these limitations, acceptable validation of the microscopic data may not be possible. In the presence of these limitations, it is recommended that comparative drug and aggregate PSD data should be submitted as supportive bioavailability and bioequivalence characterization data for suspension formulation nasal aerosols and sprays. The occurrence of drug particles and aggregates within appropriate size ranges should be tabulated for each analysis, and histograms of the drug and aggregate PSD should be provided. Count median diameter (CMD) and geometric SD (GSD) based on single-particle data (aggregates excluded) should be provided. Studies of nasal sprays should include test product placebo to provide an estimate of the occurrence of apparent drug particles (*false positives*) due to undissolved excipient. PSD by light microscopy provides supportive bioequivalence information.

Spray Pattern

Spray pattern characterizes the spray following impaction on an appropriate target (e.g., a thin-layer chromatography [TLC] plate). It provides information about the shape and density of the plume following actuation. Spray patterns should be determined on single actuations at three appropriate distances from the actuator to the target at the beginning and end of unit life. The visualization technique should preferably be specific for the drug substance. End of unit life testing is requested to ensure comparability to performance at beginning of unit life. Clear, legible photographs or photocopies of the spray patterns, not hand-drawn representations obtained by tracing the pattern, should be provided. The widest (D_{max}) and shortest (D_{min}) diameters and the ovality ratio (D_{max}/D_{min}) should be provided for each spray pattern. The SOP should include a figure describing the procedure for measurement of D_{max} and D_{min}. For bioequivalence, statistical comparisons should be based on ovality ratio and either D_{max} or D_{min} data.

Spray pattern and plume geometry (below) are recommended to assist in establishing functional equivalence of products as a result of differences in the device components of T and R products. Comparable spray pattern and plume geometry data for T and R, combined with other in vitro tests (and in vivo studies for suspensions), ensure equivalent drug deposition patterns, resulting in equivalent delivery of drug to nasal sites of action and equivalent systemic exposure or absorption.

Plume Geometry

Plume geometry describes two side views, at 90° to each other (two perpendicular planes) and relative to the axis of the plume, of the aerosol cloud when actuated into space. Plume geometry should be based on high-speed photography or other suitable methods. Photographs should be of high quality and should clearly show the dense cloud and individual large droplets or agglomerates of droplets in the vicinity of the cloud. Plume geometry may be performed only at the beginning of unit life. Plumes should be characterized at three or more times after a single actuation, chosen to characterize the plume early upon formation, as the plume has started to dissipate, and at some intermediate time. Photographs of plumes should be used to measure plume length, plume width, and plume (spray cone) angle. All photographs and data characterizing the plume dimensions in two planes should be submitted, including the scale used to indicate actual size. Comparative bioequivalence data are supportive.

Priming and Repriming

Priming and repriming data provide information to ensure delivery of the labeled dose of drug and thus are part of the in vitro bioavailability and bioequivalence assessment. Similar studies should be conducted on nasal sprays. For products approved under an NDA, priming and repriming data based on single actuations should be provided for multiple orientations.

For products approved under an ANDA, the labeling is the same as that for the reference listed drug, except for specific changes described in the regulations (21 CFR 314.94(a)(8)(iv). For nasal sprays and some nasal aerosols, the reference product labeling (package insert and/or patient package insert) describes the number of actuations necessary to prime the product on initial use and on repriming following one or more periods of nonuse (e.g., 24 h and 7 days following last dose). Comparative priming and repriming data are requested to document that priming of the test product is attained within the number of priming actuations stated in the reference product labeling. For reference product nasal aerosols lacking priming recommendations, priming studies are recommended to characterize the test product relative to the reference product. In the absence of reference product priming recommendations, an adequate number of single actuations should be studied to ensure that test and reference products have each attained an emitted dose equal to the labeling claim. Repriming studies of test products are requested only when the reference product labeling includes repriming instructions.

Priming and repriming data for the test product in multiple orientations should be provided in the CMC portion of the ANDA submission. Therefore, comparative bioequivalence studies may be based on products stored in the valve upright position. For any nasal aerosol product in which the reference product labeling recommends storage in the valve down position, additional comparative priming and repriming data should be provided for this orientation. For suspension products, the unprimed canister or bottle should be shaken for a standardized time (e.g., 5 s) and a dose should then be immediately collected. For nasal aerosols, a standardized period (e.g., 30–60 s) should be allowed between successive actuations. Doses may be collected in the same apparatus used for the dose or spray content uniformity through container life test. When priming and/or repriming information is included in the labeling, comparison of equivalence should be based on the emitted dose of the single actuation immediately following the specified number of priming or repriming actuations. The emitted dose of each earlier actuation should also be provided. When priming information is not specified, the emitted dose of each successive actuation up to and including attainment of label claim should be provided. Comparative bioequivalence data in the absence of priming are supportive.

Tail-Off Profile

Whereas dose or spray content uniformity conducted at the end of the labeled number of actuations ensures that the product delivers the labeled dose through the number of actuations stated in product labeling, the tail-off profile characterizes the decrease in emitted dose following delivery of the labeled number of actuations (i.e., from end of unit life to product exhaustion). Tail-off profile characteristics may vary as a function of valve or pump design, bottle geometry, and other factors and may be characterized in terms of uniformity of decline, rate of decline, and intercanister or interbottle variability in unit dose. For bioavailability assessment, tail-off data are noncomparative. For bioequivalence assessment, comparative tail-off profiles are requested to ensure similarity in drug delivery as the product nears exhaustion. Data should be based on the emitted dose of individual actuations. Comparative bioequivalence data are supportive; however, the test product should be no more erratic in dose delivery than the reference product, and the rate of decline in delivery should be generally similar between products.

BIOAVAILABILITY AND BIOEQUIVALENCE: CLINICAL STUDIES FOR LOCAL DELIVERY

General Information

The same adequate and well-controlled clinical trials in humans used to establish the safety and effectiveness of the drug product (21 CFR 314.126) may be used, in some cases, to establish bioavailability or, when comparative, bioequivalence (21 CFR 320.24). Although bioavailability and bioequivalence studies with a clinical endpoint are sometimes incapable of showing a dose–response relationship and may not be consistently reproducible (21 CFR 320.24(b)(4)), they are sometimes the only means available to document bioavailability and bioequivalence in drug products intended for local delivery and action. A number of FDA guidance provide information about the general conduct of clinical studies, including clinical studies to document bioavailability and bioequivalence. These include *General Considerations for Clinical Trials* (ICH E8, December 1997); *Structure and Content of Clinical Study Reports* (ICH E3, July 1996); *Good Clinical Practice: Consolidated Guideline* (ICH E6, May 1997); and *Statistical Principles for Clinical Trials* (ICH E9, May 1997).

Bioequivalence Clinical Study Endpoints

Clinical evaluations should be made at baseline and during treatment. The efficacy endpoint should be patient self-rated total nasal symptom scores (TNSS). These most often include a composite score of runny nose, sneezing, nasal itching, and, for drugs other than antihistamines, congestion. The efficacy endpoint should be expressed as change from baseline (pretreatment) of the TNSS, expressed in absolute units and percent change. In addition to the efficacy measures, all three-study designs should incorporate safety assessments.

Clinical Study Batches

The product quality bioavailability batch used for the study should be the same pivotal clinical trial batch used in the in vitro bioavailability studies. Where bioequivalence studies are needed for an NDA, the batches of test and reference products should be the same batches employed in the in vitro testing. The product quality batches used to establish the local delivery bioequivalence for an ANDA should be the test and reference batches employed in the in vitro bioequivalence testing.

Clinical Bioequivalence Study Designs and Subject Inclusion Criteria

A bioequivalence study with a clinical endpoint to establish equivalent local delivery of drug from test and reference products to the nose should document sensitivity of the study to discriminate between differing doses (i.e., show a dose–response relationship). This documentation typically relies on the inclusion of a second dose of the reference product, and preferably of the test product, that may be higher or lower, to demonstrate that the efficacy response is different between the two doses. Doses may differ by two- or fourfold, and to increase study sensitivity, the lower dose examined may be below the minimum labeled dose (e.g., one-half or one-quarter of the recommended dose, depending on the limitations of the formulation).

Although many clinical study design options may be considered to establish bioequivalence, outlined in the following are three suggested study designs for evaluating clinical responses for nasally administered drugs for SAR: (1) traditional treatment, (2) day(s) in the park, and (3) environmental exposure unit (EEU). The three-study designs use SAR patients as the study population to document bioequivalence for all indications in product labeling for nasally administered drug products are described here. Recommended studies are designed as treatment studies rather than prophylaxis studies. Depending on the time to onset of therapeutic effect of the drug being tested, the medication effect can be evaluated after a single dose (e.g., antihistamines) or after short-term treatment (e.g., corticosteroids). In all three-study designs, an assessment of onset of action and efficacy at the end of the dosing interval is recommended, because both measures are important clinically and may offer better dose discrimination.

Because specific study recommendations are not available from the FDA, a protocol for a bioequivalence study with a clinical endpoint for a specific suspension drug product should be submitted to the appropriate review division at FDA. For the three-study designs, a pilot study may be useful to determine the optimal dosing duration and doses to be used in the bioequivalence study.

Traditional Treatment Study

The recommended design for this study is a randomized, double-blind, placebo-controlled, parallel group study with a single-blind placebo lead-in period (generally 1–14 days) in which efficacy and safety of the test product are assessed for a 2-week duration. Symptom assessment should be made at least twice daily (i.e., *reflective* scores) and also at the end-of-dosing interval (i.e., *instantaneous* scores). Evaluation of both reflective and instantaneous assessments of the TNSS is critical in establishing bioequivalence with a clinical endpoint. Safety measures should include physical examination, laboratory monitoring (chemistry, liver function tests, hematology, urinalysis, serum pregnancy testing in females), monitoring of vital signs, adverse event reporting, and performance of 12 lead ECGs before and after treatment with study drug.

Day(s) in the Park (Outdoors) Study

The recommended design for this study is a randomized, double-blind, placebo-controlled, parallel group study in a park setting in which subjects are exposed to relevant outdoor allergens. On the study day, patients should undergo a baseline period of evaluation in the park setting to establish a minimum level of allergic rhinitis symptoms prior to randomization to study drug treatment. Patients should remain outdoors in the park for a prespecified length of time over one to two consecutive days. Nasal symptoms should be evaluated on a periodic basis throughout the full dosing interval to characterize onset of action and end-of-dosing interval efficacy. Safety assessment generally involves adverse event reporting.

Environmental Exposure Unit Study

The recommended design for this study is a randomized, double-blind, placebo-controlled, parallel group study in a controlled indoor environment termed an *EEU chamber*. Repeated pretreatment exposure to the relevant allergen allows screening for symptomatic responders for enrollment in the treatment phase. On the study day, patients should be exposed to the allergen in the EEU and monitored for a baseline period to ensure a minimum level of allergic rhinitis symptoms prior to randomization to study drug treatment. Patients should remain in the EEU for a prespecified length of time over 1 or 2 days. Nasal symptoms should be evaluated on a periodic basis throughout the full dosing interval to characterize onset of action and end-of-dosing interval efficacy. Safety assessment generally involves adverse event reporting.

Subjects employed in each of the three-study designs should be patients with a history of SAR and a positive allergy test for specific allergens (e.g., allergen skin test). Patients with other significant diseases should be excluded from the study. Patients should be experiencing a defined minimum level of symptom severity at the time of study enrollment.

BIOAVAILABILITY AND BIOEQUIVALENCE: PK SYSTEMIC EXPOSURE STUDIES

Plasma concentration–time profiles from bioavailability and bioequivalence studies should be used to evaluate systemic exposure for suspension drug products that produce sufficiently high drug concentrations of the active ingredient and/or active moiety after nasal administration to obtain meaningful AUC and C_{max} data. The product quality bioavailability study to characterize systemic exposure may be one of the same PK studies conducted to address clinical pharmacology and biopharmaceutics questions of regulatory interest. The bioavailability study may be conducted in healthy subjects or SAR patients. The bioavailability batch used for the PK systemic exposure study preferably should be a pivotal clinical trial batch. Alternatively, a PK batch similar to a batch used in a pivotal clinical trial may be used, in which case, any differences between the PK batch and the pivotal clinical trial batch should be discussed with appropriate CDER review staff prior to the study. If the PK batch is not one of the three batches used for the in vitro bioavailability studies, in vitro bioavailability data should be provided for the PK batch using the same protocols as for the three batches.

For an NDA or an ANDA, the in vivo bioequivalence study should be conducted with a replicate crossover or nonreplicate crossover design. The study may be single or multiple dose. The batches of test and reference product should be batches employed in the in vitro testing. For an ANDA, the batches of test and reference used for the systemic exposure study should be the same batches used for the clinical study for local delivery, and each of these batches should be one of the three batches used for the in vitro bioequivalence studies. Subjects for the study should be healthy (non-SAR patients), with exclusions primarily for reasons of safety. Several actuations from the drug product in each nostril may be needed to achieve measurable concentrations of the active ingredient and/or active moiety in an accessible biological fluid such as blood or plasma. For an ANDA, an IND in accordance with 21 CFR 320.31 will be needed when the number of doses in a single-dose or multiple-dose study exceeds the single or total daily dose specified in the labeling of the approved NDA.

Attempts should be made in the conduct of a PK systemic exposure study to minimize loss of drug due to excess fluid drainage into the nasopharynx or externally from the nasal cavity. The bioanalytical method should be validated for accuracy, precision, specificity, and sensitivity. Statistical analysis should be conducted on the log-transformed data. Average bioequivalence may be used for studies with replicate crossover or nonreplicate crossover designs. Individual bioequivalence with scaling may be used for studies with replicate crossover designs. A pilot study is recommended to assess the analytical methodology and to estimate the numbers of actuations and subjects to be used in the full-scale study.

BIOAVAILABILITY AND BIOEQUIVALENCE: PD OR CLINICAL STUDIES FOR SYSTEMIC ABSORPTION

General Information

Clinical studies for bioavailability, or bioequivalence studies with a PD or clinical endpoint, are needed to assess the systemic absorption of those suspension drug products for which PK systemic exposure studies are not feasible. Published data suggest that systemic bioequivalence of suspension formulation antihistamine nasal products may be established based on PK data. At the present time, approved nasal mast-cell stabilizer nasal spray and anticholinergic nasal spray products are solutions for which bioequivalence may be established based upon in vitro studies only. These types of studies will thus generally be needed only for corticosteroid nasal aerosols and nasal sprays. The product quality bioavailability study to characterize systemic absorption may be one of the same clinical studies conducted to establish the safety of the active ingredient and/or active moiety in the drug product. Because this section does not provide specific recommendations for clinical studies for systemic absorption, sponsors should submit a protocol for a bioequivalence study with a PD or clinical endpoint for a specific drug product to the appropriate review division at FDA.

Bioequivalence Study Endpoints for Corticosteroids

The recommended systemic absorption bioequivalence study design for nasal corticosteroids is suppression of the HPA axis. The endpoint may be either 24 h urinary free cortisol adjusted for urinary creatinine, based on a full 24 h urine collection, or serum cortisol levels collected every 4 h over a 24 h period, with exclusion of the middle of the night sample. Endpoints for placebo and test and reference treatments should be baseline adjusted prior to statistical analyses.

Clinical Study Batches

The product quality bioavailability batch used for the study should be a pivotal clinical trial batch used in the in vitro bioavailability studies. For bioequivalence studies for an NDA, the batches of T and R should be batches used in in vitro testing. For an ANDA, the batches of test and reference product used for the systemic absorption study should be the same batches used for the clinical study for local delivery. Each of these batches should be one of the three batches used for the in vitro bioequivalence studies.

Clinical Study Designs and Subject Inclusion Criteria

The study can be conducted as a placebo-controlled, randomized, multiple-dose parallel design comparing test and reference products. The study should be conducted in healthy, nonallergic volunteers not previously exposed to corticosteroids, and subjects should be domiciled within the clinical study center during the dosing days. Three treatments, test and reference products at the labeled dose (maximum labeled dose when labeling includes more than one dose) and a placebo of the test product, should be used. Each treatment period should consist of 14 days of dosing. Timed urine or serum samples for determination of 24 h urinary free cortisol or 24 h serum cortisol levels should be collected prior to dosing (baseline) and during the last 24 h of the 14 days of dosing. In addition, we recommend determining two to three interval 24 h urinary free cortisol or 24 h serum cortisol levels (e.g., performing additional assessments on days 4, 7, and/or 10) to better profile the onset of the effect of test and reference products, should detectable adrenal suppression occur.

Alternatively, the study could be conducted as a placebo-controlled, randomized, multiple-dose crossover design comparing test and reference. As in the parallel design study, the study should be conducted in healthy, nonallergic volunteers not previously exposed to corticosteroids. During the dosing days, subjects should be domiciled within the clinical study center to ensure compliance with the study protocol. Three treatments, test and reference at the labeled dose (maximum labeled dose when labeling includes more than one dose), and a placebo of the test product should be used. Each treatment period should consist of 14 days of dosing. A shorter dosing duration would be considered with adequate scientific justification. Washout periods between treatments should be adequate to eliminate the possibility of a carryover effect. Urine or serum samples for determination of 24 h urinary free cortisol or 24 h serum cortisol levels should be collected prior to each dosing period (baseline data) and during the last 24 h of each dosing period. In addition, we recommend determining two to three interval 24 h urinary free cortisol or 24 h serum cortisol levels (e.g., performing additional assessments on days 4, 7, and/or 10) to better profile the onset of the effect of test and reference products, should detectable adrenal suppression occur.

STATISTICAL ANALYSES

In vitro studies yield both profile and nonprofile data, which require different statistical analyses. Noncomparative bioavailability in vitro data analyses for both profile and nonprofile data are required. For bioequivalence studies, methods of comparison for nonprofile analyses are required.

In Vitro Bioavailability Data

The means, SDs, and percent CVs should be reported for the measures to document BA. The overall means for the formulation should be averaged over all bottles or canisters, life stages (except for priming and repriming evaluations), and batches. In addition to overall means, means at each life stage for each batch averaged over all bottles or canisters, and for each life stage averaged over all batches, are requested. For profile data, means, SDs, and percent CVs should be reported for deposition in each of Groups 1, 2, and 3 of the CI or MSLI data, as well as on the individual accessories and stage within Group 1.

In Vitro Bioequivalence Data: Nonprofile Analyses Using a Confidence Interval Approach

Nonprofile analyses should be applied to the following tests: (1) dose or spray content uniformity through container life, (2) droplet size distribution, (3) spray pattern, and (4) priming and/or repriming, when this information is specified in the labeling.

Study Protocol

Data for the bioequivalence criterion should be based on testing a suitable number of bottles or canisters from each of three batches of the T and R drug products. Each bottle or canister should be tested for the measure (parameter) of interest at beginning and end or beginning, middle, and end of unit life, as indicated in "Bioavailability and bioequivalence: in vitro studies" section and Table 11.1. Rather than evaluating performance at each life stage separately, a criterion is recommended that combines the multiple life stages. In doing so, the multiple life stages are considered as providing measures of the same underlying quantity. The recommended criterion considers deviations from uniformity across bottle or canister life stages; results are ideally uniform. Lack of uniformity between life stages should be treated as another variance component in the criterion.

For suspension formulation nasal sprays and solution formulation and suspension formulation nasal aerosols, the number of canisters or bottles (units) of product to be studied should not be fewer than 30 for each of the test and reference products (i.e., no fewer than 10 from each of three batches). For solution formulation nasal sprays, no fewer than 10 units from each of the three batches or three sublots should be studied. The number of units is a function of T to R product means and variances. Estimates of these mean differences and variances will necessitate pilot studies.

Criterion for Comparisons, Confidence Interval, and Bioequivalence Limit

The equivalence approach for nonprofile tests relies on (1) a criterion to allow the comparison, (2) a confidence interval for the criterion, and (3) a bioequivalence limit for the criterion:

Criterion for Comparison

The in vitro population bioequivalence criterion and bioequivalence limit are

$$\frac{(\mu_T - \mu_R)^2 + (\sigma_T^2 - \sigma_R^2)}{\sigma_R^2} \leq \theta \tag{11.1}$$

where

μ_T, μ_R is the *T* and *R* means (log scale)
θ_{BT}, θ_{BR} is the between batch *T* and *R* SDs (log scale)
θ_{CT}, θ_{CR} is the between canister *T* and *R* SDs (log scale)
σ_R^2 is the $\sigma_{BR}^2 + \sigma_{CR}^2 + \sigma_{LR}^2$
σ_T^2 is the $\sigma_{BT}^2 + \sigma_{CT}^2 + \sigma_{LT}^2$
σ_{LT}, σ_{LR} is the within *T* and *R* canister between life stage SD
θ is the in vitro bioequivalence (upper) limit

The overall means for the two formulations should be averaged over all bottles or canisters, life stages (except for priming and repriming evaluations), and batches.

The general approach should be to calculate a 95% upper bound for the criterion. If this upper bound is less than or equal to the upper limit, 2, the test product may be judged to be bioequivalent to the reference product at the 5% level. A population, rather than average, bioequivalence criterion is recommended in order to estimate whether the test product may be more variable than the reference product. The test product should be as or more consistent in the delivery of drug than is the reference product. An individual bioequivalence approach is not appropriate for in vitro data because there are no subjects, thus no subject-by-formulation interaction.

Determining a 95% Upper Bound

CDER recommends that a method of moments approach be used for estimating the means and variances needed to determine the population bioequivalence criterion. Approaches based on restricted maximum likelihood (REML) may be used in special cases. For determining the 95% upper bound, CDER recommends using a method analogous to one proposed for individual bioequivalence.

Specification of the Population Bioequivalence Upper Limit

The general form of the upper limit, 2, is analogous to the form of the population bioequivalence criterion, which is

$$\frac{(\text{Mean difference in natural log scale})^2 \pm \text{Variance terms}}{\text{Comparison variance}} \tag{11.2}$$

The corresponding form for the upper limit is then

$$\frac{(\text{Average bioequivalence limit in natural log scale})^2 \pm \text{Variance terms offset}}{\text{Scaling variance}} \tag{11.3}$$

This formula contains three values to be specified: (1) average bioequivalence limit, (2) variance terms offset, (3) and scaling variance. These values have not been specified by FDA.

Average Bioequivalence Limit

Due to the low variability of in vitro measurements, at the present time, CDER recommends that the limit not be larger than 90/111 (i.e., the ratio of geometric means would fall within 0.90 and 1.11). A value of 0.90 is tentatively recommended as the average bioequivalence limit. This value should be used in calculating the population bioequivalence limit.

Variance Terms Offset

This value arises to allow some difference among the total variances that may be inconsequential. In this regard, the variance terms offset is analogous to the average bioequivalence limit. The variance terms offset also helps correct for the effect on power and sample size for the need to estimate the variances. Because of the low variability of in vitro measurements, the variance terms offset, denoted ϵ_P, should be taken as 0.0. CDER is also considering ϵ_P equal to 0.01.

Scaling Variance

This value adjusts the bioequivalence criterion depending on the reference product variance. When this variance is greater than the scaling variance, σ_{T0}^2, the limit is widened. When this variance is less than the scaling variance, the limit is narrowed.

Mixed scaling should be employed for in vitro studies.

With mixed scaling, when the reference variance *in the study* is less than the scaling variance, the population bioequivalence criterion should be modified to its *constant-scaled* form:

$$\frac{(\mu_T - \mu_R)^2 + (\sigma_T^2 - \sigma_R^2)}{\sigma_{T0}^2} \tag{11.4}$$

Mixed scaling is used to avoid penalizing test products for cases with very low reference variance. It is CDER's current intent to select F_{T0} for in vitro studies so that most studies will use constant scaling and thus, that F_{T0} will be at least 0.10. The upper limit may be interpreted by reference to a population distance ratio (PDR). The PDR is the ratio of the test–reference distance (in the log scale) to the reference–reference distance. In contrast to individual bioequivalence, the distances for population bioequivalence are based on administration to separate. The population bioequivalence criterion, denoted by PBC, is related to the PDR by

$$\text{PDR} = \left(1 + \frac{\text{PBC}}{2}\right)^{1/2} \tag{11.6}$$

Substituting the bioequivalence limit 2 for PBC expresses the upper limit in the PDR scale. The specification of 0.90 for the average limit, 0.0 for the variance offset, and 0.10 for the scaling SD corresponds to an upper limit for PDR of 1.25.

In Vitro Bioequivalence Data: Supportive Nonprofile and Profile Analyses

The following tests provide supportive characterization data: (1) plume geometry; (2) tail-off profile; (3) priming data, when reference product labeling does not specify priming information; and (4) drug CMD and drug and aggregate PSD data from microscopic analyses. The comparative data requested in "Bioavailability and bioequivalence: in vitro studies" section should be provided, based upon the same number of bottles or canisters recommended in the protocol of section. Statistical criteria need not be applied.

In Vitro Bioequivalence Data: Profile Analyses Using a Confidence Interval Approach

Profile analyses apply to CI or MSLI data for nasal aerosols and nasal sprays. Analyses may rely on a criterion for comparisons of means and variances relative to a bioequivalence limit, with calculation of a 90% confidence interval. The general approach is adaptable to CIs of varying numbers of stages and accessories, or groups of stages and accessories. As discussed in "Droplet and drug particle size distribution" section, profile comparisons may be based on drug deposition within three groups.

Study Protocol

Data for the bioequivalence criterion should be based on testing a suitable number of bottles or canisters from each of three batches of the T and R drug products (or three sublots for solution formulation nasal sprays). Each canister should be tested for deposition at the beginning and end life stages. The number of canisters to be studied from each batch, which should not be less than 10, is a function of test to reference product means and variances. Estimates of these mean differences and variances will require pilot studies.

Criterion for Comparison

The criterion considered appropriate for the profile comparison is

$$Rd \leq \theta \tag{11.7}$$

where
- Rd is the in vitro bioequivalence criterion
- θ is the in vitro bioequivalence limit

Rd is derived with the following notation:

Let
- P_R, P_T be the population mean profile across the batches of the reference product and test product
- F_{BR}, F_{BT} be the distribution of deviation of batch mean from population mean profile of the reference product and test product
- F_{CR}, F_{CT} be the distribution of deviation of canister mean from batch mean profile of the reference product and test product
- P_R, P_T be the observed profile of a given puff of reference product and test product (i.e., P_R has a compound distribution of MN(100, P_R), F_{BR}, and F_{CR}, and p_t has a compound distribution of MN(100, P_T), F_{BT}, and F_{CT}, where MN(100,P) is a multinomial distribution with $n = 100$ and $P = (p_1, p_2, \ldots, p_s)$ for an impactor of S stages)

and
- d_{TR} be the observed distance between p_R and p_T, the test–reference distance
- $d_{RR'}$ be the observed distance between p_R and $p_{R'}$, the reference–reference distance (i.e., reference–reference deviation)
- rd be the $d_{TR}/d_{RR'}$, observed ratio of test–reference distance to reference–reference deviation

The in vitro bioequivalence measure is defined by

$$Rd = E(rd)$$

where
- $E(rd)$ is the expected value of rd

Further information on d_{TR}, $d_{RR'}$, rd, and the in vitro profile comparison procedure is provided in Appendix 11A.

Determining a 95% Upper Bound

Since there is no exact or asymptotic distribution of the average *rd*, the 95% upper bound should be determined by the 95th percentile of the empirical sampling distribution generated by a random sample of the matched triplet (test, reference 1, reference 2) of canisters. A description of the procedure is provided in Appendix 11B.

Specification of the Upper Limit

Reserved (simulation studies to develop specifications for the upper limit are ongoing).

MULTIPLE STRENGTHS

A small number of nasal sprays for local action are available in two strengths. Current examples are (1) ipratropium bromide nasal spray, a solution formulation, and (2) beclomethasone dipropionate nasal spray, a suspension formulation. Lower strengths of a product ordinarily would achieve the lower dose per actuation using a lower concentration formulation, without changing the actuator and metering valve or pump (other than dip tube) used in the higher strength product. The following sections describe recommended bioavailability and bioequivalence studies for low strengths of nasal sprays for which bioavailability or bioequivalence for the higher strengths has previously been established. Recommendations are also provided for cases in which bioavailability or bioequivalence is initially established on the low-strength product. No approved nasal aerosols are available in multiple strengths, thus bioavailability and bioequivalence recommendations are not considered for these products.

Solution Formulation Nasal Sprays

The bioavailability of lower or higher strength solution formulation nasal sprays should be based on conduct of all applicable in vitro tests. These studies are generally noncomparative in character. Documentation of bioequivalence between *T* and *R* products should follow the recommendations described in "Formulation and container and closure system" section regarding formulation and container and closure system. Abbreviated in vitro testing is recommended to document bioequivalence of the low-strength *T* product to the low-strength *R* product, provided bioequivalence of the high-strength product has been documented (Table 11.1).

With the exception of the reduced testing, the same protocols and acceptance criteria used to establish bioequivalence of the high-strength products should be used for the low-strength products. In vivo studies are not needed for documentation of bioavailability or bioequivalence of solution

TABLE 11.1
In Vitro Bioavailability and Bioequivalence Studies for Nasal Aerosols and Nasal Sprays

In Vitro Test	High Strength	Low Strength
Dose content uniformity	BE*	BE
Priming and repriming	Yes	Yes
Tail off	Yes	Yes
Droplet size distribution by laser diffraction	BME	B
Droplet size distribution by cascade impactor	BE	No
Spray pattern	BE	B
Plume geometry	B	No

[a] Data requested as part of the bioavailability or bioequivalence submission.

[b] Measures requested for comparative in vitro bioequivalence documentation.

* B, beginning; M, middle; E, end.

formulation nasal sprays. For cases in which bioequivalence is documented for the low-strength product, to subsequently document bioequivalence for the high-strength product, all applicable in vitro tests described in "Bioavailability and bioequivalence: in vitro studies" section should be conducted.

Suspension Formulation Nasal Sprays

The bioavailability of lower-strength suspension formulation nasal sprays should be based on conduct of all applicable in vitro tests described in "Bioavailability and bioequivalence: in vitro studies" section and systemic exposure studies, assuming availability of bioanalytical methodology to allow measurement of systemic concentrations. In the absence of this methodology, bioavailability for systemic absorption should be documented through clinical studies. Bioequivalence conditions for the lower-strength product should be the following:

- Documentation of bioequivalence for the high-strength test products and high-strength reference products, based on acceptable comparative formulations and container and closure systems, comparative in vitro data, and comparative in vivo data
- Acceptable comparative formulations and container and closure systems for the low-strength test products and low-strength reference products
- Acceptable comparative studies for low-strength test products and low-strength reference products for all applicable in vitro tests in "Bioavailability and bioequivalence: in vitro studies" section
- Proportionally similar unit dose between high- and low-dose test product and high- and low-dose reference product
- Equivalent droplet and drug PSD between high- and low-dose test product and high- and low-dose reference product

Provided the aforementioned conditions are met, in vivo studies are not needed for documentation of bioequivalence of the lower-strength products.

For cases in which an ANDA applicant initially documents bioequivalence on the low-strength product and subsequently submits an ANDA for the high-strength product, full in vitro and in vivo documentation of bioequivalence should be provided for the high-strength product. For cases in which an ANDA applicant has documented bioequivalence for its high-strength product and wishes to conduct applicable in vitro tests and in vivo study on the low-strength product, bioequivalence criteria need not include in vitro comparisons between high- and low-strength products.

SMALLER CONTAINER SIZES

Nasal aerosols and nasal sprays may be available in two container sizes. Current examples are the following:

1. Beclomethasone dipropionate nasal aerosol, a suspension formulation.
2. Fluticasone propionate nasal spray, a suspension formulation.
3. Cromolyn sodium nasal spray, a solution formulation. Smaller container sizes of nasal aerosols should be formulated with the same components and composition, metering valve, and actuator as the large container size that was studied in pivotal clinical trials (NDA) or for which bioequivalence has been documented (ANDA). Smaller container sizes of nasal sprays should be formulated with the same components and composition, pump, and actuator as the large container size that was studied in pivotal clinical trials (NDA) or for which bioequivalence has been documented (ANDA). Where this is the case, no further documentation of either bioavailability or bioequivalence is necessary. However, reestablishing proper priming, given a change in the dead volume of the pump and actuator, may in some cases be appropriate.

APPENDIX 11A: IN VITRO PROFILE COMPARISON PROCEDURE BASED ON CHI-SQUARE DIFFERENCES

This appendix describes a method of comparing CI or MSLI deposition profiles on "S" stages or accessories, or groups of stages and accessories, from droplet and/or particle sizing studies. Equivalence may be assessed by comparing the profile difference between test product and reference product canisters (nasal aerosols) or bottles (nasal sprays) to the profile variation between reference product canisters or bottles. The profile comparison is based on chi-square differences.

The following table represents the population mean profiles P_T and P_R of one test canister and one reference canister, respectively.

	Stage										
Product	**1**	**2**	**3**	**4**	**.**	**.**	***s***	**.**	**.**	***S***	**Total**
Test	P_{T1}	P_{T2}	P_{T3}	P_{T4}	.	.	P_{Ts}	.	.	P_{TS}	100
Reference	P_{R1}	P_{R2}	P_{R3}	P_{R4}	.	.	P_{Rs}	.	.	P_{RS}	100

The profile difference between test and reference product canisters is assessed by the chi-square measure as follows:

$$D_{TR} = \frac{\left(P_{T1} - P_{R1}\right)^2}{\left(\left(P_{T1} + P_{R1}\right)/2\right)} + \frac{\left(P_{T2} - P_{R2}\right)^2}{\left(\left(P_{T2} + P_{R2}\right)/2\right)} + \cdots + \frac{\left(P_{TS} - P_{RS}\right)^2}{\left(\left(P_{TS} + P_{RS}\right)/2\right)}$$

Similarly, the profile variation (i.e., difference) between any two canisters of the reference product is

$$D_{RR'} = \frac{\left(P_{R1} - P_{R'1}\right)^2}{\left(\left(P_{R1} - P_{R'1}\right)/2\right)} + \frac{\left(P_{R2} - P_{R'2}\right)^2}{\left(\left(P_{R2} + P_{R'2}\right)/2\right)} + \cdots + \frac{\left(P_{RS} - P_{R'S}\right)^2}{\left(\left(P_{RS} + P_{R'S}\right)/2\right)}$$

The approach involves a comparison of D_{TR}, the profile difference between one test canister and one reference canister, to $D_{RR'}$, the profile variation between two canisters of the reference product, where the latter is based on two randomly selected reference canisters. The comparison of profile differences is given by the ratio of D to D. A large D is one that is large relative to the variation that would be expected between two canisters of the reference product.

In order to estimate D and D, the observed data of one canister of test product and two different canisters of reference product need to be matched as a triplet. The observed profiles of the three canisters of a given triplet may be represented in the following table.

	Stage										
Product	**1**	**2**	**3**	**4**	**.**	**.**	***s***	**.**	**.**	***S***	**Total**
Test	p_{T1}	p_{T2}	p_{T3}	p_{T4}	.	.	p_{Ts}	.	.	p_{TS}	100
Reference 1	p_{R1}	p_{R2}	p_{R3}	p_{R4}	.	.	p_{Rs}	.	.	p_{RS}	100
Reference 2	$p_{R'1}$	$p_{R'2}$	$p_{R'3}$	$p_{R'4}$	.	.	$p_{R's}$	.	.	$p_{R'S}$	100

The observed profile difference d_{TR} between test and reference products is

$$D_{TR} = \frac{\left(p_{T1} - \left(p_{R1} + p_{R'1}\right)/2\right)^2}{\left(\left(p_{T1} + \left(p_{R1} + p_{R'1}\right)/2\right)/2\right)} + \frac{\left(p_{T2} - \left(p_{R2} + p_{R'2}\right)/2\right)^2}{\left(\left(p_{T2} + \left(p_{R2} + p_{R'2}\right)/2\right)/2\right)} + \cdots + \frac{\left(p_{TS} - \left(p_{RS} + p_{R'S}\right)/2\right)}{\left(\left(p_{TS} + \left(p_{RS} + p_{R'S}\right)/2\right)/2\right)}$$

The reference product canister-to-canister variation within the triplet is estimated by the profile difference between the two paired reference canisters, R and R':

$$d_{RR'} = \frac{\left(p_{R1} - p_{R'1}\right)^2}{\left(\left(p_{R1} + p_{R'1}\right)/2\right)} + \frac{\left(p_{R2} + p_{R'2}\right)^2}{\left(\left(p_{R2} + p_{R'2}\right)/2\right)} + \cdots + \frac{\left(p_{RS} - p_{R'S}\right)^2}{\left(\left(p_{RS} + p_{R'S}\right)/2\right)}$$

For a given triplet of canisters (test, reference 1, reference 2), the ratio of d_{TR} to $d_{RR'}$ may be obtained as follows:

$$rd = \frac{d_{TR}}{d_{RR'}}$$

Assuming that there are $N(T, R, R')$ triplets in the sample, the unbiased estimate of Rd [$= E(rd)$] is the sample mean of the N observed $d_{TR}/d_{RR'}$ values.

For an experiment consisting of three lots each of test and reference products, and with 10 canisters per lot, the lots can be matched into 6 different combinations of triplets with 2 different reference lots in each triplet. The 10 canisters of a test lot can be paired with the 10 canisters of each of the two reference lots in (10 factorial)^2 = (3,628,800)^2 combinations in each of the lot triplets. Hence, a random sample of the N canister-pairing of the six test–reference 1–reference 2 lot triplets is needed. Rd is estimated by the sample mean of the rd's calculated for the triplets in the selected sample of N:

$$Rd = \text{Sample mean of} \left(\frac{d_{TR}}{d_{RR'}} \right)$$

APPENDIX 11B: DETERMINATION OF THE 95% UPPER BOUND FOR IN VITRO PROFILE COMPARISONS

Assume the profile comparison is to be carried out with a random sample with no replacement of $N = 500$ matches (from the population of 6 × (10 factorial)^2 matches). The average of the 500 sample rd's ($= d_{TR}/d_{RR'}$) gives Rd. The 95% upper bound of Rd is the 95th percentile of the 500 calculated rd's (i.e., the 25th largest rd among the 500 calculated rd's).

12 Bioequivalence of Complimentary and Alternate Medicines

BACKGROUND

The U.S. government has again led the development of health-care systems by establishing a White House Commission on complementary and alternative medicines (CAM).

CAM can be defined as a group of medical, health-care, and healing systems other than those included in mainstream health care in the United States. CAM includes the worldviews, theories, modalities, products, and practices associated with these systems and their use to treat illness and promote health and well-being.

Although heterogeneous, the major CAM systems have many common characteristics, including a focus on individualizing treatments, treating the whole person, promoting self-care and self-healing, and recognizing the spiritual nature of each individual. In addition, many CAM systems have characteristics commonly found in mainstream health care, such as a focus on good nutrition and preventive practices. Unlike mainstream medicine, CAM often lacks or has only limited experimental and clinical study; however, scientific investigation of CAM is beginning to address this knowledge gap. Thus, boundaries between CAM and mainstream medicine, as well as among different CAM systems, are often blurred and are constantly changing.

Examples of the health-care systems, practices, and products typically classified as CAM in the United States are listed in Table 12.1.

Terms "mainstream," "conventional," "allopathic," and "biomedical" are used synonymously to refer to the principal form of health care and medicine available in the United States.

Many of the CAM systems of health care listed in Table 12.1 have evolved from the collective clinical experiences of many practitioners over generations of practice, such as in traditional Chinese medicine. Others have evolved from the clinical experiences of a single practitioner or small groups of practitioners who have developed a particular intervention.

Despite their diversity, there are some common threads that run among many traditional systems of health care as well as systems that have emerged more recently. These similarities include an emphasis on whole systems, the promotion of self-care and the stimulation of self-healing processes, the integration of mind and body, the spiritual nature of illness and healing, and the prevention of illness by enhancing the vital energy, or subtle forces, in the body.

The history of CAM in the United States is a long, complex story that has been shaped by scientific, economic, and social factors. A detailed rendering of this history is beyond the scope of this report. This section instead provides a brief overview of the more recent developments that have helped shape the present status of CAM in this country and its prospects for contributing to the health and well-being of our nation.

Early American health care consisted of an eclectic mix of systems. In fact, until the middle of the nineteenth century, the vast majority of primary medical care in this country was provided by botanical healers, midwives, chiropractors, homeopaths, and an assortment of other lay healers offering a variety of herbs and nostrums for a range of illnesses.

TABLE 12.1
CAM System of Health-Care, Therapies, and Products

Major Domains of CAM	Examples under Each Domain
Alternative health-care systems	Ayurvedic medicine
	Chiropractic
	Homeopathic medicine
	Native American medicine (e.g., sweat lodge, medicine wheel)
	Naturopathic medicine
	Traditional Chinese medicine (e.g., acupuncture, Chinese herbal medicine)
Mind–body interventions	Meditation
	Hypnosis
	Guided imagery
	Dance therapy
	Music therapy
	Art therapy
	Prayer and mental healing
Biological-based therapies	Herbal therapies
	Special diets (e.g., macrobiotics, extremely low-fat or high-carbohydrate diets)
	Orthomolecular medicine (e.g., megavitamin therapy)
	Individual biological therapies (e.g., shark cartilage, bee pollen)
Therapeutic massage, body work, and somatic movement therapies	Massage
	Feldenkrais
	Alexander method
Energy therapies	Qigong
	Reiki
	Therapeutic touch
Bioelectromagnetics	Magnet therapy

This began to change in the latter part of the nineteenth century, however, with the development and validation of the germ theory and significant scientific advances in antiseptic techniques, anesthesia, and surgery. Beginning in the late 1800s and lasting until the early twentieth century, there also was a major revolution in medical education that helped scientific medicine evolve into the dominant health-care system in this country.

This revolution in medical education began with the publication of William Osler's (1847–1919) textbook, *The Principles and Practice of Medicine* in 1892, which brought diagnostic clarity to medical practice. By 1905, Osler's textbook was the primary medical textbook in the vast majority of U.S. medical schools. This revolution culminated with the release of a report by Abraham Flexner in 1910 that served to crystallize the educational reform movement. After the release of the Flexner's report, many medical institutions that did not meet its standards were driven out of business or forced to implement significantly more rigorous training programs. Schools for many unorthodox healing systems either ceased to exist or became marginalized.

The isolation and elaboration of life-saving hormones, sulfa drugs, and other antibiotics in the early and middle of the twentieth century, conventional medicine cemented its place as the nation's preeminent form of health care in this country. Although most of the other health-care systems and their therapies did not disappear, they were considered by most of the public and the mainstream medical community to be unscientific relics of the past. As a result, many were practiced in relative obscurity.

With the reduced threat of infectious diseases and other acute illnesses, conventional medicine began to turn its focus to the more complex and costly problems of chronic, degenerative illnesses. As a result of public health interventions developed earlier in the twentieth century, people began living significantly longer. This gradual aging of the population began to significantly increase the prevalence of chronic conditions, such as arthritis, back pain, diabetes, hypertension, heart disease, and cancer, putting further pressure on conventional medicine to address these conditions.

As the health-care system developed more sophisticated means of diagnosing and managing chronic illnesses, the cost of health care began to rise dramatically. Between 1965 and 1975, national health-care expenditures more than tripled, rising from just over $41 billion to nearly $130 billion. Although employers and government programs covered some of these increases, out-of-pocket expenditures more than doubled during this same period. Since then, costs have continued to rise, with national health-care expenditures reaching more than $1.2 trillion in 2000, the latest year for which such figures are available, and they are expected to reach more than $2.6 trillion by 2010.

It was during this time of increasing rates of chronic illness and escalating health-care costs that medical pluralism began to reemerge in this country. This reemergence was spurred on by a number of overlapping and sometimes interrelated movements. Beginning in the 1950s, the whole foods and dietary supplement movements began to change Americans' view of food as not only something they needed to stay alive but also as potential therapeutic agents. In the late 1960s and early 1970s, Americans were increasingly exposed to a variety of traditional health care systems from foreign and indigenous cultures, many of which dated back to antiquity. *New York Times* writer James Reston's account of his emergency appendectomy in a Chinese hospital during then Secretary of State Henry Kissinger's visit to China in 1971 was particularly influential in this process. Reston's article described how his postoperative pain and discomfort were relieved by acupuncture and herbs. For most Americans, this was their first glimpse of traditional Chinese medicine and its potential uses.

During this same period, the growing "counterculture" movement in America sparked a fascination with the religious and philosophical traditions of Asian cultures. Transcendental meditation, which is derived from Hinduism, became widely known and practiced. Meanwhile, there was a growing interest in indigenous health-care traditions, such as Native American and Mexican-American health-care practices, particularly their reliance on herbs and natural substances. This movement, in turn, led to a renewed interest in "natural" health-care movements that had developed in this country in the nineteenth century but had been relegated to the background of the American health-care landscape.

The late 1970s saw the emergence of the holistic health-care movement in this country. Holistic practice (holism comes from the Greek word "holos" or "whole") emphasized an attention to the whole person, including the physical, spiritual, psychological, and ecological dimensions of healing. Holistic health care incorporates practices and concepts of Eastern philosophy and diverse cultural traditions, including acupuncture and the use of herbs, massage, and relaxation techniques as well as conventional medical practices. It gained its greatest following among nurses. However, many physicians, particularly those in the new specialty of family medicine, also became interested in this movement. The American Holistic Medical and Nurses Associations were formed, large professional and public conferences held, and a number of holistic medical clinics and holistic health centers opened.

The late 1970s and early 1980s also was a time when a variety of self-care movements emerged; they offered programs or sponsored events to help individuals and families increase wellness or reduce their risk of onset of illness through diet or lifestyle changes. The years since then have been a particularly active time for the personal fitness movement, which increasingly is making use of the techniques of other systems of healing, such as yoga, tai chi, and massage.

Today, the use of CAM approaches and therapies is more prevalent in a number of patient populations in the Unites States, no matter how narrowly or broadly it is defined. Physicians, hospitals, and other conventional health-care organizations also are showing a growing interest in CAM.

Although such prevalence of use and interest in CAM is not an indication that these practices are effective, it does suggest that those with chronic conditions and the physicians who treat them are looking for more therapeutic options than are widely available in conventional health-care settings. Indeed, for some chronic conditions, state-of-the-art conventional therapies have provided only modest gains. For example, according to a number of assessments over the years, expensive mainstream health-care approaches to managing chronic lower back pain often have not been very effective. This is perhaps why individuals with back pain are some of the most frequent users of CAM practices.

The Food and Drug Administration (FDA) has responded to the viability of CAM practices and anticipates many products approved under the guidance provided by the CAM division of FDA.

FDA PERSPECTIVE ON CAM

The history of legislation described earlier in the United States demonstrates that the U.S. agencies have been proactive in creating pathways that will lead to affordable medicines. Whereas the United States remains the largest market of pharmaceutical products with over $700 billion of market, it has not ignored the emerging needs to recognize other systems of medicines such as the herbal therapies and other nonconventional methods; in summary, any modality that is not clearly connected with a chemical structure and activity can be labeled as nontraditional in the modern understanding of pharmacology. In the United States, FDA chose a designation of CAM to encompass all herbal and nontraditional medicinal systems. This was a giant step for the U.S. FDA that had long been a promoter of a clean connection between chemistry and pharmacology. Such is not the case in the use of this broad range of drugs that fall in this category of treatment modalities.

The term "complementary and alternative medicine" (CAM) encompasses a wide array of health-care practices, products, and therapies that are distinct from practices, products, and therapies used in "conventional" or "allopathic" medicine. Some forms of CAM, such as traditional Chinese medicine and the Indian Ayurvedic medicine, have been practiced for centuries, whereas others, such as electrotherapy, are more recent in origin. In the United States, the practice of CAM has risen dramatically in recent years. In 1992, the congress established the Office of Unconventional Therapies, which later became the Office of Alternative Medicine (OAM), to explore "unconventional medical practices." In 1998, OAM became the National Center for Complementary and Alternative Medicine (NCCAM). NCCAM is a center within the National Institutes of Health. The Institute of Medicine, in its book entitled *Complementary and Alternative Medicine in the United States*, stated that more than one-third of American adults reported using some form of CAM and that visits to CAM providers each year exceed those to primary care physicians. As the practice of CAM has increased in the U.S. FDA observed increased confusion as to whether certain products used in CAM products are subject to regulation under the Food, Drug, and Cosmetic (FDC) Act or Public Health Service (PHS) Act; this is further compounded by the importation of a large number of CAM products since these are widely used, not just in the developing countries but also in many developed markets such as Germany and Japan. The FDA guidance regarding the CAM products makes two fundamental points:

1. First, depending on the CAM therapy or practice, a product used in a CAM therapy or practice *may* be subject to regulation as a biological product, cosmetic, drug, device, or food (including food additives and dietary supplements) under the Act or the PHS Act. For example, the PHS Act defines "biological product," and the Act defines (among other things) cosmetic, device, dietary supplement, drug, as well as "new drug" and "new animal drug, "food," and food additive.
2. Second, neither the Act nor the PHS Act exempts CAM products from regulation. This means, for example, if a person decides to produce and sell raw vegetable juice for use in

juice therapy to promote optimal health, that product is a food subject to the requirements for foods in the Act and FDA regulations, including the hazard analysis and critical control point (HACCP) system requirements for juices in 21 CFR part 120. If the juice therapy *is* intended for use as part of a disease treatment regimen instead of for the general wellness, the vegetable juice would also be subject to regulation as a drug under the Act.

The NCCAM defines CAM as "a group of diverse medical and health-care systems, practices, and products that are not presently considered to be part of conventional medicine." It interprets "complementary" medicine as being used together with conventional medicine, whereas "alternative" medicine is used in place of conventional medicine. NCCAM classifies CAM therapies into four categories or "domains." These are biologically based practices, energy therapies, manipulative and body-based methods, and mind–body medicine.

NCCAM once had a fifth domain, "alternative medical systems," but now considers "alternative medical systems" (now known as "whole medical systems") to be a separate category rather than another domain because alternative medical systems use practices from the four domains listed earlier. For purposes of this guidance, FDA adopts the same domains and "whole medical systems" category that NCCAM uses.

According to NCCAM, the domain called "biologically based practices" includes, but is not limited to, botanicals, animal-derived extracts, vitamins, minerals, fatty acids, amino acids, proteins, prebiotics and probiotics, whole diets, and "functional foods." Many biologically based products within this domain are subject to statutory and regulatory requirements under the Act or the PHS Act. The intended use of a product plays a central role in how it is regulated. For example:

- Botanical products, depending on the circumstances, may be regulated as drugs, cosmetics, dietary supplements, or foods. All four types of products are subject to the Act. For example, a botanical product intended for use in treating a disease would generally be regulated as a drug; a botanical product taken by mouth, labeled as a dietary supplement, and intended for use to affect the structure or function of the body would generally be regulated as a dietary supplement; a raw or dried botanical intended for use as an ingredient to flavor food would generally be regulated as a food or as a food additive, depending on whether the botanical was generally recognized as safe (GRAS) for its intended use in food; and a lotion containing botanical ingredients and intended for use in moisturizing the skin would generally be regulated as a cosmetic.
- Probiotics may be regulated as dietary supplements, foods, or drugs under the Act, depending on the product's intended use. Other factors may also affect the classification of the product, for example, whether the product contains a "dietary ingredient" as defined in section 201(ff)(1) of the Act (21 U.S.C. 321(ff)(1)), whether it is represented as a conventional food or as a meal replacement (see section 201(ff)(2)(B) of the Act), and, for probiotics used as ingredients in a conventional food, whether the ingredient is GRAS for its intended use (see section 201(s) of the Act (21 U.S.C. 321(s)). In addition to any requirements that apply based on the product's classification under the Act, probiotics may also be subject to the PHS Act's provisions concerning the prevention of communicable disease, due to potential disease-causing microorganisms that might be contained in such products. Finally, if a probiotic is a drug under the Act, it may be subject to regulation as a biological product under the PHS Act as well.
- Products that NCCAM would consider to be "functional foods" may be subject to FDA regulation as foods, dietary supplements, or drugs under the Act. As with botanicals and probiotics, the classification of a "functional food" under the Act is based primarily on the product's intended use and may also involve other factors, depending on the elements of the statutory definition of a particular product category.

NCCAM considers energy medicine to involve energy fields of two types:

1. Veritable energy fields, which can be measured and used either as mechanical vibrations (such as sound) or electromagnetic forces, including visible light, magnetism, monochromatic radiation (such as laser light), and other light rays
2. Putative energy fields (or biofields) that have defied measurement to date by reproducible methods. According to NCCAM, therapies involving putative energy fields "are based on the concept that human beings are infused with a subtle form of energy" and therapists "claim that they work with this subtle energy, see it with their own eyes, and use it to effect changes in the physical body and influence health."

In a sense, "conventional" medicine already uses various forms of "energy" medicine. For example, a magnetic resonance imaging (MRI) device uses electromagnetic waves to create images of internal body organs and tissues. As another example, an ultrasound machine uses sound waves to create images of body organs, tissues, and fetuses. Given their intended uses, we regulate these products as medical devices under the Act. The CAM products that use veritable energy fields in the diagnosis of disease or other conditions or in the cure, mitigation, treatment, or prevention of disease in man or animals or to affect the structure or any function of the body of man or animals may be medical devices under the Act. Additionally, if the product is electronic and emits radiation, it may be subject to additional requirements to ensure that there is no unnecessary exposure of people to radiation. CAM products that use putative energy fields in the diagnosis of disease or other conditions, or in the cure, mitigation, treatment, or prevention of disease in man or animals, may be medical devices under the Act. For example, FDA regulates acupuncture needles as "class II" medical devices.

Under the umbrella of manipulative and body-based practices is a heterogeneous group of CAM interventions and therapies. These include chiropractic and osteopathic manipulation, massage therapy, Tui Na, reflexology, rolfing, Brown technique, Trager bodywork, Alexander technique, Feldenkrais method, and a host of others. Manipulative and body-based practices focus primarily on the structures and systems of the body, including the bones and joints, the soft tissues, and the circulatory and lymphatic systems. To the extent that manipulative and body-based practices involve practitioners physically manipulating a patient's body, without using tools or machines, FDA does not consider that such practices are subject to regulation under the Act or the PHS Act. If, however, the manipulative and body-based practices involve the use of equipment (such as massage devices) or the application of a product (such as a lotion, cream, or oil) to the skin or other parts of the body, those products may be subject to regulation under the Act, depending on the nature of the product and its intended use.

NCCAM describes mind–body medicine as focusing on "the interactions among the brain, mind, body, and behavior, and the powerful ways in which emotional, mental, social, spiritual, and behavioral factors can directly affect health." It states that mind–body medicine "typically focuses on intervention strategies that are thought to promote health, such as relaxation, hypnosis, visual imagery, meditation, yoga, biofeedback, tai chi, qi gong, cognitive-behavioral therapies, group support, autogenic training, and spirituality." In general, CAM practices in this domain would *not* be subject to FDA jurisdiction under the Act or the PHS Act. As with the manipulative and body-based practices domain, however, any equipment or other products used as part of the practice of mind–body medicine may be subject to FDA regulation, depending on the nature of the product and its intended use. For example, biofeedback machines intended to help a patient learn to affect body functions, such as muscle activity, are regulated as class II devices.

NCCAM describes "whole medical systems" as involving "complete systems of theory and practice that have evolved independently from or parallel to allopathic (conventional) medicine." These may reflect individual cultural systems, such as traditional Chinese medicine and Ayurvedic medicine. Some elements common to whole medical systems are a belief that the body has the power to heal itself and that healing may involve techniques that use the mind, body, and spirit. Although it is unlikely that a whole medical system itself would be subject to regulation under the Act or the PHS Act, products used as *components* of whole medical systems may be subject to FDA regulation for the reasons described earlier.

To understand how the Act or the PHS Act might apply to CAM products, it is important to understand the Act's statutory definitions or, in the case of the PHS Act, our authority regarding biological products.

To illustrate how the definitions of drug or new drug might apply, consider an herbal product that is intended to treat arthritis in humans. The herbal product, which would be a "biologically based practice" insofar as CAM domains are concerned, would be a "drug" under section 201(g)(1)(B) of the Act because it is intended for use in the diagnosis, cure, mitigation, treatment, or prevention of disease (arthritis) in man. The same herbal product would also be a "new drug" under section 201(p)(1) of the Act unless it is generally recognized, among experts qualified by scientific training and experience to evaluate the safety and effectiveness of drugs, as safe and effective for use under the conditions prescribed, recommended, or suggested in the labeling. "New drug" status triggers the Act's requirements for premarket review and approval by FDA.

To illustrate how a CAM product might be a "device" under the Act, acupuncture is a CAM therapy that seeks to stimulate energy pathways ("meridians") by puncturing, pressing, heating, using electrical current, or using herbal medicines. Fine needles are often used, and these acupuncture needles are "devices" under section 201(h) of the Act because they are intended for use in the cure, mitigation, treatment, or prevention of disease in man or are intended to affect the structure or function of the body of man. We regulate acupuncture needles (see 21 CFR 880.5580), but not the practice of acupuncture itself. A detailed discussion of the Act's device provisions is beyond the scope of this guidance document. Note, however, that the Act establishes classifications for devices (class I, II, or III) that affect how they are regulated. The Act also imposes certain requirements on those who manufacture devices (including requirements pertaining to establishment registration and product listing, premarket review, labeling, postmarket reporting, and good manufacturing practices). Certain requirements also apply to device distributors.

To illustrate how a CAM practice might involve "foods," juice therapy uses juice made from vegetables and fruits. Absent any claims that would make the juice subject to the drug definition, the juice would be a "food" under section 201(f) of the Act because it is an article used for food or drink for man. A detailed discussion of the Act food provisions is beyond the scope of this guidance document. However, anyone who intends to market CAM products that might be subject to regulation under these provisions should familiarize himself or herself with the Act's requirements for foods, particularly with respect to safety and labeling. The Act and our food regulations can be found at our website at www.fda.gov/opacom/laws.

To illustrate how a CAM product might involve "food additives" under section 201(s) of the Act, some CAM practices involve dietary modifications where substances such as botanicals or enzymes are added to foods in the diet. If a manufacturer adds such a substance to a food, the substance may fall within the "food additive" definition at section 201(s) of the Act. A food additive is subject to premarket approval by FDA under section 409 of the Act (21 U.S.C. 348). Food additives that we have not approved or that do not comply with applicable FDA regulations prescribing safe conditions of use are deemed to be unsafe under section 409(a) of the Act, and foods that contain such additives are adulterated under section 402(a)(2)(C) of the Act (21 U.S.C. 342(a)(2)(C)). The Act provides that a substance is exempt from the definition of a food additive and, thus, from premarket approval if, among other reasons, it is GRAS by qualified experts under the conditions of intended use. Whether a substance added to a food is considered to be a food additive or is GRAS, any claims associating the substance with the reduction of a disease risk are "health claims" (defined in 21 CFR 101.14(a)(1)) that require premarket review by FDA. A detailed discussion of the Act's food additive provisions is beyond the scope of this guidance document. However, anyone intending to market CAM products that are or contain substances that might be subject to regulation as food additives should consult with the Act's food additive requirements.

Except for purposes of section 201(g) (of the Act), a dietary supplement shall be deemed to be a food within the meaning of this Act. To illustrate how a CAM product might be a "dietary supplement" under section 201(ff) of the Act, consider botanical products used in naturopathy.

Naturopathy is a CAM whole medical system that views disease as a manifestation of alterations in the processes by which the body heals itself. For example, naturopathic cranberry tablets might be labeled for use to maintain the health of the urinary tract. In this example, the cranberry tablets generally would be regulated as "dietary supplements" under section 201(ff)(1) of the Act if they were labeled for use to "maintain the health of the urinary tract" rather than "prevent urinary tract infections." The cranberry tablets would be regulated as "drugs" under section 201(g) of the Act if they were labeled for use to "treat urinary tract infections" even if they were labeled as dietary supplements. A detailed discussion of the Act's dietary supplement provisions is beyond the scope of this guidance document.

It is possible that certain products used in conjunction with CAM practices may be "cosmetics" under the Act. For example, if a CAM practice involves massage with a moisturizer, the moisturizer could be a "cosmetic" to the extent that it is "rubbed, poured, sprinkled, or sprayed on" the body for beautification or appearance-altering purposes. However, if the moisturizer's intended use is also for the diagnosis, cure, mitigation, treatment, or prevention of disease or to affect the structure or any function of the body, then it may also be subject to regulation as a drug. Other examples of drug/cosmetic combinations are deodorants that are also antiperspirants, moisturizers and makeup marketed with sun-protection claims, and shampoos that also treat dandruff. The Act does not require premarket approval for cosmetics, but it does prohibit the marketing of adulterated or misbranded cosmetics in interstate commerce. Anyone intending to market CAM products that might be subject to regulation as cosmetics should familiarize himself or herself with the safety and labeling requirements for these products in the Act and our regulations.

If a CAM product manufacturer attempted to use a live, disease-causing virus as a component of a CAM product, FDA could exercise its authority under section 361 of the PHS Act and 21 CFR 1240.30 to take action against the product, in addition to consider the applicability of section 351 of the PHS Act.

The bioequivalence testing of complimentary and alternate medicine products is required in accordance with how these products are classified, whether a biological, cosmetic, or a device. In February 2007, the FDA announced the availability of a draft guidance for industry entitled *Complementary and Alternative Medicine Products and Their Regulation by the Food and Drug Administration*. The term "complementary and alternative medicine" (CAM) encompasses a wide array of health-care practices, products, and therapies that are distinct from practices, products, and therapies used in "conventional" or "allopathic" medicine.

In the United States, the practice of CAM has risen dramatically in recent years. In 1992, the congress established the Office of Unconventional Therapies, which later became the OAM, to explore "unconventional medical practices." In 1998, OAM became the NCCAM. NCCAM is a center within the National Institutes of Health. The Institute of Medicine, in its book entitled *Complementary and Alternative Medicine in the United States*, stated that more than one-third of American adults reported using some form of CAM and that visits to CAM providers each year exceed those to primary care physicians.

As the practice of CAM has increased in the United States, FDA saw increased confusion as to whether certain products used in CAM (which, for convenience, we will refer to as "CAM products") are subject to regulation under the Act or the PHS Act. FDA also saw an increase in the number of CAM products imported into the United States.

Therefore, the draft guidance discusses when a CAM product is subject to the Act or the PHS Act. (When the draft guidance mentions a particular CAM therapy, practice, or product, it does so in order to provide background information or to serve as an example or illustration; any mention of a particular CAM therapy, practice, or product should not be construed as expressing FDA's support for or endorsement of that particular CAM therapy, practice, or product or, unless specified otherwise, as an agency determination that a particular product is safe and effective for its intended uses or is safe for use.) The draft guidance makes the following two fundamental points:

1. First, depending on the CAM therapy or practice, a product used in a CAM therapy or practice may be subject to regulation as a biological product, cosmetic, drug, device, or food (including food additives and dietary supplements) under the Act or the PHS Act.
2. Second, neither the Act nor the PHS Act exempts CAM products from regulation.

The term "complementary and alternative medicine" (CAM) encompasses a wide array of health-care practices, products, and therapies that are distinct from practices, products, and therapies used in "conventional" or "allopathic" medicine. Some forms of CAM, such as traditional Chinese medicine and Ayurvedic medicine, have been practiced for centuries, whereas others, such as electrotherapy, are more recent in origin. In the United States, the practice of CAM has risen dramatically in recent years. In 1992, the congress established the Office of Unconventional Therapies, which later became the OAM, to explore "unconventional medical practices." In 1998, OAM became the NCCAM. NCCAM is a center within the National Institutes of Health. The Institute of Medicine, in its book entitled *Complementary and Alternative Medicine in the United States*, stated that more than one-third of American adults reported using some form of CAM and that visits to CAM providers each year exceed those to primary care physicians.

As the practice of CAM has increased in the United States, the FDA ("we") has seen increased confusion as to whether certain products used in CAM (which, for convenience, we will refer to as "CAM products") are subject to regulation under the Federal FDC Act ("the Act") or "PHS Act." There is also an increase in the number of CAM products imported into the United States.

The FDA guidance makes two fundamental points:

1. First, depending on the CAM therapy or practice, a product used in a CAM therapy or practice *may* be subject to regulation as a biological product, cosmetic, drug, device, or food (including food additives and dietary supplements) under the Act or the PHS Act. For example, the PHS Act defines "biological product," and the Act defines (among other things)
   a. Cosmetic
   b. Device
   c. Dietary supplement
   d. Drug, as well as "new drug" and "new animal drug"
   e. Food
   f. Food additive
2. Second, neither the Act nor the PHS Act exempts CAM products from regulation. This means, for example, if a person decides to produce and sell raw vegetable juice for use in juice therapy to promote optimal health, that product is a food subject to the requirements for foods in the Act and FDA regulations, including the HACCP system requirements for juices in 21 CFR part 120. If the juice therapy is intended for use as part of a disease treatment regimen instead of for the general wellness, the vegetable juice would also be subject to regulation as a drug under the Act.

NCCAM defines CAM as "a group of diverse medical and health care systems, practices, and products that are not presently considered to be part of conventional medicine." It interprets "complementary" medicine as being used together with conventional medicine, whereas "alternative" medicine is used in place of conventional medicine. NCCAM classifies CAM therapies into four categories or "domains." These are

1. Biologically based practices
2. Energy therapies

3. Manipulative and body-based methods
4. Mind–body medicine

NCCAM once had a fifth domain, "alternative medical systems," but now considers "alternative medical systems" (now known as "whole medical systems") to be a separate category rather than another domain because alternative medical systems use practices from the four domains listed earlier. For purposes of this guidance, FDA adopts the same domains and "whole medical systems" category that NCCAM uses.

According to NCCAM, the domain called "biologically based practices" includes, but is not limited to, botanicals, animal-derived extracts, vitamins, minerals, fatty acids, amino acids, proteins, prebiotics and probiotics, whole diets, and "functional foods." Many biologically based products within this domain are subject to statutory and regulatory requirements under the Act or the PHS Act. The intended use of a product plays a central role in how it is regulated. For example:

- Botanical products, depending on the circumstances, may be regulated as drugs, cosmetics, dietary supplements, or foods. All four types of products are subject to the Act. For example, a botanical product intended for use in treating a disease would generally be regulated as a drug; a botanical product taken by mouth, labeled as a dietary supplement, and intended for use to affect the structure or function of the body would generally be regulated as a dietary supplement; a raw or dried botanical intended for use as an ingredient to flavor food would generally be regulated as a food or as a food additive, depending on whether the botanical was GRAS for its intended use in food; and a lotion containing botanical ingredients and intended for use in moisturizing the skin would generally be regulated as a cosmetic.
- Probiotics may be regulated as dietary supplements, foods, or drugs under the Act, depending on the product's intended use. Other factors may also affect the classification of the product, for example, whether the product contains a "dietary ingredient" as defined in section 201(ff)(1) of the Act (21 U.S.C. 321(ff)(1)), whether it is represented as a conventional food or as a meal replacement (see section 201(ff)(2)(B) of the Act), and, for probiotics used as ingredients in a conventional food, whether the ingredient is GRAS for its intended use (see section 201(s) of the Act (21 U.S.C. 321(s)). In addition to any requirements that apply based on the product's classification under the Act, probiotics may also be subject to the PHS Act's provisions concerning the prevention of communicable disease, due to potential disease-causing microorganisms that might be contained in such products. Finally, if a probiotic is a drug under the Act, it may be subject to regulation as a biological product under the PHS Act as well.
- Products that NCCAM would consider to be "functional foods" may be subject to FDA regulation as foods, dietary supplements, or drugs under the Act. As with botanicals and probiotics, the classification of a "functional food" under the Act is based primarily on the product's intended use and may also involve other factors, depending on the elements of the statutory definition of a particular product category.

NCCAM considers energy medicine to involve energy fields of two types:

1. Veritable energy fields, which can be measured and use either mechanical vibrations (such as sound) or electromagnetic forces, including visible light, magnetism, monochromatic radiation (such as laser light), and other light rays
2. Putative energy fields (or biofields) that have defied measurement to date by reproducible methods. According to NCCAM, therapies involving putative energy fields "are based on the concept that human beings are infused with a subtle form of energy" and therapists "claim that they work with this subtle energy, see it with their own eyes, and use it to effect changes in the physical body and influence health."

In a sense, "conventional" medicine already uses various forms of "energy" medicine. For example, an MRI device uses electromagnetic waves to create images of internal body organs and tissues. As another example, an ultrasound machine uses sound waves to create images of body organs, tissues, and fetuses. Given their intended uses, we regulate these products as medical devices under the Act. CAM products that use veritable energy fields in the diagnosis of disease or other conditions or in the cure, mitigation, treatment, or prevention of disease in man or animals or to affect the structure or any function of the body of man or animals may be medical devices under the Act. Additionally, if the product is electronic and emits radiation, it may be subject to additional requirements to ensure that there is no unnecessary exposure of people to radiation. CAM products that use putative energy fields in the diagnosis of disease or other conditions or in the cure, mitigation, treatment, or prevention of disease in man or animals may be medical devices under the Act. For example, we regulate acupuncture needles as "class II" medical devices.

According to NCCAM under the umbrella of manipulative and body-based practices is a heterogeneous group of CAM interventions and therapies. These include chiropractic and osteopathic manipulation, massage therapy, Tui Na, reflexology, rolfing, Brown technique, Trager bodywork, Alexander technique, Feldenkrais method, and a host of others. Manipulative and body-based practices focus primarily on the structures and systems of the body, including the bones and joints, the soft tissues, and the circulatory and lymphatic systems.

To the extent that manipulative and body-based practices involve practitioners physically manipulating a patient's body, without using tools or machines, we do not believe that such practices are subject to regulation under the Act or the PHS Act. If, however, the manipulative and body-based practices involve the use of equipment (such as massage devices) or the application of a product (such as a lotion, cream, or oil) to the skin or other parts of the body, those products may be subject to regulation under the Act, depending on the nature of the product and its intended use.

NCCAM describes mind–body medicine as focusing on "the interactions among the brain, mind, body, and behavior, and the powerful ways in which emotional, mental, social, spiritual, and behavioral factors can directly affect health." It states that mind–body medicine "typically focuses on intervention strategies that are thought to promote health, such as relaxation, hypnosis, visual imagery, meditation, yoga, biofeedback, tai chi, qi gong, cognitive-behavioral therapies, group support, autogenic training, and spirituality."

In general, CAM practices in this domain would *not* be subject to our jurisdiction under the Act or the PHS Act. As with the manipulative and body-based practices domain, however, any equipment or other products used as part of the practice of mind–body medicine may be subject to FDA regulation, depending on the nature of the product and its intended use. For example, biofeedback machines intended to help a patient learn to affect body functions, such as muscle activity, are regulated as class II devices.

NCCAM describes whole medical systems as involving "complete systems of theory and practice that have evolved independently from or parallel to allopathic (conventional) medicine." These may reflect individual cultural systems, such as traditional Chinese medicine and Ayurvedic medicine. Some elements common to whole medical systems are a belief that the body has the power to heal itself and that healing may involve techniques that use the mind, body, and spirit. Although it is unlikely that a whole medical system itself would be subject to regulation under the Act or the PHS Act, products used as *components* of whole medical systems may be subject to FDA regulation for the reasons described earlier.

13 Bioequivalence of Biosimilar Products

BACKGROUND

Drugs derived from biological sources such as vaccines, sera, therapeutic proteins, and antibodies are large molecules and structurally difficult to characterize, have side effects like immunogenic responses, and have stability profiles that are difficult to predict and structure–activity relationship that is ill defined, all leading to realization that the bioequivalence of these products cannot be demonstrated by the currently used methods used for chemically derived drugs (small molecule).

For small-molecule drug products, a clear pathway exists on demonstrating bioequivalence through filing of an abbreviated new drug application (ANDA) under the *Drug Price Competition and Patent Term Restoration* Act of 1984 in the United States (the Hatch and Waxman Act). By excluding all biological products from the 1984 Waxman–Hatch Act, the choice of generic biological products was eliminated since it was general consensus that a true generic form, which will be substitutable, may not exist. Also, in 1984, there was no expiry of patents of biological drugs, so the issue was not raised as intensely as it is done now.

Unlike small-molecule drug products, the *generic versions* of biological products are viewed as biosimilar and *not* as generic drug products, which are substitutable. The biosimilars as referred to by European Medicines Agency (EMA) of European Union (EU) may have different designations given by other agencies such as follow-on biologics (FOB) by the U.S. FDA and subsequent entered biologics (SEB) by the Public Health Agency (PHA) of Canada. The vocabulary used to describe these products during the period of regulatory submission is also very specific. In the first stage, the developer shows analytical similarity (not comparability) that leads to demonstration of biosimilarity (once the pharmacokinetics [PK]/pharmacodynamics [PD] and/or the clinical data are compiled); any changes made to an approved product require comparability protocol exercised; in the United States, a designation of interchangeability can be achieved by additional demonstration of safety and efficacy when the biosimilar and originator products are switched back and forth.

Biological drugs coming off patent are not called generic biologics, even though the intent is the same; instead, they are called biosimilars, FOB, or subsequent entry biologics; the designation of biosimilars, however, has gained sufficient acceptance. Biosimilars are fundamentally different from those of traditional chemical generic drugs, as these are made using a biological living system, mostly a genetically modified organism such as *E. coli* or Chinese hamster ovary cell. In establishing equivalence of chemical drugs, first we establish chemical equivalence, then pharmaceutical equivalence (where suggested), and finally the clinical equivalence. It is readily seen why this cascade of equivalence will not apply to biological drugs. First, the chemical equivalence is challenged because of the 3D structure of these molecules vis-à-vis the 2D structure of most chemical drugs. While the two products may have identical molecular weight, how the protein is folded can make significant difference in their biological properties. Compounding this observation is the realization that molecules in hundreds of thousands of Dalton range do not offer a direct structure–activity relationship, meaning that it is uncertain if one or other 3D arrangement is preferred. This poses a very serious challenge to the developers of biosimilar products. When developing a new drug, the developer has to demonstrate acceptable toxicity and efficacy without any reference to a reference listed drug (RLD). In the case of biosimilars, the developer must demonstrate similar structure, if not totally identical. Keeping in mind that the originator product has a much larger variability and the biosimilar product must fall within this range; therefore, by definition, the similarity is demonstrated by emulating a profile rather than a fixed specification as required for small molecules.

For a long time, the originators have warned the developers of biosimilar products that when it comes to biological drugs, the process controls the quality of the product—what is dubbed as "product by process," indicating that the in-process controls, to which only the originator is privy, are key to the successful manufacturing of a safe and effective biological product. While it is correct that there are many more factors that can affect the quality of a biosimilar product than a generic product, these process-related effects are not beyond the comprehension of a biosimilar product developer. For example, it is well known that the biological product if shaken vigorously can produce aggregate or degrade in its 3D structure. However, these quality aspects are readily determined by size exclusion chromatography and many other newer techniques that were not available when the originator developed the product. All warnings regarding the manufacturing of biosimilar products regarding modified structures, improper folding, and unusual 4D interaction with formulation components can be systematically studied and a reasonable assurance provided for the safety of biosimilar products. This has been ascertained in a large number of approvals in Europe, of biosimilar erythropoietins, granulocyte-colony-stimulating factor (G-CSF), growth hormone, and, most recently, the monoclonal antibodies.

Since the development of biosimilar products entails extensive scientific investigation and establishment of manufacturing under highly controlled environment—all requiring substantial initial investment and ongoing surveillance cost—a significant factor will keep the competition in the field of biosimilar products to a much smaller number of companies than what was observed for the chemical generic companies.

Summarizing the aforementioned, the biosimilar products require a different pathway of approval for the following reasons:

1. These are extremely large molecules. A monoclonal antibody may have a molecular weight of 150 KD or higher and that is 100–200 times the size of commonly used chemical drugs. The chemical structure of biological drugs is not definitive; there are limitations and uniqueness of 3D and even 4D structural variability that make the biological drugs extremely difficult to characterize; compounding this perspective is the realization that biological drugs are often not a single structure but a group of structures as found in the monoclonal antibodies where different isomers of the glycosylated products makes the entire entity of the active product.
2. The pharmaceutical equivalent of the biological drugs is not an issue since most of them fall in the injectable category (proteins are unstable otherwise), so it is not difficult to claim this level of equivalence as the exact composition of all injectable products is fully disclosed. The exact composition of the innovator product is fully disclosed so this does not create any risk.
3. The toxicological equivalence of biological drugs offers a rather unusual situation for the developers of these drugs. Since in the case of chemical drugs the toxicity is a function of the chemical structure and quantity administered, it is readily ascertained; however, biological drugs offer a challenge to developers since their side effects and toxicity is not necessarily related to quantity or the basic chemical structure. More often, subtle difference in the structure, which is inevitable, is responsible for their toxicity. Since many biological drugs are proteins and antibodies, immunological response to these is the most common side effect. The immunogenicity is a property of proteins, particularly when combined with carbohydrates, as is the case with monoclonal antibodies; however, all biological drugs have the immunogenic potential and it is this aspect of the toxicity of biological drugs that makes their bioequivalence evaluation extremely difficult. In some instances, such as the monoclonal antibodies, there appears to be no satisfactory animal toxicity model available making the choice of a toxicological study protocol redundant. Realizing this,

the regulatory agencies are open to suggestions from the developers for alternate models, such as the use of transgenic mice or even avoidance of any toxicological studies. There is also an unusual situation that arises when comparing the toxicity of biosimilar product with the originator product where toxicology models are available. The biosimilar must show similar toxicology profile, and thus for safer drugs where toxicological signs are not readily obtained, the studies can get very complicated and large.

4. The clinical equivalence, which is generally assumed to be proportional to the blood level profiles, is complicated by the realization that the relationship between the plasma level profiles and the clinical response may not be linear in the case of biological drugs, whose mechanism of action (MOA) is not always fully understood. Does the PK study reveal anything significant or not remain to be seen as surrogate tests that are developed to obviate the need for blood level studies? The PD studies are the most useful element of demonstrating similarity where a MOA is relatively easier to understand.
5. Immune responses remain the most significant consideration in the development of biosimilar products. Being proteins, these drugs are expected to have the ability to trigger immune responses, some more than others. For example, filgrastim (Neupogen) has very little immunogenicity; the originator showed about 3% response and the biosimilars 0%; on the other hand, drugs like adalimumab (Humira) may show an immunogenic response in more than 40% of patients. How are these characteristics compared remains the biggest challenge. Compounding the immunogenic response is the nature of response, something that remains to be explored in the field of proving biosimilarity. One of the most notorious stories regarding the immune response of biological drugs is the incidence of pure red cell aplasia (PRCA) in the use of erythropoietin where the formulation was altered (different surfactant) without conducting any safety trials, which results in dozens of deaths. This incidence demonstrated the continued need to handle biological drugs differently than what is the routine for small-molecule, fully characterized drugs. Interestingly, excluding the incidence listed earlier, the track record of biological drugs has been relatively very good, even better than the small-molecule drugs in the review of postmarketing surveillance and safety. And this happened despite a large number of substandard biological drugs marketed, mainly in the developing countries. But that may be an inaccurate assessment since the toxic responses to these drugs generally show up after years of their use, mostly as autoimmune responses that might trigger such disorders as diabetes and multiple sclerosis (MS). It is for this reason that the bioequivalence of biological products has been treated differently by the regulatory agencies.

As the patents for recombinant manufacturing of therapeutic proteins began to expire, there was a big effort made by generic pharmaceutical companies to enter the market of generic biological products companies, as the financial returns were highly lucrative. Interestingly, almost three-fourths of all new drugs developed have biological characteristics and this follows the great success in the use of biological therapeutics over the past three decades.

U.S. REGULATIONS

BIOSIMILARITY

The Biologics Price Competition and Innovation Act (BPCI Act) was enacted as part of the Affordable Care Act on March 23, 2010, in the United States. The BPCI Act creates an abbreviated licensure pathway for biological products demonstrated to be biosimilar to, or interchangeable with, a reference product. Biosimilar or biosimilarity means that "the biological product is highly similar to the reference product notwithstanding minor differences in clinically inactive components" and that "there are no clinically meaningful differences between the biological product and the reference product in terms of the safety, purity, and potency of the product."

In the BPCI Act, a biosimilar product is defined as a product that is *highly similar* to the reference product notwithstanding minor differences in clinically inactive components and there are no clinically meaningful differences in terms of safety, purity, and potency. Based on this definition, a biological medicine is considered biosimilar to a reference biological medicine if it is highly similar to the reference in safety, purity, (quality) and efficacy. However, little or no discussion regarding that *How similar is considered highly similar?* in the BPCI Act is given.

The BPCI Act seems to suggest that a biosimilar product should be highly similar to the reference drug product in all spectrums of good drug characteristics such as identity, strength, quality, purity, safety, and stability. In practice, however, it is almost impossible to demonstrate that a biosimilar product is highly similar to the reference product in all aspects of good drug characteristics in a single study. Thus, to ensure that a biosimilar product is highly similar to the reference product in terms of these good drug characteristics, different biosimilar studies may be required. For example, if safety and efficacy is a concern, then a clinical trial must be conducted to demonstrate that there are no clinically meaningful differences in terms of safety and efficacy. On the other hand, to ensure highly similar in quality, assay development/validation, process control/validation, and product specification of the reference product are necessarily established. In addition, test for comparability in manufacturing process between biosimilars and the reference must be performed. In some cases, if a surrogate endpoint such as PK, PD, or genomic marker is predictive of the primary efficacy/safety clinical endpoint, then a PK/PD or genomic study may be used to assess biosimilarity between biosimilars and the reference product.

It should be noted that current regulatory requirements are guided based on a case-by-case basis by the following basic principles: (1) the extent of the physicochemical and biological characterization of the product, (2) nature or possible changes in the quality and structure of the biological product due to the changes in the manufacturing process (and their unexpected outcomes), (3) clinical/regulatory experiences with the particular class of the product in question, and (4) several factors that need to be considered for biocomparability.

There are several regulatory reasons for this as well. For example, in the United States, a biosimilar product is supposed to carry a label that states the following:

- This product is approved as biosimilar to a reference product for stated indication(s) and route of administration(s).
- This product (has or has not) been determined to be interchangeable with the reference product. FDA has established a dual route of approval where a product may be listed as interchangeable.

For the purpose of establishing guidelines to prove bioequivalence of biological products, the FDA provides a specific definition of a biological product: "Biological product means a virus, therapeutic serum, toxin, antitoxin, vaccine, blood, blood component or derivative, allergenic product, protein (except any chemically synthesized polypeptide), or analogous product, or arsphenamine or derivative of arsphenamine (or any other trivalent organic arsenic compound), applicable to the prevention, treatment, or cure of a disease or condition of human beings. Protein means any alpha amino acid polymer with a specific defined sequence that is greater than 40 amino acids in size. Chemically synthesized polypeptide means any alpha amino acid polymer that is (a) made entirely by chemical synthesis and (b) is less than 100 amino acids in size."

Proving biosimilarity takes a very different path than what is generally established in proving bioequivalence of small molecule. Section 351(k) of the PHS Act (42 U.S.C. 262(k)), added by the BPCI Act, sets forth the requirements for an application for a proposed biosimilar product and an application or a supplement for a proposed interchangeable product. The interchangeable product will be more like a bioequivalent small-molecule product wherein a substitution can be made unless otherwise restricted by the prescriber. In both instances, whether it is interchangeable or not, a robust plan for demonstrating biosimilarity is required. Most of the current guidelines being developed by the FDA are more focused on demonstration of biosimilarity and the onus of proving interchangeability remain on the sponsors.

The scientific and technical investigations associated with proving biosimilarity are much more intense, even from the intensity often associated with the development of a new biological product. Proving biosimilarity requires comparing two products, both highly variable and having at times unpredictable behavior and that raises the difficulties—it is like nailing two randomly flying birds with one arrow. As an example of the associated difficulties, the specification of a reference standard needs to be established first for there are no compendia standards and even when they are present (e.g., the monographs in European Pharmacopoeia); these are not generally acceptable to the regulatory agencies for the purpose of creating specification to prove biosimilarity.

Since the U.S. FDA has recently released its guidance on developing biosimilar products and while no approvals have been granted yet, there is sufficient history of approval of biological drugs using an abbreviated path, and the examples include growth hormone, interferon beta (IFN-β), and heparin. This historic perspective is important for the developers of biosimilar products. What differentiates these approvals is the level of science that went into proving safety and efficacy without having to conduct full-scale clinical trials. The FDA continues its stance in the guidance for biosimilar products where the scientific evidence will be critical to future approvals.

Basic Understanding

There are three key elements to the advice given by FDA:

1. Providing a totality of evidence
2. Eliminating any residual uncertainly as much as possible
3. Using a stepwise approach to the development of biosimilar products in consultation with the FDA

The FDA considers the totality of the evidence provided by a sponsor to support a demonstration of biosimilarity, which can include a comparison of the proposed product and the reference product with respect to structure, function, animal toxicity, human PK and PD, clinical immunogenicity, and clinical safety and effectiveness.

Section 351(i) of the PHS Act defines biosimilarity to mean "that the biological product is highly similar to the reference product notwithstanding minor differences in clinically inactive components" and that "there are no clinically meaningful differences between the biological product and the reference product in terms of the safety, purity, and potency of the product."

The BPCI Act also amended the definition of biological product to include "protein (except any chemically synthesized polypeptide)." Under section 351(k) of the PHS Act, a proposed biological product that is demonstrated to be biosimilar to a reference product can rely on certain existing scientific knowledge about the safety, purity, and potency of the reference product to support licensure.

Analytical studies demonstrate that the biological product is highly similar to the reference product notwithstanding minor differences in clinically inactive components; the FDA has the discretion to determine that an element described earlier is unnecessary in a 351(k) application.

Unlike small-molecule drugs, whose structure can usually be completely defined and entirely reproduced, proteins are typically more complex and are unlikely to be shown to be structurally identical to a reference product. Many potential differences in protein structure can arise. Because even minor structural differences (including certain changes in glycosylation patterns) can significantly affect a protein's safety, purity, and/or potency, it is important to evaluate these differences.

Animal studies (including the assessment of toxicity) and a clinical study or studies (including the assessment of immunogenicity and PK or PD) are sufficient to demonstrate safety, purity, and potency in one or more appropriate conditions of use for which the reference product is licensed and intended to be used and for which licensure is sought for the biological product.

Scientific Basis

In general, proteins can differ in at least three ways: (1) primary amino acid sequence; (2) modification to amino acids, such as sugar moieties (glycosylation) or other side chains; and (3) higher-order structure (protein folding and protein–protein interactions). Modifications to amino acids may lead to heterogeneity and can be difficult to control. Protein modifications and higher-order structure can be affected by environmental conditions, including formulation, light, temperature, moisture, packaging materials, container closure systems, and delivery device materials. Additionally, process-related impurities may increase the likelihood and/or the severity of an immune response to a protein product, and certain excipients may limit the ability to characterize the drug substance.

Advances in analytical sciences enable some protein products to be extensively characterized with respect to their physicochemical and biological properties, such as higher-order structures and functional characteristics. These analytical methodologies have increasingly improved the ability to identify and characterize not only the drug substance of a protein product but also the excipients and product- and process-related impurities. Despite such significant improvements in analytical techniques, however, current analytical methodology may not be able to detect all relevant structural and functional differences between two proteins. Thus, data derived from analytical studies, animal studies, and a clinical study or studies are required to demonstrate biosimilarity unless FDA determines an element unnecessary.

Manufacturing Process Considerations

Different manufacturing processes may alter a protein product in a way that could affect the safety or effectiveness of the product. For example, differences in biological systems used to manufacture a protein product may cause different posttranslational modifications, which in turn may affect the safety or effectiveness of the product. Thus, when the manufacturing process for a marketed protein product is changed, the application holder must assess the effects of the change and demonstrate through appropriate analytical testing, functional assays, and/or in some cases animal and/or clinical studies that the change does not have an adverse effect on the identity, strength, quality, purity, or potency of the product as they relate to the safety or effectiveness of the product.

The International Conference on Harmonisation (ICH) guidance Q5E *Comparability of Biotechnological/Biological Products Subject to Changes in Their Manufacturing Process* describes scientific principles in the comparability assessment for manufacturing changes.

Demonstrating that a proposed product is biosimilar to a reference product typically will be more complex than assessing the comparability of a product before and after manufacturing changes made by the same manufacturer. This is because a manufacturer who modifies its own manufacturing process has extensive knowledge and information about the product and the existing process, including established controls and acceptance parameters.

In contrast, the manufacturer of a proposed biosimilar product will likely have a different manufacturing process such as different cell line, raw materials, equipment, processes, process controls, and acceptance criteria from that of the reference product and no direct knowledge of the manufacturing process for the reference product. Therefore, even though some of the scientific principles described in ICH Q5E may also apply in the demonstration of biosimilarity, in general, more data and information will be needed to establish biosimilarity than would be needed to establish that a manufacturer's postmanufacturing change product is comparable to the premanufacturing change product.

In general, a sponsor needs to provide information to demonstrate biosimilarity based on data directly comparing the proposed product with the reference product and includes required analytical studies and at least one human PK and/or PD study intended to support a demonstration of biosimilarity to the reference product licensed under section 351(a).

However, under certain circumstances, a sponsor may seek to use data derived from animal or clinical studies comparing a proposed product with a non-U.S.-licensed product to address, in part, the requirements under section 351(k)(2)(A) of the PHS Act. In such a case, the sponsor should provide adequate data or information to scientifically justify the relevance of this comparative data to an assessment of biosimilarity and to establish an acceptable bridge to the U.S.-licensed reference product.

Stepwise Approach

The FDA recommends that sponsors use a stepwise approach to developing the data and information needed to support a demonstration of biosimilarity. At each step, the sponsor should evaluate the extent to which there is residual uncertainty about the biosimilarity of the proposed product and identify the next steps to try to address that uncertainty. Where possible, studies conducted should be designed to maximize their contribution to demonstrating biosimilarity. For example, a clinical immunogenicity study may also provide other useful information about the safety profile of the proposed product.

The stepwise approach should start with extensive structural and functional characterization of both the proposed product and the reference product, which serves as the foundation of a biosimilar development program. The more comprehensive and robust the comparative structural and functional characterization—the extent to which these studies are able to identify (qualitatively or quantitatively) differences in relevant product attributes between the proposed product and reference product (including the drug substance, excipients, and impurities)—the more useful such characterization will be in determining what additional studies may be needed. For example, the more rigorous structural and functional comparisons show minimal or no difference between the proposed product and the reference product, the stronger the scientific justification for a selective and targeted approach to animal and/or clinical testing to support a demonstration of biosimilarity.

It may be useful to further quantify the similarity or differences between the two products using a meaningful fingerprint-like analysis algorithm that covers a large number of additional product attributes and their combinations with high sensitivity using orthogonal methods. Such a strategy may further reduce the possibility of undetected structural differences between the products and lead to a more selective and targeted approach to animal and/or clinical testing.

Mechanism of Action

A sufficient understanding of the MOA of the drug substance and clinical relevance of any observed structural differences, clinical knowledge of the reference product and its class indicating that the overall safety risks are low, and the availability of a clinically relevant PD measure may provide further scientific justifications for a selective and targeted approach to animal and/or clinical studies. The sponsor should then consider the role of animal data in assessing toxicity and, in some cases, in providing additional support for demonstrating biosimilarity and in contributing to the immunogenicity assessment. The sponsor should then conduct comparative human PK studies, and PD studies if there is a clinically relevant PD measure, in an appropriate study population. Sponsors should then compare the clinical immunogenicity of the two products. If there are residual uncertainties about the biosimilarity of the two products after conducting structural and functional studies, animal toxicity studies, human PK and PD studies, and clinical immunogenicity assessment, the sponsor should then consider what comparative clinical safety and effectiveness data may be adequate.

Totality of Evidence

In evaluating a sponsor's demonstration of biosimilarity, FDA will consider the totality of the data and information submitted in the application, including structural and functional characterization, nonclinical evaluation, human PK and PD data, clinical immunogenicity data, and clinical safety and effectiveness data. FDA uses a risk-based, totality-of-the-evidence approach to evaluate all available data and information submitted in support of the biosimilarity of the proposed product.

A sponsor may be able to demonstrate biosimilarity even though there are formulation or minor structural differences, provided that the sponsor provides sufficient data and information demonstrating that the differences are not clinically meaningful and the proposed product otherwise meets the statutory criteria for biosimilarity. For example, differences in certain posttranslational modifications or differences in certain excipients (e.g., human serum albumin) might not preclude a finding of biosimilarity if data and information provided by the sponsor show that the proposed product is highly similar to the reference product notwithstanding minor differences in clinically inactive components and that there are no clinically meaningful differences between the products in terms of safety, purity, and potency.

Clinically meaningful differences could include a difference in the expected range of safety, purity, and potency of the proposed and reference products. By contrast, slight differences in rates of occurrence of adverse events between the two products ordinarily are not considered clinically meaningful differences.

Product Specificity

The type and amount of analyses and testing that will be sufficient to demonstrate biosimilarity will be determined on a product-specific basis.

Because some excipients may affect the ability to characterize products, a sponsor should provide evidence that the excipients used in the reference product will not affect the ability to characterize and compare the products. This should include information that demonstrates biosimilarity based on data derived from, among other things, analytical studies that demonstrate that the biological product is highly similar to the reference product notwithstanding minor differences in clinically inactive components, unless FDA determines that an element is unnecessary in a 351(k) application.

FDA expects that a sponsor first will extensively characterize the proposed product and reference product with state-of-the-art technology, because extensive characterization of both products serves as the foundation for a demonstration of biosimilarity.

In general, FDA expects that the expression construct for a proposed product will encode the same primary amino acid sequence as the reference product. However, minor modifications such as N- or C-terminal truncations that will not affect safety and effectiveness may be justified and should be explained by the sponsor. Additionally, sponsors should consider all relevant characteristics of the proposed product (e.g., the primary, secondary, tertiary, and quaternary structure; posttranslational modifications; and biological activities) to demonstrate that the proposed product is highly similar to the reference product notwithstanding minor differences in clinically inactive components. The more comprehensive and robust the comparative structural and functional characterizations are, the stronger the scientific justification for a selective and targeted approach to animal and/or clinical testing.

Analytical Methodology

Sponsors should use an appropriate analytical methodology with adequate sensitivity and specificity for structural characterization of the proteins.

Sponsors should conduct extensive structural characterization in multiple representative lots of the proposed product and the reference product to understand the lot-to-lot variability of both drug substances in the manufacturing processes. Lots used for the analysis should support the biosimilarity of both the clinical material used in confirmatory clinical trials and the to-be-marketed proposed product.

Sponsors should justify the selection of the representative lots, including the number of lots. In addition, FDA recommends that sponsors analyze the finished dosage form of multiple lots of the proposed product and the reference product, assessing excipients and any formulation effect on purity, product- and process-related impurities, and stability. Greater emphasis is placed on evaluating the following:

1. Primary structures, such as amino acid.
2. Higher-order structures, including secondary, tertiary, and quaternary structure (including aggregation).
3. Enzymatic posttranslational modifications, such as glycosylation and phosphorylation.
4. Other potential variants, such as protein deamidation and oxidation.
5. Intentional chemical modifications, such as PEGylation sites and characteristics between the proposed product and the reference product, are among the factors that may affect the extent and nature of subsequent animal or clinical testing.

If the reference product cannot be adequately characterized with state-of-the-art technology, the sponsor should consult FDA for guidance on whether an application for such a protein product is appropriate for submission under section 351(k) of the PHS Act.

Functional Assays

The pharmacological activity of protein products can be evaluated by in vitro and/or in vivo functional assays. These assays may include, but are not limited to, bioassays, biological assays, binding assays, and enzyme kinetics. A functional evaluation comparing a proposed product to the reference product using these types of assays is also an important part of the foundation that supports a demonstration of biosimilarity and may be used to scientifically justify a selective and targeted approach to animal and/or clinical testing.

Sponsors can use functional assays to provide additional evidence that the biological activity and potency of the proposed product are highly similar to those of the reference product and/or to demonstrate that there are no clinically meaningful differences between the proposed product and the reference product. Such assays also may be used to provide additional evidence that the MOA of the two products is the same to the extent the MOA of the reference product is known.

Functional assays can be used to provide additional data to support results from structural analysis, investigate the consequences of observed structural differences, and explore structure–activity relationships. To be useful, these assays should be comparative, so they can provide evidence of similarity, or reveal differences, in the performance of the proposed product compared to the reference product, especially differences resulting from structural variations that cannot be detected using current analytical methods. FDA also recommends that sponsors discuss limitations of the assays they used when interpreting results in their submissions to the FDA.

Functional assays can also provide information that complements the animal and clinical data in assessing the potential clinical effects of minor differences in structure between the proposed product and reference product. For example, cell-based bioactivity assays can be used to detect the potential for inducing cytokine release syndrome in vivo. The available information about these assays, including sensitivity, specificity, and extent of validation, can affect the amount and type of additional animal or clinical data that may be needed to establish biosimilarity. As for the structural evaluation, appropriate lots should be used in the analysis.

Animal Data

The PHS Act also requires that a 351(k) application include information that demonstrates biosimilarity based on data derived from animal studies (including the assessment of toxicity), unless FDA determines that such studies are not necessary in a 351(k) application.

As a scientific matter, animal toxicity data are considered useful when, based on the results of extensive structural and functional characterization, uncertainties remain about the safety of the proposed product that need to be addressed before initiation of clinical studies in humans.

Animal toxicity studies are generally not useful if there is no animal species that can provide pharmacologically relevant data for the protein product (i.e., no species in which the biological activity of the protein product mimics the human response). However, there may be some instances when animal data from a pharmacologically nonresponsive species (including rodents) may be useful to support clinical studies with a proposed product that has not been previously tested in human subjects, for example, comparative PK and systemic tolerability studies.

The scope and extent of any animal toxicity studies will depend on the body of information available on the reference product, the proposed product, and the extent of known similarities or differences between the two. If animal toxicity studies are not warranted, additional comparative in vitro testing, using human cells or tissues when appropriate, may be warranted.

When animal toxicity studies are conducted, it will generally be useful to perform a comparative animal toxicology study with the proposed product and reference product (i.e., comparative bridging toxicology studies). The selection of dose, regimen, duration, and test species for these studies should provide a meaningful toxicological comparison between the two products. It is important to understand the limitations of such animal studies (e.g., small sample size, intraspecies variations) when interpreting results comparing the proposed product and the reference product.

A sponsor may be able to provide a scientific justification for a stand-alone toxicology study using only the proposed product instead of a comparative toxicology study.

In general, nonclinical safety pharmacology, reproductive and developmental toxicity, and carcinogenicity studies are not warranted when the proposed product and reference product have been demonstrated to be highly similar through extensive structural and functional characterization and animal toxicity studies.

Animal PK and PD Measures

Under certain circumstances, a single-dose study in animals comparing the proposed product and reference product using PK and PD measures may contribute to the totality of evidence that supports a demonstration of biosimilarity. Specifically, sponsors can use results from animal studies to support the degree of similarity based on PK and PD profiles of the proposed product and the reference product. PK and PD measures also can be incorporated into a single animal toxicity study, where appropriate. Animal PK and PD assessment will not negate the need for human PK and PD studies.

Animal Immunogenicity Studies

Animal immunogenicity assessments generally do not predict potential immunogenic responses to protein products in humans. However, when differences in manufacturing (e.g., impurities or excipients) between the proposed product and the reference product may result in differences in immunogenicity, the measurement of antiprotein antibody responses in animals may provide useful information relevant to patient safety. Additionally, significant differences in the immune response profile in inbred strains of mice, for example, may indicate that the proposed product and the reference product differ in one or more product attributes not captured by other analytical methods. If available, this information is of value in the design of clinical immunogenicity assessment.

Clinical Studies

The sponsor should demonstrate that "there are no clinically meaningful differences between the biological product and the reference product in terms of the safety, purity, and potency of the product." In general, the clinical program for a 351(k) application *must* include a clinical study or studies (including an assessment of immunogenicity and PK or PD) sufficient to demonstrate safety, purity, and potency in one or more appropriate conditions of use for which the reference product is licensed and intended to be used and for which licensure is sought for the biological product.

The scope and magnitude of clinical studies will depend on the extent of residual uncertainty about the biosimilarity of the two products after conducting structural and functional characterization and possible animal studies. The frequency and severity of safety risks and other safety and effectiveness concerns for the reference product may also affect the design of the clinical program. Lessening the number or narrowing the scope of any of these types of clinical studies (i.e., human PK, PD, clinical immunogenicity, or clinical safety and effectiveness) should be scientifically justified by the sponsor.

Human Pharmacology Data

Human PK and PD studies comparing a proposed product to the reference product generally are fundamental components in supporting a demonstration of biosimilarity. FDA asserts that both PK and PD studies (where there is a relevant PD measure) generally will be expected to establish biosimilarity, unless a sponsor can scientifically justify that an element is unnecessary. Human PK and PD profiles of a protein product often cannot be adequately predicted from functional assays and/or animal studies alone. Therefore, comparative human PK studies and, if clinically relevant PD measures are available, comparative human PD studies would be expected, unless a sponsor can provide a scientific justification that such studies are unnecessary. In addition, a human PK study that demonstrates similar exposure (e.g., serum concentration over time) with the proposed product and reference product can provide support for a biosimilarity demonstration. For example, a human PK study can be particularly useful when the exposure correlates to clinical safety and effectiveness. A human PD study that demonstrates a similar effect on a clinically relevant PD measure or measures related to effectiveness or specific safety concerns (except for immunogenicity, which is evaluated separately) can also provide strong support for a biosimilarity determination.

Sponsors should provide a scientific justification for the selection of the human PK and PD study population (e.g., patients vs. healthy subjects) and parameters, taking into consideration the relevance of such population and parameters, the population and parameters studied for the licensure for the reference product, and the current knowledge of the intrasubject and intersubject variability of human PK and PD for the reference product. For example, FDA recommends that, to the extent possible, the sponsor select PD measures that (1) are relevant to clinical outcomes (e.g., on mechanistic path of MOA or disease process related to effectiveness or safety); (2) can be assessed after a sufficient period of time after dosing, and with appropriate precision; and (3) have the sensitivity to detect clinically meaningful differences between the proposed product and reference product.

Sponsors should predefine and justify the criteria for PK and PD parameters for studies included in the application to demonstrate biosimilarity. Establishing a similar human PK and PD profile contributes to the demonstration of biosimilarity and may provide a scientific basis for a selective and targeted approach to subsequent clinical testing. Demonstrating that the proposed product and reference product have similar effects on a PD measure that is known to be clinically related to PK and PD studies provides quite different types of information. In simple terms, a PK study measures how the body acts on a drug, how the drug is absorbed, distributed, metabolized, and eliminated, and a PD study measures how the drug acts on the body, typically assessing a measure or measures related to the drug's biochemical and physiological effects on the body. Therefore, one type of study does not duplicate or substitute for the information provided by the other. Both PK and PD studies provide important information for assessing biosimilarity and therefore, as a scientific matter, comparative human PK and PD studies (where there is a relevant PD measure) generally will be expected.

Safety or effectiveness can provide further support for a selective and targeted approach to clinical safety/effectiveness studies. In certain circumstances, human PK and PD data may provide sufficient clinical data to support a demonstration of biosimilarity.

The factors that can affect the ability of the human PK and PD studies to support a selective and targeted approach to the clinical program and contribute to a demonstration of biosimilarity include whether the human PK and PD studies have used (1) clinically relevant PK and PD parameters (multiple PD measures that assess different domains of activities may be of value); (2) populations, dose(s), and route of administration that are the most sensitive to detect differences in PK and PD profiles; and (3) sensitive and relevant assays.

Clinical Immunogenicity Assessment

The goal of the clinical immunogenicity assessment is to evaluate potential differences between the proposed product and the reference product in the incidence and severity of human immune responses. Immune responses may affect both the safety and effectiveness of the product by, for example, altering PK, inducing anaphylaxis, or promoting development of neutralizing antibodies (NAbs) that neutralize the product as well as its endogenous protein counterpart. Thus, establishing that there are no clinically meaningful differences in immune response between a proposed product and the reference product is a key element in the demonstration of biosimilarity.

Structural, functional, and animal data are generally not adequate to predict immunogenicity in humans. Therefore, at least one clinical study that includes a comparison of the immunogenicity of the proposed product to that of the reference product will *generally* be expected. The extent and timing (e.g., premarket testing vs. pre- and postmarket testing) of a clinical immunogenicity program will vary depending on a range of factors, including the extent of analytical similarity between the proposed product and the reference product and the incidence and clinical consequences of immune responses for the reference product. For example, if the clinical consequence is severe (e.g., when the reference product is a therapeutic counterpart of an endogenous protein with a critical, nonredundant biological function or is known to provoke anaphylaxis), more extensive immunogenicity assessments will likely be needed.

If the immune response to the reference product is rare, two separate studies may be sufficient to evaluate immunogenicity: (1) a premarket study powered to detect major differences in immune responses between the two products and (2) a postmarket study designed to detect more subtle differences in immunogenicity. The overall design of immunogenicity studies will consider both the severity of consequences and the incidence of immune responses. FDA recommends use of a comparative parallel design (i.e., a head-to-head study) to assess potential differences in the risk of immunogenicity and support appropriate labeling.

It is generally only important to demonstrate that the immunogenicity of the proposed product is not increased, so a one-sided design will ordinarily be adequate to compare clinical immunogenicity of the proposed product and reference product. Acceptable differences in incidence and other immune response parameters should be discussed with the FDA in advance of the study.

Differences in immune responses between a proposed product and the reference product in the absence of observed clinical sequelae may be of concern and may warrant further evaluation to assess whether there are clinically meaningful differences between the proposed product and the reference product. The study population used to compare immunogenicity should be justified and agreed to by the agency. If a sponsor is seeking to extrapolate immunogenicity findings for one indication to other indications, the sponsor should consider using the study population and treatment regimen that are the most sensitive for detecting a difference in immune responses. Most often, this will be the population and regimen for the reference product for which development of immune responses with adverse outcomes is most likely to occur (e.g., patients with autoimmune diseases would be more likely to develop immune responses than patients with malignancies).

The selection of clinical immunogenicity endpoints or PD measures associated with immune responses to therapeutic protein products (e.g., antibody formation and cytokine levels) should take into consideration the immunogenicity issues that have emerged during the use of the reference product. Sponsors should prospectively define the clinical immune response criteria (e.g., definitions of significant clinical events), using established criteria where available, for each type of potential immune response and obtain agreement from FDA on these criteria before initiating the study. The follow-up period should be determined based on (1) the time course for the generation of immune responses (such as the development of NAbs, cell-mediated immune responses) and expected clinical sequelae (informed by experience with the reference product), (2) the time course of disappearance of the immune responses and clinical sequelae following cessation of therapy, and (3) the length of administration of the product. For example, the minimal follow-up period for chronically administered agents should be 1 year, unless the sponsor can justify a shorter duration.

As a scientific matter, it is expected that the following will be assessed in clinical immunogenicity studies:

- *NAb*: all of the aforementioned, plus neutralizing capacity to all relevant functions (e.g., uptake and catalytic activity, neutralization for replacement enzyme therapeutics)
- *Binding antibody*: titer, specificity, relevant isotype distribution, time course of development, persistence, disappearance, and association with clinical sequelae

The sponsor should develop assays capable of sensitively detecting immune responses, even in the presence of circulating drug product (proposed product and reference product). The proposed product and reference product should be assessed in the same assay with the same patient sera whenever possible. FDA recommends that immunogenicity assays be developed and validated with respect to both the proposed product and reference product early in development. Sponsors should consult with FDA on the sufficiency of assays before initiating any clinical immunogenicity study.

Clinical Safety and Effectiveness Data

As a scientific matter, comparative safety and effectiveness data *will* be necessary to support a demonstration of biosimilarity *if* there are residual uncertainties about the biosimilarity of the two products based on structural and functional characterization, animal testing, human PK and PD data, and clinical immunogenicity assessment.

A sponsor may provide a scientific justification if it believes that some or all of these comparisons on clinical safety and effectiveness are not necessary. The following are examples of factors that may influence the type and extent of the comparative clinical safety and effectiveness data needed:

- The extent to which differences in structure, function, and nonclinical pharmacology and toxicology predict differences in clinical outcomes, as well as the degree of understanding of the MOA of the reference product and disease pathology
- The extent to which human PK or PD predicts clinical outcomes (e.g., PD measures known to be clinically relevant to effectiveness)
- The extent of clinical experience with the reference product and its therapeutic class, including the safety and risk/benefit profile (e.g., whether there is a low potential for off-target adverse events) and appropriate endpoints and biomarkers for safety and effectiveness (e.g., availability of established, sensitive clinical endpoints)
- The extent of any clinical experience with the proposed product
- The nature and complexity of the reference product, the extensiveness of structural and functional characterization, and the findings and limitations of comparative structural, functional, and nonclinical testing, including the extent of observed differences

Sponsors should provide a scientific justification for how it intends to integrate these factors to determine whether and what types of clinical trials are needed and the design of any necessary trials. For example, if comparative clinical trials (using an equivalence or a noninferiority design) are needed, these factors are also relevant to determining the equivalence or noninferiority margin.

Additionally, specific safety or effectiveness concerns regarding the reference product and its class (including history of manufacturing- or source-related adverse events) may warrant more comparative clinical safety and effectiveness data.

Alternatively, if the reference product has a long, relatively safe marketing history and there have been multiple versions of the reference product on the market with no apparent differences in clinical safety and effectiveness profiles, there may be a basis for a selective and targeted approach to the clinical program.

Clinical Study Design Issues

Clinical studies should be designed such that they can demonstrate that the proposed product has neither decreased nor increased activity compared to the reference product. Decreased activity ordinarily would preclude licensure of a proposed product. Increased activity might be associated with more adverse effects or might suggest that the proposed product should be treated as an entirely different product with superior efficacy, in which case the appropriate licensure pathway would be section 351(a) of the PHS Act.

A study employing a two-sided test in which the null hypothesis is either (1) the proposed product is inferior to the reference product or (2) the proposed product is superior to the reference product based on a prespecified equivalence margin is the most straightforward study design for accomplishing this objective. The margins should be scientifically justified and adequate to enable the detection of clinically meaningful differences in effectiveness and safety between the proposed product and the reference product.

A sponsor should use clinical knowledge about the reference product and its therapeutic class to establish an appropriate equivalence margin. Although the upper (superiority) and lower (inferiority) bounds of the margin will usually be the same, there may be cases in which a different upper and lower bound may be appropriate. In some cases, a one-sided test—non-inferiority design—may be appropriate for comparing safety and effectiveness and also advantageous as it would generally allow for a smaller sample size than an equivalence (two-sided) design. For example, if it is well established that doses of the reference product higher than are recommended in its labeling do not create safety concerns, a one-sided test may be sufficient for comparing the efficacy of certain protein products (e.g., those products that pharmacodynamically saturate the target at some level and are used at or near the maximal level of clinical effect).

Because it is generally important to demonstrate that a proposed product has no more risk in terms of safety and immunogenicity compared to a reference product, a one-sided test may also be adequate in a clinical study evaluating immunogenicity or other safety endpoints as long as it is clear that lower immunogenic or other adverse events would not have implications for the effectiveness of a protein product. A noninferiority margin should also be scientifically based and prespecified.

FDA recommends that sponsors provide a scientific justification for the proposed size and length of their clinical trials to allow for (1) sufficient exposure to the proposed noninferiority clinical trials that contains a discussion on choosing the noninferiority margin (product and reference product); (2) the detection of relevant safety signals (including immunogenic responses), except for rare events or those that require prolonged exposure; and (3) the detection of clinically meaningful differences in effectiveness and safety between the two products.

The size of the clinical trials also may be influenced by the specific treatment effect(s) and the effect size of the reference product, as well as the size of the disease population. FDA recommends that sponsors consider the use of population pharmacokinetics (PPK) to explain observed differences in safety and effectiveness that may occur due to variability in PK. PPK methods are described in the guidance for industry on PPK and involve the collection of only a few blood samples per patient. PPK methods are an efficient way to quantitate the influence of covariates (e.g., age or renal function) on PK and, in some cases, PD.

Sponsors should consult the PPK guidance, in particular the discussion concerning the design of PPK studies to ensure the validity of the study results. FDA recommends that a sponsor use endpoints and study populations that will be clinically relevant and sensitive in detecting clinically meaningful differences in safety and effectiveness between the proposed product and reference product. A sponsor can use endpoints that are different from those in the reference product's clinical trials if they are scientifically justified. For example, certain endpoints (such as PD measures) are more sensitive than clinical endpoints and, therefore, may enable more precise comparisons of relevant therapeutic effects (e.g., international normalized ratio, or INR, is more sensitive to anticoagulant comparisons than the incidence of cerebral bleeds or stroke). There may be situations when multiple PD measures enhance the sensitivity of a study.

The adequacy of the endpoints also depends on the extent to which PD measures correlate with clinical outcome, the extent of structural and functional data support for biosimilarity, the understanding of MOA, and the nature or seriousness of outcome effected (risk of difference). When selecting the study population for a comparative safety and effectiveness study, a sponsor should consider, for example, whether its study population has characteristics consistent with those of the population studied for the licensure of the reference product for the same indication and whether patients have different comorbidities and disease states (e.g., immunocompetent or immunosuppressed) and receive different concomitant medications. In general, using similar study populations is essential for supporting the constancy assumption that is critical to interpreting the noninferiority finding in a one- or two-sided comparative test.

For human PK and PD studies, FDA recommends the use of a crossover design for products with a short half-life (e.g., shorter than 5 days) and low incidence of immunogenicity. For products with a longer half-life (e.g., more than 5 days), a parallel study will usually be needed.

In addition, sponsors should provide a scientific justification for the selection of study subjects (e.g., healthy volunteers or patients) and study dose (e.g., one dose). A draft guidance entitled *Non-inferiority Clinical Trials* contains a discussion on the constancy assumption. These include

- The target/receptor(s) for each relevant activity/function of the product
- The binding, dose/concentration–response, and pattern of molecular signaling upon engagement of target/receptor(s)
- The relationship between product structure and target/receptor interactions
- The location and expression of the target/receptor(s) or multiple dose(s), route of administration, and sample size

FDA recommends that sponsors consider the duration of time it takes for a PD measure or biomarker to change and the possibility of nonlinear PK caused by dose or PD. FDA also recommends consideration of the role of modeling and simulation in designing clinical studies on human PK or PD. When there are established dose–response or systemic exposure–response relationships (response may be PD measures or clinical endpoints), comparative exposure–response data can support a selective and targeted approach to clinical safety/effectiveness studies.

It is important to select, whenever possible, doses for study on the steepest part (as opposed to the plateau) of the dose–response curve for the proposed product, because even drugs with quite different potency will appear similar if the doses are studied on or near the plateau of a dose–response curve.

Sponsors should consider the limitations of the clinical trial design and results. As noted, when the administered dose is on the plateau of a dose–response curve, the clinical trial will not be sensitive in detecting PD differences between the two products. In such a case, a sponsor should use lower doses if available and appropriate (e.g., known to have the same effect or ethically acceptable to give lower doses notwithstanding differences in effect) or a sponsor could use a study subgroup whose response is not on the plateau of the dose–response curve. A low efficacy rate (e.g., ≤25%) also may reduce the sensitivity of detecting product differences in patients in a clinical trial.

Extrapolation of Clinical Data across Indications

If the proposed product meets the statutory requirements for licensure as a biosimilar product under section 351(k) of the PHS Act based on, among other things, the data derived from a clinical study sufficient to demonstrate safety, purity, and potency in an appropriate condition of use, the potential exists for the proposed product to be licensed for one or more additional conditions of use for which the reference product is licensed. However, the sponsor will need to provide sufficient scientific justification for extrapolating clinical data to support a determination of biosimilarity for each condition of use for which licensure is sought. Such scientific justification should address, for example, the following issues for the tested and extrapolated conditions of use.

The MOA(s) in each condition of use for which licensure is sought; this may include the following:

- The PK and biodistribution of the product in different patient populations.
- PD measures may provide important information on the MOA differences in expected toxicities in each condition of use and patient population (including whether expected toxicities are related to the pharmacological activity of the product or to off-target activities).
- Any other factor that may affect the safety or effectiveness of the product in each condition of use and patient population for which licensure is sought. In choosing which condition of use to study that would permit subsequent extrapolation of clinical data to other conditions of use, FDA recommends that a sponsor consider whether the tested condition of use is the most sensitive one in detecting clinically meaningful differences in safety (including immunogenicity) and effectiveness.

A sponsor should be cautious with respect to the extrapolation of safety risk profiles across indications, because patient populations for different indications may have different comorbidities and receive different concomitant medications. The sponsor of a proposed product may seek licensure only for a condition of use that has been previously licensed for the reference product.

Postmarketing Considerations

Robust postmarketing safety monitoring is an important component in ensuring the safety and effectiveness of biological products, including biosimilar therapeutic protein products. Because some aspects of postmarketing safety monitoring are product specific, FDA encourages sponsors to consult with appropriate FDA divisions to discuss the sponsors' proposed approach to postmarketing safety monitoring.

Postmarketing safety monitoring should first take into consideration any particular safety or effectiveness concerns associated with the use of the reference product and its class, as well as the proposed product in its development and clinical use (if marketed outside the United States). Postmarketing safety monitoring for a proposed product should also have adequate mechanisms in place to differentiate between the adverse events associated with the proposed product and those associated with the reference product, including the identification of adverse events associated with the proposed product that have not been previously associated with the reference product. Rare, but potentially serious, safety risks (e.g., immunogenicity) may not be detected during preapproval clinical testing because the size of the population exposed likely will not be large enough to assess rare events. In particular cases, such risks may need to be evaluated through postmarketing surveillance or studies. In addition, like any other biological products, FDA may take any appropriate action to ensure the safety and effectiveness of a proposed product, including, for example, requiring a postmarketing study to evaluate certain safety risks.

As discussed earlier, many product-specific factors can influence the components of a product development program intended to establish that a proposed product is biosimilar to a reference product. Therefore, FDA will ordinarily provide feedback on a case-by-case basis on the components of a development program for a proposed product. In addition, it may not be possible to identify in advance all the necessary components of a development program, and the assessment of one element (e.g., structural analysis) at one step can influence decisions about the type and amount of subsequent data for the next step. For these reasons, as indicated earlier, FDA recommends that sponsors use a stepwise procedure to establish the totality of the evidence that supports a demonstration of biosimilarity.

SUMMARY CONSIDERATIONS

For assessment of biosimilarity of FOB, the following questions are commonly asked. First, what endpoints should be used for assessment of biosimilarity? Second, should a clinical trial always be conducted? To address these two questions, we may revisit the definition of biosimilarity as described in the BPCI Act. A biological product that is demonstrated to be *highly similar* to an FDA-licensed biological product may rely on certain existing scientific knowledge about safety, purity (quality), and potency (efficacy) of the reference product. Thus, if one would like to show that the safety and efficacy of a biosimilar product are highly similar to that of the reference product, then a clinical trial may be required. In some cases, clinical trials for assessment of biosimilarity may be waived if there exists substantial evidence that surrogate endpoints or biomarkers are predictive of the clinical outcomes. On the other hand, clinical trials are required for assessment of drug interchangeability in order to show that the safety and efficacy between a biosimilar product and a reference product are similar in any given patient of the patient population under study.

EXTENT OF SIMILARITY

The current criteria for assessment of bioequivalence/biosimilarity are useful for determining whether a biosimilar product is similar to a reference product. However, it does not provide additional information regarding the *degree* of similarity. As indicated in the BPCI Act, a biosimilar product is defined as a product that is *highly similar* to the reference product. However, little or no discussion regarding the degree of similarity for highly similar was provided. Besides, it is also of concern to the sponsor that "what if a biosimilar product turns out to be superior to the reference product?" A simple answer to the concern is that superiority is not biosimilarity.

PRACTICAL ISSUES

Since there are many critical (quality) attributes of a potential patient's response in FOB, for a given critical attribute, valid statistical methods are necessarily developed under a valid study design and a given set of criteria for similarity, as described in the previous section. Several areas can be identified for developing appropriate statistical methodologies for the assessment of biosimilarity of FOB. These areas include, but are not limited to, the following:

Criteria for biosimilarity (in terms of average, variability, or distribution): To address the question "how similar is similar?" we suggest establishing criteria for biosimilarity in terms of average, variability, and/or distribution.

Criteria for interchangeability: In practice, it is recognized that drug interchangeability is related to the variability due to subject-by-drug interaction. However, it is not clear whether criterion for interchangeability should be based on the variability due to subject-by-drug interaction or the variability due to subject-by-drug interaction adjusted for intrasubject variability of the reference drug.

Bridging studies for assessing biosimilarity: As most biosimilar studies are conducted using a parallel design rather than a replicated crossover design, independent estimates of variance components such as the intrasubject and the variability due to subject-by-drug interaction are not possible. In this case, bridging studies may be considered.

Other practical issues include (1) the use of a percentile method for the assessment of variability, (2) comparability in biological activities, (3) assessment of immunogenicity, (4) consistency in manufacturing processes, (5) stability testing for multiple lots and/or multiple labs, (6) the potential use of sequential testing procedures and multiple testing procedures, and (7) assessing biosimilarity using a surrogate endpoint or biomarker such as genomic data.

Quantitative Evaluation of Bioequivalence

For approval of small-molecule generic drug products, the FDA requires that evidence of average bioequivalence (ABE) in drug absorption in terms of some PK parameters such as the area under the blood and/or plasma concentration–time curve (AUC) and peak concentration (C_{max}) be provided through the conduct of bioequivalence studies. In practice, we may claim that a test drug product is bioequivalent to an innovative (reference) drug product if the 90% confidence interval (CI) for the ratio of geometric means of the primary PK parameter is completely within the bioequivalence limits of (80%, 125%). The CI for the ratio of geometric means of the primary PK parameter is obtained based on log-transformed data. In what follows, study designs and statistical methods that are commonly considered in bioequivalence studies are briefly described.

As indicated earlier, we claim that a test drug product is bioequivalent to a reference (innovative) drug product if the 90% CI for the ratio of means of the primary PK parameter is totally within the bioequivalence limits of (80%, 125%). This one-size-fits-all criterion only focuses on average bioavailability and ignores heterogeneity of variability. Thus, it is not scientifically/statistically justifiable for assessment of biosimilarity of FOB. In practice, it is then suggested that appropriate criteria, which can take the heterogeneity of variability into consideration, be developed since biosimilars are known to be variable and sensitive to small variations in environmental conditions.

For assessment of bioequivalence for chemical drug products, a crossover design is often considered, except for drug products with relatively long half-lives. Since most biosimilar products have relatively long half-lives, it is suggested that a parallel-group design should be considered. However, parallel-group design does not provide independent estimates of variance components such as inter- and intrasubject variability and variability due to subject-by-product interaction. Thus, it is a major challenge for assessing biosimilars under parallel-group designs.

Although EMA of EU has published several product-specific guidance based on the concept papers, it has been criticized that there are no objective *standards* for assessment of biosimilars because it depends upon the nature of the products. Product-specific standards seem to suggest that a *flexible* biosimilarity criterion should be considered and the flexible criterion should be adjusted for variability and/or the therapeutic index of the innovative (or reference) product.

As described earlier, there are many uncertainties for assessment of biosimilarity and interchangeability of biosimilars. As a result, it is a major challenge to both clinical scientists and biostatisticians to develop valid and robust clinical/statistical methodologies for assessment of biosimilarity and interchangeability under the uncertainties. In addition, how to address the issues of quality and comparability in manufacturing process is another challenge to both the pharmaceutical scientists and biostatisticians. The proposed general approach using the biosimilarity index (derived based on the concept of reproducibility probability) may be useful. However, further research on the statistical properties of the proposed biosimilarity index is required.

Study Design

As indicated in the *Federal Register* (Vol. 42, No. 5, section 320.26(b) and section 320.27(b), 1977), a bioavailability study (single dose or multi dose) should be crossover in design, unless a parallel or other design is more appropriate for valid scientific reasons. Thus, in practice, a standard two-sequence, two-period (or 2·2) crossover design is often considered for a bioavailability or bioequivalence study. Denote by T and R the test product and the reference product, respectively. Thus, a 2·2 crossover design can be expressed as (TR, RT), where TR is the first sequence of treatments and RT denotes the second sequence of treatments. Under the (TR, RT) design, qualified subjects who are randomly assigned to sequence 1 (TR) will receive the test product T first and then crossovered to receive the reference product R after a sufficient length of washout period. Similarly, subjects who are randomly assigned to sequence 2 (RT) will receive the reference product (R) first and then receive the test product (T) after a sufficient length of washout period.

One of the limitations of the standard 2 × 2 crossover design is that it does not provide independent estimates of intrasubject variabilities since each subject will receive the same treatment only once. In the interest of assessing intrasubject variabilities, the following alternative higher-order crossover designs for comparing two drug products are often considered: (1) Balaam's design, that is, (TT, RR, RT, TR); (2) two-sequence, three-period dual design, for example, (TRR, RTT); and (3) four-sequence, four-period design, for example, (TTRR, RRTT, TRTR, RTTR).

For comparing more than two drug products, a Williams' design is often considered. For example, for comparing three drug products, such as when a U.S. and European RLD is used simultaneously, a six-sequence, three-period (6 × 3) design is usually considered, while a 4 × 4 design is employed for comparing four drug products. The designs should be capable of stabilizing variance.

In addition to the assessment of ABE, there are other types of bioequivalence assessment such as population bioequivalence (PBE), which is intended for addressing drug prescribability, and individual bioequivalence (IBE), which is intended for addressing drug switchability. For assessment IBE/PBE, the FDA recommends that a *replicated* design be considered for obtaining independent estimates of intrasubject and intersubject variabilities and variability due to subject-by-drug product interaction. A commonly considered replicate crossover design is the replicate of a 2 × 2 crossover design given by (TRTR, RTRT). In some cases, an incomplete block design or an extrareference design such as (TRR, RTR) may be considered depending upon the study objectives of the bioavailability/bioequivalence studies.

Statistical Methods

As indicated earlier, ABE is claimed if the ratio of average bioavailability between test and reference products is within the bioequivalence limit of (80%, 125%) with 90% assurance based on log-transformed data. Along this line, commonly employed statistical methods are the CI approach and the method of interval hypotheses testing. For the CI approach, a 90% CI for the ratio of means of the primary PK response such as AUC or $C_{\max}$ is obtained under an analysis of variance model. We claim bioequivalence if the obtained 90% CI is totally within the bioequivalence limit of (80%, 125%). For the method of interval hypotheses testing, the interval hypotheses are

$$H_0\text{: Bioinequivalence versus } H_a\text{: Bioequivalence}$$

Note that the aforementioned hypotheses are usually broken into two sets of one-sided hypotheses: the first set of hypotheses is to verify that the average bioavailability of the test product is not too low, whereas the second set of hypotheses is to verify that average bioavailability of the test product is not too high. Under the two one-sided hypotheses, Schuirmann's two one-sided tests procedure is commonly employed for testing ABE.

In practice, other statistical methods such as Westlake's symmetric CI approach, CI based on Fieller's theorem, Chow and Shao's joint confidence region approach, Bayesian methods and nonparametric methods such as Wilcoxon–Mann–Whitney two one-sided tests procedure, distribution-free CI based on the Hodges–Lehmann estimator, and bootstrap CI are sometimes considered.

Special Considerations

Although the assessment of ABE for generic approval has been in practice for years, it has the following limitations: (1) it focuses only on population average, (2) it ignores the distribution of the metric, and (3) it does not provide independent estimates of intrasubject variability and ignores the subject-by-formulation interaction. There is criticism that the assessment of ABE does not address the question of drug interchangeability and it may penalize drug products with lower variability.

In addition, the use of one-fits-all criterion for assessment of ABE has been criticized in the past decade. It is suggested that the one-fits-all criterion is flexible by adjusting the intrasubject variability of the reference product and therapeutic window whenever possible. This has led to the proposed scaled average bioequivalence (SAB) criterion for assessment of bioequivalence for highly variable drug products. It should be noted that the SAB criterion is a special case of the following criteria for IBE:

$$\frac{(\mu_T - \mu_R)^2 + \sigma_D^2 + (\sigma_{WT}^2 - \sigma_{WR}^2)}{\max(\sigma_{WR}^2, \sigma_{W0}^2)} \le \theta_I \tag{13.1}$$

where

σ_{WT}^2 and σ_{WR}^2 are the within-subject variances of the test drug product and the reference drug product, respectively
σ_D^2 is the variance component due to subject-by-drug interaction
σ_{W0}^2 is a constant that can be adjusted to control the probability of passing IBE
θ_I is the bioequivalence limit for IBE

As indicated by the regulatory agencies, a generic drug can be used as a substitution of the brand-name drug if it has been shown to be bioequivalent to the brand-name drug. Current regulations do not indicate that two generic copies of the same brand-name drug can be used interchangeably, even though they are bioequivalent to the same brand-name drug. Bioequivalence between generic copies of a brand-name drug is not required. Thus, one of the controversial issues is whether these approved generic drug products can be used safely and interchangeably.

Criteria, Design, and Statistical Methods for Biosimilarity

Criteria for Biosimilarity

For the comparison between drug products, some criteria for the assessment of bioequivalence, similarity (e.g., the comparison of dissolution profiles), and consistency (e.g., comparisons between manufacturing processes) are available in either regulatory guidelines/guidance or the literature. These criteria, however, can be classified into either (1) absolute change versus relative change, (2) aggregated versus disaggregated, or (3) moment based versus probability based.

In practice, we may consider assessing bioequivalence or biosimilarity by comparing average and variability separately or simultaneously. This leads to the so-called disaggregated criterion and aggregated criterion. A disaggregate criterion will provide different levels of biosimilarity. For example, the study that passes criteria of both average and variability of biosimilarity provides stronger evidence of biosimilarity as compared to those studies that pass only the average biosimilarity. On the other hand, it is not clear whether an aggregated criterion would provide a stronger evidence of biosimilarity due to potential offset (or masked) effect between the average and variability in the aggregated criterion. Further research for establishing the appropriate statistical testing procedures based on the aggregate criterion and comparing its performance with the disaggregate criterion may be needed.

Another way of assessing biosimilarity is the moment-based criterion with the probability-based criterion for assessment of bioequivalence or biosimilarity under a parallel-group design. The results indicate that the probability-based criterion is not only a much more stringent criterion but also has sensitivity to any small change in variability. This justifies the use of the probability-based criterion for assessment of biosimilarity between FOB if a certain level of precision and reliability of biosimilarity is desired.

Study Design

As indicated earlier, a crossover design is often employed for bioequivalence assessment. In a crossover study, each drug product is administered to each subject. Thus, an estimate (approximate) for within-subject variance can be used to address switch ability and interchangeability. For a parallel-group study, each drug product is administered to a different group of subjects. Thus, we can only estimate total variance (between and within-subject variances), not individual variance components. For FOB with long half-lives, crossover study would be ineffective and unethical. In this case, we need to undertake study with parallel groups. However, a parallel-group study does not provide an estimate for within-subject variation (since there is no R vs. R).

Statistical Methods

Similar to the assessment of ABE, Shuirmann's two one-sided tests procedure or the CI is recommended for assessment of biosimilarity if similar criteria are adopted. On the other hand, if similar criteria for assessment of PBE/IBE are considered, the 95% confidence upper bound can be used for assessing biosimilarity based on linearized criteria of PBE/IBE.

INTERCHANGEABILITY

As indicated in the subsection (b)(3) amended to the Public Health Act subsection 351(k)(3), the term *interchangeable* or *interchangeability* in reference to a biological product that is shown to meet the standards described in subsection (k)(4) means that the biological product may be substituted for the reference product without the intervention of the health-care provider who prescribed the reference product. Along this line, in what follows, definition and basic concepts of interchangeability (in terms of switching and alternating) are given.

Definition and Basic Concepts

As indicated in subsection (a) that amends the Public Health Act subsection 351(k)(3), a biological product is considered to be interchangeable with the reference product if (1) the biological product is biosimilar to the reference product and (2) it can be expected to produce the same clinical result in *any given patient.* In addition, for a biological product that is administered more than once to an individual, the risk in terms of safety or diminished efficacy of alternating or switching between use of the biological product and the reference product is not greater than the risk of using the reference product without such alternation or switch.

Thus, there is a clear distinction between biosimilarity and interchangeability. In other words, biosimilarity does not imply interchangeability, which is much more stringent. Intuitively, if a test product is judged to be interchangeable with the reference product, then it may be substituted, even alternated, without a possible intervention, or even notification, of the health-care provider. However, the interchangeability is expected to produce the *same* clinical result in *any given patient,* which can be interpreted as that the same clinical result can be expected in *every single patient.* In reality, conceivably, lawsuits may be filed if adverse effects are recorded in a patient after switching from one product to another.

It should be noted that when FDA declares the biosimilarity of two drug products, it may not be assumed that they are interchangeable. Therefore, labels ought to state whether for a FOB, which is biosimilar to a reference product, interchangeability has or has not been established. However, payers and physicians may, in some cases, switch products even if interchangeability has not been established.

Switching and Alternating

Unlike drug interchangeability (in terms of prescribability and switchability, the U.S. FDA has a different perception of drug interchangeability for biosimilars. From the FDA's perspectives, interchangeability includes the concept of switching and alternating between an innovative biological product (R) and its FOB (T). The concept of switching is referred to as not only the switch from "R to T" or "T to R" (narrow sense of switchability) but also "T to T" and "R to R" (broader sense of switchability). As a result, in order to assess switching, biosimilarity for "R to T," "T to R," "T to T," and "R to R" needs to be assessed based on some biosimilarity criteria under a valid study design.

On the other hand, the concept of alternating is referred to as either the switch from T to R and then switch back to T (i.e., "T to R to T") or the switch from R to T and then switch back to R (i.e., "R to T to R"). Thus, the difference between "the switch from T to R" or "the switch from R to T" and "the switch from R to T" or "the switch from T to R" needs to be assessed for addressing the concept of alternating.

Study Design

For assessment of bioequivalence for chemical drug products, a standard two-sequence, two-period (2 × 2) crossover design is often considered, except for drug products with relatively long half-lives. Since most biosimilar products have relatively long half-lives, it is suggested that a parallel-group design should be considered. However, parallel-group design does not provide independent estimates of variance components such as inter- and intrasubject variability and variability due to subject-by-product interaction. Thus, it is a major challenge for assessing biosimilars under parallel-group designs.

In order to assess biosimilarity for "R to T," "T to R," "T to T," and "R to R," the Balaam's 4 × 2 crossover design, that is, (TT, RR, TR, RT), may be useful. For addressing the concept of alternating, a two-sequence, three-period dual design, that is, (TRT, RTR), may be useful. For addressing both concepts of switching and alternating for drug interchangeability of biosimilars, a modified Balaam's crossover design, that is, (TT, RR, TRT, RTR), is then recommended.

With small-molecule drug products, bioequivalence generally reflects therapeutic equivalence. Drug prescribability, switching, and alternating are generally considered reasonable. With biological products, however, variations are often higher (other PK factors may be sensitive to small changes in conditions). Thus, often, only parallel-group design rather than crossover kinetic studies can be performed. It should be noted that very often, with FOB, biosimilarity does *not* reflect therapeutic comparability. Therefore, switching and alternating should be pursued only with substantial caution.

Biosimilarity Index

Several recent papers have proposed a reproducibility probability as an index for determining whether it is necessary to require a second trial when the result of the first clinical trial is strongly significant. Suppose that the null hypothesis H_0 is rejected if and only if $|T| > c$, where c is a positive known constant and T is a test statistic. Thus, the reproducibility probability of observing a significant clinical result when H_a is indeed true is given by

$$p = P\left(|T| > c|H_a\right) = P\left(|T| > c|\hat{\theta}\right) \tag{13.2}$$

where θ is an estimate of θ, which is an unknown parameter or vector of parameters. Following the similar idea, a reproducibility probability can also be used to evaluate biosimilarity and interchangeability between a test product and a reference product based on any prespecified criteria for biosimilarity and interchangeability. As an example, biosimilarity index proposed can be based on the well-established bioequivalence criterion by the following steps:

Step 1. Assess the average biosimilarity between the test product and the reference product based on a given biosimilarity criterion. For illustration purpose, consider bioequivalence criterion as biosimilarity criterion. That is, biosimilarity is claimed if the 90% CI of the ratio of means of a given study endpoint falls within the biosimilarity limit of (80%, 125%) based on log-transformed data.

Step 2. Once the product passes the test for biosimilarity in step 1, calculate the reproducibility probability based on the observed ratio (or observed mean difference) and *variability*. We will refer to the calculated reproducibility probability as the *biosimilarity index*.

Step 3. We then claim biosimilarity if the following null hypothesis is rejected:

$$H_0\text{: } P \le p_0 \text{ vs. } H_a\text{:} P > p_0 \tag{13.3}$$

A CI approach can be similarly applied. In other words, we claim biosimilarity if the lower 95% confidence bound of the reproducibility probability is larger than a prespecified number p_0. In practice, p_0 can be obtained based on an estimated of reproducibility probability for a study comparing a reference product to itself (the reference product). We will refer to such a study as an *R–R* study.

In an *R–R* study, define

$$P_{TR} = P\begin{pmatrix} \text{Concluding average biosimiliarity between the test and the} \\ \text{reference products in a future trial given that the average} \\ \text{biosimiliarity based on ABE criterion has been established} \\ \text{in the first trial} \end{pmatrix} \tag{13.4}$$

Alternatively, a reproducibility probability for evaluating the biosimilarity of the two same reference products based on ABE criterion is defined as

$$P_{RR} = P\begin{pmatrix} \text{Concluding average biosimiliarity of the two same reference} \\ \text{products in a future trial given that the average biosimilarity} \\ \text{based on ABE criterion has been established in the first trial} \end{pmatrix} \tag{13.5}$$

Since the idea of the biosimilarity index is to show that the reproducibility probability in a study for comparing FOB with the innovative (reference) product is higher than a reference product with the reference product, the criterion of an acceptable reproducibility probability (i.e., p_0) for assessment of biosimilarity can be obtained based on the *R–R* study. For example, if the *R–R* study suggests the reproducibility probability of 90%, that is, $P_{RR} = 90\%$, the criterion of the reproducibility probability for bioequivalence study could be chosen as 80% of the 90%, which is $p_0 = 80\% \cdot P_{RR} = 72\%$.

The aforementioned described biosimilarity index has the advantages that (1) it is robust with respect to the selected study endpoint, biosimilarity criteria, and study design; (2) it takes variability into consideration (one of the major criticisms in the assessment of ABE); (3) it allows the definition and assessment the degree of similarity (in other words, it provides partial answer to the question that "how similar is considered similar?") and (4) the use of biosimilarity index will reflect the sensitivity of heterogeneity in variance.

The biosimilarity index concept can be applied to different functional areas (domains) of biological products such as the good drug characteristics of safety (e.g., immunogenicity), purity, and potency (as described in the BPCI Act); PK; PD; biological activities; biomarkers (e.g., genomic markers); and the manufacturing process used for the assessment of *global* biosimilarity. An overall biosimilar index across domains can be obtained by the following steps:

Obtain P_i, the probability of reproducibility for the ith domain, $i = 1, \ldots, K$.
Define the global biosimilarity index

$$P = \sum_{i=1}^{K} w_i P_i,$$

where w_i is the weight for the ith domain. The weights will have to be specified a priori.

Step 1. Claim global biosimilarity if the lower 95% confidence bound of the reproducibility probability (P) is larger than a prespecified number p_0, where p_0 is a prespecified, acceptable reproducibility probability.

Step 2. Define the global biosimilarity index $P = \sum^{K} wP$, where $iii = 1$, w_i is the weight for the ith domain.

Step 3. Claim global biosimilarity if the lower 95% confidence bound of the reproducibility probability (P) is larger than a prespecified number P_0, where P_0 is a prespecified acceptable reproducibility probability.

It should be noted that biosimilarity index is sensitive to the variability associated with the reference product. The biosimilarity index decreases as the variability increases. As an example, Figure 13.1 gives reproducibility probability curves under a 2·2 crossover design with sample sizes $n_1 = n_2 = 10$, 20, 30, 40, 50, and 60 at the 0.05 level of significance and $(\theta_L, \theta_U) = (80\%, 125\%)$ when $\sigma_d = 0.2$ and 0.3, where σ_d is the standard deviation of period difference within each subject.

In practice, alternative approaches for assessment of the proposed biosimilarity index are available. The methods include maximum likelihood approach and Bayesian approach. For the Bayesian approach, let $p(\theta)$ be the power function, where θ is an unknown parameter or vector of parameters. Under this Bayesian approach, θ is random with a prior distribution assumed to be known. The reproducibility probability can be viewed as the posterior mean of the power function for the future trial

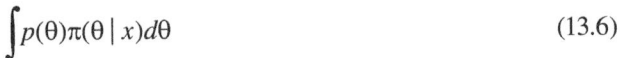

$$\int p(\theta)\pi(\theta \mid x)d\theta \tag{13.6}$$

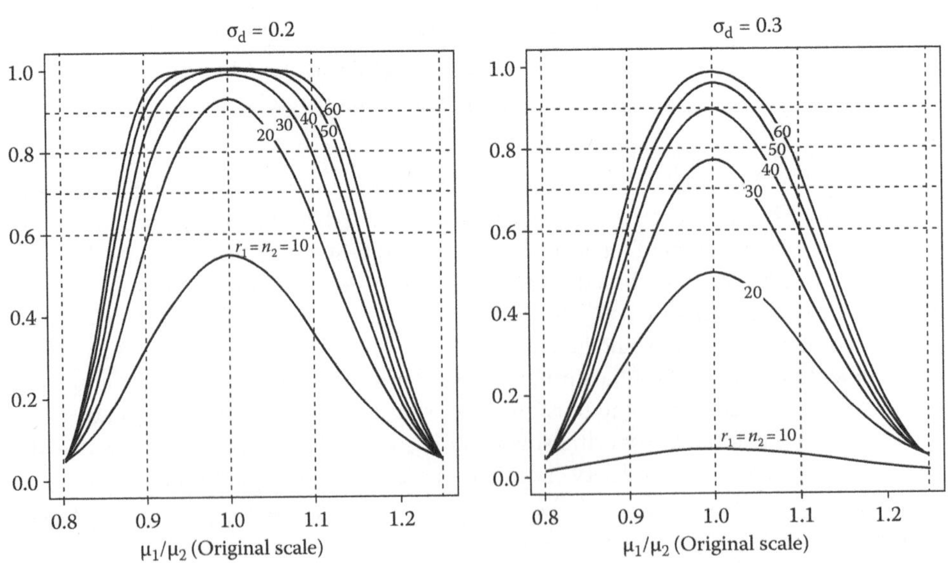

FIGURE 13.1 Impact of variability on reproducibility. (From Chow, S.C. and Liu, J.P., *J. Biopharm. Stat.* 20, 10, 2010.)

where $\pi(\theta|x)$ is the posterior density of θ, given the data set x observed for the previous trial (s). However, there may exist no explicit form for the estimation of the biosimilarity index. As a result, statistical properties of the derived biosimilarity index may not be known. In this case, the finite sample size performance of the derived biosimilarity index may only be evaluated by clinical trial simulations.

As an alternative measure for assessment of global biosimilarity across domains, we may consider

$$rd = \sum_{i=1}^{K} w_i rd_i, \text{ where } rd_i = \frac{P_{TRi}}{P_{RRi}} \tag{13.7}$$

which is the relative measure of biosimilarity between T and R as compared to that of between R and R. Based on rd_i, $i = 1,\ldots, K$, we may conduct a *profile analysis* as described in the 2003 FDA guidance on *Bioavailability and Bioequivalence Studies for Nasal Aerosols and Nasal Sprays for Local Action* (see Chapter 11 Nasal Product Bioequivalence). However, statistical properties of the profile analysis based on rd_i, $i = 1, \ldots, K$ are not fully studied. Note that the reproducibility probability decreases when μ_1/μ_2 (original scale) moves away from 1 and σ_d (log scale) is larger.

EUROPEAN PERSPECTIVE

Background

The EMA guideline lays down the nonclinical and clinical requirements for a biological medicinal product claiming to be similar to another one already marketed ("biosimilar"). The nonclinical section addresses the pharmacotoxicological assessment. The clinical section addresses the requirements for PK, PD, and efficacy studies. The section on clinical safety and pharmacovigilance addresses clinical safety studies as well as the risk management plan with special emphasis on studying the immunogenicity of the biosimilar.

The current revision of the guideline (2013) covers a risk-based approach for the design of nonclinical studies; the use of PD markers; study design, choice of appropriate patient population, and choice of surrogate endpoints in efficacy trials; design of immunogenicity studies; and extrapolation of indication.

Generally, the differences observed in the physicochemical and biological analyses will guide the planning of the nonclinical studies. Other factors that need to be taken into consideration are the mode of action of the active substance (e.g., receptor(s) involved) in all the licensed indications of the reference product and pathogenetic mechanisms involved in the disorders included in the therapeutic indications (e.g., mechanisms shared by various therapeutic indications).

The applicants should review the data from the reference product on the predictive value of in vitro assays/animal models as well as correlations between dose/exposure and PD, on one hand, and PD and clinical response, on the other hand. The availability of suitable biomarkers may abbreviate the nonclinical development.

The safety profile of the reference product will determine the focus of the safety studies both pre- and postmarketing. The Committee for Medicinal Products for Human Use (CHMP) (EMA) has issued product-class-specific guidelines to facilitate the nonclinical development of biosimilar medicinal products in certain areas. However, the applicants have to fine-tune their nonclinical studies according to the results of preceding physicochemical and in vitro biological analyses of the biosimilar and the reference product.

Nonclinical studies should be performed before initiating clinical trials. A stepwise approach should be applied to evaluate the similarity of biosimilar and reference product. In vitro studies should be conducted first and a decision then made as to the extent of what, if any, in vivo work will be required. It is important to note that design of an appropriate nonclinical study program requires a clear understanding of the reference product characteristics.

Results from the physicochemical and biological characterization studies (i.e., comparability of the biosimilar to the reference product) should be reviewed from the point of view of potential impact on efficacy and safety. The following approach may be considered and should be tailored to the product concerned on a case-by-case basis. The approach taken will need to be fully justified in the nonclinical overview:

Step 1: In vitro studies. In order to assess any difference in biological activity between the biosimilar and the reference medicinal product, data from a number of comparative in vitro studies, some of which may already be available from quality-related assays, should normally be provided. These studies could include relevant assays on

- Binding to target(s) (e.g., receptors, antigens, enzymes) known to be involved in the pharmacotoxicological effects of the reference product
- Signal transduction and functional activity/viability of cells known to be of relevance for the pharmacotoxicological effects of the reference product

The studies should be comparative in nature and should not just assess the response per se. The studies should evaluate parameters sensitive enough to detect differences.

The studies should assess the concentration–activity/binding relationship between the biosimilar and the reference medicinal product covering a concentration range where differences are most sensitively detected. They should be performed with an appropriate number of batches of product representative of that intended for clinical use. Together, these assays should broadly cover the spectrum of pharmacological/toxicological aspects known to be of relevance for the reference product and for the product class. Since in vitro assays may often be more specific and sensitive to detect differences between the biosimilar and the reference product than studies in animals, these assays can be considered as paramount for the nonclinical comparability exercise. The applicant should justify that the in vitro assays used are predictive for the in vivo situation.

If the biosimilar comparability exercise indicates early on that there are significant differences between the intended biosimilar and the reference medicinal product making it unlikely that biosimilarity will eventually be established, a stand-alone development should be considered instead.

Step 2: Determination of the need for in vivo studies. It is acknowledged that biotechnology-derived proteins may mediate in vivo effects that cannot be fully elucidated by in vitro studies. Therefore, nonclinical evaluation in in vivo studies may be necessary to provide complementary information, provided that a relevant in vivo model with regard to species or design is available. Factors to be considered when the need for in vivo nonclinical studies is evaluated include but are not restricted to

- Presence of relevant quality attributes that have not been detected in the reference product (e.g., new posttranslational modification structures)
- Significant quantitative differences in quality attributes between the intended biosimilar and the reference product
- Relevant differences in formulation, for example, use of excipients not widely used for biotechnology-derived proteins

Although each of the factors mentioned earlier do not necessarily warrant in vivo testing, these issues should be considered together to assess the level of concern and whether there is a need for in vivo testing. If the comparability exercise in the in vitro studies in step is considered satisfactory and no factors of concern are identified in step, or these factors do not block direct entrance into humans, an in vivo animal study may not be considered necessary.

If product-inherent factors that impact PK and/or biodistribution, like extensive glycosylation, cannot sufficiently be characterized on a quality and in vitro level, in vivo studies may be necessary.

Applicants should then carefully consider if these should be performed in animals or as part of the clinical testing, for example, in healthy volunteers.

If there is a need for additional in vivo information, the availability of a relevant animal species or other relevant models (e.g., transgenic animals, transplant models) should be considered. If a relevant in vivo animal model is not available, the applicant may choose to proceed to human studies taking into account principles to mitigate any potential risk.

Step 3: In vivo studies. If an in vivo evaluation is deemed necessary, the focus of the study/studies (PK and/or PD and/or safety) depends on the need for additional information. Animal studies should be designed to maximize the information obtained. The principles of the replacement, refinement, reduction (3Rs) should be considered when designing any in vivo study. Depending on the endpoints needed, it may not be necessary to sacrifice the animals at the end of the study. The duration of the study (including observation period) should be justified, taking into consideration the PK behavior of the biotechnology-derived product and its clinical use. When the model allows, the PK and PD of the biosimilar and the reference medicinal product should be quantitatively compared, including concentration–response assessment covering the therapeutic dose range in humans. For safety studies, a flexible approach should be considered, in particular if nonhuman primates are the only relevant species. The conduct of standard repeated dose toxicity studies in nonhuman primates is usually not recommended. If appropriately justified, a repeated dose toxicity study with refined design (e.g., using just one dose level of biosimilar and reference product and/or just one gender and/or no recovery animals) or an in-life evaluation of safety parameters (such as clinical signs, body weight, and vital functions) may be considered.

The conduct of toxicity studies in nonrelevant species (i.e., to assess unspecific toxicity only, based on impurities) is not recommended. Due to the different production processes used by the biosimilar and reference product manufacturers, qualitative differences of process-related impurities will occur (e.g., host cell proteins). The level of such impurities should be kept to a minimum, which is the best strategy to minimize any associated risk.

Qualitative or quantitative difference(s) of product-related variants (e.g., glycosylation patterns, charge variants) may affect biological functions of the biotechnology-derived protein and are expected to be evaluated by appropriate in vitro assays. These quality differences may have an effect on immunogenic potential and the potential to cause hypersensitivity. It is acknowledged that these effects are difficult to predict from animal studies and should be further assessed in clinical studies.

Although immunogenicity assessment in animals is generally not predictive for immunogenicity in humans, it may be needed for interpretation of in vivo studies in animals. Therefore, blood samples should be taken and stored for future evaluations if then needed. Studies regarding safety pharmacology, reproduction toxicology, and carcinogenicity are not required for nonclinical testing of biosimilars.

Studies on local tolerance are usually not required. However, if excipients are introduced for which there is no or little experience with the intended clinical route, local tolerance may need to be evaluated. If other in vivo studies are performed, evaluation of local tolerance may be part of the design of that study instead of the performance of separate local tolerance studies.

Clinical Studies

It is acknowledged that the manufacturing process of the biosimilar product will be optimized during development. However, it is recommended to generate the clinical data required for the comparability study with the test product derived from the final manufacturing process and therefore representing the quality profile of the batches to become commercialized.

Any deviation from this recommendation should be justified and supported by adequate additional bridging data. The clinical comparability exercise is normally a stepwise procedure that should begin with PK and, if feasible, PD studies followed by clinical efficacy and safety trial(s) or, in certain cases, confirmatory PK/PD studies for demonstrating clinical comparability.

Pharmacokinetic Studies

Comparative PK studies designed to demonstrate similar PK profile of the biosimilar and the reference medicinal product with regard to key PK parameters are an essential part of the biosimilar development program. The design of the study depends on various factors, including clinical context, safety, and PK characteristics of the reference product (target-mediated disposition, linear or nonlinear PK, time dependency, half-life, etc.). Furthermore, bioanalytical assays should be appropriate for their intended use and adequately validated as outlined in the guideline on bioanalytical method validation.

The criteria used in standard clinical bioequivalence studies, initially developed for chemically derived, orally administered products, may be acceptable in the absence of specific criteria for biologicals. Nevertheless, the comparability limits for the main PK parameters should be defined and justified prior to conducting the study. For the demonstration of comparable PK, it is advisable to select the most sensitive test model. Healthy volunteers lack comorbidity and comedications and are likely to have less target-mediated clearance compared to patients. A single-dose crossover study with full characterization of the PK profile, including the late elimination phase, is preferable. A parallel-group design may be necessary with substances with a long half-life and high risk of immunogenicity.

PK studies are not always possible or feasible in healthy volunteers. In this case, the PK needs to be studied in patients. The most sensitive model/population, that is, that has fewer factors that cause major interindividual or time-dependent variation, should be explored. In certain cases, such as important target-mediated clearance, highly immunogenic proteins, or highly variable PK parameters, it may be useful to collect additional PK data within the confirmatory efficacy clinical trial(s) as it allows further investigation of the clinical impact of variable PK and possible changes in the PK over time. This can be achieved by determining the PK profile in a subset of patients or by PPK.

Antidrug antibodies should be measured in parallel to PK assessment using the most appropriate sampling time points. If the reference product can be administered both intravenously (IV) and subcutaneously (SC), the evaluation of SC administration will usually be sufficient as it covers both absorption and elimination. Thus, it is possible to waive the evaluation of IV administration if comparability in both absorption and elimination has been demonstrated for the SC route. In a single-dose PK study, the primary parameters are the $AUC_{(0\text{–inf})}$ for IV administration and $AUC_{(0\text{–inf})}$ and usually C_{max} for SC administration. Secondary parameters such as t_{max}, volume of distribution, and half-life should also be estimated. In a multiple-dose study, the primary parameters should be the truncated AUC after the first administration until the second administration ($AUC_{0\text{–}t}$) and AUC over a dosage interval at steady state ($AUC\tau$). Secondary parameters are C_{max} and C_{trough} at steady state.

Pharmacodynamic Studies

It is recommended that PD markers are added to the PK studies whenever feasible. The PD markers should be selected on the basis of their clinical relevance. Normally, comparative efficacy trials are required for the demonstration of clinical comparability. In certain cases, however, comparative PK/PD studies between the test and the reference medicinal product may be sufficient to demonstrate clinical comparability, provided that all the following conditions are met where a clear dose–response relationship has been demonstrated. If not, the recommended study design is to conduct a multiple-dose–exposure–response study. This design would ensure that the biosimilar and the reference can be compared within the linear ascending part of the dose–response curve. In certain cases, a time-to-response study may be sensitive but it cannot replace dose-comparative studies.

The selected PD marker/biomarker is an accepted surrogate marker and can be related to patient outcome to the extent that demonstration of similar effect on the PD marker will ensure a similar effect on the clinical outcome. Relevant examples include absolute neutrophil count (ANC) to assess the effect of G-CSF, early viral load reduction in chronic hepatitis C to assess the effect of alpha interferons, euglycemic clamp test to compare two insulins, and magnetic resonance imaging (MRI) of disease lesions to compare two β-interferons. The evidence for a surrogacy of a PD marker/biomarker is often scanty and formal validation of surrogacy is very rare. In such cases, a combination of markers selected based on sound pharmacological principles, including dose/concentration sensitivity, may provide sufficient evidence to conclude on clinical comparability.

When evidence to establish clinical comparability will be derived from studies with PD markers/biomarkers, it is recommended to discuss such approach with regulatory authorities. This should include a proposal of the size of the proposed equivalence margin and its clinical justification.

Efficacy Trials

Usually, it is necessary to demonstrate comparable clinical efficacy of the biosimilar and the reference medicinal product in adequately powered, randomized, parallel-group comparative clinical trial(s), preferably double blind. The study population should be representative of approved therapeutic indication(s) of the reference product and be sensitive for detecting potential differences between the biosimilar and the reference.

Occasionally, changes in the clinical praxis mandate a deviation from the approved therapeutic indication, for example, in terms of concomitant medication used as combination treatment, line of therapy, or severity of the disease. Deviations need to be justified and discussed with regulatory authorities.

Study Designs

In general, an equivalence design should be used. The use of a noninferiority design may be acceptable if justified on the basis of a strong scientific rationale and taking into consideration the characteristics of the reference product, for example, safety profile/tolerability, dose range, dose–response relationship. A noninferiority trial may only be accepted where the possibility of increased efficacy can be excluded on scientific and mechanistic grounds. However, as in equivalence trials, assay sensitivity has to be considered. It is recommended to discuss the use of a noninferiority design with regulatory authorities.

Efficacy Endpoints

Efficacy trials of biosimilar medicinal products do not aim at demonstrating efficacy per se, since this has already been established with the reference product. The sole purpose of the efficacy trials is to investigate whether a clinically significant difference between the reference and biosimilar products can be detected.

The CHMP (EMA) has issued disease-specific guidelines for development of innovative medicinal products. In the development of a biosimilar medicinal product, the choice of clinical endpoints and time points of analysis of endpoints may deviate from the guidance for new active substances. Therefore, CHMP has issued product-class-specific guidelines to guide the development of biosimilar medicinal products in certain areas. In the absence of such a guideline, the applicant should select the most sensitive endpoints. Nevertheless, deviations from disease-specific guidelines need to be scientifically justified. Differences detected should always be discussed as to whether they are clinically relevant. The correlation between the "hard" clinical endpoints recommended by the guidelines for new active substances and other clinical/PD endpoints that are sensitive to detect differences may have been demonstrated in clinical trials with the reference product. In this case, it is not necessary to use the same primary efficacy endpoints as those that were used in the marketing authorization application of the reference product.

However, it is advisable to include some common endpoints (e.g., as secondary endpoints) to facilitate comparisons to the clinical trials conducted with the reference product. Clinical comparability margins should be prespecified and justified on both statistical and clinical grounds by using the data of the reference. As for all clinical comparability trial designs, assay sensitivity has to be considered.

Clinical Safety

Even if the efficacy is shown to be comparable, the biosimilar may exhibit a difference in the safety profile. Clinical safety is important throughout the clinical development program and is captured during initial PK and/or PD evaluations and also as part of the pivotal clinical efficacy study establishing comparability. Comparative safety data should normally be collected preauthorization, their amount depending on the type and severity of safety issues related to the reference product.

The duration of safety follow-up preauthorization should be justified. Care should be given to compare the type, severity, and frequency of the adverse reactions between the biosimilar and the reference product, particularly those described in the SmPC of the reference product. The applicant should provide in the application dossier an evaluation of the specific risks anticipated for the biosimilar. This includes in particular a description of possible safety concerns related to infusion-related reactions and immunogenicity of the biosimilar that may result from a manufacturing process different from that of the reference product.

The potential for immunogenicity of a biosimilar should always be investigated in a comparable manner to the reference product and should follow the principles as laid down in the aforementioned CHMP guidelines unless it can be justified that there is a need for deviation from this approach. The amount of immunogenicity data will depend on the reference product and/or the product class.

Immunogenicity testing of the biosimilar and the reference products should be conducted within the comparability exercise by using the same assay format and sampling schedule. Assays should be performed with both the reference and biosimilar molecule in parallel (in a blinded fashion) to measure the immune response against the product that was received by each patient. Usually, the incidence of antibodies and antibody titers should be measured and presented. The duration of the immunogenicity study should be justified on a case-by-case basis depending on the duration of the treatment course, disappearance of the product from the circulation (to avoid antigen interference in the assays), and the time for emergence of humoral immune response (at least 4 weeks in case of an immunosuppressive agent). The duration of follow-up should be justified based on the time course and characteristics of unwanted immune responses described for the reference medicinal product, for example, a low risk of clinically significant immunogenicity or no significant trend for increased immunogenicity over time.

In case of chronic administration, 1-year follow-up data will normally be required prelicensing. Shorter follow-up data prelicensing (e.g., months) might be justified based on the immunogenicity profile of the reference product. Immunogenicity data for the additional period, up to 1 year, could then be submitted postauthorization.

A higher immunogenicity as compared to the reference product may become an issue for the benefit/risk analysis and would question biosimilarity. However, a lower immunogenicity for the biosimilar is also possible scenario, which would not preclude approval as a biosimilar. In case of reduced development of NAbs with the biosimilar, the efficacy analysis of the entire study population could erroneously suggest that the biosimilar is more efficacious than the reference product. It is therefore recommended to prespecify an additional exploratory subgroup analysis of efficacy and safety in those patients that did not mount an antidrug antibody response during the clinical trial. This subgroup analysis could be helpful to establish that the efficacy of the biosimilar and the reference product are in principle similar if not impacted by an immune response.

For biologicals with multiple indications, immunogenicity could differ among indications and absence of immunogenicity assessment in a particular indication for the biosimilar may have to be justified.

Extrapolation of efficacy and safety from one therapeutic indication to another in case the reference medicinal product has more than one therapeutic indication, the efficacy and safety of the biosimilar has to be justified or, if necessary, demonstrated separately for each of the claimed indications. Justification will depend on, for example, clinical experience, available literature data, MOAs of the active substance of the reference product in each indication (including its degree of certainty), and receptors involved. Binding of the reference substance to the same receptors may have different effects in different target cells depending on differences in the intracellular signaling pathways, for example, due to transformation. This situation is not an argument for additional studies.

However, if there is evidence that different active sites of the reference product or different receptors of the target cells are involved in different therapeutic indications or that the safety profile of the product differs between the therapeutic indications, additional data may be needed to justify the extrapolation of safety and efficacy from the indication studied in the pivotal clinical trial. For the extrapolation of safety, the applicant should consider patient-related factors, such as different comedication, comorbidities, and immunological status, and disease-related factors, such as reactions related to the target cells, for example, lysis of tumor cells.

The extent of such data should be considered in the light of the totality of evidence derived from the biosimilar comparability exercise and the potential remaining uncertainties.

PRODUCT SPECIFIC EUROPEAN GUIDELINES

Human Follicle-Stimulating Hormone

Follicle-stimulating hormone (FSH) is a pituitary glycoprotein hormone that plays a key role in regulating reproductive function in both males and females. FSH is a heterodimeric hormone composed of two linked subunits. The alpha subunit (92 amino acids) is common to other glycoprotein hormones, whereas the beta subunit (111 amino acids) is specific. Both subunits contain oligosaccharide structures. As a consequence of carbohydrate variability, different isoforms of hFSH with different sialic acid content exist. Isoforms with a high sialic acid content remain longer in circulation. Physicochemical and biological methods are available for characterization of the protein.

Recombinant human FSH (r-hFSH) is used in assisted reproductive technologies (ART) for women to stimulate growth and recruitment of ovarian follicles and for men to induce and maintain spermatogenesis. It is administered by SC or, in some cases, intramuscular injections.

The most important side effect of FSH treatment in ovarian stimulation is the occurrence of ovarian hyperstimulation syndrome (OHSS). This possibly life-threatening condition is characterized in its most serious forms by ascites, hemoconcentration, coagulation and electrolyte disorders, and extreme ovarian enlargement. The high number of follicles recruited and high estradiol levels (released from matured follicles) are risk factors for the development of OHSS.

Immunogenicity of r-hFSH appears to be low and so far, NAbs have not been reported. Generalized hypersensitivity reactions were observed in 0.2% and <1/10,000 patients treated with two different approved r-hFSH products. Local reactions were observed more frequently (3% and >1/10 of patients treated with two different r-hFSH products).

In order to evaluate potential differences in PD properties between the biosimilar and the reference medicinal product, comparative in vitro bioassays for receptor affinity and activation should be performed (such data may already be available from bioassays submitted as part of the quality dossier). Two principal approaches exist for this purpose. First, primary granulosa cells or sertoli cells can be used. Second, permanently cultured cells (e.g., CHO) stably transfected with the human FSH receptor may be constructed. The advantage of the first approach is that the FSH receptor is investigated in its natural context. A drawback is that the number of cells is limited, which in turn limits the number of replicates and the number of different r-hFSH concentrations that can be tested to obtain reliable concentration–response relationships. The second approach, although providing enough material, relies on an artificial construct (transfected cells).

Appropriate sensitivity of the assay used for comparability testing to detect potential differences should be demonstrated, and experiments should be based on a sufficient number of dilutions per curve to characterize the whole concentration–response relationship. Binding studies including on–off kinetics should be provided as well as measures of receptor activation, that is, plasminogen activator production (only in the classical granulosa cell assay) or intracellular cAMP accumulation. Other endpoints are conceivable (e.g., reporter gene activation). The applicant should justify the approach taken.

FSH is a highly glycosylated protein and in vitro studies may not fully reflect the more complex situation in vivo. Hence, to qualify any potential differences between the biosimilar FSH and the reference product, the need for additional comparative in vivo studies should be considered.

Currently, the potency of r-hFSH-containing products is evaluated by calibration against an international standard (or an internal reference standard calibrated against the international standard; Steelman–Pohley assay). As the in vivo potency of both the biosimilar and the reference product may be evaluated in such a way, the number of different assays performed may be reduced by a study design in which the biosimilar and the reference medicinal product are compared and simultaneously calibrated against the reference standard. This reduces interassay variation and is more economical with regard to reagents and animals used. The Steelman–Pohley assay is only expected to establish biological activity but not to reveal small differences in potency between reference product and biosimilar. If feasible, an evaluation of safety endpoints, for example, body weight and local tolerance, could be included within the framework of the in vivo PD studies.

If a different bioassay—for example, an ex vivo assay such as whole follicle culture or primary granulosa cell culture—is used to compare pleiotropic effects of FSH in a natural tissue environment, this should be justified. Such an approach would further reduce the number of animals needed, circumvent interanimal variability, and would give the possibility for multiple PD readouts.

Toxicological Studies

Generally, separate repeated dose toxicity studies are not required. In specific cases, for example, when novel or less well-studied excipients are introduced, the need for additional toxicology studies should be considered.

Safety pharmacology and reproduction toxicology studies are not required for nonclinical testing of similar biological medicinal products containing r-hFSH as active substance. Studies on local tolerance are not required, unless excipients are introduced for which there is no or only little experience with the intended route of administration. If other in vivo studies are performed, evaluation of local tolerance may be evaluated as part of these studies.

Pharmacokinetic Studies

The relative PK properties of the similar biological medicinal product and the reference medicinal product should be determined in a single-dose crossover study using SC injections. Healthy female volunteers are considered appropriate. Suppression of endogenous FSH production with a gonadotropin-releasing hormone (GnRH) agonist or a combined oral contraceptive is recommended. The dose of r-hFSH should be justified, taking into account that a dose in the linear part of the dose–response curve is suitable to detect potential differences in the PK profiles of the biosimilar and the reference medicinal product. The PK parameters of interest are AUC, C_{max}, t_{max}, $t1/2$, and clearance. For the primary endpoints AUC and C_{max}, the 90% CI of the ratio test/reference should lie within 80%–125%, the conventional acceptance range for bioequivalence, unless otherwise justified. For the other parameters, descriptive statistics would be appropriate. Separate pharmacology studies for intramuscular use, if applicable, are not required.

Pharmacodynamic Studies

PD parameters should be investigated as part of the phase III trial.

Clinical Efficacy

Clinical comparability regarding efficacy between the similar and the reference biological medicinal product should be demonstrated in an adequately powered, randomized, parallel-group clinical trial.

The recommended model for the demonstration of comparability of the test and the reference product is the stimulation of multifollicular development in patients undergoing superovulation for ART such as in vitro fertilization (IVF), gamete intrafallopian transfer (GIFT), or zygote intrafallopian transfer (ZIFT). The first treatment cycle should be used for comparison of efficacy.

Double-blind trials are recommended. If the performance of a double-blind trial is not feasible, blinded assessment of study outcomes that might be particularly affected by subjective factors, such as ultrasound examinations and parameters of oocyte/embryo quality, should be carried out. The r-hFSH dose should be fixed for the first 5 days of stimulation. A GnRH agonist or GnRH antagonist protocol can be used.

"The number of oocytes retrieved" is the recommended primary endpoint. Equivalent efficacy between the test product and the reference product should be demonstrated and equivalence margins prospectively defined and justified. It should be taken into account that overstimulation as well as understimulation can result in cycle cancellation and a number of zero oocytes retrieved (primary endpoint). Thus, the data should be presented in such a way that a detailed comparison of the reasons for cancellation of ART cycles is possible.

As an alternative possibility, demonstration of noninferiority for "ongoing pregnancy rate at least 10 weeks after embryo transfer" is also an acceptable primary endpoint. In the latter case, "the number of oocytes retrieved" should be included as coprimary endpoint with an appropriate equivalence margin or as the most important secondary endpoint.

With regard to secondary endpoints, the following issues should be taken into account:

- If the number of oocytes is chosen as the primary endpoint, ongoing pregnancy rate after at least 10 weeks after embryo transfer should be evaluated as the secondary endpoint.
- In ART cycles, the dose of FSH has to be adjusted based on ovarian response, which might obscure product-specific differences. Thus, dose adjustments and possible differences between the dosages of the similar biological product and the reference product should be carefully considered. Secondary endpoints covering this issue, such as total dose of r-hFSH required, the number of days of r-hFSH stimulation, and percentage of patients with need to increase or lower the dose of r-hFSH, should be investigated. Major differences with regard to dose requirements between the similar biological product and the reference product would not be in accordance with the concept of biosimilarity.
- Parameters supporting comparable PD properties of the similar biological product and the reference product should be investigated. The respective endpoints should include the number and size distribution of follicles during treatment and at the day of ovulation induction. A further endpoint covering the initial PD effect of r-hFSH on the ovary could be the number of follicles after 5 days of FSH stimulation (before dose adjustments). In addition, serum levels of inhibin-B, estradiol, luteinizing hormone, and progesterone should be measured.
- Markers of oocyte/embryo quality should be included. The number of good-quality oocytes/embryos should be documented.

Clinical Safety

Data from the efficacy trial will usually be sufficient to characterize the adverse event profile of the biosimilar product.

An adverse reaction of special interest is OHSS. All events of OHSS should be carefully recorded, using a grading system (mild, moderate, severe) and also distinguishing between early and late onset OHSS.

Immunogenicity of a therapeutic protein is more likely when given intermittently than continuously, and the SC route of administration is more immunogenic than the IV one. Both of these factors may apply to r-hFSH as women may receive more than one ART cycle. Therefore, immunogenicity data should be provided on all women included in the efficacy trial and also on women exposed for more than one ART cycle. Immunogenicity testing should continue up to 3 months after r-hFSH treatment using validated antibody assays of adequate sensitivity and specificity. The potential impact of FSH-antibodies, if detected, on efficacy and/or safety should be assessed and the necessity for further characterization, for example, with regard to their neutralizing potential, considered.

Interferon Beta

Three different medicinal products containing recombinant IFN-β are currently approved in the EU for the first-line treatment of MS; they differ with respect to their molecular structure, injection route, recommended posology, and MS indications.

Recombinant IFN-β-1a is a single glycosylated polypeptide chain containing 166 amino acids. Two products are available, one is administered SC and the other intramuscularly.

Recombinant IFN-β-1b is produced as a single nonglycosylated polypeptide chain of 165 amino acids with no methionine at the N-terminus and an amino acid substitution at position 17 and is administered SC.

Medicinal products containing recombinant IFN-β are currently indicated for patients with relapsing MS (RMS) including those at high risk of developing MS after a single demyelinating event. The MOA of IFN-β in MS is not well established but it has been hypothesized that it acts as an immunomodulator by (1) interfering with T-cell activation in several ways, including down-regulating the expression of type II MHC molecules, inhibiting the production of proinflammatory cytokines by Th1 cells, promoting the production of anti-inflammatory cytokines by Th2 cells, activating suppressor T-cells, and (2) inhibiting permeability changes of the blood–brain barrier and the infiltration of T cells into the central nervous system (CNS).

The clinical effects of recombinant IFN-β in RMS are modest with decreases in the frequency of exacerbations by approximately 30% as compared with placebo and inconsistent results on the progression of disability.

All products are associated with similar adverse reactions, which may affect patient adherence to therapy; the most frequent are influenza-like symptoms (fever, chills, arthralgia, malaise, sweating, headache, and myalgia). Injection site reactions and asymptomatic liver and white blood cell abnormalities occur more frequently with the SC products at the recommended dose regimens. Less common adverse reactions include depression and autoimmune disorders manifested as thyroid or liver dysfunction. All products induce the development of antibodies, and in particular NAbs; in clinical trials, the incidence of NAbs has been shown to range widely, from 5% for intramuscular IFN-β-1a given weekly to 45% for SC IFN-β-1b given every other day. Most Nabs develop in the first year of therapy and they have the potential to impact clinical outcomes after 18–24 months of treatment.

In order to assess any differences in biological activity between the biosimilar and the reference medicinal product, data from a number of bioassays/pharmacological studies should be provided (e.g., receptor-binding studies; assays for characterization of antiviral, antiproliferative, and immunomodulatory effects), some of which may already be available from assays submitted as part of the quality dossier (for IFN-β-1a, the requirements of the European Pharmacopoeia monograph IFN-β-1a concentrated solution apply).

Generally, in vivo studies in animals are not recommended. Only when the outcome of the quality evaluation and/or the in vitro bioassays/pharmacological studies leave uncertainties about the comparability of the biosimilar and reference medicinal product, the need for additional studies should be considered.

In vivo studies should be designed to specifically address the remaining uncertainties identified. These could include an in vivo pharmacological study and/or a general repeated dose toxicity study in a relevant species.

Further studies in a pharmacologically responsive animal species should only be considered when it is expected that such studies would provide relevant additional information.

Clinical Studies

The clinical comparability exercise should follow a stepwise approach starting with PK and PD and continuing with efficacy and safety studies.

Pharmacokinetics

The PK properties of the biosimilar and reference products should be compared in a crossover study using the route of administration applied for. Healthy volunteers are considered an appropriate study population. The selected dose should be in the linear part of the dose–concentration curve; if available information on the reference product is too scarce, more than one dose should preferably be tested. The choice of a single or repeated dose (e.g., three doses over a week) regimen should be justified; a single dose is preferred as long as the bioanalytical method is sufficiently sensitive to characterize the full PK profile. Although antibody development is not expected after a few doses of IFN-β, their determination should be carried out before/after each treatment course in order to exclude any potential interference with the PK profile.

Serum concentrations of IFN-β are very low after the administration of therapeutic dosages and their measurement is technically difficult. Possible methods of detection include a cell-based myxovirus-resistance protein A (MxA) induction assay, which measures the biological activity of IFN-β in serum samples, and ELISA assays, which determine the IFN-β protein mass. The applicant should justify the rationale for the choice of assay.

In particular, the PK parameters of interest should include AUC, C_{max}, and also $T1/2$ or clearance. The equivalence margin has to be defined a priori and appropriately justified, especially given the high variability of the relevant PK parameters. A two-stage design may be planned in the protocol provided adjusted significance levels are used for each of the analyses.

Pharmacodynamics

PD should preferably be evaluated as part of the comparative PK studies using validated assays. There is currently no identified biological marker related to the mechanism by which IFN-β influences the clinical evolution of MS. However, a number of markers of the biological activity of IFN-β are well known, and a comprehensive comparative evaluation of some of these markers could be used to support the similarity of the biosimilar and reference medicinal products ("fingerprint approach"). MxA induction can be measured from peripheral blood leukocytes both at the protein and mRNA level; it is currently considered as one of the most sensitive markers of the biological activity of type I interferons and should be one of the selected markers. Neopterin, which was found to show a consistent and robust dose–response relationship, should also be investigated. Other possible markers include serum (2′–5′)oligoadenylate synthetase activity, interleukin 10, or TNF-related apoptosis inducing ligand (TRAIL).

MRI is a useful tool for monitoring CNS lesions in MS. Different MRI-derived parameters have been related to clinical activity, for example, gadolinium-enhancing T1-weighted lesions or new/enlarging T2-weighted lesions have been related to relapses.

Clinical Efficacy

Similar clinical efficacy between the biosimilar and reference medicinal product should be demonstrated in an adequately powered, randomized, parallel-group, equivalence clinical trial, preferably double blind. If blinding is technically not feasible, alternative measures should be applied to avoid information bias. The route of administration used in the clinical trial should be the route recommended for the reference product.

According to the guideline on medicinal products for the treatment of MS, an acceptable primary efficacy variable for a disease-modifying agent in RMS is the relapse rate, which has been used in the pivotal trials on medicinal products containing recombinant IFN-β. While in principle this would be the preferred option, such a trial may not be necessary in a biosimilar context, since the focus of this trial is to demonstrate comparable clinical activity of the biosimilar product to the reference product, which then allows bridging to the benefit/risk of the reference product. For demonstrating clinical similarity of a biosimilar and reference product, MRI of disease lesions in RMS may be sufficient (see section "Pharmacodynamics"). In addition, clinical outcomes such as relapse rate or percentage of relapse-free patients should be used as secondary endpoints in support of the MRI outcomes.

The design of the equivalence trial should ensure assay sensitivity, that is, the choice of study design, population, duration, and MRI endpoints should make it possible to detect a difference between the biosimilar and reference products, if such difference actually exists. Regarding the study design, assay sensitivity could be shown by a three-arm trial including a placebo arm for a short period of time (e.g., 4 months) sufficient to demonstrate superiority of both the biosimilar and reference products over placebo using an MRI endpoint. Patients in the placebo arm could be subsequently switched to the biosimilar product and the trial continued with the two active arms. An alternative design could be a three-arm trial with the reference product and two doses of the biosimilar product, for which differences in MRI and clinical outcomes are expected to be observed over 12 months; if the MRI curves do not differentiate the two doses over time, interpretation of the results would be difficult as the assay sensitivity of the trial would be questionable.

Whatever the design, the duration of the trial should be sufficient to show comparable efficacy on MRI endpoints and provide relevant information on clinical outcomes, that is, not less than 12 months.

The most sensitive patient population, which would enable to detect differences between the biosimilar and reference products, should be selected. This would be a homogeneous sample of patients with a confirmed diagnosis of relapsing-remitting MS (RRMS) and sufficient disease activity based on relapse frequency and/or MRI criteria to anticipate rapid changes in MRI.

MRI-based variables are acceptable primary endpoints in the context of a biosimilar comparison if backed up by relapse-related clinical outcomes; no formal equivalence test is required for clinical outcomes, which would be expected to show the same trend in effect as the MRI-based variables. A relapse should be differentiated from a pseudoexacerbation and accurately defined. Repeated MRI scans should be performed during the trial. All possible actions should be taken to ensure high-quality MRI data and maximum reliability of measurements. Updated recommendations on appropriate technical facilities and standardized procedures and training should be followed. The reading of the images should be central and blinded. The combined unique active lesions (CUA, defined as new gadolinium-enhancing T1-weighted lesions and new/enlarging T2-weighted lesions without double counting) are the most sensitive documented MRI variable and, therefore, should always be determined; a cumulative estimate over several scans may be calculated. Other MRI variables may also be used as primary endpoint if adequately justified.

The equivalence margin for the primary MRI endpoint should be prespecified and adequately justified based on MRI data for the reference medicinal product relative to placebo or, if not available, extrapolation from other IFN-β relevant data. Of note, these data are important at the planning stage of the trial but are not essential for the interpretation of the results as assay sensitivity has to be shown within the trial. It should be adequately powered with particular attention paid in the protocol to the potential dropout rate and the way of handling missing data.

Clinical Safety

Comparative safety data from the efficacy trial are usually sufficient to investigate the more frequent adverse reactions and provide an adequate preauthorization safety database for such reactions but not for rarer adverse reactions, which should be addressed postauthorization.

As IFN-β products are immunogenic, an assessment of immunogenicity by testing of sera from IFN-β-treated patients should be performed according to the principles defined in the guideline on immunogenicity assessment of therapeutic proteins. Its main objective is the comparison of the immunogenicity profile of the biosimilar and reference products over time since the antibody characteristics and effects change as a result of affinity maturation of the antibody response and/or epitope spreading. A minimum of 12-month comparative immunogenicity data should be submitted preauthorization with further assessment to be continued postapproval for at least 6 months for the biosimilar product. A strategy that includes serum sampling at baseline and at regular intervals is necessary for assessing the comparability of the dynamics of antibody development during therapy, for example, every month in the beginning of the treatment (first 3 months) followed by every 3 months.

The use of a validated, highly sensitive antibody assay, capable of detecting all antibodies (i.e., of different affinities, class, and subclass) is mandatory. Approaches that avoid specific masking of particular epitope(s) should be considered to avoid false-negative results. Following confirmation of antibody-positive samples, further characterization including determination of the ability to neutralize the biological activity of IFN-β and cross-reactivity is required. It is recommended that the standardized MxA protein NAb assay or a NAb assay that has been validated against the MxA protein NAb assay is used (EMEA/CHMP/BWP/580136/2007). The approach used to determine assay sensitivity (e.g., by using different cutoff points) should be described but the distribution of titers should also be presented at each time point for each treatment arm. Finally, patients should be categorized according to the evolution of their immune response over time using predefined criteria. For example, the patient's NAb status may be defined as antibody negative (−ve for all posttreatment samples according to predefined low/high dilutions or titers) or antibody positive, which can be categorized as "transiently positive" (1 or more posttreatment samples +ve, followed by −ve samples at all subsequent and at least two sampling time points) or "persistently positive" (two or more consecutive posttreatment samples consistently +ve). MRI activity and clinical relapses should be compared between these categories for both the biosimilar and reference product. The impact of NAbs on clinical outcomes is unlikely to be sufficiently ascertainable before 12 months of therapy and thus will need to be further evaluated postauthorization as part of the risk management plan.

The immune response to the biosimilar and reference medicinal products is expected to be comparable with regard to the incidence and titers of antibodies (neutralizing or not) as well as their impact on efficacy; although the clinical impact of binding non-NAbs is not clear, an increased frequency of such antibodies for the biosimilar product relative to the reference medicinal product would contradict the concept of biosimilarity. However, lower immunogenicity alone would have to be explained but may not preclude biosimilarity if efficacy is shown to be comparable in the various categories of patients according to their immune response (as previously defined) and provided all other data (quality, nonclinical, PK, PD, and safety) are supportive of biosimilarity.

Monoclonal Antibodies

PK study in a sufficiently sensitive and homogeneous study population (healthy volunteers or patients) normally forms an initial step of biosimilar mAb development. PK data can be helpful to extrapolate data on efficacy and safety between different clinical indications of the reference mAb. It may, on a case-by-case basis, be necessary to undertake multidose PK studies in patients or even to perform PK assessment as part of the clinical study designed to establish similar efficacy and safety.

PK studies can be combined with PD endpoints, where available. Normally, similar clinical efficacy should be demonstrated in adequately powered, randomized, parallel-group comparative clinical trial(s), preferably double-blind, normal equivalence trials. To establish comparability, deviations from disease-specific guidelines issued by the CHMP may be warranted. The guiding principle is to demonstrate similar clinical efficacy and safety compared to the reference medicinal product, not patient benefit per se, which has already been shown for the reference medicinal product. In principle, the most sensitive model and study conditions (PD or clinical) should be used in a homogeneous patient population. In cases where comparative PD studies are claimed to be most suitable to provide the pivotal evidence for similar efficacy, applicants will have to choose clinically relevant markers, justify these markers, and also provide sufficient reassurance of clinical safety, particularly immunogenicity. Extrapolation of clinical efficacy and safety data to other indications of the reference mAb, not specifically studied during the clinical development of the biosimilar mAb, is possible based on the results of the overall evidence provided from the comparability exercise and with adequate justification. As regards postauthorization follow-up, the concept to be proposed by applicants may have to exceed routine pharmacovigilance and may have to involve postauthorization safety studies (PASS).

Monoclonal antibodies have been established as a major product class of biotechnology-derived medicinal products. Different mAb products share some properties, for example, being cytotoxic to their target or neutralizing a cytokine, but differ in aspects like the MOA. They are structurally complex and may have several functional domains within a single molecule, depending on the isotype (antigen-binding region, complement-binding region, constant part interacting with Fc receptors). Each individual mAb presents a unique profile with respect to the antigen-binding region, the Fc cytotoxic effector function, and binding to Fc receptors.

Various assays have been established in the past years that allow for more in-depth characterization of complex proteins, both on a physicochemical and a functional level, for example, with potency assays, and there is experience in the assessment of minor quality differences due to changes in manufacturing processes for monoclonal antibodies. However, it may at the current stage of knowledge be difficult to interpret the relevance of minor quality differences in the physicochemical and biological characterization when comparing a biosimilar mAb to a reference mAb.

The guidance specific for monoclonal antibodies can also be applied to related substances like fusion proteins based on IgG Fc (–cept molecules).

Next-generation mAbs, defined as mAbs that are structurally and/or functionally altered (e.g., glyco-engineered mAbs with higher potency), in comparison to already licensed reference medicinal products, to achieve an improved or different clinical performance, are not biosimilars and therefore beyond the scope of this guideline.

Nonclinical studies should be performed before initiating clinical trials. In vitro studies should be conducted first and a decision then made as to the extent of what, if any, in vivo work will be required. In order to assess any difference in biological activity between the biosimilar and the reference medicinal product, data from a number of comparative in vitro studies, some of which may already be available from quality-related assays, should be provided.

In vitro nonclinical studies should be performed with an appropriate number of batches of product representative of that intended to be used in the clinical trial. These studies should include relevant assays on

- Binding to target antigen(s)
- Binding to representative isoforms of the relevant three Fc gamma receptors (FcγRI, FcγRII, and FcγRIII)
- FcRn and complement (C1q)
- Fab-associated functions (e.g., neutralization of a soluble ligand, receptor activation or blockade)
- Fc-associated functions (e.g., antibody-dependent cell-mediated cytotoxicity, ADCC; complement-dependent cytotoxicity, CDC; complement activation)

These studies should be comparative in nature and should be designed to be sensitive enough to detect differences in the concentration–activity relationship between the similar biological medicinal product and the reference medicinal product and should not just assess the response per se. It should be noted that an evaluation of ADCC and CDC is generally not needed for mAbs directed against non-membrane-bound targets. As indicated in the ICH S6 (R1) guideline, tissue cross-reactivity studies are not suitable to detect subtle changes in critical quality attributes and are thus not recommended for assessing comparability.

Together, these assays should broadly cover the functional aspects of the mAb even though some may not be considered essential for the therapeutic mode of action. As the in vitro assays may be more specific and sensitive than studies in animals, these assays can be considered paramount in the nonclinical comparability exercise.

If the comparability exercise using the aforementioned strategy indicates that the test mAb and the reference mAb cannot be considered biosimilar, it may be more appropriate to consider developing the product as a stand-alone.

It is acknowledged that some mAbs may mediate effects that cannot be fully elucidated by in vitro studies. Therefore, evaluation in an in vivo study may be necessary, provided that a relevant in vivo model with regard to species or design is available. Factors to be considered when the need for additional in vivo nonclinical studies is evaluated include, but are not restricted to, the following:

- Presence of relevant quality attributes that have not been detected in the reference product (e.g., new posttranslational modification structure).
- Presence of quality attributes in significantly different amounts than those measured in the reference product.
- Relevant differences in formulation, for example, use of excipients not widely used for mAbs.
- Although each of the factors mentioned here do not necessarily warrant in vivo testing, these issues should be considered together to assess the level of concern and whether there is a need for in vivo testing.

If the comparability exercise in the in vitro studies in step 1 is considered satisfactory and no factors of concern are identified in step 2, or these factors of concern do not block direct entrance into humans, an in vivo animal study may not be considered necessary.

If there is a need for additional information, the availability of a relevant animal species or other relevant models (e.g., transgenic animals or transplant models) should be considered. Due to the specificity of mAbs, the relevant species studied is in most cases a nonhuman primate. In all cases, the limitations of an in vivo study (such as sensitivity and variability) should be taken into account.

If a relevant in vivo animal model is not available, the applicant may choose to proceed to human studies taking into account principles to mitigate any potential risk.

In Vivo Studies

If an in vivo study is deemed necessary, the focus of the study (PK and/or PD and/or safety1) depends on the need for additional information. Animal studies should be designed to maximize the information obtained. In addition, depending on the endpoints needed, it may not be necessary to sacrifice the animals at the end of the study. The principles of the 3Rs should be considered when designing any in vivo study. The duration of the study (including observation period) should be justified, taking into consideration the PK behavior of the mAb and its clinical use.

When the model allows, the PK and PD of the similar biological medicinal product and the reference medicinal product should be quantitatively compared, including concentration–response assessment covering the therapeutic doses in humans.

The conduct of repeated dose toxicity studies in nonhuman primates is usually not recommended. Also, the conduct of toxicity studies in nonrelevant species (i.e., to assess unspecific toxicity only, based on impurities) is not recommended. Due to the different production processes used by the biosimilar and reference product manufacturers, qualitative differences of process-related impurities will occur (e.g., host cell proteins). Such impurity level should be kept to a minimum, which is the best strategy to minimize any associated risk. Qualitative or quantitative difference(s) of product-related variants (e.g., glycosylation patterns, charge variants) may affect biological functions of the mAb and are expected to be evaluated by appropriate in vitro assays. These quality differences may have an effect on immunogenic potential and potential to cause hypersensitivity. It is acknowledged that these effects are difficult to predict from animal studies and should be further assessed in clinical studies. Immunogenicity assessment in animals is generally not predictive for immunogenicity in humans, but may be needed for interpretation of in vivo studies in animals. Blood samples should be taken and stored for future evaluations if then needed.

Studies regarding safety pharmacology and reproduction toxicology are not required for nonclinical testing of biosimilar mAbs. Studies on local tolerance are usually not required. If excipients are introduced for which there is no or little experience with the intended clinical route, local tolerance may need to be evaluated. If other in vivo studies are performed, evaluation of local tolerance may be part of the design of that study instead of the performance of separate local tolerance studies.

Clinical Studies

Comparative clinical studies between the biosimilar and reference medicinal product should always be conducted. The number and type of studies might vary according to the reference product and should be justified based on a sound scientific rationale. A stepwise approach is normally recommended. Safety in this context does not usually refer to a complete repeated dose toxicity study, but rather an in-life evaluation of safety parameters such as clinical signs, body weight, and vital functions.

Throughout the development program, the extent and nature of the clinical program depends on the level of evidence obtained in the previous step(s). During the clinical development program, patients are normally enrolled commensurate with the level of evidence obtained from preceding steps, which support comparability.

Pharmacokinetics

The comparison of the PK properties of the biosimilar product and the reference medicinal product forms normally the first step of a biosimilar mAb development. The design of the study depends on various factors, including clinical context, safety, the PK characteristics of the antibody (target-mediated disposition, linear or nonlinear PK, time dependencies, half-life, etc.). Furthermore, bioanalytical assays should be appropriate for their intended use and adequately validated.

Study Design

The primary objective of the PK studies performed to support a MAA for a biosimilar is to show comparability in PK of the biosimilar with the reference medicinal product in a sufficiently sensitive and homogeneous population. This is expected to reduce variability, and thus the sample size needed to prove equivalence, and can simplify interpretation.

Healthy volunteers are likely to have less variability in PK as target-mediated clearance may be less important than in patients. Hence, if feasible, a single-dose study in healthy volunteers is recommended, which could provide important information on biosimilarity. From a PK perspective, a single-dose crossover study with full characterization of the PK profile, including the late elimination phase, is preferable. A parallel-group design may be necessary due to the long half-life of mAbs and the potential influence of immunogenicity.

A study in healthy volunteers may not be possible in case of a toxic MOA or in case the information obtained would not be sufficient to establish biosimilarity. Under these circumstances, a study in patients may be a better option. If a single-dose study is not feasible in patients, a multiple-dose study should be conducted.

It may be necessary to perform the PK study in a different population from that selected to establish similar clinical efficacy, since the most sensitive population where PK characteristics can be compared may not be the same as the most sensitive population where similar efficacy and safety can be demonstrated. In such scenarios, population PK measurements during the clinical efficacy trial are recommended since such data may add relevant data to the overall database to claim comparability.

The choice of the patient population for the PK study should be fully justified, based on a comprehensive survey of scientific literature as regards its sensitivity and also the possibility to infer PK results to other clinical indications where the reference mAb is licensed.

In case a PK study in healthy volunteers is conducted to support bioequivalence, supportive PK data from clinical studies in patients are encouraged and could provide highly supportive evidence of a similar PK behavior.

The following factors impact on the strategy of designing PK evaluations:

Disease and patient characteristics: Factors that may influence the choice of the patient population are age of usual manifestation and age range (since lower age may be less prone to presence of concomitant clinical conditions), number of previous treatments, concomitant treatments, or expression of antigen (which may be related to disease stage). For mAbs that are indicated for both monotherapy and in combination with immunosuppressant or chemotherapy, it may be sensible to study the comparative PK in the monotherapy setting in order to minimize the sources for variability. However, a first-line setting, where patients are in a better clinical condition, or an adjuvant setting in patients with early cancer, where the tumor burden is low, may be preferable; in these instances, the mAb is typically administered in combination with other therapies.

PK characteristics of the reference mAb: PK of anticancer mAbs may be time dependent, as the tumor burden may change after multiple dosing (e.g., increased half-life with multiple dosing) and this should be taken into account in the design of the study.

The existence of target-mediated clearance in addition to non-target-mediated clearance may affect the number of studies needed. In case target-mediated clearance is not relevant, one comparative PK study may be sufficient. If the reference mAb is eliminated both by target-mediated and non-target-mediated mechanisms, comparable PK should be demonstrated where each mechanism of clearance predominates: preferably one study in healthy volunteers for non-target-mediated clearance and one supportive study in patients, which can be part of the efficacy trial, to investigate comparability in target-mediated clearance.

For mAb targets that involve receptor shedding, it is advisable to measure shed receptor levels at baseline and, if relevant, during the conduct of the study, in order to verify the baseline comparability of the treatment groups. Stratification by tumor burden or receptor shedding, if possible, may help to ensure baseline comparability. An exploratory statistical analysis on postbaseline comparability at the time point relevant to the conclusion of PK equivalence could be helpful.

For mAbs licensed in several clinical indications, it is not generally required to investigate the PK profile in all of them. However, if distinct therapeutic areas are involved for one particular mAb (e.g., autoimmunity and oncology), separate PK studies may be needed if different target-mediated clearance exists for different therapeutic areas.

Doses

In principle, it is not required to test all therapeutic dosage regimens; the most sensitive dose should be selected to detect potential differences in PK between the biosimilar and the reference products. When limited data are available to know which dose is the most sensitive, it is recommended to investigate a low or the lowest recommended therapeutic dose where it is assumed that the target-mediated clearance is not yet saturated and a high or the highest therapeutic dose where it is believed that the nonspecific clearance mechanism dominates. A single-dose study with the lowest therapeutic dose in patients is considered the most adequate design to investigate the differences in target-mediated clearance, if any.

Routes of Administration

If the reference product can be administered IV and SC and if both routes are applied for, it is preferable to investigate both routes of administration. However, as the evaluation of SC administration covers both absorption and elimination, it may be possible to waive the evaluation of IV administration if comparability in both absorption and elimination has been demonstrated for the SC route using additional PK parameters such as partial AUCs.

Sampling Times

In single-dose studies, the sampling times should be selected to characterize the whole profile, including the late elimination phase. For those products administered as two (or more) consecutive doses, useful information can be obtained from both the first and last administrations since the first administration is preferred for comparative purposes and the last one can provide information on the final elimination phase that cannot be observed after the first dose.

If a multiple-dose PK study in patients is used to show similarity between the biosimilar and reference medicinal product and if elimination after the last dose cannot be characterized, sampling should normally be undertaken to characterize the concentration–time profile both after the first dose and later, preferably at steady state. Characterization of the full concentration–time profile at steady state is especially important in case of nonlinear PK of the reference mAb (e.g., many anticancer mAbs with cellular targets exhibit dose- or time-dependent PK or immunogenicity-related changes in distribution or elimination kinetics).

PK Parameters of Interest

In a single-dose study, the primary parameter should be the $AUC_{(0-inf)}$. Secondary parameters such as C_{max}, t_{max}, volume of distribution, and half-life should also be estimated. In case of SC administration, C_{max} should be a coprimary parameter. In addition, if no data are provided for the IV route, partial AUCs should be assessed to ensure comparability of both absorption and elimination.

In a multiple-dose study, the primary parameters should be the truncated AUC after the first administration until the second administration (AUC_{0-t}) and AUC over a dosage interval at steady state ($AUC\tau$). Secondary parameters are C_{max} and C_{trough} at steady state.

Antidrug antibodies should be measured in parallel of PK assessment using the most appropriate sampling time points.

Comparability margins have to be defined a priori and appropriately justified. For some reference mAbs, intersubject variability for some parameters was reported to be considerable. This may have to be accounted for in the choice of the comparability margin at least for such parameters. As a principle, any widening of the conventional equivalence margin beyond 80%–125% for the primary parameters requires thorough justification, including an estimation of potential impact on clinical efficacy and safety. For secondary parameters, CIs for ratio or differences can be presented together with descriptive statistics but no acceptance range needs to be defined. The clinical relevance of estimated differences and associated CIs should be discussed.

Timing of the PK Evaluation

Usually, proof of similar PK profiles should precede clinical efficacy trials. However, in certain scenarios, for example, for mAbs where PK is inevitably highly variable even within one clinical indication, it may, for feasibility reasons, be necessary to explore PK comparisons as part of a clinical study that is designed to establish similar clinical efficacy (as only this trial will be large enough to demonstrate PK equivalence). To start with, a comparative clinical efficacy trial that includes

PK evaluation, without a formal preceding comparative PK study, could be problematic with no former human exposure to the biosimilar mAb, together with potentially limited nonclinical in vivo data, depending on the mAb. Therefore, such a plan could only be justified on a case-by-case basis depending on the product profiles observed in the quality and nonclinical data.

Pharmacodynamics

PD parameters may contribute to the comparability exercise for certain mAbs and in certain indications. Depending on the mAb and availability of PD endpoints, the following scenarios are, theoretically, possible:

- *PD markers as support to establish comparability*: PK studies can be combined with PD endpoints, where available. This could add valuable information for the overall comparability exercise. PD markers are especially valuable if they are sensitive enough in order to detect small differences, and if they can be measured with sufficient precision. The use of multiple PD markers, if they exist, is recommended. With regard to PD evaluation, there is often a lack of specific PD endpoints. The emphasis may then have to be on nonclinical PD evaluations, for example, in vitro testing.
- *PD markers as pivotal proof of comparability*: Sponsors should always explore possibilities to study dose–concentration–response relationships or time–response relationships, since this approach, if successful, may provide strong evidence of comparability, provided that the selected doses are within the linear part of the dose–response curve.

The following prerequisites need to be met to accept that PD markers can constitute the pivotal evidence for the efficacy comparability exercise:

- A clear dose–response relationship is shown.
- At least one PD marker is an accepted surrogate marker and can be related to patient outcome to the extent that demonstration of similar effect on the PD marker will ensure a similar effect on the clinical outcome variable.

If that is not the case, then proceed to step 2 (i.e., clinical efficacy).

When PD markers are planned as pivotal evidence to establish similarity, it is recommended to discuss such approach with regulatory authorities. This should include a proposal of the size of the proposed equivalence margin and its clinical justification as regards lack of a clinical meaningful difference.

A comparative single- or repeat-dose study in the saturation part of the dose–concentration–response curve is unlikely to discriminate between different activities, should they exist, and a dose in the linear part of the dose–response curve may result in treating a patient with a too low dose. It is also acknowledged that dose–response data may not exist for the reference mAb and that exposing patients to a relatively low dose of the mAbs, in a worst case scenario, might also sensitize them to develop anti-mAb antibodies and, consequently, may make them treatment resistant. However, for some reference mAbs, clinical conditions may exist where such studies are feasible.

Clinical Efficacy

If dose-comparative and highly sensitive PD studies cannot be performed convincingly showing comparability in a clinically relevant manner, similar clinical efficacy between the similar and the reference product should be demonstrated in adequately powered, randomized, parallel-group comparative clinical trial(s), preferably double-blind, normal equivalence trials.

For most of the clinical conditions that are licensed for mAbs, specific CHMP guidance on the clinical requirements to demonstrate efficacy exists. However, to establish comparability, deviations from these guidelines (choice of endpoint, time point of analysis of endpoint, nature or dose of concomitant therapy, etc.) will be warranted in some circumstances. Such deviations need to be scientifically justified on the basis that the proposed clinical concept is designed to establish biosimilarity by employing PD markers, clinical outcomes, or both. The guiding principle is to demonstrate similar efficacy and safety compared to the reference medicinal product, not patient benefit per se, which has already been established by the reference medicinal product. Therefore, in general, the most sensitive patient population and clinical endpoint is preferred to be able to detect product-related differences, if present and, at the same time, to reduce patient and disease-related factors to a minimum in order to increase precision and to simplify interpretation. For example, patients with different disease severity and with different previous lines of treatment might be expected to respond differently, and thus differences between the study arms may be difficult to interpret, and it may remain uncertain whether such differences would be attributable to patient or disease-related factors rather than to differences between the biosimilar mAb and reference mAb.

Comparability should be demonstrated in scientifically appropriately sensitive clinical models and study conditions (whether licensed or not), and the applicant should justify that the model is relevant as regards efficacy and safety and sensitive to demonstrate comparability in the indication(s) applied for. The safety of patients should not be compromised by a comparability exercise, and patients should only be treated as medically warranted. In case there are no endpoints that are sufficiently sensitive to detect relevant differences, applicants need to implement additional measures to enable sufficient sensitivity of the overall clinical dataset obtained from the clinical study. For example, the study could be combined with a multiple-dose study or applicants could measure PD markers in addition to clinical endpoints in order to further establish comparability.

Clinical studies in special populations like the pediatric population or the elderly are normally not required since the overall objective of the development program is to establish comparability, and therefore the selection of the primary patient population is driven by the need for homogeneity and sensitivity.

The inclusion of patients from non-European countries is generally possible if there are no intrinsic differences, but it may increase heterogeneity. Knowledge of efficacy and safety of the reference mAb in a particular region may be necessary in order to prospectively define an equivalence margin. Stratification and appropriate subgroup analyses are normally expected if patients from different global regions are included in order to demonstrate consistency with the overall effect. Diagnostic and treatment strategies should be comparable in order to prevent the influence of extrinsic factors.

Clinical Studies

Clinical safety is important throughout the clinical development program and is captured during initial PK and/or PD evaluations and also as part of the pivotal clinical study establishing comparability. Care should be given to compare the type, severity, and frequency of the adverse reactions between the biosimilar mAb and the reference mAb, particularly those described for the reference product. Where no homogeneous definition exists for safety parameters (e.g., measurement of cardiotoxicity), it is recommended to use the same definitions as that used for the reference mAb in its original development program (if known) or the definitions used during postauthorization follow-up. Comparison of pharmacologically mediated adverse reactions (e.g., cardiotoxicity), that is, safety-related PD markers, could also be used as further supportive evidence for clinical comparability and could be analyzed in a similar way to that discussed for efficacy-related PD markers.

In cases where comparative and highly sensitive PD studies are suitable to provide the pivotal evidence for equivalence in clinical efficacy, applicants will have to provide sufficient reassurance of similar clinical safety, including immunogenicity. Actively controlled safety data should normally be collected preauthorization, depending on the mAb and the number of exposed patients and duration of treatment. The duration of safety follow-up preauthorization should be justified.

It might be decided to collect part of the safety data, or also additional safety data, in the post-authorization setting as described in the following. Rare events such as progressive multifocal leukoencephalopathy are unlikely to be detected in a preauthorization setting. Therefore, applicants need to propose pharmacovigilance and risk management activities for the postauthorization phase at the time of the marketing authorization application. Usually, similar pharmacovigilance activities as those of the reference medicinal product would be required, rather than a direct comparison with the reference medicinal product, since comparative data will most likely be difficult to interpret due to their rarity of occurrence and consequent lack of precision for estimated differences.

Applicants should reflect upon how retreatment of patients would be handled. Concepts should be presented at the time of marketing authorization application on how to systematically measure safety of repeat exposure of patients, for example, in oncological indications where patients undergo several treatment cycles. It is highly encouraged to extend the clinical study as a postauthorization follow-up study to a full treatment cycle, where relevant and feasible.

As regards immunogenicity assessment, applicants should refer to existing CHMP guidance. Systematic and comparative evaluation and discussion of immunogenicity is important, due to clinical consequences like loss of efficacy and also likely resistance against further treatment with the reference mAb. It may be advisable not to include patients previously treated with the reference mAb where possible or to prespecify a subgroup analysis for patient previously treated (in order to explore if pretreatment impacts immunogenicity), as previous treatment could have resulted in an antidrug antibody response that could hamper interpretation of the safety data and thus also decrease sensitivity for detecting differences. Comparative assessment of unwanted immune responses against the biosimilar and the reference mAb are normally undertaken as part of the clinical study establishing similar clinical efficacy and safety, using the same validated assay(s) (see relevant CHMP guidelines on immunogenicity assessment). A population PK approach with sparse sampling and determination of drug concentration together with antidrug antibody detection is acceptable. However, for some mAbs, antibodies can be better detected in healthy volunteers, who develop a strong immune response after a single dose within a few days. The dose of mAb administered is also an important factor to consider when investigating immunogenicity: some mAbs inhibit antibody formation when administered at high doses, and therefore studies conducted with low doses, if medically possible, are more sensitive to compare the immune response of the biosimilar and reference medicinal products.

Investigation of unwanted immunogenicity is especially important when a different expression system is employed for the biosimilar mAb compared to the reference mAb, which might, for example, yield in relevant quality attributes that have not been detected in the reference product (e.g., new posttranslational modification structure) that could result in a higher immunogenicity. This is particularly important if there is limited experience with this expression system in humans. It is recommended that such approaches are discussed in advance with regulatory authorities.

A higher immunogenicity as compared to the reference mAb may become an issue for the benefit/risk analysis and would question biosimilarity. However, also a lower immunogenicity for the biosimilar mAb is a possible scenario, which would not preclude biosimilarity. Here, the efficacy analysis of the entire patient population could suggest that the biosimilar is more efficacious (since fewer patients developed an immune response and thus more patients may show a treatment effect with the biosimilar mAb). It is therefore recommended to prespecify an additional exploratory subgroup analysis of efficacy and safety in those patients that did not mount an antidrug antibody response during the clinical trial. This subgroup analysis could be helpful to establish that the efficacy of the biosimilar and the reference mAb are in principle similar if not impacted by an immune response.

Additional long-term immunogenicity and safety data might be required postauthorization, for example, in situations where the study duration for establishing similar clinical efficacy is rather short.

Erythropoietins

Human erythropoietin is a 165 amino acid glycoprotein mainly produced in the kidneys and is responsible for the stimulation of red blood cell production. Erythropoietin for clinical use is produced by recombinant DNA technology using mammalian cells as expression system and termed epoetin.

All epoetins in clinical use have a similar amino acid sequence as endogenous erythropoietin but differ in the glycosylation pattern. Glycosylation influences PK and may affect efficacy and safety including immunogenicity. Physicochemical and biological methods are available for characterization of the protein.

Epoetin-containing medicinal products are currently indicated for several conditions such as anemia in patients with chronic renal failure and chemotherapy-induced anemia in cancer patients and for increasing the yield of autologous blood from patients in a predonation program. The MOA of epoetin is the same in all currently approved indications but the dosages required to achieve the desired response may vary considerably and are highest in the oncology indications. Epoetin can principally be administered IV or SC.

Epoetins have a relatively wide therapeutic window and are usually well tolerated provided that the stimulation of bone marrow is controlled by limiting the amount and rate of hemoglobin increase. The rate of hemoglobin increase may vary considerably between patients and is dependent not only on the dose and dosing regimen of epoetin but also other factors, such as iron stores, baseline hemoglobin and endogenous erythropoietin levels, and the presence of concurrent medical conditions such as inflammation.

Exaggerated PD response may result in hypertension and thrombotic complications. Moreover, PRCA due to neutralizing anti-epoetin antibodies has been observed, predominantly in renal anemia patients treated with SC administered epoetin. Because antibody-induced PRCA usually is a very rare event taking months to years of epoetin treatment to develop, such events are unlikely to be identified in preauthorization studies. In addition, possible angiogenic and tumor-promoting effects of epoetin might be of importance in selected populations.

Pharmacodynamics Studies

In Vitro Studies

In order to assess any alterations in reactivity between the similar biological medicinal and the reference medicinal product, data from a number of comparative bioassays (e.g., receptor-binding studies, cell proliferation assays), many of which may already be available from quality-related bioassays, should be provided.

In Vivo Studies

The erythrogenic effects of the similar biological medicinal product and the reference medicinal product should be quantitatively compared in an appropriate animal assay. Information on the erythrogenic activity may be obtained from the described repeat-dose toxicity study or from a specifically designed assay (e.g., the European Pharmacopoeia normocythemic mouse assay; data may be already available from quality-related bioassays).

Toxicological Studies

Data from at least one repeat-dose toxicity study in a relevant species (e.g., rat) should be provided. Study duration should be at least 4 weeks. Appropriate toxicokinetic measurements should be performed as part of the repeat-dose toxicity study and include a determination of antibody formation. Data on local tolerance in at least one species should be provided.

Clinical Studies

Pharmacokinetic Studies

The PK properties of the similar biological medicinal product and the reference product should be compared in single-dose crossover studies for the routes of administration applied for, usually including both SC and IV administration. Healthy volunteers are considered an appropriate study population. The selected dose should be in the sensitive part of the dose–response curve. The PK parameters of interest include AUC, C_{max}, and $T1/2$ or CL/F. Equivalence margins have to be defined a priori and appropriately justified. Differences in $T1/2$ for the IV and the SC route of administration and the dose dependence of clearance of epoetin should be taken into account when designing the studies.

Pharmacodynamic Studies

PD should preferably be evaluated as part of the comparative PK studies. The selected dose should be in the linear ascending part of the dose–response curve. In single-dose studies, reticulocyte count is the most relevant and therefore recommended PD marker for assessment of the activity of epoetin. On the other hand, reticulocyte count is not an established surrogate marker for efficacy of epoetin and therefore not a suitable endpoint in clinical trials.

Clinical Efficacy Studies

Similar clinical efficacy between the similar and the reference product should be demonstrated in adequately powered, randomized, parallel-group clinical trials. Since PK and dose requirements usually differ for IV and SC use, similar efficacy between the test and the reference product should be ensured for both routes of administration. This could be achieved by performing separate clinical trials for both routes or by performing one clinical trial for one route and providing adequate bridging data for the other route.

Confirmatory studies should preferably be double blind to avoid bias. If this is not possible, at minimum, the person(s) involved in decision making (e.g., dose adjustment) should be effectively masked to treatment allocation.

Sensitivity to the effects of epoetin is higher in erythropoietin-deficient than non-erythropoietin-deficient conditions and is also dependent on the responsiveness of the bone marrow. Patients with renal anemia and without major complications (such as severe/chronic infections or bleeding or aluminum toxicity), expected to relevantly impair the treatment response to epoetin, are therefore recommended as the study population. Other reasons for anemia should be excluded. Since epoetin doses necessary to achieve or maintain target hemoglobin levels usually differ in predialysis and dialysis patients, these two populations should not be mixed in the same study.

The following sections present different options and recommendations on how to demonstrate similar efficacy of two epoetin-containing medicinal products. A sponsor may choose from these options or modify them but should always provide sound scientific justification for the approach taken.

Demonstration of Efficacy for Both Routes of Administration

(a) Similar efficacy for both routes of administration may be demonstrated by performing two separate clinical trials.

The combination of a "correction phase" study using SC epoetin (e.g., in a predialysis population) and a "maintenance phase" study using IV epoetin (e.g., in a hemodialysis population) would be expected to provide a maximum of information on the biosimilar epoetin.

A correction phase study will determine response dynamics and dosing during the anemia correction phase and is particularly suitable to characterize the safety profile related to the PD of the similar biological medicinal product. It should include treatment-naive patients or previously treated patients after a suitably long epoetin-free and red blood cell transfusion–free period (e.g., 3 months). In case of pretreatment with long-acting erythropoiesis-stimulating agents (such as PEGylated epoetin), the treatment-free phase may need to be longer.

A maintenance phase study, on the other hand, may be more sensitive to detect differences in biological activity between the similar and the reference product, although experience suggests that correction phase studies are also likely to be sufficiently discriminatory. The study design for a maintenance phase study should minimize baseline heterogeneity and carry over effects of previous treatments. Patients included in a maintenance phase study should be optimally titrated on the reference product (stable hemoglobin in the target range on stable epoetin dose and regimen without transfusions) for a suitable duration of time (usually at least 3 months). Thereafter, study subjects should be randomized to the similar or the reference product, maintaining their prerandomization epoetin dosage, dosing regimen, and route of administration.

Alternatively, both the SC and the IV study may be performed in the maintenance setting if appropriately justified.

In the course of both studies, epoetin doses should be closely titrated to achieve (correction phase study) or maintain (maintenance phase study) target hemoglobin concentrations. The titration algorithm should be the same for both treatment groups and be in accordance with current clinical practice.

In the correction phase study, "hemoglobin responder rate" (proportion of patients achieving a prespecified hemoglobin target) or "change in hemoglobin" is the preferred primary endpoint. In the maintenance phase study, "hemoglobin maintenance rate" (proportion of patients maintaining hemoglobin levels within a prespecified range) or "change in hemoglobin" is the preferred primary endpoint. However, the fact that epoetin dose is titrated to achieve the desired response reduces the sensitivity of the hemoglobin-related endpoints to detect possible differences in the efficacy of the treatment arms. Therefore, epoetin dosage should be a coprimary endpoint in both study types.

Data for calculation of the primary efficacy endpoints should be collected during an appropriate evaluation period. A 4-week evaluation period from study month 5 to 6 in both the correction phase and the maintenance phase study has been found suitable in order to avoid potential carryover effects from baseline treatment and allow full assessment of potential differences in both endpoints in the presence of stabilized hemoglobin levels and epoetin dosages. If the primary efficacy assessment is performed at an earlier time point, the applicant will need to demonstrate that potential differences in efficacy have been fully captured.

Equivalence margins for both coprimary endpoints should be prespecified and appropriately justified and should serve as the basis for powering the studies. If change from baseline in hemoglobin is used as the primary endpoint, an equivalence margin of ±0.5 g/dL is recommended. Transfusion requirements should be included as an important secondary endpoint.

(b) Another approach to demonstrate similar efficacy for both routes of administration would be to show comparable efficacy for one route of administration in a comparative clinical trial and provide comparative single-dose and multiple-dose PK/PD bridging data in an epoetin-sensitive population (e.g., healthy volunteers) for the other route of administration. The multiple-dose PK/PD study should be at least 4 weeks in duration using a fixed epoetin dosage within the therapeutic range and change in hemoglobin as primary PD endpoint.

Since comparative immunogenicity data will always be required for SC use, if applied for, the most reasonable approach in this alternative scenario would be to perform a clinical trial using SC epoetin and to provide PK/PD bridging data for the IV route.

In this case, patients included in a SC study should be treated with test or reference ideally for a total of 12 months to obtain 12-month comparative immunogenicity data. At this point, patients on the reference medicinal product should be switched to the test product and all patients followed, for example, for another 6 months, to increase the safety and immunogenicity database of the similar medicinal product. Otherwise, regarding the design, enrolled population, and endpoints of the clinical trial, the same considerations apply as stated earlier.

Demonstration of Efficacy for One Route of Administration

If only one route of administration is intended to be applied for, a single-dose PK/PD study and either a correction phase or a maintenance phase study for the desired route should be performed. The lack of data in the other route of administration will be clearly reflected in the SmPC.

Clinical Safety

Comparative safety data from the efficacy trials are usually sufficient to provide an adequate premarketing safety database. Adverse events of specific interest include hypertension/aggravation of hypertension and thromboembolic events.

The applicant should submit at least 12-month immunogenicity data preauthorization. In the absence of standardized assays, concomitant immunogenicity data on the reference medicinal product are required for proper interpretation of results. The comparative phase should preferably cover the complete 12-month assessment period. For shorter comparative phases, the applicant will need to provide sound argument that this does not increase the uncertainty about the immunogenic potential of the biosimilar epoetin.

The use of a validated, highly sensitive antibody assay, able to detect both early (low-affinity antibodies, especially IgM class) and late (high affinity antibodies) immune responses, is mandatory. Detected antibodies need to be further characterized including their neutralizing potential. Retention samples for both correction phase and maintenance phase studies are recommended. Due to their rarity, NAbs or even PRCAs are unlikely to be captured premarketing and, if occurring, would constitute a major safety concern. Although, the relevance of binding non-NAbs is not clear, a markedly increased frequency of such antibodies for the test product would elicit a safety concern and contradict the assumption of biosimilarity.

Since the SC route of administration is usually more immunogenic than the IV route and patients with renal anemia constitute the population at risk for developing antiepoetin-antibody-induced PRCA, the immunogenicity database should include a sufficient number of SC treated patients with renal anemia, unless SC use in this population is not applied for.

LOW MOLECULAR WEIGHT HEPARINS

Heparin is a highly sulfated and heterogeneous member of the glycosaminoglycan family of carbohydrates consisting of various disaccharide units. The most common disaccharide unit is composed of a -*O*-sulfated α-L-iduronic acid and 6-*O*-sulfated, *N*-sulfated α-D-glucosamine, IdoA(S)-GlcNS(6S). Endogenous heparin is synthesized in the granules of mast cells and possesses the highest negative charge density of all known biological molecules.

Low molecular weight heparins (LMWHs) are prepared from unfractionated heparin by various chemical or enzymatic depolymerization processes. Thus, the starting material of LMWHs is of biological origin and the manufacturing process defines the characteristics of the drug substance.

The complexity of LMWH results largely from the nature of the starting material (unfractionated heparin extracted from porcine mucosa or other animal tissues), the extraction, the fractionation, and the production processes. Several state-of-the-art methods for physicochemical characterization of LMWH products are available. However, although the inhibition of activated FXa activity and the inhibition of thrombin activation reflect the main anticoagulant activities of LMWH, it is presently not clear to which extent the multiple different polysaccharides contribute to the clinical effects relevant for efficacy and safety of LMWH.

A specific LMWH differs from unfractionated heparin and may differ from other LMWHs in its PK and PD properties. As a result of the depolymerization process, LMWHs are mainly enriched in molecules with less than monosaccharide units. This reduction of molecule size is associated with a loss of thrombin inhibition activity in comparison to standard heparin and an increased inhibition of FXa. Due to difficulties in the physical detection of LMWH, conventional PK studies cannot be performed. Instead, the absorption and elimination of LMWHs are studied by using PD tests, including the measurement of anti-FXa and anti-FIIa activity. There are several authorized LMWHs that differ in their source material, manufacturing process, PK/PD properties, and therapeutic indications, which include treatment and prophylaxis of deep venous thrombosis and prevention of complications of acute coronary syndromes (unstable angina, non-ST elevation myocardial infarction [non-STEMI], and myocardial infarction with ST elevation [STEMI]). The most common adverse reactions induced by heparins are bleedings, while the most serious one is the rarely observed heparin-induced thrombocytopenia type II (HIT II). This antibody-mediated process is triggered by the induction of antibodies directed against neoantigens of platelet-factor 4 (PF4)–heparin complexes. Binding of those antibody–PF4–heparin complexes may activate platelets and generate thrombogenic platelet microaggregates. Patients developing thrombocytopenia are in danger of arterial and venous thromboembolic complications (heparin-induced thrombocytopenia and thrombosis, HITT). Although the risk of these adverse reactions appears to be reduced in comparison to unfractionated heparin, it is obligatory to monitor the platelet count regularly in all patients using LMWH and to test for PF4–heparin complex antibodies in those who develop thrombocytopenia or thromboembolic complications during heparin treatment. In conclusion, the heterogeneity of LMWH is high, the structure–effect relationship is presently not fully elucidated, and the PD markers anti-FXa and anti-FIIa activity may not fully reflect/predict efficacy. Thus, clinical trials will usually be necessary to address remaining uncertainties resulting from the physicochemical and biological comparison.

Nonclinical studies should be performed before initiating clinical trials. The studies should be comparative in nature and should be designed to detect differences in the response between the biosimilar and the reference LMWH and not just assess the response per se. The approach taken will need to be fully justified in the nonclinical overview.

Pharmacodynamic Studies

In Vitro Studies

In order to compare PD activity of the biosimilar and the reference LMWH, data from a number of comparative bioassays (based on state-of-the-art knowledge about clinically relevant PD effects of LMWH and including, at least, evaluations of anti-FXa and anti-FIIa activity) should be provided. If available, standardized assays (e.g., in accordance with the European Pharmacopoeia) should be used to measure activity. Such data may already be available from bioassays submitted as part of the quality dossier.

In Vivo Studies

If physicochemical and biological characterization of the biosimilar and the reference LMWH has been performed with a high level of resolution and convincingly demonstrated close similarity, in vivo studies are not required as part of the comparability exercise. Otherwise, the in vivo PD activity of the biosimilar and the reference LMWH should be quantitatively compared in an appropriate in vivo PD model, which takes into account state-of-the-art knowledge about clinically relevant PD effects of LMWH and includes, at least, an evaluation of anti-FXa and anti-FIIa activity and of release of tissue factor pathway inhibitor (TFPI) and/or in accordance with the intended clinical indication(s), either a suitable animal venous or an arterial.

Toxicological Studies

Generally, separate repeated dose toxicity studies are not required. In specific cases, for example, when novel or less, well-studied excipients are introduced, the need for additional toxicology studies should be considered. The conduct of toxicity studies to assess unspecific toxicity only, based on impurities is not recommended. A priori biosimilar and reference product are expected to be highly similar, which should be demonstrated with physicochemical methods. Impurities, such as proteins, should be kept at a minimum in accordance with pharmacopoeial monographs, which is the best strategy to minimize any associated risk. Studies regarding safety pharmacology and reproduction toxicology are not required for nonclinical testing of a biosimilar containing LMWH. Studies on local tolerance are not required unless excipients are introduced for which there is no or little experience with the intended route of administration. If other in vivo studies are performed, local tolerance may be evaluated as part of these studies.

Clinical Studies

Pharmacokinetic/Pharmacodynamic Studies

Due to the heterogeneity of LMWHs, conventional PK studies cannot be performed. Instead, the absorption and elimination characteristics of LMWHs should be compared by determining PD activities (including anti-FXa and anti-FIIa), as surrogate markers for their circulating concentrations. In addition, other PD tests such as TFPI activity, as well as the ratio of anti-FXa and anti-FIIa activity, should be compared. Assessment of these PD parameters will provide an important fingerprint of the polysaccharidic profile.

These PK/PD properties of the similar biological medicinal product and the reference product should be compared in a randomized, single-dose two-way crossover study in healthy volunteers using SC administration. In case the originator product is also licensed for the IV or intraarterial route, an additional comparative study should be performed via the IV route. The selected doses should be in the sensitive part of the dose–response curve and within the recommended dose ranges for the different indications. Equivalence margins should be prespecified and appropriately justified.

Clinical Efficacy

A comparative clinical efficacy trial will usually be required as part of the comparability exercise. Only if similar efficacy of the biosimilar and the reference product can be convincingly deduced from the comparison of their physicochemical characteristics, biological activity/potency and PD fingerprint profiles, based on the use of highly sensitive and specific methods, then a dedicated efficacy trial may be waived. It is expected that this is an exceptional scenario since the required amount of reassurance from analytical data and bioassays would be considerable.

Therapeutic equivalence should be demonstrated in an adequately powered, randomized, double-blind, parallel-group clinical trial. In theory, this could be done either in the setting of prevention of venous or arterial thromboembolism or in the setting of treatment of venous thromboembolism (VTE). However, the most sensitive model to detect potential differences in efficacy between the biosimilar LMWH and the reference product should be selected. Surgical patients have the highest prevalence of VTE. Furthermore, the vast majority of published trials have been performed in surgical patients with high VTE risk, especially in patients with hip and knee surgery, and thus the knowledge about influence of types of surgery, duration of trials, and risks for bleeding is the most accurate for this patient population. Therefore, it is recommended to demonstrate efficacy in the prevention of VTE in patients undergoing surgery with high VTE risk. Preferably, the trial should be conducted in major orthopedic surgery such as hip surgery.

In this clinical setting, patients with hip fracture should be well represented in the study as they have both high thrombotic risk and high perioperative bleeding risk. The posology and administration should follow European recommendations for prophylaxis with the reference product in patients requiring prolonged VTE prophylaxis. The guideline on clinical investigation of medicinal products for prophylaxis of high intra- and postoperative venous thromboembolic risk, although intended for novel medicinal products, may contain useful information for the conduct of such a trial.

However, for the purpose of investigating potential product-related differences in efficacy between the biosimilar and the reference product, the patient population should ideally be as homogenous as possible. In the VTE-prevention setting, the clinically most relevant composite endpoint consists of proximal deep vein thrombosis (DVT), pulmonary embolism (PE), and VTE-related death to demonstrate patient benefit.

However, for the purpose of biosimilarity testing, a composite endpoint consisting of total number of thromboembolic events (total DVTs, including asymptomatic distal DVT, PE, and VTE-related death) may be used. Adjudication of VTE events should be performed by a central independent and blinded committee of experts. Equivalence margins have to be defined a priori and appropriately justified, both on statistical and clinical grounds. The study should be powered to show therapeutic equivalence on one of the two composite endpoints mentioned earlier. State-of-the-art imaging technique should be used for the endpoint assessment. While proximal DVTs could be diagnosed with high specificity and sensitivity using ultrasonography, a clear assessment of distal DVT is only possible by using bilateral venography.

Thus, this invasive diagnostic procedure would be mandatory in trials including total DVT in the endpoint. The most relevant components of the primary endpoint (in particular proximal DVTs, PE, and VTE-related deaths) should favorably support the biosimilarity of the two products. Assessment of the primary endpoint should be performed at the time of occurrence of symptoms suggestive of VTE or, in asymptomatic patients, at end of treatment. The overall follow-up should be at least days to detect late thrombotic events.

Clinical Safety

Human safety data on the biosimilar will usually be needed preauthorization, even if similar efficacy can be concluded from the comparative data on physicochemical characteristics, biological activity/potency, and PD fingerprint. Comparative safety data from the efficacy trial will be sufficient to provide an adequate premarketing safety database. Care should be taken to compare the type, frequency, and severity of the adverse reactions between the similar biological medicinal product and the reference product. Major bleeding events and clinically relevant nonmajor bleeding events should be carefully assessed and documented. A consistent and clinically relevant classification of bleedings should be used. Similar to the efficacy evaluation, the adjudication of bleeding events by a central independent and blinded committee of experts, using prespecified limits, should be performed. Liver function testing is recommended. Guideline on nonclinical and clinical development of similar biological medicinal products containing LMWHs.

Sufficient reassurance will be needed that the biosimilar LMWH is not associated with excessive immunogenicity compared to the reference product. For the detection of the immune-mediated type of heparin-induced thrombocytopenia (HIT type II), monitoring of platelet count and an adequate diagnostic procedure (including determination of PF4-Heparin complex antibodies) in patients developing thrombocytopenia and/or thromboembolism (HITT) during the trial have to be performed. Monitoring of antibodies in all patients participating in the trials is not necessary. Since the frequency of immune-mediated HIT II is usually very low (<0.5%), such events are not usually expected to occur in preauthorization clinical trials.

INTERFERON ALPHA 2A OR 2B

Human interferon-alpha (IFN-α)-2a and IFN-α-2b are well-known and characterized proteins consisting of 165 amino acids. The nonglycosylated protein has a molecular weight of approx. 19,240 D. It contains two disulfide bonds, one between the cysteine residues 1 and 98 and the other between the cysteine residues 29 and 138. The sequence contains potential *O*-glycosylation sites. Physicochemical and biological methods are available for characterization of the proteins.

Recombinant IFN-α-2a or IFN-α-2b, is approved in a wide variety of conditions such as viral hepatitis B and C, leukemia, lymphoma, renal cell carcinoma, and multiple myeloma. The subtypes IFN-α-2a and IFN-α-2b have different clinical uses. IFN-α is used alone or in combination. IFN-α may have several PD effects. The relative importance of these effects in the different therapeutic indications is unknown. In general, IFN-α-2a or IFN-α-2b use in oncology indications has reduced considerably and been superseded by other treatments.

The dose and treatment regimen required to achieve the desired response vary considerably between different therapeutic indications.

IFN-α is commonly used SC although it can also be used through intramuscular or IV route. Treatment with IFN-α-2a or IFN-α-2b is associated with a variety of adverse reactions such as flu-like illness, fatigue, and myalgia. In addition, IFN-α is associated with psychiatric, hematological, and renal adverse effects.

Therapy with IFN-α-2a or IFN-α-2b may induce development of autoantibodies. A variety of immune-mediated disorders such as thyroid disease, rheumatoid arthritis, systemic lupus erythematosus, neuropathies, and vasculitis have been observed with the therapeutic use of IFN-α.

Both nonneutralizing and NAbs against the administered IFN-α have been observed.

Nonclinical Studies

Before initiating clinical development, nonclinical studies should be performed. These studies would be comparative in nature and designed to detect differences in the pharmacotoxicological response between the similar IFN-α and the reference IFN-α and not just assess the response per se. The approach taken will need to be fully justified in the nonclinical overview.

Pharmacodynamics Studies

In Vitro Studies

In order to compare differences in biological activity between the similar and the reference medicinal product, data from a number of comparative bioassays could be provided (e.g., receptor-binding studies, antiviral effects in cell culture, antiproliferative effects on human tumor cell lines), many of which may already be available from bioassays submitted as part of the quality dossier. Wherever possible, analytical methods should be standardized and validated according to relevant guidelines.

The limitations of studying antiviral effects in cell culture systems expressing hepatitis C virus (HCV), however, should be recognized, as the results do not correlate well with clinical response. Wherever possible, standardized and validated assays should be used to measure activity and potency.

In Vivo Studies

To support the comparability exercise for the sought clinical indications, the PD activity of the similar and the reference medicinal product could be quantitatively compared in

- An appropriate PD animal model (e.g., evaluating effects on PD markers as, for example, serum 2′,5′-oligoadenylate synthetase activity). If feasible, these measurements may be performed as part of the toxicological studies described in the succeeding text.
- A suitable animal tumor model (e.g., nude mice bearing human tumor xenografts).
- A suitable animal antiviral model.

Toxicological Studies

Data from at least one repeat-dose toxicity study in a relevant species should be considered (e.g., human IFN-α may show activity in the Syrian golden hamster). The study duration should be at least 4 weeks.

Appropriate toxicokinetic measurements should be performed as part of the repeat-dose toxicity study and include a determination of antibody formation.

Data on local tolerance in at least one species should be provided. If feasible, local tolerance testing can be performed as part of the described repeat-dose toxicity study.

Safety pharmacology, reproduction toxicology, mutagenicity, and carcinogenicity studies are not routine requirements for nonclinical testing of similar biological medicinal products containing recombinant human IFN-α as active substance.

Clinical Studies

Pharmacokinetic Studies

The PK properties of the similar and the reference medicinal product could be compared in single-dose crossover studies using SC and IV administration in healthy volunteers. The recommended primary PK parameter is AUC and the secondary parameters are *C* and *T*1/2 or CL/*F*.

Equivalence margins have to be defined a priori and appropriately justified.

Pharmacodynamic Studies

There are a number of PD markers, such as β2-microglobulin, neopterin, and serum 2′,5′-oligoadenylate synthetase activity, which are relevant to the interaction between IFN-α and the immune system. The selected doses should be in the linear ascending part of the dose–response curve. Whereas the relative importance of these effects in the different therapeutic indications is unknown, a comprehensive comparative evaluation of such markers following administration of test and reference products could provide useful supporting data.

Efficacy

Patient Population

The MOA of interferon comprises of several different unrelated effects. Demonstration of similar efficacy between test and reference products is required. This could be performed in treatment-naive patients with chronic hepatitis C (HCV) as delineated by the indication for the reference product. Other patient population(s) might be studied depending on the indications desired (see under "Extrapolation of evidence" section).

Study Design and Duration

A randomized, parallel-group comparison against the reference product over at least 48 weeks is recommended. If possible, the study should be double blind at least until the data to complete the primary analysis have been generated. If this is not feasible, justification should be provided and efforts to reduce/eliminate bias should be clearly identified in the protocol.

The posology (i.e., dose, route, and method of administration) should be the same as for the reference product. IFN-α should be given in line with the current standard treatment for chronic HCV infection and in accordance with the SmPC of the reference product.

The study could be designed so that the primary efficacy analysis is performed at week 12 for all enrolled patients. Preferably, a homogeneous population is recommended (e.g., one single HCV genotype). However, if a mixed population is chosen, it should be stratified based on the HCV genotype.

Endpoints

Primary: Virological response as measured by the proportion of patients with undetectable levels of HCV RNA by quantitative PCR at week 12. The assay used to measure HCV RNA and the cutoff applied should be justified. A 2-log decrease in viral load may be a coprimary endpoint.

Secondary: Virological response at week 4 and end of treatment; sustained virological response (24 weeks after completion of treatment); change in liver biochemistry including transaminase levels and morbidity.

Safety

Safety data should be collected from patients after repeated dosing in a comparative clinical trial over the treatment period plus 24 weeks of follow-up. The number of patients should be sufficient for the comparative evaluation of the adverse effect profile. Laboratory abnormalities for immune-mediated disorders should be included. The safety profile should be similar to the reference products for the common adverse events (such as flu-like illness, alopecia, myalgia, leukopenia, anemia, and thrombocytopenia).

Immunogenicity

Comparative immunogenicity data (antibody levels) should be presented during the treatment period plus 24 weeks of follow-up.

Antibodies, if present, should be further evaluated, for example, for neutralizing capacity and the resulting potential for impact on efficacy of r-IFN-α. In addition, any potential for neutralization of the effect of endogenous interferon(s) (i.e., development of autoimmunity) should be addressed. Any impact of immunogenicity should be thoroughly evaluated in those individuals:

- Not responding to treatment
- Losing response during primary treatment
- Exhibiting unexpected adverse reactions or known immune-mediated events

Extrapolation of Evidence

In principle, extrapolation from one therapeutic indication to another is appropriate where the MOA and/or the receptor is known to be the same as the condition(s) for which similarity in efficacy has been established.

If indication(s) is sought, where the MOA is not known to be the same, such extrapolation should be adequately justified.

HUMAN G-CSF

Human G-CSF is a single polypeptide chain protein of 174 amino acids with *O*-glycosylation at one threonine residue. Recombinant G-CSFs produced in *E. coli* (filgrastim) and in CHO (lenograstim) are in clinical use. Compared to the human and to the mammalian cell culture derived G-CSF, the *E. coli* protein has an additional amino-terminal methionine and no glycosylation. The recombinant G-CSF (rG-CSF) protein contains one free cysteinyl residue and two disulfide bonds. Physicochemical and biological methods are available for characterization of the protein.

Effects of G-CSF on the target cells are mediated through its transmembrane receptor that forms homooligomeric complexes upon ligand binding. Several isoforms of the G-CSF receptor arising from alternative RNA splicing leading to differences in the intracytoplasmic sequences have been isolated. One soluble isoform is known. However, the extracellular, ligand-binding domains of the known isoforms are identical. Consequently, the effects of rG-CSF are mediated via a single-affinity class of receptors.

Antibodies to the currently marketed *E. coli*–derived rG-CSF occur infrequently. These have not been described to have major consequences for efficacy or safety. rG-CSF is administered SC or IV. Possible patient-related risk factors of immune response are unknown.

Nonclinical Studies

Before initiating clinical development, nonclinical studies should be performed. These studies should be comparative in nature and should be designed to detect differences in pharmacotoxicological response between the similar biological medicinal product and the reference medicinal product—not just the response per se. The approach taken will need to be fully justified in the nonclinical overview.

Pharmacodynamic Studies

In Vitro Studies

At the receptor level, comparability of test and reference medicinal product should be demonstrated in appropriate in vitro cell-based bioassays or receptor-binding assays. Such data may already be available from bioassays that were used to measure potency in the evaluation of biological characteristics in module 3. It is important that assays used for comparability will have appropriate sensitivity to detect differences and that experiments are based on a sufficient number of dilutions per curve to fully characterize the concentration–response relationship.

In Vivo Studies

In vivo rodent models, neutropenic and nonneutropenic, should be used to compare the PD effects of the test and the reference medicinal product.

Toxicological Studies

Data from at least one repeat-dose toxicity study in a relevant species should be provided. Study duration should be at least 28 days. The study should include (1) PD measurements and (2) appropriate toxicokinetic measurements. In this context, special emphasis should be laid on the investigation of immune responses to the products.

Data on local tolerance in at least one species should be provided. If feasible, local tolerance testing can be performed as part of the described repeat-dose toxicity study.

Safety pharmacology, reproduction toxicology, mutagenicity, and carcinogenicity studies are not routine requirements for nonclinical testing of similar biological medicinal products containing recombinant G-CSF as active substance.

Clinical Studies

Pharmacokinetic Studies

The PK properties of the similar biological medicinal product and the reference medicinal product should be compared in single-dose crossover studies using SC and IV administration. The primary PK parameter is AUC and the secondary PK parameters are C_{max} and $T1/2$. The general principles for demonstration of bioequivalence are applicable.

Pharmacodynamic Studies

The ANC is the relevant PD marker for the activity of rG-CSF. The PD effect of the test and the reference medicinal products should be compared in healthy volunteers. The selected dose should be in the linear ascending part of the dose–response curve. Studies at more than one dose level may be useful. The CD34+ cell count should be reported as a secondary PD endpoint. The comparability range should be justified.

Clinical Efficacy Studies

rG-CSF can be used for several purposes such as

- Reduction in the duration of neutropenia after cancer chemotherapy or myeloablative therapy followed by bone marrow transplantation
- Mobilization of peripheral blood progenitor cells (PBPCs)
- For treatment of severe congenital, cyclic, or idiopathic neutropenia
- Treatment of persistent neutropenia in patients with advanced human immunodeficiency virus (HIV) infection

The posology varies between these conditions.

The recommended clinical model for the demonstration of comparability of the test and the reference medicinal product is the prophylaxis of severe neutropenia after cytotoxic chemotherapy in a homogenous patient group (e.g., tumor type, previous and planned chemotherapy, as well as disease stage). This model requires a chemotherapy regimen that is known to induce a severe neutropenia in patients. A two-arm comparability study is sufficient in chemotherapy models with known frequency and duration of severe neutropenia. If other chemotherapy regimens are used, a three-arm trial, including placebo, may be needed. The sponsor must justify the comparability delta for the primary efficacy variable, the duration of severe neutropenia (ANC below 0.5 × 109/l). The incidence of febrile neutropenia, infections, and the cumulative rG-CSF dose are secondary variables. The main emphasis is on the first chemotherapy cycle.

Demonstration of the clinical comparability in the chemotherapy-induced neutropenia model will allow the extrapolation of the results to the other indications of the reference medicinal product if the MOA is the same.

Alternative models, including PD studies in healthy volunteers, may be pursued for the demonstration of comparability if justified. In such cases, the sponsor should seek for scientific advice for study design and duration, choice of doses, efficacy/PD endpoints, and comparability margins.

Clinical Safety

Safety data should be collected from a cohort of patients after repeated dosing preferably in a comparative clinical trial. The total exposure should correspond to the exposure of a conventional chemotherapeutic treatment course with several cycles. The total follow-up of patients should be at least 6 months. The number of patients should be sufficient for the evaluation of the adverse effect profile, including bone pain and laboratory abnormalities.

GROWTH HORMONE

The principal bioactive human growth hormone (hGH) is a single-chain nonglycosylated 191 amino acid, 22 kD polypeptide produced in the anterior pituitary gland. Growth hormone for clinical use has an identical amino acid sequence and is produced by recombinant technology using *E. coli*, mammalian cells, or yeast cells as expression system. The structure and biological activity of somatropin can be characterized by appropriate physicochemical and biological methods. Several techniques and bioassays are available to characterize both the active substance and product-related substances/impurities such as deamidated and oxidized forms and aggregates.

Growth hormone has potent anabolic, lipolytic, and antiinsulin effects (acute insulin-like effect). The effects of GH are mediated both directly (e.g., on adipocytes and hepatocytes) and indirectly via stimulation of insulin-like growth factors (IGFs) (principally insulin-like growth factor 1 [IGF-1]). Somatropin-containing medicinal products are currently licensed for normalizing or improving linear growth and/or body composition in GH-deficient and certain non-GH-deficient states. The same receptors are thought to be involved in all currently approved therapeutic indications of rhGHs.

Somatropin has a wide therapeutic window in children during the growth phase, whereas adults may be more sensitive for certain adverse effects. Antibodies to somatropin have been described, including, very rarely, NAbs. Problems have been associated with the purity and stability of the formulations. Somatropin is administered SC; possible patient-related risk factors of immune response are unknown.

Nonclinical Studies

Before initiating clinical development, nonclinical studies should be performed. These studies should be comparative in nature and should be designed to detect differences in the pharmacotoxicological response between the similar biological medicinal product and the reference medicinal product and should not just assess the response per se. The approach taken will need to be fully justified in the nonclinical overview.

Pharmacodynamics Studies

In Vitro Studies

In order to assess any alterations in reactivity between the similar biological medicinal and the reference medicinal product, data from a number of comparative bioassays (e.g., receptor-binding studies, cell proliferation assays), many of which may already be available from quality-related bioassays, should be provided.

In Vivo Studies

An appropriate in vivo rodent model (e.g., the weight-gain assay and/or the tibia growth assay in immature hypophysectomized rats; data may already be available from quality-related bioassays) should be used to quantitatively compare the PD activity of the similar biological medicinal and the reference medicinal product.

Toxicological Studies

Data from at least one repeat-dose toxicity study in a relevant species (e.g., rat) should be provided. Study duration should be at least 4 weeks and include appropriate toxicokinetic measurements. In this context, special emphasis should be laid on the determination of immune responses.

Data on local tolerance in at least one species should be provided. If feasible, local tolerance testing can be performed as part of the described repeat-dose toxicity study.

Safety pharmacology, reproduction toxicology, mutagenicity, and carcinogenicity studies are not routine requirements for nonclinical testing of similar biological medicinal products containing rhGH as active substance.

Clinical Studies

Pharmacokinetic Studies

The relative PK properties of the similar biological medicinal product and the reference medicinal product should be determined in a single-dose crossover study using SC administration. Healthy volunteers are considered appropriate but suppression of endogenous GH production, for example, with a somatostatin analogue, should be considered. The primary PK parameter is AUC and the secondary parameters are C_{max} and $T1/2$. Comparability margins have to be defined a priori and appropriately justified.

Pharmacodynamic Studies

PD should preferably be evaluated as part of the comparative PK study. The selected dose should be in the linear ascending part of the dose–response curve. IGF-1 is the preferred PD marker for the activity of somatropin and is recommended to be used in comparative PD studies. In addition, other markers such as IGFBP-3 may be used. On the other hand, due to the lack of a clear relationship between serum IGF-1 levels and growth response, IGF-1 is not a suitable surrogate marker for the efficacy of a somatropin in clinical trials.

Clinical Efficacy Studies

Clinical comparability efficacy between the similar biological medicinal product and the reference medicinal product should be demonstrated in at least one adequately powered, randomized, parallel-group clinical trial. Clinical studies should be double blind to avoid bias. If this is not possible, at minimum, the person performing height measurements should be effectively masked to treatment allocation.

Sensitivity to the effects of somatropin is higher in GH-deficient than non-GH-deficient conditions. Treatment-naive children with GH deficiency are recommended as the target study population as this provides a sensitive and well-known model. Study subjects should be prepubertal before and during the comparative phase of the trial to avoid interference of the pubertal growth spurt with the treatment effect. This may be achieved, for example, by limiting the age/bone age at study entry. It is important that the study groups are thoroughly balanced for baseline characteristics, as this will affect the sensitivity of the trial and the accuracy of the endpoints.

Change in height velocity or change in height velocity standard deviation score from baseline to the prespecified end of the comparative phase of the trial is the recommended primary efficacy endpoint. Height standard deviation score is a recommended secondary endpoint. Adjustment for factors known to affect the growth response to somatropin should be considered.

During the comparative phase of the study, standing height should be measured at least three times per subject at each time point and the results averaged for analyses. The use of a validated measuring device is mandatory. Consecutive height measurements should be standardized and performed approximately at the same time of the day, by the same measuring device and preferably by the same trained observer. These recommendations aim to reduce measurement errors and variability.

For the determination of reliable baseline growth rates, it is important that also height measurements during the pretreatment phase are obtained in a standardized manner using a validated measuring device.

Due to significant variability in short-term growth, seasonal variability in growth, and measurement errors inherent in short-term growth measurements, the recommended duration of the comparative phase is at least 6 months and may have to be up to 12 months.

Calculation of pretreatment growth rates should be based on observation periods of no less than 6 and no more than 18 months.

Comparability margins have to be prespecified and appropriately justified, primarily on clinical grounds, and serve as the basis for powering the study.

Clinical Safety

Data from patients in the efficacy trial(s) are usually sufficient to provide an adequate premarketing safety database.

The applicant should provide comparative 12-month immunogenicity data of patients who participated in the efficacy trial(s) with sampling at 3-month intervals and testing using validated assays of adequate specificity and sensitivity.

In addition, adequate blood tests including IGF-1, IGFBP-3, fasting insulin, and blood glucose should be performed.

HUMAN INSULIN

Human insulin for therapeutic use is a nonglycosylated, disulfide-bonded heterodimer of 51 amino acids. There is extensive experience with the production of insulin for therapeutic use from animal sources, in the form of semisynthetic insulin, and through different recombinant techniques. Physicochemical and biological methods are available to characterize the primary, secondary, and tertiary structures of the recombinant insulin molecule, as well as its receptor affinity and biological activity in vitro and in vivo. Current quality guidelines on comparability provide information on the characterization and analysis of similar biological medicinal product and its comparator. For rh-insulin, attention should be given to product-related substances/impurities and process-related impurities, and in particular to desamido forms and other forms that may derive from the expression vector or arise from the conversion steps removing the C-peptide and regenerating the 3D structure.

The effects of insulin are mediated predominantly via stimulation of the insulin receptor but insulin is also a weak natural ligand of the IGF-1 receptor.

The same receptors are known to be involved in the MOA relevant for the currently approved therapeutic indications of rh-insulins.

Antibodies to rh-insulin occur frequently, mainly as cross-reacting antibodies. These have been rarely described to have major consequences for efficacy or safety. The potential for development of product-/impurity-specific antibodies needs to be evaluated. Rh-insulin is administered SC or IV. Possible patient-related risk factors of immune response are unknown.

Nonclinical Studies

Before initiating clinical development, nonclinical studies should be performed. These studies should be comparative in nature and should be designed to detect differences in the response to the similar biological medicinal product and the reference medicinal product and should not just assess the response per se. The approach taken will need to be fully justified in the nonclinical overview.

Pharmacodynamic Studies

In Vitro Studies

In order to assess any differences in properties between the similar biological medicinal product and the reference medicinal product, comparative studies such as in vitro bioassays for affinity, insulin- and IGF-1-receptor-binding assays, as well as tests for intrinsic activity should be performed. Partly, such data may already be available from bioassays that were used to measure potency in the evaluation of physicochemical characteristics. It is important that assays used for comparability testing are demonstrated to have appropriate sensitivity to detect minute differences and that experiments are based on a sufficient number of dilutions per curve to characterize the whole concentration–response relationship.

In Vivo Studies

Comparative study(ies) of PD effects would not be anticipated to be sensitive enough to detect any nonequivalence not identified by in vitro assays and is normally not required as part of the comparability exercise.

Toxicological Studies

Data from at least one repeat-dose toxicity study in a relevant species (e.g., rat) should be provided. Study duration should be at least 4 weeks. The study should include appropriate toxicokinetic measurements. In this context, special emphasis should be laid on the determination of immune responses.

Data on local tolerance in at least one species should be provided. If feasible, local tolerance testing can be performed as part of the described repeat-dose toxicity study.

Other routine toxicological studies are not required for rh-insulins developed as similar biological medicinal products.

Clinical Studies

Pharmacokinetic Studies

The relative PK properties of the similar biological medicinal product and the reference medicinal product should be determined in a single-dose crossover study using SC administration.

Comprehensive comparative data should be provided on the time–concentration profile (AUC as the primary endpoint and C_{max}, t_{max}, and $T1/2$ as secondary endpoints). Studies should be performed preferably in patients with type 1 diabetes. Factors contributing to PK variability, for example, insulin dose and site of injection/thickness of SC fat should be taken into account.

Pharmacodynamic Studies

The clinical activity of an insulin preparation is determined by its time-effect profile of hypoglycemic response, which incorporates components of PD and PK. PD data are of primary importance to demonstrate comparability of a similar rh-insulin. The double-blind, crossover hyperinsulinemic euglycemic clamp study is suitable for this characterization. Data on comparability regarding glucose infusion rate and serum insulin concentrations should be made available. The choice of study population and study duration should be justified.

Plasma glucose levels should be obtained as part of the PK study following SC administration.

Clinical Efficacy Studies

Provided that clinical comparability can be concluded from PK and PD data, there is no anticipated need for efficacy studies on intermediary or clinical variables.

Clinical Safety

Immunogenicity

The safety concerns with a similar rh-insulin relate mainly to the potential for immunogenicity. The issue of immunogenicity can only be settled through clinical trials of sufficient duration, that is, at least 12 months using SC administration. The comparative phase of this study should be at least 6 months, to be completed preapproval. Data at the end of 12 months could be presented as part of postmarketing commitment. The primary outcome measure should be the incidence of antibodies to the test and reference medicinal product.

The plans for these trials should take into account:

- Justification of study population including history of previous insulin exposure
- Definitions of prespecified analyses of the immunogenicity data with respect to effects on clinical findings (glycemic control, insulin dose requirements, local and systemic allergic reactions)

Local Reactions

If any concern is raised through nonclinical and short-term clinical studies outlined earlier, additional evaluation of local tolerability may be needed premarketing. Otherwise, such reactions should be monitored and recorded within immunogenicity trials.

14 Bioequivalence Testing
The U.S. Perspective

In the United States, the FDA approves and grants marketing authorization of generic drugs by applying the regulatory requirements provided in the Code of Federal Regulations (CFR) (Table 14.1).

A generic drug is bioequivalent to the listed drug if "the rate and extent of the absorption of the drug do not show a significant difference from the rate and extent of absorption of the listed drug when administered at the same molar dose of the therapeutic ingredient under similar experimental conditions in either a single dose or multiple doses" (21 U.S.C. 355(j)(8)(B)(i)).

The science of bioequivalence (BE) is still undergoing major changes and final rules are established after years of debate and validation of protocols. The U.S. FDA has finalized several guidelines (Table 14.2). Of greatest significance is the finalization of the Bioavailability and Bioequivalence Studies for Orally Administered Products—General Considerations that came in April 2003.

Table 14.3 lists a historical perspective of the development that led to the current guidelines in the United States.

FDA GUIDANCE ON BIOEQUIVALENT TESTING

Background

The U.S. FDA provides a comprehensive list of updated guidances; the applicants are encouraged to frequently monitor these lists available at http://www.fda.gov/Drugs/GuidanceCompliance RegulatoryInformation/Guidances/ucm064964.htm.

In September 2010, the FDA issued final regulations addressing the safety reporting requirements for investigational new drug (IND) applications found in 21 CFR part 312 and for bioavailability (BA) and BE studies found in 21 CFR part 320.

Given in the following is the current (as of end 2013) guidance according to the FDA regulations. The definition of terms throughout this chapter and others are provided in the glossary of terms. It is important to review these definitions as the regulatory agencies provide specific meanings to various terms.

PROCEDURES FOR DETERMINING THE BIOAVAILABILITY OR BIOEQUIVALENCE OF DRUG PRODUCTS

Requirements for Submission of Bioavailability and Bioequivalence Data

(a) Any person submitting a full new drug application (NDA) to the FDA shall include in the application either of the following:

  (1) Evidence measuring the in vivo BA of the drug product that is the subject of the application

  (2) Information to permit the FDA to waive the submission of evidence measuring in vivo BA

TABLE 14.1
Relevant Sections in the CFR Related to BA/BE Studies

21 CFR 314.94(a)(9)	Chemistry, manufacturing, and controls; permitted changes in inactive ingredients for parenteral, otic, ophthalmic, and topical drug products
21 CFR 320.1	Definitions of bioavailability, pharmaceutical equivalents, pharmaceutical alternatives, and bioequivalence
21 CFR 320.21	Regulatory requirements related to submission of in vivo bioavailability and bioequivalence data
21 CFR 320.22	Criteria for waiver of evidence of in vivo bioavailability or bioequivalence data
21 CFR 320.23	Basis for measuring in vivo bioavailability or demonstrating bioequivalence
21 CFR 320.24	Types of evidence to measure bioavailability or establish bioequivalence
21 CFR 320.25	Guidelines for the conduct of an in vivo bioavailability study
21 CFR 320.26	Guidelines on the design of a single dose in vivo bioavailability or bioequivalence study
21 CFR 320.27	Guidelines on the design of a multiple-dose in vivo bioavailability study
21 CFR 320.28	Correlation of bioavailability with an acute pharmacological effect or clinical evidence
21 CFR 320.29	Analytical methods for an in vivo bioavailability or bioequivalence study
21 CFR 320.30	Inquiries regarding bioavailability and bioequivalence requirements and review of protocols by the FDA
21 CFR 320.32	Procedures for establishing or amending a bioequivalence requirement
21 CFR 320.33	Criteria and evidence to assess actual or potential bioequivalence problems
21 CFR 320.36	Requirements for maintenance of records of bioequivalence testing
21 CFR 320.38	Retention of bioavailability samples
21 CFR 320.63	Retention of bioequivalence samples
21 CFR 320.24(b)	Bioequivalence Studies with Pharmacokinetic Endpoints for Drugs Submitted Under an ANDA

TABLE 14.2
Final Biopharmaceutics Guidelines of the U.S. FDA

Guideline	Date Final
Metaproterenol Sulfate and Albuterol Metered Dose Inhalers In Vitro	June 27, 1989
Topical Dermatologic Corticosteroids: In Vivo Bioequivalence	June 2, 1995
Dissolution Testing of Immediate Release Solid Oral Dosage Forms	August 1, 1997
Dissolution Testing of Immediate Release Solid Oral Dosage Forms	August 1, 1997
Extended Release Oral Dosage Forms: Development, Evaluation, and Application of In Vitro/In Vivo Correlations	September 1, 1997
Waiver of in vivo Bioavailability and Bioequivalence Studies for Immediate-Release Solid Oral Dosage Forms Based on a Biopharmaceutics Classification System	August 1, 2000
Statistical Approaches to Establishing Bioequivalence	February 1, 2001
Bioanalytical Method Validation	May 1, 2001
Food-Effect Bioavailability and Fed Bioequivalence Studies	December 1, 2002
Bioavailability and Bioequivalence Studies for Orally Administered Drug Products—General Considerations (PDF, 395KB)	March 1, 2003
Bioavailability and Bioequivalence Studies for Nasal Aerosols and Nasal Sprays for Local Action	May 3, 2003
Statistical Information from the June 1999 Draft Guidance and Statistical Information for in vitro Bioequivalence Data Posted on August 18, 1999	May 11, 2003
Bioanalytical Method Validation (Revised Final)	September 12, 2013
Bioequivalence Studies With Pharmacokinetic Endpoints for Drugs Submitted Under an Abbreviated New Drug Application	December 4, 2013

TABLE 14.3
Major BE-Related Legislative Events in the History of the United States

Year	Legislative Event
1902	Federal law for biologics, particularly vaccines, required evaluation for "safety, purity and potency." Pure Drug and Cosmetic Act.
1906	The Food and Drugs Act added drugs other than biologics.
1931	The U.S. FDA was formed.
1938	The FDCA created FDA and required safety evaluation on new drugs before marketing based on data in an NDA.
1944	Public Health Services Act.
1962	Kefauver–Harris Drug Amendments added an effectiveness requirement for the approval of NDA.
1960s	FDA permits marketing of "similars" while corresponding pioneer products undergo DESI reviews. "Similars" came into market between 1938 and 1962.
1963	The first pharmaceutical GMPs.
1970	The FDA terminates marketing of "similars" unless DESI pioneer showed safety and efficacy and "similar" manufacturer submits ANDA with formulation and manufacture information (The Supreme Court in the United States v. Generix Drug Corporation supported FDA requirement for ANDA).
	The FDA became interested in the biological availability of new drugs and a drug BE study panel was formed by the Office of Technology Assessment (OTA) to understand the chemical and therapeutic equivalent relationship of drug products. On the basis of the recommendations from OTA, the FDA formulated regulations for the submission of BA data. These regulations are currently incorporated in the 21st volume of the CFR, part 320 (21 CFR320); 75/75 (or 75/75–125) rule was originally proposed in the late 1970s as an alternative means of testing the BE of two formulations of a pharmaceutical agent. Finalized and effective regulations as CFR in 1977.
1980	Power approach for statistical analysis applied.
1983	Orphan Drug Act: in response to the spread of AIDS and need to develop drugs used in less than a population of 250,000 patients.
1984	Waxman–Hatch Act created a generic approval system for all new drugs, including those approved after 1962. The FDA finalized the BA/BE regulations (21 CFR 320) wherein the pioneer shows BA in NDA; "similars" to DESI-effective pioneers show BE leading to first U.S. generics. Several revisions to 21 CFR 320 were made including the most recent one in April 2006. The Drug Price Competition and Patent Term Restoration Act of 1984 (Pub. L. No. 98-417) (the Hatch–Waxman Amendments) created section 505(j) of the Act, which established the current ANDA approval process. The showing that must be made for an ANDA to be approved is quite different from what is required in an NDA. An NDA applicant must prove that the drug product is safe and effective. An ANDA does not have to prove the safety and effectiveness of the drug product because it relies on the finding by the FDA that the RLD is safe and effective. Instead, an ANDA applicant must demonstrate, among other things, that its drug product is bioequivalent to the RLD (2 1 U. S.C. 355(j)(2)(A)(iv)). The scientific premise underlying the Hatch–Waxman Amendments is that in most circumstances bioequivalent drug products may be substituted for each other.
1986	Discontinuation of 75/75 rule and power approach.
1986	FDA conducted public hearing due to public concern about BE.
1986–1989	The Bioequivalence Task Force formed by FDA investigated the scientific issues raised at the public hearing.
1988	The Generic Animal Drug and Patent Term Restoration Act (GADPTRA) signed into law on November 16, 1988, permits sponsors to submit an Abbreviated New Animal Drug Application (ANADA) for a generic version of any off-patent approved animal drug (with certain exceptions noted in the law) regardless of whether the drug was approved prior to 1962 and subject to the National Academy of Sciences/National Research Council/DESI (NAS/NRC/DESI) review.

(*Continued*)

TABLE 14.3 (*Continued*)
Major BE-Related Legislative Events in the History of the United States

Year	Legislative Event
	The Hatch–Waxman Act enacted in 1984 allowed generic drugs to enter the market without repeating expensive clinical trials required for their brand-name counterparts. The legislation provided accessibility to lower-cost generic drugs while still encouraging innovation and development of new drugs. The generic drug companies were allowed to market the drug after the patent and certain exclusivities expired. The effect of this legislation was quickly felt. Today, generics make up approximately 70% of the market share, a steep climb from 12% reported in 1984 by the FDA. Nevertheless, the innovator companies did not let go as easily and tried their best to slow down the entry of generic drugs as much as 44 months if the innovator drug company filed a patent infringement lawsuit against the generic company. Another limitation of the legislation, which was not intended to impede the entry of generic drugs, was the 180-day marketing exclusivity granted to the first-filed ANDA applicant. If the generic drug company is the first one to file the ANDA for the RLD, then it is granted the 180-day marketing exclusivity as the only generic drug on the market for 180 days. However, the issue with the 180-day marketing exclusivity to the first-filed ANDA is that it does not trigger until the first date of commercial marketing of the generic drug. Often times, the brand and the generic company involved in the lawsuit may agree on a settlement and the generic company may agree to delay the 180-day exclusivity beginning date as part of the settlement. As a result, the 180-day exclusivity granted to the first-filed ANDA would get delayed and halt the entry of other generic drugs for that RLD. This tactic would force generic drugs from different companies for the same RLD to wait until the end of the 180-day exclusivity period. In the end, the innovator drug company can employ such strategies to stop the entry of generic drugs into the market and maintain its power over the prescription drug market.
1989	The Bioequivalence Task Force report was released as a letter on the provision of new procedures and policies affecting the generic drug review process.
1991	Letter on the request for cooperation of regulated industry to improve the efficiency and effectiveness of the generic drug review process, by assuring the completeness and accuracy of required information and data submissions.
1992	Prescription Drug User Fee Act.
	Office of Alternative Medicine established.
	FDA issued guidance on statistical procedures for bioequivalence BE studies; letter on the provision of new information pertaining to new BE guidelines and refuse-to-file letters; two one-sided tests procedure (90% CI statistical approach).
1993	Letter to all ANDA and AADA applicants about the Generic Drug Enforcement Act of 1992 (GDEA) and the OGD intention to refuse to file incomplete submissions as required by the new law; letter to regulated industry notifying interested parties about important detailed information regarding labeling, scale-up, packaging, minor/major amendment criteria, and BE requirements.
1994	Letter on incomplete abbreviated applications, convictions under GDEA, multiple supplements, annual reports for bulk antibiotics, batch size for transdermal drugs, bioequivalence protocols, research, deviations from OGD policy.
1995	SUPAC-IR Solid Oral Dosage Forms—Scale-Up and Postapproval Changes: Chemistry, Manufacturing, and Controls, In Vitro Dissolution Testing, and In Vivo Bioequivalence Documentation.
1996	Structure and content of clinical study reports.
1997	FDA Modernization Act (Final); Dissolution Testing of Immediate Release Solid Oral Dosage Forms (Final); Extended Release Oral Dosage Forms—Development, Evaluation, and Application of In Vitro/In Vivo Correlations (Final); SUPAC-MR—Modified Release Solid Oral Dosage Forms—Scale-Up and Post approval Changes—Chemistry, Manufacturing, and Controls, In Vitro Dissolution Testing, and In Vivo Bioequivalence Documentation; SUPAC-SS—Nonsterile Semisolid Dosage Forms; Scale-Up and Postapproval Changes—Chemistry, Manufacturing, and Controls; In Vitro Release Testing and In Vivo Bioequivalence Documentation; Good Clinical Practice—Consolidated Guideline (E6); General Considerations for Clinical Trials (E8).

(*Continued*)

TABLE 14.3 (*Continued*)
Major BE-Related Legislative Events in the History of the United States

Year	Legislative Event
1998	Office of Alternative Medicine changed to National Center for Complementary and Alternative Medicine (NCCAM); Statistical Principles for Clinical Trials (E9); Ethnic Factors in the Acceptability of Foreign Clinical Data.
1999	ClinicalTrials.gov is founded to provide the public with updated information on enrollment in federally and privately supported clinical research, thereby expanding patient access to studies of promising therapies.
2000	Waiver of In Vivo Bioavailability and Bioequivalence Studies for Immediate Release Solid Oral Dosage Forms Based on a Biopharmaceutics Classification System (Final); Revising ANDA Labeling Following Revision of the RLD Labeling.
2001	Statistical Approaches to Establishing Bioequivalence (Final); Bioanalytical Method Validation (Final).
2002	Food-Effect Bioavailability and Fed Bioequivalence Studies (Final)
2003	Medicare Prescription Drug, Improvement, and Modernization Act of 2003 (MMA) amended laws, which existed under the Hatch–Waxman Act governing the FDA approval of ANDAs. It amended three major rules pertaining to listed drugs, 30-month stays, and approvals of ANDAs. The Act prohibits an ANDA applicant from amending or supplementing its application to refer to a listed drug which is different from that referred to in the application when originally submitted. The legislation clarified the types of changes, which can be submitted as amendments or supplements versus the types of changes that would require filing of a new ANDA. When the *Orange Book* identifies as a separate listed drug product with characteristics different from which the applicant is seeking approval, the applicant should submit a separate ANDA referring the corresponding listed drug. It also limited the 30-month stay per ANDA to a single term. This meant that regardless of the number of patents claimed by the innovator drug company, it would only be allowed a 30-month period to resolve any issues with the ANDA applicant. ANDA applications could be subject to no more than one 30-month stay according to the new provision. This limit of a 30-month stay per ANDA had a significant impact on the generic drug industry. The innovator drug companies could not further delay the entry of generic drugs by filing multiple patent infringement lawsuits in order to obtain multiple 30-month stays.
	Bioavailability and Bioequivalence Studies for Orally Administered Drug Products—General Considerations (Revised); Bioavailability and Bioequivalence Studies for Nasal Aerosols and Nasal Sprays for Local Action (Draft); Bioavailability and Bioequivalence Studies for Orally Administered Drug Products—General Considerations (Revised); Statistical information from the Draft Guidance and Statistical Information for In Vitro Bioequivalence Data (Draft).
2004	Handling and Retention of Bioavailability and Bioequivalence Testing Samples.
2005	Potassium Chloride Modified-Release Tablets and Capsules—In Vivo Bioequivalence and In Vitro Dissolution Testing; ANDAs—Impurities in Drug Products (Draft).
2007	FDA Amendments Act.
2009	Biologics Price Competition and Innovation Act; Submission of Summary Bioequivalence Data for Abbreviated New Drug Applications (Draft); ANDAs—impurities in Drug Substances (Final).
2010	Safety Reporting Requirements for INDs (Investigational New Drug Applications) and BA/BE (Bioavailability/Bioequivalence) Studies (Draft); Individual Product Bioequivalence Recommendations—List of Product Bioequivalence Recommendations (Revised); Guidance for Industry—Bioequivalence Recommendations for Specific Products; 12 April 2012: Bioavailability and Bioequivalence Requirements (Final).
2012	FDA Safety and Innovation Act (FDASIA): FDASIA, signed into law on July 9, 2012, gives FDA the authority to collect user fees from industry to fund reviews of innovator drugs, medical devices, generic drugs, and biosimilar biologics. It also reauthorizes two programs that encourage pediatric drug development. This is the fifth authorization of the PDUFA, first enacted in 1992, and the third authorization of the Medical Device User Fee Act, or MDUFA, first enacted in 2002. Both programs have provided steady and reliable funding to maintain and support a staff of trained reviewers who must determine whether a proposed new product is safe and effective for patients and do so within a certain time period. The new user fee programs for generic drugs and biosimilar biologics build on the successes of these two established user fee programs.

(*Continued*)

TABLE 14.3 (*Continued*)
Major BE-Related Legislative Events in the History of the United States

Year	Legislative Event
2013	Fiscal Year 2013 Regulatory Science Plan: The U.S. FDA planned to hold a public meeting in June 2013 to discuss its progress in advancing regulatory science for generic drug products and solicit feedback on its research priorities as it looks to fulfill its statutory goals under the FDASIA. The FDA is fast pushing improvements in the regulatory approval process using science as the driving force. Regulatory science is essentially the various scientific tools and methods employed to evaluate a product—in this case generic pharmaceuticals—to ensure that it is safe, effective, and manufactured to a certain standard of quality. Regulatory science has been an important subject to generic drug regulators most recently after some generic equivalents of a well-known antidepressant, Wellbutrin XR 300 mg (bupropion), were found not to be bioequivalent to the drug. In the aftermath of that revelation, the FDA has scrambled to figure out what factors were at the root cause of the BE discrepancy and how to prevent it from happening in the future. But that's just one drug. GDUFA, meanwhile, calls for the FDA to institute regulatory science initiatives on a broader, more holistic level that benefits the entire generic drug space and not just a small subset of generics of a particular product. The FDA is currently working on 13 projects that had been identified in its Fiscal Year 2013 Regulatory Science Plan, a set of projects contained within a commitment letter FDA signed during the GDUFA negotiation process. Those projects are as follows: • Bioequivalence of local acting, orally inhaled drug products • Bioequivalence of local acting topical dermatological drug products • Bioequivalence of local acting GI drug products • Quality by design of generic drug products • Modeling and simulation • Pharmacokinetic studies and evaluation of antiepileptic drugs • Excipient effects on permeability and absorption of Biopharmaceutics Classification System Class 3 drugs • Product- and patient-related factors affecting switchability of drug–device combinations • Postmarketing surveillance of generic drug usage patterns and adverse events • Evaluation of drug product physical attributes on patient acceptability • Postmarketing assessment of generic drugs and their brand-name counterparts • Physicochemical characterization of complex drug substances • Develop a risk-based understanding of potential adverse impacts to drug product quality resulting from changes in active pharmaceutical ingredients manufacturing and controls More futuristic directions include resolving the following: • What challenges currently exist that serve to limit the availability of generic pharmaceuticals? • What approaches can be taken to improve how generics are evaluated in terms of therapeutic equivalence? • How can therapeutic equivalence be monitored postapproval? • Which projects are most important for FDA to act on? • Which areas could benefit from additional draft guidance or further clarity?

(b) Any person submitting an abbreviated new drug application (ANDA) to the FDA shall include in the application either of the following:
   (1) Evidence demonstrating that the drug product that is the subject of the ANDA is bioequivalent to the reference-listed drug (RLD). A complete study report must be submitted for the BE study upon which the applicant relies for approval. For all other BE studies conducted on the same drug product formulation, the applicant must submit either a complete or summary report. If a summary report of a BE study is submitted and the

FDA determines that there may be BE issues or concerns with the product, the FDA may require that the applicant submit a complete report of the BE study to the FDA.

(2) Information to show that the drug product is bioequivalent to the RLD that would permit the FDA to waive the submission of evidence demonstrating in vivo BE as provided in paragraph (f) of this section.

(c) Any person submitting a supplemental application to the FDA shall include in the supplemental application the evidence or information set forth in paragraphs (a) and (b) of this section if the supplemental application proposes any of the following changes:

(1) A change in the manufacturing site or a change in the manufacturing process, including a change in product formulation or dosage strength, beyond the variations provided for in the approved application

(2) A change in the labeling to provide for a new indication for use of the drug product, if clinical studies are required to support the new indication for use

(3) A change in the labeling to provide for a new dosage regimen or for an additional dosage regimen for a special patient population, for example, infants, if clinical studies are required to support the new or additional dosage regimen

(d) The FDA may approve a full NDA, or a supplemental application proposing any of the changes set forth in paragraph (c) of this section, that does not contain evidence of in vivo BA or information to permit waiver of the requirement for in vivo BA data, if all of the following conditions are met:

(1) The application is otherwise approvable.

(2) The application agrees to submit, within the time specified by the FDA, either of the following:

(i) Evidence measuring the in vivo BA and demonstrating the in vivo BE of the drug product that is the subject of the application

(ii) Information to permit the FDA to waive measurement of in vivo BA

(e) Evidence measuring the in vivo BA and demonstrating the in vivo BE of a drug product shall be obtained using one of the approaches for determining BA set forth by the FDA.

(f) Information to permit the FDA to waive the submission of evidence measuring the in vivo BA or demonstrating the in vivo BE shall meet the criteria set forth by the FDA.

(g) Any person holding an approved full NDA or ANDA shall submit to the FDA a supplemental application containing new evidence measuring the in vivo BA or demonstrating the in vivo BE of the drug product that is the subject of the application if notified by the FDA that

(1) There are data demonstrating that the dosage regimen in the labeling is based on incorrect assumptions or facts regarding the pharmacokinetics (PK) of the drug product and that following this dosage regimen could potentially result in subtherapeutic or toxic levels

(2) There are data measuring significant intra-batch and batch-to-batch variability, for example, ±25%, in the BA of the drug product

(h) The requirements of this section regarding the submission of evidence measuring the in vivo BA or demonstrating the in vivo BE apply only to a full NDA or ANDA or a supplemental application for a finished dosage formulation.

CRITERIA FOR WAIVER OF EVIDENCE OF IN VIVO BIOAVAILABILITY OR BIOEQUIVALENCE

(a) Any person submitting a full NDA or ANDA, or a supplemental application proposing any of the changes set forth in 320.21(c), may request the FDA to waive the requirement for the submission of evidence measuring the in vivo BA or demonstrating the in vivo BE of the drug product that is the subject of the application. An applicant shall submit a request for waiver with the application. Except as provided in paragraph (f) of this section, the FDA shall waive the requirement for the submission of evidence of in vivo BA or BE if the drug product meets any of the provisions of paragraphs (b), (c), (d), or (e) of this section.

(b) For certain drug products, the in vivo BA or BE of the drug product may be self-evident. The FDA shall waive the requirement for the submission of evidence obtained in vivo measuring the BA or demonstrating the BE of these drug products. A drug product's in vivo BA or BE may be considered self-evident based on other data in the application if the product meets one of the following criteria:
   (1) The drug product
      (i) Is a parenteral solution intended solely for administration by injection, or an ophthalmic or otic solution
      (ii) Contains the same active and inactive ingredients in the same concentration as a drug product that is the subject of an approved full NDA or ANDA
   (2) The drug product
      (i) Is administered by inhalation as a gas, for example, a medicinal or an inhalation anesthetic
      (ii) Contains an active ingredient in the same dosage form as a drug product that is the subject of an approved full NDA or ANDA
   (3) The drug product
      (i) Is a solution for application to the skin, an oral solution, elixir, syrup, tincture, a solution for aerosolization or nebulization, a nasal solution, or similar other solubilized form
      (ii) Contains an active drug ingredient in the same concentration and dosage form as a drug product that is the subject of an approved full NDA or ANDA
      (iii) Contains no inactive ingredient or other change in formulation from the drug product that is the subject of the approved full NDA or ANDA that may significantly affect absorption of the active drug ingredient or active moiety for products that are systemically absorbed or that may significantly affect systemic or local availability for products intended to act locally

(c) The FDA shall waive the requirement for the submission of evidence measuring the in vivo BA or demonstrating the in vivo BE of a solid oral dosage form (other than a delayed-release or extended-release dosage form) of a drug product determined to be effective for at least one indication in a Drug Efficacy Study Implementation (DESI) notice or which is identical, related, or similar to such a drug product under 310.6 of this chapter unless the FDA has evaluated the drug product under the criteria set forth in 320.33, included the drug product in the *Approved Drug Products with Therapeutic Equivalence Evaluations* list, and rated the drug product as having a known or potential BE problem. A drug product so rated reflects a determination by the FDA that an in vivo BE study is required.

(d) For certain drug products, BA may be measured or BE may be demonstrated by evidence obtained in vitro in lieu of in vivo data. The FDA shall waive the requirement for the submission of evidence obtained in vivo measuring the BA or demonstrating the BE of the drug product if the drug product meets one of the following criteria:
   (1) (Reserved)
   (2) The drug product is in the same dosage form, but in a different strength, and is proportionally similar in its active and inactive ingredients to another drug product for which the same manufacturer has obtained approval and the conditions in paragraphs (d) (2) (i) through (d) (2) (iii) of this section are met:
      (i) The BA of this other drug product has been measured.
      (ii) Both drug products meet an appropriate in vitro test approved by the FDA.

(iii) The applicant submits evidence showing that both drug products are proportionally similar in their active and inactive ingredients.

(iv) Paragraph (d) of this section does not apply to delayed-release or extended-release products.

(3) The drug product is, on the basis of scientific evidence submitted in the application, shown to meet an in vitro test that has been correlated with in vivo data.

(4) The drug product is a reformulated product that is identical, except for a different color, flavor, or preservative that could not affect the BA of the reformulated product, to another drug product for which the same manufacturer has obtained approval and the following conditions are met:

(i) The BA of the other product has been measured.

(ii) Both drug products meet an appropriate in vitro test approved by the FDA.

(e) The FDA, for good cause, may waive a requirement for the submission of evidence of in vivo BA or BE if waiver is compatible with the protection of the public health. For full NDAs, the FDA may defer a requirement for the submission of evidence of in vivo BA if deferral is compatible with the protection of the public health.

(f) The FDA, for good cause, may require evidence of in vivo BA or BE for any drug product if the agency determines that any difference between the drug product and a listed drug may affect the BA or BE of the drug product.

BASIS FOR MEASURING IN VIVO BIOAVAILABILITY OR DEMONSTRATING BIOEQUIVALENCE

(a) (1) The in vivo BA of a drug product is measured if the product's rate and extent of absorption, as determined by comparison of measured parameters, for example, concentration of the active drug ingredient in the blood, urinary excretion rates, or pharmacological effects, do not indicate a significant difference from the reference material's rate and extent of absorption. For drug products that are not intended to be absorbed into the bloodstream, BA may be assessed by measurements intended to reflect the rate and extent to which the active ingredient or active moiety becomes available at the site of action.

(2) Statistical techniques used shall be of sufficient sensitivity to detect differences in rate and extent of absorption that are not attributable to subject variability.

(3) A drug product that differs from the reference material in its rate of absorption, but not in its extent of absorption, may be considered to be bioavailable if the difference in the rate of absorption is intentional, is appropriately reflected in the labeling, is not essential to the attainment of effective body drug concentrations on chronic use, and is considered medically insignificant for the drug product.

(b) Two drug products will be considered bioequivalent drug products if they are pharmaceutical equivalents or pharmaceutical alternatives whose rate and extent of absorption do not show a significant difference when administered at the same molar dose of the active moiety under similar experimental conditions, either single dose or multiple dose. Some pharmaceutical equivalents or pharmaceutical alternatives may be equivalent in the extent of their absorption but not in their rate of absorption and yet may be considered bioequivalent because such differences in the rate of absorption are intentional and are reflected in the labeling, are not essential to the attainment of effective body drug concentrations on chronic use, and are considered medically insignificant for the particular drug product studied.

TYPES OF EVIDENCE TO MEASURE BIOAVAILABILITY OR ESTABLISH BIOEQUIVALENCE

(a) BA may be measured or BE may be demonstrated by several in vivo and in vitro methods. The FDA may require in vivo or in vitro testing, or both, to measure the BA of a drug product or establish the BE of specific drug products. Information on BE requirements for specific products is included in the current edition of the FDA's publication *Approved Drug Products with Therapeutic Equivalence Evaluations* and any current supplement to the publication. The selection of the method used to meet an in vivo or in vitro testing requirement depends upon the purpose of the study, the analytical methods available, and the nature of the drug product. Applicants shall conduct BA and BE testing using the most accurate, sensitive, and reproducible approach available among those set forth in paragraph (b) of this section. The method used must be capable of measuring BA or establishing BE, as appropriate, for the product being tested.

(b) The following in vivo and in vitro approaches, in descending order of accuracy, sensitivity, and reproducibility, are acceptable for determining the BA or BE of a drug product:

(1) (i) An in vivo test in humans in which the concentration of the active ingredient or active moiety and, when appropriate, its active metabolite(s), in whole blood, plasma, serum, or other appropriate biological fluid is measured as a function of time. This approach is particularly applicable to dosage forms intended to deliver the active moiety to the bloodstream for systemic distribution within the body.

(ii) An in vitro test that has been correlated with and is predictive of human in vivo BA data.

(2) An in vivo test in humans in which the urinary excretion of the active moiety, and, when appropriate, its active metabolite(s), is measured as a function of time. The intervals at which measurements are taken should ordinarily be as short as possible so that the measure of the rate of elimination is as accurate as possible. Depending on the nature of the drug product, this approach may be applicable to the category of dosage forms described in paragraph (b)(1)(i) of this section. This method is not appropriate where urinary excretion is not a significant mechanism of elimination.

(3) An in vivo test in humans in which an appropriate acute pharmacological effect of the active moiety, and, when appropriate, its active metabolite(s), is measured as a function of time if such effect can be measured with sufficient accuracy, sensitivity, and reproducibility. This approach is applicable to the category of dosage forms described in paragraph (b)(1)(i) of this section only when appropriate methods are not available for measurement of the concentration of the moiety, and, when appropriate, its active metabolite(s), in biological fluids or excretory products but a method is available for the measurement of an appropriate acute pharmacological effect. This approach may be particularly applicable to dosage forms that are not intended to deliver the active moiety to the bloodstream for systemic distribution.

(4) Well-controlled clinical trials that establish the safety and effectiveness of the drug product, for purposes of measuring BA, or appropriately designed comparative clinical trials, for purposes of demonstrating BE. This approach is the least accurate, sensitive, and reproducible of the general approaches for measuring BA or demonstrating BE. For dosage forms intended to deliver the active moiety to the bloodstream for systemic distribution, this approach may be considered acceptable only when analytical methods cannot be developed to permit use of one of the approaches outlined in paragraphs (b)(1)(i) and (b)(2) of this section, when the approaches described in paragraphs (b)(1)(ii), (b)(1)(iii), and (b)(3) of this section are not available. This approach may also be considered sufficiently accurate for measuring BA or demonstrating BE of dosage forms intended to deliver the active moiety locally, for example, topical preparations for the skin, eye, and mucous membranes; oral dosage forms not intended to be

absorbed, for example, an antacid or radiopaque medium; and bronchodilators administered by inhalation if the onset and duration of pharmacological activity are defined.
  (5) A currently available in vitro test acceptable to the FDA (usually a dissolution rate test) that ensures human in vivo BA.
  (6) Any other approach deemed adequate by the FDA to measure BA or establish BE.
(c) The FDA may, notwithstanding prior requirements for measuring BA or establishing BE, require in vivo testing in humans of a product at any time if the agency has evidence that the product
  (1) May not produce therapeutic effects comparable to a pharmaceutical equivalent or alternative with which it is intended to be used interchangeably
  (2) May not be bioequivalent to a pharmaceutical equivalent or alternative with which it is intended to be used interchangeably
  (3) Has greater than anticipated potential toxicity related to PK or other characteristics

GUIDELINES FOR THE CONDUCT OF AN IN VIVO BIOAVAILABILITY STUDY

(a) *Guiding principles.*
  (1) The basic principle in an in vivo BA study is that no unnecessary human research should be done.
  (2) An in vivo BA study is generally done in a normal adult population under standardized conditions. In some situations, an in vivo BA study in humans may preferably and more properly be done in suitable patients. Critically ill patients shall not be included in an in vivo BA study unless the attending physician determines that there is a potential benefit to the patient.
(b) *Basic design.* The basic design of an in vivo BA study is determined by the following:
  (1) The scientific questions to be answered
  (2) The nature of the reference material and the dosage form to be tested
  (3) The availability of analytical methods
  (4) Benefit–risk considerations in regard to testing in humans
(c) *Comparison to a reference material.* In vivo BA testing of a drug product shall be in comparison to an appropriate reference material unless some other approach is more appropriate for valid scientific reasons.
(d) *Previously unmarketed active drug ingredients or therapeutic moieties.*
  (1) An in vivo BA study involving a drug product containing an active drug ingredient or therapeutic moiety that has not been approved for marketing can be used to measure the following PK data:
    (i) The BA of the formulation proposed for marketing.
    (ii) The essential PK characteristics of the active drug ingredient or therapeutic moiety, such as the rate of absorption, the extent of absorption, the half-life of the therapeutic moiety in vivo, and the rate of excretion and/or metabolism. Dose proportionality of the active drug ingredient or the therapeutic moiety needs to be established after single-dose administration and in certain instances after multiple-dose administration. This characterization is a necessary part of the investigation of the drug to support drug labeling.
  (2) The reference material in such a BA study should be a solution or suspension containing the same quantity of the active drug ingredient or therapeutic moiety as the formulation proposed for marketing.
  (3) The reference material should be administered by the same route as the formulation proposed for marketing unless an alternative or additional route is necessary to answer the scientific question under study. For example, in the case of an active drug ingredient or therapeutic moiety that is poorly absorbed after oral administration, it may be

necessary to compare the oral dosage form proposed for marketing with the active drug ingredient or therapeutic moiety administered in solution both orally and intravenously.

(e) *New formulations of active drug ingredients or therapeutic moieties approved for marketing.*
   (1) An in vivo BA study involving a drug product that is a new dosage form or a new salt or ester of an active drug ingredient or therapeutic moiety that has been approved for marketing can be used to
      (i) Measure the BA of the new formulation, new dosage form, or new salt or ester relative to an appropriate reference material
      (ii) Define the PK parameters of the new formulation, new dosage form, or new salt or ester to establish dosage recommendation
   (2) The selection of the reference material(s) in such a BA study depends upon the scientific questions to be answered, the data needed to establish comparability to a currently marketed drug product, and the data needed to establish dosage recommendations.
   (3) The reference material should be taken from a current batch of a drug product that is the subject of an approved NDA and that contains the same active drug ingredient or therapeutic moiety, if the new formulation, new dosage form, or new salt or ester is intended to be comparable to or to meet any comparative labeling claims made in relation to the drug product that is the subject of an approved NDA.

(f) *Extended-release formulations.*
   (1) The purpose of an in vivo BA study involving a drug product for which an extended-release claim is made is to determine if all of the following conditions are met:
      (i) The drug product meets the extended-release claims made for it.
      (ii) The BA profile established for the drug product rules out the occurrence of any dose dumping.
      (iii) The drug product's steady-state performance is equivalent to a currently marketed nonextended-release or extended-release drug product that contains the same active drug ingredient or therapeutic moiety and that is subject to an approved full NDA.
      (iv) The drug product's formulation provides consistent PK performance between individual dosage units.
   (2) The reference material(s) for such a BA study shall be chosen to permit an appropriate scientific evaluation of the extended-release claims made for the drug product. The reference material shall be one of the following or any combination thereof:
      (i) A solution or suspension of the active drug ingredient or therapeutic moiety
      (ii) A currently marketed noncontrolled release drug product containing the same active drug ingredient or therapeutic moiety and administered according to the dosage recommendations in the labeling of the noncontrolled release drug product
      (iii) A currently marketed extended-release drug product subject to an approved full NDA containing the same active drug ingredient or therapeutic moiety and administered according to the dosage recommendations in the labeling proposed for the extended-release drug product
      (iv) A reference material other than one set forth in paragraph (f)(2)(i), (ii), or (iii) of this section that is appropriate for valid scientific reasons

(g) *Combination drug products.*
   (1) Generally, the purpose of an in vivo BA study involving a combination drug product is to determine if the rate and extent of absorption of each active drug ingredient or therapeutic moiety in the combination drug product is equivalent to the rate and extent of absorption of each active drug ingredient or therapeutic moiety administered concurrently in separate single-ingredient preparations.
   (2) The reference material in such a BA study should be two or more currently marketed, single-ingredient drug products, each of which containing one of the active drug

ingredients or therapeutic moieties in the combination drug product. The FDA may, for valid scientific reasons, specify that the reference material shall be a combination drug product that is the subject of an approved NDA.

(3) The FDA may permit a BA study involving a combination drug product to determine the rate and extent of absorption of selected, but not all, active drug ingredients or therapeutic moieties in the combination drug product. The FDA may permit this determination if the PK and the interactions of the active drug ingredients or therapeutic moieties in the combination drug product are well known and the therapeutic activity of the combination drug product is generally recognized to reside in only one of the active drug ingredients or therapeutic moieties, for example, ampicillin in an ampicillin–probenecid combination drug product.

(h) *Use of a placebo as the reference material.* Where appropriate or where necessary to demonstrate the sensitivity of the test, the reference material in a BA study may be a placebo if

(1) The study measures the therapeutic or acute pharmacological effect of the active drug ingredient or therapeutic moiety

(2) The study is a clinical trial to establish the safety and effectiveness of the drug product.

(i) *Standards for test drug product and reference material.*

(1) Both the drug product to be tested and the reference material, if it is another drug product, shall be shown to meet all compendial or other applicable standards of identity, strength, quality, and purity, including potency and, where applicable, content uniformity, disintegration times, and dissolution rates.

(2) Samples of the drug product to be tested shall be manufactured using the same equipment and under the same conditions as those used for full-scale production.

Guidelines on the Design of a Single-Dose In Vivo Bioavailability or Bioequivalence Study

(a) *Basic principles.*

(1) An in vivo BA or BE study should be a single-dose comparison of the drug product to be tested and the appropriate reference material conducted in normal adults.

(2) The test product and the reference material should be administered to subjects in the fasting state, unless some other approach is more appropriate for valid scientific reasons.

(b) *Study design.*

(1) A single-dose study should be crossover in design, unless a parallel design or another design is more appropriate for valid scientific reasons, and should provide for a drug elimination period.

(2) Unless some other approach is appropriate for valid scientific reasons, the drug elimination period should be either

(i) At least three times the half-life of the active drug ingredient or therapeutic moiety, or its metabolite(s), measured in the blood or urine

(ii) At least three times the half-life of decay of the acute pharmacological effect

(c) *Collection of blood samples.*

(1) When comparison of the test product and the reference material is to be based on blood concentration time curves, unless some other approach is more appropriate for valid scientific reasons, blood samples should be taken with sufficient frequency to permit an estimate of both

(i) The peak concentration in the blood of the active drug ingredient or therapeutic moiety, or its metabolite(s), measured

(ii) The total area under the curve (AUC) for a time period at least three times the half-life of the active drug ingredient or therapeutic moiety, or its metabolite(s), measured

(2) In a study comparing oral dosage forms, the sampling times should be identical.

(3) In a study comparing an intravenous dosage form and an oral dosage form, the sampling times should be those needed to describe both
   (i) The distribution and elimination phase of the intravenous dosage form
   (ii) The absorption and elimination phase of the oral dosage form
(4) In a study comparing drug delivery systems other than oral or intravenous dosage forms with an appropriate reference standard, the sampling times should be based on valid scientific reasons.

(d) *Collection of urine samples.* When comparison of the test product and the reference material is to be based on cumulative urinary excretion–time curves, unless some other approach is more appropriate for valid scientific reasons, samples of the urine should be collected with sufficient frequency to permit an estimate of the rate and extent of urinary excretion of the active drug ingredient or therapeutic moiety, or its metabolite(s), measured.

(e) *Measurement of an acute pharmacological effect.*
   (1) When comparison of the test product and the reference material is to be based on acute pharmacological effect–time curves, measurements of this effect should be made with sufficient frequency to permit a reasonable estimate of the total AUC for a time period at least three times the half-life of decay of the pharmacological effect, unless some other approach is more appropriate for valid scientific reasons.
   (2) The use of an acute pharmacological effect to determine BA may further require demonstration of dose-related response. In such a case, BA may be determined by comparison of the dose–response curves as well as the total area under the acute pharmacological effect–time curves for any given dose.

Guidance on the Design of a Multiple-Dose In Vivo Bioavailability Study

(a) *Basic principles.*
   (1) In selected circumstances, it may be necessary for the test product and the reference material to be compared after repeated administration to determine steady-state levels of the active drug ingredient or therapeutic moiety in the body.
   (2) The test product and the reference material should be administered to subjects in the fasting or nonfasting state, depending upon the conditions reflected in the proposed labeling of the test product.
   (3) A multiple-dose study may be required to determine the BA of a drug product in the following circumstances:
      (i) There is a difference in the rate of absorption but not in the extent of absorption.
      (ii) There is excessive variability in BA from subject to subject.
      (iii) The concentration of the active drug ingredient or therapeutic moiety, or its metabolite(s), in the blood resulting from a single dose is too low for accurate determination by the analytical method.
      (iv) The drug product is an extended-release dosage form.

(b) *Study design.*
   (1) A multiple-dose study should be crossover in design, unless a parallel design or other design is more appropriate for valid scientific reasons, and should provide for a drug elimination period if steady-state conditions are not achieved.
   (2) A multiple-dose study is not required to be of crossover design if the study is to establish dose proportionality under a multiple-dose regimen or to establish the PK profile of a new drug product, a new drug delivery system, or an extended-release dosage form.
   (3) If a drug elimination period is required, unless some other approach is more appropriate for valid scientific reasons, the drug elimination period should be either
      (i) At least five times the half-life of the active drug ingredient or therapeutic moiety, or its active metabolite(s), measured in the blood or urine
      (ii) At least five times the half-life of decay of the acute pharmacological effect

(c) *Achievement of steady-state conditions.* Whenever a multiple-dose study is conducted, unless some other approach is more appropriate for valid scientific reasons, sufficient doses of the test product and reference material should be administered in accordance with the labeling to achieve steady-state conditions.

(d) *Collection of blood or urine samples.*

(1) Whenever comparison of the test product and the reference material is to be based on blood concentration–time curves at steady state, appropriate dosage administration and sampling should be carried out to document attainment of steady state.

(2) Whenever comparison of the test product and the reference material is to be based on cumulative urinary excretion–time curves at steady state, appropriate dosage administration and sampling should be carried out to document attainment of steady state.

(3) A more complete characterization of the blood concentration or urinary excretion rate during the absorption and elimination phases of a single dose administered at steady state is encouraged to permit estimation of the total area under concentration–time curves or cumulative urinary excretion–time curves and to obtain PK information, for example, half-life or blood clearance, that is essential in preparing adequate labeling for the drug product.

(e) *Steady-state parameters.*

(1) In certain instances, for example, in a study involving a new drug entity, blood clearances at steady state obtained in a multiple-dose study should be compared to blood clearances obtained in a single-dose study to support adequate dosage recommendations.

(2) In a linear system, the area under the blood concentration–time curve during a dosing interval in a multiple-dose steady-state study is directly proportional to the fraction of the dose absorbed and is equal to the corresponding "zero to infinity" AUC for a single-dose study. Therefore, when steady-state conditions are achieved, a comparison of blood concentrations during a dosing interval may be used to define the fraction of the active drug ingredient or therapeutic moiety absorbed.

(3) Other methods based on valid scientific reasons should be used to determine the BA of a drug product having dose-dependent kinetics (nonlinear system).

(f) *Measurement of an acute pharmacological effect.* When comparison of the test product and the reference material is to be based on acute pharmacological effect–time curves, measurements of this effect should be made with sufficient frequency to demonstrate a maximum effect and a lack of significant difference between the test product and the reference material.

CORRELATION OF BIOAVAILABILITY WITH AN ACUTE PHARMACOLOGICAL EFFECT OR CLINICAL EVIDENCE

Correlation of in vivo BA data with an acute pharmacological effect or clinical evidence of safety and effectiveness may be required if needed to establish the clinical significance of a special claim, for example, in the case of an extended-release preparation.

ANALYTICAL METHODS FOR AN IN VIVO BIOAVAILABILITY OR BIOEQUIVALENCE STUDY

(a) The analytical method used in an in vivo BA or BE study to measure the concentration of the active drug ingredient or therapeutic moiety, or its active metabolite(s), in body fluids or excretory products, or the method used to measure an acute pharmacological effect shall be demonstrated to be accurate and of sufficient sensitivity to measure, with appropriate precision, the actual concentration of the active drug ingredient or therapeutic moiety, or its active metabolite(s), achieved in the body.

(b) When the analytical method is not sensitive enough to measure accurately the concentration of the active drug ingredient or therapeutic moiety, or its active metabolite(s), in body fluids or excretory products produced by a single dose of the test product, two or more single doses may be given together to produce higher concentration if the requirements of 320.31 are met.

INQUIRIES REGARDING BIOAVAILABILITY AND BIOEQUIVALENCE REQUIREMENTS AND REVIEW OF PROTOCOLS BY THE FDA

(a) The Commissioner of Food and Drugs strongly recommends that, to avoid the conduct of an improper study and unnecessary human research, any person planning to conduct a BA or BE study submit the proposed protocol for the study to the FDA for review prior to the initiation of the study.
(b) The FDA may review a proposed protocol for a BA or BE study and will offer advice with respect to whether the following conditions are met:
  (1) The design of the proposed BA or BE study is appropriate.
  (2) The reference material to be used in the BA or BE study is appropriate.
  (3) The proposed chemical and statistical analytical methods are adequate.
(c) (1) General inquiries relating to in vivo BA requirements and methodology shall be submitted to the FDA, Center for Drug Evaluation and Research, Office of Clinical Pharmacology, 10903 New Hampshire Ave, Silver Spring, MD 20993–0002.
  (2) General inquiries relating to BE requirements and methodology shall be submitted to the FDA, Center for Drug Evaluation and Research, Division of Bioequivalence (HFD-650), 7500 Standish Pl., Rockville, MD 20855–2773.

APPLICABILITY OF REQUIREMENTS REGARDING AN "INVESTIGATIONAL NEW DRUG APPLICATION"

(a) Any person planning to conduct an in vivo BA or BE study in humans shall submit an "IND application" if
  (1) The test product contains a new chemical entity as defined in 314.108(a) of this chapter
  (2) The study involves a radioactively labeled drug product
  (3) The study involves a cytotoxic drug product
(b) Any person planning to conduct a BA or BE study in humans using a drug product that contains an already approved, non-new chemical entity shall submit an IND if the study is one of the following:
  (1) A single-dose study in normal subjects or patients where either the maximum single or total daily dose exceeds that specified in the labeling of the drug product that is the subject of an approved NDA or ANDA.
  (2) A multiple-dose study in normal subjects or patients where either the single or total daily dose exceeds that specified in the labeling of the drug product that is the subject of an approved NDA or ANDA.
  (3) A multiple-dose study on an extended-release product on which no single-dose study has been completed.
(c) The provisions of parts 50, 56, and 312 of this chapter are applicable to any BA or BE study in humans conducted under an IND.
(d) A BA or BE study in humans other than one described in paragraphs (a) through (c) of this section is exempt from the requirements of part 312 of this chapter if the following conditions are satisfied:
  (1) If the study is one described under 320.38(b) or 320.63, the person conducting the study, including any contract research organization, must retain reserve samples

of any test article and reference standard used in the study and release the reserve samples to the FDA upon request, in accordance with, and for the period specified in 320.38.

(2) An in vivo BA or BE study in humans must be conducted in compliance with the requirements for institutional review set forth in part 56 of this chapter and informed consent set forth in part 50 of this chapter.

(3) The person conducting the study, including any contract research organization, must notify the FDA and all participating investigators of any serious adverse event observed during the conduct of the study as soon as possible but in no case later than 15 calendar days after becoming aware of its occurrence. Each report must be submitted on the FDA Form 3500A or in an electronic format that the FDA can process, review, and archive. The FDA will periodically issue guidance on how to provide the electronic submission (e.g., method of transmission, media, file formats, preparation and organization of files). Each report must bear prominent identification of its contents, that is, "BA/BE safety report." The person conducting the study, including any contract research organization, must also notify the FDA of any fatal or life-threatening adverse event from the study as soon as possible but in no case later than 7 calendar days after becoming aware of its occurrence. Each notification under this paragraph must be submitted to the director, Office of Generic Drugs in the Center for Drug Evaluation and Research at the FDA. Relevant follow-up information to a BA/BE safety report must be submitted as soon as the information is available and must be identified as such, that is, "follow-up BA/BE safety report." Upon request from the FDA, the person conducting the study, including any contract research organization, must submit to the FDA any additional data or information that the agency deems necessary, as soon as possible, but in no case later than 15 calendar days after receiving the request.

PROCEDURES FOR ESTABLISHING OR AMENDING A BIOEQUIVALENCE REQUIREMENT

(a) The FDA, on its own initiative or in response to a petition by an interested person, may propose and promulgate a regulation to establish a BE requirement for a product not subject to section 505(j) of the Act if it finds there is well-documented evidence that specific pharmaceutical equivalents or pharmaceutical alternatives intended to be used interchangeably for the same therapeutic effect
   (1) Are not bioequivalent drug products
   (2) May not be bioequivalent drug products based on the criteria set forth in 320.33
   (3) May not be bioequivalent drug products because they are members of a class of drug products that have close structural similarity and similar physicochemical or PK properties to other drug products in the same class that the FDA finds are not bioequivalent drug products

(b) The FDA shall include in a proposed rule to establish a BE requirement the evidence and criteria set forth in 320.33 that are to be considered in determining whether to issue the proposal. If the rulemaking is proposed in response to a petition, the FDA shall include in the proposal a summary and analysis of the relevant information that was submitted in the petition as well as other available information to support the establishment of a BE requirement.

(c) The FDA, on its own initiative or in response to a petition by an interested person, may propose and promulgate an amendment to a BE requirement established under this subpart.

CRITERIA AND EVIDENCE TO ASSESS ACTUAL OR POTENTIAL BIOEQUIVALENCE PROBLEMS

The Commissioner of Food and Drugs shall consider the following factors, when supported by well-documented evidence, to identify specific pharmaceutical equivalents and pharmaceutical alternatives that are not or may not be bioequivalent drug products:

(a) Evidence from well-controlled clinical trials or controlled observations in patients that such drug products do not give comparable therapeutic effects
(b) Evidence from well-controlled BE studies that such products are not bioequivalent drug products
(c) Evidence that the drug products exhibit a narrow therapeutic ratio, for example, there is less than a twofold difference in median lethal dose (LD50) and median effective dose (ED50) values, or have less than a twofold difference in the minimum toxic concentrations and minimum effective concentrations in the blood, and safe and effective use of the drug products requires careful dosage titration and patient monitoring
(d) Competent medical determination that a lack of BE would have a serious adverse effect in the treatment or prevention of a serious disease or condition
(e) Physicochemical evidence that
    (1) The active drug ingredient has a low solubility in water, for example, less than 5 mg/1 mL, or, if dissolution in the stomach is critical to absorption, the volume of gastric fluids required to dissolve the recommended dose far exceeds the volume of fluids present in the stomach (taken to be 100 mL for adults and prorated for infants and children)
    (2) The dissolution rate of one or more such products is slow, for example, less than 50% in 30 min when tested using either a general method specified in an official compendium or a paddle method at 50 rpm in 900 mL of distilled or deionized water at 37°C, or differs significantly from that of an appropriate reference material such as an identical drug product that is the subject of an approved full NDA
    (3) The particle size and/or surface area of the active drug ingredient is critical in determining its BA
    (4) Certain physical structural characteristics of the active drug ingredient, for example, polymorphic forms, conforms, solvates, complexes, and crystal modifications, dissolve poorly and this poor dissolution may affect absorption
    (5) Such drug products have a high ratio of excipients to active ingredients, for example, greater than 5–1
    (6) Specific inactive ingredients, for example, hydrophilic or hydrophobic excipients and lubricants, either may be required for absorption of the active drug ingredient or therapeutic moiety or, alternatively, if present, may interfere with such absorption
(f) PK evidence that
    (1) The active drug ingredient, therapeutic moiety, or its precursor is absorbed in large part in a particular segment of the gastrointestinal (GI) tract or is absorbed from a localized site
    (2) The degree of absorption of the active drug ingredient, therapeutic moiety, or its precursor is poor, for example, less than 50%, ordinarily in comparison to an intravenous dose, even when it is administered in pure form, for example, in solution
    (3) There is rapid metabolism of the therapeutic moiety in the intestinal wall or liver during the process of absorption (first-class metabolism) so the therapeutic effect and/or toxicity of such drug product is determined by the rate as well as the degree of absorption
    (4) The therapeutic moiety is rapidly metabolized or excreted so that rapid dissolution and absorption are required for effectiveness

(5) The active drug ingredient or therapeutic moiety is unstable in specific portions of the GI tract and requires special coatings or formulations, for example, buffers, enteric coatings, and film coatings, to assure adequate absorption
(6) The drug product is subject to dose-dependent kinetics in or near the therapeutic range, and the rate and extent of absorption are important to BE

REQUIREMENTS FOR BATCH TESTING AND CERTIFICATION BY THE FDA

(a) If the commissioner determines that individual batch testing by the FDA is necessary to assure that all batches of the same drug product meet an appropriate in vitro test, he or she shall include in the BE requirement a requirement for manufacturers to submit samples of each batch to the FDA and to withhold distribution of the batch until notified by the FDA that the batch may be introduced into interstate commerce.
(b) The commissioner will ordinarily terminate a requirement for a manufacturer to submit samples for batch testing on a finding that the manufacturer has produced four consecutive batches that were tested by the FDA and found to meet the BE requirement, unless the public health requires that batch testing be extended to additional batches.

REQUIREMENTS FOR IN VITRO TESTING OF EACH BATCH

If a BE requirement specifies a currently available in vitro test or an in vitro BE standard comparing the drug product to a reference standard, the manufacturer shall conduct the test on a sample of each batch of the drug product to assure batch-to-batch uniformity.

REQUIREMENTS FOR MAINTENANCE OF RECORDS OF BIOEQUIVALENCE TESTING

(a) All records of in vivo or in vitro tests conducted on any marketed batch of a drug product to assure that the product meets a BE requirement shall be maintained by the manufacturer for at least 2 years after the expiration date of the batch and submitted to the FDA on request.
(b) Any person who contracts with another party to conduct a BE study from which the data are intended to be submitted to the FDA as part of an application submitted under part 314 of this chapter shall obtain from the person conducting the study sufficient accurate financial information to allow the submission of complete and accurate financial certifications or disclosure statements required under part 54 of this chapter and shall maintain that information and all records relating to the compensation given for that study and all other financial interest information required under part 54 of this chapter for 2 years after the date of approval of the application. The person maintaining these records shall, upon request for any properly authorized officer or employee of the FDA, at reasonable time, permit such officer or employee to have access to and copy and verify these records.

RETENTION OF BIOAVAILABILITY SAMPLES

The applicant or contract research organization shall retain the reserve samples in accordance with, and for the period specified in, 320.38 and shall release the reserve samples to the FDA upon request in accordance with 320.38. These are described in detail in the chapter on managing CROs.

(a) The applicant of an application or supplemental application submitted under section 505 of the Federal Food, Drug, and Cosmetic Act (FDCA), or, if BA testing was performed under contract, the contract research organization shall retain an appropriately identified reserve sample of the drug product for which the applicant is seeking approval (test article) and of the reference standard used to perform an in vivo BA study in accordance with and for the studies described in paragraph (b) of this section that is representative of each sample of the test article and reference standard provided by the applicant for the testing.

(b) Reserve samples shall be retained for the following test articles and reference standards and for the studies described:
   (1) If the formulation of the test article is the same as the formulation(s) used in the clinical studies demonstrating substantial evidence of safety and effectiveness for the test article's claimed indications, a reserve sample of the test article is used to conduct an in vivo BA study comparing the test article to a reference oral solution, suspension, or injection.
   (2) If the formulation of the test article differs from the formulation(s) used in the clinical studies demonstrating substantial evidence of safety and effectiveness for the test article's claimed indications, a reserve sample of the test article and of the reference standard is used to conduct an in vivo BE study comparing the test article to the formulation(s) (reference standard) used in the clinical studies.
   (3) For a new formulation, new dosage form, or a new salt or ester of an active drug ingredient or therapeutic moiety that has been approved for marketing, a reserve sample of the test article and of the reference standard is used to conduct an in vivo BE study comparing the test article to a marketed product (reference standard) that contains the same active drug ingredient or therapeutic moiety.

(c) Each reserve sample shall consist of a sufficient quantity to permit the FDA to perform five times all of the release tests required in the application or supplemental application.

(d) Each reserve sample shall be adequately identified so that the reserve sample can be positively identified as having come from the same sample as used in the specific BA study.

(e) Each reserve sample shall be stored under conditions consistent with product labeling and in an area segregated from the area where testing is conducted and with access limited to authorized personnel. Each reserve sample shall be retained for a period of at least 5 years following the date on which the application or supplemental application is approved, or, if such application or supplemental application is not approved, at least 5 years following the date of completion of the BA study in which the sample from which the reserve sample was obtained was used.

(f) Authorized FDA personnel will ordinarily collect reserve samples directly from the applicant or contract research organization at the storage site during a preapproval inspection. If authorized FDA personnel are unable to collect samples, the FDA may require the applicant or contract research organization to submit the reserve samples to the place identified in the agency's request. If the FDA has not collected or requested delivery of a reserve sample, or if the FDA has not collected or requested delivery of any portion of a reserve sample, the applicant or contract research organization shall retain the sample or remaining sample for the 5-year period specified in paragraph (e) of this section.

(g) Upon release of the reserve samples to the FDA, the applicant or contract research organization shall provide a written assurance that, to the best knowledge and belief of the individual executing the assurance, the reserve samples came from the same samples as used in the specific BA or BE study identified by the agency. The assurance shall be executed by an individual authorized to act for the applicant or contract research organization in releasing the reserve samples to the FDA.

(h) A contract research organization may contract with an appropriate, independent third party to provide storage of reserve samples provided that the sponsor of the study has been notified in writing of the name and address of the facility at which the reserve samples will be stored.

(i) If a contract research organization conducting a BA or BE study that requires reserve sample retention under this section or 320.63 goes out of business, it shall transfer its reserve samples to an appropriate, independent third party and shall notify in writing the sponsor of the study of the transfer and provide the study sponsor with the name and address of the facility to which the reserve samples have been transferred.

BIOEQUIVALENCE STUDIES WITH PHARMACOKINETIC ENDPOINTS FOR DRUGS SUBMITTED UNDER AN ANDA: NEW GUIDANCE

Introduction

In December 2013, the U.S. FDA issued additional guidance on this very important subject. This new guidance provides recommendations to applicants planning to include BE information in ANDAs and ANDA supplements. The guidance describes how to meet the BE requirements set forth in the FDCA and FDA regulations. The guidance is generally applicable to dosage forms intended for oral administration and to non-orally administered drug products in which reliance on systemic exposure measures is suitable for documenting BE (e.g., transdermal delivery systems and certain rectal and nasal drug products). The FDA believes that the guidance will also be useful when planning BE studies intended to be conducted during the postapproval period for certain changes in an ANDA.

The new guidance revises and replaces parts of two FDA guidances for industry, relating to BE and fed BE studies to be submitted in ANDAs. The new guidance does not address BA, BE, and food-effect studies in IND applications and NDAs.

Background

To receive approval for an ANDA, an applicant generally must demonstrate, among other things, that its proposed drug product is bioequivalent to the RLD (or reference product). The FDCA provides that a generic drug is bioequivalent to the listed drug if the following criteria are met:

> The rate and extent of absorption of the drug do not show a significant difference from the rate and extent of absorption of the listed drug when administered at the same molar dose of the therapeutic ingredient under similar experimental conditions in either a single dose or multiple doses.

For most products, the focus of BE studies is on the release of the drug substance from the drug product into the systemic circulation. During such BE studies, an applicant compares the systemic exposure profile of a test drug product to that of the RLD.

Establishing Bioequivalence

Under FDA regulations, an applicant must use "the most accurate, sensitive, and reproducible approach available among those set forth" in 21 CFR 320.24(b) to demonstrate BE. (See 21 CFR 320.24(a).) As noted in 21 CFR 320.24, in vivo and/or in vitro methods can be used to establish BE. In general descending order of preference, these include PK, pharmacodynamic (PD), clinical, and in vitro studies. (See 21 CFR 320.24(a).)

Pharmacokinetic Studies

General Considerations

As provided previously, the statutory definition of BE, expressed in terms of rate and extent of absorption of the active ingredient or moiety, emphasizes the use of PK endpoints in an accessible biological matrix, such as blood, plasma, and/or serum, to indicate release of the drug substance from the drug product into the systemic circulation. BE frequently relies on PK endpoints such as C_{max} (peak plasma concentration) and AUC (area under the plasma concentration time curve) that are reflective of rate and extent of absorption, respectively.

If serial measurements of the drug or its metabolites in plasma, serum, or blood cannot be accomplished, measurement of urinary excretion can be used to demonstrate BE.

Pilot Study

If the applicant chooses, a pilot study in a small number of subjects can be carried out before proceeding with a full BE study. This pilot study can be used to validate analytical methodology, assess variability, optimize sample collection time intervals, and provide other information.

Pivotal Bioequivalence Studies

General recommendations for a standard BE study based on PK measurements are provided in the attachment.

Study Designs

The FDA recommends the use of a two-period, two-sequence, two-treatment, single-dose, crossover study design, a single-dose parallel study design, or a replicate study design for BE studies. For most dosage forms that release drug intended to be systemically available, we recommend that applicants perform a two-period, two-sequence, two-treatment, single-dose, crossover study using healthy subjects. In this design, each study subject should receive each treatment (test and RLD) in random order. The crossover design may not be practical for drugs with long PK half-lives (i.e., longer than 24 h). In such cases, investigators can use a single-dose, parallel design where each treatment should be administered to a separate group of subjects with similar demographics. The general recommendations for study designs provided in the attachment should be used in designing crossover studies as well.

A replicate crossover study may be an appropriate alternative to the parallel or nonreplicate crossover study described earlier and can be conducted as either a partial (three-way) or full (four-way) replication of treatment. In this design, one or both treatments should be administered to the same subject on two separate occasions. The replicate design has the advantage of using fewer subjects although each subject should receive more treatments than in the two-treatment, crossover design. The replicate design is especially useful for highly variable drugs.

The FDA recommends that applicants use the average BE method of analysis with these study designs for establishing BE. In limited cases, applicants may use a scaled-average BE analysis approach for highly variable drugs.

For highly variable drugs (intrasubject variability >30%), applicants can conduct BE studies using a replicate-design approach. Alternatively, a single-dose, randomized, three-period reference-scaled, average BE approach is also appropriate. The reference-scaled average BE approach adjusts the BE limits of highly variable drugs by scaling to the within-subject variability of the RLD in the study and imposes a limit of 0.8–1.25 on the geometric mean ratio. The within-subject variability of RLD should be determined using a three-way modified replicate-design study in which the RLD is given twice and the test product is given once. For general information on the reference-scaled approach, investigators should refer to the published book chapter, Davit and Conner (2010).

This analysis approach is typically used with a replicate study design. Recommendations for replicate study designs and the average BE approach method can be found in the guidance for industry on Statistical Approaches to Establishing Bioequivalence. For applicants wishing to use variations of these study designs or analysis methods (e.g., a sequential design or scaled-average BE), we recommend that you submit a complete protocol for review and comment before starting the study.

Study Population

In general, unless otherwise recommended in a specific guidance,

- Subjects recruited for in vivo BE studies should be 18 years of age or older
- In vivo BE study subjects should be representative of the general population, taking into account age, sex, and race

- If a drug product is intended for use in both sexes, the applicant should include similar proportion of males and females in the study
- If the drug product is predominantly intended for use in the elderly, the applicant should include as many subjects as possible at or over age 60
- The total number of subjects in a study should be sufficient to provide adequate statistical power for BE demonstration, but the FDA does not expect that there will be sufficient power upon which to draw conclusions for each subgroup

In most cases, the FDA does not recommend statistical analysis of subgroups.

We also recommend that any restrictions on admission into a study be based primarily on safety considerations. Sometimes, safety considerations preclude the use of healthy volunteers. In such situations, applicants should attempt to enroll patients that the drug is intended to treat and whose disease process and treatments are stable for the duration of the BE study. An IND for certain BE studies may be required, for example, for cytotoxic products.

Single-Dose Studies

We usually recommend single-dose PK studies for both immediate- and modified-release (MR) drug products to demonstrate BE because these studies are generally more sensitive than a steady-state study in assessing differences in the release of the drug substance from the drug product into the systemic circulation.

Steady-State Studies

When safety considerations suggest using patients who are already receiving the medication, often the only way to establish BE without disrupting a patient's ongoing treatment is in a steady-state study. The FDA recommends that if a steady-state study is recommended, applicants carry out appropriate dosage administration and sampling to document the attainment of steady state.

Bioanalytical Methodology

The FDA recommends that applicants ensure that bioanalytical methods for BE studies are accurate, precise, selective, sensitive, and reproducible. A separate draft guidance for industry on Bioanalytical Method Validation is available to assist applicants in validating bioanalytical methods.

Pharmacokinetic Measure of Rate and Extent of Exposure

a. Rate of Absorption (Peak Exposure)

   For both single-dose and steady-state studies, we recommend that you assess the rate of absorption by measuring the peak drug concentration (C_{max}) obtained directly from the data without interpolation. The time-to-peak drug plasma concentration (t_{max}) can also provide important information regarding the rate of absorption.

b. Partial Exposure

   For orally administered immediate-release drug products, BE can generally be demonstrated by measurements of peak and total exposure. The FDA recommends the use of partial AUC as an early exposure measure under certain circumstances. The time to truncate the partial area should be related to a clinically relevant PD measure. The FDA recommends that sufficient quantifiable samples be collected to allow adequate estimation of the partial area. For further information on specific products, applicants should consult our website to determine whether a product-specific guidance for the proposed product is available.

c. Extent of Absorption (Total Exposure)
For single-dose studies, we recommend that the indicators for extent of absorption be both of the following:

- Area under the plasma/serum/blood concentration–time curve from time zero to time t (AUC_{0-t}), where t is the last time point with a measurable concentration
- Area under the plasma/serum/blood concentration–time curve from time zero to time infinity (AUC_{0-inf}), where $AUC_{0-inf} = AUC_{0-t} + C_t/\lambda z$, C_t is the last measurable drug concentration, and λz is the terminal or elimination rate constant calculated according to an appropriate method

For steady-state studies, we recommend that the indicator for extent of absorption be the area under the plasma, serum, or blood concentration–time curve over a dosing interval at steady state (AUC_{0-tau}), where tau is the length of the dosing interval.

Fed Bioequivalence Studies

Coadministration of food with oral drug products can influence BE. Therefore, fed BE studies can determine whether test and RLD products are bioequivalent when coadministered with meals. We usually recommend a single-dose, two-period, two-treatment, two-sequence, crossover study for fed BE studies. See attachment for details on study design.

When a fasting in vivo BE study is recommended for an orally administered, immediate-release product, we recommend that applicants conduct a fed study, except when the dosage and administration section of the RLD labeling states that the product should be taken only on an empty stomach (e.g., the labeling states that the product should be administered 1 h before or 2 h after a meal).

For orally administered, immediate-release products labeled to be taken only with food, fasting and fed studies are recommended, except when serious adverse events are anticipated with fasting administration. In these latter cases, we recommend that applicants conduct only a fed study; a fasting study is not recommended.

For all orally administered, MR drug products, we recommend that applicants conduct a fed BE study in addition to a fasting BE study. These studies should usually be conducted on the highest strength of the drug product, unless safety considerations preclude the use of that dose in study subjects.

Sprinkle Bioequivalence Studies

If the label of an MR RLD product states that the product can be administered sprinkled in soft foods, we recommend applicants conduct an additional BE study. For each treatment arm, the product should be sprinkled on one of the soft foods mentioned in the labeling of the RLD, normally applesauce. Aside from administration in the soft food, this additional study should follow the recommendations for the fasting BE study described in Appendix B.

Bioequivalence Studies of Products Administered in Specific Beverages

There are certain products with labeling that specifies that the product must be administered in a specific beverage. BE studies for these products should be administered mixed with one of the beverages mentioned in the labeling. If additional beverages are listed, applicants should provide evidence that using these additional beverages would not result in BE differences.

If there are questions about the use of other vehicles, or the design or analysis of such BE studies, applicants should contact the appropriate staff in the agency's OGD.

General Considerations on Other Bioequivalence Studies

In certain circumstances, other BE studies are recommended to support a demonstration of BE. In the following are some general considerations regarding these other BE studies. Sponsors should consult FDA's guidances for industry for additional information on these methods as well.

In Vitro Tests Predictive of Human In Vivo Bioavailability (In Vitro–In Vivo Correlation Studies)

In vitro–in vivo correlation (IVIVC) is a scientific approach to describe the relationship between an in vitro attribute of a dosage form (e.g., the rate or extent of drug release) and a relevant in vivo response (e.g., plasma drug concentration or amount of drug absorbed). This model relationship facilitates the rational development and evaluation of extended-release dosage forms as a surrogate for BA and/or BE testing, as well as a tool for formulation screening and setting of the dissolution/drug release acceptance criteria.

Additional information specifically on the development and validation of an IVIVC can be found in the guidance for industry on Extended Release Oral Dosage Forms: Development, Evaluation, and Application of In Vitro/In Vivo Correlations.

Pharmacodynamic

A suitably validated PD method can be used to demonstrate BE. However, the FDA does not recommend PD studies for drug products that are intended to be absorbed into the systemic circulation and for which a PK approach can be used to establish BE.

Comparative Clinical Studies

When it is not possible to use the previously described methods, well-controlled BE studies with clinical endpoints in patients can be used to establish BE.

In Vitro Studies

Under certain circumstances, BE can be evaluated using in vitro approaches (e.g., dissolution/drug release testing) under 21 CFR 320.24(b). The FDA does not recommend in vitro approaches for drug products that are intended to be systemically absorbed. Such approaches would be appropriate, however, in other circumstances (e.g., for drug products that bind bile acids in the GI tract).

Establishing Bioequivalence for Different Dosage Forms

The following sections provide recommendations for establishing BE for specific dosage forms. As explained in the following, in certain cases BE testing may be waived.

Oral Solutions

For oral solutions, elixirs, syrups, tinctures, or other solubilized forms, an in vivo BE testing requirement may be waived for certain products on the ground that in vivo BE is self-evident. In such instances, the applicant would be deemed to have complied with and fulfilled any requirement for in vivo BE data. (See 21 CFR 320.22(b)(3).) For example, BE can be waived for an oral solution if the formulation has the same active ingredient in the same concentration and dosage form as the RLD, and does not contain any excipient that significantly affects drug absorption or availability. (See 21 CFR 320.22(b)(3).)

Immediate-Release Products: Capsules and Tablets

Preapproval

For immediate-release capsule and tablet products, we recommend the following studies: (1) a single-dose, fasting study comparing the highest strength of the test and RLD products and (2) a single-dose, fed BE study comparing the highest strength of the test and RLD products (see "Fed bioequivalence studies" section).

Conducting an in vivo study on a strength other than the highest may be appropriate for reasons of safety, with concurrence by the Division of Bioequivalence, OGD, if the following conditions are met:

- Linear elimination kinetics has been documented over the therapeutic dose range.
- The higher strengths of the test and RLD products are proportionally similar to their corresponding lower strength.
- Comparative dissolution testing on the higher strength of the test and RLD products has been submitted and found to be acceptable.

An in vivo BE requirement for one or more strength(s) can be waived based on (i) acceptable BE study on the designated strength, (ii) acceptable in vitro dissolution testing of all the strengths, and (iii) proportional similarity of the formulations across all strengths.

The new guidance defines proportionally similar in the following ways:

- All active and inactive ingredients are in similar proportion between different strengths (e.g., a tablet of 50 mg strength has all the inactive ingredients—almost exactly half that of a tablet of 100 mg strength and almost twice that of a tablet of 25 mg strength).
- For high-potency drug substances (where the amount of active drug substance in the dosage form is relatively low), (1) the total weight of the dosage form remains nearly the same for all strengths (within +10% of the total weight of the strength on which a biostudy was performed), (2) the same inactive ingredients are used for all strengths, and (3) the change in any strength is obtained by altering the amount of the active ingredients and one or more of the inactive ingredients.
- Active and inactive ingredients that are not in similar proportion between different strengths can be considered proportionally similar with adequate justification (such as dosage form proportionality studies that demonstrate equivalent in vivo BA).

Under any of these scenarios, we recommend that in vivo BE studies be accompanied by in vitro dissolution profiles on all strengths of each product. We also recommend that applicants conduct the BE study comparing the test product and the RLD using the strength(s) specified in *Approved Drug Products with Therapeutic Equivalence Evaluations* (commonly referred to as the Orange Book; http://www.fda.gov/cder/orange/default.htm).

In addition, for highly soluble, highly permeable, rapidly dissolving, and orally administered immediate-release drug products, in vitro data may be acceptable to demonstrate BE based on the biopharmaceutics classification system as described in the guidance for industry on Waiver of In Vivo Bioavailability and Bioequivalence Studies for Immediate-Release Solid Oral Dosage Forms Based on a Biopharmaceutics Classification System.

For additional information on BE study design for a specific product, we recommend that applicants consult our website to determine whether a product-specific guidance for your proposed product is available.

Postapproval

Please refer to the guidance for industry on Immediate Release Solid Oral Dosage Forms, Scale-Up and Postapproval Chemistry, Manufacturing, and Controls; In Vitro Dissolution Testing and In Vivo Bioequivalence Documentation for information regarding bioequivalence testing recommended for specified types of postapproval changes. (In such instances, we anticipate that such approach will be adequate to demonstrate BE. See 21 CFR 320.24(b)(6).)

For postapproval changes, we recommend that applicants make the in vitro comparison between the prechange and postchange products. When in vivo BE studies are recommended to support a postapproval change for an ANDA product, the FDA recommends that applicants compare the postchange ANDA drug product to the RLD and not to the prechange ANDA product.

Suspensions

We generally recommend that you establish BE for a suspension in the same manner as for other solid oral dosage forms. In vivo studies and dissolution testing should be performed as described in "Immediate-release products: capsules and tablets" section or in "Modified-release products" section.

Modified-Release Products

MR products include delayed-release products and extended-release (controlled-release or sustained-release) products.

Delayed-Release Products

A delayed-release drug product is a dosage form that releases a drug at a time later than immediately after administration (e.g., the drug product exhibits a lag time in quantifiable plasma concentrations). Typically, the coatings (e.g., enteric coatings) have been designed to delay the release of medication until the dosage form has passed through the acidic medium of the stomach. In vivo tests for delayed-release drug products are similar to those for extended-release drug products. The FDA recommends that in vitro dissolution tests for these products document that they are stable under acidic conditions and that they release the drug only in a neutral medium (e.g., pH 6.8).

Extended-Release Products

An extended-release drug product is a dosage form that allows a reduction in dosing frequency and reduces fluctuations in plasma concentrations when compared to an immediate-release dosage form. Extended-release products can be formulated as capsules, tablets, granules, pellets, or suspensions. If any part of a drug product includes an extended-release component, the product should be treated as an MR dosage form for the purposes of establishing BE, as specified in the following.

Bioequivalence Studies

For MR products, we recommend the following studies: (1) a single-dose, fasting study comparing the highest strength of the test with the RLD and (2) a single-dose, fed BE study comparing the highest strength of the test with the RLD product. Because single-dose studies are considered more sensitive in addressing the primary question of BE (e.g., release of the drug substance from the drug product into the systemic circulation), multiple-dose studies are generally not recommended.

Demonstration of Bioequivalence: Additional Strengths

Additional strengths of MR products may be demonstrated to be bioequivalent to the corresponding reference product strengths under 21 CFR 320.24(b)(6) if all of the following conditions have been met:

- The additional strength is proportionally similar in its active and inactive ingredients to the test product strength that underwent acceptable in vivo studies.
- The additional strength has the same drug release mechanism as the strength of the test product that underwent an acceptable in vivo study.
- Dissolution testing of all strengths is acceptable. The FDA recommends that the drug products exhibit similar dissolution profiles between the strength on which BE testing was conducted and other strengths based on the f_2 test in at least three dissolution media (e.g., pH 1.2, 4.5, and 6.8). (In such instances, we anticipate that such approach will be adequate to demonstrate BE. See 21 CFR 320.24(b)(6).)

The FDA recommends that applicants generate dissolution profiles on the test and RLD products of all strengths.

Postapproval Changes

The applicant should refer to FDA's guidance for industry SUPAC: Modified Release Solid Oral Dosage Forms, Chemistry Manufacturing and Controls; In Vitro Dissolution Testing and In Vivo Bioequivalence Documentation for information regarding BE testing recommended for specified types of postapproval changes for MR dosage forms.

For postapproval changes, we recommend that applicants make an in vitro comparison between the approved (prechange) product and the test (postchange) product. If appropriate, we recommend that you use an f_2 test to compare dissolution profiles. An in vivo BE study may be needed if dissolution profiles are not shown to be similar. When in vivo BE studies are recommended to support a postapproval change for an ANDA product, the FDA recommends that applicants compare the postchange ANDA drug product to the RLD and not to the prechange ANDA product.

Chewable Tablets

Applicants should administer chewable tablets according to the directions on the label. If the label states that the tablet must be chewed before swallowing, the product should be chewed when administered in BE studies. If the label gives the option of either chewing the product or swallowing it whole, the product should be swallowed whole, with 240 mL of water, when administered in BE studies. We also recommend that you conduct in vitro dissolution testing on intact, whole tablets of the chewable drug product.

SPECIAL TOPICS

There are a number of topics that may call for special consideration addressed in the following sections. Additional questions should be referred to OGD.

Moieties to Be Measured

Parent Drug versus Metabolites

The parent drug in the dosage form should always be measured in the biological fluids collected in BE studies, unless accurate assay quantitation is not possible using state-of-the-art technology. We generally recommend that applicants measure only the parent drug, rather than metabolites, because the concentration–time profile of the parent drug is more sensitive to changes in formulation performance than a metabolite, which is more reflective of metabolite formation, distribution, and elimination. Primary metabolite(s), formed directly from the parent compound, should be measured if they are both (1) formed substantially through presystemic metabolism (first-pass, gut wall, or gut lumen metabolism) and (2) contribute significantly to the safety and efficacy of the product. This approach should be used for all drug products, including prodrugs. The FDA recommends that applicants analyze the parent drug measured in these BE studies using a confidence interval (CI) approach. You can use the metabolite data to provide supportive evidence of a comparable therapeutic outcome.

If the parent drug levels are too low to allow reliable analytical measurement in blood, plasma, or serum for an adequate length of time, the metabolite data obtained from these studies should be subject to the CI approach for BE demonstration.

Enantiomers versus Racemates

For BE studies, we recommend using an achiral assay to measure the racemate. We only recommend measuring individual enantiomers in BE studies when all of the following conditions have been met: (1) the enantiomers exhibit different PD characteristics, (2) the enantiomers exhibit different PK characteristics, (3) primary efficacy and safety activity reside with the minor enantiomer, and (4) nonlinear absorption is present (as expressed by a change in the enantiomer concentration ratio with change in the input rate of the drug) for at least one of the enantiomers. In such cases where all of these conditions are met, we recommend that applicants apply BE analysis to the enantiomers separately.

Drug Products with Complex Mixtures as the Active Ingredients

Certain drug products contain complex drug substances (e.g., active moieties or active ingredients that are mixtures of multiple synthetic and/or natural source components). Some or all of the components of these complex drug substances cannot be fully characterized with regard to chemical structure and/or biological activity. We do not encourage quantification of all active or potentially active components in PK studies. Rather, we recommend that applicants base BE studies on a small number of markers of rate and extent of absorption. Selection of the markers should be based on the characteristics of the drug product. Criteria for marker selection can include amount of the moiety in the dosage form, plasma, or blood levels of the moiety and biological activity of the moiety relative to other moieties in the complex mixture.

Long Half-Life Drugs

For an oral immediate-release product with a long elimination half-life drug (>24 h), applicants can conduct a single-dose, crossover study, provided an adequate washout period is used. If the crossover study is problematic, applicants should use a BE study with a parallel design. For either a crossover or parallel study, sample collection time should be adequate to ensure completion of GI transit of the drug product and absorption of the drug substance (which usually occurs within approximately 2–3 days). You can use C_{max} and a suitably truncated AUC to characterize peak and total drug exposure, respectively. For drugs that demonstrate low intrasubject variability in distribution and clearance, you can use an AUC truncated at 72 h ($AUC_{0-72\,h}$) in place of AUC_{0-t} or AUC_{0-inf}. For drugs demonstrating high intrasubject variability in distribution and clearance, AUC truncation should not be used.

First Point C_{MAX}

The first point of a concentration–time curve in a BE study, based on blood and/or plasma measurements, is sometimes the highest point, which raises questions of bias in the estimation of C_{max} because of insufficient early sampling times. A carefully conducted pilot study can enable an applicant to avoid this problem.

In the main BE study, collection of blood samples at an early time point, between 5 and 15 min after dosing, followed by additional sample collections (e.g., two to five) in the first hour after dosing is usually sufficient to assess peak drug concentrations. Failure to include early (5–15 min) sampling times leading to first time point C_{max} values may result in FDA not considering the data for affected subjects from the analysis.

Alcoholic Beverage Effects on Modified-Release Drug Products

The consumption of alcoholic beverages can affect the release of a drug substance from an MR formulation. The formulation can lose its MR characteristics, leading to more rapid drug release and altered systemic exposure. This can have deleterious effects on the drug's safety and/or efficacy.

The FDA recommends applicants developing certain extended-release solid oral dosage forms to conduct in vitro studies to determine the potential for dose dumping in alcohol in vivo. In vitro assessments of the drug release from the drug product using media with various alcohol concentrations may be recommended. An in vivo BE study of the drug product when administered with alcohol may be suggested in some cases. For information on specific products, we recommend that applicants consult the guidance for industry on Individual Product Bioequivalence Recommendations and any available relevant product-specific guidance.

Endogenous Compounds

Endogenous compounds are drugs that are already present in the body either because the body produces them or they are present in the normal diet. Because these compounds are identical to the drug that is being administered, determining the amount of drug released from the dosage form and absorbed by each subject can be difficult. The FDA recommends that applicants measure and approximate the baseline endogenous levels in blood (plasma) and subtract these levels from the total concentrations measured from each subject after the drug product has been administered. In this way, you can achieve an estimate of the actual drug availability from the drug product. Depending on whether the endogenous compound is naturally produced by the body or is present in the diet, the recommended approaches for determining BE differ as follows:

- When the body produces the compound, we recommend that you measure multiple baseline concentrations in the time period before administration of the study drug and subtract the baseline in an appropriate manner consistent with the PK properties of the drug.
- When there is dietary intake of the compound, we recommend that you strictly control the intake both before and during the study. Subjects should be housed at a clinic before the study and served standardized meals containing an amount of the compound similar to that in the meals to be served on the PK sampling day.

For both of the previous approaches, we recommend that you determine baseline concentrations for each dosing period that are period specific. If a baseline correction results in a negative plasma concentration value, the value should be set equal to 0 before calculating the baseline-corrected AUC. PK and statistical analysis should be performed on both uncorrected and corrected data. Determination of BE should be based on the baseline-corrected data.

Orally Administered Drugs Intended for Local Action

In some cases, when a drug substance produces its effects by local action in the GI tract, it may be appropriate to determine BE using PK endpoints. In other cases, it may be appropriate to determine BE using clinical endpoints, PD endpoints, and/or suitably designed and validated in vitro studies in addition to, or instead of, measuring drug plasma concentrations. For information on specific products, we recommend that applicants consult the guidance for industry on Bioequivalence Recommendations for Specific Products and any available relevant product-specific guidance.

In Vitro Dissolution Testing

The following guidances for industry provide recommendations on the development of dissolution methodology, setting specifications, and the regulatory applications of dissolution testing:

- Dissolution Testing of Immediate Release Solid Oral Dosage Forms
- Extended Release Oral Dosage Forms: Development, Evaluation, and Application of In Vitro/In Vivo Correlations

Immediate-Release Products

For immediate-release drug products, we recommend that applicants submit the method set forth in any related official *U.S. Pharmacopeia* (USP) drug product monograph. If there is not an official monograph for your proposed product, we recommend that you use the FDA-recommended methods and those described in the USP general chapter on dissolution (USP General Chapter <711> Dissolution). A dissolution methods database describing the FDA-recommended and USP methods is available to the public on the following website: http://www.accessdata.fda.gov/scripts/cder/dissolution/index.cfm.

If you choose to develop a new dissolution method, we recommend that you include the following information in the submission:

- The pH solubility profile of the drug substance.
- Dissolution profiles generated at different agitation speeds (e.g., 100–150 rpm) for USP Apparatus I (basket) or 50–100 rpm for USP Apparatus II (paddle).
- Dissolution profiles generated on all strengths in at least three dissolution media (e.g., pH 1.2, 4.5, and 6.8 buffer). Water can be used as an additional medium. If the drug being considered is poorly soluble, we recommend using appropriate concentrations of surfactants.

Modified-Release Products

For MR products, dissolution profiles using the method set forth in the official USP drug product monograph for the proposed product can be submitted. If there is not a USP drug product monograph for your proposed product, we recommend that applicants either use the FDA-recommended method (see the dissolution methods database mentioned earlier) or develop a method that is specific for your product. In addition, we recommend that you submit profiles using the methods described in the USP general chapter on dissolution or FDA methods in addition to those three described earlier (e.g., pH 1.2, 4.5 buffer, and 6.8 buffer). If you are proposing a method different from the FDA-recommended or USP method, we recommend that you submit data using the FDA-recommended or USP method in addition to your proposed method for comparison.

The applicant should select the agitation speed and medium that provide adequate discriminating ability, taking into account all the available in vitro and in vivo data.

The FDA recommends that you use dissolution data from three newly manufactured batches of test product to set dissolution specifications for MR dosage forms.

ATTACHMENT: GENERAL DESIGN AND DATA HANDLING OF BIOEQUIVALENCE STUDIES WITH PHARMACOKINETIC ENDPOINTS

For both replicate and nonreplicate in vivo PK BE studies, we recommend the following general approaches. Elements can be adjusted for certain drug substances and drug products.

Study Conduct

- The test or RLD products can be administered with about 8 ounces (240 mL) of water to an appropriate number of subjects under fasting conditions, unless the study is a fed BE study.
- Fed treatments: the FDA recommends that subjects start the recommended meal 30 min before administration of the drug product following an overnight fast of at least 10 h. Study subjects should eat this meal in 30 min or less and the drug product should be administered 30 min after start of the meal. The drug product should be administered with 8 fluid ounces (240 mL) of water.
- No food should be allowed for at least 4 h postdose. Water will be allowed as desired except for 1 h before and after drug administration. Subjects should receive standardized meals scheduled at the same time in each period of the study.
- Generally, the highest marketed strength can be administered as a single unit. If warranted to achieve sufficient bioanalytical sensitivity, multiple units of the highest strength can be administered, provided the total single dose remains within the labeled dose range and the total dose is safe for administration to the study subjects.
- An adequate washout period (e.g., more than five half-lives of the moieties to be measured) should separate each treatment.

- The lot numbers of both test and RLD products and the expiration date for the RLD product should be stated. The FDA recommends that the assayed drug content of the test product batch not differ from the RLD product by more than ±5%. The applicant should include a statement of the composition of the test product and, if possible, a side-by-side comparison of the compositions of test and RLD products. In accordance with 21 CFR 320.63, study drug test article of the test and RLD products must be retained for 5 years. For additional information, the applicant should refer to the guidance for industry on Handling and Retention of Bioavailability and Bioequivalence Testing Samples.
- Before and during each study phase, we recommend that subjects (1) be allowed water as desired, except for 1 h before and after drug administration, (2) be provided standardized meals no less than 4 h after drug administration, and (3) abstain from alcohol for 24 h before each study period and until after the last sample from each period has been collected.

Fed Studies Test Meal Composition

The FDA recommends that applicants conduct fed BE studies using meals that provide the greatest effects on GI physiology and systemic drug availability. The FDA recommends a high-fat (approximately 50% of total caloric content of the meal), high-calorie (approximately 800–1000 cal) test meal for fed BE studies. This test meal should derive approximately 150, 250, and 500–600 cal from protein, carbohydrate, and fat, respectively.

(An example test meal would be two eggs fried in butter, two strips of bacon, two slices of toast with butter, four ounces of hash brown potatoes, and eight ounces of whole milk. Substitutions in this test meal [e.g., beef or chicken instead of bacon] can be made as long as the meal provides a similar amount of calories from protein, carbohydrate, and fat and has comparable meal volume, density, and viscosity. In addition, the FDA routinely publishes guidances on BE study design for specific products; these should be consulted.)

The caloric breakdown of the test meal should be provided in the study report.

Sample Collection and Sampling Times

The FDA recommends that under normal circumstances, applicants sample blood, rather than urine or tissue. In most cases, drug or metabolites are measured in serum or plasma. However, in certain cases, whole blood may be more appropriate for analysis. The FDA recommends drawing blood samples at appropriate times to describe the absorption, distribution, and elimination phases of the drug. For most drugs, we recommend collecting 12–18 samples, including a predose sample, per subject, per dose. This sampling should continue for at least three or more terminal elimination half-lives of the drug. The exact timing for sample collection depends on the nature of the drug and the rate of input from the administered dosage form. The sample collection can be spaced in such a way that the maximum concentration of drug in the blood (C_{max}) and terminal elimination rate constant (K_{el}) can be estimated accurately. At least three to four samples should be obtained during the terminal log-linear phase to obtain an accurate estimate of λz from linear regression. The FDA recommends recording the actual clock time when samples are drawn as well as the elapsed time related to drug administration.

Subjects with Predose Plasma Drug Concentrations

If the predose concentration is ≤5% of C_{max} value in a subject with predose plasma concentration, you can include the subject's data without any adjustments in all PK measurements and calculations. The FDA recommends that if the predose value is greater than 5% of C_{max}, you drop the subject from all BE study evaluations.

Data Deletion Because of Vomiting

The FDA recommends that data from subjects who experience emesis during the course of a BE study for immediate-release products be deleted from statistical analysis if vomiting occurs at or before two times median t_{max}. For MR products, we recommend deleting data from the analysis if a subject vomits during a period of time less than or equal to the dosing interval stated in the labeling of the product.

FDA Recommends Applicants Provide the Following Pharmacokinetic Information in Their Submissions

- Plasma concentrations and time points.
- Subject, period, sequence, treatment.
- Intersubject, intrasubject, and/or total variability, if available.
- For single-dose BE studies: AUC_{0-t}, AUC_{0-inf}, and C_{max}. In addition, the applicant should report the following supportive information: t_{max}, K_{el} and $t_{1/2}$.
- For steady-state BE studies: AUC_{0-tau} and $C_{max,SS}$. In addition, the applicant should report $C_{min,SS}$ (concentration at the end of a dosing interval), C_{avSS} (average concentration during a dosing interval), degree of fluctuation $[(C_{max}-C_{min})/C_{avSS}]$, swing $[(C_{max,SS}-C_{min,SS})/C_{min,SS}]$, and t_{max}.

The FDA recommends applicants provide the following statistical information for AUC_{0-t}, AUC_{0-inf}, and C_{max}:

- Geometric means
- Arithmetic means
- Geometric mean ratios
- 90% CIs

We also recommend that you provide logarithmic transformation for measures used for BE demonstration.

Rounding Off CI Values

The FDA recommends that applicants not round off CI values; therefore, to pass a CI limit of 80%–125%, the value would be at least 80.00% and not more than 125.00%.

15 Bioequivalence Testing
European Perspective

BACKGROUND

European Legislation

Whereas most of the innovations in legislation to promote generic drugs have come first in the United States, the second largest market, Europe, has recently become more active in establishing pathways for approval of safe drugs.

The first European pharmaceutical directive, 65/65/EEC, was brought into force on January 26, 1965. It aimed to establish and maintain a high level of protection for public health and required prior approval for marketing of originator medicinal products. Much of the impetus behind Directive 65/65/EEC stemmed from determination to prevent a recurrence of the thalidomide disaster in the early 1960s, when thousands of babies were born with limb deformities as a result of their mothers taking thalidomide as a sedative during pregnancy. This experience, which shook public health authorities and the general public, made it clear that to safeguard public health, no medicinal product must ever again be marketed without prior authorization.

A decade later, two landmark directives 75/318/EEC and 75/319/EEC sought to bring the benefits of innovative pharmaceuticals to patients across the European community by introducing a procedure for mutual recognition, by member states, of their respective national marketing authorizations (MAs). To facilitate mutual recognition, Directive 75/319/EEC established a Committee for Proprietary Medicinal Products (CPMP), which first assessed whether candidate products complied with Directive 65/65/EEC.

The council adopted directives in 1992 on the wholesale distribution, classification for supply, labeling and packaging, and advertising of medicinal products for human use. The European Union (EU) also introduced pharmacovigilance (the surveillance of the safety of a medicinal product during its life on the market), requiring member states to establish national systems to collect and evaluate information on adverse reactions to medicinal products and to take appropriate action where necessary.

A new European system for authorizing medicinal products came into effect in January 1995 (via Regulation EEC/2309/93 and Directive 93/41/EEC) along with the establishment of the new European Medicines Evaluation Agency (EMEA). It offered two routes for authorizing medicinal products: a *centralized* procedure, through the EMEA (now the European Medicines Agency [EMA]), and a "mutual recognition" procedure through which applications are made to the member states selected by the applicant, and the procedure operates by mutual recognition of the national MA. Additionally, updates to the requirements relating to the placement on the market of high-technology medicinal products, particularly those derived from biotechnology, were put in place.

The newest piece of major legislation in Europe is the Falsified Medicines Directive (2011/62/EU), effective on January 2, 2013, which aims to protect European consumers against the threat of falsified medicines that might contain ingredients, including active ingredients, not indicated on the labeling, are of poor quality, or are in the incorrect dose—either too high or too low. As they have not been properly evaluated to check their quality, safety, and efficacy, they are potentially detrimental to public health and safety. The term "falsified" is used to distinguish from the infringement of intellectual property rights, so-called counterfeits. As falsifications become more sophisticated, the risk that these products reach patients in the EU increases every year.

The EMA has advanced significantly, more particularly in the field of approval of biological drugs as biosimilars, and have recently approved the first monoclonal antibody as biosimilar; the United States is far behind in its legislation and workability of a system suitable for the approval of biosimilar drugs.

The legal situation regarding authorization of pharmaceutical products in the EU is more complex than in the United States, with each member state having a competent authority in addition to the EMA, which oversees EU-wide authorization of medicines.

The EMA defines a generic medicine as a medicine that is developed to be the same as a medicine that has already been authorized (the "reference medicine"). A generics medicine contains the same active substance(s) as the reference medicine, and it is used at the same dose(s) to treat the same disease(s) as the reference medicine. However, the name of the medicine, its appearance (such as color or shape), and its packaging can be different from those of the reference medicine.

Authorization of a medicine in the EU can be done via three different routes: the centralized procedure (CP), the decentralized procedure (DCP), or the mutual recognition procedure (MRP). Additionally, national procedures (NPs) are in place in individual member states, which allow a medicine to be authorized by the competent authority in that specific member state.

The CP, which came into operation in 1995, allows applicants to obtain an MA that is valid throughout the EU. It is compulsory for medicinal products manufactured using biotechnological processes, for orphan medicinal products, and for human medicine products containing a new active substance that was not authorized in the community before May 20, 2004 (date of entry into force of Regulation (EC) No. 726/2004) and was intended for the treatment of AIDS, cancer, neurodegenerative disorder, or diabetes. The CP is also mandatory for veterinary medicinal products intended primarily for use as performance enhancers in order to promote growth of treated animals or to increase yields from treated animals. CP applications are made to, and approved by, the EMA.

To be eligible for the MRP, a medicinal product must have already received an MA in one member state. Since January 1, 1998, the MRP is compulsory for all medicinal products to be marketed in a member state other than that in which they were first authorized. Any national MA granted by an EU member state's national authority can be used to support an application for its mutual recognition by other member states. The MRP is based on the principle of mutual recognition, by EU member states, of their respective national MAs. An application for mutual recognition may be addressed to one or more member states. The applications submitted must be identical and all member states must be notified of them. As soon as one member state decides to evaluate the medicinal product (at which point it becomes the "reference member state"), it notifies this decision to other member states (which then become the "concerned member states") to whom applications have also been submitted. Concerned member states will then suspend their own evaluations and await the reference member state's decision on the product. This evaluation procedure—undertaken by the reference member state—may take up to 210 days and, if successful, results in the granting of an MA in that member state. When the assessment is completed, copies of the report are sent to all member states. The concerned member states then have 90 days to recognize the decision of the reference member state. National MAs are granted within 30 days after acknowledgment of the agreement. The DCP is similar to the MRP, but the difference lies in that it applies to medicinal products that have not received an MA at the time of application. With the DCP, an identical application for MA is submitted simultaneously to the competent authorities of the reference member state and of the concerned member states. At the end of the procedure, the product dossier, as proposed by the reference member state, is approved. The subsequent steps are identical to the MRP.

As in the United States, applicants for an MA for a generic medicine in the EU may submit an abbreviated application. According to Article 10(1) of Directive 2001/83/EC, an applicant for an authorization to market a generic medicine is not required to provide the results of preclinical and clinical trials if it can be demonstrated that the medicinal product is as follows:

- A generic medicinal product or a similar biological medicinal product of a reference medicinal product, which has been authorized under Article 6 of Directive 2001/83/EC for not less than 8 years. This type of application refers to information that is contained in the dossier of the authorization of the reference (*R*) product. This information is generally not completely available in the public domain. Authorizations for generic or similar biological medicinal products are therefore linked to the "original" authorization. This does not, however, mean that withdrawal of the authorization for the *R* product leads to the withdrawal of the authorization for the generic product (Case C-223/01, AstraZeneca, judgment of the European Court of Justice of October 16, 2003). The generic or similar biological medicinal product, once authorized, can, however, only be placed on the market 10 or 11 years after the authorization of the reference medicinal product, depending on the exclusivity period applicable for the reference medicinal product.

Generic medicine applications typically include chemical–pharmaceutical data and the results of bioequivalence studies, which demonstrate the similarity of the generic product relative to the reference medicine. As stated previously, the tolerance levels involved have been favorably compared to those acceptable for interbatch variation during production of the originator medicine. The authorizing regulatory agency(ies) is referred to the data that were established in the originator product's application for authorization for information concerning the safety and efficacy of the active molecule. This is only possible once the data exclusivity period has expired on the originator product's dossier. The majority of authorizations for generic medicines are granted through the MRP and the DCP. Since the introduction of the DCP, the MRP has mainly been used for extending the existing MA to other countries in what is known as the "repeat use" procedure.

The EU bioequivalence parameters are similar to those mandated in the United States, requiring that the test (*T*) and *R* products be contained within an acceptance interval of 80.00%–125.00% of the area under the concentration–time curve (AUC), which reflects the extent of exposure, or C_{max}, at a 90% confidence interval (CI). European guidelines, however, also provide a tightened acceptance interval of 90.00%–111.11% for narrow therapeutic index drugs (NTIDs) as well as different assessment requirements for highly variable drug products (HVDPs).

The EMA bioequivalence testing guidance began as Notice for Guideline on the Investigation of Bioavailability and Bioequivalence in December 1997 in the Joint Efficacy and Quality Working Group; several revisions and drafts later, it was issued as final on January 20, 2010, and became effective on August 1, 2010 (CPMP/EWP/QWP/1401/98 Rev. 1, London, January 20, 2010). This guideline specifies the requirements for bioequivalence assessment for immediate-release oral dosage forms with systemic action. The final guidelines provide a more clear description of topics such as bioequivalence assessment of highly variable drugs (HVDs)/HVDPs, the use of metabolite data, acceptance criteria for NTIDs, Biopharmaceutics Classification System (BCS)-based biowaivers, and dose strength to be used in case of application for MA of several strengths, yet there remain differences when compared to other guidelines of developed countries.

The EMA guideline has evolved into a more clear document on such topics as the assessment of bioequivalence for HVDs/HVDPs, the very limited use of metabolites in bioequivalence assessment, the dose strength to be used in case of application for MA of several strengths, and BCS-based and other biowaivers. The guidelines remain far from offering a globalization lead to these guidelines.

Although the guideline concerns immediate-release oral formulations with systemic action, some general recommendations on the bioequivalence requirements for specific immediate-release formulations as well as for other types of formulations are briefly discussed in its Appendix 15B, for example, oral dispersible tablets (ODTs), oral solutions, parenteral solutions, special dosage forms (liposomal, micellar, or emulsion) for intravenous (IV) use, and locally acting products. Bioequivalence requirements for fixed combination (FC) dosage forms are covered in the "Guideline on Clinical Development of Fixed Combination Medicinal Products" (February 19, 2009: www.ema.europa.eu/docs/en_GB/document_library/Scientific_guideline/2009/09/WC500003686.pdf). Recommendations for bioequivalence studies on modified-release oral medicinal products and transdermal dosage forms are described in a specific guideline, which is currently under revision.

In general, when comparing the bioavailability (BA) of a *T* formulation to that of an *R* formulation, a randomized, two-period, two-sequence, single-dose crossover design is recommended. A parallel design may exceptionally be used in case of long half-life drugs. The use of replicate crossover designs is recommended to assess the bioequivalence of HVDs/HVDPs and will be discussed in more detail later since the guideline has new recommendations regarding bioequivalence assessment of HVDs necessitating the use of a replicate design.

Conduct of a multiple-dose study in healthy volunteers to assess bioequivalence of an immediate-release oral formulation is only justified on the basis of insufficient sensitivity of the bioanalytical method. However, this justification will only be accepted in rare cases because due to recent developments in bioanalytical methodology, it is highly unlikely that the parent drug concentration in plasma cannot be measured accurately and precisely. The choice of a multiple-dose bioequivalence study for a medicinal product showing high intraindividual variability in drug plasma concentrations is no longer acceptable. Although bioequivalence studies should normally be carried out in healthy volunteers, it may be necessary, for tolerability reasons, to use patients instead. Since a single-dose bioequivalence study in patients may not be feasible, conduct of a multiple-dose study in patients is acceptable.

The guidelines recommend that subjects should be 18 years or older (previously, it was between 18 and 55 years) and preferably have a body mass index between 18.5 and 30 kg/m^2 (previously, it was "within the normal range"). The use of healthy volunteers to assess bioequivalence is now justified because this approach is considered adequate to allow extrapolation of the results to patient populations for which the reference medicinal product is approved. The guidance further states: "all treated subjects should be included in the analysis."

The drug products should normally be administered after an overnight fast, and no food is allowed for at least 4 h post dose (fasting study). However, in the case where the Summary of Product Characteristics of the *R* formulation (originator) recommends its intake with food, the study should be carried out under fed conditions. The guideline now provides more information on how to carry out a bioequivalence study under fed conditions. Since food composition (fat content) and timing of the meal relative to medicinal product ingestion are crucial for the assessment of oral drug BA, the revised guideline recommends, where a fed bioequivalence study is carried out, administration of the *T* and *R* preparation immediately after completing a high-fat meal, which represents the *worst-case scenario*. The exact composition of a high-fat meal is also now described in detail. Unlike the FDA, which generally recommends a fasting and fed study for immediate-release oral drug products, the EMA requires bioequivalence studies under both fasted and fed conditions only in exceptional cases, that is, for products with specific formulation characteristics such as microemulsions and solid dispersions, and only if these products can be taken with or without a meal.

The guideline allows a "two-stage" design in calculating the 94.12% CI instead of the usual 90% CI, corresponding to an adjusted α of 0.0294, for both analyses of the stage 1 results and the combined results from stage 1 and stage 2. With these designs, if the failure to declare the two-formulation bioequivalent appears to be due to insufficient power, that is, an insufficient number of subjects included in the study to show bioequivalence between two bioequivalent drug products, it is permitted to carry out an additional study on a number of subjects, and the results from both trials can then be combined in a final analysis. When using such a "two-stage" or "add-on" design, appropriate steps must be taken to correct for multiplicity and, therefore, guarantee an overall type I error of 5%, that is, $\alpha = 0.05$. This design is also allowed by Health Canada, the National Institute of Health Sciences of Japan, as well as the World Health Organization (WHO).

The guideline requires listing in Module 2.7.1 of the Common Technical Document, all relevant studies carried out with the product for which MA is applied, that is, bioequivalence studies comparing the *T* product (same composition and manufacturing process) with a reference medicinal product marketed in the EU. Full study reports should be provided for all bioequivalence studies, except pilot studies for which study synopses are sufficient.

BIOEQUIVALENCE METRICS

In a bioequivalence study, the following pharmacokinetic parameters should be determined: the AUC from 0 to *t*, that is, to the time of the last quantifiable plasma concentration (AUC_{0-t}), the AUC from 0 to infinity ($AUC_{0-\infty}$), the residual area ($AUC_{t-\infty}$), the maximum plasma concentration (C_{max}), and the time at which C_{max} was observed (t_{max}). The terminal plasma elimination rate constant (k_e) and the corresponding plasma half-life ($t_{½}$) also need to be determined in case the AUC has to be extrapolated to infinity. The sampling schedule should follow the plasma concentration–time curve long enough to ensure that $AUC_{(0-t)}$ covers at least 80% of $AUC_{(0-\infty)}$, hence the requirement to also determine $AUC_{t-\infty}$ and $AUC_{(0-\infty)}$. However, it is not necessary to extend blood sampling beyond 72 h following administration of *T* and *R* formulation, because for an immediate-release formulation, the oral absorption process has presumably been covered by 72 h. In that case, AUC truncated at 72 h ($AUC_{0-72\ h}$) should be estimated, and $AUC_{(0-\infty)}$ and $AUC_{t-\infty}$ do not need to be calculated. According to the guideline, AUC truncated at 72 h may be used for bioequivalence assessment irrespective of the half-life of the drug. Although the guideline accepts the use of urinary excretion data to determine the extent of absorption in case it is not possible to reliably measure the plasma concentration–time profile of the parent compound, its use to determine peak exposure should be carefully justified. The pharmacodynamic effect measurements to assess bioequivalence are not allowed.

PARENT COMPOUND VERSUS METABOLITES

In almost all cases, the evaluation of bioequivalence should be based upon the measurement of plasma concentrations of the parent compound since the concentration–time profile of the parent drug is more sensitive to changes in formulation performance than that of the metabolite, which includes the processes of metabolite formation, distribution, and elimination. The use of metabolite concentrations to assess bioequivalence can only be considered if it is adequately demonstrated that the existing analytical methods for measurement of the parent compound are not sensitive enough to accurately determine the single-dose parent drug plasma concentration–time curve, and cannot be improved by using state-of-the-art techniques. Even for inactive prodrugs, demonstration of bioequivalence based on parent compound plasma concentration is generally recommended. Identified examples of drugs that demonstrate this include clopidogrel and losartan. At the time of approval of the *R* product (Plavix®), the pharmacokinetic characteristics of clopidogrel were established based on the pharmacokinetics of the inactive carboxylic acid metabolite. With improved methods available, the guideline now requires use of plasma concentrations of the parent prodrug. The same holds for losartan. An exception is the mycophenolate mofetil where only the use of metabolite data is required.

STATISTICAL ANALYSIS AND ACCEPTANCE CRITERIA

Bioequivalence assessment is based on the "two one-sided test" procedure in which the 90% CI around the geometric mean ratio (GMR) of the *T* and *R* values of an appropriate BA measure, that is, $AUC_{(0-t)}$ (or $AUC_{0-72\ h}$) and C_{max}, is required to fall within preset bioequivalence limits, which normally are 80.00%–125.00%. These bioequivalence metrics should be analyzed using analysis of variance (ANOVA). The terms to be used in the ANOVA model are usually sequence, subject within sequence, period, and formulation. According to the bioequivalence guideline, fixed effects, rather than random effects, should be used for all terms. A nonparametric analysis is not acceptable. A statistical evaluation of t_{max} is not required, but in the case where rapid release is clinically relevant or is related to adverse events, there should be no apparent difference in median t_{max} and its variability between T and R products.

HIGHLY VARIABLE DRUGS OR DRUG PRODUCTS

The guideline recommends that bioequivalence for HVDs or HVDPs can be assessed by using a wider acceptance range for the 90% CI of C_{max}, but not for AUC, compared to the usual 80.00%–125.00% acceptance limits. The widening of the acceptance interval for C_{max} should be clinically justified, and the bioequivalence study, which should be of a replicate design, must demonstrate that the within-subject variability for C_{max} of the R product (CVWR) is >30%. The extent of widening the acceptance range for C_{max} is based on the within-subject variability of the R product, observed in the bioequivalence study, using the scaled average bioequivalence (SABE) approach.

HVDs or HVDPs are generally defined as those medicinal products showing a high within-subject variability, that is, CVw > 30%, of the bioequivalence metrics AUC and/or C_{max}. Bioequivalence parameter variability can be due to characteristics of the drug substance itself, for example, extensive presystemic metabolism, or due to drug product formulation variability. Usually C_{max} shows a higher within-subject variability than AUC. Of the 212 bioequivalence studies submitted to the FDA between 2003 and 2005, 33 studies showed a high within-subject variability in AUC and/or C_{max}. In 28 of the 33 studies, only C_{max}, but not AUC, showed a CVw > 30%. Not one study showed high variability for AUC when variability of C_{max} was moderate or low (i.e., CVWR < 30%). When the within-subject variability in AUC and/or C_{max} is high, the estimated 90% CI is very wide and will exceed the usual 80.00%–125.00% acceptance limits unless a large number of study subjects are included in the bioequivalence study. This may lead to situations where bioequivalence cannot be established even when the R formulation is tested against itself. Instead of widening the bioequivalence acceptance limits from 80%–125% to 75%–133% for all drugs/drug products showing a CVw > 30%, a scientifically more appealing approach is to widen the acceptance limits based on the within-subject variability of the R formulation CVWR, that is, the so-called SABE method. The recommendation of using reference scaling is based on the general concept that reference variability should be used as an index for setting the public standard expressed in the bioequivalence acceptance limit.

For drugs/drug products with low to moderate within-subject variability, bioequivalence is usually declared if the difference between the logarithmic means of AUC or C_{max} for T and R product (μ_T and μ_R, respectively) lies between preset bioequivalence limits θ_A. Therefore, average bioequivalence (ABE) is accepted if the following criterion is satisfied:

$$-\theta_A \le \mu_T - \mu_R \le +\theta_A \tag{15.1}$$

The limits θ_A are generally symmetrical on the logarithmic scale and usually equal to ln(1.25).

True values of the population means μ_T and μ_R are not known, and therefore, their estimates, that is, m_T and m_R, have to be used:

$$-0.223 \le m_T - m_R \le +0.233 \tag{15.2}$$

For HVDs/HVDPs when the within-subject variability of the R product exceeds a preset "switching" value (CVWR of 30% corresponding to a "switching" standard deviation σ_s of 0.294), the difference between logarithmic means, that is, $\mu_T - \mu_R$, can be normalized (scaled) to the within-subject variability of the R formulation as follows:

$$-\frac{0.233}{\sigma_0} \leq \frac{m_T - m_R}{S_{WR}} \leq +\frac{0.233}{\sigma_0} \tag{15.3}$$

or

$$-\left(\frac{0.233}{\sigma_0}\right) * S_{WR} \leq m_T - m_R \leq +\left(\frac{0.233}{\sigma_0}\right) * S_{WR} \tag{15.4}$$

where

σ_0 is the "regulatory standard deviation" that can be set by the regulatory authorities
S_{WR} represents the within-subject standard deviation of the R formulation (both on the logarithmic scale)

This means that the usual bioequivalence limits (−0.223 and +0.223) are expanded in proportion to the within-subject variability of the R formulation S_{WR}, starting from the "switching" variability, as follows:

$$-\left(\frac{0.233}{\sigma_0}\right) * S_{WR} = k * S_{WR} \tag{15.5}$$

where $0.223/\sigma_0$ is the proportionality or regulatory constant k. As mentioned earlier, HVDs/HVDPs are defined to have a within-subject coefficient of variation (CV) exceeding 30%, that is, CVWR > 30%. Because the CVWR is related to the standard deviation on the logarithmic scale S_{WR} as

$$\text{CW}_{WR} = \left[\exp\left(S_{WR}^2\right) - 1\right]^{1/2} \tag{15.6}$$

a CVWR of 30% corresponds to a "switching" variability $\sigma_s = 0.294$. The EMA has chosen a regulatory standard deviation σ_0 equal to the switching variability σ_s, that is, 0.294, and therefore, the regulatory constant k is equal to 0.760. This means that the acceptance limits for HVDs/HVDPs using the ABE approach with expanding limits (ABEL), a variant of the SABE method, are

$$\left[U, L\right] = e^{\left(\pm 0.76 S_{WR}\right)} \tag{15.7}$$

Widening of the usual bioequivalence acceptance limits of 0.80–1.25 according to this method recommended by the EMA is only acceptable for C_{max}. Provided there are no safety/efficacy concerns, the bioequivalence acceptance limits can be widened to a maximum of 69.84%–143.19%, corresponding to a CVWR of 50%. An additional constraint imposed by the EMA is that the GMR for C_{max} should lie within the conventional 80.00%–125.00% acceptance range.

The EMA regulatory limits for bioequivalence acceptance are 80.00–125.00 until a within-subject switching CV of 30% is reached. For CVWR values between 30% and 50%, the bioequivalence acceptance limits are expanded according to the within-subject variability of the R formulation (ABEL approach). For CVWR values of 50% or higher, the 90% bioequivalence acceptance limits are capped at 69.84%–143.19%. The FDA recommends scaling of the usual bioequivalence limits (80%–125%) starting at a CVWR of 30%. However, the EMA and FDA use a different regulatory standardized variability (σ_0), that is, 0.294 and 0.25, respectively. This difference in the choice of σ_0 explains why the FDA expanded limits are not only discontinuous at the switching variability but also wider than the EMA expanded limits (see Chapter 7).

The use of the ABEL approach necessitates a replicate study design allowing precise estimation of the within-subject variability of C_{max} for the *R* formulation. Either a three-period or a four-period replicate crossover design is acceptable. The EMA Questions & Answers document contains a section "Clarification on the recommended statistical method for the analysis of a bioequivalence study" in which the statistical analysis of data from a three-period (*R* administered twice, *T* administered once) and a four-period (*R* and *T* administered twice) crossover study is shown in detail.

The FDA now also recommends the reference-scaled ABE approach when the bioequivalence parameters AUC and/or C_{max} show high within-subject variability, that is, ≥30% CVWR. However, the FDA uses a regulatory standard variability, σ_0, of 0.25. As a result, since scaling starts at a CVWR of 30%, the bioequivalence limits are discontinuous at the switching variability (Figure 15.1). This is different from the ABEL procedure proposed by the EMA where scaling also starts at a "switching" variability of 30.0%, but because the EMA regulatory standard variability σ_0 is set at 0.294, the bioequivalence limits are continuous and more conservative. In addition, unlike the EMA, which only allows scaling for C_{max}, the FDA allows scaling for C_{max} and AUC. Both the EMA and the FDA apply a GMR constraint, that is, the point estimate for the *T*/*R* GMR must fall between 0.80 and 1.25.

NARROW THERAPEUTIC INDEX DRUGS

The debate as to whether or not it is necessary to apply stricter guidelines for certain drug substances has been ongoing for several decades. For NTIDs, the usual acceptance interval for AUC, and also for C_{max} if necessary for safety, efficacy, or drug level monitoring reasons, may need to be tightened to 90.00%–111.11%. However, according to the same guideline: "It is not possible to define a set of criteria to categorize drugs as narrow therapeutic index drugs (NTIDs) and it must be decided case by case if an active substance is an NTID based on clinical considerations." For two drug substances, that is, cyclosporine and tacrolimus, the EMA accepts them as drugs with a narrow therapeutic index. Consequently, based on efficacy and safety considerations, the 90.00%–111.11% acceptance limits are required for cyclosporine for both AUC and C_{max}. For tacrolimus, on the other hand, the acceptance criterion is only tightened to 90.00%–111.11% for AUC but not for C_{max} where the usual 80.00%–125.00% acceptance limits apply.

There is no consensus among the various EU member states on the issue of bioequivalence acceptance criteria as well as switchability between innovator and generic medicinal products. For example, the Danish Health and Medicines Authority considers that the 90% CI for the ratio *T* versus *R* (GMR) of both AUC and C_{max} should incorporate 100% irrespective of whether acceptance limits of 80%–125% or narrower are employed. Deviations may be accepted if they can be adequately justified not to have impact on either the overall therapeutic effect or safety profile of the product. This requirement is not part of the EMA recommendations. Like the EMA, however, the Danish Health and Medicines Authority also requires tighter acceptance limits, that is, 90.00%–111.11%, for both AUC and C_{max} for substances with a narrow therapeutic index with regard to automatic substitution for the following substances or therapeutic classes: aminophylline/theophylline, lithium, vitamin K antagonists, antiepileptics apart from levetiracetam and benzodiazepines, antiarrhythmics, centrally acting anorectics, and tricyclic antidepressants. The Danish authorities clearly accept the idea that medicinal products containing the NTIDs on their list are therapeutically equivalent because stricter bioequivalence criteria were applied, and they are therefore considered switchable, and consequently, generic substitution of NTIDs is authorized except for thyroxine and the immunosuppressants cyclosporine and tacrolimus.

The Federal Agency for Medicines and Health Products (FAMHP) of Belgium published a list with 31 drug substances considered to have a narrow therapeutic index or to be highly toxic. There is a substantial overlap between the Belgian and Danish lists. Secondly, medicinal products containing one of these 31 substances on the Belgian list are considered to be "nonswitchable," and switching from the innovator to the generic medicine, or vice versa, after initiation of therapy is, therefore, discouraged by the Belgian agency. This means that the Belgian authorities, unlike the Danish Health and Medicines Authority, do not accept therapeutic equivalence, and therefore switchability, between medicinal products containing an active substance with a narrow therapeutic index, even though bioequivalence has been demonstrated and for some of them even based on the stricter acceptance limits of 90.00%–111.11%. This example illustrates how the health authorities of two EU member states interpret and apply the bioequivalence guidelines quite differently. Moreover, with the exception of cyclosporine and tacrolimus, it is currently not known to the health practitioners for which generic medicines, authorized in member states of the EU, bioequivalence with the innovator was demonstrated on the basis of the stricter 90.00%–111.11% acceptance range.

Whether tighter acceptance criteria should be applied to certain drug substances is a controversial issue, which has received much attention in the scientific literature and has led to different recommendations by various health authorities. The Health Products and Food Branch (HPFB) of Canada issued a specific guidance for industry on the bioequivalence requirements for critical dose drugs. According to this guidance, "critical dose drugs" are defined as those drugs for which comparatively small differences in dose or concentration lead to dose- and concentration-dependent, serious therapeutic failures and/or serious adverse drug reactions. For these "critical dose drugs," the 90% CI of the relative mean AUC of the T to R formulation should lie within 90%–112%, according to Canada's HPFB guidance. In addition, the 90% CI of the relative mean C_{max} of the T to R formulation for these "critical dose drugs" should be between 80% and 125%. For "uncomplicated" drugs, Canada's HPFB does not require calculation of the 90% CI for C_{max} but requires that the GMR should lie between 80% and 125%. These requirements for "critical dose drugs" are to be met in both the fasted and fed states. In an appendix in the HPFB guidance, a list of nine drugs is given: cyclosporine, digoxin, flecainide, lithium, phenytoin, sirolimus, tacrolimus, theophylline, and warfarin. The FDA guidance for industry on Bioavailability and Bioequivalence Studies for Orally Administered Drug Products recommends that the usual bioequivalence limit of 80%–125% for non-NTIDs remains unchanged for the BA measures (AUC and C_{max}) for NTID substances unless otherwise indicated by a specific guidance. The FDA has long supported the view that stricter acceptance limits for NTIDs are not necessary for purposes of therapeutic substitution. However, recently, the FDA has restarted to debate the issue whether the bioequivalence criteria used to approve generic drugs are appropriate for all drugs and specifically whether NTIDs require special considerations. Recently, for NTIDs, the FDA Advisory Committee for Pharmaceutical Science and Clinical Pharmacology recommended the use of a replicate study design to quantify the variability of both the R and the T product and the use of a scaling approach for bioequivalence assessment.

The proposed FDA regulatory default limits for bioequivalence acceptance of NTIDs are 90%–111%. A switching variability σ_0 equal to 0.1 has been proposed by the FDA. This means that at a reference within-subject variability CVWR of 10%, the bioequivalence limits are 90%–111%. When CVWR is greater than 10%, the bioequivalence limits would expand as a function of CVWR, but the expansion would be capped at 80%–125%. This maximum expansion is reached when the CVWR is approximately 21%. When the reference variability is less than 10%, the acceptance bioequivalence limits would become narrower than the default limits of 90%–111%. This approach is similar to the EMA SABE approach for HVDs/HVDPs where the switching variability σ_0 is set by the EMA at 0.294 to give bioequivalence acceptance limits of 80%–125% at a CVWR of 30%.

Most NTIDs have a relatively small within-subject variability, ranging from approximately 5% to 25%. Similar to the scaling approach recommended by the EMA for HVDs/HVDPs, the bioequivalence limits would change as a function of the within-subject variability of the R product CVWR.

The FDA proposes for NTIDs that the default bioequivalence limits be 90%–111% and that they be scaled using a switching variability (σ_0) of 0.1 (which corresponds to a CVWR of 10.03%) to a maximum of 80%–125%. This switching variability is chosen such that the bioequivalence acceptance limits are 90%–111% when the CVWR is equal to 10%. This means that when the CVWR is less than 10%, the bioequivalence limits would narrow beyond the default limits of 90%–111% as a function of the within-subject variability of the *R* product. The maximum bioequivalence acceptance limits, that is, 80%–125%, will be reached at a CVWR of approximately 21%. The FDA is also investigating the impact of additional bioequivalence acceptance criteria for NTIDs such as point estimate limits for C_{max} and AUC and a requirement that the 90% CI around the GMR for C_{max} and AUC includes 100%.

DOSAGE STRENGTH(S) TO BIOEQUIVALENCE INVESTIGATED

When MA is requested for several strengths of a product, it may be sufficient to carry out an in vivo bioequivalence study on one or two strengths only and to apply for a biowaiver for the remaining strengths. The choice of the strength(s) at which the in vivo bioequivalence study should be carried out as well as the conditions that have to be fulfilled to qualify for a biowaiver for additional strengths are explained in much more detail in the bioequivalence guideline. In this regard, it is noteworthy that the pharmacokinetics are considered to be linear (or dose proportional) if the difference in dose-adjusted AUCs is not more than 25% between the concerned strengths, that is, the strength for which an in vivo bioequivalence study is carried out and the one for which a biowaiver is requested. The in vivo bioequivalence study should generally be conducted at the highest strength. Selection of a lower strength is acceptable for products with dose-proportional pharmacokinetics and where the drug substance is highly soluble according to the BCS. Other considerations taken into account to select the strength at which to carry out the bioequivalence study are tolerability/safety of the study subjects and insufficient sensitivity of the bioanalytical method to determine the plasma concentration–time profile. One of the conditions that have to be fulfilled to extrapolate the results of a bioequivalence study carried out with one of the strengths only is related to the composition of the various strengths, which should be proportional, that is, having a constant ratio between the active substance and the various excipients (some minor deviations of the rule are accepted). When a bioequivalence study has to be performed on more than two strengths because the condition of proportional composition is not fulfilled, the bracketing approach may be used, that is, the strengths selected for the bioequivalence studies represent the extremes in strength or in composition.

BCS-BASED BIOWAIVERS

The EMA BCS-based biowaiver guidelines are described in appendix C of the EMA guideline. Unlike the FDA, which only accepts biowaivers for BCS class I substances (high solubility, high permeability), the EMA considers biowaivers for BCS class I and III substances (high solubility, low permeability). A drug substance is considered highly soluble, according to the EMA guideline, "if the highest single dose administered as immediate release formulation(s) is completely dissolved in 250 mL of buffers within the range of pH 1.0–6.8 at 37°C ± 1°C" (9). The guideline defines permeability as a function of extent of absorption following oral administration: "Complete absorption is generally related to high permeability." When the measured extent of absorption, based on human data (absolute BA, mass-balance studies), is at least 85%, complete absorption is accepted. Although the EMA accepts in vitro permeability investigations as supportive evidence of in vivo human data, the FDA seems to attach greater importance to in vitro studies (cell cultures, intestinal tissue) and in vivo/in situ perfusion experiments in animal models. In the case of biowaivers for class III substances, the in vitro dissolution of the *T* and *R* product has to be very rapid (>85% within 15 min), and special attention must be paid to the excipients since it is known that the absorption of BCS class III substances is more susceptible to transporter-mediated excipient–drug interactions. For BCS class I substances, the EMA advises the use of similar amounts of the same excipients in the *T* product compared to the *R* product. For BCS class III substances, according to the EMA, the excipients have to be qualitatively the same and quantitatively very similar in *T* and *R* preparation.

The WHO, which is not a regulatory body but publishes technical reports and guidelines that are recommendations to national authorities, allows biowaivers not only for BCS class I and BCS class III substances but also under certain circumstances for class II substances. This lack of harmonization creates confusion, which in turn leads to suspicion by health-care providers and patients, especially since many national authorities of Third-World countries give these WHO reports regulatory status. All stakeholders in the development and registration of new drug products must balance the need for scientific rigor in assuring BA/bioequivalence (and hence product quality toward consistent therapeutic outcomes) with the time and expense of conducting in vivo bioequivalence studies and the overall impact on product costs and timely availability to patients. Ideally, these guidelines should be the same worldwide to ensure that patients all over the world can benefit from affordable and safe medicinal products.

EMA GUIDELINE

Executive Summary

This EMA guideline specifies the requirements for the design, conduct, and evaluation of bioequivalence studies for immediate-release dosage forms with systemic action.

Introduction

Background

Two medicinal products containing the same active substance are considered bioequivalent if they are pharmaceutically equivalent or pharmaceutical alternatives and their bioavailabilities (rate and extent) after administration in the same molar dose lie within acceptable predefined limits. These limits are set to ensure comparable in vivo performance, that is, similarity in terms of safety and efficacy.

In bioequivalence studies, the plasma concentration–time curve is generally used to assess the rate and extent of absorption. Selected pharmacokinetic parameters and preset acceptance limits allow the final decision on bioequivalence of the tested products. AUC reflects the extent of exposure. C_{max}, the maximum plasma concentration or peak exposure, and the time to maximum plasma concentration, t_{max}, are parameters that are influenced by absorption rate.

It is the objective of this guideline to specify the requirements for the design, conduct, and evaluation of bioequivalence studies. The possibility of using in vitro instead of in vivo studies is also addressed.

Generic Medicinal Products

In applications for generic medicinal products according to Directive 2001/83/EC, Article 10(1), the concept of bioequivalence is fundamental. The purpose of establishing bioequivalence is to demonstrate equivalence in biopharmaceutics quality between the generic medicinal product and a reference medicinal product in order to allow bridging of preclinical tests and of clinical trials associated with the reference medicinal product. The current definition for generic medicinal products is found in Directive 2001/83/EC, Article 10(2)(b), which states that a generic medicinal product is a product that has the same qualitative and quantitative composition in active substances and the same pharmaceutical form as the reference medicinal product and whose bioequivalence with the reference medicinal product has been demonstrated by appropriate BA studies. The different salts, esters, ethers, isomers, mixtures of isomers, complexes, or derivatives of an active substance are considered to be the same active substance, unless they differ significantly in properties with regard to safety and/or efficacy. Furthermore, the various immediate-release oral pharmaceutical forms shall be considered to be one and the same pharmaceutical form.

Other Types of Application

Other types of applications may also require demonstration of bioequivalence, including variations, FCs, extensions, and hybrid applications.

The recommendations on design and conduct given for bioequivalence studies in this guideline may also be applied to comparative BA studies evaluating different formulations used during the development of a new medicinal product containing a new chemical entity and to comparative BA studies included in extension or hybrid applications that are not based exclusively on bioequivalence data.

Scope

This guideline focuses on recommendations for bioequivalence studies for immediate-release formulations with systemic action. It also sets the relevant criteria under which BA studies need not be required (either waiver for additional strength [see "Strength to be investigated" section], a specific type of formulation [see Appendix 15B], or BCS-based biowaiver [see Appendix 15C]).

Specific recommendations regarding bioequivalence studies for modified-release products, transdermal products, and orally inhaled products (OIPs) are given in other guidelines (see "Legal basis" section).

The scope is limited to chemical entities. Recommendation for the comparison of biologicals to reference medicinal products can be found in guidelines on similar biological medicinal products. In case bioequivalence cannot be demonstrated using drug concentrations, in exceptional circumstances, pharmacodynamic or clinical endpoints may be needed. This situation is outside the scope of this guideline, and the reader is referred to therapeutic area–specific guidelines.

Although the concept of bioequivalence possibly could be considered applicable for herbal medicinal products, the general principles outlined in this guideline are not applicable to herbal medicinal products, for which active constituents are less well defined than for chemical entities.

Furthermore, this guideline does not cover aspects related to generic substitution as this is subject to national regulation.

Legal Basis

This guideline applies to MA applications for human medicinal products submitted in accordance with the Directive 2001/83/EC as amended, under Article 10(1) (generic applications). It may also be applicable to MA applications for human medicinal products submitted under Article 8(3) (full applications), Article 10b (FC), Article 10(3) (hybrid applications) of the same directive, and for extension and variation applications in accordance with Commission Regulation (EC) Nos. 1084/2003 and 1085/2003 as well.

This guideline should be read in conjunction with the Annex I of Directive 2001/83/EC as amended, as well as European and International Conference on Harmonisation (ICH) guidelines for conducting clinical trials, including those on

- General Considerations for Clinical Trials (ICH topic E8, CPMP/ICH/291/95)
- Guideline for Good Clinical Practice (ICH E6 (R1), CPMP/ICH/135/95)
- Statistical Principles for Clinical Trials (ICH E9, CPMP/ICH/363/96)
- Structure and Content of Clinical Study Reports (ICH E3, CPMP/ICH/137/95)

- Committee for Human Medicinal Products (CHMP) guidance for users of the CP for generics/hybrid applications (EMEA/CHMP/225411/2006)
- Pharmacokinetic studies in man (EudraLex, Volume 3, 3CC3a)
- Modified Release Oral and Transdermal Dosage Forms: Sections I and II (CPMP/QWP/604/96, CPMP/EWP/280/96)
- Fixed combination medicinal products (CPMP/EWP/240/95 Rev 1)
- Requirements for clinical documentation for OIPs including the requirements for demonstration of therapeutic equivalence between two inhaled products for use in the treatment of asthma and chronic obstructive pulmonary disease (COPD) (CPMP/EWP/4151/00 rev 1)
- Clinical requirements for locally applied, locally acting products containing known constituents (CPMP/EWP/239/95)

The guideline should also be read in conjunction with relevant guidelines on pharmaceutical quality. The *T* products used in the bioequivalence study must be prepared in accordance with GMP regulations including EudraLex Volume 4.

Bioequivalence trials conducted in the EU/EEA have to be carried out in accordance with Directive 2001/20/EC. Trials conducted outside of the union and intended for use in an MA application in the EU/EEA have to be conducted to the standards set out in Annex I of the community code, Directive 2001/83/EC, as amended.

Companies may also apply for CHMP scientific advice, via the EMEA, for specific queries not covered by existing guidelines.

Main Guideline Text

Design, Conduct, and Evaluation of Bioequivalence Studies

The number of studies and study design depend on the physicochemical characteristics of the substance, its pharmacokinetic properties, and proportionality in composition and should be justified accordingly. In particular, it may be necessary to address the linearity of pharmacokinetics, the need for studies both in fed and fasting state, the need for enantioselective analysis, and the possibility of waiver for additional strengths (see "Study conduct standardization," "Characteristics to be investigated," and "Strength to be investigated" sections). Module 2.7.1 should list all relevant studies carried out with the product applied for, that is, bioequivalence studies comparing the formulation applied for (i.e., same composition and manufacturing process) with a reference medicinal product marketed in EU. Studies should be included in the list regardless of the study outcome. Full study reports should be provided for all studies, except pilot studies for which study report synopses (in accordance with ICH E3) are sufficient. Full study reports for pilot studies should be available upon request. Study report synopses for bioequivalence or comparative BA studies conducted during formulation development should also be included in Module 2.7. Bioequivalence studies comparing the product applied for with non-EU *R* products should not be submitted and do not need to be included in the list of studies.

Study Design

The study should be designed in such a way that the formulation effect can be distinguished from other effects.

Standard Design If two formulations are compared, a randomized, two-period, two-sequence single-dose crossover design is recommended. The treatment periods should be separated by a washout period sufficient to ensure that drug concentrations are below the lower limit of bioanalytical quantification in all subjects at the beginning of the second period. Normally, at least five elimination half-lives are necessary to achieve this.

Alternative Designs Under certain circumstances, provided the study design and the statistical analyses are scientifically sound, alternative well-established designs could be considered such as parallel design for substances with very long half-life and replicate designs, for example, for substances with highly variable pharmacokinetic characteristics (see "Highly variable drugs or drug products" section).

Conduct of a multiple-dose study in patients is acceptable if a single-dose study cannot be conducted in healthy volunteers due to tolerability reasons and a single-dose study is not feasible in patients.

In the rare situation where problems of sensitivity of the analytical method preclude sufficiently precise plasma concentration measurements after single-dose administration and where the concentrations at steady state are sufficiently high to be reliably measured, a multiple-dose study may be acceptable as an alternative to the single-dose study. However, given that a multiple-dose study is less sensitive in detecting differences in C_{max}, this will only be acceptable if the applicant can adequately justify that the sensitivity of the analytical method cannot be improved and that it is not possible to reliably measure the parent compound after single-dose administration, taking into account also the option of using a supratherapeutic dose in the bioequivalence study (see also "Strength to be investigated" section). Due to the recent development in the bioanalytical methodology, it is unusual that parent drug cannot be measured accurately and precisely. Hence, the use of a multiple-dose study instead of a single-dose study, due to limited sensitivity of the analytical method, will only be accepted in exceptional cases.

In steady-state studies, the washout period of the previous treatment can overlap with the buildup of the second treatment, provided the buildup period is sufficiently long (at least five times the terminal half-life).

Reference and Test Product

Reference Product For Articles 10(1) and 10(3), MA application reference must be made to the dossier of a reference medicinal product for which an MA is or has been granted in the union on the basis of a complete dossier in accordance with Articles 8(3), 10a, 10b, or 10c of Directive 2001/83/EC, as amended. The product used as *R* product in the bioequivalence study should be part of the global MA of the reference medicinal product (as defined in Article 6(1) second subparagraph of Directive 2001/83/EC). The choice of the reference medicinal product identified by the applicant in Module 1.2 application form for which bioequivalence has been demonstrated by appropriate BA studies should be justified.

T products in an application for a generic or hybrid product or an extension of a generic/hybrid product are normally compared with the corresponding dosage form of a reference medicinal product, if available on the market.

In an application for extension of a medicinal product that has been initially approved under Article 8(3) of Directive 2001/83/EC and when there are several dosage forms of this medicinal product on the market, it is recommended that the dosage form used for the initial approval of the concerned medicinal product (and which was used in clinical efficacy and safety studies) is used as *R* product, if available on the market.

The selection of the *R* product used in a bioequivalence study should be based on assay content and dissolution data and is the responsibility of the applicant. Unless otherwise justified, the assayed content of the batch used as *T* product should not differ more than 5% from that of the batch used as *R* product determined with the test procedure proposed for routine quality testing of the *T* product. The applicant should document how a representative batch of the *R* product with regard to dissolution and assay content has been selected. It is advisable to investigate more than one single batch of the *R* product when selecting *R* product batch for the bioequivalence study.

Test Product The *T* product used in the study should be representative of the product to be marketed, and this should be discussed and justified by the applicant.

For example, for oral solid forms for systemic action, the following applies:

a. The *T* product should usually originate from a batch of at least 1/10 of production scale or 100,000 units, whichever is greater, unless otherwise justified.
b. The production of batches used should provide a high level of assurance that the product and process will be feasible on an industrial scale. In case of a production batch smaller than 100,000 units, a full production batch will be required.
c. The characterization and specification of critical quality attributes of the drug product, such as dissolution, should be established from the test batch, that is, the clinical batch for which bioequivalence has been demonstrated.
d. Samples of the product from additional pilot and/or full-scale production batches, submitted to support the application, should be compared with those of the bioequivalence study test batch and should show similar in vitro dissolution profiles when employing suitable dissolution test conditions (see Appendix 15A).
e. Comparative dissolution profile testing should be undertaken on the first three production batches.
f. If full-scale production batches are not available at the time of submission, the applicant should not market a batch until comparative dissolution profile testing has been completed.
g. The results should be provided at a competent authority's request or if the dissolution profiles are not similar together with proposed action to be taken.

For other immediate-release pharmaceutical forms for systemic action, justification of the representative nature of the test batch should be similarly established.

Packaging of Study Products The *R* and *T* products should be packed in an individual way for each subject and period, either before their shipment to the trial site or at the trial site itself. Packaging (including labeling) should be performed in accordance with good manufacturing practice, including Annex 13 of the EU guide to GMP. Where necessary and in accordance with local regulations, sites should be authorized, as provided for in Article 13(1) of Directive 2001/20/EC, except where the provisions of Article 9(2) of Directive 2005/28/EC apply. Third country sites should be able to demonstrate standards equivalent to these GMP requirements compliant with local requirements.

It should be possible to identify unequivocally the identity of the product administered to each subject at each trial period. Packaging, labeling, and administration of the products to the subjects should therefore be documented in detail. This documentation should include all precautions taken to avoid and identify potential dosing mistakes. The use of labels with a tear-off portion is recommended.

Subjects

Number of subjects: The number of subjects to be included in the study should be based on an appropriate sample size calculation. The number of evaluable subjects in a bioequivalence study should not be less than 12.

Selection of subjects: The subject population for bioequivalence studies should be selected with the aim of permitting detection of differences between pharmaceutical products. In order to reduce variability not related to differences between products, the studies should normally be performed in healthy volunteers unless the drug carries safety concerns that make this unethical. This model, in vivo healthy volunteers, is regarded as adequate in most instances to detect formulation differences and to allow extrapolation of the results to populations for which the reference medicinal product is approved (the elderly, children, patients with renal or liver impairment, etc.).

The inclusion/exclusion criteria should be clearly stated in the protocol. Subjects should be 18 years of age or older and preferably have a body mass index between 18.5 and 30 kg/m^2.

The subjects should be screened for suitability by means of clinical laboratory tests, a medical history, and a physical examination. Depending on the drug's therapeutic class and safety profile, special medical investigations and precautions may have to be carried out before, during, and after the completion of the study. Subjects could belong to either sex; however, the risk to women of childbearing potential should be considered. Subjects should preferably be nonsmokers and without a history of alcohol or drug abuse. Phenotyping and/or genotyping of subjects may be considered for safety or pharmacokinetic reasons.

In parallel design studies, the treatment groups should be comparable in all known variables that may affect the pharmacokinetics of the active substance (e.g., age, body weight, sex, ethnic origin, smoking status, extensive/poor metabolic status). This is an essential prerequisite to give validity to the results from such studies.

If the investigated active substance is known to have adverse effects, and the pharmacological effects or risks are considered unacceptable for healthy volunteers, it may be necessary to include patients instead, under suitable precautions and supervision.

Study Conduct Standardization

The test conditions should be standardized in order to minimize the variability of all factors involved except that of the products being tested. Therefore, it is recommended to standardize diet, fluid intake, and exercise.

The time of the day for ingestion should be specified. Subjects should fast for at least 8 h prior to administration of the products, unless otherwise justified. As fluid intake may influence gastric passage for oral administration forms, the *T* and *R* products should be administered with a standardized volume of fluid (at least 150 mL). It is recommended that water is allowed as desired except for 1 h before and 1 h after drug administration and no food is allowed for at least 4 h post dose. Meals taken after dosing should be standardized in regard to composition and time of administration during an adequate period of time (e.g., 12 h).

In case the study is to be performed during fed conditions, the timing of administration of the drug product in relation to food intake is recommended to be according to the SmPC of the originator product. If no specific recommendation is given in the originator SmPC, it is recommended that subjects should start the meal 30 min prior to administration of the drug product and eat this meal within 30 min.

As the BA of an active moiety from a dosage form could be dependent upon gastrointestinal transit times and regional blood flows, posture and physical activity may need to be standardized.

The subjects should abstain from food and drinks, which may interact with circulatory, gastrointestinal, hepatic, or renal function (e.g., alcoholic drinks or certain fruit juices such as grapefruit juice) during a suitable period before and during the study. Subjects should not take any other concomitant medication (including herbal remedies) for an appropriate interval before as well as during the study. Contraceptives are, however, allowed. In case concomitant medication is unavoidable and a subject is administered other drugs, for instance, to treat adverse events like headache, the use must be reported (dose and time of administration), and possible effects on the study outcome must be addressed. In rare cases, the use of a concomitant medication is needed for all subjects for safety or tolerability reasons (e.g., opioid antagonists, antiemetics). In that scenario, the risk for a potential interaction or bioanalytical interference affecting the results must be addressed.

Medicinal products that according to the originator SmPC are to be used explicitly in combination with another product (e.g., certain protease inhibitors in combination with ritonavir) may be studied either as the approved combination or without the product recommended to be administered concomitantly.

In bioequivalence studies of endogenous substances, factors that may influence the endogenous baseline levels should be controlled if possible (e.g., strict control of dietary intake).

Sampling Times A sufficient number of samples to adequately describe the plasma concentration–time profile should be collected. The sampling schedule should include frequent sampling around predicted t_{max} to provide a reliable estimate of peak exposure. In particular, the sampling schedule should be planned to avoid C_{max} being the first point of a concentration–time curve. The sampling schedule should also cover the plasma concentration–time curve long enough to provide a reliable estimate of the extent of exposure, which is achieved if $AUC_{(0-t)}$ covers at least 80% of $AUC_{(0-\infty)}$. At least three to four samples are needed during the terminal log–linear phase in order to reliably estimate the terminal rate constant (which is needed for a reliable estimate of $AUC_{(0-\infty)}$). AUC truncated at 72 h ($AUC_{(0-72\ h)}$) may be used as an alternative to $AUC_{(0-t)}$ for comparison of extent of exposure as the absorption phase has been covered by 72 h for immediate-release formulations. A sampling period longer than 72 h is therefore not considered necessary for any immediate-release formulation irrespective of the half-life of the drug.

In multiple-dose studies, the predose sample should be taken immediately before (within 5 min) dosing, and the last sample is recommended to be taken within 10 min of the nominal time for the dosage interval to ensure an accurate determination of $AUC_{(0-\infty)}$.

If urine is used as the biological sampling fluid, urine should normally be collected over no less than three times the terminal elimination half-life. However, in line with the recommendations on plasma sampling, urine does not need to be collected for more than 72 h. If rate of excretion is to be determined, the collection intervals need to be as short as feasible during the absorption phase (see also "Characteristics to be investigated" section).

For endogenous substances, the sampling schedule should allow characterization of the endogenous baseline profile for each subject in each period. Often, a baseline is determined from two to three samples taken before the drug products are administered. In other cases, sampling at regular intervals throughout 1–2 days prior to administration may be necessary in order to account for fluctuations in the endogenous baseline due to circadian rhythms (see "Characteristics to be investigated" section).

Fasting or Fed Conditions In general, a bioequivalence study should be conducted under fasting conditions as this is considered to be the most sensitive condition to detect a potential difference between formulations. For products where the SmPC recommends intake of the reference medicinal product on an empty stomach or irrespective of food intake, the bioequivalence study should hence be conducted under fasting conditions. For products where the SmPC recommends intake of the reference medicinal product only in fed state, the bioequivalence study should generally be conducted under fed conditions.

However, for products with specific formulation characteristics (e.g., microemulsions, solid dispersions), bioequivalence studies performed under both fasted and fed conditions are required unless the product must be taken only in the fasted state or only in the fed state.

In cases where information is required in both the fed and fasted states, it is acceptable to conduct either two separate two-way crossover studies or a four-way crossover study.

In studies performed under fed conditions, the composition of the meal is recommended to be according to the SmPC of the originator product. If no specific recommendation is given in the originator SmPC, the meal should be a high-fat (approximately 50% of total caloric content of the meal) and high-calorie (approximately 800–1000 kcal) meal. This test meal should derive approximately 150, 250, and 500–600 kcal from protein, carbohydrate, and fat, respectively. The composition of the meal should be described with regard to protein, carbohydrate, and fat content (specified in grams, calories, and relative caloric content [%]).

Characteristics to Be Investigated

Pharmacokinetic Parameters Actual time of sampling should be used in the estimation of the pharmacokinetic parameters. In studies to determine bioequivalence after a single dose, $AUC_{(0-t)}$, $AUC_{(0-\infty)}$, residual area, C_{max}, and t_{max} should be determined. In studies with a sampling period of 72 h, and where the concentration at 72 h is quantifiable, $AUC_{(0-\infty)}$ and residual area do not need to be reported; it is sufficient to report AUC truncated at 72 h, $AUC_{(0-72\,h)}$. Additional parameters that may be reported include the terminal rate constant, λ_z, and $t_{1/2}$.

In studies to determine bioequivalence for immediate-release formulations at steady state, $AUC_{(0-\infty)}$, $C_{max,ss}$, and $t_{max,ss}$ should be determined.

When using urinary data, $Ae_{(0-t)}$ and, if applicable, R_{max} should be determined.

Noncompartmental methods should be used for determination of pharmacokinetic parameters in bioequivalence studies. The use of compartmental methods for the estimation of parameters is not acceptable.

Parent Compound or Metabolites

General Recommendations In principle, evaluation of bioequivalence should be based upon measured concentrations of the parent compound. The reason for this is that C_{max} of a parent compound is usually more sensitive to detect differences between formulations in absorption rate than C_{max} of a metabolite.

Inactive Prodrugs Also for inactive prodrugs, the demonstration of bioequivalence for parent compound is recommended. The active metabolite does not need to be measured. However, some prodrugs may have low plasma concentrations and be quickly eliminated, resulting in difficulties in demonstrating bioequivalence for parent compound. In this situation, it is acceptable to demonstrate bioequivalence for the main active metabolite without measurement of parent compound. In the context of this guideline, a parent compound can be considered to be an inactive prodrug if it has no or very low contribution to clinical efficacy.

Use of Metabolite Data as Surrogate for Active Parent Compound The use of a metabolite as a surrogate for an active parent compound is not encouraged. This can only be considered if the applicant can adequately justify that the sensitivity of the analytical method for measurement of the parent compound cannot be improved and that it is not possible to reliably measure the parent compound after single-dose administration, taking into account also the option of using a higher single dose in the bioequivalence study (see also "Strength to be investigated" section). Due to recent developments in bioanalytical methodology, it is unusual that parent drug cannot be measured accurately and precisely. Hence, the use of a metabolite as a surrogate for active parent compound is expected to be accepted only in exceptional cases. When using metabolite data as a substitute for active parent drug concentrations, the applicant should present any available data supporting the view that the metabolite exposure will reflect parent drug and that the metabolite formation is not saturated at therapeutic doses.

Enantiomers The use of achiral bioanalytical methods is generally acceptable. However, the individual enantiomers should be measured when all the following conditions are met:

1. The enantiomers exhibit different pharmacokinetics.
2. The enantiomers exhibit pronounced difference in pharmacodynamics.
3. The exposure (AUC) ratio of enantiomers is modified by a difference in the rate of absorption.

The individual enantiomers should also be measured if the aforementioned conditions are fulfilled or are unknown. If one enantiomer is pharmacologically active and the other is inactive or has a low contribution to activity, it is sufficient to demonstrate bioequivalence for the active enantiomer.

Use of Urinary Data The use of urinary excretion data as a surrogate for a plasma concentration may be acceptable in determining the extent of exposure where it is not possible to reliably measure the plasma concentration–time profile of parent compound. However, the use of urinary data has to be carefully justified when used to estimate peak exposure. If a reliable plasma C_{max} can be determined, this should be combined with urinary data on the extent of exposure for assessing bioequivalence. When using urinary data, the applicant should present any available data supporting that urinary excretion will reflect plasma exposure.

Endogenous Substances If the substance being studied is endogenous, the calculation of pharmacokinetic parameters should be performed using baseline correction so that the calculated pharmacokinetic parameters refer to the additional concentrations provided by the treatment. Administration of supratherapeutic doses can be considered in bioequivalence studies of endogenous drugs, provided that the dose is well tolerated, so that the additional concentrations over baseline provided by the treatment may be reliably determined. If a separation in exposure following administration of different doses of a particular endogenous substance has not been previously established, this should be demonstrated, either in a pilot study or as part of the pivotal bioequivalence study using different doses of the *R* formulation, in order to ensure that the dose used for the bioequivalence comparison is sensitive to detect potential differences between formulations.

The exact method for baseline correction should be prespecified and justified in the study protocol. In general, the standard subtractive baseline correction method, meaning either subtraction of the mean of individual endogenous predose concentrations or subtraction of the individual endogenous predose AUC, is preferred. In rare cases where substantial increases over baseline endogenous levels are seen, baseline correction may not be needed.

In bioequivalence studies with endogenous substances, it cannot be directly assessed whether carryover has occurred, so extra care should be taken to ensure that the washout period is of an adequate duration.

Strength to Be Investigated

If several strengths of a *T* product are applied for, it may be sufficient to establish bioequivalence at only one or two strengths, depending on the proportionality in composition between the different strengths and other product-related issues described in the following. The strength(s) to evaluate depends on the linearity in pharmacokinetics of the active substance.

In case of nonlinear pharmacokinetics (i.e., not proportional increase in AUC with increased dose), there may be a difference between different strengths in the sensitivity to detect potential differences between formulations. In the context of this guideline, pharmacokinetics is considered to be linear if the difference in dose-adjusted mean AUCs is no more than 25% when comparing the studied strength (or strength in the planned bioequivalence study) and the strength(s) for which a waiver is considered. In order to assess linearity, the applicant should consider all data available in the public domain with regard to the dose proportionality and review the data critically. Assessment of linearity will consider whether differences in dose-adjusted AUC meet a criterion of ±25%.

If bioequivalence has been demonstrated at the strengths that are most sensitive to detect a potential difference between products, in vivo bioequivalence studies for the other strength(s) can be waived.

General Biowaiver Criteria The following general requirements must be met where a waiver for additional strength(s) is claimed:

a. The pharmaceutical products are manufactured by the same manufacturing process.
b. The qualitative composition of the different strengths is the same.
c. The composition of the strengths are quantitatively proportional, that is, the ratio between the amount of each excipient to the amount of active substance(s) is the same for all strengths (for immediate-release products, coating components, capsule shell, color agents, and flavors are not required to follow this rule). If there is some deviation from quantitatively proportional composition, condition c is still considered fulfilled if conditions

(i) and (ii) *or* (i) and (iii) in the following apply to the strength used in the bioequivalence study and the strength(s) for which a waiver is considered:

i. The amount of the active substance(s) is less than 5% of the tablet core weight, the weight of the capsule content.
ii. The amounts of the different core excipients or capsule content are the same for the concerned strengths, and only the amount of active substance is changed.
iii. The amount of a filler is changed to account for the change in amount of active substance. The amounts of other core excipients or capsule content should be the same for the concerned strengths.

d. Appropriate in vitro dissolution data should confirm the adequacy of waiving additional in vivo bioequivalence testing (see "In vitro dissolution tests" section).

Linear Pharmacokinetics For products where all the aforementioned conditions (a)–(d) are fulfilled, it is sufficient to establish bioequivalence with only one strength.

The bioequivalence study should in general be conducted at the highest strength. For products with linear pharmacokinetics and where the drug substance is highly soluble (see Appendix 15C), selection of a lower strength than the highest is also acceptable. Selection of a lower strength may also be justified if the highest strength cannot be administered to healthy volunteers for safety/tolerability reasons. Further, if problems of sensitivity of the analytical method preclude sufficiently precise plasma concentration measurements after single-dose administration of the highest strength, a higher dose may be selected (preferably using multiple tablets of the highest strength). The selected dose may be higher than the highest therapeutic dose provided that this single dose is well tolerated in healthy volunteers and that there are no absorption or solubility limitations at this dose.

Nonlinear Pharmacokinetics For drugs with nonlinear pharmacokinetics characterized by a more than proportional increase in AUC with increasing dose over the therapeutic dose range, the bioequivalence study should in general be conducted at the highest strength. As for drugs with linear pharmacokinetics, a lower strength may be justified if the highest strength cannot be administered to healthy volunteers for safety/tolerability reasons. Likewise, a higher dose may be used in case of sensitivity problems of the analytical method in line with the recommendations given for products with linear pharmacokinetics previously.

For drugs with a less than proportional increase in AUC with increasing dose over the therapeutic dose range, bioequivalence should in most cases be established both at the highest strength and at the lowest strength (or a strength in the linear range), that is, in this situation, two bioequivalence studies are needed. If the nonlinearity is not caused by limited solubility but is due to, for example, saturation of uptake transporters and provided that conditions (a)–(d) earlier are fulfilled and the T and R products do not contain any excipients that may affect gastrointestinal motility or transport proteins, it is sufficient to demonstrate bioequivalence at the lowest strength (or a strength in the linear range). Selection of other strengths may be justified if there are analytical sensitivity problems preventing a study at the lowest strength or if the highest strength cannot be administered to healthy volunteers for safety/tolerability reasons.

Bracketing Approach Where bioequivalence assessment at more than two strengths is needed, for example, because of deviation from proportional composition, a bracketing approach may be used. In this situation, it can be acceptable to conduct two bioequivalence studies, if the strengths selected represent the extremes, for example, the highest and the lowest strength or the two strengths differing most in composition, so that any differences in composition in the remaining strengths are covered by the two conducted studies.

Where bioequivalence assessment is needed both in fasting and in fed state and at two strengths due to nonlinear absorption or deviation from proportional composition, it may be sufficient to assess bioequivalence in both fasting and fed state at only one of the strengths. Waiver of either the fasting or the fed study at the other strength(s) may be justified based on previous knowledge and/or pharmacokinetic data from the study conducted at the strength tested in both fasted and fed state. The condition selected (fasting or fed) to test the other strength(s) should be the one that is most sensitive to detect a difference between products.

Fixed Combinations The conditions regarding proportional composition should be fulfilled for all active substances of FCs. When considering the amount of each active substance in an FC, the other active substance(s) can be considered as excipients. In the case of bilayer tablets, each layer may be considered independently.

Bioanalytical Methodology

The bioanalytical part of bioequivalence trials should be performed in accordance with the principles of good laboratory practice (GLP). However, as human bioanalytical studies fall outside the scope of GLP, the sites conducting the studies are not required to be monitored as part of a national GLP compliance program.

The bioanalytical methods used must be well characterized, fully validated, and documented to yield reliable results that can be satisfactorily interpreted. Within study, validation should be performed using quality control (QC) samples in each analytical run.

The main characteristics of a bioanalytical method that are essential to ensure the acceptability of the performance and the reliability of analytical results are selectivity, lower limit of quantitation, the response function (calibration curve performance), accuracy, precision, and stability.

The lower limit of quantitation should be 1/20 of C_{max} or lower, as predose concentrations should be detectable at 5% of C_{max} or lower (see "Carryover effects" section).

Reanalysis of study samples should be predefined in the study protocol (and/or SOP) before the actual start of the analysis of the samples. Normally, reanalysis of subject samples because of a pharmacokinetic reason is not acceptable. This is especially important for bioequivalence studies, as this may bias the outcome of such a study.

Analysis of samples should be conducted without information on treatment.

Evaluation

In bioequivalence studies, the pharmacokinetic parameters should in general not be adjusted for differences in assayed content of the T and R batch. However, in exceptional cases where an R batch with an assay content differing less than 5% from T product cannot be found (see "Reference and test product" section), content correction could be accepted. If content correction is to be used, this should be prespecified in the protocol and justified by inclusion of the results from the assay of the T and R products in the protocol.

Subject Accountability Ideally, all treated subjects should be included in the statistical analysis. However, subjects in a crossover trial who do not provide evaluable data for both of the T and R products (or who fail to provide evaluable data for the single period in a parallel group trial) should not be included.

The data from all treated subjects should be treated equally. It is not acceptable to have a protocol, which specifies that "spare" subjects will be included in the analysis only if needed as replacements for other subjects who have been excluded. It should be planned that all treated subjects should be included in the analysis, even if there are no dropouts.

In studies with more than two treatment arms (e.g., a three-period study including two references, one from EU and another from United States, or a four-period study including T and R in fed and fasted states), the analysis for each comparison should be conducted excluding the data from the treatments that are not relevant for the comparison in question.

Reasons for Exclusion Unbiased assessment of results from randomized studies requires that all subjects are observed and treated according to the same rules. These rules should be independent from treatment or outcome. In consequence, the decision to exclude a subject from the statistical analysis must be made before bioanalysis.

In principle, any reason for exclusion is valid provided it is specified in the protocol and the decision to exclude is made before bioanalysis. However, the exclusion of data should be avoided, as the power of the study will be reduced and a minimum of 12 evaluable subjects is required.

Examples of reasons to exclude the results from a subject in a particular period are events such as vomiting and diarrhea, which could render the plasma concentration–time profile unreliable. In exceptional cases, the use of concomitant medication could be a reason for excluding a subject.

The permitted reasons for exclusion must be prespecified in the protocol. If one of these events occurs, it should be noted in the CRF as the study is being conducted. Exclusion of subjects based on these prespecified criteria should be clearly described and listed in the study report.

Exclusion of data cannot be accepted on the basis of statistical analysis or for pharmacokinetic reasons alone, because it is impossible to distinguish the formulation effects from other effects influencing the pharmacokinetics.

The exceptions to this are the following:

1. A subject with lack of any measurable concentrations or only very low plasma concentrations for reference medicinal product. A subject is considered to have very low plasma concentrations if its AUC is less than 5% of reference medicinal product geometric mean AUC (which should be calculated without inclusion of data from the outlying subject). The exclusion of data due to this reason will only be accepted in exceptional cases and may question the validity of the trial.
2. Subjects with nonzero baseline concentrations >5% of C_{max}. Such data should be excluded from bioequivalence calculation (see "Carryover effects" section).

The aforementioned effects can, for immediate-release formulations, be the result of subject noncompliance and an insufficient washout period, respectively, and should as far as possible be avoided by mouth check of subjects after intake of study medication to ensure the subjects have swallowed the study medication and by designing the study with a sufficient washout period. The samples from subjects excluded from the statistical analysis should still be assayed and the results listed (see "Presentation of data" section).

As stated in "Study conduct standardization" section, $AUC_{(0-t)}$ should cover at least 80% of $AUC_{(0-\infty)}$. Subjects should not be excluded from the statistical analysis if $AUC_{(0-t)}$ covers less than 80% of $AUC_{(0-\infty)}$, but if the percentage is less than 80% in more than 20% of the observations, then the validity of the study may need to be discussed. This does not apply if the sampling period is 72 h or more and $AUC_{(0-72\text{ h})}$ is used instead of $AUC_{(0-t)}$.

Parameters to Be Analyzed and Acceptance Limits In studies to determine bioequivalence after a single dose, the parameters to be analyzed are $AUC_{(0-t)}$ or, when relevant, $AUC_{(0-72\text{ h})}$ and C_{max}. For these parameters, the 90% CI for the ratio of the T and R products should be contained within the acceptance interval of 80.00%–125.00%. To be inside the acceptance interval, the lower bound should be ≥80.00% when rounded to two decimal places, and the upper bound should be ≤125.00% when rounded to two decimal places.

For studies to determine bioequivalence of immediate-release formulations at steady state, $AUC_{(0-\tau)}$ and $C_{max,ss}$ should be analyzed using the same acceptance interval as stated earlier.

In the rare case where urinary data have been used, $Ae_{(0-t)}$ should be analyzed using the same acceptance interval as stated earlier for $AUC_{(0-t)}$. R_{max} should be analyzed using the same acceptance interval as for C_{max}.

A statistical evaluation of t_{max} is not required. However, if rapid release is claimed to be clinically relevant and of importance for onset of action or is related to adverse events, there should be no apparent difference in median t_{max} and its variability between T and R product.

In specific cases of products with a narrow therapeutic range, the acceptance interval may need to be tightened (see "Narrow therapeutic index drugs" section). Moreover, for HVDPs, the acceptance interval for C_{max} may in certain cases be widened (see "Highly variable drugs or drug products" section).

Statistical Analysis The assessment of bioequivalence is based upon 90% CIs for the ratio of the population geometric means (*T*/*R*) for the parameters under consideration. This method is equivalent to two one-sided tests with the null hypothesis of bioinequivalence at the 5% significance level.

The pharmacokinetic parameters under consideration should be analyzed using ANOVA. The data should be transformed prior to analysis using a logarithmic transformation. A CI for the difference between formulations on the log-transformed scale is obtained from the ANOVA model. This CI is then back-transformed to obtain the desired CI for the ratio on the original scale. A nonparametric analysis is not acceptable.

The precise model to be used for the analysis should be prespecified in the protocol. The statistical analysis should take into account sources of variation that can be reasonably assumed to have an effect on the response variable. The terms to be used in the ANOVA model are usually sequence, subject within sequence, period, and formulation. Fixed effects, rather than random effects, should be used for all terms.

Carryover Effects A test for carryover is not considered relevant, and no decisions regarding the analysis (e.g., analysis of the first period only) should be made on the basis of such a test. The potential for carryover can be directly addressed by examination of the pretreatment plasma concentrations in period 2 (and beyond if applicable).

If there are any subjects for whom the predose concentration is greater than 5% of the C_{max} value for the subject in that period, the statistical analysis should be performed with the data from that subject for that period excluded. In a two-period trial, this will result in the subject being removed from the analysis. The trial will no longer be considered acceptable if these exclusions result in fewer than 12 subjects being evaluable. This approach does not apply to endogenous drugs.

Two-Stage Design It is acceptable to use a two-stage approach when attempting to demonstrate bioequivalence. An initial group of subjects can be treated and their data analyzed. If bioequivalence has not been demonstrated, an additional group can be recruited and the results from both groups combined in a final analysis. If this approach is adopted, appropriate steps must be taken to preserve the overall type I error of the experiment, and the stopping criteria should be clearly defined prior to the study. The analysis of the first-stage data should be treated as an interim analysis and both analyses conducted at adjusted significance levels (with the CIs accordingly using an adjusted coverage probability that will be higher than 90%). For example, using 94.12% CIs for both analyses of stage 1 and the combined data from stage 1 and stage 2 would be acceptable, but there are many acceptable alternatives, and the choice of how much alpha to spend at the interim analysis is at the company's discretion. The plan to use a two-stage approach must be prespecified in the protocol along with the adjusted significance levels to be used for each of the analyses.

When analyzing the combined data from the two stages, a term for stage should be included in the ANOVA model.

Presentation of Data All individual concentration data and pharmacokinetic parameters should be listed by formulation together with summary statistics such as geometric mean, median, arithmetic mean, standard deviation, CV, minimum, and maximum. Individual plasma concentration–time curves should be presented in linear–linear and log–linear scale. The method used to derive the pharmacokinetic parameters from the raw data should be specified. The number of points of the terminal log–linear phase used to estimate the terminal rate constant (which is needed for a reliable estimate of AUC∞) should be specified.

For the pharmacokinetic parameters that were subject to statistical analysis, the point estimate and 90% CI for the ratio of the *T* and *R* products should be presented.

The ANOVA tables, including the appropriate statistical tests of all effects in the model, should be submitted.

The report should be sufficiently detailed to enable the pharmacokinetics and the statistical analysis to be repeated, for example, data on actual time of blood sampling after dose, drug concentrations, the values of the pharmacokinetic parameters for each subject in each period, and the randomization scheme should be provided.

Dropout and withdrawal of subjects should be fully documented. If available, concentration data and pharmacokinetic parameters from such subjects should be presented in the individual listings, but should not be included in the summary statistics.

The bioanalytical method should be documented in a prestudy validation report. A bioanalytical report should be provided as well. The bioanalytical report should include a brief description of the bioanalytical method used and the results for all calibration standards and QC samples. A representative number of chromatograms or other raw data should be provided covering the whole concentration range for all standard and QC samples as well as the specimens analyzed. This should include all chromatograms from at least 20% of the subjects with QC samples and calibration standards of the runs including these subjects.

If, for a particular formulation at a particular strength, multiple studies have been performed, some of which demonstrate bioequivalence and some of which do not, the body of evidence must be considered as a whole. Only relevant studies, as defined in "Design, conduct, and evaluation of bioequivalence studies" section, need be considered. The existence of a study that demonstrates bioequivalence does not mean that those that do not can be ignored. The applicant should thoroughly discuss the results and justify the claim that bioequivalence has been demonstrated. Alternatively, when relevant, a combined analysis of all studies can be provided in addition to the individual study analyses. It is not acceptable to pool together studies that fail to demonstrate bioequivalence in the absence of a study that does.

Narrow Therapeutic Index Drugs

In specific cases of products with a narrow therapeutic index, the acceptance interval for AUC should be tightened to 90.00%–111.11%. Where C_{max} is of particular importance for safety, efficacy, or drug level monitoring, the 90.00%–111.11% acceptance interval should also be applied for this parameter. It is not possible to define a set of criteria to categorize drugs as NTIDs, and it must be decided case by case if an active substance is an NTID based on clinical considerations.

Highly Variable Drugs or Drug Products

HVDPs are those whose intrasubject variability for a parameter is larger than 30%. If an applicant suspects that a drug product can be considered as highly variable in its rate and/or extent of absorption, a replicate crossover design study can be carried out.

Those HVDPs for which a wider difference in C_{max} is considered clinically irrelevant based on a sound clinical justification can be assessed with a widened acceptance range. If this is the case, the acceptance criteria for C_{max} can be widened to a maximum of 69.84%–143.19%. For the acceptance interval to be widened, the bioequivalence study must be of a replicate design where it has been demonstrated that the within-subject variability for C_{max} of the reference compound in the study is >30%. The applicant should justify that the calculated intrasubject variability is a reliable estimate and that it is not the result of outliers. The request for widened interval must be prospectively specified in the protocol.

The extent of the widening is defined based upon the within-subject variability seen in the bioequivalence study using SABE according to $[U, L] = \exp[\pm k \cdot S_{WR}]$, where U is the upper limit of the acceptance range, L is the lower limit of the acceptance range, k is the regulatory constant set to 0.760, and S_{WR} is the within-subject standard deviation of the log-transformed values of C_{max} of the R product. The following table gives examples of how different levels of variability lead to different acceptance limits using this methodology.

Within-Subject CV (%)[a]	Lower Limit	Upper Limit
30	80.00	125.00
35	77.23	129.48
40	74.62	134.02
45	72.15	138.59
≥50	69.84	143.19

[a] CV (%) = $100\sqrt{(e^{S_{WR}} - 1)}$.

The GMR should lie within the conventional acceptance range 80.00%–125.00%. The possibility to widen the acceptance criteria based on high intrasubject variability does not apply to AUC where the acceptance range should remain at 80.00%–125.00% regardless of variability. It is acceptable to apply either a three-period or a four-period crossover scheme in the replicate design study.

In Vitro Dissolution Tests

General aspects of in vitro dissolution experiments are briefly outlined in Appendix 15A including basic requirements on how to use the similarity factor (f_2 test).

In Vitro Dissolution Tests Complementary to Bioequivalence Studies

The results of in vitro dissolution tests at three different buffers (normally pH 1.2, 4.5, and 6.8) and the media intended for drug product release (QC media), obtained with the batches of T and R products that were used in the bioequivalence study, should be reported. Particular dosage forms like ODTs may require investigations using different experimental conditions. The results should be reported as profiles of percent of labeled amount dissolved versus time, displaying mean values and summary statistics.

Unless otherwise justified, the specifications for the in vitro dissolution to be used for QC of the product should be derived from the dissolution profile of the T product batch that was found to be bioequivalent to the R product (see Appendix 15A).

In the event that the results of comparative in vitro dissolution of the biobatches do not reflect bioequivalence as demonstrated in vivo, the latter prevails. However, possible reasons for the discrepancy should be addressed and justified.

In Vitro Dissolution Tests in Support of Biowaiver of Strengths

Appropriate in vitro dissolution should confirm the adequacy of waiving additional in vivo bioequivalence testing. Accordingly, dissolution should be investigated at different pH values as outlined in the previous section (normally pH 1.2, 4.5, and 6.8) unless otherwise justified. Similarity of in vitro dissolution (see Appendix 15A) should be demonstrated at all conditions within the applied product series, that is, between additional strengths and the strength(s) (i.e. batch(es)) used for bioequivalence testing.

At pH values where sink conditions may not be achievable for all strengths, in vitro dissolution may differ between different strengths. However, the comparison with the respective strength of the reference medicinal product should then confirm that this finding is drug substance rather than formulation related. In addition, the applicant could show similar profiles at the same dose (e.g., as a possibility, two tablets of 5 mg versus one tablet of 10 mg could be compared).

Study Report

Bioequivalence Study Report

The report of the bioequivalence study should give the complete documentation of its protocol, conduct, and evaluation. It should be written in accordance with the ICH E3 guideline and be signed by the investigator in accordance with Annex I of the Directive 2001/83/EC as amended.

Names and affiliations of the responsible investigator(s), the site of the study, and the period of its execution should be stated. Audit certificate(s), if available, should be included in the report.

The study report should include evidence that the choice of the reference medicinal product is in accordance with Article 10(1) and Article 10(2) of Directive 2001/83/EC as amended. This should include the *R* product name, strength, pharmaceutical form, batch number, manufacturer, expiry date, and country of purchase.

The name and composition of the *T* product(s) used in the study should be provided. The batch size, batch number, manufacturing date, and, if possible, the expiry date of the *T* product should be stated.

Certificates of analysis of *R* and *T* batches used in the study should be included in an appendix to the study report.

Concentrations and pharmacokinetic data and statistical analyses should be presented in the level of detail described earlier ("Presentation of data" section).

Other Data to Be Included in an Application

The applicant should submit a signed statement confirming that the *T* product has the same quantitative composition and is manufactured by the same process as the one submitted for authorization. A confirmation whether the *T* product is already scaled up for production should be submitted. Comparative dissolution profiles (see "In vitro dissolution tests" section) should be provided.

The validation report of the bioanalytical method should be included in Module 5 of the application. Data sufficiently detailed to enable the pharmacokinetics and the statistical analysis to be repeated, for example, data on actual times of blood sampling, drug concentrations, the values of the pharmacokinetic parameters for each subject in each period, and the randomization scheme, should be available in a suitable electronic format (e.g., as comma-separated and space-delimited text files or Excel format) to be provided upon request.

Variation Applications

If a product has been reformulated from the formulation initially approved or the manufacturing method has been modified in ways that may impact on the BA, an in vivo bioequivalence study is required, unless otherwise justified. Any justification presented should be based upon general considerations, for example, as per Appendix 15C or on whether an acceptable level A in vitro/in vivo correlation has been established (see CPMP/QWP/ 604/96).

In cases where the BA of the product undergoing change has been investigated and an acceptable level A correlation between in vivo performance and in vitro dissolution has been established, the requirements for in vivo demonstration of bioequivalence can be waived if the dissolution profile in vitro of the new product is similar to that of the already approved medicinal product under the same test conditions as used to establish the correlation (see Appendix 15A). For variations of products approved under Articles 8(3), 10a, 10b, or 10c of Directive 2001/83/EC as amended, the comparative medicinal product for use in bioequivalence and dissolution studies is usually that authorized under the currently registered formulation, manufacturing process, packaging, etc.

When variations to a generic or hybrid product are made, the comparative medicinal product for the bioequivalence study should normally be a current batch of the reference medicinal product. If a valid reference medicinal product is not available on the market, comparison to the previous formulation (of the generic or hybrid product) could be accepted, if justified. For variations that do not require a bioequivalence study, the advice and requirements stated in other published regulatory guidance should be followed.

DEFINITIONS

Pharmaceutical equivalence: Medicinal products are pharmaceutically equivalent if they contain the same amount of the same active substance(s) in the same dosage forms that meet the same or comparable standards.

Pharmaceutical equivalence does not necessarily imply bioequivalence as differences in the excipients, and/or the manufacturing process can lead to faster or slower dissolution and/or absorption.

Pharmaceutical alternatives: Pharmaceutical alternatives are medicinal products with different salts, esters, ethers, isomers, mixtures of isomers, complexes, or derivatives of an active moiety, which differ in dosage form or strength.

Pharmacokinetic parameters

- $Ae_{(0-t)}$: Cumulative urinary excretion of unchanged drug from administration until time t
- $AUC_{(0-t)}$: Area under the plasma concentration curve from administration to last observed concentration at time t
- $AUC_{(0-\infty)}$: Area under the plasma concentration curve extrapolated to infinite time
- $AUC_{(0-\infty)}$: AUC during a dosage interval at steady state
- $AUC_{(0-72\ h)}$: Area under the plasma concentration curve from administration to 72 h
- C_{max}: Maximum plasma concentration
- $C_{max,ss}$: Maximum plasma concentration at steady state; residual area/extrapolated area $(AUC_{(0-\infty)} - AUC_{(0-t)})/AUC_{(0-\infty)}$
- R_{max}: Maximal rate of urinary excretion
- t_{max}: Time until C_{max} is reached
- $t_{max,ss}$: Time until $C_{max,ss}$ is reached
- $t_{1/2}$: Plasma concentration half-life
- λ_z: Terminal rate constant
- SmPC: Summary of product characteristics

APPENDIX 15A: DISSOLUTION TESTING AND SIMILARITY OF DISSOLUTION PROFILES

General Aspects of Dissolution Testing as Related to BA

During the development of a medicinal product, a dissolution test is used as a tool to identify formulation factors that are influencing and may have a crucial effect on the BA of the drug. As soon as the composition and the manufacturing process are defined, a dissolution test is used in the QC of scale-up and of production batches to ensure both batch-to-batch consistency and that the dissolution profiles remain similar to those of pivotal clinical trial batches. Furthermore, in certain instances, a dissolution test can be used to waive a bioequivalence study. Therefore, dissolution studies can serve several purposes:

1. Testing on product quality
    - To get information on the T batches used in BA/bioequivalence studies and pivotal clinical studies to support specifications for QC
    - To be used as a tool in QC to demonstrate consistency in manufacture
    - To get information on the R product used in BA/bioequivalence studies and pivotal clinical studies

2. Bioequivalence surrogate inference
   - To demonstrate in certain cases similarity between different formulations of an active substance and the reference medicinal product (biowaivers, e.g., variations, formulation changes during development, and generic medicinal products; see "In vitro dissolution tests" section and Appendix 15C)
   - To investigate batch-to-batch consistency of the products (T and R) to be used as basis for the selection of appropriate batches for the in vivo study

Test methods should be developed based on general and/or specific pharmacopoeia requirements. In case those requirements are shown to be unsatisfactory and/or do not reflect the in vivo dissolution (i.e., biorelevance), alternative methods can be considered when justified that these are discriminatory and able to differentiate between batches with acceptable and nonacceptable performance of the product in vivo. Current state-of-the-art information including the interplay of characteristics derived from the BCS classification and the dosage form must always be considered.

Sampling time points should be sufficient to obtain meaningful dissolution profiles and at least every 15 min. More frequent sampling during the period of greatest change in the dissolution profile is recommended. For rapidly dissolving products, where complete dissolution is within 30 min, generation of an adequate profile by sampling at 5 or 10 min intervals may be necessary.

If an active substance is considered highly soluble, it is reasonable to expect that it will not cause any BA problems if, in addition, the dosage system is rapidly dissolved in the physiological pH range and the excipients are known not to affect BA. In contrast, if an active substance is considered to have a limited or low solubility, the rate-limiting step for absorption may be dosage form dissolution. This is also the case when excipients are controlling the release and subsequent dissolution of the active substance. In those cases, a variety of test conditions are recommended, and adequate sampling should be performed.

Similarity of Dissolution Profiles

Dissolution profile similarity testing and any conclusions drawn from the results (e.g., justification for a biowaiver) can be considered valid only if the dissolution profile has been satisfactorily characterized using a sufficient number of time points.

For immediate-release formulations, further to the guidance given in "Introduction" section earlier, comparison at 15 min is essential to know if complete dissolution is reached before gastric emptying.

Where more than 85% of the drug is dissolved within 15 min, dissolution profiles may be accepted as similar without further mathematical evaluation. In case more than 85% is not dissolved at 15 min but within 30 min, at least three time points are required: the first time point before 15 min, the second one at 15 min, and the third time point when the release is close to 85%.
For modified-release products, the advice given in the relevant guidance should be followed. Dissolution similarity may be determined using the f_2 statistic as follows:

$$f_2 = 50 \cdot \log \left[\frac{100}{\sqrt{1 + \frac{\sum_{t=1}^{t=n} \left[\bar{R}(t) - \bar{T}(t) \right]^2}{n}}} \right]$$

where
f_2 is the similarity factor
n is the number of time points
$\bar{R}(t)$ is the mean percent reference drug dissolved at time t after initiation of the study
$\bar{T}(t)$ is the mean percent test drug dissolved at time t after initiation of the study

For both the *R* and *T* formulations, percent dissolution should be determined.

The evaluation of the similarity factor is based on the following conditions:

- A minimum of three time points (zero excluded).
- The time points should be the same for the two formulations.
- Twelve individual values for every time point for each formulation.
- Not more than one mean value of >85% dissolved for any of the formulations.
- The relative standard deviation or CV of any product should be less than 20% for the first point and less than 10% from second to last time point.

An f_2 value between 50 and 100 suggests that the two dissolution profiles are similar.

When the f_2 statistic is not suitable, then the similarity may be compared using model-dependent or model-independent methods, for example, by statistical multivariate comparison of the parameters of the Weibull function or the percentage dissolved at different time points.

Alternative methods to the f_2 statistic to demonstrate dissolution similarity are considered acceptable, if statistically valid and satisfactorily justified.

The similarity acceptance limits should be predefined and justified and not be greater than a 10% difference. In addition, the dissolution variability of the *T* and *R* product data should also be similar; however, a lower variability of the *T* product may be acceptable. Evidence that the statistical software has been validated should also be provided.

A clear description and explanation of the steps taken in the application of the procedure should be provided, with appropriate summary tables.

APPENDIX 15B: BIOEQUIVALENCE STUDY REQUIREMENTS FOR DIFFERENT DOSAGE FORMS

Although this guideline concerns immediate-release formulations, Appendix 15B provides some general guidance on the bioequivalence data requirements for other types of formulations and for specific types of immediate-release formulations.

When the *T* product contains a different salt, ester, ether, isomer, mixture of isomers, complex, or derivative of an active substance than the reference medicinal product, bioequivalence should be demonstrated in in vivo bioequivalence studies. However, when the active substance in both *T* and *R* products is identical (or contains salts with similar properties as defined in Appendix 15C, "Drug substance" section), in vivo bioequivalence studies may in some situations not be required as described in the following and in Appendix 15C.

Oral Immediate-Release Dosage Forms with Systemic Action

For dosage forms such as tablets, capsules, and oral suspensions, bioequivalence studies are required unless a biowaiver is applicable (see Appendix 15C). For orodispersible tablets (ODTs) and oral solutions, specific recommendations apply, as detailed in the following.

Orodispersible Tablets

An ODT is formulated to quickly disperse in the mouth. Placement in the mouth and time of contact may be critical in cases where the active substance also is dissolved in the mouth and can be absorbed directly via the buccal mucosa. Depending on the formulation, swallowing of the, for example, coated substance and subsequent absorption from the gastrointestinal tract also will occur. If it can be demonstrated that the active substance is not absorbed in the oral cavity but rather must be swallowed and absorbed through the gastrointestinal tract, then the product might be considered for a BCS-based biowaiver (see Appendix 15C). If this cannot be demonstrated, bioequivalence must be evaluated in human studies.

If the ODT *T* product is an extension to another oral formulation, a three-period study is recommended in order to evaluate administration of the ODT both with and without concomitant fluid intake. However, if bioequivalence between ODT taken without water and *R* formulation with water is demonstrated in a two-period study, bioequivalence of ODT taken with water can be assumed.

If the ODT is a generic/hybrid to an approved ODT reference medicinal product, the following recommendations regarding study design apply:

- If the reference medicinal product can be taken with or without water, bioequivalence should be demonstrated without water as this condition best resembles the intended use of the formulation. This is especially important if the substance may be dissolved and partly absorbed in the oral cavity. If bioequivalence is demonstrated when taken without water, bioequivalence when taken with water can be assumed.
- If the reference medicinal product is taken only in one way (e.g., only with water), bioequivalence should be shown in this condition (in a conventional two-way crossover design).
- If the reference medicinal product is taken only in one way (e.g., only with water), and the *T* product is intended for additional ways of administration (e.g., without water), the conventional and the new method should be compared with the reference in the conventional way of administration (three-treatment, three-period, six-sequence design).

In studies evaluating ODTs without water, it is recommended to wet the mouth by swallowing 20 mL of water directly before applying the ODT on the tongue. It is recommended not to allow fluid intake earlier than 1 h after administration.

Other oral formulations such as orodispersible films, buccal tablets or films, sublingual tablets, and chewable tablets may be handled in a similar way as for ODTs. Bioequivalence studies should be conducted according to the recommended use of the product.

Oral Solutions

If the *T* product is an aqueous oral solution at time of administration and contains an active substance in the same concentration as an approved oral solution, bioequivalence studies may be waived. However, if the excipients may affect gastrointestinal transit (e.g., sorbitol, mannitol), absorption (e.g., surfactants or excipients that may affect transport proteins), in vivo solubility (e.g., cosolvents), or in vivo stability of the active substance, a bioequivalence study should be conducted, unless the differences in the amounts of these excipients can be adequately justified by reference to other data. The same requirements for similarity in excipients apply for oral solutions as for biowaivers (see Appendix 15C, "Excipients" section).

In those cases where the *T* product is an oral solution, which is intended to be bioequivalent to another immediate-release oral dosage form, bioequivalence studies are required.

Fixed Combination Dosage Forms

Bioequivalence requirements are covered in the *Guideline on Clinical Development of Fixed Combination Medicinal Products*. The possibility for a biowaiver of fixed combination medicinal products is addressed in "Fixed combinations" section in Appendix 15C.

Non-Oral Immediate-Release Dosage Forms with Systemic Action

This section applies to, for example, rectal formulations. In general, bioequivalence studies are required. A biowaiver can be considered in the case of a solution that contains an active substance in the same concentration as an approved solution and with the same qualitative and similar quantitative composition in excipients (conditions under oral solutions may apply in this case).

Parenteral Solutions

Bioequivalence studies are generally not required if the *T* product is to be administered as an aqueous IV solution containing the same active substance as the currently approved product. However, if any excipients interact with the drug substance (e.g., complex formation) or otherwise affect the disposition of the drug substance, a bioequivalence study is required unless both products contain the same excipients in very similar quantity, and it can be adequately justified that any difference in quantity does not affect the pharmacokinetics of the active substance.

In the case of other parenteral routes, for example, intramuscular (IM) or subcutaneous (SC), and when the *T* product is of the same type of solution (aqueous or oily) and contains the same concentration of the same active substance and the same excipients in similar amounts as the medicinal product currently approved, bioequivalence studies are not required. Moreover, a bioequivalence study is not required for an aqueous parenteral solution with comparable excipients in similar amounts, if it can be demonstrated that the excipients have no impact on the viscosity.

Liposomal, Micellar, and Emulsion Dosage Forms for Intravenous Use

- *Liposomal formulations*: Pharmacokinetic issues related to liposomal formulations for IV administration require special considerations that are not covered by the present guideline.
- *Emulsions*: Emulsions normally do not qualify for a biowaiver.

However, emulsion formulations may be considered eligible for a biowaiver where

a. The drug product is not designed to control release or disposition
b. The method and rate of administration are the same as the currently approved product

In these cases, the composition should be qualitatively and quantitatively the same as the currently approved emulsion, and satisfactory data should be provided to demonstrate very similar physicochemical characteristics, including size distribution of the dispersed lipid phase, and supported by other emulsion characteristics considered relevant, for example, surface properties, such as zeta potential and rheological properties.

- *Lipids for IV parenteral nutrition* may be considered eligible for a biowaiver if satisfactory data are provided to demonstrate comparable physicochemical characteristics. Differences in composition may be justified, taking into consideration the nature and the therapeutic purposes of such dosage forms.
- *Micelle-forming formulations*: Micelle solutions for IV administration may be regarded as "complex" solutions and therefore normally do not qualify for a biowaiver. However, micelle formulations may be considered eligible for a biowaiver where
  a. Rapid disassembly of the micelle on dilution occurs and the drug product is not designed to control release or disposition
  b. The method and rate of administration are the same as the currently approved product
  c. The excipients do not affect the disposition of the drug substance

In these cases, the composition of the micelle infusion, immediately before administration, should be qualitatively and quantitatively the same as that currently approved, and satisfactory data should be provided to demonstrate similar physicochemical characteristics, for example, the critical micelle concentration, the solubilization capacity of the formulation (such as maximum additive concentration), free and bound active substance, and micelle size.

This also applies in case of minor changes to the composition quantitatively or qualitatively, provided this does not include any change of amount or type of surfactants.

Modified-Release Dosage Forms with Systemic Action

Modified-Release Oral and Transdermal Dosage Forms

Bioequivalence studies are required in accordance with the guideline on Modified Release Oral and Transdermal Dosage Forms: Section II (Pharmacokinetic and Clinical Evaluation) (CPMP/EWP/280/96).

Modified-Release Intramuscular or Subcutaneous Dosage Forms

For suspensions or complexes or any kind of matrix intended to delay or prolong the release of the active substance for IM or SC administration, demonstration of bioequivalence follows the rules for extravascular modified-release formulations, for example, transdermal dosage forms as per corresponding guideline.

Locally Acting, Locally Applied Products

For products for local use (after oral, nasal, pulmonary, ocular, dermal, rectal, vaginal, etc., administration) intended to act at the site of application, recommendations can be found in other guidelines (CPMP/EWP/4151/00 rev 1, CPMP/EWP/239/95).

A waiver of the need to provide equivalence data may be acceptable in the case of solutions, for example, eye drops, nasal sprays, or cutaneous solutions, if the T product is of the same type of solution (aqueous or oily) and contains the same concentration of the same active substance as the medicinal product currently approved. Minor differences in the excipient composition may be acceptable if the relevant pharmaceutical properties of the T product and R product are identical or essentially similar. Any qualitative or quantitative differences in excipients must be satisfactorily justified in relation to their influence on therapeutic equivalence. The method and means of administration should also be the same as the medicinal product currently approved, unless otherwise justified.

Whenever systemic exposure resulting from locally applied, locally acting medicinal products entails a risk of systemic adverse reactions, systemic exposure should be measured. It should be demonstrated that the systemic exposure is not higher for the T product than for the R product, that is, the upper limit of the 90% CI should not exceed the upper bioequivalence acceptance limit 125.00.

Gases

If the product is a gas for inhalation, bioequivalence studies are not required.

APPENDIX 15C: BCS-BASED BIOWAIVER

Introduction

The BCS-based biowaiver approach is meant to reduce in vivo bioequivalence studies, that is, it may represent a surrogate for in vivo bioequivalence. In vivo bioequivalence studies may be exempted if an assumption of equivalence in in vivo performance can be justified by satisfactory in vitro data.

Applying for a BCS-based biowaiver is restricted to highly soluble drug substances with known human absorption and considered not to have a narrow therapeutic index (see "Narrow therapeutic index drugs" section). The concept is applicable to immediate-release solid pharmaceutical products for oral administration and systemic action having the same pharmaceutical form. However, it is not applicable for sublingual, buccal, and modified-release formulations. For orodispersible formulations, the BCS-based biowaiver approach may only be applicable when absorption in the oral cavity can be excluded.

BCS-based biowaivers are intended to address the question of bioequivalence between specific *T* and *R* products. The principles may be used to establish bioequivalence in applications for generic medicinal products, extensions of innovator products, variations that require bioequivalence testing, and between early clinical trial products and to-be-marketed products.

Summary Requirements

BCS-based biowaiver are applicable for an immediate-release drug product if the following apply:

- The drug substance has been proven to exhibit high solubility and complete absorption (BCS class I; for details, see "Drug substance" section).
- Either very rapid (>85% within 15 min) or similarly rapid (85% within 30 min) in vitro dissolution characteristics of the T and R product has been demonstrated, considering specific requirements (see "In vitro dissolution" section).
- Excipients that might affect BA are qualitatively and quantitatively the same. In general, the use of the same excipients in similar amounts is preferred (see "Excipients" section).

BCS-based biowaiver are also applicable for an immediate-release drug product if the following apply:

- The drug substance has been proven to exhibit high solubility and limited absorption (BCS class III; for details, see "Drug substance" section).
- Very rapid (>85% within 15 min) in vitro dissolution of the *T* and *R* product has been demonstrated, considering specific requirements (see "In vitro dissolution" section).
- Excipients that might affect BA are qualitatively and quantitatively the same and other excipients are qualitatively the same and quantitatively very similar (see "Excipients" section).

Generally, the risks of an inappropriate biowaiver decision should be more critically reviewed (e.g., site-specific absorption, risk for transport protein interactions at the absorption site, excipient composition, and therapeutic risks) for products containing BCS class III than for BCS class I drug substances.

Drug Substance

Generally, sound peer-reviewed literature may be acceptable for known compounds to describe the drug substance characteristics of importance for the biowaiver concept.

Biowaiver may be applicable when the active substances in *T* and *R* products are identical. Biowaiver may also be applicable if *T* and *R* contain different salts provided that both belong to BCS class I (high solubility and complete absorption; see "Solubility" and "Absorption" sections). Biowaiver is not applicable when the *T* product contains a different ester, ether, isomer, mixture of isomers, complex, or derivative of an active substance from that of the *R* product, since these differences may lead to different bioavailabilities not deducible by means of experiments used in the BCS-based biowaiver concept. The drug substance should not belong to the group of NTIDs (see "Narrow therapeutic index drugs" section on NTIDs).

Solubility

The pH–solubility profile of the drug substance should be determined and discussed. The drug substance is considered highly soluble if the highest single dose administered as immediate-release formulation(s) is completely dissolved in 250 mL of buffers within the range of pH 1–6.8 at 37°C ± 1°C. This demonstration requires the investigation in at least three buffers within this range (preferably at pH 1.2, 4.5, and 6.8) and in addition at the pK_a, if it is within the specified pH range. Replicate determinations at each pH condition may be necessary to achieve an unequivocal solubility classification (e.g., shake-flask method or other justified method). Solution pH should be verified prior and after addition of the drug substance to a buffer.

Absorption

The demonstration of complete absorption in humans is preferred for BCS-based biowaiver applications. For this purpose, complete absorption is considered to be established where measured extent of absorption is ≥85%. Complete absorption is generally related to high permeability.

Complete drug absorption should be justified based on reliable investigations in human. Data from absolute BA or mass-balance studies could be used to support this claim. When data from mass-balance studies are used to support complete absorption, it must be ensured that the metabolites taken into account in determination of fraction absorbed are formed after absorption. Hence, when referring to total radioactivity excreted in urine, it should be ensured that there is no degradation or metabolism of the unchanged drug substance in the gastric or intestinal fluid. Phase 1 oxidative and phase 2 conjugative metabolism can only occur after absorption (i.e., cannot occur in the gastric or intestinal fluid). Hence, data from mass-balance studies support complete absorption if the sum of urinary recovery of parent compound and urinary and fecal recovery of phase 1 oxidative and phase 2 conjugative drug metabolites account for ≥85% of the dose.

In addition, highly soluble drug substances with incomplete absorption, that is, BCS class III compounds, could be eligible for a biowaiver provided certain prerequisites are fulfilled regarding product composition and in vitro dissolution (see also "Excipients" section). The more restrictive requirements will also apply for compounds proposed to be BCS class I, but complete absorption could not convincingly be demonstrated.

Reported bioequivalence between aqueous and solid formulations of a particular compound administered via the oral route may be supportive as it indicates that absorption limitations due to (immediate-release) formulation characteristics may be considered negligible. Well-performed in vitro permeability investigations including reference standards may also be considered supportive to in vivo data.

Drug Product

In Vitro Dissolution

General Aspects

Investigations related to the medicinal product should ensure immediate-release properties and prove similarity between the investigative products, that is, *T* and *R* show similar in vitro dissolution under physiologically relevant experimental pH conditions. However, this does not establish an in vitro/in vivo correlation. In vitro dissolution should be investigated within the range of pH 1–6.8 (at least pH 1.2, 4.5, and 6.8). Additional investigations may be required at pH values in which the drug substance has minimum solubility. The use of any surfactant is not acceptable.

T and *R* products should meet requirements as outlined in "Reference and test product" section of the main guideline text. In line with these requirements, it is advisable to investigate more than one single batch of the *T* and *R* products. Comparative in vitro dissolution experiments should follow current compendial standards. Hence, thorough description of experimental settings and analytical methods including validation data should be provided. It is recommended to use 12 units of the product for each experiment to enable statistical evaluation. Usual experimental conditions include the following:

- Apparatus: Paddle or basket.
- Volume of dissolution medium: 900 mL or less.
- Temperature of the dissolution medium: 37°C ± 1°C.
- Agitation: Paddle apparatus, usually 50 rpm; basket apparatus, usually 100 rpm.
- Sampling schedule: For example, 10, 15, 20, 30, and 45 min.
- Buffer: pH 1.0–1.2 (usually 0.1 N HCl or SGF without enzymes), pH 4.5, and pH 6.8 (or SIF without enzymes) (pH should be ensured throughout the experiment; Ph. Eur. buffers recommended).
- Other conditions: No surfactant; in case of gelatin capsules or tablets with gelatin coatings, the use of enzymes may be acceptable.

Complete documentation of in vitro dissolution experiments is required including a study protocol, batch information on *T* and *R* batches, detailed experimental conditions, validation of experimental methods, individual and mean results, and respective summary statistics.

Evaluation of In Vitro Dissolution Results

Drug products are considered "very rapidly" dissolving when more than 85% of the labeled amount is dissolved within 15 min. In cases where this is ensured for the *T* and *R* products, the similarity of dissolution profiles may be accepted as demonstrated without any mathematical calculation.

Absence of relevant differences (similarity) should be demonstrated in cases where it takes more than 15 min but not more than 30 min to achieve almost complete (at least 85% of labeled amount) dissolution. F_2 testing (see Appendix 15A) or other suitable tests should be used to demonstrate profile similarity of *T* and *R*. However, the discussion of dissolution profile differences in terms of their clinical/therapeutical relevance is considered inappropriate since the investigations do not reflect any in vitro/in vivo correlation.

Excipients

Although the impact of excipients in immediate-release dosage forms on BA of highly soluble and completely absorbable drug substances (i.e., BCS class I) is considered rather unlikely, it cannot be completely excluded. Therefore, even in the case of class I drugs, it is advisable to use similar amounts of the same excipients in the composition of test like in the *R* product.

If a biowaiver is applied for a BCS class III drug substance, excipients have to be qualitatively the same and quantitatively very similar in order to exclude different effects on membrane transporters.

As a general rule, for both BCS class I and III drug substances, well-established excipients in usual amounts should be employed, and possible interactions affecting drug BA and/or solubility characteristics should be considered and discussed. A description of the function of the excipients is required with a justification whether the amount of each excipient is within the normal range. Excipients that might affect BA, like sorbitol, mannitol, sodium lauryl sulfate, or other surfactants, should be identified as well as their possible impact on

- Gastrointestinal motility
- Susceptibility of interactions with the drug substance (e.g., complexation)
- Drug permeability
- Interaction with membrane transporters

Excipients that might affect BA should be qualitatively and quantitatively the same in the *T* product and the *R* product.

Fixed Combinations

BCS-based biowaiver is applicable for immediate-release FC products if all active substances in the FC belong to BCS class I or III and the excipients fulfill the requirements outlined in "Excipients" section. Otherwise, in vivo bioequivalence testing is required.

16 Bioequivalence Testing
The ROW Perspective

BACKGROUND

The degree of assessment of bioequivalence (BE) of drug product is highly influenced by the regulatory environment of the country of marketing. Highly regulated markets have more stringent regulatory policy than countries that are not tightly regulated. The magnitude of regulatory influence is often dictated by the availability of resources, expertise, and lack of regulation or its implementation. Thus, there is a greater need to harmonize the regulatory environment globally for BE assessment so that the drug product marketed in different parts and regions of the world would have optimum drug product quality in terms of interchangeability.

In the recent years, there has been a significant progress made toward harmonization but these efforts have mostly been regional in nature and broader integration of regulations is still a long way from becoming a reality. The main efforts have come from the International Conference on Harmonisation (ICH), a consortium of regulatory authorities from Europe, Japan, and United States. Other organizations are also involved in this effort. ICH has primarily focused on developing guidelines for standardizing and harmonizing the regulatory requirements, primarily for the chemistry and manufacturing control, safety, and efficacy aspects of new drug product quality. In addition, it has developed specific documents for content and format of drug product dossier. It has not yet focused on harmonizing the requirements for approval of generic equivalents.

Other national and international organizations are also involved in this endeavor. One of the major international organizations is the World Health Organization (WHO) that has made tremendous progress specifically in developing international consensus with regard to the regulatory requirements for assessing BE for marketing authorization of multisource pharmaceutical products (MPPs) for interchangeability, selection of comparator product for BE assessment, and other related documents. There are other European and Asian organizations that are also actively involved in the harmonization efforts. Some national, regional, and international organizations are involved in the area of assessment of BE and improving the quality of pharmaceutical products globally. Some of these organizations and regulatory authorities have also developed guidance related to BE and product quality.

Harmonized BE criteria for the interchangeability of pharmaceutical products address the issue of waivers for in vivo trials, which are expensive and, generally, not always discriminating enough to form the sole basis of approval of interchangeability. As discussed in the succeeding text, the worldwide requirements to demonstrate BE vary widely, mostly as a result of the lack of expertise available at the regulatory authorities to evaluate and mostly to the ability to enforce such requirements, both from an economic and ethical perspective. The waiver for BE testing therefore becomes a topic of great interest worldwide. The issue of ethical submissions becomes a problem as often the companies have used the comparisons between the products that could not be tracked properly.

Several consortiums have debated this topic for years and a consensus has begun to develop on this topic. A large number of policy documents address this topic and include the published FDA and ICH guidelines, *Health Canada's Guideline on Preparation of DIN Submissions*, WHO document (1999) entitled *Marketing Authorization of Pharmaceutical Products with Special Reference to Multisource (Generic) Products: a Manual for Drug Regulatory Authorities* and *Multisource (Generic) Pharmaceutical Products: Guidelines on Registration Requirements to Establish Interchangeability*, Committee for Proprietary Medicinal Products (CPMP) (2001)'s *Note for Guidance on the Investigation of Bioavailability and Bioequivalence* (CPMP/EWP/QWP/98), and Pan American Network on Regulatory Harmonization: Bioavailability and Bioequivalence working group 2004.

Drug regulatory authorities must ensure that all pharmaceutical products, including generic drug products, conform to the same standards of quality, efficacy, and safety required of innovator drug products. Therefore, regulatory frameworks must be able to respond to varied and emerging drugs and dosage forms where BE demonstration is required; issues such as BE of topical products, products acting locally, endogenous therapeutic proteins, and, more recently, botanical products now need regulatory pathways, besides streamlining and reducing cost of evaluation of more traditional dosage forms where cost considerations, especially in the Third World, and often a lack of good correlation between in vivo studies and clinical response are observed. This chapter addresses these issues and provides a pathway for the prospective filers of marketing approval applications worldwide.

Whereas there is a general consensus among the West European, North American, and Japanese regulatory authorities on the BE requirements for marketing authorization of generic products, such is not the case in the rest of the world. For example, the diverse nature of the requirement in South America typifies the heterogeneity in other continents. For example, an examination of the regulatory systems of the 10 South American agencies showed that out of the 96 active ingredients, only 4 active ingredients commonly require BE studies in all 10 countries: valproic acid, carbamazepine, cyclosporine, and phenytoin. All of them are considered high health risks. The countries with the least number of active ingredient with BE study requirements are Colombia (only 5) followed by Costa Rica (only 7) and the countries with the highest number of requirements remain the United States and Canada. Chile is in the process of establishing that requirement for all active ingredients that require BE studies. Whereas the WHO has established certain guidelines, these are not widely followed in much of the Third World countries and BE studies remain haphazardly managed. The following are some of the common occurrences in the marketing approvals of generic products in the Third World countries:

- Nonvalidated test methods
- Statistically incorrect experimental designs
- Lack of authenticity of study
- Lack of assurance that the study is conducted on the manufactured batches; the MNCs routinely submitting studies from their filings in the West in support of products to be manufactured locally

The process of regulatory approval for interchangeable multisource generic drug products is not identical in all countries and the regulatory agency for each country may differ in its regulatory requirements for the demonstration of BE. Moreover, there is no universal reference-listed drug (RLD) to be used as the comparator for the proposed generic drug product. Each country establishes its own RLD product that is commonly available in its own domestic marketplace. In addition, the statistical criteria for BE may vary in each country where the brand drug product is marketed. Consequently, a generic drug product that has established BE to the RLD product in the United States may not be bioequivalent to the RLD product in another country.

Most countries appear to have adopted the basic tenets and approaches used by those countries that pioneered (Canada and United States) the introduction and application of this particular tool for the assessment of BE of different formulations of the same drug product. Apart from its general use, both by innovator and generic drug companies, there still appears to be a certain degree of naivety and ignorance associated with BE testing where the notion persist that it is specifically a tool for use only in the assessment of generic drug products.

However, some causes for concern still remain, and those relate to issues of semantics and interpretation of the main objectives such as the association between BE, therapeutic equivalence, and generic substitution. In particular, statements such as "*may or may not be interchangeable*" or "*may or may not be bioequivalent* or *therapeutically equivalent*," which appear in some formal definitions such as the following, are disconcerting.

Note: a product that has been approved based on comparison with a nondomestic comparator product *may or may not* be interchangeable with currently marketed domestic products.

MPPs. Pharmaceutically equivalent or pharmaceutically alternative products that *may or may not* be therapeutically equivalent. MPP that are therapeutically equivalent are interchangeable.

Pharmaceutical alternatives. Pharmaceutical alternatives deliver the same active moiety by the same route of administration but are otherwise not pharmaceutically equivalent. They *may or may not* be bioequivalent or therapeutically equivalent to the comparator product.

In addition, the term *essentially similar* as used in European Union (EU) guidance and documents creates confusion with respect to considerations of BE, therapeutic equivalence, and interchangeability. The issue of *pharmaceutical alternatives* remains inconsistent between countries with respect to permission for market approval with interchangeability status.

Finally, the issue and questionable use of an "acceptable reference product" or comparator product for generic substitution (AB rating) differs between countries. In the United States, it is quite clear that the reference product for use in a BE study must be an RLD as published in the Orange Book, whereas in many other countries, including the EU, a nondomestic reference product may be used to establish BE, therapeutic equivalence, and interchangeability of an MPP. In spite of the foregoing, the applications of the principles of average BE and its related methodologies have clearly served the international pharmaceutical industry very well over the years since its introduction. BE testing has proved to be an extremely valuable tool, expedient, efficient, and relatively inexpensive in comparison with the costs to conduct clinical trials. It serves both innovator/brand companies during their formulation development and postapproval changes of products initially approved on the basis of clinical safety and efficacy studies and generic manufacturers to obtain market approval for their MPP.

AUSTRALASIA (AUSTRALIA AND NEW ZEALAND)

The Australian requirements according to the Therapeutic Goods Administration (TGA) of Australia are controlled by the Therapeutic Goods Act 1989, and the New Zealand requirements as set out by Medsafe, a business unit of the New Zealand Ministry of Health (MoH), are in accordance with the Medicines Act 1981, section 20. It was initially intended that from July 1, 2006, all medicines were to receive regulatory approval from a proposed trans-Tasman joint regulatory authority. This joint agency was to have responsibility for the control of medicines in both Australia and New Zealand. However, establishment of the joint agency has been postponed and there has been no indication if and when such an agency will come into being. Currently, the Australian and New Zealand registration requirements are largely aligned with the Committee of Proprietary Medicinal Products (CPMP) guidelines of the European Medicines Agency (EMA). Prescription medicines for Australia are evaluated by the Drug Safety and Evaluation Branch (DSEB) of the TGA, and generic prescription medicines in both Australia and New Zealand are required to conform to the EU's Common Technical Document (CTD) format although, unlike the EU, an abbreviated format is acceptable.

BRAZIL

In Brazil, the *National Policy for Drug Products* published in 1998 created the National Agency for Sanitary Surveillance (Agencia Nacional de Vigilancia Sanitaria, ANVISA), which is responsible for the approval of the law and the publication of the technical guidances for the registration of generic products. The law and guidances for the registration of generic products in Brazil are based on the regulations of countries such as Canada, United States, and the EU. Brazil was the first country in South America to implement the evaluation of pharmaceutical equivalency and BE studies for the registration of generic products.

CANADA

Canadian regulators, under the Health Products and Food Branch (HPFB), Therapeutic Products Directorate (TPD), Health Canada, were the first to apply pharmacokinetics (PK) to safety and efficacy risk assessment of generic drug products as a consequence of 1969 amendments to the Patent Act (compulsory licensing). Guidelines (Reports A, B, and C) were only published in the 1990s by an Expert Advisory Committee (EAC), currently referred to as the Scientific Advisory Committee (SAC). However, Canada is governed as a Confederation of Provinces and the regulations and guidelines for BE are federal, leading to a Notice of Compliance (NOC) to sponsors for marketing in Canada. Although an application may lead to a declaration of BE for specific products, the various Canadian provinces and associated provincial formulary committees may not accept the federal decision of BE (which also evaluates quality) to be sufficient to list a particular product as interchangeable.

Revision of the Canadian guidelines was mentioned during presentations at the 2008 and 2009 annual meetings of the Canadian Society of Pharmaceutical Sciences (CSPS) held in Banff last May 2008 and in Toronto last June 2009.

EUROPEAN UNION

The EU has established requirements, which must be met by a generic drug product to receive marketing authorization. The EU offers four routes for the registration of generic drug products:

- National procedure (NP)
- Mutual recognition procedure (MRP)
- Decentralized procedure (DCP)
- Centralized procedure (CP)

The *NP* is strictly limited to medicinal products that are not authorized in more than one member state and may lead to marketing authorization of the generic drug product in the concerned member state. The *MRP* makes provision for the extension of marketing authorizations granted to one member state, the so-called reference member state (RMS), to one or more member states identified by the applicant. The *DCP* was implemented from November 2005 where a submission can be made to each of the member states, where it is intended to obtain a marketing authorization with a choice of one of them as the RMS. The *CP* has been in use since 2004 for marketing authorization of medicinal products in the EU. Here, a single application is introduced for evaluation that is carried out within the Committee for Medicinal Products for Human Use (CHMP) of the EMA and is valid throughout the EU with the same rights and obligations in each of the member states.

The first European BE guidelines were published in 1991 and assessment was based on the principles published in the scientific literature, FDA guidelines, and the first European guidelines on PK studies in man. In 2001, the EMA's CPMP published the current version of the *Note for Guidance on the Investigation of Bioavailability and Bioequivalence*. A draft version of revised BE guidelines, entitled *Guideline on the Investigation of Bioequivalence*, was made publicly available in August 2008 on the EMA website and a modified version is due to come into effect in 2010.

INDIA

BE assessment for generic medicines in India was instituted by the incorporation of Schedule Y of the Drugs and Cosmetics Act in 1988, followed by subsequent amendments of Schedule Y in 1989 and 2005.

In India, generic medicines are those medicines that are labeled with their generic names. There is no separate law for registering generic medicines. Drugs and drug products are classified as either "new drugs" or drugs other than new drugs. However, in general, "generic drugs" refer to those drugs that are no longer subject to patent protection and are being marketed by their generic name. In fact, it's interesting to note that India only started observing product patents from January 1, 2005.

India has a two-tier regulatory system, the Central Drugs Standard Control Organization (CDSCO) under the Government of India, and each state has its own drug regulatory system having certain powers.

A few years ago, India became a popular destination for the contract research organizations (CROs) to conduct BE studies; however, the bureaucracy in India has literally killed this industry as the length of time it takes to secure approval to conduct BE studies is unreasonable. Even animal toxicology studies require government approval and this has significantly hurt the CRO industry in India.

JAPAN

In Japan, data from 2006 indicated that generic drug products accounted for as little as 16.9% of the market by volume 5.7% by value. In the following year, the Japanese government announced a specific program to increase the use of generic products to more than 30% by the year 2012 and a system for generic substitution was targeted for implementation in April 2008.

Approval to manufacture and market generic drug products in accordance with the Pharmaceutical Affairs Law in Japan is granted by the Minister of Health, Labour and Welfare (MHLW). An independent administrative organization, the Pharmaceutical and Medical Devices Agency (PMDA), under the auspices of the MHLW, undertakes the review and assessment of BE data as part of the "Equivalence and Compliance Review."

The first guideline for BE studies for generic drugs was released in 1971, in which large animals, such as dogs and rabbits, could be used in BE studies, but humans were not required. Subsequently, in 1997, the MHLW published a revised guideline *The Guideline for Bioequivalence Studies of Generic Products* (1997). Several newer guidelines and revisions have been published such as the *Guideline for Bioequivalence Studies for Different Strengths of Oral Solid Dosage Forms* (2000), *Guideline for Bioequivalence Studies for Additional Dosage Forms of Oral Solid Dosage Forms* (2001), *Guideline for Bioequivalence Studies of Generic Products for Topical Dermal Application* (2003), *Guideline for Bioequivalence Studies for Formulation Changes of Oral Solid Dosage Forms* (2000), *Draft Guideline for Bioequivalence Studies for Changes in Manufacturing of Oral Solid Dosage Forms: Conventional, Enteric Coated and Prolonged Release Products* (2003), and *Revision of Guideline for Bioequivalence Studies of Generic Products* (2006).

For oral drug products, in Japan, dissolution testing of these dosage forms is required together with BE studies and plays an important role in selecting appropriate subjects for the in vivo study. In fact, it is noteworthy that the use of dissolution testing for bioavailability (BA)/BE in Japan clearly marks a distinct difference between Japanese BA/BE guidelines and those of most other countries.

Given the importance of the Japanese submission and significant differences with the U.S. and EMA guidelines, an English summary of these guidelines is provided here.

SOUTH AFRICA

The Medicines Control Council (MCC) in South Africa is a statutory body that was established in terms of the Medicines and Related Substances Control Act (MRSCA), 101 of 1965, to oversee the regulation of medicines in South Africa. To facilitate the registration process for generic medicines, guidelines have been prepared to serve as a recommendation to applicants wishing to submit data in support of the registration of such medicines.

In the early 2000s, the requirements for the registration and market approval of generic medicines were published as official guidelines. Prior to that time, proof of safety and efficacy of generic medicines were based on requirements described in "official" notices or circulars issued by the MCC. In many instances, only in vitro dissolution testing was required based upon Circular 14/95 that was first issued in the early 1990s and subsequently updated in 1995.

During 2002, legislation was introduced to make provision for generic substitution whereby dispensers of medicines were mandated to inform patients of the benefits of the substitution for a branded medicine by an interchangeable multisource medicine and to recommend accordingly. The final decision, however, has been left in the hands of the patient.

SOUTH AMERICA (EXCLUDING BRAZIL): PAN AMERICAN HEALTH ORGANIZATION

In South American countries (Brazil has been dealt with previously), harmonization efforts have been carried out by different economic integration groups, such as *Tratado de Libre Comercio de Norteamérica* (TLCN) in Canada, United States, and Mexico; MERCOSUR in Argentina, Brazil, Paraguay, and Uruguay (Chile and Bolivia participate without being members); *Sistema de la Integración Centroamericana* (SICA) in Guatemala, El Salvador, Honduras, Nicaragua, and Costa Rica; *Comunidad Andina de Naciones* (CAN) in Bolivia, Colombia, Ecuador, Peru, and Venezuela; Caribbean Community (CARICOM) in Caribbean Islands; and by Pan American Health Organization (PAHO) through Pan American Network for Drug Regulatory Harmonization (PANDRH). Generally, most South American countries, apart from Brazil, do not have regulations for the registration of generic products as such. However, proof of BE and an inference of therapeutic equivalence and/or declaration of interchangeability (through either in vitro or in vivo methodology) are required as a condition either for registration or commercialization of generic (i.e., noninnovator) products.

BRAZIL

ANVISA is responsible for drug registration and licenses to pharmaceutical laboratories and to other companies inside the pharmaceutical production flow. The agency is also responsible for establishing regulations applicable to clinical trials and drug pricing, which is carried out by the Chamber of Drug Market Regulation (CMED).

Together with states and municipalities, the agency inspects factories, monitors the quality of drugs, exercises postmarketing surveillance, takes pharmacovigilance actions, and regulates drug promotion and marketing.

Moreover, ANVISA is in charge of analyzing patent requests related to pharmaceutical processes and products, in partnership with the National Industrial Property Institute (INPI). The International CROs that intend to be certificated by ANVISA must contact a national company to represent them at ANVISA. The selected national company will be responsible for the submission of specific application forms for first certification or certification renewal.

Brazil's ANVISA has suspended processing of new applications for primary certification of BA or BE for medical products by research centers located outside the country.

Applications filed prior to July 31, 2013, the date the suspension entered into force (RDC 37/13), will still be processed. The purpose of the suspension is to facilitate the review of existing rules governing foreign centers and the development of a new certification model that will be implemented later. The new certification model is being prepared now by the agency and will be released for public consultation after review and approval of the Board of ANVISA.

TAIWAN

The Bureau of Pharmaceutical Affairs (BPA) within the Department of Health (DOH) is responsible for the regulation of medicinal products in Taiwan. In 1984, the BPA outsourced the drafting of BA/BE guidelines and Taiwan guidelines on BA/BE for generic medicines were issued in 1987. Generic products are reviewed and approved within the BPA at the DOH. The applicant needs to certify to the DOH that the patent in question is not infringed by the generic product. Once the BPA staff have completed the filing review of the submission and have verified that the application has met all the necessary regulatory requirements, the application is then assigned to technical review, which focuses on BE data and chemistry and manufacture quality.

TURKEY

Marketing authorization for medicinal products for humans are issued by the MoH in Turkey, but the entire procedure is managed by the General Directorate of Drug and Pharmacy (GDDP). Turkish licensing regulations for all pharmaceutical products, innovator/brand, and generics came into force on June 30, 2005 (37). This legislation brought Turkish law in line with that of the EU and covered all aspects of the drug registration procedure and all new Turkish regulations are intended to be, as far as possible, compatible with those of the EU. However, BE became compulsory for a generic drug product to receive a marketing license after the publication of a regulation on May 27, 1994 (38). In general, the design and conduct of a BE study should follow Turkish and/or EU regulations on Good Clinical Practice (GCP).

WORLD HEALTH ORGANIZATION

The WHO has developed a document entitled *937, WHO Expert Committee on Specifications for Pharmaceutical Preparations*, the 40th report in which annex 7 relates to BE. WHO, which is the directing and coordinating authority for health within the United Nations system, is responsible for providing leadership on global health matters, shaping the health research agenda, setting norms and standards, articulating evidence-based policy options, providing technical support to countries, and monitoring and assessing health trends. In particular, the intention is to help national and regional authorities (in particular drug regulatory authorities) and procurement agencies, as well as major international bodies and institutions, to combat problems of substandard medicines and underpin important initiatives. Importantly, since the overall tendency is that resource-constrained or resource-poor countries are less likely to control the quality of products on the market, due to the absence of properly resourced and functioning regulatory authorities, WHO publications and guidelines are intended to fill an important void by providing appropriate information in the quest to ensure medicinal product quality.

INTERPRETATION OF TECHNICAL TERMS

Several issues prevail in the official documents and guidelines published by various regulatory authorities, which appear to have resulted in confusion and misinterpretation of criteria and conditions for the declaration of BE and consequently establishing the interchangeability/switchability of a generic product for an innovator/brand product.

The definition of the term "bioequivalence" itself gives rise to a degree of misunderstanding when related to generic substitution. For example, the Orange Book provides the following definition for BE, namely, "the absence of a significant difference in the rate and extent to which the active ingredient or active moiety in pharmaceutical equivalents or pharmaceutical alternatives becomes available at the site of drug action when administered at the same molar dose under similar conditions in an appropriately designed study."

Importantly, in this definition, provision is made for "pharmaceutical alternatives" to be declared bioequivalent to the listed reference product in the Orange Book. However, pharmaceutical alternatives are not considered as therapeutic equivalents and therefore are neither interchangeable nor switchable.

The following is stated in the Orange Book:

Drug products are considered to be therapeutic equivalents only if they are pharmaceutical equivalents and if they can be expected to have the same clinical effect and safety profile when administered to patients under the conditions specified in the labeling.

Consequently, in terms of the FDA's requirements, only products classified as therapeutically equivalent can be substituted with the full expectation that the substituted product will produce the same clinical effect and safety profile as the prescribed product. FDA specifically excludes a "pharmaceutical alternative" dosage form from attaining therapeutic equivalent status. In contrast, this specific requirement for therapeutic equivalence is either overlooked or deemed unnecessary in several countries where pharmaceutical alternatives, once shown to be bioequivalent, are considered to be interchangeable. Reference to the WHO document includes the following definitions/descriptions where the term pharmaceutical alternatives is mentioned. For example, a *multisource (generic) pharmaceutical product* is described as pharmaceutically equivalent or pharmaceutically alternative products that may or may not be therapeutically equivalent. This intention is unclear since what would be the basis of market approval for a pharmaceutically equivalent or pharmaceutically alternative generic product that has not been shown to be therapeutically equivalent.

COMPARATOR (REFERENCE) PRODUCTS

A further confusing issue also arises from some of the requirements stated in the BE WHO document where, specifically, mention is made in that document of a "comparator product" under section 3 entitled *Choice of Comparator Products*. This section describes how to identify the correct comparator product against which the proposed MPP (generic) must be compared. Generally, the comparator or reference product for use in a BE study needs to be the "nationally authorized innovator," usually the innovator/brand product that has been approved and available on that domestic market. However, the WHO document makes provision for some options of choice regarding the comparator product by including the permitted use of an innovator product currently available on the market in a well-regulated country. This brings into consideration the use of a foreign or nondomestic reference product. Although such a choice is permitted only when a nationally authorized innovator may not be available (i.e., countries where no innovator/brand product is available on their domestic market), this constraint appears to have been dismissed by some countries where, in spite of the availability of a nationally authorized innovator, use of a foreign reference product is permitted. The WHO comparator product was instituted specifically as an effort to assist national drug regulatory authorities and pharmaceutical companies in selecting appropriate comparator products from a list of comparator products derived from information collected from drug regulatory authorities and the pharmaceutical industry. The WHO document includes the following:

Note: a product that has been approved based on comparison with a nondomestic comparator product may or may not be interchangeable with currently marketed domestic products.

Hence, it is important to be aware that should a nationally authorized innovator product not be available as the comparator product, products approved on the basis of comparison to the chosen comparator product may or may not be interchangeable with other products currently available within that particular market. In other words, "generic substitution" per se cannot be recommended unconditionally.

GENERIC SUBSTITUTION (INTERCHANGEABILITY)

In some countries, where the market share of a new generic product may be relatively small, sponsors tend to submit BE studies where the reference product from another country (foreign/nondomestic) was used. Therefore, BE studies that were performed in compliance with the requirements of one country are submitted in support of approval of that generic in another country. The intent for such practice is to reduce costs and avoid the necessity of performing an additional BE study. However, concerns have been raised about such practice in view of the fact that different formulations of the same product are often used in different countries and may have significant consequences with respect to BA.

A unique situation currently exists in South Africa. The government, a few years ago, decided that when a generic medicine has been approved for marketing by the national medicines regulatory authority, the MCC, prescribers and dispensers of medicines must inform the patient that such prescribed medicines may be available at a lower price than the innovator/brand product and recommend accordingly. In other words, where an approved generic medicine exists, the generic medicine will be substituted for the innovator/brand product unless the dispenser is prohibited to do so by the patient. The mandatory instruction is based on the premise that approved generic medicines have been assessed by the MCC and deemed to be interchangeable. However, in terms of the national guidelines, most generic medicines approved and marketed in South Africa do not comply with the usual internationally accepted requirements for interchangeability since most have not been assessed by comparison with the innovator/brand product available on the South African market. The reason for the "noninterchangeability" notion is that provision has been made in the local guidelines to permit BE assessment to be conducted using a "foreign" reference product. This means that the "foreign" reference product, although being supplied by the same innovator/brand company, may not be the same (identical formulation, manufacture, etc.) as the innovator/brand product being sold on the South African market. There are many instances where the innovator/brand products are formulated differently for different markets. For example, Tegretol XR^R tablets, a prolonged action carbamazepine product, is marketed in the United States as a nondisintegrating dosage form using the OROS^R mechanism. The same innovator is listed as the manufacturer of prolonged action carbamazepine dosage forms in various other countries where those dosage forms are also tablets but which disintegrate in aqueous fluid and are marketed as Tegretol CR^R in South Africa. The release mechanisms and formulation of the United States reference-listed product and the product marketed by the same innovator in South Africa are clearly different. In some instances, this is done intentionally due to patents. In other cases, there may be unintended differences in release of the active ingredient(s), due to various factors such as the manufacturing process. BA differences between products can be due to factors such as sources of raw material and synthesis (nature) of the API including particle size and crystal forms (polymorphs, crystal shapes and degree of hydration or salvation, etc.), use of different methods of manufacture and manufacturing equipment, among others. All of these factors can have significant effects on BA with consequent implications for their BE. Hence, in the absence of specific confirmatory data, a nondomestic innovator/brand product used as the reference product in a BE study involving a generic medicine intended for a particular domestic market cannot be assumed to be bioequivalent to the domestic innovator/brand product. In spite of the foregoing, the only data required by the MCC to show that the "foreign" reference product is the "same" as the reference product marketed in South Africa are dissolution profiles comparing f_2 values between the foreign and domestic reference product conducted in three different dissolution media at pH 1.5, 4.5, and 6.8. These comparisons are not constrained to any particular class of medicine or properties such as the BCS or drug use, potency, therapeutic index (e.g., narrow), among others. Risk assessment has apparently neither been done nor is required.

Whereas it should be noted that a generic product that has been shown to be bioequivalent to a reference product purchased in a nondomestic market will likely provide a spectrum of safety/efficacy associated with the included active ingredient, it is important to stress that such a generic product cannot be deemed to be equivalent to the innovator/rand product available on another (domestic) market and vice versa, unless appropriate data have been obtained to show BE. In other words, a generic product that has been shown to be bioequivalent to a nondomestic reference product may well be usable or "prescribable" but in the absence of the necessary comparative BA data showing BE between that same generic product and the domestic reference product, that generic product cannot be considered to be interchangeable.

From the foregoing discussion it appears that the MCC, by making provision for a foreign reference product to be used in a BE study (i.e., using a nondomestic innovator/brand product as the reference), has inadvertently and naively created a two-tiered system for the approval of generic medicines in South Africa. The top tier can therefore be considered to consist of generic products approved on the basis of comparison with the domestic innovator/brand product as the reference, whereas another (second or lower) tier includes those generic products approved on the basis of a comparison of the generic with a nondomestic innovator/brand as the reference in a BE study. In addition, the latter tier probably also includes all other generic medicines that have been approved on the basis of in vitro testing only, apart from those products which incorporate an API classified as Class 1 according to the BCS.

It also appears that a similar situation of ill-conceived interchangeability for approved generic controlled-/modified-release dosage forms and also non-oral dosage forms such as topical products for local use, inhalation products, and various other such generic products that are not intended to be absorbed into the systemic circulation that these exists.

In summary, it should be emphasized that only when a generic medicine has been shown to be bioequivalent with the domestic innovator/brand product used as reference will substitution be acceptable and appropriate. On the basis of "similar" BA, as would the case be when the generic has been shown to be bioequivalent to a nondomestic reference product, a clinical decision may be justified to declare a "second-tier" generic product "prescribable" for a patient who is naive to that particular medicine, but certainly, interchangeability of such a product is highly questionable.

Although the current BE guidelines and recommendations of the major regional and national health authorities show a fair degree of consistency, a number of outstanding BE issues and concerns remain to be resolved. The most obvious of these controversial issues, such as the BE acceptance limits for NTI drugs and HVDs/HVDPs, the use of metabolites for BE assessment, and conditions to grant biowaivers, are not always dealt with in the same way by the various health authorities. Global harmonization should therefore be the next logical step in the continuing process to improve the BE guidelines as a means to guarantee safe and efficacious drug products for the consumer in all parts of the world. Global harmonization efforts by the ICH and the WHO should be stepped up in collaboration with the regulatory agencies of the western world as more nations throughout the world have come to rely on low-cost, good-quality multisource (generic) pharmaceutical products to provide lower health-care costs without sacrificing important public health goals. However, as pointed out earlier, consensus on a number of BE issues have not been reached at this point in time among international regulatory agencies. In addition, differing levels of commitment and resources by the various countries and regions constitute another formidable barrier that has to be overcome to harmonize BE approaches to ensure development of optimally performing and affordable drug products for use by health practitioners and patients in the global community.

Due to significant recognition of the BA/BE concept all over the world, tremendous advancements have been made by the FDA as well as various national, international, and supranational regulatory authorities. In parallel, pharmaceutical industry and academia are also contributing exclusively in the area of assessment of BE. Currently available approaches to determine BE of generic products are largely standardized due to discussion and consensus reached among various stakeholders at numerous national and international meetings, conferences, and workshops (e.g., American Association of Pharmaceutical Scientists, Federation Internationale Pharmaceutique). Thus, the currently available excellent scientific and regulatory guidance documents are due to the combined efforts of industry, academia, and regulatory scientists.

GLOBAL DIVERSITY

The adaptation of the BE concept worldwide for over the years has enabled the production and approval of quality generic products through profound scientific, technical, and regulatory advances (especially through replicate designs, application of BCS, scaled average BE) by various approaches to assess BE for various complex and special groups of drugs. This continuing success story of BE is based on the contribution to efficacy, safety, and quality by international regulatory authorities, pharma industry researchers, academic researchers, and indeed the efforts from ICH, WHO, and various international conferences. However, a lot remains to be done, especially to promote global harmonization of BE approaches, which should focus on uniformity, standardization of nomenclature, agreement on general concepts, alternative approaches for locally acting drug products, choice of test procedures, outlier challenge, and consideration of BE criteria and objectives, all of which reflect regulatory decision-making standards, as well as ensuring product quality over time for both innovator and generic drugs. To achieve these objectives, efforts should continue from international health organizations, pharmaceutical industries, researchers, and regulatory authorities to understand and to develop more efficient and scientifically valid approaches to assess BE and develop generic drugs in a cost-effective manner.

The magnitude of regulatory influence is often dictated by the availability of resources, expertise, and lack of regulation or its implementation. Thus, there is a greater need to harmonize the regulatory environment globally for BE assessment as far as practicable so that the drug product marketed in different parts and regions of the world would have optimum drug product quality in terms of interchangeability. In the recent years, some significant progress has been made toward harmonization; in addition, some regulatory authorities are also in the process of cooperating with their counterparts from other countries to harmonize the regulatory requirements while streamlining their own regulatory requirements.

Tremendous work toward harmonization was initiated and completed by some organizations, especially the ICH and the WHO. ICH is a consortium of regulatory authorities from Europe, Japan, and the United States that focused primarily on developing guidelines for standardizing and harmonizing the regulatory requirements, mainly for aspects of chemistry and manufacturing control, safety, and efficacy of new drug product quality. In addition, it developed specific documents for the content and format of drug product dossiers. It has not yet focused on harmonizing the requirements for approval of generic equivalents. On the other hand, the WHO has made remarkable progress specifically in developing international consensus on the regulatory requirements for assessing BE for marketing authorization of MPPs for interchangeability, selection of comparator product for BE assessment, and other related regulatory documents. Apart from the ICH and WHO, other European and Asian organizations (national and international) are actively involved in harmonization efforts for assessing of BE and improving the quality of pharmaceutical products globally.

GLOBAL AGENCIES

Every country now has its own individual regulatory authority as well as regulatory guidance for BE studies, and the regulatory environment of the respective country of marketing influences the magnitude of assessment of BE of drug product. The regulatory authorities of various countries and international organizations are listed and briefly described in Table 16.1.

GENERAL ASSESSMENT OF BIOEQUIVALENCE

The global paradigm for the assessment of BE of different drug products is based on the fundamental assumption that two products are equivalent when the rate and extent of absorption of the test/generic drug does not show a significant difference from the rate and extent of absorption of the reference/brand drug under similar experimental conditions as defined. Global agencies classify BE studies in the descending order of preference as

1. PK endpoint studies
2. Pharmacodynamic (PD) endpoint studies
3. Clinical endpoint studies
4. In vitro endpoint studies

TABLE 16.1
Global Regulatory Agencies and Organizations

Country	Agency	Web Address
Armenia	Scientific Center of Drug and Medical Technologies Expertise (SCDMTE)	http://www.pharm.am/
ASEAN	Association of Southeast Asian Nations Consultative Committee for Standards and Quality	http://www.aseansec.org/
Australia	Therapeutic Goods Administration (TGA)	http://www.tga.gov.au/
Belgium	Pharmaceutical Inspectorate	http://afigp.fgov.be/
Brazil	National Health Surveillance Agency (ANVISA)	http://www.anvisa.gov.br/
Bulgaria	Drug Agency	http://www.bda.bg/
Canada	Health Canada	http://www.hc-sc.gc.ca/
China, People's Republic of	National Institute for the Control of Pharmaceutical and Biological Products	http://www.nicpbp.org.cn/cmsweb/
Colombia	Instituto Nacional de Vigilancia de Medicamentos Y Alimentos (INVIMA)	http://web.invima.gov.co/
Czech Republic	State Institute for Drug Control	http://www.sukl.cz/
Europe	European Medicines Agency (EMA)	http://www.ema.europa.eu/
European Union	European Commission and EMA	http://www.ema.europa.eu/
Fiji	Ministry of Health	http://www.health.gov.fj/
Finland	National Agency for Medicines	http://www.nam.fi/
France	Agence Francaise de Securite Sanitaire des Produits de Sante (AFSSAPS)	http://www.afssaps.fr/
Germany	Federal Institute for Drugs and Medical Devices	http://www.bfarm.de/
Japan	Global GMP Harmonization	http://www.nihs.go.jp/drug/section3/hiyama070518-3.pdf
Global Harmonization Task Force	GHTF	http://www.ghtf.org/
Greece	National Organization for Medicines	http://www.eof.gr/
Hong Kong	Department of Health	http://www.dh.gov.hk/
Iceland	Icelandic Medicines Agency (IMA)	http://www.imca.is/
India	Central Drugs Standard Control Organization (CDSCO)	http://cdsco.nic.in/
Indonesia	Ministry of Health	http://www.depkes.go.id/
International Conference on Harmonization	ICH	http://www.ich.org/
Ireland	Medicines Board	http://www.imb.ie/
Israel	Ministry of Health	http://www.health.gov.il/
Italy	National Institute of Health	http://www.iss.it/
Japan	Pharmaceuticals and Medical Devices Agency (PMDA)	http://www.pmda.go.jp/
Kenya	Ministry of Health	http://www.publichealth.go.ke/
Korea	Korea Food and Drug Administration (K-FDA)	http://www.kfda.go.kr/
Malaysia	National Pharmaceutical Control Bureau	http://portal.bpfk.gov.my/
Mexico	Ministry of Health	http://www.salud.gob.mx/
Namibia	Ministry of Health and Social Services	http://www.healthforall.net/grnmhss/
Netherlands	Medicines Evaluation Board	http://www.cbg-meb.nl/
New Zealand	Medicines and Medical Devices Safety Authority (MEDSAFE)	http://www.medsafe.govt.nz/
Norway	Norwegian Medicines Agency	http://www.legemiddelverket.no/

(*Continued*)

TABLE 16.1 (*Continued*)
Global Regulatory Agencies and Organizations

Country	Agency	Web Address
Poland	Drug Institute	http://www.il.waw.pl/
Saudi Arabia	Ministry of Health	http://www.moh.gov.sa/
Singapore	Health Sciences Authority (HSA)	http://www.hsa.gov.sg
South Africa	Medicines Control Council (MCC)	http://www.mccza.com/
Spain	Spanish Drug Agency	http://www.msc.es/
Sri Lanka	Ministry of Health	http://www.health.gov.lk/
Sweden	Medical Products Agency	http://www.lakemedelsverket.se/
Switzerland	Swiss Agency for Therapeutic Products	http://www.swissmedic.ch/
Taiwan	Department of Health (DOH)	http://www.doh.gov.tw/
Tanzania	Ministry of Health	http://wwww.tanzania.go.tz/
United Arab Emirates	Federal Department of Pharmacies	http://www.uae.gov.ae/
United Kingdom	Medicines and Healthcare Products Regulatory Agency (MHRA)	http://www.mhra.gov.uk/
Unites States	U.S. Food and Drug Administration (FDA)	http://www.fda.gov/
World Health Organization	WHO	http://www.who.int/
Zimbabwe	Ministry of Health and Child Welfare	http://www.gta.gov.zw/health.html

Pharmacokinetic Endpoint Studies

These studies are most widely preferred to assess BE for drug products, where drug level can be determined in an easily accessible biological fluid (such as plasma, blood, urine) and drug level is correlated with the clinical effect. The statutory definition of BA and BE, expressed in rate and extent of absorption of the active moiety or ingredient to the site of action, emphasizes the use of PK measures to indicate release of the drug substance from the drug product with absorption into the systemic circulation. Regulatory guidance recommends that measures of systemic exposure be used to reflect clinically important differences between test and reference products in BA and BE studies. These measures include (1) total exposure AUC_{0-t} or $AUC_{0-\infty}$ for single-dose studies and $AUC_{0-\tau}$ for steady-state studies, (2) peak exposure (C_{max}), and (3) early exposure (partial AUC to peak time of the reference product of an immediate-release drug product). Reliance on systemic exposure measures will reflect comparable rate and extent of absorption, which, in turn, will achieve the underlying goal of assuring comparable therapeutic effects. Single-dose studies to document BE were preferred because they are generally more sensitive in assessing in vivo release of the drug substance from the drug product when compared to multiple-dose studies. Table 16.2 describes the general PK parameters (primary and secondary) for single-dose, multiple-dose, and urinary data.

The following are the circumstances that demand multiple-dose study/steady-state PK:

- Dose- or time-dependent PK
- For modified-release products for which the fluctuation in plasma concentration over a dosage interval at steady state needs to be assessed
- If problems of sensitivity preclude sufficiently precise plasma concentration measurements after single-dose administration
- If the intraindividual variability in the plasma concentration or disposition precludes the possibility of demonstrating BE in a reasonably sized single-dose study and this variability is reduced at steady state
- When a single-dose study cannot be conducted in healthy volunteers due to tolerability reasons and a single-dose study is not feasible in patients

TABLE 16.2
Primary Pharmacokinetic Parameters Used in Bioavailability and Bioequivalence Testing

Study Type	Primary Parameters	Secondary Parameters
Single dose	C_{max}, AUC_{0-t}, $AUC_{0-\infty}$	t_{max}, AUC% extrapolation, MRT, Kel, t1/2
Steady state	$C_{max(ss)}$, $C_{min(ss)}$, $AUC_{0-\tau}$	$T_{min(ss)}$, $t_{max(ss)}$, % swing, % fluctuation
Urinary based	$Ae_{(0-t)}$, $Ae_{(0-\infty)}$, R_{max}	t_{lag}

- If the medicine has a long terminal half-life and blood concentrations after a single dose cannot be followed for a sufficient time
- For those medicines that induce their own metabolism or show large intraindividual variability
- For combination products for which the ratio of plasma concentration of the individual substances is important
- If the medicine is likely to accumulate in the body
- For enteric-coated preparations in which the coating is innovative

Under normal circumstances, blood should be the biological fluid sampled to measure drug concentrations. Most drugs may be measured in serum or plasma; however, in some drugs, whole blood (e.g., tacrolimus) may be more appropriate for analysis. If the blood concentrations are too minute to be detected and a substantial amount (40%) of the drug is eliminated unchanged in the urine, the urine may serve as the biological fluid to be sampled (e.g., alendronic acid).

Table 16.2 lists the primary PK parameters used in BA and BE studies.

Pharmacodynamic Endpoint Studies

PK studies measure systemic exposure but are generally inappropriate to document local delivery BA and BE. In such cases, BA may be measured, and BE may be established, based on a pharmacodynamic study, providing an appropriate pharmacodynamic endpoint is available. Pharmacodynamic evaluation is measurement of the effect on a pathophysiological process, such as a function of time, after administration of two different products to serve as a basis for BE assessment. Regulatory authorities request justification from the applicant for the use of pharmacodynamic effects/parameters for the establishment of BE criteria. These studies generally become necessary under two conditions: (1) if the drug and/or metabolite(s) in plasma or urine cannot be analyzed quantitatively with sufficient accuracy and sensitivity and (2) if drug concentration measurement cannot be used as surrogate endpoints for the demonstration of efficacy and safety of the particular pharmaceutical product. The other important specifications for pharmacodynamic studies include (1) a dose–response relationship should be demonstrated; (2) sufficient measurements should be taken to provide an appropriate pharmacodynamic response profile; (3) the complete dose–effect curve should remain below the maximum physiological response; (4) all pharmacodynamic measurements/methods should be validated for specificity, accuracy, and reproducibility. Examples of these pharmacodynamic studies include locally acting drug products and oral inhalation drug products, such as metered-dose inhalers and dry powder inhalers, and topically applied dermatologic drug products, such as creams and ointments. Bronchodilator drug products, such as albuterol metered-dose inhalers, produce relaxation of smooth muscle of the airways. For these drug products, a pharmacodynamic endpoint, based either on increase in forced expiratory volume in 1 s (FEV1) or on measurement of PD20 or PC20 (the dose or concentration, respectively, of a challenge agent), is clinically relevant and may be used for BA and BE studies.

Clinical Endpoint Studies or Comparative Clinical Trials

In the absence of PK and pharmacodynamic approaches, adequate and well-controlled clinical trials may be used to establish BE. Several international regulatory authorities provide general information about the conduct of clinical studies to establish BE.

In Vitro Endpoint Studies

More recently, a Biopharmaceutics Classification System (BCS) has categorized drug substances as having either high or low solubility and permeability and drug products as exhibiting rapid dissolution. According to this approach, drug substances may be classified into four primary groups:

1. Highly soluble and highly permeable
2. Highly permeable and poorly soluble
3. Highly soluble and poorly permeable
4. Poorly soluble and poorly permeable

Using this BCS approach, a highly permeable, highly soluble drug substance formulated into a rapidly dissolving drug product may need only in vitro dissolution studies to establish BE. In addition, in vitro approaches to document BE for drugs with no known BA problems and approved before 1962 remain acceptable as per FDA regulations. Dissolution tests can also be used to reduce the number of in vivo studies in other circumstances and to (1) assess batch-to-batch quality and support batch release, (2) provide process control and quality assurance, and (3) assess the need for further BE studies relative to minor postapproval changes, where they function as a signal of bioinequivalence.

DESIGN AND ANALYSIS

The general considerations for the advancement of conducting BE studies are as follows:

- Study design and protocol
- Bioanalysis
- Selection of appropriate analyte(s)
- BE metrics and data treatment
- Statistical approaches and analysis
- Acceptance criteria for BE

STUDY DESIGN

Successfully determining the BE of generic drugs to their respective reference drugs depends mostly on design and managing the conduct of study such that the highest quality samples are obtained. Some regulatory authorities provide specific information on RLDs to be used to demonstrate BE (Table 16.3).

Attention should also be paid to sizing the study properly (to achieve sufficient statistical power to demonstrate BE), enrolling subjects as per relevant inclusion and exclusion criteria, ensuring that the appropriate overall design (simple two-period crossover, replicate design to gain direct information on within-subject variability for both test and reference product or parallel design) can adequately address the question at hand, standardization of the environmental conditions (such as, fasting, fed, ambulatory, supine), and ensuring that GCPs are strictly adhered to and documented. All of these should be planned a priori and embodied in the overall protocol and study plan for the smooth execution of BE studies.

TABLE 16.3
Agencies Providing Specific Information on Drugs to Conduct Bioequivalence Studies

Agency	URL
United States	http://www.accessdata.fda.gov/scripts/cder/drugsatfda/index.cfm http://www.fda.gov/Drugs/GuidanceComplianceRegulatoryInformation/Guidances/ucm075214.htm
Canada	http://webprod.hc-sc.gc.ca/dpd-bdpp/index-eng.jsp
Europe	http://www.medicines.org.uk/EMC/browsedocuments.aspx
Australia	Australia: https://www.ebs.tga.gov.au/

Generally, the study design and number of studies (single dose and/or multiple dose and/or fasting and/or fed) depend on the RLD product, physicochemical properties of the drug, its PK properties, and proportionality in composition with justification along with respective regulatory guidance and specifications. Table 16.4 describes various study designs generally used for BE studies.

Genetic variations among ethnic and/or racial background can alter the drug disposition (e.g., white persons who predominantly express less P-glycoprotein in intestinal epithelial cells than black persons) and thus lead to potential sources of variability in PK parameters apart from geographical, food habits, and metabolic variations. For BE studies, these problems will be minimized using crossover designs, and hence, U.S. and Europe regulatory agencies (but not Japan, Korea, and Mexico, for example) are accepting BE studies from other countries also, as these factors mostly do not have much effect on test and reference products. BE studies should be generally performed on a healthy population unless safety warranties (patient population should be preferred, if the risk associated with the drug is more in healthy population; e.g., anticancer drugs) as they facilitate the provision of adequate information to detect formulation differences and allow extrapolation of this information to populations for which the brand drug is approved.

The regulatory specifications on strength to be investigated, demographics, sample size, number of studies required, fasting and/or fed requirements, standardization of experimental conditions (fluid intake, posture, and physical activity), add-on design, and sampling and washout criteria are briefly described in Tables 16.5 through 16.12.

As a result of random variation or a larger than expected relative difference, there is no guarantee that the sample size as calculated will pass the standards. If the study is run with the appropriate size and the standards are not met, the sponsor may add more subjects, and this approach is generally referred to as an "add-on" study (Table 16.11).

BIOANALYSIS

In a general prospective of BA/BE studies, bioanalysis should be the subsequent step following clinical operations of the study, and it should be executed with strict adherence to good laboratory practices, standard operating procedures, and specific regulatory requirements. Bioanalysis is a term generally used to describe the quantitative measurement of a compound (drug) or its metabolite in biological fluids, primarily blood, plasma, serum, urine, or tissue extracts. Bioanalysis typically consists of two important components (1) sample preparation and (2) detection of the desired compound using a validated method. Excellent scientific and regulatory guidance documents are available that outline the requirements for a fully validated method. The application of validated methodology presupposes that the most appropriate analyte is monitored to attest to the question of BE.

TABLE 16.4
Brief Description of Bioavailability and Bioequivalence Testing Designs

Design	Significance	Advantages	Disadvantages
Crossover	• When intrasubject CV (approx. 15%) is usually substantially smaller than that intersubject CV (approx. 30%). • Generally recommended by all regulatory authorities.	• Since the treatments are compared on the same subject, the intersubject variability does not contribute to the error variability. • Subject randomization causes unbiased determination of treatment effects. • Large information based on minimum sample size. • Straightforward statistical analysis.	• Carryover effects and period effects are possible due to inappropriate washout period. • Long duration. • Possibility of more dropouts leads to insufficient power. • Not suitable for long half-life drugs. • Not optimal for studies in patients and highly variable drugs.
Parallel	• If the drug has a very long terminal elimination half-life. • Duration of the washout time for the two-period crossover study is so long (if >1 month). • If the intrasubject CV is higher with crossover design.	• Design is simple and robust. • Dropouts will be comparatively less. • Duration of the study is less than crossover study. • Study with patients is possible. • Straightforward statistical analysis.	• Subjects cannot serve as their own controls for intrasubject comparisons. • Large sample size is required. • Lower statistical power than crossover. • Phenotyping mandatory for drugs showing polymorphism.
Replicate	• Useful for the highly variable drugs (intrasubject CV ≥ 30%).	• Allows comparisons of within-subject variances for the test and reference products. • Indicates whether a test product exhibits higher or lower within-subject variability in the bioavailability measures when compared to the reference product. • Provides more information about the intrinsic factors underlying formulation performance. • Reduces the number of subjects needed in the bioequivalence study. • The number of subjects required to demonstrate bioequivalence can be reduced by up to about 50%. • Design increases the power of the study when the variability in the systemic exposure of the test drug and formulation is high.	• Involves larger volume of blood withdrawn from each subject. • Longer duration of the entire study. • Increased possibility of subject dropouts. • Expensive.
Variance balanced design	• For more than two formulations. • Desirable to estimate the pairwise effects with the same degree of precision.	• Allows to choose between two more candidate test formulations. • Comparison of test formulation with several reference formulations. • Standard design for the establishment of dose proportionality.	• Statistical analysis is more complicated (especially when dropout rate is high). • May need measures against multiplicity (increasing the sample size).

TABLE 16.5
Brief Description of the Criteria on Strength to be Investigated in Bioequivalence Studies

Regulatory Authority	Linear Pharmacokinetics	Nonlinear Pharmacokinetics
Europe and Australia	General: The bioequivalence study should in general be conducted at the highest strength. Highly soluble drug and any safety concern: Lower strength acceptable. Problems of sensitivity of the analytical method: Highest strength acceptable.	For drugs with nonlinear pharmacokinetics characterized by a more than proportional increase in AUC with increasing dose over the therapeutic dose range, the bioequivalence study should in general be conducted at the highest strength. As for drugs with linear pharmacokinetics, a lower strength may be justified if the highest strength cannot be administered to healthy volunteers for safety/tolerability reasons. Likewise, a higher dose may be used in case of sensitivity problems of the analytical method in line with the recommendations given for products with linear pharmacokinetics earlier. For drugs with a less than proportional increase in AUC with increasing dose over the therapeutic dose range, bioequivalence should in most cases be established both at the highest strength and at the lowest strength (or strength in the linear range), i.e., in this situation, two bioequivalence studies are needed. If the nonlinearity is not caused by limited solubility but is due to, e.g., saturation of uptake transporters, and provided that (1) same manufacturing process, (2) qualitative composition of the different strengths is the same, (3) composition of the strengths are quantitatively proportional, and (4) appropriate in vitro dissolution data should confirm the adequacy of waiving additional in vivo bioequivalence testing and the test and reference products do not contain any excipients that may affect gastrointestinal motility or transport protein, it is sufficient to demonstrate bioequivalence sport proteins at the lowest strength (or strength in the linear range).

United States	Reference-listed drug in the Orange Book; usually the highest strength if formulations are proportionally similar. For an ANDA, conducting an in vivo study on a strength that is not the highest may be appropriate for reasons of safety, subject to approval by the Division of Bioequivalence, Office of Generic Drugs, and provided that the following conditions are met: (1) Linear elimination kinetics has been shown over the therapeutic dose range; (2) the higher strengths of the test and reference products are proportionally similar to their corresponding lower strength; (3) comparative dissolution testing on the higher strength of the test and reference products is submitted and found to be appropriate.	Not specified.
Saudi Arabia	For conventional solid oral drug products, in vivo bioequivalence studies are conducted on the highest strength. This requirement for the lower strengths can be waived provided (1) in vivo bioequivalence is demonstrated on the highest strengths; (2) in vitro dissolution testing is acceptable; and (3) the formulation for the lower strengths are proportionally similar to the strength that has undergone in vivo bioequivalence testing (i.e., the ratio of active ingredients and excipients between the strengths is essentially the same).	Not specified.
Canada	Generally, use strength with largest sensitivity to identify differences in formulation. Reference product is (1) a drug product that has been issued a notice of compliance pursuant to section C.08.004 of the Food and Drug Regulations and is currently marketed in Canada by the innovator or (2) a drug product acceptable to the director.	
Asia	Test products in an application for a generic product are normally compared with the corresponding dosage form of an innovator medicinal product (reference product). The choice of reference product should be justified by the applicant and agreed upon by the regulatory authority. If the innovator product is not available, an alternative comparator product approved by drug regulatory authority of the country can be used.	
New Zealand	When the drug product is the first market entry of that type of dosage form, the reference product should normally be the innovator's prompt-release formulation. The comparison should be between a single dose of the drug formulation and doses of the prompt-release formulation, which it is intended to replace.	
Korea	Reference drug product is an approved drug product (or an approved imported drug product), the safety and efficacy of which have beenestablished or recognized by the commissioner of the KFDA.	

TABLE 16.6
Regulatory Criteria on Subject Demographics for Bioequivalence Studies

Regulatory Authority	Sex	Age (Years)	Body Mass Index (BMI) (kg/m²)
India	Male or female	Healthy adult volunteers	Not specified
Asia	Either sex	18–55	18–30; Asians: 18–25
United States	Both sexes	18	Not specified
Europe	Either sex	18	18.5–30
Canada	Both sexes	18–55	Height/weight ratio for healthy volunteer subjects should be within 15% of the normal range
Australia	Either sex	18–55	Accepted normal BMI.
South Africa	Either sex	18–55	Accepted normal BMI or within 15% of the ideal body mass or any other recognized reference
Russia	Both sexes	19–45	Weight of body does not fall outside the limits ±15% on Ketle total-height index
Korea	Healthy adult	19–55	Not specified
Japan	Healthy adult	Not specified	Not specified
People's Republic of China	Both sexes	18–40	Standard weight range
Mexico	Avoiding pharmacokinetic differences between sexes is well documented; volunteers of just one sex must be included	18–55	Weight 10% from the ideal weight
Saudi Arabia	If females are included in the study, the effects of gender differences and menstrual cycle (if applicable) are examined statistically	18–50	Within 15% of ideal body weight, height, and body build
New Zealand	Both sexes	Age range prior to the onset of age-related physiological changes (usually 18–60)	Average weight (e.g., within ±15% of their ideal weight as given in the current Metropolitan Life insurance Company Height and Mass Tables)

Selection of Appropriate Analyte(s)

Each regulatory authority has its own specifications for selection of an appropriate analyte to be measured as well as consideration for BE. Most commonly, the investigator should consult the relevant regulatory agency for guidance on a particular therapeutic agent. The general considerations are discussed in the following sections.

Parent Drug versus Metabolite(s)

BE based on test/reference comparisons of PK measures serves two purposes: (1) to act as a surrogate for therapeutic equivalence and (2) to provide in vivo evidence of pharmaceutical quality. The overall objective of BE is to ensure that generic products have efficacy and safety characteristics

TABLE 16.7
Regulatory Criteria on Sample Size for Bioequivalence Studies

Regulatory Authority	Minimum	Sample Size Specifications
India	Should not be <16 unless justified for ethical reasons	The number of subjects required for a study should be statistically significant and should be sufficient to allow for possible withdrawals or removals (dropouts) from the study.
Asia	Should not be <12	The number of subjects required is determined by (1) the error variance associated with the primary characteristic to be studied as estimated from a pilot experiment, from previous studies, or from published data; (2) the significance level desired; (3) the expected deviation from the reference product compatible with bioequivalence (delta, i.e., percentage difference from 100%); and (4) the required power.
United States	12	A sufficient number of subjects should complete the study to achieve adequate power for a statistical assessment.
Europe	Should not be <12	The number of subjects to be included in the study should be based on an appropriate sample size calculation.
Canada	12	(1) Obtain an estimate of the intrasubject CV from the literature or from a pilot study, (2) choose one of Figures 3.1 through 3.3 (mentioned in bioequivalence guidance document) by determining which one has the closest rounded-up CV to that estimated in (1), given earlier, (3) choose an expected true ratio of test over reference means (usually 100%) and move up the graph to the 0.90 probability of acceptance, and (4) a linear extrapolation between given sample sizes is adequate. This sample size calculation must be provided in the study protocol. More subjects than the sample size calculation required should be recruited into the study. This strategy allows for possible dropouts and withdrawals.
Australia	Should not be <16 unless justified	Same as that of Asian guidelines.
South Africa	Should not be <12 (general); 20 subjects (for modified-release oral dosage forms)	The number of subjects should be justified on the basis of providing at least 80% power of meeting the acceptance criteria; alternatively, the sample size can be calculated using appropriate power equations, which should be presented in the protocol.
Russia	18	In quantity sufficient for ensuring statistical importance of study. Thus, capacity of the statistical test for bioequivalence study must be supported at a level of not less than 80% for revealing 20% distinctions between comparison parameter.
Korea	12	The number of subjects should meet the requirements for statistical validity. The number of subjects can be determined based on the characteristics of the active component of the pertinent drug product.
Japan	20	A sufficient number of subjects for assessing bioequivalence should be included. If bioequivalence cannot be demonstrated because of an insufficient number, an add-on subject study can be performed using not less than half the number of subjects in the initial study. A sample size of 20 (n = 10/group) for the initial study and pooled size of 30 for initial plus add-on subject study may suffice if test and reference products are equivalent in dissolution and similar in average AUC and C_{max}.

(Continued)

TABLE 16.7 ***(Continued)***
Regulatory Criteria on Sample Size for Bioequivalence Studies

Regulatory Authority	Minimum	Sample Size Specifications
Saudi Arabia	A number of subjects of less than 24 may be accepted (with a minimum of 12 subjects) when statistically justifiable	Generally recommends a number of 24 normal healthy subjects. Should enroll a number of subjects sufficient to ensure adequate statistical results, which is based on the power function of the parametric statistical test procedure applied. The number of subjects should be determined using appropriate methods taking into account the error variance associated with the primary parameters to be studied (as estimated for a pilot experiment, from previous studies or from published data), the significance level desired ($\alpha = 0.05$), and the deviation from the reference product compatible with bioequivalence (±20%) and compatible with safety and efficacy.
New Zealand	12	The number of subjects should provide the study with a sufficient statistical power (usually ≥80%) to detect the allowed difference (usually 20%) between the test and reference medicines for AUC and C_{max}. This number (n) may, in many cases, be estimated in advance from published or pilot study data using formula. If the calculated number of subjects appears to be higher than is ethically justifiable, it may be necessary to accept a statistical power that is less than desirable. Normally, it is not practical to use more than about 40 subjects in a bioavailability study.
Mexico	Sample size must not be <24 subjects considering both sequences or it must meet the requirement related to a difference to be detected of ±20% for the reference product's mean, associated with a type I error of 0.05 and a minimal potency of (1-*) of 0.8 for this kind of design. A sample size of <24 subjects must be scientifically justified.	
Brazil	The number of healthy volunteers shall at all times assure an adequate statistical power to guarantee reliability of bioequivalence study results.	

TABLE 16.8
Regulatory Criteria on Number of Studies Required for Conducting Bioequivalence Studies

Regulatory Authority	Immediate-Release Formulations	Modified-Release Formulations
India	Generally, a single-dose, nonreplicate, fasting study. Food-effect studies are required (1) when it is recommended that the study drug should be taken with food (as would be in routine clinical practice) and (2) when fasting state studies make assessment of C_{max} and t_{max} difficult. If multiple-study design is important, appropriate dosage administered and sampling be carried out to document attainment of steady state.	Should conduct fasting as well as food-effect studies. If multiple-study design is important, appropriate dosage administered and sampling carried out to document attainment of steady state.
United States	Generally, two studies • A single-dose, nonreplicate fasting study. • A food-effect, nonreplicate study. Food-effect study can be excepted in the following cases: (1) when both test product and RLD are rapidly dissolving, have similar dissolution profiles, and contain a drug substance with high solubility and high permeability (BCS Class I); (2) when the dosage and administration section of the RLD label states that the product should be taken only on an empty stomach; or (3) when the RLD label does not make any statements about the effect of food on absorption or administration. If food effect mentioned in RLD label and if multiple-study design is important, appropriate dosage administered and sampling be carried out to document attainment of steady state	Should conduct fasting as well as food-effect studies. If multiple-study design is important, appropriate dosage administered and sampling be carried out to document attainment of steady state.
Europe and Australia	Generally, a single-dose, nonreplicate, fasting study. Food-effect studies are required if the summary of product characteristics of the reference product contains specific recommendations in relation with food interaction.	Should conduct fasting, food-effect, as well as steady-state studies.
Canada	Generally, comparative BA studies conducted in the fasting state. Fed study is acceptable if there is a documented serious safety risk to subjects from single-dose administration of the drug or drug product in the absence of food, then an appropriately designed study conducted in the presence of only a sufficient quantity of food to prevent the toxicity may be acceptable for purposes of bioequivalence assessment. For complicated IR formulations (narrow therapeutic range drugs, highly toxic drugs, and nonlinear drugs), both fasted and fed studies.	Usual requirement is for both fasted and fed studies. If multiple-study design is important, appropriate dosage administered and sampling be carried out to document attainment of steady state.
South Africa	Should be done under fasting conditions unless food effects affect bioavailability of drug or reference product dosage recommended.	Both fed and fasted studies are required. If multiple-study design is important, it should be carried out as per regulatory specifications.
Korea	Generally, a single-dose, nonreplicate, fasting study.	Should conduct fasting, food-effect, as well as steady-state studies.

(Continued)

TABLE 16.8 (*Continued*)
Regulatory Criteria on Number of Studies Required for Conducting Bioequivalence Studies

Regulatory Authority	Immediate-Release Formulations	Modified-Release Formulations
Japan	Both fasting and food-effect studies.	Should conduct fasting, food-effect, as well as steady-state studies.
Saudi Arabia	Generally, a single-dose, nonreplicate, fasting study is required. Food-effect studies are required (1) if documented evidence of effect of food on drug absorption, (2) the drug is recommended to be administered with food, and (3) the drug may produce gastric irritation under fasting conditions, thus may be taken with food.	Should conduct fasting as well as food-effect studies.
New Zealand	Generally, a single-dose fasting study is required. Fed study is required when it is recommended that the drug be given with food or fasted studies make assessment of C_{max} and t_{max} difficult.	Should conduct fasting as well as food-effect studies. Steady-state studies are generally required if the drugs are likely to accumulate along with single-dose studies.

TABLE 16.9
Regulatory Criteria for Conducting Fasting and Fed Bioequivalence Studies

Regulatory Authority	Fasting Requirements	Fed Study Requirement
India	Overnight fast (at least 10 h), with a subsequent fast of 4 h following dosing. For multiple-dose fasting studies, when an evening dose must be given, 2 h before and after the dosing.	950–1000 kcal of high-fat breakfast approximately 15 min before dosing (at least 50% of calories must come from fat, 15%–20% from proteins and rest from carbohydrates). The vast ethnic and cultural restrictions of the Indian subcontinent preclude the recommendation by a single standard high fat; in this case, protocol should specify the appropriate and suitable diet.
United States	Following an overnight fast of at least 10 h, with a subsequent fast of 4 h postdose.	A high-fat (approximately 50% of total caloric content of the meal), high-calorie (approximately 800–1000 cal) meal is recommended. This test meal should derive approximately 150, 250, and 500–600 cal from protein, carbohydrate, and fat, respectively. The caloric breakdown of the test meal should be provided in the study report. If the caloric breakdown of the meal is significantly different from the one described earlier, it should require a scientific rationale for this difference. Following an overnight fast of at least 10 h, subjects should start the recommended meal 30 min prior to dosing. Study subjects should eat this meal in 30 min or less; however, the drug product should be administered 30 min after start of the meal.
Europe and Australia	Should fast for at least 8 h prior to dosing, unless otherwise justified and no food is allowed for at least 4 h postdose.	The composition of the meal is recommended to be according to the SPC of the originator product. If no specific recommendation is given in the originator SPC, the meal should be a high-fat (approximately 50% of total caloric content of the meal) and high-calorie (approximately 800–1000 kcal) meal. This test meal should derive approximately 150, 250, and 500–600 kcal from protein, carbohydrate, and fat, respectively. The composition of the meal should be described in terms of protein, carbohydrate, and fat content (specified in grams, calories, and relative caloric content (%)).
Canada	Following an overnight fast of at least 10 h, with a subsequent fast of 4 h postdose.	Should be a representative meal in which sufficient food is given to allow potential perturbation of systemic BA of the drug from the drug product. The sponsor should justify the choice of meal and relate the specific components and timing of food administration. Example: two eggs fried in butter, two strips of bacon, two slices of toast with butter, 120 g of hash browns, and 240 mL of whole milk.
South Africa	Fasting prior to dosing and after dosing should be standardized.	Use of high-calorie and high-fat meals is recommended.
Korea	Should be fasted for at least 10 h before and up to 4 h after the drug administration.	High-fat diet should be taken within 20 min in at least a 10 h fasting state. The drug products should be administered 30 min after the meal starts.
Saudi Arabia	Following an overnight fast of at least 10 h, with a subsequent fast of 4 h postdose.	A high-fat (approximately 50% of total caloric content of the meal), high-calorie (approximately 1000 cal) breakfast. Alternative meals with equivalent nutritional content can be used.
New Zealand	After an overnight fast of at least 10 h, with a subsequent fast of 2–4 h following dose administration.	The meal should contain approximately 30–40 g of fat.

TABLE 16.10
Regulatory Criteria on Fluid Intake, and Posture and Physical Activity for Bioequivalence Studies

Regulatory Authority	Fluid Intake	Posture and Physical Activity
India	Standardization of fluid intake and physical activity is required and it should be stated in protocol.	
United States	Subjects should be administered the drug product with 240 mL (8 fluid ounces) of water; water is not allowed as desired except for 1 h before and 1 h after the drug administration.	Standardized
Asia, Europe, and Australia	The drug products should be administered with a standardized volume of fluid (at least 150 mL). Prior to and during each study phase, subjects should be allowed water as desired except for 1 h before and after drug administration.	As the bioavailability of an active moiety from a dosage form could be dependent upon gastrointestinal transit times, and regional blood flows, posture and physical activity may need to be standardized.
Canada	On the morning of the study, up to 250 mL of water may be permitted up to 2 h before drug administration. The dose should be taken with water of a standard volume (e.g., 150 mL) and at a standard temperature. Two hours after drug administration, 250 mL of xanthine-free fluids is permitted.	For most drugs, subjects should not be allowed to recline until at least 2 h after drug ingestion. Physical activity and posture should be standardized as much as possible to limit effects on gastrointestinal blood flow and motility. The same pattern of posture and activity should be maintained for each study day.
South Africa	The volume of fluid administered at the time of dosing should be constant (e.g., 200 mL); fluids taken after dosing should also be standardized.	Should be standardized.
Korea	Drug products should be administered with 240 mL of water; drinking water 1 h before and after the administration of drug products is not allowed.	Subjects should not be in a supine position at least 2 h after the administration of drug products and should maintain a posture and do only activities that would minimize the effects on their gastrointestinal blood flow rate and motility.
Saudi Arabia	The test or reference products should be administered with about 8 fluid ounces (240 mL) of water; water allowed as desired except for 1 h before and after drug administration.	Appropriate restrictions on fluid intake and physical activities should be made.
New Zealand	The quantity, type, and timing of food and fluid taken concurrently with the medicine should be stated and should be controlled.	Standardization of posture and physical activity is important. Subjects should not be allowed to recline until at least 2 h after oral administration of the medicine.

TABLE 16.11
Regulatory "Add-On Criteria" for Conducting Bioequivalence Studies

Regulatory Authority	
Europe and Australia	It is acceptable to use a two-stage approach when attempting to demonstrate bioequivalence. An initial group of subjects can be treated and their data analyzed. If bioequivalence has not been demonstrated, an additional group can be recruited and the results from both groups combined in a final analysis. If this approach is adopted, appropriate steps must be taken to preserve the overall type I error of the experiment and the stopping criteria should be clearly defined prior to the study. The analysis of the first-stage data should be treated as an interim analysis and both analyses conducted at adjusted significance levels.
South Africa	If the bioequivalence study was performed with the appropriate size but bioequivalence cannot be demonstrated because of a result of a larger than expected random variation or a relative difference, an add-on subject study can be performed using not less than half the number of subjects in the initial study. Combining is acceptable only if the same protocol was used and preparations from the same batches were used. Add-on designs must be carried out strictly according to the study protocol and standard operating procedures and must be given appropriate statistical treatment, including consideration of consumer risk.
Canada	As a result of random variation or a larger than expected relative difference, there is no guarantee that the sample size as calculated will pass the standards. If the study is run with the appropriate size and the standards are not met, the sponsor may add more subjects (a minimum of 12). The same protocol should be used (i.e., same formulations, same lots, same blood sampling times, a minimum number of 12 subjects). The choice to use this strategy, as with all designs, should be declared and justified a priori. The level of confidence should be adjusted using the Bonferroni procedure. The t-value should be that for $P = 0.025$ instead of 0.05.
Japan	Also for add-on study, an additional 10 subjects is recommended along with initial subjects.

similar to those of the corresponding reference product. For the most part, traditional BE studies have been carried out on the basis of measurement of only the parent drug in body fluids such as plasma or serum. In some cases, however, monitoring a metabolite, or the parent and metabolite(s), may be more appropriate. A number of reasons for use of metabolite data have been put forward, such as (1) the parent is an inactive prodrug, (2) plasma concentrations of the parent drug are too low to monitor because of inadequate assay sensitivity, (3) the parent drug is metabolized rapidly to an active metabolite, and (4) the parent drug and a metabolite both have therapeutic activities but the metabolite is present in higher concentrations when the parent drug is rapidly and extensively metabolized such that only metabolite(s) data are available.

Enantiomers versus Racemates

For BA/BE studies, measurement of both enantiomers may be important. For BE studies, measurement of the racemate using an achiral assay has been recommended, without measurement of individual enantiomers except when (1) the enantiomers exhibit different pharmacodynamic characteristics, (2) the enantiomers exhibit different PK, (3) the primary activity resides with the minor enantiomers, and (4) nonlinear absorption is present (as expressed by a change in the enantiomers concentration ratio with change in the input rate of the drug) for at least one of the enantiomers.

Drug Products with Complex Mixtures

Certain drug products may contain complex drug substances, that is, active moiety or active ingredient(s), which are mixtures of multiple synthetic and/or natural source components.

TABLE 16.12
Regulatory Criteria on Sampling and Washout Period for Conducting Bioequivalence Studies

Regulatory Authority	Sampling Criteria	Washout Criteria
India	*Blood sampling*: Should be extended to at least 3 elimination half-lives, at least 3 sampling points during absorption phase, 3–4 at the projected t_{max}, and 4 points during elimination phase; sampling should be continued for a sufficient period to ensure that AUC_{0-t} to $AUC_{0-\infty}$ is only a small percentage (normally <20%) of the total AUC. Truncated AUC is undesirable except in the presence of enterohepatic recycling.	Adequate and ideally, it should be ≥5 half-lives of the moieties to be measured.
United States	*Urinary sampling*: Collect urine samples for 7 or more half-lives Blood samples should be drawn at appropriate times to describe the absorption, distribution, and elimination phases of the drug; 12–18 samples, including a predose sample, should be collected per subject per dose and should continue for at least 3 or more terminal half-lives of the drug.	An adequate washout period (e.g., more than 5 half-lives of the moieties to be measured).
Europe	*Single-dose blood sampling*: Sufficient sampling is required; frequent sampling around predicted t_{max}; avoid C_{max} to be the first point; accommodate reliable estimate (AUC_{0-t} covers at least 80% of $AUC_{0-\infty}$); at least 3–4 points during the terminal log-linear phase; AUC truncated at 72 h ($AUC_{0-72\,h}$) may be used as an alternative to AUC_{0-t} or comparison of extent of exposure. *Multiple-dose blood sampling*: Predose sample should be taken immediately before (within 5 min) dosing and the last sample is recommended to be taken within 10 min of the nominal time for the dosage interval to ensure an accurate determination of $AUC_{0-\tau}$. *Urinary sampling*: Urine should normally be collected over no less than three times the terminal elimination half-life.	Sufficient washout period (usually at least 5 terminal half-lives).
Australia	*Single-dose blood sampling*: Should provide adequate estimation of C_{max}; cover plasma concentration–time curve long enough to provide a reliable estimation of the extent of absorption; 3–4 samples during the terminal log-linear phase. AUC truncated at 72 h is permitted for long half-life drugs. *Multiple-dose blood sampling*: When differences between morning and evening or nightly dosing are known, sampling should be carried out over a full 24 h cycle.	Adequate washout period.
Canada	*Blood sampling*: Sampling should be sufficient to account for at least 80% of the known $AUC_{0-\infty}$, C_{max}, and terminal disposition; 3 times the terminal half-life of the drug; 12–18 samples should be collected per each subject per dose; and 4 or more points be determined during the terminal log-linear phase. *Urine sampling*: Urine should be collected over no less than 3 times the terminal elimination half-life. For a 24 h study, sampling times of 0–2, 2–4, 4–8, 8–12, and 12–24 h are usually appropriate.	Normally, it should be not less than 10 times the mean terminal half-life of the drug. Normally, the interval between study days should not exceed 3–4 weeks.

South Africa	*Blood sampling*: Sampling should be sufficient to account for at least 80% of the known $AUC_{0-\infty}$ and C_{max}; collecting at least 3–4 samples above the LOQ during the terminal log-linear phase; sampling period is approximately three terminal half-lives of the drug; AUC truncated at 72 h is permitted for long half-life drugs; 12–18 samples should be collected per each subject per dose; at least 3–4 samples above LOQ should be obtained during the terminal log-linear phase. *Urine sampling*: Sufficient urine should be collected over an extended period and generally no less than 7 times the terminal elimination half-life; for a 24 h study, sampling times of 0–2, 2–4, 4–8, 8–12, and 12–24 h postdose are usually appropriate.	Adequate washout period.
Korea	*Blood sampling*: Sampling should be sufficient to estimate all the required parameters for BA; cover 3 or more times the terminal half-life; at least 2 points before t_{max}; sufficient to account for at least 80% of the known $AUC_{0-\infty}$; number of blood samples should be 12; AUC truncated at 72 h is permitted for long half-life drugs. *Urine sampling*: Adequate number of urine samples should be covered to estimate the amount and excretory rate.	Adequate and should be >5 times the half-life of the active ingredients.
Saudi Arabia	Sufficient samples are collected to estimate all the required parameters during absorption and elimination for bioequivalence assessment. A sampling period extending to at least 4–5 terminal elimination half-lives of the drug or 4–5 the longest half-life of the pertinent analyte (if more than 1 analyte) is usually sufficient.	An adequate washout period (e.g., more than 5 half-lives of the moieties to be measured).
New Zealand	*Single-dose blood sampling*: Sampling should be sufficient to account for at least 80% of the known $AUC_{0-\infty}$ and should extend to at least 3 elimination half-lives of the drug; truncated AUC is undesirable except in unavoidable circumstances like the presence of enterohepatic recycling. *Multiple-dose blood sampling*: Sampling should be carried out over a full 24 h cycle so that any effects of circadian rhythms may be detected, unless these rhythms can be argued not to have practical significance. *Urine sampling*: Adequate number of urine samples should be covered to estimate the amount and excretory rate. For a 24 h study, sampling times of 0–2, 2–4, 4–8, 8–12, and 12–24 h are usually appropriate. Where urinary excretion is measured in a single-dose study, it is necessary to collect urine for 7 or more half-lives.	An adequate washout period (at least 3 times the dominating half-life).

Some or all of the components of these complex drug substances may not be characterized by chemical structure and/or biological activity. In this circumstance, BA and BE studies may be based on selected markers of peak and total exposure.

BIOEQUIVALENCE METRICS AND DATA TREATMENT

The most frequent data treatment involves analysis of variance using a suitable program such as SAS® (Statistical Analysis System, SAS Institute, Cary, NC) or WinNonlin® (Pharsight Corporation, St. Louis, MO) so that contributions from subject, period, product/formulation, and interactions between these can be examined. Geometric mean ratios and log transformed data are examined to test the hypothesis that the 90% confidence interval of extent (AUC_{0-t}) and $AUC_{0-\infty}$ and the maximum concentration (C_{max}) fall within the acceptance limits of 80%–125%. More recently, other data treatments have been popular, which include partial area measurements and exposure metrics including C_{max}/AUC, especially with highly variable drugs (HVDs), and with drugs having a long terminal t1/2, specialized dosage forms, and/or whose time to C_{max} is considered important (e.g., certain analgesics). In all of these cases, the objective has been to err on the side of protecting the consumer while at times increasing risk to the manufacturer. Hence, over the last 15 years, considerable debate has occurred globally about the fundamental scientific rationale used to establish BE for some of these "special" cases, in an effort to solve these issues associated with harmonization of drug equivalence approaches.

STATISTICAL APPROACHES

Considerable debate has ensued over the past 20 years on statistical testing and BE studies. After protracted, wide-ranging, and in-depth discussion among various experts from different locations, specific statistical regulatory guidance is available to investigators conducting BE studies. The various PK parameters derived from the plasma concentration–time curve are subjected to ANOVA in which the variance is partitioned into components due to subjects, periods, and treatments. The classical null hypothesis test is the hypothesis of equal means, H_0: $\mu_T = \mu_R$ (i.e., products are bioequivalent), where μ_T and μ_R represent the expected mean bioavailabilities of the test and reference products, respectively. The alternate hypothesis therefore is H_1: $\mu_T \neq \mu e_R$ (i.e., products are bioinequivalent).

The detection of the difference becomes simply a function of sample size, and since the probable magnitude of the difference is the critical factor, this gives rise to two anomalies: (1) a large difference between two formulations that is nevertheless not statistically significant if error variability is high and/or sample size not large enough and (2) a small difference, probably of no therapeutic importance whatsoever, which is shown to be statistically significant if error variability is minimal and/or sample size adequately large.

The first case suggests a lack of sensitivity in the analysis, and the second an excess of it. Consequently, any practice that increases the variability of the study (sloppy design, assay variability, and within-formulation variability) would reduce the chances of finding a significant difference and hence improve the chances of concluding BE. The FDA therefore recognized that a finding of no statistical significance in the first case was not necessarily evidence of BE and consequently asked for a retrospective examination of the power of the test of null hypothesis.

Adequate statistical approaches should be considered to establish the BE of generic product to that of reference product. Much worldwide discussion and interaction has focused on facilitating the appropriate statistical approaches to establish interchangeability between generic drug and reference drug. The pertinent statistical approaches include (1) study power, (2) 75/75 rule, and (3) 90% confidence interval.

Study Power

The conduct of a BE study should require some prior knowledge of the performance of the products (generic and brand drugs) in the human body so that an appropriate number of test subjects can be enrolled and provide adequate power to test the hypothesis with a reasonable likelihood (i.e., at least 80%) that the two products are indeed bioequivalent. In fact, the alternative hypothesis that two products (generic and brand drugs) are statistically different leads to the conclusion that they are not bioequivalent. The two criteria considered most important to understand are the inherent variability of the drug and the geometric mean ratio between the test and reference product. Both of these parameters can be determined through the conduct of a pilot study ($n = 6–12$) to determine the proper sample size required for the pivotal study to establish BE as well as to minimize the possibility of undersizing the study.

75/75 Rule

This approach was the first application wherein individual bioequivalence (IBE) was being tested. The biomedical community felt that unless the change in the biological system was greater than 20%–25%, it would really not pose a significant clinical risk of invalidating the use of one therapeutic strategy versus another. This formed the basis for the 75/75 rule, which states that two products are equivalent if, and only if, at least 75% of the individuals being tested had ratios (of the various PK parameters obtained from the individual results) between the 75% and 125% limits and the study conducted has the statistical power to detect a 20% difference between the two products.[1] This approach was sound until the arrival of the 90% confidence interval. Later, the 75/75 rule lost most of its appeal when it was noted that both the test and reference products each have their own variability, and, therefore, a 90% confidence interval approach was more appropriate for giving some consideration to the differential variability between the test and reference products.

90% Confidence Interval

Westlake was the first to suggest the use of confidence intervals as a BE test to evaluate whether the mean amount of drug absorbed using the test formulation was close to the mean amount absorbed of the reference product. Subsequently, in July 1992, the guidance on *Statistical Procedures for Bioequivalence Studies Using a Standard Two-treatment Crossover Design* was released by the FDA. It was revised in 2001 and is available as *Statistical Approaches to Establishing Bioequivalence* (http://www.fda.gov/downloads/Drugs/GuidanceComplianceRegulatoryInformation/Guidances/ucm070244.pdf). This is based primarily on average BE (ABE), wherein the average values for the PK parameters were determined for the test and reference products and compared using a 90% confidence interval for the ratio of the averages using a two one-sided *t*-tests procedure. The abioequivalence approach for BE, however, has limitations for addressing drug switchability, since it focuses only on the comparison of population averages between the test and reference formulations. This concept was really based on the fact that if the ratios of the two PK parameters of clinical interest (such as AUC, C_{max}) are to be compared, each with their own variability that may or may not be randomly distributed, then such a comparison can truly be done only through a confidence interval approach. This concept is well accepted by almost all regulatory authorities to establish the BE.

ACCEPTANCE CRITERIA FOR BIOEQUIVALENCE

An equivalence approach is generally recommended, which usually relies on (1) a criterion to allow the comparison, (2) a confidence interval for the criterion, and (3) a BE limit. To compare measures in these studies, data are generally analyzed by using an average BE criterion with some considerations allowed for special category drugs.

GENERAL

To establish BE, the calculated 90% confidence interval should fall within a BE limit of 80%–125% using logarithm transformed data (adopted since the concentration parameters C_{max} and AUC may or may not be normally distributed). Currently, the BE limits of 80%–125% have been applied to almost all drug products by regulatory authorities. More detailed information on acceptance criteria for BE is given in Table 16.13.

FOR HIGHLY VARIABLE DRUGS

In the context of BE, HVDs are considered to be drugs and drug products exhibiting intrasubject variability greater than 30% coefficient of variation in the PK measures, AUC, and/or C_{max}. Due to this high variability, large sample size may be needed in BE studies to give adequate statistical power to meet FDA BE limits, and thus designing BE studies for HVDs is challenging. Consequently, development of generic products for HVDs is a major concern for the generic

TABLE 16.13
Regulatory Acceptance Criteria for Bioequivalence

	90% Confidence Interval on Log Transformed Data					
	Single-Dose Study			Steady-State Study		
Regulatory Authority	C_{max}	AUC_{0-t}	$AUC_{0-\infty}$	C_{max}	C_{min}	AUC_{τ}
India	80–125	80–125	80–125	80–125	80–125	80–125
Asia	80–125	80–125	80–125	80–125	80–125	80–125
United States	80–125	80–125	80–125	80–125	80–125	80–125
Europe	80–125	80–125	Not applicable	80–125		80–125
Canada	Ratio must be 80–125. Need to pass also on potency corrected data. Add-on studies may be allowed if intra-CV greater than expected	80–125	Not applicable	80–125	80–125	80–125
Australia	80–125	80–125	Not applicable	80–125	80–125	80–125
South Africa	75–133	80–125	Not applicable	75–133	75–133	80–125 (including% swing and% fluctuation)
Russia	75–133	80–125	80–125	75–133	75–133	80–125
Korea	80–125	80–125	80–125	80–125	80–125	80–125
Mexico	80–125	80–125	Not applicable	80–125	80–125	80–125
Saudi Arabia	80–125	80–125	80–125	80–125	80–125	80–125 (including% swing and% fluctuation)
New Zealand	80–125	80–125	80–125	80–125	80–125	80–125

TABLE 16.14
Regulatory Bioequivalence Acceptance Criteria for Special Class Drugs

Regulatory Authority	Highly Variable Drugs 90% Confidence Interval Log Transformed Data		Narrow Therapeutic Index Drugs 90% Confidence Interval Log Transformed Data	
	C_{max}	AUC	C_{max}	AUC_{0-t}
Asia	The interval must be prospectively defined, e.g., 0.75–1.33 and justified for addressing in particular any safety or efficacy concerns for patients switched between formulations	In rare cases, a wider acceptance range may be acceptable if it is based on sound clinical justification	Acceptance interval may need to be tightened	Acceptance interval may need to be tightened
United States	GMR (80–125) 95% upper bound for $(\mu_T - \mu_R)/\delta 2$ WR # 0.7976 (using scaled average approach)	GMR (80–125) 95% upper bound or $(\mu_T - \mu_R)/\delta 2$ WR # 0.7976 (using scaled average approach)	80–125	80–125
Europe	—	—	90.00–111.11	90.00–111.11
Canada	GMR (80–125)	GMR (80–125) 90% CI (80–125)	—	—
Saudi Arabia	75–133	Wider acceptance range may be acceptable and this should be justified clinically	90–111	—
Japan	—	—	90.00–111.11	90.00–111.11

drugs industry. Major regulatory agencies also considered different approaches for evaluating BE of HVDs. From 2004 onward, the FDA started looking for alternative approaches to resolve this issue and eventually found that replicate crossover design and scaled ABE provides a good approach for evaluating the BE of HVDs and drug products as it would effectively decrease sample size, without increasing patient risk. Recently, the FDA has issued *Method for Statistical Analysis Using the Reference-Scaled Average Bioequivalence Approach for Progesterone Capsules*, which clearly states how to perform statistical analysis for HVDs, such as progesterone using the replicate crossover design and reference-scaled abioequivalence approach (more information is available at http://www.fda.gov/downloads/Drugs/GuidanceCompliance RegulatoryInformation/Guidances/UCM209294.pdf). The various regulatory agency acceptance criteria for HVDs are given in Table 16.14.

For Narrow Therapeutic Index Drugs

Narrow therapeutic index drugs (NTIDs) can be defined as drugs that require therapeutic drug concentration or pharmacodynamic monitoring and/or drugs for which drug product labeling indicates a narrow therapeutic range designation. Perhaps, tighter restrictions on these drugs would aid in the establishment of truly bioequivalent drug products within this class. Thus, additional testing and controls may be needed to ensure the quality of these drug products. The regulatory acceptance criterion for NTIDs is given in Table 16.14. A list of these drugs is provided in Table 16.15.

TABLE 16.15
Narrow Therapeutic Index Drugs (FDA)

Aminophylline tablets, ER tablets	Carbamazepine tablets, oral suspension
Clindamycin hydrochloride capsules	Clonidine hydrochloride tablets
Clonidine transdermal patches	Dyphylline tablets
Disopyramide phosphate capsules, ER capsules	Ethinyl estradiol/progestin oral contraceptive tablets
Guanethidine sulfate tablets	Isoetharine mesylate inhalation aerosol
Isoproterenol sulfate tablets	Lithium carbonate capsules, tablets, ER tablets
Metaproterenol sulfate tablets	Minoxidil tablets
Oxtriphylline tablets, DR tablets, ER tablets	Phenytoin, sodium capsules (prompt or extended), oral suspension
Prazosin hydrochloride capsules	Primidone tablets, oral suspension
Procainamide hydrochloride, capsules, tablets, ER tablets	Quinidine sulfate capsules, tablets, ER tablets
Quinidine gluconate tablets, ER tablets	Theophylline capsules, ER capsules, tablets, ER tablets
Valproic acid capsules, syrup	Divalproex, sodium DR capsules, DR tablets
Warfarin, sodium tablets	

WORLD HEALTH ORGANIZATION GUIDELINES

The WHO interchangeability requirement include providing a BE study report that comprises, in the case of multisource (generic) preparations, BE study based on the WHO guidelines. BE data are required from all oral preparations except aqueous solutions at the time of administration. Orally or parenterally administered aqueous solutions will be assessed by chemical–pharmaceutical characteristics only. Also, BE study is required from preparations indicated for serious conditions requiring assured therapeutic response. All compounds in the present list correspond to this characteristic. Instead of BE trial, comparative clinical trial using clinical or pharmacodynamic endpoints can be presented. These endpoints should be justified and validated for the compound and trial should be designed to show equivalence. Trial showing the absence of significant difference cannot be accepted.

BE study report should contain at least the following items:

- Description of study design. The most appropriate study type is a two-period randomized crossover study. If other study types were used (e.g., parallel group design), these should be justified by the applicant. In general, single-dose study with sufficiently long period for blood samples collection is acceptable.
- Information about investigators, study site, and study dates.
- Data about preparations used: manufacturer, place of manufacture, and batch number.
- Reference preparation in BE study should be well-known preparation used in most countries of the world. The best acceptable reference is innovator preparation or product from WHO list of international comparator products if listed.
- Characterization of study subjects. BE study should be normally performed in healthy volunteers. If patients were used, the applicant should justify this. Number of subjects should not be smaller than 12. Study report should contain inclusion and exclusion criteria and listing of demographic data of all subjects.
- Description of study procedures. Administration of test products, meals, times of blood sampling, or urine collection periods should be described in the clinical report.
- Description and validation of drug determination methods in investigated material.

- Analytical method should be validated over the measured drug concentration range. Validation should contain methodology and results of sensitivity, specificity, accuracy, precision, and repeatability determination.
- All measured drug concentrations should be presented.
- Calculation methodology of PK parameters. Preferred is noncompartmental analysis. If modeled parameters were used, these models should be validated for the compound. All measured and calculated PK parameters should be presented in the report.
- Description of statistical methodology and results of statistical calculations. Statistical calculations should be based on the equivalence evaluation. The statistical method of choice is the two one-sided t-tests procedure and the calculation of 90% confidence intervals of the test/reference.
- The main parameters to assess the BE are area under the plasma concentration–time curve (AUC) and maximum concentrations (C_{max}) ratios.
- The 90% confidence interval for the AUC ratio should lie within a BE range of 80%–125%. In some specific cases of drugs with a narrow therapeutic range, the acceptance range may need to be tightened.
- The 90% confidence interval for the C_{max} ratio should lie within a BE range of 80%–125%. In some specific cases of drugs with a narrow therapeutic range, the acceptance range may need to be tightened. In certain cases for drugs with an inherently high intrasubject variability, a wider acceptance range (e.g., 75%–133%) may be acceptable. The range used must be defined prospectively and should be justified, taking into account safety and efficacy considerations.
- Summary of pharmacology, toxicology, and efficacy of the product. In case of products containing new active ingredients and new combinations of active ingredients, provide full information on safety and efficacy as defined in guidelines by the EU, the U.S. Food and Drug Administration, or the Japanese Ministry of Health and Welfare.

JAPANESE PERSPECTIVE ON BIOEQUIVALENCE STUDIES OF GENERIC PRODUCTS

Introduction

The Japanese guidance suggests that if the objective of the study to assure therapeutic equivalence of generic products to innovator products is not feasible, then pharmacological effects supporting therapeutic efficacy or therapeutic effectiveness in major indications should be compared. (These comparative tests are hereafter called pharmacodynamic studies and clinical studies, respectively.) For oral products, dissolution tests should be performed, since they provide important information concerning BE.

Terminology

Terms used in the guideline are defined as follows:

Bioavailability: The rate and extent of absorption of active ingredients or active metabolites from a product into the systemic circulation.

Bioequivalent products: Drug products having the equivalent BA.

Therapeutically equivalent products: Drug products having the equivalent therapeutic efficacies.

Innovator products: Drug products that have been approved as a new drug or a drug that corresponds to one.

Generic products: Products of which active ingredients, strengths, dosage forms, and dosage regimens are the same as those of innovator products.

Tests

Oral Immediate-Release Products and Enteric-Coated Products

Reference and Test Products In principle, dissolution tests should be performed using the following test solution (1) or (2), using six vessels or more for three lots of innovator products by the paddle method at 50 rpm. Among the three lots, the one that shows intermediate dissolution should be selected as the reference product. When the average dissolutions of the three lots reach 85% within 15 min, any lots can be used as the reference product.

1. The specification test solution when the dissolution specifications are established in the specifications and test procedures.
2. Among the test solutions described in the dissolution conditions, when the average dissolution of at least one lot reaches 85%, the test solution providing the slowest dissolution should be selected. When the average dissolution of any of the lots does not reach 85%, the test solution providing the fastest dissolution should be used.

If a reference product cannot be appropriately selected for the drug product by dissolution testing as described earlier, the reference product should be the innovator product lot that shows intermediate characteristics when either a dissolution (release) test appropriate for the characteristics of the drug product or a substitute physicochemical test is performed. If the drug is administered as a liquid where the active ingredient dissolves, an appropriate lot can be used as a reference product without performing dissolution tests.

It is recommended to use a lot manufactured at the same lot size as the full-scale production. However, a lot manufactured at a scale of not less than 1/10 of a full-scale production also can be used. If the product is a homogeneous liquid where the active ingredient dissolves, a lot of which manufacturing scale is less than the 1/10 can be used. Manufacturing method of the test product and commercial products should be similar, and quality and BA of both products should be equivalent.

A reference product whose content or potency is as close as possible to the labeled claim should be used. Furthermore, it is preferable that the difference between the content or potency of the test product and that of the reference product be within 5% of the labeled claim.

Bioequivalence Studies

Test Methods

Appropriate study protocol including the required number of subjects and sampling intervals should be determined according to preliminary studies and previously reported data. The rationale of the protocol should be described.

Design

In principle, crossover studies should be employed with random assignment of individual subjects to each group. Parallel designs can be employed for drugs with extremely long half-lives.

Number of Subjects

A sufficient number of subjects for assessing BE should be included. If BE cannot be demonstrated because of an insufficient number, an add-on subject study can be performed using not less than half the number of subjects in the initial study.

The add-on subject study should include at least one half of the number of subjects in the initial study. If the number of subjects in the initial study is 20 or more (10 subjects per group) or the total number of subjects in the initial study and add-on study is 30 or more, BE may be assessed based on the difference between the average BA of the test product and that of the reference product and the results of dissolution testing, without depending on confidence intervals, as is explained in the following.

Multiple-dose studies or studies with stable isotopes may be useful for HVDs that require large sample sizes.

Selection of Subjects

In principle, healthy adult volunteers should be employed.

If the use of the drug is limited to a specific population and test and reference products show a significant difference in dissolution *a under one or more of conditions of the dissolution test, the bioequivalent studies should be performed using subjects from the specific population (Figure 16.1).

If the use of the drug is not limited to a specific population and test and reference products showed a specific significant difference in dissolution *b at around pH 6.8 by the dissolution test or between pH 3.0 and 6.8 for products containing basic drugs, subjects with low gastric acidity (achlorhydric subjects) should be employed unless the application of the drug is limited to a specific population. This rule is not applied to enteric-coated products.

When it is unfavorable to use healthy subjects because of potent pharmacological action or adverse (side) effects, patients receiving the medication should be employed. If the clearance of drug differs to a large extent among subjects due to genetic polymorphism, subjects with higher clearance should be employed.

Before, during, and after studies, subjects' health condition should be monitored with close attention, especially, to adverse (side) effects.

(a) Significant difference in dissolution means the following two cases:
   1. The average dissolution of the slower-dissolution product is 50% or less at the time when the average dissolution of the faster-dissolution product reaches 80%. Also, when the average dissolution of the faster product is 85% or more within 15 min, the average dissolution of the slower product is not more than 60% of that of the faster product. However, this rule is not applied when the average dissolution of both products is 85% or more within 15 min after lag time (defined as the time when 5% of the drug dissolves) and the difference in the mean lag time in dissolution between test and reference products is within 10 min.
   2. The average dissolution of the slower product is not more than 60% of that of the faster product at the final testing time when the average dissolution of either product does not reach 80% within the specified testing time. The rule is also not applied when the average dissolution of both products at the final testing time specified in section 3.V.3. (2 h in pH 1.2 medium and 6 h in others) is 20% or less, because of difficulty of appropriate comparison of their dissolution.

(b) Specific significant difference in dissolution means that when test and reference products showed a significant difference in dissolution around pH 6.8 (between pH 3.0 and 6.8 for products containing basic drugs), and they do not show a significant difference in other test conditions. This rule is not applied when test and reference products show a significant difference in dissolution around pH 6.8 (between pH 3.0 and 6.8 for products containing basic drugs), and also, they show the same degree of or more significant difference in all other pH conditions.

Drug Administration

1. *Dose*: One dose unit or a clinical usual dose should generally be employed. A higher dose, which does not exceed the maximal dose of the dosage regimen, may be employed when analytical difficulties exist, such as high detection limit.

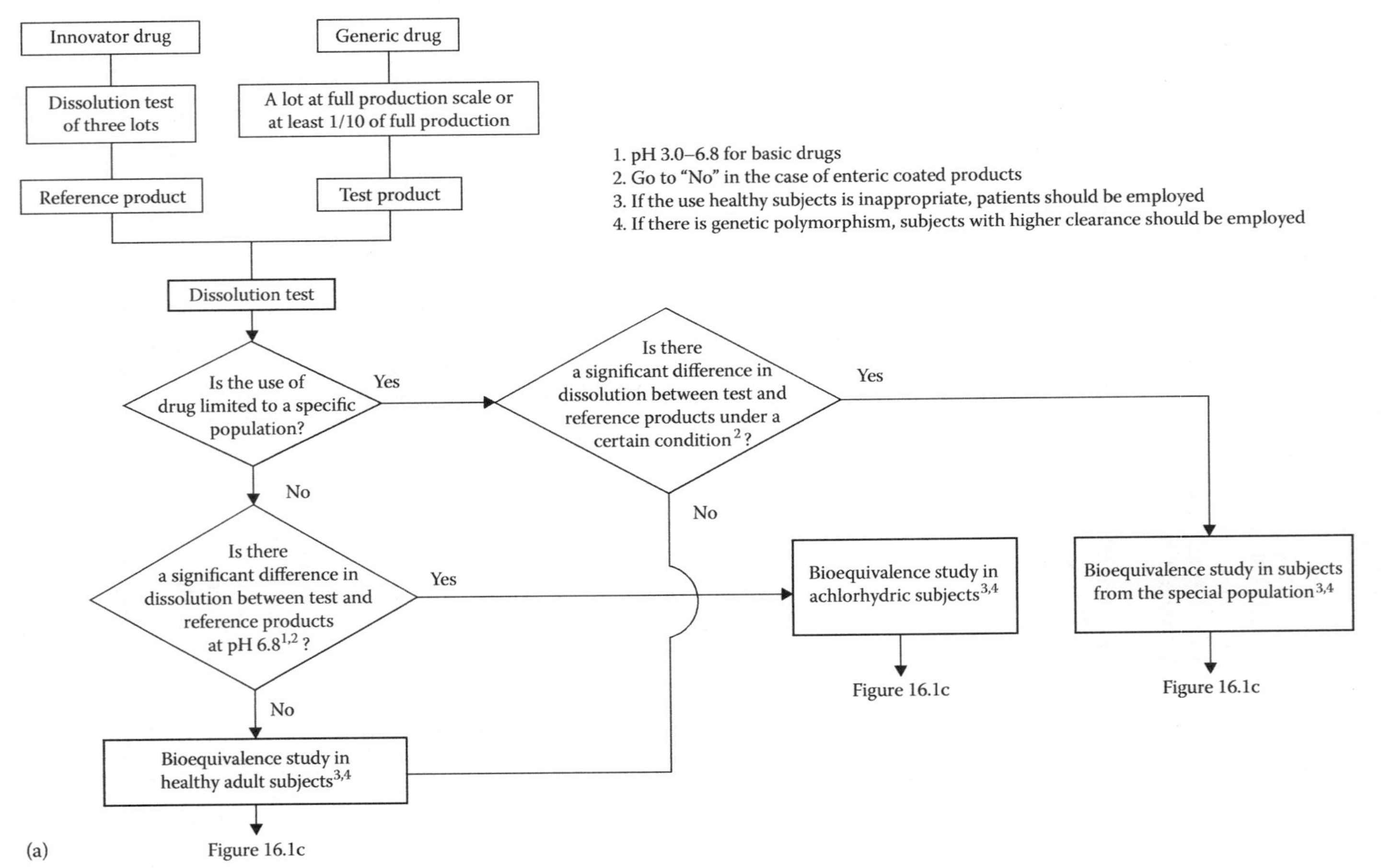
Innovator drug
Dissolution test of three lots
Reference product
Generic drug
A lot at full production scale or at least 1/10 of full production
Test product
1. pH 3.0–6.8 for basic drugs
2. Go to "No" in the case of enteric coated products
3. If the use healthy subjects is inappropriate, patients should be employed
4. If there is genetic polymorphism, subjects with higher clearance should be employed
Dissolution test
Is the use of drug limited to a specific population?
Yes
Is there a significant difference in dissolution between test and reference products under a certain condition2?
Yes
No
No
Is there a significant difference in dissolution between test and reference products at pH 6.81,2?
Yes
Bioequivalence study in achlorhydric subjects3,4
Bioequivalence study in subjects from the special population3,4
Figure 16.1c
Figure 16.1c
No
Bioequivalence study in healthy adult subjects3,4
(a)
Figure 16.1c

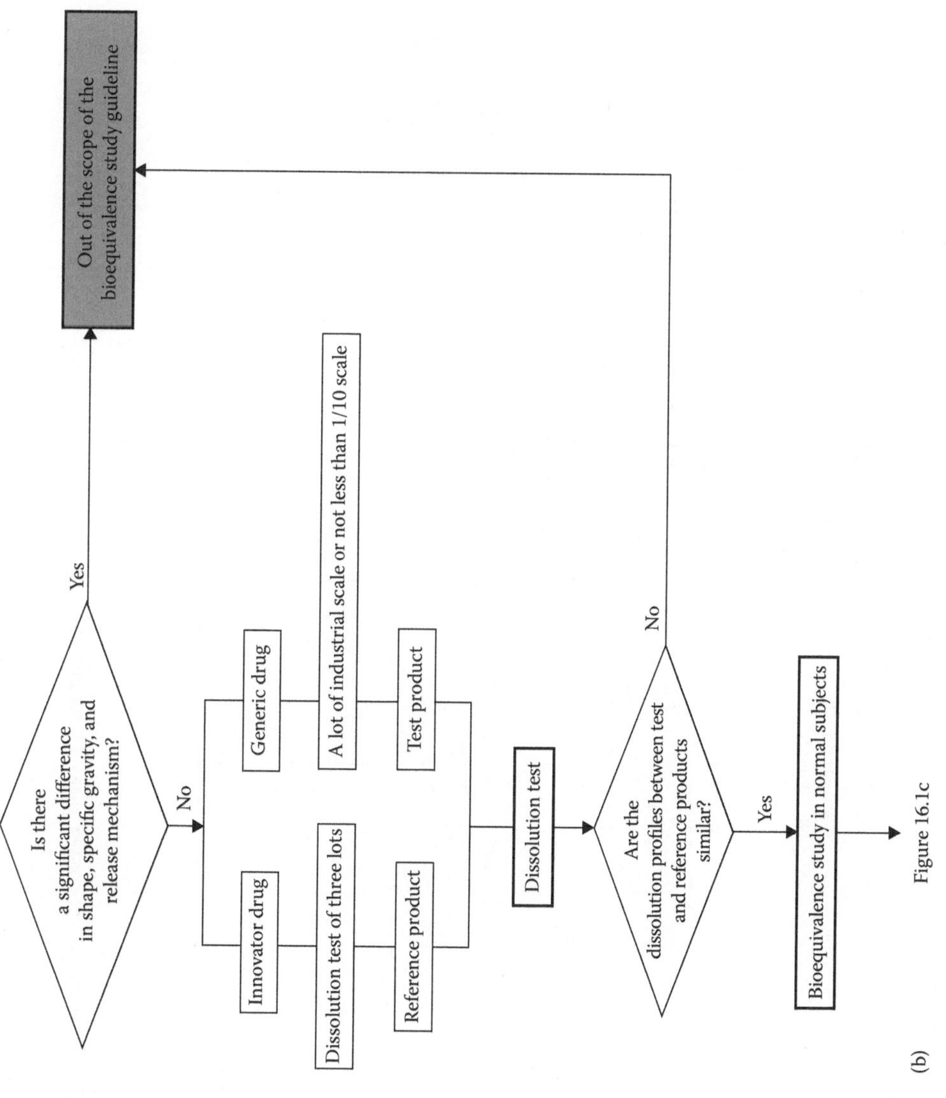

FIGURE 16.1 ***(Continued)*** BE study of oral dosage forms. (a) BE of oral immediate-release products and enteric-coated products. (b) BE for oral extended release products.

(Continued)

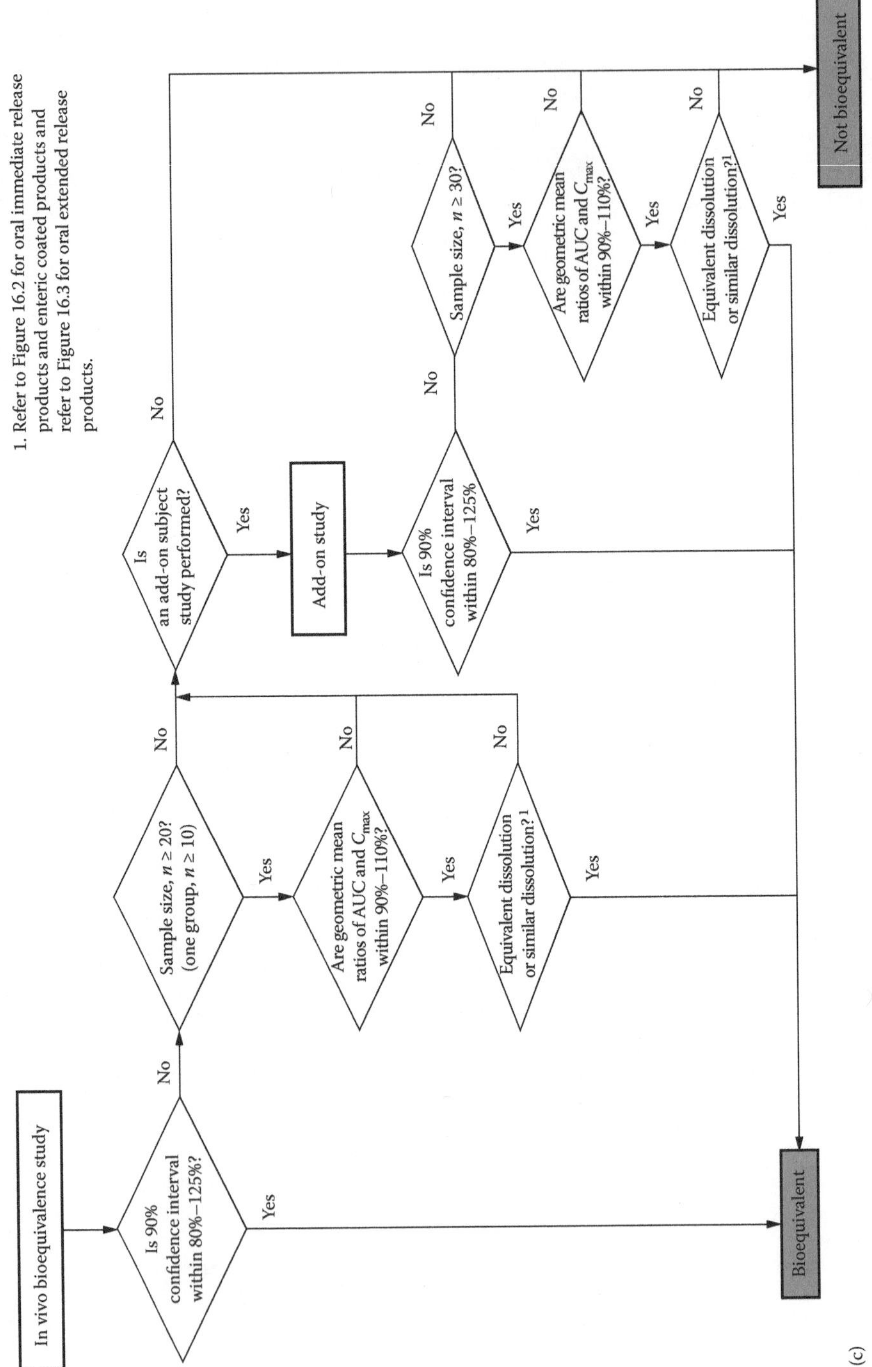

FIGURE 16.1 *(Continued)* BE study of oral dosage forms. (c) Judgment of BE.

2. *Single versus multiple-dose studies*: In principle, BE studies should be performed by single-dose studies. Multiple-dose studies may be employed for drugs that are repeatedly administered to patients.
   a. *Single-dose studies*: Drugs are usually given to subjects with 100–200 mL water (normally 150 mL) after fasting for more than 10 h. Fasting lasts for at least 4 h post-dose. If the postprandial dose is specified in the dosage regimen, and if the BA in fasting state is very poor, or high incidence of severe adverse events is anticipated, drugs may be given after food. In the fed study, a low-fat diet of 700 kcal or less containing not more than 20% by energy of the lipid should be employed. The meal should be eaten within 20 min, and drugs are administered according to the dosing regimen or 30 min after the meal, if the dosing time is not indicated in the regimen.
   b. *Multiple-dose studies*: Drugs should, in principle, be administered to subjects under fasting conditions as in the single-dose studies when biological fluids are sampled for the assessment of BA. In the time period before fluids are sampled, drugs should repeatedly be given between meals (drugs should be administered more than 2 h after a meal) at constant intervals.

Measurement of Biological Samples

1. *Biological fluids to be sampled*: Blood samples should generally be employed. Urine samples may be used if there is a rationale.
2. *Sampling schedule*: Blood samples should be taken at a frequency sufficient for assessing C_{max}, AUC, and other parameters. Sampling points should be at least 7, including zero time, 1 point before C_{max}, 2 points around C_{max}, and 3 points during the elimination phase. Sampling should be continued until AUC_t is over 80% of AUC_∞ (normally, more than three times the elimination half-life after t_{max}). However, when the elimination half-life of unchanged active ingredient or active metabolites to be measured is extremely long, blood samples should be collected for at least 72 h. When urine samples are used, they should be collected in the same manner as blood samples. When F is evaluated by deconvolution, body fluid must be collected until completion of absorption, but collection of body fluid over a long period of time is not necessarily required.
3. *Substances to be measured*: As a general rule, the unchanged active ingredient should be measured. Major active metabolites may be measured instead of the unchanged active ingredient, if it is rational. Stereoselective assay is not generally required. However, when it is indicated that there exist stereoisomers with different activities for the main pharmacological effect, and stereoselective absorption or elimination dependent on the absorption rate is noticeable, the enantiomer with higher activity should be measured.
4. *Analytical method*: Analytical methods should be fully validated regarding specificity, accuracy, precision, linearity, quantitation limit, and stability of substances in samples and so forth.

Washout Period

Washout periods in crossover studies between administrations of test and reference products should usually be more than five times the elimination half-life of the unchanged active ingredient or active metabolites to be measured.

Assessment of Bioequivalence

Parameters to Be Assessed

If blood is the sampled body fluid, AUC_t and C_{max} should be used as the parameters for evaluation of BE in a single-dose study, and AUC_τ and C_{max} should be used in a multiple-dose study. The measured value should be used for C_{max}, and the value calculated by the trapezoidal integration for AUC. If F can be calculated by deconvolution, F may be used instead of AUC.

AUC_∞, t_{max}, MRT, kel, and so on should be used as reference parameters. In multiple-dose studies, C_τ is also used as a reference parameter. However, if differences in the time required for manifestation of the effect of the drug could affect its clinical usefulness, t_{max} should also be used as a parameter for evaluation of equivalence.

If urine is the sampled body fluid, the parameters Ae_t, Ae_τ, Ae_∞, U_{max}, and U_τ should be used instead of AUC_t, AUC_τ, AUC_∞, C_{max}, and C_τ. If differences in the time required for manifestation of the effect of the drug could affect its clinical usefulness, urine cannot be used as the sampled body fluid.

Bioequivalence Range

If AUC and C_{max} are lognormally distributed, the BE acceptance range for each parameter is 0.80–1.25 when expressed as the ratio of the parameter's population means for the test product and reference product. If AUC and C_{max} are normally distributed, the acceptance range for each parameter is –0.20 to +0.20 when expressed as the ratio of the difference between the population mean for the test product and that for the reference product to the population mean for the reference product. If the drug's effect is not strong, a wider BE acceptance range than those mentioned earlier is sometimes set for C_{max}. If t_{max} is used as a parameter to evaluate equivalence, an appropriate BE acceptance range should be set beforehand.

Statistical Analysis

As a general rule, the parameters other than t_{max} are usually distributed lognormally, so those parameters should be logarithmically transformed before the statistical analysis. The 90% shortest confidence interval or two one-sided *t*-tests with the significance level of 5% should be used. Other reasonable statistical methods also can be used. When add-on subject study is performed and there are no fundamental differences between the two studies in formulation, design, assay, and subjects, data from the initial and add-on subject studies can be pooled and statistically analyzed. In the pooled analysis, the study must be added as a source of variation.

Acceptance Criteria

Products are considered to be bioequivalent, if the 90% confidence interval of difference in the average values of logarithmic parameters to be assessed between test and reference products is within the acceptable range of log(0.80)–log(1.25). However, even though the confidence interval is not in the aforementioned range, test products are accepted as bioequivalent, if the following three conditions are satisfied: (1) the total sample size of the initial BE study is not less than 20 (n = 10/group) or pooled sample size of the initial and add-on subject studies is not less than 30, (2) the differences in average values of logarithmic parameters to be assessed between two products are between log(0.90) and log(1.11), and (3) dissolution rates of test and reference products are evaluated to be similar. Reference parameters should be subjected to statistical assessment. If a significant difference is detected in the parameters between reference and test products, effects of this difference on therapeutic equivalence should be explained.

Pharmacodynamic Studies

A pharmacodynamic study is one in which therapeutic equivalence is demonstrated by using the pharmacological effect in humans as an index. Pharmacodynamic studies are used for drugs whose unchanged active ingredient or active metabolite is difficult to measure in blood or urine and for drugs whose BA does not indicate their therapeutic effect. In pharmacodynamic studies, it is advisable to compare the pharmacological effect over time. For antacids or digestive enzymes, suitable in vitro efficacy tests can be employed.

The acceptance criteria of equivalence in pharmacodynamic study should be established by considering the pharmacological activity of each drug.

Clinical Studies

Clinical studies are performed to establish the therapeutic equivalence of drugs using clinical effectiveness as an index. If BE studies and pharmacodynamic studies are impossible or inappropriate, clinical study is applied.

The acceptance criteria of equivalence in clinical study should be established by considering the pharmacological characteristics and activity of the respective drug.

Dissolution Tests

Dissolution tests should be performed, using a suitably validated dissolution system and assay.

1. *Number of vessels*: 12 vessels or more under each testing condition.
2. *Testing time*: 2 h in pH 1.2 medium and 6 h in other test fluids. The test can be stopped at the time when the average dissolution of reference product reaches 85%.
3. *Testing conditions*: The test should be carried out under the following conditions.
   Apparatus: JP paddle apparatus.
   Volume of test solution: Basically 900 mL.
   Temperature: 37°C ± 0.5°C
   Test solutions: The first and second fluids for the dissolution test (JP16) are used as pH 1.2 and 6.8 test solutions, respectively. Diluted McIlvaine buffers (pH is adjusted by 0.05 mol/L disodium hydrogen phosphate and 0.025 mol/L citric acid) are used for other pH solutions. If the average dissolution rate of the reference product does not reach 85% by 6 h under any of the aforementioned dissolution test conditions but does reach 85% in another suitable dissolution media, a test using the other dissolution media may be added.

   a. Products containing acidic drugs

Agitation (rpm)	pH
50[a]	(1) 1.2
	(2) 5.5–6.5[b]
	(3) 6.8–7.5[b]
	(4) Water
100[c]	(1), (2), or (3)[b]

[a] The paddle method at 75 rpm or the basket method at 100 rpm can be used instead of the paddle method at 50 rpm, when coning phenomenon of disintegrates in the bottom of vessel is observed.

[b] In the test solutions where the average dissolution of reference product reaches 85% within the testing time specified, the test solution where the dissolution is the slowest should be selected. When the average dissolution of the reference product does not reach 85% within the specified time in any of test fluids, the test solution should be selected where the dissolution is the fastest.

[c] If, in a dissolution test by the paddle method at 50 or 75 rpm, both the reference product and the test product dissolve an average of 85% or more within 30 min in dissolution medium that could be used with the paddle method at 100 rpm, the dissolution test by the paddle method at 100 rpm may be omitted.

b. Products containing neutral or basic drugs and coated products

Agitation (rpm)	pH
50[a]	(1) 1.2
	(2) 3.0–5.0[b]
	(3) 6.8
	(4) Water
100[c]	(1), (2), or (3)[b]

[a] The paddle method at 75 rpm or the basket method at 100 rpm can be used instead of the paddle method at 50 rpm, when coning phenomenon of disintegrates in the bottom of vessel is observed.

[b] In the test solutions where the average dissolution of reference product reaches 85% within the testing time specified, the test solution where the dissolution is the slowest should be selected. When the average dissolution of the reference product does not reach 85% within the specified time in any of test fluids, the test solution should be selected where the dissolution is the fastest.

[c] If, in a dissolution test by the paddle method at 50 or 75 rpm, both the reference product and the test product dissolve an average of 85% or more within 30 min in dissolution medium that could be used with the paddle method at 100 rpm, the dissolution test by the paddle method at 100 rpm may be omitted.

c. Products containing poorly soluble drugs
A drug product containing a poorly soluble drug is a drug product for which, when the test is performed at 50 rpm, the average dissolution rate of the reference product does not reach 85% within the designated test time in any of the dissolution media specified in (1) or in (2) with no surfactant in the medium.

Agitation (rpm)	pH	Surfactant
50[a]	(1) 1.2	None
	(2) 4.0	None
	(3) 6.8	None
	(4) Water	None
	(5) 1.2	Polysorbate 80[b]
	(6) 4.0	Polysorbate 80[b]
	(7) 6.8	Polysorbate 80[b]
100[c]	(5), (6), or (7)[d]	Polysorbate 80[e]

[a] The paddle method at 75 rpm or the basket method at 100 rpm can be used instead of the paddle method at 50 rpm, when coning phenomenon of disintegrates in the bottom of vessel is observed.

[b] Investigate polysorbate 80 concentrations of 0.01%, 0.1%, 0.5%, and 1.0% (W/V). Determine the lowest polysorbate 80 concentration necessary for the reference product to dissolve an average of 85% or more in at least one of dissolution media (5), (6), and (7) within the designated test time and add that concentration of polysorbate 80 to dissolution medium (5), (6), or (7). If the reference product does not dissolve an average of 85% in any of the dissolution media within the designated test time, choose the polysorbate 80 concentration at which dissolution is fastest. In the event that polysorbate 80 affects the dissolution behavior of the drug by interacting with the drug or with excipients, or in another such event, it is permissible to replace potassium dihydrogen phosphate with sodium dihydrogen phosphate as the buffering agent and use sodium lauryl sulfate. However, if sodium lauryl sulfate is used, the solubility of the drug may not exceed the solubility maximum concentration specified for polysorbate 80.

[c] If, in a dissolution test by the paddle method at 50 or 75 rpm, both the reference product and the test product dissolve an average of 85% or more within 30 min in dissolution medium that could be used with the paddle method at 100 rpm, the dissolution test by the paddle method at 100 rpm may be omitted.

[d] In the test solutions where the average dissolution of reference product reaches 85% within the testing time specified, the test solution where the dissolution is the slowest should be selected. When the average dissolution of the reference product does not reach 85% within the specified time in any of test fluids, the test solution should be selected where the dissolution is the fastest.

[e] The same concentration as that in the case of 50 rpm.

d. Enteric-coated products

Agitation (rpm)	pH
50[a]	(1) 1.2
	(2) 6.0
	(3) 6.8
100[b]	(2)

[a] The paddle method at 75 rpm or the basket method at 100 rpm can be used instead of the paddle method at 50 rpm, when coning phenomenon of disintegrates in the bottom of vessel is observed.

[b] If, in a dissolution test by the paddle method at 50 or 75 rpm, both the reference product and the test product dissolve an average of 85% or more within 30 min in dissolution medium that could be used with the paddle method at 100 rpm, the dissolution test by the paddle method at 100 rpm may be omitted.

Enteric-coated products containing poorly soluble drugs should be tested by adding polysorbate 80 to the test fluids (2) and (3) according to the dissolution test method for products containing poorly soluble drugs as described earlier.

4. *Acceptance criteria for similarity of dissolution profiles*: The average dissolution rate of the test product is compared with the average dissolution rate of the reference product. If dissolution of the reference product or test product has a lag time, the dissolution curve can be adjusted with the dissolution lag time (Appendix 16B). Criteria a. through c. in the succeeding text are applied after the lag time. However, when dissolution curves are corrected, the difference between the average dissolution lag times of the test product and reference product must be not more than 10 min. The time points for comparing dissolution rates when assessment is performed by the f_2 function are specified in Appendix 16B.

When any of the criteria in the following are met under all the sets of dissolution test conditions, the dissolution behavior is judged as similar. However, the average dissolution rate of the reference product must reach 85% or more within the designated test time under at least one set of dissolution test conditions.

If the comparison time point is to be less than 15 min, dissolution behavior may be evaluated using a comparison time point of 15 min. If correction for lag time is performed, the comparison time point of 15 min is the time before correction.

If the pH of the dissolution test medium is 1.2 for an enteric-coated product, dissolution behavior may be evaluated using only the dissolution rate at the designated test time (after 2 h).

A judgment of similarity in dissolution does not mean BE (Figure 16.2).

1. When the average dissolution of the reference product reaches 85% within 15 min: the average dissolution of the test product reaches 85% within 15 min or is within that of the reference product ±15% at 15 min.
2. When the average dissolution of the reference product reaches 85% at between 15 and 30 min: the average dissolution of the test product are within that of the reference product ±15% at two appropriate time points when the average dissolution of the reference product are around 60% and 85% or f_2 value is not less than 42.
3. When the average dissolution of the reference product does not reach 85% within 30 min: the results meet one of the following criteria.
   a. When the average dissolution of the reference product reaches 85% within the testing time specified, the average dissolution of the test product are within that of the reference product ±15% at two appropriate time points when the average dissolution of the reference product are around 40% and 85% or f_2 value is not less than 42.

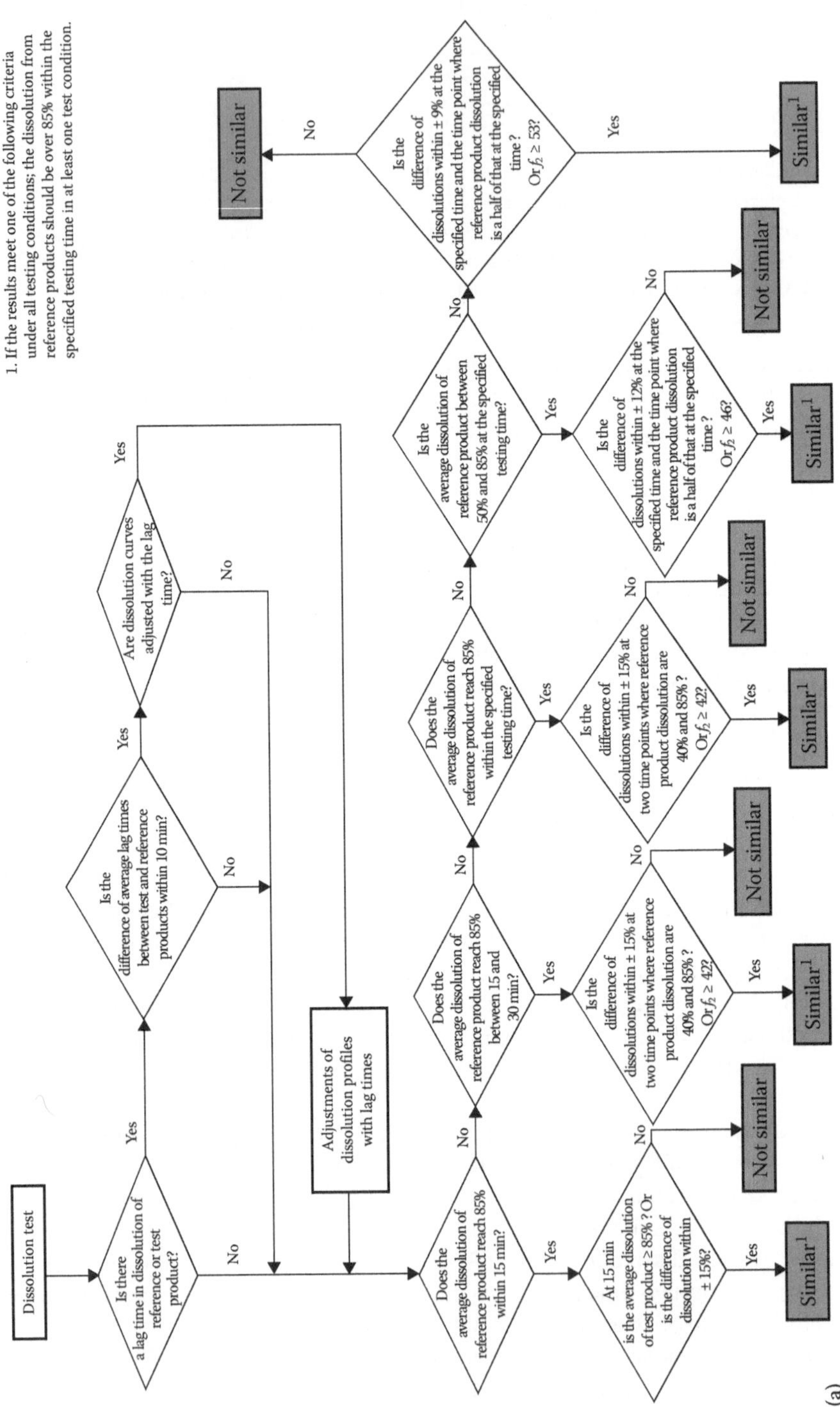
Dissolution test
Is there a lag time in dissolution of reference or test product?
Yes
No
Is the difference of average lag times between test and reference products within 10 min?
Yes
No
Are dissolution curves adjusted with the lag time?
Yes
No
Adjustments of dissolution profiles with lag times
Does the average dissolution of reference product reach 85% within 15 min?
Yes
No
At 15 min is the average dissolution of test product ≥ 85% ? Or is the difference of dissolution within ± 15%?
Yes
Similar1
No
Not similar
Does the average dissolution of reference product reach 85% between 15 and 30 min?
Yes
No
Is the difference of dissolutions within ± 15% at two time points where reference product dissolution are 40% and 85% ? Or f2 ≥ 42?
Yes
Similar1
No
Not similar
Does the average dissolution of reference product reach 85% within the specified testing time?
Yes
No
Is the difference of dissolutions within ± 15% at two time points where reference product dissolution are 40% and 85% ? Or f2 ≥ 42?
Yes
Similar1
No
Not similar
Is the average dissolution of reference product between 50% and 85% at the specified testing time?
Yes
No
Is the difference of dissolutions within ± 12% at the specified time and the time point where reference product dissolution is a half of that at the specified time ? Or f2 ≥ 46?
Yes
Similar1
No
Not similar
Is the difference of dissolutions within ± 9% at the specified time and the time point where reference product dissolution is a half of that at the specified time ? Or f2 ≥ 53?
Yes
Similar1
No
Not similar
1. If the results meet one of the following criteria under all testing conditions; the dissolution from reference products should be over 85% within the specified testing time in at least one test condition.
(a)

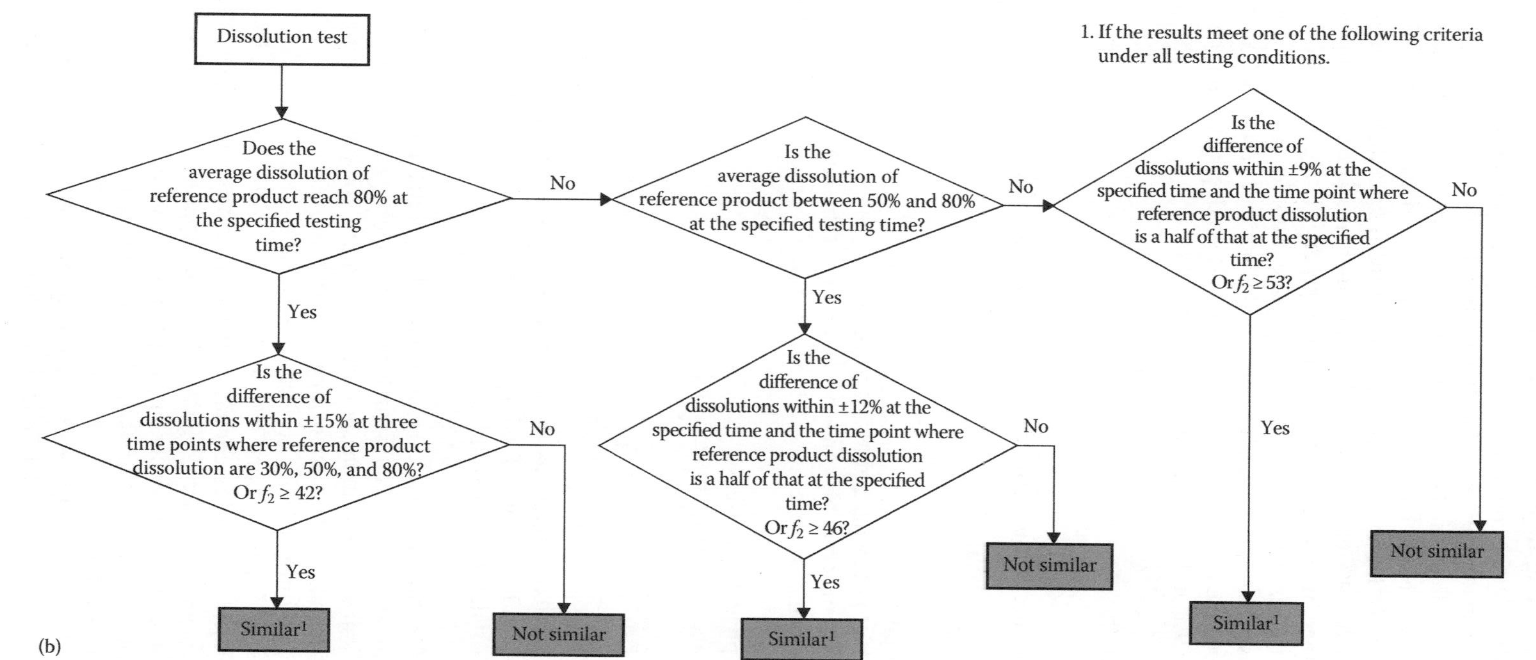

FIGURE 16.2 (*Continued*) Judgment of dissolution similarity. (a) Oral immediate-release products and enteric-coated products. (b) Oral extended release products.

b. When the average dissolution of the reference products reaches 50% and does not reaches 85% within the testing time specified, the average dissolution of the test product are within that of the reference product ±12% at the testing time specified and at an appropriate time point when the average dissolution of the reference product reaches about a half of the average dissolution at the testing time specified or f_2 value is not less than 46.
c. When the average dissolution of the reference product does not reach 50% within the testing time specified, the average dissolution of the test product are within that of the reference product ±9% at the testing time specified and at an appropriate time point when the average dissolution of the reference product is about a half of the average dissolution at the testing time specified or f_2 value is not less than 53. However, when the average dissolution of the reference product is not more than 10% at the stipulated dissolution time, the average dissolution of the test product is within that of the reference product ±9% at the testing time specified only.

Reporting of Test Results

Samples

1. Brand name and lot no. of the reference product. Code no. or name, lot no., and lot size of the test product
2. Type of dosage form
3. Name of drug substances
4. Labeled claims or potencies
5. Measured contents or potencies and assay procedures
6. Solubility of drugs at different pHs and in water used for dissolution tests
7. Particle size or specific surface area for poorly soluble drugs and their measurement procedures
8. Types of polymorph and solubility
9. Others (e.g., pKa and physicochemical stability)

Results

1. Summary
2. Dissolution tests:
   a. List of test conditions (apparatus, stirring speed, types and volumes of test solutions)
   b. Assay: method and summary of validation
   c. Summary of validation of dissolution tests
   d. Results
      i. Results of preliminary tests performed to select a reference product. Tables listing dissolution rate of individual sample under each testing condition, average dissolution, and standard deviations of each lot. Figures comparing average dissolution curves of each lot under each testing condition.
      ii. Results of preliminary tests performed to select test media.
      iii. Comparison of reference and test products tables listing dissolved% of individual sample under each testing condition, the average dissolution, and standard deviations of test and reference products. Figures comparing average dissolution curves of test and reference products under each testing condition.
3. BE studies. The following should be described. Preliminary test items should also be reported.
   a. Experimental conditions
      i. Subjects: Age, sex, body weight, and other data obtained by laboratory tests are described. Individual gastric acidity should be reported if necessary or otherwise available.
      ii. Drug administration.

Duration of fasting, coadministered water volume, and time of food ingestion after drug administration are described. In the case of postprandial administration, menu, content of meal (protein, fat, carbohydrate, calories, and others), and time from food ingestion to drug administration are described.

iii. Assay: procedure and summary of validation.

b. Results

i. Individual subject data tables showing drug concentration in biological fluids at each sampling time, C_{max}, C_{τ}, AUC_t, AUC_{τ}, AUC_{∞}, kel, t_{max}, and MRT. For all items, provide the untransformed data. The correlation coefficient for determining kel should be reported together with time points used. The ratios of C_{max} and AUC_t of test product to those of reference product in each individual should be reported. Figures comparing individual drug concentration–time profiles of the two products drawn on a linear/linear scale.

ii. Averages and standard deviations tables showing averages and standard deviations of raw data of drug levels in biological fluids at each time point, C_{max}, C_{τ}, AUC_t, AUC_{τ}, AUC_{∞}, kel, t_{max}, and MRT. The ratios of average of C_{max} and AUC_t of test product to those of reference product should be reported. Figures comparing average drug level–time profiles of the two products drawn on a linear/linear scale.

iii. Statistical analysis and equivalence assessment analysis of variance tables for C_{max}, C_{τ}, AUC_t, AUC_{τ}, AUC_{∞}, kel, t_{max}, and MRT that are logarithmically transformed when required. The statistical results for C_{max}, AUC_t, and AUC_{τ}. For other parameters, statistical testing results of the null hypotheses should be reported where the average values of test and reference products are assumed to be equivalent.

iv. Analysis of PK parameters. If deconvolution is used, the program, algorithm, PK models, and fitting information should be listed.

v. Other information on dropouts (data, reasons), monitoring records of health status of subjects.

4. Pharmacodynamic studies. Reporting of results should follow the description of BE studies.
5. Clinical studies. Reporting of results should follow the description of BE studies.

Oral Extended Release Products

Reference and Test Products

In principle, dissolution tests should be performed using the following test solution (1) or (2), using six vessels or more for three lots of innovator products by the paddle method at 50 rpm. Among the three lots, the one that shows intermediate dissolution should be selected as the reference product:

1. The specification test solution when the dissolution specifications are established in the specifications and test procedures.
2. Among the test solutions described in the dissolution conditions, when the average dissolution of at least one lot reaches 85%, the test solution providing the slowest dissolution should be selected. When the average dissolution of any of the lots does not reach 85%, the test solution providing the fastest dissolution should be used.

If a reference product cannot be appropriately selected for the drug product by dissolution testing as described earlier, the reference product should be the innovator product lot that shows intermediate characteristics when either a dissolution (release) test appropriate for the characteristics of the drug product or a substitute physicochemical test is performed.

The test generic product must not differ markedly from the innovator product in size, shape, specific gravity, or release mechanism. For the test product lot size, and content or potency, follow the "Oral Immediate-Release Products and Enteric-Coated Products" standards earlier in this section. The dissolution behavior of the test product must be similar to that of the reference product. Assess the similarity of dissolution behavior.

Bioequivalence Studies

Test Method

BE studies should be performed by single-dose studies in both the fasted and fed states. In the case of postprandial administration, a high-fat diet of 900 kcal or more containing 35% lipid content should be used. The meal should be eaten within 20 min, and drugs administered within 10 min thereafter.

When a high incidence of severe adverse events is indicated after dosing in the fasting state, the fasting dose studies can be replaced with postprandial dose studies with the low-fat meal employed in the study for oral immediate-release products and enteric-coated products. Other testing conditions should follow those of oral immediate-release products and enteric-coated products.

Assessment of Bioequivalence

1. BE range, parameters, data transformation, and statistical analysis. These are the same as those of oral immediate-release products and enteric-coated products.
2. Acceptance criteria

Products are considered to be bioequivalent, if the 90% confidence interval of difference in the average values of logarithmic parameters to be assessed between test and reference products is within the acceptable range of log(0.80)–log(1.25). However, even though the confidence interval is not in the aforementioned range, test products are accepted as bioequivalent, if the following three conditions are satisfied: (1) the total sample size of the initial BE study is not less than 20 (n = 10/group) or pooled sample size of the initial and add-on subject studies is not less than 30, (2) the difference in average values of logarithmic AUC and C_{max} between two products is between log(0.90) and log(1.11), and (3) the dissolution characteristics of the test product are equivalent to those of the reference product. The dissolution profiles are judged to be equivalent following the standards outlined earlier in this section. The assessment of reference parameters follows that of oral immediate-release products and enteric-coated products.

Pharmacodynamic and Clinical Studies

If BE studies cannot be performed, pharmacodynamic or clinical studies should be carried out to evaluate therapeutic equivalence according to the studies for oral immediate-release products and enteric-coated products.

Dissolution Tests

1. *Number of units*: 12 units or more under each testing conditions.
2. *Testing time*: The testing time is normally 24 h, but at pH 1.2, the test may be concluded after 2 h. The test can be stopped at the time when the average dissolution of reference product reaches 85%.
3. *Test conditions*: The test should be carried out under the following conditions.

   Apparatus: Paddle apparatus, rotating basket, and disintegration testing apparatus can be selected, the reason for which should be stated.

Volume of test solution, temperature, and test solutions should follow the description of oral immediate-release products and enteric-coated products.

Apparatus	Agitation (rpm)	pH	Other Conditions
Paddle	50	(1) 1.2	
		(2) 3.0–5.0[a]	
		(3) 6.8–7.5[a]	
		(4) Water	
		(3)	Polysorbate 80 (1.0% W/V)
	100	(3)	
	200	(3)	
Basket	100	(3)	
	200	(3)	
Disintegration	30[b]	(3)	Without disk
	30[b]	(3)	With disk

[a] In the test solutions where the average dissolution of reference product reaches 80% within 24 h, the test solution where the dissolution is the slowest should be selected. When the average dissolution of the reference product does not reach 80% within 24 h in any of test fluids, the test solution where the dissolution is the fastest should be selected.

[b] Strokes/min.

4. *Acceptance criteria for similarity and equivalence of dissolution profiles*: If the results meet one of the following criteria shown in (1) under all testing conditions, the dissolution profile of the test product is judged to be similar to that of the reference product.

If the average dissolution of the reference product reaches 80% within the testing time point specified in at least one test condition, and the results meet one of the following criteria shown in (2) under all testing conditions, the dissolution profile of the test product is judged to be equivalent to that of the reference product.

When similarity factor, f_2, is used, Appendix 16A (2) time points for f_2 should be employed. A judgment of similarity or equivalence in dissolution does not mean BE (Figure 16.3).

1. *Similarity*
   a. When the average dissolution of the reference product reaches 80% within the testing time specified, the average dissolution of the test product are within that of the reference product ±15% at three appropriate time points when the average dissolution of the reference product are around 30%, 50%, and 80% or f_2 value is not less than 42.
   b. When the average dissolution of the reference product reaches 50% and does not reach 80% within the testing time point specified, the average dissolution of the test product are within that of the reference product ±12% at the testing time specified and at an appropriate time point when the average dissolution of the reference product reaches about a half of the average dissolution at the testing time specified or f_2 value is not less than 46.
   c. When the average dissolution of the reference product does not reach 50% within the testing time specified, the average dissolution of the test product are within that of the reference product ±9% at the testing time specified and at an appropriate time point when the average dissolution of the reference product is about a half of the average dissolution at the testing time specified or f_2 value is not less than 53. However, when the average dissolution of the reference product is not more than 10% within the testing time specified, the average dissolution of the test product is within that of the reference product ±9% at the testing time specified only.

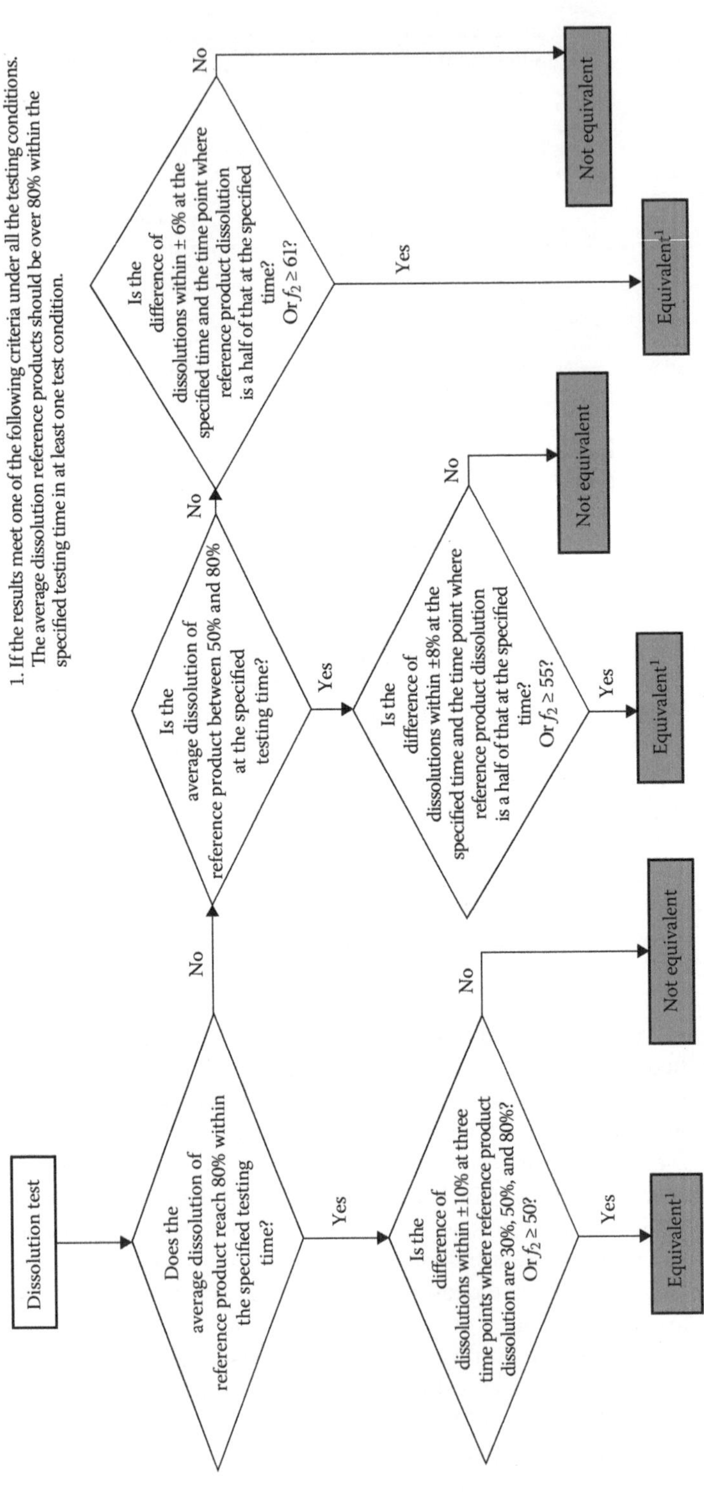

FIGURE 16.3 Judgment of dissolution equivalence.

2. *Equivalence*
   a. When the average dissolution of the reference product reaches 80% within the testing time specified, the average dissolution of the test product are within that of the reference product ±10% at three appropriate time points when the average dissolution of the reference product are around 30%, 50%, and 80% or f_2 value is not less than 50.
   b. When the average dissolution of the reference product reaches 50% and does not reach 80% within the testing time point specified, the average dissolution of the test product are within that of the reference product ±8% at the testing time specified and at an appropriate time point when the average dissolution of the reference product reaches about a half of the average dissolution at the testing time specified or f_2 value is not less than 55.
   c. When the average dissolution of the reference product does not reach 50% within the testing time specified, the average dissolution of the test product are within that of the reference product ±6% at the testing time specified and at an appropriate time point when the average dissolution of the reference product is about a half of the average dissolution at the testing time specified or f_2 value is not less than 61. However, when the average dissolution of the reference product is not more than 10% within the testing time specified, the average dissolution of the test product is within that of the reference product ±6% at the testing time specified only.

Reporting of Test Results

The shape, specific gravity, and release mechanism of the test product should be described, which do not differ significantly from those of the innovator product. The description of other results is the same as that for oral immediate products and enteric-coated products.

Non-Oral Dosage Forms

The test for the products for topical use should be following the *Guideline for Bioequivalence Studies of Generic Products for Topical Use*, an attachment of Division-Notification No. 1124004 of the Pharmaceutical and Food Safety Bureau, amendments to the *Guideline for Bioequivalence Studies of Generic Products* and other guidelines, dated November 24, 2006.

For other non-oral dosage forms, the test should be performed following the description in the succeeding text.

Reference and Test Products

Suitable release tests or alternative physicochemical tests should be performed for three lots of an innovator product from which one lot providing intermediate characteristics should be selected as a reference product. If the drug is administered as a liquid where the active ingredient dissolves, an appropriate lot can be used as a reference product. The lot size and drug content or potency follows the description for oral immediate-release products and enteric-coated products.

Bioequivalence Studies

The test should follow the BE test for oral immediate-release products and enteric-coated products but the results of release or physicochemical tests are not used as supportive data for the assessment of BE.

Pharmacodynamic and Clinical Studies

Follow the description of the oral immediate-release products and enteric-coated products. In pharmacodynamic study, it is desirable to compare the efficacy–time profiles. Appropriate animal studies will be allowed for products for topical use (skin) of which pharmacological effects on the surface of the skin and can be evaluated appropriately, for example, hemostatic agent, disinfecting agent, and intention promoters, which do not need to penetrate the stratum corneum for exerting their pharmacological effects. For bactericides for external use, appropriate in vitro efficacy tests can be employed.

Dissolution (Release) Tests or Physicochemical Tests

Release or physicochemical characteristics should be compared between test and reference products by appropriate tests that will vary depending on the product.

Reporting of Test Results

Follow the description of the oral immediate-release products and enteric-coated products.

Dosage Forms of Which Bioequivalence Studies Are Waived

Injections for intravenous administration, administered as an aqueous solution.

APPENDIX 16A: f_2 (SIMILARITY FACTOR) AND TIME POINTS FOR COMPARISONS

Definition of f_2

The following equation defines f_2. T_i and R_i show the average dissolutions of the test and reference products at the time point (i), respectively, and n is the number of time points at which the average dissolution are compared:

$$f_2 = 50\log\left[\frac{100}{\sqrt{1+\frac{\sum_{i=1}^{n}(T_i - R_i)^2}{n}}}\right]$$

Time Points for f_2

1. When the average dissolution of the reference product reaches 85% (80% for extended release products) between 15 and 30 min: 15, 30, 45 min
2. When the average dissolution of the reference product reaches 85% (80% for extended release products) between 30 min and the testing time point specified: at Ta/4, 2Ta/4, 3Ta/4 and Ta, where Ta is the time point at which average dissolution of the reference product reaches approximately 85% (80% for extended release products)
3. When the average dissolution of the reference product does not reach 85% (80% for extended release products) within the testing time point specified: at Ta/4, 2Ta/4, 3Ta/4 and Ta, where Ta is the time point at which average dissolution of the reference product reaches approximately 85% (80% for extended release products) of the amount dissolved at the testing time point specified

APPENDIX 16B: ADJUSTING DISSOLUTION CURVES WITH LAG TIMES

The lag time is defined as the time when 5% of the labeled claim of the active ingredient dissolves from the product. A lag time should be determined for respective product by linear interpolation, and then respective dissolution curve is obtained by adjusting dissolution curve with the lag time. Average dissolution curves of the test and reference products are obtained, which can be used for the assessment of similarity and equivalence in dissolution curves.

APPENDIX 16C: LIST OF ABBREVIATIONS OF PARAMETERS

Ae_t	Cumulative amount of drug excreted in the urine from zero to the final sampling time t
Ae	Cumulative amount of drug excreted in the urine from zero to infinity
AUC	Area under drug concentration in blood–time curves
AUC_t	AUC from zero to the final sampling time t
AUC_τ	AUC over one dose interval (τ) at steady state
AUC_∞	AUC from zero to infinity
C_{max}	The maximum drug concentration in blood
C_τ	Drug level in blood at the time τ after dosing at steady state
F	Ratio of bioavailability of a test product to that of the standard preparation (intravenous dose or oral aqueous dose)
Kel	Elimination rate constant
MRT	Mean residence time
t_{max}	Time to the maximum drug concentration in blood or time to the maximum urinary excretion rate
U_{max}	The maximum urinary excretion rate of drug
U_t	Urinary excretion rate of drug at the final sampling time over one dose interval (t) at steady state

17 Bioequivalence Testing Protocols

To receive approval for an abbreviated new drug application (ANDA), an applicant generally must demonstrate, among other things, that its product has the same active ingredient, dosage form, strength, route of administration, and conditions of use as the listed drug and that the proposed drug product is bioequivalent to the reference-listed drug (RLD) (21 U.S.C. 355(j)(2)(A); 21 CFR 314.94(a)). Bioequivalent drug products show no significant difference in the rate and extent of absorption of the therapeutic ingredient (21 U.S.C. 355(j)(8); 21 CFR 320.1(e)). Bioequivalence (BE) studies are undertaken in support of ANDA submissions with the goal of demonstrating BE between a proposed generic drug product and its RLD. The regulations governing BE are provided at 21 CFR in part 320. The U.S. Food and Drug Administration (FDA) has recently begun to promulgate individual BE requirements. To streamline the process for making guidance available to the public on how to design product-specific BE studies, the U.S. FDA will be issuing product-specific BE recommendations (www.fda.gov/cder/ogd/index.htm). Given later are the current recommendations for more significant products.

ClinicalTrials.gov is a web-based resource that provides patients, their family members, health-care professionals, researchers, and the public with easy access to information on publicly and privately supported clinical studies on a wide range of diseases and conditions. The website is maintained by the National Library of Medicine (NLM) at the National Institutes of Health (NIH). Information on ClinicalTrials.gov is provided and updated by the sponsor or principal investigator of the clinical study. Studies are generally submitted to the website (i.e., registered) when they begin, and the information on the site is updated throughout the study. In some cases, results of the study are submitted after the study ends. This website and database of clinical studies is commonly referred to as a "registry" and "results database."

ClinicalTrials.gov includes information about medical studies in human volunteers. Most of the records in ClinicalTrials.gov describe clinical trials (also called interventional studies). A clinical trial is a research study in which human volunteers are assigned to interventions (e.g., a medical product, behavior, or procedure) based on a protocol (or plan) and are then evaluated for effects on biomedical or health outcomes. ClinicalTrials.gov also includes records describing observational studies and programs providing access to investigational drugs outside of clinical trials (expanded access). Studies listed in the database are conducted in all 50 states and in 185 countries. ClinicalTrials.gov does not contain all clinical studies conducted in the United States because not all studies are required by law to be registered. However, the number of studies registered each year has increased over time as more policies and laws requiring registration have been enacted and as more sponsors and investigators voluntarily register their studies.

ClinicalTrials.gov was created as a result of the Food and Drug Administration Modernization Act of 1997 (FDAMA). FDAMA required the U.S. Department of Health and Human Services, through NIH, to establish a registry of clinical trials information for both federally and privately funded trials conducted under investigational new drug (IND) applications to test the effectiveness of experimental drugs for serious or life-threatening diseases or conditions. NIH and the FDA worked together to develop the site, which was made available to the public in February 2000.

The ClinicalTrials.gov registration requirements were expanded after Congress passed the Food and Drug Administration Amendments Act of 2007 (FDAAA). Section 801 of FDAAA (FDAAA 801) requires more types of trials to be registered and additional trial registration information to be

TABLE 17.1
BE Reported on www.clinicaltrials.gov in October 2013

Region	Studies
World	1532
United States	626
South Asia	198
Canada	171
Europe	141
East Asia	58
Southeast Asia	29
Japan	27
South America	26
Africa	11
Mexico	11
Middle East	10
Pacifica	6
Central America	4
North Asia	3

submitted. The law also requires the submission of results for certain trials. This led to the development of the ClinicalTrials.gov results database, which contains information on study participants and a summary of study outcomes, including adverse events. The results database was made available to the public in September 2008. FDAAA 801 also established penalties for failing to register or submit the results of trials. See the History, Policies, and Laws page for more information about the development of ClinicalTrials.gov.

Because ClinicalTrials.gov is a government website, it does not host, or receive funding from, advertising or the display of commercial content.

One significant source of information on protocols that used BE testing is the website www.clinicaltrials.gov. Table 17.1 shows a geographical distribution of BE trials reported as of October 2013. A majority of trials are conducted in North America, and as a result, the U.S. FDA guidelines are of great importance.

These data provide a very good idea about the types of drug development in progress. Table 17.2 shows the breakdown of the type of conditions for which drug BE are conducted.

PRODUCT-SPECIFIC GUIDELINES

Introduction

This guidance describes FDA's process for making available to the public FDA guidance on how to design BE studies for specific drug products to support ANDAs. Under this process, applicants planning to carry out such studies in support of their ANDAs will be able to access BE study guidance on the FDA website, rather than having to request this information from the agency and wait for the agency to respond, as has been the case in the past. The FDA believes that making this information available on the Internet will streamline the guidance process, making it more efficient than the previous process. This process also will provide a meaningful opportunity for the public to consider and comment on BE study recommendations for specific drug products.

FDA's guidance documents, including this guidance, do not establish legally enforceable responsibilities. Instead, guidances describe the agency's current thinking on a topic and should be viewed only as recommendations, unless specific regulatory or statutory requirements are cited. The use of the word should in agency guidances mean that something is suggested or recommended, but not required.

TABLE 17.2
Conditions and Frequency of Studies Reported

Condition	Study(ies) Reported	Condition	Study(ies) Reported
Abnormalities, multiple	1	Blood platelet disorders	1
Acne vulgaris	9	Blood protein disorders	1
Acquired hemophilia	1	Body temperature changes	1
Acquired immunodeficiency syndrome	17	Body weight	2
Acute coronary syndrome	1	Bone diseases	10
Acute lymphoblastic leukemia	2	Bone diseases, developmental	4
Addison disease	1	Bone diseases, endocrine	3
Adnexal diseases	4	Bone diseases, metabolic	5
Adrenal gland diseases	1	Bone marrow diseases	4
Adrenal insufficiency	1	Brain diseases	40
Affective disorders, psychotic	8	Breast diseases	13
Agoraphobia	1	Breast neoplasms	14
Alopecia	2	Bronchial diseases	11
Alopecia areata	2	Bronchial neoplasms	4
Alzheimer disease	3	Calcium metabolism disorders	1
Alzheimer disease, familial	3	Candidiasis	2
Anemia	2	Candidiasis, vulvovaginal	2
Angina pectoris	4	Carcinoma	5
Angina pectoris, variant	2	Carcinoma, bronchogenic	4
Angina, unstable	2	Carcinoma, non-small cell lung	2
Anxiety disorders	7	Cardiovascular abnormalities	1
Arbovirus infections	1	Cataract	3
Arnold–Chiari malformation	2	Central nervous system diseases	42
Arrhythmias, cardiac	1	Central nervous system infections	2
Arterial occlusive diseases	4	Cervical dystonia	1
Arteriosclerosis	4	Chagas disease	1
Arthritis	8	Chest pain	5
Arthritis, psoriatic	1	Chromosome aberrations	3
Arthritis, rheumatoid	6	Chromosome disorders	3
Asthma	11	Chronic lymphocytic leukemia	1
Atherosclerosis	1	Chronic myeloid leukemia	3
Atrial fibrillation	1	Chronic myeloproliferative disorders	3
Atrophy	3	Coagulation protein disorders	4
Attention deficit and disruptive behavior disorders	2	Cognition disorders	3
		Colitis	3
Attention deficit disorder with hyperactivity	2	Colitis, ulcerative	3
Autoimmune diseases	16	Colonic diseases	4
Autoimmune diseases of the nervous system	2	Colorectal neoplasms	1
Autonomic nervous system diseases	1	Congenital abnormalities	3
Bacterial infections	7	Congenital hypothyroidism	1
Bacterial meningitis	1	Connective tissue diseases	7
Basal ganglia diseases	3	Constipation	1
Behavioral symptoms	23	Coronary artery disease	4
Bipolar disorder	8	Coronary disease	1
Blood coagulation disorders	6	Cystic fibrosis	3
Blood coagulation disorders, inherited	4	Delirium	3

(*Continued*)

TABLE 17.2 (*Continued*)
Conditions and Frequency of Studies Reported

Condition	Study(ies) Reported	Condition	Study(ies) Reported
Delirium, dementia, amnestic, cognitive disorders	3	Eye infections	1
		Facial dermatoses	6
Dementia	3	Facies	7
Demyelinating autoimmune diseases, CNS	2	Failure to thrive	1
Demyelinating diseases	2	Fever	1
Depression	22	Fibrosis	3
Depressive disorder	22	Flavivirus infections	1
Depressive disorder, major	1	Foot diseases	1
Dermatitis	4	Foot ulcer	3
Dermatitis, allergic contact	1	Gambling	3
Dermatitis, atopic	1	Ganglion cysts	3
Dermatitis, contact	2	Gastroenteritis	3
Dermatomycoses	3	Gastrointestinal diseases	18
Diabetes complications	3	Gastrointestinal neoplasms	1
Diabetes mellitus	78	Genetic diseases, inborn	11
Diabetes mellitus, type 1	7	Genetic diseases, X-linked	2
Diabetes mellitus, type 2	31	Genital diseases, female	10
Diabetic angiopathies	3	Genital diseases, male	15
Diabetic foot	3	Genital dwarfism	1
Diabetic neuropathies	3	Genital neoplasms, female	3
Diarrhea	1	Genital neoplasms, male	2
Digestive system diseases	18	Glanzmann thrombasthenia	1
Digestive system neoplasms	1	Glucose metabolism disorders	47
Disorders of sex development	1	Gonadal disorders	4
DNA virus infections	6	Gonadal dysgenesis	1
Duodenal diseases	2	Gram-positive bacterial infections	3
Dwarfism	4	Growth disorders	4
Dwarfism, pituitary	2	Growth hormone deficiency	2
Dyskinesias	3	Habits	3
Dyslipidemias	11	Headache	7
Dyssomnias	4	Headache disorders	6
Dystonia	1	Headache disorders, primary	6
Dystonic disorders	1	Heart defects, congenital	1
Ear diseases	2	Heart diseases	6
Edema	2	Heart valve diseases	1
Encephalitis	1	Heartburn	3
Encephalitis, Japanese	1	Helminthiasis	1
Endocrine gland neoplasms	3	Hematologic diseases	11
Endocrine system diseases	59	Hemophilia	4
Enterovirus infections	9	Hemophilia A	4
Enuresis	1	Hemophilia A, congenital	4
Epilepsy	25	Hemophilia B	2
Epilepsy, generalized	1	Hemorrhage	3
Epilepsy, tonic–clonic	1	Hemorrhagic disorders	6
Erectile dysfunction	3	Hemostatic disorders	6
Exanthema	6	Hepadnaviridae infections	2
Eye diseases	3	Hepatitis	9

(*Continued*)

TABLE 17.2 (*Continued*)
Conditions and Frequency of Studies Reported

Condition	Study(ies) Reported	Condition	Study(ies) Reported
Hepatitis A	9	Intestinal diseases	6
Hepatitis B	2	Intestinal neoplasms	1
Hepatitis B, chronic	1	Ischemia	4
Hepatitis C	6	Japanese encephalitis	1
Hepatitis C, chronic	4	Joint diseases	6
Hepatitis, chronic	5	Keratoacanthoma	1
Hepatitis, viral, human	9	Keratosis	6
Herpes genitalis	2	Keratosis, actinic	6
Herpes labialis	1	Kidney diseases	2
Herpes simplex	4	Kidney failure, chronic	2
Herpes zoster	3	Lamellar ichthyosis	1
Herpesviridae infections	4	Leg ulcer	3
Hirsutism	2	Lens diseases	2
HIV infections	17	Lentivirus infections	13
Hot flashes	1	Leukemia	5
Hypercholesterolemia	6	Leukemia, B-cell	1
Hyperglycemia	2	Leukemia, B-cell, chronic	1
Hyperkinesis	2	Leukemia, lymphocytic, chronic, B-cell	1
Hyperlipidemias	8	Leukemia, lymphoid	3
Hyperparathyroidism	1	Leukemia, myelogenous, chronic, BCR–ABL positive	3
Hyperparathyroidism, secondary	1		
Hyperphosphatemia	2	Leukemia, myeloid	3
Hyperplasia	9	Lipid metabolism disorders	11
Hypersensitivity	24	Liver diseases	9
Hypersensitivity, delayed	1	Lung diseases	21
Hypersensitivity, immediate	17	Lung diseases, obstructive	12
Hypertension	34	Lung neoplasms	4
Hypertension, pulmonary	1	Lymphatic diseases	4
Hypertrichosis	2	Lymphoma	3
Hypoadrenalism	1	Lymphoma, non-Hodgkin	1
Hypocalcemia	1	Lymphoma, small cleaved-cell, diffuse	1
Hypoglycemia	1	Lymphoproliferative disorders	5
Hypopituitarism	2	Malaria	5
Hypotension	1	Malaria, falciparum	4
Hypotension, orthostatic	1	Malnutrition	283
Hypothyroidism	2	Melanoma	1
Hypotrichosis	1	Meningitis	1
Ichthyosis	1	Meningitis, bacterial	1
Idiopathic pulmonary hypertension	1	Menopause	9
Immunologic deficiency syndromes	14	Menopause, premature	1
Immunoproliferative disorders	5	Mental disorders	63
Infant, newborn, diseases	4	Mental disorders diagnosed in childhood	3
Inflammation	8	Mental retardation	1
Inflammatory bowel diseases	3	Metabolic diseases	62
Influenza, human	1	Metrorrhagia	1
Intermittent claudication	2	Migraine disorders	5

(*Continued*)

TABLE 17.2 (*Continued*)
Conditions and Frequency of Studies Reported

Condition	Study(ies) Reported	Condition	Study(ies) Reported
Mitral valve prolapse	1	Ovarian diseases	4
Mood disorders	32	Ovarian epithelial cancer	2
Mouth diseases	7	Ovarian neoplasms	3
Movement disorders	4	Overnutrition	2
Multiple myeloma	1	Overweight	2
Multiple sclerosis	2	Pain	34
Musculoskeletal diseases	16	Pain, postoperative	1
Mycobacterium infections	3	Pancreatic diseases	3
Mycoses	6	Panic disorder	1
Myelodysplastic syndromes	2	Paraproteinemias	1
Myelofibrosis	1	Parasitic diseases	8
Myeloproliferative disorders	3	Parasomnias	4
Myocardial ischemia	4	Parathyroid diseases	1
Nail diseases	2	Parkinson disease	3
Nausea	8	Parkinsonian disorders	3
Neoplasm metastasis	4	Pathological conditions, anatomical	3
Neoplasms, germ cell and embryonal	1	Peptic ulcer	2
Neoplasms, glandular and epithelial	2	Peripheral nervous system diseases	1
Neoplasms, nerve tissue	1	Philadelphia chromosome	1
Neoplasms, plasma cell	1	Phosphorus metabolism disorders	2
Neuralgia	1	Picornaviridae infections	9
Neurobehavioral manifestations	1	Pituitary diseases	2
Neurocirculatory asthenia	1	Pneumonia	1
Neurodegenerative diseases	6	Polyuria	1
Neuroectodermal tumors	1	Postoperative complications	3
Neuroendocrine tumors	1	Postoperative nausea and vomiting	1
Neuroepithelioma	1	Prader–Willi syndrome	1
Neurologic manifestations	19	Precancerous conditions	8
Neuromuscular diseases	1	Precursor cell lymphoblastic leukemia–lymphoma	2
Nevus	1		
Nevus, pigmented	1	Preleukemia	2
Nocturnal enuresis	1	Primary dysautonomias	1
Nose diseases	5	Primary myelofibrosis	1
Nutrition disorders	9	Primary ovarian insufficiency	1
Obesity	2	Prinzmetal's variant angina	2
Obesity, morbid	1	Prostatic diseases	10
Onychomycosis	2	Prostatic hyperplasia	8
Opioid-related disorders	1	Prostatic neoplasms	2
Orthomyxoviridae infections	1	Protozoan infections	7
Orthostatic intolerance	1	Pruritus	1
Osteoarthritis	1	Psoriasis	1
Osteoporosis	5	Psychomotor agitation	2
Osteoporosis, postmenopausal	1	Psychotic disorders	63
Otitis	2	Pulmonary disease, chronic obstructive	1
Otitis externa	2	Rectal diseases	1
Otorhinolaryngologic diseases	7	Recurrence	2
Ovarian cancer	3	Renal insufficiency	2

(*Continued*)

TABLE 17.2 (*Continued*)
Conditions and Frequency of Studies Reported

Condition	Study(ies) Reported	Condition	Study(ies) Reported
Renal insufficiency, chronic	2	Stomatognathic diseases	7
Respiration disorders	2	Substance-related disorders	5
Respiratory aspiration	5	Syndrome	20
Respiratory hypersensitivity	16	Synovial cyst	3
Respiratory tract diseases	33	Tauopathies	3
Respiratory tract infections	12	Tension-type headache	1
Respiratory tract neoplasms	4	Thoracic neoplasms	4
Restless legs syndrome	2	Thrombasthenia	1
Retroviridae infections	13	Thyroid diseases	2
Rheumatic diseases	7	Tic disorders	1
Rhinitis	5	Tics	1
Rhinitis, allergic, perennial	2	Tinea	3
Rhinitis, allergic, seasonal	3	Tinea pedis	1
RNA virus infections	23	Tobacco use disorder	6
Rosacea	1	Torticollis	1
Salivary gland diseases	6	Translocation, genetic	1
Schizophrenia	10	Trichomonas infections	1
Schizophrenia and disorders with psychotic features	9	Trypanosomiasis	1
		Tuberculosis	3
Sclerosis	2	Tuberculosis, central nervous system	1
Sebaceous gland diseases	6	Tuberculosis, meningeal	1
Seizures	14	Tuberculous meningitis	1
Sex chromosome disorders of sex development	1	Turner syndrome	1
Sexual dysfunction, physiological	3	Tylosis	6
Sexual dysfunctions, psychological	3	Ulcer	9
Sexually transmitted diseases	15	Urinary bladder diseases	2
Sexually transmitted diseases, viral	15	Urinary bladder, overactive	2
Signs and symptoms, digestive	12	Urinary incontinence	1
Signs and symptoms, respiratory	1	Urinary incontinence, stress	1
Skin abnormalities	1	Urination disorders	2
Skin diseases	43	Urogenital abnormalities	1
Skin diseases, eczematous	3	Urogenital neoplasms	5
Skin diseases, genetic	1	Urologic diseases	6
Skin diseases, infectious	7	Urological manifestations	4
Skin diseases, papulosquamous	1	Uterine diseases	1
Skin manifestations	1	Uterine hemorrhage	1
Skin ulcer	3	Vaginal diseases	5
Sleep disorders	4	Vaginitis	5
Sleep disorders, intrinsic	4	Vaginosis, bacterial	1
Sleep initiation and maintenance disorders	2	Vascular diseases	45
Small-cell lung carcinoma	2	Virilism	2
Smoking	9	Virus diseases	28
Spinal diseases	1	Vomiting	8
Spondylarthritis	1	Vulvar diseases	2
Spondyloarthropathies	1	Vulvovaginitis	2
Spondylitis	1	Water–electrolyte imbalance	1
Stomach diseases	2	Xerostomia	6

Background

What Are BE Studies?

To receive approval for an ANDA, an applicant generally must demonstrate, among other things, that its product has the same active ingredient, dosage form, strength, route of administration, and conditions of use as the listed drug and that the proposed drug product is bioequivalent to the RLD (21 U.S.C. 355(j)(2)(A); 21 CFR 314.94(a)). If a drug acts through absorption into the bloodstream, bioequivalent drug products are those that show no significant difference in the rate and extent of absorption of the therapeutic ingredient (21 U.S.C. 355(j)(8)(B); 21 CFR 320.1(e)). For a drug that is not intended to be absorbed into the bloodstream, FDA may establish alternative methods to show BE that may be expected to detect a significant difference between the drug and the listed drug in safety and therapeutic effect (21 U.S.C. 355(j)(8)(C); 21 CFR 320.24). BE studies are undertaken in support of ANDA submissions with the goal of demonstrating BE between a proposed generic drug product and its RLD. The regulations governing BE are provided at 21 CFR part 320.

How Did the Agency Make This Information Available in the Past?

Previously, the Office of Generic Drugs (OGD) provided guidance on how to design BE studies for specific products only when asked for assistance by individual applicants. We had determined that making recommendations to applicants about how to design BE studies would help the generic drug industry, the innovator drug industry, contract research organizations, academia, and others understand the agency's expectations with regard to demonstrating BE. In most cases, the requested information was not available anywhere else, and, in some cases, OGD performed its own research before responding to an applicant's request for product-specific information. In many cases, OGD responded to individual requests for information on BE studies in letter format after specific recommendations were prepared within the Center for Drug Evaluation and Research (CDER). This meant that information about BE studies was only being provided to those specifically requesting such information. In addition, the staff developing the recommendations and responding to requests for information have been the same individuals who are responsible for reviewing the BE data in ANDAs. With the increase in the number of ANDA submissions and in the requests for BE information during the last few years, the process of providing BE recommendations has become extremely time-consuming for the agency.

In October 2000, to help address this growing problem, FDA issued the guidance Bioavailability and Bioequivalence Studies for Orally Administered Drug Products—General Considerations, which describes general recommendations for demonstrating BE. These general recommendations were helpful, but many individuals have continued to seek assistance from the agency in designing their product-specific BE studies, as certain drug products may raise BE issues not squarely addressed in more general guidance. As a result, after exploring various mechanisms that would allow us to conserve our resources while responding to the needs of industry and other interested persons, OGD has developed a new approach to making guidance available on product-specific BE studies. As before, the agency intends to develop BE recommendations based on its understanding of the characteristics of the listed drug, information derived from published literature, agency research, and consultations within different offices in CDER as needed based upon the novelty or complexity of the BE considerations. Once developed, the agency intends to make BE recommendations for specific drug products available through the process described here.

FDA is not required to publish draft or final product-specific BE recommendations before it approves an ANDA for the drug product. If the agency determines that, as required by the statute and regulations (21 U.S.C. 355(j)(2); 21 CFR 314.94), an ANDA contains sufficient evidence that the proposed generic drug product is bioequivalent to its RLD and the application meets the other requirements for approval, FDA will approve the ANDA. In assessing whether an ANDA contains adequate evidence of BE, the agency considers available relevant information, which may include information submitted by the public to dockets for citizen petitions and product-specific BE recommendations.

Procedures for Making Recommendations Available

To streamline the process for making guidance available to the public on how to design product-specific BE studies, the agency intends to use the following process:

- Product-specific BE recommendations will be developed and posted on the Internet on the FDA Drugs guidance page (http://www.fda.gov/Drugs/GuidanceComplianceRegulatoryInformation/Guidances/default.htm, Individual Product Bioequivalence Recommendations) in draft to facilitate public comment. Users can also search for a specific product BE recommendation using the search tool on the guidance page.
- Product-specific BE recommendations may contain differing amounts of detail and background information regarding the specific basis for the recommendations, depending upon the novelty and complexity of the scientific considerations.
- Newly posted draft and final BE recommendations will be announced in the New/Revised/Withdrawn list, which is posted on the FDA Drugs guidance page.
- The agency will issue a notice in the Federal Register (FR) announcing the availability on the FDA website of new and revised product-specific draft and final BE recommendations. The notice will identify a comment period for the draft recommendations.
- Comments on product-specific draft BE recommendations will be considered in developing final BE recommendations.
- The BE recommendations will be revised as appropriate to ensure that the most up-to-date BE information is available to the public.
- On occasion, the appropriate methodology for establishing BE for a specific drug product will be the subject of a citizen petition or other correspondence and of a product-specific BE recommendation. This situation is particularly likely when the drug product raises novel or complex BE issues, as may be posed by certain topical and nonsystemically absorbed products. When the same BE issue is under consideration in different contexts, the agency will take into account the status of related matters in determining how to best address the scientific issues. This may involve, for example, coordinating the consideration of a pending citizen petition with the development and publication of a product-specific BE recommendation or identifying in the notice of availability for the draft BE recommendation the docket number of a related citizen petition.

Abacavir Sulfate; Lamivudine, Tablets/Oral. Recommended studies: two studies. (1) Type of study: fasting design—single-dose, two-treatment, two-period crossover in vivo. Strength: 600 mg/300 mg. Subjects: normal healthy males and females, general population. Additional comments: (2) Type of study: fed design—single-dose, two-treatment, two-period crossover in vivo. Strength: 600 mg/300 mg. Subjects: normal healthy males and females, general population. Additional comments: Analytes to measure (in appropriate biological fluid): abacavir and lamivudine in plasma. BE based on (90% confidence interval [CI]): abacavir and lamivudine. Waiver request of in vivo testing: not applicable.

Abacavir Sulfate, Tablets/Oral. Recommended studies: two studies. (1) Type of study: fasting design—single-dose, two-treatment, two-period crossover in vivo. Strength: 300 mg. Subjects: normal healthy males and females, general population. Additional comments: (2) Type of study: fed design—single-dose, two-treatment, two-period crossover in vivo. Strength: 300 mg. Subjects: normal healthy males and females, general population. Additional comments: Analytes to measure (in appropriate biological fluid): abacavir in plasma. BE based on (90% CI): abacavir. Waiver request of in vivo testing: not applicable.

Abacavir Sulfate; Lamivudine; Zidovudine, Tablets/Oral. Recommended studies: two studies. (1) Type of study: fasting design—single-dose, two-way crossover in vivo. Strength: 300 mg/150 mg/300 mg. Subjects: normal healthy males and females, general population. Additional comments: (2) Type of study: fed design—single-dose, two-way crossover in vivo. Strength: 300 mg/150 mg/300 mg. Subjects: normal healthy males and females, general population. Additional comments: Analytes to measure (in appropriate biological fluid): abacavir, lamivudine, and zidovudine in plasma. BE based on (90% CI): abacavir, lamivudine, and zidovudine. Waiver request of in vivo testing: not applicable.

Acamprosate Calcium, Delayed-Release Tablet/Oral. Recommended studies: two studies. (1) Type of study: fasting design—single-dose, two-way crossover in vivo. Strength: 333 mg. Subjects: normal healthy males and females, general population. Females should not be pregnant and, if applicable, should practice abstention or contraception during the study. Additional comments: (2) Type of study: fed design—single-dose, two-way crossover in vivo. Strength: 333 mg. Subjects: normal healthy males and females, general population. Females should not be pregnant and, if applicable, should practice abstention or contraception during the study. Additional comments: Analytes to measure: acamprosate in plasma. Acamprosate exists completely dissociated in plasma. Therefore, BE measures may be reported in terms of acetylhomotaurine. BE based on (90% CI): acamprosate. Waiver request of in vivo testing: not applicable.

Acitretin, Capsules/Oral. Recommended studies: two studies. (1) Type of study: fasting design—single-dose, two-way crossover in vivo. Strength: 25 mg. Subjects: normal healthy males and females, general population. Additional comments: Pregnant female subjects should be excluded from the BE studies. (2) Type of study: fed design—single-dose, two-way crossover in vivo. Strength: 25 mg. Subjects: normal healthy males and females, general population. Additional comments: Please see comment previously. Analytes to measure: All-*trans*-acitretin and 13-*cis*-acitretin in plasma. Since acitretin undergoes extensive presystemic metabolism and interconversion by isomerization to 13-*cis*-acitretin, measurement of all-*trans*-acitretin and 13-*cis*-acitretin in plasma is recommended. The pharmacokinetic (PK) parameters for all-*trans*-acitretin should meet the current BE criteria. The 13-*cis*-acitretin data will be used as supportive evidence. For the metabolite, the following data should be submitted: individual and mean concentrations, individual and mean PK parameters, and geometric means and ratios of means for AUC and C_{max}. BE based on (90% CI): all-*trans*-acitretin. Waiver request of in vivo testing: 10 mg based on (i) acceptable BE studies on the 25 mg strength, (ii) proportional similarity across all strengths, and (iii) acceptable in vitro dissolution testing of all strengths.

Acyclovir, Tablet/Oral. Recommended studies: two studies. (1) Type of study: fasting design—single-dose, two-treatment, two-period crossover in vivo. Strength: 800 mg. Subjects: normal healthy males and females, general population. Additional comments: (2) Type of study: fed design—single-dose, two-treatment, two-period crossover in vivo. Strength: 800 mg. Subjects: normal healthy males and females, general population. Additional comments: Analytes to measure: acyclovir in plasma. BE based on (90% CI): acyclovir. Waiver request of in vivo testing: 400 mg based on (i) acceptable BE studies on the 800 mg strength, (ii) proportional similarity of the formulations across all strengths, and (iii) acceptable in vitro dissolution testing of all strengths. Please conduct comparative dissolution testing on 12 dosage units of all strengths of the test and reference products.

Alendronate Sodium, Tablets/Oral. Recommended studies: one study. Type of study: fasting design: single-dose, two-way crossover in vivo. Strength: 70 mg. Subjects: normal healthy males and females, general population. Additional comments: Analytes to measure (in appropriate biological fluid): alendronate in urine. BE based on (90% CI): alendronate. The BE study should be based on urinary excretion data. The following PK parameters should be calculated: Ae (amount of drug excreted during each collection interval), total Ae (0–48) (total amount of drug excreted over the entire period of sample collection), Re (rate of drug excretion), R_{max} (maximum excretion rate), and t_{max} (time of the maximum excretion rate). All parameters should be calculated using a noncompartmental model. The statistical analysis using analysis of variance (ANOVA) should be performed on total Ae (0–48) and R_{max}. The 90% CI criteria should be applied to these parameters and should be within the limits of 80%–125%. Waiver request of in vivo testing: 5, 10, 35, and 40 mg based on (i) acceptable BE study on the 70 mg strength, (ii) proportional similarity across all strengths, and (iii) acceptable in vitro dissolution testing of all strengths.

Alfuzosin Hydrochloride, Extended-Release Tablets/Oral. Recommended studies: two studies. (1) Type of study: fasting design—single-dose, two-treatment, two-period crossover in vivo. Strength: 10 mg. Subjects: normal healthy males and females, general population. Additional comments: (2) Type of study: fed design—single-dose, two-treatment, two-period crossover in vivo. Strength: 10 mg. Subjects: normal healthy males and females, general population. Additional comments: Analytes to measure: alfuzosin. BE based on (90% CI): alfuzosin. Waiver request of in vivo testing: not applicable. In addition to the method earlier, for modified-release products, dissolution profiles on 12 dosage units each of test and reference products generated using USP Apparatus I at 100 rpm and/or Apparatus II at 50 rpm in at least three dissolution media (pH 1.2, 4.5, and 6.8 buffer) should be submitted in the application. Agitation speeds may have to be increased if appropriate. It is acceptable to add a small amount of surfactant, if necessary. Please include early sampling times of 1, 2, and 4 h and continue every 2 h until at least 80% of the drug is released, to provide assurance against premature release of drug (dose dumping) from the formulation.

Almotriptan Malate, Tablets/Oral. Recommended studies: two studies. (1) Type of study: fasting design—single-dose, two-way crossover in vivo. Strength: 12.5 mg. Subjects: normal healthy males and females, general population. Additional comments: (2) Type of study: fed design—single-dose, two-way crossover in vivo. Strength: 12.5 mg. Subjects: normal healthy males and females, general population. Additional comments: Analytes to measure (in appropriate biological fluid): almotriptan in plasma. BE based on (90% CI): almotriptan. Waiver request of in vivo testing: 6.25 mg based on (i) acceptable BE studies of the 12.5 mg strength, (ii) proportional similarity across all strengths, and (iii) acceptable in vitro dissolution testing of all strengths.

Alosetron Hydrochloride, Tablets/Oral. Recommended studies: two studies. (1) Type of study: fasting design—single-dose, two-way crossover in vivo. Strength: 1 mg (base). Subjects: normal healthy females, general population. Additional comments: (2) Type of study: fed design—single-dose, two-way crossover in vivo. Strength: 1 mg (base). Subjects: normal healthy females, general population. Additional comments: Analytes to measure (in appropriate biological fluid): Alosetron in plasma. BE based on (90% CI): alosetron. Waiver request of in vivo testing: 0.5 mg (base) based on (i) acceptable BE studies on the 1 mg strength, (ii) proportional similarity across all strengths, and (iii) acceptable in vitro dissolution testing of all strengths.

Alprazolam, Extended-Release Tablets/Oral. Recommended studies: two studies. (1) Type of study: fasting design—single-dose, two-treatment, two-period crossover in vivo. Strength: 3 mg. Subjects: normal healthy males and females, general population. Additional comments: (2) Type of study: fed design—single-dose, two-treatment, two-period crossover in vivo. Strength: 3 mg. Subjects: normal healthy males and females, general population. Additional comments: Analytes to measure: alprazolam in plasma. BE based on (90% CI): alprazolam. Waiver request of in vivo testing: 0.5, 1, and 2 mg based on (i) acceptable BE studies on the 3 mg strength, (ii) proportional similarity of the formulations across all strengths, and (iii) acceptable in vitro dissolution testing of all strengths. In addition to the aforementioned method, for modified-release products, dissolution profiles on 12 dosage units each of test and reference products generated using USP Apparatus I at 100 rpm and/or Apparatus II at 50 rpm in at least three dissolution media (pH 1.2, 4.5, and 6.8 buffer) should be submitted in the application. Agitation speeds may have to be increased if appropriate. It is acceptable to add a small amount of surfactant, if necessary. Please include early sampling times of 1, 2, and 4 h and continue every 2 h until at least 80% of the drug is released, to provide assurance against premature release of drug (dose dumping) from the formulation.

Alprazolam, Tablet/Oral. Recommended studies: one study. Type of study: fasting design—single-dose, two-treatment, two-period crossover in vivo. Strength: 1 mg. Subjects: normal healthy males and females, general population. Additional comments: Analytes to measure: alprazolam in plasma. BE based on (90% CI): alprazolam. Waiver request of in vivo testing: 0.25, 0.5, and 2 mg based on (i) acceptable BE studies on the 1 mg strength, (ii) proportional similarity of the formulations across all strengths, and (iii) acceptable in vitro dissolution testing of all strengths. Please conduct comparative dissolution testing on 12 dosage units of all strengths of the test and reference products.

Amlodipine Besylate, Tablets/Oral. Recommended studies: two studies. (1) Type of study: fasting design—single-dose, two-way crossover in vivo. Strength: 10 mg. Subjects: normal healthy males and females, general population. Additional comments: (2) Type of study: fed design—single-dose, two-way crossover in vivo. Strength: 10 mg Subjects: normal healthy males and females, general population. Additional comments: Analytes to measure: amlodipine in plasma. BE based on (90% CI): amlodipine. Waiver request of in vivo testing: 2.5 and 5 mg based on (i) acceptable BE studies on the 10 mg strength, (ii) proportional similarity across all strengths, and (iii) acceptable in vitro dissolution testing of all strengths.

Amlodipine Besylate; Benazepril Hydrochloride, Capsules/Oral. Recommended studies: two studies. (1) Type of study: fasting design—single-dose, two-way crossover in vivo. Strength: 10 mg/40 mg. Subjects: normal healthy males and females, general population. Additional comments: Female subjects should be excluded from the BE studies if they are pregnant. (2) Type of study: fed design—single-dose, two-way crossover in vivo. Strength: 10 mg/40 mg. Subjects: normal healthy males and females, general population. Additional comments: Please see comment earlier. Analytes to measure: amlodipine, benazepril, and active metabolite benazeprilat in plasma. BE based on (90% CI): amlodipine and benazepril. Please submit the metabolite data as supportive evidence of comparable therapeutic outcome. For the metabolite, the following data should be submitted: individual and mean concentrations, individual and mean PK parameters, and geometric means and ratios of means for AUC and C_{max}. Waiver request of in vivo testing: 2.5 mg/10 mg, 5 mg/10 mg, 5 mg/20 mg, 5 mg/40 mg, and 10 mg/20 mg, based on (i) acceptable BE studies on the 10 mg/40 mg strength, (ii) proportional similarity across all strengths, and (iii) acceptable in vitro dissolution testing of all strengths.

Please include early sampling times of 1, 2, and 4 h and continue every 2 h until at least 80% of the drug is released, to provide assurance against premature release of drug (dose dumping) from the formulation. Due to concerns of dose dumping from this drug product when taken with alcohol, please conduct additional dissolution testing using various concentrations of ethanol in the dissolution medium, as follows: testing conditions, 900 mL, 0.1 N HCl, Apparatus II (paddle) at 50 rpm, with and without the alcohol (see later). Test 1: 12 units tested according to the proposed method (with 0.1 N HCl), with data collected every 15 min for a total of 2 h. Test 2: 12 units analyzed by substituting 5% (v/v) of test medium with Alcohol USP and data collection every 15 min for a total of 2 h. Test 3: 12 units analyzed by substituting 20% (v/v) of test medium with Alcohol USP and data collection every 15 min for a total of 2 h. Test 4: 12 units analyzed by substituting 40% (v/v) of test medium with Alcohol USP and data collection every 15 min for a total of 2 h. Both test and RLD products must be tested accordingly, and data must be provided on individual unit, means, range, and percentage of coefficient of variation (%CV) on both strengths.

Amoxicillin; Clavulanate Potassium, Chewable Tablets/Oral. Recommended studies: two studies. (1) Type of study: fasting design—single-dose, two-treatment, two-period crossover in vivo. Strength: 250 mg/62.5 mg. Subjects: Normal healthy males and females, general population. Additional comments: (2) Type of study: fed design—single-dose, two-treatment, two-period crossover in vivo. Strength: 250 mg/62.5 mg. Subjects: normal healthy males and females, general population. Additional comments: Analytes to measure (in appropriate biological fluid): amoxicillin and clavulanic acid in plasma. BE based on (90% CI): amoxicillin and clavulanic acid. Waiver request of in vivo testing: 125 mg/31.25 mg, based on (i) acceptable BE studies on the 250 mg/62.5 mg strength, (ii) formulation proportionality across all strengths, and (iii) acceptable in vitro dissolution testing of all strengths.

Amoxicillin; Clavulanate Potassium, Chewable Tablets/Oral. Recommended studies: two studies. (1) Type of study: fasting design—single-dose, two-treatment, two-period crossover in vivo. Strength: 400 mg/57 mg. Subjects: normal healthy males and females, general population. Additional comments: (2) Type of study: fed design—single-dose, two-treatment, two-period crossover in vivo. Strength: 400 mg/57 mg. Subjects: normal healthy males and females, general population. Additional comments: Analytes to measure (in appropriate biological fluid): amoxicillin and clavulanic acid in plasma. BE based on (90% CI): amoxicillin and clavulanic acid. Waiver request of in vivo testing: 200 mg/28.5 mg, based on (i) acceptable BE studies on the 400 mg/57 mg strength, (ii) proportional similarity of the formulations, and (iii) acceptable in vitro dissolution testing of all strengths.

Amoxicillin; Clavulanate Potassium, Suspension/Oral. Recommended studies: three studies. (1) Type of study: fasting design—single-dose, two-way crossover in vivo. Strength: 600 mg/EQ 42.9 mg (base)/5 mL. Subjects: normal healthy males and females, general population. Additional comments: (2) Type of study: fed design—single-dose, two-way crossover in vivo. Strength: 600 mg/EQ 42.9 mg (base)/5 mL. Subjects: normal healthy males and females, general population. Additional comments: (3) Type of study: fasting design—single-dose, two-way crossover in vivo. Strength: 400 mg/EQ 57 mg (base)/5 mL. Subjects: normal healthy males and females, general population. Additional comments: Analytes to measure: amoxicillin and clavulanate potassium in plasma. BE based on (90% CI): amoxicillin and clavulanate potassium. Waiver request of in vivo testing: 200 mg/EQ 28.5 mg (base)/5 mL based on (i) acceptable BE studies on the 400 mg/EQ 57 mg (base)/5 mL strength, (ii) proportional similarity of the 200 mg/EQ 28.5 mg (base)/5 mL and 400 mg/EQ 57 mg (base)/5 mL strengths, and (iii) acceptable in vitro dissolution testing of the 200 mg/EQ 28.5 mg (base)/5 mL and 400 mg/EQ 57 mg (base)/5 mL strengths.

Amprenavir, Capsules/Oral. Recommended studies: two studies. (1) Type of study: fasting design—single-dose, two-way crossover in vivo. Strength: single dose of 200 mg (4 × 50 mg). Subjects: normal healthy males and females, general population. Additional comments: Females should not be pregnant or lactating and, if applicable, should practice abstention or contraception during the study. (2) Type of study: fed design—single-dose, two-way crossover in vivo. Strength: single dose of 200 mg (4 × 50 mg). Subjects: normal healthy males and females, general population. Additional comments: Please see comment earlier. Analytes to measure (in appropriate biological fluids): amprenavir in plasma. BE based on (90% CI): amprenavir. Waiver request of in vivo testing: not applicable.

Anagrelide Hydrochloride, Capsules/Oral. Recommended studies: two studies. (1) Type of study: fasting design—single-dose, two-way crossover in vivo. Strength: 1 mg. Subjects: normal healthy males and females, general population. Additional comments: (2) Type of study: fed design—single-dose, two-way crossover in vivo. Strength: 1 mg. Subjects: normal healthy males and females, general population. Additional comments: Analytes to measure: anagrelide in plasma. BE based on (90% CI): anagrelide. Waiver request of in vivo testing: 0.5 mg based on (i) acceptable BE studies on the 1 mg strength, (ii) proportional similarity across all strengths, and (iii) acceptable in vitro dissolution testing of all strengths.

Anastrozole, Tablets/Oral. Recommended studies: two studies. (1) Type of study: fasting design—single-dose, two-way crossover in vivo. Strength: 1 mg. Subjects: please conduct the studies in postmenopausal subjects or surgically sterile females. Additional comments: Please do not include subjects who are using female hormone replacement therapies, thyroid hormone replacement therapies, or antihypertensive therapies in the study population. Anastrozole has a long terminal elimination half-life. Please ensure adequate washout periods between treatments in the crossover studies. You may also consider using a parallel study design due to anastrozole's long half-life. For long half-life drug products, an AUC truncated to 72 h may be used in place of AUC0-t or AUC0-8. Please collect sufficient blood samples in the BE studies to adequately characterize the peak concentration (C_{max}) and time to reach peak concentration (t_{max}). (2) Type of study: fed design—single-dose, two-way crossover in vivo. Strength: 1 mg. Subjects: please see comments earlier. Additional comments: Please see comments earlier. Analytes to measure (in appropriate biological fluid): anastrozole in plasma. BE based on (90% CI): anastrozole. Waiver request of in vivo testing: not applicable.

Aprepitant, Capsules/Oral. Recommended studies: two studies. (1) Type of study: fasting design—single-dose, two-way crossover in vivo. Strength: 125 mg. Subjects: normal healthy males and females, general population. Additional comments: (2) Type of study: fed design—single-dose, two-way crossover in vivo. Strength: 125 mg. Subjects: normal healthy males and females, general population. Additional comments: Analytes to measure (in appropriate biological fluid): aprepitant in plasma. BE based on (90% CI): aprepitant. Waiver request of in vivo testing: 40 and 80 mg based on (i) acceptable BE studies on the 125 mg strength, (ii) proportional similarity to the 125 mg strength, and (iii) acceptable in vitro dissolution testing.

Aripiprazole, Tablets/Oral. Recommended studies: three studies. (1) Type of study: fasting design—single-dose, two-treatment, two-period crossover in vivo. Dose and tablet strength: 10 mg. Subject: normal healthy males and females, general population. Additional comments: (2) Type of study: fed design—single-dose, two-treatment, two-period crossover in vivo. Dose and tablet strength: 10 mg. Subjects: normal healthy males and females, general population. Additional comments: (3) Type of study: fasting design—single-dose, two-treatment, two-period crossover in vivo. Dose and tablet strength: 5 mg; if adequate exposure is not possible with a 5 mg dose, you may consider using a 10 mg dose (2 × 5 mg). Subjects: normal healthy males and females, general population. Additional comments: Notes: life-threatening adverse events attributed to acute laryngeal dystonia have been reported following administration of a single dose of 30 mg aripiprazole to healthy volunteers in BE studies. Although such events have not been reported at doses lower than 30 mg, because of the life-threatening nature of these events and because the dose response relationship is not known for this event, the following safety precautions are recommended for healthy volunteer studies of aripiprazole at all doses: Study protocols should specify standard procedures to diagnose and treat dystonic reactions should they occur. Subjects younger than 45 years of age should be excluded. There appears to be an inverse linear relationship between age and the incidence of acute dystonic reactions. Adults under 35 years of age were reported to have a 15-fold higher rate of neuroleptic-induced dystonia compared to a group of patients 60–80 years of age. The occurrence of dystonias appears to be rare at ages of approximately 45 years and higher. Protocols should include stringent drug screening procedures to ensure that subjects are free of illicit drugs at the time of administration of each study drug dose. The screening interview should include specific questions to exclude subjects with a prior personal or family history of dystonic reactions to medications. Prospective study subjects should also be specifically questioned about prior neuroleptic drug exposures. Aripiprazole has been poorly tolerated by healthy volunteers in some BE studies, particularly at the 15 and 30 mg dose levels. In several cases, adverse events have resulted in a high incidence of dropouts. Adverse events in aripiprazole studies have included nausea, vomiting, dizziness, syncope, insomnia, headache, fatigue, hypotension, hot flashes, weakness, diaphoresis, and confusion. To minimize the occurrence of adverse events and to ensure the safety of healthy volunteer subjects in clinical trials of aripiprazole, the following is recommended: Subjects should be monitored in-house for at least 3 days after dosing and until adverse events have resolved. Subjects should be kept supine for at least 8 h starting no longer than 15 min after each dose. Subjects should be asked to use the bathroom soon before dosing. Subjects should be encouraged to use urinals or bedpans during the first 8 h after dosing and at any time after dosing if the subject is experiencing adverse events such as nausea, dizziness, or hypotension. If subjects do use the bathroom during the first 8 h after dosing or while experiencing adverse events such as nausea, dizziness, or hypotension, they should be assisted to and from the bathroom by study personnel. At a minimum, routine 12-lead ECGs should be performed at 3–5 h after dosing and at 8–12 h after dosing. Continuous ECG monitoring during those time periods may be considered as an alternative. Vital signs monitoring should continue post dosing throughout the period that subjects are housed, commencing no later than 30 min following dosing. Vital signs should be monitored frequently (at least every 0.5–1 h) for at least the first 8 h after dosing and the first hour after subjects are allowed to rise from the supine position. Prespecified limits should be defined for reporting adverse events related to vital signs (e.g., hypotension, bradycardia). Vital sign readings that meet these predefined limits should be reported as adverse events, even if they are not performed during a scheduled assessment (e.g., vital signs performed as part of an assessment of an adverse event). The protocol should include standard procedures for the assessment and management of potential adverse events, including vital signs and ECG monitoring as appropriate for adverse events possibly associated with hypotension. Women of childbearing potential should be enrolled only if they are using effective contraceptives. A negative pregnancy test is needed within 24 h prior to each dose. These subjects should also be informed of the potential teratogenicity of the study drug as part of the informed consent process. Nursing women should also be excluded. The protocol should include measures to prevent relative dehydration at the time of dosing, such as encouragement of water intake whenever possible prior to dosing. Consideration should be made to providing a standard meal just prior to the standard fasting period before dosing. During the informed consent process, subjects should be advised of the high incidence of adverse events that have occurred in some healthy volunteer studies of aripiprazole. Aripiprazole has a long terminal elimination half-life. Please ensure adequate washout periods between treatments in the crossover studies. You may also consider using a parallel study design due to aripiprazole's long half-life. For long half-life drug products, an AUC truncated to 72 h may be used in place of AUC0-t or AUC0-inf. Please collect sufficient blood samples in the BE studies to adequately characterize the peak concentration (C_{max}) and time to reach peak concentration (t_{max}). Analytes to measure (in appropriate biological fluid): aripiprazole in plasma. BE based on (90% CI): aripiprazole. Waiver request of in vivo testing (assuming conduct of the three in vivo studies earlier): 2, 15, 20, and 30 mg, based on (i) acceptable BE studies on the 5 and 10 mg strengths, (ii) proportional similarity across all strengths, and (iii) acceptable in vitro dissolution testing of all strengths.

Armodafinil, Tablets/Oral. Recommended studies: two studies. (1) Type of study: fasting design—single-dose, two-treatment, two-period crossover in vivo. Strength: 250 mg. Subjects: normal healthy males and females, general population. Additional comments: (2) Type of study: fed design—single-dose, two-treatment, two-period crossover in vivo. Strength: 250 mg. Subjects: normal healthy males and females, general population. Additional comments: Analytes to measure (in appropriate biological fluid): armodafinil in plasma. BE based on (90% CI): armodafinil. Waiver request of in vivo testing: 50 and 150 mg, based on (i) acceptable BE studies on the 250 mg strength, (ii) proportional similarity of the formulations, and (iii) acceptable in vitro dissolution testing of all strengths.

Atazanavir Sulfate, Capsules/Oral. Recommended studies: two studies. (1) Type of study: fasting design—single-dose, two-way crossover in vivo. Strength: 300 mg. Subjects: normal healthy males and females, general population. Additional comments: (2) Type of study: fed design—single-dose, two-way crossover in vivo. Strength: 300 mg. Subjects: normal healthy males and females, general population. Additional comments: Analytes to measure (in appropriate biological fluid): atazanavir in plasma. BE based on (90% CI): atazanavir. Waiver request of in vivo testing: 100, 150, and 200 mg based on (i) acceptable BE studies on the 300 mg strength, (ii) proportional similarity across all strengths, and (iii) acceptable in vitro dissolution testing of all strengths.

Atomoxetine Hydrochloride, Capsules/Oral. Recommended studies: two studies. (1) Type of study: fasting design—single-dose, two-way crossover in vivo. Strength: 60 mg. Subjects: normal healthy males and females, general population. Additional comments: 60 mg is studied because higher doses may cause unacceptable side effects in normal healthy subjects. (2) Type of study: fed design—single-dose, two-way crossover in vivo. Strength: 60 mg. Subjects: normal healthy males and females, general population. Additional comments: Please see comment earlier. Analytes to measure (in appropriate biological fluid): atomoxetine in plasma. BE based on (90% CI): atomoxetine. Waiver request of in vivo testing: 5, 10, 18, 25, 40, 80, and 100 mg based on (i) acceptable BE studies on the 60 mg strength, (ii) proportional similarity across all strengths, and (iii) acceptable in vitro dissolution testing of all strengths. The 5 mg strength of STRATTERA™ is currently not marketed. If a firm is interested in seeking approval for this strength, please submit a citizen petition requesting the U.S. FDA make a determination that this particular strength was not withdrawn for reasons of safety or effectiveness or check the FR for a previously submitted citizen petition. Submission of the citizen petition to the FDA should be done prior to an ANDA submission.

Atorvastatin Calcium, Tablets/Oral. Recommended studies: two studies. (1) Type of study: fasting design—single-dose, two-way crossover in vivo. Strength: 80 mg. Subjects: normal healthy males and females, general population. Additional comments: (2) Type of study: fed design—single-dose, two-way crossover in vivo. Strength: 80 mg. Subjects: normal healthy males and females, general population. Additional comments: Analytes to measure: atorvastatin, *ortho*- and *para*hydroxylated metabolites of atorvastatin. The *ortho*- and *para*hydroxylated metabolites of atorvastatin are formed by presystemic metabolism and contribute meaningfully to efficacy. For the metabolites, the following data should be submitted: individual and mean concentrations, individual and mean PK parameters, and geometric means and ratios of means for AUC and C_{max}. BE based on (90% CI): atorvastatin. Waiver request of in vivo testing: 10, 20, and 40 mg based on (i) acceptable BE studies on the 80 mg strength, (ii) proportional similarity across all strengths, and (iii) acceptable in vitro dissolution testing of all strengths.

Atovaquone, Tablets/Oral. Recommended studies: two studies. (1) Type of study: fasting design—single-dose, two-treatment, two-period crossover in vivo. Strength: 250 mg. Subjects: normal healthy males and females, general population. Additional comments: You may also consider using a parallel study design due to atovaquone's long half-life. For long half-life drug products, an AUC truncated to 72 h may be used in place of AUC0-t or AUC0-8. (2) Type of study: fed design—single-dose, two-treatment, two-period crossover in vivo. Strength: 250 mg. Subjects: normal healthy males and females, general population. Additional comments: Please see comment earlier. Analytes to measure (in appropriate biological fluid): atovaquone in plasma. BE based on (90% CI): atovaquone. Waiver request of in vivo testing: not applicable. Atovaquone is known to be practically insoluble in both water and 0.1 M HCl (<0.0002 mg/mL at 25°C). Use of conventional aqueous dissolution media with and without surfactant has been found unsuccessful and not reproducible in some laboratories working with atovaquone tablet products. If encountering the same difficulty, you may consider developing a dissolution method similar to the method available in the dissolution database. Although the use of the high alcoholic medium is not considered conventional, it has been found justifiable by the FDA for this drug substance. You may develop an alternate dissolution testing method for the drug product and submit the dissolution testing results when the application is filed.

Azithromycin, Tablets/Oral. Recommended studies: two studies. (1) Type of study: fasting design—single-dose, two-way crossover in vivo. Strength: 600 mg. Subjects: normal healthy males and females, general population. Additional comments: (2) Type of study: fed design—single-dose, two-way crossover in vivo. Strength: 600 mg. Subjects: normal healthy males and females, general population. Additional comments: Analytes to measure: azithromycin. BE based on (90% CI): azithromycin. Waiver request of in vivo testing: 250 and 500 mg based on (i) acceptable BE studies on the 600 mg strength, (ii) proportional similarity across all strengths, and (iii) acceptable in vitro dissolution testing of all strengths. Since azithromycin tablets, 250, 500, and 600 mg, are the subject of three separate new drug applications (NDA's), three separate ANDAs must be submitted. You may request a waiver of in vivo BE testing of the 250 mg and the 500 mg strengths if you meet the criteria. In addition, please cross-reference the in vivo BE studies conducted on the higher strength along with your waiver request.

Balsalazide Disodium, Capsule/Oral. Recommended studies: two studies. (1) Type of study: fasting design—single-dose, two-way crossover in vivo. Strength: 2250 mg dose (3 × 750 mg). Subjects: normal healthy males and females, general population. Females should not be pregnant and, if applicable, should practice abstention or contraception during the study. Additional comments: (2) Type of study: fed design—single-dose, two-way crossover in vivo. Strength: 2250 mg dose (3 × 750 mg). Subjects: normal healthy males and females, general population. Additional comments: Please see comment earlier. Analytes to measure (in appropriate biological fluid): balsalazide and mesalamine in plasma. BE based on (90% CI): balsalazide and mesalamine. Waiver request of in vivo testing: not applicable. In vitro dissolution testing under the following conditions should be submitted to support documentation of BE: Apparatus and rotation speed, USP Apparatus I (basket), at 100 rpm; medium, (1) 0.1 N HCl, (2) pH 4.5 buffer, (3) pH 6.8 buffer, (4) pH 7.4 buffer; volume, 900 mL; temperature, 37°C; and sample times, 5, 10, 15, 20, 30, 45, and 60 min and until at least 80% of the labeled content is dissolved.

Benzonatate, Capsule/Oral. Recommended studies: benzonatate capsules, 100 and 200 mg, may be considered for waiver of in vivo BE testing pursuant to 21 CFR 320.22(c) provided the in vitro dissolution profiles of your benzonatate capsules, 100 and 200 mg, and the RLDs are comparable. Analytes to measure (in appropriate biological fluid): not applicable. BE based on (90% CI): not applicable. Waiver request of in vivo testing: not applicable.

Benzphetamine Hydrochloride, Tablet/Oral. Recommended studies: A benzphetamine hydrochloride tablet is a DESI-effective drug product without known BE problems. Therefore, in vivo BE testing is not requested. Comparative dissolution testing on 12 dosage units of all strengths of the test and reference products is requested. You may request a waiver of in vivo BE study requirements on this product under 21 CFR 320.22(c). Analytes to measure: not applicable. BE based on (90% CI): not applicable. Waiver request of in vivo testing: 50 mg.

Bicalutamide, Tablets/Oral. Recommended studies: two studies. (1) Type of study: fasting design—single-dose, two-way crossover in vivo. Strength: 50 mg. Subjects: normal healthy males and females, general population. Additional comments: Female subjects should be excluded from the BE studies if they are pregnant. Bicalutamide has a long terminal elimination half-life. Please ensure adequate washout periods between treatments in the crossover studies. You may also consider using a parallel study design due to bicalutamide's long half-life. For long half-life drug products, an AUC truncated to 72 h may be used in place of AUC0-t or AUC0-8. (2) Type of study: fed design—single-dose, two-way crossover in vivo. Strength: 50 mg. Subjects: normal healthy males and females, general population. Additional comments: Please see comments earlier. Analytes to measure: bicalutamide, using an achiral assay. BE based on (90% CI): bicalutamide. Waiver request of in vivo testing: not applicable.

Bisoprolol Fumarate, Tablets/Oral. Recommended studies: two studies. (1) Type of study: fasting design—single-dose, two-way crossover in vivo. Strength: 10 mg. Subjects: normal healthy males and females, general population. Additional comments: (2) Type of study: fed design—single-dose, two-way crossover in vivo. Strength: 10 mg. Subjects: normal healthy males and females, general population. Additional comments: Analytes to measure: Bisoprolol in plasma. BE based on (90% CI): bisoprolol. Waiver request of in vivo testing: 5 mg based on (i) acceptable BE studies on the 10 mg strength, (ii) proportional similarity across all strengths, and (iii) acceptable in vitro dissolution testing of all strengths. Please conduct comparative dissolution testing on 12 dosage units of all strengths of the test and reference products.

Bisoprolol Fumarate; Hydrochlorothiazide, Tablet/Oral. Recommended studies: two studies. (1) Type of study: fasting design—single-dose, two-way crossover in vivo. Strength: 10 mg/6.25 mg. Subjects: normal healthy males and females, general population. Additional comments: (2) Type of study: fed design—single-dose, two-way crossover in vivo. Strength: 10 mg/6.25 mg. Subjects: normal healthy males and females, general population. Additional comments: Analytes to measure: bisoprolol and hydrochlorothiazide in plasma. BE based on (90% CI): bisoprolol and hydrochlorothiazide. Waiver request of in vivo testing: 2.5 mg/6.25 mg and 5 mg/6.25 mg based on (i) acceptable BE studies on the 10 mg/6.25 mg strength, (ii) proportional similarity across all strengths, and (iii) acceptable in vitro dissolution testing of all strengths.

Bupropion Hydrochloride, Extended-Release Tablets/Oral. Recommended studies: two studies. (1) Type of study: fasting design—single-dose, two-treatment, two-period crossover in vivo. Strength: 150 mg. Subjects: normal healthy males and females, general population. Additional comments: (2) Type of study: fed design—single-dose, two-treatment, two-period crossover in vivo. Strength: 150 mg. Subjects: normal healthy males and females, general population. Additional comments: Analytes to measure (in appropriate biological fluid): bupropion and hydroxybupropion (active metabolite of bupropion) in plasma. BE based on (90% CI): bupropion. Waiver request of in vivo testing: 300 mg based on (i) acceptable BE studies on the 150 mg strength, (ii) proportional similarity across all strengths, and (iii) acceptable in vitro dissolution testing of all strengths. In addition to the previous method, for modified-release products, dissolution profiles on 12 dosage units each of test and reference products generated using USP Apparatus I at 100 rpm and/or Apparatus II at 50 rpm in at least three dissolution media (pH 1.2, 4.5, and 6.8 buffer) should be submitted in the application. Agitation speeds may have to be increased if appropriate. It is acceptable to add a small amount of surfactant, if necessary. Please include early sampling times of 1, 2, and 4 h and continue every 2 h until at least 80% of the drug is released, to provide assurance against premature release of drug (dose dumping) from the formulation. Due to concerns of dose dumping from this drug product when taken with alcohol, please conduct additional dissolution testing using various concentrations of ethanol in the dissolution medium, as follows: testing conditions, 900 mL, 0.1 N HCl, Apparatus I (basket) at 75 rpm, with and without the alcohol (see later). Test 1: 12 units tested according to the proposed method (with 0.1 N HCl), with data collected every 15 min for a total of 2 h. Test 2: 12 units analyzed by substituting 5% (v/v) of test medium with Alcohol USP and data collection every 15 min for a total of 2 h. Test 3: 12 units analyzed by substituting 20% (v/v) of test medium with Alcohol USP and data collection every 15 min for a total of 2 h. Test 4: 12 units analyzed by substituting 40% (v/v) of test medium with Alcohol USP and data collection every 15 min for a total of 2 h. Both test and RLD products must be tested accordingly, and data must be provided on individual unit, means, range, and %CV on both strengths.

Candesartan Cilexetil, Tablets/Oral. Recommended studies: two studies. (1) Type of study: fasting design—single-dose, two-way crossover in vivo. Strength: 32 mg. Subjects: normal, healthy males and females, general population. Additional comments: Females should not be pregnant and, if applicable, should practice abstention or contraception during the study. (2) Type of study: fed design—single-dose, two-way crossover in vivo. Strength: 32 mg. Subjects: normal, healthy males and females, general population. Additional comments: Please see comments earlier. Analytes to measure (in appropriate biological fluid): candesartan in plasma. BE based on (90% CI): candesartan. Requests for waivers of in vivo testing: 4, 8, and 16 mg based on (i) acceptable BE studies on the 32 mg strength, (ii) acceptable dissolution testing of all strengths, and (iii) proportional similarity in the formulations of all strengths.

Candesartan Cilexetil; Hydrochlorothiazide, Tablets/Oral. Recommended studies: two studies. (1) Type of study: fasting design—single-dose, two-way crossover in vivo. Strength: 32 mg/12.5 mg. Subjects: normal, healthy males and females, general population. Additional comments: Female subjects should be excluded from the BE studies if they are pregnant. (2) Type of study: fed design—single-dose, two-way crossover in vivo. Strength: 32 mg/12.5 mg. Subjects: normal, healthy males and females, general population. Additional comments: Female subjects should be excluded from the BE studies if they are pregnant. Analytes to measure (in appropriate biological fluid): candesartan and hydrochlorothiazide in plasma. BE based on (90% CI): candesartan and hydrochlorothiazide. Requests of waivers of in vivo. Testing: 16 mg/12.5 mg, based on (i) acceptable BE studies on the 32 mg/12.5 mg strength, (ii) formulation proportionality across all strengths, and (iii) acceptable in vitro dissolution testing of all strengths.

Carbamazepine, Extended-Release Capsules/Oral. Recommended studies: three studies. (1) Type of study: fasting design—single-dose, two-treatment, two-period crossover in vivo. Strength: 300 mg. Subjects: normal healthy males and females, general population. Additional comments: Female subjects should not be enrolled in BE studies of carbamazepine, if they are pregnant. Only females who are either surgically sterile or practicing a recognized safe method of contraception should be included in a study. You should clearly define in the study protocol what is considered a "safe method of contraception." BE studies conducted for this product may be referenced to support a request for a waiver of evidence of in vivo BE for generic products referencing EQUETRO. Please submit separate applications for each RLD. (2) Type of study: fed design—single-dose, two-treatment, two-period crossover in vivo. Strength: 300 mg. Subjects: normal healthy males and females, general population. Additional comments: Please see previous comment. (3) Type of study: fasting (capsule compared to RLD, sprinkled on a spoonful of applesauce) design—single-dose, two-treatment, two-period crossover in vivo. Strength: 300 mg. Subjects: normal healthy males and females, general population. Additional comments: Please see previous comment. Analytes to measure (in appropriate biological fluid): carbamazepine in plasma. BE based on (90% CI): carbamazepine. Waiver request of in vivo testing: 100 and 200 mg based on (i) acceptable BE studies on the 300 mg strength, (ii) proportional similarity across all strengths, and (iii) acceptable in vitro dissolution testing of all strengths. In addition to the method earlier, for modified-release products, dissolution profiles on 12 dosage units each of test and reference products generated using USP Apparatus I at 100 rpm and/or Apparatus II at 50 rpm in at least three dissolution media (pH 1.2, 4.5, and 6.8 buffer) should be submitted in the application. Agitation speeds may have to be increased if appropriate. It is acceptable to add a small amount of surfactant, if necessary. Please include early sampling times of 1, 2, and 4 h and continue every 2 h until at least 80% of the drug is released, to provide assurance against premature release of drug (dose dumping) from the formulation.

Carbamazepine, Extended-Release Tablets/Oral. Recommended studies: two studies. (1) Type of study: fasting design—single-dose, two-treatment, two-period crossover in vivo. Strength: 400 mg. Subjects: normal healthy males and females, general population. Additional comments: Females must have a negative baseline pregnancy test within 24 h prior to receiving the drug. Females should not be pregnant or lactating and, if applicable, should practice abstention or contraception during the study. (2) Type of study: fed design—single-dose, two-treatment, two-period crossover in vivo. Strength: 400 mg. Subjects: normal healthy males and females, general population. Additional comments: Please see comments earlier. Analytes to measure (in appropriate biological fluid): carbamazepine in plasma. BE based on (90% CI): carbamazepine. Waiver request of in vivo testing: 100 and 200 mg, based on (i) acceptable BE studies on the 400 mg tablet, (ii) proportional similarity of the formulations, and (iii) acceptable in vitro dissolution testing of all strengths. In addition to the method earlier, for modified-release products, dissolution profiles on 12dosage units each of test and reference products generated using USP Apparatus I at 100 rpm and/or Apparatus II at 50 rpm in at least three dissolution media (pH 1.2, 4.5, and 6.8 buffer) should be submitted in the application. Agitation speeds may have to be increased if appropriate. It is acceptable to add a small amount of surfactant, if necessary. Please include early sampling times of 1, 2, and 4 h and continue every 2 h until at least 80% of the drug is released, to provide assurance against premature release of drug (dose dumping) from the formulation.

Carbamazepine, Suspension/Oral. Recommended studies: two studies. (1) Type of study: fasting design—single-dose, two-way crossover in vivo. Strength: 100 mg/5 mL. Subjects: normal healthy males and females, general population. Additional comments: (2) Type of study: fed design—single-dose, two-way crossover in vivo. Strength: 100 mg/5 mL. Subjects: normal healthy males and females, general population. Additional comments: Analytes to measure (in appropriate biological fluid): carbamazepine in plasma. BE based on (90% CI): carbamazepine. Waiver request of in vivo testing: not applicable.

Carbidopa; Entacapone; Levodopa, Tablets/Oral. Recommended studies: two studies. (1) Type of study: fasting design—single-dose, two-way crossover in vivo. Strength: (37.5, 200, 150 mg) carbidopa, entacapone, levodopa. Subjects: normal healthy males and females, general population. Females should not be pregnant and, if applicable, should practice abstention or contraception during the study. Additional comments: (2) Type of study: fasting design—single-dose, two-way crossover in vivo. Strength: (12.5, 200, 50 mg) carbidopa, entacapone, levodopa. Subjects: normal healthy males and females, general population. Females should not be pregnant and, if applicable, should practice abstention or contraception during the study. Additional comments: Analytes to measure (in appropriate biological fluid): carbidopa, entacapone, and levodopa in plasma. BE based on (90% CI): carbidopa, entacapone, and levodopa. Waiver request of in vivo testing: (25, 200, 100 mg) carbidopa, entacapone, and levodopa tablets, based on (i) acceptable BE study on the 37.5, 200, and 150 mg tablet, (ii) formulation proportionality across all strengths, and (iii) acceptable in vitro dissolution testing of all strengths.

Carvedilol, Tablets/Oral. Recommended studies: two studies. (1) Type of study: fasting design—single-dose, two-way crossover in vivo. Strength: 12.5 mg. Subjects: normal healthy males and females, general population. Additional comments: Due to safety concerns, the OGD recommends that you conduct the BE studies using carvedilol tablets, 12.5 mg, instead of the 25 mg strength. (2) Type of study: fed design—single-dose, two-way crossover in vivo. Strength: 12.5 mg. Subjects: normal healthy males and females, general population. Additional comments: Please see previous comment. Analytes to measure: carvedilol and 4-hydroxyphenyl-carvedilol metabolite of carvedilol in plasma. BE based on (90% CI): carvedilol. Please submit the metabolite data as supportive evidence of comparable therapeutic outcome. For the metabolite, the following data should be submitted: individual and mean concentrations, individual and mean PK parameters, and geometric means and ratios of means for AUC and C_{max}. Waiver request of in vivo testing: 3.125, 6.25, and 25 mg based on (i) acceptable BE studies on the 12.5 mg strength, (ii) proportional similarity across all strengths, and (iii) acceptable in vitro dissolution testing of all strengths.

Cefdinir, Capsules/Oral. Recommended studies: two studies. (1) Type of study: fasting design—single-dose, two-treatment, two-period crossover in vivo. Strength: 300 mg. Subjects: normal healthy males and females, general population. Additional comments: (2) Type of study: fed design—single-dose, two-treatment, two-period crossover in vivo. Strength: 300 mg. Subjects: normal healthy males and females, general population. Additional comments: Analytes to measure (in appropriate biological fluid): cefdinir in plasma. BE based on (90% CI): cefdinir. Waiver request of in vivo testing: not applicable.

Cefditoren Pivoxil, Tablets/Oral. Recommended studies: two studies. (1) Type of study: fasting design—single-dose, two-way crossover in vivo. Strength: 200 mg. Subjects: normal healthy males and females, general population. Additional comments: (2) Type of study: fed design—single-dose, two-way crossover in vivo. Strength: 200 mg. Subjects: normal healthy males and females, general population. Additional comments: Analytes to measure (in appropriate biological fluid): cefditoren (not the prodrug cefditoren pivoxil) in plasma. BE based on (90% CI): cefditoren. Waiver request of in vivo testing: not applicable.

Cefixime, Suspension/Oral. Recommended studies: two studies. (1) Type of study: fasting design—single-dose, two-way crossover in vivo. Strength: 200 mg/5 mL. Subjects: normal healthy males and females, general population. Additional comments: Females should not be pregnant or lactating and, if applicable, should practice abstention or contraception during the study. (2) Type of study: fed design—single-dose, two-way crossover in vivo. Strength: 200 mg/5 mL. Subjects: normal healthy males and females, general population. Additional comments: Please see previous comment. Analytes to measure (in appropriate biological fluid): cefixime in plasma. BE based on (90% CI): cefixime. Waiver request of in vivo testing: 100 mg/5 mL based on (i) acceptable BE studies on the 200 mg/5 mL strength, (ii) proportional similarity of the formulations across all strengths, and (iii) acceptable in vitro dissolution testing of all strengths. A dosage unit for a suspension is the labeled strength (5 mL). A total of 12 units from 12 different bottles should be used.

Celecoxib, Capsules/Oral. Recommended studies: two studies. (1) Type of study: fasting design—single-dose, two-way crossover in vivo. Strength: 400 mg. Subjects: normal healthy males and females, general population. Additional comments: (2) Type of study: fed design—single-dose, two-way crossover in vivo. Strength: 400 mg. Subjects: normal healthy males and females, general population. Additional comments: Analytes to measure: celecoxib in plasma. BE based on (90% CI): celecoxib. Waiver request of in vivo testing: 100 and 200 mg based on (i) acceptable BE studies on the 400 mg strength, (ii) proportional similarity across all strengths, and (iii) acceptable in vitro dissolution testing of all strengths.

Cetirizine Hydrochloride, Chewable Tablets/Oral. Recommended studies: two studies. (1) Type of study: fasting design—single-dose, two-way crossover in vivo. Strength: 10 mg. Subjects: normal healthy males and females, general population. Additional comments: (2) Type of study: fed design—single-dose, two-way crossover in vivo. Strength: 10 mg. Subjects: normal healthy males and females, general population. Additional comments: Analytes to measure (in appropriate biological fluid): cetirizine in plasma. BE based on (90% CI): cetirizine. Waiver request of in vivo testing: 5 mg based on (i) acceptable BE studies on the 10 mg strength, (ii) proportional similarity of the formulations across all strengths, and (iii) acceptable in vitro dissolution testing of all strengths.

Cetirizine Hydrochloride; Pseudoephedrine Hydrochloride, Extended-Release Tablets/Oral. Recommended studies: two studies. (1) Type of study: fasting design—single-dose, two-way crossover in vivo. Strength: 5 mg/120 mg. Subjects: normal healthy males and females, general population. Additional comments: (2) Type of study: fed design—single-dose, two-way crossover in vivo. Strength: 5 mg/120 mg. Subjects: normal healthy males and females, general population. Additional comments: Analytes to measure: cetirizine and pseudoephedrine in plasma. BE based on (90% CI): cetirizine and pseudoephedrine. Waiver request of in vivo testing: not applicable. For modified-release products, dissolution profiles generated using USP Apparatus I at 100 rpm and/or Apparatus II at 50 rpm in at least three dissolution media (pH 1.2, 4.5, and 6.8 buffer, water) should be submitted in the application. Agitation speeds may have to be increased if appropriate. It is acceptable to add a small amount of surfactant, if necessary. The following sampling times are recommended: 1, 2, and 4 h and every 2 h thereafter, until at least 80% of the drug is dissolved. Comparative dissolution profiles should include individual tablet data as well as the mean, range, and standard deviation at each time point for 12 tablets.

Cevimeline Hydrochloride, Capsule/Oral. Recommended studies: two studies. (1) Type of study: fasting design—single-dose, two-way crossover in vivo. Strength: 30 mg. Subjects: normal healthy males and females, general population. Females should not be pregnant and, if applicable, should practice abstention or contraception during the study. Additional comments: (2) Type of study: fed design—single-dose, two-way crossover in vivo. Strength: 30 mg. Subjects: normal healthy males and females, general population. Females should not be pregnant and, if applicable, should practice abstention or contraception during the study. Additional comments: Analytes to measure (in appropriate biological fluid): cevimeline in plasma. BE based on (90% CI): cevimeline. Waiver request of in vivo testing: not applicable.

Cilostazol, Tablets/Oral. Recommended studies: two studies. (1) Type of study: fasting design—single-dose, two-way crossover in vivo. Strength: 100 mg. Subjects: normal healthy males and females, general population. Additional comments: Patients should be advised to take cilostazol at least one-half hour before or two hours after food. Therefore, a fed study is not recommended. (2) Type of study: fasting design—single-dose, two-way crossover in vivo. Strength: 50 mg. Subjects: normal healthy males and females, general population. Additional comments: Please see previous comments. Analytes to measure: Cilostazol in plasma. BE based on (90% CI): cilostazol. Waiver request of in vivo testing: not applicable.

Cinacalcet Hydrochloride, Tablets/Oral. Recommended studies: two studies. (1) Type of study: fasting design—single-dose, two-way crossover in vivo. Strength: 90 mg. Subjects: normal healthy males and females, general population. Females should not be pregnant and, if applicable, should practice abstention or contraception during the study. Additional comments: (2) Type of study: fed design—single-dose, two-way crossover in vivo. Strength: 90 mg .Subjects: normal healthy males and females, general population. Females should not be pregnant and, if applicable, should practice abstention or contraception during the study. Additional comments: Analytes to measure (in appropriate biological fluid): cinacalcet in plasma. BE based on (90% CI): cinacalcet. Waiver request of in vivo testing: 60 and 30 mg based on (i) acceptable BE studies on the 90 mg strength, (ii) proportional similarity across all strengths, and (iii) acceptable in vitro dissolution testing of all strengths.

Ciprofloxacin Hydrochloride, Extended-Release Tablets/Oral. Recommended studies: two studies. (1) Type of study: fasting design—single-dose, two-treatment, two-period crossover in vivo. Strength: 500 mg. Subjects: normal healthy males and females, general population. Additional comments: (2) Type of study: fed design—single-dose, two-treatment, two-period crossover in vivo. Strength: 500 mg. Subjects: normal healthy males and females, general population. Additional comments: Analytes to measure (in appropriate biological fluid): ciprofloxacin in plasma. BE based on (90% CI): ciprofloxacin. Waiver request of in vivo testing: Not Applicable. In addition to the method earlier, for modified-release products, dissolution profiles on 12 dosage units each of test and reference products generated using USP Apparatus I at 100 rpm and/or Apparatus II at 50 rpm in at least three dissolution media (pH 1.2, 4.5, and 6.8 buffer) should be submitted in the application. Agitation speeds may have to be increased if appropriate. It is acceptable to add a small amount of surfactant, if necessary. Please include early sampling times of 1, 2, and 4 h and continue every 2 h until at least 80% of the drug is released, to provide assurance against premature release of drug (dose dumping) from the formulation.

Ciprofloxacin; Ciprofloxacin Hydrochloride, Extended-Release Tablets/Oral. Recommended studies: three studies. (1) Type of study: fasting design—single-dose, two-way crossover in vivo. Strength: 1000 mg (425.2 mg, EQ 574.9 mg base). Subjects: normal healthy males and females, general population. Additional comments: (2) Type of study: fed design—single-dose, two-way crossover in vivo. Strength: 1000 mg (425.2 mg, EQ 574.9 mg base). Subjects: normal healthy males and females, general population. (3) Type of study: fasting design—single-dose, two-way crossover in vivo. Strength: 500 mg (212.6 mg, EQ 287.5 mg base). Subjects: normal healthy males and females, general population. Analytes to measure: ciprofloxacin. BE based on (90% CI): ciprofloxacin. Waiver request of in vivo testing: The 500 mg strength of ciprofloxacin extended-release tablets is *not* eligible for (a) Waiver of in vivo testing based on an acceptable in vivo BE study of the 1000 mg strength. For modified-release products, dissolution profiles generated using USP Apparatus I at 100 rpm and/or Apparatus II at 50 rpm in at least three dissolution media (pH 1.2, 4.5, and 6.8 buffer, water) should be submitted in the application. Agitation speeds may have to be increased if appropriate. It is acceptable to add a small amount of surfactant, if necessary. The following sampling times are recommended: 1, 2, and 4 h and every 2 h thereafter, until at least 80% of the drug is dissolved. Comparative dissolution profiles should include individual tablet data as well as the mean, range, and standard deviation at each time point for 12 tablets.

Clarithromycin, Extended-Release Tablets/Oral. Recommended studies: two studies. (1) Type of study: fasting design—single-dose, two-way crossover in vivo. Strength: 500 mg. Subjects: normal healthy males and females, general population. Additional comments: (2) Type of study: fed design—single-dose, two-way crossover in vivo. Strength: 500 mg. Subjects: normal healthy males and females, general population. Additional comments: Analytes to measure (in appropriate biological fluid): clarithromycin in plasma. BE based on (90% CI): clarithromycin. Waiver request of in vivo testing: not applicable. For modified-release products, dissolution profiles generated using USP Apparatus I at 100 rpm and/or Apparatus II at 50 rpm in at least three dissolution media (pH 1.2, 4.5, and 6.8 buffer, water) should be submitted in the application. Agitation speeds may have to be increased if appropriate. It is acceptable to add a small amount of surfactant, if necessary. The following sampling times are recommended: 1, 2, and 4 h and every 2 h thereafter, until at least 80% of the drug is dissolved. Comparative dissolution profiles should include individual tablet data as well as the mean, range, and standard deviation at each time point for 12 tablets.

Clarithromycin Granules for Suspension/Oral. Recommended studies: two studies. (1) Type of study: fasting design—single-dose, two-way crossover in vivo. Strength: 250 mg/5 mL. Subjects: normal healthy males and females, general population. Additional comments: (2) Type of study: fed design—single-dose, two-way crossover in vivo. Strength: 250 mg/5 mL. Subjects: normal healthy males and females, general population. Additional comments: Analytes to measure: clarithromycin in plasma. BE based on (90% CI): clarithromycin. Waiver request of in vivo testing: 125 mg/5 mL based on (i) acceptable BE studies on the 250 mg/5 mL strength, (ii) proportional similarity of the formulations across all strengths, and (iii) acceptable in vitro dissolution testing of all strengths.

Clonidine Hydrochloride, Tablets/Oral. Recommended studies: one study. Type of study: fasting design—single-dose, two-way crossover in vivo. Strength: 0.3 mg. Subjects: normal healthy males and females, general population. Additional comments: Analytes to measure: clonidine in plasma. BE based on (90% CI): clonidine. Waiver request of in vivo testing: 0.1 and 0.2 mg based on (i) acceptable BE study on the 0.3 mg strength, (ii) proportional similarity of the formulations across all strengths, and (iii) acceptable in vitro dissolution testing of all strengths. Please conduct comparative dissolution testing on 12 dosage units of all strengths of the test and reference products.

Clopidogrel Bisulfate, Tablets/Oral. Recommended studies: two studies. (1) Type of study: fasting design—single-dose, two-way crossover in vivo. Strength: 75 mg. Subjects: normal healthy males and females, general population. Additional comments: (2) Type of study: fed design—single-dose, two-way crossover in vivo. Strength: 75 mg. Subjects: normal healthy males and females, general population. Additional comments: Analytes to measure: clopidogrel in plasma. BE based on (90% CI): clopidogrel. Waiver request of in vivo testing: not applicable. Please conduct comparative dissolution testing on 12 dosage units of all strengths of the test and reference products.

Danazol, Capsules/Oral. Recommended studies: two studies. (1) Type of study: fasting design—single-dose, two-way crossover in vivo. Strength: 200 mg. Subjects: normal healthy males and females, general population. Additional comments: Due to teratogenicity concerns, females in these studies should not be pregnant. (2) Type of study: fed design—single-dose, two-way crossover in vivo. Strength: 200 mg. Subjects: normal healthy males and females, general population. Additional comments: Due to teratogenicity concerns, females in these studies should not be pregnant. Analytes to measure: danazol in plasma. BE based on (90% CI): danazol. Waiver request of in vivo testing: 50 and 100 mg based on (i) acceptable BE studies on the 200 mg strength, (ii) proportional similarity of the formulations across all strengths, and (iii) acceptable in vitro dissolution testing of all strengths. Please conduct comparative dissolution testing on 12 dosage units of all strengths of the test and reference products.

Dantrolene Sodium, Capsules/Oral. Recommended studies: one study. Type of study: fasting design—single-dose, two-way crossover in vivo. Strength: 100 mg. Subjects: normal healthy males and females, general population. Additional comments: Analytes to measure: dantrolene in plasma. BE based on (90% CI): dantrolene. Waiver request of in vivo testing: 25 and 50 mg based on (i) acceptable BE studies on the 100 mg strength, (ii) proportional similarity across all strengths, and (iii) acceptable in vitro dissolution testing of all strengths.

Darifenacin Hydrobromide, Extended-Release Tablets/Oral. Recommended studies: two studies. (1) Type of study: fasting design—single-dose, two-way crossover in vivo. Strength: 15 mg. Subjects: normal healthy males and females, general population. Additional comments: Females should not be pregnant and, if applicable, should practice abstention or contraception during the study. (2) Type of study: fed design—single-dose, two-way crossover in vivo. Strength: 15 mg. Subjects: normal healthy males and females, general population. Additional comments: Please see previous comment. Analytes to measure (in appropriate biological fluid): darifenacin in plasma. BE based on (90% CI): darifenacin. Waiver request of in vivo testing: 7.5 mg based on (i) acceptable BE studies on the 15 mg strength, (ii) proportional similarity across all strengths, and (iii) acceptable in vitro dissolution testing of all strengths.

Darunavir Ethanolate, Tablet/Oral. Recommended studies: two studies. (1) Type of study: fasting design—single-dose, two-treatment, two-period crossover in vivo. Strength: single dose of 600 mg (2 × 300 mg). Subjects: normal healthy males and females, general population. Additional comments: (2) Type of study: fed design—single-dose, two-treatment, two-period crossover in vivo. Strength: single dose of 600 mg (2 × 300 mg). Subjects: normal healthy males and females, general population. Additional comments: Analytes to measure (in appropriate biological fluid): darunavir in plasma. BE based on (90% CI): darunavir. Waiver request of in vivo testing: not applicable.

Deferasirox, Tablets for Oral Suspension. Recommended studies: one study. Type of study: fasting design—single-dose, two-way crossover in vivo. Strength: 500 mg. Subjects: normal healthy males and females, general population. Additional comments: The following passage is reproduced from the Dosage and Administration section of the labeling: Tablets should be completely dispersed by stirring in water, orange juice, or apple juice until a fine suspension is obtained. Doses of <1 g should be dispersed in 3.5 oz of liquid and doses of >1 g in 7.0 oz of liquid. After swallowing the suspension, any residue should be resuspended in a small volume of liquid and swallowed. Tablets should not be chewed or swallowed whole. Analytes to measure (in appropriate biological fluid):

deferasirox in plasma. BE based on (90% CI): deferasirox. Waiver request of in vivo testing: 250 and 125 mg tablets based on (i) acceptable BE studies on the 500 mg strength, (ii) proportional similarity across all strengths, and (iii) acceptable in vitro dissolution testing of all strengths.

Delavirdine Mesylate, Tablets/Oral. Recommended studies: two studies. (1) Type of study: fasting design—single-dose, two-treatment, two-period crossover in vivo. Strength: 200 mg. Subjects: normal healthy males and females, general population. Females should not be pregnant or lactating and, if applicable, should practice abstention or contraception during the study. (2) Type of study: fed design—single-dose, two-treatment, two-period crossover in vivo. Strength: 200 mg. Subjects: normal healthy males and females, general population. Females should not be pregnant or lactating and, if applicable, should practice abstention or contraception during the study. Analytes to measure (in appropriate biological fluid): delavirdine in plasma. BE based on (90% CI): delavirdine. Waiver request of in vivo testing: 100 mg based on (i) acceptable BE studies on the 200 mg strength, (ii) proportional similarity of the 100 mg formulation to the 200 mg strength, and (iii) acceptable in vitro dissolution testing of all strengths.

Desloratadine, Orally Disintegrating Tablets/Oral. Recommended studies: two studies. (1) Type of study: fasting design—single-dose, two-way crossover in vivo. Strength: 5 mg. Subjects: normal healthy males and females, general population. Additional comments: (2) Type of study: fed design—single-dose, two-way crossover in vivo. Strength: 5 mg. Subjects: normal healthy males and females, general population. Additional comments: Analytes to measure: desloratadine and the active metabolite, 3-hydroxydesloratadine in plasma. Please submit the metabolite data as supportive evidence of the comparable therapeutic outcome. For the metabolite, the following data should be submitted: individual and mean concentrations, individual and mean PK parameters, and geometric means and ratios of means for AUC and C_{max}. BE based on (90% CI): desloratadine. Waiver request of in vivo testing: 2.5 mg based on (i) acceptable BE studies on the 5 mg strength, (ii) proportional similarity across all strengths, and (iii) acceptable in vitro dissolution testing of all strengths.

Desloratadine, Tablets/Oral. Recommended studies: two studies. (1) Type of study: fasting design—single-dose, two-way crossover in vivo. Strength: 5 mg. Subjects: normal healthy males and females, general population. Additional comments: (2) Type of study: fed design—single-dose, two-way crossover in vivo. Strength: 5 mg. Subjects: normal healthy males and females, general population. Additional comments: Analytes to measure: desloratadine and its metabolite, 3-hydroxydesloratadine. BE based on (90% CI): desloratadine. Please submit the metabolite data as supportive evidence of comparable therapeutic outcome. For the metabolite, the following data should be submitted: individual and mean concentrations, individual and mean PK parameters, and geometric means and ratios of means for AUC and C_{max}. Waiver request of in vivo testing: not applicable.

Dexmethylphenidate, Tablets/Oral. Recommended studies: two studies. (1) Type of study: fasting design—single-dose, two-way crossover in vivo. Strength: 10 mg. Subjects: normal healthy males and females, general population. Additional comments: (2) Type of study: fed design—single-dose, two-way crossover in vivo. Strength: 10 mg. Subjects: normal healthy males and females, general population. Additional comments: Analytes to measure: dexmethylphenidate in plasma. BE based on (90% CI): dexmethylphenidate. Waiver request of in vivo testing: 2.5 and 5 mg based on (i) acceptable BE studies on the 10 mg strength, (ii) proportional similarity across all strengths, and (iii) acceptable in vitro dissolution testing of all strengths.

Dextromethorphan Polistirex, Extended-Release Oral Suspension/Oral. Recommended studies: two studies. (1) Type of study: fasting design—single-dose, two-way crossover in vivo. Strength: 30 mg/5 mL. Subjects: normal healthy males and females, general population. Additional comments: (2) Type of study: fed design—single-dose, two-way crossover in vivo. Strength: 30 mg/5 mL. Subjects: normal healthy males and females, general population. Additional comments: Analytes to measure (in appropriate biological fluid): dextromethorphan and its metabolite dextrorphan in plasma. BE based on (90% CI): dextromethorphan. Please submit the metabolite data as supportive evidence of comparable therapeutic outcome. For the metabolite, the following data should be submitted: individual and mean concentrations, individual and mean PK parameters, and geometric means and ratios of means for AUC and C_{max}. Waiver request of in vivo testing: not applicable. A dosage unit for a suspension is the labeled strength (5 mL). A total of 12 units from 12 different bottles should be used. In addition to the previous method, for modified-release products, dissolution profiles on 12 dosage units each of test and reference products generated using USP Apparatus I at 100 rpm and/or Apparatus II at 50 rpm in at least three dissolution media (pH 1.2, 4.5, and 6.8 buffer) should be submitted in the application. Agitation speeds may have to be increased if appropriate. It is acceptable to add a small amount of surfactant, if necessary. Please include early sampling times of 1, 2, and 4 h and continue every 2 h until at least 80% of the drug is released, to provide assurance against premature release of drug (dose dumping) from the formulation.

Diclofenac Sodium; Misoprostol, Delayed Tablets/Oral. Recommended studies: two studies. (1) Type of study: fasting design—single-dose, two-way crossover in vivo. Strength: 75 mg/0.2 mg. Subjects: normal healthy males and females, general population. Additional comments: Female subjects should be excluded from the BE study if they are pregnant. (2) Type of study: fed design—single-dose, two-way crossover in vivo. Strength: 75 mg/0.2 mg. Subjects: normal healthy males and females, general population. Additional comments: Female subjects should be excluded from the BE study if they are pregnant. Analytes to measure: diclofenac and misoprostol's metabolite, misoprostol acid in plasma. BE based on (90% CI): diclofenac and misoprostol's metabolite, misoprostol acid. Please submit the metabolite data as supportive evidence of comparable therapeutic outcome. For the metabolite, the following data should be submitted: individual and mean concentrations, individual and mean PK parameters, and geometric means and ratios of means for AUC and C_{max}. Waiver request of in vivo testing: 50 mg/0.2 mg based on (i) acceptable BE studies on the 75 mg/0.2 mg strength, (ii) proportional similarity of the formulations across all strengths, and (iii) acceptable in vitro dissolution testing of all strengths.

Dicloxacillin Sodium, Capsules/Oral. Recommended studies: one study. Type of study: fasting design—single-dose, two-way crossover in vivo. Strength: 500 mg. Subjects: normal healthy males and females, general population. Additional comments: Analytes to measure: dicloxacillin in plasma. BE based on (90% CI): dicloxacillin. Waiver request of in vivo testing: 125 and 250 mg based on (i) acceptable BE study on the 500 mg, (ii) proportional similarity of the formulations across all strengths, and (iii) acceptable in vitro dissolution testing of all strengths. Please conduct comparative dissolution testing on 12 dosage units of all strengths of the test and reference products.

Didanosine, Chewable Tablets/Oral. Recommended studies: one study. Type of study: fasting design—single-dose, two-treatment, two-period crossover in vivo Strength: 2 × 200 mg (400 mg dose). Subjects: normal healthy males and females, general population. Additional comments: Analytes to measure (in appropriate biological fluid): didanosine in plasma using an achiral method. BE based on (90% CI): didanosine. Waiver request of in vivo testing: 25, 50, 100, and 150 mg, based on (i) acceptable BE study on the 200 mg strength, (ii) proportional similarity of the formulations across all strengths, and (iii) acceptable in vitro dissolution testing of all strengths.

Didanosine, Delayed-Release Capsules, Enteric-Coated Beadlets/Oral. Recommended studies: one study. Type of study: fasting design—single-dose, two-treatment, two-period crossover in vivo. Strength: 400 mg. Subjects: normal healthy males and females, general population. Additional comments: Analytes to measure (in appropriate biological fluid): didanosine in plasma using an achiral method. BE based on (90% CI): didanosine. Waiver request of in vivo testing: 125, 200, and 250 mg, based on (i) acceptable BE studies on the 400 mg strength, (ii) proportional similarity of the formulations across all strengths, and (iii) acceptable in vitro dissolution testing of all strengths.

Digoxin, Tablets/Oral. Recommended studies: two studies. (1) Type of study: fasting design—single-dose, two-way crossover in vivo. Strength: 0.25 mg. Subjects: normal healthy males and females, general population. Additional comments: If reliable blood drug levels cannot be obtained using a 1 × 0.25 mg dose, you may use a single dose of 2 × 0.25 mg tablets. Please carefully monitor the study subjects for adverse events. A washout period of about 2 weeks is suggested. Please continue sample collection for approximately 6 days, that is, at least three or more terminal half-lives of the drug. (2) Type of study: fed design—single-dose, two-way crossover in vivo. Strength: 0.25 mg. Subjects: normal healthy males and females, general population. Additional comments: Please see previous comments. Analytes to measure (in appropriate biological fluid): digoxin in plasma. BE based on (90% CI): digoxin. Waiver request of in vivo testing: 0.125 mg based on (i) acceptable BE studies on the 0.25 mg strength, (ii) proportional similarity of the formulations across all strengths, and (iii) acceptable in vitro dissolution testing of all strengths. Please conduct comparative dissolution testing on 12 dosage units of all strengths of the test and reference products.

Diltiazem Hydrochloride, Extended-Release Capsules/Oral. Recommended studies: two studies. (1) Type of study: fasting design—single-dose, two-way, crossover in vivo. Strength: 360 mg. Subjects: normal healthy males and females, general population. Additional comments: (2) Type of study: fed design—single-dose, two-way, crossover in vivo. Strength: 360 mg. Subjects: normal healthy males and females, general population. Additional comments: Analytes to measure: diltiazem and the active metabolites desacetyldiltiazem and desmethyldiltiazem in plasma. Please submit the metabolite data as supportive evidence of comparable therapeutic outcome. For the metabolite, the following data should be submitted: individual and mean concentrations, individual and mean PK parameters, and geometric means and ratios of means for AUC and C_{max}. BE based on (90% CI): diltiazem. Waiver request of in vivo testing: 120, 180, 240, and 300 mg based on (i) acceptable BE studies on the 360 mg strength, (ii) proportional similarity across all strengths, and (iii) acceptable in vitro dissolution testing of all strengths. For modified-release products, dissolution profiles generated using USP Apparatus I at 100 rpm and/or Apparatus II at 50 rpm in at least three dissolution media (pH 1.2, 4.5, and 6.8 buffer, water) should be submitted in the application. Agitation speeds may have to be increased if appropriate. It is acceptable to add a small amount of surfactant, if necessary. The following sampling times are recommended: 1, 2, and 4 h and every 2 h thereafter, until at least 80% of the drug is dissolved. Comparative dissolution profiles should include individual tablet data as well as the mean, range, and standard deviation at each time point for 12 capsules.

Diltiazem Hydrochloride, Extended-Release Capsules/Oral. Recommended studies: two studies. (1) Type of study: fasting design—single-dose, two-treatment, two-period crossover in vivo. Strength: 240 mg. Subjects: normal healthy males and females, general population. Additional comments: Females must have a negative baseline pregnancy test within 24 h prior to receiving the drug. Females should not be pregnant or lactating and, if applicable, should practice abstention or contraception during the study. (2) Type of study: fed design—single-dose, two-treatment, two-period crossover in vivo. Strength: 240 mg. Subjects: normal healthy males and females, general population. Additional comments: Please see comments earlier. Analytes to measure (in appropriate biological fluid): diltiazem and the active metabolites desacetyldiltiazem and desmethyldiltiazem in plasma. Please submit the metabolite data as supportive evidence of comparable therapeutic outcome. For the metabolites, the following data should be submitted: individual and mean concentrations, individual and mean PK parameters, and geometric means and ratios of means for AUC and C_{max}. BE based on (90% CI): diltiazem. Waiver request of in vivo testing: 120 and 180 mg based on acceptable (i) BE studies on the 240 mg strength, (ii) proportional similarity of the formulations, and (iii) acceptable in vitro dissolution testing of all strengths. For modified-release products, dissolution profiles generated using USP Apparatus I at 100 rpm and/or Apparatus II at 50 rpm in at least three dissolution media (pH 1.2, 4.5, and 6.8 buffer, water) should be submitted in the application. Agitation speeds may have to be increased if appropriate. It is acceptable to add a small amount of surfactant, if necessary. The following sampling times are recommended: 1, 2, and 4 h and every 2 h thereafter, until at least 80% of the drug is dissolved. Comparative dissolution profiles should include individual tablet data as well as the mean, range, and standard deviation at each time point for 12 capsules.

Diltiazem Hydrochloride, Extended-Release Capsules/Oral. Recommended studies: three studies. (1) Type of study: fasting design—single-dose, two-treatment, two-period crossover in vivo. Strength: 420 mg. Subjects: normal healthy males and females, general population.

Additional comments: Females must have a negative baseline pregnancy test within 24 h prior to receiving the drug. Females should not be pregnant or lactating and, if applicable, should practice abstention or contraception during the study. (2) Type of study: fed design—single-dose, two-treatment, two-period crossover in vivo. Strength: 420 mg. Subjects: normal healthy males and females, general population. Additional comments: Please see previous comments. (3) Type of study: fasting sprinkle-in-applesauce design—single-dose, two-treatment, two-period crossover in vivo. Strength: 420 mg. Subjects: normal healthy males and females, general population. Additional comments: Please administer the dose after sprinkling the entire contents of the capsule on a teaspoonful of applesauce in accordance with the approved labeling of the RLD. Analytes to measure (in appropriate biological fluid): diltiazem and the active metabolites desacetyldiltiazem and desmethyldiltiazem in plasma. Please submit the metabolite data as supportive evidence of comparable therapeutic outcome. For the metabolites, the following data should be submitted: individual and mean concentrations, individual and mean PK parameters, and geometric means and ratios of means for AUC and C_{max}. BE based on (90% CI): diltiazem. Waiver request of in vivo testing: 120, 180, 240, 300, and 360 mg based on (i) acceptable BE studies on the 420 mg strength, (ii) proportional similarity of the formulations, and (iii) acceptable in vitro dissolution testing of all strengths. For modified-release products, dissolution profiles generated using USP Apparatus I at 100 rpm and/or Apparatus II at 50 rpm in at least three dissolution media (pH 1.2, 4.5, and 6.8 buffer, water) should be submitted in the application. Agitation speeds may have to be increased if appropriate. It is acceptable to add a small amount of surfactant, if necessary. The following sampling times are recommended: 1, 2, and 4 h and every 2 h thereafter, until at least 80% of the drug is dissolved. Comparative dissolution profiles should include individual tablet data as well as the mean, range, and standard deviation at each time point for 12 units.

Diltiazem Hydrochloride, Extended-Release Tablets/Oral. Recommended studies: two studies. (1) Type of study: fasting design—single-dose, two-way, crossover in vivo. Strength: 420 mg. Subjects: normal healthy males and females, general population. Additional comments: (2) Type of study: fed design—single-dose, two-way, crossover in vivo. Strength: 420 mg. Subjects: normal healthy males and females, general population. Additional comments: Analytes to measure: diltiazem and the active metabolites desacetyldiltiazem and desmethyldiltiazem in plasma. Please submit the metabolite data as supportive evidence of comparable therapeutic outcome. For the metabolite, the following data should be submitted: individual and mean concentrations, individual and mean PK parameters, and geometric means and ratios of means for AUC and C_{max}. BE based on (90% CI): diltiazem. Waiver request of in vivo testing: 120, 180, 240, 300, and 360 mg based on (i) acceptable BE studies on the 420 mg strength, (ii) proportional similarity across all strengths, and (iii) acceptable in vitro dissolution testing of all strengths. For modified-release products, dissolution profiles generated using USP Apparatus I at 100 rpm and/or Apparatus II at 50 rpm in at least three dissolution media (pH 1.2, 4.5, and 6.8 buffer, water) should be submitted in the application. Agitation speeds may have to be increased if appropriate. It is acceptable to add a small amount of surfactant, if necessary. The following sampling times are recommended: 1, 2, and 4 h and every 2 h thereafter, until at least 80% of the drug is dissolved. Comparative dissolution profiles should include individual tablet data as well as the mean, range, and standard deviation at each time point for 12 tablets.

Dipyridamole, Tablets/Oral. Recommended studies: one study. Type of study: fasting design—single-dose, two-way crossover in vivo. Strength: 75 mg. Subjects: normal healthy males and females, general population. Additional comments: Analytes to measure: dipyridamole in plasma. BE based on (90% CI): dipyridamole. Waiver request of in vivo testing: 25 and 50 mg based on (i) acceptable BE study on the 75 mg strength, (ii) proportional similarity of the formulations across all strengths, and (iii) acceptable in vitro dissolution testing of all strengths. Please conduct comparative dissolution testing on 12 dosage units of all strengths of the test and reference products.

Divalproex Sodium, Delayed-Release Pellets Capsule/Oral. Recommended studies: three studies. (1) Type of study: fasting design—single-dose, two-way crossover in vivo. Strength: 125 mg.

Subjects: normal healthy males and females, general population. Additional comments: Normal liver function test should be required prior to dosing with divalproex sodium in BE studies. Females should not be pregnant and, if applicable, should practice abstention or contraception during the study. (2) Type of study: fed design—single-dose, two-way crossover in vivo. Strength: 125 mg. Subjects: normal healthy males and females, general population. Additional comments: Please see previous comment. (3) Type of study: fasting sprinkle-in-applesauce design—single-dose, two-way crossover in vivo. Strength: 125 mg. Subjects: normal healthy males and females, general population. Additional comments: Please see comment earlier. Analytes to measure: valproic acid in plasma. It is not necessary to measure plasma concentrations of the metabolites. BE based on (90% CI): valproic acid. Waiver request of in vivo testing: not applicable.

Divalproex Sodium, Extended-Release Tablets/Oral. Recommended studies: two studies. (1) Type of study: fasting design—single-dose, two-way, crossover in vivo. Strength: 500 mg. Subjects: normal healthy males and females, general population. Additional comments: Females should not be pregnant or lactating and, if applicable, should practice abstention or contraception during the study. (2) Type of study: fed design—single-dose, two-way, crossover in vivo. Strength: 500 mg. Subjects: normal healthy males and females, general population. Additional comments: Please see previous comments. Analytes to measure: valproic acid in plasma. BE based on (90% CI): valproic acid. Waiver request of in vivo testing: 250 mg based on (i) acceptable BE studies on the 500 mg strength, (ii) proportional similarity across all strengths, and (iii) acceptable in vitro dissolution testing of all strengths. For modified-release products, dissolution profiles generated using USP Apparatus I at 100 rpm and/or Apparatus II at 50 rpm in at least three dissolution media (pH 1.2, 4.5, and 6.8 buffer, water) should be submitted in the application. Agitation speeds may have to be increased if appropriate. It is acceptable to add a small amount of surfactant, if necessary. The following sampling times are recommended: 1, 2, and 4 h and every 2 h thereafter, until at least 80% of the drug is dissolved. Comparative dissolution profiles should include individual tablet data as well as the mean, range, and standard deviation at each time point for 12 tablets.

Dofetilide, Capsules/Oral. Recommended studies: two studies. (1) Type of study: fasting design—single dose, two-way crossover, in vivo. Strength: 0.5 mg. Subjects: normal healthy males and females, general population. Females should not be pregnant and, if applicable, should practice abstention or contraception during the study. Additional comments: A black box warning concerns the risk of drug-induced arrhythmia. The study should be conducted in a facility that can provide continuous cardiac monitoring in the presence of personnel trained in management of serious ventricular arrhythmias. Any subject that develops a prolonged QTc interval should be monitored until the QTc is within normal limits. (2) Type of study: fed design—single dose, two-way crossover, in vivo. Strength: 0.5 mg. Subjects: normal healthy males and females, general population. Females should not be pregnant and, if applicable, should practice abstention or contraception during the study. Additional comments: Please see previous comments. Analytes to measure (in appropriate biological fluid): dofetilide in plasma. BE based on (90% CI): dofetilide. Waiver request of in vivo testing: 0.25 and 0.125 mg based on (i) acceptable BE studies on the 0.5 mg strength, (ii) proportional similarity to the 0.5 mg strength, and (iii) acceptable in vitro dissolution testing.

Donepezil Hydrochloride, Orally Disintegrating Tablet/Oral. Recommended studies: two studies. (1) Type of study: fasting design—single-dose, two-treatment, two-period crossover in vivo. Strength: 10 mg. Subjects: normal healthy males and females, general population. Additional comments: (2) Type of study: fed design—single-dose, two-treatment, two-period crossover in vivo. Strength: 10 mg. Subjects: normal healthy males and females, general population. Additional comments: Analytes to measure: donepezil in plasma. BE based on (90% CI): donepezil. Waiver request of in vivo testing: 5 mg based on (i) acceptable BE studies on the 10 mg strength, (ii) proportional similarity of the formulations across all strengths, and (iii) acceptable in vitro dissolution testing of all strengths.

Donepezil Hydrochloride, Tablets/Oral. Recommended studies: two studies. (1) Type of study: fasting design—single-dose, two-way, crossover in vivo. Strength: 10 mg. Subjects: normal healthy males and females, general population. Additional comments: (2) Type of study: fed design—single-dose, two-way, crossover in vivo. Strength: 10 mg. Subjects: normal healthy males and females, general population. Additional comments: Analytes to measure: donepezil in plasma. BE based on (90% CI): donepezil. Waiver request of in vivo testing: 5 mg based on (i) acceptable BE studies on the 10 mg strength, (ii) proportional similarity across all strengths, and (iii) acceptable in vitro dissolution testing of all strengths.

Doxazosin Mesylate, Extended-Release Tablets/Oral. Recommended studies: two studies. (1) Type of study: fasting design—single-dose, two-way crossover in vivo. Strength: 8 mg. Subjects: normal healthy males and females, general population. Females should not be pregnant and, if applicable, should practice abstention or contraception during the study. Additional comments: (2) Type of study: fed design—single-dose, two-way crossover in vivo. Strength: 8 mg. Subjects: normal healthy males and females, general population. Females should not be pregnant and, if applicable, should practice abstention or contraception during the study. Additional comments: Analytes to measure (in appropriate biological fluid): doxazosin in plasma. BE based on (90% CI): doxazosin. Waiver request of in vivo testing: 4 mg based on (i) acceptable BE studies on the 8 mg strength, (ii) proportional similarity across both strengths, and (iii) acceptable in vitro dissolution testing of both strengths. In addition to the method earlier, for modified-release products, dissolution profiles on 12 dosage units each of test and reference products generated using USP Apparatus I at 100 rpm and/or Apparatus II at 50 rpm in at least three dissolution media (pH 1.2, 4.5, and 6.8 buffer) should be submitted in the application. Agitation speeds may have to be increased if appropriate. It is acceptable to add a small amount of surfactant, if necessary. Please include early sampling times of 1, 2, and 4 h and continue every 2 h until at least 80% of the drug is released, to provide assurance against premature release of drug (dose dumping) from the formulation.

Doxycycline, Delayed-Release Capsules/Oral. Recommended studies: two studies. (1) Type of study: fasting design—single-dose, two-way crossover in vivo. Strength: 40 mg. Subjects: normal healthy males and females, general population. Additional comments: Females should not be pregnant or lactating and, if applicable, should practice abstention or contraception during the study. (2) Type of study: fed design—single-dose, two-way crossover in vivo. Strength: 40 mg. Subjects: normal healthy males and females, general population. Additional comments: Please see previous comment. Analytes to measure (in appropriate biological fluid): doxycycline in plasma. BE based on (90% CI): doxycycline. Waiver request of in vivo testing: not applicable.

Doxycycline, Tablets/Oral. Recommended studies: two studies. (1) Type of study: fasting design—single-dose, two-way, crossover in vivo. Strength: 150 mg. Subjects: normal healthy males and females, general population. Additional comments: (2) Type of study: fed design—single-dose, two-way, crossover in vivo. Strength: 150 mg. Subjects: normal healthy males and females, general population. Additional comments: Analytes to measure (in appropriate biological fluid): doxycycline in plasma. BE based on (90% CI): doxycycline. Waiver request of in vivo testing: 50, 75, and 100 mg based on (i) acceptable BE studies on the 150 mg strength, (ii) proportional similarity across all strengths, and (iii) acceptable in vitro dissolution testing of all strengths.

Drospirenone; Estradiol, Tablets/Oral. Recommended studies: two studies. (1) Type of study: fasting design—single-dose, two-way crossover in vivo. Strength: 0.5 and 1 mg. Subjects: normal healthy postmenopausal women. Additional comments: (2) Type of study: fed design—single-dose, two-way crossover in vivo. Strength: 0.5 and 1 mg. Subjects: normal healthy postmenopausal women. Additional comments: Analytes to measure (in appropriate biological fluid): drospirenone and unconjugated estradiol, unconjugated estrone, and total estrone in plasma. BE based on (90% CI): drospirenone and baseline-adjusted total estrone. Statistical analysis should be performed on data both with and without baseline adjustment. BE acceptance criteria will be based on baseline-adjusted results only. Baseline adjustment: data of each subject and period should be adjusted for the mean of −1 and −0.5 h and predose levels for that same subject and period. If, after adjustment, any negative concentrations result, they should be set equal to zero. Waiver request of in vivo testing: not applicable.

Duloxetine Hydrochloride, Delayed-Release Pellets Capsule/Oral. Recommended studies: two studies. (1) Type of study: fasting design—single-dose, two-treatment, two-period crossover in vivo. Strength: 60 mg. Subjects: normal healthy males and females, general population. Additional comments: Females should not be pregnant and, if applicable, should practice abstention or contraception during the study. Due to the need to maintain the enteric coating, the subjects in a BE study should be advised not to crush or chew the enteric-coated pellets. (2) Type of study: fed design—single-dose, two-treatment, two-period crossover in vivo. Strength: 60 mg. Subjects: normal healthy males and females, general population. Additional comments: Please see previous. Analytes to measure (in appropriate biological fluid): duloxetine in plasma. BE based on (90% CI): duloxetine. Waiver request of in vivo testing: 20 and 30 mg based on (i) acceptable BE studies on the 60 mg strength, (ii) proportional similarity of the formulations across all strengths, and (iii) acceptable in vitro dissolution testing of all strengths.

Dutasteride, Capsules/Oral. Recommended studies: two studies. (1) Type of study: fasting design—single-dose, two-way crossover in vivo. Strength: 0.5 mg. Subjects: normal healthy males. Additional comments: (2) Type of study: fed design—single-dose, two-way crossover in vivo. Strength: 0.5 mg. Subjects: normal healthy males. Additional comments: Note: as an option, due to the relatively long half-life, the firm may wish to conduct these studies using a parallel design. As an additional option for either the crossover or parallel design, the firm may wish to truncate the AUC at 72 h. Analytes to measure (in appropriate biological fluid): dutasteride in plasma. BE based on (90% CI): dutasteride. Waiver request of in vivo testing: not applicable.

Efavirenz, Capsules/Oral. Recommended studies: one study. Type of study: fasting design—single-dose, two-treatment, two-period crossover in vivo. Strength: 200 mg. Subjects: normal healthy males and females, general population. Additional comments: Analytes to measure (in appropriate biological fluid): efavirenz in plasma. BE based on (90% CI): efavirenz. Waiver request of in vivo testing: 50 and 100 mg based on (i) acceptable BE study on the 200 mg strength, (ii) proportional similarity of the formulations across all strengths, and (iii) acceptable in vitro dissolution testing of all strengths.

Efavirenz, Tablets/Oral. Recommended studies: one study. Type of study: fasting design—single-dose, two-way crossover in vivo. Strength: 600 mg. Subjects: normal healthy males and females, general population. Additional comments: Analytes to measure: efavirenz in plasma. BE based on (90% CI): efavirenz. Waiver request of in vivo testing: not applicable.

Emtricitabine, Capsules/Oral. Recommended studies: two studies. (1) Type of study: fasting design—single-dose, two-way crossover in vivo. Strength: 200 mg. Subjects: normal healthy males and females, general population. Additional comments: Females should not be pregnant and, if applicable, should practice abstention or contraception during the study. (2) Type of study: fed design—single-dose, two-way crossover in vivo. Strength: 200 mg. Subjects: normal healthy males and females, general population. Additional comments: Please see previous comments. Analytes to measure: Emtricitabine in plasma. BE based on (90% CI): emtricitabine. Waiver request of in vivo testing: not applicable.

Entacapone, Tablets/Oral. Recommended studies: two studies. (1) Type of study: fasting design—single-dose, two-way, crossover in vivo. Strength: 200 mg. Subjects: normal healthy males and females, general population. Additional comments: Due to the high inter- and intrasubject variability observed with this product, you may want to consider using a replicate study design. Since the drug product is to be used predominantly in the elderly, please include as many subjects of 60 years of age or older as possible. (2) Type of study: fed design—single-dose, two-way, crossover in vivo. Strength: 200 mg. Subjects: normal healthy males and females, general population. Additional comments: Please see previous comments. Analytes to measure: entacapone in plasma. BE based on (90% CI): entacapone. Waiver request of in vivo testing: not applicable.

Entecavir, Tablets/Oral. Recommended studies: one study. Type of study: fasting design—single-dose, two-way crossover in vivo. Strength: 1 mg. Subjects: normal healthy males and females, general population. Additional comments: As an option, due to the relatively long half-life, the firm may wish to conduct this study using a parallel design. As an additional option for either the crossover or parallel design, the firm may wish to truncate the AUC at 72 h. Analytes to measure (in appropriate biological fluid): entecavir in plasma. BE based on (90% CI): entecavir. Waiver request of in vivo testing: 0.5 mg based on (i) acceptable BE studies on the 1 mg strength, (ii) proportional similarity across all strengths, and (iii) acceptable in vitro dissolution testing of all strengths.

Eplerenone, Tablets/Oral. Recommended studies: two studies. (1) Type of study: fasting design—single-dose, two-way crossover in vivo. Strength: 50 mg. Subjects: normal healthy males and females, general population. Additional comments: (2) Type of study: fed design—single-dose, two-way crossover in vivo. Strength: 50 mg. Subjects: normal healthy males and females, general population. Additional comments: Analytes to measure (in appropriate biological fluid): eplerenone in plasma. BE based on (90% CI): eplerenone. Waiver request of in vivo testing: 25 mg based on (i) acceptable BE studies on the 50 mg strength, (ii) proportional similarity to the 50 mg strength, and (iii) acceptable in vitro dissolution testing.

Eprosartan Mesylate; Hydrochlorothiazide, Tablets/Oral. Recommended studies: two studies. (1) Type of study: fasting design—single-dose, two-way crossover in vivo. Strength: 600 mg/25 mg. Subjects: normal healthy males and females, general population. Additional comments: Females should not be pregnant or lactating and, if applicable, should practice abstention or contraception during the study. Please include provisions for appropriate monitoring and intervention in the case of possible drug-related adverse events (e.g., subjects complaining of dizziness/lightheadedness should have blood pressure/heart rate assessed). (2) Type of study: fed design—single-dose, two-way crossover in vivo. Strength: 600 mg/25 mg. Subjects: normal healthy males and females, general population. Additional comments: Please see comment earlier. Analytes to measure (in appropriate biological fluid): eprosartan and hydrochlorothiazide in plasma. BE based on (90% CI): eprosartan and hydrochlorothiazide. Waiver request of in vivo testing: 600 mg/12.5 mg, based on (i) acceptable BE studies on the 600 mg/25 mg strength, (ii) proportional similarity of the formulations across all strengths, and (iii) acceptable in vitro dissolution testing of all strengths.

Erlotinib Hydrochloride, Tablets/Oral. Recommended studies: one study. Type of study: fasting design—single-dose, two-way crossover in vivo. Strength: 150 mg. Subjects: normal healthy males and females, general population. Additional comments: Females should not be pregnant or lactating and, if applicable, should practice abstention or contraception during the study. Any subject experiencing an adverse event should be followed until the adverse event has completely resolved. Analytes to measure (in appropriate biological fluid): erlotinib in plasma. BE based on (90% CI): erlotinib. Waiver request of in vivo testing: 100 and 25 mg based on (i) acceptable BE studies on the 150 mg strength, (ii) proportional similarity of the formulations across all strengths, and (iii) acceptable in vitro dissolution testing of all strengths.

Escitalopram Oxalate, Tablets/Oral. Recommended studies: two studies. (1) Type of study: fasting design—single-dose, two-way crossover in vivo. Strength: 20 mg. Subjects: normal healthy males and females, general population. Additional comments: (2) Type of study: fed design—single-dose, two-way crossover in vivo. Strength: 20 mg. Subjects: normal healthy males and females, general population. Additional comments: Analytes to measure: escitalopram, using an achiral assay. BE based on (90% CI): escitalopram. Waiver request of in vivo testing: 5 and 10 mg based on (i) acceptable BE studies on the 20 mg strength, (ii) proportional similarity across all strengths, and (iii) acceptable in vitro dissolution testing of all strengths.

Esomeprazole Magnesium, Delayed-Release Capsules/Oral. Recommended studies: three studies. (1) Type of study: fasting design—single-dose, two-way crossover in vivo. Strength: 40 mg. Subjects: normal healthy males and females, general population. Additional comments: (2) Type of study: fed design—single-dose, two-way crossover in vivo. Strength: 40 mg. Subjects: normal healthy males and females, general population. Additional comments: (3) Type of study: sprinkle design—single-dose, two-way crossover in vivo. Strength: 40 mg. Subjects: normal healthy males and females, general population. Additional comments: Fasting study, with treatments sprinkled over a spoonful of applesauce. Analytes to measure: esomeprazole using an achiral assay. BE based on (90% CI): esomeprazole. Waiver request of in vivo testing: 20 mg based on (i) acceptable BE studies on the 40 mg strength, (ii) proportional similarity of the formulations across all strengths, and (iii) acceptable in vitro dissolution testing of all strengths. For dissolution method development, please refer to USP, "Delayed-Release (Enteric-Coated) Articles—General Drug Release Standard." Esomeprazole is an acid labile drug substance; therefore, please measure esomeprazole from the beadlets of the EC capsule and not from the dissolution medium (0.1 N HCl) during the acid stage; using 12 additional capsules of the test and reference products, proceed to the buffer stage.

Esterified Estrogens, Tablets/Oral. Recommended studies: one study. Type of study: fasting design—single-dose, two-way crossover in vivo. Strength: 2.5 mg. Subjects: normal healthy postmenopausal or surgically sterile females. Additional comments: Analytes to measure (in appropriate biological fluid): estrone sulfate and equilin sulfate in plasma. (1) Please provide baseline correction for endogenous estrone sulfate in the analysis. Please measure baseline estrone sulfate levels at −1, −0.5, and 0 h. The mean of the predose estrone sulfate levels should be used for the baseline adjustment of the postdose levels. Any negative values obtained from baseline correction should be designated as zero (0), and any subject with baseline-adjusted predose concentrations (at time 0 h) greater than 5% of their C_{max} should be excluded from the BE statistical analysis and the 90% CI based on the remaining subjects. (2) The selected blood sampling schedule should include sufficient time points around t_{max} for the best estimate of C_{max} and should be sufficiently long for the best characterization of the elimination phase of both analytes (at least 96 h). (3) The analytical assay method selected should be sufficiently sensitive and specific to measure estrone sulfate and equilin sulfate concentrations in plasma and should have a lower limit of quantitation (LLOQ) of 50 pg/mL or less for both analytes. (4) Based on the estimated half-life of the two analytes, the washout duration should be greater than five times the half-life; therefore, at least a 2-week washout period between doses is recommended. BE based on (90% CI): estrone sulfate and equilin sulfate. Waiver request of in vivo testing: 0.3, 0.625, and 1.25 mg based on (i) acceptable BE studies on the 2.5 mg strength, (ii) proportional similarity of the formulations across all strengths, and (iii) acceptable in vitro dissolution testing of all strengths.

Eszopiclone, Tablets/Oral. Recommended studies: two studies. (1) Type of study: fasting design—single-dose, two-treatment, two-period crossover in vivo. Strength: 3 mg. Subjects: normal healthy males and females, general population. Additional comments: (2) Type of study: fed design—single-dose, two-treatment, two-period crossover in vivo. Strength: 3 mg. Subjects: normal healthy males and females, general population. Additional comments: Analytes to measure (in appropriate biological fluid): eszopiclone in plasma. BE based on (90% CI): eszopiclone. Waiver request of in vivo testing: 1 and 2 mg, based on (i) acceptable BE studies on the 3 mg tablet, (ii) proportional similarity of the formulations, and (iii) acceptable in vitro dissolution testing of all strengths.

Ethambutol Hydrochloride, Tablets/Oral. Recommended studies: two studies. (1) Type of study: fasting design—single-dose, two-treatment, two-period crossover in vivo. Strength: 400 mg. Subjects: normal healthy males and females, general population. Additional comments: (2) Type of study: fed design—single-dose, two-treatment, two-period crossover in vivo. Strength: 400 mg. Subjects: normal healthy males and females, general population. Additional comments: Analytes to measure (in appropriate biological fluid): ethambutol in plasma. BE based on (90% CI): ethambutol. Waiver request of in vivo testing: 100 mg, based on (i) acceptable BE studies on the 400 mg strength, (ii) proportional similarity of the formulations, and (iii) acceptable in vitro dissolution testing of all strengths.

Ethinyl Estradiol and Levonorgestrel, Tablets/Oral. Recommended studies: two studies. (1) Type of study: fasting design—single-dose, two-treatment, two-period crossover in vivo. Strength: 0.03 mg/0.15 mg tablet of ethinyl estradiol and levonorgestrel. Subjects: normal healthy males and females, general population. Additional comments: (2) Type of study: fasting design—single-dose, two-treatment, two-period crossover in vivo. Strength: 0.01 mg tablet of ethinyl estradiol. Subjects: normal healthy males and females, general population. Additional comments: Analytes to measure (in appropriate biological fluid): ethinyl estradiol and levonorgestrel in plasma for the combination tablets. Only ethinyl estradiol for the single component tablet. BE based on (90% CI): ethinyl estradiol and levonorgestrel. Waiver request of in vivo testing: BE studies conducted on Seasonique® (ethinyl estradiol and levonorgestrel) tablets, 0.03 mg/0.15 mg, may be referenced to support a request for a waiver of evidence of in vivo BE for Seasonale® (ethinyl estradiol and levonorgestrel) tablets, 0.03 mg/0.15 mg. Please submit separate applications for each RLD.

Ethinyl Estradiol and Levonorgestrel, Tablets/Oral. Recommended studies: one study. Type of study: fasting design—single-dose, two-treatment, two-period crossover in vivo. Strength: 0.03 mg/0.15 mg. Subjects: normal healthy males and females, general population. Additional comments: Analytes to measure (in appropriate biological fluid): ethinyl estradiol and levonorgestrel in plasma. Only ethinyl estradiol for the single component tablet in Seasonique. BE based on (90% CI): ethinyl estradiol and levonorgestrel. Waiver request of in vivo testing: BE studies conducted on Seasonique® (ethinyl estradiol and levonorgestrel) tablets, 0.03 mg/0.15 mg, may be referenced to support a request for a waiver of evidence of in vivo BE for Seasonale® (ethinyl estradiol and levonorgestrel) tablets, 0.03 mg/0.15 mg. Please submit separate applications for each RLD.

Etidronate Disodium, Tablets/Oral. Recommended studies: one study. Type of study: fasting design—single-dose, two-way crossover in vivo. Strength: 400 mg. Subjects: normal healthy males and females, general population. Additional comments: Analytes to measure (in appropriate biological fluid): etidronate in plasma. BE based on (90% CI): etidronate. Waiver request of in vivo testing: 200 mg based on (i) acceptable BE study on the 400 mg strength, (ii) proportional similarity across all strengths, and (iii) acceptable in vitro dissolution testing of all strengths. Please conduct comparative dissolution testing on 12 dosage units of all strengths of the test and reference products.

Exemestane, Tablets/Oral. Recommended studies: two studies. (1) Type of study: fasting design—single-dose, two-way, crossover in vivo. Strength: 25 mg. Subjects: normal healthy postmenopausal women, general population. Additional comments: This product is indicated for use in postmenopausal women. Because of teratogenicity concerns with this product, females in these studies should not be of childbearing potential. We recommended that you attempt to include as many postmenopausal women as possible. (2) Type of study: fed design—single-dose, two-way, crossover in vivo. Strength: 25 mg. Subjects: normal healthy postmenopausal women, general population. Additional comments: Please see previous comments. Analytes to measure (in appropriate biological fluid): exemestane in plasma. BE based on (90% CI): exemestane. Waiver request of in vivo testing: not applicable.

Famotidine, Orally Disintegrating Tablets/Oral. Recommended studies: two studies. (1) Type of study: fasting design—single-dose, two-way crossover in vivo. Strength: 40 mg. Subjects: normal healthy males and females, general population. Additional comments: (2) Type of study: fed design—single-dose, two-way crossover in vivo. Strength: 40 mg. Subjects: normal healthy males and females, general population. Additional comments: Analytes to measure: famotidine in plasma. BE based on (90% CI): famotidine. Waiver request of in vivo testing: 20 mg based on (i) acceptable BE studies on the 40 mg strength, (ii) proportional similarity across all strengths, and (iii) acceptable in vitro dissolution testing of all strengths.

Famotidine, Tablets/Oral. Recommended studies: two studies. (1) Type of study: fasting design—single-dose, two-way, crossover in vivo. Strength: 40 mg. Subjects: normal healthy males and females, general population. Additional comments: (2) Type of study: fed design—single-dose, two-way, crossover in vivo. Strength: 40 mg. Subjects: normal healthy males and females, general population. Additional comments: Analytes to measure: famotidine in plasma. BE based on (90% CI): famotidine. Waiver request of in vivo testing: 10 and 20 mg based on (i) acceptable BE studies on the 40 mg strength, (ii) proportional similarity across all strengths, and (iii) acceptable in vitro dissolution testing of all strengths. Note: Separate applications should be submitted for the prescription (Rx) and over-the-counter (OTC) products. You may request a waiver of in vivo BE testing for the OTC product; if you conduct the studies on the Rx product, submit acceptable dissolution data on all strengths and the formulations of the products are proportional. Please cross-reference in the OTC application the studies conducted for the Rx product. Please conduct comparative dissolution testing on 12 dosage units of all strengths of the test and reference products.

Felbamate, Oral Suspension/Oral. Recommended studies: one study. Type of study: fasting design—multiple-dose, two-way steady-state crossover in vivo. Strength: 600 mg/5 mL. Subjects: male and nonpregnant female epilepsy patients. Additional comments: Please also consider the following additional safety monitoring: (a) If any evidence of bone marrow (hematologic) depression occurs, felbamate treatment should be discontinued and a hematologist consulted to ensure appropriate medical care. (b) Additional criteria for exclusion from the study relative to baseline be practiced including (i) twofold increase in the highest, 2-day pre-study seizure frequency, (ii) single generalized, tonic–clonic seizure if none occurred during pretreatment screening, and/or (iii) significant prolongation of generalized, tonic–clonic seizures. Analytes to measure: felbamate in plasma. (1) Measurements of felbamate are requested on at least 2 consecutive days immediately prior to PK analysis days 7 and 14 to confirm steady-state concentrations of felbamate (i.e., additional consecutive measures on days 5, 6 and 12, 13). (2) Because felbamate is rapidly absorbed and reaches a peak plasma concentration within 1–3 h post consumption, please also include blood sampling at 0.25 h after drug dosing to accurately measure the absorption/distribution phases of the felbamate PK profile. (3) Patients who receive multiples of 600 mg of felbamate per day (1200–4800 mg/day) would be eligible for the study by continuing their established maintenance dose. Because patients will be administered different dosing regimens, the dose needs to be included in the ANOVA statistical model. Dose normalization is not advised. (4) No washout period is necessary between treatment periods. (5) You are encouraged to submit protocols for the in vivo BE studies to be conducted at steady state in patients already taking the RLD at a therapeutic dose for review prior to initiating the studies. BE based on (90% CI): felbamate. Waiver request of in vivo testing: not applicable. A dosage unit for a suspension is the labeled strength (5 mL). A total of 12 units from 12 different bottles should be used.

Felbamate, Tablets/Oral. Recommended studies: one study. Type of study: fasting design—multiple-dose, two-way steady-state crossover in vivo. Strength: 600 mg. Subjects: male and nonpregnant female epilepsy patients. Additional comments: Please also consider the following additional safety monitoring: (a). If any evidence of bone marrow (hematologic) depression occurs, felbamate treatment should be discontinued and a hematologist consulted to ensure appropriate medical care. (b) Additional criteria for exclusion from the study relative to baseline be practiced including (i) twofold increase in the highest, 2-day pre-study seizure frequency, (ii) single generalized, tonic–clonic seizure if none occurred during pretreatment screening, and/or (iii) significant prolongation of generalized, tonic–clonic seizures. Analytes to measure: felbamate in plasma. (1) Measurements of felbamate are requested on at least 2 consecutive days immediately prior to PK analysis days 7 and 14 to confirm steady-state concentrations of felbamate (i.e., additional consecutive measures on days 5, 6 and 12, 13). (2) Because felbamate is rapidly absorbed and reaches a peak plasma concentration within 1–3 h post consumption, please also include blood sampling at 0.25 h after drug dosing to accurately measure the absorption/distribution phases of the felbamate PK profile. (3) Patients who receive multiples of 600 mg tablets of felbamate per day (1200–4800 mg/day) would be eligible for the study by continuing their established maintenance dose. Because patients will be administered different dosing regimens, the dose needs to be included in the ANOVA statistical model. Dose normalization is not advised. (4) No washout period is necessary between treatment periods. (5) You are encouraged to submit protocols for the in vivo BE studies to be conducted at steady state in patients already taking the RLD at a therapeutic dose for review prior to initiating the studies. BE based on (90% CI): felbamate. Waiver request of in vivo testing: 400 mg based on (i) acceptable BE studies on the 600 mg strength, (ii) proportional similarity of the formulations across all strengths, and (iii) acceptable in vitro dissolution testing of all strengths.

Fenofibrate, Capsules/Oral. Recommended studies: two studies. (1) Type of study: fasting design—single-dose, two-way crossover in vivo. Strength: 150 mg. Subjects: normal healthy males and females, general population. Additional comments: Females should not be pregnant or lactating and, if applicable, should practice abstention or contraception during the study. (2) Type of study: fed design—single-dose, two-way crossover in vivo. Strength: 150 mg. Subjects: normal healthy males and females, general population. Additional comments: Please see previous comment. Analytes to measure (in appropriate biological fluid): fenofibric acid, the active metabolite of fenofibrate in plasma. BE based on (90% CI): fenofibric acid. Waiver request of in vivo testing: 50 and 100 mg based on (i) acceptable BE studies on the 150 mg strength, (ii) proportional similarity of the formulations across all strengths, and (iii) acceptable in vitro dissolution testing of all strengths.

Fenofibrate, Capsules/Oral. Recommended studies: two studies. (1) Type of study: fasting design—single-dose, two-way crossover in vivo. Strength: 130 mg. Subjects: normal healthy males and females, general population. Additional comments: (2) Type of study: fed design—single-dose, two-way crossover in vivo. Strength: 130 mg. Subjects: normal healthy males and females, general population. Additional comments: Analytes to measure: fenofibric acid in plasma. BE based on (90% CI): fenofibric acid. Waiver request of in vivo testing: 43 mg based on (i) acceptable BE studies on the 130 mg strength, (ii) proportional similarity across all strengths, and (iii) acceptable in vitro dissolution testing of all strengths.

Fenofibrate, Tablets/Oral. Recommended studies: two studies. (1) Type of study: fasting design—single-dose, two-way crossover in vivo. Strength: 145 mg. Subjects: normal healthy males and females, general population. Additional comments: (2) Type of study: fed design—single-dose, two-way crossover in vivo. Strength: 145 mg. Subjects: normal healthy males and females, general population. Additional comments: Analytes to measure: due to the difficulties with the fenofibrate assay, only the metabolite, fenofibric acid, should be measured. BE based on (90% CI): fenofibric acid. Waiver request of in vivo testing: 48 mg based on (i) acceptable BE studies on the 145 mg strength, (ii) proportional similarity of the formulations 48 and 145 mg strengths, and (iii) acceptable in vitro dissolution testing of 48 and 145 mg strengths.

Fexofenadine Hydrochloride, Capsules/Oral. Recommended studies: one study. Type of study: fasting design—single-dose, two-way crossover in vivo. Strength: 60 mg. Subjects: normal healthy males and females, general population. Additional comments: Analyte to measure: fexofenadine in plasma. BE based on (90% CI): fexofenadine. Waiver request of in vivo testing: not applicable.

Fexofenadine Hydrochloride, Tablets/Oral. Recommended studies: one study. Type of study: fasting design—single-dose, two-way crossover in vivo. Strength: 180 mg. Subjects: normal healthy males and females, general population. Additional comments: Analytes to measure: fexofenadine in plasma. BE based on (90% CI): fexofenadine. Waiver request of in vivo testing: 30 and 60 mg based on (i) acceptable BE study on the 180 mg strength, (ii) proportional similarity of the formulations across all strengths, and (iii) acceptable in vitro dissolution testing of all strengths.

Flavoxate Hydrochloride, Tablets/Oral. Recommended studies: one study. Type of study: fasting design—single-dose, two-way crossover in vivo. Strength: 100 mg. Subjects: normal healthy males and females, general population. Additional comments: Analytes to measure (in appropriate biological fluid): flavoxate and the metabolite, 3 methyl-flavone-8-carboxylic acid in plasma. BE based on (90% CI): flavoxate or the metabolite, 3 methyl-flavone-8-carboxylic acid. If flavoxate can be reliably measured, a CI approach for BE determination should be used for flavoxate. If flavoxate cannot be reliably measured, a CI approach for BE determination should be used for 3 methyl-flavone-8-carboxylic acid. Waiver request of in vivo testing: not applicable.

Fluconazole, Tablet/Oral. Recommended studies: one study. Type of study: fasting design—single-dose, two-treatment, two-period crossover in vivo. Strength: 200 mg. Subjects: normal healthy males and females, general population. Additional comments: Analytes to measure: fluconazole in plasma. BE based on (90% CI): fluconazole. Waiver request of in vivo testing: 50, 100, and 150 mg based on (i) acceptable BE study on the 200 mg strength, (ii) proportional similarity of the formulations across all strengths, and (iii) acceptable in vitro dissolution testing of all strengths.

Fluoxetine Hydrochloride; Olanzapine, Capsules/Oral. Recommended studies: two studies. (1) Type of study: fasting design—single-dose, two-way crossover in vivo. Strength: 50 mg/6 mg. Subjects: normal healthy males and females, general population. Additional comments: (2) Type of study: fed design—single-dose, two-way crossover in vivo. Strength: 50 mg/6 mg. Subjects: normal healthy males and females, general population. Additional comments: Analytes to measure: fluoxetine and olanzapine in plasma. BE based on (90% CI): fluoxetine and olanzapine. Waiver request of in vivo testing: 25 mg/3 mg, 25 mg/6 mg, 25 mg/12 mg, and 50 mg/12 mg based on (i) acceptable BE studies on the 50 mg/6 mg strength, (ii) proportional similarity across all strengths, and (iii) acceptable in vitro dissolution testing of all strengths.

Fluvastatin Sodium, Capsules/Oral. Recommended studies: two studies. (1) Type of study: fasting design—single-dose, two-way crossover in vivo. Strength: 40 mg. Subjects: normal healthy males and females, general population. Additional comments: Due to teratogenicity concerns, female subjects enrolled in these studies should not be pregnant. (2) Type of study: fed design—single-dose, two-way crossover in vivo. Strength: 40 mg. Subjects: normal healthy males and females, general population. Additional comments: Please see previous comment. Analytes to measure (in appropriate biological fluid): fluvastatin in plasma (achiral assay). BE based on (90% CI): fluvastatin. Waiver request of in vivo testing: 20 mg based on (i) acceptable BE studies on the 40 mg strength, (ii) proportional similarity of the formulations across all strengths, and (iii) acceptable in vitro dissolution testing of all strengths.

Fluvastatin Sodium, Extended-Release Tablets/Oral. Recommended studies: two studies. (1) Type of study: fasting design—single-dose, two-way crossover in vivo Strength: 80 mg. Subjects: normal healthy males and females, general population. Additional comments: Due to the teratogenicity concerns with fluvastatin sodium, female subjects enrolled in these studies should not be pregnant. (2) Type of study: fed design—single-dose, two-way, crossover in vivo. Strength: 80 mg. Subjects: normal healthy males and females, general population. Additional comments: Please see comments earlier. Analytes to measure (in appropriate biological fluid): fluvastatin in plasma (achiral assay). BE based on (90% CI): fluvastatin. Waiver request of in vivo testing: not applicable. For modified-release products, dissolution profiles generated using USP Apparatus I at 100 rpm and/or Apparatus II at 50 rpm in at least three dissolution media (pH 1.2, 4.5, and 6.8 buffer, water) should be submitted in the application. Agitation speeds may have to be increased if appropriate. It is acceptable to add a small amount of surfactant, if necessary. The following sampling times are recommended: 1, 2, and 4 h and every 2 h thereafter, until at least 80% of the drug is dissolved. Comparative dissolution profiles should include individual tablet data as well as the mean, range, and standard deviation at each time point for 12 tablets.

Fosamprenavir Calcium, Suspension/Oral. Recommended studies: two studies. (1) Type of study: fasting design—single-dose, two-treatment, two-period crossover in vivo. Strength: EQ 50 mg base/mL (dose = 28 mL corresponding to a dose of 1400 mg). Subjects: normal healthy males and females, general population. Additional comments: Females should not be pregnant or lactating and, if applicable, should practice abstention or contraception during the study. Bottle should be shaken well before drug administration. (2) Type of study: fed design—single-dose, two-treatment, two-period crossover in vivo. Strength: EQ 50 mg base/mL (dose = 28 mL corresponding to a dose of 1400 mg). Subjects: normal healthy males and females, general population. Additional comments: Please see previous comment. Analytes to measure (in appropriate biological fluid): amprenavir, the active metabolite of fosamprenavir, in plasma. BE based on (90% CI): amprenavir. Waiver request of in vivo testing: not applicable.

Fosamprenavir Calcium, Tablets/Oral. Recommended studies: two studies. (1) Type of study: fasting design—single-dose, two-way crossover in vivo. Strength: 700 mg. Subjects: normal healthy males and females, general population. Additional comments: (2) Type of study: fed design—single-dose, two-way crossover in vivo. Strength: 700 mg. Subjects: normal healthy males and females, general population. Additional comments: Analytes to measure (in appropriate biological fluid): amprenavir (not the prodrug fosamprenavir) in plasma. BE based on (90% CI): amprenavir. Waiver request of in vivo testing: not applicable.

Fosinopril Sodium, Tablets/Oral. Recommended studies: two studies. (1) Type of study: fasting design—single-dose, two-way crossover in vivo. Strength: 40 mg. Subjects: normal healthy males and females, general population. Additional comments: Females should not be pregnant and, if applicable, should practice abstention or contraception during the study. (2) Type of study: fed design—single-dose, two-way crossover in vivo. Strength: 40 mg. Subjects: normal healthy males and females, general population. Additional comments: Please see previous comment. Analytes to measure: metabolite fosinoprilat in plasma. BE based on (90% CI): metabolite fosinoprilat. Waiver request of in vivo testing: 10 and 20 mg based on (i) acceptable BE studies on the 40 mg strength, (ii) proportional similarity across all strengths, and (iii) acceptable in vitro dissolution testing of all strengths.

Fosinopril Sodium; Hydrochlorothiazide, Tablets/Oral. Recommended studies: two studies. (1) Type of study: fasting design—single-dose, two-way crossover in vivo. Strength: 20 mg/12.5 mg. Subjects: normal healthy males and females, general population. Additional comments: Female subjects should be excluded from the studies if they are pregnant. (2) Type of study: fed design—single-dose, two-way crossover in vivo. Strength: 20 mg/12.5 mg. Subjects: normal healthy males and females, general population. Additional comments: Female subjects should be excluded from the studies if they are pregnant. Analytes to measure (in appropriate biological fluid): the metabolite of fosinopril, fosinoprilat, and hydrochlorothiazide in plasma. BE based on (90% CI): fosinoprilat and hydrochlorothiazide. Waiver request of in vivo testing: 10 mg/12.5 mg based on (i) acceptable BE studies on the 20 mg/12.5 mg strength, (ii) proportional similarity of the formulations across all strengths, and (iii) acceptable in vitro dissolution testing of all strengths.

Gabapentin, Capsule/Oral. Recommended studies: two studies. (1) Type of study: fasting design—single-dose, two-way crossover in vivo. Strength: 400 mg. Subjects: normal healthy males and females, general population. Additional comments: (2) Type of study: fed design—single-dose, two-way crossover in vivo. Strength: 400 mg. Subjects: normal healthy males and females, general population. Additional comments: Analytes to measure: gabapentin in plasma. BE based on (90% CI): gabapentin. Waiver request of in vivo testing: 100 and 300 mg based on (i) acceptable BE studies on the 400 mg strength, (ii) proportional similarity of the formulations across all strengths, and (iii) acceptable in vitro dissolution testing of all strengths.

Gabapentin, Tablets/Oral. Recommended studies: two studies. (1) Type of study: fasting design—single-dose, two-way, crossover in vivo. Strength: 800 mg. Subjects: normal healthy males and females, general population. Additional comments: (2) Type of study: fed design—single-dose, two-way, crossover in vivo. Strength: 800 mg. Subjects: normal healthy males and females, general population. Additional comments: Analytes to measure: gabapentin in plasma. BE based on (90% CI): gabapentin. Waiver request of in vivo testing: 100, 300, 400, and 600 mg based on (i) acceptable BE studies on the 800 mg strength, (ii) proportional similarity across all strengths, and (iii) acceptable in vitro dissolution testing of all strengths.

Galantamine Hydrobromide, Extended-Release Capsule/Oral. Recommended studies: two studies. (1) Type of study: fasting design—single-dose, two-way crossover in vivo. Strength: 8 mg. Subjects: normal healthy males and females, general population. Females should not be pregnant and, if applicable, should practice abstinence or contraception during the study. Additional comments: The most frequent adverse events leading to drug discontinuation are nausea, vomiting, dizziness, and syncope. Please include appropriate safety precautions in your protocols. These include adequate monitoring of vital signs and adverse events, stopping criteria in the event of an unacceptable degree of hypotension or bradycardia and appropriate evaluation and management of adverse events. Please assure that the investigator(s) will be vigilant in recognizing and managing any unacceptable clinical or laboratory findings. (2) Type of study: fed design—single-dose, two-way crossover in vivo. Strength: 8 mg. Subjects: normal healthy males and females, general population. Females should not be pregnant and, if applicable, should practice abstinence or contraception during the study. Additional comments: Please see comments earlier. Analytes to measure (in appropriate biological fluid): galantamine in plasma. BE based on (90% CI): galantamine. Waiver request of in vivo testing: 16 and 24 mg based on (i) acceptable BE studies on the 8 mg strength, (ii) proportional similarity across all strengths, and (iii) acceptable in vitro dissolution testing of all strengths.

In addition to the method earlier, for modified-release products, dissolution profiles on 12 dosage units each of test and reference products generated using USP Apparatus I at 100 rpm and/or Apparatus II at 50 rpm in at least three dissolution media (pH 1.2, 4.5, and 6.8 buffer) should be submitted in the application. Agitation speeds may have to be increased if appropriate. It is acceptable to add a small amount of surfactant, if necessary. Please include early sampling times of 1, 2, and 4 h and continue every 2 h until at least 80% of the drug is released, to provide assurance against premature release of drug (dose dumping) from the formulation.

Ganciclovir, Capsules/Oral. Recommended studies: two studies. (1) Type of study: fasting design—single-dose, two-way crossover in vivo. Strength: 500 mg. Subjects: due to safety concerns with the use of healthy subjects, the study population should be patients with advanced HIV+ infection, who are at risk for developing cytomegalovirus disease. Additional comments: (2) Type of study: fed design—single-dose, two-way crossover in vivo. Strength: 500 mg. Subjects: there are safety concerns with using healthy subjects. Therefore, the study population should be patients with advance HIV+ infection who are at risk for developing cytomegalovirus disease. Additional comments: Analytes to measure: ganciclovir in plasma. BE based on (90% CI): ganciclovir. Waiver request of in vivo testing: 250 mg based on (i) acceptable BE studies on the 500 mg strength, (ii) proportional similarity of the formulations across all strengths, and (iii) acceptable in vitro dissolution testing of all strengths.

Gemifloxacin Mesylate, Tablets/Oral. Recommended studies: two studies. (1) Type of study: fasting design—single-dose, two-way crossover in vivo. Strength: 320 mg (base equivalent). Subjects: normal healthy males and females, general population. Females should not be pregnant and, if applicable, should practice abstention or contraception during the study. Additional comments: Females should not be lactating. Subjects should not have a history of prolongation of the QTc interval or ongoing proarrhythmic conditions such as clinically significant bradycardia or acute myocardial ischemia. (2) Type of study: fed design—single-dose, two-way crossover in vivo. Strength: 320 mg. Subjects: normal healthy males and females, general population. Females should not be pregnant and, if applicable, should practice abstention or contraception during the study. Additional comments: Females should not be lactating. Subjects should not have a history of prolongation of the QTc interval or ongoing proarrhythmic conditions such as clinically significant bradycardia or acute myocardial ischemia. Analytes to measure (in appropriate biological fluid): gemifloxacin in plasma. BE based on (90% CI): gemifloxacin. Waiver request of in vivo testing: not applicable.

Glimepiride, Tablets/Oral. Recommended studies: two studies. (1) Type of study: fasting design—single-dose, two-way crossover in vivo. Strength: 1 mg. Subjects: normal healthy males and females, general population. Additional comments: Because of the potential for hypoglycemia from using a dose of 4 mg of glimepiride tablets, you should conduct the BE studies using the 1 mg dose. Each dose in the studies should be administered with 240 mL of 20% glucose solution to minimize hypoglycemic effects. After dosing, 60 mL of 20% glucose solution should be given to each subject every 15 min for the following 4 h. (2) Type of study: fed design—single-dose, two-way crossover in vivo. Strength: 1 mg. Subjects: normal healthy males and females, general population. Additional comments: Please see previous comment. Analytes to measure: glimepiride in plasma. BE based on (90% CI): glimepiride. Waiver request of in vivo testing: 2 and 4 mg based on (i) acceptable BE studies on the 1 mg strength, (ii) proportional similarity across all strengths, and (iii) acceptable in vitro dissolution testing of all strengths.

Glimepiride/Rosiglitazone Maleate, Tablets/Oral. Recommended studies: two studies. (1) Type of study: fasting design—single-dose, two-treatment, two-period crossover in vivo. Strength: 1 mg/4 mg. Subjects: normal healthy males and females, general population. Females must have a negative baseline pregnancy test within 24 h prior to receiving the drug. Females should not be pregnant or lactating and, if applicable, should practice abstention or contraception during the study. Additional comments: Because of the potential for hypoglycemia from BE studies using the 4 mg dose of glimepiride tablets, in vivo BE study of the 1 mg glimepiride/4 mg rosiglitazone maleate tablets is recommended. In addition, each dose in the study should be administered with 240 mL of 20% glucose solution to minimize hypoglycemic effects. After dosing, 60 mL of 20% glucose solution should be given to each subject every 15 min for the following 4 h. (2) Type of study: fed design—single-dose, two-treatment, two-period crossover in vivo. Strength: 1 mg/4 mg. Subjects: normal healthy males and females, general population. Females must have a negative baseline pregnancy test within 24 h prior to receiving the drug. Females should not be pregnant or lactating and, if applicable, should practice abstention or contraception during the study. Additional comments: Please see previous additional comments. Analytes to measure (in appropriate biological fluid): glimepiride and rosiglitazone in plasma. BE based on (90% CI): glimepiride and rosiglitazone. Waiver request of in vivo testing: 2 mg/4 mg, 4 mg/4 mg, 2 mg/8 mg, and 4 mg/8 mg tablets, based on (i) acceptable BE studies on the 1 mg/4 mg tablet, (ii) proportional similarity of the formulations, and (iii) acceptable in vitro dissolution testing of all strengths.

Glipizide; Metformin Hydrochloride, Tablet/Oral. Recommended studies: two studies. (1) Type of study: fasting design—single-dose, two-treatment, two-period crossover in vivo. Strength: 5 mg/500 mg. Subjects: normal healthy males and females, general population. Additional comments: Since the drug product causes hypoglycemia, it is recommended that subjects receive 60 mL of 20% glucose solution in water after each dose and every 15 min for 4 h during fasting and fed BE studies. (2) Type of study: fed design—single-dose, two-treatment, two-period crossover in vivo. Strength: 5 mg/500 mg. Subjects: normal healthy males and females, general population. Additional comments: Please see previous comment. Analytes to measure: glipizide and metformin in plasma. BE based on (90% CI): glipizide and metformin. Waiver request of in vivo testing: 2.5 mg/250 mg and 2.5 mg/500 mg based on (i) acceptable BE studies on the 5 mg/500 mg strength, (ii) proportional similarity of the formulations across all strengths, and (iii) acceptable in vitro dissolution testing of all strengths.

Glyburide; Metformin Hydrochloride, Tablets/Oral. Recommended studies: two studies. (1) Type of study: fasting design—single-dose, two-way crossover in vivo. Strength: 5 mg/500 mg. Subjects: normal healthy males and females, general population. Additional comments: The drug products should be administered with 240 mL of 20% glucose solution in water, followed by 60 mL of the glucose solution administered every 15 min for up to 4 h after dosing. (2) Type of study: fed design—single-dose, two-way crossover in vivo. Strength: 5 mg/500 mg. Subjects: normal healthy males and females, general population. Additional comments: The drug products should be administered with 240 mL of 20% glucose solution in water, followed by 60 mL of the glucose solution administered every 15 min for up to 4 h after dosing. Analytes to measure: glyburide and metformin. BE based on (90% CI): glyburide and metformin. Waiver request of in vivo testing: 1.25 mg/250 mg and 2.5 mg/500 mg based on (i) acceptable BE studies on the 5 mg/500 mg strength, (ii) proportional similarity across all strengths, and (iii) acceptable in vitro dissolution testing of all strengths.

Granisetron Hydrochloride, Tablets/Oral. Recommended studies: two studies. (1) Type of study: fasting design—single-dose, two-way crossover in vivo. Strength: 1 mg Subjects: normal healthy males and females, general population. Additional comments: (2) Type of study: fed design—single-dose, two-way crossover in vivo. Strength: 1 mg. Subjects: normal healthy males and females, general population. Additional comments: Analytes to measure: granisetron in plasma. BE based on (90% CI): granisetron. Waiver request of in vivo testing: not applicable.

Hydrochlorothiazide, Capsule/Oral. Recommended studies: two studies. (1) Type of study: fasting design—single-dose, two-way crossover in vivo. Strength: 12.5 mg. Subjects: normal healthy males and females, general population. Additional comments: (2) Type of study: fed design—single-dose, two-way crossover in vivo. Strength: 12.5 mg. Subjects: normal healthy males and females, general population. Additional comments: Analytes to measure: hydrochlorothiazide in plasma. BE based on (90% CI): hydrochlorothiazide. Waiver request of in vivo testing: not applicable.

Hydrochlorothiazide; Irbesartan, Tablets/Oral. Recommended studies: two studies. (1) Type of study: fasting design—single-dose, two-way crossover in vivo. Strength: 25 mg/300 mg. Subjects: normal healthy males and females, general population. Additional comments: Females should not be pregnant and, if applicable, should practice abstention or contraception during the study.

(2) Type of study: fed design—single-dose, two-way crossover in vivo. Strength: 25 mg/300 mg. Subjects: normal healthy males and females, general population. Additional comments: Please see previous comment. Analytes to measure: hydrochlorothiazide and irbesartan in plasma. BE based on (90% CI): hydrochlorothiazide and irbesartan. Waiver request of in vivo testing: 12.5 mg/150 mg and 12.5 mg/300 mg based on (i) acceptable BE studies on the 25 mg/300mg strength, (ii) proportional similarity across all strengths, and (iii) acceptable in vitro dissolution testing of all strengths.

Hydrochlorothiazide; Losartan Potassium, Tablets/Oral. Recommended studies: two studies. (1) Type of study: fasting design—single-dose, two-way crossover in vivo. Strength: 25 mg/100 mg. Subjects: normal healthy males and females, general population. Additional comments: Females should not be pregnant and, if applicable, should practice abstention or contraception during the study. (2) Type of study: fed design—single-dose, two-way crossover in vivo. Strength: 25 mg/100 mg. Subjects: normal healthy males and females, general population. Additional comments: Please see previous comments. Analytes to measure: hydrochlorothiazide, losartan, and its carboxylic metabolite in plasma. For the carboxylic acid metabolite, the following data should be submitted: (1) individual and mean concentration, (2) individual and mean PK parameters, and (3) geometric means and ratios of means for AUC and C_{max}. BE based on (90% CI): hydrochlorothiazide and losartan. Waiver request of in vivo testing: 12.5 mg/50 mg and 12.5 mg/100 mg based on (i) acceptable BE studies on the 25 mg/100 mg strength, (ii) proportional similarity across all strengths, and (iii) acceptable in vitro dissolution testing of all strengths.

Hydrochlorothiazide; Olmesartan Medoxomil, Tablets/Oral. Recommended studies: two studies. (1) Type of study: fasting design—single-dose, two-way crossover in vivo. Strength: 25 mg/40 mg. Subjects: normal healthy males and females, general population. Females should not be pregnant and, if applicable, should practice abstention or contraception during the study. Additional comments: The labeling for this drug contains a black box regarding pregnancy and fetal/neonatal morbidity and mortality. (2) Type of study: fed design—single-dose, two-way crossover in vivo. Strength: 25 mg/40mg. Subjects: normal healthy males and females, general population. Females should not be pregnant and, if applicable, should practice abstention or contraception during the study. Additional comments: Please see previous comment. Analytes to measure (in appropriate biological fluid): hydrochlorothiazide and olmesartan in plasma. BE based on (90% CI): hydrochlorothiazide and olmesartan. Waiver request of in vivo testing: 12.5 mg/40 mg and 12.5 mg/20 mg strengths based on (i) acceptable BE studies on the 25 mg/40 mg strength, (ii) proportional similarity of the formulations across all strengths, and (iii) acceptable in vitro dissolution testing of all strengths.

Hydrochlorothiazide; Valsartan, Tablet/Oral. Recommended studies: two studies. (1) Type of study: fasting design—randomized, single-dose, two-treatment, two-period crossover in vivo. Strength: 25 mg/160 mg. Subjects: normal healthy males and females, general population. Additional comments: Females should not be pregnant and, if applicable, should practice abstention or contraception during the study. (2) Type of study: fed design—randomized, single-dose, two-treatment, two-period crossover in vivo. Strength: 25 mg/160 mg. Subjects: normal healthy males and females, general population. Additional comments: Please see comment earlier. Analytes to measure (in appropriate biological fluid): valsartan and hydrochlorothiazide in plasma. BE based on (90% CI): valsartan and hydrochlorothiazide. Waiver request of in vivo testing: 12.5 mg/80 mg and 12.5 mg/160 mg based on (i) acceptable BE studies on the 25 mg/160 mg strength, (ii) proportional similarity of the formulations across all strengths, and (iii) acceptable in vitro dissolution testing of all strengths.

Ibandronate Sodium, Tablets/Oral. Recommended studies: two studies. (1) Type of study: fasting design—single-dose, parallel design, or two-way crossover in vivo. Strength: 2.5 mg. Subjects: normal healthy males and females, general population. Additional comments: Please include as many postmenopausal women as possible in the studies. (2) Type of study: fasting design—single-dose, parallel design, or two-way crossover in vivo. Strength: 150 mg. Subjects: normal healthy males and females, general population. Additional comments: Please include as many postmenopausal women as possible in the studies. Analytes to measure: ibandronate in plasma. BE based on (90% CI): ibandronate. Waiver request of in vivo testing: not applicable.

Ibuprofen and Pseudoephedrine Hydrochloride, Suspension/Oral. Recommended studies: two studies. (1) Type of study: fasting design—single-dose, two-way crossover in vivo. Strength: 100 mg/5 mL and 15 mg/5 mL. Subjects: normal healthy males and females, general population. Additional comments: (2) Type of study: fed design—single-dose, two-way crossover in vivo. Strength: 100 mg/5 mL and 15 mg/5 mL Subjects: normal healthy males and females, general population. Additional comments: Analytes to measure: ibuprofen and pseudoephedrine in plasma. BE based on (90% CI): ibuprofen and pseudoephedrine. Waiver request of in vivo testing: not applicable. A dosage unit for a suspension is the labeled strength (5 mL). A total of 12 units from 12 different bottles should be used. In addition to the method earlier, for modified-release products, dissolution profiles on 12 dosage units each of test and reference products generated using USP Apparatus I at 100 rpm and/or Apparatus II at 50 rpm in at least three dissolution media (pH 1.2, 4.5, and 6.8 buffer) should be submitted in the application. Agitation speeds may have to be increased if appropriate. It is acceptable to add a small amount of surfactant, if necessary. Please include early sampling times of 1, 2, and 4 h and continue every 2 h until at least 80% of the drug is released, to provide assurance against premature release of drug (dose dumping) from the formulation.

Indinavir Sulfate, Capsule/Oral. Recommended studies: two studies. (1) Type of study: fasting design—single-dose, two-treatment, two-period crossover in vivo. Strength: 400 mg. Subjects: normal healthy males and females, general population. Additional comments: (2) Type of study—fed design—single-dose, two-treatment, two-period crossover in vivo. Strength: 400 mg. Subjects: normal healthy males and females, general population. Additional comments: Analytes to measure (in appropriate biological fluid): indinavir in plasma. BE based on (90% CI): indinavir. Waiver request of in vivo testing: 100, 200, and 333 mg based on (i) acceptable BE studies on the 400 mg strength, (ii) proportional similarity of the formulations across all strengths, and (iii) acceptable in vitro dissolution testing of all strengths.

Irbesartan, Tablet/Oral. Recommended studies: two studies. (1) Type of study: fasting design—single-dose, two-treatment, two-period crossover in vivo. Strength: 300 mg. Subjects: normal healthy males and females, general population. Additional comments: Female subjects should be excluded from the BE studies if they are pregnant. (2) Type of study: fed design—single-dose, two-treatment, two-period crossover in vivo. Strength: 300 mg. Subjects: normal healthy males and females, general population. Additional comments: Please see comment earlier. Analytes to measure: irbesartan in plasma: BE based on (90% CI): irbesartan. Waiver request of in vivo testing: 75 and 150 mg based on (i) acceptable BE studies on the 300 mg strength, (ii) proportional similarity of the formulations across all strengths, and (iii) acceptable in vitro dissolution testing of all strengths.

Isosorbide Mononitrate, Extended-Release Tablets/Oral. Recommended studies: two studies. (1) Type of study: fasting design—single-dose, two-way, crossover in vivo. Strength: 120 mg. Subjects: normal healthy males and females, general population. Additional comments: (2) Type of study: fed design—single-dose, two-way, crossover in vivo. Strength: 120 mg. Subjects: normal healthy males and females, general population. Additional comments: Analytes to measure: isosorbide mononitrate in plasma. BE based on (90% CI): isosorbide mononitrate. Waiver request of in vivo testing: 30 and 60 mg based on (i) acceptable BE studies on the 120 mg strength, (ii) proportional similarity across all strengths, and (iii) acceptable in vitro dissolution testing of all strengths. In addition to the method earlier, for modified-release products, dissolution profiles on 12 dosage units each of test and reference products generated using USP Apparatus I at 100 rpm and/or Apparatus II at 50 rpm in at least three dissolution media (pH 1.2, 4.5, and 6.8 buffer) should be submitted in the application. Agitation speeds may have to be increased if appropriate. It is acceptable to add a small amount of surfactant, if necessary. Please include early sampling times of 1, 2, and 4 h and continue every 2 h until at least 80% of the drug is released, to provide assurance against premature release of drug (dose dumping) from the formulation.

Isradipine, Capsule/Oral. Recommended studies: two studies. (1) Type of study: fasting design—single-dose, two-way crossover in vivo. Strength: 5 mg. Subjects: normal healthy males and females, general population. Additional comments: (2) Type of study: fed design—single-dose, two-way crossover in vivo. Strength: 5 mg. Subjects: normal healthy males and females, general population. Additional comments: Analytes to measure (in appropriate biological fluid): isradipine in plasma. BE based on (90% CI): isradipine. Waiver request of in vivo testing: 2.5 mg based on (i) acceptable BE studies on the 5 mg strength, (ii) proportional similarity of the formulations across all strengths, and (iii) acceptable in vitro dissolution testing of all strengths.

Isradipine, Extended-Release Tablets/Oral. Recommended studies: two studies. (1) Type of study: fasting design—single-dose, two-way, crossover in vivo. Strength: 10 mg. Subjects: normal healthy males and females, general population. Additional comments: (2) Type of study: fed design—single-dose, two-way, crossover in vivo. Strength: 10 mg. Subjects: normal healthy males and females, general population. Additional comments: Analytes to measure (in appropriate biological fluid): isradipine in plasma. BE based on (90% CI): isradipine. Waiver request of in vivo testing: 5 mg based on (i) acceptable BE studies on the 10 mg strength, (ii) proportional similarity across all strengths, and (iii) acceptable in vitro dissolution testing of all strengths. In addition to the method earlier, for modified-release products, dissolution profiles on 12 dosage units each of test and reference products generated using USP Apparatus I at 100 rpm and/or Apparatus II at 50 rpm in at least three dissolution media (pH 1.2, 4.5, and 6.8 buffer) should be submitted in the application. Agitation speeds may have to be increased if appropriate. It is acceptable to add a small amount of surfactant, if necessary. Please include early sampling times of 1, 2, and 4 h and continue every 2 h until at least 80% of the drug is released, to provide assurance against premature release of drug (dose dumping) from the formulation.

Itraconazole, Capsule/Oral. Recommended studies: two studies. (1) Type of study: fasting design—single-dose, two-treatment, two-period crossover in vivo. Strength: 100 mg. Subjects: normal healthy males and females, general population. Additional comments: (2) Type of study: fed design—single-dose, two-treatment, two-period crossover in vivo. Strength: 100 mg. Subjects: normal healthy males and females, general population. Additional comments: Analytes to measure (in appropriate biological fluid): itraconazole and its active metabolite, hydroxyitraconazole, in plasma. BE based on (90% CI): itraconazole. Waiver request of in vivo testing: not applicable.

Lamivudine, Tablets/Oral. Recommended studies: two studies. (1) Type of study: fasting design—single-dose, two-treatment, two-period crossover in vivo. Strength: 300 mg. Subjects: normal healthy males and females, general population. Additional comments: (2) Type of study: fed design—single-dose, two-treatment, two-period crossover in vivo. Strength: 300 mg. Subjects: normal healthy males and females, general population. Additional comments: Analytes to measure (in appropriate biological fluid): lamivudine in plasma. BE based on (90% CI): lamivudine. Waiver request of in vivo testing: 150 mg based on (i) acceptable BE studies on the 300 mg strength, (ii) proportional similarity of the formulations across all strengths, and (iii) acceptable in vitro dissolution testing of all strengths.

Lamivudine, Tablets/Oral. Recommended studies: two studies. (1) Type of study: fasting design—single-dose, two-treatment, two-period crossover in vivo. Strength: 100 mg. Subjects: normal healthy males and females, general population. Additional comments: (2) Type of study: fed design—single-dose, two-treatment, two-period crossover in vivo. Strength: 100 mg. Subjects: normal healthy males and females, general population. Additional comments: Analytes to measure (in appropriate biological fluid): lamivudine in plasma. BE based on (90% CI): lamivudine. Waiver request of in vivo testing: not applicable.

Lamivudine; Zidovudine, Tablet/Oral. Recommended studies: two studies. (1) Type of study: fasting design—single-dose, two-treatment, two-period crossover in vivo. Strength: 150 mg/300 mg. Subjects: normal healthy males and females, general population. Additional comments: (2) Type of study: fed design—single-dose, two-treatment, two-period crossover in vivo.

Strength: 150 mg/300 mg. Subjects: normal healthy males and females, general population. Additional comments: Analytes to measure: lamivudine and zidovudine in plasma. BE based on (90% CI): lamivudine and zidovudine. Waiver request of in vivo testing: not applicable.

Lamotrigine, Chewable Dispersible Tablet/Oral. Recommended studies: two studies. (1) Type of study: fasting design—single-dose, two-treatment, two-period crossover in vivo. Strength—single dose of 50 mg (2 × 25 mg). Subjects: normal healthy males and females, general population. Additional comments: (2) Type of study: fed design—single-dose, two-treatment, two-period crossover in vivo. Strength: single dose of 50 mg (2 × 25 mg). Subjects: normal healthy males and females, general population. Additional comments: Analytes to measure (in appropriate biological fluid): lamotrigine in plasma. Please utilize a validated analytical method such as LC–MS/MS to reliably measure plasma lamotrigine concentrations. An LLOQ of 10 ng/mL is recommended to adequately characterize the PKs at 50 mg study dose. BE based on (90% CI): lamotrigine. Waiver request of in vivo testing: 2 and 5 mg based on (i) acceptable BE studies on the 25 mg strength, (ii) proportional similarity of the formulations across all strengths, and (iii) acceptable in vitro dissolution testing of all strengths.

Lamotrigine, Tablet/Oral. Recommended studies: two studies. (1) Type of study: fasting design—single-dose, two-treatment, two-period crossover in vivo. Strength: single dose of 50 mg (2 × 25 mg). Subjects: normal healthy males and females, general population. Additional comments: Due to safety concerns, studies on the highest strength are not recommended. (2) Type of study: fed design—single-dose, two-treatment, two-period crossover in vivo. Strength: single dose of 50 mg (2 × 25 mg). Subjects: normal healthy males and females, general population. Additional comments: See previous comment. Analytes to measure (in appropriate biological fluid): lamotrigine in plasma. Please utilize a validated analytical method such as LC–MS/MS to reliably measure plasma lamotrigine concentrations. An LLOQ of 10 ng/mL is recommended to adequately characterize the PKs at 50 mg study dose. BE based on (90% CI): lamotrigine. Waiver request of in vivo testing: 100, 150, and 200 mg based on (i) acceptable BE studies on the 25 mg strength, (ii) proportional similarity of the formulations across all strengths, and (iii) acceptable in vitro dissolution testing of all strengths. Please find the dissolution information for this product at this website and conduct comparative dissolution testing on 12 dosage units of all strengths of the test and reference products.

Leflunomide, Tablets/Oral. Recommended studies: three studies. (1) Type of study: fasting design—single-dose, two-way crossover in vivo. Strength: 100 mg. Subjects: normal healthy males and females, general population. Additional comments: (2) Type of study: fasting design—single-dose, two-way crossover in vivo. Strength: 20 mg. Subjects: normal healthy males and females, general population. Additional comments: Only female subjects who are unable to bear children should be included in the study, and male subjects wishing to father a child during the study should be excluded from the study. Since the half-life of the metabolite A77 1726 is very long, you may consider BE studies with parallel designs. (3) Type of study: fed design—single-dose, two-way crossover in vivo. Strength: 20 mg. Subjects: normal healthy males and females, general population. Additional comments: Please see previous comment. Analytes to measure: Please measure only the leflunomide's metabolite, A77 1726, in plasma. BE based on (90% CI): the metabolite of leflunomide, A77 1726. Waiver request of in vivo testing: 10 mg based on (i) acceptable BE studies on the 20 mg strength, (ii) proportional similarity across all strengths, and (iii) acceptable in vitro dissolution testing of all strengths.

Levonorgestrel, Tablets/Oral. Recommended studies: one study. Type of study: fasting design—single-dose, two-way crossover in vivo. Strength: 0.75 mg. Subjects: normal healthy females, general population. Additional comments: Analytes to measure: levonorgestrel in plasma. BE based on (90% CI): levonorgestrel. Waiver request of in vivo testing: not applicable.

Lidocaine, Patch/Topical. Recommended studies: two studies. (1) Type of study: fasting design—single dose, in vivo, using three topical patches. Strength: 5%, 700 mg/patch. Subjects:

normal healthy males and females, general population. Additional comments: Apply three topical patches (2100 mg total dose) simultaneously over a 12 h period. You may use a smaller number of patches provided the plasma concentrations of lidocaine are measurable to adequately characterize the PK profile of lidocaine for BE assessment based on the 90% CI criteria. Please include a 24 h post dose sampling time in the BE study. In addition to PK data, please report the "apparent dose" delivered. The apparent dose can be determined by subtracting the remaining amount of lidocaine in each patch (used patch) from the manufactured amount. The amount of adhesive residue from each patch left on the skin should be analyzed and included in the calculation. Analytes to measure: lidocaine in plasma. Please utilize a validated analytical method such as LC–MS/MS to reliably measure plasma lidocaine concentrations. An LLOQ of 0.20 ng/mL is recommended to adequately characterize the PKs at the 2100 mg study dose. BE based on (90% CI): lidocaine. (2) Type of study: skin irritation/sensitization study design—single dose, in vivo (preceded by an induction phase and a rest period). Strength: 5%, 700 mg/patch. Subjects: normal healthy males and females, general population. Additional comments: Specific recommendations are provided later for the skin irritation/sensitization/adhesion study. General comments: Please note that the name of RLD is designated as lidocaine topical patch, 5%. This designation is based on the concentration of lidocaine in the adhesive, which is 5%. Please formulate your product to contain 5% of lidocaine in the adhesive, to have the same surface area and the same total amount of lidocaine in the patch as the RLD.

Linezolid, Tablets/Oral. Recommended studies: two studies. (1) Type of study: fasting design—single-dose, two-treatment, two-period crossover in vivo. Strength: 600 mg. Subjects: normal healthy males and females, general population. Additional comments: (2) Type of study: fed design—single-dose, two-treatment, two-period crossover in vivo. Strength: 600 mg. Subjects: normal healthy males and females, general population. Additional comments: Analytes to measure (in appropriate biological fluid): linezolid in plasma. BE based on (90% CI): linezolid. Waiver request of in vivo testing: not applicable.

Liothyronine Sodium, Tablets/Oral. Recommended studies: one study. (1) Type of study: fasting design—single-dose, two-way crossover in vivo. Dose and strength: 100 mcg (2 × 50 mcg). Subjects: normal healthy males and females, general population. Additional comments: Baseline levels of liothyronine should be measured at 3 predose time points (–30, –15, 0 min). The mean of the three predose samples should be subtracted from each measured postdose concentration. Analytes to measure (in appropriate biological fluid): total (free + bound) liothyronine in plasma. BE based on (90% CI): total (free + bound) liothyronine in plasma after baseline correction. Waiver request of in vivo testing: 25 and 5 mcg based on (i) acceptable BE studies on the 50 mcg strength, (ii) proportional similarity of the formulations across all strengths, and (iii) acceptable in vitro dissolution testing of all strengths. Please conduct comparative dissolution testing on 12 dosage units of all strengths of the test and reference products.

Lisinopril, Tablets/Oral. Recommended studies: two studies. (1) Type of study: fasting design—single-dose, two-treatment, two-period crossover in vivo. Strength: 40 mg. Subjects: normal healthy males and females, general population. Additional comments: Females must have a negative baseline pregnancy test within 24 h prior to receiving the drug. Females should not be pregnant or lactating and, if applicable, should practice abstention or contraception during the study. (2) Type of study: fed design—single-dose, two-treatment, two-period crossover in vivo. Strength: 40 mg. Subjects: normal healthy males and females, general population. Additional comments: Please see comments earlier. Analytes to measure (in appropriate biological fluid): lisinopril in plasma. BE based on (90% CI): lisinopril. Waiver request of in vivo testing: 2.5, 5, 10, 20, and 30 mg, based on (i) acceptable BE studies on the 40 mg strength, (ii) proportional similarity of the formulations, and (iii) acceptable in vitro dissolution testing of all strengths.

Lisinopril; Hydrochlorothiazide, Tablet/Oral. Recommended studies: two studies. (1) Type of study: fasting design—single-dose, two-treatment, two-period crossover in vivo. Strength: 25 mg/20 mg. Subjects: normal healthy males and females, general population. Additional comments: Female subjects enrolled in the BE studies should not be pregnant and, if applicable, should practice abstention or contraception during the study. (2) Type of study: fed design—single-dose, two-treatment, two-period crossover in vivo. Strength: 25 mg/20 mg. Subjects: normal healthy males and females, general population. Additional comments: Please see previous comment. Analytes to measure: lisinopril and hydrochlorothiazide in plasma. BE based on (90% CI): lisinopril and hydrochlorothiazide. Waiver request of in vivo testing: 12.5 mg/10 mg and 12.5 mg/20 mg based on (i) acceptable BE studies on the 25 mg/20 mg strength, (ii) proportional similarity of the formulations across all strengths, and (iii) acceptable in vitro dissolution testing of all strengths.

Lopinavir; Ritonavir, Tablets/Oral. Recommended studies: two studies. (1) Type of study: fasting design—single-dose, two-treatment, two-period crossover in vivo. Strength: 200 mg/50 mg (400 mg/100 mg dose). Subjects: normal healthy males and females, general population. Additional comments: Pregnant and lactating women should be excluded from participation in studies. Women must have a negative baseline pregnancy test prior to receiving the drug. (2) Type of study: fed design—single-dose, two-treatment, two-period crossover in vivo. Strength: 200 mg/50 mg (400 mg/100 mg dose). Subjects: normal healthy males and females, general population. Additional comments: Please see previous comments. Analytes to measure (in appropriate biological fluid): lopinavir and ritonavir in plasma. BE based on (90% CI): lopinavir and ritonavir. Waiver request of in vivo testing: not applicable.

Loratadine, Orally Disintegrating Tablets/Oral. Recommended studies: two studies. (1) Type of study: fasting design—single-dose, two-way crossover in vivo. Strength: 10 mg. Subjects: normal healthy males and females, general population. Additional comments: (2) Type of study: fed design—single-dose, two-way crossover in vivo. Strength: 10 mg. Subjects: normal healthy males and females, general population. Additional comments: Analytes to measure: loratadine and the active metabolite, descarboethoxyloratadine, in plasma. Please submit the metabolite data as supportive evidence of the comparable therapeutic outcome. For the metabolite, the following data should be submitted: individual and mean concentrations, individual and mean PK parameters, and geometric means and ratios of means for AUC and C_{max}. BE based on (90% CI): loratadine. Waiver request of in vivo testing: not applicable.

Losartan Potassium, Tablet/Oral. Recommended studies: two studies. (1) Type of study: fasting design—single-dose, two-treatment, two-period crossover in vivo. Strength: 100 mg. Subjects: normal healthy males and females, general population. Additional comments: Pregnant women should be excluded from participation in the BE studies. (2) Type of study: fed design—single-dose, two-treatment, two-period crossover in vivo. Strength: 100 mg. Subjects: normal healthy males and females, general population. Additional comments: See previous comment. Analytes to measure (in appropriate biological fluid): losartan and the metabolite carboxylic acid in plasma. BE based on (90% CI): losartan. Please submit the metabolite data as supportive evidence of comparable therapeutic outcome. For the carboxylic acid metabolite, the following data should be submitted: individual and mean concentrations, individual and mean PK parameters, and geometric means and ratios of means for AUC and C_{max}. Waiver request of in vivo testing: 25 and 50 mg based on (i) acceptable BE studies on the 100 mg strength, (ii) proportional similarity of the formulations across all strengths, and (iii) acceptable in vitro dissolution testing of all strengths.

Mefloquine Hydrochloride, Tablets/Oral. Recommended studies: one study. Type of study: fed design—single-dose, parallel design in vivo. Strength: 250 mg. Subjects: normal healthy males and females, general population. Additional comments: Mefloquine has been shown to cause esophagitis/gastritis when administered under fasting conditions. A fasting BE study is not recommended. Analytes to measure: mefloquine in plasma. BE based on (90% CI): mefloquine. Waiver request of in vivo testing: not applicable.

Meloxicam, Suspension/Oral. Recommended studies: two studies. (1) Type of study: fasting design—single-dose, two-way crossover in vivo. Dose and suspension strength: 5 mL of 7.5 mg/5 mL. Subjects: normal healthy males and females, general population. Additional comments: Females should not be pregnant and, if applicable, should practice abstinence or contraception during the study. (2) Type of study: fed design—single-dose, two-way crossover in vivo. Dose and suspension strength: 5 mL of 7.5 mg/5 mL. Subjects: normal healthy males and females, general population. Additional comments: Please see comment earlier. Analytes to measure (in appropriate biological fluid): meloxicam in plasma. BE based on (90% CI): meloxicam. Waiver request of in vivo testing: not applicable.

Meloxicam, Tablets/Oral. Recommended studies: two studies. (1) Type of study: fasting design—single-dose, two-way, crossover in vivo. Strength: 15 mg. Subjects: normal healthy males and females, general population. Additional comments: (2) Type of study: fed design—single-dose, two-way, crossover in vivo. Strength: 15 mg. Subjects: normal healthy males and females, general population. Additional comments: Analytes to measure: meloxicam in plasma. BE based on (90% CI): meloxicam. Waiver request of in vivo testing: 7.5 mg based on (i) acceptable BE studies on the 15 mg strength, (ii) proportional similarity across all strengths, and (iii) acceptable in vitro dissolution testing of all strengths. Memantine Hydrochloride, Tablets/Oral. Recommended studies: two studies. (1) Type of study: fasting design—single-dose, two-way crossover in vivo. Strength: 10 mg. Subjects: normal healthy males and females, general population. Females should not be of childbearing potential. (2) Type of study: fed design—single-dose, two-way crossover in vivo. Strength: 10 mg. Subjects: normal healthy males and females, general population. Females should not be of childbearing potential. Additional comments: Analytes to measure (in appropriate biological fluid): memantine in plasma. BE based on (90% CI): memantine. Waiver request of in vivo testing: 5 mg, based on (i) acceptable BE studies on the 10 mg strength, (ii) formulation proportionality of 10 and 5 mg strengths, and (iii) acceptable dissolution testing on both strengths.

Mercaptopurine, Tablet/Oral. Recommended studies: one study. Submission of an IND application is required prior to the conduct of a BE study for a cytotoxic drug product such as mercaptopurine (see 21 CFR §320.31). Type of study: steady-state study in patients. Strength: 50 mg. Studies may be conducted at steady state in patients receiving therapeutic doses (usually 100–200 mg/day in the average adult) or maintenance daily doses (usually 50–100 mg/day in the average adult). Patients should be in a stable regimen using the same dosage unit (multiples of the same strength). Additional comments: Patients with inherited deficiency of the enzyme thiopurine methyl transferase must be excluded from these studies. The protocol may exclude concomitant chemotherapy and should exclude prior exposure to doxorubicin. The informed consent should include a description of the known genotoxicity of 6-mercaptopurine in human cells and animal models. Analytes to measure (in appropriate biological fluid): mercaptopurine in plasma. BE based on (90% CI): mercaptopurine. Waiver request of in vivo testing: not applicable.

Mesalamine, Enema/Rectal. Recommended studies: one study. The following study is recommended to establish BE of mesalamine rectal enema provided that the test product is qualitatively (Q1) and quantitatively (Q2) the same as the RLD. Type of study: fasting design—single-dose, two-way crossover in vivo or replicate design. Strength: 4 gm/60 mL. Subjects: normal healthy males and females, general population. Additional comments: The proposed generic and RLD formulations should have comparable particle size. Analytes to measure (in appropriate biological fluid: mesalamine (5-ASA) in plasma. BE based on (90% CI): mesalamine (5-ASA). Waiver request of in vivo testing: not applicable. In vitro dissolution testing under the following conditions should be submitted to support documentation of BE: Please conduct comparative dissolution testing on 12 dosage units of the test and reference products using 900 mL of the following media: 0.1 N HCl and buffers at pH 4.5, 6.8, and 7.2 using Apparatus II (paddle) at 25 and 50 rpm. Please ensure that the dissolution method is adequate to distinguish mesalamine dissolved in dissolution media from drug particles. You may modify the filtration method in the dissolution testing, if necessary.

Mesalamine, Suppository/Rectal. Recommended studies: three studies. (1) Type of study: BE study with clinical endpoints. Design: Parallel design, three arms (test, reference, and placebo) in vivo. Strengths: 500 and 1000 mg. Subjects: patients with ulcerative proctitis. Additional comments: Please submit a protocol to the clinical review team for recommendations on study design.

(2) Type of study: BE studies with PK endpoints (fasting) design—single-dose, two-way crossover in vivo. Strengths: 500 and 1000 mg, comparing to the respective strengths of the RLD. Subjects: normal healthy males and females, general population. Additional comments: Because the 500 and 1000 mg strengths are not proportionally similar, a BE study with clinical endpoints and a BE study with PK endpoints (fasting) will be needed for each strength product, if you wish to develop each strength. Analytes to measure (PK study): mesalamine in plasma. BE (PK study) based on (90% CI): mesalamine. Waiver request of in vivo testing: not applicable.

Metaxalone, Tablet/Oral. Recommended studies: two studies. (1) Type of study: fasting design—single-dose, two-treatment, two-period crossover in vivo. Strength: 800 mg. Subjects: normal healthy males and females, general population. Additional comments: (2) Type of study: fed design—single-dose, two-treatment, two-period crossover in vivo. Strength: 800 mg. Subjects: normal healthy males and females, general population. Additional comments: Analytes to measure (in appropriate biological fluid): metaxalone in plasma. BE based on (90% CI): metaxalone. Waiver request of in vivo testing: 400 mg based on (i) acceptable BE studies on the 800 mg strength, (ii) proportional similarity of the formulations across all strengths, and (iii) acceptable in vitro dissolution testing of all strengths. Please note that metaxalone tablets, 400 mg, have been discontinued from the market. If you would like to market the 400 mg strength, please submit a citizen petition pursuant to 21 CFR 314.122, requesting that the FDA determine whether this strength was discontinued due to safety and/or effectiveness reasons. Please follow the citizen petition format outlined in 21 CFR 10.20 and 10.30.

Metformin Hydrochloride, Extended-release Tablet/Oral. Recommended studies: two studies. (1) Type of study: fasting design—single-dose, two-treatment, two-period crossover in vivo. Strength: 750 mg. Subjects: normal healthy males and females, general population. Additional comments: The drug products should be administered with 240 mL of a 20% glucose solution in water, followed by 60 mL of the glucose solution administered every 15 min for up to 4 h after dosing. (2) Type of study: fed design—single-dose, two-treatment, two-period crossover in vivo. Strength: 750 mg. Subjects: normal healthy males and females, general population. Additional comments: Please see previous comment. Analytes to measure (in appropriate biological fluid): metformin in plasma. BE based on (90% CI): metformin. Waiver request of in vivo testing: 500 mg based on (i) acceptable BE studies on the 750 mg strength, (ii) proportional similarity of the formulations across all strengths, and (iii) acceptable in vitro dissolution testing of all strengths.

Metformin Hydrochloride; Pioglitazone Hydrochloride, Tablets/Oral. Recommended studies: two studies. (1) Type of study: fasting design—single-dose, two-way crossover in vivo. Strength: 850 mg metformin HCl and 15 mg pioglitazone HCl (as the base). Subjects: normal healthy males and females, general population. Females must have a negative baseline pregnancy test within 24 h prior to receiving the drug. Females should not be pregnant or lactating and, if applicable, should practice abstention or contraception during the study. Additional comments: To avoid hypoglycemic episodes in healthy volunteers, the drug products should be administered with 240 mL of a 20% glucose solution in water, followed by 60 mL of the glucose solution administered every 15 min for up to 4 h after dosing. (2) Type of study: fed design—single-dose, two-way crossover in vivo. Strength: 850 mg metformin HCl and 15 mg pioglitazone HCl (as the base). Subjects: normal healthy males and females, general population. Females must have a negative baseline pregnancy test within 24 h prior to receiving the drug. Females should not be pregnant or lactating and, if applicable, should practice abstention or contraception during the study. Additional comments: To avoid hypoglycemic episodes in healthy volunteers, the drug products should be administered with 240 mL of a 20% glucose solution in water, followed by 60 mL of the glucose solution administered every 15 min for up to 4 h after dosing. Analytes to measure (in appropriate biological fluid): metformin, pioglitazone, and hydroxypioglitazone (M-IV) in plasma. BE based on (90% CI): metformin and pioglitazone. Waiver request of in vivo testing: (500, 15 mg) metformin HCl and pioglitazone HCl tablets, based on (i) acceptable BE study on the (850, 15 mg) tablet and (ii) acceptable in vitro dissolution testing of all strengths.

Metoprolol Succinate, Extended-Release Tablets/Oral. Recommended studies: three studies. (1) Type of study: fasting design—single-dose, two-way crossover in vivo. Strength: 200 mg. Subjects: normal healthy males and females, general population. Additional comments: (2) Type of study: fed design—single-dose, two-way crossover in vivo. Strength: 200 mg. Subjects: normal healthy males and females, general population. (3) Type of study: fasting design—single-dose, two-way crossover in vivo. Strength: 50 mg. Subjects: normal healthy males and females, general population. Analytes to measure (in appropriate biological fluid): metoprolol in plasma. BE based on (90% CI): metoprolol. Waiver request of in vivo testing: 25 and 100 mg tablets, based on (i) acceptable BE studies on the 50 and 200 mg strengths, (ii) proportional similarity across all strengths, and (iii) acceptable in vitro dissolution testing of all strengths. In addition to the method earlier, for modified-release products, dissolution profiles on 12 dosage units each of test and reference products generated using USP Apparatus I at 100 rpm and/or Apparatus II at 50 rpm in at least three dissolution media (pH 1.2, 4.5, and 6.8 buffer) should be submitted in the application. Agitation speeds may have to be increased if appropriate.

Miglustat, Capsule/Oral. Recommended studies: two studies. (1) Type of study: fasting design—single-dose, two-way crossover in vivo. Strength: 100 mg. Subjects: normal healthy males and females, general population. Females should not be pregnant and, if applicable, should practice abstention or contraception during the study. Additional comments: Pregnancy category X. Miglustat may cause fetal harm when administered to a pregnant woman. The drug is contraindicated in women who are or may become pregnant. (2) Type of study: fed design—single-dose, two-way crossover in vivo. Strength: 100 mg. Subjects: normal healthy males and females, general population. Females should not be pregnant and, if applicable, should practice abstention or contraception during the study. Additional comments: Please see previous comments. Analytes to measure (in appropriate biological fluid): miglustat in plasma. BE based on (90% CI): miglustat. Waiver request of in vivo testing: not applicable.

Minocycline Hydrochloride, Extended-Release Tablets/Oral. Recommended studies: two studies. (1) Type of study: fasting design—single-dose, two-way crossover in vivo. Strength: 135 mg. Subjects: normal healthy males and females, general population. Females should not be pregnant and, if applicable, should practice abstention or contraception during the study. (2) Type of study: fed design—single-dose, two-way crossover in vivo. Strength: 135 mg. Subjects: normal healthy males and females, general population. Females should not be pregnant and, if applicable, should practice abstention or contraception during the study. Analytes to measure (in appropriate biological fluid): minocycline in plasma. BE based on (90% CI): minocycline. Waiver request of in vivo testing: 45 and 90 mg based on (i) acceptable BE studies on the 135 mg strength, (ii) proportional similarity 45 and 90 mg formulations to the 135 mg strength, and (iii) acceptable in vitro dissolution testing of all strengths. In addition to the previous method, for modified-release products, dissolution profiles on 12 dosage units each of test and reference products generated using USP Apparatus I at 100 rpm and/or Apparatus II at 50 rpm in at least three dissolution media (pH 1.2, 4.5, and 6.8 buffer) should be submitted in the application. Agitation speeds may have to be increased if appropriate. It is acceptable to add a small amount of surfactant, if necessary. Please include early sampling times of 1, 2, and 4 h and continue every 2 h until at least 80% of the drug is released, to provide assurance against premature release of drug (dose dumping) from the formulation.

Mirtazapine, Orally Disintegrating Tablets/Oral. Recommended studies: two studies. (1) Type of study: fasting design—single-dose, two-treatment, two-period crossover in vivo. Strength: 15 mg. Subjects: normal healthy males and females, general population. Additional comments: Due to safety concerns, studies on the lower strength are recommended. (2) Type of study: fed design—single-dose, two-treatment, two-period crossover in vivo. Strength: 15 mg. Subjects: normal healthy males and females, general population. Additional comments: Please see previous comment. Analytes to measure: mirtazapine in plasma. BE based on (90% CI): mirtazapine. Waiver request of in vivo testing: 30 and 45 mg based on (i) acceptable BE studies on the 15 mg strength, (ii) proportional similarity of the formulations across all strengths, and (iii) acceptable in vitro dissolution testing of all strengths.

Modafinil, Tablet/Oral. Recommended studies: two studies. (1) Type of study: fasting design—single-dose, two-treatment, two-period crossover in vivo. Strength: 200 mg. Subjects: normal healthy males and females, general population. Additional comments: (2) Type of study: fed design—single-dose, two-treatment, two-period crossover in vivo. Strength: 200 mg. Subjects: normal healthy males and females, general population. Additional comments: Analytes to measure: modafinil using an achiral assay. BE based on (90% CI): modafinil. Waiver request of in vivo testing: 100 mg based on (i) acceptable BE studies on the 200 mg strength, (ii) proportional similarity of the formulations across all strengths, and (iii) acceptable in vitro dissolution testing of all strengths.

Moexipril Hydrochloride, Tablet/Oral. Recommended studies: one study. Type of study: fasting design—single-dose, two-treatment, two-period crossover in vivo. Strength: 15 mg. Subjects: normal healthy males and females, general population. Additional comments: Pregnant women should be excluded from participation in the BE study. Analytes to measure (in appropriate biological fluid): moexipril in plasma. BE based on (90% CI): moexipril. Waiver request of in vivo testing: 7.5 mg based on (i) acceptable BE studies on the 15 mg strength, (ii) proportional similarity of the formulations across all strengths, and (iii) acceptable in vitro dissolution testing of all strengths.

Montelukast Sodium, Chewable Tablet/Oral. Recommended studies: two studies. (1) Type of study: fasting design—single-dose, two-treatment, two-period crossover in vivo. Strength: 5 mg. Subjects: normal healthy males and females, general population. Additional comments: (2) Type of study: fed design—single-dose, two-treatment, two-period crossover in vivo. Strength: 5 mg. Subjects: normal healthy males and females, general population. Additional comments: Analytes to measure (in appropriate biological fluid): montelukast in plasma. BE based on (90% CI): montelukast. Waiver request of in vivo testing: 4 mg based on (i) acceptable BE studies on the 5 mg strength, (ii) proportional similarity of the formulations across all strengths, and (iii) acceptable in vitro dissolution testing of all strengths.

Morphine Sulfate, Extended-Release Capsules/Oral. Recommended studies: three studies. (1) Type of study: fasting design—single-dose, two-way, crossover in vivo. Strength: 100 mg. Subjects: normal healthy males and females, general population. Additional comments: Please use a narcotic antagonist such as naltrexone if the study involves healthy subjects. You should consult a physician who is an expert in the administration of opioids for an appropriate dose of narcotic antagonist. (2) Type of study: fed design—single-dose, two-way, crossover in vivo. Strength: 100 mg. Subjects: normal healthy males and females, general population. Additional comments: Please use a narcotic antagonist such as naltrexone if the study involves healthy subjects. You should consult a physician who is an expert in the administration of opioids for an appropriate dose of narcotic antagonist. (3) Type of study: sprinkle design—single-dose, two-way, crossover in vivo. Strength: 100 mg. Subjects: normal healthy males and females, general population. Additional comments: Please use a narcotic antagonist such as naltrexone if the study involves healthy subjects. You should consult a physician who is an expert in the administration of opioids for an appropriate dose of narcotic antagonist. Analytes to measure: morphine and morphine-6-glucuronide. BE based on (90% CI): morphine. Waiver request of in vivo testing: 20, 30, 50, and 60 mg based on (i) acceptable BE studies on the 100 mg strength, (ii) proportional similarity across all strengths, and (iii) acceptable in vitro dissolution testing of all strengths. In addition to the aforementioned method, for modified-release products, dissolution profiles on 12 dosage units each of test and reference products generated using USP Apparatus I at 100 rpm and/or Apparatus II at 50 rpm in at least three dissolution media (pH 1.2, 4.5, and 6.8 buffer) should be submitted in the application. Agitation speeds may have to be increased if appropriate. It is acceptable to add a small amount of surfactant, if necessary. Please include early sampling times of 1, 2, and 4 h and continue every 2 h until at least 80% of the drug is released, to provide assurance against premature release of drug (dose dumping) from the formulation. Due to concerns of dose dumping from this drug product when taken with alcohol, please conduct additional dissolution testing using various concentrations of ethanol in the dissolution medium, as follows: testing conditions 900 mL, 0.1 N HCl, Apparatus I (basket) at 100 rpm, with and without the alcohol (see later). Test 1: 12 units tested according to the proposed method (with 0.1 N HCl), with data collected every 15 min for a total of 2 h. Test 2: 12 units analyzed by substituting 5% (v/v) of test medium with Alcohol USP and data collection every 15 min for a total of 2 h. Test 3: 12 units analyzed by substituting 20% (v/v) of test medium with Alcohol USP and data collection every 15 min for a total of 2 h. Test 4: 12 units analyzed by substituting 40% (v/v) of test medium with Alcohol USP and data collection every 15 min for a total of 2 h. Both test and RLD products must be tested accordingly, and data must be provided on individual unit, means, range, and %CV on both strengths.

Mycophenolate Mofetil, Capsules/Oral. Recommended studies: two studies. (1) Type of study: fasting design—single-dose, randomized, two-treatment, two-period, two-sequence crossover in vivo. Strength: 250 mg. Subjects: normal healthy males and females, general population. Additional comments: (2) Type of study: fed design—single-dose, randomized, two-treatment, two-period, two-sequence crossover in vivo. Strength: 250 mg. Subjects: normal healthy males and females, general population. Additional comments: Analytes to measure (in appropriate biological fluid): mycophenolate mofetil and the active metabolite, mycophenolic acid (MPA), in plasma. BE based on (90% CI): mycophenolate mofetil. If mycophenolate mofetil plasma concentrations can be reliably measured and its PKs accurately determined, please analyze the data for the parent compound using the CI approach. The data for the active metabolite can be used as supportive evidence. However, if you can demonstrate that it is not possible to measure mycophenolate mofetil in plasma accurately and reliably, please analyze the metabolite using the CI approach. Waiver request of in vivo testing: not applicable.

Mycophenolate Mofetil Hydrochloride, Tablets/Oral. Recommended studies: two studies. (1) Type of study: fasting design—single-dose, two-treatment, two-period crossover in vivo. Strength: 500 mg. Subjects: normal healthy males and females, general population. Additional comments: (2) Type of study: fed design—single-dose, two-treatment, two-period crossover in vivo. Strength: 500 mg. Subjects: normal healthy males and females, general population. Additional comments: Analytes to measure (in appropriate biological fluid): mycophenolate mofetil and the active metabolite, MPA, in plasma. BE based on (90% CI): mycophenolate mofetil. If mycophenolate mofetil plasma concentrations can be reliably measured and its PKs accurately determined, please analyze the data for the parent compound using the CI approach. The data for the active metabolite can be used as supportive evidence. However, if you can demonstrate that it is not possible to measure mycophenolate mofetil in plasma accurately and reliably, please analyze the metabolite using the CI approach. Waiver request of in vivo testing: not applicable.

Nabumetone, Tablets/Oral. Recommended studies: two studies. (1) Type of study: fasting design—single-dose, two-way crossover in vivo. Strength: 750 mg. Subjects: normal healthy males and females, general population. Additional comments: (2) Type of study: fed design—single-dose, two-way crossover in vivo. Strength: 750 mg. Subjects: normal healthy males and females, general population. Additional comments: Analytes to measure: 6-methoxy-2-naphthyl-acetic acid (6-MNA). BE based on (90% CI): 6-MNA. Waiver request of in vivo testing: 500 mg based on (i) acceptable BE studies on the 750 mg strength, (ii) proportional similarity of the formulations across all strengths, and (iii) acceptable in vitro dissolution testing of all strengths. Please conduct comparative dissolution testing on 12 dosage units of all strengths of the test and reference products.

Nateglinide, Tablets/Oral. Recommended studies: two studies. (1) Type of study: fasting design—single-dose, two-way crossover in vivo. Strength: 120 mg. Subjects: normal, healthy males and females, general population. Additional comments: All subjects should fast overnight for at least 10 h prior to dosing and for 4 h after dosing. A single oral dose (120 mg) should be administered with 240 mL of 20% glucose solution. Since, multiple plasma concentration peaks were often observed under fasting conditions, please ensure that the same sampling schedule is followed during the study for both test and reference drug administration. Females should not be pregnant or lactating and, if applicable, should practice abstention or contraception during the study. (2) Type of study: fed design—single-dose, two-way crossover in vivo. Strength: 120 mg. Subjects: normal, healthy males and females, general population. Additional comments: A single oral dose (120 mg) should be administered with 240 mL of water 30 min after start of a standard high-fat FDA breakfast. Subjects should start the recommended meal 30 min prior to administration of the drug product. Study subjects should eat this meal in 30 min or less; however, the drug product should be administered 30 min after start of the meal. Females should not be pregnant or lactating and, if applicable, should practice abstention or contraception during the study. Analytes to measure (in appropriate biological fluid): nateglinide in plasma. BE based on (90% CI): nateglinide. Waiver request of in vivo testing: 60 mg, based on (i) acceptable BE studies on the 120 mg strength, (ii) proportional similarity across all strengths, and (iii) acceptable in vitro dissolution testing of all strengths.

Nelfinavir Mesylate, Suspension/Oral. Recommended studies: two studies. (1) Type of study: fasting design—single-dose, two-treatment, two-period crossover in vivo. Strength: 50 mg/scoopful. Subjects: normal healthy males and females, general population. Additional comments: (2) Type of study: fed design—single-dose, two-treatment, two-period crossover in vivo. Strength: 50 mg/scoopful. Subjects: normal healthy males and females, general population. Additional comments: Analytes to measure (in appropriate biological fluid): nelfinavir in plasma. BE based on (90% CI): nelfinavir. Waiver request of in vivo testing: not applicable.

Nelfinavir Mesylate, Tablets/Oral. Recommended studies: three studies. (1) Type of study: fasting design—randomized, single-dose, two-treatment, two-period crossover in vivo. Strength: 625 mg. Subjects: normal healthy males and females, general population. Additional comments: High PK variability has been observed with nelfinavir when administered to fasting subjects. Thus, it is the firm's responsibility to enroll an adequate number of subjects to demonstrate BE. Since nelfinavir appears to be a highly variable drug when administered under fasting conditions, conducting a replicate design study as an alternative to a two-way crossover study may be considered. A replicate study design has the advantage that fewer subjects can be used than in a two-way crossover study. The FDA recommends that a replicate design BE study uses the following two sequences: ABAB (test, reference, test, reference) and BABA (reference, test, reference, test). (2) Type of study: fasting design—randomized, single-dose, two-treatment, two-period crossover in vivo. Strength: 250 mg. Subjects: normal healthy males and females, general population. Additional comments: Please see previous. (3) Type of study: fed design—randomized, single-dose, two-treatment, two-period crossover in vivo. Strength: 625 mg. Subjects: normal healthy males and females, general population. Additional comments: Please see previous. Analytes to measure (in appropriate biological fluid): nelfinavir in plasma. Please develop a method of adequate sensitivity to accurately measure nelfinavir concentrations in plasma. If it is not possible to accurately measure nelfinavir plasma concentrations following administration of a single dosage unit, it is acceptable to administer a higher dose. A single dose as high as 1250 mg may be safely administered to healthy normal subjects. BE based on (90% CI): nelfinavir. Waiver request of in vivo testing: not applicable.

Nevirapine, Suspension/Oral. Recommended studies: two studies. (1) Type of study: fasting design—single dose, one period, parallel in vivo. Strength: 50 mg/5mL. Subjects: normal healthy males and females, general population. Additional comments: Due to safety concerns of sever life-threatening skin reactions and hepatotoxicity, single-dose parallel study designs in normal healthy subjects are recommended. (2) Type of study: fed design—single dose, one period, parallel in vivo. Strength: 50 mg/5mL. Subjects: normal healthy males and females, general population. Additional comments: Please see previous comments. Analytes to measure (in appropriate biological fluid): nevirapine in plasma. BE based on (90% CI): nevirapine. Waiver request of in vivo testing: not applicable.

Nevirapine, Tablet/Oral. Recommended studies: two studies. (1) Type of study: fasting design—single dose, randomized, two treatments, one period, parallel, open label in vivo, Strength: 200 mg. Subjects: normal healthy males and females, general population. Additional comments: Due to safety concerns of severe life-threatening skin reactions and hepatotoxicity, single-dose parallel study designs in healthy volunteers are recommended. (2) Type of study: fed design—single dose, randomized, two treatments, one period, parallel, open label in vivo. Strength: 200 mg. Subjects: normal healthy males and females, general population. Additional comments: Please see previous comment. Analytes to measure (in appropriate biological fluid): nevirapine in plasma. BE based on (90% CI): nevirapine. Waiver request of in vivo testing: not applicable.

Olanzapine, Orally Disintegrating Tablet/Oral. Recommended studies: two studies. (1) Type of study: fasting design—single-dose, two-treatment, two-period crossover in vivo. Strength: 5 mg. Subjects: normal healthy males and females, general population. Additional comments: Due to safety concerns, studies should be conducted using the 5 mg strength. (2) Type of study: fed design—single-dose, two-treatment, two-period crossover in vivo. Strength: 5 mg. Subjects: normal healthy males and females, general population. Additional comments: Please see previous comment. Analytes to measure: olanzapine in plasma. BE based on (90% CI): olanzapine. Waiver request of in vivo testing: 10, 15, and 20 mg based on (i) acceptable BE studies on the 5 mg strength, (ii) proportional similarity of the formulations across all strengths, and (iii) acceptable in vitro dissolution testing of all strengths.

Olmesartan Medoxomil, Tablets/Oral. Recommended studies: two studies. (1) Type of study: fasting design—single-dose, two-way crossover in vivo. Strength: 40 mg. Subjects: normal healthy males and females, general population. Females should not be pregnant and, if applicable, should practice abstention or contraception during the study. Additional comments: Labeling for this drug contains a black box regarding pregnancy and fetal/neonatal morbidity and mortality. (2) Type of study: fed design—single-dose, two-way crossover in vivo. Strength: 40 mg. Subjects: normal healthy males and females, general population. Females should not be pregnant and, if applicable, should practice abstention or contraception during the study. Additional comments: Please see previous comments. Analytes to measure (in appropriate biological fluid): olmesartan in plasma. BE based on (90% CI): olmesartan. Waiver request of in vivo testing: 20 and 5 mg based on (i) acceptable BE studies on the 40 mg strength, (ii) proportional similarity across all strengths, and (iii) acceptable in vitro dissolution testing of all strengths.

Olsalazine Sodium, Capsule/Oral. Recommended studies: one study. Type of study: fed design—single-dose, two-treatment, two-period crossover in vivo. Strength: 250 mg. Subjects: normal healthy males and females, general population. Additional comments: Please use the lowest single dose possible to obtain accurate PK parameters for both olsalazine and mesalamine. Please enroll enough subjects to achieve adequate statistical power to demonstrate BE to the RLD. A pilot study may be necessary to assist in the determination of the appropriate number of subjects to enroll in the pivotal study. The number of subjects should be sufficient to allow for dropouts. You may also refer to Appendix C of the guidance for industry, "Statistical Approaches to Establishing Bioequivalence" at http://www.fda.gov/cder/guidance/index.htm. Analytes to measure (in appropriate biological fluid): olsalazine and mesalamine in plasma. BE based on (90% CI): olsalazine and mesalamine. Waiver request of in vivo testing: not applicable. In addition, please perform dissolution testing over a range of pH values comparing the test and reference products. Varying pH conditions should be studied to approximate the pH conditions that olsalazine sodium capsules will be subjected to in the GI tract. Therefore, the following pH conditions should be used using 12 dosage units of the test and reference products: apparatus, USP Apparatus I (basket); speed, 100 rpm; medium, 0.1 N HCl, pH 4.5 buffer, pH 6.8 buffer; volume, 900 mL; and sampling times, 5, 10, 15, 20, 30, 45, and 60 min and until at least 80% of the labeled content is dissolved.

Omeprazole, Delayed-Release Capsule/Oral. Recommended studies: four studies (1) Type of study: fasting design—single-dose, two-treatment, two-period crossover in vivo. Strength: 40 mg. Subjects: normal healthy males and females, general population. Additional comments: (2) Type of study: fasting, sprinkle design—single-dose, two-treatment, two-period crossover in vivo. Strength: 40 mg. Subjects: normal healthy males and females, general population. Additional comments: Please administer the dose after sprinkling the entire contents of the capsule on a teaspoonful of applesauce in accordance with the approved labeling of the RLD. (3) Type of study: fed design—single-dose, two-treatment, two-period crossover in vivo. Strength: 40 mg. Subjects: normal healthy males and females, general population. Additional comments: (4) Type of study: fasting design—single-dose, two-treatment, two-period crossover in vivo. Strength: 20 mg. Subjects: normal healthy males and females, general population. Additional comments: Analytes to measure (in appropriate biological fluid): omeprazole in plasma. BE based on (90% CI): omeprazole. Waiver request of in vivo testing: 10 mg based on (i) acceptable BE studies on the 20 mg strength, (ii) proportional similarity of the formulations on 10 and 20 mg strengths, and (iii) acceptable in vitro dissolution testing of 10 and 20 mg strengths. Please conduct comparative dissolution testing on 12 dosage units of all strengths of the test and reference products.

Omeprazole Magnesium, Delayed-Release Tablets/Oral. Recommended studies: two studies. (1) Type of study: fasting design—single-dose, two-treatment, two-period crossover in vivo. Strength: 20 mg. Subjects: normal healthy males and females, general population. Additional comments: (2) Type of study: fed design—single-dose, two-treatment, two-period crossover in vivo. Strength: 20 mg. Subjects: normal healthy males and females, general population. Additional comments: Analytes to measure: omeprazole in plasma. BE based on (90% CI): omeprazole. Waiver request of in vivo testing: not applicable.

Omeprazole, Powder for Suspension/Oral. Recommended studies: one study. Type of study: fasting design—single-dose, two-treatment, two-period crossover in vivo. Strength: 40 mg/packet. Subjects: normal healthy males and females, general population. Additional comments: Analytes to measure: omeprazole in plasma. BE based on (90% CI): omeprazole. Waiver request of in vivo testing: 20 mg/packet based on (i) acceptable BE study on the 40 mg strength, (ii) proportional similarity of the formulations across all strengths, and (iii) acceptable in vitro dissolution testing of all strengths. Since omeprazole powder for oral suspension, 20 and 40 mg/packet, is subject to two separate NDAs, two separate ANDA must be submitted. A waiver of in vivo BE testing is available.

Omeprazole, Sodium Bicarbonate, and Magnesium Hydroxide, Chewable Tablets/Oral. Recommended studies: one study. Type of study: fasting design—single-dose, two-way crossover in vivo. Strength: 40 mg/600 mg/700 mg. Subjects: normal healthy males and females, general population. Additional comments: Females should not be pregnant or lactating and, if applicable, should practice abstention or contraception during the study. Analytes to measure (in appropriate biological fluid): omeprazole in plasma. BE based on (90% CI): omeprazole. Waiver request of in vivo testing: 20 mg/600 mg/700 mg based on (i) acceptable BE studies on the 40 mg/600 mg/700 mg strength, (ii) proportional similarity of the formulations across all strengths, and (iii) acceptable in vitro dissolution testing of all strengths.

Ondansetron Hydrochloride, Tablet/Oral. Recommended studies: two studies. (1) Type of study: fasting design—single-dose, two-treatment, two-period crossover in vivo. Strength: 24 mg. Subjects: normal healthy males and females, general population. Additional comments: (2) Type of study: fed design—single-dose, two-treatment, two-period crossover in vivo. Strength: 24 mg. Subjects: normal healthy males and females, general population. Additional comments: Analytes to measure: ondansetron in plasma. BE based on (90% CI): ondansetron. Waiver request of in vivo testing: 4 and 8 mg based on (i) acceptable BE studies on the 24 mg strength, (ii) proportional similarity of the formulations across all strengths, and (iii) acceptable in vitro dissolution testing of all strengths.

Ondansetron, Orally Disintegrating Tablet. Recommended studies: two studies. (1) Type of study: fasting design—single-dose, two-treatment, two-period crossover in vivo. Strength: 8 mg. Subjects: normal healthy males and females, general population. Additional comments: (2) Type of study: fed design—single-dose, two-treatment, two-period crossover in vivo. Strength: 8 mg. Subjects: normal healthy males and females, general population. Additional comments: Analytes to measure: ondansetron in plasma. BE based on (90% CI): ondansetron. Waiver request of in vivo testing: 4 mg based on (i) acceptable BE studies on the 8 mg strength, (ii) proportional similarity of the formulations across all strengths, and (iii) acceptable in vitro dissolution testing of all strengths.

Oxcarbazepine, Suspension/Oral. Recommended studies: two studies. (1) Type of study: fasting design—single-dose, two-treatment, two-period crossover in vivo. Strength: 300 mg/5 mL (600 mg dose). Subjects: normal healthy males and females, general population. Additional comments: (2) Type of study: fed design—single-dose, two-treatment, two-period crossover in vivo. Strength: 300 mg/5 mL (600 mg dose). Subjects: normal healthy males and females, general population.

Additional comments: Analytes to measure (in appropriate biological fluid): oxcarbazepine and its 10-hydroxy metabolite (monohydroxy derivative [MHD]) in plasma using an achiral assay. BE based on (90% CI): oxcarbazepine. Please submit the metabolite data as supportive evidence of comparable therapeutic outcome. For the metabolite, the following data should be submitted: individual and mean concentrations, individual and mean PK parameters, and geometric means and ratios of means for AUC and C_{max}. Waiver request of in vivo testing: not applicable. Please note that a dosage unit for a suspension is the labeled strength (5 mL). A total of 12 units from 12 different bottles should be used.

Oxcarbazepine, Tablet/Oral. Recommended studies: two studies. (1) Type of study: fasting design—single-dose, two-treatment, two-period crossover in vivo. Strength: 600 mg. Subjects: normal healthy males and females, general population. Additional comments: (2) Type of study: fed design—single-dose, two-treatment, two-period crossover in vivo. Strength: 600 mg. Subjects: normal healthy males and females, general population. Additional comments: Analytes to measure (in appropriate biological fluid): oxcarbazepine and active metabolite 10-MHD in plasma using an achiral assay. BE based on (90% CI): oxcarbazepine. Please submit the metabolite data as supportive evidence of comparable therapeutic outcome. For the metabolite, the following data should be submitted: individual and mean concentrations, individual and mean PK parameters, and geometric means and ratios of means for AUC and C_{max}. Waiver request of in vivo testing: 150 and 300 mg based on (i) acceptable BE studies on the 600 mg strength, (ii) proportional similarity of the formulations across all strengths, and (iii) acceptable in vitro dissolution testing of all strengths.

Oxymorphone Hydrochloride, Extended-Release Tablets/Oral. Recommended studies: two studies. (1) Type of study: fasting design—single-dose, two-treatment, two-period crossover in vivo. Strength: 40 mg. Subjects: normal healthy males and females, general population. Additional comments: Please use a narcotic antagonist such as naltrexone if the study involves healthy subjects. You should consult a physician who is an expert in the administration of opioids for an appropriate dose of narcotic antagonist. (2) Type of study: fed design—single-dose, two-treatment, two-period crossover in vivo. Strength: 40 mg. Subjects: normal healthy males and females, general population. Additional comments: Please see previous comments. Analytes to measure (in appropriate biological fluid): oxymorphone and its metabolite, 6-OH-oxymorphone, in plasma. BE based on (90% CI): oxymorphone. For 6-OH-oxymorphone, please submit individual and mean concentrations, individual and mean PK parameters, and geometric means and ratios of means for AUC and C_{max}. Waiver request of in vivo testing: 5, 10, and 20 mg based on (i) acceptable BE studies of the 40 mg strength, (ii) proportional similarity across all strengths, and (iii) acceptable in vitro dissolution testing of all strengths. For modified-release products, dissolution profiles generated using USP Apparatus I at 100 rpm and/or Apparatus II at 50 rpm in at least three dissolution media (pH 1.2, 4.5, and 6.8 phosphate buffer, water) should be submitted in the application. Agitation speeds may have to be increased if appropriate. It is acceptable to add a small amount of surfactant, if necessary. The following sampling times are recommended: 1, 2, 4, and every 2 h thereafter, until at least 80% of the labeled content is dissolved. Comparative dissolution profiles should include individual tablet data as well as the mean, range, and standard deviation at each time point for 12 tablets. Due to concerns of dose dumping from this drug product when taken with alcohol, please conduct additional dissolution testing using various concentrations of ethanol in the dissolution medium, as follows: testing conditions, 900 mL, 0.1 N HCl, Apparatus I (basket) at 75 rpm, with and without the alcohol (see later). Test 1: 12 units tested according to the proposed method (with 0.1 N HCl), with data collected every 15 min for a total of 2 h. Test 2: 12 units analyzed by substituting 5% (v/v) of test medium with Alcohol USP and data collection every 15 min for a total of 2 h. Test 3: 12 units analyzed by substituting 20% (v/v) of test medium with Alcohol USP and data collection every 15 min for a total of 2 h. Test 4: 12 units analyzed by substituting 40% (v/v) of test medium with Alcohol USP and data collection every 15 min for a total of 2 h. Both test and RLD products must be tested accordingly, and data must be provided on individual unit, means, range, and %CV on both strengths.

Oxymorphone Hydrochloride, Tablets/Oral. Recommended studies: one study. Type of study: fasting design—single-dose, two-treatment, two-period crossover in vivo. Strength: 10 mg. Subjects: normal healthy males and females, general population. Additional comments: Please use a narcotic antagonist such as naltrexone if the study involves healthy subjects. You should consult a physician who is an expert in the administration of opioids for an appropriate dose of narcotic antagonist. Analytes to measure (in appropriate biological fluid): oxymorphone and its metabolite, 6-OH-oxymorphone, in plasma. For 6-OH-oxymorphone, the following data should be submitted: individual and mean concentrations, individual and mean PK parameters, and geometric means and ratios of means for AUC and C_{max}. BE based on (90% CI): oxymorphone. Waiver request of in vivo testing: 5 mg, based on (i) acceptable BE studies on the 10 mg strength, (ii) proportional similarity of the formulations, and (iii) acceptable in vitro dissolution testing of all strengths.

Paliperidone, Extended-Release Tablets/Oral. Recommended studies: two studies. (1) Type of study: fasting design—single-dose, two-treatment, two-period crossover in vivo. Strength: 6 mg. Subjects: normal healthy males and females, general population. Additional comments: Females must have a negative baseline pregnancy test within 24 h prior to receiving the drug. Females should not be pregnant or lactating and, if applicable, should practice abstention or contraception during the study. (2) Type of study: fed design—single-dose, two-treatment, two-period crossover in vivo. Strength: 6 mg. Subjects: normal healthy males and females, general population. Additional comments: Please see previous comment. Analytes to measure (in appropriate biological fluid): paliperidone in plasma. BE based on (90% CI): paliperidone. Waiver request of in vivo testing: 3 and 9 mg based on (i) acceptable BE studies of the 6 mg strength, (ii) proportional similarity across all strengths, and (iii) acceptable in vitro dissolution testing of all strengths. In addition to the aforementioned method, for modified-release products, dissolution profiles on 12 dosage units each of test and reference products generated using USP Apparatus I at 100 rpm and/or Apparatus II at 50 rpm in at least three dissolution media (pH 1.2, 4.5, and 6.8 buffer, water) should be submitted in the application. Agitation speeds may have to be increased if appropriate. It is acceptable to add a small amount of surfactant, if necessary. Please include early sampling times of 1, 2, and 4 h and continue every 2 h until at least 80% of the drug is released. Comparative dissolution profiles should include individual tablet data as well as the mean, range, and standard deviation at each time point for 12 tablets.

Pantoprazole Sodium, Delayed-Release Tablet/Oral. Recommended studies: two studies. (1) Type of study: fasting design—single-dose, two-treatment, two-period crossover in vivo. Strength: 40 mg. Subjects: normal healthy males and females, general population. Additional comments: (2) Type of study: fed design—single-dose, two-treatment, two-period crossover in vivo. Strength: 40 mg. Subjects: normal healthy males and females, general population. Additional comments: Analytes to measure (in appropriate biological fluid): pantoprazole in plasma. BE based on (90% CI): pantoprazole. Waiver request of in vivo testing: 20 mg based on (i) acceptable BE studies on the 40 mg strength, (ii) proportional similarity of the formulations across all strengths, and (iii) acceptable in vitro dissolution testing of all strengths.

Paricalcitol, Capsule/Oral. Recommended studies: two studies. (1) Type of study: fasting design—single-dose, two-treatment, two-period crossover in vivo. Strength: 4 mcg. Subjects: normal healthy males and females, general population. Additional comments: Females must have a negative baseline pregnancy test within 24 h prior to receiving the drug. Females should not be pregnant or lactating and, if applicable, should practice abstention or contraception during the study. (2) Type of study: fed design—single-dose, two-treatment, two-period crossover in vivo. Strength: 4 mcg. Subjects: normal healthy males and females, general population. Additional comments: Please see previous comment. Analytes to measure (in appropriate biological fluid): paricalcitol in plasma. BE based on (90% CI): paricalcitol. Waiver request of in vivo testing: 2 and 1 mcg tablets, based on (i) acceptable BE studies of the 4 mcg strength, (ii) proportional similarity across all strengths, and (iii) acceptable in vitro dissolution testing of all strengths.

Perindopril Erbumine, Tablet/Oral. Recommended studies: two studies. (1) Type of study: fasting design—single-dose, two-treatment, two-period crossover in vivo. Strength: 8 mg. Subjects: normal healthy males and females, general population. Additional comments: Female subjects enrolled in the BE studies should not be pregnant and, if applicable, should practice abstention or contraception during the study. (2) Type of study: fed design—single-dose, two-treatment, two-period crossover in vivo. Strength: 8 mg. Subjects: normal healthy males and females, general population. Additional comments: Please see previous comments. Analytes to measure (in appropriate biological fluid): perindopril and the active metabolite, perindoprilat, in plasma. BE based on (90% CI): perindopril. Please submit the metabolite data as supportive evidence of comparable therapeutic outcome. For the metabolite, the following data should be submitted: individual and mean concentrations, individual and mean PK parameters, and geometric means and ratios of means for AUC and C_{max}. Waiver request of in vivo testing: 2 and 4 mg based on (i) acceptable BE studies on the 8 mg strength, (ii) proportional similarity of the formulations across all strengths, and (iii) acceptable in vitro dissolution testing of all strengths.

Phenytoin, Chewable Tablets/Oral. Recommended studies: two studies. (1) Type of study: fasting design—single-dose, two-treatment, two-period crossover in vivo. Strength: 300 mg dose (6 × 50 mg) and use a washout period of at least 14 days. Subjects: normal healthy males and females, general population. Additional comments: The tablets should be swallowed whole. (2) Type of study: fed design—single-dose, two-treatment, two-period crossover in vivo. Strength: 300 mg dose (6 × 50 mg) and use a washout period of at least 14 days. Subjects: normal healthy males and females, general population. Additional comments: The tablets should be swallowed whole. Analytes to measure (in appropriate biological fluid): phenytoin in plasma. BE based on (90% CI): phenytoin. Waiver request of in vivo testing: not applicable.

Phenytoin Sodium, Extended Capsule/Oral. Recommended studies: two studies. (1) Type of study: fasting design—single-dose, two-way crossover in vivo. Strength: single-dose 300 mg. Subjects: normal healthy males and females, general population. Additional comments: Washout period of at least 14 days. The single-dose studies for fasting and fed can be conducted as single-dose, two-treatment, four-period, replicated design. The strength(s) designated in the *Orange Book* as the RLD should be used in the studies. (2) Type of study: fed design—single-dose, two-way crossover in vivo. Strength: single-dose 300 mg. Subjects: normal healthy males and females, general population. Additional comments: Please see previous comments. Analytes to measure: phenytoin in plasma. BE based on (90% CI): phenytoin. Waiver request of in vivo testing: 200 mg based on (i) acceptable BE studies on the 300 mg strength, (ii) proportional similarity across all strengths, and (iii) acceptable in vitro dissolution testing of all strengths. Please conduct comparative dissolution testing on 12 dosage units of all strengths of the test and reference products using the USP method. In addition to the aforementioned method, dissolution profiles on 12 dosage units each of test and reference products generated using USP Apparatus I at 100 rpm and/or Apparatus II at 50 rpm in at least three dissolution media (pH 1.2, 4.5, and 6.8 buffer) should be submitted in the application. Agitation speeds may have to be increased if appropriate. It is acceptable to add a small amount of surfactant, if necessary. Please include early sampling times of 1, 2, and 4 h and continue every 2 h until at least 80% of the drug is released, to provide assurance against premature release of drug (dose dumping) from the formulation.

Phenytoin Sodium, Extended Capsule/Oral. Recommended studies: four studies (1) Type of study: fasting design—single-dose, two-way crossover in vivo. Strength: single dose of 300 mg (3 × 100 mg). Subjects: normal healthy males and females, general population. Additional comments: Washout period of at least 14 days. The single-dose studies for fasting and fed can be conducted as single-dose, two-treatment, four-period, replicated design. The strength(s) designated in the *Orange Book* as the RLD should be used in the studies. (2) Type of study: fed design—single-dose, two-way crossover in vivo. Strength: single dose of 300 mg (3 × 100 mg). Subjects: normal healthy males and females, general population. Additional comments: Please see previous comments. (3) Type of study: fasting design—single-dose, two-way crossover in vivo. Strength: single dose of

300 mg (10 × 30 mg). Subjects: normal healthy males and females, general population. Additional comments: Please see previous comments. (4) Type of study: fed design—single-dose, two-way crossover in vivo. Strength: single dose of 300 mg (10 × 30 mg). Subjects: normal healthy males and females, general population. Additional comments: Please see previous comments. Analytes to measure: phenytoin in plasma. BE based on (90% CI): phenytoin. Waiver request of in vivo testing: not applicable. Please conduct comparative dissolution testing on 12 dosage units of all strengths of the test and reference products using the USP method. In addition to the aforementioned method, dissolution profiles on 12 dosage units each of test and reference products generated using USP Apparatus I at 100 rpm and/or Apparatus II at 50 rpm in at least three dissolution media (pH 1.2, 4.5, and 6.8 buffer) should be submitted in the application. Agitation speeds may have to be increased if appropriate. It is acceptable to add a small amount of surfactant, if necessary. Please include early sampling times of 1, 2, and 4 h and continue every 2 h until at least 80% of the drug is released, to provide assurance against premature release of drug (dose dumping) from the formulation.

Phenytoin, Suspension/Oral. Recommended studies: two studies. (1) Type of study: fasting design—single-dose, two-way crossover in vivo. Strength: 125 mg/5 mg (dose 300 mg). Subjects: normal healthy males and females, general population. Additional comments: Washout period of at least 14 days. The single-dose studies for fasting and fed can be conducted as single-dose, two-treatment, four-period, replicated design. The strength(s) designated in the *Orange Book* as the RLD should be used in the studies. (2) Type of study: fed design—single-dose, two-way crossover in vivo. Strength: 125 mg/5 mg (dose 300 mg). Subjects: normal healthy males and females, general population. Additional comments: Please see previous comments. Analytes to measure: phenytoin in plasma. BE based on (90% CI): phenytoin. Waiver request of in vivo testing: not applicable. Please conduct comparative dissolution testing on 12 dosage units of all strengths of the test and reference products using the USP method. A dosage unit for a suspension is the labeled strength (5 mL). A total of 12 units from 12 different bottles should be used.

Pilocarpine Hydrochloride, Tablet/Oral. Recommended studies: two studies. (1) Type of study: fasting design—single-dose, two-treatment, two-period crossover in vivo. Strength: 7.5 mg. Subjects: normal healthy males and females, general population. Additional comments: (2) Type of study: fed design—single-dose, two-treatment, two-period crossover in vivo. Strength: 7.5 mg. Subjects: normal healthy males and females, general population. Additional comments: Analytes to measure (in appropriate biological fluid): pilocarpine and the metabolite, pilocarpic acid, in plasma. Pilocarpine has been shown to be unstable in heparinized plasma and convert to pilocarpic acid during storage. Therefore, you should pay attention to the stabilization of pilocarpine and separation of the drug from its metabolites in the assay development and validation. Recent literature states that the use of EDTA as an anticoagulant during blood sampling may be helpful in stabilizing pilocarpine. The stability of pilocarpine in plasma samples and the assay specificity of pilocarpine, especially in relation to its metabolites and plasma endogenous components, should be clearly demonstrated in the assay method validation report submitted to the FDA. BE based on (90% CI): pilocarpine. If pilocarpine can be reliably measured, a CI approach for BE determination should be used for pilocarpine. If pilocarpine cannot be reliably measured, a CI approach for BE determination should be used for pilocarpic acid. Waiver request of in vivo testing: 5 mg based on (i) acceptable BE studies on the 7.5 mg strength, (ii) proportional similarity of the formulations across all strengths, and (iii) acceptable in vitro dissolution testing of all strengths.

Pimozide, Tablets/Oral. Recommended studies: one study. Type of study: fasting design—single-dose, two-way crossover in vivo. Strength: 2 mg. Subjects: normal healthy males and females, general population. Additional comments: Females should not be pregnant or lactating and, if applicable, should practice abstention or contraception during the study. Analytes to measure (in appropriate biological fluid): pimozide in plasma. Pimozide has a long terminal elimination half-life. Please ensure adequate washout periods between treatments in the crossover studies. You may also consider using a parallel study design due to pimozide's long half-life. For long half-life drug products, an AUC truncated to 72 h may be used in place of AUC0-t or AUCinf. Please collect sufficient blood samples in the BE study to adequately characterize the peak concentration (C_{max}) and time to reach peak concentration (t_{max}). BE based on (90% CI): pimozide. Waiver request of in vivo testing: 1 mg based on (i) acceptable BE studies on the 2 mg strength, (ii) proportional similarity of the formulations across all strengths, and (iii) acceptable in vitro dissolution testing of all strengths.

Posaconazole, Suspension/Oral. Recommended studies: two studies. (1) Type of study: fasting design—single-dose, two-treatment, two-period crossover in vivo. Strength: 40 mg/mL (dose 400 mg). Subjects: normal healthy males and females, general population. Additional comments: Females must have a negative baseline pregnancy test within 24 h prior to receiving the drug. Females should not be pregnant or lactating and, if applicable, should practice abstention or contraception during the study. (2) Type of study: fed design—single-dose, two-treatment, two-period crossover in vivo. Strength: 40 mg/mL (dose 400 mg). Subjects: normal healthy males and females, general population. Additional comments: Please see previous comment. Analytes to measure (in appropriate biological fluid): posaconazole in plasma. BE based on (90% CI): posaconazole. Waiver request of in vivo testing: not applicable. Please note that a dosage unit for a suspension is the labeled strength (ml). A total of 12 units from 12 different bottles should be used. Pravastatin, Sodium Tablet/Oral. Recommended studies: two studies. (1) Type of study: fasting design—single-dose, two-treatment, two-period crossover in vivo. Strength: 80 mg. Subjects: normal healthy males and females, general population. Additional comments: (2) Type of study: fed design—single-dose, two-treatment, two-period crossover in vivo. Strength: 80 mg. Subjects: normal healthy males and females, general population. Additional comments: Analytes to measure (in appropriate biological fluid): pravastatin in plasma. BE based on (90% CI): pravastatin. Waiver request of in vivo testing: 10, 20, and 40 mg based on (i) acceptable BE studies on the 80 mg strength, (ii) proportional similarity of the formulations across all strengths, and (iii) acceptable in vitro dissolution testing of all strengths.

Quetiapine Fumarate, Tablet/Oral. Recommended studies: three studies. (1) Type of study: fasting design—single dose, in vivo. Strength: 25 mg. Subjects: normal healthy males and females, general population. Additional comments: Please include careful safety precautions in your protocols, including adequate monitoring of vital signs and adverse events, stopping criteria in the event of an unacceptable degree of hypotension or tachycardia, and appropriate evaluation and management of adverse events. Please assure that the investigator(s) will be vigilant in recognizing and managing any unacceptable clinical or laboratory findings. It is recommended that a study protocol be submitted for review before initiating a BE study for this product. (2) Type of study: fed design—single dose, in vivo. Strength: 25 mg. Subjects: normal healthy males and females, general population. Additional comments: Please see previous comments. (3) Type of study and design: steady state, in vivo. Strength: 300 mg. Subjects: schizophrenic patients already receiving quetiapine in a stable regimen. Additional comments: Please see comments earlier. Analytes to measure: quetiapine in plasma. BE based on (90% CI): quetiapine. Waiver request of in vivo testing: 50, 100, 150, 200, and 400 mg based on (i) acceptable BE studies on the 25 and 300 mg strength, (ii) proportional similarity of the formulations across all strengths, and (iii) acceptable in vitro dissolution testing of all strengths. Please conduct comparative dissolution testing on 12 dosage units each of all strengths of the test and reference products.

Quinapril Hydrochloride, Tablets/Oral. Recommended studies: two studies. (1) Type of study: fasting design—single-dose, two-way, crossover in vivo. Strength: 40 mg. Subjects: normal healthy males and females, general population. Additional comments: Pregnant women should be excluded from participation in BE studies with ACE inhibitors. (2) Type of study: fed design—single-dose, two-way, crossover in vivo. Strength: 40 mg. Subjects: normal healthy males and females, general population. Additional comments: Pregnant women should be excluded from participation in BE studies with ACE inhibitors. Analytes to measure: quinapril and the metabolite, quinaprilat, in plasma. BE based on (90% CI): quinapril. Please submit the metabolite data as supportive evidence of comparable therapeutic outcome. For the metabolite, the following data should be submitted:

individual and mean concentrations, individual and mean PK parameters, and geometric means and ratios of means for AUC and C_{max}. Waiver request of in vivo testing: 5, 10, and 20 mg based on (i) acceptable BE studies on the 40 mg strength, (ii) proportional similarity across all strengths, and (iii) acceptable in vitro dissolution testing of all strengths.

Quinine Sulfate, Capsules/Oral. Recommended studies: two studies. (1) Type of study: fasting design—single-dose, two-way crossover in vivo. Strength: 324 mg. Subjects: normal healthy males and females, general population. Additional comments: Females should not be pregnant or lactating and, if applicable, should practice abstention or contraception during the study. Subjects with a QTc interval of >480 ms by ECG should also be excluded. (2) Type of study: fed design—single-dose, two-way crossover in vivo. Strength: 324 mg. Subjects: normal healthy males and females, general population. Additional comments: Please see previous comments. Analytes to measure (in appropriate biological fluid): quinine in plasma. BE based on (90% CI): quinine. Waiver request of in vivo testing: not applicable.

Raloxifene Hydrochloride, Tablet/Oral. Recommended studies: two studies. (1) Type of study: fasting design—single-dose, two-treatment, two-period crossover in vivo. Strength: 60 mg. Subjects: normal healthy males and females, general population. Additional comments: (2) Type of study: fed design—single-dose, two-treatment, two-period crossover in vivo. Strength: 60 mg. Subjects: normal healthy males and females, general population. Additional comments: Analytes to measure (in appropriate biological fluid): raloxifene and the metabolites, raloxifene-4′-glucuronide and raloxifene-6′-glucuronide, in plasma. BE based on (90% CI): raloxifene. Please submit the metabolite data as supportive evidence of comparable therapeutic outcome. For the metabolite, the following data should be submitted: individual and mean concentrations, individual and mean PK parameters, and geometric means and ratios of means for AUC and C_{max}. Waiver request of in vivo testing: not applicable.

Ramipril, Capsule/Oral. Recommended studies: two studies. (1) Type of study: fasting design—single-dose, two-treatment, two-period crossover in vivo. Strength: 10 mg. Subjects: normal healthy males and females, general population. Additional comments: Female subjects enrolled in the BE studies should not be pregnant and, if applicable, should practice abstention or contraception during the study. (2) Type of study: fed design—single-dose, two-treatment, two-period crossover in vivo. Strength: 10 mg. Subjects: normal healthy males and females, general population. Additional comments: Please see previous comment. Analytes to measure (in appropriate biological fluid): ramipril and the metabolite, ramiprilat, in plasma. BE based on (90% CI): ramipril. If ramipril can be reliably measured, a CI approach for BE determination should be used for ramipril. If ramipril cannot be reliably measured, a CI approach for BE determination should be used for ramiprilat. Waiver request of in vivo testing: 1.25, 2.5, and 5 mg based on (i) acceptable BE studies on the 10 mg strength, (ii) proportional similarity of the formulations across all strengths, and (iii) acceptable in vitro dissolution testing of all strengths.

Ribavirin, Capsule/Oral. Recommended studies: two studies. (1) Type of study: fasting design—single-dose, two-treatment, two-period crossover in vivo. Strength: 200 mg. Subjects: normal healthy males and females, general population. Additional comments: (2) Type of study: fed design—single-dose, two-treatment, two-period crossover in vivo. Strength: 200 mg. Subjects: normal healthy males and females, general population. Additional comments: Analytes to measure (in appropriate biological fluid): ribavirin in plasma. BE based on (90% CI): Ribavirin. Waiver request of in vivo testing: not applicable.

Ribavirin, Tablet/Oral. Recommended studies: two studies. (1) Type of study: fasting design—single-dose, two-treatment, two-period crossover in vivo. Strength: 600 mg. Subjects: normal healthy males and females, general population. Additional comments: (2) Type of study: fed design—single-dose, two-treatment, two-period crossover in vivo. Strength: 600 mg. Subjects: normal healthy males and females, general population. Additional comments: Analytes to measure (in appropriate biological fluid): ribavirin in plasma. BE based on (90% CI): ribavirin. Waiver request of in vivo testing: 200 and 400 mg strengths based on (i) acceptable BE studies on the 600 mg strength, (ii) proportional similarity of the formulations across all strengths, and (iii) acceptable in vitro dissolution testing of all strengths.

Rifampin, Capsule/Oral. Recommended studies: one study. Type of study: fasting design—single-dose, two-treatment, two-period crossover in vivo. Strength: 300 mg Subjects: normal healthy males and females, general population. Additional comments: Analytes to measure (in appropriate biological fluid): rifampin in plasma. BE based on (90% CI): rifampin. Waiver request of in vivo testing: 150 mg based on (i) acceptable BE studies on the 300 mg strength, (ii) proportional similarity of the formulations across all strengths, and (iii) acceptable in vitro dissolution testing of all strengths. Please submit separate applications for each strength. You may cross-reference the study submitted in the application for the higher strength to request waivers of in vivo testing for the lower strength. Please conduct comparative dissolution testing on 12 dosage units of all strengths of the test and reference products using the USP method.

Riluzole, Tablets/Oral. Recommended studies: one study. (1) Type of study: fasting design—single-dose, two-way crossover in vivo. Strength: 50 mg. Subjects: normal, healthy males and females, general population. Additional comments: Analytes to measure (in appropriate biological fluid): riluzole in plasma. BE based on (90% CI): riluzole. Waiver request of in vivo testing: not applicable.

Risedronate Sodium, Tablet/Oral. Recommended studies: two studies. (1) Type of study: fasting design—single-dose, two-way crossover in vivo. Strength: 75 mg. Subjects: normal healthy males and females, general population. Additional comments: (2) Type of study: fasting design—single-dose, two-treatment, two-period crossover in vivo. Strength: 35 mg. Subjects: normal healthy males and females, general population. Additional comments: As an option, due to the relatively long half-life, the firm may wish to conduct this study using a parallel design. As an additional option for either the crossover or parallel design, the firm may wish to truncate the AUC at 72 h. Analytes to measure (in appropriate biological fluid): risedronate in plasma. BE based on (90% CI): risedronate. Waiver request of in vivo testing: 5 and 30 mg based on (i) acceptable BE study on the 35 and 75 mg strengths, (ii) proportional similarity of the formulations across all strengths, and (iii) acceptable in vitro dissolution testing of all strengths.

Risedronate Sodium; Calcium Carbonate, Tablets/Oral (co-packaged). Recommended studies: one study. Type of study: fasting design—single-dose, two-way crossover in vivo. Strength: 35 mg (risedronate sodium tablet). Subjects: normal healthy males and females, general population. Additional comments: As an option, due to the relatively long half-life, the firm may wish to conduct this study using a parallel design. As an additional option for either the crossover or parallel design, the firm may wish to truncate the AUC at 72 h. Analytes to measure (in appropriate biological fluid): risedronate in plasma. BE based on (90% CI): risedronate. Waiver request of in vivo testing: not applicable. For calcium carbonate tablet, please conduct comparative dissolution testing on 12 dosage units.

Risperidone, Tablet/Oral. Recommended studies: two studies. (1) Type of study: fasting design—single-dose, two-treatment, two-period crossover in vivo. Strength: 1 mg. Subjects: normal healthy males and females, general population. Additional comments: Due to safety concerns, BE studies should be conducted using the 1 mg strength. (2) Type of study: fed design—single-dose, two-treatment, two-period crossover in vivo. Strength: 1 mg. Subjects: normal healthy males and females, general population. Additional comments: Please see previous comment. Analytes to measure: risperidone in plasma. BE based on (90% CI): risperidone. Waiver request of in vivo testing: 0.25, 0.5, 2, 3, and 4 mg based on (i) acceptable BE studies on the 1 mg strength, (ii) proportional similarity of the formulations across all strengths, and (iii) acceptable in vitro dissolution testing of all strengths.

Ritonavir, Capsules/Oral. Recommended studies: two studies. (1) Type of study: fasting design—single-dose, two-treatment, two-period crossover in vivo. Strength: 100 mg. Subjects: normal healthy males and females, general population. Additional comments: (2) Type of study: fed design—single-dose, two-treatment, two-period crossover in vivo. Strength: 100 mg. Subjects: normal healthy males and females, general population. Additional comments: Analytes to measure (in appropriate biological fluid): ritonavir in plasma. BE based on (90% CI): ritonavir. Waiver request of in vivo testing: not applicable.

Rizatriptan Benzoate, Tablets/Oral. Recommended studies: two studies. (1) Type of study: fasting design—single-dose, two-way crossover in vivo. Strength: 10 mg. Subjects: normal healthy males and females, general population. Additional comments: (2) Type of study: fed design—single-dose, two-way crossover in vivo. Strength: 10 mg. Subjects: normal healthy males and females, general population. Additional comments: Analytes to measure (in appropriate biological fluid): rizatriptan in plasma. BE based on (90% CI): rizatriptan. Waiver request of in vivo testing: 5mg, based on (i) acceptable BE studies on the 10 mg strength, (ii) proportional similarity across all strengths, and (iii) acceptable in vitro dissolution testing of all strengths.

Rosiglitazone Maleate, Tablets/Oral. Recommended studies: two studies. (1) Type of study: fasting design—single-dose, two-way crossover in vivo. Strength: 8 mg. Subjects: normal, healthy males and females, general population. Additional comments: (2) Type of study: fed design—single-dose, two-way crossover in vivo. Strength: 8 mg. Subjects: normal, healthy males and females, general population. Additional comments: Analytes to measure (in appropriate biological fluid): rosiglitazone in plasma. BE based on (90% CI): rosiglitazone. Waiver request of in vivo testing: 2 and 4 mg, based on (i) acceptable BE studies on the 8 mg strength, (ii) proportional similarity across all strengths, and (iii) acceptable in vitro dissolution testing of all strengths.

Rosuvastatin Calcium, Tablets/Oral. Recommended studies: two studies. (1) Type of study: fasting design—single-dose, two-way crossover in vivo. Strength: 40 mg. Subjects: normal healthy males and females, general population. Additional comments: (2) Type of study: fed design—single-dose, two-way crossover in vivo. Strength: 40 mg. Subjects: normal healthy males and females, general population. Additional comments: Analytes to measure (in appropriate biological fluid): rosuvastatin in plasma. BE based on (90% CI): rosuvastatin. Waiver requests of in vivo testing: 5, 10, and 20 mg strengths based on (i) acceptable BE studies on the 40 mg strength, (ii) proportional similarity across all strengths, and (iii) acceptable dissolution testing of all strengths.

Saquinavir Mesylate, Capsules/Oral. Recommended studies: two studies. (1) Type of study: fasting design—single-dose, two-treatment, two-period crossover in vivo. Strength: 200 mg. Subjects: normal healthy males and females, general population. Additional comments: (2) Type of study: fed design—single-dose, two-treatment, two-period crossover in vivo. Strength: 200 mg. Subjects: normal healthy males and females, general population. Additional comments: Analytes to measure (in appropriate biological fluid): saquinavir in plasma. BE based on (90% CI): saquinavir. Waiver request of in vivo testing: not applicable.

Saquinavir Mesylate, Tablet/Oral. Recommended studies: two studies. (1) Type of study: fasting design—single-dose, two-treatment, two-period crossover in vivo. Strength: 500 mg. Subjects: normal healthy males and females, general population. Additional comments: (2) Type of study: fed design—single-dose, two-treatment, two-period crossover in vivo. Strength: 500 mg. Subjects: normal healthy males and females, general population. Additional comments: Analytes to measure (in appropriate biological fluid): saquinavir in plasma. BE based on (90% CI): saquinavir. Waiver request of in vivo testing: not applicable.

Sertraline Hydrochloride, Tablets/Oral. Recommended studies: two studies. (1) Type of study: fasting design—single-dose, two-treatment, two-period crossover in vivo. Strength: 100 mg. Subjects: normal healthy males and females, general population. Additional comments: Due to safety concerns, BE studies should be conducted on the 100 mg strength. (2) Type of study: fed design—single-dose, two-treatment, two-period crossover in vivo. Strength: 100 mg. Subjects: normal healthy males and females, general population. Additional comments: Analytes to measure (in appropriate biological fluid): sertraline in plasma. BE based on (90% CI): sertraline. Waiver request of in vivo testing: 25, 50, 150, and 200 mg based on (i) acceptable BE studies on the 100 mg strength, (ii) proportional similarity of the formulations across all strengths, and (iii) acceptable in vitro dissolution testing of all strengths.

Sibutramine Hydrochloride, Capsule/Oral. Recommended studies: two studies. (1) Type of study: fasting design—single-dose, two-treatment, two-period crossover in vivo. Strength: 15 mg. Subjects: normal healthy males and females, general population. Additional comments: Due to safety concerns, studies should not be conducted using doses higher than 15 mg. (2) Type of study: fed design—single-dose, two-treatment, two-period crossover in vivo. Strength: 15 mg. Subjects: normal healthy males and females, general population. Additional comments: Please see previous comment. Analytes to measure: sibutramine and the major first-generation active (desmethyl) metabolites M1 and M2, using an achiral assay. BE based on (90% CI): sibutramine. If sibutramine can be reliably measured, a CI approach for BE determination should be used for sibutramine. If sibutramine cannot be reliably measured, a CI approach for BE determination should be used for major first-generation active (desmethyl) metabolites M1 and M2. Waiver request of in vivo testing: 5 and 10 mg based on (i) acceptable BE studies on the 15 mg strength, (ii) proportional similarity of the formulations across all strengths, and (iii) acceptable in vitro dissolution testing of all strengths.

Sildenafil Citrate, Tablets/Oral. Recommended studies: two studies. (1) Type of study: fasting design—single-dose, two-way crossover in vivo. Strength: 100 mg. Subjects: normal, healthy males. Additional comments: (2) Type of study: fed design—single-dose, two-way crossover in vivo. Strength: 100 mg. Subjects: normal, healthy males. Additional comments: Analytes to measure (in appropriate biological fluid): sildenafil and active metabolite piperazine *N*-desmethylsildenafil in plasma. BE based on (90% CI): sildenafil. Please submit the metabolite data as supportive evidence of comparable therapeutic outcome. For the metabolite, the following data should be submitted: individual and mean concentrations, individual and mean PK parameters, and geometric means and ratios of means for AUC and C_{max}. Waiver request of in vivo testing: 25 and 50 mg based on (i) acceptable BE studies on the 100 mg strength, (ii) proportional similarity across all strengths, and (iii) acceptable in vitro dissolution testing of all strengths.

Simvastatin, Tablets/Oral. Recommended studies: two studies. (1) Type of study: fasting design—single-dose, two-way crossover in vivo. Strength: 80 mg. Subjects: normal healthy males and females, general population. Additional comments: (2) Type of study: fed design—single-dose, two-way crossover in vivo. Strength: 80 mg. Subjects: normal healthy males and females, general population. Additional comments: Analytes to measure: simvastatin and its beta-hydroxy acid metabolite in plasma. Please submit the metabolite data as supportive evidence of comparable therapeutic outcome. For the beta-hydroxy metabolite of simvastatin, the following data should be submitted: individual and mean concentrations, individual and mean PK parameters, and geometric means and ratios of means for AUC and C_{max}. BE based on (90% CI): simvastatin. Waiver request of in vivo testing: 5, 10, 20, and 40 mg based on (i) acceptable BE studies on the 80 mg strength, (ii) proportional similarity of the formulations across all strengths, and (iii) acceptable in vitro dissolution testing of all strengths. Please conduct comparative dissolution testing on 12 dosage units of all strengths of the test and reference products using the method specified in the USP method.

Sirolimus, Tablets/Oral. Recommended studies: two studies. (1) Type of study: fasting design—single-dose, two-treatment, two-period crossover in vivo. Strength: 2 mg. Subjects: normal healthy males and females, general population. Additional comments: (2) Type of study: fed design—single-dose, two-treatment, two-period crossover in vivo. Strength: 2 mg. Subjects: normal healthy males and females, general population. Additional comments: Analytes to measure (in appropriate biological fluid): sirolimus in plasma. BE based on (90% CI): sirolimus. Waiver request of in vivo testing: 1 mg based on (i) acceptable BE studies on the 2 mg strength, (ii) proportional similarity of the formulations across all strengths, and (iii) acceptable in vitro dissolution testing of all strengths.

Solifenacin Succinate, Tablet/Oral. Recommended studies: two studies. (1) Type of study: fasting design—single dose, parallel in vivo. Strength: 10 mg. Subjects: normal healthy males and females, general population. Additional comments: Females should not be pregnant or lactating and, if applicable, should practice abstention or contraception during the study. Note: As an option, you may conduct this study using a single dose, two-way crossover design. As an additional option for either the crossover or parallel design, you may truncate the AUC at 72 h, provided the drug demonstrates low intrasubject variability in distribution and clearance. (2) Type of study: fed design—single dose, parallel in vivo. Strength: 10 mg. Subjects: normal healthy males and females, general population. Additional comments: Please see previous comments. Analytes to measure (in appropriate biological fluid): solifenacin in plasma. BE based on (90% CI): solifenacin. Waiver request of in vivo testing: 5 mg based on (i) acceptable BE studies on the 10 mg strength, (ii) proportional similarity of the formulations across all strengths, and (iii) acceptable in vitro dissolution testing of all strengths.

Stavudine, Capsules/Oral. Recommended studies: two studies. (1) Type of study: fasting design—single-dose, two-treatment, two-period crossover in vivo. Strength: 40 mg. Subjects: normal healthy males and females, general population. Additional comments: (2) Type of study: fed design—single-dose, two-treatment, two-period crossover in vivo. Strength: 40 mg. Subjects: normal healthy males and females, general population. Additional comments: Analytes to measure (in appropriate biological fluid): stavudine in plasma. BE based on (90% CI): stavudine. Waiver request of in vivo testing: 15, 20, and 30 mg based on (i) acceptable BE studies on the 40 mg strength, (ii) acceptable dissolution testing of the 15, 20, 30, and 40 mg strengths, and (iii) proportional similarity in the formulations of the 15, 20, 30, and 40 mg strengths.

Sulfamethoxazole; Trimethoprim, Suspension/Oral. Recommended studies: one study. Type of study: fasting design—single-dose, two-treatment, two-period crossover in vivo. Strength: 200 mg/40 mg per 5 mL. Subjects: normal healthy males and females, general population. Additional comments: Analytes to measure (in appropriate biological fluid): sulfamethoxazole and trimethoprim in plasma. BE based on (90% CI): sulfamethoxazole and trimethoprim. Waiver request of in vivo testing: not applicable.

Sumatriptan Succinate, Tablets/Oral. Recommended studies: two studies. (1) Type of study: fasting design—single-dose, two-way crossover in vivo. Strength: 100 mg. Subjects: normal, healthy males and females, general population. Additional comments: (2) Type of study: fed design—single-dose, two-way crossover in vivo. Strength: 100 mg. Subjects: normal, healthy males and females, general population. Additional comments: Analytes to measure (in appropriate biological fluid): sumatriptan in plasma. BE based on (90% CI): sumatriptan. Waiver request of in vivo testing: 25 and 50 mg, based on (i) acceptable BE studies on the 100 mg strength, (ii) proportional similarity across all strengths, and (iii) acceptable in vitro dissolution testing of all strengths.

Tacrolimus, Capsule/Oral. Recommended studies: two studies. (1) Type of study: fasting design—single-dose, two-treatment, two-period crossover in vivo. Strength: 5 mg. Subjects: normal healthy males and females, general population. Additional comments: Females must have a negative baseline pregnancy test within 24 h prior to receiving the drug. Females should not be pregnant or lactating and, if applicable, should practice abstention or contraception during the study. (2) Type of study: fed design—single-dose, two-treatment, two-period crossover in vivo. Strength: 5 mg. Subjects: normal healthy males and females, general population. Additional comments: Please see comments earlier. Analytes to measure (in appropriate biological fluid): tacrolimus in whole blood. BE based on (90% CI): tacrolimus. Waiver request of in vivo testing: 0.5 and 1 mg, based on (i) acceptable BE studies on the 5 mg strength, (ii) proportional similarity of the formulations across all strengths, and (iii) acceptable in vitro dissolution testing of all strengths.

Tadalafil, Tablets/Oral. Recommended studies: two studies. (1) Type of study: fasting design—single-dose, two-way crossover in vivo. Strength: 20 mg. Subjects: normal healthy males, general population. Additional comments: (2) Type of study: fed design—single-dose, two-way crossover in vivo. Strength: 20 mg. Subjects: normal healthy males, general population. Additional comments: Analytes to measure (in appropriate biological fluid): tadalafil in plasma. BE based on (90% CI): tadalafil. Waiver request of in vivo testing: 5 and 10 mg based on (i) acceptable BE studies on the 20 mg strength, (ii) proportional similarity across all strengths, and (iii) acceptable in vitro dissolution testing of all strengths.

Tamsulosin Hydrochloride, Capsules/Oral. Recommended studies: two studies. (1) Type of study: fasting design—single-dose, two-way crossover in vivo. Strength: 0.4 mg. Subjects: normal, healthy males and females, general population. Additional comments: (2) Type of study: fed design—single-dose, two-way crossover in vivo. Strength: 0.4 mg. Subjects: normal, healthy males and females, general population. Additional comments: Analytes to measure (in appropriate biological fluid): tamsulosin in plasma. BE based on (90% CI): tamsulosin. Waiver request of in vivo testing: not applicable.

Telithromycin, Tablets/Oral. Recommended studies: two studies. (1) Type of study: fasting design—single-dose, two-way crossover in vivo. Strength: 400 mg. Subjects: normal healthy males and females, general population. Females should not be pregnant and, if applicable, should practice abstention or contraception during the study. Additional comments: The study design should include a screen for signs and symptoms of possible hepatotoxicity prior to administering each subsequent dose of telithromycin in a crossover or replicate crossover design. In order to minimize the risk of hepatotoxicity, please do not exceed a 400 mg dose in the BE study. Subjects who consume alcohol should be excluded from BE studies of telithromycin. (2) Type of study: fed design—single-dose, two-way crossover in vivo. Strength: 400 mg. Subjects: normal healthy males and females, general population. Females should not be pregnant and, if applicable, should practice abstention or contraception during the study. Additional comments: Please see previous comments. Analytes to measure (in appropriate biological fluid): Telithromycin in plasma. BE based on (90% CI): telithromycin. Waiver request of in vivo testing: 300 mg based on (i) acceptable BE studies on the 400 mg strength, (ii) proportional similarity across both strengths, and (iii) acceptable in vitro dissolution testing of both strengths.

Telmisartan, Tablets/Oral. Recommended studies: two studies. (1) Type of study: fasting design—single-dose, two-treatment, two-period crossover in vivo. Strength: 80 mg. Subjects: normal healthy males and females, general population. Additional comments: Females should not be pregnant and, if applicable, should practice abstention or contraception during the study. (2) Type of study: fed design—single-dose, two-treatment, two-period crossover in vivo. Strength: 80 mg. Subjects: normal healthy males and females, general population. Additional comments: Please see previous comment. Analytes to measure (in appropriate biological fluid): telmisartan in plasma. BE based on (90% CI): telmisartan. Waiver request of in vivo testing: 20 and 40 mg, based on (i) acceptable BE studies on the 80 mg strength, (ii) proportional similarity of the formulations across all strengths, and (iii) acceptable in vitro dissolution testing of all strengths.

Tenofovir Disoproxil Fumarate, Tablet/Oral. Recommended studies: two studies. (1) Type of study: fasting design—single-dose, two-treatment, two-period crossover in vivo. Strength: 300 mg. Subjects: normal healthy males and females, general population. Additional comments: (2) Type of study: fed design—single-dose, two-treatment, two-period crossover in vivo. Strength: 300 mg. Subjects: normal healthy males and females, general population. Additional comments: Analytes to measure (in appropriate biological fluid): tenofovir in serum. BE based on (90% CI): tenofovir. Waiver request of in vivo testing: not applicable.

Terazosin Hydrochloride, Capsules/Oral. Recommended studies: one study. Type of study: fasting design—single-dose, two-way crossover in vivo. Strength: 2 mg. Subjects: normal healthy males and females, general population. Additional comments: Due to safety concerns, the studies should be conducted using the 2 mg strength. Analytes to measure: terazosin in plasma. BE based on (90% CI): terazosin. Waiver request of in vivo testing: 1, 5, and 10 mg based on (i) acceptable BE studies on the 2 mg strength, (ii) proportional similarity across all strengths, and (iii) acceptable in vitro dissolution testing of all strengths.

Terbinafine Hydrochloride, Tablet/Oral. Recommended studies: two studies. (1) Type of study: fasting design—randomized, single-dose, two-treatment, two-period crossover in vivo. Strength: 250 mg. Subjects: normal healthy males and females, general population. Additional comments: (2) Type of study: fed design—randomized, single-dose, two-treatment, two-period crossover in vivo. Strength: 250 mg. Subjects: normal healthy males and females, general population. Additional comments: Analytes to measure: terbinafine in plasma. BE based on (90% CI): terbinafine. Waiver request of in vivo testing: not applicable.

Testosterone, Extended-Release Tablets/Buccal. Recommended studies: two studies. (1) Type of study: fasting design—single-dose, two-way crossover in vivo. Strength: 30 mg. Subjects: testosterone-deficient (hypogonadal) males. Additional comments: Subjects should not currently be receiving any treatment for their hypogonadism. The inclusion criterion for testosterone-deficient (hypogonadal) males is serum testosterone levels below 2.5 ng/mL. At least three predose levels will serve as baseline. A "fed" BE study is not recommended because the product is a buccal adhesive, not to be ingested. This obviates the need for oral dose dumping assessment due to food. (2) Type of study: In vitro adhesion comparative performance testing study design—a tensiometry study is recommended to compare the peak detachment force for test and reference products. Water is recommended between the buccal tablets and the base plate of the tensiometer. The loading weight and length of time the loading weight is applied to press the buccal tablet into contact with the base plate should be specified. Following removal of the weight, the rate at which the buccal tablet is pulled away from the base plate should be specified. The peak detachment force should be measured as the force required to detach the buccal tablet from the base plate. The comparative adhesion test should be conducted using 12 individual units of the test and reference products. Prior to conducting studies for submission to the ANDA, the firm should determine appropriate loading weight, length of time the loading weight is applied to press the buccal tablet into contact with the base plate of the tensiometer, and the rate at which the buccal tablet is pulled away from the base plate. These studies should be conducted to assure the appropriateness of the test conditions to the test and reference products. Analytes to measure (in appropriate biological fluid): total testosterone in plasma. BE based on (90% CI): Baseline-adjusted testosterone. Waiver request of in vivo testing: not applicable.

Ticlopidine Hydrochloride, Tablet/Oral. Recommended studies: two studies. (1) Type of study: fasting design—single-dose, two-treatment, two-period crossover in vivo. Strength: 250 mg. Subjects: normal healthy males and females, general population. Additional comments: (2) Type of study: fed design—single-dose, two-treatment, two-period crossover in vivo. Strength: 250 mg. Subjects: normal healthy males and females, general population. Additional comments: Analytes to measure: ticlopidine in plasma. BE based on (90% CI): ticlopidine. Waiver request of in vivo testing: not applicable.

Tinidazole, Tablet/Oral. Recommended studies: one study. Type of study: fed design—single-dose, two-treatment, two-period crossover in vivo. Strength: 500 mg. Subjects: normal healthy males and females, general population. Additional comments: Analytes to measure (in appropriate biological fluid): tinidazole in plasma. BE based on (90% CI): tinidazole. Waiver request of in vivo testing: 250 mg based on (i) acceptable BE studies of the 40 mg strength, (ii) proportional similarity across all strengths, and (iii) acceptable in vitro dissolution testing of all strengths. Dissolution testing to document BE: Apparatus, USP Apparatus I (basket); rotation speed, 100 rpm; medium, 0.1 N HCl (or 0.1 N HCl with NaCl) at pH 1.2, pH 4.5 acetate buffer, pH 6.8 phosphate buffer and water; volume, 900 mL; temperature, 37°C; sample times, 5, 10, 15, 20, 25, 30, and 40 min or as needed for profile comparisons. Additional comments: All raw data (test and reference products) should be submitted with means at each sampling point, the range (minimum and maximum values), the %CV, and f2 value tabulated (if appropriate). The dissolution testing should be conducted on 12 units from the same lot numbers that are used in the in vivo BE study.

Tipranavir, Capsules/Oral. Recommended studies: two studies. (1) Type of study: fasting design—single-dose, two-treatment, two-period crossover in vivo. Strength: 250 mg (please administer a 500 mg dose, 2 × 250 mg). Subjects: normal healthy males and females, general population. Additional comments: Females must have a negative baseline pregnancy test within 24 h prior to receiving the drug. Females should not be pregnant or lactating and, if applicable, should practice abstention or contraception during the study. (2) Type of study: fed design—single-dose, two-treatment, two-period crossover in vivo. Strength: 250 mg (please administer a 500 mg dose, 2 × 250 mg). Subjects: normal healthy males and females, general population. Additional comments: Please see comment earlier. Analytes to measure (in appropriate biological fluid): tipranavir in plasma. BE based on (90% CI): tipranavir. Waiver request of in vivo testing: not applicable.

Tizanidine Hydrochloride, Capsule/Oral. Recommended studies: two studies. (1) Type of study: fasting design—single-dose, two-treatment, two-period crossover in vivo. Strength: 6 mg. Subjects: normal healthy males and females, general population. Additional comments: (2) Type of study: fed design—single-dose, two-treatment, two-period crossover in vivo. Strength: 6 mg.

Subjects: normal healthy males and females, general population. Additional comments: Analytes to measure (in appropriate biological fluid): tizanidine in plasma. BE based on (90% CI): tizanidine. Waiver request of in vivo testing: 2 and 4 mg based on (i) acceptable BE studies on the 6 mg strength, (ii) proportional similarity of the formulations across all strengths, and (iii) acceptable in vitro dissolution testing of all strengths.

Tolterodine Tartrate, Extended-Release Capsules/Oral. Recommended studies: two studies. (1) Type of study: fasting design—single-dose, two-treatment, two-period crossover in vivo. Strength: 4 mg. Subjects: normal healthy males and females, general population. Additional comments: (2) Type of study: fed design—single-dose, two-treatment, two-period crossover in vivo. Strength: 4 mg. Subjects: normal healthy males and females, general population. Additional comments: Analytes to measure (in appropriate biological fluid): tolterodine and the 5-hydroxymethyl tolterodine (5-OHM) metabolite in plasma. For the metabolite, the following data should be submitted: individual and mean concentrations, individual and mean PK parameters, and geometric means and ratios of means for AUC and C_{max}. BE based on (90% CI): tolterodine. Waiver request of in vivo testing: 2 mg, based on (i) acceptable BE studies on the 4 mg strength, (ii) proportional similarity of the formulations across all strengths, and (iii) acceptable in vitro dissolution testing of all strengths. For modified-release products, dissolution profiles generated using USP Apparatus I at 100 rpm and/or Apparatus II at 50 rpm in at least three dissolution media (pH 1.2, 4.5, and 6.8 buffer, water) should be submitted in the application. Agitation speeds may have to be increased if appropriate. It is acceptable to add a small amount of surfactant, if necessary. The following sampling times are recommended: 1, 2, and 4 h and every 2 h thereafter, until at least 80% of the drug is dissolved. Comparative dissolution profiles should include individual tablet data as well as the mean, range, and standard deviation at each time point for 12 capsules.

Tolterodine Tartrate, Tablet/Oral. Recommended studies: two studies. (1) Type of study: fasting design—randomized, single-dose, two-treatment, two-period, two-sequenced crossover in vivo. Strength: 2 mg. Subjects: normal healthy males and females, general population. Additional comments: (2) Type of study: fed design—randomized, single-dose, two-treatment, two-period, two-sequenced crossover in vivo. Strength: 2 mg. Subjects: normal healthy males and females, general population. Additional comments: Analytes to measure (in appropriate biological fluid): tolterodine and the active metabolite, 5-OHM, in plasma. BE based on (90% CI): tolterodine. Please submit the metabolite data as supportive evidence of comparable therapeutic outcome. For the metabolite, the following data should be submitted: individual and mean concentrations, individual and mean PK parameters, and geometric means and ratios of means for AUC and C_{max}. Waiver request of in vivo testing: 1 mg based on (i) acceptable BE studies on the 2 mg strength, (ii) proportional similarity of the formulations across all strengths, and (iii) acceptable in vitro dissolution testing of all strengths.

Topiramate Sprinkle, Capsule/Oral. Recommended studies: two studies. (1) Type of study: fasting design—single-dose, two-treatment, two-period crossover in vivo. Strength: 25 mg. Subjects: normal healthy males and females, general population. Additional comments: Females must have a negative baseline pregnancy test within 24 h prior to receiving the drug. Females should not be pregnant or lactating and, if applicable, should practice abstention or contraception during the study. (2) Type of study: fed design—single-dose, two-treatment, two-period crossover in vivo. Strength: 25 mg. Subjects: normal healthy males and females, general population. Additional comments: Please see previous comments. Analytes to measure (in appropriate biological fluid): topiramate in plasma. BE based on (90% CI): topiramate. Waiver request of in vivo testing: 15 mg based on (i) acceptable BE studies on the 25 mg strength, (ii) proportional similarity of the formulations across all strengths, and (iii) acceptable in vitro dissolution testing of all strengths.

Topiramate, Tablet/Oral. Recommended studies: two studies. (1) Type of study: fasting design—single-dose, two-treatment, two-period crossover in vivo. Strength: 25 mg. Subjects: normal healthy males and females, general population. Additional comments: Due to safety concerns, studies should be conducted on the 25 mg strength. Animal studies with topiramate have demonstrated selective developmental toxicity, including teratogenicity. Although no studies have been conducted in pregnant women taking topiramate, in postmarketing experience, cases of hypospadias have been reported in male infants exposed in utero to topiramate, with or without other anticonvulsants; however, a causal relationship with topiramate has not been established. Therefore, the following precautions are recommended for the BE study: pregnant women should be excluded from the study, and a negative pregnancy test should be required within 24 h before dosing for all women of childbearing potential. Women of childbearing potential should be enrolled only if using an effective method of contraception. Written informed consent must include the finding of birth defects in animal studies and the unknown risk to a human fetus if exposed to this drug. (2) Type of study: fed design—single-dose, two-treatment, two-period crossover in vivo. Strength: 25 mg. Subjects: normal healthy males and females, general population. Additional comments: Please see comments earlier. Analytes to measure (in appropriate biological fluid): topiramate in plasma. BE based on (90% CI): topiramate. Waiver request of in vivo testing: 50, 100, and 200 mg based on (i) acceptable BE studies on the 25 mg strength, (ii) proportional similarity of the formulations across all strengths, and (iii) acceptable in vitro dissolution testing of all strengths.

Torsemide, Tablets/Oral. Recommended studies: two studies. (1) Type of study: fasting design—single-dose, two-treatment, two-period crossover in vivo. Strength: 20 mg. Subjects: normal healthy males and females, general population. Additional comments: Due to safety concerns associated with administering torsemide tablets, 100 mg, to healthy subjects, in vivo BE studies should be conducted on the 20 mg strength. (2) Type of study: fed design—single-dose, two-treatment, two-period crossover in vivo. Strength: 20 mg. Subjects: normal healthy males and females, general population. Additional comments: Please see comment earlier. Analytes to measure (in appropriate biological fluid): torsemide in plasma. BE based on (90% CI): torsemide. Waiver request of in vivo testing: 5, 10, and 100 mg, based on (i) acceptable BE studies on the 20 mg strength, (ii) proportional similarity of the formulations across all strengths, and (iii) acceptable in vitro dissolution testing of all strengths.

Tramadol, Extended-Release Tablets/Oral. Recommended studies: two studies. (1) Type of study: fasting design—single-dose, two-treatment, two-period crossover in vivo. Strength: 100 mg. Subjects: normal healthy males and females, general population. Additional comments: (2) Type of study: fed design—single-dose, two-treatment, two-period crossover in vivo. Strength: 100 mg. Subjects: normal healthy males and females, general population. Additional comments: Analytes to measure (in appropriate biological fluid): tramadol in plasma by achiral assay (nonstereospecific method). BE based on (90% CI): tramadol. Waiver request of in vivo testing: 200 and 300 mg based on (i) acceptable BE studies on the 100 mg strength, (ii) proportional similarity of the formulations across all strengths, and (iii) acceptable in vitro dissolution testing of all strengths. In addition to the aforementioned method, for modified-release products, dissolution profiles on 12 dosage units each of test and reference products generated using USP Apparatus I at 100 rpm and/or Apparatus II at 50 rpm in at least three dissolution media (pH 1.2, 4.5, and 6.8 buffer) should be submitted in the application. Agitation speeds may have to be increased if appropriate. It is acceptable to add a small amount of surfactant, if necessary. Please include early sampling times of 1, 2, and 4 h and continue every 2 h until at least 80% of the drug is released, to provide assurance against premature release of drug (dose dumping) from the formulation. Due to concerns of dose dumping from this drug product when taken with alcohol, please conduct additional dissolution testing using various concentrations of ethanol in the dissolution medium, as follows: testing conditions, 900 mL, 0.1 N HCl, Apparatus I (basket) at 75 rpm, with and without the alcohol (see later). Test 1: 12 units tested according to the proposed method (with 0.1 N HCl), with data collected every 15 min for a total of 2 h. Test 2: 12 units analyzed by substituting 5% (v/v) of test medium with Alcohol USP and data collection every 15 min for a total of 2 h. Test 3: 12 units analyzed by substituting 20% (v/v) of test medium with Alcohol USP and data collection every 15 min for a total of 2 h. Test 4: 12 units analyzed by substituting 40% (v/v) of test medium with Alcohol USP and data collection every 15 min for a total of 2 h. Both test and RLD products must be tested accordingly, and data must be provided on individual unit, means, range, and %CV on both strengths.

Tramadol Hydrochloride, Tablets/Oral. Recommended studies: two studies. (1) Type of study: fasting design—single-dose, two-way crossover in vivo. Strength: 50 mg. Subjects: normal healthy males and females, general population. Additional comments: (2) Type of study: fed design—single-dose, two-way crossover in vivo. Strength: 50 mg. Subjects: normal healthy males and females, general population. Additional comments: Analytes to measure: tramadol in plasma using an achiral assay. BE based on (90% CI): tramadol. Waiver request of in vivo testing: not applicable.

Tramadol Hydrochloride; Acetaminophen, Tablets/Oral. Recommended studies: two studies. (1) Type of study: fasting design—single-dose, two-treatment, two-period crossover in vivo. Strength: 37.5 mg/325 mg. Subjects: normal healthy males and females, general population. Additional comments: (2) Type of study: fed design—single-dose, two-treatment, two-period crossover in vivo. Strength: 37.5 mg/325 mg. Subjects: normal healthy males and females, general population. Additional comments: Analytes to measure (in appropriate biological fluid): tramadol using an achiral assay and acetaminophen. BE based on (90% CI): tramadol and acetaminophen. Waiver request of in vivo testing: not applicable.

Trandolapril, Tablet/Oral. Recommended studies: two studies. (1) Type of study: fasting design—single-dose, two-treatment, two-period crossover in vivo. Strength: 4 mg. Subjects: normal healthy males and females, general population. Additional comments: Females should not be pregnant and, if applicable, should practice abstention or contraception during the study. (2) Type of study: fed design—single-dose, two-treatment, two-period crossover in vivo. Strength: 4 mg. Subjects: normal healthy males and females, general population. Additional comments: Please see previous comment. Analytes to measure: trandolapril and its active metabolite, trandolaprilat, in plasma. BE based on (90% CI): trandolapril. Please submit the metabolite data as supportive evidence of comparable therapeutic outcome. For the metabolite, the following data should be submitted: individual and mean concentrations, individual and mean PK parameters, and geometric means and ratios of means for AUC and C_{max}. Waiver request of in vivo testing: 1 and 2 mg based on (i) acceptable BE studies on the 4 mg strength, (ii) proportional similarity of the formulations across all strengths, and (iii) acceptable in vitro dissolution testing of all strengths.

Triamterene, Capsule/Oral. Recommended studies: one study. Type of study: fasting design—single-dose, two-treatment, two-period crossover in vivo. Strength: 100 mg. Subjects: normal healthy males and females, general population. Additional comments: Analytes to measure (in appropriate biological fluid): triamterene in plasma. BE based on (90% CI): triamterene. Waiver request of in vivo testing: 50 mg based on (i) acceptable BE study on the 100 mg strength, (ii) proportional similarity of the formulations across all strengths, and (iii) acceptable in vitro dissolution testing of all strengths. Please conduct comparative dissolution testing on 12 dosage units of all strengths of the test and reference products using the USP method.

Trospium Chloride, Tablet/Oral. Recommended studies: one study. Type of study: fasting design—single-dose, two-treatment, two-period crossover in vivo. Strength: 20 mg. Subjects: normal healthy males and females, general population. Additional comments: Females should not be pregnant or lactating and, if applicable, should practice abstention or contraception during the study. Analytes to measure (in appropriate biological fluid): trospium in plasma. BE based on (90% CI): trospium. Waiver request of in vivo testing: not applicable.

Valacyclovir Hydrochloride, Tablets/Oral. Recommended studies: two studies. (1) Type of study: fasting design—single-dose, two-treatment, two-period crossover in vivo. Strength: 1000 mg. Subjects: normal healthy males and females, general population. Additional comments: (2) Type of study: fed design—single-dose, two-treatment, two-period crossover in vivo. Strength: 1000 mg. Subjects: normal healthy males and females, general population. Additional comments: Analytes to measure (in appropriate biological fluid): valacyclovir and its metabolite, acyclovir, in both studies. If valacyclovir plasma concentrations can be reliably measured and its PK parameters accurately determined, you should analyze the valacyclovir data using the CI approach. The acyclovir data can be used to provide supportive evidence of comparable therapeutic outcome. BE based on (90% CI): valacyclovir. If valacyclovir cannot be reliably measured, you should analyze the acyclovir data obtained from these studies using the CI approach. Waiver request of in vivo testing: 500 mg based on (i) acceptable BE studies on the 1000 mg strength, (ii) proportional similarity of the formulations across all strengths, and (iii) acceptable in vitro dissolution testing of all strengths.

Valsartan, Tablets/Oral. Recommended studies: two studies. (1) Type of study: fasting design—single-dose, two-treatment, two-period crossover in vivo. Strength: 320 mg. Subjects: normal healthy males and females, general population. Additional comments: A dose of 320 mg can be safely administered to healthy subjects. Please include provisions for appropriate monitoring and intervention in the case of possible drug-related adverse events (e.g., subjects complaining of dizziness/lightheadedness should have blood pressure/heart rate assessed). Females should not be pregnant and, if applicable, should practice abstention or contraception during the study. (2) Type of study: fed design—single-dose, two-treatment, two-period crossover in vivo. Strength: 320 mg. Subjects: normal healthy males and females, general population. Additional comments: Please see previous comment. Analytes to measure (in appropriate biological fluid): valsartan in plasma. BE based on (90% CI): valsartan. Waiver request of in vivo testing: 40, 80, and 160 mg based on (i) acceptable BE studies on the 320 mg strength, (ii) proportional similarity of the formulations across all strengths, and (iii) acceptable in vitro dissolution testing of all strengths.

Vardenafil Hydrochloride, Tablets/Oral. Recommended studies: two studies. (1) Type of study: fasting design—single-dose, two-way crossover in vivo. Strength: 20 mg. Subjects: normal healthy males, general population. Additional comments: (2) Type of study: fed design—single-dose, two-way crossover in vivo. Strength: 20 mg. Subjects: normal healthy males, general population. Additional comments: Analytes to measure (in appropriate biological fluid): vardenafil in plasma. BE based on (90% CI): vardenafil. Waiver request of in vivo testing: 2.5, 5, and 10 mg, based on (i) acceptable BE studies on the 20 mg strength, (ii) proportional similarity across all strengths, and (iii) acceptable in vitro dissolution testing of all strengths.

Varenicline Tartrate, Tablet/Oral. Recommended studies: two studies. (1) Type of study: fasting design—single-dose, two-treatment, two-period crossover in vivo. Strength: 1.0 mg. Subjects: normal healthy males and females, general population; smokers and nonsmokers may be used. Additional comments: Females should not be pregnant and, if applicable, should practice abstention or contraception during the study. (2) Type of study: fed design—single-dose, two-treatment, two-period crossover in vivo. Strength: 1.0 mg. Subjects: normal healthy males and females, general population; smokers and nonsmokers may be used. Additional comments: Please see previous comments. Analytes to measure (in appropriate biological fluid): varenicline in plasma. BE based on (90% CI): varenicline. Waiver request of in vivo testing: 0.5 mg, based on (i) acceptable BE studies of the 1 mg strength, (ii) proportional similarity across all strengths, and (iii) acceptable in vitro dissolution testing of all strengths.

Venlafaxine Hydrochloride, Extended-Release Capsule/Oral. Recommended studies: two studies. (1) Type of study: fed design—single-dose, two-treatment, two-period crossover in vivo. Strength: 150 mg. Subjects: normal healthy males and females, general population. Additional comments: Due to safety concerns, BE studies under fasting conditions are not recommended. (2) Type of study: fed sprinkle design—single-dose, two-treatment, two-period crossover in vivo. Strength: 150 mg. Subjects: normal healthy males and females, general population. Additional comments: Please administer the dose after sprinkling the entire contents of the capsule on a teaspoonful of applesauce in accordance with the approved labeling of the reference product under fed conditions. Please see previous comment. Analytes to measure: venlafaxine, and its metabolite *O*-desmethylvenlafaxine, in plasma. BE based on (90% CI): venlafaxine. Waiver request of in vivo testing: 37.5 and 75 mg based on (i) acceptable BE studies on the 150 mg strength, (ii) proportional similarity of the formulations across all strengths, and (iii) acceptable in vitro dissolution testing of all strengths. In addition to the aforementioned method, for modified-release products, dissolution profiles on 12 dosage units each of test and reference products generated using USP Apparatus I at 100 rpm and/or Apparatus II at 50 rpm in at least three dissolution media (pH 1.2, 4.5, and 6.8 buffer) should be submitted in the application. Agitation speeds may have to be increased if appropriate.

It is acceptable to add a small amount of surfactant, if necessary. Please include early sampling times of 1, 2, and 4 h and continue every 2 h until at least 80% of the drug is released, to provide assurance against premature release of drug (dose dumping) from the formulation.

Verapamil Hydrochloride, Extended-Release Capsules/Oral. Recommended studies: three studies. (1) Type of study: fasting design—single-dose, two-way crossover in vivo. Strength: 360 mg. Subjects: normal healthy males and females, general population. Additional comments: (2) Type of study: fed design—single-dose, two-way crossover in vivo. Strength: 360 mg. Subjects: normal healthy males and females, general population. Additional comments: (3) Type of study: fasting, sprinkled-over-applesauce design—single-dose, two-way crossover in vivo. Strength: 360 mg. Subjects: normal healthy males and females, general population. Additional comments: Analytes to measure: verapamil and its metabolite, norverapamil, in plasma. Please submit the metabolite data as supportive evidence of the comparable therapeutic outcome. For the metabolite, the following data should be submitted: individual and mean concentrations, individual and mean PK parameters, and geometric means and ratios of means for AUC and C_{max}. BE based on (90% CI): verapamil. Waiver request of in vivo testing: 120, 180, and 240 mg based on (i) acceptable BE studies on the 360 mg strength, (ii) proportional similarity across all strengths, and (iii) acceptable in vitro dissolution testing of all strengths. For modified-release products, dissolution profiles generated using USP Apparatus I at 100 rpm and/or Apparatus II at 50 rpm in at least three dissolution media (pH 1.2, 4.5, and 6.8 phosphate buffer, water) should be submitted in the application. Agitation speeds may have to be increased if appropriate. It is acceptable to add a small amount of surfactant, if necessary. The following sampling times are recommended: 1, 2, 4, and every 2 h thereafter, until at least 80% of the labeled content is dissolved. Comparative dissolution profiles should include individual tablet data as well as the mean, range, and standard deviation at each time point for 12 capsules.

Verapamil Hydrochloride, Extended-Release Capsules/Oral. Recommended studies: three studies. (1) Type of study: fasting, bedtime (PM)-dosing design—single-dose, two-way crossover in vivo. Strength: 300 mg. Subjects: normal healthy males and females, general population. Additional comments: PM dosing. (2) Type of study: fed, morning (AM)-dosing design—single-dose, two-way crossover in vivo. Strength: 300 mg. Subjects: normal healthy males and females, general population. Additional comments: AM dosing. (3) Type of study: fasting, sprinkled-over-applesauce, AM-dosing design—single-dose, two-way crossover in vivo. Strength: 300 mg. Subjects: normal healthy males and females, general population. Additional comments: AM dosing. Analytes to measure: verapamil and its metabolite, norverapamil, in plasma utilizing a validated LC–MS/MS method. For norverapamil, the following data should be submitted: individual and mean concentrations, individual and mean PK parameters, and geometric means and ratios of means for AUC and C_{max}. The following sampling times are recommended: predose and 2, 3, 4, 6, 7, 8, 9, 10, 12, 14, 16, 20, 24, 30, 36, and 48 h post dose. BE based on (90% CI): verapamil. Waiver request of in vivo testing: 100 and 200 mg based on (i) acceptable BE studies on the 300 mg strength, (ii) proportional similarity across all strengths, and (iii) acceptable in vitro dissolution testing of all strengths. For modified-release products, dissolution profiles generated using USP Apparatus I at 100 rpm and/or Apparatus II at 50 rpm in at least three dissolution media (pH 1.2, 4.5, and 6.8 phosphate buffer, water) should be submitted in the application. Agitation speeds may have to be increased if appropriate. It is acceptable to add a small amount of surfactant, if necessary. The following sampling times are recommended: 1, 2, 4, and every 2 h thereafter, until at least 80% of the labeled content is dissolved. Comparative dissolution profiles should include individual tablet data as well as the mean, range, and standard deviation at each time point for 12 capsules.

Voriconazole, Suspension/Oral. Recommended studies: one study. Type of study: fasting design—single-dose, two-way crossover in vivo. Strength: 200 mg/5 mL. Subjects: normal healthy males and females, general population. Females should not be pregnant and, if applicable, should practice abstention or contraception during the study. Additional comments: Analytes to measure (in appropriate biological fluid): voriconazole in plasma. BE based on (90% CI): voriconazole. Waiver request of in vivo testing: not applicable.

Zafirlukast, Tablets/Oral. Recommended studies: one study. Type of study: fasting design—single-dose, two-treatment, two-period crossover in vivo. Strength: 20 mg. Subjects: normal healthy males and females, general population. Additional comments: Analytes to measure (in appropriate biological fluid): zafirlukast in plasma. BE based on (90% CI): zafirlukast. Waiver request of in vivo testing: 10 mg based on acceptable (i) BE studies on the 20 mg tablet, and (ii) proportional similarity of the formulations and (iii) acceptable in vitro dissolution testing of all strengths.

Zalcitabine, Tablets/Oral. Recommended studies: two studies. (1) Type of study: fasting design—single-dose, two-treatment, two-period crossover in vivo. Strength: 0.75 mg. Subjects: normal healthy males and females, general population. Additional comments: (2) Type of study: fed design—single-dose, two-treatment, two-period crossover in vivo. Strength: 0.75 mg. Subjects: normal healthy males and females, general population. Additional comments: Analytes to measure (in appropriate biological fluid): zalcitabine in plasma. BE based on (90% CI): zalcitabine. Waiver request of in vivo testing: 0.375 mg based on (i) acceptable BE studies on the 0.75 mg strength, (ii) proportional similarity of the formulations across all strengths, and (iii) acceptable in vitro dissolution testing of all strengths.

Zaleplon, Tablets/Oral. Recommended studies: two studies. (1) Type of study: fasting design—single-dose, two-way crossover in vivo. Strength: 10 mg. Subjects: normal healthy males and females, general population. Additional comments: Patients should be advised not to drive if they are experiencing drowsiness and/or dizziness at the end of the study. (2) Type of study: fed design—single-dose, two-way crossover in vivo. Strength: 10 mg. Subjects: normal healthy males and females, general population. Additional comments: Analytes to measure: zaleplon in plasma. BE based on (90% CI): zaleplon. Waiver request of in vivo testing: 5 mg based on (i) acceptable BE studies on the 10 mg strength, (ii) proportional similarity across all strengths, and (iii) acceptable in vitro dissolution testing of all strengths.

Zidovudine, Capsules/Oral. Recommended studies: two studies. (1) Type of study: fasting design—single-dose, two-treatment, two-period crossover in vivo. Strength: 100 mg. Subjects: normal healthy males and females, general population. Additional comments: (2) Type of study: fed design—single-dose, two-treatment, two-period crossover in vivo. Strength: 100 mg. Subjects: normal healthy males and females, general population. Additional comments: Analytes to measure (in appropriate biological fluid): zidovudine in plasma. BE based on (90% CI): zidovudine. Waiver request of in vivo testing: not applicable. Please conduct dissolution testing on 12 dosage units each of the test and reference products using the USP method.

Zidovudine, Tablets/Oral. Recommended studies: two studies. (1) Type of study: fasting design—single-dose, two-treatment, two-period crossover in vivo. Strength: 300 mg. Subjects: normal healthy males and females, general population. Additional comments: (2) Type of study: fed design—single-dose, two-treatment, two-period crossover in vivo. Strength: 300 mg. Subjects: normal healthy males and females, general population. Additional comments: Analytes to measure (in appropriate biological fluid): zidovudine in plasma. BE based on (90% CI): zidovudine. Waiver request of in vivo testing: not applicable. Please conduct comparative dissolution testing on 12 dosage units of all strengths of the test and reference products.

Zileuton, Tablets/Oral. Recommended studies: two studies. (1) Type of study: fasting design—single-dose, two-treatment, two-period crossover in vivo. Strength: 600 mg. Subjects: normal healthy males and females, general population. Additional comments: (2) Type of study: fed design—single-dose, two-treatment, two-period crossover in vivo. Strength: 600 mg. Subjects: normal healthy males and females, general population. Additional comments: Analytes to measure (in appropriate biological fluid): zileuton in plasma. BE based on (90% CI): zileuton. Waiver request of in vivo testing: not applicable.

Ziprasidone, Hydrochloride Capsules/Oral. Recommended studies: two studies. (1) Type of study: fasting design—single-dose, two-way crossover in vivo. Strength: 20 mg. Subjects: normal healthy males and females, general population. Additional comments: Given that the risk of QT prolongation is associated with higher doses and little, if any, such effect is expected with a

20 mg dose, a screening ECG to exclude subjects with prolonged QT or other ECG abnormality is recommended, along with monitoring of vital signs and adverse events. Females should not be pregnant or lactating and, if applicable, should practice abstention or contraception during the study. (2) Type of study: fed design—single-dose, two-way crossover in vivo. Strength: 20 mg. Subjects: normal healthy males and females, general population. Additional comments: Please see previous comment. Analytes to measure: ziprasidone in plasma. BE based on (90% CI): ziprasidone. Waiver request of in vivo testing: 40, 60, and 80 mg based on (i) acceptable BE studies on the 20 mg strength, (ii) proportional similarity across all strengths, and (iii) acceptable in vitro dissolution testing of all strengths.

Zolmitriptan, Orally Disintegrating Tablets/Oral. Recommended studies: two studies. (1) Type of study: fasting design—single-dose, two-treatment, two-period crossover in vivo. Strength: 5 mg. Subjects: normal healthy males and females, general population. Additional comments: The whole tablet should be placed on the tongue and allowed to disintegrate for 30 s. After 30 s, all subjects should consume 240 mL of water. (2) Type of study: fed design—single-dose, two-treatment, two-period crossover in vivo. Strength: 5 mg. Subjects: normal healthy males and females, general population. Additional comments: Please see previous comment. Analytes to measure (in appropriate biological fluid): zolmitriptan in plasma. BE based on (90% CI): zolmitriptan. Waiver request of in vivo testing: 2.5 mg based on (i) acceptable BE studies on the 5 mg strength, (ii) proportional similarity of the formulations, and (iii) acceptable in vitro dissolution testing of all strengths.

Zolpidem, Tablets/Oral. Recommended studies: two studies. (1) Type of study: fasting design—single-dose, two-way crossover in vivo. Strength: 10 mg. Subjects: normal healthy males and females, general population. Additional comments: Patients should be advised not to drive if they are experiencing drowsiness and/or dizziness at the end of the study. (2) Type of study: fed design—single-dose, two-way crossover in vivo. Strength: 10 mg. Subjects: normal healthy males and females, general population. Additional comments: Analytes to measure: zolpidem in plasma. BE based on (90% CI): zolpidem. Waiver request of in vivo testing: 5 mg based on (i) acceptable BE studies on the 10 mg strength, (ii) proportional similarity across all strengths, and (iii) acceptable in vitro dissolution testing of all strengths.

Zonisamide, Capsules/Oral. Recommended studies: two studies. (1) Type of study: fasting design—single-dose, two-treatment, two-period crossover in vivo. Strength: 100 mg. Subjects: normal healthy males and females, general population. Females must have a negative baseline pregnancy test within 24 h prior to receiving the drug. Females should not be pregnant or lactating and, if applicable, should practice abstention or contraception during the study. Additional comments: Since zonisamide has a long half-life, you can consider performing a parallel design study, truncating the AUC at 72 h. If you choose to do a crossover design study, the washout period should be adequate to provide for drug elimination. Please verify that zonisamide's clearance has low intrasubject variability. (2) Type of study: fed design—single-dose, two-treatment, two-period crossover in vivo. Strength: 100 mg. Subjects: normal healthy males and females, general population. Females must have a negative baseline pregnancy test within 24 h prior to receiving the drug. Females should not be pregnant or lactating and, if applicable, should practice abstention or contraception during the study. Additional comments: Please see previous additional comments. Analytes to measure (in appropriate biological fluid): zonisamide in serum. BE based on (90% CI): zonisamide. Waiver request of in vivo testing: 25 and 50 mg, based on (i) acceptable BE studies on the 100 mg capsule, (ii) proportional similarity of the formulations, and (iii) acceptable in vitro dissolution testing of all strengths.

18 Bioequivalence Documentation

BACKGROUND

Sponsors would do well if they require their contract research organizations (CROs) to produce these reports in a clear, comprehensive format that allows easy access to all the information submitted. Whereas the Food and Drug Administration (FDA) does not require any specific format for the submission of the report, the layout described here has been successful in achieving fast approval of bioequivalence (BE) studies and it is recommended that sponsors check their CRO's standard format against the recommendations made here for the purpose of assuring completeness, accuracy, ready accessibility, and compliance (Tables 18.1 and 18.2).

SUBMISSION OF SUMMARY BE DATA FOR ANDAS FOR FDA

This guidance represents FDA's current thinking on this topic. Applicants can use an alternative approach if the approach satisfies the requirements of the applicable statutes and regulations.

INTRODUCTION

This guidance is intended to assist applicants who are submitting abbreviated new drug applications (ANDAs) in complying with FDA's requirements for the submission of BE data. FDA's final rule on "Requirements for Submission of Bioequivalence Data" (the BE data rule) requires an ANDA applicant to submit data from *all* BE studies the applicant conducts on a drug product formulation submitted for approval, including studies that do not demonstrate that the generic product meets the current BE criteria. All BE studies conducted on the same drug product formulation must be submitted to the agency as either a complete study report or a summary report of the BE data. The amended regulations include a definition of *same drug product formulation* (section 320.1(g)).

This guidance provides information on the following subjects:

- Types of ANDA submissions covered by the BE data rule
- Recommended format for summary reports of BE studies
- Types of formulations the agency considers to be the same drug product formulation for different dosage forms based on differences in composition

The guidance is applicable to BE studies conducted for ANDAs during both preapproval and post-approval periods.

BACKGROUND

The Federal Food, Drug, and Cosmetic Act (the FDCA) and FDA regulations require that an ANDA applicant submit, among other things, information showing that the applicant's drug

TABLE 18.1
Document Types and Management for CRO Monitoring

Document Type	Description	Control System
Cover page	Single page listing the document number (describe in the GDP SOP), study number, full title of the study including the sponsor's name, generic name of drug, the brand names of sponsor as well as reference-listed drug (RLD) (including lot numbers, date of manufacture, and date of expiry), dosage strength (and all the strengths the application supports), basic statistical design used, single or multiple dosing, fed or unfed study, route of administration, duration of study, place where study was completed, and date of completion. This page will also bear signatures of at least the following: 1. Written by 2. Reviewed by 3. Approved by (CRO) 4. Approval by sponsor 5. Certification of compliance with GDP, GLP, and GCP by head of quality assurance (QA) at CRO 6. Certification of compliance with GDP 7. Certification of compliance with GLP 8. Certificate of compliance with GCP (institutional review board [IRB]) 9. Certificate of compliance with 21 CFR various parts on use of computers and software 10. Certification of current calibration of instruments by quality control (QC) head	Good document practice SOP describing the convention used for naming and numbering the documents and numbering of studies that can be traced to location where the original documents are stored.
Table of content	Extensive listing of the entire document including all tables, figures, charts, references, and enclosures is required. Preparation of this table of content is simplified if a predetermined convention of naming the categories is used; the preferred form is 1.2.3 succession. While quoting references, do not use footnote function, but instead use a superscript; this allows separation of documents at the regulatory submission without losing control of document. There should be a single document file compiled and one table of content prepared; however, it is possible that the various sections of the study may be separated and reviewed by different reviewers; to expedite that, separate each section by a divider and provide a complete table of content before each section (a redundant function).	Good document practice.
Study summary	A tabulated document summarizing all the aspects of the study, the results, and conclusions reached including composite plasma level curves; the purpose of this section is to prepare the reviewer for specific exploration of the report.	
Abbreviations and glossary of terms	Whereas sponsors and CROs may create their own abbreviations, a glossary of all terms used should be provided making use of standard terms only; all references quoted are provided here.	

(Continued)

TABLE 18.1 (*Continued*)
Document Types and Management for CRO Monitoring

Document Type	Description	Control System
Section I: BE Report	This is the actual report of BE testing and includes reference to any deviations made from the approved protocol; full description of the drug used and its pharmacokinetic and other disposition properties that might affect the time course of drug in the body. More specifically, this report will include 1. Ethics 1.1 IRB 1.2 Ethical conduct of the study 1.3 Informed consent 1.4 Justification of the study 2. Introduction 2.1 History and source of drug 2.2 Chemistry 2.3 Pharmacology 2.4 Pharmacokinetics 2.5 Therapeutic uses 2.6 Adverse events 3. Investigation 3.1 Investigators and study Administrative structure 3.2 Study objectives 3.3 Investigational plan 3.4 Rationale of study design 3.5 Selection of study population 3.5.1 Study subjects demography 3.5.2 Inclusion criteria 3.5.3 Exclusion criteria 3.6 Subjects identification 3.7 Case report form (CRF) note 3.8 Confinement 3.9 Removal of subjects from study 3.10 Dietary restrictions, standardized diet, and fluid intake 3.11 Study drug administration 3.12 Identity of study medications 3.13 Assignment of study's subjects and randomization 3.14 Times of dosing 3.15 Treatment compliance 3.16 Physical activities after drug intake 3.17 Prior and concurrent medication 3.18 Clinical laboratory 3.19 Description of study facilities 3.20 Collection and handling of blood samples for analysis 3.22 Bioanalytical drug determination methodology 3.23 Data QA 3.24 Pharmacokinetic calculations	The BE protocol vis-à-vis the report includes the history of how the document was developed, the source documents used (e.g., information from published sources, protocols submitted to the FDA and available through FOI, RLD's promotional and label material used, guidelines of FDA, ICH, WHO, and others used in preparing the protocol), all correspondence (final forms) with the regulatory authorities and a justification of the dosage form, strength, and the manner of sourcing used. This protocol should be all inclusive of acceptance and rejection criteria, study design, analytical methods (to be developed or where developed their sensitivity), statistical analysis, and certification for GLP and GCP compliance. The purpose of this section is to lay out the road map that was supposed to be followed in conducting the study; later in the study report, any deviations from the agreed and approved protocol will be highlighted and reasons for deviation described and justified.The bioanalytical methodology and its validation is generally the proprietary property of the CRO and is not allowed for publication and mostly the portion redacted from the FOI-based documentation; also redacted are some critical inclusion and exclusion criteria in the study.

(*Continued*)

TABLE 18.1 (*Continued*)
Document Types and Management for CRO Monitoring

Document Type	Description	Control System
	3.25 Statistical analysis *3.25.1 Confidence intervals (CIs)* *3.25.2 Analysis of variance (ANOVA)* *3.25.3 Sample size determination* 3.26 Data tabulation, descriptive statistics, and diagrammatic data presentation 4. Study subjects 4.1 Disposition of subjects 4.2 Withdrawals and exclusions 4.3 Demographic characteristics 4.4 Variations from the study protocol 5. Safety evaluation 5.1 Benefit–risk ratio 5.2 Extent of exposure 5.3 Adverse events 5.3.1 Brief summary of adverse events 5.3.2 Display of adverse events 5.4 Clinical laboratory evaluation 5.5 Vital signs, physical assessment, and other clinical observations 5.6 Safety and tolerance 6. Results and BE evaluation 6.1 Data sets from study subjects 6.2 Adjustment due to anomalies *6.2.1 Adjustment due to collection anomalies* *6.2.2 Adjustment due to analytical anomalies* *6.2.3 Adjustment due to pharmacokinetic anomalies* *6.2.4 Nonzero predose concentrations* 6.3 Handling of withdrawals and missing data 6.4 Pharmacokinetic parameters 6.5 Statistical inferences *6.5.1 BE conclusion* *6.5.2 ANOVA* 7. Discussion and conclusions 8. References 9. APPENDIX 1 Appendix 1.1 Approved protocol Appendix 1.2 IRB approval with signatures Appendix 1.3 Sample CRFs Appendix 1.4 Sample informed consent form Appendix 1.5 Curriculum vitae (CV) of investigators Appendix 1.6 Randomization plan Appendix 1.7 Pharmacokinetic and statistic output Appendix 1.8 Bioanalytical report 10. Appendix II (not included for brevity) Appendix 2.1 Withdrawals	

(*Continued*)

TABLE 18.1 (*Continued*)
Document Types and Management for CRO Monitoring

Document Type	Description	Control System
	Appendix 2.2 Impact of the variations from the study protocol on the study outcomeAppendix 2.3 Screening examination data	
	Appendix 2.4 Adverse events	
	Appendix 2.5 Drug administration times for all subjects	
	Appendix 2.6 Diet composition	
	Appendix 2.7 Laboratory results values	
	Appendix 2.8 Clinical assessment for all subjects	

product is bioequivalent to the approved product designated by the FDA as the RLD (section 505(j)(2)(A)(iv) of the FDCA (21 U.S.C. 355(j)(2)(A)(iv); sections 314.94(a)(7) and 320.21(b) 1)). In the past, ANDA applicants submitted only the BE studies that demonstrated that a generic product met BE criteria, but they did not typically submit additional BE studies conducted on the same drug product formulation, including studies that did not show the product met BE criteria.

The BE data rule amended FDA's regulations to require that an ANDA applicant submit data from all BE studies that the applicant conducts on the drug product formulation submitted for approval (sections 314.81(b)(2)(vi), 314.94(a)(7), 314.96(a)(i), and 320.21(b)(1) and (c)). We believe that data from any additional BE studies may be important in our determination of whether a product is bioequivalent to the RLD and are relevant to our evaluation of generic products in general. These data will increase our understanding of generic drug development and how changes in components and composition may affect formulation performance, as well as promote further development of science-based BE policies.

Submission of All BE Studies

FDA regulations, as amended by and clarified in the BE data rule, require that a complete report be submitted for the BE studies upon which the applicant relies for approval and either a complete or summary report be submitted for each additional study conducted on the same drug product formulation (sections 314.81(b)(2)(vi), 314.94(a)(7), 314.96(a)(i), and 320.21(b)(1) and (c)). This requirement includes both in vivo and in vitro testing conducted to demonstrate BE. The regulations also provide that, if a summary report is submitted—and the agency believes that there may be BE issues or concerns with the drug product—the agency may request that a complete report be prepared and submitted to the FDA.

What Types of ANDA Submissions Must Include All BE Studies?

Under the BE data rule, ANDA applicants are required to submit information from all BE studies conducted on the same formulation of the drug product contained in the following submissions:

- ANDAs (section 314.94)
- ANDA amendments (section 314.96(a))
- ANDA supplements that require BE studies under section 320.21(c)
- ANDAs submitted under a suitability petition (section 314.93)
- ANDA annual reports (section 314.81(b)(2)(vi))

TABLE 18.2
List of Standard SOPs

No.	Name of SOP
1.	Conduct of BE study
2.	Archiving and retrieval of documents related to BE study
3.	QA of the BE study; audits of clinical and bioanalytical part of the study and the study report
4.	Study files/study report
5.	Preparation and review of the protocol for the study
6.	Amendment to the protocol for the study, protocol deviations/violation recording and reporting
7.	Sponsor/CRO QA agreement in conducting the BE study
8.	Study approval process by ethical committee
9.	BA/BE report
10.	Written informed consent (obtaining written informed consent for screening from study volunteers)
11.	Allotment of identification numbers to volunteers at various stages in BE study
12.	CRF, preparation of CRF, review and completion, data collection, and CRF completion
13.	Adverse/serious adverse event monitoring, recording, and reporting
14.	Organization chart of the study (flow chart)
15.	Training of the personnel
16.	Responsibilities of the members of the research team
17.	Monitoring of the study by the sponsor
18.	Conduct of prestudy meeting
19.	Study start-up
20.	Eligibility criteria for registration and registration of individuals into volunteer bank
21.	Handling of volunteer withdrawal
22.	Allotment of identification numbers to volunteers at various stages in biostudy
23.	Screening of enrolled volunteers for the study and frequency
24.	Payments to research subjects for BA/BE studies
25.	Procedures for entry into and exit from clinical unit
26.	Handling of subject check-in and checkout
27.	Housekeeping at clinical unit
28.	Planning, preparation, evaluation, and service of standardized meals for biostudies
29.	Distribution of meals to study subjects
30.	Administration of oral solid dosage forms of the drug to human subjects during BA/BE study
31.	Cannulation of study subjects
32.	Collection of blood samples from study subjects
33.	System for number of biosamples
34.	Recording of vital signs of subjects
35.	Operation and verification of fire alarm system if present
36.	Oxygen administration to subject from medical oxygen cylinder, emergency care of subjects during BA/BE study
37.	Availability of ambulance during BA/BE study (SOP for calling ambulance when necessary)
38.	Centrifugation and separation of blood samples
39.	Storage of plasma/serum samples
40.	Segregation of biosamples
41.	Transfer of plasma/serum samples to bioanalytical laboratory (handling and delivery)
42.	Procedures for washing glassware
43.	Recording temperature and relative humidity of rooms
44.	Instruction on operation and maintenance procedures for all the equipment in the clinical unit
45.	Numbering the equipment and log books for use in the clinical unit (system of numbering)
46.	Control of access to pharmacy
47.	Pharmacy area requirements

(*Continued*)

TABLE 18.2 (*Continued*)
List of Standard SOPs

No.	Name of SOP
48.	Authorization related to drug storage, dispensing, and retrieval from storage for BE study
49.	Study drug receipt, return, and accountability documentation (forms)
50.	Study drug receipt and return procedures
51.	Storage of drugs
52.	Line clearance before and after dispensing
53.	Documentation of line clearance and dispensing; packaging records and release of dispensed drugs (certificates and forms)
54.	Retention of samples of study drugs
55.	Disposal of archived study drugs and storage procedures
56.	Disposal of biological materials
57.	Procedures for bioanalytical laboratory (SOPs for the all equipment, analytical methods, reagent preparation)
58.	Out-of-specification (OOS) situation in the laboratory
59.	Acceptance criteria for analytical runs: acceptance of calibration curves, acceptance of the runs based on operating characteristics (OC) samples results
60.	Chromatographic acceptance criteria, chromatogram integration
61.	Sample reassay (how many times?)
62.	Pharmacokinetic data from bioanalytical data and how it is edited
63.	Statistics in the BE study (programs that are acceptable to be used)
64.	Equipment calibration

What Format Should Be Used for a Summary Report?

For a suggested format for summary reports, go online to FDA's GenericDrugs:InformationforIndu stry web page. The Division of Bioequivalence has developed model data summary tables in a concise format consistent with a common technical document (CTD) formatted application. The tables are located under the heading "Generic Drug Development, Abbreviated New Drug Application (ANDA) Submissions, and Review Information." The agency recommends that these table formats be used to organize the data for summary reports required by the BE data rule.

Same Drug Product Formulation

The FDA amended the regulations to require an applicant to submit data from all BE studies conducted on the same formulation of the drug product submitted for approval. In section 320.1(g), the FDA added a definition of the term *same drug product formulation*:

> *Same drug product formulation* means the formulation of the drug product submitted for approval and any formulations that have minor differences in composition or method of manufacture from the formulation submitted for approval, but are similar enough to be relevant to the FDA's determination of bioequivalence.

The definition of *same drug product formulation* in section 320.1(g) applies regardless of whether the products are manufactured at the same or different manufacturing sites.

In the following sections, we discuss differences in composition to consider when comparing drug product formulations. For immediate-release (IR) and extended-release (ER) drug products, we discuss the following:

- Minor differences in composition that are unlikely to have any detectable impact on formulation quality and performance between the formulations being compared. These differences would result in formulations that meet the definition of *same drug product*

formulation and for which BE studies must be submitted (sections 314.81(b)(2)(vi), 314.94(a)(7), 314.96(a)(i), and 320.21(b)(1) and (c)).
- Differences in composition that are likely to result in a significant difference in formulation quality and performance between the formulations being compared. These differences would result in formulations that do not meet the definition of *same drug product formulation* and for which BE studies need not be submitted.

Immediate-Release Drug Products

Immediate-Release Formulations Considered to Be the Same

Minor differences that result in product formulations that are considered to be the same include the following:

- A difference in an ingredient intended to affect the color or flavor of the drug product
- A different approved ingredient of the printing ink
- A difference in the technical grade and/or specification of an excipient (e.g., Avicel PH102 vs. Avicel PH200)
- A difference in particle size of the drug substance or excipients
- Formulations with different amounts of excipients are considered to be the same drug product formulation if
  - For an individual excipient, the difference in weight between the formulations being compared is less than or equal to the percentage shown in Table 18.3
  - The cumulative total of all excipient weight differences is less than or equal to 10%

Illustrative examples of IR formulations considered to be the same are provided as follows:

- If the amount of a filler excipient in an experimental formulation (A) is 105 mg and the same filler excipient in the formulation proposed to be marketed (B) is 100 mg, the difference in the excipient weight is 5%. These two formulations would be considered the same because the difference in weight of the filler excipient is less than 10%.

TABLE 18.3
IR Formulations: Differences in Excipient Weights

Excipient	Difference (≤) in Excipient Weights between Two Formulations[a]
Filler	10
Disintegrant	
Starch	6
Others	2
Binder	3
Lubricant	
Calcium or magnesium stearate	0.5
Others	2
Glidant	
Talc	2
Others	0.2
Film coat	2

[a] Percentage of difference between the formulation proposed for marketing and another experimental.

- In the case of multiple excipient changes, if an experimental formulation (A) contains 95 mg of a filler excipient and 103 mg of a disintegrant and the formulation proposed for marketing (B) contains 100 mg of the same filler excipient and 100 mg of the same disintegrant, the difference in weight for the filler excipient is 5% and the difference in weight for the disintegrant is 3%. The cumulative change is 8%, which is less than 10% for all excipient differences. Therefore, these formulations would be considered the same.

Immediate-Release Formulations Considered Not to Be the Same

A difference that results in product formulations that are considered not to be the same would include the addition or deletion of an excipient (with the exception of a difference in an ingredient intended to affect the color or flavor of the drug product or a difference in an ingredient of the printing ink).

Formulations with different amounts of the same excipients are considered not to be the same drug product formulation if

- For an individual excipient, the difference in excipient weight between the formulations being compared exceeds the percentages shown in Table 18.3
- The cumulative total of all excipient weight differences exceeds 10%
- Illustrative examples of IR formulations considered not to be the same are provided as follows:
  - If the amount of a filler excipient in an experimental formulation (A) is 115 mg and the filler excipient in the formulation proposed for marketing (B) is 100 mg, the difference in the excipient weight would be 15%. These two formulations would not be considered the same because the difference in weight of the filler excipient is greater than 10%.
  - In the case of multiple excipient changes, if an experimental formulation (A) contains 90 mg of a filler excipient and 106 mg of a disintegrant and the formulation proposed for marketing (B) contains 100 mg of the filler excipient and 100 mg of the disintegrant, the difference in weight for the excipient is 10% and the difference in weight for the disintegrant is 6%. The cumulative change would be 16%. Therefore, these formulations would not be considered the same, and any studies conducted with formulation A would not need to be submitted.

Immediate-Release Formulations: Other

Studies performed on experimental formulations containing different polymorphic forms of the drug substance also should be submitted.

Extended-Release Drug Products: Nonrelease Controlling Excipients

Extended-Release Formulations Considered to Be the Same (Nonrelease Controlling Excipients)

Minor differences that result in product formulations that are considered to be the same include the following:

- A difference in an ingredient intended to affect the color or flavor of the drug product
- A different approved ingredient of the printing ink
- A difference in the technical grade and/or specification of a nonrelease controlling excipient (e.g., Avicel PH102 vs. Avicel PH200)
- A difference in particle size of the drug substance or excipients
- Formulations with different amounts of the same nonrelease controlling excipients are considered to be the same drug product formulation if
  - For an individual excipient, the difference in excipient weight between the formulations being compared is less than or equal to the percentages listed in Table 18.4
  - The cumulative total of all excipient weight differences is less than or equal to 10%

TABLE 18.4
ER Formulations: Differences in Excipient Weights

Nonrelease Controlling Excipient	Difference (≤) in Excipient Weights between Two Formulations[a]
Filler	10
Disintegrant	
Starch	6
Others	2
Binder	1
Lubricant	
Calcium or magnesium stearate	0.5
Others	2
Glidant	
Talc	2
Others	0.2
Film coat	2

[a] Percentage of difference between the formulation proposed for marketing and another experimental formulation.

Extended-Release Formulations Considered Not to Be the Same (Nonrelease Controlling Excipients)

Examples of differences that result in product formulations that are considered not to be the same include the following:

- The addition or deletion of an excipient (except for a difference in an ingredient intended to affect the color or flavor of the drug product or a difference in an ingredient of the printing ink).
- A difference in weight of a nonrelease controlling excipient between the formulations being compared that exceeds the percentage listed in Table 18.4.
- The cumulative total difference in weights of all nonrelease controlling excipients exceeds 10%.

Extended-Release Formulations: Other (Nonrelease Controlling Excipients)

Studies performed on experimental formulations containing different polymorphic forms of the drug substance also should be submitted.

Extended-Release Drug Products: Release Controlling Excipients

Extended-Release Formulations Considered to Be the Same (Release Controlling Excipients)

Examples of minor differences that result in product formulations that are considered to be the same include the following:

- A difference in the technical grade and/or specification of the release controlling excipient(s) (e.g., Eudragit RS 100 vs. Eudragit RL 100)
- A difference in particle size of the drug substance or excipients
- A difference in the amount of release controlling excipient(s), expressed as the difference in weight of the release controlling excipient(s) in the experimental formulation compared to the formulation proposed for marketing, of less than or equal to 10%

Extended-Release Formulations Considered Not to Be the Same (Release Controlling Excipients)

Examples of differences that result in product formulations that are considered not to be the same include the following:

- The addition or deletion of a release controlling excipient
- A difference in the amount of release controlling excipient(s), expressed as the difference in weight of the release controlling excipient(s) in the experimental formulation compared to the formulation proposed for marketing, of greater than 10%

Extended-Release Formulations: Other (Release Controlling Excipients)

Studies performed on experimental formulations containing different polymorphic forms of the drug substance also should be submitted.

Semisolid Dosage Forms

For the purposes of this guidance, formulations of semisolid dosage form products are considered to be the same if the experimental formulation is in the same category as the formulation proposed for marketing (e.g., the formulations being compared are both for creams) and any differences between formulations are as described in the following:

- If the difference in the amount of an individual excipient between the experimental formulation and the formulation intended to be marketed is less than or equal to 5%, the two formulations are considered to be the same.
- If more than one excipient amount is changed, and the cumulative total of differences in the amount of all excipients is less than or equal to 7%, the two formulations are considered to be the same.
- Formulations with differences in particle size distribution of the drug substance, if the drug is in suspension, are considered to be the same.
- Formulations with differences in technical grade of a structure forming excipient are considered not to be the same.

Other Complex Dosage Forms

For other complex dosage forms (e.g., transdermals, injectable suspensions, and suppositories), limited information is available regarding quantitative and qualitative changes that could have a significant impact on the bioavailability (BA) of the product. Because of this lack of information, we consider all experimental formulations that are pharmaceutically equivalent to the formulation of the complex dosage form product intended to be marketed to be the same as the RLD. Therefore, the agency requests submission of either a summary report or a complete report of all BA or BE studies conducted during the development of the drug product. This information will increase our understanding of the development of the generic product and how changes in components, composition, and methods of manufacture have affected formulation performance. Access to this information will promote further development of science-based BE policies for complex dosage forms.

INTERNATIONAL SUBMISSIONS

While most regulatory agencies follow either U.S. or EMEA format, many have simplified presentation realizing that the drugs being filed in their countries have already been approved and the filing is accompanied by a certificate of pharmaceutical products (CPP).

A typical example will be the submission requirement in Egypt.

Format and Content of BE Study Report to be Submitted to the Central Administration of Pharmaceutical Affairs

Title Page

- Study title
- Name of the product
- Name of active ingredient(s)($API_{(S)}$)
- Name of sponsor
- Name and address of BE center
- Name, affiliation, and signature (accompanied with date) of the chairman of the board
- Name, affiliation, and signature (accompanied with date) of the principal investigator
- Date of the study (start, completion)

Report Contents

I. Study objective
II. Study resume
   - Drug review

Pharmacokinetic characteristics, pharmacodynamics, indications, pharmacological actions, and preparations

- Product information (include enclosed leaflet)

Item	Test Product	Reference Product
1. Product name		
2. $API_{(S)}$		
3. Molecular formula		
4. Dosage form		
5. Type of the product (IR, sustained-release, etc.)		
6. Dosage regimen		
7. Strength		
8. Batch size		
9. Batch number		
10. Manufacture date		
11. Expiry date		
12. Storage conditions		
13. Quantitative formulation	(to be attached)	(if available)

- Summary of BE study
  - Study design
  - Number of subjects
  - Treatment conditions (fasting or with food)
  - Type of obtained biological samples
  - Analytical procedure
  - Pharmacokinetic parameters
  - Statistical methods
    - Summary of BE data
    - Figure of mean plasma concentration–time profile
    - Figure of mean cumulative urinary excretion (if applicable)
    - Figure of mean urinary excretion rates (if applicable)
  - Results and conclusion (table of mean parameters t_{max}, C_{max}, $AUC_{0-\infty}$, AUC_{0-t})

In Vitro Testing

- Potency determination
- Uniformity of dosage unit (weight variation or content uniformity)
- Dissolution testing
  - Dissolution assay methodology
  - Dissolution assay validation
  - Comparative dissolution profile

Protocol, Approvals, and Details of In Vivo Study

- Protocol with justification
- Subjects
  - Number of subjects participated in the study (the number of subjects should not be less than 24 healthy subjects; in case of high variability, a larger number may be required)
  - Exclusion and inclusion criteria
  - Deviation from protocol
- Study design
  - Subjects assignment in the study
  - Number of periods
  - Sequence (randomization plan)
  - Treatments
  - Washout periods
  - Dosage form administration (fasting, with food, fluid intake with product, time, type of food and fluids throughout the study, etc.)
  - Procedures to minimize risk
  - Time and frequency of sampling
  - Date of drug administration and samples analysis of periods 1 and 2
- Collection of biological samples
  - Sufficient number of biological samples should be collected during the absorption phase (not less than 3 points).
  - Intensive sampling should be carried out around the time of the expected peak concentration.
  - Sufficient number of samples should be collected in the log-linear elimination phase of the drug. (A sampling period extending to at least four to five half lives of the drug is usually sufficient.)
  - Storage conditions of biological samples.
- Data analysis
- Letter of acceptance of protocol from MOH regulatory authorities (submitted upon request)
- Informed consent form
- Letter of approval of IRB and/or ethics committee

Clinical Study

- Summary of the study
- CRF including
  - Demographic characteristics of the subjects (sex, age, weight, and height)
  - The clinical evaluation data of subjects:
    - Hematological (complete blood count and blood group)
    - Biochemical (glucose and lipid profile and liver and kidney function)
    - Serological (HIV, HCV)
    - Urine analysis
- Vital signs of subjects (blood pressure, chest examination, abdomen examination, pulse rate, etc.)
- Adverse reactions report

Assay Method and Validation

- Assay method description (with reference(s) if applicable)
  - Equipment, materials, solvents, and their sources
  - Internal standard (IS) (name, concentration, and molecular formula)
  - Preparation of stock and standard solutions (in details)
- Validation procedure in terms of calibration curve, linearity, selectivity, lower limit of quantitation (LOQ), precision (inter- and intraday), accuracy, range, stability (freeze and thaw), and QC samples
- Data of the previously mentioned parameters
- Figures of calibration curve(s) and chromatograms of standard and QC samples (blank, spiked, and actual subject samples should be included)
- Sample back calculation

Pharmacokinetic Parameters

- Definitions
- Drug levels at each sampling time
- Pharmacokinetic parameters, calculation, and tables ($AUC_{0\rightarrow t}$, $AUC_{0\rightarrow\infty}$, C_{max}, t_{max})
- Figure of mean plasma concentration–time profile
- Figures of individual subjects plasma concentration–time profile
- Figure of mean cumulative urinary excretion (if applicable)
- Figures of individual subject cumulative urinary excretion (if applicable)
- Figure of mean urinary excretion rates (if applicable)
- Figures of individual subject urinary excretion rates (if applicable)
- Tables of individual subjects data arranged by drug, drug/period, drug/sequence

Statistical Analysis

- Any acceptable statistical program can be used.
- In case of using ANOVA, two–way ANOVA for crossover design should be performed for the pharmacokinetic parameters.

  The ANOVA should include factors accounting for the following sources of variation:
  - Treatments (drugs or formulations)
  - Periods (phases)
  - Sequence (group or order)
  - Subjects within sequence
- Logarithmic transformation of the pharmacokinetic parameters: C_{max}, $AUC_{0\rightarrow t}$, and $AUC_{0\rightarrow\infty}$ should be performed before data analysis.
- The pharmacokinetic parameter, t_{max}, should be expressed as median values and analyzed on untransformed data.
- The two one-sided hypotheses at the alpha = 0.05 level of significance should be performed for AUC(s) and C_{max} by constructing the 90% CI for the ratio between the test and the reference averages. For $AUC_{0\rightarrow t}$ and $AUC_{0\rightarrow\infty}$, CI should be between 80% and 125%. For C_{max}, the CIs should be between 70% and 143% depending on the drug under investigation.
- Summary of statistical significance.
- Summary of statistical parameters.

Appendices

- Randomization schedule
- Chromatograms of at least 20% of subjects

- Clinical facilities' description
- Analytical facilities' description
- CV of the investigators
- Table containing names, affiliations, and signatures (accompanied with date) of the principal, clinical, and analytical investigator(s)
- Table of team names', responsibilities and signatures
- List of IRB and/or ethics committee and signatures

References

For any details not mentioned in the previously included items, WHO guidelines should be followed "until further notification."

Provided here are examples of blank forms that may be used in the conduct of BE studies:

Screening Record

This page must be filed separately.

Subject's Identification Data

First name ______________________________

Middle name ______________________________

Last name ______________________________

Street address ______________________________

City ______________________________

Country ______________________________

Occupation ______________________________

Telephone number ______________________________

Sex: ❒ **Male** ❒ **Female**

Date of birth |_____|_____|_____| (dd | mm | yy)

Age |_____| years

Interviewer: ____________________ Date: ______/______/______

Confirmation by the Principal Investigator

I confirm that the data included here is in complete accordance with the source documents of this subject.

Name of the principal investigator: ________________________

Signature:________________________

Date |______|______|______| (dd | mm | yy)

Subject's Identification Data (*Continued*)

Height |______| cm

Weight |______| kg

Body frame ❒ Small ❒ Medium ❒ Large ❒ N/A

Refer to the study master file for more details about the body frame.

Date of birth |_____|______|______| (dd|mm|yy)

Age |______| years

Recorded by:____________________________ Date:______________________

(See height–weight ranges in the master file of the study.)

Vital Signs

Study Day		**Screening**	
Date (dd | mm | yy)		|____|____|____|	
Actual Time (hh|mm)		|________|____|	
Blood Pressure	**After at least 3 min sitting**	|________| |________|	
Pulse	**After at least 3 min sitting**	|________________|	
Respiration rate (/min)		|________________|	
Temperature (°C)		|________|	o Oral o Rectal o Axillary

Performed by: ___________________________________ Date: _______/______/______

Clinical investigator signature: _____________________ Date: _______/______/______

Medical History

			Notes
1. Family history	❒ Normal	❒ Abnormal	________
2. Allergy (*including drug allergy*)	❒ Normal	❒ Abnormal	________
3. Cardiovascular	❒ Normal	❒ Abnormal	________
4. Respiratory	❒ Normal	❒ Abnormal	________
5. Renal	❒ Normal	❒ Abnormal	________
6. Hepatic	❒ Normal	❒ Abnormal	________
7. Gastrointestinal	❒ Normal	❒ Abnormal	________
8. Special diet in the last 30 days (***Excessive vitamin intake, popular diets, significant weight gain or loss, vegetarian or psychological eating disorders***)	❒ Normal	❒ Abnormal	________
9. Surgery	❒ No	❒ Yes	(mm/yy) |____|____| |____|____| |____|____|
10. Psychiatric disease	❒ No	❒ Yes	________
11. Epilepsy or other seizures	❒ No	❒ Yes	________
12. G-6-PD deficiency	❒ No	❒ Yes	________
13. Bleeding/coagulation disorders or severe anemia	❒ No	❒ Yes	________
14. Acute infection within the last week	❒ No	❒ Yes	________
15. Thyroid gland abnormalities	❒ No	❒ Yes	________
16. Skin abnormalities	❒ No	❒ Yes	________
17. Eye/ear/nose/throat abnormalities	❒ No	❒ Yes	________
18. Diabetes	❒ No	❒ Yes	________
19. Neurological abnormalities	❒ No	❒ Yes	________
20. Musculoskeletal disorders	❒ No	❒ Yes	________

Interviewer: ____________________ Date: ____/____/____

Clinical investigator signature: ____________________ Date: ____/____/____

Medical History (*Continued*)

Notes

21. Latest blood donation ❐ No ❐ Yes QTY:________________ Date:________

22. Latest participation in another clinical trial ❐ No ❐ Yes ____________________________________

23. Planned hospitalization (*within the next 3 months*) ❐ No ❐ Yes ____________________________________

24. Tobacco use ❐ No ❐ Yes QTY:____________________________
❐ Ex-consumer ____________________________________

25. Caffeine use ❐ No ❐ Yes ____________________________________

Consumer of
❐ Coke ❐ No ❐ Yes QTY______________________________
❐ Tea ❐ No ❐ Yes QTY______________________________
❐ Coffee ❐ No ❐ Yes QTY______________________________
❐ Ex-consumer ____________________________________

26. Alcohol consumption ❐ No ❐ Yes QTY______________________________
❐ Ex-consumer ____________________________________

27. Drug abuse ❐ No ❐ Yes ____________________________________
____
____

28. Others ❐ No ❐ Yes ____________________________________

Interviewer: _________________________ Date: _______/_______/______

Clinical investigator signature: ___________________________ Date: _______/_______/______

Alcohol Screen (If Applicable)

Alcohol Screen Result ❐ Negative ❐ Positive Date |____|____|____| (DD.MM.YY)

Comments:

____

Screening for Drugs of Abuse (If Applicable)

Drug Name	Drugs of Abuse	
	❐ Negative	❐ Positive
	❐ Negative	❐ Positive
	❐ Negative	❐ Positive
	❐ Negative	❐ Positive
	❐ Negative	❐ Positive
	❐ Negative	❐ Positive

This form can be expanded depending on the number of drugs that will be screened.

Recorded by: ______________________________ Date: ______/______/______

Clinical investigator signature: ___________________ Date: ______/______/______

Physical Examination

			Notes
1. Appearance	❐ Normal	❐ Abnormal	____________
2. Skin	❐ Normal	❐ Abnormal	____________
3. Eye	❐ Normal	❐ Abnormal	____________
4. Nose	❐ Normal	❐ Abnormal	____________
5. Throat	❐ Normal	❐ Abnormal	____________
6. Mouth	❐ Normal	❐ Abnormal	____________
7. Neck	❐ Normal	❐ Abnormal	____________
8. Breast	❐ Normal	❐ Abnormal	____________
9. Lungs	❐ Normal	❐ Abnormal	____________
10. Heart	❐ Normal	❐ Abnormal	____________
11. Abdomen	❐ Normal	❐ Abnormal	____________
12. Kidney	❐ Normal	❐ Abnormal	____________
13. Spine	❐ Normal	❐ Abnormal	____________
14. Lymph nodes	❐ Normal	❐ Abnormal	____________
15. Extremities	❐ Normal	❐ Abnormal	____________
16. Neurological and reflex	❐ Normal	❐ Abnormal	____________
17. Genitalia	❐ Normal	❐ Abnormal	____________
18. Rectum	❐ Normal	❐ Abnormal	____________
19. Mental status	❐ Normal	❐ Abnormal	____________
20. Others	❐ No	❐ Yes	____________

If "others," please specify ____

____

Result of ECG Examination

See attached ECG record ❐ Normal ❐ Abnormal ❐ NA

(*if applicable*)

____

____

____

____

Recorded by: ______________________________ Date: _____/_____/_____

Clinical investigator signature: ____________________ Date: _____/_____/_____

Selection Criteria

A. Inclusion Criteria

			Notes
1. Male, age 18–45 years, inclusive.	❐ Yes	❐ No	__________
2. Body weight within the limits for height (see "Height–weight ranges" study master file).	❐ Yes	❐ No	__________
3. Medical history, vital signs, physical examination (including neurological assessment), ECG (if applicable) without evidence of clinically significant deviation from normal medical condition, performed not longer than 2 weeks before the initiation of the clinical study.	❐ Yes	❐ No	__________
4. Clinical laboratory tests are performed no longer than 1 month from the initiation of the clinical study.	❐ Yes	❐ No	__________
5. Results of blood and urine examinations are within the normal range or deviation is not considered clinically significant by the clinical investigator.	❐ Yes	❐ No	__________
6. Subject does not have allergy to the drugs under investigation.	❐ Yes	❐ No	__________

B. Exclusion Criteria

			Notes
Medical history, physical examination, and laboratory tests with evidence of clinically significant deviation from normal medical condition.	❐ Yes	❐ No	__________
Evidence for any abnormality in carbohydrate metabolism.	❐ Yes	❐ No	__________
3. Results of blood and/or urine examinations, which are clinically significant.	❐ Yes	❐ No	__________
4. Acute infection within 1 week preceding the first study drug administration.	❐ Yes	❐ No	__________

5. History of drug or alcohol abuse. ❐ Yes ❐ No ____________
6. Subject is a heavy smoker. ❐ Yes ❐ No ____________
7. Subject does not agree not to take any prescription or nonprescription drug within 1 week before the first study drug administration. ❐ Yes ❐ No ____________
8. Subject is on a special diet (e.g., subject is vegetarian). ❐ Yes ❐ No ____________

Checked by: ____________ Date: ___/___/___
Clinical investigator signature: ____________ Date: ___/___/___

B. Exclusion Criteria (*Continued*)

Notes

9. Subject consumes large quantities of alcohol or beverages containing methylxanthines, e.g., caffeine. ❐ Yes ❐ No
10. Subject does not agree not to consume any beverages or foods containing alcohol 24 h prior to the first study drug administration until the end of the study. ❐ Yes ❐ No ____________
11. Subject does not agree not to consume any beverages or foods containing methylxanthines, e.g., caffeine (coffee, tea, cola, chocolate, etc.) 24 h (or as specified in the study protocol) prior to the study drug administration of either study phase until the last blood sample of the respective study phase was collected. ❐ Yes ❐ No ____________
12. Subject has a history of severe diseases which have direct impact on the study. ❐ Yes ❐ No ____________
13. Participation in a BE study within the last month before the first study drug administration. ❐ Yes ❐ No ____________
14. Participation in a clinical study within the last 3 months before the first study drug administration. ❐ Yes ❐ No ____________
15. Subject intends to be hospitalized within 3 months after the first study drug administration. ❐ Yes ❐ No ____________
16. Subjects who, through completion of this study, would have donated more than 500 mL of blood in 14 days, 750 mL in 30 days, 1000 mL in 90 days, 1250 mL in 120 days, 1500 mL in 180 days, 2000 mL in 270 days, and 2500 mL in 1 year. ❐ Yes ❐ No ____________

Checked by: ____________ Date: ___/___/___
Subject can be included ❐ Yes ❐ No ____________
If no (reason for exclusion)____________
____
____
____

Clinical investigator signature: ____________ Date: ___/___/___

Phase I Clinical Record

Study Phase: Phase I

1. Admission time |________|________| Date: _____/_____/_____

(hh:mm) (DD.MM.YY)

2. *Drugs taken within the last week:*____

____

3. *Alcohol consumption within the last 24 h:*______________________________________

4. Alcohol screening result: o Negative o Positive o N/A

5. Caffeine consumption within the last 24 h (or as specified in the study protocol) (coffee, tea, cola, chocolate, etc.):____

____

6. *Meals leftovers* (see meal composition in the master file of the study)

Meal	*Date (DD.MM.YY)*	*Time (h) from–to*	*Meals leftovers*
Dinner (day −1)			
Lunch (day 1)			
Dinner (day 1)			

Vital Signs Measurement Form

Study Phase: Phase I

Time from Drug Administration	**Date**	**Actual Time (hh/mm)**	**Blood Pressure (mmHg)**	**Pulse (/min)**	**Respiration Rate (/min)**	**Temperature (°C)**		***Performed by/Date***	***Clinical Inves-tigator Signature/ Date***
			(after at least 3 min sitting)						
Before study drug administration.		|_____|___|	|______|/ |______|	|______	|______	|____|	o Oral o Rectal o Axillary		
4.00 h		|_____|___|	|______|/ |______|	|______|	|______|	|____|	o Oral o Rectal o Axillary		
10.00 h		|_____|___|	|______|/ |______|	|______|	|______|	|____|	o Oral o Rectal o Axillary		

Drug Administration Form

Study Phase: Phase I

Treatment Given

❒ Treatment A ❒ Treatment B ❒ Treatment C

	Time (*hh.mm*)	Date (*DD.MM.YY*)
Administration of Study Drug	|________|________|	____/____/____

Volume of water:________ mL. Temperature of water:________ °C

Hand check ❒ ❒ N/A *Others* ❒ ❒ N/A

Mouth check ❒ ❒ N/A Specify: ________________

Comments:____

____

Administered by: ____________________________ Date: _____/_____/_____
Checked by: ____________________________ Date: _____/_____/_____

Blood Glucose Strip (Applicable for Antidiabetic Drugs Only)

	Actual Test Time	**Value (Normal <100 mg/dL)**	**Performed by/Date**
Immediately before study drug administration **N.A**	|______|______| (hh:mm)	|________________| (mg/dL)	________________
() Hour after study drug administration	|______|______| (hh:mm)	|________________| (mg/dL)	________________
() Hour after study drug administration	|______|______| (hh:mm)	|________________| (mg/dL)	________________
() Hour after study drug administration	|______|______| (hh:mm)	|________________| (mg/dL)	________________
() Hour after study drug administration	|______|______| (hh:mm)	|________________| (mg/dL)	________________
() Hour after study drug administration	|______|______| (hh:mm)	|________________| (mg/dL)	________________
() Hour after study drug administration	|______|______| (hh:mm)	|________________| (mg/dL)	________________

Blood Sampling Form

Study Phase: Phase I

Stopwatch code no.: ___________

Date (DD.MM.YY)	Sample ID	Sample Collection Time			Signature	Comments (if necessary)
		Time (hours)	Theoretical	Actual		
	Predose Sample (BK)		NA			
	Study Drug					
	CGL (I) 1	0.25				
	CGL (I) 2	0.50				
	CGL (I) 3	0.75				
	CGL (I) 4	1.00				
	CGL (I) 5	1.25				
	CGL (I) 6	1.50				
	CGL (I) 7	1.75				
	CGL (I) 8	2.00				
	CGL (I) 9	2.50				
	CGL (I) 10	3.00				
	CGL (I) 11	4.00				
	CGL (I) 12	6.00				
	CGL (I) 13	8.00				
	CGL (I) 14	12.00				
	CGL (I) 15	16.00				
	CGL (I) 16	24.00				

Check for Adverse Events

Study Phase: Phase I

Clinical Assessment

1. No adverse event occurred during Phase I (treatment was well tolerated by the subject):

❒ Yes ❒ No If no, specify: ____

____

____

____

Interviewer: _________________________________ Date: |______|______|______|

Clinical investigator signature: ____________________ Date: |______|______|______|

2. The subject left the study without any changes of the baseline condition:

☐ Yes ☐ No If no, specify: ________________________________

____

____

____

Clinical investigator signature: ____________________ Date: |_____|_____|_____|

For clinical investigator: Please refer to the adverse events registration record and use an adverse events form to record any adverse events as per CLP-010.

Comments

Any clinically important observation, not reported in other parts of the CRF, including observations toward a possible activity outside the studied indication.

☐ **Yes** ☐ **No**

Comments

Interviewer: ______________________________ Date: |_____|_____|_____|

Clinical investigator signature: ____________________ Date: |_____|_____|_____|

Phase II Clinical Record

Study Phase: Phase II

1. Admission time |_______|_______| Date: _____/_____/_____
(hh:mm) (DD.MM.YY)
2. *Drugs taken within the wash out period:* ________________________________

____

____

3. *Alcohol consumption within the last 24 h:*________________________________
4. *Alcohol screening result:* ○ *Negative* ○ *Positive* ○ *N/A*
5. *Caffeine consumption within the last 24 h (or as specified in the study protocol) (coffee, tea, cola, chocolate, etc.):*________________________________

____

6. *Meals leftovers* (see meal composition in the master file of the study)

Meal	*Date (DD. MM.YY)*	*Time (h)* from–to	*Meal leftovers*
Dinner (day −1)			
Lunch (day 1)			
Dinner (day 1)			

Vital Signs Measurement Form

Study Phase: Phase II

Time from Drug Administration	Date	Actual Time (hh/mm)	Blood Pressure (mmHg) (after at least 3 min sitting)	Pulse (/min)	Respiration Rate (/min)	Temper-ature (°C)	*Performed by/Date*	*Clinical Investigator Signature/ Date*
Before study drug administration		|____|__|	|_____|/ |_____|	|_____|	|_____|	|_____|	o Oral o Rectal o Axillary	
4.00 h		|____|__|	|_____|/ |_____|	|_____|	|_____|	|_____|	o Oral o Rectal o Axillary	
10.00 h		|____|__|	|_____|/ |_____|	|_____|	|_____|	|_____|	o Oral o Rectal o Axillary	

Drug Administration Form

Study Phase: Phase II

Treatment Given

❐ Treatment A ❐ Treatment B ❐ Treatment C

	Time (*hh.mm*)	Date (*DD.MM.YY*)
Administration of Study Drug	|______|______|	____/____/____

Volume of water:________ mL. Temperature of water:________ °C

Hand check ❐ ❐ N/A *Others* ❐ ❐ N/A

Mouth check ❐ ❐ N/A Specify ________________

Comments:____

____

Administered by: __________________________ Date: _____/_____/_____

Checked by: ___________________________ Date: _____/_____/_____

Blood Glucose Strip (Applicable for Antidiabetic Drugs Only)

	Actual Test Time	Value (Normal <100 mg/dl)	Performed by/Date
Immediately before study drug administration **N.A**	|___|___| (hh:mm)	|______| (mg/dL)	______
() Hour after study drug administration	|___|___| (hh:mm)	|______| (mg/dL)	______
() Hour after study drug administration	|___|___| (hh:mm)	|______| (mg/dL)	______
() Hour after study drug administration	|___|___| (hh:mm)	|______| (mg/dL)	______
() Hour after study drug administration	|___|___| (hh:mm)	|______| (mg/dL)	______
() Hour after study drug administration	|___|___| (hh:mm)	|______| (mg/dL)	______
() Hour after study drug administration	|___|___| (hh:mm)	|______| (mg/dL)	______

Blood Sampling Form

Study Phase: Phase II

Stopwatch code no.: __________

Date (DD. MM.YY)	Sample ID	Sample Collection Time			Signature	Comments (if necessary)
		Time (Hours)	Theoretical	Actual		
	Predose Sample (BK)		NA			
	Study Drug Administration					
	CGL (II) 1	0.25				
	CGL (II) 2	0.50				
	CGL (II) 3	0.75				
	CGL (II) 4	1.00				
	CGL (II) 5	1.25				
	CGL (II) 6	1.50				
	CGL (II) 7	1.75				
	CGL (II) 8	2.00				
	CGL (II) 9	2.50				
	CGL (II) 10	3.00				
	CGL (II) 11	4.00				
	CGL (II) 12	6.00				
	CGL (II) 13	8.00				
	CGL (II) 14	12.00				
	CGL (II) 15	16.00				
	CGL (II) 16	24.00				

Check for Adverse Events

Study Phase: Phase II

Clinical assessment

No adverse event occurred during Phase II (treatment was well tolerated by the subject):

❒ Yes ❒ No If no, specify: ______________________________
____
____
____

Interviewer: ______________________ Date: |______|______|______|

Clinical investigator signature: ______________ Date: |______|______|______|

The subject left the study without any changes of the baseline condition:

❒ Yes ❒ No If no, specify: ______________________________
____
____
____

Clinical investigator signature: ______________ Date: |______|______|______|

For clinical investigator: Please refer to the adverse events registration record and use an adverse events form to record any adverse events as per CLP-010.

Comments

Any clinically important observation, not reported in other parts of the case report form, including observations toward a possible activity outside the studied indication.

❒ Yes ❒ No

Comments

Interviewer: ______________________ Date: |______|______|______|

Clinical investigator signature: ______________ Date: |______|______|______|

Subject Evaluation Form (Closeout)

❒ **Used after completion of the study (24 h after donating all the samples**[a])

The CRF **of this subject has been reviewed and the subject was asked for any adverse event (see relevant forms) that occurred during the study. The subject completed the study and he or she**

○ **Left the study without any changes of the baseline condition**

○ **Left the study with some changes of the baseline condition**

Specify:

____

____

____

____

❒ **Used in case of withdrawal**

Reason for withdrawal:

____

____

____

The following measures were taken to evaluate the previous case (see attached examinations):

____

____

____

(In case of vital signs measurement, use additional vital signs form.)

Clinical investigator signature: ____________________ Date: |______|______|______|

[a] *Unless otherwise specified in the study protocol.*

Follow-Up Record

1. Physical Examination

			Notes
1. Appearance	❒ Normal	❒ Abnormal	______
2. Skin	❒ Normal	❒ Abnormal	______
3. Eye	❒ Normal	❒ Abnormal	______
4. Nose	❒ Normal	❒ Abnormal	______
5. Throat	❒ Normal	❒ Abnormal	______
6. Mouth	❒ Normal	❒ Abnormal	______
7. Neck	❒ Normal	❒ Abnormal	______
8. Breast	❒ Normal	❒ Abnormal	______
9. Lungs	❒ Normal	❒ Abnormal	______
10. Heart	❒ Normal	❒ Abnormal	______
11. Abdomen	❒ Normal	❒ Abnormal	______
12. Kidney	❒ Normal	❒ Abnormal	______
13. Spine	❒ Normal	❒ Abnormal	______
14. Lymph nodes	❒ Normal	❒ Abnormal	______
15. Extremities	❒ Normal	❒ Abnormal	______
16. Neurological and reflex	❒ Normal	❒ Abnormal	______
17. Genitalia	❒ Normal	❒ Abnormal	______
18. Rectum	❒ Normal	❒ Abnormal	______
19. Mental status	❒ Normal	❒ Abnormal	______
20. Others	❒ No	❒ Yes	______

21. If "others," please specify ______________________________
____
____

2. Result of ECG Examination ❒ Normal ❒ Abnormal ______________

- See attached ECG record

(*if applicable*)

____
____
____
____

Recorded by: ______________________ Date: |______|______|______|

Clinical investigator signature: ______________________ Date: |______|______|______|

FORMATTING OF BE SUMMARY TABLES

1. Please provide these tables as pdf files and in MSWord. Place the MSWord format of all the tables in Module 2.7 and the pdf files in the appropriate eCTD/CTD locations.
2. Margins for the paper should be 1" for the top and bottom and 1.25" for the left and right sides.
3. All text should be Times New Roman 10.
4. Please use the Default Table Style when creating the tables when they are created in Microsoft® Word. (Select Menu Table-Table Auto Format-Table Normal)
5. Tables 18.5 through 8.15 should be in PORTRAIT orientation.
6. Tables 18.16 through 18.20 should be in LANDSCAPE orientation.

TABLE 18.5
Submission Summary

Drug product name
Strength(s)
Applicant name
Address
Point of contact
  Name
  Address
  Telephone number
  Fax number

Or, please provide an electronic copy of Form 356H.
This information is needed for a complete BE review and, although required for the archival copy submitted to the Agency, it is frequently not readily available in the Bioequivalence Submission. The Division of Bioequivalence prefers that this information be submitted as a electronic Form 356H. If this is not possible, then please complete Table 18.5.

TABLE 18.6
Bioanalytical Method Validation

Information Requested	Data
Bioanalytical method validation report location	Provide the volume(s) and page(s)
Analyte	Provide the name(s) of the analyte(s)
IS	Identify the IS used
Method description	Brief description of extraction method; analytical method
LOQ	LOQ units
Average recovery of drug (%)	%
Average recovery of IS (%)	(%)
Standard curve concentrations (units/mL)	Standard curve range and appropriate concentrations units
QC concentrations (units/mL)	List all the concentrations used
QC intraday precision range (%)	Range or per QC
QC intraday accuracy range (%)	Range or per QC
QC intraday precision range (%)	Range or per QC
QC intraday accuracy range (%)	Range or per QC
Benchtop stability (h)	Hours at room temperature
Stock stability (days)	Days at 4°C
Processed stability (h)	Hours at room temperature; hours @ 4°C
Freeze–thaw stability (cycles)	Number of cycles
Long-term storage stability (days)	17 days at –20°C (or others)
Dilution integrity	Concentration diluted X-fold
Selectivity	No interfering peaks noted in blank plasma surplus

Please include table for each analyte.
Please submit all method validation SOPs.

TABLE 18.7
Demographic Profile of Subjects Completing the BE Study

Study No.

		Treatment Groups	
		Test Product $N =$	**Reference Product** $N =$
Age (years)	Mean ± SD Range	50 ± 15 21–64	
Age groups	<18	*N*(%)	*N*(%)
	18–40	*N*(%)	*N*(%)
	41–64	*N*(%)	N(%)
	65–75	*N*(%)	*N*(%)
	>75	*N*(%)	*N*(%)
Sex	Male	*N*(%)	*N*(%)
	Female	*N*(%)	*N*(%)
Race	Asian	*N*(%)	*N*(%)
	Black	*N*(%)	*N*(%)
	Caucasian	*N*(%)	*N*(%)
	Hispanic	*N*(%)	*N*(%)
	Others	*N*(%)	*N*(%)
BMI	Mean ± SD Range		
Other factors			

Please provide a separate table for each BE study.

TABLE 18.8
Incidence of Adverse Events in Individual Studies

Body System/Adverse Event	**Reported Incidence by Treatment Groups Fasted/Fed BE Study Study No.**	
	Test	**Reference**
Body as a whole		
Dizziness	*N*(%)	*N*(%)
etc.	*N*(%)	*N*(%)
Cardiovascular		
Hypotension		
etc.		
Gastrointestinal		
Constipation		
etc.		
Other organ sys.		
Total	*N*(%)	*N*(%)

Provide separate table for each BE study.

TABLE 18.9
Study Information

Study number	
Study title	
Study type	☐ In vivo BE ☐ In vitro BE ☐ Permeability ☐ Others (specify)
Submission location:Study report	Location, e.g., 5.3.1.2
Validation report	Location, e.g., 5.3.1.2
Bioanalytical report	Location, e.g., 5.3.1.4
Clinical site (name, address, phone #, fax #)	
Principal clinical investigator (name, e-mail)	
Dosing dates	
Analytical site (name, address, phone #, fax #)	
Analysis dates	
Principal analytical investigator (name, e-mail)	
Storage period of biostudy samples (no. of days from the first day of sample collection to the last day of sample analysis)	

Please provide separate table for each BE study.

TABLE 18.10
Product Information

Product	Test	Reference
Treatment ID		
Manufacturer		
Batch/lot no.		
Manufacturer date		N/A
Expiration date	N/A	
Strength		
Bio-batch size		N/A
Production batch size		N/A
Potency		
Content uniformity (mean, %CV)		N/A
Dose administered		
Route of administration		

TABLE 18.11
Dropout Information

Study No.

Subject No	Reason for Dropout/Replacement[a]	Period	Replaced?	Replaced with

[a] Please provide time, treatment (test or reference), and cause of dropout, if reason of dropout is other than "personal reasons."

Please provide separate table for each BE study.

TABLE 18.12
Protocol Deviations

Study No.

Type	Subject #s (Test)	Subject #s (Ref.)

Please provide a separate table for each BE study.

TABLE 18.13
Summary of Standard Curve and QC Data for BE Sample Analyses[a]

BE Study No. Analyte Name

Parameter	Standard Curve Samples
Concentration (ng, mcg/mL)	
Inter day precision (%CV)	
Inter day accuracy (%actual)	
Linearity	(Range of R^2 values)
Linearity range (ng, mcg/mL)	
Sensitivity/LOQ (ng, mcg/mL)	

BE Study No.
Analyte Name

Parameter	QC Samples
Concentrating (ng, mcg/mL)	
Inter day precision (%CV)	
Inter day accuracy (%Actual)	

[a] If applicable, please provide separate tables for the parent drug and metabolite(s).

TABLE 18.14
SOP's Dealing with Bioanalytical Repeats of Study Samples[a]

SOP No.	Effective Date of SOP	SOP Title

[a] Please include the SOP for bioanalytical repeats in your submission.

TABLE 18.15
Composition of Meal Used in Fed BE Study[a]

Composition	Percent of Total Kcal	Kcal
Fat		
Carbohydrate		
Protein		
Total		

[a] If the standard meal referenced in the CDER Guidance for Industry Food-Effect Bioavailability and Fed Bioequivalence Studies is used, then it is not necessary to complete the table. In that case, please add a statement in the fed BE study report indicating that the "FDA standard meal" was used. If an alternative meal is used, then please complete the summary table earlier.

TABLE 18.16
Summary of BA Studies

Study Reg. No.	Study Objective	Study Design	Treatments (Doses, Dosage Form, Route) [Product ID]	Subjects No. (M/F)Type Age: Mean (Range)	Mean Parameters (±SD)						Study Report Location
					C_{max} (units/mL)	t_{max} (h)	AUC_{0-t} (units)	AUC_{∞} (units)	$T½$ (h)	Kel (h^{-1})	
			Test product strength Tab./Cap./Susp p.o. [Batch #]	# completing (#M/#F) Healthy subjects or patients mean age (range)	M(%CV)	Median (Range)	M(%CV)	M(%CV)	M(%CV)	M(%CV)	
Study #	Fasting Study title	Randomized single-dose crossover	Ref. product strength Tab./Cap./Susp p.o. [Batch #]		M(%CV)	Median (Range)	M(%CV)	M(%CV)	M(%CV)	M(%CV)	Vol. # p.#
			Test product strength Tab./Cap./Susp p.o. [Batch #]	# completing (#M/#F) Healthy subjects or patients mean age (range)	M(%CV)	Median (Range)	M(%CV)	M(%CV)	M(%CV)	M(%CV)	
Study #	Fed Study title	Randomized single-dose crossover	Ref. product strength Tab./Cap./Susp p.o. [Batch #]		M(%CV)	Median (Range)	M(%CV)	M(%CV)	M(%CV)	M(%CV)	Vol. # p.#

TABLE 18.17
Statistical Summary of the Comparative BA Data

Drug
Dose (# × mg)
Least Squares Geometric Means, Ratio of Means, and 90% CIs
Fasted BE Study (Study No.)

Parameter	Test	Reference	Ratio	90% CI
AUC_{0-t}				
AUC_{∞}				
C_{max}				

Fasted BE Study (Study No.)

Parameter	Test	Reference	Ratio	90% CI
AUC_{0-t}				
AUC_{∞}				
C_{max}				

TABLE 18.18
Summary of In Vitro Dissolution Studies

Dissolution Condition Apparatus:
Speed of Rotation:
Medium:
Volume:
Temperature:

Firm's Proposed Specification

Dissolution Testing Site (Name, Address)

Study Ref No.	Testing Date	Product ID\Batch No. (Test—Manufacture Date) (Reference—Expiration Date)	Dosage Strength and Form	No. of Dosage Units		Collection Times (min or h)	Study Report Location
Study Report #		Test Product	mg Tablet Capsule	12	Mean		
					Range		
					%CV		
Study Report #		Reference Product	mg Tablet Capsule	12	Mean		
					Range		
					%CV		

Provide dissolution data for all strengths (test and reference).

TABLE 18.19
Formulation Data

Ingredient	Amount (mg)/Tablet		Amount (mg)/Tablet	
	Strength 1	Strength 2	Strength 1	Strength 2
Cores				
Coating				
Total			100.00	100.00

Please include the formulation of all strength.

TABLE 18.20
Reanalysis of Study Samples

Study No.

Additional Information in Volume(s), Page(s)

Reason Why Assay Was Repeated	Number of Samples Reanalyzed				Number of Recalculated Values Used after Reanalysis			
	Actual Number		% of Total Assays		Actual Number		% of Total Assays	
	T	*R*	*T*	*R*	*T*	*R*	*T*	*R*
Pharmacokinetic[a]								
Reason A (e.g., below LOQ)								
Reason B								
Reason C								
Etc.								
Total								

[a] If no repeats were performed for pharmacokinetic reasons, insert "0.0."
Please provide separate table for each analyte measured for each in vivo study.

19 Good Laboratory Practices

BACKGROUND

Title 21, part 58 of the Code of Federal Regulations (CFR) describes the required practices for nonclinical laboratory testing as it applies to analytical methods development and analysis of biological samples. Note that this is separate from the good laboratory practices (GLPs) required for clinical trials, which are described elsewhere in the book. Compliance with these GLP guidelines is crucial to meet the audit requirement of the Food and Drug Administration (FDA). Given in the following are the overall principles for GLP compliance and the details of documents required for GLP audits.

The area of GLP compliance is less well understood than the GCP or the analytical method validation; thus, given in the following as a preamble to this subject are the most often asked questions about the GLP compliance:

1. The GLPs do not apply to validation trials conducted to confirm the analytical methods used to determine the concentration of test article in animal tissues and drug dosage forms. However, the GLPs apply to the chemical procedures used to characterize the test article, to determine the stability of the test article and its mixtures, and to determine the homogeneity and concentration of test article mixtures. Likewise, the GLPs apply to the chemical procedures used to analyze specimens (e.g., clinical chemistry, urinalysis). The GLPs do not apply to the work done to develop chemical methods of analysis or to establish the specifications of a test article.
2. GLP compliance applies to all those laboratories that intend to submit data to FDA whether they are located in the United States or anywhere in the world. Photocopies of raw data, which are dated and verified by signature of the copier, are considered to be "exact" copies of the raw data. Are records of instrument calibration considered to be raw data?
3. A computer printout derived from data transferred to computer media from laboratory data sheets is not considered to be raw data.
4. The GLPs do not require that a sponsor approve the study director for a contracted study? It is the responsibility of the testing facility management.
5. If a firm functions as a primary contractor for nonclinical laboratory studies and the actual studies are then subcontracted to nonclinical laboratories, who is considered a "sponsor" will depend on the specific provisions of the contract.
6. The responsibility for test article characterization is not specifically assigned in the GLP; it is a subject of the specific contractual arrangement between the sponsor and the contractor.
7. The contract laboratories do not have to show the sponsor's name on the master schedule sheet and this information can be coded but the code must be revealed to the FDA investigator on request.
8. A contractor need not include in the final report information on test article characterization and stability if such information has been collected by the sponsor. The contractor should identify in its final report which information will be subsequently supplied by the sponsor.
9. The FDA-483 is the written notice of objectionable practices or deviations from the regulations that is prepared by the FDA investigator at the end of the inspection. The items listed on the form serve as the basis for the exit discussion with laboratory management at which time management can either agree or disagree with the items and can offer possible

corrective actions to be taken. Management may also respond to the district office in writing after it has had sufficient time to properly study the FDA-483. The FDA investigator prepares an establishment inspection report (EIR), which summarizes the observations made at the laboratory and which contains exhibits concerning the studies audited (protocols, standard operating procedures [SOPs], curriculum vitae [CVs], etc.). The EIR is then reviewed by district personnel as well as headquarters personnel. This review may reveal additional GLP deviations that should be and are communicated to laboratory management.

10. The GLP investigators should not comment on the scientific merits of a protocol or the scientific interpretation given in the final report. Their function is strictly a noting of observations and verification. Scientific judgments are made by the respective headquarters review units that deal with the test article.
11. A GLP EIR cannot be reviewed by laboratory management prior to its issuance. The GLP EIR is an internal agency document, which reflects the observations and findings of the FDA investigator. It cannot be released to anyone outside the agency until agency action has been completed and the released copy is purged of all trade secret information. Laboratories that disagree with portions of the EIR should write a letter, which contains the areas of disagreement to the local FDA District Office. The laboratories can ask that their letters accompany the EIR whenever it is requested under the Freedom of Information Act.
12. The FDA investigators can take photographs of objectionable practices and conditions.
13. Overseas laboratories are scheduled for inspection on the basis of having submitted to FDA the results of significant studies on important products; there is no 2-year cycle of inspections.
14. The following background materials are used by agency investigators to prepare for a GLP inspection:
    a. The GLP regulations
    b. The management briefings postconference report
    c. Assorted memoranda and policy issuances
    d. The GLP Compliance Program
    e. The protocol of an ongoing study, if available
    f. The final report of a completed study, if available
    g. The inspection report of the most recent inspection
15. If the results of an inspection reveal that significant deviations from the GLPs exist, the laboratory will be sent a regulatory letter that lists the major deviations and that requests a response within 10 days. The response should describe those actions that the laboratory has taken or plans to take to effect correction. The response should also encompass items that were listed on the FDA-483 and those that were discussed during the exit discussion with laboratory management. A specific timetable should be given for accomplishing the planned actions. The reasonableness of the timetable will be determined by FDA compliance staff, based on the needs of the particular situation. For less significant deviations, the laboratory will be sent a notice of adverse findings letter that also lists the deviations but that requests a response within 30 days. Again, the reasonableness of the response will be determined by FDA staff. The FDA-483 lists observations of violative conditions that have the capability to adversely affect nonclinical laboratory studies. Corrective actions should be instituted as soon as possible. Laboratory management is informed of all routine GLP inspections prior to the inspection, but special compliance or investigative inspections need not be preannounced.
16. The study director cannot be the chief executive of a nonclinical laboratory. The GLPs require that there be a separation of function between the study director and the QAU director. In the example, the QAU director would be reporting to the study director. The GLPs do permit the designation of an "acting" or "deputy" study director to be responsible

for a study when the study director is on leave. Should study records identify the designated "deputy" or "acting" study director?

17. The study director is responsible for adherence to the GLPs. The QAU is not expected to perform a scientific evaluation of a study or to "second-guess" the scientific procedures that are used. QAU inspections are made to ensure that the GLPs, SOPs, and protocols are being followed and that the data summarized in the final report accurately reflect the results of the study. A variety of procedures can be used to do this, but certainly the procedures should include an examination and correlation of the raw data records. The QAU must keep copies of all protocols as currently amended. The only SOPs that the QAU are required to keep are those concerned with the operations and procedures of the QAU.
18. The QAU is not required to monitor compliance with regulations promulgated by other government agencies.
19. An individual who is involved in a nonclinical laboratory study cannot perform QAU functions for portions of the study that the individual is not involved with. However, the individual can perform QAU functions for a study that he or she is not involved with. The QAU does review amendments to the final report.
20. The master schedule sheet should list all nonclinical laboratory studies conducted on FDA-regulated products and intended to support an application for a research or marketing permit.
21. The QAU may in its periodic reports to management and the study director recommend actions to solve existing problems.
22. The QAU should assure that the computer-formatted data accurately reflect the raw data. For the statistical analyses that would comprise a report from a participating scientist, it should be checked by QAU and appended to the final report. The QAU is also responsible for maintaining the laboratory archives. The QAU can be constituted as a single person, provided that the workload is not excessive and other duties do not prevent the person from doing an adequate job. It would be prudent to designate an alternate in case of disability, vacations, etc.
23. The GLPs do not isolate responsibility for defining study phases and designating critical study phases; logically, the task should be done by the study director and the participating scientists working in concert with the QAU and laboratory management. It can be covered by an SOP.
24. The agency has not established guidelines for the frequency of calibration of balances used in nonclinical laboratory studies. This would be a large undertaking in part due to the wide variety of equipment that is available and to the differing workloads that would be imposed on the equipment. It is suggested that you work with the equipment manufacturers and your study directors to arrive at a suitable calibration schedule. The key point is that the calibration should be frequent enough to assure data validity. The maintenance and calibration schedules should be part of the SOPs for each instrument.
25. When an equipment manufacturer performs the routine equipment maintenance, do the equipment manufacturer's maintenance procedures have to be described in the facilities' SOPs? The facilities' SOPs would have to state that maintenance was being performed by the equipment manufacturer according to their own procedures.
26. The GLPs do not specify the amount of detail to be included in the SOPs. The SOPs are intended to minimize the introduction of systematic error into a study by ensuring that all personnel will be familiar with and use the same procedures. The adequacy of the SOPs is a key responsibility of management. A guideline of adequacy that could be used is to determine whether the SOPs are understood and can be followed by trained laboratory personnel.
27. The study director cannot authorize changes in the SOPs. Each workstation should have access to the SOPs applicable to the work performed at the station. A complete set of the

SOPs, including authorized amendments, should be maintained in the archives. The SOPs are approved by the laboratory management and not QAU. The GLPs do not specify the contents of individual SOPs, but the SOP that deals with computerized data acquisition should include the purpose of the program, the specifications, the procedures, the end products, the language, the interactions with other programs, the procedures for assuring authorized data entry and access, the procedures for making and authorizing changes to the program, the source listing of the program, and perhaps even a flow chart. The laboratory's computer specialists should determine what other characteristics need to be described in the SOP.

28. All reagents used in a nonclinical laboratory have to be labeled to indicate identity, titer or concentration, storage requirements, and expiration date. Purchased reagents usually carry all these items except for the expiration date, so the laboratory should label the reagent containers with an expiration date. The expiration date selected should be in line with laboratory experience and need not require specific stability testing.
29. The procedures for confirming the quality of incoming reagents used in nonclinical laboratory studies are left to laboratory management decision but the SOPs should document the actual procedures used.
30. The GLPs do not require the use of product accountability procedures for reagents and chemicals used in a nonclinical laboratory study.
31. The study director or the QAU is permitted to request analysis of reserved samples; sufficient reserve sample should be retained so that the sample is not exhausted. Physical and chemical tests conducted on test articles are required to be done under the GLPs.
32. An analytical method need not be totally contained in the protocol. The protocol must state the type and frequency of tests to be made. Type can be connoted by reference to literature citations or the SOPs as applicable.
33. Each nonclinical laboratory study requires a sponsor-approved specific protocol; however, the laboratory that conducts the study can also qualify as the sponsor of the study. Unforeseen circumstances, which have only a one-time effect (different date of sample collection), need to be reported only in the raw data and the final report. However, such circumstances, which result in a systematic change, for example, in the SOPS or in the protocol, should also be made by a protocol amendment. The protocol amendment need not be made in advance but should be made as rapidly as possible.
34. The protocol must list the type and frequency of tests, analyses, and measurements to be made in the study. Where these are covered by SOPS, they should be listed in the protocol.
35. Raw data collected in nonclinical laboratory studies need not be cosigned by a second individual, and there is no requirement for maintaining bound copies of data recorded.
36. The GLP requirements that are applicable to computerized data acquisition systems include the following criteria:
    a. Only authorized individuals can make data entries.
    b. Data entries may not be deleted, but changes may be made in the form of dated amendments, which provide the reason for data change.
    c. The database must be made as tamperproof as possible.
    d. The SOPs should describe the procedures used for ensuring the validity of the data.
    e. Either the magnetic media or hardcopy printouts are considered to be raw data.
37. It is acceptable to manually transcribe raw data into notebooks if it is verified accurate by signature and date; technically, the GLPs do not preclude such an approach. It is not a preferred procedure, however, since the chance of transcription errors would exist. Accordingly, such an approach should be used only when necessary, and in this event, the raw data should also be retained.
38. All circumstances that may have affected the quality of the data have to be described in the final report. The GLPs do not address the issue of approval of the final report.

According to the GLPs, the final report is official when it is signed and dated by the study director. If persons reviewing the final report request changes, then such changes must be made by way of a formal amendment. The final report need identify only the name of the study director, the names of other participating scientists, and the names of all supervisory personnel.

39. Certain raw data records that are not study specific such as instrument calibration need not be filed in the archives in each study file. These can be filed in a retrievable fashion such as chronological in the archive. At the completion of a study, QAU records and inspection reports should be retained in the archives.
40. At the termination of a nonclinical laboratory study, whether a contractor can send all of the raw data, study records, and specimens to the sponsor of the study is not specifically addressed in GLPs. section 58.195(g) requires contract laboratories that go out of business to transfer all raw data and records to the sponsor. Likewise, section 58.190(b) permits raw data and study records to be stored elsewhere (other than the contract laboratory location) provided that the contract laboratory's archives have reference to the other locations and provided that the final study report identifies the other locations as directed by section 58.195(a)(13). Consequently, it is permissible for the sponsor to retain all raw data and records from the date of termination of the nonclinical laboratory study. Common sense dictates, however, that the contract laboratory keep copies of the material that has been forwarded to the sponsor.
41. For the blood and urine specimens, which are analyzed for both labile and stable constituents, it is necessary to retain the specimen for the term required by the regulations or for as long as their quality permits meaningful reevaluation, whichever is shorter.
42. The preparation of the conforming amendment statement is the responsibility of the product sponsor and the statement should be submitted as part of the application for a research or marketing permit. The contractor, however, should identify for the sponsor those non-GLP practices, which were used in each nonclinical laboratory study so that a proper conforming amendment statement can be prepared. This can be signed by the same individual in the firm who signs the official application for a research or marketing permit.
43. The FDA does not necessarily reject nonclinical laboratory studies that have not been conducted in full compliance with the GLPs. The GLP Compliance Program provides guidance on the issue. For FDA to reject a study, it is necessary to find that there were deviations from the GLPs and that these deviations were of such a nature as to compromise the quality and integrity of the study covered by the agency inspection.
44. The of the SOPs need not be submitted along with an application for a research or marketing permit.

ORGANIZATION AND PERSONNEL

Personnel

a. Each individual engaged in the conduct of or responsible for the supervision of a nonclinical laboratory study shall have education, training, and experience, or a combination thereof, to enable that individual to perform the assigned functions.
b. Each testing facility shall maintain a current summary of training and experience and job description for each individual engaged in or supervising the conduct of a nonclinical laboratory study.
c. There shall be a sufficient number of personnel for the timely and proper conduct of the study according to the protocol.
d. Personnel shall take necessary personal sanitation and health precautions designed to avoid contamination of test and control articles and test systems.

e. Personnel engaged in a nonclinical laboratory study shall wear clothing appropriate for the duties they perform. Such clothing shall be changed as often as necessary to prevent microbiological, radiological, or chemical contamination of test systems and test and control articles.
f. Any individual found at any time to have an illness that may adversely affect the quality and integrity of the nonclinical laboratory study shall be excluded from direct contact with test systems, test and control articles, and any other operation or function that may adversely affect the study until the condition is corrected. All personnel shall be instructed to report to their immediate supervisors any health or medical conditions that may reasonably be considered to have an adverse effect on a nonclinical laboratory study.

Testing Facility Management

For each nonclinical laboratory study, testing facility management shall

a. Designate a study director as described in section 58.33, before the study is initiated
b. Replace the study director promptly if it becomes necessary to do so during the conduct of a study
c. Assure that there is a quality assurance unit (QAU) as described in section 58.35
d. Assure that test and control articles or mixtures have been appropriately tested for identity, strength, purity, stability, and uniformity, as applicable
e. Assure that personnel, resources, facilities, equipment, materials, and methodologies are available as scheduled
f. Assure that personnel clearly understand the functions they are to perform
g. Assure that any deviations from these regulations reported by the QAU are communicated to the study director and corrective actions are taken and documented

Study Director

For each nonclinical laboratory study, a scientist or other professional of appropriate education, training, and experience, or combination thereof, shall be identified as the study director. The study director has overall responsibility for the technical conduct of the study, as well as for the interpretation, analysis, documentation, and reporting of results, and represents the single point of study control. The study director shall assure that

a. The protocol, including any change, is approved as provided by section 58.120 and is followed
b. All experimental data, including observations of unanticipated responses of the test system, are accurately recorded and verified
c. Unforeseen circumstances that may affect the quality and integrity of the nonclinical laboratory study are noted when they occur, and corrective action is taken and documented
d. Test systems are as specified in the protocol
e. All applicable GLP regulations are followed
f. All raw data, documentation, protocols, specimens, and final reports are transferred to the archives during or at the close of the study

Quality Assurance Unit

a. A testing facility shall have a QAU, which shall be responsible for monitoring each study to assure management that the facilities, equipment, personnel, methods, practices, records, and controls are in conformance with the regulations in this part. For any given study, the QAU shall be entirely separate from and independent of the personnel engaged in the direction and conduct of that study.

b. The QAU shall
   1. Maintain a copy of a master schedule sheet of all nonclinical laboratory studies conducted at the testing facility indexed by test article and containing the test system, nature of study, date study was initiated, current status of each study, identity of the sponsor, and name of the study director
   2. Maintain copies of all protocols pertaining to all nonclinical laboratory studies for which the unit is responsible
   3. Inspect each nonclinical laboratory study at intervals adequate to assure the integrity of the study and maintain written and properly signed records of each periodic inspection showing the date of the inspection, the study inspected, the phase or segment of the study inspected, the person performing the inspection, findings and problems, action recommended and taken to resolve existing problems, and any scheduled date for reinspection. Any problems found during the course of an inspection which are likely to affect study integrity shall be brought to the attention of the study director and management immediately.
   4. Periodically submit to management and the study director written status reports on each study, noting any problems and the corrective actions taken
   5. Determine that no deviations from approved protocols or SOPs were made without proper authorization and documentation
   6. Review the final study report to assure that such report accurately describes the methods and SOPs and that the reported results accurately reflect the raw data of the nonclinical laboratory study
   7. Prepare and sign a statement to be included with the final study report, which shall specify the dates inspections were made and findings reported to management and to the study director

c. The responsibilities and procedures applicable to the QAU, the records maintained by the QAU, and the method of indexing such records shall be in writing and shall be maintained. These items including inspection dates, the study inspected, the phase or segment of the study inspected, and the name of the individual performing the inspection shall be made available for inspection to authorized employees of the FDA.

d. A designated representative of the FDA shall have access to the written procedures established for the inspection and may request testing facility management to certify that inspections are being implemented, performed, documented, and followed up in accordance with this paragraph.

FACILITIES

General

Each testing facility shall be of suitable size and construction to facilitate the proper conduct of nonclinical laboratory studies. It shall be designed so that there is a degree of separation that will prevent any function or activity from having in adverse effect on the study.

Facilities for Handling Test and Control Articles

a. As necessary to prevent contamination or mix-ups, there shall be separate areas for receipt and storage of the test and control articles.

Laboratory Operation Areas

Separate laboratory space shall be provided, as needed, for the performance of the routine and specialized procedures required by nonclinical laboratory studies.

Specimen and Data Storage Facilities

Space shall be provided for archives, limited to access by authorized personnel only, for the storage and retrieval of all raw data and specimens from completed studies.

EQUIPMENT

Equipment Design

Equipment used in the generation, measurement, or assessment of data and equipment used for facility environmental control shall be of appropriate design and adequate capacity to function according to the protocol and shall be suitably located for operation, inspection, cleaning, and maintenance.

Maintenance and Calibration of Equipment

a. Equipment shall be adequately inspected, cleaned, and maintained. Equipment used for the generation, measurement, or assessment of data shall be adequately tested, calibrated, and/or standardized.
b. The written SOPs required under section 58.81(b)(11) shall set forth in sufficient detail the methods, materials, and schedules to be used in the routine inspection, cleaning, maintenance, testing, calibration, and/or standardization of equipment and shall specify, when appropriate, remedial action to be taken in the event of failure or malfunction of equipment. The written SOPs shall designate the person responsible for the performance of each operation.
c. Written records shall be maintained of all inspection, maintenance, testing, calibration, and/or standardizing operations. These records, containing the date of the operation, shall describe whether the maintenance operations were routine and followed the written SOPs. Written records shall be kept of nonroutine repairs performed on equipment as a result of failure and malfunction. Such records shall document the nature of the defect, how and when the defect was discovered, and any remedial action taken in response to the defect.

TESTING FACILITIES OPERATION

Standard Operating Procedures

a. A testing facility shall have SOPs in writing setting forth nonclinical laboratory study methods that management is satisfied and are adequate to ensure the quality and integrity of the data generated in the course of a study. All deviations in a study from SOPs shall be authorized by the study director and shall be documented in the raw data. Significant changes in established SOPs shall be properly authorized in writing by the management.
b. SOPs shall be established for, but not limited to, the following:
   Receipt, identification, storage, handling, mixing, and method of sampling of the test and control articles
   - Test system observations
   - Laboratory tests
   - Data handling, storage, and retrieval
   - Maintenance and calibration of equipment
   - Transfer, proper placement, and identification of animals
c. Each laboratory area shall have immediately available laboratory manuals and SOPs relative to the laboratory procedures being performed. Published literature may be used as a supplement to SOPs.
d. A historical file of SOPs, and all revisions thereof, including the dates of such revisions, shall be maintained.

Reagents and Solutions

All reagents and solutions in the laboratory areas shall be labeled to indicate identity, titer or concentration, storage requirements, and expiration date. Deteriorated or outdated reagents and solutions shall not be used.

TEST AND CONTROL ARTICLES

Test and Control Article Characterization

a. The identity, strength, purity, and composition or other characteristics which will appropriately define the test or control article shall be determined for each batch and shall be documented. Methods of synthesis, fabrication, or derivation of the test and control articles shall be documented by the sponsor or the testing facility. In those cases where marketed products are used as control articles, such products will be characterized by their labeling.
b. The stability of each test or control article shall be determined by the testing facility or by the sponsor either (1) before study initiation or (2) concomitantly according to written SOPs, which provide for periodic analysis of each batch.
c. Each storage container for a test or control article shall be labeled by name, chemical abstract number or code number, batch number, expiration date, if any, and, where appropriate, storage conditions necessary to maintain the identity, strength, purity, and composition of the test or control article. Storage containers shall be assigned to a particular test article for the duration of the study.
d. For studies of more than 4 weeks duration, reserve samples from each batch of test and control articles shall be retained for the period of time provided by section 58.195.

Test and Control Article Handling

Procedures shall be established for a system for the handling of the test and control articles to ensure that

a. There is proper storage
b. Distribution is made in a manner designed to preclude the possibility of contamination, deterioration, or damage
c. Proper identification is maintained throughout the distribution process
d. The receipt and distribution of each batch is documented (such documentation shall include the date and quantity of each batch distributed or returned)

PROTOCOL FOR AND CONDUCT OF A NONCLINICAL LABORATORY STUDY

Protocol

a. Each study shall have an approved written protocol that clearly indicates the objectives and all methods for the conduct of the study. The protocol shall contain, as applicable, the following information:
    1. A descriptive title and statement of the purpose of the study
    2. Identification of the test and control articles by name, chemical abstract number, or code number
    3. The name of the sponsor and the name and address of the testing facility at which the study is being conducted
    4. The procedure for identification of the test system
    5. A description of the experimental design, including the methods for the control of bias

6. The type and frequency of tests, analyses, and measurements to be made
7. The records to be maintained
8. The date of approval of the protocol by the sponsor and the dated signature of the study director
9. A statement of the proposed statistical methods to be used

b. All changes in or revisions of an approved protocol and the reasons therefore shall be documented, signed by the study director, dated, and maintained with the protocol.

Conduct of a Nonclinical Laboratory Study

a. The nonclinical laboratory study shall be conducted in accordance with the protocol.
b. The test systems shall be monitored in conformity with the protocol.
c. Specimens shall be identified by test system, study, nature, and date of collection. This information shall be located on the specimen container or shall accompany the specimen in a manner that precludes error in the recording and storage of data.
d. All data generated during the conduct of a nonclinical laboratory study, except those that are generated by automated data collection systems, shall be recorded directly, promptly, and legibly in ink. All data entries shall be dated on the date of entry and signed or initialed by the person entering the data. Any change in entries shall be made so as not to obscure the original entry, shall indicate the reason for such change, and shall be dated and signed or identified at the time of the change. In automated data collection systems, the individual responsible for direct data input shall be identified at the time of data input. Any change in automated data entries shall be made so as not to obscure the original entry, shall indicate the reason for change, and shall be dated, and the responsible individual shall be identified.

RECORDS AND REPORTS

Reporting of Nonclinical Laboratory Study Results

a. A final report shall be prepared for each nonclinical laboratory study and shall include, but not necessarily be limited to, the following:
   1. Name and address of the facility performing the study and the dates on which the study was initiated and completed
   2. Objectives and procedures stated in the approved protocol, including any changes in the original protocol
   3. Statistical methods employed for analyzing the data
   4. The test and control articles identified by name, chemical abstracts number or code number, strength, purity, and composition or other appropriate characteristics
   5. Stability of the test and control articles under the conditions of administration
   6. A description of the methods used
   7. A description of the test system used
   8. A description of all circumstances that may have affected the quality or integrity of the data
   9. The name of the study director, the names of other scientists or professionals, and the names of all supervisory personnel involved in the study
   10. A description of the transformations, calculations, or operations performed on the data, a summary and analysis of the data, and a statement of the conclusions drawn from the analysis

11. The signed and dated reports of each of the individual scientists or other professionals involved in the study
12. The locations where all specimens, raw data, and the final report are to be stored
13. The statement prepared and signed by the QAU

b. The final report shall be signed and dated by the study director.
c. Corrections or additions to a final report shall be in the form of an amendment by the study director. The amendment shall clearly identify that part of the final report that is being added to or corrected and the reasons for the correction or addition and shall be signed and dated by the person responsible.

Storage and Retrieval of Records and Data

a. All raw data, documentation, protocols, final reports, and specimens generated as a result of a nonclinical laboratory study shall be retained.
b. There shall be archives for orderly storage and expedient retrieval of all raw data, documentation, protocols, specimens, and interim and final reports. Conditions of storage shall minimize deterioration of the documents or specimens in accordance with the requirements for the time period of their retention and the nature of the documents or specimens. A testing facility may contract with commercial archives to provide a repository for all material to be retained. Raw data and specimens may be retained elsewhere provided that the archives have specific reference to those other locations.
c. An individual shall be identified as responsible for the archives.
d. Only authorized personnel shall enter the archives.
e. Material retained or referred to in the archives shall be indexed to permit expedient retrieval.

Retention of Records

a. Record retention requirements require a period of at least 2 years following the date on which an application for a research or marketing permit, in support of which the results of the nonclinical laboratory study were submitted, is approved by the FDA. This requirement does not apply to studies supporting investigational new drug (IND) applications or applications for investigational device exemptions (IDEs), records of which shall be kept for a period of at least 5 years following the date on which the results of the nonclinical laboratory study are submitted to the FDA in support of an application for a research or marketing permit.
b. The master schedule sheet, copies of protocols, and records of quality assurance inspections as required shall be maintained by the QAU as an easily accessible system of records.
c. Summaries of training and experience and job descriptions may be retained along with all other testing facility employment records.
d. Records and reports of the maintenance and calibration and inspection of equipment.
e. Records required by this part may be retained either as original records or as true copies such as photocopies, microfilm, microfiche, or other accurate reproductions of the original records.
f. If a facility conducting nonclinical testing goes out of business, all raw data, documentation, and other material specified in this section shall be transferred to the archives of the sponsor of the study. The Food and Drug Administration shall be notified in writing of such a transfer.

AUDIT OF FACILITIES FOR GLP COMPLIANCE

The FDA routinely conducts audits of facilities submitting data to the FDA for approval of applications for marketing authorization. The objective of this audit is

- To verify the quality and integrity of data submitted in a research or marketing application
- To inspect (approximately every 2 years) nonclinical laboratories conducting safety studies that are intended to support applications for research or marketing of regulated products
- To audit safety studies and determine the degree of compliance with GLP regulations

The types of inspections include the following:

- Surveillance inspections: Surveillance inspections are periodic, routine determinations of a laboratory's compliance with GLP regulations. These inspections include a facility inspection and audits of ongoing and/or recently completed studies.
- Directed inspections: Directed inspections are assigned to achieve a specific purpose, such as
  - Verifying the reliability, integrity, and compliance of critical safety studies being reviewed in support of pending applications
  - Investigating issues involving potentially unreliable safety data and/or violative conditions brought to the FDA's attention
  - Reinspecting laboratories previously classified official action indicated (OAI) (usually within 6 months after the firm responds to a warning letter)
  - Verifying the results from third-party audits or sponsor audits submitted to the FDA for consideration in determining whether to accept or reject questionable or suspect studies

General Instructions to Investigators

1. The investigator will determine the current state of GLP compliance by evaluating the laboratory facilities, operations, and study performance.
2. Organization chart—If the facility maintains an organization chart, obtain a current version of the chart for use during the inspection and submit it in the EIR.
3. Facility floor-plan diagram—Obtain a diagram of the facility. The diagram may identify areas that are not used for GLP activities. If it does not, request that appropriate facility personnel identify any areas that are not used for GLP activities. Use it during the inspection and submit it in the EIR.
4. Master schedule sheet—Obtain a copy of the firm's master schedule sheet for all studies listed since the last GLP inspection or last 2 years and select studies as defined in 21 CFR 58.3(d). If the inspection is the first inspection of the facility, review the entire master schedule. If studies are identified as non-GLP, determine the nature of several studies to verify the accuracy of this designation. See 21 CFR 58.1 and 58.3(d). In contract laboratories, determine who decides if a study is a GLP study.
5. Identification of studies
   a. Directed inspections—Inspection assignments will identify studies to be audited.
   b. Surveillance inspections—Inspection assignments may identify one or more studies to be audited. If the assignment does not identify a study for coverage, or if the referenced study is not suitable to assess all portions of current GLP compliance, the investigator will select studies as necessary to evaluate all areas of laboratory operations. When additional studies are selected, first priority should be given to FDA studies for submission to the assigning center.

Note: Studies performed for submission to other government agencies, for example, Environmental Protection Agency, National Toxicology Program, and National Cancer Institute, will not be audited without authorization from the Bioresearch Monitoring Program coordinator (HFC-230). However, this authorization is not necessary to briefly look at one of these studies to assess the ongoing operations of a portion of the facility.

Establishment Inspections

The facility inspection should be guided by the GLP regulations. The following areas should be evaluated and described as appropriate:

1. **Organization and personnel (21 CFR 58.29, 58.31, 58.33)**
   a. *Purpose*: To determine whether the organizational structure is appropriate to ensure that studies are conducted in compliance with GLP regulations and to determine whether management, study directors, and laboratory personnel are fulfilling their responsibilities under the GLPs.
   b. *Management responsibilities (21 CFR 58.31)*: Identify the various organizational units, their role in carrying out GLP study activities, and the management responsible for these organizational units. This includes identifying personnel who are performing duties at locations other than the test facility and identifying their line of authority. If the facility has an organization chart, much of this information can be determined from the chart. Determine if management has procedures for assuring that the responsibilities in 58.31 can be carried out. Look for evidence of management involvement, or lack thereof, in the following areas:
      1. Assigning and replacing study directors
      2. Control of study director workload (use the Master Schedule to assess workload)
      3. Establishment and support of the QAU, including assuring that deficiencies reported by the QAU are communicated to the study directors and acted upon
      4. Assuring that test and control articles or mixtures are appropriately tested for identity, strength, purity, stability, and uniformity
      5. Assuring that all study personnel are informed of and follow any special test and control article handling and storage procedures
      6. Providing required study personnel, resources, facilities, equipment, and materials
      7. Reviewing and approving protocols and SOPs
      8. Providing GLP or appropriate technical training
   c. *Personnel (21 CFR 58.29)*: Identify key laboratory and management personnel, including any consultants or contractors used, and review personnel records, policies, and operations to determine if the following apply:
      1. Summaries of training and position descriptions are maintained and are current for selected employees.
      2. Personnel have been adequately trained to carry out the study functions that they perform.
      3. Personnel have been trained in GLPs.
      4. Practices are in place to ensure that employees take necessary health precautions, wear appropriate clothing, and report illnesses to avoid contamination of the test and control articles and test systems.
      5. If the firm has computerized operations, determine the following:
         a. Who was involved in the design, development, and validation of the computer system?
         b. Who is responsible for the operation of the computer system, including inputs, processing, and output of data?

c. Whether computer system personnel have training commensurate with their responsibilities, including professional training and training in GLPs.
d. Whether some computer system personnel are contractors who are present on-site full-time, or nearly full-time. The investigation should include these contractors as though they were employees of the firm. Specific inquiry may be needed to identify these contractors, as they may not appear on organization charts.
e. Interview and observe personnel using the computerized systems to assess their training and performance of assigned duties.

d. *Study director (21 CFR 58.33)*
   1. Assess the extent of the study director's actual involvement and participation in the study. In those instances when the study director is located off-site, review any correspondence/records between the testing facility management and QAU and the off-site study director. Determine that the study director is being kept immediately apprised of any problems that may affect the quality and integrity of the study.
   2. Assess the procedures by which the study director
      a. Assures the protocol and any amendments have been properly approved and are followed
      b. Assures that all data are accurately recorded and verified
      c. Assures that data are collected according to the protocol and SOPs
      d. Documents unforeseen circumstances that may affect the quality and integrity of the study and implements corrective action
      e. Assures that study personnel are familiar with and adhere to the study protocol and SOPs
      f. Assures that study data are transferred to the archives at the close of the study

e. *EIR documentation and reporting*: Collect exhibits to document deficiencies. This may include SOPs, organizational charts, position descriptions, and CVs, as well as study-related memos, records, and reports for the studies selected for review. The use of outside or contract facilities must be noted in the EIR. The assigning center should be contacted for guidance on inspection of these facilities.

2. **Quality assurance unit (QAU; 21 CFR 58.35)**
   a. *Purpose*: To determine if the test facility has an effective, independent QAU that monitors significant study events and facility operations, reviews records and reports, and assures management of GLP compliance
   b. *QAU operations (21 CFR 58.35(b-d))*: Review QAU SOPs to assure that they cover all methods and procedures for carrying out the required QAU functions and confirm that they are being followed. Verify that SOPs exist and are being followed for QAU activities including, but not limited to, the following:
      1. Maintenance of a master schedule sheet
      2. Maintenance of copies of all protocols and amendments
      3. Scheduling of its in-process inspections and audits
      4. Inspection of each nonclinical laboratory study at intervals adequate to assure the integrity of the study and maintenance of records of each inspection
      5. Immediately notify the study director and management of any problems that are likely to affect the integrity of the study
      6. Submission of periodic status reports on each study to the study director and management
      7. Review of the final study report
      8. Preparation of a statement to be included in the final report that specifies the dates inspections were made and findings reported to management and to the study director
      9. Inspection of computer operations

Verify that, for any given study, the QAU is entirely separate from and independent of the personnel engaged in the conduct and direction of that study. Evaluate the time QAU personnel spend in performing in-process inspection and final report audits. Determine if the time spent is sufficient to detect problems in critical study phases and if there are adequate personnel to perform the required functions.

Note: The investigator may request the firm's management to certify in writing that inspections are being implemented, performed, documented, and followed up in accordance with this section (See 58.35(d)).

c. *EIR documentation and reporting*: Obtain a copy of the master schedule sheet dating from the last routine GLP inspection or covering the past 2 years. If the master schedule is too voluminous, obtain representative pages to permit headquarters review. When master schedule entries are coded, obtain the code key. Deficiencies should be fully reported and documented in the EIR. Documentation to support deviations may include copies of QAU SOPs, list of QAU personnel, their CVs or position descriptions, study-related records, protocols, and final reports.

3. **Facilities (21 CFR 58.41-51)**
   a. *Purpose*: To assess whether the facilities are of adequate size and design
   b. *Facility inspection*
      1. Review environmental controls and monitoring procedures for critical areas (i.e., animal rooms, test article storage areas, laboratory areas, handling of biohazardous material) and determine if they appear adequate and are being followed.
      2. Review the SOPs that identify materials used for cleaning critical areas and equipment and assess the facility's current cleanliness.
      3. Determine whether there are appropriate areas for the receipt, storage, mixing, and handling of the test and control articles.
      4. Determine whether separation is maintained in rooms where two or more functions requiring separation are performed.
      5. Determine that computerized operations and archived computer data are housed under appropriate environmental conditions (e.g., protected from heat, water, and electromagnetic forces).
   c. *EIR documentation and reporting*: Identify which facilities, operations, SOPs, etc., were inspected. Only significant changes in the facility from previous inspections need be described. Facility floor plans may be collected to illustrate problems or changes. Document any conditions that would lead to contamination of test articles or to unusual stress of test systems.
4. **Equipment (21 CFR 58.61-63)**
   a. *Purpose*: To assess whether equipment is appropriately designed and of adequate capacity and is maintained and operated in a manner that ensures valid results.
   b. *Equipment inspection*: Assess the following:
      1. The general condition, cleanliness, and ease of maintenance of equipment in various parts of the facility.
      2. The heating, ventilation, and air-conditioning system design and maintenance, including documentation of filter changes and temperature/humidity monitoring in critical areas.
      3. Whether equipment is located where it is used and that it is located in a controlled environment, when required.
      4. Non-dedicated equipment for preparation of test and control article carrier mixtures is cleaned and decontaminated to prevent cross contamination.

5. For representative pieces of equipment, check the availability of the following:
    a. SOPS and/or operating manuals
    b. Maintenance schedule and log
    c. Standardization/calibration procedure, schedule, and log
    d. Standards used for calibration and standardization
6. For computer systems, assess that the following procedures exist and are documented (see also attachment A):
    a. Validation study, including validation plan and documentation of the plan's completion
    b. Maintenance of equipment, including storage capacity and backup procedures
    c. Control measures over changes made to the computer system, which include the evaluation of the change, necessary test design, test data, and final acceptance of the change
    d. Evaluation of test data to assure that data are accurately transmitted and handled properly when analytical equipment is directly interfaced to the computer
    e. Procedures for emergency backup of the computer system (e.g., backup battery system and data forms for recording data in the event of a computer failure or power outage)

c. *EIR documentation and reporting*: The EIR should list which equipment, records, and procedures were inspected and the studies to which they are related. Detail any deficiencies that might result in contamination of test articles, uncontrolled stress to test systems, and/or erroneous test results.

5. **Testing facility operations (21 CFR 58.81)**
    a. *Purpose*: To determine if the facility has established and follows written SOPs necessary to carry out study operations in a manner designed to ensure the quality and integrity of the data
    b. *SOP evaluation*
        1. Review the SOP index and representative samples of SOPs to ensure that written procedures exist to cover at least all of the areas identified in 58.81(b).
        2. Verify that only current SOPs are available at the personnel workstations.
        3. Review key SOPs in detail and check for proper authorization signatures and dates and general adequacy with respect to the content (i.e., SOPs are clear and complete and can be followed by a trained individual).
        4. Verify that changes to SOPs are properly authorized and dated and that a historical file of SOPs is maintained.
        5. Ensure that there are procedures for familiarizing employees with SOPs.
        6. Determine that there are SOPs to ensure the quality and integrity of data, including input (data checking and verification), output (data control), and an audit trail covering all data changes.
        7. Verify that a historical file of outdated or modified computer programs is maintained. If the firm does not maintain old programs in digital form, ensure that a hard copy of all programs has been made and stored.
        8. Verify that SOPs are periodically reviewed for current applicability and that they are representative of the actual procedures in use.
        9. Review selected SOPs and observe employees performing the operation to evaluate SOP adherence and familiarity.
    c. *EIR documentation and reporting*: Submit SOPs, data collection forms, and raw data records as exhibits that are necessary to support and illustrate deficiencies.

6. **Reagents and solutions (21 CFR 58.83)**
   a. *Purpose*: To determine that the facility ensures the quality of reagents at the time of receipt and subsequent use
      1. Review the procedures used to purchase, receive, label, and determine the acceptability of reagents and solutions for use in the studies.
      2. Verify that reagents and solutions are labeled to indicate identity, titer or concentration, storage requirements, and expiration date.
      3. Verify that for automated analytical equipment, the profile data accompanying each batch of control reagents are used.
      4. Check that storage requirements are being followed.
7. **Test and control articles (21 CFR 58.105-113)**
   a. *Purpose*: To determine that procedures exist to assure that test and control articles and mixtures of articles with carriers meet protocol specifications throughout the course of the study and that accountability is maintained
   b. *Characterization and stability of test articles (21 CFR 58.105)*: The responsibility for carrying out appropriate characterization and stability testing may be assumed by the facility performing the study or by the study sponsor. When test article characterization and stability testing is performed by the sponsor, verify that the test facility has received documentation that this testing has been conducted.
      1. Verify that procedures are in place to ensure that
         a. The acquisition, receipt and storage of test articles, and means used to prevent deterioration and contamination are as specified
         b. The identity, strength, purity, and composition, (i.e., characterization) to define the test and control articles are determined for each batch and are documented
         c. The stability of test and control articles is documented
         d. The transfer of samples from the point of collection to the analytical laboratory is documented
         e. Storage containers are appropriately labeled and assigned for the duration of the study
         f. Reserve samples of test and control articles for each batch are retained for studies lasting more than 4 weeks
   c. *Test and control article handling (21 CFR 58.107)*
      1. Determine that there are adequate procedures for
         a. Documentation for receipt and distribution
         b. Proper identification and storage
         c. Precluding contamination, deterioration, or damage during distribution
      2. Inspect test and control article storage areas to verify that environmental controls, container labeling, and storage are adequate.
      3. Observe test and control article handling and identification during the distribution and administration to the test system.
      4. Review a representative sample of accountability records and, if possible, verify their accuracy by comparing actual amounts in the inventory. For completed studies, verify documentation of final test and control article reconciliation.
8. **Protocol and conduct of nonclinical laboratory study (21 CFR 58.120 130)**
   a. *Purpose*: To determine if study protocols are properly written and authorized and that studies are conducted in accordance with the protocol and SOPs
   b. *Study protocol (21 CFR 58.120)*
      1. Review SOPs for protocol preparation and approval and verify they are followed.
      2. Review the protocol to determine if it contains required elements.

3. Review all changes, revisions, or amendments to the protocol to ensure that they are authorized, signed, and dated by the study director.
4. Verify that all copies of the approved protocol contain all changes, revisions, or amendments.

c. *Conduct of the Nonclinical Laboratory Study (21 CFR 58.130)*: Evaluate the following laboratory operations, facilities, and equipment to verify conformity with protocol and SOP requirements for
   1. Test system monitoring
   2. Recording of raw data (manual and automated)
   3. Corrections to raw data (corrections must not obscure the original entry and must be dated, initialed, and explained)
   4. Randomization of test systems
   5. Collection and identification of specimens
   6. Authorized access to data and computerized systems

d. *EIR reporting and documentation*: Identify the study(ies) inspected and, if available, the associated FDA research or marketing permit numbers. Report and document any deficiencies observed. Submit, as exhibits, a copy of all protocols and amendments that were reviewed.

9. **Records and reports (21 CFR 58.185-195)**
   a. *Purpose*: To assess how the test facility stores and retrieves raw data, documentation, protocols, final reports, and specimens
   b. *Reporting of study results (21 CFR 58.185)*: Determine if the facility prepares a final report for each study conducted. For selected studies, obtain the final report and verify that it contains the following:
      1. The required elements in 21 CFR 58.185(a)(1-14), including the identity (name and address) of any subcontractor facilities and portion of the study contracted and a description of any computer program changes
      2. Dated signature of the study director (21 CFR 58.185(b))
      3. Corrections or additions to the final report made in compliance with 21 CFR 58.185(c)
   c. *Storage and retrieval of records and data (21 CFR 58.190)*
      1. Verify that raw data, documentation, protocols, final reports, and specimens have been retained.
      2. Identify the individual responsible for the archives. Determine if delegation of duties to other individuals in maintaining the archives has occurred.
      3. Verify that archived material retained or referred to in the archives is indexed to permit expedient retrieval. It is not necessary that all data and specimens be in the same archive location. For raw data and specimens retained elsewhere, the archives index must make specific reference to those other locations.
      4. Verify that access to the archives is controlled and determine that environmental controls minimize deterioration.
      5. Ensure that there are controlled procedures for adding or removing material. Review archive records for the removal and return of data and specimens. Check for unexplained or prolonged removals.
      6. Determine how and where computer data and backup copies are stored, that records are indexed in a way to allow access to data stored on electronic media, and that environmental conditions minimize deterioration.
      7. Determine to what electronic media such as tape cassettes or ultrahigh-capacity portable disks the test facility has the capacity of copying records in electronic form. Report names and identifying numbers of both copying equipment type and electronic medium type to enable agency personnel to bring electronic media to future inspections for collecting exhibits.

   d. *EIR documentation and reporting*: Provide a brief summary of the facility's report preparation procedures and their retention and retrieval of records, reports, and specimens. If records are archived off-site, obtain a copy of documentation of the records which were transferred and where they are located. Describe and document deficiencies.

Data Audit

In addition to the procedures outlined previously for evaluating the overall GLP compliance of a firm, the inspection should include the audit of at least one completed study. Studies for audit may be assigned by the center or selected by the investigator as described in part III, A. The audit will include a comparison of the protocol (including amendments to the protocol), raw data, records, and specimens against the final report to substantiate that protocol requirements were met and that findings were fully and accurately reported.

1. For each study audited, the study records should be reviewed for quality to ensure that data are as follows:
   a. Attributable—The raw data can be traced, by signature or initials and date to the individual observing and recording the data. Should more than one individual observe or record the data, that fact should be reflected in the data.
   b. Legible—The raw data are readable and recorded in a permanent medium. If changes are made to original entries, the changes
      1. Must not obscure the original entry
      2. Indicate the reason for change
      3. Must be signed or initialed and dated by the person making the change
   c. Contemporaneous—The raw data are recorded at the time of the observation.
   d. Original—The first recording of the data.
   e. Accurate—The raw data are true and complete observations. For data entry forms that require the same data to be entered repeatedly, all fields should be completed or a written explanation for any empty fields should be retained with the study records.
2. General
   a. Determine if there were any significant changes in the facilities, operations, and QAU functions other than those previously reported.
   b. Determine whether the equipment used was inspected, standardized, and calibrated prior to, during, and after use in the study. If equipment malfunctioned, review the remedial action and ensure that the final report addresses whether the malfunction affected the study.
   c. Determine if approved SOPs existed during the conduct of the study.
   d. Compare the content of the protocol with the requirements in 21 CFR.
   e. Review the final report for the study director's dated signature and the QAU statement as required in 21 CFR 58.35(b)(7).
3. Protocol vs. final report: Study methods described in the final report should be compared against the protocol and the SOPs to confirm those requirements were met.
4. Final report vs. raw data: The audit should include a detailed review of records, memorandum, and other raw data to confirm that the findings in the final report completely and accurately reflect the raw data. Representative samples of raw data should be audited against the final report.
5. Specimens vs. final report: The audit should include examination of a representative sample of specimens in the archives for confirmation of the number and identity of specimens in the final report.
6. EIR documentation and reporting

7. Full reporting

   A full report will be prepared and submitted in the following situations:
   a. The initial GLP inspection of a facility
   b. All inspections that may result in an OAI classification
   c. Any assignment specifically requesting a full report
8. Abbreviated reporting
   a. Field investigators may use abbreviated reporting for the following types of assignments:
      1. Surveillance inspections (except for initial inspections) of a facility when it is apparent from the findings that the inspection may result in a final classification of no action indicated (NAI) or voluntary action indicated (VAI). These reports must include enough documented information to support the final classification.
      2. Directed inspections and data audits provided the report *fully covers* all aspects of the specific topic of the inspection (i.e., operations, past deficiencies, assigned studies, etc.) and documents significant adverse findings to support the final classification.

COMPUTERIZED SYSTEMS

Computer systems and operations are thoroughly covered during inspection of any facility. No additional reporting is required under this attachment.

In August 1997, the agency's regulation on electronic signatures and electronic recordkeeping became effective. The regulation, at 21 CFR part 11, describes the technical and procedural requirements that must be met if a firm chooses to maintain records electronically and/or use electronic signatures. part 11 works in conjunction with other FDA regulations and laws that require recordkeeping. Those regulations and laws ("predicate rules") establish requirements for record content, signing, and retention.

Certain older electronic systems may not have been in full compliance with part 11 by August 1997 and modification to these so-called legacy systems may take more time. part 11 does not grandfather legacy systems and FDA expects that firms using legacy systems are taking steps to achieve full compliance with part 11.

If a firm is keeping electronic records or using electronic signatures, determine if they are in compliance with 21 CFR part 11. Determine the depth of part 11 coverage on a case-by-case basis, in light of initial findings and program resources. At a minimum, ensure that (1) the firm has prepared a corrective action plan for achieving full compliance with part 11 requirements and is making progress toward completing that plan in a timely manner; (2) accurate and complete electronic and human readable copies of electronic records, suitable for review, are made available; and (3) employees are held accountable and responsible for actions taken under their electronic signatures. If initial findings indicate the firm's electronic records and/or electronic signatures may not be trustworthy and reliable, or when electronic recordkeeping systems inhibit meaningful FDA inspection, a more detailed evaluation may be warranted.

Personnel: Part III, C.1.c. (21 CFR 58.29)

Determine the following:

- Who was involved in the design, development, and validation of the computer system?
- Who is responsible for the operation of the computer system, including inputs, processing, and output of data?
- If computer system personnel have training commensurate with their responsibilities, including professional training and training in GLPs.
- Whether some computer system personnel are contractors who are present on-site full-time or nearly full-time. The investigation should include these contractors as though they were employees of the firm. Specific inquiry may be needed to identify these contractors, as they may not appear on organization charts.

QAU Operations: Part III, C.2 (21 CFR 58.35(b-d))

- Verify SOPs exist and are being followed for QAU inspections of computer operations.

Facilities: Part III, C.3 (21 CFR 58.41-51)

- Determine that computerized operations and archived computer data are housed under appropriate environmental conditions.

Equipment: Part III, C.4 (21 CFR 58.61-63)

For computer systems, check that the following procedures exist and are documented:

- Validation study, including validation plan and documentation of the plan's completion
- Maintenance of equipment, including storage capacity and backup procedures
- Control measures over changes made to the computer system, which include the evaluation of the change, necessary test design, test data, and final acceptance of the change
- Evaluation of test data to assure that data are accurately transmitted and handled properly when analytical equipment is directly interfaced to the computer
- Procedures for emergency backup of the computer system (e.g., backup battery system and data forms for recording data in case of a computer failure or power outage)

Testing Facility Operations: Part III, C.5 (21 CFR 58.81)

- Verify that a historical file of outdated or modified computer programs is maintained.

Records and Reports: Part III, C.10.b. (21 CFR 58.185–195)

- Verify that the final report contains the required elements in 58.185(a)(1-14), including a description of any computer program changes.

Storage and Retrieval of Records and Data: Part III, C.10.c. (21 CFR 58.190)

- Assess archive facilities for degree of controlled access and adequacy of environmental controls with respect to computer media storage conditions.
- Determine how and where computer data and backup copies are stored, that records are indexed in a way to allow access to data stored on electronic media, and that environmental conditions minimize deterioration.
- Determine how and where original computer data and backup copies are stored.

GOOD LABORATORY PRACTICE QUESTIONS AND ANSWERS

Since June 20, 1979, the FDA has been asked many questions on the GLP regulations (GLPs, 21 CFR 58). This information given in the following is routinely reviewed by field investigators prior to making GLP inspections and by headquarters personnel involved in the GLP program. Therefore, the applicants should understand the various nuances of the answers provided by FDA regarding GLPs. If there are any specific questions that are not covered here, the applicants are encouraged to write to Bioresearch Monitoring Program coordinator (HFC-130), FDA, 5600 Fishers Lane Rockville, MD 20857.

Subpart A: General Provisions

Section 58.1: Scope

1. Do the GLPs apply to validation trials conducted to confirm the analytical methods used to determine the concentration of test article in animal tissues and drug dosage forms?

   No.

2. Do the GLPs apply to the following studies on animal health products: overdosage studies in the target species, animal safety studies in the target species, tissue residue accumulation and depletion studies, and udder irritation studies?
   Yes.
3. Do the GLPs apply to safety studies on cosmetic products?
   No. Such studies are not carried out in support of a marketing permit. However, the GLPs represent good quality control a goal that all testing facilities should strive to attain.
4. Do safety studies done to determine the potential drug-abuse characteristics of a test article have to be done under the GLPs?
   Yes they do, but only when the studies are required to be submitted to the FDA as part of an application for a research or marketing permit.
5. Do the GLPs apply to the organoleptic evaluation of processed foods?
   No.
6. Do the GLPs apply to all of the analytical support work conducted to provide supplementary data to a safety study?
   The GLPs apply to the chemical procedures used to characterize the test article, to determine the stability of the test article and its mixtures, and to determine the homogeneity and concentration of test article mixtures. Likewise, the GLPs apply to the chemical procedures used to analyze specimens (e.g., clinical chemistry, urinalysis). The GLPs do not apply to the work done to develop chemical methods of analysis or to establish the specifications of a test article.
7. Is it possible to obtain an exemption from specific provisions of the GLPs for special nonclinical laboratory studies?
   Yes. The GLPs were written with the aim of being applicable to a broad variety of studies, test articles, and test systems. Nonetheless, the FDA realizes that not all of the GLP provisions apply to all studies and, indeed, for some special studies certain of the GLP provisions may compromise proper science. For this reason, laboratories may petition the FDA for exemption for certain studies from some of the GLP provisions. The petition should contain sufficient facts to justify granting the exemption.
8. Are subcontractor laboratories that furnish a particular service such as ophthalmology exams; reading of animal ECGs, EEGs, and EMGs; preparation of blocks and slides from tissues; statistical analysis; and hematology covered by the GLPs?
   Yes, to the extent that they contribute to a study that is subject to the GLPs.

Section 58.3: Definitions

1. Are animal cage cards considered to be raw data?
   Raw data is defined as "any laboratory worksheets, records, memorandum, notes that are the result of original observations and activities and are necessary for the reconstruction and evaluation of-the report of that study." Cage cards are not raw data if they contain information like animal number, study number, study dates, and cage number (information that is not the result of original observations and that is not necessary for study reconstruction). However, if an original observation is put on the cage cards, then all cards must be saved as raw data.
2. Are photocopies of raw data, which are dated and verified by signature of the copier, considered to be "exact" copies of the raw data?
   Yes.
3. Are records of quarantine, animal receipt, environmental monitoring, and instrument calibration considered to be raw data?
   Yes.

4. A laboratory conducts animal studies to establish a baseline set of data for a different test species/strain. No test article is administered but the toxicology laboratory facilities and procedures will be used and the resulting data may eventually be submitted do the FDA as part of a research or marketing permit. Are the studies considered to be nonclinical laboratory studies that are covered by the GLPs?

   Generally, a nonclinical laboratory study involves a test article studied under laboratory conditions for the purpose of determining its safety. The cited example does not fit the definition so it would not be covered by the GLPs. Since the data from the baseline studies may be used to interpret the results of a nonclinical laboratory study, it is recommended, but not required, that the study be conducted in accord with GLPs in order to ensure valid baseline data.

5. The definition of "nonclinical laboratory study" excludes field trials in animals. What is a field trial in animals?

   A field trial in animals is similar to a human clinical trial. It is conducted for the purpose of obtaining data on animal drug efficacy and it is excluded from coverage under the GLPs.

6. Necropsies are done by prosectors trained by and working under the supervision of a pathologist. The necropsy data are recorded by the prosector on data sheets, and when making the final report, the pathologist summarizes the data collected by the prosector as well as by him or herself. What constitutes the raw data in this example?

   Both the prosector's data sheets and the signed and dated report of the pathologist would be considered raw data.

7. Is a computer printout derived from data transferred to computer media from laboratory data sheets considered to be raw data?

   No.

8. Are the assay plates used in the 10t1/2 mammalian cell transformation assay considered to be specimens?

   Yes.

9. If a firm uses parapathologists to screen tissue preparations, are the parapathologists' data sheets considered to be raw data?

   Yes.

Section 58.10: Applicability to Studies Performed under Grants and Contracts

1. Certain contracts specify that a series of nonclinical laboratory studies be done on a single test article. Do the GLPs permit the designation of different study directors for each study under the contract?
2. Do the GLPs require that a sponsor approve the study director for a contracted study?
3. A firm functions as a primary contractor for nonclinical laboratory studies. The actual studies are then subcontracted to nonclinical laboratories. Is the firm considered to be a "sponsor?"

   Yes.

   No. Testing facility management designates the study director.

   The GLPs define "sponsor" as a person who initiates and supports a nonclinical laboratory study. Sponsorship in the cited example would be determined by the specific provisions of the contract.

4. Who is responsible for test article characterization—the sponsor or the contractor?

   The GLPs do not assign the responsibility in this area. The matter is a subject of the specific contractual arrangement between the sponsor and the contractor.

5. Do contract laboratories have to show the sponsor's name on the master schedule sheet or can this information be coded?

   The information can be coded but the code must be revealed to the FDA investigator on request.

6. A sponsor desires to contract for a nonclinical laboratory study to be conducted in a foreign laboratory. Must the sponsor notify the foreign laboratory that compliance with the U.S. GLPs is required?

   Yes.

7. Must a contractor include in the final report information on test article characterization and stability when such information has been collected by the sponsor?

   No. The contractor should identify in its final report which information will be subsequently supplied by the sponsor.

8. Must a sponsor reveal toxicology data already collected on a test article to a contract laboratory?

   No. If use of the test article involves a potential danger to laboratory personnel, the contract laboratory should be advised so that appropriate precautions can be taken.

Section 58.15: Inspection of a Testing Facility

1. What is the usual procedure for the issuance of a form FDA-483?

   The FDA-483 is the written notice of objectionable practices or deviations from the regulations that is prepared by the FDA investigator at the end of the inspection. The items listed on the form serve as the basis for the exit discussion with laboratory management at which time management can either agree or disagree with the items and can offer possible corrective actions to be taken. Management may also respond to the district office in writing after it has had sufficient time to properly study the FDA-483.

2. Will a laboratory subsequently be notified of GLP deviations not listed on the FDA-483?

   This does happen. The FDA investigator prepares an EIR, which summarizes the observations made at the laboratory and which contains exhibits concerning the studies audited (protocols, SOPs, CVs, etc.). The EIR is then reviewed by district personnel as well as headquarters personnel. This review may reveal additional GLP deviations that should be and are communicated to laboratory management.

3. What kinds of domestic toxicology laboratory inspections does FDA perform and how frequently are they done?

   FDA performs four kinds of inspections related to the GLPs and nonclinical laboratory studies. These include the following: a GLP inspection, an inspection undertaken as a periodic, routine determination of a laboratory's compliance with the GLPs, which includes examination of an ongoing study as well as a completed study; a data audit, an inspection made to verify that the information contained in a final report submitted to the FDA is accurate and reflected by the raw data; a directed inspection from any of the series of inspections conducted for various compelling reasons (questionable data in a final report, tips from informers, etc.); and a follow-up inspection, an inspection made sometime after a GLP inspection that revealed objectionable practices and conditions. The purpose of the follow-up inspection is to assure that proper corrective actions have been taken. GLP inspections are scheduled once every 2 years, whereas the other kinds of inspections are scheduled as needed.

4. Should GLP investigators comment on the scientific merits of a protocol or the scientific interpretation given in the final report?

   No. Their function is strictly a noting of observations and verification. Scientific judgments are made by the respective headquarters' review units that deal with the test article.

5. Can a GLP EIR be reviewed by laboratory management prior to issuance?

   No. The GLP EIR is an internal FDA document, which reflects the observations and findings of the FDA investigator. It cannot be released to anyone outside the FDA until

FDA action has been completed and the released copy is purged of all trade secret information. Laboratories that disagree with portions of the EIR should write a letter, which contains the areas of disagreement to the local FDA District Office. The laboratories can ask that their letters accompany the EIR whenever it is requested under the Freedom of Information Act.

6. Can FDA investigators take photographs of objectionable practices and conditions?

   It is the FDA position that photographs can be taken as a part of the inspection and this position has been sustained by a district court decision.

7. The GLP Compliance Program requires the FDA investigator to select an ongoing study in order to inspect current laboratory operations. What criteria are used to select the study?

   The studies are selected in accord with FDA priorities, that is, the longest-term study on the most significant product.

8. Does FDA inspect international nonclinical laboratories once every 2 years?

   No. Overseas laboratories are scheduled for inspection on the basis of having submitted to the FDA the results of significant studies on important products.

9. What background materials are used by FDA investigators to prepare for a GLP inspection?

   Prior to an inspection, the following materials are usually reviewed:

   a. *The GLP regulations*
   b. *The management briefings postconference report*
   c. *Assorted memoranda and policy issuances*
   d. *The GLP Compliance Program*
   e. *The protocol of an ongoing study, if available*
   f. *The final report of a completed study, if available*
   g. *The inspection report of the most recent inspection*

10. How long does FDA allow a laboratory to effect corrective actions after an inspection has been made?

    If the results of an inspection reveal that significant deviations from the GLPs exist, the laboratory will be sent a regulatory letter that lists the major deviations and that requests a response within 10 days. The response should describe those actions that the laboratory has taken or plans to take to effect correction. The response should also encompass items that were listed on the FDA-483 and those that were discussed during the exit discussion with laboratory management. A specific timetable should be given for accomplishing the planned actions. The reasonableness of the timetable will be determined by FDA compliance staff, based on the needs of the particular situation.

    For less significant deviations, the laboratory will be sent a notice of adverse findings letter that also lists the deviations but that requests a response within 30 days. Again, the reasonableness of the response will be determined by the FDA staff.

11. Does a laboratory's responsibility for corrective action listed on a FDA-483 begin at the conclusion of an inspection or upon receipt of correspondence from the originating bureau in which corrective action is requested?

    The FDA-483 lists observations of violative conditions that have the capability to adversely affect nonclinical laboratory studies. Corrective actions should be instituted as soon as possible.

12. Does FDA preannounce all GLP inspections?

    Laboratory management is informed of all routine GLP inspections prior to the inspection, but special compliance or investigative inspections need not be preannounced.

Subpart B: Organization and Personnel

Section 58.29: Personnel

1. For what sequence in the supervisory chain should position descriptions be available?
   Position descriptions should be available for each individual engaged in or supervising the conduct of the study.
2. Should current summaries of training and experience list attendance at scientific and technical meetings?
   Yes. The FDA considers such attendance as a valuable adjunct to the other kinds of training received by laboratory personnel.
3. If certain specialists (pathologists, statisticians, ophthalmologists, etc.) are contracted to conduct certain aspects of a study, need they be identified in the final report?
   Yes.
4. Does the QAU have to be composed of technical personnel?
   No. Management is however responsible for assuring that "personnel clearly understand the functions they are to perform" (section 58.31(f)) and that each individual engaged in the study has the appropriate combination of education, training, and experience (section 58.29(a)).

Section 58.31: Testing Facility Management

1. Can the study director be the chief executive of a nonclinical laboratory?
   No. The GLPs require that there be a separation of function between the study director and the QAU director. In the example, the QAU director would be reporting to the study director.

Section 58.33: Study Director

1. The GLPs permit the designation of an "acting" or "deputy" study director to be responsible for a study when the study director is on leave. Should study records identify the designated "deputy" or "acting" study director?
   Yes.
2. Is the study director responsible for adherence to the GLPs?
   Yes.

Section 58.35: Quality Assurance Unit

1. As a QAU person, I have no expertise in the field of pathology. How do I audit pathology findings?
   The QAU is not expected to perform a scientific evaluation of a study or to "second-guess" the scientific procedures that are used. QAU inspections are made to ensure that the GLPs, SOPs, and protocols are being followed and that the data summarized in the final report accurately reflect the results of the study. A variety of procedures can be used to do this but certainly the procedures should include an examination and correlation of the raw data records.
2. Must the QAU keep copies of all protocols and amendments and SOPs and amendments?
   The QAU must keep copies of all protocols as currently amended. The only SOPs that the QAU are required to keep are those concerned with the operations and procedures of the QAU.
3. Does the QAU have to monitor compliance with regulations promulgated by other government agencies?
   The GLPs do not require this.
4. Can an individual who is involved in a nonclinical laboratory study perform QAU functions for portions of the study that the individual is not involved with?

5. Does the QAU review amendments to the final report?
   No. However, the individual can perform QAU functions for a study that he or she is not involved with.
   Yes.
6. What studies are required to be listed on the master schedule sheet?
   The master schedule sheet should list all nonclinical laboratory studies conducted on FDA-regulated products and intended to support an application for a research or marketing permit.
7. May the QAU in its periodic reports to management and the study director recommend actions to solve existing problems?
   Yes.
8. If raw data are transcribed and sent to the sponsor for (a) preparing the data in computer format or (b) performing a statistical analysis, what are the responsibilities of the QAU?
   For (a), the QAU should assure that the computer-formatted data accurately reflect the raw data. For (b), the statistical analyses would comprise a report from a participating scientist; therefore, it should be checked by QAU and appended to the final report.
9. Can the QAU also be responsible for maintaining the laboratory archives?
   Yes.
10. Can a QAU be constituted as a single person?
    Yes, provided that the workload is not excessive and other duties do not prevent the person from doing an adequate job. It would be prudent to designate an alternate in case of disability, vacations, etc.
11. Who is responsible for defining study phases and designating critical study phases and can these be covered in the SOP?
    The GLPs do not isolate this responsibility. Logically, the task should be done by the study director and the participating scientists working in concert with the QAU and laboratory management. It can be covered by an SOP.

Subpart C: Facilities

Section 58.41: General

No questions were asked on the subject.

Section 58.43: Animal Care Facilities

1. Do the GLPs require clean/dirty separation for the animal care areas?
   No. They do require adequate separation of species and studies.
2. Do the GLPs require that separate animal rooms be used to house test systems and conduct different studies?
   No. The GLPs require separate areas adequate to assure proper separation of test systems, isolation of individual projects, animal quarantine, and routine or specialized housing of animals, as necessary to achieve the study objectives.
3. Do the GLPs require that access to animal rooms be limited only to authorized individuals?
   No. However, undue stresses and potentially adverse influences on the test system should be minimized.

Section 58.45: Animal Supply Facilities

No questions were asked on the subject.

Section 58.47: Facilities for Handling Test and Control Articles

1. Do test and control articles have to be maintained in locked storage units?
   No, but accurate records of test and control article accountability must be maintained.

Section 58.49: Laboratory Operation Areas

No questions were asked on the subject.

Section 58.51: Specimen and Data Storage Facilities

1. What do the GLPs require with regard to facilities for the archives?
   Space should be provided for archives limited to access by authorized personnel. Storage conditions should minimize deterioration of documents and specimens.

Section 58.53: Administrative and Personnel Facilities

No questions were asked on the subject.

Subpart D: Equipment

Section 58.61: Equipment Design

No questions were asked on the subject.

Section 58.63 Maintenance and Calibration of Equipment

1. Has the FDA established guidelines for the frequency of calibration of equipment (balances) used in nonclinical laboratory studies?
   The FDA has not established guidelines for the frequency of calibration of balances used in nonclinical laboratory studies. This would be a large undertaking in part due to the wide variety of equipment that is available and to the differing workloads that would be imposed on the equipment. It is suggested that you work with the equipment manufacturers and your study directors to arrive at a suitable calibration schedule. The key point is that the calibration should be frequent enough to assure data validity. The maintenance and calibration schedules should be part of the SOPs for each instrument.
2. When an equipment manufacturer performs the routine equipment maintenance, do the equipment manufacturer's maintenance procedures have to be described in the facilities' SOPs?
   No. The facilities' SOPs would have to state that maintenance was being performed by the equipment manufacturer according to their own procedures.

Subpart E: Testing Facilities Operation

Section 58.81: Standard Operating Procedures

1. What amount of detail should be included in the SOPs?
   The GLPs do not specify the amount of detail to be included in the SOPs. The SOPs are intended to minimize the introduction of systematic error into a study by ensuring that all personnel will be familiar with and use the same procedures. The adequacy of the SOPs is a key responsibility of management. A guideline of adequacy that could be used is to determine whether the SOPs are understood and can be followed by trained laboratory personnel.
2. Can the study director authorize changes in the SOPS?
   No. Approval of the SOPs and changes thereto is a function of laboratory management.
3. How many copies of the complete laboratory SOPs are needed?
   Each workstation should have access to the SOPs applicable to the work performed at the station. A complete set of the SOPs, including authorized amendments, should be maintained in the archives.
4. Who approves the SOPs of the QAU?
   Laboratory management
5. To what extent are computer programs to be documented as SOPs?
   The GLPs do not specify the contents of individual SOPs, but the SOP that deals with computerized data acquisition should include the purpose of the program, the

specifications, the procedures, the end products, the language, the interactions with other programs, the procedures for assuring authorized data entry and access, the procedures for making and authorizing changes to the program, the source listing of the program, and perhaps even a flow chart. The laboratory's computer specialists should determine what other characteristics need to be described in the SOP.

Section 58.83: Reagents and Solutions

1. What are the GLP requirements for labeling of reagents purchased directly from manufacturers?

   All reagents used in a nonclinical laboratory have to be labeled to indicate identity, titer or concentration, storage requirements, and expiration date. Purchased reagents usually carry all these items except for the expiration date, so the laboratory should label the reagent containers with an expiration date. The expiration date selected should be in line with laboratory experience and need not require specific stability testing.
2. How extensive should the procedures be for confirming the quality of incoming reagents used in nonclinical laboratory studies?

   Laboratory management should make this decision but the SOPs should document the actual procedures used.
3. Do the procedures used for preparing the S9 activator fraction (liver microsomal challenge) are to be performed in accord with the GLPs?

   No. The GLPs consider the S9 activator fraction to be a reagent. Therefore, it must be labeled properly, stored properly, and tested prior to use in accord with adequate SOPs, and it cannot be used if its potency is below established specifications.
4. Do the GLPs require the use of product accountability procedures for reagents and chemicals used in a nonclinical laboratory study?

   No.

Section 58.90: Animal Care

1. Can diseased animals received from a supplier be diagnosed, treated, certified "well," and then entered into a nonclinical laboratory study?

   The GLPs provide for this procedure by including provisions directed toward animal quarantine and isolation. The question of whether such animals can be entered into a study, however, is a scientific one that should be answered by the veterinarian in charge and the study director and other scientists involved in the study.
2. Do the GLPs prohibit the use of primates for multiple nonclinical laboratory studies?

   No. Again, the question is a scientific one and the potential impact of multiple use on study interpretation should be carefully assessed.
3. Is a photocopy of an animal purchase order, which has been signed and dated by the individual receiving the shipment, sufficient proof of animal receipt?

   Yes, but actual shipping tickets are also acceptable.
4. Does the FDA have guidelines for animal bedding?

   No, but the GLPs prohibit the use of bedding which can interfere with the objectives of the study.
5. Does the FDA permit the sterilization of animal feed with ethylene oxide?

   No.
6. For certain test systems (timed-pregnant rodents), it is not possible to use long quarantine periods. Do the GLPs specify quarantine periods for each test system?

   No. The quarantine period can be established by the veterinarian in charge of animal care and should be of sufficient length to permit evaluation of health status.

7. How are feed and water contaminants to be dealt with?
   The protocol should include a positive statement as to the need for conducting feed analysis for contaminants. If analysis is necessary, the identities and specifications for the contaminants should be listed. The need for analysis as well as the specifications should be determined by the study scientists. Water contaminants can be handled similarly.
8. How is the adequacy of bedding materials to be handled?
   This can be handled, as are the analyses for possible contaminants in feed and water. The study director and associated scientists should consider the bedding and its possible impact on the study. The results of this consideration should appear in the protocol.
9. What do the GLPs require in regard to assuring the genetic quality of animals used in a nonclinical laboratory study?
   This is a scientific issue that is not specifically addressed by the GLPs. Suitability of the test system for use in a study is a protocol matter and any required testing procedure should be arrived at by the study scientists.
10. Do the GLPs require specific procedures for the microbiological monitoring of animals used in nonclinical laboratory studies?
    The procedures used should be in accord with acceptable veterinary medical practice.
11. The Japanese are preparing animal care guidelines, which are similar but not identical to the U.S. guidelines prepared by the National Institutes of Health (NIH). Would these be acceptable?
    Japanese guidelines that are similar, but no less stringent, in the important particulars with the NIH guidelines would be acceptable to the FDA.
12. What is the frequency of feed contaminant analysis?
    If contaminant analyses are required by the protocol, then the GLPs require periodic analysis of the feed to ensure that the contaminant level is at or below that judged to be acceptable. Statistical procedures should be used to determine the frequency of analysis since this is dependent on the specific chemical characteristics of the interfering contaminant.
13. It is necessary to use "official" methods of analysis to determine the levels of interfering contaminants?
    No. The methods should be appropriate for the analysis and the FDA reserves the right to examine the raw data supporting the analytical results.
    1. Do the GLPs require production facilities to be dedicated to the manufacture of specific animal feeds used in nonclinical laboratory studies?
    2. Is a separate room required for animal necropsy?
       No.
       No. The GLPs require separate areas and/or rooms as necessary to prevent any activity from having an adverse effect on the study. If the necropsy is done in an animal room, precautions should be taken to minimize disturbances that may interfere with the study.

Subpart F: Test and Control Articles

Section 58.105: Test and Control Article Characterization

1. Is it necessary to retain samples of feed from nonclinical laboratory studies in which the feed serves as the control article?
   Yes. It is not necessary, however, to retain reserve samples of feed from studies that involves test article administration by routes other than feed.
2. What expiration date is placed on the label of test articles whose stability is being assessed concurrently with the conduct of the study?
   In this situation, the stability of the test article is unknown, but periodic analysis data exist. The label should contain a statement such as "see protocol" or "see periodic

analysis results" so that test article users will know that current analytical data should be examined prior to continued use of the test article.

3. If analysis of the reserve samples is required by the study director or the QAU, is it permitted?
   Yes, but sufficient reserve sample should be retained so that the sample is not exhausted.
4. Are physical and chemical tests conducted on test articles required to be done under the GLPs?
   According to section 58.105, such tests conducted to characterize the specific batch of test article used in the nonclinical laboratory study are covered.

Section 58.107: Test and Control Article Handling

1. With regard to safety studies in large animals (cattle, horses, etc.), must test article accountability be maintained and can the animals be used for food purposes?
   Test article accountability must be maintained. For guidance on whether the treated animals can be used for food, you should contact the appropriate individuals in the Center for Veterinary Medicine.

Section 58.113: Mixtures of Articles with Carriers

1. Do the GLPs require tests for homogeneity, concentration, and stability on mixtures of control articles used as positive controls?
   Yes.
2. Do test or control article concentration assays have to be performed on each batch of test or control article carrier mixture?
   No. The GLPs require only periodic analysis of test or control article carrier mixtures.
3. What is the purpose of periodic analysis requirement for test or control article mixtures?
   This requirement provides additional assurance that the test system is being exposed to protocol-specified quantities of test article. Whereas, in most instances, proper assurance is obtained through adequate uniformity-of-mixing studies, adequate SOPs, and trained personnel, occasionally, the mixing equipment can malfunction or other uncontrollable events can occur that lead to improper dosages. These events can be recognized through periodic analysis.
4. For acute studies, does the test article carrier mixture have to be analyzed (single-dose studies)?
   Yes, but the analysis need not be done prior to the study provided the mixture is stable in storage.
5. For liquid dosing studies where the test article mixture is made by dilution of the highest dose, which dose should be analyzed?
   The lowest dose would be appropriate since it would confirm the efficacy of the dilution process; however, the GLPs do not prohibit the analysis of any of the other doses.
6. Do homogeneity studies need to be done on solutions and suspensions of test articles used in acute nonclinical laboratory studies?
   The answers to these questions are yes for suspensions of test articles and no for true solutions of test articles.
7. The analysis of test article mixtures that are used in acute studies is problematic. Usually at the stage of product development, the analytical method is not fully developed. Also, getting the analytical department to schedule the analysis is difficult. Stability is not a problem since fresh solutions are used. In view of the fact that acute studies are not pivotal in gaining approval of a research or marketing permit, is it necessary to analyze test article mixtures?
   Yes. Although acute studies may be of lesser importance in assessing the safety of human drugs, they are important for animal drugs, biological products, and certain food additives. For this reason, there must be some assurance that the test system was dosed with protocol-specified quantities of test article. The GLPs do not require that the analysis be done prior to the use of the test article mixture provided that the mixture is stable on storage.

Subpart G: Protocol for the Conduct of a Nonclinical Laboratory Study

Section 58.120: Protocol

1. What are the proposed starting and completion dates for a nonclinical laboratory study?

   There is a good deal of confusion on these dates and proper interpretation impacts on several GLP areas. Accordingly, the following clarification is offered: At the time of protocol development, the study director is to propose to management the approximate time frame of the study. Section 58.120(a)(4), therefore, requires that the protocol contain the proposed starting and completion date of the study. These dates are somewhat discretionary provided that they are identified in the protocol. Suitable identification can be the date of first dosing of the test system to the date of last dosing, the date of allocation of the test system to the experimental units to the date of necropsy of the last animal on test, the date of receipt of the test system to the date of final histopathological examination, or any combination of these or any other logical starting and completion dates. After this, the protocol is signed by the study director and forwarded for approval to management. Management approves, if indicated, signs and dates, and at this point, the study becomes a regulated study and must be entered on the master schedule sheet. The study is carried on the master schedule sheet until the study director submits a signed and dated final report. Thus, for master schedule sheet purposes, the starting date of the study is the date of protocol approval by management and the completion date of the study is the date of signature of the final report by the study director. Neither of the foregoing timeframes need be used to define the study terms described in section 58.35(b)(3) and section 58.105(d). For these sections, the traditional terms found in the toxicology literature may be used.

2. Must an analytical method be totally contained in the protocol?

   No. The protocol must state the type and frequency of tests to be made. Type can be connoted by reference to literature citations or the SOPs as applicable.

   1. Does each nonclinical laboratory study require a sponsor-approved specific protocol?
   2. Do unforeseen circumstances that occur during a study and that necessitate minor operational changes have to be reported as protocol amendments?

      Yes. However, the laboratory that conducts the study can also qualify as the sponsor of the study.

      Unforeseen circumstances, which have only a one-time effect (different date of sample collection, animal weighings) need to be reported only in the raw data and the final report. However, such circumstances, which result in a systematic change, for example, in the SOPS or in the protocol, should also be made by a protocol amendment. The protocol amendment need not be made in advance but should be made as rapidly as possible.

5. Pathologists at a firm would like to take tissues from animals in a nonclinical study, which would be used to conduct exploratory research studies. The tissues would not be part of the nonclinical laboratory study design and the results would not necessarily pertain to the study objectives. What would the GLPs require in this case?

   The protocol should state that tissues are to be taken from the experimental animals and that the tissues would be used for exploratory research purposes. If any effects were observed in the exploratory research studies, which would influence the interpretation of the results of the nonclinical laboratory study, these effects must be reported in the final report.

6. Does the protocol have to list the SOPs used in a specific study?

   The protocol must list the type and frequency of tests, analyses, and measurements to be made in the study. Where these are covered by SOPs, they should be listed in the protocol.

7. Do the GLPs require that absorption studies be done on each test article?
   No. The GLPs require that, if absorption studies are needed to achieve the scientific objectives of the study plan, the protocol should describe the methods to be used to determine absorption. Whether or not absorption studies are required is a scientific issue to be decided by the study scientists.
8. Who assesses protocol validity (number of animals, test article dosage, test system, etc.)?
   This is done by the study scientists using the scientific literature, published guidelines, advice from regulatory agencies, and prior experimental work.

Section 58.130: Conduct of a Nonclinical Laboratory Study

1. Do raw data collected in nonclinical laboratory studies have to be cosigned by a second individual?
2. What are the GLP requirements that are applicable to computerized data acquisition systems?
   No. An acceptable system must satisfy the following criteria:
   a. Only authorized individuals can make data entries.
   b. Data entries may not be deleted, but changes may be made in the form of dated amendments, which provide the reason for data change.
   c. The database must be made as tamperproof as possible.
   d. The SOPs should describe the procedures used for ensuring the validity of the data.
   e. Either the magnetic media or hardcopy printouts are considered to be raw data.
3. In Japan, employees do not sign raw data records but rather they use an official seal, which is unique to the employee. Is this an acceptable procedure?
   Yes.
4. Do tissue slides have to carry the complete sample labeling information stated in the GLPs?
   No, accession numbers are permitted providing that these numbers can be translated into the information required under section 58.130(c).
5. Is a positive notation (a statement of what was done in the raw data) required for routine laboratory operations such as the following?
   a. Identifying animals
   b. Shaving or abrading rabbits
   c. Specific dosing procedures
   d. Fasting of animals
   Yes.
6. Do the GLPs require the entry of raw data into bound notebooks?
   No.
7. Is it acceptable to manually transcribe raw data into notebooks if it is verified accurate by signature and date?
   Technically, the GLPs do not preclude such an approach. It is not a preferred procedure, however, since the chance of transcription errors would exist. Accordingly, such an approach should be used only when necessary, and in this event, the raw data should also be retained.

Subpart J: Records and Reports

Section 58.185: Reporting of Nonclinical Laboratory Study Results

1. Do contributing scientist's reports have to be prepared and appended to final reports or can the contributing scientist's report be included in the final report prepared by the study director and signed by each contributing scientist?
   The signed reports of contributing scientists should be appended to the final report.

2. Does section 58:115(a) describe the format for submission of a final report?
   The cited section describes the information that has to be submitted in a final report but the specific format is left up to the laboratory.
3. Do all circumstances that may have affected the quality of the data have to be described in the final report?
   Yes.
4. Who approves the final report of a nonclinical laboratory study?
   The GLPs do not address the issue of approval of the final report. According to the GLPs, the final report is official when it is signed and dated by the study director. If persons reviewing the final report request changes, then such changes must be made by way of a formal amendment.
5. Can the chemistry information required by section 58.185(a)(4) be located elsewhere in the application for a research or marketing permit?
   Yes. The final report should, however, reference the location of the chemistry information.
6. Does everyone who participated in a study have to be identified in the final report?
   No. The final report need identify only the name of the study director, the names of other participating scientists, and the names of all supervisory personnel.
7. Does the phase of the study, which has been inspected, need to be identified in the QAU statement in the final report?
   No.
   1. How are protocol deviations, which are discovered after the completion of the study, to be handled?
   2. How does the FDA view interim reports of nonclinical laboratory studies?
      The deviations should be described in the final report and in the study records.
      Interim reports are to be treated the same as final reports, that is, they are to be reviewed by the QAU so that the summarized data accurately reflect the raw data.

Section 58.190: Storage and Retrieval of Records and Data

1. Certain raw data records are not study specific (pest control, instrument calibration). Must these be filed in the archives in each study file?
   No. These can be filed in a retrievable fashion such as chronological in the archives.
2. Where should the QAU records be retained?
   At the completion of a study, QAU records and inspection reports should be retained in the archives.
3. At the termination of a nonclinical laboratory study, can a contractor send all of the raw data, study records, and specimens to the sponsor of the study?
   The regulations do not specifically address this issue. Section 58.195(g) requires contract laboratories that go out of business to transfer all raw data and records to the sponsor. Likewise, section 58.190(b) permits raw data and study records to be stored elsewhere (other than the contract laboratory location) provided that the contract laboratory's archives have reference to the other locations and provided that the final study report identifies the other locations as directed by section 58.195(a)(13).

   Consequently, it is permissible for the sponsor to retain all raw data and records from the date of termination of the nonclinical laboratory study. Common sense dictates, however, that the contract laboratory keep copies of the material that has been forwarded to the sponsor.
4. Can a study director or a pathologist be responsible for storing and retaining specimens and raw data?
   Yes, the GLPs permit multiple archival locations provided that these locations are identified in the central archives and that they provide adequate storage conditions and authorized access features.

Section 58.195: Retention of Records

1. With regard to blood and urine specimens, which are analyzed for both labile and stable constituents, is it necessary to retain the specimen until the most stable constituent deteriorates?

   All specimens should be retained for the term required by the regulations or for as long as their quality permits meaningful reevaluation, whichever is shorter.

2. For a GLP-regulated metabolism study, whole tissues are homogenized and aliquots thereof are used for analysis. Is it necessary to retain all of the remaining homogenate as a reserve sample?

   No, it is only necessary to retain a representative sample large enough to repeat the original measurements.

3. If animals used in acute studies are subjected to necropsy, is it necessary to retain the organs as study specimens.

   Yes.

Conforming Amendments

1. Do acute studies not done in conformity with the GLPs have to be identified in the conforming amendment statement?

   Yes.

2. How extensive should the conforming amendment statement be for preliminary exploratory studies that are exempt from GLP coverage?

   The statement should be brief and indicate the GLP-exempt status of the study.

3. For contracted nonclinical laboratory studies, who is responsible for preparing the GLP compliance statement required by the conforming amendments?

   The preparation of the conforming amendment statement is the responsibility of the product sponsor and the statement should be submitted as part of the application for a research or marketing permit. The contractor, however, should identify for the sponsor those non-GLP practices, which were used in each nonclinical laboratory study so that a proper conforming amendment statement can be prepared.

4. Who signs the conforming amendment statement?

   This can be the same individual in the firm who signs the official application for a research or marketing permit.

5. Is a specific conforming amendment statement as required by part 314(f)(7) to be prepared for each nonclinical laboratory study?

   Yes. GLP deviations have to be identified for all nonclinical laboratory studies. This can be done by preparing a single comprehensive statement, which includes all safety studies in the respective official filing. The conforming amendment statement in the official filing should be located in proximity to the animal safety studies section.

General

1. Have any nonclinical laboratories been disqualified since June 20, 1979?

   No.

2. Does the FDA reject nonclinical laboratory studies that have not been conducted in full compliance with the GLPs?

   Not necessarily. The GLP Compliance Program provides guidance on the issue. For FDA to reject a study, it is necessary to find that there were deviations from the GLPs and that these deviations were of such a nature as to compromise the quality and integrity of the study covered by the FDA inspection.

3. Must copies of the SOPs be submitted along with an application for a research or marketing permit?

   No.
4. What should be done about nonclinical laboratory studies that are stopped prior to completion?

   *The FDA recognizes that a variety of circumstances *(disease outbreak, power failures, etc.) can lead to the premature termination of a nonclinical laboratory study. In these cases, a short final report should be prepared that describes the reasons for study termination.*
5. Has the FDA established permissible limits for environmental controls (temperature, humidity, and lighting) for the animal facilities?

   No. These are scientific matters that should be described in the protocol and/or the SOPs. Of course, accurate records should be maintained.

20 Bioanalytical Method Validation

BACKGROUND

The assessment of the quality of an analytical method is far more than a statistical challenge; it is a matter of good ethics and good business practices. Many regulatory documents have been released in the pharmaceutical industry to address quality issues. These are primarily International Conference on Harmonisation (ICH) and Food and Drug Administration (FDA) documents. Those that are related to analytical and bioanalytical method validation (ICH, 1995, 1997; FDA, 2001) suggest that analytical methods must comply with specific acceptance criteria to be recognized as validated procedures. The primary aim of these documents is to require evidence that the analytical methods are suitable for their intended use. Unfortunately, discrepancies exist among these documents with respect to the definition of acceptance criteria, and limited guidance is provided for estimating the performance criteria. Appendix A to this book contains the definitions of various terms used in this chapter.

Table 20.1 describes the overall requirements of submission of bioanalytical validation.

The concepts and inconsistencies in the general methodologies used for establishing validation of analytical methods are an evolving field and require a keen understanding of the statistical principles individual to the test methods to establish reliable estimates of the performance criteria. Whereas a large number of CFR 21 part 11 compliant statistical analysis software are available, one of the most reliable and comprehensive software is offered by SAS (www.sas.com). Several references are provided in this chapter for the use of SAS software.

Method Classification Based on Data Types

The ultimate goal of an analytical method or procedure is to measure accurately a quantity, such as the concentration of an analyte, or to measure a specific activity, as, for example, for a biomarker. However, many assays such as cell-based and enzyme activity biomarker assays may not be very sensitive, may lack precision, and/or may not offer definitive reference standards. Assays based on physicochemical (such as chromatographic methods) or biochemical (such as ligand-binding assays [LBAs]) properties of an analyte assume that these quantifiable characteristics are reflective of the quantities, concentration, or biological activity of the analyte. For the purpose of analytical validation, these are classified as follows:

Qualitative methods generate data, which do not have a continuous proportionality relationship with the amount of analyte in a sample; the data are categorical in nature. Data may be nominal such as a present/absent call for a gene or gene product. Alternatively, data might be ordinal in nature, with discrete scoring scales (e.g., 1 to 5 or –, +, +++) such as for immunohistochemistry assays.

Quantitative methods are assays where the response signal has a continuous relationship with the quantity or activity of the analyte. These responses can therefore be described by a mathematical function. Inclusion of reference standards at discrete concentrations allows the quantification of sample responses by interpolation. The availability of a well-defined reference standard may

TABLE 20.1
EMA Bioanalytical Method Validation

Analytical validation report	<Study code>	
Location(s)	<Vol/page, link>	
This analytical method was used in the following studies:	<Study IDs>	
Short description of the method	<e.g., HPLC–MS–MS, GC–MS, ligand binding>	
Biological matrix	<e.g., plasma, whole blood, urine>	
Analyte	<Name>	
Location of product certificate	<Vol/page, link>	
IS[a]	<Name>	
Location of product certificate	<Vol/page, link>	
Calibration concentrations (units)		
Lower limit of quantification (units)	<LLOQ>, <accuracy%>, <precision%>	
QC concentrations (units)		
Between-run accuracy	<Range or by QC>	
Between-run precision	<Range or by QC>	
Within-run accuracy	<Range or by QC>	
Within-run precision	<Range or by QC>	
Matrix factor (MF) (all QC)[a]	Low QC	High QC
IS normalized MF (all QC)[a]	<Mean>	<Mean>
CV% of IS normalized MF (all QC)[a]	<Mean>	<Mean>
% of QCs with >85% < 115% n.v.[a,b]	<CV%>	<CV%>
% matrix lots with mean <80% or >120% n.v.[a,b]	<%>	<%>
	<%>	<%>
Long-term stability of the stock solution and working solutions[c] (observed change %)	Confirmed up to <time> at <°C><%, range or by QC>	
Short-term stability in biological matrix at room temperature or at sample processing temperature (observed change %)	Confirmed up to <time><%, range or by QC>	
Location	<Vol/page, link>	
Autosampler storage stability (observed change %)	Confirmed up to <time><%, range or by QC>	
Post-preparative stability (observed change %)	Confirmed up to <time><%, range or by QC>	
Freeze and thaw stability (observed change %)	<–temperature °C, # cycles,><range or by QC>	
Dilution integrity	Concentration diluted <X-fold> accuracy <%> precision <%>	
Partial validation[d]	<Describe shortly the reason of revalidation(s)>	
Location(s)	<Vol/page, link>	
Cross validation(s)[d]	<Describe shortly the reason of cross validations>	
Location(s)	<Vol/page, link>	

[a] Might not be applicable for the given analytical method.
[b] n.v. = nominal value.
[c] Report short-term stability results if no long-term stability on stock and working solution are available.
[d] These rows are optional. Report any validation study that was completed after the initial validation study.

be limited or may not be representative of the in vivo presentation, so quantification may not be absolute. To that end, three types of quantitative methods have been defined:

- A *definitive quantitative assay* uses calibrators fit to a known model to provide absolute quantitative values for unknown samples. Typically, such assays are only possible where the analyte is not endogenous. An example of this is a small molecule drug.

- A *relative quantitative assay* is similar in approach, but generally involves the measurement of endogenously occurring analytes. In this case, even a *zero* or blank calibrator may contain some amount of analyte, and quantification can only be done relative to this *zero* level. Examples of this include immunoassays for cytokines, such as sTNFRII, or gene expression assays, for example, reverse transcriptase polymerase chain reaction (RT-PCR).
- A *quasi-quantitative assay* does not involve the use of calibrators, mostly due to the lack of suitable reference material, so the analytical result for a test sample is reported only in terms of the assay signal (e.g., optical density in ELISA).

Objective of an Analytical Method

The objective of a definite and relative quantitative analytical method is to be able to quantify as accurately as possible *each* of the unknown quantities that the laboratory will have to determine. In other words, what all analysts expect from an analytical procedure is that the difference between the measurement or observation (X) and the unknown "true value" μ_T of the test sample be small or inferior to an acceptance limit λ:

$$-\lambda < X - \mu_T < \lambda \Leftrightarrow \left|X - \mu_T\right| < \lambda \tag{20.1}$$

The acceptance limit λ can be different depending on the requirements of the analyst and the objective of the analytical procedure. The objective is linked to the requirements usually admitted by the practice (e.g., 1% or 2% on bulk, 5% on pharmaceutical specialties, 15% for biological samples). Acceptance limits vary in clinical applications depending on factors such as the physiological variability and the intent of use.

Objective of the Pre-Study Validation Phase

The aim of the pre-study validation phase is to generate information to guarantee that the analytical method will provide, in routine use, measurements close to the true value without being affected by other elements present in the sample. In other words, the validation phase should demonstrate that the inequality described in Equation 20.1 holds for a certain proportion of the sample population.

The difference between the measurement X and its true value is a sum of a systematic error (bias or trueness) and a random error (variance or precision). The true values of these parameters are unknown, but they can be estimated based on the validation experiments. The reliability of these estimates depends on the adequacy of these experiments (design, size).

Consequently, the objective of the validation phase is to evaluate whether, given the estimates of bias and variance, the expected proportion of measures that will fall within the acceptance limits is greater than a predefined level, say, β:

$$E_{\hat{\mu},\hat{\sigma}}\left[P\left(\left|X - \mu_T\right| < \lambda \middle| \hat{\mu}_M, \hat{\sigma}_M\right)\right] \geq \beta \tag{20.2}$$

Although Equation 20.2 cannot be solved exactly within a frequentist framework, approximate solutions that can be used in practice will be discussed elsewhere in this book.

Classical Design in Pre-Study Validation

Experiments performed during pre-study validation are designed to mimic the processes and practices to be followed during routine application of a method. All aspects of the analytical method should be taken into account, such as the lot of a solvent, operator, and preparation of samples. If measurements generated under these "simulated" conditions are acceptable, then the method will

TABLE 20.2
Minimal Sample Size for *r* Runs and *s* Replicates per Run for 10% Acceptance Limits

	Within-Run Variance									
	1%		2%		3%		4%		5%	
Between-Run Variance	*r*	*s*	*r*	*s*	*r*	*s*	*r*	*s*	*r*	*s*
1%	4	3	4	3	4	3	4	4	5	9
	5	3	5	3	5	3	5	4	6	7
2%	4	3	4	3	4	3	4	6	8	9
	5	3	5	3	5	3	5	6	9	7
3%	4	4	4	6	5	6	7	10		
	5	3	5	3	6	5	8	9		
4%	7	10	9	8						
	8	7	10	6						

be declared valid for routine use. Usually, two sets of samples will be prepared for simulating the real process: calibration and validation samples (CS and VS).

- CS must be prepared according to the protocol that will be followed during routine use, that is, the same operational mode, the same number of concentration levels for the standard curve, and the same number of repetitions at each level.
- VS must be prepared in the sample matrix when applicable. In the validation phase, they mimic the unknown samples that the analytical procedure will have to quantify in routine use. Each validation standard should be prepared independently, in order to have realistic estimates of the variance components.

The minimum design of a pre-study validation phase is at least two replicates per run or series in a minimum of three runs. However, it is highly recommended to consider at least six runs in order to have a good estimate of the between-run variance. The number of runs and replicates to perform at each concentration level to demonstrate that an analytical procedure is valid could be estimated (by simulations) and depends, of course, on the inherent but unknown properties of the analytical procedure itself. The more variable the method, the more experiments are necessary.

Table 20.2 displays the minimal sample size for *r* runs and *s* replicates per run for 10% acceptance limits (the table was computed via simulations, assuming a potential small bias of 2%). It is clear that the number of runs increases with increasing between-run variance. The higher number of runs can be compensated by more replicates per runs, but this leads to a larger total number of experiments (*rs*). Also as expected when the sum of bias (2%), the within-run and between-run variances, becomes greater than 10%, it becomes unlikely that the method will ever be validated for such acceptance limits. More development in the laboratory is required to achieve this objective. The reproducibility that requires between-laboratory experiments will not be discussed in this chapter.

VALIDATION CRITERIA

Analytical methods used in the testing of biological fluids must be validated, and whereas any approach that establishes compliance with the requirements given later is acceptable to regulatory authorities, it is a good idea to follow a formal protocol to the development of these methods. The bioanalytical testing procedures involve such methods as gas chromatography (GC), high-pressure liquid chromatography (LC), and combined GC and LC mass spectrometric (MS) procedures such as LC–MS, LC–MS–MS, GC–MS, and GC–MS–MS performed for the quantitative determination of drugs and/or metabolites in biological matrices such as blood, serum, plasma, or urine. Selective and sensitive analytical methods for the quantitative evaluation of drugs and their metabolites (analytes) are critical for the successful conduct of preclinical and/or biopharmaceutics and clinical pharmacology studies. Bioanalytical method validation includes all of the procedures that demonstrate that a particular method used for quantitative measurement of analytes in a given biological matrix, such as blood, plasma, serum, or urine, is reliable and reproducible for the intended use.

The main validation criteria widely recommended by various regulatory documents (ICH, FDA, European Union) and commonly used in analytical laboratories are

- Specificity–selectivity
- Response function (calibration curve)
- Linearity
- Precision (repeatability and intermediate precision)
- Accuracy (trueness)
- Measurement error (TE)
- Limit of detection (LOD)
- Limit of quantification (LOQ)
- Assay range
- Sensitivity

In addition, according to the domains concerned, other specific criteria can be required:

- Analyte stability
- Recovery
- Effect of the dilution

Validation involves documenting, through the use of specific laboratory investigations, that the performance characteristics of the method are suitable and reliable for the intended analytical applications. The acceptability of analytical data corresponds directly to the criteria used to validate the method. A full validation is necessary for an analytical procedure to pass from the development phase to the phase of routine analysis. The validation step is not only necessary but also required at the time specifications (tests and acceptance limits) that are set up for an active ingredient or a finished product.

The validation criteria mentioned earlier must be established, insofar as possible, in the same matrix as that of samples to be analyzed. Every new analytical procedure will have to be validated for each type of matrix (e.g., for each type of biological fluid and for each animal species). Nevertheless, the definition of a matrix depends on analyst responsibility. Some matrix regrouping, generally admitted by the profession for an application domain given, can be performed.

Published methods of analysis are often modified to suit the requirements of the laboratory performing the assay. These modifications should be validated to ensure suitable performance of the analytical method. When changes are made to a previously validated method, the analyst should exercise judgment as to how much additional validation is needed. During the course of a typical drug development program, a defined bioanalytical method undergoes many modifications. The evolutionary changes to support specific studies and different levels of validation demonstrate the validity of an assay's performance. Different types and levels of validation are defined and characterized as follows.

Pre-Validation

The classical design of a pre-study validation consists in performing different runs with replicates in each run. Let's note p the number of runs, n_i the number of replicates in the ith run, $N = n_1 + \cdots + n_p$ the total number of measurements, and x_{ij} the jth measurement in the ith run.

Validation criteria such as accuracy and precision are estimated for each concentration level by statistical analysis of the back-calculated quantities. Computationally, a one-factor random effects analysis of variance (ANOVA) model is fit to the back-calculated values at each level with run as the random effects factor:

$$x_{ij} = \mu + \alpha_i + \varepsilon_{ij}$$

where
- μ is the mean of calculated concentrations
- α_i and ε_{ij} are normally distributed with mean 0 and variances σ_B^2 and σ_W^2, respectively

Here, σ_B^2 is the run-to-run variance and σ_W^2 is the within-run variance. The estimates of μ, σ_B^2, and σ_W^2 are given by

$$\hat{\mu} = \bar{x}_{..} = \frac{1}{N}\sum_{i=1}^{p} n_i \bar{x}_{i\cdot}, \quad \hat{\sigma}_W^2 = \frac{1}{N-p}\sum_{i=1}^{p}\sum_{j=1}^{n_i}\left(x_{ij} - \bar{x}_{i\cdot}\right)^2$$
$$\hat{\sigma}_B^2 = \frac{p-1}{N-\bar{n}}\left\{\left[\frac{1}{p-1}\sum_{i=1}^{p}\left(\bar{x}_{i\cdot} - \bar{x}_{..}\right)^2\right] - \hat{\sigma}_W^2\right\} \tag{20.3}$$

where $\bar{x}_i = n_i^{-1}\Sigma_{j=1}nix_{ij}$ and $\bar{n} = N^{-1}\Sigma_{i=1}pn_i$. By definition, the estimate of the within-run variance corresponds to the variance of repeatability, and the sum of the within-run and run-to-run components corresponds to the intermediate precision. That is, the intermediate precision variance is $\sigma_B^2 + \sigma_W^2$.

The relative error (RE), the coefficient of variation of the intermediate precision (CV_{IP}), and the total error (TE) are calculated as follows:

$$RE = 100\frac{\hat{\mu} - \mu}{\mu}, \quad CV_{IP} = 100\frac{\sqrt{\hat{\sigma}_W^2 + \hat{\sigma}_B^2}}{\mu}, \quad TE = RE + CV_{IP} \tag{20.4}$$

Full Validation

Full validation is important when developing and implementing a bioanalytical method for the first time and is particularly important for a new drug entity or where metabolites are added to an existing assay for quantification.

Partial validation: Partial validations are modifications of already validated bioanalytical methods. Partial validation can range from as little as one intra-assay accuracy and precision determination to a nearly full validation. Typical bioanalytical method changes that fall into this category include, but are not limited to, the following:

- Bioanalytical method transfers between laboratories or analysts
- Change in analytical methodology (e.g., change in detection systems)
- Change in anticoagulant in harvesting biological fluid
- Change in matrix within species (e.g., human plasma to human urine)
- Change in sample processing procedures
- Change in species within matrix (e.g., rat plasma to mouse plasma)
- Change in relevant concentration range
- Changes in instruments and/or software platforms
- Limited sample volume (e.g., pediatric study)

- Rare matrices
- Selectivity demonstration of an analyte in the presence of concomitant medications
- Selectivity demonstration of an analyte in the presence of specific metabolites

Cross validation: Cross validation is a comparison of validation parameters when two or more bioanalytical methods are used to generate data within the same study or across different studies. An example of cross validation would be a situation where an original validated bioanalytical method serves as the *reference* and the revised bioanalytical method is the *comparator*. The comparisons should be done both ways.

When sample analyses within a single study are conducted at more than one site or more than one laboratory, cross validation with spiked matrix standards and subject samples should be conducted at each site or laboratory to establish interlaboratory reliability. Cross validation should also be considered when data generated using different analytical techniques (e.g., LC–MS–MS vs. ELISA) in different studies are included in a regulatory submission.

All modifications should be assessed to determine the recommended degree of validation. The analytical laboratory conducting pharmacology/toxicology and other preclinical studies for regulatory submissions should adhere to FDA's good laboratory practices (GLPs) (21 CFR part 58) and to sound principles of quality assurance throughout the testing process. The bioanalytical method for human bioavailability (BA), bioequivalence (BE), pharmacokinetic (PK), and drug interaction studies must meet the criteria in 21 CFR 320.29. The analytical laboratory should have a written set of standard operating procedures (SOPs) to ensure a complete system of quality control (QC) and assurance. The SOPs should cover all aspects of analysis from the time the sample is collected and reaches the laboratory until the results of the analysis are reported. The SOPs also should include record keeping, security and chain of sample custody (accountability systems that ensure integrity of test articles), sample preparation, and analytical tools such as methods, reagents, equipment, instrumentation, and procedures for QC and verification of results.

The process by which a specific bioanalytical method is developed, validated, and used in routine sample analysis can be divided into (1) reference standard preparation, (2) bioanalytical method development and establishment of assay procedure, and (3) application of validated bioanalytical method to routine drug analysis and acceptance criteria for the analytical run and/or batch. These three processes are described later in this chapter.

Reference Standard

Analysis of drugs and their metabolites in a biological matrix is carried out using samples spiked with calibration (reference) standards and using QC samples. The purity of the reference standard used to prepare spiked samples can affect study data. For this reason, an authenticated analytical reference standard of known identity and purity should be used to prepare solutions of known concentrations. If possible, the reference standard should be identical to the analyte. When this is not possible, an established chemical form (free base or acid, salt, or ester) of known purity can be used. Three types of reference standards are usually used: (1) certified reference standards (e.g., USP compendial standards), (2) commercially supplied reference standards obtained from a reputable commercial source, and/or (3) other materials of documented purity custom synthesized by an analytical laboratory or other noncommercial establishment. The source and lot number, expiration date, certificates of analyses when available, and/or internally or externally generated evidence of identity and purity should be furnished for each reference standard.

Method Development

The method development and establishment phase defines the chemical assay. The fundamental parameters for a bioanalytical method validation are accuracy, precision, selectivity, sensitivity, reproducibility, and stability. Measurements for each analyte in the biological matrix should be validated. In addition, the stability of the analyte in spiked samples should be determined. Typical method development and establishment for a bioanalytical method include determination of (1) selectivity; (2) accuracy, precision, and recovery; (3) calibration curve; and (4) stability of analyte in spiked samples.

Selectivity is the ability of an analytical method to differentiate and quantify the analyte in the presence of other components in the sample. For selectivity, analyses of blank samples of the appropriate biological matrix (plasma, urine, or other matrix) should be obtained from at least six sources. Each blank sample should be tested for interference, and selectivity should be ensured at the lower limit of quantification (LLOQ). Potential interfering substances in a biological matrix include endogenous matrix components, metabolites, decomposition products, and, in the actual study, concomitant medication and other exogenous xenobiotics. If the method is intended to quantify more than one analyte, each analyte should be tested to ensure that there is no interference.

The *accuracy* of an analytical method describes the closeness of mean test results obtained by the method to the true value (concentration) of the analyte. Accuracy is determined by replicate analysis of samples containing known amounts of the analyte. Accuracy should be measured using a minimum of five determinations per concentration. A minimum of three concentrations in the range of expected concentrations is recommended. The mean value should be within 15% of the actual value except at LLOQ, where it should not deviate by more than 20%. The deviation of the mean from the true value serves as the measure of accuracy.

The *precision* of an analytical method describes the closeness of individual measures of an analyte when the procedure is applied repeatedly to multiple aliquots of a single homogeneous volume of biological matrix. Precision should be measured using a minimum of five determinations per concentration. A minimum of three concentrations in the range of expected concentrations is recommended. The precision determined at each concentration level should not exceed 15% of the coefficient of variation (CV) except for the LLOQ, where it should not exceed 20% of the CV. Precision is further subdivided into within-run, intra-batch precision or repeatability, which assesses precision during a single analytical run, and between-run, inter-batch precision or repeatability, which measures precision with time, and may involve different analysts, equipment, reagents, and laboratories.

The *recovery* of an analyte in an assay is the detector response obtained from an amount of the analyte added to and extracted from the biological matrix, compared to the detector response obtained for the true concentration of the pure authentic standard. Recovery pertains to the extraction efficiency of an analytical method within the limits of variability. Recovery of the analyte need not be 100%, but the extent of recovery of an analyte and of the internal standard (IS) should be consistent, precise, and reproducible. Recovery experiments should be performed by comparing the analytical results for extracted samples at three concentrations (low, medium, and high) with unextracted standards that represent 100% recovery.

Calibration

The response function for an analytical procedure is the existing relationship, within a specified range, between the response (signal, e.g., area under the curve, peak height, absorption) and the concentration (quantity) of the analyte in the sample. The calibration curve should be described preferably by a simple monotonic response function that gives accurate measurements. Note that the response function is frequently confused with the linearity criteria. However, the later criterion refers to the relationship between the quantity introduced and the quantity back-calculated from the calibration curve. Because of the confusion, it is common to see laboratory analysts try very hard to ensure that the response function is linear in the classical sense, that is, a straight line. Not only is this not required, but it is often irrelevant and can lead to large errors in measured results (e.g., for LBAs). A significant source of bias and imprecision in analytical measurements can be the choice of the statistical model for the calibration curve.

Computational Aspects

Statistical models for calibration curves can be either linear or nonlinear in their parameter(s). The choice between these two families of models will depend on the type of method and/or the range of concentrations of interest. If the range is very narrow, locally, an unweighted linear model may suffice, while a larger range may require a more advanced and weighted model. High-performance liquid chromatography (HPLC) methods are usually linear while immunoassays are typically non-linear. Weighting may be important for both methods because a common feature for many analytical methods is that the variance of the signal is a function of the level or quantity to be measured.

Methodologies for fitting linear and nonlinear models generally require different SAS procedures. For both model types, curves are fit by finding values for the model parameters that minimize the sum of squares of the distances between observations and the fitted curve. For linear models, parameter estimates can be derived analytically while this is not the case for many nonlinear models. Consequently, iterative procedures are often required to estimate the parameters of a nonlinear model. In this section, both linear and nonlinear models will be considered.

In case of heterogeneous variances of the signal across the concentration range, it is recommended that observations be weighted when fitting a curve. If observations are not weighted, an observation more distant to the curve than others has more influence on the curve fit. As a consequence, the curve fit may not be good where the variances are smaller. Weighting each term of the sum of squares is frequently used to solve this problem, where this can be viewed as minimizing the relative distances instead of minimizing the actual distances. When replicates are present at each concentration level, it is often better to fit the model to their average/median response values. Regardless of model type, it is assumed that all observations fit to a model are completely independent. In reality, replicates are often not independent for many analytical procedures because of the steps followed in preparation and analysis of samples. In such cases, replicates should not be used separately.

Models are typically applied on either a linear scale or log scale of the assay signal and/or the calibrator concentrations. The linear scale is used in case of homogeneous variance across the concentration range, and the log scale is often more appropriate when variance increases with increasing response.

Linear and Polynomial Models

Most commonly used types of polynomial models include simple linear regression (with or without an intercept) and quadratic regression models.

Nonlinear Models (PROC NLIN)

To fit a nonlinear model, one needs to rely on iterative methods. These methods begin with an initial set of parameter values for the model of interest and update the parameter values at each step in order to improve the fit. The iterative process is stopped when the fit can no longer be improved.

Nonlinear models frequently used in curve calibration include the 4-parameter logistic regression, 5-parameter logistic regression, and power model. Consider, for example, the 4-parameter logistic regression model:

$$y = f(x) = \beta_1 + \frac{\beta_2 - \beta_1}{1 + \left(x/\beta_3\right)^{\beta_4}} \tag{20.5}$$

where β_1, β_2, β_3, and β_4 are the top asymptote, bottom asymptote, concentration corresponding to half distance between β_1 and β_2, and the slope, respectively.

Both the NLIN and NLMIXED procedures can be used to fit such models. Except for the fact that they both optimize a function of interest, they do not work in the same manner. PROC NLIN fits nonlinear models by minimizing the error sum of squares, and it can handle only models with fixed effects. PROC NLMIXED enables us to fit models with fixed and random effects by maximizing an approximation to the likelihood function integrated over the random effects. In this context, PROC NLMIXED is used only in models with fixed effect, and thus, the problem of integration is avoided.

Nonlinear Models (PROC NLMIXED)

PROC NLMIXED supports a large number of iterative methods for fitting nonlinear models. Unfortunately, there is no general rule for choosing the most appropriate method. The choice is problem dependent, and, most of the time, one needs to select the iterative method by trial and error (see PROC NLMIXED documentation for general recommendations).

Note that the METHOD option in PROC NLMIXED does not specify the optimization method as in PROC NLIN but rather the method for approximating the integral of the likelihood function over the random effects. The TECHNIQUE option is used in PROC NLMIXED to select the optimization method.

Precision Profile for Immunoassays

After a calibration curve and weighting model have been chosen, a precision profile may be employed to characterize the precision of the back-calculated concentrations for unknown test samples using this calibration curve. The precision profile is a plot of the CV of the calibrated concentration versus the true concentration on a log scale. Ideally, the calculated standard error of the calibrated concentration must take into account both the variability in the calibration curve and variability in the assay response. Wald's method is generally recommended for computing these standard errors, and the resulting CV is given by

$$CV(x_0)=\frac{100}{x_0}\left\{\left[\frac{\partial f^{-1}(y_0,\hat{\beta})}{\partial y}\right]\frac{\hat{\sigma}^2 y_0^{2\hat{\theta}}}{m}+\left[\frac{\partial f^{-1}(y_0,\hat{\beta})}{\partial y}\right]'\Sigma(\hat{\beta})\left[\frac{\partial f^{-1}(y_0,\hat{\beta})}{\partial y}\right]\right\}^{1/2} \tag{20.6}$$

where

m is the number of replicates

$\Sigma(\hat{\beta})$ is the covariance matrix of the parameter estimates $\hat{\beta}$

As an illustration, *Program 4.4* computes a precision profile for a 5-parameter logistic model:

$$y=f(x)=\beta_1+\frac{\beta_2-\beta_1}{[1+(x/\beta_3)^{\beta_4}]^{\gamma}} \tag{20.7}$$

The γ parameter is known as the asymmetry factor, and, when it is set to 1, this model is equivalent to a 4-parameter logistic model.

The estimates of quantification limits from precision profiles are "optimistic" because they are based on only the calibration curve data themselves. These limits do not take into account matrix interference, cross-reactivity, operational factors, etc. However, these limits serve as a useful screening tool before beginning the pre-study validation exercise. Since the pre-study validation package encompasses several other sources of variability as well, if the quantification limits from a precision profile are not satisfactory, then almost definitely, the quantification limits derived from a rigorous pre-study validation package will not be satisfactory. In this case, it will be worth going back to the drawing board and further optimizing the assay protocol before proceeding to the pre-study validation phase.

TABLE 20.3
Inverse Functions for Widely Used Response Functions

Response Function	Back-Calculated Value x^* Using the Inverse Function
$Y = \beta X$	$x^* = y / \hat{\beta}$
$Y = \beta_0 + \beta_1 X$	$x^* = (y - \hat{\beta}_0)/\hat{\beta}_1$
$Y = \beta_0 + \beta_1 X + \beta_2 X^2$	$x^* = \left[-\hat{\beta}_1 + \sqrt{\hat{\beta}_1^2 - 4\hat{\beta}_2(\hat{\beta}_0 - y)}\right] / 2\hat{\beta}_2$
$Y = \beta_1 + \dfrac{\beta_2 - \beta_1}{1 + (X / \beta_3)^{\beta_4}}$	$x^* = \hat{\beta}_3 \left(\dfrac{\hat{\beta}_2 - \hat{\beta}_1}{y - \hat{\beta}_1} - 1\right)^{-1/\hat{\beta}_4}$

Back-Calculated Quantities or Inverse Predictions

Once a calibration curve is fitted, concentrations of the samples of interest are calculated by inverting the estimated calibration function. In a pre-study validation, the calibration curves are fitted separately for each run, and the VSs are calculated using the calibration curve for the same run. The resulting data set consists of different concentration levels, and, at each level, there are multiple runs and replicates within each run. Most of the time, the number of runs and the number of replicates are the same for all concentration levels. Inverse functions for widely used response functions are shown in Table 20.3.

Limits of Quantification and Range of the Assay

The upper and lower limits of quantification (ULOQ and LLOQ) of an analytical procedure are the lowest and highest amounts of the targeted substance in the sample that can be quantitatively determined under the prescribed experimental conditions. As a consequence, the range of an analytical procedure is the range between the lower and upper limits of quantification for which the analytical procedure was demonstrated to have a suitable level of measurement error.

In practice, the information needed to establish the limits of quantitation and the associated range is already available in the measurement profile plot. The limits of quantitation are the most extreme (low, high) concentrations (quantities) at which the tolerance interval is still within the acceptance limits, should the tolerance limits cross the acceptance limits. If all tolerance limits lie within the acceptance limits, the limits of quantitation are defined as the most extreme quantities tested in the study.

Limit of Detection

The LOD of an analytical procedure is the lowest amount of the targeted substance in the sample that can be detected reliably, but not necessarily quantified as an accurate value using the experimental conditions prescribed. A variety of methods to estimate the LOD have been proposed in the literature, generally based on the calibration information and estimates. None of them are really satisfactory and are sensitive to various assumptions related to the model, the design of experiment, and modeling of heterogeneity of variances. Based on our experience, the "best" and most consistent estimate proposed for bioanalytical methods aiming at covering a large range of concentrations is based on dividing the LLOQ by 3 or 3.33. This is justified by the fact that it is largely accepted that the LOD is 3 times the noise of the signal and the LLOQ is 10 times the same noise.

A calibration (standard) curve is the relationship between instrument response and known concentrations of the analyte. A calibration curve should be generated for each analyte in the sample. A sufficient number of standards should be used to adequately define the relationship between concentration and response. A calibration curve should be prepared in the same biological matrix as the samples in the intended study by spiking the matrix with known concentrations of the analyte. The number of standards used in constructing a calibration curve will be a function of the anticipated range of analytical values and the nature of the analyte/response relationship. Concentrations of standards should be chosen on the basis of the concentration range expected in a particular study. A calibration curve should consist of a blank sample (matrix sample processed without IS), a zero sample (matrix sample processed with IS), and six to eight nonzero samples covering the expected range, including LLOQ.

- *Lower limit of quantification (LLOQ)*: The lowest standard on the calibration curve should be accepted as the LOQ if the following conditions are met:
  - The analyte response at the LLOQ should be at least five times the response compared to blank response.
  - Analyte peak (response) should be identifiable, discrete, and reproducible with a precision of 20% and accuracy of 80–120%.
- *Calibration curve/standard curve/concentration–response:* The simplest model that adequately describes the concentration–response relationship should be used. Selection of weighting and use of a complex regression equation should be justified. The following conditions should be met in developing a calibration curve:
  - 20% deviation of the LLOQ from nominal concentration
  - 15% deviation of standards other than LLOQ from nominal concentration

At least four out of six nonzero standards should meet the aforementioned criteria, including the LLOQ and the calibration standard at the highest concentration. Excluding the standards should not change the model used.

Specificity–Selectivity

The specificity of an analytical procedure is the ability to unequivocally assess the analyte in the presence of components that may be expected to be present. Usually, the analyst must demonstrate that the measured result is directly related to the analyte or product of interest and that other aspects in the sample, such as the matrix, do not interfere with the signal or measurement. For example, selectivity of a chromatographic method is verified typically by showing that the product of interest is clearly separated from all other products.

Linearity

The linearity of an analytical procedure is defined in terms of its ability to obtain results directly proportional to the concentrations (quantities) of the analyte in the sample within a defined range (ICH, 1995). It is important to note that the linearity criterion is applied to the results, that is, back-calculated quantities or concentrations, rather than the response signals or instrument response as a function of the dose or quantities.

Accuracy and Precision

Accuracy (Trueness)

The accuracy (the preferred term is trueness), of an analytical procedure, according to ICH and related documents, expresses the closeness of agreement between the mean value obtained from a series of measurements and the value that is accepted either as a conventional true value or an accepted reference value (e.g., international standard, standard from a pharmacopoeia). It is a measure of the systematic error of test results obtained by the analytical method from its theoretical true/reference value. The measure of trueness is generally expressed in terms of recovery and absolute/relative bias.

Note that "accuracy" is synonymous with bias or trueness only within the pharmaceutical set of regulations covered by ICH (and related national documents implementing ICH Q2A and Q2B). Outside the pharmaceutical industry, for example, in industries covered by the International Organization for Standardization or National Committee for Clinical Laboratory Standards guidelines (food, chemistry, clinical biology, and other industries), "accuracy" refers to the TE, that is, the sum of trueness and precision.

Precision

The precision of an analytical procedure is defined by the closeness of agreement (usually expressed as standard deviation or CV) between a series of measurements obtained from multiple sampling of the same homogeneous sample (independent assays) under the prescribed conditions. The term "independent results" means that the results are obtained and prepared the same way that unknown samples will be quantified and prepared.

Precision provides information on random errors and can be evaluated at three levels: repeatability, intermediate precision (within laboratory), and reproducibility (between laboratories). The precision only represents the distribution of the random errors and is not related to the true or specified value. A measure of precision is calculated from the standard deviation of the results.

Quantitative measures of precision depend in a critical manner on stipulated conditions. One can distinguish among the following three conditions:

- *Repeatability.* Repeatability expresses the precision under conditions where the results of independent assays are obtained by the same analytical procedure on identical samples in the same laboratory, with the same operator, using the same equipment and during a short interval of time. It is estimated by the within-series variance component.
- *Intermediate precision.* Intermediate precision expresses the precision under conditions where the results of independent assays are obtained by the same analytical procedure on identical samples in the same laboratory, with different operators, using different equipment and during a given time interval. It is estimated by the sum of within-series and between-series variance components. Intermediate precision is representative of the total random error for a single measurement within a laboratory, whatever the day or series.
- *Reproducibility.* Reproducibility expresses the precision under conditions where the results are obtained by the same analytical procedure on identical samples in different laboratories, with different operators and using different equipment. It is estimated by the sum of within-series, between-series, and between-laboratories variance components.

Total Error or Measurement Error

The measurement error of an analytical procedure is related to the closeness of agreement between the value found and the value that is accepted either as a conventional true value or an accepted reference value. The closeness of agreement observed is based on the sum of the systematic and random errors; in other words, the TE linked to the result. Consequently, the measurement error is the expression of the sum of the trueness and precision, that is, the TE.

As shown later, the observation X is a result of the true sample value μ_T, the method's Bias (estimated by the mean of many results) and Precision (estimated by the standard deviation or, in most cases, the intermediate precision). Equivalently, the difference between an observation X and the true value is the sum of the systematic and random errors, that is, total error or measurement error.

$$X = \mu_T + \text{Bias} + \text{Precision}$$

$$\Leftrightarrow X - \mu_T = \text{Bias} + \text{Precision}$$

$$\Leftrightarrow X - \mu_T = \text{Total error}$$

$$\Leftrightarrow X - \mu_T = \text{Measurement error}$$

Decision Rule

Equation 20.2, which describes the main objective of an analytical method, cannot be solved exactly. A simple way to resolve this problem and make a reliable decision relies on computing the β-expectation tolerance intervals:

$$E_{\hat{\mu}_M,\hat{\sigma}_M}\left[P_X\left(\hat{\mu}_M - k\hat{\sigma}_M < X < \hat{\mu}_M + k\hat{\sigma}_M \middle| \hat{\mu}_M,\hat{\sigma}_M\right)\right] = \beta \tag{20.8}$$

where the k factor is determined so that the expected proportion of the population falling within the interval is equal to β. If the β-expectation tolerance interval is totally included within the acceptance limits $[-\lambda, +\lambda]$, that is, if $\hat{\mu}_M - k\hat{\sigma}_M > -\lambda$ and $\hat{\mu}_M + \text{k}\hat{\mu}_M < \lambda$, the expected proportion of measurements within the same acceptance limits is greater than or equal to β. Note that the opposite statement is not true, that is, if either $\hat{\mu}_M - \text{k}\hat{\mu}_M < -\lambda$ or $\hat{\mu}_M + \text{k}\hat{\mu}_M > \lambda$, the expected proportion is not necessarily smaller than β.

Most of the time, an analytical procedure is intended to quantify over a range of quantities or concentrations. Consequently, during the validation phase, samples are prepared to adequately cover this range, and a β-expectation tolerance interval is calculated at each level.

A measurement error profile is obtained, on one hand, by connecting the lower limits and, on the other hand, by connecting the upper limits. A procedure is valid over a certain range of values if the measurement error profile is included within the acceptance limits $[-\lambda, +\lambda]$.

A measurement error profile gives the analyst a sense of what a procedure will be able to produce over the intended range. The interpretation of a measurement error profile is that it shows where 100β% of the measures provided by this analytical method will lie, which is directly connected to the objective of the analytical method (produce measures close to the unknown true values).

In practice, the β-expectation tolerance interval is obtained as follows (on a relative scale):

$$\left[\text{RE} - Q_t\left(v, \frac{1+\beta}{2}\right)\sqrt{1+\frac{1}{pnB^2}}\text{CV}_{IP}, \text{RE} + Q_t\left(v, \frac{1+\beta}{2}\right)\sqrt{1+\frac{1}{pnB^2}}\text{CV}_{IP}\right] \tag{20.9}$$

where

p is the number of runs

n is the number of replicates within each run

$Q_t(a,b)$ is the 100b% quantile of the t distribution with a degrees of freedom, and

$$R = \frac{\hat{\sigma}_B^2}{\hat{\sigma}_W^2}, \quad B = \sqrt{\frac{R+1}{nR+1}}, \quad v = (R+1)^2\left[\frac{(R+1/n)^2}{p-1} + \frac{1-1/n}{pn}\right]^{-1} \tag{20.10}$$

Stability

Drug stability in a biological fluid is a function of the storage conditions, the chemical properties of the drug, the matrix, and the container system. The stability of an analyte in a particular matrix and container system is relevant only to that matrix and container system and should not be extrapolated to other matrices and container systems. Stability procedures should evaluate the stability of the analytes during sample collection and handling, after long-term (frozen at the intended storage temperature) and short-term (benchtop, room temperature) storage, and after going through freeze and thaw cycles and the analytical process. Conditions used in stability experiments should reflect situations likely to be encountered during actual sample handling and analysis. The procedure should also include an evaluation of analyte stability in stock solution.

All stability determinations should use a set of samples prepared from a freshly made stock solution of the analyte in the appropriate analyte-free, interference-free biological matrix. Stock solutions of the analyte for stability evaluation should be prepared in an appropriate solvent at known concentrations.

- *Freeze and thaw stability:* Analyte stability should be determined after three freeze and thaw cycles. At least three aliquots at each of the low and high concentrations should be stored at the intended storage temperature for 24 h and thawed unassisted at room temperature. When completely thawed, the samples should be refrozen for 12–24 h under the same conditions. The freeze–thaw cycle should be repeated two more times, then analyzed on the third cycle. If an analyte is unstable at the intended storage temperature, the stability sample should be frozen at –70°C during the three freeze and thaw cycles.
- *Short-term temperature stability:* Three aliquots of each of the low and high concentrations should be thawed at room temperature and kept at this temperature from 4 to 24 h (based on the expected duration that samples will be maintained at room temperature in the intended study) and analyzed.
- *Long-term stability:* The storage time in a long-term stability evaluation should exceed the time between the date of first sample collection and the date of last sample analysis. Long-term stability should be determined by storing at least three aliquots of each of the low and high concentrations under the same conditions as the study samples. The volume of samples should be sufficient for analysis on three separate occasions. The concentrations of all the stability samples should be compared to the mean of back-calculated values for the standards at the appropriate concentrations from the first day of long-term stability testing.
- *Stock solution stability:* The stability of stock solutions of drug and the IS should be evaluated at room temperature for at least 6 h. If the stock solutions are refrigerated or frozen for the relevant period, the stability should be documented. After completion of the desired storage time, the stability should be tested by comparing the instrument response with that of freshly prepared solutions.
- *Post-preparative stability:* The stability of processed samples, including the resident time in the autosampler, should be determined. The stability of the drug and the IS should be assessed over the anticipated run time for the batch size in VSs by determining concentrations on the basis of original calibration standards.

Although the traditional approach of comparing analytical results for stored samples with those for freshly prepared samples has been referred to in this chapter, other statistical approaches based on confidence limits for evaluation of an analyte's stability in a biological matrix can be used. SOPs should clearly describe the statistical method and rules used. Additional validation may include investigation of samples from dosed subjects.

Principles of Bioanalytical Method Validation and Establishment

The fundamental parameters to ensure the acceptability of the performance of a bioanalytical method validation are accuracy, precision, selectivity, sensitivity, reproducibility, and stability. A specific, detailed description of the bioanalytical method should be written. This can be in the form of a protocol, study plan, report, and/or SOP. Each step in the method should be investigated to determine the extent to which environmental, matrix, material, or procedural variables can affect the estimation of analyte in the matrix from the time of collection of the material up to and including the time of analysis. It may be important to consider the variability of the matrix due to the physiological nature of the sample. In the case of LC–MS–MS-based procedures, appropriate steps should be taken to ensure the lack of matrix effects throughout the application of the method, especially if the nature of the matrix changes from the matrix used during method validation. A bioanalytical method should be validated for the intended use or application. All experiments used to make claims or draw conclusions about the validity of the method should be presented in a report (method validation report). Whenever possible, the same biological matrix as the matrix in the intended samples should be used for validation purposes. (For tissues of limited availability, such as bone marrow, physiologically appropriate proxy matrices can be substituted.)

The stability of the analyte (drug and/or metabolite) in the matrix during the collection process and the sample storage period should be assessed, preferably prior to sample analysis. For compounds with potentially labile metabolites, the stability of analyte in matrix from dosed subjects (or species) should be confirmed. The accuracy, precision, reproducibility, response function, and selectivity of the method for endogenous substances, metabolites, and known degradation products should be established for the biological matrix. For selectivity, there should be evidence that the substance being quantified is the intended analyte. The concentration range over which the analyte will be determined should be defined in the bioanalytical method, based on evaluation of actual standard samples over the range, including their statistical variation. This defines the *standard curve*. A sufficient number of standards should be used to adequately define the relationship between concentration and response. The relationship between response and concentration should be demonstrated to be continuous and reproducible. The number of standards used should be a function of the dynamic range and nature of the concentration–response relationship. In many cases, six to eight concentrations (excluding blank values) can define the standard curve. More standard concentrations may be recommended for nonlinear than for linear relationships. The ability to dilute samples originally above the upper limit of the standard curve should be demonstrated by accuracy and precision parameters in the validation.

In consideration of high-throughput analyses, including but not limited to multiplexing, multicolumn, and parallel systems, sufficient QC samples should be used to ensure control of the assay. The number of QC samples to ensure proper control of the assay should be determined based on the run size. The placement of QC samples should be judiciously considered in the run. For a bioanalytical method to be considered valid, specific acceptance criteria should be set in advance and achieved for accuracy and precision for the validation of QC samples over the range of the standards.

Specific Recommendations for Method Validation

The matrix-based standard curve should consist of a minimum of six standard points, excluding blanks, using single or replicate samples. The standard curve should cover the entire range of expected concentrations. Standard curve fitting is determined by applying the simplest model that adequately describes the concentration–response relationship using appropriate weighting and statistical tests for *goodness of fit*. LLOQ is the lowest concentration of the standard curve that can be measured with acceptable accuracy and precision. The LLOQ should be established using at least five samples independent of standards and determining the CV and/or appropriate confidence interval. The LLOQ should serve as the lowest concentration on the standard curve and should not be confused with the LOD and/or the low QC sample. The highest standard will define the upper limit of quantification (ULOQ) of an analytical method.

For validation of the bioanalytical method, accuracy and precision should be determined using a minimum of five determinations per concentration level (excluding blank samples). The mean value should be within ±15% of the theoretical value, except at LLOQ, where it should not deviate by more than ±20%. The precision around the mean value should not exceed 15% of the CV, except for LLOQ, where it should not exceed 20% of the CV. Other methods of assessing accuracy and precision that meet these limits may be equally acceptable.

The accuracy and precision with which known concentrations of analyte in biological matrix can be determined should be demonstrated. This can be accomplished by analysis of replicate sets of analyte samples of known concentrations (QC samples) from an equivalent biological matrix. At a minimum, three concentrations representing the entire range of the standard curve should be studied: one within 3× the LLOQ (low QC sample), one near the center (middle QC), and one near the upper boundary of the standard curve (high QC).

Reported method validation data and the determination of accuracy and precision should include all outliers; however, calculations of accuracy and precision excluding values that are statistically determined as outliers can also be reported.

The stability of the analyte in biological matrix at intended storage temperatures should be established. The influence of freeze–thaw cycles (a minimum of three cycles at two concentrations in triplicate) should be studied. The stability of the analyte in matrix at ambient temperature should be evaluated over a time period equal to the typical sample preparation, sample handling, and analytical run times.

Reinjection reproducibility should be evaluated to determine if an analytical run could be reanalyzed in the case of instrument failure. The specificity of the assay methodology should be established using a minimum of six independent sources of the same matrix. For hyphenated MS-based methods, however, testing six independent matrices for interference may not be important. In the case of LC–MS- and LC–MS–MS-based procedures, matrix effects should be investigated to ensure that precision, selectivity, and sensitivity will not be compromised. Method selectivity should be evaluated during method development and throughout method validation and can continue throughout application of the method to actual study samples.

Acceptance/rejection criteria for spiked, matrix-based calibration standards and validation QC samples should be based on the nominal (theoretical) concentration of analytes. Specific criteria can be set up in advance and achieved for accuracy and precision over the range of the standards, if so desired.

Microbiological and Ligand-Binding Assays

Many of the bioanalytical validation parameters and principles discussed earlier are also applicable to microbiological and LBAs. However, these assays possess some unique characteristics that should be considered during method validation.

Selectivity issues: As with chromatographic methods, microbiological and LBAs should be shown to be selective for the analyte. The following recommendations for dealing with two selectivity issues should be considered:

- *Interference from substances physiochemically similar to the analyte*
  - Cross-reactivity of metabolites, concomitant medications, or endogenous compounds should be evaluated individually and in combination with the analyte of interest.
  - When possible, the immunoassay should be compared with a validated reference method (such as LC–MS) using incurred samples and predetermined criteria for agreement of accuracy of immunoassay and reference method.

    - The dilutional linearity to the reference standard should be assessed using study (incurred) samples.
    - Selectivity may be improved for some analytes by incorporation of separation steps prior to immunoassay.
- *Matrix effects unrelated to the analyte*
    - The standard curve in biological fluids should be compared with standard in buffer to detect matrix effects.
    - Parallelism of diluted study samples should be evaluated with diluted standards to detect matrix effects.
    - Nonspecific binding should be determined.

Quantification issues: Microbiological and immunoassay standard curves are inherently nonlinear, and, in general, more concentration points may be recommended to define the fit over the standard curve range than for chemical assays. In addition to their nonlinear characteristics, the response–error relationship for immunoassay standard curves is a nonconstant function of the mean response (heteroscedasticity). For these reasons, a minimum of six nonzero calibrator concentrations, run in duplicate, is recommended. The concentration–response relationship is most often fitted to a 4- or 5-parameter logistic model, although others may be used with suitable validation. The use of *anchoring points* in the asymptotic high- and low-concentration ends of the standard curve may improve the overall curve fit. Generally, these anchoring points will be at concentrations that are below the established LLOQ and above the established ULOQ. Whenever possible, calibrators should be prepared in the same matrix as the study samples or in an alternate matrix of equivalent performance. Both ULOQ and LLOQ should be defined by acceptable accuracy, precision, or confidence interval criteria based on the study requirements.

For all assays, the key factor is the accuracy of the *reported results*. This accuracy can be improved by the use of replicate samples. In the case where replicate samples should be measured during the validation to improve accuracy, the same procedure should be followed as for unknown samples. The following recommendations apply to quantification issues:

- If separation is used prior to assay for study samples but not for standards, it is important to establish recovery and use it in determining results. Possible approaches to assess efficiency and reproducibility of recovery are (1) the use of radiolabeled tracer analyte (quantity too small to affect the assay), (2) the advance establishment of reproducible recovery, and (3) the use of an IS that is not recognized by the antibody but can be measured by another technique.
- Key reagents, such as antibody, tracer, reference standard, and matrix, should be characterized appropriately and stored under defined conditions.
- Assessments of analyte stability should be conducted in true study matrix (e.g., should not use a matrix stripped to remove endogenous interferences).
- Acceptance criteria: At least 67% (four out of six) of QC samples should be within 15% of their respective nominal value; 33% of the QC samples (not all replicates at the same concentration) may be outside 15% of nominal value. In certain situations, wider acceptance criteria may be justified.
- Assay reoptimization or validation may be important when there are changes in key reagents, as follows:
    - Labeled analyte (tracer)
        - Binding should be reoptimized.
        - Performance should be verified with standard curve and QCs.
    - Antibody
        - Key cross-reactivities should be checked.
        - Tracer experiments earlier should be repeated.
    - Matrix
        - Tracer experiments earlier should be repeated.

Method development experiments should include a minimum of six runs conducted over several days, with at least four concentrations (LLOQ, low, medium, and high) analyzed in duplicate in each run.

Application of Validated Method to Routine Analysis

Assays of all samples of an analyte in a biological matrix should be completed within the time period for which stability data are available. In general, biological samples can be analyzed with a single determination without duplicate or replicate analysis if the assay method has acceptable variability as defined by validation data. This is true for procedures where precision and accuracy variabilities routinely fall within acceptable tolerance limits. For a difficult procedure with a labile analyte where high precision and accuracy specifications may be difficult to achieve, duplicate or even triplicate analyses can be performed for a better estimate of analyte.

A calibration curve should be generated for each analyte to assay samples in each analytical run and should be used to calculate the concentration of the analyte in the unknown samples in the run. The spiked samples can contain more than one analyte. An analytical run can consist of QC samples, calibration standards, and either (1) all the processed samples to be analyzed as one batch or (2) a batch composed of processed unknown samples of one or more volunteers in a study. The calibration (standard) curve should cover the expected unknown sample concentration range in addition to a calibrator sample at LLOQ. Estimation of concentration in unknown samples by extrapolation of standard curves below LLOQ or above the highest standard is not recommended. Instead, the standard curve should be redefined or samples with higher concentration should be diluted and reassayed. It is preferable to analyze all study samples from a subject in a single run.

Once the analytical method has been validated for routine use, its accuracy and precision should be monitored regularly to ensure that the method continues to perform satisfactorily. To achieve this objective, a number of QC samples prepared separately should be analyzed with processed test samples at intervals based on the total number of samples. The QC samples in duplicate at three concentrations (one near the LLOQ [i.e., $\pm 3 \times$ LLOQ], one in midrange, and one close to the high end of the range) should be incorporated in each assay run. The number of QC samples (in multiples of three) will depend on the total number of samples in the run. The results of the QC samples provide the basis of accepting or rejecting the run. At least four of every six QC samples should be within ±15% of their respective nominal value. Two of the six QC samples may be outside the ±15% of their respective nominal value, but not both at the same concentration.

The following recommendations should be noted in applying a bioanalytical method to routine drug analysis:

- A matrix-based standard curve should consist of a minimum of six standard points, excluding blanks (either single or replicate), covering the entire range.
- Response function: Typically, the same curve fitting, weighting, and goodness of fit determined during pre-study validation should be used for the standard curve within the study. Response function is determined by appropriate statistical tests based on the actual standard points during each run in the validation. Changes in the response function relationship between pre-study validation and routine run validation indicate potential problems.
- The QC samples should be used to accept or reject the run. These QC samples are matrix spiked with analyte.
- System suitability: Based on the analyte and technique, a specific SOP (or sample) should be identified to ensure optimum operation of the system used.
- Any required sample dilutions should use like matrix (e.g., human to human) obviating the need to incorporate actual within-study dilution matrix QC samples.
- Repeat analysis: It is important to establish an SOP or guideline for repeat analysis and acceptance criteria. This SOP or guideline should explain the reasons for repeating sample analysis. Reasons for repeat analyses could include repeat analysis of clinical or preclinical

samples for regulatory purposes, inconsistent replicate analysis, samples outside of the assay range, sample processing errors, equipment failure, poor chromatography, and inconsistent PK data. Reassays should be done in triplicate if sample volume allows. The rationale for the repeat analysis and the reporting of the repeat analysis should be clearly documented.
- Sample data reintegration: An SOP or guideline for sample data reintegration should be established. This SOP or guideline should explain the reasons for reintegration and how the reintegration is to be performed. The rationale for the reintegration should be clearly described and documented. Original and reintegration data should be reported.

Acceptance criteria for the run: The following acceptance criteria should be considered for accepting the analytical run:

- Standards and QC samples can be prepared from the same spiking stock solution, provided the solution stability and accuracy have been verified. A single source of matrix may also be used, provided selectivity has been verified.
- Standard curve samples, blanks, QCs, and study samples can be arranged as considered appropriate within the run.
- Placement of standards and QC samples within a run should be designed to detect assay drift over the run.
- Matrix-based standard CSs: 75%, or a minimum of six standards, when back-calculated (including ULOQ), should fall within 15%, except for LLOQ, when it should be 20% of the nominal value. Values falling outside these limits can be discarded, provided they do not change the established model.
- "Specific Recommendation for Method Validation," should be provided for both the intra-day and intra-run experiment.
- QC Samples: QC samples replicated (at least once) at a minimum of three concentrations (one within 3× of the LLOQ [low QC], one in the midrange [middle QC], and one approaching the high end of the range [high QC]) should be incorporated into each run. The results of the QC samples provide the basis of accepting or rejecting the run. At least 67% (four out of six) of the QC samples should be within 15% of their respective nominal (theoretical) values; 33% of the QC samples (not all replicates at the same concentration) can be outside the 15% of the nominal value. A confidence interval approach yielding comparable accuracy and precision is an appropriate alternative.
- The minimum number of samples (in multiples of three) should be at least 5% of the number of unknown samples or six total QCs, whichever is greater.
- Samples involving multiple analytes should not be rejected based on the data from one analyte failing the acceptance criteria.
- The data from rejected runs need not be documented, but the fact that a run was rejected and the reason for failure should be recorded.

DOCUMENTATION

The validity of an analytical method should be established and verified by laboratory studies, and documentation of successful completion of such studies should be provided in the assay validation report. General and specific SOPs and good record keeping are an essential part of a validated analytical method. The data generated for bioanalytical method establishment and the QCs should be documented and available for data audit and inspection. Documentation for submission to the agency should include (1) summary information, (2) method development and establishment, (3) bioanalytical reports of the application of any methods to routine sample analysis, and (4) other information applicable to method development and establishment and/or to routine sample analysis.

Summary information: Summary information should include the following:

- Summary table of validation reports, including analytical method validation, partial revalidation, and cross validation reports. The table should be in chronological sequence and include assay method identification code, type of assay, and the reason for the new method or additional validation (e.g., to lower the limit of quantification).
- Summary table with a list, by protocol, of assay methods used. The protocol number, protocol title, assay type, assay method identification code, and bioanalytical report code should be provided.
- A summary table allowing cross-referencing of multiple identification codes should be provided (e.g., when an assay has different codes for the assay method, validation reports, and bioanalytical reports, especially when the sponsor and a contract laboratory assign different codes).

Documentation for method establishment: Documentation for method development and establishment should include

- An operational description of the analytical method
- Evidence of purity and identity of drug standards, metabolite standards, and ISs used in validation experiments
- A description of stability studies and supporting data
- A description of experiments conducted to determine accuracy, precision, recovery, selectivity, LOQ, calibration curve (equations and weighting functions used, if any), and relevant data obtained from these studies
- Documentation of intra- and inter-assay precision and accuracy
- In NDA submissions, information about cross validation study data, if applicable
- Legible annotated chromatograms or mass spectrograms, if applicable
- Any deviations from SOPs, protocols, or GLPs (if applicable) and justifications for deviations

Application to routine drug analysis: Documentation of the application of validated bioanalytical methods to routine drug analysis should include the following:

- Evidence of purity and identity of drug standards, metabolite standards, and ISs used during routine analyses.
- Summary tables containing information on sample processing and storage. Tables should include sample identification, collection dates, storage prior to shipment, information on shipment batch, and storage prior to analysis. Information should include dates, times, sample condition, and any deviation from protocols.
- Summary tables of analytical runs of clinical or preclinical samples. Information should include assay run identification, date and time of analysis, assay method, analysts, start and stop times, duration, significant equipment and material changes, and any potential issues or deviation from the established method.
- Equations used for back-calculation of results.
- Tables of calibration curve data used in analyzing samples and calibration curve summary data.
- Summary information on intra- and inter-assay values of QC samples and data on intra- and inter-assay accuracy and precision from calibration curves and QC samples used for accepting the analytical run. QC graphs and trend analyses in addition to raw data and summary statistics are encouraged.
- Data tables from analytical runs of clinical or preclinical samples. Tables should include assay run identification, sample identification, raw data and back-calculated results, integration codes, and/or other reporting codes.

- Complete serial chromatograms from 5% to 20% of subjects, with standards and QC samples from those analytical runs. For pivotal BE studies for marketing, chromatograms from 20% of serially selected subjects should be included. In other studies, chromatograms from 5% of randomly selected subjects in each study should be included. Subjects whose chromatograms are to be submitted should be defined prior to the analysis of any clinical samples.
- Reasons for missing samples.
- Documentation for repeat analyses. Documentation should include the initial and repeat analysis results, the reported result, assay run identification, the reason for the repeat analysis, the requestor of the repeat analysis, and the manager authorizing reanalysis. Repeat analysis of a clinical or preclinical sample should be performed only under a predefined SOP.
- Documentation for reintegrated data. Documentation should include the initial and repeat integration results, the method used for reintegration, the reported result, assay run identification, the reason for the reintegration, the requestor of the reintegration, and the manager authorizing reintegration. Reintegration of a clinical or preclinical sample should be performed only under a predefined SOP.
- Deviations from the analysis protocol or SOP, with reasons and justifications for the deviations.

Other information: Other information applicable both to method development and establishment and to routine sample analysis could include

- Lists of abbreviations and any additional codes used, including sample condition codes, integration codes, and reporting codes
- Reference lists and legible copies of any references
- SOPs or protocols covering the following areas:
    - Calibration standard acceptance or rejection criteria
    - Calibration curve acceptance or rejection criteria
    - QC sample and assay run acceptance or rejection criteria
    - Acceptance criteria for reported values when all unknown samples are assayed in duplicate
    - Sample code designations, including clinical or preclinical sample codes and bioassay sample code
    - Assignment of clinical or preclinical samples to assay batches
    - Sample collection, processing, and storage
    - Repeat analyses of samples
    - Reintegration of samples

CONCLUSION

To understand, estimate, and interpret various criteria required for assessing the validity of an analytical method, these must be reported, and the code used to compute the criteria must be documented according to best practices. Regardless of the complexity of computations and models, the objective of an analytical method (are the measurement errors acceptable?) should never be forgotten and should remain the primary focus. The information needed to make a decision is contained in the measurement error profile. All performance criteria—linearity, accuracy, precision, limits, and measurement errors—can be assessed using a graphical profile and can be easily understood and interpreted by an analyst.

Another important remark related to the decision is that, if the tolerance intervals are within the acceptance limits, all of the required criteria are guaranteed to be met. The opposite, however, is not true, that is, even when all of the performance criteria are satisfied, the measurement errors will not necessarily be acceptable. Although it is common to assume that good methods will always produce good results and most regulatory documents were written in this spirit, it is important to remember that only the opposite statement holds true: good results can only be obtained with a good method.

New Advisory from FDA

In September 2013, the FDA issued additional advisory information on bioanalytical method validation. In all likelihood, these advisory will become the final guideline; given as follows is a synopsis of this advice.

Introduction

This guidance provides assistance to sponsors of investigational new drug applications (INDs), new drug applications (NDAs), abbreviated new drug applications (ANDAs), biological license applications (BLAs), and supplements in developing bioanalytical method validation information used in human clinical pharmacology, BA, and BE studies that require PK or biomarker concentration evaluation. This guidance also applies to bioanalytical methods used for nonclinical pharmacology/toxicology studies. For studies related to the veterinary drug approval process (investigational new animal drug applications [INADs], new animal drug applications [NADAs], and abbreviated new animal drug applications [ANADAs]), this guidance may apply to blood and urine BA, BE, and PK studies.

The information in this guidance generally applies to bioanalytical procedures, such as GC; high-pressure LC; combined GC and LC MS procedures, such as LC–MS, LC–MS–MS, GC–MS, and GC–MS–MS; and LBAs, and immunological and microbiological procedures that are performed for the quantitative determination of drugs and/or metabolites and therapeutic proteins in biological matrices, such as blood, serum, plasma, urine, tissue, and skin.

This guidance provides general recommendations for bioanalytical method validation. The recommendations can be modified depending on the specific type of analytical method used.

Originally, issued in 2001, this guidance has been revised to reflect advances in science and technology related to validating bioanalytical methods.

Background

This guidance was originally developed based on the deliberations following two workshops: Analytical Methods Validation: Bioavailability, Bioequivalence, and Pharmacokinetic Studies (December 3–5, 1990) and Bioanalytical Methods Validation: A Revisit With a Decade of Progress (January 12–14, 2000). Since publication of the guidance in May 2001, additional workshops have been held that have helped guide the current revisions to the guidance: the Quantitative Bioanalytical Methods Validation and Implementation: Best Practices for Chromatographic and Ligand Binding Assays (May 1–3, 2006) and the AAPS/FDA Workshop on Incurred Sample Reanalysis (February 2008).

Validating bioanalytical methods includes performing all of the procedures that demonstrate that a particular method used for quantitative measurement of analytes in a given biological matrix (e.g., blood, plasma, serum, or urine) is reliable and reproducible for the intended use. Fundamental parameters for this validation include the following:

- Accuracy
- Precision
- Selectivity
- Sensitivity
- Reproducibility
- Stability

Validation involves documenting, through the use of specific laboratory investigations, that the performance characteristics of a method are suitable and reliable for the intended analytical applications. The acceptability of analytical data corresponds directly to the criteria used to validate the method. For pivotal studies that require regulatory action for approval or labeling, such as BE or PK studies, the bioanalytical methods should be fully validated. For exploratory methods used for the sponsor's internal decision making, less validation may be sufficient. When changes are made to a previously validated method, additional validation may be needed. For example, published methods of analysis are often modified to suit the requirements of the laboratory performing the assay, and during the course of a typical drug development program, a defined bioanalytical method often undergoes many modifications. These modifications should be validated to ensure suitable performance of the analytical method. The evolutionary changes needed to support specific studies call for different levels of validation to demonstrate the validity of method performance. The following define and characterize the different types and levels of methods validation:

Full validation
Full validation of bioanalytical methods is important

- During development and implementation of a novel bioanalytical method
- For analysis of a new drug entity
- For revisions to an existing method that add metabolite quantification

Partial validation
Partial validations evaluate modifications of already validated bioanalytical methods. Partial validation can range from as little as one intra-assay accuracy and precision determination to a nearly full validation. Typical bioanalytical method modifications or changes that fall into this category include, but are not limited to, the following:

- Bioanalytical method transfers between laboratories or analysts
- Change in analytical methodology (e.g., change in detection systems)
- Change in anticoagulant in harvesting biological fluid (e.g., heparin to EDTA)
- Change in matrix within species (e.g., human plasma to human urine)
- Change in sample processing procedures
- Change in species within matrix (e.g., rat plasma to mouse plasma)
- Change in relevant concentration range
- Changes in instruments and/or software platforms
- Modifications to accommodate limited sample volume (e.g., pediatric study)
- Rare matrices
- Selectivity demonstration of an analyte in the presence of concomitant medications

Cross validation
Cross validation is a comparison of validation parameters when two or more bioanalytical methods are used to generate data within the same study or across different studies. An example of cross validation would be a situation in which an original validated bioanalytical method serves as the reference and the revised bioanalytical method is the comparator. The comparisons should be done both ways.

When sample analyses within a single study are conducted at more than one site or more than one laboratory, cross validation with spiked matrix standards and subject samples should be conducted at each site or laboratory to establish interlaboratory reliability. Cross validation should also be considered when data generated using different analytical techniques (e.g., LC–MS–MS vs. ELISA) in different studies are included in a regulatory submission. All modifications to an existing method should be assessed to determine the recommended degree of validation.

The analytical laboratory conducting nonclinical pharmacology/toxicology studies for regulatory submissions should adhere to FDA's GLP requirements (21 CFR 135 part 58). The bioanalytical method for human BA, BE, PK, and drug interaction studies must meet the criteria specified in 21 CFR 320.29.

Analytical laboratories should have written SOPs to ensure a complete system of QC and assurance. SOPs should cover all aspects of analysis from the time the sample is collected and reaches the laboratory until the results of the analysis are reported. The SOPs also should include record keeping, security and chain of sample custody (accountability systems that ensure integrity of test articles), sample preparation, and analytical tools such as methods, reagents, equipment, instrumentation, and procedures for QC, and verification of results.

The following sections discuss in more detail chromatographic methods, LBAs, incurred sample reanalysis (ISR), and other issues that should be considered and how best to document validation methods.

Chromatographic Methods

Reference Standards

Analysis of drugs and their metabolites in a biological matrix is performed using calibration standards and QC samples (QCs) spiked with reference standards. The purity of the reference standard used to prepare spiked samples can affect study data. For this reason, authenticated analytical reference standards of known identity and purity should be used to prepare solutions of known concentrations. If possible, the reference standard should be identical to the analyte. When this is not possible, an established chemical form (free base or acid, salt or ester) of known purity can be used.

Three types of reference standards are usually used: (1) certified reference standards (e.g., USP compendial standards), (2) commercially supplied reference standards obtained from a reputable commercial source, and/or (3) other materials of documented purity custom synthesized by an analytical laboratory or other noncommercial establishment. The source and lot number, expiration date, certificates of analyses when available, and/or internally or externally generated evidence of identity and purity should be furnished for each reference and IS used. If the reference or internal standard expires, stock solutions made with this lot of standard should not be used unless purity is reestablished.

Bioanalytical Method Development and Validation

A specific, detailed, written description of the bioanalytical method should be established a priori. This can be in the form of a protocol, study plan, report, and/or SOP. Each step in the method should be investigated to determine the extent to which environmental, matrix, or procedural variables could affect the estimation of analyte in the matrix from the time of collection of the samples to the time of analysis.

Appropriate steps should be taken to ensure the lack of matrix effects throughout the application of the method, especially if the matrix used for production batches is different from the matrix used during method validation. Matrix effects on ion suppression or enhancement or on extraction efficiency should be addressed. A bioanalytical method should be validated for the intended use or application. All experiments used to make claims or draw conclusions about the validity of the method should be presented in a report (method validation report), including a description of validation runs that failed.

Measurements for each analyte in the biological matrix should be validated. Method development and validation for a bioanalytical method should include demonstrations of selectivity; accuracy, precision, and recovery; the calibration curve; sensitivity; reproducibility; and stability of analyte in spiked samples.

1. **Selectivity**

   Selectivity is the ability of an analytical method to differentiate and quantify the analyte in the presence of other components in the sample. Evidence should be provided that the substance quantified is the intended analyte. Analyses of blank samples of the appropriate biological matrix (plasma, urine, or other matrix) should be obtained from at least six sources. Each blank sample should be tested for interference, and selectivity should be ensured at the LLOQ.

Potential interfering substances in a biological matrix include endogenous matrix components, metabolites, decomposition products, and, in the actual study, concomitant medication and other xenobiotics. If the method is intended to quantify more than one analyte, each analyte should be tested to ensure that there is no interference.

2. **Accuracy, precision, and recovery**

The accuracy of an analytical method describes the closeness of mean test results obtained by the method to the actual value (concentration) of the analyte. Accuracy is determined by replicate analysis of samples containing known amounts of the analyte (i.e., QCs). Accuracy should be measured using a minimum of five determinations per concentration. A minimum of three concentrations in the range of expected study sample concentrations is recommended. The mean value should be within 15% of the nominal value except at LLOQ, where it should not deviate by more than 20%. The deviation of the mean from the nominal value serves as the measure of accuracy.

The precision of an analytical method describes the closeness of individual measures of an analyte when the procedure is applied repeatedly to multiple aliquots of a single homogeneous volume of biological matrix. Precision should be measured using a minimum of five determinations per concentration. A minimum of three concentrations in the range of expected study sample concentrations is recommended. The precision determined at each concentration level should not exceed 15% of the CV except for the LLOQ, where it should not exceed 20% of the CV. Precision is further subdivided into within-run and between-run precision. Within-run precision (intra-batch precision or within-run repeatability) is an assessment of precision during a single analytical run. Between-run precision (inter-batch precision or between-run repeatability) is an assessment of precision over time and may involve different analysts, equipment, reagents, and laboratories.

Sample concentrations above the upper limit of the standard curve should be diluted. The accuracy and precision of these diluted samples should be demonstrated in the method validation.

The recovery of an analyte in an assay is the detector response obtained from an amount of the analyte added to and extracted from the biological matrix, compared to the detector response obtained for the true concentration of the analyte in solvent. Recovery pertains to the extraction efficiency of an analytical method within the limits of variability. Recovery of the analyte need not be 100%, but the extent of recovery of an analyte and of the IS should be consistent, precise, and reproducible. Recovery experiments should be performed by comparing the analytical results for extracted samples at three concentrations (low, medium, and high) with unextracted standards that represent 100% recovery.

3. **Calibration curve**

A calibration (standard) curve is the relationship between instrument response and known concentrations of the analyte. The relationship between response and concentration should be continuous and reproducible. A calibration curve should be generated for each analyte in the sample. The calibration standards can contain more than one analyte. A calibration curve should be prepared in the same biological matrix as the samples in the intended study by spiking the matrix with known concentrations of the analyte. In rare cases, matrices may be difficult to obtain (e.g., cerebrospinal fluid). In such cases, calibration curves constructed in surrogate matrices should be justified. Concentrations of standards should be chosen on the basis of the concentration range expected in a particular study. A calibration curve should consist of a blank sample (matrix sample processed without analyte or IS), a zero sample (matrix sample processed without analyte but with IS), and at least six nonzero samples (matrix samples processed with analyte and IS) covering the expected range, including LLOQ.

Method validation experiments should include a minimum of six runs conducted over several days, with at least four concentrations (including LLOQ, low, medium, and high) analyzed in duplicate in each run.

(a) LLOQ

The lowest standard on the calibration curve should be accepted as the LLOQ if the following conditions are met:

- The analyte response at the LLOQ should be at least five times the response compared to blank response.
- Analyte peak (response) should be identifiable, discrete, and reproducible, and the back-calculated concentration should have precision that does not exceed 20% of the CV and accuracy within 20% of the nominal concentration. The LLOQ should not be confused with the LOD and/or the low QC sample.
- The LLOQ should be established using at least five samples and determining the CV, and/or appropriate confidence interval should be determined.

(b) ULOQ

The highest standard will define the ULOQ of an analytical method.

- Analyte peak (response) should be reproducible, and the back-calculated concentration should have precision that does not exceed 15% of the CV and accuracy within 15% of the nominal concentration.

(c) Calibration curve/standard curve/concentration–response

- The simplest model that adequately describes the concentration–response relationship should be used. Selection of weighting and use of a complex regression equation should be justified. Standards/calibrators should not deviate by more than 15% of nominal concentrations, except at LLOQ where the standard/calibrator should not deviate by more than 20%.
- The acceptance criterion for the standard curve is that at least 75% of nonzero standards should meet the previous criteria, including the LLOQ. Excluding an individual standard should not change the model used. Exclusion of calibrators for reasons other than failing to meet acceptance criteria and assignable causes is discouraged.

(d) QC Samples

- At least three concentrations of QCs in duplicate should be incorporated into each run as follows: one within three times the LLOQ (low QC), one in the midrange (middle QC), and one approaching the high end (high QC) of the range of the expected study concentrations.
- The QCs provide the basis of accepting or rejecting the run. At least 67% (e.g., at least four out of six) of the QCs concentration results should be within 15% of their respective nominal (theoretical) values. At least 50% of QCs at each level should be within 15% of their nominal concentrations. A confidence interval approach yielding comparable accuracy and precision in the run is an appropriate alternative.
- The minimum number of QCs should be at least 5% of the number of unknown samples or six total QCs, whichever is greater.
- It is recommended that calibration standards and QCs be prepared from separate stock solutions. However, standards and QCs can be prepared from the same spiking stock solution, provided the stability and accuracy of the stock solution have been verified. A single source of blank matrix may also be used, provided absence of matrix effects on extraction recovery and detection has been verified. At least one demonstration of precision and accuracy of calibrators and QCs prepared from separate stock solutions is expected.

Acceptance/rejection criteria for spiked, matrix-based calibration standards and QCs should be based on the nominal (theoretical) concentration of analytes.

4. **Sensitivity**
   Sensitivity is defined as the lowest analyte concentration that can be measured with acceptable accuracy and precision (i.e., LLOQ).
5. **Reproducibility**
   Reproducibility of the method is assessed by replicate measurements using the assay, including QCs and possibly incurred samples. Reinjection reproducibility should be evaluated to determine if an analytical run could be reanalyzed in the case of instrument interruptions.
6. **Stability**

The chemical stability of an analyte in a given matrix under specific conditions for given time intervals is assessed in several ways. Pre-study stability evaluations should cover the expected sample handling and storage conditions during the conduct of the study, including conditions at the clinical site, during shipment, and at all other secondary sites.

Drug stability in a biological fluid is a function of the storage conditions, the physicochemical properties of the drug, the matrix, and the container system. The stability of an analyte in a particular matrix and container system is relevant only to that matrix and container system and should not be extrapolated to other matrices and container systems.

Stability testing should evaluate the stability of the analytes during sample collection and handling, after long-term (frozen at the intended storage temperature) and short-term (benchtop, room temperature) storage, and after freeze and thaw cycles and the analytical process. Conditions used in stability experiments should reflect situations likely to be encountered during actual sample handling and analysis. If, during sample analysis for a study, storage conditions changed and/or exceeded the sample storage conditions evaluated during method validation, stability should be established under these new conditions.

The procedure should also include an evaluation of analyte stability in stock solution. All stability determinations should use a set of samples prepared from a freshly made stock solution of the analyte in the appropriate analyte-free, interference-free biological matrix. Stock solutions of the analyte for stability evaluation should be prepared in an appropriate solvent at known concentrations. Stability samples should be compared to freshly made calibrators and/or freshly made QCs. At least three replicates at each of the low and high concentrations should be assessed. Stability sample results should be within 15% of nominal concentrations.

(a) *Freeze and thaw stability*
During freeze–thaw stability evaluations, the freezing and thawing of stability samples should mimic the intended sample handling conditions to be used during sample analysis. Stability should be assessed for a minimum of three freeze–thaw cycles.

(b) *Benchtop stability*
Benchtop stability experiments should be designed and conducted to cover the laboratory handling conditions that are expected for study samples.

(c) *Long-term stability*
The storage time in a long-term stability evaluation should equal or exceed the time between the date of first sample collection and the date of last sample analysis.

(d) *Stock solution stability*
The stability of stock solutions of drug and IS should be evaluated. When the stock solution exists in a different state (solution vs. solid) or in a different buffer composition (generally the case for macromolecules) from the certified reference standard, the stability data on this stock solution should be generated to justify the duration of stock solution storage stability.

(e) *Processed sample*
The stability of processed samples, including the resident time in the autosampler, should be determined.

Validated Method: Use, Data Analysis, and Reporting

This section describes the expectations for the use of a validated bioanalytical method for routine drug analysis.

- System suitability: If system suitability is assessed, a specific SOP should be used. Apparatus conditioning and instrument performance should be determined using spiked samples independent of the study calibrators, QCs, or study samples. Data should be maintained with the study records.
- Calibration curves and QCs should be included in all analytical runs.
- An analytical run should consist of QCs, calibration standards, and one or more batches of processed samples. A batch may consist of all of the processed unknown samples of one or more subjects in a study and QCs. If the bioanalytical method necessitates separation of the overall analytical run into distinct processing batches (e.g., capacity limit of 100-well plates or solid phase extraction manifold, extraction by multiple analysts), each distinct processing batch should be processed at least in duplicates QCs at all QC levels (e.g., low, middle, high) along with the study samples. In such cases, acceptance criteria should be established for the analytical run as a whole as well as the distinct processing batches.
- The calibration (standard) curve should cover the expected study sample concentration range.
- Accuracy and precision as outlined in "Bioanalytical method development and validation" section should be provided for both the inter-run and intra-run experiments and tabulated for all runs (passed and failed).
- Concentrations in unknown samples should not be extrapolated below the LLOQ or above the ULOQ of the standard curve. Instead, the standard curve should be extended and revalidated, or samples with higher concentration should be diluted and reanalyzed. Concentrations below the LLOQ should be reported as zeros.
- Any required sample dilutions should use like matrix (e.g., human to human).
- Assays of all samples of an analyte in a biological matrix should be completed within the time period for which stability data are available.
- Response function: Typically, the same curve fitting, weighting, and goodness of fit determined during pre-study validation should be used for the calibration curve within the study. Response function should be determined by appropriate statistical tests based on the actual standard points during each run in the validation. Changes in the response function relationship between pre-study validation and routine run validation indicate potential problems. IS response should be monitored for drift. An SOP should be developed a priori to address issues related to variability of the IS response.
- The QCs should be used to accept or reject the run. Runs should be rejected if the calibration standards or QCs fall outside the acceptance criteria stated earlier.
- QCs should be interspersed with study samples during processing and analysis. The minimum number of QCs to ensure proper control of the assay should be at least 5% of the number of unknown samples or a total of six QCs, whichever is greater.
- If the study sample concentrations are clustered in a narrow range of the standard curve, additional QCs should be added to cover the sample range. Accuracy and precision of the additional QCs should be validated before continuing with the analysis. If the partial validation is acceptable, samples that have already been analyzed do not require reanalysis.
- All study samples from a subject should be analyzed in a single run.
- Carryover should be assessed and monitored during analysis. If carryover occurs, it should be mitigated or reduced.
- ISR should be performed (see "Incurred sample reanalysis" section).
- Repeat analysis: It is important to establish an SOP or guideline for repeat analysis and acceptance criteria. This SOP or guideline should explain the reasons for repeating sample analysis. Reasons for repeat analyses could include samples outside of the assay range,

sample processing errors, equipment failure, and poor chromatography. Reassays should be done in triplicate if sample volume allows. The rationale, approach, and all data for the repeat analysis and reporting should be clearly documented.

- Samples involving multiple analytes should not be rejected based on the data from one analyte failing the acceptance criteria.
- The data from rejected runs should be documented but need not be reported; however, the fact that a run was rejected and the reason for failure should be reported.
- If a unique or disproportionately high concentration of a metabolite is discovered in human studies, a fully validated assay may need to be developed for the metabolite, depending upon its activity (refer to the FDA guidance for industry Safety Testing of Drug Metabolites).
- Reported method validation data and the determination of accuracy and precision should include all outliers; however, calculations of accuracy and precision excluding values that are determined as outliers should also be reported.
- Sample data reintegration: An SOP or guideline for sample data reintegration should be established a priori. This SOP or guideline should define the criteria for reintegration and how the reintegration is to be performed. The rationale for the reintegration should be clearly described and documented. Audit trails should be maintained. Original and reintegration data should be reported.

Ligand-Binding Assays

Many of the bioanalytical validation parameters and principles discussed earlier are also applicable to microbiological assays and LBAs. These types of assays have a variety of design configurations that possess some unique characteristics that should be considered during method validation.

Key Reagents

Key reagents, such as reference standards, antibodies, tracers, and matrices, should be characterized appropriately and stored under defined conditions. Assay reoptimization or validation may be important when there are changes in key reagents. For example,

Labeled analytes (tracers)

- Binding should be reoptimized.
- Performance should be verified with standard curve and QCs.

Antibodies

- Key cross-reactivities should be checked.
- Tracer experiments earlier should be repeated.

Matrices

- Tracer experiments earlier should be repeated.

Bioanalytical Method Development and Validation

A specific, detailed, written description of the bioanalytical method should be established a priori. This can be in the form of a protocol, study plan, report, and/or SOP. Each step in the method should be investigated to determine the extent to which environmental, matrix, or procedural variables can affect the estimation of analyte in the matrix from the time of collection of the samples to the time of analysis.

It may be important to consider the variability of the matrix. Appropriate steps should be taken to ensure the lack of matrix effects throughout the application of the method, especially if the nature of the matrix changes from the matrix used during method validation. A bioanalytical method should be validated for the intended use or application. All experiments used to make claims or draw conclusions about the validity of the method should be presented in a report (method validation report).

Measurements for each analyte in the biological matrix should be validated. Method development and validation for a bioanalytical method should include demonstrations of selectivity, accuracy, precision, recovery, the calibration curve, sensitivity, reproducibility, and stability of analyte in spiked samples.

1. **Selectivity**
   As with chromatographic methods (described in "Chromatographic methods" section), LBAs should be shown to be selective for the analyte. The following recommendations for dealing with two selectivity issues should be considered:
   (a) Interference from substances physiochemically similar to the analyte
   - Cross-reactivity of metabolites, concomitant medications and their significant metabolites, or endogenous compounds should be evaluated individually and in combination with the analyte of interest.
   - When possible, the LBA should be compared with a validated reference method (such as LC–MS) using incurred samples and predetermined criteria to assess the accuracy of the LBA method.

   (b) Matrix effects
   Matrix effects should be evaluated. For example,
   - The calibration curve in biological fluids should be compared with calibrators in buffer to detect matrix effects using at least 10 sources of blank matrix
   - Parallelism of diluted study samples should be evaluated with diluted standards to detect matrix effects
   - Nonspecific binding should be determined
2. **Accuracy, precision, and recovery**
   Accuracy is determined by replicate analysis of samples containing known amounts of the analyte (QCs). Accuracy should be measured using a minimum of five determinations per concentration. A minimum of three concentrations in the range of expected study sample concentrations is recommended. The mean value should be within 20% of the actual value except at LLOQ, where it should not deviate by more than 25%.

   The precision should be measured using a minimum of five determinations per concentration. A minimum of three concentrations in the range of expected study sample concentrations is recommended. The precision determined at each concentration level should not exceed 20% of the CV except for the LLOQ, where it should not exceed 25% of the CV. Precision is further subdivided into within-run and between-run precision. Within-run (also known as intra-batch precision or repeatability) is an assessment of the precision during a single analytical run. Between-run precision (also known as inter-batch precision or repeatability) is a measurement of the precision with time and may involve different analysts, equipment, reagents, and laboratories.

   Samples with concentrations over the ULOQ should be diluted with the same matrix as used for the study samples, and accuracy and precision should be demonstrated.

   For LBAs that employ sample extraction, the recovery of an analyte is the measured concentration relative to the known amount added to the matrix. Recovery experiments should be performed for extracted samples at three concentrations.
3. **Calibration curve**
   Most LBA calibration (standard) curves are inherently nonlinear, and in general, more concentration points may be recommended to define the fit over the standard curve range than for chromatographic assays. In addition to their nonlinear characteristics, the response–error relationship for immunoassay standard curves is a variable function of the mean response (heteroscedasticity). For these reasons, the standard curve should consist of a

minimum of six, duplicate nonzero calibrator concentrations covering the entire range including LLOQ and excluding blanks (either single or replicate). The concentration–response relationship is most often fitted to a 4- or 5-parameter logistic model, although other models may be used with suitable validation. Calibrators should be prepared in the same matrix as the study samples.

If an alternate matrix is used, proper justification should be provided. A calibration curve should be generated for each analyte in the sample.

Method validation experiments should include a minimum of six runs conducted over several days, with at least six concentrations (including LLOQ, low, medium, and high) analyzed in duplicate in each run.

(a) LLOQ
- The lowest concentration on the calibration curve should be the LLOQ if the following conditions are met:
- Analyte peak (response) should be identifiable, discrete, and reproducible, and back-calculated concentration should have precision that does not exceed 25% CV and accuracy within 25% of the nominal concentration.
- The LLOQ should not be confused with the LOD and/or the low QCs. The LLOQ should be established using at least five samples and determining CV and/or appropriate confidence intervals.

(b) ULOQ

The highest standard will define the ULOQ of an analytical method.
- Analyte response should be reproducible, and the back-calculated concentration should have precision that does not exceed 20% CV and accuracy within 20% of the nominal concentration.

(c) Calibration curve/standard curve/concentration–response
- The simplest model that adequately describes the concentration–response relationship should be used. Selection of weighting and use of a complex regression equation should be justified. The standard calibrator concentrations should be within 25% of the nominal concentration at LLOQ and within 20% of the nominal concentration at all other concentrations.
- The acceptance criterion for the standard curve is that at least 75% of nonzero standards should meet the aforementioned criteria, including the LLOQ. Excluding an individual standard should not change the model used. Exclusion of calibrators for reasons other than failing to meet acceptance criteria and assignable causes is discouraged.
- TE (accuracy and precision) should not exceed 30%. Values falling outside these limits should be discarded, provided they do not change the established model.

(d) QC samples
- At least three concentrations of QCs in duplicate should be incorporated into each run as follows: one within three times the LLOQ (low QC), one in the midrange (middle QC), and one approaching the high end (high QC) of the range of the expected study sample concentrations.
- The results of the QCs provide the basis of accepting or rejecting the run. At least 67% (e.g., at least four out of six) of the QC concentration results should be within 20% of their respective nominal (theoretical) values. At least 50% of QCs at each level should be within 20% of their nominal concentrations. A confidence interval approach yielding comparable accuracy and precision in the run is an appropriate alternative.
- The minimum number of QCs should be at least 5% of the number of unknown samples or six total QCs, whichever is greater.

- It is recommended that calibration standards and QCs be prepared from separate stock solutions. However, standards and QCs can be prepared from the same spiking stock solution, provided the stability and accuracy of the stock solution have been verified. A single source of blank matrix may also be used, provided absence of matrix effects on extraction recovery and detection has been verified. At least one demonstration of precision and accuracy of calibrators and QCs prepared from separate stock solutions is expected.

Acceptance/rejection criteria for spiked, matrix-based calibration standards and QCs should be based on the nominal (theoretical) concentration of analytes.

4. **Sensitivity**
Sensitivity is defined as the lowest analyte concentration that can be measured with acceptable accuracy and precision.
5. **Reproducibility**
Reproducibility of the method is assessed by replicate measurements using the assay, including QCs and possibly incurred samples. Reinjection reproducibility should be evaluated to determine if an analytical run could be reanalyzed in the case of instrument interruptions.
6. **Stability**
The chemical stability of an analyte in a given matrix under specific conditions for given time intervals is assessed in several ways. Pre-study stability evaluations should cover the expected sample handling and storage conditions during the conduct of the study, including conditions at the clinical site, during shipment, and at all other secondary sites.

Stability samples should be compared to freshly made calibrators and/or freshly made QCs. At least three replicates at each of the low and high concentrations should be assessed. Assessments of analyte stability should be conducted in the same matrix as that of the study samples. All stability determinations should use samples prepared from a freshly made stock solution. Conditions used in stability experiments should reflect situations likely to be encountered during actual sample handling and analysis (e.g., long-term, benchtop, and room temperature storage and freeze–thaw cycles). If, during sample analysis for a study, storage conditions changed and/or exceed the sample storage conditions evaluated during method validation, stability should be established under the new conditions. Stock solution stability also should be assessed. Stability sample results should be within 15% of nominal concentrations.

(a) *Freeze and thaw stability*
During freeze–thaw stability evaluations, the freezing and thawing of stability samples should mimic the intended sample handling conditions to be used during sample analysis. Stability should be assessed for a minimum of three freeze–thaw cycles.

(b) *Benchtop stability*
Benchtop stability experiments should be designed and conducted to cover the laboratory handling conditions that are expected for study samples.

(c) *Long-term stability*
The storage time in a long-term stability evaluation should equal or exceed the time between the date of first sample collection and the date of last sample analysis.

(d) *Stock solution stability*
The stability of stock solutions of drug should be evaluated. When the stock solution exists in a different state (solutions vs. solid) or in a different buffer composition (generally the case for macromolecules) from the certified reference standard, the stability data on this stock solution should be generated to justify the duration of stock solution storage stability.

(e) *Processed sample stability*
The stability of processed samples, including the time until completion of analysis, should be determined.

Validated Method: Use, Data Analysis, and Reporting

This section describes the expectations for the use of a validated bioanalytical method for routine drug analysis.

- Standard curves and QCs should be included in all analytical runs.
- The calibration (standard) curve should cover the expected study sample concentration range.
- Accuracy and precision as outlined in "Bioanalytical method development and validation" section should be provided for both the inter-run and intra-run experiments and tabulated for all runs (passed and failed).
- Concentrations in unknown samples should not be extrapolated below the LLOQ or above the ULOQ of the standard curve. Instead, the standard curve should be extended and revalidated, or samples with higher concentrations should be diluted and reanalyzed. Concentrations below the LLOQ should be reported as zeros. Any required sample dilutions should use like matrix (e.g., human to human).
- Assays of all samples of an analyte in a biological matrix should be completed within the time period for which stability has been demonstrated.
- Response function: Typically, the same curve fitting, weighting, and goodness of fit determined during pre-study validation should be used for the standard curve within the study. Response function is determined by appropriate statistical tests based on the actual standard points during each run in the validation. Any changes in the response function relationship between pre-study validation and routine run validation indicate potential problems. An SOP should be developed a priori to address such issues.
- The QCs should be used to accept or reject the run. Runs should be rejected if the calibration standards or QCs fall outside the acceptance criteria stated earlier.
- QCs should be interspersed with study samples during processing and analysis. The minimum number of QCs to ensure proper control of the assay should be at least 5% of the number of unknown samples or a total of six QCs, whichever is greater.
- If the study sample concentrations are clustered in a narrow range of the standard curve, additional QCs should be added in the sample range. Accuracy and precision of the additional QCs should be validated before continuing with the analysis. If the partial validation is acceptable, samples that have already been analyzed do not require reanalysis.
- All study samples from a subject should be analyzed in a single run.
- Carryover should be assessed and monitored during analysis. If carryover occurs, it should be mitigated or reduced.
- ISR should be performed (see "Incurred sample reanalysis" section).
- Repeat analysis: It is important to establish an SOP or guideline for repeat analysis and acceptance criteria. This SOP or guideline should explain the reasons for repeating sample analysis. Reasons for repeat analyses could include samples outside of the assay range, sample processing errors, and equipment failure. The rationale, approach, and all data for the repeat analysis and reporting should be clearly documented.
- Samples involving multiple analytes should not be rejected based on the data from one analyte failing the acceptance criteria.
- The data from rejected runs should be documented, but need not be reported; however, the fact that a run was rejected and the reason for failure should be reported.
- If a unique or disproportionately high concentration of a metabolite is discovered in human studies, a fully validated assay may need to be developed for the metabolite depending on its activity (see guidance for industry Safety Testing of Drug Metabolites).
- Reported method validation data and the determination of accuracy and precision should include all outliers; however, calculations of accuracy and precision, excluding values that are determined as outliers, should also be reported.

Incurred Sample Reanalysis

ISR is a necessary component of bioanalytical method validation and is intended to verify the reliability of the reported subject sample analyte concentrations. ISR is conducted by repeating the analysis of a subset of subject samples from a given study in separate runs on different days to critically support the precision and accuracy measurements established with spiked QCs; the original and repeat analysis is conducted using the same bioanalytical method procedures. ISR samples should be compared to freshly prepared calibrators. ISR is expected for all in vivo human BE studies and all pivotal PK or pharmacodynamic (PD) studies. For nonclinical safety studies, the performing laboratory should conduct ISR at least once for each method and species.

For regulatory submissions containing only a few studies, it may be advantageous to incorporate ISR into the method development and validation stage by conducting a pilot study prior to the pivotal study. This approach allows for the remediation of methodological issues prior to conduct of the pivotal study. For applications with a greater number of pivotal PK or PD studies, ISR should be monitored in a larger number and variety of studies.

SOPs should be established and followed to address the following points:

- The total number of ISR samples should be 7% of the study sample size.
- In selecting samples for reanalysis, adequate coverage of the PK profile in its entirety should be provided and should include assessments around C_{max} and in the elimination phase for all study subjects.
- Two-thirds (67%) of the repeated sample results should be within 20% for small molecules and 30% for large molecules. The percentage difference of the results is determined with the following equation: ([repeat – original]* 100)/mean.

Written procedures should be in place to guide an investigation in the event of ISR failure for the purpose of resolving the lack of reproducibility. All aspects of ISR evaluations should be documented to reconstruct the study conduct as well as any investigations thereof. ISR results should be included in the final report of the respective study.

Additional Issues

Endogenous Compounds

For analytes that are also endogenous compounds, the accuracy of the measurement of the analytes poses a challenge when the assay cannot distinguish between the therapeutic and the endogenous counterpart. In such situations, the following approaches are recommended to validate and monitor assay performance. Other approaches, if justified by scientific principles, may also be considered.

- The biological matrix used to prepare calibration standards should be the same as the study samples and free of the endogenous analyte. To address the suitability of an analyte-free biological matrix, the matrix should be demonstrated to have (1) no measurable endogenous analyte and (2) no matrix effect or interference when compared to the biological matrix. The use of alternate matrices (e.g., buffers, dialyzed serum) for the preparation of calibration standards is generally not recommended unless an analyte-free biological matrix is not readily available or cannot be prepared. In such cases, use of an alternate analyte-free matrix should be justified, and the calibration standard in the alternate matrix should be demonstrated to have no matrix effect when compared to the actual biological matrix of the study samples.
- The QCs should be prepared by spiking known quantities of analyte(s) in the same biological matrix as the study samples. The endogenous concentrations of the analyte in the biological matrix should be evaluated prior to QC preparation (e.g., by replicate analysis). The concentrations for the QCs should account for the endogenous concentrations in the biological matrix (i.e., additive) and be representative of the expected study concentrations.

Biomarkers

The recommendations in this guidance pertain only to the validation of assays to measure in vivo biomarker concentrations in biological matrices such as blood or urine. Considerable effort also goes into defining the biological function of biomarkers, and confusion may arise regarding terminology. Information about defining the biological role of a biomarker is available on the FDA Drug Development Tools website.

Biomarkers are increasingly used to assess the effects of new drugs and therapeutic biological products in patient populations. Because of the important roles biomarkers can play in evaluating the safety and/or effectiveness of a new medical product, it is critical to ensure the integrity of the data generated by assays used to measure them. Biomarkers can be used for a wide variety of purposes during drug development; therefore, a fit-for-purpose approach should be used when evaluating the extent of method validation that is appropriate. When biomarker data will be used to support a regulatory action, such as the pivotal determination of safety and/or effectiveness or to support labeled dosing instructions, the assay should be fully validated.

For assays intended to support early drug development (e.g., candidate selection, go-no-go decisions, proof of concept), the sponsor should incorporate the extent of method validation they deem appropriate.

Method validation for biomarker assays should address the same questions as method validation for PK assays. The accuracy, precision, selectivity, range, reproducibility, and stability of a biomarker assay are important characteristics that define the method. The approach used for PK assays should be the starting point for validation of biomarker assays, although FDA realizes that some characteristics may not apply or that different considerations may need to be addressed.

Diagnostic Kits

Diagnostic kits are sometimes codeveloped with new drug or therapeutic biological products. The recommendations in this section of the guidance do not apply to commercial diagnostic kits that are intended for point-of-care patient diagnosis, but rather to analytical methods that are used during the development of new drugs and therapeutic biologics. The reader should refer to the appropriate Center for Devices and Radiological Health (CDRH) guidance documents regarding FDA expectations for commercial diagnostic kits. Furthermore, these recommendations do not apply to Clinical Laboratory Improvements Amendments (CLIA)-regulated entities or to assays designed to quantify or identify genes or genetic polymorphisms.

If a sponsor uses a commercially available diagnostic kit to measure a biomarker, drug, or therapeutic biological concentration during the development of a novel drug or therapeutic biological product, FDA makes the following recommendations.

LBA kits with various detection platforms are sometimes used to determine analyte concentrations in PK or PD studies when the reported results must exhibit sufficient precision and accuracy. Because such kits are generally developed for use as clinical diagnostic tools, their suitability for use in PK or PD studies should be demonstrated.

Diagnostic kit validation data provided by the manufacturer may not ensure reliability of the kit method for drug development purposes. The performance of diagnostic kits should be assessed in the facility conducting the sample analysis. Validation considerations for kit assays include, but are not limited to, the following examples:

- Site-specific validation should be performed. Specificity, accuracy, precision, and stability should be demonstrated under actual conditions of use. Modifications from kit processing instructions should be validated completely.
- Kits that use sparse calibration standards (e.g., one- or two-point calibration curves) should include in-house validation experiments to establish the calibration curve with a sufficient number of standards across the calibration range.

- Actual QC concentrations should be known. Concentrations of QCs expressed as ranges are not sufficient for quantitative applications. In such cases, QCs with known concentrations should be prepared and used, independent of the kit-supplied QCs.
- Standards and QCs should be prepared in the same matrix as the subject samples. Kits with standards and QCs prepared in a matrix different from the subject samples should be justified, and appropriate cross validation experiments should be performed. Refer to the endogenous compounds section of this guidance for additional discussion (see "Endogenous compounds" section).
- If the analyte source (reference standard) in the kit differs from that of the subject samples (e.g., protein isoform variation), testing should evaluate differences in immunological activity with the kit reagents.
- If multiple kit lots are used within a study, lot-to-lot variability and comparability should be addressed for critical reagents.
- Individual batches using multiple assay plates (e.g., 100-well ELISA plates) should include sufficient replicate QCs on each plate to monitor accuracy. Acceptance criteria should be established for the individual plates and overall analytical run.

New Technologies

FDA encourages the development and use of new bioanalytical technologies. Generally, the use and submission of data based on new technologies should be supported with data generated by established technology, until the new approaches become accepted practice.

Although the dried blood spot (DBS) methodology has been successful in individual cases, the method has not yet been widely accepted. Benefits of DBS include reduced blood sample volumes collected for drug analysis and ease of collection, storage, and transportation. A comprehensive validation will be essential prior to using DBS in regulated studies. This validation should address, at a minimum, the effects of the following issues: storage and handling temperature, homogeneity of sample spotting, hematocrit, stability, carryover, and reproducibility including ISR. Correlative studies with traditional sampling should be conducted during drug development. Sponsors are encouraged to seek feedback from the appropriate FDA review division early in drug development.

Documentation

General and specific SOPs and good record keeping are essential to a properly validated analytical method. The validity of an analytical method should be established and verified by laboratory studies, and the documentation of successful completion of such studies should be provided in the assay validation report. The data generated for bioanalytical method establishment and the QCs should be documented and available for data audit and inspection. Documentation for submission to FDA should include the following:

- Method development and validation data and reports
- Bioanalytical reports of the application of any methods to study sample analysis
- Overall summary information including limitations to use

All relevant documentation necessary for reconstructing the study as it was conducted and reported should be maintained in a secure environment. Relevant documentation includes, but is not limited to, source data; protocols and reports; records supporting procedural, operational, and environmental concerns; and correspondence records between the involved parties.

Regardless of the documentation format (i.e., paper or electronic), records should be contemporaneous with the event, and subsequent alterations should not obscure the original data. The basis for changing or reprocessing data should be documented with sufficient detail, and the original record should be maintained. Electronic audit trails should be available for all chromatography acquisition and data processing software and other means of electronic data capture. Information related to each bioanalytical run should be maintained at the laboratory and should include the analysts performing the run, start and stop times (duration), raw data, integration codes, and/or other reporting codes.

System Suitability/Equilibration

System suitability is routinely assessed before an analytical run. Data generated from system suitability checks should be maintained in a specific file on-site and should be available for inspection. System suitability samples should be different from the study samples, standards, and QCs to be analyzed in the run. Therefore, study samples, standards, or QCs should not be used as their own system suitability samples within the analytical run.

Summary Information

Summary information should include the following:

- A summary of assay methods used for each study protocol. Each summary should provide the protocol number, protocol title, assay type, assay method identification code, bioanalytical report code, and effective date of the method.
- For each analyte, a summary table of all the relevant method validation reports should be provided including partial validation and cross validation reports. The table should include assay method identification code, type of assay, the reason for the new method or additional validation (e.g., to lower the LOQ), and the dates of final reports. Changes made to the method should be clearly identified.
- A summary table cross-referencing multiple identification codes should be provided when an assay has different codes for the assay method, validation reports, and bioanalytical reports.

Documentation for Method Validation

Documentation for method validation should include the following:

- An operational description of the analytical method used in the study.
- A detailed description of the assay procedure (analyte, IS, sample pretreatment, method of extraction, and analysis).
- A description of the preparation of the calibration standards and QCs including blank matrix, anticoagulant if applicable, dates of preparation, and storage conditions.
- Evidence of purity and identity of drug, metabolites, and IS used at the time of the validation experiments. The chromatography of the analyte should be interference-free.
- The batch/lot numbers and storage conditions of the reference standards used to prepare the calibration standards and QCs of each assay should be provided.
- A description of potential interferences for the drug or metabolites in LBAs.
- A description of experiments conducted to determine accuracy, precision, recovery, selectivity, stability, limits of quantification, calibration curve (equations and weighting functions used), and a summary of the results including intra- and inter-assay precision and accuracy. QCs results that fail to meet the acceptance criteria should not be excluded from calculations of accuracy and precision unless there is an assignable cause.
- A description of cross validation or partial validation experiments and supporting study data, if applicable.
- Legible annotated chromatograms or mass spectrograms, if applicable.
- Description and supporting data of significant investigations of unexpected results if applicable.
- Tabulated data including, but not limited to, the following:
  - All validation experiments with analysis dates, whether the experiments passed or failed and the reason for the failure

- Results of calibration standards from all validation experiments, including calibration range, response function, back-calculated concentrations, accuracy, and precision
- QC results from all validation experiments (within- and between-run precision and accuracy)
- Data from all stability experiments, that is, storage temperatures, duration of storage, dates of analysis, and dates of preparation of QCs and calibration standards used in the stability experiments
- Data on selectivity, LLOQ, carryover, extraction recovery, matrix effect if applicable, dilution integrity, and anticoagulant effect if applicable

All measurements with the individual calculated concentrations should be presented in the validation report.

Documentation for Bioanalytical Report

Documentation of the application of validated bioanalytical methods to routine drug analysis should include the following:

- Evidence of purity at the time of use and identity of drug standards, metabolite standards, and ISs used during routine analyses and expiration or retest dates.
- Step-by-step description of procedures for preparation of QCs and calibrators.
- Sample identification, collection dates, storage prior to shipment, information on shipment batch, and storage prior to analysis. Information should include dates, times, and sample condition.
- Any deviations from the validated method, significant equipment and material changes, SOPs, protocols, and justifications for deviations.
- Equations and regression methods for calculation of concentration results.
- Complete serial chromatograms from 5% to 20% of subjects, with standards and QCs from those analytical runs. For pivotal BE studies used to support approval, chromatograms from 20% of serially selected subjects should be included. In other studies, chromatograms from 5% of randomly selected subjects in each study should be included. Subjects whose chromatograms are to be submitted should be defined prior to the analysis of any clinical samples.
- Reasons for missing samples.
- Repeat analyses should be documented with the reason(s) for the repeat analysis, the initial and repeat analysis results, the reported result, assay run identification, and the manager authorizing reanalysis. Repeat analysis of a clinical or nonclinical sample should be performed only under a predefined SOP.
- Data from reintegrated chromatograms should be documented with the reason for reintegration, initial and repeat integration results, the method used for reintegration, the reported result, assay run identification, and the manager authorizing reintegration. Reintegration of a clinical or nonclinical sample should be performed only under a predefined SOP.

The following tables should be included:

- Summary of intra- and inter-assay values of QCs and calibration curve standards used for accepting the analytical run. QC graphs and trend analyses are encouraged.
- A table listing all of the accepted and rejected analytical runs of clinical or nonclinical samples. The table should include assay run identification, assay method, and the subjects that were analyzed in each run. Tables with the individual back-calculated results for all study samples should be submitted.
- Examples of tabular listings of analytical data for reports are given in Tables 20.4 and 20.5); summary tables should be included in eCTD Module 2.

TABLE 20.4
Example of an Overall Summary Table for a Method Validation Report[a]

	Results	Hyperlink[b]	Comments
Methodology	LC–MS–MS	01-SOP-001	
Method validation report number	MVR-001	MVR-001	
Biological matrix	Human plasma	MVR-001	
Anticoagulant (if applicable)	EDTA	MVR-001	
Calibration curve range	XXX–YYY ng/mL	Summary tables 001MVR-01/CCTables Report text 001MVR-01/ CCText	
Analyte of interest	Compound A	NA	
IS	Compound A IS	NA	
Inter-run accuracy (for each QC concentration)	Low QC (AA ng/mL):X% Medium QC (e.g., BB ng/mL):Y% High QC (e.g., CC ng/mL):Z%	Summary tables 001MVR-01/APTables Report text 001MVR-01/ APText	
Inter-run precision (for each QC concentration)	Low QC (AA ng/mL):X% Medium QC (e.g., BB ng/mL):Y% High QC (e.g. CC ng/mL):Z%		
Dilution integrity (specify dilution factors and QC concentrations and matrix that were evaluated)	Dilution QC:CC ng/mL (dilution factor: X) Accuracy: Y% Precision: Z%	Summary tables 001MVR-01/DIL tables Report text 001MVR-01/DIL text	
Selectivity	<20% of the LLOQ-list drugs tested	Summary tables 001MVR-01/SELTables Report text 001MVR-01?SElText	
Short-term or benchtop temperature stability	Demonstrated for X hours at Y°C	Summary tables 001MVR-01/STSTables Report text 001MVR-01/ STSText	
Long-term stability	Demonstrated for X days at Y°C	Summary tables 001MVR-01/PSSTables Report text 001MVR-01/ PSSText	
Freeze–thaw stability	Demonstrated for Y cycles at Z°C	Summary tables 001MVR-01/FTSTables Report text 001MVR-01/ FTSText	
Stock solution stability	Demonstrated for Y weeks at Y°C	Summary tables 001MVR-01/SSSTables Report text 001MVR-01/ SSSText	
Processed sample stability	Demonstrated for Y cycles at Z°C	Summary tables 001MVR-01/PSSTables Report text 001MVR-01/ PSSText	

(Continued)

TABLE 20.4 (*Continued*)
Example of an Overall Summary Table for a Method Validation Report[a]

	Results	Hyperlink[b]	Comments
ISR	>67% of samples acceptable	Summary tables 001MVR-01/ISRTables Report text 001MVR-01/ISRText	
Recovery extraction efficiency		Summary tables 001MVR-01/EXTTables Report text 001MVR-01/EXTText	
Matrix effects		Summary tables 001MVR-01/MATTables Report text 001MVR-01/MATText	

Note: This table contains fictitious information, which serves illustrative purposes only.

[a] Failed method validation experiments should be listed, and data may be requested.

[b] For eCTD submissions, a hyperlink should be provided for the summary tables and report text.

TABLE 20.5
Example of Information for Reference Standards for Method Validation Conducted in Plasma Matrix[a]

Reference Standard	Retest/ Expiration Date	Lot Numbers	Validation Experiment	Dates of Analysis	Evidence of Purity (Hyperlink)	Comments
Compound A	MM/DD/YY	RS01	Rums 1–3 (accuracy and precisions) Rum 3 (selectivity experiment)	MM/DD/YY	001MVR-01/RS01	
Compound A IS	MM/DD/YY	RS02	Rums 1–3 (accuracy and precisions) Rum 3 (selectivity experiment)	MM/DD/YY	001MVR-01/RS02	

Note: This table contains fictitious information, which serves illustrative purposes only.

[a] A similar table would be included in the bioanalytical study report linking the use of reference standards to specific batches or analytical runs.

21 Good Clinical Practice

Good Clinical Practice (GCP) is an international ethical and scientific quality standard for designing, conducting, recording and reporting trials that involve the participation of human subjects. Compliance with this standard provides public assurance that the rights, safety and well-being of trial subjects are protected, consistent with the principles that have their origin in the Declaration of Helsinki, and that the clinical trial data are credible.

Declaration of Helsinki (World Medical Organization, 1996)

BASIC PRINCIPLES

1. Biomedical research involving human subjects must conform to generally accepted scientific principles and should be based on adequately performed laboratory and animal experimentation and on a thorough knowledge of the scientific literature.
2. The design and performance of each experimental procedure involving human subjects should be clearly formulated in an experimental protocol that should be transmitted to a specially appointed independent committee for consideration, comment, and guidance.
3. Biomedical research involving human subjects should be conducted only by scientifically qualified persons and under the supervision of a clinically competent medical person. The responsibility for the human subject must always rest with a medically qualified person and never rest on the subject of the research, even though the subject has given his or her consent.
4. Biomedical research involving human subjects cannot legitimately be carried out unless the importance of the objective is in proportion to the inherent risk to the subject.
5. Every biomedical research project involving human subjects should be preceded by careful assessment of predictable risks in comparison with foreseeable benefits to the subject or to others. Concern for the interests of the subject must always prevail over the interests of science and society.
6. The right of the research subject to safeguard his or her integrity must always be respected. Every precaution should be taken to respect the privacy of the subject and to minimize the impact of the study on the subject's physical and mental integrity and on the personality of the subject.
7. Physicians should abstain from engaging in research projects involving human subjects unless they are satisfied that the hazards involved are believed to be predictable. Physicians should cease any investigation if the hazards are found to outweigh the potential benefits.
8. In publication of the results of his or her research, the physician is obliged to preserve the accuracy of the results. Reports of experimentation not in accordance with the principles laid down in this declaration should not be accepted for publication.
9. In any research on human beings, each potential subject must be adequately informed of the aims, methods, anticipated benefits, and potential hazards of the study and the discomfort it may entail. He or she should be informed that he or she is at liberty to abstain from participation in the study and that he or she is free to withdraw visor his or her consent to participation at any time. The physician should then obtain the subject's freely given informed consent, preferably inheriting.

10. When obtaining informed consent for the research project, the physician should be particularly cautious if the subject is in dependent relationship to him or her or may consent under duress. In that case, the informed consent should be obtained by a physician who isn't engaged in the investigation and who is completely independent of this official relationship.
11. In case of legal incompetence, informed consent should be obtained from the legal guardian in accordance with national legislation. Where physical or mental incapacity makes it impossible to obtain informed consent, or when the subject is a minor, permission from the responsible relative replaces that of the subject in accordance with national legislation. Whenever the minor child is in fact able to give a consent, the minor's consent must be obtained in addition to the consent of the minor's legal guardian.
12. The research protocol should always contain a statement of the ethical considerations involved and should indicate that the principles enunciated in the present declaration are complied with.

NONTHERAPEUTIC BIOMEDICAL RESEARCH INVOLVING HUMAN SUBJECTS: NONCLINICAL BIOMEDICAL RESEARCH

1. In the purely scientific application of medical research carried out on a human being, it is the duty of the physician to remain the protector of the life and health of that person on whom biomedical research is being carried out.
2. The subjects should be volunteers—either healthy persons or patients for whom the experimental design is not related to the patient's illness.
3. The investigator or the investigating team should discontinue the research if in his or her or their judgment, it may, if continued, be harmful to the individual.
4. In research on man, the interest of science and society should never take precedence over considerations related to the well-being of the subject.

PRINCIPLES OF ICH GCP

1. Clinical trials should be conducted in accordance with the ethical principles that have their origin in the Declaration of Helsinki and that are consistent with GCP and the applicable regulatory requirement(s).
2. Before a trial is initiated, foreseeable risks and inconveniences should be weighed against the anticipated benefit for the individual trial subject and society. A trial should be initiated and continued only if the anticipated benefits justify the risks.
3. The rights, safety, and well-being of the trial subjects are the most important considerations and should prevail over interests of science and society.
4. The available nonclinical and clinical information on an investigational product should be adequate to support the proposed clinical trial.
5. Clinical trials should be scientifically sound and described in a clear, detailed protocol.
6. A trial should be conducted in compliance with the protocol that has received prior institutional review board (IRB)/independent ethics committee (IEC) approval/favorable opinion.
7. The medical care given to, and medical decisions made on behalf of, subjects should always be the responsibility of a qualified physician or, when appropriate, of a qualified dentist.
8. Each individual involved in conducting a trial should be qualified by education, training, and experience to perform his or her respective task(s).
9. Freely given informed consent should be obtained from every subject prior to clinical trial participation.

10. All clinical trial information should be recorded, handled, and stored in a way that allows its accurate reporting, interpretation, and verification.
11. The confidentiality of records that could identify subjects should be protected, respecting the privacy and confidentiality rules in accordance with the applicable regulatory requirement(s).
12. Investigational products should be manufactured, handled, and stored in accordance with applicable good manufacturing practice (GMP). They should be used in accordance with the approved protocol.
13. Systems with procedures that assure the quality of every aspect of the trial should be implemented.

INSTITUTIONAL REVIEW BOARD/INDEPENDENT ETHICS COMMITTEE

Responsibilities

1. An IRB/IEC should safeguard the rights, safety, and well-being of all trial subjects. Special attention should be paid to trials that may include vulnerable subjects.
2. The IRB/IEC should obtain the following documents: trial protocol(s)/amendment(s), written informed consent form(s), and consent form updates that the investigator proposes for use in the trial, subject recruitment procedures (e.g., advertisements), written information to be provided to subjects, Investigator's Brochure (IB), available safety information, information about payments and compensation available to subjects, the investigator's current curriculum vitae and/or other documentation evidencing qualifications, and any other documents that the IRB/IEC may need to fulfill its responsibilities. The IRB/IEC should review a proposed clinical trial within a reasonable time and document its views in writing, clearly identifying the trial, the documents reviewed, and the dates for the following:
    - Approval/favorable opinion
    - Modifications required prior to its approval/favorable opinion
    - Disapproval/negative opinion
    - Termination/suspension of any prior approval/favorable opinion
3. The IRB/IEC should consider the qualifications of the investigator for the proposed trial, as documented by a current curriculum vitae and/or by any other relevant documentation the IRB/IEC requests.
4. The IRB/IEC should conduct continuing review of each ongoing trial at intervals appropriate to the degree of risk to human subjects, but at least once per year.
5. The IRB/IEC may request more information than is outlined in item 10 under section "Informed Consent of Trial Subjects" be given to subjects when, in the judgment of the IRB/IEC, the additional information would add meaningfully to the protection of the rights, safety, and/or well-being of the subjects.
6. When a nontherapeutic trial is to be carried out with the consent of the subjects' legally acceptable representative (see items 12 and 14 under section "Informed Consent of Trial Subjects"), the IRB/IEC should determine that the proposed protocol and/or other document(s) adequately addresses relevant ethical concerns and meets applicable regulatory requirements for such trials.
7. Where the protocol indicates that prior consent of the trial subject or the subjects' legally acceptable representative is not possible (see item 15 under section "Informed Consent of Trial Subjects"), the IRB/IEC should determine that the proposed protocol and/or other document(s) adequately addresses relevant ethical concerns and meets applicable regulatory requirements for such trials (i.e., in emergency situations).

8. The IRB/IEC should review both the amount and method of payment to subjects to assure that neither presents problems of coercion or undue influence on the trial subjects. Payments to a subject should be prorated and not wholly contingent on completion of the trial by the subject.
9. The IRB/IEC should ensure that information regarding payment to subjects, including the methods, amounts, and schedule of payment to trial subjects, is set forth in the written informed consent form and any other written information to be provided to subjects. The way payment will be prorated should be specified.

Composition, Functions, and Operations

1. The IRB/IEC should consist of a reasonable number of members, who collectively have the qualifications and experience to review and evaluate the science, medical aspects, and ethics of the proposed trial. It is recommended that the IRB/IEC should include
   (a) At least five members
   (b) At least one member whose primary area of interest is in a nonscientific area
   (c) At least one member who is independent of the institution/trial site
   Only those IRB/IEC members who are independent of the investigator and the sponsor of the trial should vote/provide opinion on a trial-related matter.
   A list of IRB/IEC members and their qualifications should be maintained.
2. The IRB/IEC should perform its functions according to written operating procedures, should maintain written records of its activities and minutes of its meetings, and should comply with GCP and with the applicable regulatory requirement(s).
3. An IRB/IEC should make its decisions at announced meetings at which at least a quorum, as stipulated in its written operating procedures, is present.
4. Only members who participate in the IRB/IEC review and discussion should vote/provide their opinion and/or advice.
5. The investigator may provide information on any aspect of the trial, but should not participate in the deliberations of the IRB/IEC or in the vote/opinion of the IRB/IEC.
6. An IRB/IEC may invite nonmembers with expertise in special areas for assistance.

Procedures

The IRB/IEC should establish, document in writing, and follow its procedures, which should include

1. Determining its composition (names and qualifications of the members) and the authority under which it is established
2. Scheduling, notifying its members of, and conducting its meetings
3. Conducting initial and continuing review of trials
4. Determining the frequency of continuing review, as appropriate
5. Providing, according to the applicable regulatory requirements, expedited review and approval/favorable opinion of minor change(s) in ongoing trials that have the approval/favorable opinion of the IRB/IEC
6. Specifying that no subject should be admitted to a trial before the IRB/IEC issues its written approval/favorable opinion of the trial
7. Specifying that no deviations from, or changes of, the protocol should be initiated without prior written IRB/IEC approval/favorable opinion of an appropriate amendment, except when necessary to eliminate immediate hazards to the subjects or when the change(s) involves only logistical or administrative aspects of the trial (e.g., change of monitor[s], telephone number[s]) (see item 2 under section "Compliance with Protocol")

8. Specifying that the investigator should promptly report to the IRB/IEC: (a) deviations from, or changes of, the protocol to eliminate immediate hazards to the trial subjects (see item 7 under section "Procedures" and items 2 and 4 under section "Compliance with Protocol"); (b) changes increasing the risk to subjects and/or affecting significantly the conduct of the trial (see item 2 under section "Progress Reports"); (c) all adverse drug reactions (ADRs) that are both serious and unexpected; and (d) new information that may affect adversely the safety of the subjects or the conduct of the trial
9. Ensuring that the IRB/IEC promptly notify in writing the investigator/institution concerning
   (a) Its trial-related decisions/opinions
   (b) The reasons for its decisions/opinions
   (c) Procedures for appeal of its decisions/opinions

Records

The IRB/IEC should retain all relevant records (e.g., written procedures, membership lists, lists of occupations/affiliations of members, submitted documents, minutes of meetings, and correspondence) for a period of at least 3 years after completion of the trial and make them available upon request from the regulatory authority(ies).

The IRB/IEC may be asked by investigators, sponsors, or regulatory authorities to provide its written procedures and membership lists.

INVESTIGATOR

Investigator's Qualifications and Agreements

1. The investigator(s) should be qualified by education, training, and experience to assume responsibility for the proper conduct of the trial, should meet all the qualifications specified by the applicable regulatory requirement(s), and should provide evidence of such qualifications through up-to-date curriculum vitae and/or other relevant documentation requested by the sponsor, the IRB/IEC, and/or the regulatory authority(ies).
2. The investigator should be thoroughly familiar with the appropriate use of the investigational product(s), as described in the protocol, in the current IB, in the product information, and in other information sources provided by the sponsor.
3. The investigator should be aware of, and should comply with, GCP and the applicable regulatory requirements.
4. The investigator/institution should permit monitoring and auditing by the sponsor and inspection by the appropriate regulatory authority(ies).
5. The investigator should maintain a list of appropriately qualified persons to whom the investigator has delegated significant trial-related duties.

Adequate Resources

1. The investigator should be able to demonstrate (e.g., based on retrospective data) a potential for recruiting the required number of suitable subjects within the agreed recruitment period.
2. The investigator should have sufficient time to properly conduct and complete the trial within the agreed trial period.
3. The investigator should have available an adequate number of qualified staff and adequate facilities for the foreseen duration of the trial to conduct the trial properly and safely.
4. The investigator should ensure that all persons assisting with the trial are adequately informed about the protocol, the investigational product(s), and their trial-related duties and functions.

Medical Care of Trial Subjects

1. A qualified physician (or dentist, when appropriate), who is an investigator or a sub-investigator for the trial, should be responsible for all trial-related medical (or dental) decisions.
2. During and following a subject's participation in a trial, the investigator/institution should ensure that adequate medical care is provided to a subject for any adverse events (AEs), including clinically significant laboratory values, related to the trial. The investigator/institution should inform a subject when medical care is needed for intercurrent illness(es) of which the investigator becomes aware.
3. It is recommended that the investigator inform the subject's primary physician about the subject's participation in the tsrial if the subject has a primary physician and if the subject agrees to the primary physician being informed.
4. Although a subject is not obliged to give his/her reason(s) for withdrawing prematurely from a trial, the investigator should make a reasonable effort to ascertain the reason(s), while fully respecting the subject's rights.

Communication with Institutional Review Board/Independent Ethics Committee

1. Before initiating a trial, the investigator/institution should have written and dated approval/favorable opinion from the IRB/IEC for the trial protocol, written informed consent form, consent form updates, subject recruitment procedures (e.g., advertisements), and any other written information to be provided to subjects.
2. As part of the investigator's/institution's written application to the IRB/IEC, the investigator/institution should provide the IRB/IEC with a current copy of the IB. If the IB is updated during the trial, the investigator/institution should supply a copy of the updated IB to the IRB/IEC.
3. During the trial, the investigator/institution should provide to the IRB/IEC all documents subject to review.

Compliance with Protocol

1. The investigator/institution should conduct the trial in compliance with the protocol agreed to by the sponsor and, if required, by the regulatory authority(ies) and which was given approval/favorable opinion by the IRB/IEC. The investigator/institution and the sponsor should sign the protocol, or an alternative contract, to confirm agreement.
2. The investigator should not implement any deviation from, or changes of, the protocol without agreement by the sponsor and prior review and documented approval/favorable opinion from the IRB/IEC of an amendment, except where necessary to eliminate an immediate hazard(s) to trial subjects or when the change(s) involves only logistical or administrative aspects of the trial (e.g., change in monitor[s], change of telephone number[s]).
3. The investigator, or person designated by the investigator, should document and explain any deviation from the approved protocol.
4. The investigator may implement a deviation from, or a change of, the protocol to eliminate an immediate hazard(s) to trial subjects without prior IRB/IEC approval/favorable opinion. As soon as possible, the implemented deviation or change, the reasons for it, and, if appropriate, the proposed protocol amendment(s) should be submitted
   (a) To the IRB/IEC for review and approval/favorable opinion
   (b) To the sponsor for agreement
   (c) To the regulatory authority(ies), if required

Investigational Product(s)

1. Responsibility for investigational product(s) accountability at the trial site(s) rests with the investigator/institution.
2. Where allowed/required, the investigator/institution may/should assign some or all of the investigator's/institution's duties for investigational product(s) accountability at the trial site(s) to an appropriate pharmacist or another appropriate individual who is under the supervision of the investigator/institution.
3. The investigator/institution and/or a pharmacist or other appropriate individual, who is designated by the investigator/institution, should maintain records of the product's delivery to the trial site, the inventory at the site, the use by each subject, and the return to the sponsor or alternative disposition of unused product(s). These records should include dates, quantities, batch/serial numbers, expiration dates (if applicable), and the unique code numbers assigned to the investigational product(s) and trial subjects. Investigators should maintain records that document adequately that the subjects were provided the doses specified by the protocol and reconcile all investigational product(s) received from the sponsor.
4. The investigational product(s) should be stored as specified by the sponsor (see item 2 under section "Manufacturing, Packaging, Labeling, and Coding Investigational Product(s)" and item 3 under section "Supplying and Handling Investigational Product(s)") and in accordance with applicable regulatory requirement(s).
5. The investigator should ensure that the investigational product(s) are used only in accordance with the approved protocol.
6. The investigator, or a person designated by the investigator/institution, should explain the correct use of the investigational product(s) to each subject and should check, at intervals appropriate for the trial, that each subject is following the instructions properly.

Randomization Procedures and Unblinding

The investigator should follow the trial's randomization procedures, if any, and should ensure that the code is broken only in accordance with the protocol. If the trial is blinded, the investigator should promptly document and explain to the sponsor any premature unblinding (e.g., accidental unblinding, unblinding due to a serious adverse event [SAE]) of the investigational product(s).

Informed Consent of Trial Subjects

1. In obtaining and documenting informed consent, the investigator should comply with the applicable regulatory requirement(s) and should adhere to GCP and to the ethical principles that have their origin in the Declaration of Helsinki. Prior to the beginning of the trial, the investigator should have the IRB/IEC's written approval/favorable opinion of the written informed consent form and any other written information to be provided to subjects.
2. The written informed consent form and any other written information to be provided to subjects should be revised whenever important new information becomes available that may be relevant to the subjects' consent. Any revised written informed consent form and written information should receive the IRB/IEC's approval/favorable opinion in advance of use. The subject or the subjects' legally acceptable representative should be informed in a timely manner if new information becomes available that may be relevant to the subjects' willingness to continue participation in the trial. The communication of this information should be documented.

3. Neither the investigator, nor the trial staff, should coerce or unduly influence a subject to participate or to continue to participate in a trial.
4. None of the oral and written information concerning the trial, including the written informed consent form, should contain any language that causes the subject or the subject's legally acceptable representative to waive or to appear to waive any legal rights or that releases or appears to release the investigator, the institution, the sponsor, or their agents from liability for negligence.
5. The investigator, or a person designated by the investigator, should fully inform the subject or, if the subject is unable to provide informed consent, the subject's legally acceptable representative, of all pertinent aspects of the trial including the written information and the approval/favorable opinion by the IRB/IEC.
6. The language used in the oral and written information about the trial, including the written informed consent form, should be as nontechnical as practical and should be understandable to the subject or the subject's legally acceptable representative and the impartial witness, where applicable.
7. Before informed consent may be obtained, the investigator, or a person designated by the investigator, should provide the subject or the subject's legally acceptable representative ample time and opportunity to inquire about details of the trial and to decide whether or not to participate in the trial. All questions about the trial should be answered to the satisfaction of the subject or the subject's legally acceptable representative.
8. Prior to a subjects' participation in the trial, the written informed consent form should be signed and personally dated by the subject or by the subject's legally acceptable representative and by the person who conducted the informed consent discussion.
9. If a subject is unable to read or if a legally acceptable representative is unable to read, an impartial witness should be present during the entire informed consent discussion. After the written informed consent form and any other written information to be provided to subjects are read and explained to the subject or the subjects' legally acceptable representative, and after the subject or the subjects' legally acceptable representative has orally consented to the subjects' participation in the trial and, if capable of doing so, has signed and personally dated the informed consent form, the witness should sign and personally date the consent form. By signing the consent form, the witness attests that the information in the consent form and any other written information was accurately explained to, and apparently understood by, the subject or the subject's legally acceptable representative and that informed consent was freely given by the subject or the subjects' legally acceptable representative.
10. Both the informed consent discussion and the written informed consent form and any other written information to be provided to subjects should include explanations of the following:
    (a) That the trial involves research.
    (b) The purpose of the trial.
    (c) The trial treatment(s) and the probability for random assignment to each treatment.
    (d) The trial procedures to be followed, including all invasive procedures.
    (e) The subject's responsibilities.
    (f) Those aspects of the trial that are experimental.
    (g) The reasonably foreseeable risks or inconveniences to the subject and, when applicable, to an embryo, fetus, or nursing infant.
    (h) The reasonably expected benefits. When there is no intended clinical benefit to the subject, the subject should be made aware of this.
    (i) The alternative procedure(s) or course(s) of treatment that may be available to the subject and their important potential benefits and risks.

(j) The compensation and/or treatment available to the subject in the event of trial-related injury.
(k) The anticipated prorated payment, if any, to the subject for participating in the trial.
(l) The anticipated expenses, if any, to the subject for participating in the trial.
(m) That the subject's participation in the trial is voluntary and that the subject may refuse to participate or withdraw from the trial, at any time, without penalty or loss of benefits to which the subject is otherwise entitled.
(n) That the monitor(s), the auditor(s), the IRB/IEC, and the regulatory authority(ies) will be granted direct access to the subject's original medical records for verification of clinical trial procedures and/or data, without violating the confidentiality of the subject, to the extent permitted by the applicable laws and regulations and that, by signing a written informed consent form, the subject or the subject's legally acceptable representative is authorizing such access.
(o) That records identifying the subject will be kept confidential and, to the extent permitted by the applicable laws and/or regulations, will not be made publicly available. If the results of the trial are published, the subjects' identity will remain confidential.
(p) That the subject or the subject's legally acceptable representative will be informed in a timely manner if information becomes available that may be relevant to the subject's willingness to continue participation in the trial.
(q) The person(s) to contact for further information regarding the trial and the rights of trial subjects and whom to contact in the event of trial-related injury.
(r) The foreseeable circumstances and/or reasons under which the subject's participation in the trial may be terminated.
(s) The expected duration of the subject's participation in the trial.
(t) The approximate number of subjects involved in the trial.

11. Prior to participation in the trial, the subject or the subject's legally acceptable representative should receive a copy of the signed and dated written informed consent form and any other written information provided to the subjects. During a subjects' participation in the trial, the subject or the subjects' legally acceptable representative should receive a copy of the signed and dated consent form updates and a copy of any amendments to the written information provided to subjects.
12. When a clinical trial (therapeutic or nontherapeutic) includes subjects who can only be enrolled in the trial with the consent of the subjects' legally acceptable representative (e.g., minors or patients with severe dementia), the subject should be informed about the trial to the extent compatible with the subjects' understanding, and, if capable, the subject should sign and personally date the written informed consent.
13. Except as described in item 14 under section "Informed Consent of Trial Subjects", a nontherapeutic trial (i.e., a trial in which there is no anticipated direct clinical benefit to the subject) should be conducted in subjects who personally give consent and who sign and date the written informed consent form.
14. Nontherapeutic trials may be conducted in subjects with consent of a legally acceptable representative provided the following conditions are fulfilled:
    (a) The objectives of the trial cannot be met by means of a trial in subjects who can give informed consent personally.
    (b) The foreseeable risks to the subjects are low.
    (c) The negative impact on the subjects' well-being is minimized and low.
    (d) The trial is not prohibited by law.
    (e) The approval/favorable opinion of the IRB/IEC is expressly sought on the inclusion of such subjects, and the written approval/favorable opinion covers this aspect.

Such trials, unless an exception is justified, should be conducted in patients having a disease or condition for which the investigational product is intended. Subjects in these trials should be particularly closely monitored and should be withdrawn if they appear to be unduly distressed.

15. In emergency situations, when prior consent of the subject is not possible, the consent of the subject's legally acceptable representative, if present, should be requested. When prior consent of the subject is not possible and the subjects' legally acceptable representative is not available, enrollment of the subject should require measures described in the protocol and/or elsewhere, with documented approval/favorable opinion by the IRB/IEC, to protect the rights, safety, and well-being of the subject and to ensure compliance with applicable regulatory requirements. The subject or the subject's legally acceptable representative should be informed about the trial as soon as possible, and consent to continue and other consent as appropriate (see item 10 under section "Informed Consent of Trial Subjects") should be requested.

Records and Reports

1. The investigator should ensure the accuracy, completeness, legibility, and timeliness of the data reported to the sponsor in the case report forms (CRFs) and in all required reports.
2. Data reported on the CRF, which are derived from source documents, should be consistent with the source documents or the discrepancies should be explained.
3. Any change or correction to a CRF should be dated, initialed, and explained (if necessary) and should not obscure the original entry (i.e., an audit trail should be maintained); this applies to both written and electronic changes or corrections (see item 4 (n) under section "Monitoring"). Sponsors should provide guidance to investigators and/or the investigators' designated representatives on making such corrections. Sponsors should have written procedures to assure that changes or corrections in CRFs made by sponsor's designated representatives are documented, are necessary, and are endorsed by the investigator. The investigator should retain records of the changes and corrections.
4. The investigator/institution should maintain the trial documents as specified in Essential Documents for the Conduct of a Clinical Trial (see section "Essential Documents for the Conduct of a Clinical Trial") and as required by the applicable regulatory requirement(s). The investigator/institution should take measures to prevent accidental or premature destruction of these documents.
5. Essential documents should be retained until at least 2 years after the last approval of a marketing application in an International Conference on Harmonization (ICH) region and until there are no pending or contemplated marketing applications in an ICH region or at least 2 years have elapsed since the formal discontinuation of clinical development of the investigational product. These documents should be retained for a longer period however if required by the applicable regulatory requirements or by an agreement with the sponsor. It is the responsibility of the sponsor to inform the investigator/institution as to when these documents no longer need to be retained (see item 12 under section "Trial Management, Data Handling, and Record Keeping").
6. The financial aspects of the trial should be documented in an agreement between the sponsor and the investigator/institution.
7. Upon request of the monitor, auditor, IRB/IEC, or regulatory authority, the investigator/institution should make available for direct access all requested trial-related records.

Progress Reports

1. The investigator should submit written summaries of the trial status to the IRB/IEC annually, or more frequently, if requested by the IRB/IEC.
2. The investigator should promptly provide written reports to the sponsor, the IRB/IEC (see item 8 under section "Procedures"), and, where applicable, the institution on any changes significantly affecting the conduct of the trial and/or increasing the risk to subjects.

Safety Reporting

1. All SAEs should be reported immediately to the sponsor except for those SAEs that the protocol or other document (e.g., IB) identifies as not needing immediate reporting. The immediate reports should be followed promptly by detailed, written reports. The immediate and follow-up reports should identify subjects by unique code numbers assigned to the trial subjects rather than by the subjects' names, personal identification numbers, and/or addresses. The investigator should also comply with the applicable regulatory requirement(s) related to the reporting of unexpected serious ADRs to the regulatory authority(ies) and the IRB/IEC.
2. AEs and/or laboratory abnormalities identified in the protocol as critical to safety evaluations should be reported to the sponsor according to the reporting requirements and within the time periods specified by the sponsor in the protocol.
3. For reported deaths, the investigator should supply the sponsor and the IRB/IEC with any additional requested information (e.g., autopsy reports and terminal medical reports).

Premature Termination or Suspension of a Trial

If the trial is prematurely terminated or suspended for any reason, the investigator/institution should promptly inform the trial subjects, should assure appropriate therapy and follow-up for the subjects, and, where required by the applicable regulatory requirement(s), should inform the regulatory authority(ies). In addition:

1. If the investigator terminates or suspends a trial without prior agreement of the sponsor, the investigator should inform the institution where applicable, and the investigator/institution should promptly inform the sponsor and the IRB/IEC and should provide the sponsor and the IRB/IEC a detailed written explanation of the termination or suspension.
2. If the sponsor terminates or suspends a trial (see section "Premature Termination or Suspension of a Trial"), the investigator should promptly inform the institution where applicable, and the investigator/institution should promptly inform the IRB/IEC and provide the IRB/IEC a detailed written explanation of the termination or suspension.
3. If the IRB/IEC terminates or suspends its approval/favorable opinion of a trial (see item 2 under section "Responsibilities" and item 9 under section "Procedures"), the investigator should inform the institution where applicable and the investigator/institution should promptly notify the sponsor and provide the sponsor with a detailed written explanation of the termination or suspension.

Final Report(s) by Investigator

Upon completion of the trial, the investigator, where applicable, should inform the institution; the investigator/institution should provide the IRB/IEC with a summary of the trial's outcome and the regulatory authority(ies) with any reports required.

SPONSOR

Quality Assurance and Quality Control

1. The sponsor is responsible for implementing and maintaining quality assurance and quality control systems with written SOPs to ensure that trials are conducted and data are generated, documented (recorded), and reported in compliance with the protocol, GCP, and the applicable regulatory requirement(s).
2. The sponsor is responsible for securing agreement from all involved parties to ensure direct access to all trial-related sites, source data/documents, and reports for the purpose of monitoring and auditing by the sponsor and inspection by domestic and foreign regulatory authorities.
3. Quality control should be applied to each stage of data handling to ensure that all data are reliable and have been processed correctly.
4. Agreements, made by the sponsor with the investigator/institution and any other parties involved with the clinical trial, should be in writing, as part of the protocol or in a separate agreement.

Contract Research Organization

1. A sponsor may transfer any or all of the sponsor's trial-related duties and functions to a contract research organization (CRO), but the ultimate responsibility for the quality and integrity of the trial data always resides with the sponsor. The CRO should implement quality assurance and quality control.
2. Any trial-related duty and function that is transferred to and assumed by a CRO should be specified in writing.
3. Any trial-related duties and functions not specifically transferred to and assumed by a CRO are retained by the sponsor.
4. All references to a sponsor in this guideline also apply to a CRO to the extent that a CRO has assumed the trial-related duties and functions of a sponsor.

Medical Expertise

The sponsor should designate appropriately qualified medical personnel who will be readily available to advise on trial-related medical questions or problems. If necessary, outside consultant(s) may be appointed for this purpose.

Trial Design

1. The sponsor should utilize qualified individuals (e.g., biostatisticians, clinical pharmacologists, and physicians) as appropriate, throughout all stages of the trial process, from designing the protocol and CRFs and planning the analyses to analyzing and preparing interim and final clinical trial reports.
2. For further guidance, see Clinical Trial Protocol and Protocol Amendment(s) (see section "Clinical Trial Protocol and Protocol Amendment(s)"), the ICH Guideline for Structure and Content of Clinical Study Reports, and other appropriate ICH guidance on trial design, protocol, and conduct.

Trial Management, Data Handling, and Record Keeping

1. The sponsor should utilize appropriately qualified individuals to supervise the overall conduct of the trial, to handle the data, to verify the data, to conduct the statistical analyses, and to prepare the trial reports.

2. The sponsor may consider establishing an independent data-monitoring committee (IDMC) to assess the progress of a clinical trial, including the safety data and the critical efficacy endpoints at intervals, and to recommend to the sponsor whether to continue, modify, or stop a trial. The IDMC should have written operating procedures and maintain written records of all its meetings.
3. When using electronic trial data handling and/or remote electronic trial data systems, the sponsor should
   (a) Ensure and document that the electronic data processing system(s) conforms to the sponsor's established requirements for completeness, accuracy, reliability, and consistent intended performance (i.e., validation)
   (b) Maintain SOPs for using these systems
   (c) Ensure that the systems are designed to permit data changes in such a way that the data changes are documented and that there is no deletion of entered data (i.e., maintain an audit trail, data trail, edit trail)
   (d) Maintain a security system that prevents unauthorized access to the data
   (e) Maintain a list of the individuals who are authorized to make data changes (see item 5 under section "Investigator's Qualifications and Agreements" and item 3 under section "Records and Reports")
   (f) Maintain adequate backup of the data
   (g) Safeguard the blinding, if any (e.g., maintain the blinding during data entry and processing)
4. If data are transformed during processing, it should always be possible to compare the original data and observations with the processed data.
5. The sponsor should use an unambiguous subject identification code that allows identification of all the data reported for each subject.
6. The sponsor, or other owners of the data, should retain all of the sponsor-specific essential documents pertaining to the trial (see section "Essential Documents for the Conduct of a Clinical Trial").
7. The sponsor should retain all sponsor-specific essential documents in conformance with the applicable regulatory requirement(s) of the country(ies) where the product is approved and/or where the sponsor intends to apply for approval(s).
8. If the sponsor discontinues the clinical development of an investigational product (i.e., for any or all indications, routes of administration, or dosage forms), the sponsor should maintain all sponsor-specific essential documents for at least 2 years after formal discontinuation or in conformance with the applicable regulatory requirement(s).
9. If the sponsor discontinues the clinical development of an investigational product, the sponsor should notify all the trial investigators/institutions and all the regulatory authorities.
10. Any transfer of ownership of the data should be reported to the appropriate authority(ies), as required by the applicable regulatory requirement(s).
11. The sponsor-specific essential documents should be retained until at least 2 years after the last approval of a marketing application in an ICH region and until there are no pending or contemplated marketing applications in an ICH region or at least 2 years have elapsed since the formal discontinuation of clinical development of the investigational product. These documents should be retained for a longer period however if required by the applicable regulatory requirement(s) or if needed by the sponsor.
12. The sponsor should inform the investigator(s)/institution(s) in writing of the need for record retention and should notify the investigator(s)/institution(s) in writing when the trial-related records are no longer needed.

Investigator Selection

1. The sponsor is responsible for selecting the investigator(s)/institution(s). Each investigator should be qualified by training and experience and should have adequate resources (see sections "Investigator's Qualifications and Agreements" and "Adequate Resources") to properly conduct the trial for which the investigator is selected. If organization of a coordinating committee and/or selection of coordinating investigator(s) is to be utilized in multicenter trials, their organization and/or selection is the sponsor's responsibility.
2. Before entering an agreement with an investigator/institution to conduct a trial, the sponsor should provide the investigator(s)/institution(s) with the protocol and an up-to-date IB and should provide sufficient time for the investigator/institution to review the protocol and the information provided.
3. The sponsor should obtain the investigator's/institution's agreement
   - (a) To conduct the trial in compliance with GCP, with the applicable regulatory requirement(s) (see item3 under section "Investigator's Qualifications and Agreements"), and with the protocol agreed to by the sponsor and given approval/favorable opinion by the IRB/IEC (see item 1 under section "Compliance with Protocol")
   - (b) To comply with procedures for data recording/reporting
   - (c) To permit monitoring, auditing, and inspection (see item 4 under section "Investigator's Qualifications and Agreements")
   - (d) To retain the trial-related essential documents until the sponsor informs the investigator/institution these documents are no longer needed (see item 4 under section "Records and Reports" and item 12 under section "Trial Management, Data Handling, and Record Keeping")

   The sponsor and the investigator/institution should sign the protocol, or an alternative document, to confirm this agreement.

Allocation of Responsibilities

Prior to initiating a trial, the sponsor should define, establish, and allocate all trial-related duties and functions.

Compensation to Subjects and Investigators

1. If required by the applicable regulatory requirement(s), the sponsor should provide insurance or should indemnify (legal and financial coverage) the investigator/institution against claims arising from the trial, except for claims that arise from malpractice and/or negligence.
2. The sponsor's policies and procedures should address the costs of treatment of trial subjects in the event of trial-related injuries in accordance with the applicable regulatory requirement(s).
3. When trial subjects receive compensation, the method and manner of compensation should comply with applicable regulatory requirement(s).

Financing

The financial aspects of the trial should be documented in an agreement between the sponsor and the investigator/institution.

Notification/Submission to Regulatory Authority(ies)

Before initiating the clinical trial(s), the sponsor (or the sponsor and the investigator, if required by the applicable regulatory requirement[s]) should submit any required application(s) to the appropriate authority(ies) for review, acceptance, and/or permission (as required by the applicable regulatory requirement[s]) to begin the trial(s). Any notification/submission should be dated and contain sufficient information to identify the protocol.

Confirmation of Review by Institutional Review Board/Independent Ethics Committee

1. The sponsor should obtain from the investigator/institution
   (a) The name and address of the investigator's/institution's IRB/IEC
   (b) A statement obtained from the IRB/IEC that it is organized and operates according to GCP and the applicable laws and regulations
   (c) Documented IRB/IEC approval/favorable opinion and, if requested by the sponsor, a current copy of protocol, written informed consent form(s) and any other written information to be provided to subjects, subject recruiting procedures, and documents related to payments and compensation available to the subjects and any other documents that the IRB/IEC may have requested
2. If the IRB/IEC conditions its approval/favorable opinion upon change(s) in any aspect of the trial, such as modification(s) of the protocol, written informed consent form and any other written information to be provided to subjects, and/or other procedures, the sponsor should obtain from the investigator/institution a copy of the modification(s) made and the date approval/favorable opinion was given by the IRB/IEC.
3. The sponsor should obtain from the investigator/institution documentation and dates of any IRB/IEC reapprovals/reevaluations with favorable opinion and of any withdrawals or suspensions of approval/favorable opinion.

Information on Investigational Product(s)

1. When planning trials, the sponsor should ensure that sufficient safety and efficacy data from nonclinical studies and/or clinical trials are available to support human exposure by the route, at the dosages, for the duration, and in the trial population to be studied.
2. The sponsor should update the IB as significant new information becomes available (see section "Investigator's Brochure").

Manufacturing, Packaging, Labeling, and Coding Investigational Product(s)

1. The sponsor should ensure that the investigational product(s) (including active comparator(s) and placebo, if applicable) is characterized as appropriate to the stage of development of the product(s), is manufactured in accordance with any applicable GMP, and is coded and labeled in a manner that protects the blinding, if applicable. In addition, the labeling should comply with applicable regulatory requirement(s).
2. The sponsor should determine, for the investigational product(s), acceptable storage temperatures, storage conditions (e.g., protection from light), storage times, reconstitution fluids and procedures, and devices for product infusion, if any. The sponsor should inform all involved parties (e.g., monitors, investigators, pharmacists, storage managers) of these determinations.
3. The investigational product(s) should be packaged to prevent contamination and unacceptable deterioration during transport and storage.
4. In blinded trials, the coding system for the investigational product(s) should include a mechanism that permits rapid identification of the product(s) in case of a medical emergency, but does not permit undetectable breaks of the blinding.

5. If significant formulation changes are made in the investigational or comparator product(s) during the course of clinical development, the results of any additional studies of the formulated product(s) (e.g., stability, dissolution rate, bioavailability) needed to assess whether these changes would significantly alter the pharmacokinetic profile of the product should be available prior to the use of the new formulation in clinical trials.

Supplying and Handling Investigational Product(s)

1. The sponsor is responsible for supplying the investigator(s)/institution(s) with the investigational product(s).
2. The sponsor should not supply an investigator/institution with the investigational product(s) until the sponsor obtains all required documentation (e.g., approval/favorable opinion from IRB/IEC and regulatory authority[ies]).
3. The sponsor should ensure that written procedures include instructions that the investigator/institution should follow for the handling and storage of investigational product(s) for the trial and documentation thereof. The procedures should address adequate and safe receipt, handling, storage, dispensing, retrieval of unused product from subjects, and return of unused investigational product(s) to the sponsor (or alternative disposition if authorized by the sponsor and in compliance with the applicable regulatory requirement[s]).
4. The sponsor should
   (a) Ensure timely delivery of investigational product(s) to the investigator(s)
   (b) Maintain records that document shipment, receipt, disposition, return, and destruction of the investigational product(s) (see section "Essential Documents for the Conduct of a Clinical Trial")
   (c) Maintain a system for retrieving investigational products and documenting this retrieval (e.g., for deficient product recall, reclaim after trial completion, expired product reclaim)
   (d) Maintain a system for the disposition of unused investigational product(s) and for the documentation of this disposition
5. The sponsor should consider the following:
   (a) Take steps to ensure that the investigational product(s) are stable over the period of use.
   (b) Maintain sufficient quantities of the investigational product(s) used in the trials to reconfirm specifications, should this become necessary, and maintain records of batch sample analyses and characteristics. To the extent stability permits, samples should be retained either until the analyses of the trial data are complete or as required by the applicable regulatory requirement(s), whichever represents the longer retention period.

Record Access

1. The sponsor should ensure that it is specified in the protocol or other written agreement that the investigator(s)/institution(s) provide direct access to source data/documents for trial-related monitoring, audits, IRB/IEC review, and regulatory inspection.
2. The sponsor should verify that each subject has consented, in writing, to direct access to his or her original medical records for trial-related monitoring, audit, IRB/IEC review, and regulatory inspection.

Safety Information

1. The sponsor is responsible for the ongoing safety evaluation of the investigational product(s).

2. The sponsor should promptly notify all concerned investigator(s)/institution(s) and the regulatory authority(ies) of findings that could affect adversely the safety of subjects, impact the conduct of the trial, or alter the IRB/IEC's approval/favorable opinion to continue the trial.

Adverse Drug Reaction Reporting

1. The sponsor should expedite the reporting to all concerned investigator(s)/institutions(s), to the IRB(s)/IEC(s), where required, and to the regulatory authority(ies) of all ADRs that are both serious and unexpected.
2. Such expedited reports should comply with the applicable regulatory requirement(s) and with the ICH Guideline for Clinical Safety Data Management: Definitions and Standards for Expedited Reporting.
3. The sponsor should submit to the regulatory authority(ies) all safety updates and periodic reports, as required by applicable regulatory requirement(s).

Monitoring

1. *Purpose*
   The purposes of trial monitoring are to verify that
   (a) The rights and well-being of human subjects are protected
   (b) The reported trial data are accurate, complete, and verifiable from source documents
   (c) The conduct of the trial is in compliance with the currently approved protocol/amendment(s), with GCP, and with the applicable regulatory requirement(s)
2. *Selection and Qualifications of Monitors*
   (a) Monitors should be appointed by the sponsor.
   (b) Monitors should be appropriately trained and should have the scientific and/or clinical knowledge needed to monitor the trial adequately. A monitor's qualifications should be documented.
   (c) Monitors should be thoroughly familiar with the investigational product(s), the protocol, written informed consent form and any other written information to be provided to subjects, the sponsor's SOPs, GCP, and the applicable regulatory requirement(s).
3. *Extent and Nature of Monitoring*
   The sponsor should ensure that the trials are adequately monitored. The sponsor should determine the appropriate extent and nature of monitoring. The determination of the extent and nature of monitoring should be based on considerations such as the objective, purpose, design, complexity, blinding, size, and endpoints of the trial. In general, there is a need for on-site monitoring, before, during, and after the trial; however, in exceptional circumstances, the sponsor may determine that central monitoring in conjunction with procedures such as investigators' training and meetings and extensive written guidance can assure appropriate conduct of the trial in accordance with GCP. Statistically controlled sampling may be an acceptable method for selecting the data to be verified.
4. *Monitor's Responsibilities*
   The monitor(s) in accordance with the sponsor's requirements should ensure that the trial is conducted and documented properly by carrying out the following activities when relevant and necessary to the trial and the trial site:
   (a) Acting as the main line of communication between the sponsor and the investigator.
   (b) Verifying that the investigator has adequate qualifications and resources (see sections "Investigator's Qualifications and Agreements", "Adequate Resources", and "Investigator Selection") and remain adequate throughout the trial period and that facilities, including laboratories, equipment, and staff, are adequate to safely and properly conduct the trial and remain adequate throughout the trial period.

(c) Verifying, for the investigational product(s), (i) that storage times and conditions are acceptable and that supplies are sufficient throughout the trial; (ii) that the investigational product(s) are supplied only to subjects who are eligible to receive it and at the protocol specified dose(s); (iii) that subjects are provided with necessary instruction on properly using, handling, storing, and returning the investigational product(s); (iv) that the receipt, use, and return of the investigational product(s) at the trial sites are controlled and documented adequately; and (v) that the disposition of unused investigational product(s) at the trial sites complies with applicable regulatory requirement(s) and is in accordance with the sponsor.
(d) Verifying that the investigator follows the approved protocol and all approved amendment(s), if any.
(e) Verifying that written informed consent was obtained before each subject's participation in the trial.
(f) Ensuring that the investigator receives the current IB, all documents, and all trial supplies needed to conduct the trial properly and to comply with the applicable regulatory requirement(s).
(g) Ensuring that the investigator and the investigator's trial staff are adequately informed about the trial.
(h) Verifying that the investigator and the investigator's trial staff are performing the specified trial functions, in accordance with the protocol and any other written agreement between the sponsor and the investigator/institution, and have not delegated these functions to unauthorized individuals.
(i) Verifying that the investigator is enrolling only eligible subjects.
(j) Reporting the subject recruitment rate.
(k) Verifying that source documents and other trial records are accurate, complete, kept up-to-date, and maintained.
(l) Verifying that the investigator provides all the required reports, notifications, applications, and submissions and that these documents are accurate, complete, timely, legible, and dated and identify the trial.
(m) Checking the accuracy and completeness of the CRF entries, source documents, and other trial-related records against each other. The monitor specifically should verify the following:
    (i) The data required by the protocol are reported accurately on the CRFs and are consistent with the source documents. (ii) Any dose and/or therapy modifications are well documented for each of the trial subjects. (iii) AEs, concomitant medications, and intercurrent illnesses are reported in accordance with the protocol on the CRFs. (iv) Visits that the subjects fail to make, tests that are not conducted, and examinations that are not performed are clearly reported as such on the CRFs. (v) All withdrawals and dropouts of enrolled subjects from the trial are reported and explained on the CRFs.
(n) Informing the investigator of any CRF entry error, omission, or illegibility. The monitor should ensure that appropriate corrections, additions, or deletions are made, dated, explained (if necessary), and initialed by the investigator or by a member of the investigator's trial staff who is authorized to initial CRF changes for the investigator. This authorization should be documented.
(o) Determining whether all AEs are appropriately reported within the time periods required by GCP, the protocol, the IRB/IEC, the sponsor, and the applicable regulatory requirement(s).

(p) Determining whether the investigator is maintaining the essential documents (see section "Essential Documents for the Conduct of a Clinical Trial").
(q) Communicating deviations from the protocol, SOPs, GCP, and the applicable regulatory requirements to the investigator and taking appropriate action designed to prevent recurrence of the detected deviations.

5. *Monitoring Procedures*
The monitor(s) should follow the sponsor's established written SOPs as well as those procedures that are specified by the sponsor for monitoring a specific trial.
6. *Monitoring Report*
(a) The monitor should submit a written report to the sponsor after each trial-site visit or trial-related communication.
(b) Reports should include the date, site, name of the monitor, and name of the investigator or other individual(s) contacted.
(c) Reports should include a summary of what the monitor reviewed and the monitor's statements concerning the significant findings/facts, deviations and deficiencies, conclusions, actions taken or to be taken, and/or actions recommended to secure compliance.
(d) The review and follow-up of the monitoring report with the sponsor should be documented by the sponsor's designated representative.

Audit

If or when sponsors perform audits, as part of implementing quality assurance, they should consider the following:

1. *Purpose*
The purpose of a sponsor's audit, which is independent of and separate from routine monitoring or quality control functions, should be to evaluate trial conduct and compliance with the protocol, SOPs, GCP, and the applicable regulatory requirements.
2. *Selection and Qualification of Auditors*
(a) The sponsor should appoint individuals, who are independent of the clinical trials/systems, to conduct audits.
(b) The sponsor should ensure that the auditors are qualified by training and experience to conduct audits properly. An auditor's qualifications should be documented.
3. *Auditing Procedures*
(a) The sponsor should ensure that the auditing of clinical trials/systems is conducted in accordance with the sponsor's written procedures on what to audit, how to audit, the frequency of audits, and the form and content of audit reports.
(b) The sponsor's audit plan and procedures for a trial audit should be guided by the importance of the trial to submissions to regulatory authorities, the number of subjects in the trial, the type and complexity of the trial, the level of risks to the trial subjects, and any identified problem(s).
(c) The observations and findings of the auditor(s) should be documented.
(d) To preserve the independence and value of the audit function, the regulatory authority(ies) should not routinely request the audit reports. Regulatory authority(ies) may seek access to an audit report on a case-by-case basis when evidence of serious GCP noncompliance exists or in the course of legal proceedings.
(e) When required by applicable law or regulation, the sponsor should provide an audit certificate.

Noncompliance

1. Noncompliance with the protocol, SOPs, GCP, and/or applicable regulatory requirement(s) by an investigator/institution or by member(s) of the sponsor's staff should lead to prompt action by the sponsor to secure compliance.
2. If the monitoring and/or auditing identifies serious and/or persistent noncompliance on the part of an investigator/institution, the sponsor should terminate the investigator's/institution's participation in the trial. When an investigator's/institution's participation is terminated because of noncompliance, the sponsor should notify promptly the regulatory authority(ies).

Premature Termination or Suspension of a Trial

If a trial is prematurely terminated or suspended, the sponsor should promptly inform the investigators/institutions and the regulatory authority(ies) of the termination or suspension and the reason(s) for the termination or suspension. The IRB/IEC should also be informed promptly and provided the reason(s) for the termination or suspension by the sponsor or by the investigator/institution, as specified by the applicable regulatory requirement(s).

Clinical Trial/Study Reports

Whether the trial is completed or prematurely terminated, the sponsor should ensure that the clinical trial reports are prepared and provided to the regulatory agency(ies) as required by the applicable regulatory requirement(s). The sponsor should also ensure that the clinical trial reports in marketing applications meet the standards of the ICH Guideline for Structure and Content of Clinical Study Reports. (*Note*: The ICH Guideline for Structure and Content of Clinical Study Reports specifies that abbreviated study reports may be acceptable in certain cases.)

Multicenter Trials

For multicenter trials, the sponsor should ensure the following:

1. All investigators conduct the trial in strict compliance with the protocol agreed to by the sponsor and, if required, by the regulatory authority(ies) and given approval/favorable opinion by the IRB/IEC.
2. The CRFs are designed to capture the required data at all multicenter trial sites. For those investigators who are collecting additional data, supplemental CRFs should also be provided that are designed to capture the additional data.
3. The responsibilities of coordinating investigator(s) and the other participating investigators are documented prior to the start of the trial.
4. All investigators are given instructions on following the protocol, on complying with a uniform set of standards for the assessment of clinical and laboratory findings, and on completing the CRFs.
5. Communication between investigators is facilitated.

CLINICAL TRIAL PROTOCOL AND PROTOCOL AMENDMENT(S)

The contents of a trial protocol should generally include the following topics. However, site-specific information may be provided on separate protocol page(s) or addressed in a separate agreement, and some of the information listed as follows may be contained in other protocol referenced documents, such as an IB.

General Information

1. Protocol title, protocol identifying number, and date. Any amendment(s) should also bear the amendment number(s) and date(s).
2. Name and address of the sponsor and monitor (if other than the sponsor).
3. Name and title of the person(s) authorized to sign the protocol and the protocol amendment(s) for the sponsor.
4. Name, title, address, and telephone number(s) of the sponsor's medical expert (or dentist when appropriate) for the trial.
5. Name and title of the investigator(s) who is (are) responsible for conducting the trial and the address and telephone number(s) of the trial site(s).
6. Name, title, address, and telephone number(s) of the qualified physician (or dentist, if applicable), who is responsible for all trial-site-related medical (or dental) decisions (if other than investigator).
7. Name(s) and address(es) of the clinical laboratory(ies) and other medical and/or technical department(s) and/or institutions involved in the trial.

Background Information

1. Name and description of the investigational product(s)
2. A summary of findings from nonclinical studies that potentially have clinical significance and from clinical trials that are relevant to the trial
3. Summary of the known and potential risks and benefits, if any, to human subjects
4. Description of and justification for the route of administration, dosage, dosage regimen, and treatment period(s)
5. A statement that the trial will be conducted in compliance with the protocol, GCP, and the applicable regulatory requirement(s)
6. Description of the population to be studied
7. References to literature and data that are relevant to the trial and that provide background for the trial

Trial Objectives and Purpose

A detailed description of the objectives and the purpose of the trial.

Trial Design

The scientific integrity of the trial and the credibility of the data from the trial depend substantially on the trial design. A description of the trial design should include the following:

1. A specific statement of the primary endpoints and the secondary endpoints, if any, to be measured during the trial.
2. A description of the type/design of trial to be conducted (e.g., double-blind, placebo-controlled, parallel design) and a schematic diagram of trial design, procedures, and stages.
3. A description of the measures taken to minimize/avoid bias, including (a) randomization and (b) blinding.
4. A description of the trial treatment(s) and the dosage and dosage regimen of the investigational product(s). Also include a description of the dosage form, packaging, and labeling of the investigational product(s).
5. The expected duration of subject participation and a description of the sequence and duration of all trial periods, including follow-up, if any.

6. A description of the *stopping rules* or *discontinuation criteria* for individual subjects, parts of trial, and entire trial.
7. Accountability procedures for the investigational product(s), including the placebo(s) and comparator(s), if any.
8. Maintenance of trial treatment randomization codes and procedures for breaking codes.
9. The identification of any data to be recorded directly on the CRFs (i.e., no prior written or electronic record of data) and to be considered to be source data.

Selection and Withdrawal of Subjects

1. Subject inclusion criteria
2. Subject exclusion criteria
3. Subject withdrawal criteria (i.e., terminating investigational product treatment/trial treatment) and procedures specifying (a) when and how to withdraw subjects from the trial/investigational product treatment, (b) the type and timing of the data to be collected for withdrawn subjects, (c) whether and how subjects are to be replaced, and (d) the follow-up for subjects withdrawn from investigational product treatment/trial treatment

Treatment of Subjects

1. The treatment(s) to be administered, including the name(s) of all the product(s), the dose(s), the dosing schedule(s), the route/mode(s) of administration, and the treatment period(s), including the follow-up period(s) for subjects for each investigational product treatment/trial treatment group/arm of the trial
2. Medication(s)/treatment(s) permitted (including rescue medication) and not permitted before and/or during the trial
3. Procedures for monitoring subject compliance

Assessment of Efficacy

1. Specification of the efficacy parameters
2. Methods and timing for assessing, recording, and analyzing of efficacy parameters

Assessment of Safety

1. Specification of safety parameters
2. The methods and timing for assessing, recording, and analyzing safety parameters
3. Procedures for eliciting reports of and for recording and reporting AE and intercurrent illnesses
4. The type and duration of the follow-up of subjects after AEs

Statistics

1. A description of the statistical methods to be employed, including timing of any planned interim analysis(ses).
2. The number of subjects planned to be enrolled. In multicenter trials, the numbers of enrolled subjects projected for each trial site should be specified. Reason for choice of sample size includes reflections on (or calculations of) the power of the trial and clinical justification.
3. The level of significance to be used.

4. Criteria for the termination of the trial.
5. Procedure for accounting for missing, unused, and spurious data.
6. Procedures for reporting any deviation(s) from the original statistical plan (any deviation(s) from the original statistical plan should be described and justified in protocol and/or in the final report, as appropriate).
7. The selection of subjects to be included in the analyses (e.g., all randomized subjects, all dosed subjects, all eligible subjects, evaluable subjects).

Direct Access to Source Data/Documents

The sponsor should ensure that it is specified in the protocol or other written agreement that the investigator(s)/institution(s) will permit trial-related monitoring, audits, IRB/IEC review, and regulatory inspection(s), providing direct access to source data/documents.

Quality Control and Quality Assurance

Protocols and SOPs.

Ethics

Description of ethical considerations relating to the trial.

Data Handling and Record Keeping

Financing and Insurance

Financing and insurance if not addressed in a separate agreement.

Publication Policy

Publication policy, if not addressed in a separate agreement.

Supplements

(*Note*: Since the protocol and the clinical trial/study report are closely related, further relevant information can be found in the ICH Guideline for Structure and Content of Clinical Study Reports.)

INVESTIGATOR'S BROCHURE

Introduction

The IB is a compilation of the clinical and nonclinical data on the investigational product(s) that are relevant to the study of the product(s) in human subjects. Its purpose is to provide the investigators and others involved in the trial with the information to facilitate their understanding of the rationale for, and their compliance with, many key features of the protocol, such as the dose, dose frequency/interval, methods of administration, and safety monitoring procedures. The IB also provides insight to support the clinical management of the study subjects during the course of the clinical trial. The information should be presented in a concise, simple, objective, balanced, and nonpromotional form that enables a clinician, or potential investigator, to understand it and make his or her own unbiased risk–benefit assessment of the appropriateness of the proposed trial. For this reason, a medically qualified person should generally participate in the editing of an IB, but the contents of the IB should be approved by the disciplines that generated the described data.

This guideline delineates the minimum information that should be included in an IB and provides suggestions for its layout. It is expected that the type and extent of information available will vary with the stage of development of the investigational product. If the investigational product is marketed and its pharmacology is widely understood by medical practitioners, an extensive IB may not be necessary. Where permitted by regulatory authorities, a basic product information brochure, package leaflet, or labeling may be an appropriate alternative, provided that it includes current, comprehensive, and detailed information on all aspects of the investigational product that might be of importance to the investigator. If a marketed product is being studied for a new use (i.e., a new indication), an IB specific to that new use should be prepared. The IB should be reviewed at least annually and revised as necessary in compliance with a sponsor's written procedures. More frequent revision may be appropriate depending on the stage of development and the generation of relevant new information. However, in accordance with GCP, relevant new information may be so important that it should be communicated to the investigators and possibly to the IRBs/IECs and/or regulatory authorities before it is included in a revised IB.

Generally, the sponsor is responsible for ensuring that an up-to-date IB is made available to the investigator(s) and the investigators are responsible for providing the up-to-date IB to the responsible IRBs/IECs. In the case of an investigator sponsored trial, the sponsor–investigator should determine whether a brochure is available from the commercial manufacturer. If the investigational product is provided by the sponsor–investigator, then he or she should provide the necessary information to the trial personnel. In cases where preparation of a formal IB is impractical, the sponsor–investigator should provide, as a substitute, an expanded background information section in the trial protocol that contains the minimum current information described in this guideline.

General Considerations

The IB should include the following:

1. *Title Page*
   This should provide the sponsor's name, the identity of each investigational product (i.e., research number, chemical or approved generic name, and trade name(s) where legally permissible and desired by the sponsor), and the release date. It is also suggested that an edition number, and a reference to the number and date of the edition it supersedes, be provided. An example is given in Appendix 21A.
2. *Confidentiality Statement*
   The sponsor may wish to include a statement instructing the investigator/recipients to treat the IB as a confidential document for the sole information and use of the investigator's team and the IRB/IEC.

Contents of the Investigator's Brochure

The IB should contain the following sections, each with literature references where appropriate:

1. *Table of Contents*
   An example of the Table of Contents is given in Appendix 21B.
2. *Summary*
   A brief summary (preferably not exceeding two pages) should be given, highlighting the significant physical, chemical, pharmaceutical, pharmacological, toxicological, pharmacokinetic, metabolic, and clinical information available that is relevant to the stage of clinical development of the investigational product.

3. *Introduction*
   A brief introductory statement should be provided that contains the chemical name (and generic and trade name[s] when approved) of the investigational product(s), all active ingredients, the investigational product's/products' pharmacological class and its expected position within this class (e.g., advantages), the rationale for performing research with the investigational product(s), and the anticipated prophylactic, therapeutic, or diagnostic indication(s). Finally, the introductory statement should provide the general approach to be followed in evaluating the investigational product.
4. *Physical, Chemical, and Pharmaceutical Properties and Formulation*
   A description should be provided of the investigational product substance(s) (including the chemical and/or structural formula[e]), and a brief summary should be given of the relevant physical, chemical, and pharmaceutical properties.

   To permit appropriate safety measures to be taken in the course of the trial, a description of the formulation(s) to be used, including excipients, should be provided and justified if clinically relevant. Instructions for the storage and handling of the dosage form(s) should also be given.

   Any structural similarities to other known compounds should be mentioned.
5. *Nonclinical Studies*

   Introduction

   The results of all relevant nonclinical pharmacology, toxicology, pharmacokinetic, and investigational product metabolism studies should be provided in summary form. This summary should address the methodology used, the results, and a discussion of the relevance of the findings to the investigated therapeutic and the possible unfavorable and unintended effects in humans.

   The information provided may include the following, as appropriate, if known/available:

   - Species tested
   - Number and sex of animals in each group
   - Unit dose (e.g., milligram/kilogram [mg/kg])
   - Dose interval
   - Route of administration
   - Duration of dosing
   - Information on systemic distribution
   - Duration of postexposure follow-up
   - Results, including the following aspects:
     - Nature and frequency of pharmacological or toxic effects
     - Severity or intensity of pharmacological or toxic effects
     - Time to onset of effects
     - Reversibility of effects
     - Duration of effects
     - Dose response

Tabular format/listings should be used whenever possible to enhance the clarity of the presentation.

The following sections should discuss the most important findings from the studies, including the dose response of observed effects, the relevance to humans, and any aspects to be studied in humans. If applicable, the effective and nontoxic dose findings in the same animal species should be compared (i.e., the therapeutic index should be discussed). The relevance of this information to the proposed human dosing should be addressed. Whenever possible, comparisons should be made in terms of blood/tissue levels rather than on a mg/kg basis.

(a) *Nonclinical Pharmacology*
A summary of the pharmacological aspects of the investigational product and, where appropriate, its significant metabolites studied in animals should be included. Such a summary should incorporate studies that assess potential therapeutic activity (e.g., efficacy models, receptor binding, and specificity) as well as those that assess safety (e.g., special studies to assess pharmacological actions other than the intended therapeutic effect[s]).

(b) *Pharmacokinetics and Product Metabolism in Animals*
A summary of the pharmacokinetics and biological transformation and disposition of the investigational product in all species studied should be given. The discussion of the findings should address the absorption and the local and systemic bioavailability of the investigational product and its metabolites and their relationship to the pharmacological and toxicological findings in animal species.

(c) *Toxicology*
A summary of the toxicological effects found in relevant studies conducted in different animal species should be described under the following headings where appropriate:
- Single dose
- Repeated dose
- Carcinogenicity
- Special studies (e.g., irritancy and sensitization)
- Reproductive toxicity
- Genotoxicity (mutagenicity)

6. *Effects in Humans*

Introduction
A thorough discussion of the known effects of the investigational product(s) in humans should be provided, including information on pharmacokinetics, metabolism, pharmacodynamics, dose response, safety, efficacy, and other pharmacological activities. Where possible, a summary of each completed clinical trial should be provided. Information should also be provided regarding results of any use of the investigational product(s) other than from in clinical trials, such as from experience during marketing.

(a) *Pharmacokinetics and Product Metabolism in Humans*
A summary of information on the pharmacokinetics of the investigational product(s) should be presented, including the following, if available:
- Pharmacokinetics (including metabolism, as appropriate, and absorption, plasma protein binding, distribution, and elimination)
- Bioavailability of the investigational product (absolute, where possible, and/or relative) using a reference dosage form
- Population subgroups (e.g., gender, age, and impaired organ function)
- Interactions (e.g., product–product interactions and effects of food)
- Other pharmacokinetic data (e.g., results of population studies performed within clinical trial(s)

(b) *Safety and Efficacy*
A summary of information should be provided about the investigational product's/products' (including metabolites, where appropriate) safety, pharmacodynamics, efficacy, and dose response that were obtained from preceding trials in humans (healthy volunteers and/or patients). The implications of this information should be discussed. In cases where a number of clinical trials have been completed, the use of summaries of safety and efficacy across multiple trials by indications in subgroups may provide

a clear presentation of the data. Tabular summaries of ADRs for all the clinical trials (including those for all the studied indications) would be useful. Important differences in ADR patterns/incidences across indications or subgroups should be discussed.

The IB should provide a description of the possible risks and ADRs to be anticipated on the basis of prior experiences with the product under investigation and with related products. A description should also be provided of the precautions or special monitoring to be done as part of the investigational use of the product(s).

(c) *Marketing Experience*
The IB should identify countries where the investigational product has been marketed or approved. Any significant information arising from the marketed use should be summarized (e.g., formulations, dosages, routes of administration, and adverse product reactions). The IB should also identify all the countries where the investigational product did not receive approval/registration for marketing or was withdrawn from marketing/registration.

7. *Summary of Data and Guidance for the Investigator*
This section should provide an overall discussion of the nonclinical and clinical data and should summarize the information from various sources on different aspects of the investigational product(s), wherever possible. In this way, the investigator can be provided with the most informative interpretation of the available data and with an assessment of the implications of the information for future clinical trials. Where appropriate, the published reports on related products should be discussed. This could help the investigator to anticipate ADRs or other problems in clinical trials. The overall aim of this section is to provide the investigator with a clear understanding of the possible risks and adverse reactions and of the specific tests, observations, and precautions that may be needed for a clinical trial. This understanding should be based on the available physical, chemical, pharmaceutical, pharmacological, toxicological, and clinical information on the investigational product(s). Guidance should also be provided to the clinical investigator on the recognition and treatment of possible overdose and ADRs that is based on previous human experience and on the pharmacology of the investigational product.

APPENDIX 21A

Title page (Example)
Sponsor's name
Product:
Research Number:
Name(s): Chemical, Generic (if approved)
Trade Name(s) (if legally permissible and desired by the sponsor)

Investigator's Brochure
Edition Number: Release Date:
Replaces Previous Edition Number: Date:

APPENDIX 21B

Table of Contents of Investigator's Brochure (Example)
- Confidentiality Statement (optional)
- Signature Page (optional)
1. Table of Contents
2. Summary

3. Introduction
4. Physical, Chemical, and Pharmaceutical Properties and Formulation
5. Nonclinical Studies
5.1 Nonclinical Pharmacology
5.2 Pharmacokinetics and Product Metabolism in Animals
5.3 Toxicology
6. Effects in Humans
6.1 Pharmacokinetics and Product Metabolism in Humans
6.2 Safety and Efficacy
6.3 Marketing Experience
7. Summary of Data and Guidance for the Investigator

NB: References on

1. Publications
2. Reports These references should be found at the end of each chapter Appendices (if any)

ESSENTIAL DOCUMENTS FOR THE CONDUCT OF A CLINICAL TRIAL

Introduction

Essential documents are those documents that individually and collectively permit evaluation of the conduct of a trial and the quality of the data produced. These documents serve to demonstrate the compliance of the investigator, sponsor, and monitor with the standards of good clinical practice (GCP) and with all applicable regulatory requirements.

Essential documents also serve a number of other important purposes. Filing essential documents at the investigator/institution and sponsor sites in a timely manner can greatly assist in the successful management of a trial by the investigator, sponsor, and monitor. These documents are also the ones that are usually audited by the sponsor's independent audit function and inspected by the regulatory authority(ies) as part of the process to confirm the validity of the trial conduct and the integrity of data collected.

The minimum list of essential documents that has been developed follows. The various documents are grouped in three sections according to the stage of the trial during which they will normally be generated: (1) before the clinical phase of the trial commences, (2) during the clinical conduct of the trial, and (3) after completion or termination of the trial. A description is given of the purpose of each document, and whether it should be filed in either the investigator/institution or sponsor files, or both. It is acceptable to combine some of the documents, provided the individual elements are readily identifiable.

Trial master files should be established at the beginning of the trial, both at the investigator/institution's site and at the sponsor's office. A final closeout of a trial can only be done when the monitor has reviewed both investigator/institution and sponsor files and confirmed that all necessary documents are in the appropriate files.

Any or all of the documents addressed in this guideline may be subject to, and should be available for, audit by the sponsor's auditor and inspection by the regulatory authority(ies).

Before the Clinical Phase of the Trial Commences

During this planning stage, the following documents should be generated and should be on file before the trial formally starts:

1. Investigator's Brochure (IB)
2. Signed protocol and amendments, if any, and sample case report form (CRF)

3. Information given to trial subject
    - Informed consent form (including all applicable translations)
    - Any other written information
    - Advertisement for subject recruitment (if used)
4. Financial aspects of the trial
5. Insurance statement (where required)
6. Signed agreement between involved parties, for example,
    - Investigator/institution and sponsor
    - Investigator/institution and contract research organization (CRO)
    - Sponsor and CRO
    - Investigator/institution and authority(ies) (where required)
7. Dated, documented approval/favorable opinion of IRB/IEC of the following:
    - Protocol and any amendments
    - CRF (if applicable)
    - Informed consent form(s)
    - Any other written information to be provided to the subject(s)
    - Advertisement for subject recruitment (if used)
    - Subject compensation (if any)
    - Any other documents given approval/favorable opinion
8. IRB/IEC composition
9. Regulatory authority(ies) authorization/approval/notification of protocol (where required)
10. Curriculum vitae and/or other relevant documents evidencing qualifications of investigator(s) and subinvestigator(s)
11. Normal value(s)/range(s) for medical/laboratory/technical procedure(s) and/or test(s) included in the protocol
12. Medical/laboratory/technical procedures/tests
    - Certification
    - Accreditation
    - Established quality control and/or external quality assessment or other validation (where required)
13. Sample of label(s) attached to investigational product container(s)
14. Instructions for handling of investigational product(s) and trial-related materials (if not included in protocol or IB)
15. Shipping records for investigational product(s) and trial-related materials
16. Certificate(s) of analysis of investigational product(s) shipped
17. Decoding procedures for blinded trials
18. Master randomization list
19. Pretrial monitoring report to document that the site is suitable for the trial X
20. Trial initiation monitoring report

During the Clinical Conduct of the Trial

In addition to having on file the previous documents, the following should be added to the files during the trial as evidence that all new relevant information is documented as it becomes available:

1. IB updates
2. Any revision to
    - Protocol/amendment(s) and CRF
    - Informed consent form
    - Any other written information provided to subjects
    - Advertisement for subject recruitment (if used)

3. Dated, documented approval/favorable opinion of institutional review board (IRB)/independent ethics committee (IEC) of the following:
    - Protocol amendment(s)
    - Revision(s) of
        - Informed consent form
        - Any other written information to be provided to the subject
        - Advertisement for subject recruitment (if used)
        - Any other documents given approval/favorable opinion
        - Continuing review of trial (where required)
4. Regulatory authority(ies) authorizations/approvals/notifications where required for protocol amendment(s) and other documents
5. Curriculum vitae for new investigator(s) and/or subinvestigator(s)
6. Updates to normal value(s)/range(s) for medical/laboratory/technical procedure(s)/test(s) included in the protocol
7. Updates of medical/laboratory/technical procedures/tests
    - Certification
    - Accreditation
    - Established quality control and/or external
    - Quality assessment or other validation (where required)
8. Documentation of investigational product(s) and trial-related materials shipment
9. Certificate(s) of analysis for new batches of investigational products
10. Monitoring visit reports
11. Relevant communications other than site visits
    - Letters
    - Meeting notes
    - Notes of telephone calls
12. Signed informed consent forms
13. Source documents
14. Signed, dated, and completed CRF
15. Documentation of CRF corrections
16. Notification by originating investigator to sponsor of serious adverse events and related reports
17. Notification by sponsor and/or investigator, where applicable, to regulatory authority(ies) and IRB(s)/IEC(s) of unexpected serious adverse drug reactions and of other safety information
18. Notification by sponsor to investigators of safety information
19. Interim or annual reports to IRB/IEC and authority(ies)
20. Subject screening log
21. Subject identification code list
22. Subject enrollment log
23. Investigational products accountability at the site
24. Signature sheet
25. Record of retained body fluids/tissue samples (if any)

After Completion or Termination of the Trial

After completion or termination of the trial, all of the documents identified in the sections "Before the Clinical Phase of the Trial Commences" and "During the Clinical Conduct of the Trial" should be in the file together with the following:

1. Investigational product(s) accountability at site
2. Documentation of investigational product destruction

3. Completed subject identification code list
4. Audit certificate (if available)
5. Final trial closeout monitoring report
6. Treatment allocation and decoding documentation
7. Final report by investigator to IRB/IEC where required, and where applicable, to the regulatory authority(ies)
8. Clinical study report

22 Computer and Software Validation

BACKGROUND

Over the past 30 years, data collection, recording, analysis, reporting, and regulatory submissions have become greatly dependent on electronic computerized systems. Regulatory agencies worldwide have begun accepting submissions electronically even allowing these applications being signed off electronically. This change in the traditional paper trail system requires significant changes to data handling and greater emphasis on validating the regulatory submissions. In laboratories conducting bioequivalence studies, the following instances arise where validation of computer systems is required:

1. Record keeping systems including patient databases
2. Software controlling operation of analytic equipment
3. Software used to evaluate data statistics and store data

The current systems proposed to validate the use of computers and software have an interesting historical background that is important to review. Back in the 1970s, there was a reported error in matrix conversion because of numeric overflow; in the 1980s, the erroneous use of "n" instead of "$n - 1$" for degrees of freedom threw the automated analysis out; in the 1990s, credibility of a chip maker was questioned when it was shown that the division by 3 does not yield result that is 3× the value. All of these software bugs have prompted greater emphasis on the commercial-off-the-shelf (COTS) products, which are fully validated. There is also greater emphasis today on collaborative research resulting in such projects as Human Genome Project; cancer Biomedical Information Grid (caBIG™) (an open source, open access, voluntary information network); and the Gates Group requirement of data sharing for the $287 million funding in AIDS research; the conduct of these projects requires robust hardware and software systems across many platforms. Back in the 1960s, sponsors submitted FORTRAN code and Food and Drug Administration (FDA) reviewers poured over each line of code that became so onerous that the U.S. government funded the development of the statistical analytical system (SAS) software at the University of North Carolina. The regulatory requirement for *validation and verification* has a bias toward the COTS and the recent CDRF draft guidance on Bayesian mentions WinBUGS and CDRH has a LINUX cluster. The Bayesian inference using Gibbs sampling (BUGS) project is concerned with flexible software for the Bayesian analysis of complex statistical models using Markov chain Monte Carlo (MCMC) methods. The project began in 1989 in the MRC Biostatistics Unit and led initially to the "classic" BUGS program and then onto the WinBUGS software developed jointly with the Imperial College School of Medicine at St Mary's, London. The development now also includes the OpenBUGS project in the University of Helsinki, Finland. There are now a number of versions of BUGS, which can be confusing. WinBUGS 1.4.1 features a graphical user interface and online monitoring and convergence diagnostics. The OpenBUGS project is based at the University of Helsinki. Open source version of the core BUGS code with a variety of interfaces and running under Linux as LinBUGS. OpenBUGS is the main development platform and is currently experimental, but will eventually become the standard version. Just another Gibbs sampler (JAGS) by Martyn Plummer is an open source software and not really a version of BUGS: JAGS uses essentially the same model description language but it has been completely rewritten. Use of all of this software requires good understanding of Bayesian statistical principles. The available software can be classified into three categories; the open source software are programs distributed freely with source code and anyone can modify them and redistribute without any licensing; generally, these programs are technology neutral and include such examples as OpenBUGS and libraries; there are no regulations prohibiting the use of open source software. The General Public License software executables include noncommercial "freeware" or "shareware" and examples include the WinBUGS. Finally, there are custom-code and open source compilers such as SAS. The CRF title 21 section 11.10 Controls for closed systems have the following requirements:

a. Validation to ensure accuracy, reliability, consistent intended performance, and the ability to discern invalid or altered records
b. Accurate and complete copies of records in both human readable and electronic form suitable for inspection, review, and copying by the agency
c. Protection of records throughout the record retention period
d. Limiting system access to authorized individuals
e. Use of secure, computer-generated, time-stamped audit trails
f. Use of operational systems checks, authority checks, device checks
g. Education, training, and experience of operators and holding individuals accountable
h. Systems documentation

Software validation principles include the following:

a. Good software engineering to support final conclusion that the software is validated.
b. Approach based on the intended use and the safety risk associated with the software.
c. Software validation and verification conducted throughout the entire software life cycle.
d. Party with regulatory responsibility needs to establish that the software is validated for the intended use.
e. Software validation is a matter of developing a level of confidence.

The computerized systems that are used to create, modify, maintain, archive, retrieve, or transmit clinical data are required to be maintained and/or submitted to the FDA regarding the safety and effectiveness of new human and animal drugs, biological products, and medical devices, and certain food and color additives are subject to 21 CFR part 11 requirements for validation and integrity.

The FDA has the authority to inspect all records relating to clinical investigations, which include bioequivalence testing, regardless of how they were created or maintained. The FDA established the Bioresearch Monitoring (BIMO) program of inspections and audits to monitor the conduct and reporting of clinical trials to ensure that supporting data from these trials meet the highest standards of quality and integrity and conform to FDA's regulations. FDA's acceptance of data from clinical trials for decision-making purposes depends on FDA's ability to verify the quality and integrity of the data during FDA on-site inspections and audits. To be acceptable, the data should meet certain fundamental elements of quality whether collected or recorded electronically or on paper. For example, data should be attributable, legible, contemporaneous, original, and accurate. Persons using the data from computerized systems should have confidence that the data are no less reliable than data in paper form.

The procedures described in the following may be applicable to data or source documents that are created (1) in hard copy and later entered into a computerized system, (2) by direct entry by a human into a computerized system, and (3) automatically by a computerized system.

DATA HANDLING AND STORAGE PRINCIPLES

The following general principles with regard to computerized systems that are used to create, modify, maintain, archive, retrieve, or transmit clinical data required to be maintained and/or submitted to the FDA are recommended:

1. Each study protocol identifies at which steps a computerized system will be used to create, modify, maintain, archive, retrieve, or transmit data.
2. For each study, the documentation must identify what software and hardware are to be used in computerized systems that create, modify, maintain, archive, retrieve, or transmit data. This documentation should be retained as part of the study records.
3. The computerized systems be designed (1) so that all requirements assigned to these systems in a study protocol are satisfied (e.g., data are recorded in metric units, the study blinded) and (2) to preclude errors in data creation, modification, maintenance, archiving, retrieval, or transmission.
4. It is important to design a computerized system in such a manner so that all applicable regulatory requirements for record keeping and record retention in clinical trials are met with the same degree of confidence as is provided with paper systems.
5. The clinical investigator must retain records required to be maintained for a period of time specified in these regulations. Retaining the original source document or a certified copy of the source document at the site where the investigation was conducted can assist in meeting these regulatory requirements. It can also assist in the reconstruction and evaluation of the trial throughout and after the completion of the trial.
6. When original observations are entered directly into a computerized system, the electronic record is the source document.
7. Records relating to an investigation must be adequate and accurate in the case of investigational new drug applications (INDs), complete in the case of investigational new animal drug applications (INADs), and accurate, complete, and current in the case of investigational device exemptions (IDEs). An audit trail that is electronic or consists of other physical, logical, or procedural security measures to ensure that only authorized additions, deletions, or alterations of information in the electronic record have occurred may be needed to facilitate compliance with applicable records regulations.
8. It is recommended that data be retrievable in such a fashion that all information regarding each individual subject in a study is attributable to that subject.
9. To ensure the authenticity and integrity of electronic records, it is important that security measures be in place to prevent unauthorized access to the data in the electronic record and to the computerized system.

It is recommended that standard operating procedures (SOPs) pertinent to the use of the computerized system be available on site. It is recommended that SOPs be established for the following:

- System setup/installation
- Data collection and handling
- System maintenance
- Data backup, recovery, and contingency plans

- Security
- Change control
- Alternative recording methods (in the case of system unavailability)

Computer Access Controls

To ensure that individuals have the authority to proceed with data entry, data entry systems must be designed to limit access so that only authorized individuals are able to input data. Examples of methods for controlling access include using combined identification codes/passwords or biometric-based identification at the start of a data entry session. Controls and procedures must be in place that are designed to ensure the authenticity and integrity of electronic records created, modified, maintained, or transmitted using the data entry system. Therefore, it is recommended that each user of the system have an individual account into which the user logs in at the beginning of a data entry session, inputs information (including changes) on the electronic record, and logs out at the completion of data entry session.

It is recommended that individuals work only under their own password or other access key and not share these with others. It is recommended that individuals not be allowed to log onto the system to provide another person access to the system. It is recommended that passwords or other access keys be changed at established intervals.

When someone leaves a workstation, it is recommended that the SOP require that person to log off the system. Alternatively, an automatic log off may be appropriate for long idle periods. For short periods of inactivity, it is recommended that some kind of automatic protection be installed against unauthorized data entry. An example could be an automatic screen saver that prevents data entry until a password is entered.

Audit Trails or Other Security Measures

Persons who use electronic record systems to maintain an audit trail as one of the procedures to protect the authenticity, integrity, and, when appropriate, the confidentiality of electronic records. As clarified in the part 11, *Scope and Application* guidance, however, the FDA intends to exercise enforcement discretion regarding specific part 11 requirements related to computer-generated, time-stamped audit trails and any corresponding requirements. Persons must still comply with all applicable predicate rule requirements for clinical trials, for example, records related to the conduct of the study must be adequate and accurate. It is therefore important to keep track of all changes made to information in the electronic records that document activities related to the conduct of the trial. Computer-generated, time-stamped audit trails or information related to the creation, modification, or deletion of electronic records may be useful to ensure compliance with the appropriate predicate rule.

In addition, clinical investigators must, upon request by the FDA, at reasonable times, permit agency employees to have access to and copy and verify any required records or reports made by the investigator. In order for the FDA to review and copy this information, FDA personnel should be able to review audit trails or other documents that track electronic record activities both at the study site and at any other location where associated electronic study records are maintained. To enable FDA's review, information about the creation, modification, or deletion of electronic records should be created incrementally and in a chronological order. To facilitate FDA's inspection of this information, it is recommended that clinical investigators retain either the original or a certified copy of any documentation created to track electronic record activities.

Even if there are no applicable predicate rule requirements, it may be important to have computer-generated, time-stamped audit trails or other physical, logical, or procedural security measures to ensure the trustworthiness and reliability of electronic records. It is recommended that any decision on whether to apply computer-generated audit trails or other appropriate security measures be based on the need to comply with predicate rule requirements, a justified and documented risk assessment, and a determination of the potential effect on data quality and record integrity. Firms should determine and document the need for audit trails based on a risk assessment that takes into consideration circumstances surrounding system use, the likelihood that information might be compromised, and any system vulnerabilities.

If you determine that audit trails or other appropriate security measures are needed to ensure electronic record integrity, it is recommended that personnel who create, modify, or delete electronic records not be able to modify the documents or security measures used to track electronic record changes. It is recommended that audit trails or other security methods used to capture electronic record activities document who made the changes and when and why changes were made to the electronic record.

Some examples of methods for tracking changes to electronic records include as follows:

- Computer-generated, time-stamped electronic audit trails.
- Signed and dated printed versions of electronic records that identify what, when, and by whom changes were made to the electronic record. When using this method, it is important that appropriate controls be utilized that ensure the accuracy of these records (e.g., sight verification that the printed version accurately captures all of the changes made to the electronic record).
- Signed and dated printed standard electronic file formatted versions (e.g., pdf, XML, or SGML) of electronic records that identify what, when, and by whom changes were made to the electronic record.
- Procedural controls that preclude unauthorized personnel from creating, modifying, or deleting electronic records or the data contained therein.

Date/Time Stamps

It is recommended that controls be put in place to ensure that the system's date and time are correct. The ability to change the date or time should be limited to authorized personnel, and such personnel should be notified if a system date or time discrepancy is detected. It is recommended that someone always document changes to date or time. It is not expected that documentation of time changes that systems make automatically to adjust to daylight saving time conventions be made available.

It is also recommended that dates and times include the year, month, day, hour, and minute. The FDA encourages establishments to synchronize systems to the date and time provided by trusted third parties.

Clinical study computerized systems are likely be used in multicenter trials and may be located in different time zones. For systems that span different time zones, it is better to implement time stamps with a clear understanding of the time zone reference used. It is recommended that system documentation explain time zone references as well as zone acronyms or other naming conventions.

Systems Features

It is recommended that a number of computerized system features be available to facilitate the collection, inspection, review, and retrieval of quality clinical data. Key features are described here.

Systems Used for Direct Entry of Data

It is recommended that prompts, flags, or other help features be incorporated into the computerized system to encourage consistent use of clinical terminology and to alert the user to data that are out of acceptable range. It is recommended against the use of features that automatically enter data into a field when the field is bypassed.

Retrieval of Data and Record Retention

The FDA expects to be able to reconstruct a clinical study submitted to the agency. This means that documentation should fully describe and explain how data were obtained and managed and how electronic records were used to capture data. It is suggested that your decision on how to maintain records be based on predicate rule requirements and that this documented decision be based on a justified risk assessment and a determination of the value of the records over time. The FDA does not object to required records that are archived in electronic format; nonelectronic media such as microfilm, microfiche, and paper; or a standard electronic file format (such as pdf, XML, or SGML). Persons must still comply with all predicate rule requirements, and the records themselves and any copies of required records should preserve their original content and meaning. Paper and electronic record and signature components can coexist (i.e., as a hybrid system) as long as the predicate requirements are met, and the content and meaning of those records are preserved.

It is not necessary to reprocess data from a study that can be fully reconstructed from available documentation. Therefore, actual application software, operation systems, and software development tools involved in processing of data or records do not need to be retained.

System Security

In addition to internal safeguards built into the computerized system, external safeguards should be put in place to ensure that access to the computerized system and to the data is restricted to authorized personnel. It is recommended that staff be kept thoroughly aware of system security measures and the importance of limiting access to authorized personnel.

SOPs should be developed and implemented for handling and storing the system to prevent unauthorized access. Controlling system access can be accomplished through the following provisions:

- Operational system checks
- Authority checks
- Device (e.g., terminal) checks
- The establishment of and adherence to written policies that hold individuals accountable for actions initiated under their electronic signatures

It is recommended that access to data be restricted and monitored through the system's software with its required log-on, security procedures, and audit trail (or other selected security measures to track electronic record activities). It is recommended that procedures and controls be implemented to prevent the data from being altered, browsed, queried, or reported via external software applications that do not enter through the protective system software.

It is recommended that a cumulative record be available that indicates, for any point in time, the names of authorized personnel, their titles, and a description of their access privileges. It is recommended that the record be kept in the study documentation, accessible at the site.

If a sponsor supplies computerized systems exclusively for clinical trials, it is recommended that the systems remain dedicated to the purpose for which they were intended and validated. If a computerized system being used for a clinical study is part of a system normally used for other purposes, it is recommended that efforts be made to ensure that the study software be logically and physically isolated as necessary to preclude unintended interaction with nonstudy software. If any of the software programs are changed, it is recommended that the system be evaluated to determine the effect of the changes on logical security.

It is recommended that controls be implemented to prevent, detect, and mitigate effects of computer viruses, worms, or other potentially harmful software code on study data and software.

System Dependability

It is recommended that sponsors ensure and document that all computerized systems conform to their own established requirements for completeness, accuracy, reliability, and consistent intended performance.

It is recommended that systems documentation be readily available at the site where clinical trials are conducted and provide an overall description of the computerized systems and the relationships among hardware, software, and physical environment.

As noted in the part 11, *Scope and Application* guidance, the FDA intends to exercise enforcement discretion regarding specific part 11 requirements for validation of computerized systems. It is suggested that your decision to validate computerized systems and the extent of the validation take into account the impact the systems have on your ability to meet predicate rule requirements. You should also consider the impact those systems might have on the accuracy, reliability, integrity, availability, and authenticity of required records and signatures. Even if there is no predicate rule requirement to validate a system, it may still be important to validate the system, based on criticality and risk, to ensure the accuracy, reliability, integrity, availability, and authenticity of required records and signatures.

It is recommended that the sponsor base its approach on a justified and documented risk assessment and determination of the potential of the system to affect data quality and record integrity. For example, in the case where data are directly entered into electronic records and the business practice is to rely on the electronic record, validation of the computerized system is important. However, when a word processor is used to generate SOPs for use at the clinical site, validation would not be important.

If validation is required, the FDA may ask to see the regulated company's documentation that demonstrates software validation. The study sponsor is responsible for making any such documentation available if requested at the time of inspection at the site where software is used. Clinical investigators are not generally responsible for validation unless they originated or modified software.

Legacy Systems

As noted in the part 11, *Scope and Application* guidance, the FDA intends to exercise enforcement discretion with respect to all part 11 requirements for systems that otherwise were fully operational prior to August 20, 1997, the effective date of part 11, under the circumstances described in the following. These systems are also known as legacy systems. The FDA does not intend to take enforcement action to enforce compliance with any part 11 requirements if all the following criteria are met for a specific system:

- The system was in operation before the part 11 effective date.
- The system met all applicable predicate rule requirements prior to the part 11 effective date.
- The system currently meets all applicable predicate rule requirements.
- There is documented evidence and justification that the system is fit for its intended use.

If a system has changed since August 20, 1997, and if the changes would prevent the system from meeting predicate rule requirements, part 11 controls should be applied to part 11 records and signatures pursuant to the enforcement policy expressed in the part 11 guidance.

Off-the-Shelf Software

While the FDA has announced that it intends to exercise enforcement discretion regarding specific part 11 requirements for validation of computerized systems, persons must still comply with all predicate rule requirements for validation. It was suggested in the guidance for industry on part 11 that the impact of computerized systems on the accuracy, reliability, integrity, availability, and authenticity of required records and signatures be considered when you decide whether to validate, and noted that even absent a predicate rule requirement to validate a system, it might still be important to validate in some instances.

For most off-the-shelf (OTS) software, the design level validation will have already been done by the company that wrote the software. Given the importance of ensuring valid clinical trial data, the FDA suggests that the sponsor or contract research organization (CRO) have documentation (either original validation documents or on-site vendor audit documents) of this design level validation by the vendor and would itself have performed functional testing (e.g., by use of test data sets) and researched known software limitations, problems, and defect corrections. Detailed documentation of any additional validation efforts performed by the sponsor or CRO will preserve the findings of these efforts.

In the special case of database and spreadsheet software that is (1) purchased OTS, (2) designed for and widely used for general purposes, (3) unmodified, and (4) not being used for direct entry of data, the sponsor or CRO may not have documentation of design level validation. The FDA suggests that the sponsor or CRO perform functional testing (e.g., by use of test data sets) and research known software limitations, problems, and defect corrections.

In the case of OTS software, it is recommended that the following be available to the FDA on request:

- A written design specification that describes what the software is intended to do and how it is intended to do it
- A written test plan based on the design specification, including both structural and functional analyses
- Test results and an evaluation of how these results demonstrate that the predetermined design specification has been met

Change Control

The FDA recommends that written procedures be put in place to ensure that changes to the computerized system, such as software upgrades, including security and performance patches, equipment, or component replacement, or new instrumentation, will maintain the integrity of the data and the integrity of protocols. It is recommended that the effects of any changes to the system be evaluated and a decision made regarding whether, and if so, what level of validation activities related to those changes would be appropriate. It is recommended that validation be performed for those types of changes that exceed previously established operational limits or design specifications. Finally, it is recommended that all changes to the system be documented.

Systems Control

It is recommended that appropriate system control measures be developed and implemented.

- Software version control: It is recommended that measures be put in place to ensure that versions of software used to generate, collect, maintain, and transmit data are the versions that are stated in the systems documentation.
- Contingency plans: It is recommended that written procedures describe contingency plans for continuing the study by alternate means in the event of failure of the computerized system.
- Backup and recovery of electronic records: When electronic formats are the only ones used to create and preserve electronic records, it is recommended that backup and recovery procedures be outlined clearly in the SOPs and be sufficient to protect against data loss. It is recommended that records be backed up regularly in a way that would prevent a catastrophic loss and ensure the quality and integrity of the data. It is recommended that records be stored at a secure location specified in the SOPs. Storage is

typically off-site or in a building separate from the original records. It is recommended that backup and recovery logs be maintained to facilitate an assessment of the nature and scope of data loss resulting from a system failure. Firms that rely on electronic and paper systems should determine the extent to which backup and recovery procedures are needed based on the need to meet predicate rule requirements, a justified and documented risk assessment, and a determination of the potential effect on data quality and record integrity.

Training of Personnel

Firms using computerized systems must determine that persons who develop, maintain, or use electronic systems have the education, training, and experience to perform their assigned tasks. It is recommended that training be provided to individuals in the specific operations with regard to computerized systems that they are to perform. It is recommended that training be conducted by qualified individuals on a continuing basis, as needed, to ensure familiarity with the computerized system and with any changes to the system during the course of the study. It is further recommended that employee education, training, and experience be documented.

Copies of Records

The FDA has the authority to inspect all records relating to clinical investigations, regardless of how the records were created or maintained. Therefore, you should provide the FDA investigator with reasonable and useful access to records during an FDA inspection and supply copies of electronic records by

- Producing copies of records held in common portable formats when records are maintained in these formats
- Using established automated conversion or export methods, where available, to make copies available in a more common format (e.g., pdf, XML, or SGML formats)

Regardless of the method used to produce copies of electronic records, it is important that the copying process used produces copies that preserve the content and meaning of the record. For example, if you have the ability to search, sort, or trend records, copies given to the FDA should provide the same capability if it is reasonable and technically feasible. The FDA expects to inspect, review, and copy records in a human readable form at your site, using your hardware and following your established procedures and techniques for accessing records.

Electronic Signature Certification

Persons using electronic signatures to meet an FDA signature requirement must, prior to or at the time of such use, certify to the FDA that the electronic signatures in their system, used on or after August 20, 1997, are intended to be the legally binding equivalent of traditional handwritten signatures.

GENERAL PRINCIPLES OF SOFTWARE VALIDATION

Planning, verification, testing, traceability, configuration management, and many other aspects of good software engineering are important activities that together help to support a final conclusion that software is validated. It is recommended to integrate the software life cycle management and risk management activities. Based on the intended use and the safety risk associated with the software to be developed, the software developer should determine the specific approach, the combination of techniques to be used, and the level of effort to be applied. While this guidance does not recommend any specific life cycle model or any specific technique or method, it does recommend that software validation and verification activities be conducted throughout the entire software life cycle.

Since the software operating a device such as HPLC equipment or LC/MS/MS, data storage systems and analysis using statistical software is of utmost importance, it must be made parts of the QA function of the bioequivalence testing laboratory. It is unlikely that the user would develop any software on its own to perform the functions incumbent in the operation of a bioequivalence laboratory. Since the software is developed by someone other than the user (e.g., OTS software), the software developer may not be directly responsible for compliance with FDA regulations. In that case, the party with regulatory responsibility (i.e., the user) needs to assess the adequacy of the OTS software developer's activities and determine what additional efforts are needed to establish that the software is validated for the device manufacturer's intended use. The FDA believes in reducing burden and encourages firms to suggest an alternative approach that would be less burdensome.

Quality System Regulations

While the quality system regulation states that design input requirements must be documented and that specified requirements must be verified, the regulation does not further clarify the distinction between the terms "requirement" and "specification." A *requirement* can be any need or expectation for a system or for its software. Requirements reflect the stated or implied needs of the customer, and may be market-based, contractual, or statutory, as well as an organization's internal requirements. There can be many different kinds of requirements (e.g., design, functional, implementation, interface, performance, or physical requirements). Software requirements are typically derived from the system requirements for those aspects of system functionality that have been allocated to software. Software requirements are typically stated in functional terms and are defined, refined, and updated as a development project progresses. Success in accurately and completely documenting software requirements is a crucial factor in successful validation of the resulting software.

A *specification* is defined as "a document that states requirements." It may refer to or include drawings, patterns, or other relevant documents and usually indicates the means and the criteria whereby conformity with the requirement can be checked. There are many different kinds of written specifications, for example, system requirement specification, software requirement specification, software design specification, software test specification, and software integration specification. All of these documents establish "specified requirements" and are design outputs for which various forms of verification are necessary.

Verification and Validation

The quality system regulation is harmonized with *ISO 8402*:1994, which treats "verification" and "validation" as separate and distinct terms. On the other hand, many software engineering journal articles and textbooks use the terms "verification" and "validation" interchangeably or in some cases refer to software "verification, validation, and testing (VV&T)" as if it is a single concept, with no distinction among the three terms.

Software verification provides objective evidence that the design outputs of a particular phase of the software development life cycle meet all of the specified requirements for that phase. Software verification looks for consistency, completeness, and correctness of the software and its supporting documentation, as it is being developed, and provides support for a subsequent conclusion that software is validated. Software testing is one of many verification activities intended to confirm that software development output meets its input requirements. Other verification activities include various static and dynamic analyses, code and document inspections, walk-throughs, and other techniques.

Software validation is a part of the design validation for a finished device, but is not separately defined in the quality system regulation. For purposes of this guidance, the FDA considers software validation to be "confirmation by examination and provision of objective evidence that software specifications conform to user needs and intended uses, and that the particular requirements implemented through software can be consistently fulfilled." In practice, software validation activities may occur both during and at the end of the software development life cycle to ensure that all requirements have been fulfilled. Since software is usually part of a larger hardware system, the validation of software typically includes evidence that all software requirements have been implemented correctly and completely and are traceable to system requirements. A conclusion that software is validated is highly dependent upon comprehensive software testing, inspections, analyses, and other verification tasks performed at each stage of the software development life cycle.

Software verification and validation are difficult because a developer cannot test forever, and it is hard to know how much evidence is enough. In large measure, software validation is a matter of developing a "level of confidence" that the device meets all requirements and user expectations for the software automated functions and features of the device. Measures such as defects found in specification documents, estimates of defects remaining, testing coverage, and other techniques are all used to develop an acceptable level of confidence before shipping the product. The level of confidence, and therefore the level of software validation, verification, and testing effort needed, will vary depending upon the safety risk (hazard) posed by the automated functions of the device.

IQ/OQ/PQ

For many years, both the FDA and regulated industry have attempted to understand and define software validation within the context of process validation terminology. For example, industry documents and other FDA validation guidance sometimes describe user site software validation in terms of installation qualification (IQ), operational qualification (OQ), and performance qualification (PQ). While IQ/OQ/PQ terminology has served its purpose well and is one of many legitimate ways to organize software validation tasks at the user site, this terminology may not be well understood among many software professionals, and it is not used elsewhere in this document.

While software shares many of the same engineering tasks as hardware, it has some very important differences. For example,

- The vast majority of software problems are traceable to errors made during the design and development process. While the quality of a hardware product is highly dependent on design, development, and manufacture, the quality of a software product is dependent primarily on design and development with a minimum concern for software manufacture. Software manufacturing consists of reproduction that can be easily verified. It is not difficult to manufacture thousands of program copies that function exactly the same as the original; the difficulty comes in getting the original program to meet all specifications.
- One of the most significant features of software is branching, that is, the ability to execute alternative series of commands, based on differing inputs. This feature is a major contributing factor for another characteristic of software—its complexity. Even short programs can be very complex and difficult to fully understand.
- Typically, testing alone cannot fully verify that software is complete and correct. In addition to testing, other verification techniques and a structured and documented development process should be combined to ensure a comprehensive validation approach.
- Unlike hardware, software is not a physical entity and does not wear out. In fact, software may improve with age, as latent defects are discovered and removed. However, as software

is constantly updated and changed, such improvements are sometimes countered by new defects introduced into the software during the change.

- Unlike some hardware failures, software failures occur without advanced warning. The software's branching that allows it to follow differing paths during execution may hide some latent defects until long after a software product has been introduced into the marketplace.
- Another related characteristic of software is the speed and ease with which it can be changed. This factor can cause both software and nonsoftware professionals to believe that software problems can be corrected easily. Combined with a lack of understanding of software, it can lead managers to believe that tightly controlled engineering is not needed as much for software as it is for hardware. In fact, the opposite is true. Because of its complexity, the development process for software should be even more tightly controlled than for hardware, in order to prevent problems that cannot be easily detected later in the development process.
- Seemingly insignificant changes in software code can create unexpected and very significant problems elsewhere in the software program. The software development process should be sufficiently well planned, controlled, and documented to detect and correct unexpected results from software changes.
- Given the high demand for software professionals and the highly mobile workforce, the software personnel who make maintenance changes to software may not have been involved in the original software development. Therefore, accurate and thorough documentation is essential.
- Historically, software components have not been as frequently standardized and interchangeable as hardware components. However, medical device software developers are beginning to use component-based development tools and techniques. Object-oriented methodologies and the use of OTS software components hold promise for faster and less expensive software development. However, component-based approaches require very careful attention during integration. Prior to integration, time is needed to fully define and develop reusable software code and to fully understand the behavior of OTS components.

For these and other reasons, software engineering needs an even greater level of managerial scrutiny and control than does hardware engineering.

Software validation is a critical tool used to assure the quality of control and output; software validation can increase the usability and reliability and increased robustness of the data obtained. Software validation can also reduce long-term costs by making it easier and less costly to reliably modify software and revalidate software changes. Software maintenance can represent a very large percentage of the total cost of software over its entire life cycle. An established comprehensive software validation process helps to reduce the long-term cost of software by reducing the cost of validation for each subsequent release of the software.

Principles of Software Validation

Whereas it is unlikely that a BE laboratory personnel will be involved in the validation of software more than testing it at the IQ/OQ and PQ levels, it is important to understand how software is validated when requiring this certification from the vendors.

Software quality assurance needs to focus on preventing the introduction of defects into the software development process and not on trying to "test quality into" the software code after it is written. Software testing is very limited in its ability to surface all latent defects in software code. For example, the complexity of most software prevents it from being exhaustively tested.

Software testing is a necessary activity. However, in most cases, software testing by itself is not sufficient to establish confidence that the software is fit for its intended use. In order to establish that confidence, software developers should use a mixture of methods and techniques to prevent software errors and to detect software errors that do occur. The "best mix" of methods depends on many factors including the development environment, application, size of project, language, and risk.

To build a case that the software is validated requires time and effort. Preparation for software validation should begin early, that is, during design and development planning and design input. The final conclusion that the software is validated should be based on evidence collected from planned efforts conducted throughout the software life cycle.

Software validation takes place within the environment of an established software life cycle. The software life cycle contains software engineering tasks and documentation necessary to support the software validation effort. In addition, the software life cycle contains specific verification and validation tasks that are appropriate for the intended use of the software. This guidance does not recommend any particular life cycle models—only that they should be selected and used for a software development project.

The software validation process is defined and controlled through the use of a plan. The software validation plan defines "what" is to be accomplished through the software validation effort. Software validation plans are a significant quality system tool. Software validation plans specify areas such as scope, approach, resources, schedules, and the types and extent of activities, tasks, and work items.

The software validation process is executed through the use of procedures. These procedures establish "how" to conduct the software validation effort. The procedures should identify the specific actions or sequence of actions that must be taken to complete individual validation activities, tasks, and work items.

Due to the complexity of software, a seemingly small local change may have a significant global system impact. When any change (even a small change) is made to the software, the validation status of the software needs to be reestablished. Whenever software is changed, a validation analysis should be conducted not just for validation of the individual change but also to determine the extent and impact of that change on the entire software system. Based on this analysis, the software developer should then conduct an appropriate level of software regression testing to show that unchanged but vulnerable portions of the system have not been adversely affected. Design controls and appropriate regression testing provide the confidence that the software is validated after a software change.

Validation coverage should be based on the software's complexity and safety risk—not on firm size or resource constraints. The selection of validation activities, tasks, and work items should be commensurate with the complexity of the software design and the risk associated with the use of the software for the specified intended use. For lower-risk devices, only baseline validation activities may be conducted. As the risk increases, additional validation activities should be added to cover the additional risk. Validation documentation should be sufficient to demonstrate that all software validation plans and procedures have been completed successfully.

Validation activities should be conducted using the basic quality assurance precept of "independence of review." Self-validation is extremely difficult. When possible, an independent evaluation is always better, especially for higher-risk applications. Some firms contract out for a third-party independent verification and validation, but this solution may not always be feasible. Another approach is to assign internal staff members that are not involved in a particular design or its implementation, but who have sufficient knowledge to evaluate the project and conduct the verification and validation activities. Smaller firms may need to be creative in how tasks are organized and assigned in order to maintain internal independence of review.

Specific implementation of these software validation principles may be quite different from one application to another. The user has flexibility in choosing how to apply these validation principles, but retains ultimate responsibility for demonstrating that the software has been validated.

Software is designed, developed, validated, and regulated in a wide spectrum of environments and for a wide variety of devices with varying levels of risk. FDA-regulated medical device applications include software that

- Is a component, part, or accessory of a medical device
- Is itself a medical device
- Is used in manufacturing, design and development, or other parts of the quality system

In each environment, software components from many sources may be used to create the application (e.g., in-house-developed software, OTS software, contract software, shareware). In addition, software components come in many different forms (e.g., application software, operating systems, compilers, debuggers, and configuration management tools). The validation of software in these environments can be a complex undertaking; therefore, it is appropriate that all of these software validation principles be considered when designing the software validation process. The resultant software validation process should be commensurate with the safety risk associated with the system, device, or process.

Software validation activities and tasks may be dispersed, occurring at different locations and being conducted by different organizations. However, regardless of the distribution of tasks, contractual relations, source of components, or the development environment, the user or specification developer retains ultimate responsibility for ensuring that the software is validated.

Software validation is accomplished through a series of activities and tasks that are planned and executed at various stages of the software development life cycle. These tasks may be onetime occurrences or may be iterated many times, depending on the life cycle model used and the scope of changes made as the software project progresses.

For each of the software life cycle activities, there are certain "typical" tasks that support a conclusion that the software is validated. However, the specific tasks to be performed, their order of performance, and the iteration and timing of their performance will be dictated by the specific software life cycle model that is selected and the safety risk associated with the software application. For very-low-risk applications, certain tasks may not be needed at all. However, the software developer should at least consider each of these tasks and should define and document which tasks are or are not appropriate for their specific application. The following discussion is generic and is not intended to prescribe any particular software life cycle model or any particular order in which tasks are to be performed.

Typical Tasks

Quality Planning

- Risk (hazard) management plan
- Configuration management plan
- Software quality assurance plan
- Software verification and validation plan:
  - Verification and validation tasks and acceptance criteria
  - Schedule and resource allocation (for software verification and validation activities)
  - Reporting requirements:
    - Formal design review requirements
    - Other technical review requirements
- Problem reporting and resolution procedures
- Other support activities

Requirements

- Preliminary risk analysis
- Traceability analysis:
  - Software requirements to system requirements (and vice versa)
  - Software requirements to risk analysis
- Description of user characteristics
- Listing of characteristics and limitations of primary and secondary memory
- Software requirement evaluation
- Software user interface requirement analysis
- System test plan generation
- Acceptance test plan generation
- Ambiguity review or analysis

Design

- Updated software risk analysis
- Traceability analysis—design specification to software requirements (and vice versa)
- Software design evaluation
- Design communication link analysis
- Module test plan generation
- Integration test plan generation
- Test design generation (module, integration, system, and acceptance)

Construction or Coding

- Traceability analyses:
  - Source code to design specification (and vice versa)
  - Test cases to source code and to design specification
- Source code and source code documentation evaluation
- Source code interface analysis
- Test procedure and test case generation (module, integration, system, and acceptance)

Testing by the Software Developer

- Test planning
- Structural test case identification
- Functional test case identification
- Traceability analysis—testing:
  - Unit (module) tests to detailed design
  - Integration tests to high-level design
  - System tests to software requirements
- Unit (module) test execution
- Integration test execution
- Functional test execution
- System test execution
- Acceptance test execution
- Test result evaluation
- Error evaluation/resolution
- Final test report

User Site Testing

- Acceptance test execution
- Test result evaluation

- Error evaluation/resolution
- Final test report

VALIDATION OF AUTOMATED PROCESS EQUIPMENT AND QUALITY SYSTEM SOFTWARE

The quality system regulation requires that "when computers or automated data processing systems are used as part of production or the quality system, the (device) manufacturer shall validate computer software for its intended use according to an established protocol." (See 21 CFR 820.70(i).) This has been a regulatory requirement of FDA's medical device good manufacturing practice (GMP) regulations since 1978.

In addition to the earlier validation requirement, computer systems that implement part of a device manufacturer's production processes or quality system (or that are used to create and maintain records required by any other FDA regulation) are subject to the Electronic Records; Electronic Signatures regulation. (See 21 CFR part 11.) This regulation establishes additional security, data integrity, and validation requirements when records are created or maintained electronically. These additional part 11 requirements should be carefully considered and included in system requirements and software requirements for any automated record keeping systems. System validation and software validation should demonstrate that all part 11 requirements have been met.

Computers and automated equipment are used extensively throughout all aspects of medical device design, laboratory testing and analysis, product inspection and acceptance, production and process control, environmental controls, packaging, labeling, traceability, document control, complaint management, and many other aspects of the quality system. Increasingly, automated plant floor operations can involve extensive use of embedded systems in

- Programmable logic controllers
- Digital function controllers
- Statistical process control
- Supervisory control and data acquisition
- Robotics
- Human–machine interfaces
- Input/output devices
- Computer operating systems

Software tools are frequently used to design, build, and test the software that goes into an automated medical device. Many other commercial software applications, such as word processors, spreadsheets, databases, and flowcharting software, are used to implement the quality system. All of these applications are subject to the requirement for software validation, but the validation approach used for each application can vary widely.

Whether production or quality system software is developed in-house by the device manufacturer, developed by a contractor, or purchased OTS, it should be developed using the basic principles outlined earlier. The user has latitude and flexibility in defining how validation of that software will be accomplished, but validation should be a key consideration in deciding how and by whom the software will be developed or from whom it will be purchased. The software developer defines a life cycle model. Validation is typically supported by

- Verifications of the outputs from each stage of that software development life cycle
- Checking for proper operation of the finished software in the device manufacturer's intended use environment

The level of validation effort should be commensurate with the risk posed by the automated operation. In addition to risk other factors, such as the complexity of the process software and the degree to which the user is dependent upon that automated process to produce a safe and effective device, determine the nature and extent of testing needed as part of the validation effort. Documented requirements and risk analysis of the automated process help to define the scope of the evidence needed to show that the software is validated for its intended use. For example, an automated milling machine may require very little testing if the user can show that the output of the operation is subsequently fully verified against the specification before release. On the other hand, extensive testing may be needed for

- A plant-wide electronic record and electronic signature system
- An automated controller for a sterilization cycle
- An automated test equipment used for inspection and acceptance of finished circuit boards in a life-sustaining/life-supporting device

Numerous commercial software applications may be used as part of the quality system (e.g., a spreadsheet or statistical package used for quality system calculations, a graphics package used for trend analysis, or a commercial database used for recording device history records or for complaint management). The extent of validation evidence needed for such software depends on the device manufacturer's documented intended use of that software. For example, a user who chooses not to use all the vendor-supplied capabilities of the software only needs to validate those functions that will be used and for which the user is dependent upon the software results as part of production or the quality system. However, high-risk applications should not be running in the same operating environment with nonvalidated software functions, even if those software functions are not used. Risk mitigation techniques such as memory partitioning or other approaches to resource protection may need to be considered when high-risk applications and lower-risk applications are to be used in the same operating environment. When software is upgraded or any changes are made to the software, the user should consider how those changes may impact the "used portions" of the software and must reconfirm the validation of those portions of the software that are used. (See 21 CFR 820.70(i).)

User Requirements

A very important key to software validation is a documented user requirement specification that defines

- The "intended use" of the software or automated equipment
- The extent to which the user is dependent upon that software or equipment for production of a quality medical device

The user needs to define the expected operating environment including any required hardware and software configurations, software versions, and utilities. The user also needs to

- Document requirements for system performance, quality, error handling, startup, shutdown, security, etc.
- Identify any safety-related functions or features, such as sensors, alarms, interlocks, logical processing steps, or command sequences
- Define objective criteria for determining acceptable performance

The validation must be conducted in accordance with a documented protocol, and the validation results must also be documented. (See 21 CFR 820.70(i).) Test cases should be documented that will exercise the system to challenge its performance against the predetermined criteria, especially for its most critical parameters. Test cases should address error and alarm conditions, startup, shutdown, all applicable user functions and operator controls, potential operator errors, maximum and minimum ranges of allowed values, and stress conditions applicable to the intended use of the equipment. The test cases should be executed, and the results should be recorded and evaluated to determine whether the results support a conclusion that the software is validated for its intended use.

A user may conduct a validation using their own personnel or may depend on a third party such as the equipment/software vendor or a consultant. In any case, the user retains the ultimate responsibility for ensuring that the production and quality system software

- Is validated according to a written procedure for the particular intended use
- Will perform as intended in the chosen application

The user should have a documentation including

- Defined user requirements
- Validation protocol used
- Acceptance criteria
- Test cases and results
- A validation summary

that objectively confirms that the software is validated for its intended use.

VALIDATION OF OTS SOFTWARE AND AUTOMATED EQUIPMENT

Most of the automated equipment and systems used by device manufacturers are supplied by third-party vendors and are purchased OTS. The user is responsible for ensuring that the product development methodologies used by the OTS software developer are appropriate and sufficient for the device manufacturer's intended use of that OTS software. For OTS software and equipment, the user may or may not have access to the vendor's software validation documentation. If the vendor can provide information about their system requirements, software requirements, validation process, and the results of their validation, the medical user can use that information as a beginning point for their required validation documentation. The vendor's life cycle documentation, such as testing protocols and results, source code, design specification, and requirement specification, can be useful in establishing that the software has been validated. However, such documentation is frequently not available from commercial equipment vendors, or the vendor may refuse to share their proprietary information.

Where possible and depending upon the risk involved, the user should consider auditing the vendor's design and development methodologies used in the construction of the OTS software and should assess the development and validation documentation generated for the OTS software. Such audits can be conducted by the user or by a qualified third party. The audit should demonstrate that the vendor's procedures and results of the verification and validation activities performed for the OTS software are appropriate and sufficient for the safety and effectiveness requirements of the medical device to be produced using that software.

Some vendors who are not accustomed to operating in a regulated environment may not have a documented life cycle process that can support the device manufacturer's validation requirement. Other vendors may not permit an audit. Where necessary validation information is not available from the vendor, the user will need to perform sufficient system-level "black box" testing to establish that the software meets their "user needs and intended uses." For many applications, black box testing alone is not sufficient. Depending upon the risk of the device produced, the role of the OTS software in the process, the ability to audit the vendor, and the sufficiency of vendor-supplied information, the use of OTS software or equipment may or may not be appropriate, especially if there are suitable alternatives available. The user should also consider the implications (if any) for continued maintenance and support of the OTS software should the vendor terminate their support.

For some OTS software development tools, such as software compilers, linkers, editors, and operating systems, exhaustive black box testing by the user may be impractical. Without such testing—a key element of the validation effort—it may not be possible to validate these software tools. However, their proper operation may be satisfactorily inferred by other means. For example, compilers are frequently certified by independent third-party testing, and commercial software products may have "bug lists," system requirements, and other operational information available from the vendor that can be compared to the device manufacturer's intended use to help focus the "black box" testing effort. OTS operating systems need not be validated as a separate program. However, system-level validation testing of the application software should address all the operating system services used, including maximum loading conditions, file operations, handling of system error conditions, and memory constraints that may be applicable to the intended use of the application program.

23 Outsourcing and Monitoring of Bioequivalence Studies

BACKGROUND

The generics drug market has grown substantially over the last decades and this expansion is expected to continue. In the United States, almost 70% of all scripts are written generically, though they still represent 30% of the total cost of scripts. While Europe has continued to enjoy the benefits of generic biological drugs, the United States is yet to approve its first product as of the end of 2013. Sponsors are increasing the level of abbreviated new drug application (ANDA) filings and, in support, adopting a business model that increasingly relies on outsourcing to optimize efficiencies and contain costs.

As a business model, outsourcing has strategic and competitive benefits well recognized by the generic pharmaceutical industry processors and suppliers that comprise the bulk of the industry. For instance, the outsourcing of bioequivalence (BE) studies and the manufacture of active pharmaceutical ingredients (APIs) and finished dosage forms (FDFs) is widely practiced.

A majority of sponsors outsource facets of their BE studies to less-regulated regions, such as India, China, Russia, and Eastern Europe, with India being the preferred outsourcing destination and China and Russia being the less commonly used outsourcing destinations. However, recently, India has begin to slide in a number of studies as a result of regulatory changes that require long approval times and onerous compliance requirements such as approval of the Ministry of Health even to conduct toxicology studies in animals.

Outsourcing BE has its challenges, particularly regarding all matters concerning Food and Drug Administration (FDA) quality assurance requirements and regulatory issues. Despite these challenges, outsourcing is very popular: commonly outsourced activities include clinical and sample analysis, results reporting, and data and statistical analysis. The most frequently outsourced BE function is clinical activity, with almost 65% of all sponsors doing this.

A majority of the sponsors in the study currently outsource fewer than one-quarter of their BE studies to contract research organizations (CROs) in less-regulated markets (such as India), while one-third outsource most studies to those regions. Approximate distribution of BE studies outsourced to CROs in less-regulated markets (such as India or Eastern Europe) is given in Table 23.1.

A majority of companies are expected to increase their use of CROs in less-regulated regions during the next 2 years. More than two-thirds of companies expect to move more BE studies to India and almost half of companies are planning less work in Eastern Europe. India has recently run into serious regulatory hurdles where the Ministry of Health has created a highly complicated and difficult process of commencing BE trials; even the toxicology studies in animals now require Ministry of Health scrutiny and months of delays in securing approvals to commence the studies.

On average, participating sponsors conduct 3.27 BE studies (including both pilot and pivotal work) for each submitted ANDA. The median number of studies is three. In some instances, there are over 20 studies conducted for each ANDA filing. The cost of each patient enrolled on the average is around $1200. While the cost is the lowest in India and the highest in the United States, for studies, this may be the opposite.

TABLE 23.1
Percentage of Companies and Extent of Sourcing

Percent of Companies	Percentage of Studies Outsourced
22	76–100
11	51–75
11	26–50
56	1–25

For most companies, key benefits of using pilot studies in BE testing are improved design of pivotal studies and faster screening of experimental formulations. Saving time and/or money are additional benefits of using pilots. In addition, more than two-thirds of sponsors have dedicated clinical quality assurance (CQA) groups.

Outsourcing activities also commonly include working with CROs, investigators, and clinical sites, which may or may not be in a contract relationship with the CROs. The selection of a CRO is based on a best-practice selection process involving bids from more than one CRO and framed so that outsourced activities are generally contractually delineated with deliverables on a set time frame and the ability to terminate the outsourcing relationship.

Outsourced activities are amenable to auditing, and sponsors audit the CROs, investigators, and clinical sites frequently, with the majority of the audits occurring before the BE studies begin. Multiple deciding factors are followed to select which studies need on-site monitoring. Common factors often include whether or not a study is pilot or pivotal, possesses certain "uniqueness" factors and previous study experience with the CRO.

Quality assurance, which is a critical step in the BE study process, is seldom outsourced. Most sponsors house a dedicated and centralized CQA function and for good reason. The CQA function plays a key role in prestudy assessment of clinical sites and the CROs and is viewed as an essential component in improving BE study quality, reducing deficiency letters, accelerating the review process, and improving the relationship with the FDA.

Maintaining a regulatory department is a critical element of the study process and, similar to QA, is seldom outsourced. The regulatory department plays a key role in ensuring that documentation is complete and up to date to ensure necessary repeat analysis of clinical, medical, and analytical studies and internal regulatory compliance for acceptance of BE data in support of ANDA and to promote regulatory transparency.

To ensure excellence, as well as cost containment and efficiencies, while managing the risk, partner selection is a critical process and the key to BE outsourcing excellence. Efficient organizations have in place best-practice procedures for monitoring the outsourcing partner and the progress of the BE studies underway. Vigilance, due diligence, and communication create an atmosphere of success and will always support the effective management of these relationships.

Selection criteria should be carefully considered—standards that focus on solid business success and reputation, as well as financial stability, are important, as is a demonstrated expertise in BE study execution. Does the CRO provide a dedicated team knowledgeable in drug characteristics biology and pharmaceutics? Does it have established, successful relationships with key regulatory agencies?

When it comes to managing a BE study outsourcing relationship effectively, best practices include setting up standard operating procedures and conducting pre- and posttrial audits. Correspondingly, on-site monitoring and regular quality assessments for different groups involved in the BE study are elementary as well. Finally, maintaining strict compliance regimes while maintaining good communication between the various organizational facets is fundamental to winning value from an outsourcing partnership.

Best practices in the progress of the BE study include instituting efficiencies in the testing itself, by optimizing the use of the outsourcing partner through negotiating agreements that save time and money in terms of utilization by experiment and optimization of employee productivity.

Risk management strategies that include holding the outsourcing partner to higher-quality standards in terms of managing documentation will always serve the interests of both parties. It's understood that with BE studies, complete documentation is essential to ensure the necessary repeat analysis of clinical, medical, and analytical studies. It is also essential for the ability to convert from paper to electronic formats and for regulatory compliance and transparency and FDA filings.

Contributing as well to BE study success is the active involvement of the generic company outside of the relationship, with the CROs active in the design of pilot studies to improve design of pivotal studies and faster screening of experimental formulations; design and FDA acceptance of the protocols are also key components.

The Generic Drug User Fee Act of 2012 (GDUFA) is a new entry and is likely to have an impact on costs associated with the outsourcing of manufacturing of materials. While GDUFA is designed to speed access to safe and effective generic drugs to the public and reduce costs to industry, it requires that domestic and foreign FDF and API manufacturing facilities be inspected and pay user fees.

GDUFA is likely to require the generic company to audit their outsourcing partners as the higher probability of an FDA inspection will mean a greater risk of supply interruption due to compliance failures. The increased inspection activity is likely to reduce the numbers of outsourcing partners and raise prices, which are likely to be passed to the generic company. However, it is equally likely that the quality and consistency of APIs and FDFs will improve, which would benefit sponsors.

Because GDUFA requires the payment of user fees, which are lower for domestic manufacturing facilities than for foreign manufacturing facilities, these costs are likely to be passed on to the generic company.

It is expected that sponsors will continue to outsource BE studies and manufacturing of APIs and FDFs. In order to maximize efficiencies and cost containment from outsourcing, sponsors need to evaluate their in-house activities versus their outsourcing activities.

Generic companies also need to carefully select their outsourcing partner, holding both core competencies and cost containment as key components. Following selection, instituting best practices to successfully manage the outsourcing relationship are key factors for outsourcing excellence. CRO industry revenue was estimated at $33.6 billion for 2012 and is expected to reach $37.4 billion in 2013.

According to the independent Tufts Center for the Study of Drug Development, clinical trials conducted by CROs are completed an average of 30% more quickly than those conducted in-house. This results in an average time saving of some 4–5 months, translating to $120 million to $150 million in increased revenue potential.

The American Council of Research Organizations (ACRO) member companies employ approximately 95,000 people worldwide. Each year, our members conduct more than 11,000 clinical trials in 115 countries involving nearly two million research participants. ACRO member companies have contributed to the development of all of the top 50 selling biopharmaceutical products globally and then participate in the development of the vast majority of new treatments and therapies approved globally each year. The top five therapeutic areas for CROs are oncology, CNS, infectious disease, metabolic disorders, and cardiovascular disease. Vaccine development is another growing area of research for CROs.

ACRO member companies manage nearly one million square feet of laboratory space, process more than 16 million samples each year, and deliver more than 60 million individual test results. Approximately two-thirds of CRO business is from the pharmaceutical industry, 27% from biotech, and the remainder funded by the medical device, foundation, and government sectors.

Approximately 46% of clinical trials are conducted in the United States, 30% in Europe, and the remainder in Asia, Latin America, Africa, and the Middle East.

HANDLING AND RETENTION OF TESTED SAMPLES

Introduction

There are specific guidelines available from the U.S. FDA for study sponsors and/or drug manufacturers, CROs, site management organizations (SMOs), clinical investigators (CIs), and independent third parties regarding the procedure for handling reserve samples from relevant bioavailability (BA) and BE studies, as required by 21 CFR 320.38 and 320.63. The highlights of the guidance are on

- How the test article and reference standard for ability and BE studies should be distributed to the testing facilities
- How testing facilities should randomly select samples for testing and material to maintain as reserve samples
- How the reserve samples should be retained

Background

The Generic Drug Enforcement Act of 1992 (GDEA) was enacted in response to what the Congress concluded was a corrupt approval process for generic drugs through ANDAs. The GDEA created numerous and far-reaching powers that are being asserted by the U.S. FDA. It authorized the FDA to debar businesses and individuals and also permit the FDA to deny approval of ANDAs, withdraw approval of ANDAs, suspend the distribution of drugs produced by entities under investigation, and impose civil penalties. The rationale for enactment of the GDEA was that the Congress had found that "there is substantial evidence that significant corruption occurred in the Food and Drug Administration's process of approving drugs under ANDAs." In 1984, the Congress had enacted legislation that authorized the FDA to approve certain generic drug products without requiring an applicant to go through the rigorous process mandated for innovator drug products. Based on an investigation that began in 1988, the House Energy and Commerce Committee's Subcommittee on Oversight and Investigations concluded that various companies that sold generic drugs had in return for gratuities to FDA employees to obtain preferential treatment in obtaining approval of ANDAs.

On April 10, 1992, the legislation passed the Senate, and on April 28, 1992, HR 2454, in its amended form, passed the House. President Bush signed it into law on May 13, 1992. Given in Table 23.2 is a listing of individuals who are currently (end of 2013) debarred by the FDA. It is extremely important that the sponsors check the currency of this list and ask the potential CROs to provide an ongoing certification that none of these people will be engaged in any work conducted for submission to the FDA.

FDA Debarment

For putting unlawful profit ahead of consumer safety, 38 drug industry employees are facing a lifetime bar on practicing their livelihood. They were convicted, under the Federal Food, Drug, and Cosmetic Act, of felonies that included submitting false data to the FDA, lying to FDA investigators, paying or accepting bribes, and selling prescription drug samples. As a result, the 38 were "debarred" by the FDA from working for a drug company.

Most worked for generic drug firms (firms that make drugs that are equivalent to the first, brand-name versions) in positions like "vice president for quality control" or "director for research and development."

The word "debar" means to shut out or exclude. The FDA's authority to debar people from the drug industry comes from the GDEA, often called the "debarment act" because it authorizes, and sometimes even requires, the FDA to forbid people (or firms) convicted of certain crimes—basically, crimes related to the FDA's regulation of drugs—from participating in the drug industry. Other parts of this Act give the agency additional authorities.

TABLE 23.2
FDA Debarment List

Firms			
Name of Firm	**Effective Date**	**End/Term of Debarment**	**FR DATE.txt (MM/DD/YY)**
None as of this date			

Persons			
Name of Person	**Effective Date**	**End/Term of Debarment**	**FR DATE.txt (MM/DD/YY)**
Aiache, Adrien E.	05/27/2011	5 years%	05/27/2011
Akhigbe, Ehigiator O.	12/17/2010	25 years%	12/17/2010
Albanese, Anthony W.	11/23/2009	Permanent^	11/23/2009
Anthony, James Michael	11/07/1997	Permanent^	11/07/1997
Azeem, Mohammed	04/26/1993	Permanent^	04/26/1993
		FR correction	05/05/1993
Bae, Kun Chae	12/30/1993	Permanent^*	12/30/1993
Banks, Norma D.	08/28/1997	Permanent^	08/28/1997
Bansal, Padam C.	11/29/1993	Permanent^*	11/29/1993
Berman, David E.	30/09/2011	3 years%	03/09/2011
o	03/11/1997	Sp.Trmnation+	03/11/1997
Bhutani, Baldev Raj	12/02/2004	Permanent^	12/02/2004
Borison, Richard L.	09/30/2003	10 years%	09/30/2003
Brancato, David J.	01/06/1994	Permanent^*	01/06/1994
Bushlow, John W.	03/22/1996	Permanent^*	03/22/1996
Butkovitz, Anne L.	10/17/2006	Permanent^	10/17/2006
Campbell, Maria Anne Kirkman	09/02/2008	Permanent ^	09/02/2008
Caro Acevedo, Eduardo	03/24/2005	5 years%	03/24/2005
Chang, Charles Y.	03/08/1993	Permanent^	03/08/1993
		FR correction	04/26/1993
Charpentier, Laverne M.	12/02/2002	5 years%*	12/02/2002
Chatman, Cathryn Lyn a.k.a. Cathryn Lyn Garcia	04/07/2011	5 years%	04/07/2011
Chatterji, Dulal C.	11/01/1995	Permanent^#	01/22/1997
Choi, Andrew K.	08/08/2011	4 years%	08/08/2011
	06/11/1998	Sp.Trmnation+	06/11/1998
Cioffi, Albert Ronald	10/25/2011	5 years%	10/25/2011
Colton, Steven F.	06/17/1993	Permanent^*	06/17/1993
Concepcion, Jose	04/05/2012	5 years%	04/05/2012
Copanos, John D.	03/11/1996	Permanent^*	03/11/1996
		FR correction	04/18/1996
Courtney, Robert Ray	10/20/2003	Permanent^	10/20/2003
Desai, Kanubhai C.	10/06/1993	Permanent^	10/06/1993
Diamond, Bruce I.	11/14/2013	10 years%	11/14/2013
Dicola, Charles G.	11/05/1993	Permanent^*	11/05/1993
**Donnelly, Mary	11/29/1993	Permanent^	11/29/1993
Elbert, Robert	04/03/1997	Permanent^*	04/03/1997
Elsharaiha, Rami	09/29/2000	Permanent^	09/29/2000
Feuer, Scott	06/02/1998	5 years%	06/10/1998

(Continued)

TABLE 23.2 (*Continued*)
FDA Debarment List

Persons			
Name of Person	**Effective Date**	**End/Term of Debarment**	**FR DATE.txt (MM/DD/YY)**
Fiddes, Robert A.	11/06/2002	20 years%	11/06/2002
	01/21/2003	FR correction	01/21/2003
Finelli, Gena R.	04/21/1993	Permanent^	04/21/1993
Fogari, Robert A.	07/08/1993	Permanent^*	07/08/1993
Foyle, Ashley Brandon	05/01/2013	5 years%	03/18/2013
Freeman, David	05/08/2013	5 years%	05/08/2013
Garfinkel, Barry D.	04/02/1997	Permanent^*	04/02/1997
Girdhari, Premchand	01/21/2000	Permanent^*	01/21/2000
Gonsalves, Jr., Wallace	10/21/2009	Permanent^	10/21/2009
Hendrick, Kim C.	08/04/2009	Permanent^	08/04/2009
Herman, Hedviga	10/17/1997	Permanent^	10/17/1997
Holland, James A.	03/04/2010	5 years%*	03/04/2010
Hossain, Liaquat	11/05/1993	Permanent^*	11/05/1993
Islam, Amirul	08/27/1997	Permanent^#	08/27/1997
	10/09/2003	Sp. Trmnation+	10/09/2003
Justice, Glen R.	07/09/2012	25 years%	07/09/2012
Kalidindi, Sanyasi Raju	04/21/1993	Permanent^	04/21/1993
Kane, Niaja	11/23/2009	Permanent^	11/23/2009
Kimball, James T.	01/30/2007	Permanent ^	01/30/2007
Kindness, George J.	11/24/2008	Permanent ^	11/24/2008
Kletch, Walter S.	11/29/1994	Permanent^*	11/29/1994
Knott, Susan F.	12/11/2012	2 years%	12/11/2012
Kokes, Edwin	04/30/2003	Permanent^	04/30/2003
Kornak, Paul H.	08/04/2009	Permanent^	08/04/2009
Kostas, Constantine I.	06/25/1998	Permanent^*	06/10/1998
Lai, Elaine Yee-Ling	11/13/2002	5 years%	11/13/2002
		FR correction	11/23/2003
Lais, Patrick J.	02/23/2010	Permanent^	02/23/2010
Lentini, Jerome	05/09/2012	Permanent^*	05/09/2012
Liang, Cheng Yi	03/06/2013	Permanent^	03/06/2013
Long, Susan M.	12/23/1993	Permanent^*	12/23/1993
Macwan, Ashish	04/05/2012	5 years%	04/05/2012
Mannan, Muhammad Z.	04/06/1993	Permanent^	04/06/1993
Marcus, Jay	09/29/2000	5 years%	09/29/2000
Marks, Stephen L.	02/22/2012	Permanent^	02/22/2012
Matkari, Rajaram K.	10/20/1993	Permanent^*	10/20/1993
	06/13/2000	Withdrawn++	06/13/2000
Mays, Gary D.	01/24/1997	5 years%	01/24/1997
Mehlmauer, Marilyn A.	4/4/2011	4 years%	4/4/2011
Mendell, Arnold S.	12/21/1994	Permanent^*	12/21/1994
Morris, Andrew	05/16/1994	Permanent^#	01/11/1995
Nathan, Ray	08/09/2011	Permanent^	08/09/2011
Norman, Allyn M.	06/12/2009	5 years%	06/12/2009
One person removed from list	11/06/2002	Rescission!!!	01/16/2003
Page, Roy C.	11/24/2008	Permanent ^	11/24/2008

(*Continued*)

TABLE 23.2 (*Continued*)
FDA Debarment List

Persons			
Name of Person	**Effective Date**	**End/Term of Debarment**	**FR DATE.txt (MM/DD/YY)**
Pai, Daphne (a.k.a. Lau, Daphne)	11/05/1993	Permanent^*	11/05/1993
Palazzo, Maria Carmen	04/06/2011	Permanent^	04/06/2011
Panagotacos, Daphne I.	05/09/2012	5 years%*	05/09/2012
Pappas, Anastasios	03/09/2011	5 years%	03/09/2011
***Patel, Ashok	11/08/1994	Permanent^*	11/08/1994
Parikh, Jyotin	04/06/2012	5 years%	04/06/2012
Perkal, Mark B.	11/29/1993	Permanent^	11/29/1993
	09/11/1998	Sp.Trmnation+	09/11/1998
Petrik, Craig H.	04/30/2002	Permanent^	04/30/2002
Peugeot, Renee	01/13/2003	Permanent^	01/13/2003
Poet, Albert	03/03/2011	Permanent^	03/03/2011
Prasad, Kumar	06/30/1993	Permanent^	06/30/1993
Quamruzzaman, Abu	11/18/1993	Permanent^#	12/30/1994
Rana, Nandlal	10/20/1997	Permanent^	10/20/1997
Reuben, Scott S.	11/16/2011	Permanent^	11/16/2011
Rivers, Jacob H.	04/12/1993	Permanent^	04/12/1993
Rodgers, Jr., Thomas M.	07/28/2005	5 years%	07/28/2005
Rodriguez, Juan Manuel	07/26/1993	Permanent^	07/26/1993
Rosio, Timothy J.	06/24/2011	4 years%	06/24/2011
Rothenberg, Gayle	11/08/2011	Permanent^	11/08/2011
Ruetschi, Maja S.	04/04/2011	5 years%	04/04/2011
Ryan, Patrick T.	11/29/1994	Permanent^	11/29/1994
		FR correction	01/17/1995
Sacher, Robert E.	08/11/1997	Permanent^	08/11/1997
Sardesai, Suhas V.	09/30/2003	Permanent^	09/30/2003
Sawaya, Mary E. a.k.a. Marty Sawaya	06/12/2009	Permanent^	06/12/2009
Schetlick, Gloria H.	04/26/1993	Permanent^	04/26/1993
Martinez-Seldon, Deborah	10/05/2011	Permanent^	10/05/2011
Seldon, Stephen Lee	05/27/2011	Permanent^	05/27/2011
Shah, Atul	12/05/1994	Permanent^*	12/05/1994
	03/11/1997	Sp.Trmnation+	03/11/1997
Shah, Dilip	08/31/1993	Permanent^*	08/31/1993
Shah, Satish R.	08/01/1994	Permanent^*	08/01/1994
Shah, Shashikant	04/05/2012	5 years%	04/05/2012
Shainfeld, Frederick Jay	03/10/1995	Permanent^#	02/28/1996
Sharp, Lisa Jean	08/23/2012	Permanent^	08/23/2012
****Shulman, Robert *NMI*	08/27/1993	Permanent^	08/27/1993
Shrum, Kelly Dean	08/23/2012	Permanent^	08/23/2012
Snyder, Jr., Harry W.	01/13/2003	Permanent^	01/13/2003
Spencer, Wayne E.	09/13/2012	Permanent^	09/13/2012
Striefsky, Edmund J.	09/30/2003	Permanent^	09/30/2003
Sturm, Jan T.	06/22/1993	Permanent^	06/22/1993
		FR correction	07/06/1993

(*Continued*)

TABLE 23.2 (*Continued*)
FDA Debarment List

Persons			
Name of Person	**Effective Date**	**End/Term of Debarment**	**FR DATE.txt (MM/DD/YY)**
Theodore, Thomas Ronald	08/05/2003	Permanent^	08/05/2003
Torgerson, Michelle Lynn	11/13/2009	Permanent^	11/13/2009
Uddin, Mohammad *NMI*	09/29/2000	Permanent^	09/29/2000
Ullom, Brian	02/02/2010	Permanent^	02/02/2010
Vale, Jason	01/12/2010	Permanent^	01/12/2010
Van Wormer, Mark, E.	03/10/2011	Permanent^	03/10/2011
Vegesna, Raju	04/12/1993	Permanent^	04/12/1993
Wells, Ivyl W.	04/19/2011	Permanent^	04/19/2011
Xu, Kevin	04/08/2010	Permanent^	04/08/2010
Yanikian, Sami Arshak	07/02/2012	10 years%	07/02/2012
Yoser, Seth M.	05/20/2010	Permanent^	08/18/2010

Notations:

^ Mandatory debarment (section 306(a)).

% Permissive debarment (section 306(b)).

* Hearing requested and denied.

Acquiesced to debarment.

+ Special termination of debarment (section 306(d)(4)(C) and (d)(4)(D)).

++ Order to withdraw order of debarment (debarment terminated) (section 306(d)(3)(B)(i)).

!!! Rescission of debarment order.

a.k.a. Also known as.

NMI No middle initial known to be used.

** Ms. Mary L. Donnelly, of Green Bay City, Michigan, United States (FDA Docket No. 93N-0264), is not to be confused with Dr. Mary Sandra Donnelly, currently affiliated with St. Michael's Hospital of Toronto, Ontario, Canada.

*** Ashok Patel of Upper Saddle River, New Jersey, United States (FDA Docket No. 92N-0417), is not to be confused with Dr. Ashokkumar Bhailaibhai Patel (a.k.a. Ashok B. Patel) currently affiliated with Radiant Research, St. Louis, Missouri.

**** Robert Shulman, of Centerport, NY, United States (FDA Docket No. 93N-0069), is not to be confused with Dr. Robert J. Shulman, currently affiliated with Baylor College of Medicine, Children's Nutrition Research Center, Texas Children's Hospital. Following the generic drug scandal in the 1980s, the FDA issued an interim rule in the *Federal Register* of November 8, 1990, on the retention of BA and BE testing samples. The intent of the interim rule was to deter possible bias and fraud in BA and BE testing by study sponsors and/or drug manufacturers. Following public comments, a final rule was issued in the *Federal Register.*

When a person is debarred, the FDA notifies the public by publishing a notice in the Federal Register. Also, the FDA keeps an up-to-date debarment list. A copy of the list can be obtained by contacting the FDA's Office of Enforcement (HFC-230), 12720 Twinbrook Parkway, Rockville, MD 20852; phone: (301) 827-0410; fax: (301) 827-0482; e-mail, tchin@fdaem.ssw.dhhs.gov.

Each time a company—any drug company, not just a generic drug maker—applies for approval of a drug, it must submit to the FDA a signed statement that no debarred people worked on the application. If a drug firm employs a debarred person, even as a consultant or contractor, it can be fined up to $1 million. The person illegally working in the industry can be fined up to $250,000.

The law is broader than its title implies, affecting in large part brand-name as well as generic drug companies. It is called the "Generic Drug Enforcement Act" because the Congress passed the public protection measure in response to the confidence-shaking discovery in 1989 of widespread corruption in the generic drug industry.

That year, the FDA learned that some generic drug companies had committed illegal acts—things like falsifying data on drug formulations and illegally giving money to FDA chemists reviewing their drug applications—to gain preferential treatment.

"The data falsifications weren't just honest mistakes by generic companies," says David Read, chair of the FDA's Debarment Task Force, which meets as necessary to make debarment policy decisions. "They were calculated attempts to circumvent FDA's regulations. They constituted serious violations of the public trust."

To restore confidence in generic drugs, the FDA fired agency employees who had taken bribes and reinspected drug manufacturing facilities. The FDA also tightened its regulatory processes, for better verification of data used to support approval decisions.

Even with improved procedures, the FDA lacks the resources to audit every piece of data in every drug application. "The drug approval process is based on a system of trust," Read says. "The agency receives hundreds of drug applications a year, each consisting of many volumes. When the Congress created the new drug approval process, it was relying on drug companies and the agency to be fundamentally honest in their dealings with each other."

Debarment supplements the FDA's existing compliance instruments—injunctions, seizures, recalls, civil penalties, and criminal sanctions. By rooting out dishonest people, the debarment act bolstered the FDA's efforts to clean up the generic industry.

Under the law, a debarred person can't work for a drug firm "in any capacity." According to the U.S. Court of Appeals for the District of Columbia, even a job as a cook in a drug firm's cafeteria would be forbidden because of the opportunity for close contact between the debarred person and the drug firm's management. "All direct employment by a drug company, whether in the board room or the cafeteria or somewhere in between, is forbidden," the court said.

Besides direct employment, some jobs for a contractor that provides services to a drug firm are also off-limits.

"Debarment is a serious measure, but it's not intended as punishment," says Read. "It protects the public by ensuring that people with a history of dishonest conduct in the drug approval process will no longer be participants in that process."

Some debarees have claimed that the law is unconstitutional—that it is an ex post facto law—because it applies to crimes committed before the Generic Drug Enforcement Act was even passed in 1992. Because the law is a public protection measure and not punishment, courts have found that it is not illegal to debar people for conduct occurring before the law existed.

For the same reason, the Act doesn't violate the constitution's double jeopardy clause because it doesn't punish someone twice for the same offense.

Termination

Even the permanence of so-called mandatory debarment doesn't make debarment a punishment. All debarments except one so far have been mandatory debarments, meaning that the types of convictions—federal felony convictions relating to the development, approval, or regulation of a drug product—compelled the FDA to exclude the people from the industry.

Unlike "permissive debarment," which lasts up to 5 years, mandatory debarments of individuals are considered permanent. Although the mandatory debarments are imposed for a lifetime, the label of permanent may be misleading because the law provides a way for debarees to apply for "termination" of the debarment. They may be allowed to return to the drug industry if they substantially assist in the investigations or prosecutions of others in drug-related cases and submit persuasive evidence that they are rehabilitated and are no longer a threat to the drug approval process.

So far, no debarments have been terminated and a small number of the people debarred so far may be eligible to have their debarments terminated.

In the preamble to the final rule, the agency stated that the study sponsor and/or drug manufacturer should not separate out the reserve samples of the test article and reference standard before sending the drug product to the testing facility. This is to ensure that the reserve samples are in fact representative of the batches provided by the study sponsor and/or drug manufacturer for the testing. The study sponsor and/or drug manufacturer should send to the testing facility batches of the test article and reference standard so that the testing facility can *randomly select* samples for testing and material to maintain as reserve samples. The drug product should also be maintained in the sponsor's and/or manufacturer's original container.

Also in the preamble to the final rule, the agency noted that reserve sample retention is the responsibility of the organization that conducts the BA or BE study. The intent is to eliminate the possibility of sample substitution by the study sponsor and/or drug manufacturer or prevent the alteration of any reserve samples from a study conducted by a contractor before release of drug product samples to the FDA.

The FDA's Division of Scientific Investigations (DSI) and field investigators from the Office of Regulatory Affairs (ORA) conduct inspections of clinical and analytical sites that perform BA and BE studies for study sponsors and/or drug manufacturers seeking approval of generic and new drug products. A frequent finding from these inspections is the absence of reserve samples at the testing facilities where the studies are conducted. In many cases, DSI finds that testing facilities return reserve samples to the study sponsors and/or drug manufacturers, against the direction of the regulations in 21 CFR 320.38 and 320.63. In other cases, study sponsors and/or drug manufacturers, SMOs, or contract packaging facilities designate the study test article and reference standard for each subject and preclude the testing facilities from randomly selecting representative reserve samples from the supplies. DSI also finds that deviations from the regulations more often occur in BE studies with pharmacodynamic or clinical endpoints in which the studies are confused with clinical safety or efficacy studies. The pharmacodynamic or clinical endpoint BE studies are usually multisite, blinded studies conducted under contract (either directly with the study sponsor or drug manufacturer or through an SMO) by physicians or CIs who use their own clinics or offices to conduct the studies. Moreover, some CIs believe that they are not CROs and are not required to retain reserve samples. This guidance clarifies the responsibilities for retention of reserve samples.

Sampling Techniques

The FDA recommends that the study sponsor and/or drug manufacturer send to the testing facility batches of the test article and reference standard packaged in such a way that the testing facility can randomly select samples for BE testing and samples to maintain as reserve samples. This will ensure that the reserve samples are in fact representative of the batches provided by the study sponsor and/or drug manufacturer and that they are retained in the study sponsor's original container. Because the study sponsor and/or drug manufacturer may provide a testing facility with a variety of container sizes and packaging, the FDA is flexible in applying the representativeness requirement described in 21 CFR 320.38. For example, any of the following random sampling techniques might be used by the testing facility for the container size and packaging described (bolded text is particularly relevant).

Single container—If a single container of the test article and reference standard is provided to the testing facility, the testing facility should remove a sufficient quantity of the test article and reference standard from their respective containers to conduct the study; the remainder in each container should be retained as reserve samples in the original containers.

Multiple containers—If multiple containers of the test article and reference standard are provided to the testing facility, the testing facility should ***randomly select*** enough containers of the test article and reference standard to conduct the study; the remaining containers of the test article and reference standard should be retained as the reserve sample in the original containers. Generally, multiple open bottles are discouraged. The FDA encourages testing facilities to limit the number of open containers retained as study reserves.

Unit dose—If the test article and reference standard are provided to the testing facility in unit dose packaging, the testing facility should ***randomly select*** a sufficient quantity of unit doses of the test article and reference standard to conduct the study; the remaining unit doses of the test article and of the reference standard should be retained as the reserve samples in the original unit dose packaging. ***Therefore, it would be inappropriate to provide the study medications in unit dose packaging and all the reserve samples in bulk containers.***

Blinded study—If the study is to be blinded and the test article and reference standard are provided to the testing facility in unit dose packaging with each unit dose labeled with a randomization code, ***the study sponsor and/or drug manufacturer should provide the testing facility with a labeled set of the test article and reference standard sufficient to conduct the study and with additional, identically labeled sets sufficient to retain the "five times quantity" (see "Quantity of reserve samples" section). The testing facility should randomly select a labeled set to conduct the study; the remaining labeled sets would be retained in their unit dose packaging as the reserve samples.*** For a blinded study, we recommend that the study sponsor and/or drug manufacturer also provide to the testing facility a sealed code for use by the FDA should it be necessary to break the code. The sealed code should be maintained at the testing facility.

Retention for Multiple Studies and Shipments

If the same batches of the test articles and reference standards initially provided to the testing facility are used in performing more than one study, only one reserve sample of the test article and reference standard in sufficient quantity need to be retained. The reserve samples should be identified as having come from the same batches as used in each study. However, if additional supplies of the test article and reference standard will be used by a testing facility to perform the same study or additional studies, the testing facility should retain a sufficient quantity of reserve samples from the subsequent shipment. If a CRO with multiple testing facilities conducts more than one BE study (e.g., fed and fasted studies) for the same drug product, and the study test article and reference standard are sent to the testing facilities in different shipments, we recommend that sufficient quantity of reserve samples be kept for each study at each testing facility. These approaches are to ensure that the reserve samples are in fact representative of the batch provided by the study sponsor and/or drug manufacturer to the testing facility.

Quantity of Reserve Samples

The quantity of reserve samples should be sufficient to permit the agency to perform five times all of the release tests required in the application or supplemental application. The rationale for requiring the *five times quantity* is provided in the final rule. The CI can obtain the amount that constitutes the five times quantity from the sponsor and/or drug manufacturer. For solid oral dosage forms (e.g., tablets, capsules), an upper limit of 300 units each for the test article and reference standard can be considered sufficient to meet the five times quantity. Because the agency has limited experience with the retention and testing of nonsolid oral dosage forms, the agency is unable to recommend an upper limit for the retention of nonsolid oral dosage forms at this time. In the case of a reference standard that is an extemporaneously compounded solution or suspension or a reconstitutable powder, we recommend that the pure active ingredient and the unconstituted powder be retained. For a multisite BA or BE study, we recommend that the total amount of reserve samples to be retained across ***all*** testing facilities satisfy the five times quantity requirement. Each site is asked to retain a reasonable amount of test article and reference standard to be determined by considering (1) the total number of testing facilities participating in the study, (2) the number of subjects expected to be enrolled at each testing facility, and (3) a minimum limit (e.g., 5 dose units) for each of the test articles and reference standards. If the reserve samples from more than one testing facility are transferred to an independent third party for storage, we recommend that the independent third party segregate the reserve samples from the various testing facilities so that any given reserve sample can be unambiguously associated with the testing facility from which it came.

Responsibilities in Various Study Settings

Because of the variety of study settings potentially involved in conducting BA and BE studies, several examples are provided here. These examples are not the only possible study settings. However, in ***all*** instances, the chain of custody of the reserve samples used in the study should be preserved. The sponsor and/or manufacturer and any storage facility should document and maintain the transfer records for agency verification.

Studies Conducted at CROs, Universities, Hospitals, or Physicians' Offices

CROs are the most common study sites. Many BA/BE studies of oral dosage forms are conducted at CROs to support approval of ANDAs, new drug applications (NDAs), and NDA supplements. CROs typically conduct single-site, open-label, crossover design studies with healthy volunteers as participants.

Study sponsors and drug manufacturers sometimes conduct BA and BE studies through a CRO, university faculty, hospitals, or CIs in private practice. The testing facilities are usually clinical study units in universities, hospitals, or clinics run by physicians.

The responsibilities of the study sponsor and/or drug manufacturer include

- Packaging, distributing, and shipping the test article and reference standard to the testing facility
- Monitoring the study if it is conducted under an investigational new drug (IND) application (rarely needed for most ANDA studies)

The responsibilities of the testing facility are as follows:

- The CI or designee (such as the study coordinator or research pharmacist of the testing facility) should randomly select sufficient test article and reference standard to conduct the study from the supplies received from the sponsor and/or drug manufacturer and retain the remaining study samples as study reserves.
- The testing facility or the pharmacy of the testing facility should retain the reserve samples.
- If the testing facility does not have adequate storage, or goes out of business, the reserve samples can be transferred to an independent third party with an adequate facility for storage under conditions consistent with product labeling.

Note: When studies are conducted at universities, hospitals, or physicians' offices, the CI or physician conducting the study should ***not*** send the reserve samples back to the study sponsor and/or drug manufacturer. The goal is to eliminate the possibility for sample substitution by the study sponsor and/or drug manufacturer and to preclude the alteration of a reserve sample from a study conducted by another entity before the release of the reserve sample to the FDA.

Studies Involving SMOs

When BA or BE studies are conducted by an SMO, they are frequently multisite, open-label studies of oral dosage forms in patients or multisite, open-label studies of non-oral dosage forms with pharmacodynamic or clinical endpoints. Often, the study sponsor and/or drug manufacturer contracts with an SMO to recruit CIs and to monitor a study. The SMO is involved directly or indirectly (i.e., by subcontracting to another party) in packaging and shipping of study test articles and reference standards to the testing facilities. The testing facilities are usually the clinical study units of CROs, universities, hospitals, or clinics run by physicians.

The responsibility of the study sponsor or drug manufacturer is to ship the test article and reference standard to the SMO under contract or to the packaging facility under subcontract to the SMO.

The responsibilities of the SMO include

- Packaging, distributing, and shipping the test article and reference standard to all testing facilities (or subcontracting a packaging facility to perform this function)
- Monitoring the study at different sites if it is conducted under an IND (rarely needed for most ANDA studies)

The SMO should ***not*** randomly select and retain reserve study samples. As explained in the preamble to the final rule, the agency intended that sufficient test article and reference standard to conduct the study should be randomly selected at each testing facility and that each testing facility should retain the remaining study samples as reserves.

The responsibilities of the testing facilities are as follows:

- The CI or designee (such as the study coordinator or the research pharmacist of each testing facility) should randomly select sufficient test article and reference standard to conduct the study from the supplies received from the SMO under contract, or from the packaging facility under subcontract with the SMO, and retain the remaining study samples as study reserves.
- Each testing facility or the pharmacy of each testing facility should retain the reserve samples.
- Following the completion of the study, if one or more of the testing facilities do not have adequate storage, reserve samples can be transferred to an independent third party with an adequate facility for storage under conditions consistent with product labeling. The reserve samples should not be transferred back to an SMO or any other organization that deals with packaging the test articles and reference standard for storage. This is to eliminate the possibility of commingling reserve samples from packaging activities (21 CFR 211.84 and 211.170) and BE studies (21 CFR 320.38 and 320.63). As stated in "Studies conducted at CROs, universities, hospitals, or physicians' offices" section, the reserve samples should ***not*** be shipped back to the sponsor or manufacturer.

Blinded Studies with Pharmacodynamic or Clinical Endpoints Involving an SMO

Blinded BE studies are often conducted at multiple sites and involve non-oral dosage forms with pharmacodynamic or clinical endpoints.

In multisite, blinded BE studies, the sponsor and/or drug manufacturer needs to consider whether the study design will allow for selection and retention of reserve samples in accordance with 21 CFR 320.38 and 320.63 and the final rule. If the study design is too complex to meet the regulatory requirements for reserve samples, the study design may need to be reconsidered.

- Packaging, distributing, and shipping test article and reference standard to all testing facilities (or subcontracting a packaging facility to perform this function). The FDA recommends that the SMO provide the testing facilities with enough code-labeled sets to conduct the study and to retain the five times quantity. Based on inspection experience, DSI does not recommend that test article and reference standard be prenumbered for subjects, because assigning unit doses to a designated subject number precludes the random selection of drug used for dosing and drug used for reserve samples.
- Monitoring the study at different sites if it is conducted under an IND (rarely needed for most ANDA studies)

Note: The SMO should not select reserve samples. In addition, the reserve samples should not be transferred by the testing facility back to an SMO or any other organization that deals with packaging the test articles and reference standard for storage.

- The CI or designee (such as the study coordinator or the research pharmacist) of each testing facility should randomly select sufficient test article and reference standard to conduct the study from the supplies received from the SMO under contract, or from the packaging facility under subcontract with the SMO, and retain the remaining study samples as study reserves. The CI should be aware of the sampling techniques used for blinded studies.
- Each testing facility or the pharmacy of each testing facility should retain the reserve samples. Please note that if a placebo is used in blinded BE studies, reserve samples for the placebo should be retained along with the test article and reference standard reserves.
- The sealed treatment code of the study should be kept at the testing facility. This is applicable even if the reserve samples are forwarded to an independent third party.

  If one or more of the testing facilities do not have adequate storage, or go out of business, the reserve samples can be forwarded to an independent third party with an adequate facility for storage under conditions consistent with product labeling.

Given in the following is a suggested packaging and random selection plan for a blinded, multisite study of a dermatological cream product involving an SMO: the study enrolls 300 subjects with approximately 60 subjects at five testing facilities. The five times quantity for the test article and reference standard is 50 tubes for each product. In preparation for conducting the study, the SMO prepares 200 boxes that contain one code-labeled tube of test article and one code-labeled tube of reference standard in each box. The SMO randomly distributes 40 boxes to each clinical testing facility. The clinical facility randomly selects 30 of the boxes to dose 60 subjects. The remaining 10 boxes serve as the reserve samples. In this example, staff (e.g., a pharmacist) not involved with the study may be recommended to ensure the study remains blinded. This packaging system ensures that an equal number of test article and reference standard are administered to the subjects at each site and that an equal number of test article and reference standard will be maintained as reserve samples. Since 10 boxes are kept at each of five testing facilities, 50 tubes each of test article and reference standard are retained and the five times quantity reserve sample requirement is met. In addition, the requirement of random selection by each testing facility is also met.

In-House Studies Conducted by a Study Sponsor and/or Drug Manufacturer

Only about 7% of all sites inspected by DSI from 1997 to 2002 conducted in-house BA and BE studies. If a study sponsor and/or drug manufacturer conducts such a study, manufacturing reserve samples (21 CFR 211.170) and BE study reserve samples (21 CFR 320.38 and 320.63) should be separated. The in-house clinical research unit should operate as an independent unit for the purposes of sample retention. All matters (e.g., manufacturing, purchasing, packaging, transfer records) concerning the test article and reference standard should be clearly documented and available to FDA investigators during an inspection. Standard procedures concerning security and accountability of the test article and reference standard for each study should be established to eliminate the possibility of sample substitution. Sponsors conducting in-house studies can engage an independent third party to store reserve samples. If an independent third party is not used, there should be (1) a totally segregated and fully compliant in-house storage area, (2) procedures and policies in place to show that adequate test article and reference standard are retained, (3) controlled access to the reserve samples, and (4) a rigorous and unbroken chain of custody for the reserve samples.

The study sponsor and/or drug manufacturer (clinical research department) should be responsible for packaging and transferring the test article and reference standard to the in-house clinical study unit.

The testing facility (in-house clinical study unit) should be responsible for the following:

- Documentation of all matters concerning the transfer and receipt of the test article and reference standard.
- Random selection of sufficient test article and reference standard to conduct the study and retention of the remaining study samples as reserves. The selection is generally made by the CI, study coordinator, or research pharmacist (if available) in the clinical study unit. The FDA recommends that a staff member (e.g., a study nurse) witness the random selection process and dosing.
- Retention of reserve samples in a secure area. To ensure the authenticity of the reserve samples, access to this area should be limited. The FDA encourages maintenance of an entry log to the storage area.
- Preparation for adequate storage of reserve samples. If the in-house testing facilities do not have adequate storage, or go out of business, the reserve samples can be forwarded to an independent third party with an adequate facility for secure storage under conditions consistent with product labeling.

In Vitro Bioequivalence Studies

21 CFR 320.63 states:

> The applicant of an abbreviated application or a supplemental application submitted under section 505 of the Federal Food, Drug, and Cosmetic Act, or, if bioequivalence testing was performed under contract, the contract research organization shall retain reserve samples of any test article and reference standard used in conducting an in vivo or in vitro bioequivalence study required for approval of the abbreviated application or supplemental application.

Thus, the regulations for reserve samples apply to in vitro BE studies. The in vitro BE studies required for approval of nasal aerosols and nasal sprays for local action are an example of this. Note that in vitro studies conducted to compare dissolution rates for different strengths of the same formulation are not subject to the reserve sample regulations. For an in vitro BE study, the roles of the study sponsor and/or drug manufacturer and the testing facility are similar to those described for in vivo BE studies conducted by CROs and in the examples of in vivo BE studies conducted in-house by a study sponsor and/or drug manufacturer.

Exception for Inhalant Products

As stated in 21 CFR 320.38(c), each reserve sample shall consist of a sufficient quantity of samples to permit the FDA to perform five times all of the release tests required in the application or supplemental application. Dose content uniformity or spray content uniformity release tests alone usually take 30 units (canisters or bottles) per batch. Performance of other release tests can suggest a need for additional units. The number of reserve sample units to be retained for three batches of test article and reference standard could exceed 1000 units (up to 250 units for each batch of the test article and reference standard) based on the five times quantity requirement. The agency has determined that in lieu of the "five times quantity" requirement, the quantity of inhalant (nasal aerosol or nasal spray) test article and reference standard retained for testing and analyses should be at least 50 units for each batch (see the preamble to the final rule).

For NDAs, at least 50 units of each of the three batches of nasal aerosol or nasal spray needed for BA studies should be retained. However, where the reference standard is another nasal aerosol or nasal spray, at least 50 units of that batch should also be retained. For ANDAs, at least 50 units of each of three batches should be retained for each of the test articles and reference standards used for in vivo or in vitro BE studies. If multiple testing facilities are used in a BA or BE study, the total amount of reserves for each product across ***all*** testing facilities would be at least 50 units, and each testing facility should retain a reasonable amount of test articles and reference standards. For NDAs and ANDAs, if the in vivo or in vitro studies include placebo aerosols or sprays, at least 50 units of each placebo batch should also be retained. These recommendations apply only to nasal aerosol and nasal sprays for local action that are to be marketed as multiple dose products, typically labeled to deliver 30 or more actuations per canister or bottle.

RISK-BASED MONITORING OF CROs

Introduction

Monitoring the work conducted by the CROs is an onerous task with high risk to sponsors in failing the regulatory compliance of the studies submitted. This can result in missing the deadlines for filing ANDAs and the related financial benefits. The U.S. FDA has recently provided additional advise on monitoring CROs on a risk-based strategy to reduce the cost of monitoring and assuring that the studies shall remain in regulatory compliance and enhance human subject protection and the quality of clinical trial data by focusing sponsor oversight on the most important aspects of study conduct and reporting.

Sponsors can use a variety of approaches to fulfill their responsibilities for monitoring CI conduct and performance in IND studies conducted under 21 CFR part 312, investigational device exemption (IDE) studies conducted under 21 CFR part 812 as suggested by the U.S. FDA recently (August 2013); however, these guidance are equally applicable to the data collected in the filing of ANDAs. It is noteworthy that the regulatory agencies may still ask for phase III trials, abbreviated phase III trials, or even phase IV trials to support an ANDA filing. So, while the FDA is treading gradually into suggesting a risk-based approach, the sponsors will be wise to consider this equally applicable to BE studies.

The FDA guidance describes strategies for monitoring activities that reflect a modern, risk-based approach that focuses on critical study parameters and relies on a combination of monitoring activities to oversee a study effectively. For example, the guidance specifically encourages greater use of centralized monitoring methods where appropriate.

Background

Effective monitoring of clinical investigations by sponsors is critical to the protection of human subjects and the conduct of high-quality studies (21 CFR part 312, subpart D) generally (Responsibilities of Sponsors and Investigators) and 21 CFR part 812, subpart C generally (Responsibilities of Sponsors). Sponsors of clinical investigations involving human drugs, biological products, medical devices, and combinations thereof are required to provide oversight to ensure adequate protection of the rights, welfare, and safety of human subjects and the quality of the clinical trial data submitted to the FDA (21 CFR 312.50 requires a sponsor to, among other things, ensure "proper monitoring of the investigation(s)" and "that the investigation(s) is conducted in accordance with the general investigational plan and protocols contained in the IND"; 21 CFR 812.40 states that sponsors are responsible for, among other things, "ensuring proper monitoring of the investigation"; see also 21 CFR 312.53(d), 312.56(a), 812.43(d), and 812.46). The FDA's regulations require sponsors to monitor the conduct and progress of their clinical investigations. The regulations are not specific about how sponsors are to conduct such monitoring and are therefore compatible with a range of approaches to monitoring that will vary depending on multiple factors.

During the past two decades, the number and complexity of clinical trials have grown dramatically. These changes create new challenges to clinical trial oversight, particularly increased variability in CI experience, site infrastructure, treatment choices, and standards of health care, as well as challenges related to geographic dispersion. At the same time, increasing use of electronic systems and records and improvements in statistical assessments present opportunities for alternative monitoring approaches (e.g., centralized monitoring) that can improve the quality and efficiency of sponsor oversight of clinical investigations. The FDA encourages sponsors to develop monitoring plans that manage important risks to human subjects and data quality and address the challenges of oversight in part by taking advantage of the innovations in modern clinical trials. A risk-based approach to monitoring does not suggest any less vigilance in oversight of clinical investigations. Rather, it focuses sponsor oversight activities on preventing or mitigating important and likely risks to data quality and to processes critical to human subject protection and trial integrity. Moreover, a risk-based approach is dynamic, more readily facilitating continual improvement in trial conduct and oversight. For example, monitoring findings should be evaluated to determine whether additional actions (e.g., training of CI and site staff, clarification of protocol requirements) are necessary to ensure human subject protection and data quality across sites.

This guidance focuses principally on monitoring, which is one aspect of the processes and procedures needed to ensure clinical trial quality and subject safety. Monitoring is a quality control tool for determining whether study activities are being carried out as planned, so that deficiencies can be identified and corrected. Monitoring, or oversight, alone cannot ensure quality. Rather, quality is an overarching objective that must be built into the clinical trial enterprise. The FDA recommends a *quality risk management* approach to clinical trials and is considering the need for additional guidance describing this approach. The FDA realizes that the term *monitoring* is used in different ways in the clinical trial context. It can refer to the assessment of CI conduct, oversight, and reporting of findings of a clinical trial; to the ongoing evaluation of safety data and the emerging benefit–risk profile of an investigational product; and to the monitoring of internal sponsor and CRO processes and systems integral to proposing, designing, performing, recording, supervising, reviewing, or reporting clinical investigations.

Therefore, for purposes of this guidance, *monitoring* refers to the methods used by sponsors of investigational studies, or CROs delegated responsibilities for the conduct of IND/ANDA studies, to oversee the conduct of, and reporting of data from, clinical investigations, including appropriate CI supervision of study site staff and third-party contractors. Monitoring activities include communication with the CI and study site staff; review of the study site's processes, procedures, and records; and verification of the accuracy of data submitted to the sponsor.

Current Monitoring Practices and FDA Guidance

A survey conducted through the Clinical Trials Transformation Initiative (CTTI) (CTTI is a public–private partnership involving the FDA, academia, industry representatives, patient and consumer representatives, professional societies, investigator groups, and other government agencies, initiated in 2008; CTTI's mission is to identify practices that will increase the quality and efficiency of clinical trials) indicated that a range of practices has been used to monitor the conduct of clinical trials. These practices vary in intensity, focus, and methodology and include centralized monitoring of clinical data by statistical and data management personnel; targeted on-site visits to higher-risk CIs (e.g., where centralized monitoring suggests problems at a site); and frequent, comprehensive on-site visits to all CI sites by sponsor personnel or representatives (e.g., clinical monitors or clinical research associates).

Although survey participants reported a range of monitoring methods, periodic, frequent visits to each CI site to evaluate study conduct and review data for each enrolled subject remain the predominant mechanism by which pharmaceutical, biotechnology, and medical device companies monitor the progress of clinical investigations. For major efficacy trials, companies typically conduct on-site monitoring visits at approximately 4- to 8-week intervals, at least partly because of the perception that the frequent on-site monitoring visit model, with 100% verification of all data, historically has been the FDA's preferred way for sponsors to meet their monitoring obligations. In contrast, academic coordinating centers, cooperative groups, and government organizations use on-site monitoring less extensively. For example, some government agencies and oncology cooperative groups typically visit sites only once every 2 or 3 years to qualify or certify clinical study sites to ensure they have the resources, training, and safeguards to conduct clinical trials. The FDA also recognizes that regulators and practitioners have relied on data from critical outcome studies (e.g., many National Institutes of Health-sponsored trials, Medical Research Council-sponsored trials in the United Kingdom, International Study of Infarct Survival [ISIS] trials, and GISSI), which had no regular on-site monitoring and used primarily centralized and other alternative monitoring methods. These examples suggest that use of alternative monitoring approaches should be considered by all sponsors, including commercial sponsors, when developing risk-based monitoring strategies and plans.

The 1996 International Conference on Harmonisation of Technical Requirements for Registration of Pharmaceuticals for Human Use (ICH) guidance on good clinical practice (ICH E6) and the 2011 International Standards Organization (ISO) Clinical investigation of medical devices for human subjects—good clinical practice (ISO 14155:2011) address monitoring. Both ICH E6 and ISO 14155:2011 specifically provide for flexibility in how trials are monitored. ICH E6 and ISO 14155:2011 advise sponsors to consider the objective, design, complexity, size, and endpoints of a trial in determining the extent and nature of monitoring for a given trial. The ISO standard further states that a sponsor's assessment of these factors should be used to develop a monitoring plan, a recommendation consistent with the FDA's recommendation for monitoring plan development in this guidance. Although the ICH guidance and ISO standard specifically provide for the possibility of reduced, or even no, on-site monitoring, they also make clear that it would be appropriate to rely entirely on centralized monitoring only in exceptional circumstances.

The FDA has communicated the goals of, and recommendations for, risk-based monitoring to FDA staff in review, inspection, and compliance functions. The FDA's bioresearch monitoring compliance program guidance manuals (CPGMs) for sponsors, CROs, and monitors (CPGM 7348.810) and for CIs and sponsor-investigators (CPGM 7348.811) are compatible with the approaches described in this guidance. For example, CPGM 7348.810 informs the FDA field staff that the regulations do not prescribe a specific monitoring technique. While CPGM 7348.810 refers to site visits and does not discuss centralized monitoring, the focus is on the review of monitoring activities through documentation and whether these activities were carried out in accordance with the sponsor's (or CRO's) monitoring procedures.

FDA's Rationale for Risk-Based Monitoring

The FDA believes that risk-based monitoring could improve sponsor oversight of clinical investigations. This guidance is therefore intended to make it clear that risk-based monitoring, including the appropriate use of centralized monitoring and reliance on technological advances (e.g., e-mail, webcasts, online training modules), can meet statutory and regulatory requirements under appropriate circumstances.

There is a growing consensus that risk-based approaches to monitoring, focused on risks to the most critical data elements and processes necessary to achieve study objectives, are more likely than routine visits to all clinical sites and 100% data verification to ensure subject protection and overall study quality. For example, incorporation of centralized monitoring practices, where appropriate, should improve a sponsor's ability to ensure the quality of clinical trial data. Several publications suggest that certain data anomalies (e.g., fraud, including fabrication of data, and other nonrandom data distributions) may be more readily detected by centralized monitoring techniques than by on-site monitoring. It has been suggested that a statistical approach to central monitoring can "help improve the effectiveness of on-site monitoring by prioritizing site visits and by guiding site visits with central statistical data checks," an approach that is supported by illustrative examples using actual trial data sets. A recent review of on-site monitoring findings collected during a multicenter international trial also suggests that centralized monitoring can identify the great majority of on-site monitoring findings. The review determined that centralized monitoring activities could have identified more than 90% of the findings identified during on-site monitoring visits.

The FDA encourages sponsors to tailor monitoring plans to the needs of the trial. The FDA recognizes that this guidance places greater emphasis on centralized monitoring than appeared feasible at the time ICH E6 was finalized. However, the FDA considers the approach to monitoring described in this guidance to be consistent with ICH E6 and ISO 14155:2011. The FDA believes it is reasonable to conclude that the flexibility described in ICH E6 and ISO 14155:2011 was intended to permit innovative approaches to improve the effectiveness of monitoring. Notably, the advancement in electronic systems and increasing use of electronic records (i.e., electronic data capture [EDC] systems) facilitate remote access to electronic data and, increasingly, to some source data. Additionally, statistical assessments using data submitted on paper case report forms (CRFs) or via EDC may permit timely identification of clinical sites that require additional training, monitoring, or both. We expect that the pharmaceutical and device industries will, for the foreseeable future, continue to use some amount of on-site monitoring, but we anticipate decreased use of on-site monitoring with evolving monitoring methods and technological capabilities.

The following sections reflect the FDA's current thinking on monitoring and include recommendations on how to develop and implement a study-specific monitoring plan as well as how to document monitoring activities. The FDA acknowledges that there are limited empirical data to support the utility of the various methods employed to monitor clinical investigations (e.g., superiority of one method vs. another), including data to support on-site monitoring.[26] As a result, the recommendations are based, in part, on the FDA's experience from the review of protocols during the IND/ANDA or IDE phase, data submitted in preapproval applications, results of inspections conducted to ensure human subject protection and data integrity, and information obtained from public outreach efforts conducted under the auspices of the CTTI.

Overview of Monitoring Methods

On-Site and Centralized Monitoring

This section is intended to assist sponsors in identifying and designing monitoring practices appropriate to a given clinical trial. It describes some of the capabilities of on-site and centralized monitoring processes and factors to consider in determining which monitoring practices may be appropriate for a given clinical trial. See the succeeding text for a discussion of factors to consider when determining the types, frequency, and extent of monitoring activities and for examples of events or results that would trigger a change in planned monitoring activities.

On-Site Monitoring

On-site monitoring is an in-person evaluation carried out by sponsor personnel or representatives at the sites at which the clinical investigation is being conducted. On-site monitoring can identify data entry errors (e.g., discrepancies between source records and CRFs) and missing data in source records or CRFs; provide assurance that study documentation exists; assess the familiarity of the site's study staff with the protocol and required procedures; and assess compliance with the protocol and investigational product accountability. On-site monitoring can also provide a sense of the quality of the overall conduct of the trial at a site (e.g., attention to detail, thoroughness of study documentation, appropriate delegation of study tasks, appropriate CI supervision of site staff performing critical study functions). On-site monitoring can therefore be particularly helpful early in a study, especially if the protocol is complex and includes novel procedures with which CIs may be unfamiliar. Findings at the site may lead to training efforts at both the site visited and elsewhere.

Centralized Monitoring

Centralized monitoring is a remote evaluation carried out by sponsor personnel or representatives (e.g., clinical monitors, data management personnel, or statisticians) at a location other than the sites at which the clinical investigation is being conducted. Centralized monitoring processes can provide many of the capabilities of on-site monitoring as well as additional capabilities.

The FDA encourages greater use of centralized monitoring practices, where appropriate, than has been the case historically, with correspondingly less emphasis on on-site monitoring. The types of monitoring activities and the extent to which centralized monitoring practices can be employed depend on various factors, including the sponsor's use of electronic systems; the sponsor's access to subjects' electronic records, if applicable; the timeliness of data entry from paper CRF, if applicable; and communication tools available to the sponsor and study site. These may vary by study and by site. Sponsors who plan to use centralized monitoring processes should ensure that the processes and expectations for site record keeping, data entry, and reporting are well-defined and ensure timely access to clinical trial data and supporting documentation. If sponsors intend to rely heavily on centralized monitoring practices, they should identify, in the monitoring plan, when one or more on-site monitoring visits would be indicated.

Examples of Alternative Monitoring Techniques

As discussed in "Background" section, monitoring activities broadly include communication with the CI and study site staff; review of the study site's processes, procedures, and records; and verification of the accuracy of data submitted to the sponsor. This section highlights areas for which centralized monitoring techniques could be considered. For certain monitoring activities, centralized monitoring techniques can be considered in place of, or to complement, traditional monitoring techniques. Specific techniques used should be prospectively included in the monitoring plan and should be informed by the risk assessment:

- Centralized monitoring techniques should be used to the extent appropriate and feasible to supplement or reduce the frequency and extent of on-site monitoring with monitoring activities that can be done as well or better remotely or with monitoring activities that can be accomplished using centralized processes only. Examples include the following:
  - Monitor data quality through routine review of submitted data to identify and follow-up on missing data, inconsistent data, data outliers, and potential protocol deviations that may be indicative of systemic or significant errors in data collection and reporting at a site.
  - Conduct statistical analyses to identify data trends not easily detected by on-site monitoring, such as
    - Standard checks of range, consistency, and completeness of data
    - Checks for unusual distribution of data within and between study sites, such as too little variance
  - Analyze site characteristics, performance metrics (e.g., high screen failure or withdrawal rates, high frequency of eligibility violations, delays in reporting data), and clinical data to identify trial sites with characteristics correlated with poor performance or noncompliance.
  - Verify critical source data remotely as described in the monitoring plan, in cases where such source data are accessible or where CRF data are, according to the protocol, source data.

- Complete administrative and regulatory tasks. Such tasks include, for example, verifying continuous institutional review board (IRB) approval by reviewing electronic IRB correspondence, if available; performing portions of investigational product accountability, such as comparison of randomization and CRF data, to preliminarily assess whether the subject was administered or dispensed the assigned product and to evaluate consistency between investigational product receipt, use, and disposition records; and verifying whether previously requested CRF corrections were made.

Centralized techniques, including routine review of submitted data and statistical and other analyses, may also be used to identify significant concerns (e.g., need for clarification of a protocol procedure, indications of data fabrication) with noncritical data that may not have otherwise been a focus of monitoring (e.g., source document verification).

Target on-site monitoring by identifying higher-risk clinical sites (e.g., sites with data anomalies or a higher frequency of errors, protocol violations, or dropouts relative to other sites), through the activities described earlier. Such findings, whether related to critical or noncritical data, may warrant more intensive and consideration of on-site monitoring.

Communication with Study Site Staff

Communication between the monitor and the study site staff is an essential component of monitoring. Various modes of communication (e.g., teleconferences, videoconferencing, e-mail) could be considered for specific study time points (e.g., study initiation) and activities (e.g., to discuss findings of a monitor's e-CRF review, training of new site staff).

Review of Site's Processes, Procedures, and Records

Techniques for monitoring informed consent and site records are included here as examples of approaches to monitoring site's processes, procedures, and records.

Informed Consent

Verification of subjects' informed consent is a critical activity that should be monitored. Alternatives to the traditional approach (monitors verifying the original signature on the consent form for each subject at the site) may be more effective in identifying inadequacies in the consent process and may be more efficient. For example, the study site electronically sends (e.g., fax, e-mail) the signed page(s) of consent forms to the monitor, or the monitor performs remote comparison of dates of study procedures and documentation of informed consent on CRFs. An internet portal that enables the site staff to upload signed consent forms and enables access by designated monitors is a tool that can be considered. Use of electronic informed consent may also facilitate sponsor oversight of human subject protection. We recognize that sponsors must attend to privacy and confidentiality concerns when considering techniques for monitoring informed consent remotely.

Site's Records

A growing portion of source documents (e.g., laboratory and radiology reports, source documents submitted by the CI for other purposes such as health records documenting serious adverse events or adjudicated events) are electronic and may be available to the sponsor remotely. Furthermore, consistent with ICH E6 and ISO 14155:2011, original observations can be entered directly into the e-CRF or transmitted to the e-CRF from various locations, devices, or instruments. The FDA recognizes that sponsors may not have remote access to electronic health records maintained by hospitals, universities, and other institutions because of data privacy and security concerns as well as technological challenges. Sponsors should consider risk-based approaches to monitoring using the format of study information (i.e., electronic, paper, or combination of electronic and paper), tools, and other resources available to them.

A variety of centralized monitoring techniques can be used to replace, supplement, and target on-site monitoring activities. The majority of these techniques (e.g., checks for completeness of data, sites with a higher frequency of protocol violations relative to other sites, sites with high screen failure rates) can be performed regardless of the extent of use of electronic records in the study. For example, the majority of these techniques can be performed using CRF data collected either using electronic data capture systems or entered into a database from a paper CRF collected by the sponsor. A recent publication discusses statistical techniques for identifying various types of data errors (Reference section). The statistical techniques described in this guidance may not be routinely used by all sponsors and may not be appropriate for every trial, but they are included in this guidance as examples of monitoring techniques that may be considered by sponsors.

Additional monitoring techniques, such as routine review of data as they are submitted, are possible for studies that use electronic CRFs. Although not a monitoring technique, another method of ensuring data quality routinely implemented in e-CRFs is the use of electronic prompts in the e-CRF to minimize errors and omissions at the time of data entry, particularly if data are entered directly into the e-CRF.

Source Data Verification and Corroboration

The sponsor should consider the quantity and types of source data that need to be verified against CRFs or corroborated against other records (e.g., review of medical record to corroborate a subject's response of "no hospitalizations" since the previous visit on a CRF) during the sponsor's identification of critical data and processes or in the risk assessment, or both. The sponsor should include a description of the quantity and types of source records to verify or corroborate in the monitoring plan. The sponsor should consider which source records are likely to provide the most meaningful information about a subject's participation and the CI's conduct and oversight. For example, for a particular study, there may be minimal benefit in comparing 100% of the source data for each subject to the CRFs for each study visit. Rather, it may be sufficient to compare the most critical data points for a sample of subjects and study visits as an indicator of data accuracy. Similarly, for a particular study, although collection of all concomitant medications, body temperature, and body weight are required by the protocol and are documented in the medical record and transcribed to a CRF, they may not be identified by the sponsor as critical data, because a small error rate in those variables would not affect the outcome of the trial. In the absence of information indicating potential concerns with the data (e.g., sites with data anomalies, inconsistent data), source document verification or corroboration of these noncritical data may not provide significantly useful information to the sponsor.

Risk-Based Monitoring

No single approach to monitoring is appropriate or necessary for every clinical trial. The FDA recommends that each sponsor design a monitoring plan that is tailored to the specific human subject protection and data integrity risks of the trial. Ordinarily, such a risk-based plan would include a mix of centralized and on-site monitoring practices. The monitoring plan should identify the various methods intended to be used and the rationale for their use.

Monitoring activities should focus on preventing or mitigating important and likely sources of error in the conduct, collection, and reporting of critical data and processes necessary for human subject protection and trial integrity. Sponsors should prospectively identify critical data and processes, then perform a risk assessment to identify and understand the risks that could affect the collection of critical data or the performance of critical processes, and then develop a monitoring plan that focuses on the important and likely risks to critical data and processes.

Identify Critical Data and Processes to Be Monitored

Sponsors should prospectively identify critical data and processes that if inaccurate, not performed, or performed incorrectly would threaten the protection of human subjects or the integrity of the study results. As examples, the following types of data and processes should ordinarily be identified as critical:

- Verification that informed consent was obtained appropriately.
- Adherence to protocol eligibility criteria designed to exclude individuals for whom the investigational product may be less safe than the protocol intended and to include only subjects from the targeted study population for whom the test article is most appropriate.
- Procedures for documenting appropriate accountability and administration of the investigational product (e.g., ensuring the integrity of randomization at the site level, where appropriate).
- Conduct and documentation of procedures and assessments related to
  - (a) Study endpoints
  - (b) Protocol-required safety assessments
  - (c) Evaluating, documenting, and reporting serious adverse events and unanticipated adverse device effects, subject deaths, and withdrawals, especially when a withdrawal may be related to an adverse event
- Conduct and documentation of procedures essential to trial integrity, such as ensuring the study blind is maintained, both at the site level and at the sponsor level, as appropriate, referring to specified events for adjudication, and allocation concealment. Other types of data (e.g., covariates such as concomitant treatments or demographic characteristics, routine laboratory tests performed as part of subject monitoring that do not address protocol-specified safety or efficacy endpoints) and processes (e.g., a hospital pharmacy's storage of an investigational product with no specific critical handling instructions) identified by the sponsor as noncritical often may be monitored less intensively.
- There is increasing recognition that some types of errors in a clinical trial are more important than others. For example, a low, but nonzero rate of errors in capturing certain baseline characteristics of enrolled subjects (e.g., age, concomitant treatment, or concomitant illness) will not, in general, have a significant effect on study results if the errors are distributed randomly. In contrast, a small number of errors related to study endpoints (e.g., not following protocol-specified definitions) can profoundly affect study results, as could failure to report rare but important adverse events. Based on the FDA's inspection and review experience, infrequent errors in noncritical data are unlikely to alter the FDA's conclusions about whether a product is safe and effective and whether participants' safety was appropriately monitored.

Risk Assessment

This guidance discusses the risk assessment, a component of risk management, as applied in the context of clinical monitoring. Risk assessment generally involves identifying risks, analyzing risks, and then determining whether risks need to be modified by implementing controls (e.g., processes, policies, or practices). The risk assessment recommended in this guidance to inform development of a monitoring plan may also support efforts to manage risks across a clinical trial (e.g., through modifying the protocol design or implementation) or development program. This guidance does not provide comprehensive detail on how to perform a risk assessment. There are many risk assessment methodologies and tools from a variety of industries that can be applied to clinical trials.

Following the identification of critical data and processes (see "Identify critical data and processes to be monitored" section), sponsors should perform a risk assessment to identify and understand the nature, sources, and potential causes of risks that could affect the collection of critical data or the performance of critical processes. Risks to critical data and processes most merit consideration during risk assessment, to ensure that monitoring efforts are focused on preventing or mitigating important and likely sources of error in their conduct, collection, and reporting.

Risk identification for monitoring purposes should generally consider the types of data to be collected, the specific activities required to collect these data, and the range of potential safety and other human subject protection concerns that are inherent to the clinical investigation (e.g., based on trial design or investigational product).

The identified risks should be assessed and prioritized by considering the following:

- The likelihood of errors occurring
- The impact of such errors on human subject protection and trial integrity
- The extent to which such errors would be detectable

Sponsors should use the results of the risk assessment in developing the monitoring plan (e.g., determining which risks may be addressed through monitoring, determining the types and intensity of monitoring activities best suited to addressing these risks). Sponsors may also determine that some risks are better managed through activities other than monitoring, for example, modifying the protocol to remove the source of the risk. Sponsors should periodically evaluate emerging risks and whether monitoring activities require modification to effectively oversee the risks.

Factors to Consider When Developing a Monitoring Plan

A monitoring plan ordinarily should focus on preventing or mitigating important and likely risks, identified by the risk assessment, to critical data and processes. The types (e.g., on-site, centralized), frequency (e.g., early, for initial assessment and training vs. throughout the study), and extent (e.g., comprehensive [100% data verification] vs. targeted or random review of certain data [less than 100% data verification]) of monitoring activities will depend to some degree on a range of factors, considered during the risk assessment, including the following:

- *Complexity of the study design*
  More intensive monitoring (e.g., increased frequency and extent of review) may be necessary as study design complexity increases. Examples may include studies with adaptive designs, stratified designs, complex dose titrations, or multiple device placement studies.
- *Types of study endpoints*
  Endpoints that are more interpretative or subjective may require on-site visits to assess the totality of subject records and to review application of protocol definitions with the CI. More objective endpoints (e.g., death, hospitalization, or clinical laboratory values and standard measurements) may be more suitable for remote verification. Endpoints for which inappropriate subject withdrawal or lack of follow-up may impede study evaluation are likely to need more intensive monitoring to identify the reason(s) subjects are withdrawing and to determine whether follow-up can be improved.
- *Clinical complexity of the study population*
  A study that involves a population that is seriously ill or vulnerable may require more intensive monitoring and consideration of on-site monitoring visits to be sure appropriate protection is being provided.
- *Geography*
  Sites in geographic areas where there are differences in standards of medical practice or subject demographics or where there is a less established clinical trial infrastructure may require more intensive monitoring and consideration of on-site monitoring visits.
- *Relative experience of the CI and of the sponsor with the CI*
  CIs who lack significant experience in conducting and overseeing investigations, using a novel or innovative medical device, or with the surgical procedure associated with medical device use may benefit from more intensive monitoring and frequent communication to ensure CI understanding of responsibilities. In addition, the relative experience of a sponsor with the CI may be a factor in determining an appropriate monitoring plan.

Electronic Data Capture

Use of EDC systems with the capability to assess quality metrics (e.g., missing data, data error rates, protocol violations) in real time could help identify potentially higher-risk sites for the purpose of targeting sites in need of more intensive monitoring:

- *Relative safety of the investigational product*
  A study of a product that has significant safety concerns or for which there is no prior experience in human clinical trials (e.g., a phase 1 pharmaceutical investigation or a device feasibility study) may require more intensive monitoring and consideration of on-site monitoring visits to ensure appropriate CI oversight of subject safety.
- *Stage of the study*
  A tapered approach to monitoring may be used where appropriate, with more intensive monitoring at initiation and during early stages of a trial. For example, a tapered approach could be used for a complex study where more intensive and on-site monitoring might be required early but where, once procedures are established, less intensive monitoring might suffice. Similarly, a tapered approach could be used for relatively inexperienced CIs.
- *Quantity of data*
  Some centralized monitoring tools may be more useful as the quantity of data (e.g., size or duration of trial, number of sites) collected increases.

Monitoring Plan

For each clinical trial, the sponsor should develop a monitoring plan that describes the monitoring methods, responsibilities, and requirements for the trial. The monitoring plan should include a brief description of the study, its objectives, and the critical data and study procedures, with particular attention to data and procedures that are unusual in relation to clinical routine and require training of study site staff. The plan should also communicate the specific risks to be addressed by monitoring and should provide those involved in monitoring with adequate information to effectively carry out their duties. A monitoring plan may reference existing policies and procedures (e.g., standard operating procedure describing general monitoring processes or issue investigation and resolution). All sponsor and CRO personnel involved with monitoring, including those who review or determine appropriate action regarding potential issues identified through monitoring, should review the monitoring plan and associated documents (e.g., standard operating procedures or other documents referenced in the monitoring plan).

The components of a monitoring plan might include the following:

Description of Monitoring Approaches

- A description of each monitoring method to be employed during the study and how it will be used to address important risks and ensure the validity of critical data
- Criteria for determining the timing, frequency, and extent of planned monitoring activities
- Specific activities required for each monitoring method employed during the study, including reference to required tools, logs, or templates
- Definitions of events or results (e.g., findings from central monitoring activities) that would trigger changes in planned monitoring activities for a particular CI

  For example, if it is determined that a CI differs markedly from other CIs in making safety-related findings or other key safety metrics, in rate of enrollment, in the number of protocol deviations, or in the rate of missing CRFs, the CI's site should be considered for targeted on-site visits. The establishment of acceptable variation for particular critical data and processes would facilitate identification of significant deviations.

- Identification of possible deviations or failures that would be critical to study integrity and how these are to be recorded and reported

  For example, sponsors may wish to establish a specific mechanism for tracking and notifying key study personnel of deviations related to collection or reporting of data necessary to interpret the primary endpoint, regardless of which monitoring method identified a concern.

The study monitoring plan should also describe how various monitoring activities will be documented, regardless of whether they are conducted on-site or centrally.

Communication of Monitoring Results

- Format, content, timing, and archiving requirements for reports and other documentation of monitoring activities
- Process for appropriate communication—of routine monitoring results to management and other stakeholders (e.g., CRO, data management)
  - (a) Of immediate reporting of significant monitoring issues to appropriate parties (e.g., sponsor management, CI and site staff, IRB, FDA), as necessary
  - (b) From study management and other stakeholders to monitors

For example, data management personnel may provide monitors with routine reports of outstanding CRFs or of common data queries at or across sites that may enable effective targeting of monitoring activities.

Management of Noncompliance

- Processes for addressing unresolved or significant issues (e.g., significant noncompliance with the investigational plan, suspected or confirmed data falsification) identified by monitoring, whether at a particular site or across study sites
- Processes to ensure that root cause analyses are conducted where important deviations are discovered and that appropriate corrective and preventive actions (e.g., additional training on a study or site level) are implemented to address issues identified by monitoring
- Other quality management practices applicable to the clinical investigation (e.g., reference to any other written documents describing appropriate actions regarding noncompliance)

Ensuring Quality Monitoring

- Description of any specific training required for personnel carrying out monitoring activities, including personnel conducting internal data monitoring, statistical monitoring, or other centralized review activities. Training should include principles of clinical investigations and human subject protection. In addition, study-specific training should include discussion of the trial design, protocol requirements, the study monitoring plan, applicable standard operating procedures, appropriate monitoring techniques, and applicable electronic systems.
- Planned audits of monitoring to ensure that sponsor and CRO staff conduct monitoring activities in accordance with the monitoring plan, applicable regulations, guidance, and sponsor policies, procedures, templates, and other study plans. Auditing is a quality assurance tool that can be used to evaluate the effectiveness of monitoring to ensure human subject protection and data integrity.

- Many sponsors have successfully implemented on-site comonitoring visits (i.e., monitoring visits performed by both a study monitor and the monitor's supervisor or another evaluator designated by the sponsor or CRO) to evaluate whether monitors are effectively carrying out visit activities, in compliance with the study monitoring plan. These visits may be conducted either for randomly selected monitors or may be targeted to specific monitors, based upon questions arising from review of monitoring visit documentation. See ICH E6, section 5.19, and ISO 14155:2011, section 6.11, for additional information on audits.

Monitoring Plan Amendments

Sponsors should consider what events would indicate a need for review and revision of the monitoring plan and establish processes to permit timely updates where necessary. For example, a protocol amendment, change in the definition of significant protocol deviations, or identification of new risks to study integrity could result in a change to the monitoring plan.

Documenting Monitoring Activities

Documentation of monitoring activities should generally include the following:

- The date of the activity and the individual(s) conducting and participating in it
- A summary of the data or activities reviewed
- A description of any noncompliance, potential noncompliance, data irregularities, or other deficiencies identified
- A description of any actions taken, to be taken, or recommended, including the person responsible for completing actions and the anticipated date of completion

Documentation of monitoring should include sufficient detail to allow verification that the monitoring plan was followed. Monitoring documentation should be provided to appropriate management in a timely manner for review and follow-up, as indicated.

Additional Strategies to Ensure Study Quality

Although the focus of this guidance is on monitoring the oversight and conduct of, and reporting of data from, clinical investigations, the FDA considers monitoring to be just one component of a multifactor approach to ensuring the quality of clinical investigations. Many other factors contribute to the quality of a clinical investigation. This section highlights additional areas that complement monitoring and can affect study quality.

A fundamental component of ensuring quality monitoring is a sponsor's compliance with monitoring plans and any accompanying procedures.

Protocol and Case Report Form Design

The most important tool for ensuring human subject protection and high-quality data is a well-designed and articulated protocol. A poorly designed or ambiguous protocol may introduce systemic errors that can render a clinical investigation unreliable despite rigorous monitoring. Additionally, the complexity of the trial design and the type and amount of data collected may influence data quality. The CRF, which captures the data required by the protocol, is another critical tool for which design directly affects the quality of trial data. Care should be taken to ensure that the CRF captures data accurately (e.g., as required by the protocol) and that the CRF design and instructions facilitate consistent data collection across CI sites.

Clinical Investigator Training and Communication

Clinical trial monitors conducting on-site visits have historically played an important role in training the CI and site staff during a study. On-site visits also have served as a primary means of providing feedback to CIs and study personnel on study conduct. Without meaningful training prior to the conduct of a study and of appropriate instruction during the study (e.g., when changes are made to the protocol), CIs and their staff may have difficulty carrying out a trial correctly. Sponsors who plan less frequent or limited on-site monitoring should consider the following:

- Monitoring activities should include sufficient time for discussion of CI's and site staff's responsibilities, feedback, and additional training, if needed, during the conduct of the study.
- It may be necessary to implement alternative training (e.g., teleconferences, webcasts, online training modules) and communication methods (see "Communication with study site staff" section) for providing and documenting ongoing, timely training and feedback, as well as to provide notification of significant changes to study conduct or other important information.

Delegation of Monitoring Responsibilities to a CRO

If a sponsor of an IND study delegates the responsibility for ensuring proper monitoring to a CRO, FDA regulations (21 CFR 312.52) require the written transfer of any obligations from a sponsor to a CRO and require the CRO to comply with the regulations.[8] Although sponsors can transfer responsibilities for monitoring to a CRO(s), they retain responsibility for oversight of the work completed by the CRO(s) that assume this responsibility. Sponsors should evaluate CRO compliance with regulatory requirements and contractual obligations in an ongoing manner. For example, sponsor oversight of monitoring performed by a CRO may include the sponsor's periodic review of monitoring reports and vendor performance or quality metrics and documented communication between the sponsor and CRO regarding monitoring progress and findings.

Sponsors and CROs should consider additional factors when a sponsor transfers responsibilities for monitoring to a CRO. Sponsors and CROs should prospectively establish a clear understanding of both parties' responsibilities and of the expectations for the conduct of the transferred obligations. Sponsors should share information with a CRO that may inform decisions a CRO may make regarding the monitoring practices for a trial (e.g., findings of a risk assessment). Sponsors should prospectively evaluate monitoring procedures and monitoring plans developed by a CRO to ensure the monitoring approach is consistent with applicable aspects of the trial. In addition, sponsors and CROs should have processes in place for timely exchange of relevant information (e.g., significant monitoring findings, significant changes in risk for a trial).

Clinical Investigator and Site Selection and Initiation

In addition to regulatory requirements for CI selection, sponsors should consider factors such as sponsor's previous experience with the CI or site, workload of the CI and study staff, and resource availability at the study site during CI and site selection.

Site initiation is a critical study activity that often involves sponsor personnel from a range of disciplines, including monitors. Key components of site initiation include ensuring the CIs and site staff understand their responsibilities, including applicable regulatory requirements as well as study processes and procedures, including the sponsor's processes for monitoring the investigation. Communication and documentation tools for monitoring discussed in this guidance can also be used for site selection and initiation activities.

CRO SELECTION

Table 23.3 lists a few CROs without making any endorsement of any of these companies.

TABLE 23.3
List of a Few Potential CROs

Algorithme Pharma (http://www.algopharm.com)
Anapharm (http://www.inventivhealthclinics.com)
Argint International (www.argintinternational.com)
Aster Cephac (www.aster-cephac.com)
Auriga Research Limited (http://www.aurigaresearch.com)
AXIS Clinicals Ltd. (http://www.axisclinicals.com)
Azidus Laboratories Ltd. (http://www.azidus.com)
Bio Analytical Research Corporation (BARC) (www.barclab.com)
BioArc Research Solutions (http://www.bioarcresearch.com)
BioReliance (www.bioreliance.com)
Bioscience Labs (http://biosciencelabs.com/)
Biovail Contract Research (http://www.biovail-cro.com)
Boston Biostatistics, Inc. (www.bostonbio.com)
Cato Research Ltd. (www.cato.com)
Centre of Pharmacokinetics Research (http://www.filab.com.pl/filab2/index.php)
CEPHA (www.cepha.cz)
CEPHA s.r.o. (http://www.ccpha.cz)
Certus International, Inc. (www.certusintl.com)
Clinical and Analytical Research Center (http://www.cro.vimspectrum.ro/index.php/en/)
Clinical Trial Management Services, Inc. (CTMS) (www.ctmsinc.com)
CMIC Co., Ltd. (www.cmic-holdings.co.jp/e)
Covance (www.covance.com)
CroMedica (www.cromedica.com)
CTSI (www.ctsi-cro.com)
Exodon (http://www.exodon.com/)
GRC Genuine Research Organization (http://www.grc-me.com)
Hill Top Research (www.hill-top.com)
ICON (Nasdaq:ICLR) (www.iconclinical.com)
ICON (www.iconplc.com)
INC Research (www. Incresearch.com)
Info Kinetics (http://info-kinetics.com)
International Drug Development Institute (IDDI) (www.iddi.com)
International Pharmaceutical Research Center (http://iprc.com.jo)
Jordan Center for Pharmaceutical Research (JPRC) (http://www.japm.com)
Kiecana Clinical Research (www.kiecana.com.pl)
Kynetyx HT—CRO (http://www.kynetyxht.ro)
Lambda Therapeutic Research Inc. (http://www.lambda-cro.pl)
MCT—Pharmaceutical Development Company (http://www.mct-cro.com/regional-reach/cro-jordan/)
MDS Pharma Services (www.mdsps.com)
OCT—Clinical Trials (www.oct-clinicaltrials.com)
Omnicare Clinical Research (www.omnicarecr.com)
Parexel (http://www.parexel.com)
Parexel (www.parexel.com)
Pharm PlanNet (http://www.crs-group.de)
Pharma Medica Research Inc. (http://www.pharmamedica.com)
Pharma Professional Services (PPS) (http://www.phaps.com)

(*Continued*)

TABLE 23.3 (*Continued*)
List of a Few Potential CROs

Pharmaceutical Product Development (PPDI) (www.ppdi.com)
Pharmaceutical Research Unit (PRU) (http://www.pru.com.jo)
PPD (www.ppdl.com)
PRA (www.praintl.com)
PRA International (www.prainternational.com)
Protech Pharmaservices Corporation (PPC) (http://www.ppccro.com)
QuailCRO Clinical Research (www.quailcro.com)
Quinta-Analytica (http://www.quinta.cz)
Quintiles (www.quintiles.com)
Rho, Inc. (www.RhoWorld.com)
Synchron Research Services Pvt. Ltd. (http://synchronresearch.com)
Synteract, Inc. (www.synteract.com)
TherImmune Research Corporation
TKL Research (www.tklresearch.com)
Valid-Trio (www.valid-trio.com)
VIMTA Labs Ltd. (http://www.vimta.com)

Selecting a CRO requires a lot of investigation about the capability, record of performance, area of specialty, integrity, and timeliness of producing the data. For BE studies, there are three distinct functions that a CRO performs:

1. Recruit subjects and conduct trials
2. Bioanalytical services
3. Data analysis and reporting

In some instances, all of these services are provided from in-house resources, in others, these may be outsourced. The cost of studies is a significant factor but the timeliness of study reporting would often supersede any cost considerations particularly where the filing date is critical to secure any available market exclusivity. However, like many other industries, the low-cost centers are appearing in China and India. In China, if the reference-listed drug (RLD) is already approved, then it only takes an internal board review to approve the trials; however, in India, the bureaucracy has created significant barriers, and where timeliness is important, India will be a poor choice.

The experience of a CRO in having conducted similar studies is an extremely crucial factor to consider. From assisting in the writing of the protocol to suggesting an appropriate number of subjects, the CRO can be very helpful.

In all instances, the sponsor must build an in-house team to continuously monitor the course of the studies, assure that the data are correctly reported and recorded, and provide an almost real-time vigilance of the study through electronic data rooms.

24 Epilog
Future of Bioequivalence Testing

BACKGROUND

Bioequivalence is defined in 21 CFR 320.1 as "the absence of a significant difference in the rate and extent to which the active ingredient or active moiety in pharmaceutical equivalents or pharmaceutical alternatives becomes available at the site of drug action when administered at the same molar dose under similar conditions in an appropriately designed study." This definition has been emulated globally by every regulatory agency worldwide resulting in a creation of protocols and experimental designs requiring pharmacokinetic testing in humans. One assumption involved in all of these models is that since in most instances the site of action may not be accessible for sampling, the indirect testing through study of the pharmacokinetic profiles allows simulation of the profile of drug concentration at the site of action.

Establishing bioequivalence between two products is required in several situations:

- Prototype formulations during early development and pivotal clinical trial formulations
- Mapping a process that relates critical manufacturing variables (CMVs), including formulation, processes, and equipment variables that can significantly affect drug release from the product
- Innovator formulations that differ from their new drug application (NDA) formulations as a result of scale-up and postapproval changes (SUPAC)
- Implementation of improved manufacturing technologies
- Scale-up changes in manufacturing locations
- Multisource products filed for approval under abbreviated new drug application (ANDA) as allowed under the Hatch–Waxman law

The current guidelines proposed by global regulatory agencies require comparison of the pharmacokinetic profiles of the test and the reference drug product with specific ranges of acceptability of data to declare bioequivalence. The classical conventional human pharmacokinetic in vivo bioequivalence study employs a single-dose, two-period, two-treatment, two-sequence, open-label, randomized crossover design comparing equal doses of the test and reference products in fasted, adult, healthy volunteers (e.g., $n = 24$). These studies are an indirect measure of bioequivalence and supposed to represent a study of the drug concentration profile at the site of action.

21 CFR 320.25(a) further codifies the universal belief that "no unnecessary human testing should be performed" and goes on to suggest: "The basic principle in an in vivo bioavailability study is that no unnecessary human research should be done." The nature of exposure to healthy humans in conducting bioequivalence testing is exacerbated when highly toxic drugs or drugs requiring multiple dosing are administered. In an effort to reduce unnecessary exposure to humans, the regulatory agencies have made biowaivers (except in Japan) available for certain class of drugs that are not likely to have problem in achieving bioequivalence. This is provided under a biopharmaceutical classification system (BCS) wherein some classes of drugs are exempted from bioequivalence testing.

The in vitro bioequivalence testing has a long history of use. 21 CFR 320.33 has provided criteria to assess actual or potential bioequivalence problems. In the late 1970s, drug products that had met these criteria were deemed not *bioproblem* drug products. In vitro studies were expected to correctly assess bioequivalence for products that were not "bioproblem" drug products. For instant or immediate-release (IR) products not containing a "bioproblem" drug, the Food and Drug Administration (FDA) allowed drug efficacy study implementation (DESI)-effective drugs to be assessed for bioequivalence through in vitro studies alone. Since 1979, such products that passed bioequivalence testing were assigned an AA rating in FDA's "Approved Drug Products with Therapeutic Equivalence Ratings." 21 CFR 320.24 also describes situations when in vitro studies can be used alone to document bioequivalence. The fact that the U.S. FDA allows developers to challenge the bioequivalence testing opens the doors to more creative applications of the in vitro tests.

The use of dissolution testing in place of pharmacokinetic studies is generally allowed for those drugs that dissolve rapidly and where permeability of drug molecules across biological barriers is not problematic. This includes, for example, the drugs that dissolve at least 85% in 15 min (very rapidly dissolving) or in 30 min (rapidly dissolving) or less in pH 1.2, 4.5, and 6.8 dissolution media. These are the drugs where bioequivalence is self-evident making bioequivalence studies redundant.

There is however no global concurrence on a BCS for biowaivers; developers are not generally able to make global filings for bioequivalence testing for their products. There is a dire need to reevaluate the scientific rationale behind the current methods of testing bioequivalence with the aim to reduce or eliminate this testing where possible.

PITFALLS IN BIOEQUIVALENCE TESTING

The existing methodology used globally to demonstrate bioequivalence has many pitfalls that need to be reviewed prior to suggesting an alternate method of bioequivalence testing.

Site of Action Requirement

The official definition of bioequivalence as enumerated earlier requires demonstration of equivalence of concentration at the *site of action*. The fact that the site of action of most drugs is not known and not possible to sample makes this universal definition questionable. When a new chemical entity (NCE) is developed, the developer is required to provide data on the pharmacokinetic (what the body does to the drug) profile as well as the pharmacodynamic (what the drug does to the body) profile to establish safe dosing. The developer is not required to identify the site of action nor the mechanism of action, and the dosing is never based on any evaluation of the concentration at the site of action. The bioequivalence testing under the conditions enumerated earlier is conducted to assure that the two products are equally effective and thus interchangeable. Requiring an assessment at the site of action enhances the requirements beyond what is required for the NCE despite the established demonstration of dosing and safety of the active ingredient. The use of pharmacokinetic profiles to simulate the concentration at the site of action is also flawed since it assumes a linear dose–response making the pharmacokinetic profile a poor choice to meet the basic requirement of bioequivalence.

In brief, the official definition of bioequivalence requiring comparisons between two products to evaluate any difference between them based on the concentration achieved at the site of action is at best impractical and at worse irrelevant to the goal of demonstrating bioequivalence. There is a need to revise the definition of bioequivalence. One choice will be to require statistically insignificant pharmacokinetic profiles, but that too will be unacceptable as an indirect test of the difference in the drug products; the purpose of bioequivalence testing is to decipher any differences in the two drug products, not the safety and efficacy that have already been established.

Statistical Modeling Errors

The bioequivalence evaluation based on pharmacokinetic profiling is achieved through the use of complex statistical models to account for the high variability (intersubject, intrasubject) of the data

H_0: Products are not bioequivalent
H_1: Products are bioequivalent

Result of testing	Truth: H_0 is true	Truth: H_0 is false
Fail to reject H_0	Good	Type II (producer risk, β)
Reject H_0	Type I (consumer risk, α)	Good

FIGURE 24.1 Types of errors in hypothesis testing.

collected. The studies are powered to various levels of confidence intervals, and the number of subjects enrolled depends on the coefficient of variation in the data collected. As a result, many developers conduct a pilot study to establish this variability to correctly power their final study. Despite years of developing suitable statistical models, errors remain evident in the bioequivalence testing making these studies less useful.

Two types of errors are common in bioequivalence studies: type I and type II (Figure 24.1).

H_0: Products are not bioequivalent.

In bioequivalence testing, the null hypothesis states that products are not bioequivalent (H_0), while the alternate hypothesis states that products are bioequivalent (H_1). Type I error occurs when products are erroneously concluded to be bioequivalent when they are not bioequivalent. Type I error represents a risk to the consumer (i.e., a health risk to the patient). Type II error occurs when products are erroneously concluded to be not bioequivalent when they are bioequivalent. Type II error represents a risk to the producer as a good product is rejected.

Assuming that the conventional human pharmacokinetic in vivo bioequivalence testing is a perfect indication of whether products are bioequivalent, the extent that products pass class I (BCS) with rapid dissolution but fail in vivo bioequivalence testing is analogous to the type I error rate of in vitro testing.

Highly variable drugs (HVDs) are drugs with high within-subject variability (ANOVA-CV $\geq$ 30%) in C_{max} and/or AUC. HVDs typically have flat dose–response curves and large therapeutic windows, such that clinically important adverse drug reactions (ADRs) occur at much higher doses than those required for efficacy. Currently, in the United States, the same conventional bioequivalence statistical analysis (i.e., AUC and C_{max}; log-transformed data; ANOVA model with period, sequence, subject (seq), and treatment; and 90% confidence intervals must fit between 80% and 125%) is applied to HVDs, as well as non-HVDs. It is well appreciated that HVDs often require a greater number of subjects than non-HVDs, in order to avoid type II errors when products are erroneously concluded to be not bioequivalent when they are in fact bioequivalent. High variability is a frequent basis for low in vivo bioequivalence study power, necessitating larger number of subject to achieve sufficient power. Drug formulations associated with an intraindividual variability of 35% or more generally fail to meet bioequivalence criteria at an astronomic rate of over 85%. In spite of this pattern of high in vivo bioequivalence testing failure for HVDs, evidence indicates that high variability is frequently not due to poor product quality, an aspect that is primarily investigated in bioequivalence testing but is mostly confounded within the biological variability.

In addition to being subjected to type I and type II errors, the conventional in vivo bioequivalence testing also is imperfectly designed. In traditional bioequivalence testing, the residual variance is composed of (1) analytical variability, (2) within-subject variability in absorption, distribution, metabolism and excretion (ADME), (3) within-formulation variability, (4) subject-by-formulation interaction, and (5) unexplained variability. The conventional two-period designs cannot separate these variance components. Hence, passing the traditional bioequivalence test assumes that the two products have sufficient product quality in that within-formulation variability and subject-by-formulation interaction are small. Traditional bioequivalence testing does not consider differences in within-subject variability between test and reference. Replicate designs where each product is administered twice allow partitioning of the subject-by-formulation interaction from residual variance and estimation of within-subject variability of each of the test and reference. Conventional in vivo bioequivalence testing is not sensitive to detecting a subject-by-formulation interaction effect or a reference that is a highly variable drug product (HVDP), except of course that such increase in variability will necessitate an increase in subject numbers to establish bioequivalence.

While a few theoretical scenarios provide a basis for a subject-by-formulation interaction effect, it perhaps is surprising that conventional in vivo bioequivalence testing is not sensitive to detecting a reference that is an HVDP. The term HVDP differs from HVD. HVDs are drugs with high within-subject variability (ANOVA-CV ≥ 30%) in C_{max} and/or AUC. HVDs are typically associated with high first pass. An HVDP is a formulation of poor pharmaceutical quality where the drug itself is not highly variable and where within-formulation variability (e.g., capsule to capsule variability) is large. Conventional in vivo bioequivalence testing is not sensitive to detecting a reference that is an HVDP.

The weakest reason to favor in vivo bioequivalence over in vitro testing is that the two approaches can provide different results. When products are truly bioequivalent (or truly not bioequivalent), both tests can be correct, both incorrect, or one correct and the other incorrect. The lack of concordance between in vitro and in vivo results reflects type I and type II errors of each approach. This does not reduce the importance of in vitro testing, only the demonstration of the lack of correlation.

Situations where in vitro test should be viewed as preferred include class I drugs with rapid dissolution, class III drugs with very rapid dissolution, and HVDs with rapid dissolution and that are not bio (equivalence) problem drugs. These situations represent a substantial majority of drugs. Class I and III drugs make up about 50% of all marketed oral solid drug products. Upwards of 31% of drugs are HVDs. Since most HVDs show high first-pass metabolism and since many such drugs may be expected to be highly permeable, it can be estimated that a substantial majority of drugs are candidates for in vitro bioequivalence testing as the preferred bioequivalence test.

The current method of statistical modeling of pharmacokinetic studies to elucidate difference in the quality of drug products is inappropriate and wasteful.

Waiver System

The BCS classifies active substances into four different groups (Table 24.1) according to their aqueous solubility at the highest dose in a volume of 250 mL and intestinal permeability. This system of classification was proposed by Gordon Amidon in the mid-1990s. The idea was that when a drug formulated as an IR solid oral drug product has an in vivo high solubility into the gastrointestinal tract and has a high permeability (class 1 drugs), the rate and extent of drug absorption are expected

TABLE 24.1
BCS and Correlation with IVIVC Distribution of Drugs in Various BCS Classes

Class	% Prevalence	Solubility	Permeability	Likelihood of IVIVC
I	30–36	High	High	IVIVC expected (if dissolution is rate-limiting step)
II	30–34	Low	High	IVIVC expected
III	19–28	High	Low	Little or no IVIVC
IV	3–7	Low	Low	Little or no IVIVC

to be equivalent to the in vitro dissolution test, being 85% dissolved in less than 30 min with three different buffers, making in vivo bioequivalence demonstration unnecessary. This approach has been accepted by the FDA and European Medicines Agency (EMA) but is still rarely used, at least in Europe. Nevertheless, the EMA has gone one step further and has released new bioequivalence guidelines that include recommendations on BCS-based biowaivers. Biowaivers would be applicable to products containing BCS class 1 drugs that exhibit high solubility, complete absorption, and rapid in vitro dissolution and also to pharmaceutical products containing substances exhibiting high solubility, limited absorption (BCS class 3), and very rapid in vitro dissolution, once the product has been previously justified for the biowaiver and has demonstrated other specific requirements concerning active substances and excipients. Also, the new EMA guideline said that "excipients that might affect bioavailability are qualitatively and quantitatively the same. In general, the use of the same excipients in similar amounts is preferred." More than that, the WHO had relaxed the solubility ratio and permeability criterion for class 1 and class 3 allowing too some bioexceptions in pharmaceutical products containing BCS class 2 drugs that are weak acids with dose solubility ratio of 250 mL or lower at 37°C over a pH range of 1.2–6.8. Such waivers would increase the speed and decrease the cost of bringing orally administered multisource therapeutics to market.

Recent literature analyzing the bioequivalence outcomes and their BCS classification has shown that the basic premise of awarding waivers based on solubility and permeability does not hold. It has taken decades to accumulate enough data to prove the uselessness of the BCS waivers. Comparisons between pharmaceutical products with active substances from the four BCS classes do not show any differential characteristics of each class in terms of *n*-, inter-, and intrasubject variability for C_{max} or AUC. Despite the usually employed test dissolution methodology proposed as quality control, pharmaceutical products with active substances from the four classes of BCS showed nonbioequivalent studies.

Table 24.1 also shows the relative prevalence of the various classes of drugs in the BCS. It shows that more than 50% (class I and III) of drugs fall in the high-solubility range where the testing may not be required. This means that most drugs currently waived for bioequivalence testing may not be meeting the quality requirements.

The most common criterion for classifying these drugs, in vitro–in vivo correlation (IVIVC), has little value because the dissolution profiles (the in vitro profile) and blood levels (in vivo profile) can be totally independent of each other. There has been too much emphasis given to IVIVC in the past without realizing that combining a biological process with a chemical process defeats the purpose of the exercise.

In Vitro–In Vivo Correlation

The basic premise in the BCS is based on two parameters of the chemical entity: solubility and permeability, both of which are inherent properties of the chemical entity and have nothing to do with differences in the drug products. These two parameters are also interdependent wherein permeability is a function of both aqueous and nonaqueous solubility. Using a thermodynamic property of the active ingredient to establish the quality of a drug product is deeply flawed. Drugs having different permeability potentials behave differently, and drugs that have little permeability issues are generally granted biowaivers. These biowaivers can be questioned as being a carte blanch to a certain class of drugs allowing variations in the drug products that may be relevant to the quality of the drug products.

Biological Variability

For drugs that have problems in crossing the biological barriers, blood level studies are required assuming that somehow the release of drugs at the site of administration can be monitored through blood level monitoring. Reality is that in the case of drugs with problem in absorption, the blood levels are less likely to discriminate between the drug products since the variability introduced by the drug products is much smaller compared to the variability in blood levels resulting from the variations in the absorption properties of the drug.

Postabsorption events such as metabolism and enterohepatic recycling further add to the variability in the pharmacokinetic profiles. Recognition of this inherent variability has resulted in bioequivalence testing requirements that vary across various regulatory agencies to account for the confounding factors such as within-subject variability in ADME and enterohepatic recirculation. For example, the Canadian agency does not require a confidence interval for C_{max}, but corrects for drug content; FDA requirements differ. The Committee for Proprietary Medicinal Products (CPMP)/European Medicines Evaluation Agency (EMEA) guideline allows broadening the bioequivalence limits (e.g., 75%–133%) under certain situations. There are also proposals to broaden the bioequivalence limits according to the within-subject variability of the reference. Additionally, in vivo bioequivalence testing is subject to metric issues, where C_{max} is not viewed as an ideal metric for rate, such that early exposure may sometimes need to be measured. These limitations of in vivo bioequivalence testing have been frequently discussed, resulting in a range of different criteria to assess bioequivalence from pharmacokinetic data.

The differences between two drug products are confounded by the biological variability related to drug absorption and disposition making the current method of testing bioequivalence irrelevant. This observation is clearly demonstrated in the range of pharmacokinetic data considered acceptable for bioequivalence demonstration. Most bioequivalence studies allow a window of acceptability that is mostly 80%–125% or allowing almost 50% variability for the products to be bioequivalent. It is easily understood that the variability in the delivery of drug to the site of administration is likely to be much smaller than the window of variability allowed when blood levels are used as a surrogate for the concentration of drug at the site of administration.

Multiple Dosing

Besides the high variability of the pharmacokinetic profiles because of biological reasons, more complications in using in vivo bioequivalence testing result where multiple dosing is required. First, the drug levels achieved in multiple dosing confound the variations in each dose resulting in a pharmacokinetic profile less capable of differentiating between drug products; second, since the exposure to healthy subjects can be large, the studies are often conducted in patients where the drug clearance mechanisms may be compromised. This also adds to the cost of the study substantially. In multiple-dose studies, each subsequent dose also has the ability to nonlinearly affect the pharmacokinetic profile resulting in extremely complex pharmacokinetic and pharmacodynamic profiles that may have little correlation with the quality of the drug product.

Dissolution Testing

The current methodologies for dissolution testing as used in awarding biowaivers and to monitor the quality of drug products are based on a choice of dissolution medium that is supposed to emulate the site of drug administration. Recent studies have shown that this approach of dissolution testing to establish IVIVC to secure biowaivers do not correlate with bioequivalence outcomes. The fact is that this milieu for the drug delivery systems is not well established and is also highly variable as the drug product passes through a course of anatomy. The correct approach should be to study dissolution rates under not a single but a variety of conditions regardless of their relevance to physiological conditions; once a variety of dissolution profiles are established under different conditions, a comparison between a reference and a test product should suffice declaration of bioequivalence. The rationale behind this proposition is that the two forces within a drug product, interparticle and intermolecular, create the differences in the release rates, and these can be studied only under a variety of conditions of dissolution. A comparable change in the dissolution profile induced by a variety of conditions will point to similarity of the interparticle and intermolecular bonding and thus the quality of drug products.

There is no single universal dissolution media that a priori predicts in vivo drug dissolution. There is no single vitro (and in situ) permeability test condition that mimics the complex intestinal mucosa that a drug can experience over the course of its passage through the gastrointestinal lumen. To overcome this, multicondition dissolution testing and multicondition permeability testing involving mainly various pH levels have been recommended. All of these studies and suggestions continue to force an IVIVC that does not truly exist.

Complex Systems

Whereas detailed guidelines are available for instant or IR drug products, complex systems where controlled or sustained release is intended or where blood levels are not a predictor of the response and toxicity and extremely complex bioequivalence testing are mandated, and in some cases, this means conducting limited clinical trials. This defeats the basic purpose of simplifying the process of approval of generic products or to validate change controls. This complexity arises from the way bioequivalence is defined; if the two drug products have reproducible patterns of release and the underlying factors that produce these profiles are similar, then the need for any complex testing will be obviated.

Onetime Testing

Bioequivalence testing is currently required to be conducted only once when the needs arise as specified earlier. This is the most serious flaw in the current requirements. In most instances, the sample tested might present an early change or development. The phase I materials are generally modified when reaching full commercial production, and these changes are likely missed out in the evaluation of bioequivalence. In the 1980s, the FDA got hit with several fraudulent filings of ANDAs where the developers used the reference drug products for the test substance as well; it took years to decipher this. This would not have been possible if there were a requirement of continuous comparison of the test and reference products. Minor process changes do not require bioequivalence retesting, but over a period of time, several minor changes may have brought significant changes that are currently overlooked.

Time

One of the basic purposes of initiating waivers of clinical trials in ANDAs was to allow faster development of generic products. Whereas waiving clinical trials did reduce the time considerably, the remaining in vivo requirements in bioequivalence trials extend the product development time; this can be significantly curtailed if the purpose of these studies is reexamined to conclude that none of these testing is required to assure a lifetime of quality of products.

Cost

The cost of bioequivalence studies, particularly when fed and multiple dose studies are required, remains very high and keeps out of market many developers; additionally, most developers engage lower-cost vendors where the quality of data may not be as dependable. Eliminating the requirements of bioequivalence testing will increase the number of multisource products and reduce the cost of these drugs to consumers.

Interchangeability

The basic tenet of the Hatch–Waxman law was to provide a generic interchangeable alternate at a lower cost. The awarding interchangeability of products has the assumption that the reference and the test products continue to be similar over the lifetime of the product. Nothing can be farther from truth. Without continuous monitoring of the quality of drug products against the innovator or the earlier product (where it can be based on a predetermined criteria), the lifetime quality of drug products cannot be guaranteed. Additionally, the innovators, against whose product is the interchangeability granted, may change its process of manufacturing and specification and sourcing of API, create a new dissolution profile, and alter the formulation—none of which is confided with the generic drug product manufacturers—and even when this is disclosed, the generic manufacturer is not required to adopt these changes. This creates a serious problem in keeping the two products interchangeable and may be a cause of the reported failures of the generic products. By requiring the generic manufacturers to continuously compare their products with the innovator product should obviate this problem. However, repeated bioequivalence testing is not justified, and in vitro testing on a continuous basis comparing the innovator and the generic product is required to assure that the generic products remain interchangeable.

Global Harmonization

Global regulatory agencies have different opinions on biowaivers. For example, no biowaivers are allowed in Japan. Other agencies use a variety of methods of classifying the drugs for biowaivers and the type of in vitro testing required; this makes global filing of generic products difficult and an expensive task for developers. Even when there is a concordance on the protocol of in vivo bioequivalence testing among the agencies, the requirements for statistical treatment of data differ. These differences point out to the weaknesses inherent in the utilization of bioequivalence testing. A better global harmonization can be achieved by eliminating the need for bioequivalence testing and replacing it with appropriate in vitro tests that can be readily replicated.

New Approach to Bioequivalence Demonstration

Bioequivalence testing is supposed to identify any substantial differences in drug products, not to evaluate the clinical efficacy and toxicity. The chemical drug products used in drug products can be thoroughly analyzed and compared, and the regulatory agencies must demand better comparison profiles of the chemical drug products including impurities, crystalline structure, solvates, and many other properties. Once that is established, the products are declared chemically equivalent. Whether a developer chooses the same formulation of its generic product is neither necessary nor possible—this may well be a pharmaceutical alternate rather than pharmaceutical equivalent. From this point forward, a comparison of the quality of two products is mainly governed by the efficacy of the drug delivery system.

If an identical molecule is released in an equal concentration and rate at the site of administration from two drug products, then these should be declared bioequivalent. To fully understand and exploit this suggestion, we need to examine and appreciate the course of a drug molecule from the drug product to the site of action as given in the following:

Intact drug delivery system ⇒ dispersed drug delivery system ⇒ drug substance in solution at the site of administration ⇒ **membrane transport ⇒ drug substance in the biosystem (e.g., plasma) ⇒ drug substance at the pharmacological action site ⇒ clinical or toxic response.**

The steps in bold represent the processes that are independent of the quality of the drug product; these are also the processes that are highly variable and should not be made part of the evaluation of the drug product, whose only job is to deliver the active pharmaceutical ingredient at the site of action; when the concentration and the rate of delivery at the site of administration are identical, the drug products should be declared bioequivalent. However, determining the concentration and the rate of availability at the site of administration itself poses problems; unless we can provide a direct sampling, which is not possible, a surrogate test needs to be developed. A surrogate test need not be correlating the properties of the drug delivery system at the site of administration but assure that the two drug products will behave similarly at the site of administration. This concept of comparison is novel and does not replicate the same errors that were committed in developing the use of pharmacokinetic profiling to simulate the drug concentration at the site of action.

Fortunately, sound scientific principles can be applied to providing an assurance of the quality of drug products if the mechanism of delivery of active pharmaceutical ingredient at the site of administration is better understood. Generally, there are two physical and chemical forces, interparticle and intramolecular bonding that must be overcome to yield a monomolecular solution. The key to evaluating these bonding forces comes from dissolution profiles obtained under a variety of conditions that can change the ease with which these forces are overcome during dissolution. If the two products demonstrate a similar shift in dissolution rate, the assumption that the bonding forces are the same or can be overcome identically can be assumed.

Drug molecules reaching the site of action face a multitude of inevitable barriers—the biobarriers—the classical bipoidal layers that are present in our body to protect us from the entry of foreign chemicals that can harm us. These barriers cascade throughout the body leading all the way to the cellular level of the site of pharmacological response. The only function of a drug product is to deliver the active drug at the site of administration. From this point forward, the disposition of drug molecules is controlled by the body physiology independently of the drug product.

There is a need to fully exploit dissolution testing to correlate with the release characteristics at the site of administration rather than correlate it with the pharmacokinetic profile.

The new and novel method proposed here is to test the dissolution rate of the drug products under conditions that may have little to do with the physiological environment and more to do with the chemical potential of the drug products. The proposed testing differs from the current dissolution methods that are limited to a single condition, more particularly, a physiological condition that fails to differentiate between the drug products as they might behave at the site of administration.

THERMODYNAMIC POTENTIAL

Thermodynamic forces drive the factors that cause drug molecules to be released from a drug product: solubility, surface area, temperature, the strength of interparticle binding, intermolecular binding forces, and a complex interdependence of these factors.

Chemical reactions proceed with the evolution or absorption of heat. This heat flow represents differences in chemical energy associated with the rearrangement of atoms in molecules, the making and breaking of bonds to form new substances. When measured at constant pressure, this is the enthalpy change (ΔH) for the reaction.

Physical changes can also involve heat. Typically, the dissolving of a solid in water will involve measurable heat. There is no absolute agreement on whether dissolving itself should be categorized as wholly physical, partly chemical, etc., but intermolecular forces are certainly involved (at the very least). Regardless of the appropriate "label" for dissolving as a process, there is an overall energy term. It is known as the heat of solution, ΔH_{soln}.

It may not be possible to know the precise mechanism for a particular dissolution, but in a hypothetical generalized scenario, there would be at least three energy changes involved:

- Solute particles are separated from the solid mass (energy is absorbed, ΔH_1).
- Solvent particles move apart to make space for dissolved solute (energy is absorbed, ΔH_2).
- Solute and solvent particles are attracted to one another (energy is released, ΔH_3).

For most solids dissolving in water, the sum of the first two terms is greater than the third and thus dissolving is frequently endothermic ($\Delta H_{soln} = +$) and solubility generally increases with increasing temperature. When heats of solution become very highly positive, it is often because the solute and solvent are dissimilar and, in the extreme case, immiscible. The old rule of "like dissolves like" is an approximation, but an important one.

The force that determines the penetration of drug molecules across biological barriers is dependent on the chemical potential of the drug at the site of absorption; while concentration may be subject to several variants, the chemical potential is the correct measure of the effectiveness of a drug product. According to Scheme 1, the free energy of the system changes toward negative free energy, as the drug product transforms from a single drug product to a monomolecular solution. Based on the Gibbs–Helmholtz equation,

$$\Delta G = \Delta H - T\Delta S \tag{24.1}$$

The spontaneous conversion of a highly aggregated drug product into highly dispersed monomolecular solution is entropy driven that supersedes the effects of any enthalpy-driven factors leading to a negative free energy. Once a maximum dispersion of drug molecules is achieved, the chemical potential of the molecular drug products is maximized.

Dissolution profiles represent the thermodynamic transformation of drug products; this phenomenon can be used to test the differences between two drug products if an orthogonal testing approach is adopted where the purpose is to decipher the thermodynamic potential differences between the tested drug products.

A comprehensive approach to compare the thermodynamic potential of the two drug products requires a systematic approach:

1. Assure chemical equivalence. This means that the API has exactly the same chemical structure as the innovator or another product that is compared with. This includes the profiling of the impurities, chemical variants such as stereoisomers, chiral forms, crystalline structure, amorphous component, solvates, and hydrates.
2. Assure pharmaceutical equivalence. This means the design of the drug delivery system and the physical characteristics of all components (e.g., particle size), including excipients. Also assured at this stage is the stability profile of the API in the compared drug delivery systems. Pharmaceutical equivalence is not required and alternates can be used.
3. Assure in vitro drug release equivalence. Dissolution profiles derived under conditions that are capable of capturing any differences in the ability of the drug product to release the drug are established. Currently, the U.S. FDA database on dissolution testing conditions is limited to a set of specific conditions that do not provide adequate comparison opportunity. The new testing method involves a variety of dissolution conditions that show a definite shift in the dissolution profile; a similar shift in the dissolution profile under various thermodynamic stressors (dissolution conditions) establishes bioequivalence and a method for lifetime assurance of drug product quality.

This surrogate approach labeled as thermodynamic equivalence surrogate test (TEST) for bioequivalence has significant advantages over the current methods of testing. This includes (1) reduced costs to developer, (2) assessment of product performance more directly and routinely, (3) obviating exposure to human subjects on an ethical ground, (4) providing a monitoring of drug product quality throughout the life cycle of the product, and (5) universal application to all types of dosage forms, from IR to timed, programmed, sustained, and long-term release.

Thermodynamic Equivalence Surrogate Test for Bioequivalence

The thermodynamic potential (chemical potential) of a drug product is the key determinant of its effectiveness, and a comparison of this potential between a test and a reference drug product should suffice to establish their bioequivalence. The strength of this testing depends on the robustness of the testing conditions developed as these will be highly dependent on the nature of the drug and the dosage form. As a result, the given approach can be equally applicable to all types of dosage forms including immediate, controlled, timed, sustained, or any other type of release profile since the approach is independent of the functionality of the drug and its evaluation and the comparisons are limited to the ability of the drug product to demonstrate identical thermodynamic potential.

Thermodynamic equivalence can be readily established if the dissolution profiles of the test and reference drug products are compared under conditions that yield a variance in the dissolution profile; the premise being that if both reference and test drug products show a similar shift in the dissolution profile, then the underlying thermodynamic potential will also be the same. Providing multiple profiles in a variety of conditions of dissolution will pick out even smaller yet relevant differences between the test and the reference drug products, not possible to be identified by the current methodology.

Currently, the most common dissolution testing is done at fixed conditions:

- The U.S. FDA suggests 50 rpm (paddle) or 100 rpm (basket), 900 mL, U.S. Pharmacopoeia (USP) buffer, 37°C.
- The WHO suggests 75 rpm (paddle) or 100 rpm (basket), 900 mL, USP buffer, 37°C.
- The EU: nothing explicit.

No surfactants and other additives are allowed. There is a need to substantially expand the scope of dissolution testing. In brief, conditions of dissolution that alter the thermodynamic stress during dissolution should be developed to create a matrix of dissolution data, and the overall comparability of these data between the reference and the test drug product would establish thermodynamic equivalence.

The fundamental principle of creating thermodynamic stress involves a clear understanding of the types of forces that come into play in the dissolution of drugs. The first is the interparticle binding in a solid drug product and the other is intermolecular interaction; both of these should be overcome to allow a monomolecular dispersion of the active drug at the site of absorption.

The thermodynamic stress conditions that modify the release characteristics of the drug product may include temperature, pH, dielectric constant, polarity (and bipolarity), osmolality, electrical field, and various permutations and combinations of these conditions. The selection of the stress condition should be based on the ability to alter the dissolution profile significantly for both the test and the reference drug product leading to a possibility of comparison and thus establishment of bioequivalence if the dissolution profiles are superimposable.

The dissolution profiles can be tested using such parameters as concentration at various time intervals and the partial and cumulative AUC under the concentration–time curve and then comparing the dissolution profile of the test and reference drug product and proving comparability by superimposable dissolution profiles.

When comparing the test and reference products, the dissolution profiles are currently compared using a similarity factor (f_2). The similarity factor is a logarithmic reciprocal square root transformation of the sum of squared error and is a measurement of the similarity in the percent (%) of dissolution between the two curves:

$$f_2 = 50 + \log\left\{\left[1 + \left(\frac{1}{n}\right)\sum_{t=1}^{n}\left(R_t - T_t\right)^2\right]^{-0.5} \cdot 100\right\} \tag{24.2}$$

Two dissolution profiles are considered similar when the f_2 value is ≥50. To allow the use of mean data, the coefficient of variation should not be more than 20% at the earlier time points (e.g., 10 min) and should not be more than 10% at other time points. Note that when both the test and the reference products dissolve 85% or more of the label amount of the drug in less than 15 min using all three dissolution media recommended earlier, the profile comparison with an f_2 test is unnecessary.

The use of TEST requires developing a new method of declaring similarity where the drug product is not allowed to dissolve as fast so that the differences can be studied. The thermodynamic stressors listed in the succeeding text are more typical than most relevant; there can be a stressor highly specific to the drug of choice and in all instances, a permutation and combination of these stressors can be studied to provide a matrix of tests that will allow an objective direct measure of the differences between the drug products and thus a comparison of their quality in delivering the active drug products at the site of absorption.

Temperature

The temperature of 37°C is most commonly used with intent to connect this condition to a physiological milieu; in TEST application, the objective is to identify the differences in two drug products in environment that causes a shift in the dissolution rates. Any relevance to physiological conditions can be ancillary but not of any other value. For thermodynamic stress purpose, the temperature needs to be varied significantly, from close the freezing point of the solution to high temperature just short of causing any significant degradation of the active drug. It should be noted that the depression of freezing point and the elevation of boiling point in an electrolyte solution could be substantial and taken into account in deciding the range of temperature studied. Generally, a temperature range of 5°C–50°C should suffice. However, if a change in the temperature of the dissolution media does not produce a statistically significant shift in the dissolution profile, then this stress is not suitable to discern the differences between the test and the reference product. This may happen in the case of highly soluble drugs, in which case dissolution media should be selected to moderate the dissolution rates.

Dissolution profiles obtained at different temperatures reflect the interparticle and intermolecular interactions within the drug product and form an excellent basis of matching the thermodynamic potential of the drug products.

A significant iteration of this exercise can be a programmed heating of the dissolution solution allowing a single profile starting with a low temperature and then reaching to the highest limit selected. A statistically superimposable profile between drug products should indicate high level of interparticle binding and interactions of drug product with the excipients.

To avoid any limitations due to solubility considerations, the current dissolution testing is conducted under sink conditions. However, in establishing a thermodynamic equivalence, this may not be necessary to maintain. Again, the purpose of thermodynamic equivalence demonstration is a comparison of the test product with a reference product, and whatever testing condition that allows it without degrading the product is acceptable.

Dielectric Properties

Drug solubility plays a significant role in the BCS for eligibility for waivers from bioequivalence testing. The changes in dielectric constant of the medium have a dominant effect on the solubility of the ionizable solute in which higher dielectric constant can cause more ionization of the solute and results in more solubilization. As an example, water (*DW*, 298 = 78.5) has higher dissociation strength on ions in comparison with ethanol (*DE*, 298 = 24.2), which results in more solubilization power of ions in water (Figure 24.1).

The following equation is a theoretical model for solubility correlation in two different media or phases:

$$\text{Log}\left(\frac{S_1}{S_2}\right) = \left(0.4343 \times \frac{e^2}{2rkT}\right)\left(\frac{1}{D_2}\right) - \left(\frac{I}{D_1}\right) \tag{24.3}$$

where

S_1 and S_2 are the solubilities of the solute in media 1 and 2
e is the charge of an electron
r is the effective radius of the ion in the medium
k is the Boltzmann constant
T is the absolute temperature
D_1 and D_2 are the dielectric constants of the media 1 and 2, respectively

Unfortunately, by using this equation, the predicted solubility values (when r values are known) or predicted r values (when solubility values are known) based on experimental data do not seem to be meaningful. However, one can consider the constant value of ($0.4343 \times e^2/2rk$) as AT for a specific solute and obtain

$$\text{Log}\left(\frac{S_1}{S_2}\right)=\left(\frac{AT}{T}\right)\left(\frac{1}{D_2}\right)-\left(\frac{1}{D_1}\right) \tag{24.4}$$

where AT is a slope that can be calculated using two experimental solubility data points (e.g., solubility values in water and ethanol). The resulted AT values show indirect relation with temperature. This is expected, as it has indirect correlation with r of Equation 24.3, which has direct correlation with temperature. Also, it seems that AT values are not mainly affected by the structure of the solutes under study.

The main advantage of the proposed model is that it does not require any experimental solubility data in mixed solvents. Just two experimental solubility data points in monosolvents and dielectric constants of solvent systems under consideration are employed in the prediction process. It almost provides good results, which might show its applicability in solubility prediction.

The main disadvantage of the proposed prediction method is that it is applicable only for the solubility prediction of electrolytes or zwitterions in which the ionization is the dominant parameter and the phenomenon could be represented.

The rate of release of drug products into a solution is determined by the kinetic energy as studied earlier in the variation of the temperature, but it is also affected by the dielectric properties of the dissolution medium; the dielectric constant also affects the solubility and thus the driving force for the release of drug molecules from the drug products. Given in the following is a listing of the dielectric constants of various media that may be suitable for this testing. One reason why these have not been tested is because the focus of most of the research in the past has been to emulate a physiological model rather than examine the drug product under conditions to yield differentiating dissolution rates (Table 24.2).

Surfactants

Surfactants alter the solubility of drug by micellization; enhanced wettability of the surface of the drug product can additionally affect the dissolution rates by reducing the coating of particles by air and also by affecting the energy of interparticle and intermolecular bonds. If the addition of a surfactant changes the dissolution profile and the test and reference product show similar changes, it will indicate equality of the chemical potential at the site of administration. The choice of surfactants can include both polar or nonpolar surfactants and a combination of surfactants.

pH

The effect of dissolution medium pH on dissolution of drugs has been well studied, and the current biowaiver guidelines include demonstrating dissolution at various pH as a prerequisite to requesting waivers from bioequivalence testing. The premise behind the current requirements is to correlate these profiles in a physiological milieu. Given that the excipients as well as the active drug is highly affected in its state of ionization based on pH, the real test of similarity will involve studying dissolution throughout the available pH range, perhaps from 1 to 14, and then comparing the drug products based on the superimposability of these profiles. Subtle differences in the physicochemical characteristics would readily become evident under these conditions.

TABLE 24.2
Dielectric Constant of Various Liquids

Compound	Dielectric Constant
Vacuum	1.0
Pyridine	1.1
Butane	1.4
Methane	1.7
Pentane	1.8
Heptane	1.9
Chlorine	2.0
Decane	2.0
Dodecane	2.0
Hexane	2.0
Octane	2.0
Turpentine (wood)	2.2
Carbon tetrachloride	2.2
Toluene	2.3
Benzene	2.3
Palmitic acid	2.3
Stearic acid	2.3
Cumene	2.4
Styrene	2.4
Naphthalene	2.5
Caproic acid	2.6
Carbon disulfide	2.6
Linoleic acid	2.7
Pinene	2.7
Terpinene	2.7
Furan	3.0
Bromine	3.1
Cotton seed oil	3.1
Olive oil	3.1
Resorcinol	3.2
Linseed oil	3.3
Ether	4.3
Phenol	4.3
Castor oil	4.7
Chloroform	4.8
Acetic acid	6.2
Ethylamine	6.3
Aniline	7.3
Cresol	10.6
Propylene	11.9
Hexanol	13.3
Ammonia (aqua)	16.5
Acetone	20.7
Alcohol, propyl	21.8
Ethanol	24.3

(Continued)

TABLE 24.2 (*Continued*)
Dielectric Constant of Various Liquids

Compound	Dielectric Constant
Methanol	33.1
Ethylene glycol	37.0
Furfural	42.0
Glycerol	42.5
Hydrazine	52.0
Glycerin	55.0
Water	80.4

Osmolality

Whereas buffers are used to adjust the pH for dissolution testing, the buffer capacity and thus the osmolality of the dissolution medium have not been the focus of any significant studies. The osmolality plays a significant role in determining the electrostatic interactions between the drug product, the released molecules, and the components of the dissolution medium. The range of osmolality should be wide, likely between 100 and 1000 mOsm/L and even higher to cause a significant change in the dissolution profile of the reference test product. There is no need to select an osmolality based on any physiological relevance as it is used merely as a thermodynamic stressor. If suitable osmolality to induce changes in the dissolution profile is not observed, then other more suitable thermodynamic stressor should be chosen.

Lipophilicity

While the stress of dielectric characteristics determines the polarity of the medium, the lipophilicity of the dissolution medium can play a significant role of thermodynamic stressor. In many instances, selecting appropriate lipophilicity of the dissolution medium can substantially slow down or enhance the dissolution rates. The goal of including this stressor is to provide a highly valuable thermodynamic stressor to create a matrix of dissolution profiles.

Bipolarity

A biphasic system with both polar and nonpolar characteristics can provide an excellent thermodynamic stressor when other stressors may not be so effective. Ideally, an emulsion (o/w type) dissolution medium should be used to determine if this constitutes a thermodynamic stressor. Alternately, a w/o type dissolution medium can also be used. An interesting biphasic system can be the milk containing various percentages of fats; other biphasic systems such as coacervates can also be very useful.

Electrical Field

An electrical field applied to the dissolution medium can induce significant movement of charged species, both that come from the drug product and those that are present in the dissolution medium. Voltage ranges from 1 to 20 DC V can be very useful in creating an electrical field that can modify the release characteristics of the drug products.

Physical Stress

The normal physical stress provided in a typical dissolution apparatus is a paddle moving the liquid as seen in the most widely used USP apparatus. A variety of other mixing means may have variable effect on the release profile of the active drug. The purpose of choosing alternates is not to simulate any physiological conditions but to provide an environment where the dissolution profiles are shifted and then observe the differences between the test and the reference drug products. The physical stress may involve using apparatus that allow rolling of the drug product, mechanical vibrations, ultrasonic vibrations, orbital motion, linear motion vertically or horizontally, and any other type that can result in a change in the dissolution profile.

Sink Condition

Most dissolution testing is conducted under sink conditions to obviate the effect of drug saturation solubility on the dissolution rates; however, in reality, the site of administration may not have sufficient volume and thus not emulate a sink condition. The dissolution testing should be conducted at both sink and nonsink conditions making the solubility parameter as another thermodynamic stress. In those cases where the drug is readily dissolved, dissolution media should be chosen to reduce the dissolution rates to allow reasonable ranges of comparisons. Under sink conditions, the dissolution profile will change but not the plateau level; under nonsink conditions, both can change providing better opportunities of comparing two drug products.

Duration of Testing

Since the solubility of the active drug is constant under specific conditions, every dissolution test will eventually lead to a plateau effect as it is entirely driven by the solubility. How fast a dissolution profile reaches the plateau is determined by the structural differences in the drug products. An appropriate test method will avoid reaching a plateau, and generally, short-term testing will reveal greater information about the differences between drug products; studies conducted for less than 60 min will be most relevant.

Proposed Protocols

TEST™ for bioequivalence will require establishing a matrix of tests, using at least two independent thermodynamic stressors such as temperature and polarity. The next step will be to identify at least three conditions for each stressor that demonstrate a clear shift in the dissolution profile and then generating a dissolution matrix; for example, in the case of two stressors and three conditions, it will yield a matrix of nine test profiles. Applying the same conditions of testing to both the reference and test drug products, a statistical analysis of the reproducibility of effects between the test and reference drug product would establish thermodynamic equivalence. An almost unlimited set of conditions can be applied to force a variation at three levels for every drug.

SUMMARY

In summary, there are several concrete reasons for challenging the current methodology of testing bioequivalence:

- The purpose of bioequivalence testing should be to identify any differences in two drug products, regardless of their projected impact on clinical efficacy. This is necessary since the correlation between formulation differences and clinical efficacy will never be fully understood.

- The pharmacokinetic profile of a drug product does not constitute a good surrogate for the concentration of drug at the site of action; this is mostly presumptuous.
- The ultimate test of bioequivalence should be the rate and extent of delivery of drug at the site of administration, not the site of action, as currently required; this is a significant step to consider since the site of action for most drugs remains elusive and almost never accessible for sampling.
- The potential for a drug to be absorbed in the biological system is proportional to the chemical potential of the drug at the site of administration, which is in turn proportional to the thermodynamic potential of the drug product.
- The thermodynamic potential of drug products can be measured by subjecting them to conditions wherein the dissolution profiles are altered; observing similar changes between the test and the reference drug product should establish thermodynamic equivalence.
- The thermodynamic equivalence can be applied to every type of dosage form regardless of its release characteristics since the test establishes that under all conditions, the chemical potential will be identical to the innovator product or a reference product.
- The thermodynamic potential test can be used as a routine test to monitor the quality of the product throughout the lifetime of the product.
- There is no rationale for exposing healthy humans to drug testing when the results show that these data are inconclusive.
- Significant cost savings and reduction in the development time can be achieved by eliminating all in vivo testing.
- The use of thermodynamic equivalence will eliminate the exposure to humans, reduce the cost of development, and allow faster regulatory approvals—all needed to bring the cost of drugs on a global basis.

Appendix A: Glossary of Terms

Accepted reference value: An accepted reference value is a value used as a reference, agreed for a comparison and derived from the following: a theoretical or established value, based on scientific reasons; an assigned or certified value, based on experimental data from a national or international organization; a consensus value, based on a collaborative experimental work; and the mathematical expectation of the (measurable) quantity, in case where the previous points are not applicable, that is, the mean of a specified population of determinations (cf. NF ISO 5725-1).

Accuracy: The degree of closeness of the determined value to the nominal or known true value under prescribed conditions. This is sometimes termed trueness.

Accuracy: (IEEE) (1) A qualitative assessment of correctness or freedom from error. (2) A quantitative measure of the magnitude of error. Contrast with precision (CDRH). (3) The measure of an instrument's capability to approach a true or absolute value. It is a function of precision and bias. See bias, precision, calibration. The degree of closeness of the determined value to the nominal or known true value under prescribed conditions. This is sometimes termed trueness.

Act: The Federal Food, Drug, and Cosmetic Act, as amended (sections 201-902, 52 Stat. 1040 et seq., as amended [21 U.S.C. 321-392]).

Adverse drug reaction (ADR): In the preapproval clinical experience with a new medicinal product or its new usages, particularly as the therapeutic dose(s) may not be established, all noxious and unintended responses to a medicinal product related to any dose should be considered adverse drug reactions. The phrase responses to a medicinal product means that a causal relationship between a medicinal product and an adverse event is at least a reasonable possibility, that is, the relationship cannot be ruled out. Regarding marketed medicinal products: a response to a drug that is noxious and unintended and that occurs at doses normally used in man for prophylaxis, diagnosis, or therapy of diseases or for modification of physiological function (see the *ICH Guideline for Clinical Safety Data Management: Definitions and Standards for Expedited Reporting*).

Adverse event (AE): Any untoward medical occurrence in a patient or clinical investigation subject administered with a pharmaceutical product and who does not necessarily have a causal relationship with this treatment. An adverse event (AE) can therefore be any unfavorable and unintended sign (including an abnormal laboratory finding), symptom, or disease temporally associated with the use of a medicinal (investigational) product, whether or not related to the medicinal (investigational) product (see the *ICH Guideline for Clinical Safety Data Management: Definitions and Standards for Expedited Reporting*).

Algorithm analysis: (IEEE) A software V&V task to ensure that the algorithms selected are correct, appropriate, and stable and meet all accuracy, timing, and sizing requirements.

Algorithm: (IEEE) (1) A finite set of well-defined rules for the solution of a problem in a finite number of steps. (2) Any sequence of operations for performing a specific task.

Analysis: (1) To separate into elemental parts or basic principles so as to determine the nature of the whole. (2) A course of reasoning showing that a certain result is a consequence of assumed premises. (3) (ANSI) The methodical investigation of a problem, and the separation of the problem into smaller related units for further detailed study.

Analyte or activity: The analyte (or activity when relevant) is the matter of the analytical procedure. The analyte is a physical entity (e.g., water activity), chemical entity (e.g., active substance alone or in a pharmaceutical formulation, total lipids, aspartame, lead), or biological entity (e.g., ATP-metric activity). In the case of quantitative analytical procedures of farm and food products, "analyte" is equivalent to "measurand."

Analyte: A specific chemical moiety being measured; it can be an intact drug, a biomolecule or its derivative, a metabolite, and/or a degradation product in a biologic matrix.

Analytical laboratory: A facility used by a pharmaceutical sponsor or contract research organization to determine the nature and proportionate quantities of the constituents of a compound for an in vivo BE study. An analytical laboratory typically completes an assay to determine the drug concentration in body fluids.

Analytical procedure (or method or assay): Written procedure that describes all the means and the operating procedures required to perform the analysis of the analyte. That is, field of application, principle and/or reactions, definitions, reactants, equipments, operating procedures, expression of results, suitability tests, and test reports.

Analytical run (or batch): A complete set of analytical and study samples with appropriate number of standards and QCs for their validation. Several runs (or batches) may be completed in 1 day, or one run (or batch) may take several days to complete.

Analytical run: A complete set of analytical and study samples with appropriate number of standards and QCs for their validation. Several runs may be completed in 1 day, or one run may take several days to complete.

ANDA: Abbreviated drug application.

Anomaly: (IEEE) Anything observed in the documentation or operation of software that deviates from expectations based on previously verified software products or reference documents.

ANSI: American National Standards Institute.

API: Active pharmaceutical ingredients.

Applicable regulatory requirement(s): Any law(s) and regulation(s) addressing the conduct of clinical trials of investigational products.

Approval (in relation to institutional review boards): The affirmative decision of the institutional review board (IRB) that the clinical trial has been reviewed and may be conducted at the institution site within the constraints set forth by the IRB, the institution, good clinical practice (GCP), and the applicable regulatory requirements.

ASCII: American Standard Code for Information Interchange.

Attributable data: Attributable data are those that can be traced to individuals responsible for observing and recording the data. In an automated system, attributability could be achieved by a computer system designed to identify individuals responsible for any input.

AUC_{0-inf}: Area under the concentration–time curve extrapolated to infinity.

AUC_{0-t}: Area under the concentration–time curve from time zero to the last measurable time point.

AUC_{0-tau}: Area under the concentration–time curve for one dosing interval at steady state.

Audit: A systematic and independent examination of trial-related activities and documents to determine whether the evaluated trial-related activities were conducted, and the data were recorded, analyzed, and accurately reported according to the protocol, sponsor's standard operating procedures (SOPs), good clinical practice (GCP), and the applicable regulatory requirement(s).

Audit certificate: A declaration of confirmation by the auditor that an audit has taken place.

Audit report: A written evaluation by the sponsor's auditor of the results of the audit.

Audit trail: Documentation that allows reconstruction of the course of events. (1) (ISO) Data in the form of a logical path linking a sequence of events, used to trace the transactions that have affected the contents of a record. (2) A chronological record of system activities that is sufficient to enable the reconstruction, reviews, and examination of the sequence of environments and activities surrounding or leading to each event in the path of a transaction from its inception to output of final results. An audit trail is a secure, computer-generated, time-stamped electronic record that allows reconstruction of the course of events relating to the creation, modification, and deletion of an electronic record.

Audit: (1) (IEEE) An independent examination of a work product or set of work products to assess compliance with specifications, standards, contractual agreements, or other criteria. (2) (ANSI) To conduct an independent review and examination of system records and activities in order to test the adequacy and effectiveness of data security and data integrity procedures, to ensure compliance with established policy and operational procedures, and to recommend any necessary changes.

BA: Bioavailability.

Baseline: (NIST) A specification or product that has been formally reviewed and agreed upon, that serves as the basis for further development, and that can be changed only through formal change control procedures.

Batch formula (composition): A complete list of the ingredients and their amounts to be used for the manufacture of a representative batch of the drug product. All ingredients should be included in the batch formula whether or not they remain in the finished product.

Batch: A specific quantity or lot of a test or control article that has been characterized according to section 58.105(a). (IEEE) Pertaining to a system or mode of operation in which inputs are collected and processed all at one time, rather than being processed as they arrive, and a job, once started, proceeds to completion without additional input or user interaction. Contrast with conversational, interactive, online, real time; A batch is a number of unknown samples from one or more patients in a study and QCs that are processed at one time; A specific quantity of a drug or other material produced according to a single manufacturing order during the same cycle of manufacture and intended to have uniform character and quality, within specified limits (21 CFR 210.3(b)(2)).

Batch processing: Execution of programs serially with no interactive processing. Contrast with real-time processing.

BCS: Biopharmaceutics classification system.

BE: Bioequivalence.

Benchmark: A standard against which measurements or comparisons can be made.

Bias: A measure of how closely the mean value in a series of replicate measurements approaches the true value.

Bioavailability: The rate and extent to which the active ingredient or active moiety is absorbed from a drug product and becomes available at the site of action. For drug products that are not intended to be absorbed into the bloodstream, bioavailability may be assessed by measurements intended to reflect the rate and extent to which the active ingredient or active moiety becomes available at the site of action.

Biobatch: The lot of drug product formulated for purposes of pharmacokinetic evaluation in a bioavailability/bioequivalency study. For modified release solid oral, this batch should be 10% or greater than the proposed commercial production batch or at least 100,000 units, whichever is greater.

Bioequivalence (BE): The absence of a significant difference in the rate and extent to which the active ingredient or active moiety in pharmaceutical equivalents or pharmaceutical alternatives becomes available at the site of drug action when administered at the same molar dose under similar conditions in an appropriately designed study.

Bioequivalence conditions: "A drug shall be considered to be bioequivalent to a listed drug if— (1) the rate and extent of absorption of the drug do not show a significant difference from the rate and extent of absorption of the listed drug when administered at the same molar dose of the therapeutic ingredient under similar experimental conditions in either a single dose or multiple doses; or (2) the extent of absorption of the drug does not show a significant difference from the extent of absorption of the listed drug when administered at the same molar dose of the therapeutic ingredient under similar experimental conditions in either a single dose or multiple doses and the difference from the listed drug in the rate of absorption of the drug is intentional, is reflected in its proposed labeling, is not essential to the attainment of effective body drug concentrations on chronic use, and is considered medically insignificant for the drug." (section 505 (j)[7](B) of the Federal Food, Drug, and Cosmetic Act.)

Bioequivalence: The absence of a significant difference in the rate and extent to which the active ingredient or active moiety in pharmaceutical equivalents or pharmaceutical alternatives becomes available at the site of drug action when administered at the same molar dose under similar conditions in an appropriately designed study. Where there is an intentional difference in rate (e.g., in certain extended release dosage forms), certain pharmaceutical equivalents or alternatives may be considered bioequivalent if there is no significant difference in the extent to which the active ingredient or moiety from each product becomes available at the site of drug action. This applies only if the difference in the rate at which the active ingredient or moiety becomes available at the site of drug action is intentional and is reflected in the proposed labeling, is not essential to the attainment of effective body drug concentrations on chronic use, and is considered medically insignificant for the drug.

Bioequivalence requirement: A requirement imposed by the Food and Drug Administration for in vitro and/or in vivo testing of specified drug products that must be satisfied as a condition of marketing.

Bioequivalence Studies for Modified Release Drug Product: The OGD Guidance (3). The bioequivalence study should be conducted using the reference-listed drug (RLD) product and/or the innovator drug product as the reference and the test product should be the product (generic or innovator) that has undergone postapproval change.

Bioequivalent drug products: (1) Pharmaceutical equivalent or pharmaceutical alternative products that display comparable bioavailability when studied under similar experimental conditions. The regulatory authorities describes one set of conditions under which a test and reference-listed drug shall be considered bioequivalent: the rate and extent of absorption of the test drug do not show a significant difference from the rate and extent of absorption of the reference drug when administered at the same molar dose of the therapeutic ingredient under similar experimental conditions in either a single dose or multiple doses or the extent of absorption of the test drug does not show a significant difference from the extent of absorption of the reference drug when administered at the same molar dose of the therapeutic ingredient under similar experimental conditions in either a single dose or multiple doses and the difference from the reference drug in the rate of absorption of the drug is intentional, is reflected in its proposed labeling, is not essential to the achievement of effective body drug concentrations on chronic use, and is considered medically insignificant for the drug. (2) This term describes pharmaceutical equivalent or pharmaceutical alternative products that display comparable bioavailability when studied under similar experimental conditions.

Biological matrix: A discrete material of biological origin that can be sampled and processed in a reproducible manner. Examples are blood, serum, plasma, urine, feces, saliva, sputum, and various discrete tissues.

Bioresearch monitoring (BIMO) goal date (or inspection summary goal date): The date by which the OGD Division of Bioequivalence (DBE) project manager anticipates a review of the inspection results from Division of Scientific Investigations (DSI). This date is determined in consultation with DSI staff and includes time to assess the inspection results.

Blank: (1) A sample of a biological matrix to which no analytes have been added and is used to assess the specificity of the bioanalytical method. (2) Test performed in the absence of the matrix (blank reactant) or in a matrix without analyte (blank matrix). By extension, the instrument response in the absence of the analyte is used (blank instrumental).

Blinding/masking: A procedure in which one or more parties to the trial are kept unaware of the treatment assignment(s). Single blinding usually refers to the subject(s) being unaware, and double blinding usually refers to the subject(s), investigator(s), monitor, and, in some cases, data analyst(s) being unaware of the treatment assignment(s).

Bomb: A Trojan horse that attacks a computer system upon the occurrence of a specific logical event (logic bomb), the occurrence of a specific time-related logical event (time bomb), or is hidden in electronic mail or data and is triggered when read in a certain way (letter bomb). See Trojan horse, virus, worm.

Boolean: Pertaining to the principles of mathematical logic developed by George Boole, a nineteenth century mathematician. Boolean algebra is the study of operations such as *add*, *subtract*, and *multiply* carried out on variables that can have only one of two possible values, that is, 1 (true) and 0 (false).

Bootstrap: (IEEE) A short computer program that is permanently resident or easily loaded into a computer and whose execution brings a larger program, such as an operating system or its loader, into memory.

Boundary value analysis: (NBS) A selection technique in which test data are chosen to lie along "boundaries" of the input domain [or output range] classes, data structures, procedure parameters, etc. Choices often include maximum, minimum, and trivial values or parameters. This technique is often called stress testing. See testing, boundary value.

Boundary value: (1) (IEEE) A data value that corresponds to a minimum or maximum input, internal, or output value specified for a system or component. (2) A value that lies at, or just inside or just outside, a specified range of valid input and output values.

Brand-name drug: A brand-name drug is a drug marketed under a proprietary, trademark-protected name.

Bulk drug substance: Any substance represented for use in a drug and when in the manufacturing, processing, or packaging of a drug becomes an active ingredient of a finished dosage form. This does not include intermediates used in the synthesis of such substances.

Calibration standard (or calibration sample): Calibration standards are samples of known concentrations, with or without matrix, that allow drawing the calibration curve.

Calibration standard: A biological matrix to which a known amount of analyte has been added. Calibration standards are used to construct calibration curves from which the concentrations of analytes in quality control samples and in unknown study samples are determined.

Calibration: Ensuring continuous adequate performance of sensing, measurement, and actuating equipment with regard to specified accuracy and precision requirements. See accuracy, bias, precision.

Case report form (CRF): A printed, optical, or electronic document designed to record all of the protocol required information to be reported to the sponsor on each trial subject.

CavSS: Average plasma concentration at steady state.

Certified copy: A copy of original information that has been verified, as indicated by dated signature, as an exact copy having all of the same attributes and information as the original.

Change control: The processes, authorities for, and procedures to be used for all changes that are made to the computerized system and/or the system's data. Change control is a vital subset of the quality assurance (QA) program within an establishment and should be clearly described in the establishment's SOPs.

Clinical facility: A site where patients or subjects are examined and observed during an in vivo BE study.

Clinical investigator: An individual who actually conducts a clinical investigation (i.e., under whose immediate direction the drug is administered or dispensed to a subject) (21 CFR 312.3(b)). In this guidance, when a clinical investigation involves BA or BE studies, the clinical investigator has the responsibility of retaining the reserve samples at the testing facility or through an independent third party.

Clinical trial/study: Any investigation in human subjects intended to discover or verify the clinical, pharmacological, and/or other pharmacodynamic effects of an investigational product(s) and/or to identify any adverse reactions to an investigational product(s) and/or to study absorption, distribution, metabolism, and excretion of an investigational product(s) with the object of ascertaining its safety and/or efficacy. The terms clinical trial and clinical study are synonymous.

Clinical trial/study report: A written description of a trial/study of any therapeutic, prophylactic, or diagnostic agent conducted in human subjects, in which the clinical and statistical description, presentations, and analyses are fully integrated into a single report (see the *ICH Guideline for Structure and Content of Clinical Study Reports*).

C_{max}**:** Peak concentration.

$C_{max,SS}$**:** Peak concentrations during the dosing interval at steady state.

$C_{min,SS}$**:** Minimum or trough concentrations at steady state.

Code of Federal Regulations CPMP: Commission for Proprietary Medicinal Products.

Comparator (product): An investigational or marketed product (i.e., active control), or placebo, used as a reference in a clinical trial.

Compliance (in relation to trials): Adherence to all the trial-related requirements, good clinical practice (GCP) requirements, and the applicable regulatory requirements.

Compliance classification: The compliance status of an inspection.

Contract research organization (CRO): (1) A person or an organization (commercial, academic, or other) contracted by the sponsor to perform one or more of a sponsor's trial-related duties and functions. (2) An independent contractor of the sponsor or manufacturer that assumes one or more of the obligations of a sponsor (e.g., design of a protocol, selection or monitoring of investigations, evaluation of reports, and preparation of materials to be submitted to the FDA) (21 CFR 312.3(b)). This guidance addresses BA and BE studies submitted to support approvals of new and generic drugs. These studies are usually conducted by CROs under contract to study sponsors and/or drug manufacturers. Many CROs have their own testing facilities, with physicians (to serve as clinical investigators) and clinical support staff (e.g., nurses, medical technologists) to conduct the BA and BE studies.

Contract: A written, dated, and signed agreement between two or more involved parties that sets out any arrangements on delegation and distribution of tasks and obligations and, if appropriate, on financial matters. The protocol may serve as the basis of a contract.

Control article: Any food additive, color additive, drug, biological product, electronic product, medical device for human use, or any article other than a test article, feed, or water that is administered to the test system in the course of a nonclinical laboratory study for the purpose of establishing a basis for comparison with the test article.

Coordinating committee: A committee that a sponsor may organize to coordinate the conduct of a multicenter trial.

Coordinating investigator: An investigator assigned the responsibility for the coordination of investigators at different centers participating in a multicenter trial.

Correctness: (IEEE) The degree to which software is free from faults in its specification, design, and coding. The degree to which software, documentation, and other items meet specified requirements. The degree to which software, documentation, and other items meet user needs and expectations, whether specified or not.

Critical control point: (QA) A function or an area in a manufacturing process or procedure, the failure of which, or loss of control over, may have an adverse effect on the quality of the finished product and may result in a unacceptable health risk.

Critical design review: (IEEE) A review conducted to verify that the detailed design of one or more configuration items satisfy specified requirements; to establish the compatibility among the configuration items and other items of equipment, facilities, software, and personnel; to assess risk areas for each configuration item; and, as applicable, to assess the results of producibility analyses, review preliminary hardware product specifications, evaluate preliminary test planning, and evaluate the adequacy of preliminary operation and support documents.

Critical drugs: "Critical dose drugs" are defined as those drugs where comparatively small differences in dose or concentration lead to dose- and concentration-dependent, serious therapeutic failures and/or adverse drug reactions that may be persistent, irreversible, slowly reversible, or life-threatening events.

Critical equipment variable: A specific design, operating principle, or automation of equipment that can affect a specific performance variable critical to the ultimate and predictable performance of the dosage form and its drug.

Critical manufacturing variable: Includes those manufacturing materials (critical composition variable), methods, equipment, and processes that significantly affect drug release, from the formulation (e.g., coating thickness, particle size, crystal form, excipient type, concentrations and distribution, and tablet hardness).

Critical processing variable: A specific step, unit process, or condition of a unit process that can affect a specific performance variable critical to the ultimate and predictable performance of the dosage form and its drug.

Criticality analysis: (IEEE) Analysis that identifies all software requirements that have safety implications, and assigns a criticality level to each safety-critical requirement based upon the estimated risk.

Criticality: (IEEE) The degree of impact that a requirement, module, error, fault, failure, or other item has on the development or operation of a system. Syn: severity.

Cross-validation: Comparison validation parameters of two bioanalytical methods.

Data analysis: (IEEE) (1) Evaluation of the description and intended use of each data item in the software design to ensure the structure and intended use will not result in a hazard. Data structures are assessed for data dependencies that circumvent isolation, partitioning, data aliasing, and fault containment issues affecting safety and the control or mitigation of hazards. (2) Evaluation of the data structure and usage in the code to ensure each is defined and used properly by the program. Usually performed in conjunction with logic analysis.

Data validation: (1) (ISO) A process used to determine if data are inaccurate, incomplete, or unreasonable. The process may include format checks, completeness checks, check key tests, reasonableness checks, and limit checks. (2) The checking of data for correctness or compliance with applicable standards, rules, and conventions.

Data: Representations of facts, concepts, or instructions in a manner suitable for communication, interpretation, or processing by humans or by automated means.

Delayed release: For enteric-coated drug products, drug-release procedures described in USP 23 NF 18, sections <711> and <724>, should be followed. When the guidance refers to dissolution testing in addition to application/compendia release requirements, the dissolution test should be performed in 0.1 N HCl for 2 h (acid stage) followed by testing in USP buffer media, in the range of pH 4.5–7.5 (buffer stage) under standard (application/compendia) test conditions and increased agitation speeds using the application/compendia test apparatus. For the rotating basket method (Apparatus 1), a rotation speed of 50, 100, and 150 rpm may be used, and for the rotating paddle method (Apparatus 2), a rotation speed of 50, 75, and 100 rpm may be studied. Multipoint dissolution profiles should be obtained during the buffer stage of testing. Adequate sampling should be performed, for example, at 15, 30, 45, 60, and 120 min (following the time from which the dosage form is placed in the buffer) until either 80% of the drug is released or an asymptote is reached. The aforementioned dissolution testing should be performed using the changed drug product and the biobatch or marketed batch (unchanged drug product).

Delayed release: Release of a drug (or drugs) at a time other than immediately following oral administration.

Direct access: Permission to examine, analyze, verify, and reproduce any records and reports that are important to evaluation of a clinical trial. Any party (e.g., domestic and foreign regulatory authorities, sponsor's monitors, and auditors) with direct access should take all reasonable precautions within the constraints of the applicable regulatory requirement(s) to maintain the confidentiality of subjects' identities and sponsor's proprietary information.

Direct entry: Recording data where an electronic record is the original capture of the data. Examples are the keying by an individual of original observations into the system or automatic recording by the system of the output of a balance that measures subject's body weight.

Directed inspection: An inspection based on substantive information suggesting scientific misconduct, major human subject protection violations, or compromised BE data.

Dissolution testing extended release: Dissolution testing should be conducted on 12 individual dosage units for the changed drug product and the biobatch or marketed batch (unchanged drug product). The potential for pH dependence of drug release from a modified release drug product is well recognized. Multipoint dissolution profiles should be obtained using discriminating agitation speed and medium. A surfactant may be used with appropriate justification. Early sampling times of 1, 2, and 4 h should be included in the sampling schedule to provide assurance against premature release of the drug (dose dumping) from the formulation. Differing sampling times should be justified to prevent premature drug release. See current USP or NF.

DMF: Drug master file.

Documentation: All records, in any form (including, but not limited to, written, electronic, magnetic, and optical records, and scans, x-rays, and electrocardiograms), that describe or record the methods, conduct, and/or results of a trial, the factors affecting a trial, and the actions taken.

Documentation plan: (NIST) A management document describing the approach to a documentation effort. The plan typically describes what documentation types are to be prepared, what their contents are to be, when this is to be done and by whom, how it is to be done, and what are the available resources and external factors affecting the results.

Documentation, level of: (NIST) A description of required documentation indicating its scope, content, format, and quality. Selection of the level may be based on project cost, intended usage, extent of effort, or other factors; for example, level of concern.

Documentation, software: (NIST) Technical data or information, including computer listings and printouts, in human-readable form, that describe or specify the design or details, explain the capabilities, or provide operating instructions for using the software to obtain desired results from a software system. See specification; specification, requirements; specification; design; software design description; test plan, test report, user's guide.

Documentation: (ANSI) The aids provided for the understanding of the structure and intended uses of an information system or its components, such as flowcharts, textual material, and user manuals.

Dosage form: The form of the completed pharmaceutical product, for example, tablet, capsule, injection, elixir, and suppository.

DRA: Drug regulatory authority.

Driver: A program that links a peripheral device or internal function to the operating system and provides for activation of all device functions. Syn: device driver. Contrast with test driver.

Drug: Any substance or pharmaceutical product for human or veterinary use that is intended to modify or explore physiological systems or pathological states for the benefit of the recipient.

Drug master file: A drug master file (DMF) is a master file that provides a full set of data on an API. In some countries, the term may also comprise data on an excipient or a component of a product such as a container.

Drug product: A drug product is a finished dosage form (e.g., tablet and capsule) that contains a drug substance, generally, but not necessarily, in association with one or more other ingredients (21 CFR 314.3(b)). A solid oral dosage form includes but is not limited to tablets, chewable tablets, enteric-coated tablets, capsules, caplets, encapsulated beads, and gelcaps.

Drug regulatory authority: A national body that administers the full spectrum of drug regulatory activities, including at least all of the following functions: marketing authorization of new products and variation of existing products, quality control laboratory testing, and adverse drug reaction monitoring.

Drug substance: An active ingredient that is intended to furnish pharmacological activity or other direct effect in the diagnosis, cure, mitigation, treatment, or prevention of a disease or to affect the structure of any function of the human body, but does not include intermediates used in the synthesis of such ingredient (21 CFR 314.3(b)).

Electronic record: Any combination of text, graphics, data, audio, pictorial, or other information representation in digital form that is created, modified, maintained, archived, retrieved, or distributed by a computer system.

Electronic Signature: A computer data compilation of any symbol or series of symbols executed, adopted, or authorized by an individual to be the legally binding equivalent of the individual's handwritten signature.

Embedded computer: A device that has its own computing power dedicated to specific functions, usually consisting of a microprocessor and firmware. The computer becomes an integral part of the device as opposed to devices that are controlled by an independent, stand-alone computer. It implies software that integrates operating system and application functions.

Embedded software: (IEEE) Software that is part of a larger system and performs some of the requirements of that system; for example, software used in an aircraft or rapid transit system. Such software does not provide an interface with the user. See firmware.

Enantiomers: Two stereoisomers (molecules that are identical in atomic constitution and bonding but differ in the 3D arrangement of the atoms) that are related to each other by a reflection: they are mirror images of each other, which are nonsuperimposable. Every stereocenter in one has the opposite configuration in the other. Two compounds that are enantiomers of each other have the same physical properties, except for the direction in which they rotate the polarized light and how they interact with different optical isomers of other compounds.

End user: (ANSI) (1) A person, device, program, or computer system that uses an information system for the purpose of data processing in information exchange. (2) A person whose occupation requires the use of an information system but does not require any knowledge of computers or computer programming. See user.

Enteric coated: Intended to delay the release of the drug (or drugs) until the dosage form has passed through the stomach. Enteric-coated products are delayed release dosage forms.

Equipment: Automated or nonautomated, mechanical or nonmechanical equipment used to produce the drug product, including equipment used to package the drug product.

Essential documents: Documents that individually and collectively permit evaluation of the conduct of a study and the quality of the data produced.

Essential drugs: Essential drugs are those that satisfy the health-care needs of the majority of the population. As indicated by the Expert Committee on the Use of Essential Drugs (5), each country may generate its own list of essential drugs.

Establishment evaluation request (EER): A request made to evaluate establishments listed in an application.

Event table: A table that lists events and the corresponding specified effect[s] of or reaction[s] to each event.

Excipient: Any component of a finished dosage form other than the claimed therapeutic ingredient or ingredients.

Extended release: Extended release products are formulated to make the drug available over an extended period after ingestion. This allows a reduction in dosing frequency compared to a drug presented as a conventional dosage form (e.g., as a solution or an immediate release dosage form).

Failure analysis: Determining the exact nature and location of a program error in order to fix the error, to identify and fix other similar errors, and to initiate corrective action to prevent future occurrences of this type of error. Contrast with debugging.

FDA Compliance Program Guidance Manual (CPGM), Compliance Program 7348.001: Bioresearch Monitoring—In Vivo Bioequivalence: The program describing the procedures used by FDA staff in performing inspections of BE studies.

FDA: Food and Drug Administration.

Feasibility study: Analysis of the known or anticipated need for a product, system, or component to assess the degree to which the requirements, designs, or plans can be implemented.

File maintenance: (ANSI) The activity of keeping a file up to date by adding, changing, or deleting data.

File transfer protocol: (1) Communications protocol that can transmit binary and ASCII data files without loss of data. (2) TCP/IP protocol that is used to log onto the network, list directories, and copy files. It can also translate between ASCII and EBCDIC. See TCP/IP.

File: (1) (ISO) A set of related records treated as a unit; for example, in stock control, a file could consist of a set of invoices. (2) The largest unit of storage structure that consists of a named collection of all occurrences in a database of records of a particular record type. Syn: data set.

Finished product: A product that has undergone all stages of production, including packaging in its final container and labeling.

Firmware: (IEEE) The combination of a hardware device, for example, an IC and computer instructions and data that reside as read-only software on that device. Such software cannot be modified by the computer during processing. See embedded software.

Flowchart or flow diagram: (2) (ISO) A graphical representation in which symbols are used to represent such things as operations, data, flow direction, and equipment for the definition, analysis, or solution of a problem. (2) (IEEE) A control flow diagram in which suitably annotated geometrical figures are used to represent operations, data, or equipment and arrows are used to indicate the sequential flow from one to another. Syn: flow diagram.

Formulation: A listing of the ingredients and composition of the dosage form.

Full validation: Establishment of all validation parameters that apply to sample analysis for the bioanalytical method for each analyte.

GCP: Good clinical practices.

Generic drug: A generic drug is the same as a brand-name drug in dosage, safety, strength, how it is taken, quality, performance, and intended use.

Generic products: Generic products may be marketed either under the approved nonproprietary name or under a brand (proprietary) name. They may be marketed in dosage forms and/or strengths different from those of the innovator products. Where the term generic product is used, it means a pharmaceutical product, usually intended to be interchangeable with the innovator product, which is usually manufactured without a license from the innovator company and marketed after expiry of the patent or other exclusivity rights. The term should not be confused with generic names for active pharmaceutical ingredients (APIs). The term generic product has somewhat different meanings in different jurisdictions. The use of this term is therefore avoided as much as possible, and the term multisource pharmaceutical product is used instead.

GMP: Good manufacturing practice.

Good clinical practice (GCP): A standard for the design, conduct, performance, monitoring, auditing, recording, analyses, and reporting of clinical trials that provides assurance that the data and reported results are credible and accurate and that the rights, integrity, and confidentiality of trial subjects are protected.

Good Laboratory Practice and Bioequivalence Investigations Branch (GBIB): The unit within the Division of Scientific Investigations responsible for assigning and/or performing inspections of facilities conducting BE and nonclinical studies.

Hard copy: Printed, output on paper.

Immediate release: Allows the drug to dissolve in the gastrointestinal contents, with no intention of delaying or prolonging the dissolution or absorption of the drug.

Immediate release dosage form: A dosage form that is intended to release the entire active ingredient on administration with no enhanced, delayed, or extended release effect.

Immediate release solid oral dosage forms: A significant body of information on the stability of the drug product is likely to exist after 5 years of commercial experience for new molecular entities or 3 years of commercial experience for new dosage forms.

In vitro–in vivo correlation: A predictive mathematical model describing the relationship between an in vitro property of an oral dosage form (usually the rate or extent of drug dissolution or release) and a relevant in vivo response (e.g., plasma drug concentration or amount of drug absorbed). For modified release dosage forms, changes in release-controlling excipients in the formulation should be within the range of release-controlling excipients of the established correlation. In the presence of an established in vitro/in vivo correlation, only application/compendia dissolution testing need be performed.

Incurred sample reanalysis (ISR): A repeated measurement of analyte concentration from study samples to demonstrate reproducibility.

IND: Investigational new drug.

Independent data-monitoring committee (IDMC) (data and safety monitoring board, monitoring committee, data-monitoring committee): An independent data-monitoring committee that may be established by the sponsor to assess at intervals the progress of a clinical trial, the safety data, and the critical efficacy endpoints and to recommend to the sponsor whether to continue, modify, or stop a trial.

Independent ethics committee (IEC): An independent body (a review board or a committee, institutional, regional, national, or supranational), constituted of medical professionals and non-medical members, whose responsibility it is to ensure the protection of the rights, safety, and well-being of human subjects involved in a trial and to provide public assurance of that protection, by, among other things, reviewing and approving/providing favorable opinion on the trial protocol, the suitability of the investigator(s), facilities, and the methods and material to be used in obtaining and documenting informed consent of the trial subjects. The legal status, composition, function, operations, and regulatory requirements pertaining to independent ethics committees may differ among countries but should allow the independent ethics committee to Act in agreement with GCP as described in this guideline.

Independent third party: A person that has no affiliation with the study sponsor and/or drug manufacturer; An independent third party indicates a person that has no affiliation with the study sponsor and/or drug manufacturer.

Informed consent: A process by which a subject voluntarily confirms his or her willingness to participate in a particular trial, after having been informed of all aspects of the trial that are relevant to the subject's decision to participate. Informed consent is documented by means of a written, signed, and dated informed consent form.

Innovator pharmaceutical product: A pharmaceutical product that was first authorized for marketing (normally as a patented drug) based on documentation of its safety, efficacy, and pharmaceutical quality (according to contemporary regulatory requirements). When drugs have been available in the marketplace for many years, it may not be possible to identify an innovator pharmaceutical product. In these cases, an innovator product may be defined as a medicinal authorized and marketed on the basis of a full dossier, that is, including chemical, biological, pharmacological–toxicological, and clinical data.

Inspection: The Act by a regulatory authority(ies) of conducting an official review of documents, facilities, records, and any other resources that are deemed by the authority(ies) to be related to the clinical trial and that may be located at the site of the trial, at the sponsor's and/or contract research organizations (CROs) facilities, or at other establishments deemed appropriate by the regulatory authority(ies).

Installation: (ANSI) The phase in the system life cycle that includes assembly and testing of the hardware and software of a computerized system. Installation includes installing a new computer system, new software or hardware, or otherwise modifying the current system.

Institutional review board (IRB): An independent body constituted of medical, scientific, and nonscientific members, whose responsibility is to ensure the protection of the rights, safety, and well-being of human subjects involved in a trial by, among other things, reviewing, approving, and providing continuing review of trial protocol and amendments and of the methods and material to be used in obtaining and documenting informed consent of the trial subjects.

Interchangeability: An interchangeable pharmaceutical product is one that is therapeutically equivalent to a comparator (reference) product.

Interim clinical trial/study report: A report of intermediate results and their evaluation based on analyses performed during the course of a trial.

Internal standard (IS): Test compound(s) (e.g., structurally similar analog, stable labeled compound) added to both calibration standards and samples at known and constant concentration to facilitate quantification of the target analyte(s).

Investigational product: A pharmaceutical form of an active ingredient or placebo being tested or used as a reference in a clinical trial, including a product with a marketing authorization when used or assembled (formulated or packaged) in a way different from the approved form or when used for an unapproved indication or when used to gain further information about an approved use.

Investigator/institution: An expression meaning "the investigator and/or institution, where required by the applicable regulatory requirements."

Investigator: A person responsible for the conduct of the clinical trial at a trial site. If a trial is conducted by a team of individuals at a trial site, the investigator is the responsible leader of the team and may be called the principal investigator. See also subinvestigator.

Investigator's brochure: A compilation of the clinical and nonclinical data on the investigational product(s) that is relevant to the study of the investigational product(s) in human subjects.

IR: Immediate release.

IS: Internal standards.

ISO: International Organization for Standardization.

Justification: Reports containing scientific data and expert professional judgment to substantiate decisions.

Key element: (QA) An individual step in a critical control point of the manufacturing process.

Legally acceptable representative: An individual or other body authorized under applicable law to consent, on behalf of a prospective subject, to the subject's participation in the clinical trial.

Life cycle methodology: The use of any one of several structured methods to plan, design, implement, test, and operate a system from its conception to the termination of its use.

Limit of detection (LOD): The lowest concentration of an analyte that the bioanalytical procedure can reliably differentiate from background noise.

Lot: A batch or a specific identified portion of a batch having uniform character and quality within specified limits or, in the case of a drug product produced by continuous process, a specific identified amount produced in a unit of time or quantity in a manner that assures its having uniform character and quality within specified limits (21 CFR 210.3(b)(10)).

Lower limit of quantification (LLOQ): The lowest amount of an analyte in a sample that can be quantitatively determined with suitable precision and accuracy.

Matrix effect: The direct or indirect alteration or interference in response due to the presence of unintended analytes (for analysis) or other interfering substances in the sample.

Matrix: All constituents of the laboratory sample other than the analyte. A type of matrix is defined as a group of materials or of products considered by the analyst as having a homogeneous behavior with regard to the analytical method used.

Method: A comprehensive description of all procedures used in sample analysis.

Methods validation: The analytical process of actual use testing of the applicant's proposed regulatory method(s) in an FDA laboratory.

Methods verification: The process of testing a compendia ANDA drug substance or drug product by compendia procedures in an FDA laboratory for purposes of ensuring compliance with compendia specifications and evaluating the appropriateness of a particular formulation for analysis by the compendia methods.

Modified release dosage forms: Dosage forms whose drug-release characteristics of time course and/or location are chosen to accomplish therapeutic or convenience objectives not offered by conventional dosage forms such as a solution or an immediate release dosage form. Modified release solid oral dosage forms include both delayed and extended release drug products.

Modified release solid oral dosage forms: A significant body of information should include, for "modified release solid oral dosage forms," a product-specific body of information. This product-specific body of information is likely to exist after 5 years of commercial experience for the original modified release solid oral drug product or 3 years of commercial experience for any subsequent modified release solid oral drug product utilizing similar drug-release mechanism.

Monitoring report: A written report from the monitor to the sponsor after each site visit and/or other trial-related communication according to the sponsor's SOPs.

Monitoring: The Act of overseeing the progress of a clinical trial and of ensuring that it is conducted, recorded, and reported in accordance with the protocol, standard operating procedures (SOPs), good clinical practice (GCP), and the applicable regulatory requirement(s).

Multitasking: (IEEE) A mode of operation in which two or more tasks are executed in an interleaved manner. Syn: parallel processing.

Multicenter trial: A clinical trial conducted according to a single protocol but at more than one site and, therefore, carried out by more than one investigator.

Multisource and single-source drug products: In most instances, it refers to those pharmaceutical equivalents available from more than one manufacturer that may or may not be therapeutically equivalent. Multisource pharmaceutical products that are therapeutically equivalent are interchangeable.

Narrow therapeutic index: It means that there is less than a twofold difference in median lethal dose (LD 50) and median effective does (ED 50) values or have less than twofold difference in the minimum toxic concentration and minimum effective concentration in the blood; steep dose–response.

Nonclinical laboratory study: In vivo or in vitro experiments in which test articles are studied prospectively in test systems under laboratory conditions to determine their safety. The term does not include studies utilizing human subjects or clinical studies or field trials in animals. The term does not include basic exploratory studies carried out to determine whether a test article has any potential utility or to determine physical or chemical characteristics of a test article.

Nonclinical study: Biomedical studies not performed on human subjects.

Nonrelease-controlling excipient (Noncritical composition variable): An excipient in the final dosage form whose primary function does not include modifying the duration of release of the active drug substance from the dosage form.

Octal: The base 8 number system. Digits are 0, 1, 2, 3, 4, 5, 6, and 7.

OEM: Original equipment manufacturer.

Official action indicated (OAI): Objectionable conditions or practices that were found that represented significant departures from the regulations and could require administrative or regulatory sanctions.

Operating principles: Rules or concepts governing the operation of the system.

Operating system: (ISO) Software that controls the execution of programs and that provides services such as resource allocation, scheduling, input/output control, and data management. Usually, operating systems are predominantly software, but partial or complete hardware implementations are possible.

Operation and maintenance phase: (IEEE) The period of time in the software life cycle during which a software product is employed in its operational environment, monitored for satisfactory performance, and modified as necessary to correct problems or to respond to changing requirements.

Opinion (in relation to independent ethics committee): The judgment and/or the advice provided by an independent ethics committee (IEC).

Original data: Original data are those values that represent the first recording of study data. FDA is allowing original documents and the original data recorded on those documents to be replaced by certified copies provided the copies are identical and have been verified as such.

Partial validation: Modification of validated bioanalytical methods that do not necessarily call for full revalidation.

Perfective maintenance: (IEEE) Software maintenance performed to improve the performance, maintainability, or other attributes of a computer program. Contrast with adaptive maintenance and corrective maintenance.

Pharmaceutical alternatives: Drug products that contain the identical therapeutic moiety, or its precursor, but not necessarily in the same amount or dosage form or as the same salt or ester. Each such drug product individually meets either the identical or its own respective compendial or other applicable standard of identity, strength, quality, and purity, including potency and, where applicable, content uniformity, disintegration times, and/or dissolution rates.

Pharmaceutical equivalents: Drug products in identical dosage forms that contain identical amounts of the identical active drug ingredient, that is, the same salt or ester of the same therapeutic moiety, or, in the case of modified release dosage forms that require a reservoir or overage or such forms as prefilled syringes where residual volume may vary, that deliver identical amounts of the active drug ingredient over the identical dosing period; do not necessarily contain the same inactive ingredients; and meet the identical compendial or other applicable standard of identity, strength, quality, and purity, including potency and, where applicable, content uniformity, disintegration times, and/or dissolution rates.

Pharmaceutical equivalents: Drug products are considered to be pharmaceutical equivalents if they contain the same active ingredient(s), have the same dosage form and route of administration, and are identical in strength or concentration.

Pharmaceutical product: Any preparation for human or veterinary use that is intended to modify or explore physiological systems or pathological states for the benefit of the recipient. Note: In many countries in Latin America, multisource drug products are referred to as "productos similares" and are marketed under an approved new brand proprietary. However, when they are marketed under the nonproprietary name (unbranded), they are usually known as "generic products." These products cannot be considered interchangeable until appropriate evidence has been submitted to show interchangeability.

Pilot scale: The manufacture of either drug substance or drug product by a procedure fully representative of and simulating that used for full manufacturing scale. For solid oral dosage forms, this is generally taken to be, at a minimum, one-tenth that of full production, or 100,000 tablets or capsules, whichever is larger (4).

Platform: The hardware and software that must be present and functioning for an application program to run [perform] as intended. A platform includes, but is not limited to, the operating system or executive software, communication software, microprocessor, network, input/output hardware, any generic software libraries, database management, user interface software, and the like.

Precision: The closeness of agreement (i.e., degree of scatter) among a series of measurements obtained from multiple sampling of the same homogenous sample under the prescribed conditions.

Predicate rule: This term refers to underlying requirements set forth in the Federal Food, Drug, and Cosmetic Act, the PHS Act, and FDA regulations (other than 21 CFR part 11). Regulations governing good clinical practice and human subject protection can be found at 21 CFR parts 50, 56, 312, 511, and 812.

Process: A series of operations, actions, and controls used to manufacture a drug product.

Processed sample: The final extract (prior to instrumental analysis) of a sample that has been subjected to various manipulations (e.g., extraction, dilution, concentration).

Processed: The final extract (prior to instrumental analysis) of a sample that has been subjected to various manipulations (e.g., extraction, dilution, concentration).

Project plan: (NIST) A management document describing the approach taken for a project. The plan typically describes work to be done, resources required, methods to be used, the configuration management and quality assurance procedures to be followed, the schedules to be met, the project organization, etc. Project in this context is a generic term. Some projects may also need integration plans, security plans, test plans, quality assurance plans, etc. See documentation plan, software development plan, test plan, software engineering.

Proof of correctness: (NBS) The use of techniques of mathematical logic to infer that a relation between program variables assumed true at program entry implies that another relation between program variables holds at program exit.

Protocol: A document that describes the objective(s), design, methodology, statistical considerations, and organization of a trial. The protocol usually also gives the background and rationale for the trial, but these could be provided in other protocol-referenced documents. Throughout the ICH GCP guideline, the term protocol refers to protocol and protocol amendments.

QA: Quality assurance.

QC: Quality control.

Qualification, installation: (FDA) Establishing confidence that process equipment and ancillary systems are compliant with appropriate codes and approved design intentions and that manufacturer's recommendations are suitably considered.

Qualification, operational: (FDA) Establishing confidence that process equipment and subsystems are capable of consistently operating within established limits and tolerances.

Qualification, process performance: (FDA) Establishing confidence that the process is effective and reproducible.

Qualification, product performance: (FDA) Establishing confidence through appropriate testing that the finished product produced by a specified process meets all release requirements for functionality and safety.

Quality assurance (QA): All those planned and systematic actions that are established to ensure that the trial is performed and the data are generated, documented (recorded), and reported in compliance with good clinical practice (GCP) and the applicable regulatory requirement(s).

Quality assurance unit: Any person or organizational element, except the study director, designated by testing facility management to perform the duties relating to quality assurance of nonclinical laboratory studies.

Quality assurance, software: (IEEE) (1) A planned and systematic pattern of all actions necessary to provide adequate confidence that an item or product conforms to established technical requirements. (2) A set of activities designed to evaluate the process by which products are developed or manufactured.

Quality assurance: (1) (ISO) The planned systematic activities necessary to ensure that a component, module, or system conforms to established technical requirements. (2) All actions that are taken to ensure that a development organization delivers products that meet performance requirements and adhere to standards and procedures. (3) The policy, procedures, and systematic actions established in an enterprise for the purpose of providing and maintaining some degree of confidence in data integrity and accuracy throughout the life cycle of the data, which includes input, update, manipulation, and output. (4) (QA) The actions, planned and performed, to provide confidence that all systems and components that influence the quality of the product are working as expected individually and collectively.

Quality control (QC): The operational techniques and activities undertaken within the quality assurance system to verify that the requirements for quality of the trial-related activities have been fulfilled.

Quality control sample (QC): (1) A spiked sample used to monitor the performance of a bioanalytical method and to assess the integrity and validity of the results of the unknown samples analyzed in an individual batch. (2) A sample with a known quantity of analyte that is used to monitor the performance of a bioanalytical method and to assess the integrity and validity of the results of the unknown samples analyzed in an individual run.

Quantification range: The range of concentrations, including ULOQ and LLOQ, that can be reliably and reproducibly quantified with accuracy and precision through the use of a concentration–response relationship.

Racemate: A racemate is optically inactive. Because the two isomers rotate plane-polarized light in opposite directions, they cancel out; therefore, a racemic mixture does not rotate plane-polarized light. In contrast to the two separate enantiomers, which generally have identical physical properties, a racemate often has different properties compared to either one of the pure enantiomers. Different melting points and solubilities are very common, but differing boiling points are also possible. Pharmaceuticals can be available as a racemate or as a pure enantiomer, which might have different potencies.

Randomization: The process of assigning trial subjects to treatment or control groups using an element of chance to determine the assignments in order to reduce bias.

Ranges: The extent to which or the limits between which acceptable variation exists.

Raw data: Any laboratory worksheets, records, memoranda, notes, or exact copies thereof that are the result of original observations and activities of a nonclinical laboratory study and are necessary for the reconstruction and evaluation of the report of that study. In the event that exact transcripts of raw data have been prepared (e.g., tapes which have been transcribed verbatim, dated, and verified accurate by signature), the exact copy or exact transcript may be substituted for the original source as raw data. Raw data may include photographs, microfilm or microfiche copies, computer printouts, magnetic media, including dictated observations, and recorded data from automated instruments.

Recovery: The extraction efficiency of an analytical process, reported as a percentage of the known amount of an analyte carried through the sample extraction and processing steps of the method.

Reference-listed drug (RLD): A reference-listed drug is an approved drug product to which new generic versions are compared to show that they are bioequivalent.

Reference product: The pharmaceutical product with which the "new" product is intended to be interchangeable in clinical practice. The reference product is usually the innovators product for which safety, efficacy, and quality have been documented.

Reference standard: The reference product used in a BE study. It is usually the innovator's product or a marketed product of the drug under investigation. For BA studies, the reference standard can be an oral solution of the drug under investigation.

Regulatory authorities: Bodies having the power to regulate. In the ICH GCP guideline, the expression regulatory authorities includes the authorities that review submitted clinical data and those that conduct inspections (see 1.29). These bodies are sometimes referred to as competent authorities.

Regulatory methods: The analytical procedures proposed by the applicant and agreed upon by the agency to determine whether the drug substance or drug product meets its established specifications. For drug substances and drug products having monographs in the USP, the USP analytical methods are considered regulatory by definition.

Relational database: Database organization method that links files together as required. Relationships between files are created by comparing data such as account numbers and names. A relational system can take any two or more files and generate a new file from the records that meet the matching criteria. Routine queries often involve more than one data file; for example, a customer file and an order file can be linked in order to ask a question that relates to information in both files, such as the names of the customers that purchased a particular product. Contrast with network database, flat file.

Release-controlling excipient (critical composition variable): An excipient in the final dosage form whose primary function is to modify the duration of release of the active drug substance from the dosage form.

Release mechanism: The process by which the drug substance is released from the dosage form. Typically, the definition contains the energy source or pictorially describes the way the drug is released.

Representative: Corresponding to or replacing some other species or the like; exemplifying a group or kind; typical.

Reproducibility: The precision between two laboratories. It also represents precision of the method under the same operating conditions over a short period of time.

Requirement: (IEEE) (1) A condition or capability needed by a user to solve a problem or achieve an objective. (2) A condition or capability that must be met or possessed by a system or system component to satisfy a contract, standard, specification, or other formally imposed documents. (3) A documented representation of a condition or capability as in (1) or (2).

Reserve samples. Reserve samples and retention samples are used interchangeably.

Retention period: (ISO) The length of time specified for data on a data medium to be preserved.

Retrospective trace: (IEEE) A trace produced from historical data recorded during the execution of a computer program. Note: this differs from an ordinary trace, which is produced cumulatively during program execution.

Revalidation: Relative to software changes, revalidation means validating the change itself, assessing the nature of the change to determine potential ripple effects, and performing the necessary regression testing.

Risk: (IEEE) A measure of the probability and severity of undesired effects. Often taken as the simple product of probability and consequence.

Robustness: The degree to which a software system or component can function correctly in the presence of invalid inputs or stressful environmental conditions. See software reliability.

Routine inspection: An inspection to determine the compliance of a clinical facility or analytical laboratory with U.S. regulations. Typically, there is no prior indication of misconduct, human subject protection problems, or suspect data.

Routine: (IEEE) A subprogram that is called by other programs and subprograms. Note: This term is defined differently in various programming languages.

Safety: (DOD) Freedom from those conditions that can cause death, injury, occupational illness, or damage to or loss of equipment or property or damage to the environment.

Same drug product formulation: The formulation of the drug product submitted for approval and any formulations that have minor differences in composition or method of manufacture from the formulation submitted for approval but are similar enough to be relevant to the agency's determination of bioequivalence.

Same: Agreeing in kind, amount; unchanged in character or condition.

Sample: A generic term encompassing controls, blanks, unknowns, and processed samples.

Satisfactory current good manufacturing practice (cGMP) inspection: A satisfactory cGMP inspection is one during which (1) no objectionable conditions or practices were found during an inspection or (2) objectionable conditions were found, however, corrective action is left to the firm to take voluntarily and the objectionable conditions do not justify further administrative or regulatory actions.

Scale-down: The process of decreasing the batch size. Significant body of information.

Scale-up: The process of increasing the batch size.

Selectivity: The ability of the bioanalytical method to measure and differentiate the analytes in the presence of components that may be expected to be present. These could include metabolites, impurities, degradants, or matrix components.

Selectivity/specificity: The ability of the bioanalytical method to measure and differentiate the analytes in the presence of components that may be expected to be present. These could include metabolites, impurities, degradants, or matrix components.

Sensitivity: Defined as the lowest analyte concentration that can be measured with acceptable accuracy and precision (i.e., LLOQ).

Serious adverse event (SAE) or serious adverse drug reaction (serious ADR): Any untoward medical occurrence that at any dose, results in death, is life threatening, requires inpatient hospitalization or prolongation of existing hospitalization, and results in persistent or significant disability/incapacity.

Side effect: An unintended alteration of a program's behavior caused by a change in one part of the program, without taking into account the effect the change has on another part of the program.

Similar: Having a general likeness.

Site management organization (SMO): (1) An organization that manages clinical study sites on behalf of the sponsor and/or drug manufacturer. (2) In this guidance, site management organization refers to an organization that manages clinical study sites on behalf of the sponsor and/or drug manufacturer.

Software documentation: (NIST) Technical data or information, including computer listings and printouts, in human-readable form, that describe or specify the design or details, explain the capabilities, or provide operating instructions for using the software to obtain desired results from a software system. See specification; specification, requirements; specification, design; software design description; test plan, test report, user's guide.

Software reliability: (IEEE) (1) The probability that a software will not cause the failure of a system for a specified time under specified conditions. The probability is a function of the inputs to and use of the system in the software. The inputs to the system determine whether existing faults, if any, are encountered. (2) The ability of a program to perform its required functions accurately and reproducibly under stated conditions for a specified period of time.

Software safety change analysis: (IEEE) Analysis of the safety-critical design elements affected directly or indirectly by the change to show the change does not create a new hazard, does not impact on a previously resolved hazard, does not make a currently existing hazard more severe, and does not adversely affect any safety-critical software design element.

Software safety code analysis: (IEEE). Verification that the safety-critical portions of the design are correctly implemented in the code.

Software safety design analysis: (IEEE) Verification that the safety-critical portion of the software design correctly implements the safety-critical requirements and introduces no new hazards. See logic analysis, data analysis, interface analysis, constraint analysis, functional analysis, software element analysis, timing and sizing analysis, reliability analysis, software hazard analysis, system safety.

Software safety requirements analysis: (IEEE) Analysis evaluating software and interface requirements to identify errors and deficiencies that could contribute to a hazard. See criticality analysis, specification analysis, timing and sizing analysis, different software systems analyses, software hazard analysis, system safety.

Software safety test analysis: (IEEE) Analysis demonstrating that safety requirements have been correctly implemented and that the software functions safely within its specified environment. Tests may include unit level tests, interface tests, software configuration item testing, system level testing, stress testing, and regression testing. See software hazard analysis, system safety.

Software validation: Confirmation by examination and provision of objective evidence that software specifications conform to user needs and intended uses and that the particular requirements implemented through the software can be consistently fulfilled. Design level validation is that portion of the software validation that takes place in parts of the software life cycle before the software is delivered to the end user.

Software: (ANSI) Programs, procedures, rules, and any associated documentation pertaining to the operation of a system. Contrast with hardware.

SOPs: Standard operating procedures.

Source code: (1) (IEEE) Computer instructions and data definitions expressed in a form suitable for input to an assembler, compiler, or other translator. (2) The human-readable version of the list of instructions [program] that cause a computer to perform a task. Contrast with object code.

Source data: All information in original records and certified copies of original records of clinical findings, observations, or other activities in a clinical trial necessary for the reconstruction and evaluation of the trial. Source data are contained in source documents (original records or certified copies).

Source documents: Original documents, data, and records (e.g., hospital records, clinical and office charts, laboratory notes, memoranda, subjects' diaries or evaluation checklists, pharmacy dispensing records, recorded data from automated instruments, copies or transcriptions certified after verification as being accurate copies, microfiches, photographic negatives, microfilm or magnetic media, x-rays, subject files, and records kept at the pharmacy, at the laboratories, and at medicotechnical departments involved in the clinical trial).

Specification, formal: (NIST) (1) A specification written and approved in accordance with established standards. (2) A specification expressed in a requirements specification language. Contrast with requirement.

Specification, functional: (NIST) A specification that documents the functional requirements for a system or system component. It describes what the system or component is to do rather than how it is to be built. Often part of a requirements specification. Contrast with requirement.

Specification, interface: (NIST) A specification that documents the interface requirements for a system or system component. Often part of a requirements specification. Contrast with requirement.

Specification, performance: (IEEE) A document that sets forth the performance characteristics that a system or component must possess. These characteristics typically include speed, accuracy, and memory usage. Often part of a requirements specification. Contrast with requirement.

Specification, product: (IEEE) A document that describes the as-built version of the software.

Specification, programming: (NIST) See specification, design.

Specification, requirements: (NIST) A specification that documents the requirements of a system or system component. It typically includes functional requirements, performance requirements, interface requirements, design requirements (attributes and constraints), and development (coding) standards. Contrast with requirement.

Specification: (IEEE) A document that specifies, in a complete, precise, verifiable manner, the requirements, design, behavior, or other characteristics of a system or component and, often, the procedures for determining whether these provisions have been satisfied. Contrast with requirement. See specification, formal; specification, requirements; specification, functional; specification, performance; specification, interface; specification, design; coding standards; design standards.

Specimen: Any material derived from a test system for examination or analysis.

Sponsor–investigator: An individual who both initiates and conducts, alone or with others, a clinical trial and under whose immediate direction the investigational product is administered to, dispensed to, or used by a subject. The term does not include any person other than an individual (e.g., it does not include a corporation or an agency). The obligations of a sponsor–investigator include both those of a sponsor and those of an investigator.

Sponsor: An individual, company, institution, or organization that takes responsibility for the initiation, management, and/or financing of a clinical trial.

Stability: The chemical stability of an analyte in a given matrix under specific conditions for given time intervals.

Standard curve: The relationship between the experimental response value and the analytical concentration (also called a calibration curve).

Standard operating procedures (SOPs): Detailed, written instructions to achieve uniformity of the performance of a specific function.

Study completion date: The date the final report is signed by the study director.

Study director: The individual responsible for the overall conduct of a nonclinical laboratory study.

Study initiation date: The date the protocol is signed by the study director.

Subinvestigator: Any individual member of the clinical trial team designated and supervised by the investigator at a trial site to perform critical trial-related procedures and/or to make important trial-related decisions (e.g., associates, residents, research fellows). See also investigator.

Subject identification code: A unique identifier assigned by the investigator to each trial subject to protect the subject's identity and used in lieu of the subject's name when the investigator reports adverse events and/or other trial-related data.

Subject/trial subject: An individual who participates in a clinical trial, either as a recipient of the investigational product(s) or as a control.

SUPAC: Scale-up and postapproval changes.

System life cycle: The course of developmental changes through which a system passes from its conception to the termination of its use; for example, the phases and activities associated with the analysis, acquisition, design, development, test, integration, operation, maintenance, and modification of a system.

System suitability: Determination of instrument performance (e.g., sensitivity and chromatographic retention) by analysis of a set of reference standards conducted prior to the analytical run.

Technical grade: Technical grades of excipients may differ in (1) specifications and/or functionality, (2) impurities, and (3) impurity profiles.

Test article: Any food additive, color additive, drug, biological product, electronic product, medical device for human use, or any other article subject to regulation under the Act or under sections 351 and 354–360F of the Public Health Service Act.

Test system: Any animal, plant, microorganism, or subparts thereof to which the test or control article is administered or added for study. Test system also includes appropriate groups or components of the system not treated with the test or control articles.

Test: (IEEE) An activity in which a system or component is executed under specified conditions; the results are observed or recorded and an evaluation is made of some aspect of the system or component.

Testability: (IEEE) (1) The degree to which a system or component facilitates the establishment of test criteria and the performance of tests to determine whether those criteria have been met. (2) The degree to which a requirement is stated in terms that permit establishment of test criteria and performance of tests to determine whether those criteria have been met.

Testing facility: (1) The entity performing the BA or BE (in vivo or in vitro) study. The testing facility can be a CRO, university, hospital, clinic of a clinical investigator, or in-house clinical study unit of a study sponsor and/or drug manufacturer, where dosing and sampling (i.e., blood, urine, or clinical endpoints) are performed. In issuing the final rule, the agency intended that reserve samples should be kept at the testing facility. (2) A person who actually conducts a nonclinical laboratory study, that is, actually uses the test article in a test system. Testing facility includes any establishment required to register under section 510 of the Act that conducts nonclinical laboratory studies and any consulting laboratory described in section 704 of the Act that conducts such studies. Testing facility encompasses only those operational units that are being or have been used to conduct nonclinical laboratory studies.

Testing, boundary value: A testing technique using input values at, just below, and just above the defined limits of an input domain and with input values causing outputs to be at, just below, and just above the defined limits of an output domain. See boundary value analysis; testing, stress.

Testing, compatibility: The process of determining the ability of two or more systems to exchange information. In a situation where the developed software replaces an already working program, an investigation should be conducted to assess possible comparability problems between the new software and other programs or systems.

Therapeutic equivalents: Drug products are considered to be therapeutic equivalents only if they are pharmaceutical equivalents and if they can be expected to have the same clinical effect and safety profile when administered to patients under the conditions specified in the labeling. The FDA classifies as therapeutically equivalent those products that meet the following general criteria: (1) they are approved as safe and effective; (2) they are pharmaceutical equivalents in that they (a) contain identical amounts of the same active drug ingredient in the same dosage form and route of administration and (b) meet compendial or other applicable standards of strength, quality, purity, and identity; (3) they are bioequivalent in that (a) they do not present a known or potential bioequivalence problem, and they meet an acceptable in vitro standard, or (b) if they do present such a known or potential problem, they are shown to meet an appropriate bioequivalence standard; (4) they are adequately labeled; and (5) they are manufactured in compliance with current good manufacturing practice regulations. The FDA considers drug products to be therapeutically equivalent if they meet the criteria outlined earlier, even though they may differ in certain other characteristics such as shape, scoring configuration, release mechanisms, packaging, excipients (including colors, flavors, preservatives), expiration date/time, and minor aspects of labeling (e.g., the presence of specific pharmacokinetic information) and storage conditions.

When such differences are important in the care of a particular patient, it may be appropriate for the prescribing physician to require that a particular brand be dispensed as a medical necessity. With this limitation, however, FDA believes that products classified as therapeutically equivalent can be substituted with the full expectation that the substituted product will produce the same clinical effect and safety profile as the prescribed product.

Thermodynamic equivalence: A term coined by the author. A demonstration of equal chemical equivalence at the site of absorption between the innovator and the generic product.

Trial site: The location(s) where trial-related activities are actually conducted.

Trojan horse: A method of attacking a computer system, typically by providing a useful program that contains code intended to compromise a computer system by secretly providing for unauthorized access, the unauthorized collection of privileged system or user data, the unauthorized reading or altering of files, the performance of unintended and unexpected functions, or the malicious destruction of software and hardware.

Unambiguous: (1) Not having two or more possible meanings. (2) Not susceptible to different interpretations. (3) Not obscure, not vague. (4) Clear, definite, certain.

Unexpected adverse drug reaction: An adverse reaction, the nature or severity of which is not consistent with the applicable product information (e.g., investigator's brochure for an unapproved investigational product or package insert/summary of product characteristics for an approved product) (see the *ICH Guideline for Clinical Safety Data Management: Definitions and Standards for Expedited Reporting*).

Unknown: A biological sample that is the subject of the analysis.

Upper limit of quantification (ULOQ): The highest amount of an analyte in a sample that can be quantitatively determined with precision and accuracy.

User: (ANSI) Any person, organization, or functional unit that uses the services of an information processing system. See end user.

USP: The current edition of the *United States Pharmacopeia* and its supplements.

Validate: To prove to be valid.

Validation protocol: (FDA) A written plan stating how validation will be conducted, including test parameters, product characteristics, production equipment, and decision points on what constitutes acceptable test results.

Validation standard (or validation sample): Validation standards are samples reconstituted in the matrix or in any other reference material with true values set by consensus and used to validate the analytical procedure.

Validation, process: (FDA) Establishing documented evidence that provides a high degree of assurance that a specific process will consistently produce a product meeting its predetermined specifications and quality characteristics.

Validation, prospective: (FDA) Validation conducted prior to the distribution of either a new product or product made under a revised manufacturing process, where the revisions may affect the product's characteristics.

Validation, retrospective: (FDA) (1) Validation of a process for a product already in distribution based upon accumulated production, testing, and control data. (2) Retrospective validation can also be useful to augment initial premarket prospective validation for new products or changed processes. Test data are useful only if the methods and results are adequately specific. Whenever test data are used to demonstrate conformance to specifications, it is important that the test methodology be qualified to assure that the test results are objective and accurate.

Validation, software: (NBS) Determination of the correctness of the final program or software produced from a development project with respect to the user needs and requirements. Validation is usually accomplished by verifying each stage of the software development life cycle. See verification, software.

Validation, verification, and testing: (NIST) Used as an entity to define a procedure of review, analysis, and testing throughout the software life cycle to discover errors, determine functionality, and ensure the production of quality software.

Validation: Establishing through documented evidence a high degree of assurance that a specific process will consistently produce a product that meets its predetermined specifications and quality attributes. A validated manufacturing process is one that has been proven to do what it purports or is represented to do. The proof of validation is obtained through collection and evaluation of data, preferably beginning from the process development phase and continuing through into the production phase. Validation necessarily includes process qualification (the qualification of materials, equipment, systems, buildings, and personnel), but it also includes the control of entire processes for repeated batches or runs.

Variable: A name, label, quantity, or data item whose value may be changed many times during processing. Contrast with constant.

Vendor: A person or an organization that provides software and/or hardware and/or firmware and/or documentation to the user for a fee or in exchange for services. Such a firm could be a medical device manufacturer.

Verifiable: Can be proved or confirmed by examination or investigation. See measurable.

Vulnerable subjects: Individuals whose willingness to volunteer in a clinical trial may be unduly influenced by the expectation, whether justified or not, of benefits associated with participation or of a retaliatory response from senior members of a hierarchy in case of refusal to participate. Examples are members of a group with a hierarchical structure, such as medical, pharmacy, dental, and nursing students; subordinate hospital and laboratory personnel; employees of the pharmaceutical industry; members of the armed forces; and persons kept in detention. Other vulnerable subjects include patients with incurable diseases, persons in nursing homes, unemployed or impoverished persons, patients in emergency situations, ethnic minority groups, homeless persons, nomads, refugees, minors, and those incapable of giving consent.

Well-being (of the trial subjects): The physical and mental integrity of the subjects participating in a clinical trial.

WHO: World Health Organization.

Appendix B: Dissolution Testing Requirements for U.S. FDA Submission

How to use this table:

1. The entries are listed in the following order separated by a comma. If nothing appears after the date entry, that means that there are no waivers allowed and all dosage forms must be tested.

   Drug Name, Dosage Form, USP Apparatus, Speed (RPMs), Medium, Volume (mL), Recommended Sampling Times (minutes), Date Updated, Waiver.
2. If it states Refer to USP, in that case, no conditions of dissolution will be provided.

Abacavir Sulfate, Tablet, II (Paddle), 75, 0.1 N HCl, 900, 5, 10, 15, and 30, 03/22/2006.

Abacavir Sulfate/Lamivudine, Tablet, II (Paddle), 75, 0.1 N HCl, 900, 10, 20, 30, and 45, 01/03/2007.

Abacavir Sulfate/Lamivudine/Zidovudine, Tablet, II (Paddle), 75, 0.1 N HCl, 900, 5, 10, 15, 30, and 45, 01/03/2007.

Abiraterone Acetate, Tablet, II (Paddle), 50, 0.25% SLS in 56.5 mM phosphate buffer, pH 4.5, 900, 10, 20, 30, 45, and 60, 02/28/2013.

Acamprosate Calcium, Tablet (Delayed Release), I (Basket), 180, Acid Stage: 0.1 N HCl Buffer Stage: "Citrate–sodium hydroxide" buffer pH 6.8 (150 mL of 2 N NaOH, 21.014 g of citric acid and ultrapure water to 1000 mL) (Method B), 1000, 120 (Acid) 30, 60, 90, 120, and 180 (buffer), 12/20/2005.

Acarbose, Tablet, II (Paddle), 75, Water (deaerated), 900, 10, 15, 20, 30, and 45, 03/22/2006, 50 and 100 mg based on (1) acceptable in vivo bioequivalence study on the 25 mg strength, (2) acceptable in vitro dissolution testing of all strengths, and (3) proportional similarity of the formulations.

Acetaminophen, Suppository, II (Paddle), 50, Phosphate buffer, pH 5, 900, 15, 30, 45, 60, and 90, 08/17/2006.

Acetaminophen, Tablet (Extended Release), Refer to USP, 03/03/2011, Acetaminophen Extended Release Caplet or Gelcap, 650 mg based on (1) acceptable bioequivalence studies on the 650 mg strength Geltab or Caplet, (2) acceptable in vitro dissolution testing of both dosage forms, and (3) proportional similarity of both formulations.

Acetaminophen/Butalbital, Tablet, II (Paddle), 50, Water (deaerated), 900, 15, 30, 45, 60, and 90, 01/03/2007, 325 mg/50 mg and 650 mg/50 mg pursuant to 21 CFR 320.22(c) provided that the in vitro dissolution profiles of the proposed product are comparable to those of the reference product.

Acetaminophen/Butalbital/Caffeine, Tablet, Refer to USP, 01/14/2008, 325 mg/50 mg/40 mg, 500 mg/50 mg/40 mg, and 750 mg/50 mg/40 mg pursuant to 21 CFR 320.22(c) provided that the in vitro dissolution profiles of the proposed product are comparable to those of the reference product.

Acetaminophen/Butalbital/Caffeine/Codeine Phosphate, Capsule, II (Paddle), 50, Water (deaerated), 900, 10, 20, 30, 45, and 60, 03/04/2006.

Acetaminophen/Caffeine/Dihydrocodeine Bitartrate, Capsule, I (Basket), 100, Water, 900, 10, 20, 30, 45, and 60, 01/03/2007.

Acetaminophen/Caffeine/Dihydrocodeine Bitartrate, Tablet, II (Paddle), 50, Water, 900, 10, 15, 30, 45, and 60, 07/25/2007, 712.8 mg/60 mg/32 mg and 356.4 mg/30 mg/16 mg pursuant to 21 CFR 320.22(c) provided that the in vitro dissolution profiles of your proposed product are comparable to those of the reference product.

Acetaminophen/Hydrocodone Bitartrate, Tablet, Refer to USP, 07/25/2007, 325 mg/5 mg, 325 mg/7.5 mg, 325 mg/10 mg, 500 mg/2.5 mg, 500 mg/5 mg, 500 mg/7.5 mg, 500 mg/10 mg, 650 mg/7.5 mg, 650 mg/10 mg, 660 mg/10 mg, 750 mg/7.5 mg, and 750 mg/10 mg pursuant to 21 CFR 320.22(c) provided that the in vitro dissolution profiles of the proposed product are comparable to those of the reference product.

Acetaminophen/Oxycodone, Tablet, Refer to USP, 01/14/2008.

Acetaminophen/Pentazocine HCl, Tablet, I (Basket), 100, Water (deaerated), 900, 10, 20, 30, 45, and 60, 01/12/2004.

Acetaminophen/Propoxyphene HCl, Tablet, Refer to USP, 01/15/2010.

Acetaminophen/Propoxyphene Napsylate, Tablet, Refer to USP, 01/15/2010.

Acetaminophen/Tramadol HCl, Tablet, II (Paddle), 50, 0.1 N HCl, 900, 5, 10, 15, 20, and 30, 03/04/2006.

Acetazolamide, Capsule (Extended Release), II (Paddle), 75, Acetate Buffer, pH 4.5 with 2.2% Tween 20, 900, 1, 2, 5, 7, 9, 12, and 14 h, 01/15/2010, 125 mg based on (1) acceptable bioequivalence studies on the 250 mg strength, (2) acceptable dissolution testing across all strengths, and (3) proportional similarity in the formulations across all strengths.

Acetazolamide, Tablet, Refer to USP, 07/21/2011.

Acetazolamide, Tablet, Refer to USP, 07/14/2008.

Acitretin, Capsule, Refer to USP, 09/22/2011, 10 mg based on (1) acceptable bioequivalence studies on the 25 mg strength, (2) proportionally similar across all strengths, and (3) acceptable in vitro dissolution testing of all strengths.

Acrivastine/Pseudoephedrine HCl, Capsule, II (Paddle), 50, 0.01 N HCl, 900, 5, 10, 15, and 30, 01/12/2004.

Acyclovir, Suspension, II (Paddle), 50, 0.1 N HCl, 900, 10, 20, 30, 45, and 60, 02/20/2004.

Acyclovir, Tablet, Refer to USP, 06/18/2007, 400 mg based on (1) acceptable bioequivalence studies on the 800 mg strength, (2) proportional similarity of the formulations across all strengths, and (3) acceptable in vitro dissolution testing of all strengths.

Acyclovir, Capsule, Refer to USP, 01/05/2012.

Adefovir Dipivoxil, Tablet, II (Paddle), 50, 0.01 N HCl, 600, 10, 20, 30, 45, and 60, 04/10/2008.

Albendazole, Tablet, Refer to USP, 08/15/2013.

Albuterol Sulfate, Tablet, Refer to USP, 09/03/2008, 2 mg based on (1) acceptable bioequivalence studies on the 20 mg strength, (2) acceptable dissolution testing across all strengths, and (3) proportional similarity in the formulations across all strengths.

(Continued)

Albuterol Sulfate, Tablet (Extended Release), II (Paddle), 50, 0.1 N HCl, 900, 1, 2, 4, 6, 9, and 12 h, 04/09/2007.

Alendronate Sodium, Tablet, Refer to USP, 01/14/2008, 5, 10, 35, and 40 mg based on (1) an acceptable bioequivalence study on the 70 mg strength, (2) proportional similarity of the formulations across all strengths, and (3) acceptable in vitro dissolution testing of all strengths.

Alendronate Sodium/Cholecalciferol, Tablet, II (Paddle), For Alendronate: 50; For Cholecalciferol: 75; For Alendronate: Deaerated Water; For Cholecalciferol: 0.3% SDS in USP Water; For Alendronate: 900; For Cholecalciferol: 500, 10, 15, 20, 30, and 45, 11/25/2008, 70 mg/2800 IU based on (1) acceptable bioequivalence studies on the 70 mg/5600 mg strength, (2) proportional similarity of the formulations across all strengths, and (3) acceptable in vitro dissolution testing of all strengths.

Alfuzosin HCl, Tablet (Extended Release), II (Paddle), 100, 0.01 N HCl, 900, 1, 2, 12, 20 h, 06/18/2007.

Aliskiren Hemifumarate, Tablet, I (Basket), 100, 0.01 N HCl, 500, 10, 20, 30, and 45, 09/03/2008, 150 mg based on (1) acceptable in vivo bioequivalence study(ies) on the 300 mg strength, (2) proportional similarity of the formulations across all strengths, and (3) acceptable in vitro dissolution testing on all strengths.

Aliskiren Hemifumarate/Amlodipine Besylate, Tablet, I (Basket), 100, 0.01 N HCl, pH 2.0, 500, 10, 15, 20, 30, and 45, 06/07/2012, Eq 300 mg base; Eq 5 mg base, Eq 150 mg base; Eq 10 mg base, and Eq 150 mg base; Eq 5 mg base based on (1) acceptable bioequivalence studies on the 300 mg/10 mg strength, (2) acceptable in vitro dissolution testing of all strengths, and (3) proportional similarity of the formulations across all strengths.

Aliskiren Hemifumarate/Amlodipine Besylate/Hydrochlorothiazide, Tablet, I (Basket), 100, 0.01 N HCl, 900, 10, 15, 20, 30, and 45, 06/07/2012.

Aliskiren Hemifumarate/Hydrochlorothiazide, Tablet, I (Basket), 100, 0.1 N HCl, 900, 10, 15, 20, 30, and 45, 10/08/2009, 150 mg/25 mg and 150 mg/12.5 mg based on (1) acceptable bioequivalence studies on the 300 mg/25 mg strength, (2) acceptable dissolution testing across all strengths, and (3) proportional similarity in the formulations across all strengths.

Aliskiren Hemifumarate/Valsartan, Tablet, I (Basket), 100, Phosphate Buffer, pH 6.8, 1000, 5, 10, 15, 20, 30, and 45, 12/23/2010, Eq. 150 mg base/160 mg based on (1) acceptable bioequivalence studies on the Eq. 300 mg base/320 mg strength, (2) acceptable in vitro dissolution testing of all strengths, and (3) proportional similarity of the formulations across all strengths. Please refer to the Mirtazapine Tablet Draft Guidance for additional information regarding waivers of in vivo testing.

Allopurinol, Tablet, Refer to USP, 07/25/2007, Eq. 150 mg base/160 mg based on (1) acceptable bioequivalence studies on the Eq. 300 mg base/320 mg strength, (2) acceptable in vitro dissolution testing of all strengths, and (3) proportional similarity of the formulations across all strengths. Please refer to the Mirtazapine Tablet Draft Guidance for additional information regarding waivers of in vivo testing.

Almotriptan Malate, Tablet, II (Paddle), 50, 0.1 N HCl, 900, 5, 10, 15, and 30, 01/20/2006, 6.25 mg based on (1) acceptable bioequivalence studies of the 12.5 mg strength, (2) proportionally similar across all strengths, and (3) acceptable in vitro dissolution testing of all strengths.

Alosetron HCl, Tablet, II (Paddle), 50 (for 1 mg) and 75 (for 0.5 mg), Water (deaerated), 500, 10, 20, 30, and 45, 01/26/2006, 0.5 mg (base) based on (1) acceptable bioequivalence studies on the 1 mg strength, (2) proportionally similar across all strengths, and (3) acceptable in vitro dissolution testing of all strengths.

Alprazolam, Tablet (Extended Release), I (Basket), 100, 1% Phosphate Buffer, pH 6.0, 500, 1, 4, 8, 12, and 16 h, 02/08/2007, 0.5, 1, and 2 mg based on (1) acceptable bioequivalence studies on the 3 mg strength, (2) proportional similarity of the formulations across all strengths, and (3) acceptable in vitro dissolution testing of all strengths.

Alprazolam, Tablet, Refer to USP, 06/18/2007.

Alprazolam, Tablet (Orally Disintegrating), II (Paddle), 50, 70 mM Potassium Phosphate Buffer, pH 6.0, 500, 2, 5, 10, 15, and 20, 10/06/2008.

Altretamine, Capsule, Refer to USP, 01/29/2010.

Alvimopan, Capsule, II (Paddle), 50, 0.1 N HCl, 900, 5, 10, 15, 20, 30, and 45, 10/21/2010.

Amantadine HCl, Capsule, Refer to USP, 12/23/2010.

Amantadine HCl, Tablet, II (Paddle), 50, Water (deaerated), 500, 10, 20, 30, 45, and 60, 01/12/2004.

Ambrisentan, Tablet, II (Paddle), 75, 0.05 M Acetate Buffer, pH 5.0, 900, 5, 10, 15, 30, and 45, 05/20/2009, 5 mg based on (1) acceptable bioequivalence studies on the 10 mg strength, (2) acceptable in vitro dissolution testing of all strengths, and (3) proportional similarity of the formulations across all strengths. Please refer to the Mirtazapine Tablet Draft Guidance for additional information regarding waivers of in vivo testing.

Amiloride HCl, Tablet, Refer to USP, 06/07/2012.

Amiloride HCl/Hydrochlorothiazide, Tablet, Refer to USP, 06/07/2012.

(Continued)

Aminosalicylic, Granule (Delayed Release), II (Paddle), 100, Acid Stage: 0.1 N HCl; Buffer Stage 2: pH 7.5 Phosphate Buffer, 1000, Acid Stage: 2 h; Buffer Stage: 1, 2, 3, and 4 h, 07/14/2008.

Amiodarone HCl (Test 1), Tablet, II (Paddle), 100, 1% SLS in water, 1000, 10, 20, 30, 45, 60, and 90, 01/12/2004, 100, 300, and 400 mg based on (1) acceptable bioequivalence studies on the 200 mg strength, (2) acceptable in vitro dissolution testing of all strengths, and (3) proportional similarity of the formulations across all strengths. Please refer to the Mirtazapine Tablet Draft Guidance for additional information regarding waivers of in vivo testing.

Amiodarone HCl (Test 2), Tablet, I (Basket), 50, Acetate Buffer, pH 4.0, with 1% Tween 80, 900, 10, 20, 30, 45, 60, and 90, 01/12/2004.

Amitriptyline HCl, Tablet, Refer to USP, 01/14/2008, 10, 50, 75, 100, and 150 mg based on (1) acceptable bioequivalence studies on the 25 mg strength, (2) acceptable in vitro dissolution testing of all strengths, and (3) proportional similarity of the formulations across all strengths.

Amlodipine Besylate, Tablet, II (Paddle), 75, 0.01 N HCl, 500, 10, 20, 30, 45, and 60, 01/14/2004, 2.5 and 5 mg based on (1) acceptable bioequivalence studies on the 10 mg strength, (2) proportionally similar across all strengths, and (3) acceptable in vitro dissolution testing of all strengths.

Amlodipine Besylate, Tablet (Orally Disintegrating), II (Paddle), 50, 0.01 M HCl, 500, 5, 10, 15, and 20, 10/06/2008.

Amlodipine Besylate/Atorvastatin Calcium, Tablet, II (Paddle), 75, Phosphate Buffer, pH 6.8, 900, 5, 10, 15, and 30, 04/02/2009.

Amlodipine Besylate/Benazepril HCl, Capsule, I (Basket), 100, 0.01 N HCl, 500, 10, 20, 30, 45, and 60, 06/20/2007.

Amlodipine Besylate/Hydrochlorothiazide/Olmesartan Medoxomil, Tablet, II (Paddle), 50, Phosphate Buffer, pH 6.8, 900, 5, 10, 15, 20, 30, and 45, 07/21/2011.

Amlodipine Besylate/Hydrochlorothiazide/Valsartan, Tablet, II (Paddle), 50, Phosphate Buffer, pH 6.8, 900, 10, 20, 30, and 45, 03/25/2010.

Amlodipine Besylate/Hydrochlorothiazide/Valsartan (10/25/320 mg), Tablet, II (Paddle), 55, Phosphate Buffer, pH 6.8, 900, 10, 20, 30, and 45, 03/25/2010.

Amlodipine Besylate/Olmesartan Medoxomil, Tablet, II (Paddle), 50, Phosphate Buffer, pH 6.8, 900, 10, 20, 30, and 45, 08/11/2008.

Amlodipine Besylate/Telmisartan, Tablet, II (Paddle), 75, Telmisartan: Phosphate Buffer, pH 7.5; Amlodipine: 0.01 N HCl, pH 2, Telmisartan: 900; Amlodipine: 500, Telmisartan: 10, 15, 20, 30, and 45; Amlodipine: 10, 15, 20, 30, and 45, 08/05/2010.

Amlodipine Besylate/Valsartan, Tablet, II (Paddle), 75, Phosphate Buffer, pH 6.8, 1000, 5, 10, 15, 20, 30, and 45, 07/21/2011.

Amoxicillin, Tablet (Extended Release), II (Paddle), 100, 3-Stage dissolution: 50 mM potassium phosphate monobasic buffer at pH 4.0 (0–2 h), 6.0 (2–4 h), and 6.8 (4 hand beyond), 900, 0.25, 0.5, 1, 2, 2.25, 2.5, 3, 4, 4.25, 4.5, 5, and 6 h, 10/21/2010.

Amoxicillin, Capsule, Refer to USP, 01/31/2013.

Amoxicillin, Tablet, Refer to USP, 01/31/2013.

Amoxicillin, For Oral Suspension, II (Paddle), 50, Water (degassed), 900, 5, 10, 15, 20, 30, and 45, 06/06/2013.

Amoxicillin/Clarithromycin/Lansoprazole, Capsule/Tablet/Capsule (Copackage), Refer to USP for monographs of Amoxicillin Capsules, Clarithromycin Tablets, and Lansoprazole Delayed-Release Capsules, 02/28/2013.

Amoxicillin/Clarithromycin/Omeprazole, Capsule/Tablet/Capsule (Copackage), Refer to USP for monographs of Amoxicillin Capsules, Clarithromycin Tablets, and Omeprazole Delayed-Release Capsules, 02/28/2013.

Amoxicillin/Clavulanate Potassium, Tablet, Refer to USP, As appropriate, 0, 0.5, 1, 2, 3, 4, and 5 h, 10/04/2012.

Amoxicillin/Clavulanate Potassium, Tablet (Chewable), Refer to USP, 01/14/2008.

Amoxicillin/Clavulanate Potassium, Suspension, II (Paddle), 75, Water (deaerated), 900, 5, 10, 15, and 30, 01/14/2004.

Amphetamine Aspartate/Amphetamine Sulfate/Dextroamphetamine Saccharate/Dextroamphetamine Sulfate, Capsule (Extended Release), II (Paddle), 50, Dilute HCl, pH 1.1 for first 2 h, then add 200 mL of 200 mM Phosphate Buffer and adjust to pH 6.0 for the remainder, 0–2 h: 750 mL. After 2 h: 950 mL, 0.5, 1, 2, 3, and 4 h, 07/25/2007.

Amphetamine Aspartate/Amphetamine Sulfate/Dextroamphetamine Saccharate/Dextroamphetamine Sulfate, Tablet, I (Basket), 100, Deionized Water, 500, 10, 20, 30, and 45, 11/25/2008.

Amphetamine ER, Capsule, II (Paddle), 50, 750 mL of dilute HCl, pH 1.1 for the first 2 h, then add 200 mL of 200 mM phosphate buffer, and adjust to pH 6 (w/HCl or NaOH) for the remainder, 750 mL of dilute HCl, 200 mL of phosphate buffer, 1, 2, 3, 4, and 6 h, 08/17/2006.

Ampicillin/Ampicillin Trihydrate, For Oral Suspension, II (Paddle), 25, Water (deaerated), 900, 5, 10, 15, 20, 01/03/2007.

Amprenavir, Capsule, II (Paddle), 50, 0.1 N HCl, 900, 10, 15, 30, and 45, 02/19/2008.

(*Continued*)

Anagrelide HCl, Capsule, I (Basket), 100, 0.1 N HCl, 900, 5, 10, 15, 30, and 45, 01/14/2004.

Anastrozole, Tablet, II (Paddle), 50, Water, 900, 5, 10, 15, and 30, and 45, 01/03/2007.

Apixaban Tablets, Tablet, II (Paddle), 75, 0.05 M Sodium Phosphate Buffer with 0.05% SLS, pH 6.8, 900, 5, 10, 20, 30, and 45, 05/09/2013.

Aprepitant, Capsule, II (Paddle), 100, 2.2% sodium dodecyl sulfate in distilled water, 900, 10, 15, 20, 30, and 45, 01/20/2006, 40 and 80 mg based on (1) acceptable bioequivalence studies on the 125 mg strength, (2) proportionally similar to the 125 mg strength, and (3) acceptable in vitro dissolution testing.

Aripiprazole, Tablet (Orally Disintegrating), II (Paddle), 75, Acetate Buffer, pH 4.0, 1000, 10, 20, 30, and 45, 08/11/2008, 15, 20, and 30 mg based on (1) acceptable bioequivalence studies on the 10 mg, (2) proportional similarity of all strengths, and (3) acceptable in vitro dissolution testing of all strengths.

Aripiprazole, Tablet, II (Paddle), 60, pH 1.2 USP Buffer (Hydrochloric Acid), 900, 10, 20, 30, and 45, 12/20/2005, 2, 15, 20, and 30 mg, based on (1) acceptable bioequivalence studies on the 5 and 10 mg strengths (2) proportionally similar across all strengths, and (3) acceptable in vitro dissolution testing of all strengths.

Armodafinil, Tablet, II (Paddle), 50, 0.1 N HCl, 900, 10, 20, 30, and 45, 01/14/2008, 50 and 150 mg, based on acceptable (1) bioequivalence studies on the 250 mg strength, and (2) proportional similarity of the formulations and (3) acceptable in vitro dissolution testing of all strengths.

Asenapine Maleate, Tablet (Sublingual), II (Paddle), 50, Acetate Buffer, pH 4.5, 500, 1, 2, 3, 4, and 5, 05/09/2013.

Aspirin/Butalbital/Caffeine, Capsule, Refer to USP, 06/24/2010, 325 mg/50 mg/40 mg pursuant to 21 CFR 320.22(c) provided that the in vitro dissolution profiles of the proposed product are comparable to those of the reference product.

Aspirin/Butalbital/Caffeine, Tablet, Refer to USP, 06/24/2010, 325 mg/50 mg/40 mg pursuant to 21 CFR 320.22(c) provided that the in vitro dissolution profiles of the proposed product are comparable to those of the reference product.

Aspirin/Butalbital/Caffeine/Codeine Phosphate, Capsule, Refer to USP, 08/27/2009.

Aspirin/Caffeine/Orphenadrine Citrate, Tablet, I (Basket), 75, Water (deaerated), 900, 10, 20, 30, 45, and 60, 01/15/2004.

Aspirin/Dipyridamole, Capsule, I (Basket), 100, 0.01 N HCl for first hour, 0.1 M Phosphate Buffer, pH 5.5, thereafter, 0–1 h: 900 mL. 900 mL thereafter, Acid: 10, 20, 30, 45, and 60 min; Buffer: 1, 2, 5, and 7 h, 10/09/2007.

Aspirin/Hydrocodone Bitartrate, Tablet, II (Paddle), 75, Acetate Buffer, pH 4.5, 900, 10, 20, 30, 45, 60, and 90, 01/15/2004.

Aspirin/Meprobamate, Tablet, I (Basket), 100, Water (deaerated), 900, 10, 20, 30, 45, 60, and 90, 01/15/2004.

Aspirin/Methocarbamol, Tablet, II (Paddle), 50, Water (deaerated), 900, 10, 20, 30, 45, 60, and 90, 01/15/2004.

Aspirin/Oxycodone HCl, Tablet, Refer to USP, 01/15/2010, 325 mg/4.8355 mg and 325 mg/4.5 mg/0.38 mg pursuant to 21 CFR 320.22(c) provided that the in vitro dissolution profiles of the proposed product are comparable to those of the reference product.

Atazanavir Sulfate, Capsule, II (Paddle), 50, 0.025 N HCl, 1000, 10, 20, 30, and 45, 01/20/2006, 100, 150, and 200 mg based on (1) acceptable bioequivalence studies on the 300 mg strength, (2) proportionally similar across all strengths, and (3) acceptable in vitro dissolution testing of all strengths.

Atenolol, Tablet, Refer to USP, 07/25/2007, 25 and 50 mg based on (1) acceptable bioequivalence studies on the 100 mg strength, (2) acceptable in vitro dissolution testing of all strengths, and (3) proportional similarity of the formulations across all strengths.

Atomoxetine HCl, Capsule, II (Paddle), 50, 0.1 N HCl, 1000, 10, 20, 30, and 45, 12/20/2005, 5, 10, 18, 25, 40, 80, and 100 mg based on (1) acceptable bioequivalence studies on the 60 mg strength, (2) proportionally similar across all strengths, and (3) acceptable in vitro dissolution testing of all strengths.

Atorvastatin Calcium, Tablet, II (Paddle), 75, 0.05 M Phosphate buffer, pH 6.8, 900, 5, 10, 15, and 30, 01/15/2004, Eq 10, 20, and 40 mg base based on (1) acceptable bioequivalence studies on the Eq 80 mg base strength, (2) acceptable in vitro dissolution testing of all strengths, and (3) proportional similarity of the formulations across all strengths.

Atovaquone, Oral Suspension, Develop a dissolution method, 07/21/2009.

Atovaquone, Tablet, II (Paddle), 50, 40% isopropanol buffered to pH 8.0 with potassium dihydrogen phosphate, 900, 10, 20, 30, 45, 60, and 90, 06/18/2007.

Atovaquone/Proguanil HCl, Tablet, II (Paddle) with PEAK vessels, 50, 40% isopropanol buffered to pH 8.0 with potassium dihydrogen phosphate, 900, 15, 30, 45, and 60, 08/17/2006.

Auranofin, Capsule, II (Paddle), 50, Water (deaerated), 900, 10, 20, 30, and 45, 01/15/2004.

Azacitidine, Injectable Suspension, Develop a dissolution method, 09/03/2008.

Azathioprine, Tablet, Refer to USP, 04/08/2010, 25, 75, and 100 mg based on (1) acceptable bioequivalence studies on the 50 mg strength, (2) proportional similarity of the formulations across all strengths, and (3) acceptable in vitro dissolution testing of all strengths.

(Continued)

Azilsartan Kamedoxomil, Tablet, II (Paddle), 50, Phosphate Buffer, pH 7.8 (deaerated), 900, 5, 10, 15, 20, 30, and 45, 05/09/2013.

Azilsartan Kamedoxomil/Chlorthalidone, Tablet, II (Paddle), 50, Phosphate Buffer, pH 6.8 containing 1.0% Tween 80, 900, 5, 10, 15, 20, 30, and 45, 05/09/2013.

Azithromycin, Tablet, II (Paddle), 75, 0.1 M Phosphate Buffer, pH 6.0, 900, 10, 20, 30, and 45, 01/14/2008, 250 and 500 mg based on (1) acceptable bioequivalence studies on the 600 mg strength, (2) proportionally similar across all strengths, and (3) acceptable in vitro dissolution testing of all strengths.

Azithromycin, Oral Suspension, II (Paddle), 50, Phosphate buffer, pH 6.0, 900, 10, 20, 30, and 45, 08/17/2006, 100 mg/5 mL based on (1) acceptable bioequivalence studies on the 200 mg/5 mL strength, (2) acceptable dissolution testing across all strengths, and (3) proportional similarity in the formulations across all strengths.

Azithromycin, Suspension (Extended Release), II (Paddle), 50, Phosphate Buffer, pH 6.0, 900, 15, 30, 45, 60, 120, and 180, 04/15/2008.

Baclofen, Tablet, Refer to USP, 12/15/2009, 10 mg based on (1) acceptable bioequivalence studies on the 20 mg strength, (2) proportional similarity of the formulations across all strengths, and (3) acceptable in vitro dissolution testing of all strengths.

Baclofen, Tablet (Orally Disintegrating), II (Paddle), 25, 50 mM Acetate Buffer, pH 4.5, 500 mL (10 mg) or 1000 mL (20 mg), 5, 10, 15, and 30, 07/14/2008, 10 mg based on (1) acceptable bioequivalence study on the 20 mg strength, (2) proportional similarity in the formulations, and (3) acceptable in vitro dissolution testing of all strengths.

Balsalazide Disodium, Capsule, II (Paddle) with sinker, 50, pH 6.8 buffer, 900, 10, 20, 30, and 45, 01/26/2006.

Balsalazide Disodium, Tablet, II (Paddle), 100, Water (degassed), 1000, 10, 20, 30, 45, 60, 75, 90, and 120, 07/31/2013.

Bedaquiline Fumarate, Tablet, I (Basket), 150, 0.01 N HCl, 900, 10, 15, 20, 30, and 45, 06/06/2013.

Benazepril HCl, Tablet, II (Paddle), 50, Water (deaerated), 500, 10, 20, 30, and 45, 01/16/2004, 5, 10, and 20 mg based on (1) acceptable bioequivalence studies on the 40 mg strength, (2) acceptable in vitro dissolution testing of all strengths, and (3) proportional similarity of the formulations across all strengths.

Benazepril HCl/Hydrochlorothiazide, Tablet, I (Basket), 100, 0.1 N HCl, 500, 10, 20, 30, and 45, 01/16/2004.

Bendroflumethiazide/Nadolol, Tablet, Refer to USP, 07/25/2007.

Benzonatate, Capsule, Refer to USP.

Benzphetamine HCl, Tablet, II (Paddle), 50, Water, 900, 10, 20, 30, and 45, 06/20/2007, 50 mg.

Bepridil HCl, Tablet, I (Basket), 100, 0.1 N HCl, 900, 10, 20, 30, 45, and 60, 01/16/2004.

Betamethasone Acetate/Betamethasone Sodium Phosphate, Injectable Suspension, IV (Flow-through cell), Flow at 8 mL/min, 0.05% SLS, pH 3.0 or develop an in vitro release method using USP IV (Flow-Through Cell), and, if applicable, Apparatus II (Paddle) or any other appropriate method, for comparative evaluation by the agency, 5, 10, 15, 30, 45, 60, 90, 120, 180, 240, 300, and 360, 04/08/2010.

Bethanechol Chloride, Tablet, Refer to USP, 10/06/2008, 5, 10, 25, and 50 mg pursuant to 21 CFR.320.22(c) provided the in vitro dissolution profiles of your product are comparable to those of the reference product.

Bexarotene, Capsule, II (Paddle), 50, Tier 1 Medium: 0.5% HDTMA in 0.05 M phosphate buffer, pH 7.5 Tier 2 Medium: 0.5% HDTMA in 0.05 M phosphate buffer, pH 7.5 with 0.05 g/L pancreatin enzyme, 900, 15, 30, 45, and 60, 08/17/2006.

Bicalutamide, Tablet, II (Paddle), 50, 1% SLS in water, 1000, 10, 20, 30, 45, and 60, 12/15/2005.

Bismuth Subcitrate Potassium/Metronidazole/Tetracycline HCl, Capsule, II (Paddle), 75, Tetracycline and Metronidazole: 0.1 N HCl; Bismuth Subcitrate Potassium: Water, 900, 5, 15, 20, 30, and 45, 10/06/2008.

Bisoprolol Fumarate, Tablet, Refer to USP, 06/18/2007, 5 mg based on (1) acceptable bioequivalence studies on the 10 mg strength, (2) proportionally similar across all strengths, and (3) acceptable in vitro dissolution testing of all strengths.

Bisoprolol Fumarate/Hydrochlorothiazide, Tablet, II (Paddle), 75, 0.1 N HCl, 900, 5, 10, 20, 30, and 45, 01/20/2004, 2.5 mg/6.25 mg and 5 mg/6.25 mg based on (1) acceptable bioequivalence studies on the 10 mg/6.25 mg strength, (2) proportionally similar across all strengths, and (3) acceptable in vitro dissolution testing of all strengths.

Boceprevir, Capsule, II (Paddle) with sinker, 50, 50 mM phosphate buffer, pH 6.8 with 0.1% sodium dodecyl sulfate, 900, 10, 20, 30, 45, 60, and 75, 01/31/2013.

Bosentan, Tablet, II (Paddle), 50, 1% SLS in water, 900, 15, 30, 45, and 60, 09/02/2010, 62.5 mg based on (1) acceptable bioequivalence studies on the 125 mg strength, (2) acceptable in vitro dissolution testing of all strengths, and (3) proportional similarity of the formulations across all strengths. Please refer to the Mirtazapine Tablet Draft Guidance for additional information regarding waivers of in vivo testing.

Brinzolamide, Ophthalmic Suspension, Develop a method to characterize in vitro release, 09/01/2011.

Bromocriptine Mesylate, Tablet, Refer to USP, 07/25/2007.

(*Continued*)

Budesonide, Capsule, II (Paddle) with sinker, 75, Acid stage: 0.1 N HCl; Buffer stage: Phosphate Buffer, pH 7.5, Acid stage: 1000; Buffer stage: 1000, Acid stage: 2 h; Buffer stage: 1, 2, 4, 6, and 8 h, 05/20/2009.

Bumetanide, Tablet, Refer to USP, 07/14/2008, 0.5 and 1 mg based on (1) acceptable bioequivalence studies on the 2 mg strength, (2) acceptable dissolution testing across all strengths, and (3) proportional similarity in the formulations across all strengths.

Buprenorphine, Film, Transdermal (Extended Release), VI (Cylinder) with adapter, if needed, 50, 0.9% Sodium Chloride at 32°C, 600, 0.5, 1, 2, 4, 6, 8, 12, 16, and 24 h, 05/09/2013.

Buprenorphine HCl, Tablet (Sublingual), I (Basket), 100, Water, 500, 2, 5, 8, 10, 15, and until at least 80% of the labeled content is dissolved, 04/09/2007, 2 mg strength based on (1) acceptable bioequivalence study on the 8 mg strength, (2) acceptable in vitro dissolution testing of all strengths, and (3) proportional similarity of the formulations across all strengths.

Buprenorphine HCl/Naloxone HCl, Film (Sublingual), V (Paddle over Disk) with 56 mm, 40 mesh stainless steel disk, 100, Acetate Buffer, pH 4.0, 900, 1, 2, 3, 5, 7, and 10, 05/09/2013, 2 mg/0.5 mg (base) strength based on (1) acceptable bioequivalence study on the 8 mg/2 mg (base) strength, (2) acceptable in vitro dissolution testing of all strengths, and (3) proportional similarity of the formulations across all strengths.

Buprenorphine HCl/Naloxone HCl, Tablet (Sublingual), I (Basket), 100, Water, 500, 1, 3, 5, 7.5, 10, 15, and 20, 07/01/2010.

Bupropion HCl, Tablet, Refer to USP, 08/15/2013, Bupropion Hydrochloride Tablet, 75 mg based on (1) acceptable bioequivalence studies on the 100 mg strength, (2) acceptable in vitro dissolution testing of all strengths, and (3) proportional similarity of the formulations across all strengths. Please refer to the Mirtazapine Tablet Draft Guidance for additional information regarding waivers of in vivo testing.

Bupropion HCl, Tablet (Extended Release), Refer to USP, 07/25/2007.

Bupropion Hydrobromide, Tablet (Extended Release), I (Basket), 75, 0.1 N HCl, 900, 1, 2, 4, 6, 8, and 10 h, 06/10/2009.

Buspirone Hydrochloride, Tablet, Refer to USP, 07/21/2009.

Busulfan, Tablet, II (Paddle), 50, Water (deaerated), 500, 5, 10, 15, and 30, 07/14/2008.

Cabergoline, Tablet, II (Paddle), 50, 0.1 N HCl, 500, 5, 10, 15, and 30, 01/20/2004.

Calcitriol, Capsule, Develop a quantitative rupture test, 06/03/2008, 0.25 mcg based on (1) acceptable bioequivalence studies on the 0.5 mcg strength, (2) proportionally similar across all strengths, and (3) acceptable in vitro dissolution testing of all strengths.

Calcium Acetate, Capsule, II (Paddle), 50, Water, 900, 5, 10, 15, 20, and 30, 07/21/2009.

Calcium Acetate, Tablet, Refer to USP, 01/14/2008.

Candesartan Cilexetil (16, 8, and 4 mg), Tablet, II (Paddle), 50, 0.35% Polysorbate 20 in 0.05 M Phosphate Buffer, pH 6.5, 900, 10, 20, 30, 45, and 60, 06/20/2007.

Candesartan Cilexetil (32 mg), Tablet, II (Paddle), 50, 0.70% Polysorbate 20 in 0.05 M Phosphate Buffer, pH 6.5, 900, 10, 20, 30, 45, and 60, 06/20/2007, 4, 8, and 16 mg based on (1) acceptable bioequivalence studies on the 32 mg strength, (2) acceptable dissolution testing of all strengths, and (3) proportional similarity in the formulations of all strengths.

Candesartan Cilexetil/Hydrochlorothiazide (16/12.5 mg), Tablet, II (Paddle), 50, 0.35% Polysorbate 20 in phosphate buffer pH 6.5, 900, 10, 20, 30, 45, and 60, 01/29/2010.

Candesartan Cilexetil/Hydrochlorothiazide (32/12.5 and 32/25 mg), Tablet, II (Paddle), 50, 0.70% Polysorbate 20 in phosphate buffer pH 6.5, 900, 15, 20, 30, 45, and 60, 01/29/2010, 16 mg/12.5 mg and 32 mg/12.5 mg based on (1) acceptable bioequivalence studies on the 32 mg/25 mg strength, (2) formulation proportionality across all strengths, and (3) acceptable in vitro dissolution testing of all strengths.

Capecitabine, Tablet, II (Paddle), 50, Water (deaerated), 900, 10, 20, 30, and 45, 01/23/2004, 150 mg based on (1) acceptable bioequivalence study on the 500 mg strength, (2) acceptable in vitro dissolution testing all strengths, and (3) proportional similarity of the formulations across all strengths.

Carbamazepine, Suspension, II (Paddle), 50, Water (deaerated), 900, 10, 20, 30, 45, and 60, 01/20/2004.

Carbamazepine, Capsule (Extended Release), II (Paddle), 75, First 4 h: Dilute Acid, pH 1.1. After 4 h: Phosphate Buffer, pH 7.5 with 0.1% sodium lauryl sulfate (SLS), First 4 h: 900. After 4 h: 900, 1, 2, 4, 6, 8, 10, and 12 h, 09/01/2011, 100 and 200 mg based on (1) acceptable bioequivalence studies on the 300 mg strength, (2) proportionally similar across all strengths, and (3) acceptable in vitro dissolution testing of all strengths.

(Continued)

Carbamazepine, Tablet, Refer to USP, If an applicant desires to develop the entire product line (100, 200, 300, and 400 mg), in vivo bioequivalence studies should compare test, Carbamazepine Tablets, 400 mg with 2 × 200 mg of the reference listed drug (RLD). In vivo bioequivalence study requirements for the 100, 200, and 300 mg strengths may be waived based on (1) acceptable bioequivalence studies on the 400 mg strength, (2) acceptable in vitro dissolution testing of all strengths, and (3) proportional similarity of the formulations across all strengths. Please refer to the Mirtazapine Tablet Draft Guidance for additional information regarding waivers of in vivo testing.

Carbamazepine, Tablet (Chewable), II (Paddle), 75, 1% SLS in Water, 900, 15, 30, 45, 60, and 90, 12/23/2010.

Carbamazepine, Tablet (Extended Release), Refer to USP, 01/14/2008, 100 and 200 mg, based on acceptable (1) bioequivalence studies on the 400 mg tablet, (2) proportional similarity of the formulations, and (3) acceptable in vitro dissolution testing of all strengths.

Carbidopa/Entacapone/Levodopa, Tablet, I (Basket), Carbidopa and Levodopa: 50; Entacapone: 125; For both Carbidopa and Levodopa: 0.1 N HCl; For Entacapone: Phosphate buffer pH 5.5, Carbidopa and Levodopa: 750 mL. Entacapone: 900 mL, 10, 20, 30, 45, and 60, 01/03/2007, (37.5; 200; 150 mg), (31.25; 200; 125 mg), and (25; 200; 100 mg) based on (1) acceptable bioequivalence studies on the (50; 200; 200 mg) strength, (2) acceptable in vitro dissolution testing of the aforementioned strengths, and (3) proportional similarity in the formulations across the aforementioned strengths. Please refer to Mirtazapine Tablet Draft Guidance for additional information regarding waiver of in vivo testing. (18.75; 200; 75 mg) based on (1) acceptable bioequivalence studies on the (12.5; 200; 50 mg) strength, (2) acceptable in vitro dissolution testing of the aforementioned strengths, and (3) proportional similarity in the formulations across the aforementioned strengths. Please refer to Mirtazapine Tablet Draft Guidance for additional information regarding waiver of in vivo testing.

Carbidopa/Levodopa, Tablet (Orally Disintegrating), II (Paddle), 50, 0.1 N HCl, 750, 5, 10, 15, 30, and 45, 07/25/2007, 10 mg/100 mg and 25 mg/100 mg based on (1) acceptable bioequivalence studies on the 25 mg/250 mg strength, (2) acceptable in vitro dissolution testing of all strengths, and (3) proportional similarity of the formulations across all strengths.

Carbidopa/Levodopa, Tablet, Refer to USP, 01/14/2008.

Carbidopa/Levodopa, Tablet (Extended Release), II (Paddle), 50, 0.1 N HCl, 900, 0.5, 0.75, 1, 1.5, 2, 2.5, 3, and 4 h, 08/15/2013, 25 mg/100 mg based on (1) acceptable bioequivalence studies on the 50 mg/200 mg strength, (2) acceptable in vitro dissolution testing of all strengths, and (3) proportional similarity of the formulations across all strengths. Please refer to the Mirtazapine Tablet Draft Guidance for additional information regarding waivers of in vivo testing.

Carglumic Acid, Tablet, II (Paddle), 100, 0.05 M Phosphate Buffer, pH 6.8, 750, 5, 10, 15, 20, and 30, 08/15/2013.

Carisoprodol, Tablet, Refer to USP, 01/29/2010, 350 mg strength of the test drug product, based on acceptable formulation data and in vitro dissolution testing under the criteria set forth in 21 CFR 320.22(c). Carisoprodol 350 mg Tablet is a DESI1—effective drug for which there are no known or suspected bioequivalence problems and as such is rated "AA" in FDA/CDER's Approved Drug Products with Therapeutic Equivalence Evaluations ("Orange Book").

Carvedilol, Tablet, II (Paddle), 50, SGF without enzyme, 900, 10, 20, 30, and 45, 01/21/2004, 3.125, 6.25, and 25 mg based on (1) acceptable bioequivalence studies on the 12.5 mg strength, (2) proportionally similar across all strengths, and (3) acceptable in vitro dissolution testing of all strengths.

Carvedilol Phosphate, Capsule (Extended Release), II (Paddle), 100, 0.1 N HCl, 900, 1, 4, 8, 12, 18, and 24 h, 04/02/2009, 10, 20, and 80 mg based on (1) acceptable bioequivalence studies on the 40 mg strength, (2) proportionally similar across all strengths, and (3) acceptable in vitro dissolution testing of all strengths.

Cefaclor, Tablet (Chewable), Refer to USP, 03/03/2011.

Cefaclor, Tablet (Extended Release), Refer to USP, 03/03/2011.

Cefaclor, Capsule, Refer to USP, 03/03/2011, Cefaclor Capsule, 250 mg based on (1) acceptable bioequivalence studies on the 500 mg strength, (2) acceptable in vitro dissolution testing of all strengths, and (3) proportional similarity of the formulations across all strengths. Please refer to the Mirtazapine Tablet Draft Guidance for additional information regarding waivers of in vivo testing.

Cefadroxil, Tablet, Refer to USP, 09/02/2010.

Cefadroxil, Capsule, Refer to USP, 09/02/2010.

Cefadroxil, Suspension, II (Paddle), 25, Water, 900, 5, 10, 15, 30, and 45, 07/25/2007.

Cefdinir, Capsule, II (Paddle), 50, Phosphate Buffer, pH 6.8, 900, 5, 10, 15, 30, and 45, 07/25/2007.

Cefdinir, Suspension, II (Paddle), 50, 0.05 M Phosphate buffer, pH 6.8, 900, 10, 20, 30, and 45, 04/09/2007.

Cefditoren Pivoxil, Tablet, II (Paddle), 75, Simulated Gastric Fluid without enzyme, 900, 5, 10, 15, 20, and 30, 01/15/2010, 200 mg, based on (1) acceptable bioequivalence studies on the 400 mg strength tablets, (2) proportional similarity of formulations across all strengths, and (3) acceptable in vitro drug release testing of all strengths.

(Continued)

Cefixime, Tablet, Refer to USP, 12/23/2010.

Cefixime, Tablet (Chewable), II (Paddle), 25, Phosphate Buffer, pH 7.2, 900, 10, 15, 20, 30, and 45, 12/23/2010.

Cefixime, Capsule, I (Basket), 100, 0.05 M Phosphate Buffer, pH 7.2, 900, 10, 20, 30, 45, and 60, 08/15/2013.

Cefixime, Suspension, II (Paddle), 50, 0.05 M Phosphate buffer, pH 7.2, 900, 10, 20, 30, and 45, 04/09/2007, 100 mg/5 mL based on (1) acceptable bioequivalence studies on the 200 mg/5 mL strength, (2) acceptable in vitro dissolution testing of both strengths, and (3) proportional similarity of the formulations across all strengths. Please refer to the Mirtazapine Tablet Draft Guidance for additional information regarding dose proportionality.

Cefpodoxime Proxetil, Tablet, Refer to USP, 07/25/2007.

Cefpodoxime Proxetil, Suspension, II (Paddle), 50, Glycine Buffer (0.04 M) pH 3.0, 900, 10, 20, 30, and 45, 12/20/2005, Eq 50 mg base/5 mL based on (1) acceptable bioequivalence studies on the Eq 100 mg base/5 mL strength, (2) acceptable in vitro dissolution testing of all strengths, and (3) proportional similarity of the formulations across all strengths. Please refer to the Mirtazapine Tablet Draft Guidance for additional information regarding waivers of in vivo testing.

Cefprozil, Tablet, Refer to USP, 10/04/2012.

Cefprozil, For Oral Suspension, II (Paddle), 25, Water, 900, 5, 10, 15, 20, and 30, 10/04/2012, 125 mg/5 mL based on (1) acceptable bioequivalence studies on the 250 mg/5 mL strength, (2) acceptable in vitro dissolution testing of all strengths, and (3) proportional similarity of the formulations across all strengths. Please refer to the Mirtazapine Tablet Draft Guidance for additional information regarding waivers of in vivo testing.

Cefprozil, Tablet, Refer to USP, 07/25/2007, 250 mg based on (1) acceptable bioequivalence studies on the 500 mg strength, (2) acceptable in vitro dissolution testing of all strengths, and (3) proportional similarity of the formulations across all strengths. Please refer to the Mirtazapine Tablet Draft Guidance for additional information regarding waivers of in vivo testing.

Cefprozil Monohydrate, Suspension, II (Paddle), 25, Water (deaerated), 900, 5, 10, 15, and 30, 01/21/2004.

Ceftibuten Dihydrate, Suspension, II (Paddle), 50, 0.05 M Phosphate Buffer, pH 7.0, 1000, 10, 20, 30, and 45, 01/21/2004.

Cefuroxime Axetil, Tablet, Refer to USP, 07/25/2007, 125 mg/5 mL based on (1) acceptable bioequivalence studies on the 250 mg/5 mL strength, (2) acceptable in vitro dissolution testing of all strengths, and (3) proportional similarity of the formulations across all strengths.

Celecoxib, Capsule, II (Paddle), 50, 100, and 200 mg: 50 rpm; 400 mg: 75 rpm, Tier 1 Medium: 0.04 M tribasic sodium phosphate (pH 12) with 1% SLS. Tier 2 Initial Medium: 750 mL of simulated gastric fluid, USP (includes pepsin); At 20 min, while stirring, add 180 mL of appropriate concentrations of SLS solution (for a final concentration of 1% SLS). Add about 70 mL of 1.2 N NaOH to adjust the pH to 12, Tier 1: 1000 mL Tier 2: 750 mL (initial) 1000 mL (final), 10, 20, 30, 45, and 60, 07/01/2010, 100 and 200 mg based on (1) acceptable bioequivalence studies on the 400 mg strength, (2) proportionally similar across all strengths, and (3) acceptable in vitro dissolution testing of all strengths.

Cephalexin, Capsule, Refer to USP, 04/02/2009, 250, 333, and 500 mg based on (1) acceptable bioequivalence studies on the 750 mg strength, (2) acceptable dissolution testing across all strengths, and (3) proportional similarity in the formulations across all strengths.

Cephalexin, Suspension, II (Paddle), 25, Water, 900, 5, 10, 20, and 30, 07/25/2007.

Cetirizine HCl, Tablet (Regular and Chewable), II (Paddle), 50, Water (deaerated), 900, 10, 20, 30, and 45, 03/04/2006, 5 mg based on (1) acceptable bioequivalence studies on the 10 mg strength, (2) proportional similarity of the formulations across all strengths, and (3) acceptable in vitro dissolution testing of all strengths.

Cetirizine HCl/Pseudoephedrine HCl, Tablet (Extended Release), I (Basket), 100, 0.1 N HCl, 500, 0.17, 0.25, 0.5, 1, 2, 6, and 8 h, 06/18/2007.

Cevimeline HCl, Capsule, II (Paddle) with option to use a sinker, 50, 0.1 N HCl, 900, 5, 10, 15, and 30, 01/26/2006.

Chlorambucil, Tablet, II (Paddle), 75, 0.1 N HCl, 900, 10, 20, 30, and 45, 08/17/2006.

Chlorpheniramine Maleate, Tablet (Extended Release), III (Reciprocating Cylinder), 27 dpm, Row 1: Test Fluid 1 (0.1 N HCl) for the first hour. Row 2: Test fluid 2 (Phosphate Buffer, pH 7.5) for fifth hour, Row 1: 250 mL. Row 2: 250 mL, 1 h for test fluid 1 and 4 h for test fluid 2, 07/25/2007.

Chlorpheniramine Maleate/Ibuprofen/Pseudoephedrine HCl, Tablet, II (Paddle), 50, 0.05 M Phosphate Buffer, pH 6.5, 900, 10, 20, 30, and 45, 02/20/2004.

Chlorpheniramine Polistirex/Hydrocodone Polistirex, Extended Release Oral Suspension, II (Paddle), 50, Simulated Gastric Fluid (SGF) at 37°C ± 0.5°C, 495, 1, 2, 3, 6, 8, 12, 16, and 24 h, 06/30/2011.

(*Continued*)

Chlorpheniramine Polistirex/Hydrocodone Polistirex, Capsule (Extended Release), II (Paddle), 50, Simulated Intestinal Fluid without enzyme, 500, 1, 4, 12, and 24 h, 11/25/2008, 4 mg/5 mg based on (1) acceptable bioequivalence studies on the 8 mg/10 mg strength, (2) acceptable dissolution testing across all strengths, and (3) proportional similarity in the formulations across all strengths.

Chlorpromazine HCl, Tablet, Refer to USP, 01/05/2012, 50 and 200 mg based on (1) acceptable bioequivalence studies on the 100 mg strength, (2) acceptable in vitro dissolution testing of all strengths, and (3) proportional similarity of the formulations across all strengths. 10 mg based on (1) acceptable bioequivalence studies on the 25 mg strength, (2) acceptable in vitro dissolution testing of both strengths, and (3) proportional similarity of the formulations across both strengths. Please refer to the Mirtazapine Tablet Draft Guidance for additional information regarding waivers of in vivo testing.

Chlorthalidone, Tablet, Refer to USP, 04/15/2008, 15 and 25 mg based on (1) acceptable bioequivalence studies on the 50 mg strength (2) proportional similarity of the formulations across all strengths, and (3) acceptable in vitro dissolution testing of all strengths.

Chlorzoxazone, Tablet, Refer to USP, 01/14/2008.

Choline Fenofibrate, Capsule (Delayed Release), II (Paddle), 50, Acid Stage: 0.05 M Sodium Phosphate, pH 3.5 ± 0.05; Buffer Stage: 0.05 M Sodium Phosphate, pH 6.8 ± 0.05, Acid stage: 500; Buffer stage: 900, Acid stage: 120; Buffer stage: 15, 30, 60, 90, 120, 240, and 360, 07/01/2010, Eq 45 mg based on (1) acceptable bioequivalence studies on the Eq 135 mg strength, (2) proportionally similar across all strengths, and (3) acceptable in vitro dissolution testing on all strengths.

Ciclopirox, Topical Suspension, Develop a method to characterize in vitro release, 03/25/2010.

Cilostazol, Tablet, II (Paddle), 75, 0.3% SLS in water, 900, 15, 30, 45, 60, and 90, 08/17/2006.

Cinacalcet HCl, Tablet, II (Paddle), 75, 0.05 N HCl, 900, 10, 20, 30, and 45, 01/26/2006, 60 and 30 mg based on (1) acceptable bioequivalence studies on the 90 mg strength, (2) proportionally similar across all strengths, and (3) acceptable in vitro dissolution testing of all strengths.

Ciprofloxacin, Oral suspension, II (Paddle), 100, 0.05 M Acetate Buffer with 0.025% Brij35 (polyoxyethylene lauryl ether), pH 4.5, 900, 10, 20, 30, and 45, 03/25/2010, 250 mg/5 mL based on (1) acceptable bioequivalence studies on the 500 mg/5 mL strength, (2) acceptable dissolution testing across all strengths, and (3) proportional similarity in the formulations across all strengths.

Ciprofloxacin HCl, Tablet, Refer to USP, 09/02/2010, Eq. 100, 250, and 750 mg base, based on (1) acceptable bioequivalence studies on the Eq. 500 mg base strength, (2) acceptable in vitro dissolution testing of all strengths, and (3) proportional similarity of the formulations across all strengths.

Ciprofloxacin HCl, Tablet (Extended Release), I (Basket), 100, 0.1 N HCl, 900, 1, 2, 4, and 7 h or until at least 80% released, 01/14/2008.

Ciprofloxacin HCl/Hydrocortisone, Otic Suspension, Develop a method to characterize in vitro release, 09/01/2011.

Ciprofloxacin/Ciprofloxacin HCl (AB), Tablet (Extended Release), II (Paddle), 50, 0.1 N HCl, 900, 15, 30, 60, and 120, 01/14/2008, The 500 mg strength of ciprofloxacin extended-release tablets is NOT eligible for a waiver of in vivo testing based on an acceptable in vivo bioequivalence study of the 1000 mg strength.

Citalopram HBr, Tablet, Refer to USP, 01/14/2008, 20 and 10 mg based on (1) acceptable bioequivalence studies on the 40 mg strength, (2) acceptable dissolution testing across all strengths, and (3) proportional similarity in the formulations across all strengths.

Citalopram Hydrobromide, Capsule, II (Paddle), 50, 0.1 N HCl, 900, 10, 20, 30, and 45, 10/06/2008, 20 and 10 mg based on (1) acceptable bioequivalence studies on the 40 mg strength, (2) acceptable dissolution testing across all strengths, and (3) proportional similarity in the formulations across all strengths.

Clarithromycin, Tablet (Extended Release), Refer to USP, 900, 10/06/2008.

Clarithromycin, Tablet, Refer to USP, 07/25/2007.

Clarithromycin, Suspension, II (Paddle), 50, 0.05 M Phosphate Buffer, pH 6.8, 900, 10, 20, 30, 45, and 60, 01/23/2004, 125 mg/5 mL based on (1) acceptable bioequivalence studies on the 250 mg/5 mL strength, (2) proportional similarity of the formulations across all strengths, and (3) acceptable in vitro dissolution testing of all strengths.

Clindamycin HCl, Capsule, Refer to USP, 09/01/2011, 75 and 150 mg based on (1) acceptable bioequivalence studies on the 300 mg strength, (2) acceptable in vitro dissolution testing of all strengths, and (3) proportional similarity in the formulations across all strengths. Please refer to the Mirtazapine Tablet Draft Guidance for additional information regarding waivers of in vivo testing.

Clobazam, Tablet, II (Paddle), 75, 0.1 N HCl (degassed), 900, 5, 10, 20, 30, 45, and 60, 07/31/2013.

Clobazam, Oral Suspension, II (Paddle), 75, 0.1 N HCl (degassed), 900, 5, 10, 15, 20, 25, and 30, 07/31/2013.

Clomiphene Citrate, Tablet, Refer to USP, 08/15/2013.

(*Continued*)

Clonazepam, Tablet, Refer to USP, 04/08/2010, 0.5 and 2 mg based on (1) acceptable bioequivalence studies on the 1 mg strength, (2) proportional similarity of the formulations across all strengths, and (3) acceptable in vitro dissolution testing of all strengths.

Clonazepam, Tablet (Orally Disintegrating), II (Paddle), 50, Water, 900, 5, 10, 15, 30, and 45, 07/25/2007.

Clonidine, Transdermal, Refer to USP, 02/18/2009.

Clonidine (0.1 mg), Tablet (Extended Release), II (Paddle) with sinker, 50, Acid stage: 0.01 N HCl; Buffer stage: Phosphate Buffer, pH 7.0, Acid stage: 500; Buffer stage: 500, Acid stage: 1 and 2 h; Buffer stage: 1, 2, 4, 6, 10, 14, and 16 h, 01/26/2012.

Clonidine (Eq. 0.17 mg and Eq. 0.26 mg), Tablet (Extended Release), II (Paddle), 50, 500 mL 0.1 N HCl for the first hour, then add 400 mL 0.27 M Sodium Phosphate (Dibasic) buffer solution, Acid stage: 500; Buffer stage: 900, 1, 2, 3, 6, 9, 12, 16, 20, and 24 h, 07/01/2010.

Clonidine HCl, Tablet, Refer to USP, 06/18/2007.

Clopidogrel Bisulfate, Tablet, Refer to USP, 07/25/2007.

Clorazepate Dipotassium, Tablet, Refer to USP, 01/31/2013.

Clotrimazole, Tablet (Vaginal), II (Paddle), 50, 0.1 N HCl, 900, 10, 20, 30, and 45, 01/24/2004.

Clotrimazole, Lozenge, Refer to USP, 10/06/2008.

Clozapine, Tablet, Refer to USP, 07/21/2011.

Clozapine, Tablet (Orally Disintegrating), II (Paddle), 50 rpm (12.5, 25, and 100 mg); 75 rpm (150 and 200 mg), pH 4.5 Acetate Buffer, 900, 5, 10, 15, 20, and 30, 06/09/2011.

Codeine Sulfate, Tablet, Refer to USP, 09/01/2011.

Colchicine, Tablet, Refer to USP, 08/05/2010.

Colesevelam HCl, Tablet, Disintegration Testing as per USP <701> in various media such as deionized water, simulated gastric fluid, and simulated intestinal fluid, 10/28/2010.

Cyclobenzaprine, Capsule (Extended Release), II (Paddle), 50, 0.1 N HCl, 900, 2, 4, 6, 8, 12, and 16 h, 09/03/2008.

Cyclobenzaprine HCl, Tablet, Refer to USP, 07/25/2007.

Cyclophosphamide, Tablet, I (Basket), 100, Water (deaerated), 900, 10, 20, 30, 45, and 60, 01/24/2004.

Cycloserine, Capsule, Refer to USP, 05/09/2013.

Cyclosporine (100 mg) (AB1), Capsule (Liquid filled), II (Paddle), 75, 0.1 N HCl containing 4 mg of *N,N*-dimethyldodecylamine-*N*-oxide per mL, 1000, 10, 20, 30, 45, 60, and 90, 01/14/2008.

Cyclosporine (25 mg) (AB1), Capsule (Liquid filled), II (Paddle), 75, 0.1 N HCl containing 4 mg of *N,N*-dimethyldodecylamine-*N*-oxide per mL, 500, 10, 20, 30, 45, 60, and 90, 01/14/2008.

Cysteamine Bitartrate, Capsule, I (Basket), 75, 0.1 N HCl, 900, 10, 20, 30, and 45, 01/24/2004.

Dabigatran Etexilate Mesylate, Capsule, I (Basket) for 75 strength: I (Basket with modified diameter of 24.5 mm) for 150 mg strength:, 100, 0.01 N HCl (pH 2.0), 900, 10, 20, 30, and 45, 09/22/2011, 75 mg based on (1) acceptable bioequivalence studies on the 150 mg strength, (2) acceptable in vitro dissolution testing of all strengths, and (3) proportional similarity of the formulations across all strengths.

Dalfampridine, Tablet (Extended Release), II (Paddle), 50, Phosphate Buffer, pH 6.8, 900, 0.5, 1, 2, 4, 6, 8, 10, and 12 h, 06/07/2012.

Danazol, Capsule, Refer to USP, 06/18/2007, 50 and 100 mg based on (1) acceptable bioequivalence studies on the 200 mg strength, (2) proportional similarity of the formulations across all strengths, and (3) acceptable in vitro dissolution testing of all strengths.

Dantrolene Sodium, Capsule, I (Basket), 100, 0.5% Hyamine 10× in water, adjust to pH 6.8 with 0.1 N KOH or 0.1 N HCl, 900, 10, 20, 30, 40, and 60, 01/27/2004, 25 and 50 mg based on (1) acceptable bioequivalence studies on the 100 mg strength, (2) proportionally similar across all strengths, and (3) acceptable in vitro dissolution testing of all strengths.

Dapsone, Tablet, Refer to USP, 12/23/2010, 25 mg based on (1) acceptable bioequivalence studies on the 100 mg strength, (2) acceptable in vitro dissolution testing of all strengths, and (3) proportional similarity in the formulations across all strengths. Please refer to the Mirtazapine Tablet Draft Guidance for additional information regarding waivers of in vivo testing.

Darifenacin Hydrobromide, Tablet (Extended Release), I (Basket), 100, 0.01 M HCl Comparative dissolution data should also be provided in 900 mL pH 4.5 buffer, pH 6.8 buffer, and water using Apparatus I (Basket) at 100 rpm, 900, 1, 4, 8, 12, 16, 20, and 24 h, 01/20/2006, 7.5 mg based on (1) acceptable bioequivalence studies on the 15 mg strength, (2) proportionally similar across all strengths, and (3) acceptable in vitro dissolution testing of all strengths.

(*Continued*)

Darunavir Ethanolate, Tablet, II (Paddle), 75, 2% Tween 20 in 0.05 M Sodium Phosphate Buffer, pH 3.0, 900, 10, 20, 30, and 45, 09/13/2007, Darunavir Ethanolate Tablet, 75, 150, 300, 400, and 600 mg (base), based on (1) acceptable bioequivalence studies on the 800 mg (base) strength, (2) acceptable in vitro dissolution testing across all strengths, and (3) proportional similarity of the formulations across all strengths. Please refer to the Mirtazapine Tablet Draft Guidance for additional information regarding waivers of in vivo testing.

Dasatinib, Tablet, II (Paddle), 60, pH 4.0 Acetate buffer containing 1% Triton X-100, 1000, 10, 15, 30, and 45, 10/30/2009, 20, 50, and 70 mg tablet, based on (1) acceptable bioequivalence studies on the 100 mg tablet, (2) proportional similarity of the formulations across all strengths and (3) acceptable in vitro dissolution testing of all strengths.

Deferasirox, Tablet (for Oral Suspension), II (Paddle), 50, Phosphate buffer pH 6.8 with 0.5% Tween 20, 900, 10, 20, 30, and 45, 06/21/2006, 250 and 125 mg tablets based on (1) acceptable bioequivalence studies on the 500 mg strength, (2) proportionally similar across all strengths, and (3) acceptable in vitro dissolution testing of all strengths.

Delavirdine Mesylate, Tablet, II (Paddle), 50, 0.05 M Phosphate Buffer, pH 6.0 containing 0.6% w/v SDS, 900, 10, 20, 30, 45, and 60, 12/03/2007, 100 mg based on (1) acceptable bioequivalence studies on the 200 mg strength, (2) proportional similarity of the 100 mg formulation to the 200 mg strength, and (3) acceptable in vitro dissolution testing of all strengths.

Demeclocycline HCl, Capsule, Refer to USP, 07/25/2007.

Demeclocycline HCl, Tablet, Refer to USP, 07/25/2007, 150 mg based on (1) acceptable bioequivalence studies on the 300 mg strength, (2) acceptable dissolution testing across all strengths, and (3) proportional similarity in the formulations across all strengths.

Desipramine HCl, Tablet, Refer to USP, 01/31/2013, 10, 25, 50, 75, and 150 mg based on (1) acceptable bioequivalence studies on the 100 mg strengths, (2) acceptable in vitro dissolution testing of all strengths, and (3) proportional similarity of the formulations across all strengths. Please refer to the Mirtazapine Tablet Draft Guidance for additional information regarding waivers of in vivo testing.

Desloratadine, Tablet (Orally Disintegrating), II (Paddle), 50, 0.1 N HCl, 900, 3, 6, 10, 15, 06/18/2007, 2.5 mg based on (1) acceptable bioequivalence studies on the 5 mg strength, (2) proportionally similar across all strengths, and (3) acceptable in vitro dissolution testing of all strengths.

Desloratadine, Tablet, II (Paddle), 50, 0.1 N HCl, 500, 15, 20, 30, and 45, 03/04/2006.

Desloratadine/Pseudoephedrine Sulfate (2.5 mg/120 mg), Tablet (Extended Release), II (Paddle), 50, First hour: 0.1 N HCl; After 1 h: 0.1 M Potassium Phosphate Buffer pH 7.5, 1000, For Desloratadine: 10, 20, 30, and 45; For Pseudoephedrine Sulfate: 1, 2, 6, and 8 h, 04/02/2009.

Desloratadine/Pseudoephedrine Sulfate (5 mg/240 mg), Tablet (Extended Release), II (Paddle), 50, First hour: 0.1 N HCl; After 1 h: 0.1 M Potassium Phosphate Buffer pH 7.5, 1000, For Desloratadine: 10, 20, 30, and 45; For Pseudoephedrine Sulfate: 1, 2, 4, 8, 16, and 24 h, 04/02/2009.

Desmopressin Acetate, Tablet, II (Paddle), 75, Water (deaerated), 500, 10, 20, 30, and 45, 12/15/2005, 0.1 mg based on (1) acceptable bioequivalence studies on the 0.2 mg strength, (2) acceptable dissolution testing across all strengths, and (3) proportional similarity of all strengths.

Desogestrel/Ethinyl Estradiol, Tablet, Refer to USP, 11/04/2008, 0.125 mg/0.025 mg and 0.1 mg/0.025 mg based on (1) acceptable bioequivalence studies on the 0.15 mg/0.025 mg strength, (2) proportional similarity of the formulations across all strengths, and (3) acceptable in vitro dissolution testing of all strengths.

Desvenlafaxine Succinate, Tablet (Extended Release), I (Basket), 100, 0.9% NaCl in water, 900, 1, 2, 4, 8, 12, 16, 20, and 24 h, 04/02/2009.

Dexamethasone, Implant (Intravitreal), VII (with reciprocating 50 mesh baskets), 30 cycles/min, Phosphate Buffered Saline containing 0.05 g/L sodium dodecyl sulfate at 45°C ± 0.5°C, 30, 12, 24, 48, 72, 96, 120, 144, 168, 192, 216, and 24 h, 10/21/2010.

Dexamethasone, Tablet, Refer to USP, 04/02/2009, Dexamethasone Tablets, 0.5, 0.75, 1, 1.5, 2, and 4 mg based on (1) acceptable in vivo bioequivalence study on the 6 mg strength, (2) acceptable dissolution testing of all strengths, and (3) proportional similarity in the formulations of all strengths.

Dexamethasone/Tobramycin, Ophthalmic Suspension, Develop a method to characterize in vitro release, 04/02/2009.

Dexlansoprazole, Capsule (Delayed Release), I (Basket), 100, Acid Stage: 0.1 N HCl, Buffer Stage: pH 7.0 Phosphate Buffer with 5 mM SLS, Acid stage: 500; Buffer stage: 900, Acid stage: 120; Buffer Stage: 10, 20, 40, 50, 60, 75, 105, and 120, 08/05/2010, 30 mg based on (1) acceptable bioequivalence studies on the 60 mg strength, (2) acceptable in vitro dissolution testing of all strengths, and (3) proportional similarity of the formulations across all strengths. Please refer to the Mirtazapine Draft Guidance for additional information regarding waivers of in vivo testing.

(Continued)

Dexmethylphenidate HCl, Capsule (Extended Release), I (Basket), 100, First 2 h: 0.01 N HCl, Hours 2–10: Phosphate Buffer, pH 6.8, Acid: 500, Buffer: 500, 0.5, 1, 2, 4, 6, and 10 h, 01/14/2008, 5, 10, 15, 20, 25, 30, and 35 mg based on (1) acceptable bioequivalence studies on the 40 mg strength, (2) acceptable in vitro dissolution testing of all strengths, and (3) proportional similarity of the formulations across all strengths. Please refer to the Mirtazapine Tablet Draft Guidance for additional information regarding waivers of in vivo testing under 1.1CFR § 320.24(b)(6).

Dexmethylphenidate HCl, Tablet, I (Basket), 100, Water, 900, 10, 15, 30, and 45, 06/18/2007.

Dextroamphetamine Sulfate, Capsule (Extended Release), I (Basket), 100, 0.1 N HCl, 500, 1, 4, 8, and 12 h, 11/25/2008, 5 and 10 mg based on (1) acceptable bioequivalence studies on the 15 mg strength, (2) acceptable dissolution testing across all strengths, and (3) proportional similarity in the formulations across all strengths.

Dextroamphetamine Sulfate, Tablet, I (Basket), 100, Water, 500, 10, 20, 30, 45, and 60, 01/31/2013.

Dextromethorphan HBr/Guaifenesin, Tablet (Extended Release), I (Basket), 50, 0.01 N HCl, 900, 1, 2, 6, and 12 h, 11/25/2008, 30 mg/600 mg based on (1) acceptable bioequivalence studies on the 60 mg/1200 strength, (2) acceptable dissolution testing across all strengths, and (3) proportional similarity in the formulations across all strengths.

Dextromethorphan Hydrobromide/Quinidine Sulfate, Capsule, I (Basket), 100, pH 1.2, Simulated Gastric Fluid without enzyme, 900, 5, 10, 15, 20, and 30, 01/05/2012.

Dextromethorphan Polistirex, Suspension (Extended Release), II (Paddle), 50, 0.1 N HCl, 500, 30, 60, 90, and 180, 10/06/2008.

Diazepam, Tablet, Refer to USP, 07/25/2007.

Diazepam, Gel (Rectal), II (Paddle), 50, 0.05 M Phosphate Buffer, pH 6.8, 500, 5, 10, 15, 30, and 45, 04/02/2009.

Diclofenac Epolamine, Topical patch, V (Paddle over Disk) with a watch dish (a diameter of 6 cm), 50, pH 6.8 phosphate buffer at 32°C ± 0.5°C, 500, 15, 30, 45, 60, 90, 120, and 180, 10/21/2010.

Diclofenac Potassium, Powder for Oral Solution, II (Paddle), 75, 0.05 M phosphate buffer (Trisodium Phosphate Dodecahydrate in 0.1 N HCl and pH adjusted to 6.8), 400, 2.5, 5, 7.5, and 10, 10/21/2010.

Diclofenac Potassium, Capsule, II (Paddle), 50, 50 mM Phosphate buffer pH 6.8, 900, 10, 15, 20, 30, and 45, 10/21/2010.

Diclofenac Potassium, Tablet, II (Paddle), 50, SIF without enzyme, 900, 10, 20, 30, 45, 60, and 90, 01/27/2004.

Diclofenac Sodium, Tablet (Extended Release), Refer to USP, 06/10/2009.

Diclofenac Sodium, Tablet (Delayed Release), Refer to USP, 06/10/2009.

Diclofenac Sodium/Misoprostol Enteric Coated, Tablet (Delayed Release), II (Paddle) (diclo) II (Paddle) (miso), 100 (diclo) 50 (miso), Diclofenac: Acid Stage: 0.1 N HCl Buffer Stage: 750 mL 0.1 N HCl + 250 mL 0.2 M phos. buffer, pH 6.8 (Method A) Misoprostol: Water (deaerated), Diclo: Acid: 750 Buffer:1000 Miso: 500, Diclo.: 120 (acid) 15, 30, 45, and 60 (Buffer). Miso: 10, 20, and 30, 12/15/2005.

Dicloxacillin Sodium, Capsule, Refer to USP, 06/18/2007, 125 and 250 mg based on (1) acceptable bioequivalence study on the 500 mg, (2) proportional similarity of the formulations across all strengths, and (3) acceptable in vitro dissolution testing of all strengths.

Didanosine, Tablet (Chewable), II (Paddle), 75, Water (deaerated), 900, 10, 20, 30, and 45, 01/26/2004.

Didanosine, Capsule (Delayed-Release Pellets), I (Basket), 100, Acid stage: 0.1 N HCl; Buffer stage: 0.1 N HCl:0.2 M Tribasic Sodium Phosphate (3:1), pH 6.8, 1000, Acid stage: 60, 90, and 120; Buffer stage: 10, 20, 30, 45, and 60, 01/26/2004, 125, 200, and 250 mg, based on (1) acceptable bioequivalence studies on the 400 mg strength, (2) proportional similarity of the formulations across all strengths, and (3) acceptable in vitro dissolution testing of all strengths.

Dienogest/Estradiol Valerate, Tablet, II (Paddle), 50, 0.4% SLS in water, 900, 10, 15, 20, 30, and 45, 06/07/2012, Dienogest; Estradiol valerate Tablets, 2 mg/2 mg based on (1) acceptable bioequivalence studies on the 3 mg/2 mg strength, (2) the formulations are proportionally similar, and (3) acceptable in vitro dissolution testing of both strengths. Estradiol Valerate Tablets, 1 mg based on (1) acceptable bioequivalence studies on the 3 mg strength, (2) the formulations are proportionally similar, and (3) acceptable in vitro dissolution testing of both strengths.

Diethylpropion HCl, Tablet (Extended Release), I (Basket), 100, Water (deaerated), 900, 1, 3, 5, 7, and 9 h, 05/20/2009.

Diflunisal, Tablet, Refer to USP, 04/15/2008.

Digoxin, Tablet, Refer to USP, 06/18/2007, 0.125 mg based on (1) acceptable bioequivalence studies on the 0.25 mg strength, (2) proportional similarity of the formulations across all strengths, and (3) acceptable in vitro dissolution testing of all strengths.

Diltiazem HCl, Tablet (Extended Release), II (Paddle), 100, Phosphate Buffer, pH 5.8, 900, 2, 8, 14, and 24 h, 02/19/2008, 120, 180, 240, and 300 mg based on (1) acceptable bioequivalence studies on the 360 mg strength, (2) proportionally similar across all strengths, and (3) acceptable in vitro dissolution testing of all strengths.

(Continued)

Diltiazem HCl (AB2), Capsule (Extended Release), Refer to USP, 02/19/2008, 120 and 180 mg based on acceptable (1) bioequivalence studies on the 240 mg strength, and (2) proportional similarity of the formulations, and (3) acceptable in vitro dissolution testing of all strengths.

Diltiazem HCl (AB3), Capsule (Extended Release), Refer to USP, 02/19/2008, 120, 180, 240, 300, and 360 mg based on acceptable (1) bioequivalence studies on the 420 mg strength, (2) proportional similarity of the formulations and (3) acceptable in vitro dissolution testing of all strengths.

Diltiazem HCl (AB4), Capsule (Extended Release), Refer to USP, 02/19/2008.

Dinoprostone, Vaginal Suppository, Develop a method to characterize in vitro release, 10/04/2012.

Dinoprostone, Vaginal Insert (Extended Release), II (Paddle), 50, Deionized Water, 500, 0.25, 0.5, 1, 2, 2.5, 3, 3.5, 4, and 5 h, 09/01/2011.

Diphenhydramine Citrate/Ibuprofen, Tablet, II (Paddle), 50, 50 mM Phosphate Buffer, pH 6.5, 900, 10, 20, 30, and 45, 01/14/2008.

Diphenhydramine HCl/Ibuprofen, Capsule, I (Basket), 100, 200 mM Phosphate Buffer, pH 7.2, 900, 10, 20, 30, and 45, 01/14/2008.

Dipyridamole, Tablet, Refer to USP, 06/18/2007, 25 and 50 mg based on (1) acceptable bioequivalence studies on the 75 mg strength, (2) acceptable in vitro dissolution testing of all strengths, and (3) proportional similarity of the formulations across all strengths.

Disopyramide Phosphate, Capsule (Extended Release), Refer to USP, 11/04/2008, 100 mg based on (1) acceptable bioequivalence studies on the 150 mg strength, (2) acceptable dissolution testing across all strengths, and (3) proportional similarity in the formulations across all strengths.

Disopyramide Phosphate, Capsule, Refer to USP, 09/03/2008, 100 mg based on (1) acceptable bioequivalence studies on the 150 mg strength, (2) acceptable dissolution testing across all strengths, and (3) proportional similarity in the formulations across all strengths.

Disulfiram, Tablet, II (Paddle), 100, 2% SDS, 900, 15, 30, 45, 60, 75, 90, 105, and 120, 06/18/2007, Disulfiram Tablet, 250 mg based on (1) acceptable bioequivalence studies on the 500 mg strength, (2) acceptable in vitro dissolution testing of all strengths, and (3) proportional similarity of the formulations across all strengths. Please refer to the Mirtazapine Tablet Draft Guidance for additional information regarding waivers of in vivo testing.

Divalproex Sodium, Tablet (Delayed Release), Refer to USP, 07/25/2007, 125 and 250 mg based on (1) acceptable bioequivalence studies on the 500 mg strength, (2) acceptable in vitro dissolution testing for all strengths, and (3) proportional similarity in the formulations across all strengths.

Divalproex sodium, Tablet (Extended Release), II (Paddle), 100, Acid phase: 0.1 N HCl for 45 min; Drug Release: (after 45 min) 0.05 M Phosphate Buffer with 75 mM SDS, pH 5.5, Acid phase:500 mL; Drug release: 900 mL, 1.5, 3, 6, 9, 12, 15, 18, 21, and 24 h, 10/28/2010, 250 mg based on (1) acceptable bioequivalence studies on the 500 mg strength, (2) proportionally similar across all strengths, and (3) acceptable in vitro dissolution testing of all strengths.

Divalproex Sodium, Capsule (Delayed-Release Pellet), II (Paddle), 50, 0.05 M Phosphate Buffer, pH 7.5, 500, 2, 4, 6, 8, and 10 h, 10/06/2008.

Dofetilide, Capsule, I (Basket), 100, 0.001 M HCl, 900, 10, 15, 30, and 45, 01/20/2006, 0.25, 0.125 mg based on (1) acceptable bioequivalence studies on the 0.5 mg strength, (2) proportionally similar to the 0.5 mg strength, and (3) acceptable in vitro dissolution testing.

Dolasetron Mesylate, Tablet, Refer to USP, 07/01/2010, 50 mg (1) acceptable bioequivalence studies on the 100 mg strength, (2) acceptable in vitro dissolution testing for all strengths, and (3) proportional similarity in the formulations across all strengths.

Donepezil HCl, Tablet (Orally Disintegrating [ODT]), II (Paddle), 50, 0.1 N HCl, 900, 10, 20, 30, and 45, 03/04/2006, 5 mg based on (1) acceptable bioequivalence studies on the 10 mg strength, (2) acceptable in vitro dissolution testing of all strengths, and (3) proportional similarity of the formulations across all strengths.

Donepezil HCl, Tablet, II (Paddle), 50, 0.1 N HCl, 900, 10, 20, 30, and 40, 01/27/2004.

Donepezil HCl (23 mg), Tablet, II (Paddle), 50, 0.05 M Phosphate Buffer, pH 6.8, 900, 1, 2, 3, 4, 6, 8, and 10 h, 12/23/2010.

Doxazosin Mesylate, Tablet, II (Paddle), 50, 0.01 N HCl, 900, 10, 20, 30, 45, and 60, 01/27/2004, 2, 4, and 8 mg based on (1) acceptable bioequivalence studies on the 1 mg strength, (2) proportional similarity across both strengths, and (3) acceptable in vitro dissolution testing of all strengths.

Doxazosin Mesylate, Tablet (Extended Release), II (Paddle), 75, SGF without enzyme, 900, 1, 2, 4, 6, 8, 12, and 16 h, 01/03/2007, 4 mg based on (1) acceptable bioequivalence studies on the 8 mg strength, (2) proportionally similar across both strengths, and (3) acceptable in vitro dissolution testing of both strengths.

(*Continued*)

Doxepin HCl, Capsule, Refer to USP, 08/05/2010.

Doxepin HCl, Tablet, II (Paddle), 50, Simulated Gastric Fluid w/o enzyme (pH 1.1–1.3), 900, 5, 10, 15, 20, 30, and 45, 09/02/2010, Eq 3 mg base based on (1) acceptable bioequivalence studies on the Eq 6 mg base strength, (2) acceptable in vitro dissolution testing of both strengths, and (3) proportional similarity of the formulations across all strengths.

Doxercalciferol, Capsule, Develop a quantitative rupture test, 06/03/2008, Eq 10 mg base, Eq 50 mg base, Eq 75 mg base, Eq 100 mg base, and Eq 150 mg base based on (1) acceptable bioequivalence studies on the Eq 25 mg base strength, (2) acceptable in vitro dissolution testing of all strengths, and (3) proportional similarity of the formulations across all strengths.

Doxorubicin HCl, Injectable (Liposomal), Develop a method to characterize in vitro release, starting at pH 6.00 ± 0.05 and at 47°C ± 0.5°C. Replicate for 12 dosage vials, 10/04/2012.

Doxycycline, Suspension, II (Paddle), 25, 0.01 N HCl, 900, 5, 10, 15, and 20, 09/03/2008.

Doxycycline, Tablet, II (Paddle), 75, 0.01 N HCl, 900, 15, 30, 45, 60, and 90, 01/14/2008, 50, 75, and 100 mg based on (1) acceptable bioequivalence studies on the 150 mg strength, (2) proportionally similar across all strengths, and (3) acceptable in vitro dissolution testing of all strengths.

Doxycycline, Capsule (Delayed Release), II (Paddle), 75, Dilute HCl, pH 1.1 for 2 h and then add 200 mL of 0.1 N NaOH in 200 mM Phosphate Buffer. Adjust pH to 6.0 using 2 N HCl and/or 2 N NaOH, Acid stage: 750; Buffer stage: 950, 1, 2, 2.5, 3, and 4 h, 10/06/2008.

Doxycycline Hyclate, Capsule, Refer to USP, 07/14/2008, 50 mg based on (1) acceptable bioequivalence studies on the 100 mg strength, (2) acceptable dissolution testing across all strengths, and (3) proportional similarity in the formulations across all strengths.

Doxycycline Hyclate, Tablet (Delayed Release), I (Basket), 50, Acid stage: 0.06 N HCl; Buffer stage: Neutralized Phthalate Buffer, pH 5.5, Acid stage: 900; Buffer stage: 1000, Acid stage: 5, 10, 15, 20, and 30; Buffer stage: 10, 20, 30, and 45, 04/02/2009, 100 mg base Eq and 75 mg base Eq strengths based on (1) acceptable bioequivalence studies on the 150 mg base Eq strength, (2) acceptable dissolution testing across all strengths, and (3) proportional similarity in the formulations across all strengths.

Dronabinol, Capsule, II (Paddle), 100 and 150, 10% Labrasol in Water; (in addition, the USP capsule rupture test should also be conducted), 500, 5, 10, 15, 30, 45, 60, and until at least 80% of the labeled content is released, 01/31/2007, 2.5 and 10 mg based on (1) acceptable bioequivalence studies on the 5 mg strength, (2) acceptable in vitro dissolution testing for all strengths, and (3) proportional similarity in the formulations across all strengths.

Dronedarone HCl, Tablet, II (Paddle), 75, pH 4.5 Phosphate buffer, 1000, 10, 15, 20, 30, 45, 60, 90, and 120, 08/05/2010.

Drospirenone/Estradiol, Tablet, II (Paddle), 50, Water, 900, 10, 20, 30, and 45, 01/03/2007, 0.25 mg/0.5 mg based on (1) acceptable bioequivalence studies on the 0.5 mg/1 mg strength, (2) acceptable dissolution testing on all strengths, and (3) proportional similarity in the formulations across all strengths. If only the lower strength, 0.25 mg/0.5 mg, is to be marketed first, then the fasting and fed studies should be conducted on this lower strength, comparing it with the equal strength of the reference listed drug (RLD) product. However, if the higher strength, 0.5 mg/1 mg, is to be marketed following the in vivo studies of the lower strength, then an additional fasting study will be requested for the higher strength.

Drospirenone/Ethinyl Estradiol, Tablet, II (Paddle), 50, Water, 900, 10, 20, 30, and 45, 09/22/2011.

Drospirenone/Ethinyl Estradiol/Levomefolate Calcium, Tablet, II (Paddle), 50, Phosphate Buffered Saline pH 6.8 containing 0.03% ascorbic acid, 900, 5, 10, 15, 20, 30, and 45, 01/05/2012.

Duloxetine HCl, Capsule (Delayed-Release Pellets), I (Basket), 100, [A] Gastric Challenge: 0.1 N HCl [B] Buffer Medium: pH 6.8 phosphate buffer (USP), 1000, 120 min (For A) 15, 30, 45, 60, and 90 min (For B), 03/22/2006, 20, 30 mg based on (1) acceptable bioequivalence studies on the 60 mg strength, (2) proportional similarity of the formulations across all strengths, and (3) acceptable in vitro dissolution testing of all strengths.

Dutasteride, Capsule (Soft Gelatin), II (Paddle), 50, Tier I: Dissolution Medium: 0.1 N HCl with 2% (w/v) sodium dodecyl sulfate (SDS) (900 mL) Tier II: Dissolution Medium: 0.1 N HCl with pepsin (as per USP) (450 mL) for the first 25 min, followed by addition of 0.1 N HCl with SDS (4% w/v) (450 mL) for the remainder of the dissolution test, 900, 15, 30, 45, and 60, 08/05/2010.

Dutasteride/Tamsulosin HCl, Capsule, Dutasteride: II (Paddle) with sinker. Tamsulosin: II (Paddle), Dutasteride: 75 Tamsulosin: 50, Dutasteride::Tier I: Dissolution Medium: 1%w/v cetyltrimethylammonium bromide (CTAB) in 0.1 N HCl. Tier II: Dissolution Medium: 1% w/v CTAB in 0.1 N HCl with 0.16% w/v pepsin. Tamsulosin Acid Stage (0–2 h): 0.1 N HCl. Buffer stage: Add 250 mL of 0.2 M Sodium Phosphate Tribasic, Dodecahydrate pH 6.8, Dutasteride: 900. Tamsulosin: Acid stage: 750; Buffer stage: 1000, Dutasteride: 15, 30, 45, and 60 min. Tamsulosin: Acid Stage: 2 h Buffer stage: 0.5, 1, 2, 3, 5, 7, and 10 h, 01/26/2012.

(Continued)

Efavirenz, Capsule, II (Paddle) A sinker may be used with justification if necessary, 50, 1% Sodium Lauryl Sulfate in water, 900, 15, 30, 45, and 60, 03/22/2006, 50 and 100 mg based on (1) acceptable bioequivalence study on the 200 mg strength, (2) proportional similarity of the formulations across all strengths, and (3) acceptable in vitro dissolution testing of all strengths.

Efavirenz, Tablet, II (Paddle), 50, 2% SLS in water, 1000, 10, 15, 30, 45, 60, 06/18/2007.

Efavirenz 600 mg; Emtricitabine 200 mg; Tenofovir Disoproxil Fumarate 300 mg, Tablet, II (Paddle), 100, 2% SLS in water, 1000, 10, 20, 30, and 45, 01/03/2007.

Eletriptan Hydrobromide, Tablet, I (Basket), 100, 0.1 N HCl, 900, 5, 10, 15, and 30, 04/02/2009, 20 mg base Eq based on (1) acceptable bioequivalence studies on the 40 mg base Eq strength, (2) acceptable dissolution testing across all strengths, and (3) proportional similarity in the formulations across all strengths.

Eltrombopag Olamine, Tablet, II (Paddle), 50, 0.5% Polysorbate 80 in Phosphate Buffer, pH 6.8, 900, 10, 15, 20, 30, 45, and 60, 06/07/2012.

Emtricitabine, Capsule, II (Paddle), 50, Tier 1: 0.1 N HCl Tier 2: 0.1 N HCl containing Pepsin 750,000 USP units/L. Tier 2 is used after failure of Tier 1 testing, 900, 10, 20, 30, and 45, 12/16/2005.

Emtricitabine/Tenofovir Disoproxil Fumarate, Tablet, II (Paddle), 50, 0.01 N HCl, 900, 5, 10, 15, 30, and 45, 01/03/2007.

Enalapril Maleate, Tablet, Refer to USP, 09/03/2008, 10, 5, and 2.5 mg based on (1) acceptable bioequivalence studies on the 20 mg strength, (2) acceptable dissolution testing across all strengths, and (3) proportional similarity in the formulations across all strengths.

Entacapone, Tablet, II (Paddle), 50, Phosphate Buffer, pH 5.5, 900, 10, 20, 30, and 45, 01/29/2004.

Entecavir, Tablet, II (Paddle), 50, Phosphate buffer pH 6.8 (50 mM), 1000, 10, 20, 30, and 45, 06/21/2006, 0.5 mg based on (1) acceptable bioequivalence studies on the 1 mg strength, (2) proportionally similar across all strengths, and (3) acceptable in vitro dissolution testing of all strengths.

Eplerenone, Tablet, II (Paddle), 50, 0.1 N HCl, 1000, 10, 20, 30, and 45, 12/19/2005, 25 mg based on (1) acceptable bioequivalence studies on the 50 mg strength, (2) proportionally similar to the 50 mg strength, and (3) acceptable in vitro dissolution testing.

Eprosartan Mesylate, Tablet, II (Paddle), 75, 0.2 M Phosphate Buffer, pH 7.5, 1000, 15, 30, 45, and 60, 07/14/2008, 400 mg based on (1) acceptable bioequivalence studies on the 600 mg strength, (2) acceptable dissolution testing across all strengths, and (3) proportional similarity in the formulations across all strengths.

Eprosartan Mesylate/Hydrochlorothiazide, Tablet, II (Paddle), 75, 0.2 M Phosphate Buffer, pH 7.5, 1000, 10, 20, 30, and 45, 02/19/2008, 600 mg/12.5 mg, based on (1) acceptable bioequivalence studies on the 600 mg/25 mg strength, (2) proportional similarity of the formulations across all strengths, and (3) acceptable in vitro dissolution testing of all strengths.

Ergocalciferol, Capsule, II (Paddle), 100, 0.5 N NaOH with 10% Triton X-100, 500, 15, 30, 45, 60, and 90, 08/05/2010.

Erlotinib HCl, Tablet, II (Paddle), 75, 0.1 N HCl containing 1% SDS, 1000, 15, 30, 45, and 60, 03/22/2006, 100 and 25 mg based on (1) acceptable bioequivalence studies on the 150 mg strength, (2) proportional similarity of the formulations across all strengths, and (3) acceptable in vitro dissolution testing of all strengths.

Erythromycin Ethylsuccinate, Suspension, II (Paddle), 75, Monobasic Sodium Phosphate, pH 6.8 Buffer with 1% SLS Buffer w/1% SLS, 900, 10, 20, 30, 45, and 60, 01/27/2004.

Erythromycin Ethylsuccinate/Sulfisoxazole Acetyl, Granules for Oral suspension, Develop a dissolution method, 09/02/2010.

Escitalopram Oxalate, Tablet, II (Paddle), 75, 0.1 N HCl, 900, 10, 20, 30, and 45, 02/20/2004, 5 and 10 mg based on (1) acceptable bioequivalence studies on the 20 mg strength, (2) proportionally similar across all strengths, and (3) acceptable in vitro dissolution testing of all strengths.

Escitalopram Oxalate, Capsule, II (Paddle), 50, 0.1 N HCl, 900, 10, 20, 30, and 45, 10/06/2008, 5 and 10 mg based on (1) acceptable bioequivalence studies on the 20 mg strength, (2) acceptable dissolution testing across all strengths, and (3) proportional similarity in the formulations across all strengths.

Esomeprazole Magnesium, Capsule (Delayed-Release Pellets), II (Paddle), 100, Acid stage: 0.1 N HCl; Buffer stage: Sodium Phosphate Buffer, pH 6.8, Acid stage: 300; Buffer stage: 1000, Acid stage: 120; Buffer stage: 10, 20, 30, 45, and 60, 02/26/2004, 20 mg based on (1) acceptable bioequivalence studies on the 40 mg strength, (2) proportional similarity of the formulations across all strengths, and (3) acceptable in vitro dissolution testing of all strengths. Please refer to the Mirtazapine Tablet Guidance for additional information regarding waivers of in vivo testing.

Esomeprazole Magnesium, For Oral Suspension (Delayed Release), II (Paddle), 100, Acid stage: 0.1 N HCl; Buffer stage: Sodium Phosphate Buffer, pH 6.8, Acid stage: 300; Buffer stage: 1000, Acid stage: 120; Buffer stage: 10, 20, 30, 45, and 60, 09/02/2010, Eq 10 mg base/packet and Eq 20 mg base/packet based on (1) acceptable bioequivalence studies on the 40 mg strength, (2) acceptable in vitro dissolution testing of all strengths, and (3) proportional similarity in the formulations across all strengths.

(*Continued*)

Estazolam, Tablet, II (Paddle), 50, Water (deaerated), 900, 10, 20, 30, and 45, 01/27/2004.

Esterified Estrogens, Tablet, II (Paddle), 50, Water, 900, 15, 30, 45, 60, 90, 120, and 180, 02/19/2008, 0.3, 0.625, and 1.25 mg based on (1) acceptable bioequivalence studies on the 2.5 mg strength, (2) proportional similarity of the formulations across all strengths, and (3) acceptable in vitro dissolution testing of all strengths.

Estradiol, Vaginal Tablet, I (Basket), 40, Phosphate Buffer, pH 4.75 ± 0.05, 500, 1, 2, 3, 5, 8, 10, and 12 h, 07/21/2009, BE with clinical endpoint study on the 25 mcg based on (1) acceptable BE with PK endpoints studies on the 10 and 25 mcg strengths, (2) acceptable BE with clinical endpoint study on the 10 mcg strength, (3) proportional similarity of the 10 mcg formulation to the 25 mcg strength, and (4) acceptable in vitro dissolution testing of both strengths.

Estradiol, Vaginal Ring, Incubator shaker, 130, 0.9% Saline, 250, 1, 9, 16, 17, 18, 19, 45 days, 01/03/2007.

Estradiol (0.014 mg/24 h), Film, Transdermal (Extended Release), Develop a method to characterize in vitro release, 10/28/2010, 0.025 mg/24 h, 0.0375 mg/24 h, 0.05 mg/24 h and 0.075 mg/24 h may be considered for a waiver of in vivo bioequivalence testing based on (1) acceptable bioequivalence studies on the 0.1 mg/24 h strength, (2) acceptable dissolution testing of all strengths, and (3) proportional similarity in the formulations across all strengths.

Estradiol (0.025 mg/24 h, 0.0375 mg/24 h, 0.05 mg/24 h, 0.06 mg/24 h, 0.075 mg/24 h and 0.1 mg/24 h), Film, Transdermal (Extended Release), Develop a method to characterize in vitro release, 10/28/2010, 0.014 mg/24 h, 0.025 mg/24 h, 0.0375 mg/24 h, 0.05 mg/24 h, mg/24 h, and 0.075 mg/24 h may be considered for a waiver of in vivo bioequivalence testing based on (1) acceptable bioequivalence studies on the 0.1 mg/24 h strength, (2) acceptable dissolution testing of all strengths, and (3) proportional similarity in the formulations across all strengths.

Estradiol (Test 1) (0.025 mg/24 h, 0.0375 mg/24 h, 0.05 mg/24 h, 0.075 mg/24 h, and 0.1 mg/24 h), Film, Transdermal (Extended Release), VI (Cylinder) attach the patch to a disk at the bottom of the cylinder, 50, Water at 32°C ± 0.5°C, 0.025 mg/24 h and 0.0375 mg/24 h: 500 mL; 0.05 mg/24 h, 0.075 mg/24 h, and 0.1 mg/24 h: 900 mL, 1, 2, 4, 6, 8, 10, and 12 h, 10/28/2010, 0.014 mg/24 h, 0.025 mg/24 h, 0.0375 mg/24 h, 0.05 mg/24 h, mg/24 h, and 0.075 mg/24 h may be considered for a waiver of in vivo bioequivalence testing based on (1) acceptable bioequivalence studies on the 0.1 mg/24 h strength, (2) acceptable dissolution testing of all strengths, and (3) proportional similarity in the formulations across all strengths.

Estradiol (Test 2) (0.05 mg/24 h and 0.1 mg/24 h), Film, Transdermal (Extended Release), V (Paddle over Disk) with a stainless steel disk, 50, Water at 32°C ± 0.5°C, 900, 6, 12, 18, 24, 36, 48, 60, 72, and 96 h, 10/28/2010, 0.05 mg/24 h may be considered for a waiver of in vivo bioequivalence testing based on (1) acceptable bioequivalence studies on the 0.1 mg/24 h strength, (2) acceptable dissolution testing of both strengths, and (3) proportional similarity in the formulations.

Estradiol/Norethindrone Acetate, Tablet, Refer to USP, 01/05/2012.

Estradiol/Norgestimate (1 mg/0.09 mg), Tablet, II (Paddle), 50, 0.3% SLS in water, 500, 10, 20, 30, and 45, 07/09/2004.

Estramustine Phosphate Sodium, Capsule, I (Basket), 100, Water, 900, 10, 20, 30, and 45, 07/15/2009.

Estrogens Conjugated Synthetic A, Tablet, I (Basket), 50, Water, 900, 1, 2, 3, 5, 8, 10, and 12 h, 09/02/2010, 0.3, 0.45, 0.625, and 0.9 mg based on (1) acceptable in vivo bioequivalence study on the 1.25 mg strength, (2) formulations of all strengths are proportionally similar, and (3) acceptable dissolution testing for all strengths.

Estrogens, Conjugated Synthetic B, Tablet, II (Paddle), 50, Water, 900, 2, 5, 8, and 12 h, 10/06/2008.

Eszopiclone, Tablet, II (Paddle), 50, 0.1 N HCl, 500, 10, 20, 30, and 45, 09/13/2007, 1 and 2 mg, based on acceptable (1) bioequivalence studies on the 3 mg tablet, (2) proportional similarity of the formulations, and (3) acceptable in vitro dissolution testing of all strengths.

Ethacrynic Acid, Tablet, Refer to USP, 12/23/2010.

Ethambutol HCl, Tablet, Refer to USP, 01/14/2008, 100 mg, based on acceptable (1) bioequivalence studies on the 400 mg strength, (2) proportional similarity of the formulations, and (3) acceptable in vitro dissolution testing of all strengths.

Ethinyl Estradiol, Tablet, Refer to USP, 09/22/2011, 0.035 mg/0.215 mg, 0.035 mg/0.180 mg, 0.025 mg/0.25 mg, 0.025 mg/0.215 mg, and 0.025 mg/0.18 mg based on (1) adequate bioequivalence studies on the 0.035 mg/0.25 mg strength, (2) proportional similarity in the formulations of all strengths, and (3) adequate in vitro dissolution testing of all strengths.

Ethinyl Estradiol/Ethynodiol Diacetate, Tablet, II (Paddle), 75, 0.25% Sodium Lauryl Sulfate (SLS) in Water, 600, 10, 20, 30, and 45, 07/14/2008.

Ethinyl Estradiol/Etonogestrel, Vaginal Ring, Develop a method to characterize in vitro release, 01/31/2013.

Ethinyl Estradiol/Levonorgestrel, Tablet, Refer to USP, 02/19/2008, 0.03 mg/0.05 mg based on (1) acceptable bioequivalence studies on the 0.03 mg/0.125 mg and 0.04 mg/0.075 mg strengths, (2) proportional similarity in the formulations of all strengths, and (3) acceptable in vitro dissolution testing of all strengths.

Ethinyl Estradiol/Levonorgestrel (AB), Tablet, Refer to USP, 02/19/2008.

Ethinyl Estradiol/Levonorgestrel (AB2), Tablet, Refer to USP, 11/04/2008.

(*Continued*)

Ethinyl Estradiol/Norethindrone, Tablet, Refer to USP, 07/15/2009, 0.035 mg/0.5 mg and 0.035 mg/0.75 mg, and 0.035 mg/1 mg based on (1) acceptable bioequivalence studies on the 0.035 mg/1 mg strength, (2) acceptable dissolution testing across all strengths, and (3) proportional similarity in the formulations across all strengths.

Ethinyl Estradiol/Norethindrone, Tablet (Chewable), II (Paddle), 75, 0.09% Sodium Lauryl Sulfate in 0.1 N HCl, 500, 10, 20, 30, and 45, 01/14/2008.

Ethinyl Estradiol/Norethindrone Acetate, Tablet, Refer to USP, 07/15/2009, 20 mcg/1 mg and 30 mcg/1 mg based on (1) acceptable bioequivalence studies on the 35 mcg/1 mg strength, (2) acceptable dissolution testing across all strengths, and (3) proportional similarity in the formulations across all strengths.

Ethinyl Estradiol/Norgestimate, Tablet, II (Paddle), 75, 0.05% Tween 20 in water, 600, 5, 10, 20, and 30, 01/14/2008.

Ethinyl Estradiol/Norgestimate (AB), Tablet, II (Paddle), 75, 0.05% Tween 20 in water, 600, 10, 20, 30, and 45, 01/14/2008.

Ethinyl Estradiol/Norgestrel, Tablet, II (Paddle), 75, Water with 5 ppm of Tween 80, 500, 10, 20, 30, 45, 60, and 90, 01/28/2004.

Ethinyl Estradiol; Norelgestromin, Film, Transdermal, Modified USP Type V (Paddle over disk), 50, 0.1% Hydroxypropyl-beta-cyclodextrin at 32°C, 900, 0.25, 0.5, 1, 2, 4, 8, 12, 16, 20, and 24 h, 05/20/2009.

Ethionamide, Tablet, I (Basket), 75, 0.1 N HCl, 900, 10, 20, 30, 45, and 60, 01/31/2013.

Ethosuximide, Capsule, Refer to USP, 04/15/2008.

Etidronate Disodium, Tablet, Refer to USP, 06/18/2007, 200 mg based on (1) acceptable bioequivalence study on the 400 mg strength, (2) proportionally similar across all strengths, and (3) acceptable in vitro dissolution testing of all strengths.

Etodolac, Tablet, Refer to USP, 01/14/2008.

Etodolac, Tablet (Extended Release), Refer to USP, 06/24/2010, 400 and 500 mg based on (1) acceptable bioequivalence studies on the 600 mg strength, (2) acceptable in vitro dissolution testing of all strengths, and (3) proportional similarity of the formulations across all strengths.

Etoposide, Capsule, Refer to USP, 06/24/2010.

Etravirine (100 mg), Tablet, II (Paddle), 50, 1.0% Sodium lauryl sulfate (SLS) in 0.01 M HCl in two phases: Phase 1: 500 mL of degassed 0.01 M HCl for 10 min. Phase 2: Add 400 mL of 2.25% SLS in 0.01 M HCl, 500 (phase 1): 900 (phase 2), Phase 1: No Sampling. Phase 2: 5, 10, 20, 30, 45, 60, and 90, 06/30/2011.

Etravirine (200 mg), Tablet, II (Paddle), 70, 1.0% Sodium lauryl sulfate (SLS) in 0.01 M HCl in two phases: Phase 1: 1000 mL of degassed 0.01 M HCl for 10 min. Phase 2: Add 800 mL of 2.25% SLS in 0.01 M HCl, 1000 (phase 1): 1800 (phase 2), Phase 1: No Sampling. Phase 2: 5, 10, 20, 30, 45, 60, and 90, 06/30/2011, 25 and 100 mg based on (1) acceptable fasting and fed BE studies on the 200 mg strength, (2) proportional similarity in the formulations across both strengths, and (3) acceptable dissolution testing across both strengths.

Everolimus, Tablet, II (Paddle), 50, Water with 0.4% sodium dodecyl sulfate, 500, 10, 20, 30, and 45, 07/01/2010, 2.5, 5, and 7.5 mg based on (1) acceptable bioequivalence studies on the 10 mg strength, (2) proportional similarity of all strengths, and (3) acceptable in vitro dissolution testing across all strengths.

Exemestane, Tablet, I (Basket), 100, 0.5%(w/v) SLS Solution, 900, 10, 20, 30, and 45, 08/17/2006.

Ezetimibe, Tablet, II (Paddle), 50, 0.45% SLS in 0.05 M Acetate Buffer, pH 4.5, 500, 10, 20, 30, and 45, 01/14/2008.

Ezetimibe/Simvastatin, Tablet, II (Paddle), 50, 0.01 M Sodium Phosphate, pH 7.0/0.5% SDS, 900, 5, 10, 20, and 30, 01/03/2007, 10 mg/10 mg, 10 mg/20 mg, and 10 mg/40 mg based on (1) acceptable bioequivalence studies on the 10 mg/80 mg strength, (2) acceptable dissolution testing across all strengths, and (3) proportional similarity in the formulations across all strengths.

Ezogabine, Tablet, II (Paddle), 75, 0.01 N HCl, 1000, 5, 10, 15, 20, and 30, 08/15/2013, 50, 200, and 300 mg based on (1) acceptable bioequivalence studies on the 400 mg strength, (2) acceptable dissolution testing across all strengths, and (3) proportional similarity in the formulations across all strengths. Please refer to the Mirtazapine Tablet Guidance for additional information regarding waivers of in vivo testing.

Famciclovir, Tablet, II (Paddle), 50, 0.1 N HCl, 900, 10, 20, 30, and 45, 04/09/2007, 125 and 250 mg based on (1) acceptable bioequivalence studies on the 500 mg strength, (2) acceptable dissolution testing across all strengths, and (3) proportional similarity in the formulations across all strengths.

Famotidine, Tablet (Orally Disintegrating), II (Paddle), 50, 0.1 M Phosphate Buffer, pH 4.5, 900, 2, 5, 10, 15, and 20, 10/06/2008, 20 mg based on (1) acceptable bioequivalence studies on the 40 mg strength, (2) proportionally similar across all strengths, and (3) acceptable in vitro dissolution testing of all strengths.

Famotidine, Tablet (Chewable), II (Paddle), 50, 0.1 M Phosphate Buffer, pH 4.5, 900, 10, 20, 30, 45, and 60, 01/29/2004.

Famotidine, Tablet, Refer to USP, 06/18/2007, 10 and 20 mg based on (1) acceptable bioequivalence studies on the 40 mg strength, (2) proportionally similar across all strengths, and (3) acceptable in vitro dissolution testing of all strengths.

(Continued)

Famotidine, Suspension, II (Paddle), 25 and 50, 0.1 M Phosphate buffer, pH 4.5, 900, 10, 15, 30, and 45, 11/25/2008.

Famotidine/Calcium Carbonate/Magnesium Hydroxide, Tablet (Chewable), Develop a dissolution method, 12/15/2009.

Famotidine/Ibuprofen, Tablet, II (Paddle), 50, 0.05 M Phosphate Buffer, pH 7.2, 900, 5, 10, 15, 20, 30, and 45, 08/15/2013.

Febuxostat, Tablet, II (Paddle), 75, 0.05 M Phosphate Buffer, pH 6.0, 900, 5, 10, 15, 20, and 30, 08/15/2013, 40 mg based on (1) acceptable bioequivalence studies on the 80 mg strength, (2) acceptable in vitro dissolution testing across all strengths, and (3) proportional similarity of all strengths.

Felbamate, Tablet, Refer to USP, 08/15/2013, 400 mg based on (1) acceptable bioequivalence studies on the 600 mg strength, (2) acceptable in vitro dissolution testing of all strengths, and (3) proportional similarity of the formulations across all strengths.

Felbamate, Suspension, II (Paddle), 50, Water (deaerated), 900, 5, 10, 15, and 30, 01/28/2004.

Felodipine, Tablet (Extended Release), Refer to USP, 01/14/2008, 2.5 and 5 mg based on (1) acceptable bioequivalence studies on the 10 mg strength, (2) acceptable in vitro dissolution testing for all strengths, and (3) proportional similarity of the formulations across all strengths.

Fenofibrate, Capsule, II (Paddle), 75, Phosphate Buffer w/2% Tween 80 and 0.1% pancreatin, pH 6.8, 900, 15, 30, 45, 60, 90, and 120, 02/19/2008, 50 and 100 mg based on (1) acceptable bioequivalence studies on the 150 mg strength, (2) proportional similarity of the formulations across all strengths, and (3) acceptable in vitro dissolution testing of all strengths.

Fenofibrate, Capsule (Micronized), II (Paddle), 75, 0.025 M SLS in water, 1000, 10, 20, 30, 40, and 60, 06/03/2008.

Fenofibrate (40 and 120 mg), Tablet, II (Paddle), 75, 0.75% Sodium lauryl sulfate in water, 900, 5, 10, 20, 30, 45, and 60, 10/21/2010, 40 mg based on (1) acceptable bioequivalence studies on the 120 mg strength, (2) acceptable dissolution testing across all strengths, and (3) proportional similarity in the formulations across all strengths.

Fenofibrate (48 and 145 mg), Tablet, II (Paddle), 50, 25 mM Sodium lauryl sulfate in water, 1000, 5, 10, 20, 30, 45, and 60, 10/21/2010, 48 mg based on (1) acceptable bioequivalence studies on the 145 mg strength, (2) proportional similarity of the formulations 48 and 145 mg strengths, and (3) acceptable in vitro dissolution testing of 48 and 145 mg strengths.

Fenofibrate (54 and 160 mg), Tablet, II (Paddle), 50, 0.05 M Sodium lauryl sulfate in water, 1000, 5, 10, 20, 30, 45, and 60, 10/21/2010, 50 mg based on (1) acceptable bioequivalence studies on the 160 mg strength, (2) acceptable dissolution testing across all strengths, and (3) proportional similarity in the formulations across all strengths.

Fenofibric Acid, Tablet, II (Paddle), 75, Phosphate buffer, pH 6.8, 900, 5, 15, 30, 45, and 60, 08/05/2010, 35 mg based on (1) acceptable bioequivalence studies on the 105 mg strength, (2) acceptable in vitro dissolution testing of all strengths, and (3) proportional similarity of the formulations across all strengths.

Fenoprofen Calcium, Capsule, Refer to USP, 11/25/2008.

Fentanyl, Transdermal, VII (Reciprocating holder) cylinder, 30 cycles/min amplitude of about 2 m, Equimolar mixture of 0.005 M phosphoric acid solution, and 0.005 M sodium phosphate, monobasic monohydrate (pH ~ 2.6) at 32°C. Change the test samples into fresh preequilibrated release medium at the time points indicated. Remove the protective liner and place the film onto a piece of nylon netting with adhesive facing the net. Secure the netting and transdermal system using nylon tie wraps at the top and bottom of the cylinder on the holder. The adhesive side faces toward the media, 250 mL for the 75 and 100 mcg/h, 200 mL for the 50 mcg/h, and 150 mL for the 25 and 12.5 mcg/h dosage strength, 0.5, 1, 2, 4, and 24 h, 06/09/2011, 12.5, 50, 75, and 100 µg/h based on (1) acceptable bioequivalence studies on the 25 µg/h strength, (2) proportionally similar 12.5, 50, 75, and 100 µg/h formulations to the 25 µg/h strength, and (3) acceptable in vitro dissolution testing of all strengths.

Fentanyl Citrate, Lozenges, II (Paddle), 175, 0.1 M Phosphate Buffer, pH 4.5, 500, 5, 10, 20, 30, and 40, 05/20/2009, 0.2, 0.6, 0.8, 1.2, and 1.6 mg based on (1) acceptable bioequivalence study on the 0.4 mg strength, (2) acceptable in vitro dissolution testing of all strengths, and (3) proportional similarity of the formulations across all strengths.

Fentanyl Citrate, Tablet (Sublingual), II (Paddle), 50, Phosphate Buffer, pH 6.8, 500, 1, 3, 5, 7, 10, 15, and 20, 08/15/2013, Eq 0.1 mg base, Eq 0.2 mg base, Eq 0.3 mg base, Eq 0.6 mg base, and Eq 0.8 mg base based on (1) acceptable bioequivalence studies on the Eq 0.4 mg base strength, (2) acceptable in vitro dissolution testing of all strengths, and (3) proportional similarity of the formulations across all strengths.

Fentanyl Citrate (0.1 and 0.4 mg), Tablet (Buccal), II (Paddle) small volume dissolution apparatus, 100, Phosphate Buffered Saline solution, pH 7.0, 100, 3, 5, 7.5, 10, 15, and 20, 11/20/2009, 0.1, 0.2, 0.6, and 0.8 mg (base equivalent) based on (1) acceptable bioequivalence study on the 0.4 mg strength, (2) acceptable in vitro dissolution testing of all strengths, and (3) proportional similarity of the formulations across all strengths.

Fentanyl Citrate (0.2, 0.3, 0.6, and 0.8 mg), Tablet (Buccal), II (Paddle) small volume dissolution apparatus, 100, Phosphate Buffered Saline solution, pH 7.0, 200, 3, 5, 7.5, 10, 15, and 20, 11/20/2009.

(*Continued*)

Fentanyl Citrate (0.2, 0.4, 0.6, and 0.8 mg), Film (Buccal), I (Basket) 100 mL dissolution vessel, 100, 25-mM Phosphate Buffer, pH 6.4, 60, 5, 10, 15, 20, 30, and 45, 12/15/2009.

Fentanyl Citrate (1.2 mg), Film (Buccal), I (Basket) 100 mL dissolution vessel, 100, 25-mM Phosphate Buffer, pH 6.4, 100, 5, 10, 15, 20, 30, and 45, 12/15/2009.

Fesoterodine Fumarate, Tablet (Extended Release), II (Paddle) with sinker, 75, Phosphate Buffer, pH 6.8, 900, 1, 2, 4, 6, 8, 10, 12, 16, and 20 h, 08/15/2013, 4 mg based on (1) acceptable bioequivalence studies on the 8 mg strength, (2) acceptable in vitro dissolution testing of all strengths, and (3) proportional similarity of the formulations across all strengths.

Fexofenadine HCl, Tablet (Orally Disintegrating), II (Paddle), 50, 0.001 N HCl, 500, 5, 10, 15, 30, and 45, 09/03/2008.

Fexofenadine HCl, Suspension, II (Paddle), 50, 0.001 M HCl, 900, 10, 20, 30, and 45, 11/25/2008.

Fexofenadine HCl, Tablet, II (Paddle), 50, 0.001 N HCl, 900, 5, 10, 20, 30, and 45, 02/19/2004, 30 and 60 mg based on (1) acceptable bioequivalence study on the 180 mg strength, (2) proportional similarity of the formulations across all strengths, and (3) acceptable in vitro dissolution testing of all strengths.

Fexofenadine HCl, Capsule, II (Paddle), 50, Water (deaerated), 900, 10, 20, 30, 45, and 60, 01/29/2004.

Fexofenadine HCl/Pseudoephedrine HCl, Tablet (Extended Release), Refer to USP, 04/02/2009.

Finasteride, Tablet, Refer to USP, 07/25/2007, 1 mg based on (1) acceptable bioequivalence studies on the 5 mg strength, (2) acceptable in vitro dissolution testing of all strengths, and (3) proportional similarity of the formulations across all strengths. Please refer to the Mirtazapine Tablet Guidance for additional information regarding waivers of in vivo testing.

Fingolimod, Capsule, I (Basket), 100, 0.1 N HCl with 0.2% SDS (sodium dodecyl sulfate), 500, 5, 10, 15, 20, and 30, 08/15/2013.

Flavoxate HCl, Tablet, I (Basket), 100, 0.1 N HCl, 900, 5, 10, 20, and 30, 01/29/2004.

Flecainide Acetate, Tablet, Refer to USP, 12/15/2009, 50 and 100 mg based on (1) acceptable bioequivalence studies on the 150 mg strength, (2) proportional similarity of the formulations across all strengths, and (3) acceptable in vitro dissolution testing of all strengths.

Fluconazole, Tablet, II (Paddle), 50, Water (deaerated), 900 (For 150, 200, 300, and 400 mg tabs) 500 (For 50 and 100 mg tabs), 10, 20, 30, 45, and 60, 03/04/2006, 50, 100, and 150 mg based on (1) acceptable bioequivalence study on the 200 mg strength, (2) proportional similarity of the formulations across all strengths, and (3) acceptable in vitro dissolution testing of all strengths.

Fluconazole (200 mg/5 mL), Suspension, II (Paddle), 50, Water (deaerated), 900, 10, 20, 30, and 45, 01/30/2004.

Fluconazole (50 mg/5 mL), Suspension, II (Paddle), 50, Water (deaerated), 500, 10, 20, 30, and 45, 01/30/2004.

Flucytosine, Capsule, Refer to USP, 06/24/2010, 250 mg based on (1) acceptable bioequivalence study on the 500 mg strength, (2) acceptable in vitro dissolution testing of all strengths, and (3) proportional similarity of the formulations across all strengths.

Fludarabine Phosphate, Tablet, II (Paddle), 50, Water, 900, 5, 10, 15, 20, and 30, 06/07/2012.

Fludrocortisone Acetate, Tablet, Refer to USP, 05/20/2009.

Fluoxetine, Capsules (Delayed Release), Refer to USP, 07/25/2007.

Fluoxetine HCl, Capsule, Refer to USP, 09/02/2010.

Fluoxetine HCl, Tablet, I (Basket), 100, 0.1 N HCl, 1000, 5, 10, 15, and 30, 01/03/2007, Eq 10 and 15 mg base, based on (1) acceptable bioequivalence studies on the Eq 20 mg base strength, (2) acceptable in vitro dissolution testing of all strengths, and (3) proportional similarity in the formulations across all strengths. Please refer to the Mirtazapine Tablet Guidance for additional information regarding waivers of in vivo testing.

Fluoxetine/Olanzapine, Capsule, II (Paddle), 50, 0.1 N HCl, 900, 10, 20, 30, and 45, 08/17/2006.

Flutamide, Capsule, Refer to USP, 01/31/2013.

Fluvastatin Sodium, Capsule, Refer to USP, 01/14/2008, 20 mg based on (1) acceptable bioequivalence studies on the 40 mg strength, (2) proportional similarity of the formulations across all strengths, and (3) acceptable in vitro dissolution testing of all strengths.

Fluvastatin Sodium, Tablet (Extended Release), I (Basket), 50, Water (deaerated), 1000, 0.5, 2, 4, 6, and 8 h, 09/22/2011.

Fluvoxamine Maleate, Capsule (Extended Release), II (Paddle), 50, Phosphate Buffer, pH 6.8, 900, 1, 2, 4, 6, 8, and 12 h, 01/15/2010, 100 mg based on (1) acceptable bioequivalence studies on the 150 mg strength, (2) acceptable in vitro dissolution testing of 100 and 150 mg strengths, and (3) proportional similarity of the formulations of all strengths.

Fluvoxamine Maleate, Tablet, II (Paddle), 50, Water (deaerated), 900, 10, 20, 30, and 45, 01/03/2007.

Fosamprenavir Calcium, Tablet, II (Paddle), 75, 250 mM Sodium Acetate/Acetic acid buffer pH 3.5, 900, 10, 20, 30, and 45, 12/16/2005.

(Continued)

Fosamprenavir Calcium, Oral Suspension, II (Paddle), 25, 10 mM HCl, 900, 5, 10, 15, and 20, 12/03/2007.

Fosinopril Sodium, Tablet, II (Paddle), 50, Water (deaerated), 900, 10, 20, 30, and 45, 01/30/2004, 10 and 20 mg based on (1) acceptable bioequivalence studies on the 40 mg strength, (2) proportionally similar across all strengths, and (3) acceptable in vitro dissolution testing of all strengths.

Fosinopril Sodium/Hydrochlorothiazide, Tablet, Refer to USP, 08/11/2008, 10 mg/12.5 mg based on (1) acceptable bioequivalence studies on the 20/12.5 mg strength, (2) proportional similarity of the formulations across all strengths, and (3) acceptable in vitro dissolution testing of all strengths.

Frovatriptan succinate, Tablet, II (Paddle), 50, Phosphate Buffer pH 5.5, 900, 5, 10, 15, 20, and 30, 11/04/2008.

Furosemide, Tablet, Refer to USP, 08/05/2010, 20 and 40 mg based on (1) acceptable bioequivalence studies on the 80 mg strength, (2) acceptable in vitro dissolution testing of all strengths, and (3) proportional similarity of the formulations across all strengths.

Gabapentin, Tablet, Refer to USP, 06/03/2008, 100, 300, 400, and 600 mg based on (1) acceptable bioequivalence studies on the 800 mg strength, (2) proportionally similar across all strengths, and (3) acceptable in vitro dissolution testing of all strengths.

Gabapentin, Capsule, Refer to USP, 06/03/2008, 100 and 300 mg based on (1) acceptable bioequivalence studies on the 400 mg strength, (2) proportional similarity of the formulations across all strengths, and (3) acceptable in vitro dissolution testing of all strengths.

Gabapentin Enacarbil, Tablet (Extended Release), II (Paddle), 50, 10 mM Phosphate buffer at pH 7.4 with 1.0% SLS, 500 (for 300 mg); 900 (for 600 mg), 0.5, 1, 2, 4, 6, 8, 12, and 24 h, 01/31/2013, 300 mg based on (1) acceptable bioequivalence studies on the 600 mg strength, (2) acceptable in vitro dissolution testing of all strengths, and (3) proportional similarity of the formulations across all strengths.

Galantamine HBr, Tablet, Refer to USP, 08/11/2008, 8, 12 mg based on (1) acceptable bioequivalence studies on the 4 mg strength, (2) proportional similarity across all strengths, and (3) acceptable in vitro dissolution testing of all strengths.

Galantamine HBr, Capsule (Extended Release), II (Paddle), 50, 50 mM potassium dihydrogen phosphate buffer pH 6.5 Comparative dissolution data should also be provided in 900 mL pH 0.1 HCl, pH 4.5 buffer, and water using Apparatus II (Paddle) at 50 rpm, 900, 1, 4, 10, and 12 h, 01/20/2006, 16, 24 mg based on (1) acceptable bioequivalence studies on the 8 mg strength, (2) proportionally similar across all strengths, and (3) acceptable in vitro dissolution testing of all strengths.

Ganciclovir, Capsule, II (Paddle), 60, Water (deaerated), 900, 10, 20, 30, 45, and 60, 02/02/2004, 250 mg based on (1) acceptable bioequivalence studies on the 500 mg strength, (2) proportional similarity of the formulations across all strengths, and (3) acceptable in vitro dissolution testing of all strengths.

Gefitinib, Tablet, II (Paddle), 50, Tween 80 (5% v/v) in water, 1000, 10, 20, 30, 45, and 60, 10/28/2010.

Gemfibrozil, Tablet, Refer to USP, 07/25/2007.

Gemifloxacin Mesylate, Tablet, II (Paddle), 50, 0.01 N HCl, 900, 10, 20, 30, and 45, 01/03/2007.

Glimepiride, Tablet, II (Paddle), 75, Phosphate Buffer, pH 7.8, 900, 5, 10, 15, and 30, 07/23/2004, 2 and 4 mg based on (1) acceptable bioequivalence studies on the 1 mg strength, (2) proportionally similar across all strengths, and (3) acceptable in vitro dissolution testing of all strengths.

Glimepiride/Pioglitazone HCl, Tablet, II (Paddle), 75, For Pioglitazone: pH 2.0, HCl Buffer; For Glimepiride: pH 6.8, Sodium Phosphate Buffer with 0.2% sodium dodecyl sulfate, 900; For Pioglitazone: 10, 15, 20, 30, and 45; For Glimepiride: 10, 15, 20, and 30, 04/02/2009, 4 mg/30 mg based on (1) acceptable bioequivalence studies on the 2 mg/30 mg strength, (2) acceptable dissolution testing across all strengths, and (3) proportional similarity in the formulations across all strengths.

Glimepiride/Rosiglitazone Maleate, Tablet, II (Paddle), 75, 0.01 M HCl with 0.5% Sodium Dodecyl Sulfate, 900, 5, 10, 15, 30, 45, and 60, 01/03/2007, 2 mg/4 mg; 4 mg/4 mg; 2 mg/8 mg; and 4 mg/8 mg tablets, based on acceptable (1) bioequivalence studies on the 1 mg/4 mg tablet, (2) proportional similarity of the formulations, and (3) acceptable in vitro dissolution testing of all strengths.

Glipizide, Tablet (Extended Release), II (Paddle), 50, Simulated Intestinal Fluid without pancreatin, pH 7.5, 900, 1, 2, 4, 8, 16 h and until at least 80% dissolved, 04/10/2008, 2.5 and 5 mg based on (1) acceptable bioequivalence studies on the 10 mg strength, (2) proportionally similar across all strengths, and (3) acceptable in vitro dissolution testing of all strengths.

Glipizide, Tablet, Refer to USP, 08/05/2010, 5 mg based on (1) acceptable bioequivalence studies on the 10 mg strength, (2) acceptable in vitro dissolution testing of all strengths, and (3) proportional similarity of the formulations across all strengths.

Glipizide/Metformin HCl, Tablet, Refer to USP, 12/18/2008, 2.5 mg/250 mg and 2.5 mg/500 mg based on (1) acceptable bioequivalence studies on the 5 mg/500 mg strength, (2) proportional similarity of the formulations across all strengths, and (3) acceptable in vitro dissolution testing of all strengths.

(Continued)

Glyburide (Micronized), Tablet, II (Paddle), 50, 0.05 M Phosphate Buffer, pH 7.5, 900, 10, 20, 30, 45, and 60, 02/02/2004.

Glyburide (Nonmicronized), Tablet, II (Paddle), 75, 0.05 M Borate Buffer, pH 9.5, 500, 10, 20, 30, 45, and 60, 02/02/2004.

Glyburide/Metformin HCl, Tablet, Refer to USP, 01/14/2008, 1.25 mg/250 mg and 5 mg/500 mg based on (1) acceptable bioequivalence studies on the 2.5 mg/500 mg strength, (2) acceptable in vitro dissolution testing across all strengths, and (3) proportional similarity in the formulations across all strengths.

Glycopyrrolate, Tablet, Refer to USP, 07/25/2007, 1 and 2 mg strengths of the test drug products, based on acceptable formulation data and in vitro dissolution testing under the criteria set forth in 21 CFR.320.22(c).

Goserelin Acetate, Implant, Prior to sampling, the jar is removed from incubation and mechanically swirled with digital orbital shaker, Swirl orbit of 50 mm at 205 rpm for 6 s, Each implant should be incubated in 50 mL of phosphate buffered saline, pH 7.4, at 39°C (warmed overnight before the implants are added), in a 120 mL Wheaton jar, 50, 3, 14, 35, 56, and 84 days (10.8 mg strength); 7, 14, 17, 21, and 28 days (3.6 mg strength), 11/04/2008.

Granisetron, Film, Transdermal (Extended Release), VI (Cylinder), 50, 80 μL/L phosphoric acid (85%) at 32°C ± 0.5°C, 1000, 2, 6, 12, 24, 36, 48, 60, 72, and 96 h, 03/03/2011.

Granisetron HCl, Tablet, II (Paddle), 50, Phosphate buffer, pH 6.5, 500, 10, 20, 30, 45, and 60, 06/05/2006.

Griseofulvin, Oral Suspension, II (Paddle), 25, and 50, 0.54% Sodium Lauryl Sulfate (SLS) in Water, 1000, 10, 20, 30, and 45, 10/28/2010.

Griseofulvin (Microcrystalline), Oral Suspension, II (Paddle), 25 and 50, 0.54% Sodium Lauryl Sulfate (SLS) in Water, 1000, 10, 20, 30, and 45, 10/28/2010.

Griseofulvin (Microcrystalline), Tablet, Refer to USP, 01/15/2010, 125 mg strength based on: (1) acceptable in vivo bioequivalence studies on the 500 mg strength, (2) proportional similarity of the formulations across all strengths, and (3) acceptable in vitro dissolution testing on all strengths.

Griseofulvin (Ultramicrocrystalline), Tablet, Refer to USP, 11/04/2008, 125 mg based on (1) acceptable bioequivalence studies on the 250 mg strength, (2) acceptable dissolution testing across all strengths, and (3) proportional similarity in the formulations across all strengths.

Guaifenesin, Tablet (Extended Release), I (Basket), 75, 0.1 N HCl, 900, 1, 2, 4, 6, and 12 h, 01/03/2007.

Guaifenesin/Pseudoephedrine Hydrochloride, Tablet (Extended Release), I (Basket), 50, 0.01 N HCl, 900, 1, 2, 6, and 12 h, 11/25/2008.

Guanfacine, Tablet (Extended Release), II (Paddle), 75, HCl Buffer, pH 2.2, 900, 1, 2, 4, 6, 8, 10, 12, 16, 20, and 24 h, 07/01/2010.

Haloperidol, Tablet, Refer to USP, 11/25/2008, 0.5, 1, 5, 10, and 20 mg based on (1) acceptable bioequivalence studies on the 2 mg strength, (2) acceptable dissolution testing across all strengths, and (3) proportional similarity in the formulations across all strengths.

Homatropine Methylbromide/Hydrocodone Bitartrate, Tablet, Refer to USP, 10/30/2009, 1.5 mg/5 mg based on acceptable formulation data and in vitro dissolution testing under the criteria set forth in 21 CFR 320.22(c).

Hydralazine HCl, Tablet, Refer to USP, 04/10/2008, 10, 25, 50, and 100 mg strengths of the test product, based on acceptable in vitro dissolution testing under 21 CFR § 320.22(c).

Hydrochlorothiazide, Capsule, I (Basket), 100, 0.1 N HCl, 900, 10, 20, 30, and 45, 02/03/2004.

Hydrochlorothiazide, Tablet, Refer to USP, 07/25/2007, 12.5 and 25 mg based on (1) acceptable bioequivalence studies on the 50 mg strength, (2) acceptable in vitro dissolution testing of all strengths, and (3) proportional similarity of the formulations across all strengths.

Hydrochlorothiazide/Irbesartan, Tablet, II (Paddle), 50, 0.1 N HCl, 1000, 10, 20, 30, and 45, 09/24/2008, 12.5 mg/150 mg based on (1) acceptable bioequivalence studies on the 12.5 mg/300 mg strength, (2) acceptable in vitro dissolution testing of all strengths, and (3) proportional similarity of the formulations across all strengths. Please refer to the Mirtazapine Tablet Guidance for additional information regarding waivers of in vivo testing.

Hydrochlorothiazide/Lisinopril, Tablet, II (Paddle), 50, 0.1 N HCl, 900, 10, 20, 30, 45, and 60, 02/03/2004, 12.5 mg/10 mg based on (1) acceptable bioequivalence studies on the 25 mg/20 mg strength, (2) acceptable in vitro dissolution testing of all strengths, and (3) proportional similarity of the formulations across all strengths.

Hydrochlorothiazide/Losartan Potassium, Tablet, I (Basket), 100, Water (deaerated), 900, 10, 20, 30, 45, and 60, 02/03/2004, 12.5 mg/50 mg and 12.5 mg/100 mg based on (1) acceptable bioequivalence studies on the 25 mg/100 mg strength, (2) acceptable in vitro dissolution testing of all strengths, and (3) proportional similarity of the formulations across all strengths.

Hydrochlorothiazide/Metoprolol Tartrate, Tablet, Refer to USP, 01/05/2012, 50 mg/25 mg based on (1) acceptable bioequivalence studies on the 100 mg/25 mg strength, (2) acceptable in vitro dissolution testing of all strengths, and (3) proportional similarity of the formulations across all strengths.

(*Continued*)

Hydrochlorothiazide/Moexipril HCl, Tablet, II (Paddle), 50, 0.1 N HCl, 900, 5, 10, 15, and 30, 02/10/2004, 12.5 mg/7.5 mg, and 12.5 mg/15 mg based on (1) acceptable bioequivalence study on the 15 mg/25 mg strength, (2) proportional similarity of the formulations across all strengths, and (3) acceptable in vitro dissolution testing of all strengths.

Hydrochlorothiazide/Olmesartan Medoxomil, Tablet, II (Paddle), 50, 0.05 M Phosphate Buffer, pH 6.8, 900, 5, 10, 15, 20, 30, 45, and 60, 07/09/2007, 12.5 mg/40 mg and 12.5 mg/20 mg strengths based on (1) acceptable bioequivalence studies on the 25 mg/40 mg strength, (2) proportional similarity of the formulations across all strengths, and (3) acceptable in vitro dissolution testing of all strengths.

Hydrochlorothiazide/Quinapril HCl, Tablet, I (Basket), 100, Water (deaerated), 900, 5, 10, 20, and 30, 02/03/2004, 12.5 mg/20 mg and 12.5 mg/10 mg based on (1) acceptable bioequivalence studies on the 25 mg/20 mg strength, (2) acceptable in vitro dissolution testing of all strengths, and (3) proportional similarity of the formulations across all strengths.

Hydrochlorothiazide/Spironolactone, Tablet, Refer to USP, 08/27/2009, 25 mg/25 mg based on (1) acceptable bioequivalence studies on the 50 mg/50 mg strength, (2) proportional similarity of the formulations across all strengths, and (3) acceptable in vitro dissolution testing of all strengths.

Hydrochlorothiazide/Telmisartan, Tablet, II (Paddle), 75, Phosphate Buffer, pH 7.5, 900, 10, 15, 20, 30, 45, and 60, 04/10/2008.

Hydrochlorothiazide/Triamterene, Tablet, Refer to USP, 07/31/2013, 12.5 mg/40 mg and 12.5 mg/80 mg based on (1) acceptable bioequivalence studies on the 25 mg/80 mg strength, (2) acceptable dissolution testing across all strengths, and (3) proportional similarity in the formulations across all strengths.

Hydrochlorothiazide/Valsartan, Tablet, II (Paddle), 50, Phosphate Buffer pH 6.8, 1000, 10, 20, 30, and 45, 02/03/2004, 12.5 mg/80 mg, 12.5 mg/160 mg, 12.5 mg/320 mg, 25 mg/160 mg based on (1) acceptable bioequivalence studies on the 25 mg/320 mg strength, (2) proportional similarity of the formulations across all strengths, and (3) acceptable in vitro dissolution testing of all strengths.

Hydrochlorothiazide/Triamterene, Capsule, Refer to USP, 10/06/2008, 25 mg/37.5 mg strength based on (1) acceptable bioequivalence studies on the 25 mg/50 mg strength; (2) proportional similarity across all strengths and (3) acceptable dissolution testing on all strengths.

Hydrocodone Bitartrate/Ibuprofen, Tablet, II (Paddle), 50, Phosphate Buffer, pH 7.2, 900, 5, 10, 15, and 30, 02/04/2004, 2.5 mg/200 mg, 5 mg/200 mg, and 10 mg/200 mg based on: (1) acceptable bioequivalence study on the 7.5 mg/200 mg, (2) proportional similarity across all strengths, and (3) acceptable in vitro dissolution testing of all strengths.

Hydrocortisone, Tablet, Refer to USP, 05/09/2013, 5 and 10 mg based on (1) acceptable bioequivalence studies on the 20 mg strength, (2) acceptable in vitro dissolution testing of all strengths, and (3) proportional similarity of the formulations across all strengths. Please refer to the Mirtazapine Tablet Guidance for additional information regarding waivers of in vivo testing.

Hydromorphone HCl, Tablet (Extended Release), VII (Reciprocating holder) (Sample holder Cage), 30 cycles/min, Water, 50, 1, 2, 4, 6, 8, 10, 12, 16, 20, and 24 h, 05/09/2013.

Hydromorphone HCl, Tablet, Refer to USP, 07/25/2007, 2 and 4 mg based on (1) acceptable bioequivalence studies on the 8 mg strength, (2) acceptable in vitro dissolution testing of all strengths, and (3) proportional similarity of the formulations across all strengths.

Hydroxyurea, Capsule, Refer to USP, 09/03/2008, 200, 250, 300, and 400 mg based on (1) acceptable bioequivalence studies on the 500 mg strength, (2) acceptable dissolution testing across all strengths, and (3) proportional similarity in the formulations across all strengths.

Hydroxyzine HCl, Tablet, Refer to USP, 07/25/2007, 10, 25, and 50 mg pursuant to 21 CFR 320.22(c) provided that the in vitro dissolution profiles of the proposed product are comparable to those of the reference product.

Hydroxyzine Pamoate, Capsule, Refer to USP, 04/02/2009, 25 mg based on (1) acceptable bioequivalence studies on the 50 mg strength, (2) proportional similarity of the formulations across all strengths, and (3) acceptable in vitro dissolution testing of all strengths.

Hydroxyzine Pamoate, Suspension, Develop a dissolution method, 04/02/2009.

Ibandronate Sodium, Tablet, II (Paddle), 50, Water, 500, 5, 10, 15, 30, and 45, 01/03/2007.

Ibuprofen, Tablet (Chewable), II (Paddle), 50, 0.05 M Phosphate Buffer, pH 7.2, 900, 10, 20, 30, and 45, 02/04/2004.

Ibuprofen, Suspension, Refer to USP, 11/04/2008.

Ibuprofen, Suspension/Drop, II (Paddle), 50, Phosphate Buffer, pH 7.2, 900, 5, 10, 15, and 20, 11/04/2008.

Ibuprofen, Tablet, Refer to USP, 07/25/2007, 400 and 600 mg strengths, based on (1) acceptable bioequivalence studies on the 800 mg strength, (2) proportional similarity of the formulations across all strengths, and (3) acceptable in vitro dissolution testing of all strengths.

(Continued)

Ibuprofen, Capsule (Soft Gelatin/Liquid Fill), I (Basket), 150, 50 mM Phosphate Buffer, pH 7.2, 900, 5, 10, 20, 30, and 45, 05/09/2013.

Ibuprofen Potassium, Capsule (Soft Gelatin/Liquid Fill), I (Basket), 150, Phosphate Buffer, pH 7.2, 900, 5, 10, 20, and 30, 02/04/2004.

Ibuprofen/Diphenhydramine, Capsule, I (Basket), 100, Phosphate Buffer (200 mM), pH 7.2, 900, 10, 20, 30, and 45, 01/03/2007.

Ibuprofen/Oxycodone HCl, Tablet, I (Basket), 100, Phosphate buffer, pH 7.2, 500, 10, 20, 30, and 45, 04/09/2007.

Ibuprofen/Phenylephrine HCl, Tablet, II (Paddle), 50, 50 mM Potassium Phosphate Buffer, pH 6.5, (degassed), 900, 10, 15, 20, 30, and 45, 01/05/2012.

Ibuprofen/Pseudoephedrine HCl, Capsule, I (Basket), 150, Tier 1: 0.05 M phosphate buffer, pH 7.2 Tier 2: 0.05 M phosphate buffer, pH 7.2 with NMT 1750 USP protease units/L of 1 x USP pancreatin, 900, 10, 20, 30, and 45, 03/04/2006.

Ibuprofen/Pseudoephedrine HCl, Suspension, II (Paddle), 50, 0.05 M Phosphate Buffer, pH 7.2, 900, 5, 10, 15, and 30, 02/04/2004.

Icosapent Ethyl, Capsule, Develop an in vitro release method using USP IV (Flow-Through Cell), and, if applicable, Apparatus II (Paddle) or any other appropriate pharmacopoeial apparatus, for comparative evaluation by the agency, 08/15/2013.

Iloperidone, Tablet, II (Paddle), 50, 0.1 N HCl, 500, 5, 10, 15, 30, 45, and 60, 08/05/2010, 2, 4, 6, 8, 10, and 12 mg, based on (1) acceptable bioequivalence study using the 1 mg tablets, (2) proportional similarity in the formulations of the 1, 2, 4, 6, 8, 10, and 12 mg strengths, and (3) acceptable comparable in vitro dissolution testing on all strengths.

Imatinib Mesylate, Tablet, II (Paddle), 50, 0.1 N HCl, 1000, 5, 10, 15, 20, and 30, 09/22/2011, 100 mg based on (1) acceptable bioequivalence studies on the 400 mg strength, (2) acceptable in vitro dissolution testing of all strengths, and (3) proportional similarity of the formulations across all strengths.

Imipramine HCl, Tablet, Refer to USP, 01/14/2008.

Imipramine Pamoate, Capsule, I (Basket), 100, 0.1 N HCl without pepsin and with 0.3% pepsin (addition of pepsin is recommended only when significant slow dissolution is observed), 900, 30, 60, 90, 120, 150, and 180, 01/14/2008, Eq 100, 125, and 150 mg HCl based on (1) acceptable bioequivalence studies on the Eq 75 mg HCl strength, (2) acceptable in vitro dissolution testing of all strengths, and (3) proportional similarity in the formulations across all strengths. Please refer to the Mirtazapine Tablet Guidance for additional information regarding waivers of in vivo testing.

Indapamide, Tablet, Refer to USP, 04/15/2008, 1.25 mg may be considered for a waiver of in vivo bioequivalence testing based on (1) acceptable bioequivalence studies on the 2.5 mcg strength, (2) proportionally similar across all strengths, and (3) acceptable in vitro dissolution testing of all strengths.

Indinavir Sulfate, Capsule, II (Paddle), 50, 0.1 M Citrate Buffer, pH 3.8, 900, 10, 15, 20, and 30, 02/04/2004, 100, 200, and 333 mg based on (1) acceptable bioequivalence studies on the 400 mg strength, (2) proportional similarity of the formulations across all strengths, and (3) acceptable in vitro dissolution testing of all strengths.

Indomethacin, Capsule (Extended Release), Refer to USP, 07/25/2007.

Indomethacin, Capsule, Refer to USP, 12/15/2009, 25 mg based on (1) acceptable bioequivalence studies on the 50 mg strength, (2) proportional similarity of the formulations across all strengths, and (3) acceptable in vitro dissolution testing of all strengths.

Irbesartan, Tablet, Refer to USP, 08/11/2008, 75 and 150 mg based on (1) acceptable bioequivalence studies on the 300 mg strength, (2) proportional similarity of the formulations across all strengths, and (3) acceptable in vitro dissolution testing of all strengths.

Isocarboxazid, Tablet, II (Paddle), 50, 0.1 N HCl, 900, 10, 20, 30, 45, and 60, 02/04/2004.

Isoniazid, Tablet, Refer to USP, 04/15/2008, 100 and 300 mg strengths based on acceptable in vitro dissolution testing under 21 CFR § 320.22(c).

Isosorbide Dinitrate/Hydralazine HCl, Tablet, I (Basket), 100, 0.05 N HCl, 900, 10, 15, 20, 25, 30, and 45, 06/10/2009.

Isosorbide Mononitrate, Tablet, II (Paddle), 50, Water (deaerated), 900, 5, 10, 15, and 30, 02/04/2004.

Isosorbide Mononitrate, Tablet (Extended Release), Refer to USP, 11/25/2008, 30, 60 mg based on (1) acceptable bioequivalence studies on the 120 mg strength, (2) proportionally similar across all strengths, and (3) acceptable in vitro dissolution testing of all strengths.

Isotretinoin, Capsule, I (Basket, with 20 mesh), 100, 0.05 M Potassium Phosphate Buffer, dibasic, pH 7.8, containing 0.5% solid LDAO, 900, 15, 30, 45, 60, and 90, 10/06/2008, 10 and 30 mg based on (1) acceptable bioequivalence studies on the 20 and 40 mg strengths, (2) proportional similarity of the formulations across all strengths, and (3) acceptable in vitro dissolution testing of all strengths.

(*Continued*)

Isradipine, Capsule, II (Paddle), 50, 0.1% Lauryl Dimethylamine Oxide (LDAO) in water, 500, 10, 20, 30, 45, and 60, 02/25/2004.

Isradipine (10 mg), Tablet (Extended Release), II (Paddle), 50, 0.2% Lauryl Dimethylamine Oxide (LDAO) in water, 1000, 2, 4, 8, 12, 16, and 24 h, 02/25/2004, 5 mg based on (1) acceptable bioequivalence studies on the 10 mg strength, (2) proportionally similar across all strengths, and (3) acceptable in vitro dissolution testing of all strengths.

Isradipine (5 mg), Tablet (Extended Release), II (Paddle), 50, 0.2% Lauryl Dimethylamine Oxide (LDAO) in water, 500, 2, 4, 8, 12, 16, and 24 h, 02/25/2004, 2.5 mg based on (1) acceptable bioequivalence studies on the 5 mg strength, (2) proportional similarity of the formulations across all strengths, and (3) acceptable in vitro dissolution testing of all strengths.

Itraconazole, Capsule, II (Paddle), 100, SGF without Enzyme, 900, 10, 20, 30, 45, 60, and 90, 02/04/2004.

Itraconazole, Tablet, II (Paddle), 75, 0.1 N HCl, 900, 5, 15, 30, 45, 60, 75, and 90, 08/15/2013.

Ivermectin, Tablet, II (Paddle), 50, 0.5% SDS in 0.01 M Monobasic Sodium Phosphate, pH 7.0, 900, 10, 20, 30, 45, and 60, 02/04/2004.

Ketoconazole, Tablet, I (Basket), 100, Simulated gastric fluid w/o pepsin, 800, 15, 30, 45, 60, and 90, 01/03/2007.

Ketoprofen, Tablet, II (Paddle), 50, SIF Buffer without enzyme, pH 7.4, 900, 10, 20, 30, 45, and 60, 02/05/2004.

Ketoprofen, Capsule, II (Paddle), 50, 0.05 M Phosphate Buffer pH 7.4, 1000, 10, 20, 30, and 45, 07/25/2007.

Ketorolac Tromethamine, Tablet, Refer to USP, 04/15/2008.

Labetalol HCl, Tablet, Refer to USP, 08/27/2009, 100 and 300 mg based on (1) acceptable bioequivalence studies on the 200 mg strength, (2) acceptable dissolution testing across all strengths, and (3) proportional similarity in the formulations across all strengths.

Lacosamide, Tablet, II (Paddle), 50, 0.1 N HCl, 900, 10, 15, 20, 30, and 45, 06/07/2012, 50, 100, and 150 mg based on (1) acceptable bioequivalence studies on the 200 mg strength, (2) acceptable in vitro dissolution testing of all strengths, and (3) proportional similarity of the formulations across all strengths.

Lamivudine (for 100 and 150 mg), Tablet, II (Paddle), 50, Water (deaerated), 900, 10, 20, 30, and 45, 03/22/2006.

Lamivudine (for 300 mg only), Tablet, II (Paddle), 75, 0.1 N HCl, 900, 5, 10, 15, and 30, 03/22/2006, 150 mg based on (1) acceptable bioequivalence studies on the 300 mg strength, (2) proportional similarity of the formulations across all strengths, and (3) acceptable in vitro dissolution testing of all strengths.

Lamivudine 150 mg/Zidovudine 300 mg Tablets and Abacavir Sulfate 300 mg Tablets copackaged, Tablet, II (Paddle), 75, 0.1 N HCl, 900, 5, 10, 15, 20, 30, and 40, 01/03/2007.

Lamivudine/Stavudine/Nevirapine, Tablet, II (Paddle), 75, 0.1 N HCl, 900, 10, 20, 30, 45, and 60, 01/03/2007.

Lamivudine/Zidovudine, Tablet, II (Paddle), 75, 0.1 N HCl, 900, 10, 20, 30, and 45, 02/20/2004.

Lamivudine/Zidovudine + Efavirenz, Tablet (Copackage), II (Paddle), Lamivudine and Zidovudine: 75 Efavirenz: 50, Lamivudine and Zidovudine: 0.1 N HCl Efavirenz: 2% SLS in water, Lamivudine and Zidovudine: 1000 Efavirenz: 900, 10, 20, 30, and 45, 01/03/2007.

Lamivudine/Zidovudine + Nevirapine, Tablet (Copackage), II (Paddle), 50, Lamivudine and Zidovudine: water Nevirapine: 0.06 M HCl (pH 1.2), 900, 10, 15, 30, 45, and 60, 01/03/2007.

Lamivudine/Zidovudine/Nevirapine, Tablet, II (Paddle), 50, 0.01 N HCl, 900, 10, 15, 30, 45, and 60, 01/03/2007.

Lamotrigine, Tablet (Extended Release), II (Paddle), 50, Acid Stage: 0.01 M HCl; Buffer Stage: Phosphate Buffer, pH 6.8 + 0.5% SLS (Add 200 mL of 0.0205 M tribasic sodium phosphate (pH 12) solution containing 2.25% w/v SLS to 700 mL of HCl), Acid Stage: 700; Buffer Stage: 900, Acid stage: 120; Buffer stage: 1, 2, 3, 5, 7, 10, 12, and 15 h, 03/25/2010, 25, 100, 200, and 300 mg based on (1) acceptable bioequivalence studies on the 50 mg strength, (2) acceptable in vitro dissolution testing of all strengths, and (3) proportional similarity of the formulations across all strengths.

Lamotrigine, Tablet (Regular), II (Paddle), 50, 0.1 N HCl, 900, 5, 10, 15, 20, and 30, 03/04/2006, 100, 150, and 200 mg based on (1) acceptable bioequivalence studies on the 25 mg strength, (2) proportional similarity of the formulations across all strengths, and (3) acceptable in vitro dissolution testing of all strengths.

Lamotrigine, Tablet (Chewable dispersible), II (Paddle), 50, 0.1 N HCl, 900, 5, 10, 15, 20, and 30, 01/14/2008, 2 and 5 mg based on (1) acceptable bioequivalence studies on the 25 mg strength, (2) proportional similarity of the formulations across all strengths, and (3) acceptable in vitro dissolution testing of all strengths.

Lansoprazole, Capsule (Delayed Release), Refer to USP, 11/04/2008, 15 mg based on (1) acceptable bioequivalence studies on the 30 mg strength, (2) acceptable dissolution testing across all strengths, and (3) proportional similarity in the formulations across all strengths.

(*Continued*)

Lansoprazole, Tablet (Delayed Release, Orally Disintegrating), II (Paddle), 75, Acid Stage: 0.1 N HCl; Buffer Stage: Phosphate Buffer, pH 6.8 with 5 mM Sodium Dodecyl Sulfate, 500 (Acid), 900 (Buffer), 60 (Acid), 10, 20, 30, and 45 (Buffer), 11/04/2008, 15 mg based on (1) acceptable bioequivalence studies on the 30 mg strength, (2) acceptable dissolution testing across all strengths, and (3) proportional similarity in the formulations across all strengths.

Lanthanum Carbonate, Chewable Tablet, Reciprocating Cylinder (Apparatus 3 modified), 10 dpm (dip rate per minute), 0.25 N HCl, 900 (modified from the standard apparatus 3 vessel to achieve sink condition), 10, 20, 30, and 45, 01/03/2007, 500 and 750 mg based on (1) acceptable in vitro bioequivalence studies on the 1000 mg strength, (2) proportional similarity in the formulations across all strengths, and (3) acceptable in vitro dissolution of all strengths.

Lapatinib Ditosylate, Tablet, II (Paddle), 55, 2% Polysorbate 80 in 0.1 N HCl, 900, 10, 15, 30, and 45, 10/30/2009.

Leflunomide, Tablet, II (Paddle), 100, Water (deaerated), 1000, 10, 20, 30, and 45, 02/05/2004.

Leflunomide (100 mg), Tablet, II (Paddle), 100, Water (deaerated) + 0.6% Polyoxyethylene Lauryl Ether, 1000, 10, 20, 30, and 45, 05/31/2007, 10 mg based on (1) acceptable bioequivalence studies on the 20 mg strength, (2) proportionally similar across all strengths, and (3) acceptable in vitro dissolution testing of all strengths.

Lenalidomide, Capsule, II (Paddle), 50, 0.01 N HCl, 900, 10, 15, 20, 30, and 45, 04/15/2008, 2.5, 5, 10, and 15 mg based on (1) acceptable fasting and fed bioequivalence studies on the 25 mg strength, (2) proportional similarity of the formulations across all strengths, and (3) acceptable in vitro dissolution testing of all strengths.

Letrozole, Tablet, Refer to USP, 04/10/2008.

Leucovorin Calcium, Tablet, Refer to USP, 07/14/2008, 5, 10, and 15 mg based on (1) acceptable bioequivalence studies on the 25 mg strength, (2) acceptable dissolution testing across all strengths, and (3) proportional similarity in the formulations across all strengths.

Leuprolide Acetate, Injectable (Extended Release), Develop a dissolution method using USP IV (Flow-Through Cell), and, if applicable, Apparatus II (Paddle) or any other appropriate method, for comparative evaluation by the agency, 01/15/2010, 11.25 mg/vial–3 month and 22.5 mg/vial based on (1) an acceptable bioequivalence study on 30 mg/vial strength, (2) acceptable dissolution testing across all strengths, and (3) qualitative (Q1) and quantitative (Q2) sameness to the respective RLD strength.

Levetiracetam, Tablet (Extended Release), I (Basket), 100, 0.05 M Phosphate Buffer, pH 6.0, 900, 1, 2, 4, 6, 8, and 12 h, 04/02/2009, 500 mg strength based on (1) acceptable bioequivalence studies on the 750 mg strength, (2) acceptable in vitro dissolution testing of all strengths, and (3) proportional similarity of the formulations across all strengths.

Levetiracetam, Tablet, II (Paddle), 50, Water (deaerated), 900, 5, 10, 15, and 30, 02/05/2004, 250, 500, and 750 mg based on (1) acceptable bioequivalence studies on the 1000 mg strength, (2) proportional similarity of the formulations across all strengths, and (3) acceptable in vitro dissolution testing of all strengths.

Levocetirizine Dihydrochloride, Tablet, II (Paddle), 50, Water, 900, 10, 20, 30, and 45, 08/11/2008.

Levofloxacin, Tablet, I (Basket), 100, 0.1 N HCl, 900, 10, 20, 30, and 45, 06/18/2007, 250 and 500 mg based on (1) acceptable bioequivalence studies on the 750 mg strength, (2) acceptable in vitro dissolution testing of all strengths, and (3) proportional similarity in the formulations across all strengths.

Levonorgestrel, Tablet, II (Paddle), 75, 0.1 N HCl with 0.1% SLS, 1000, 10, 20, 30, 45, 60, and 90, 02/05/2004, 0.75 mg based on (1) acceptable fasting and fed BE studies on the 1.5 mg strength, (2) proportional similarity in the formulations across all strengths, and (3) acceptable dissolution testing across all strengths.

Levothyroxine Sodium, Tablet, Refer to USP, 07/25/2007.

Lidocaine, Topical Patch, Paddle over Disk (Apparatus 5), 50, Acetic acid/sodium acetate buffer, pH 4.0 at 32°C, 500, 10, 20, 30, 60, 120, and 180, 01/03/2007.

Linagliptin, Tablet, I (Basket), 50, 0.1 N HCl, 900, 5, 10, 15, 20, 30, and 45, 08/15/2013.

Linezolid, Tablet, II (Paddle), 50, 0.05 M Phosphate Buffer, pH 6.8, 900, 5, 10, 20, 30, and 45, 01/14/2008.

Linezolid, Suspension, II (Paddle), 50, 0.05 M Phosphate Buffer, pH 6.8, 900, 10, 20, 30, and 45, 01/14/2008.

Liothyronine Sodium, Tablet, Refer to USP, 06/18/2007, 5 and 25 mcg based on (1) acceptable bioequivalence studies on the 50 mcg strength, (2) acceptable in vitro dissolution testing of all strengths, and (3) proportional similarity of the formulations across all strengths.

Lisdexamfetamine Dimesylate, Capsule, II (Paddle), 50, 0.1 N HCl, 900, 5, 10, 15, and 20, 10/06/2008, 20, 30, 40, 50, and 60 mg based on (1) acceptable bioequivalence studies on the 70 mg strength, (2) acceptable dissolution testing across all strengths, and (3) proportional similarity in the formulations across all strengths.

Lisinopril, Tablet, Refer to USP, 01/14/2008, 2.5, 5, 10, 20, and 30 mg, based on acceptable (1) bioequivalence studies on the 40 mg strength, (2) proportional similarity of the formulations and (3) acceptable in vitro dissolution testing of all strengths.

Lithium Carbonate, Tablet (Extended Release), Refer to USP, 01/14/2008.

(*Continued*)

Lithium Carbonate, Capsule, Refer to USP, 07/25/2007, 150 and 300 mg based on (1) acceptable bioequivalence studies on the 600 mg strength, (2) proportionally similar across all strengths, and (3) acceptable in vitro dissolution testing of all strengths.

Lithium Carbonate, Tablet, Refer to USP, 04/10/2008.

Lomefloxacin HCl, Tablet, II (Paddle), 50, 0.01 N HCl, 900, 10, 20, 30, and 45, 02/05/2004.

Lomustine, Capsule, Develop a dissolution method.

Lopinavir/Ritonavir, Capsule (Soft Gelatin), II (Paddle), 50, Tier 1: 0.05 M Polyoxyethylene 10 Lauryl Ether with 10 mM Sodium Phosphate monobasic (pH 6.8); Tier II: same as mentioned earlier with NMT 1750 USP units/L of Pancreatin, 900, 10, 15, 30, and 45, 06/18/2007.

Lopinavir/Ritonavir, Tablet (Combination), II (Paddle), 75, 0.06 M Polyoxyethylene 10 Lauryl Ether, 900, 15, 30, 60, 90, and 120, 09/13/2007, 100 mg/25 mg based on (1) acceptable bioequivalence studies on the 200 mg/50 mg strength, (2) acceptable dissolution testing across both strengths, and (3) proportional similarity in the formulations across both strengths. Please refer to the Mirtazapine Tablet Guidance for additional information regarding waivers of in vivo testing.

Loratadine, Capsule (Soft Gelatin), II (Paddle) with sinker, 75, Tier I: 0.1 N HCl with 0.1% Tween 20. Tier II: 0.1 N HCl with 0.1% Tween 20 with addition of pepsin (as per USP), 900, 10, 20, 30, 45, and 60, 02/28/2013.

Loratadine, Tablet (Chewable), II (paddle), 50, 0.1 N HCl, 500, 15, 30, 45, and 60, 07/14/2008.

Loratadine, Tablet (Orally Disintegrating), I (Basket), 50, SGF without enzyme, 900, 2, 4, 6, and 10, 07/14/2008.

Loratadine/Pseudoephedrine Sulfate (10 mg/240 mg), Tablet (Extended Release), I (Basket), 75, 900 mL 0.1 N HCl for 1 h, then replace the medium with 900 mL 0.05 M phosphate buffer at pH 6.8 containing 0.01% sodium lauryl sulfate, 900, Loratadine: 10, 15, 20, 30, and 45; Pseudoephedrine: 1, 2, 4, 8, 12, 16, 18, and 24 h, 08/05/2010.

Loratadine/Pseudoephedrine Sulfate (5 mg/120 mg), Tablet (Extended Release), II (Paddle), 50, 900 mL 0.1 N HCl for 1 h, then replace with 900 mL 0.05 M phosphate buffer at pH 8.2 containing 0.01% sodium lauryl sulfate, 900, Loratadine:15, 20, 30, 45, 60, and 90; Pseudoephedrine: 1, 2, 4, 8, 12, and 16 h, 08/05/2010.

Lorazepam, Tablet, Refer to USP, 01/14/2008, 0.5 and 1 mg based on (1) acceptable bioequivalence studies on the 2 mg strength, (2) proportional similarity of the formulations across all strengths, and (3) acceptable in vitro dissolution testing of all strengths.

Losartan Potassium, Tablet, II (Paddle), 50, Water (deaerated), 900, 10, 20, 30, and 45, 02/06/2004, 25 and 50 mg based on (1) acceptable bioequivalence studies on the 100 mg strength, (2) proportional similarity of the formulations across all strengths, and (3) acceptable in vitro dissolution testing of all strengths.

Losartan Potassium, Tablet, Refer to USP, 01/05/2012.

Loteprednol Etabonate/Tobramycin, Ophthalmic Suspension, Develop a method to characterize in vitro release, 01/31/2013.

Lovastatin/Niacin, Tablet (Extended Release), I (Basket), 100, For Niacin: Water; For Lovastatin: 0.05 M phosphate buffer, pH 7.0 with 0.5% sodium dodecyl sulfate, 900; For Niacin: 0.5, 1, 2, 3, 6, 9, 12, 20, and 24 h; For Lovastatin: 15, 30, 45, and 60 min, 01/14/2008.

Lubiprostone, Capsule (Soft Gelatin), II (Paddle), 50, 0.1 N HCl/1% HCO-40 (Polyoxyl 40 hydrogenated castor oil), 900, 15, 30, 45, 60, 90, and 120, 08/19/2010, 8 mcg based on (1) acceptable bioequivalence studies on the 24 mcg strength, (2) proportional similarity of the formulations across all strengths, and (3) acceptable in vitro quantitative capsule rupture (dissolution) testing all strengths.

Lurasidone HCl, Tablet, II (Paddle), 50, McIlvaine buffer, pH 3.8 [0.0.025 M Citric acid Solution + 0.05 M Na_2HPO_4 solution (3:2)] Measure the pH and adjust to 3.8, if necessary. Degas before use, 900, 5, 10, 15, 20, and 30, 01/31/2013.

Magnesium Hydroxide/Omeprazole/Sodium Bicarbonate, Tablet (Chewable), II (Paddle), 150, 0.029 M sodium phosphate buffer w/0.5% SDS, pH 7.4, 900, 15, 30, 45, and 60, 02/19/2008.

Magnesium Hydroxide/Omeprazole/Sodium Bicarbonate, Tablet (Chewable), II (Paddle), 150, pH 7.4 Phosphate Buffer with 0.5% SDS, 900, 15, 30, 45, 60, and 90, 10/06/2008.

Maraviroc, Tablet, I (Basket), 100, 0.01 N HCl, 900, 10, 15, 20, 30, and 45, 10/21/2010, 150 mg based on (1) acceptable bioequivalence studies on the 300 mg strength, (2) proportional similarity of the 150 and 300 mg strengths, and (3) acceptable in vitro dissolution testing of the 150 and 300 mg strengths.

Mebendazole, Tablet (Chewable), II (Paddle), 75, 0.1 N HCl containing 1% Sodium Lauryl Sulfate, 900, 15, 30, 45, 60, 90, and 120, 10/06/2008.

Meclizine HCl, Tablet (Chewable), I (Basket), 100, 0.01 N HCl, 900, 10, 20, 30, 45, and 60, 04/08/2010, 25 mg pursuant to 21 CFR § 320.22(c) provided that the in vitro dissolution profiles of the proposed product are comparable to those of the reference product.

(Continued)

Meclizine HCl, Tablet, I (Basket), 100, 0.01 N HCl, 900, 10, 20, 30, 45, and 60, 08/27/2009, 12.5, 25, and 50 mg strengths of the test product, based on acceptable formulation data and in vitro dissolution testing under 21 CFR § 320.22(c).

Mefenamic Acid, Capsule, Refer to USP, 12/15/2009.

Mefloquine HCl, Tablet, I (Basket), 100, SGF without enzyme, 900, 10, 20, 30, 45, and 60, 02/06/2004.

Megestrol Acetate, Oral Suspension, Refer to USP, 12/15/2009.

Meloxicam, Suspension, II (Paddle), 25, Phosphate buffer at pH 7.5, 900, 5, 10, 15, and 30, 01/26/2006.

Meloxicam, Tablet, II (Paddle), 75, Phosphate Buffer, pH 7.5, 900, 10, 20, 30, 45, and 60, 02/20/2004, 7.5 mg based on (1) acceptable bioequivalence studies on the 15 mg strength, (2) proportionally similar across all strengths, and (3) acceptable in vitro dissolution testing of all strengths.

Melphalan, Tablet, Refer to USP, 07/14/2008.

Memantine HCl, Capsule (Extended Release), I (Basket), 100, pH 1.2 Buffer, Simulated Gastric Fluid without enzyme, 900, 1, 2, 4, 6, 8, 10, 12, and 16 h, 10/28/2010, 7, 14, and 21 mg based on (1) acceptable bioequivalence studies on the 28 mg strength, (2) acceptable in vitro dissolution testing of all strengths, and (3) proportional similarity of the formulations across all strengths. Please refer to the Mirtazapine Tablet Guidance for additional information regarding waivers of in vivo testing.

Memantine HCl, Tablet, I (Basket), 100, 0.1 N HCl with NaCl (12 g NaCl in 6 L water adjust pH to 1.2 with HCl), 900, 10, 20, 30, and 45, 12/16/2005, 5 mg based on (1) acceptable bioequivalence studies on the 10 mg strength, (2) acceptable dissolution testing across all strengths, and (3) proportional similarity of the formulations across all strengths.

Menthol/Methyl Salicylate, Topical Patch, VI (Cylinder), 50, Neutralized phthalate buffered solution (0.2 M potassium biphthalate) with pH of 5.0 at 32°C ± 0.5°C, 900, 10, 20, 30, 60, 120, 150, and 180, 01/31/2013.

Meprobamate, Tablet, Refer to USP, 11/25/2008, 200 and 400 mg strengths of the test product, based on acceptable in vitro dissolution testing under 21 CFR § 320.22(c).

Mercaptopurine, Tablet, II (Paddle), 50, 0.1 N HCl, 900, 20, 30, 45, 60, 90, and 120, 02/06/2004.

Mesalamine, Suppository, II (Paddle) with option to use a sinker, 75 (for 500 mg) and 125 (for 1000 mg), For 500 mg strength: 0.2 M Phosphate buffer, pH 7.5 at 37°C; For 1000 mg strength: 0.2 M Phosphate buffer, pH 7.5 at 40°C, 900, 30, 60, 90, 120, and 150, 01/30/2006.

Mesalamine (1.2 g), Tablet (Delayed Release), II (Paddle), 100, Acid stage (A): 100 mM HCl Buffer stage (B): Phosphate Buffer, pH 6.4 Buffer stage (C): Phosphate Buffer, pH 7.2, Acid stage (A): 750 mL; Buffer stage (B): 950 mL; Buffer stage (C): 960 mL, Acid stage (A): 2 h; Buffer stage (B): 1 h; Buffer stage (C): 1, 2, 4, 6, and 8 h, 06/10/2009.

Mesalamine (250 and 500 mg), Capsule (Extended Release), Refer to USP, 06/10/2009, Mesalamine extended-release capsules, 250 mg, may be considered for a waiver of in vivo bioequivalence testing based on (1) acceptable bioequivalence studies on the 500 mg strength, (2) acceptable dissolution testing of the 250 and 500 mg strengths, and (3) proportional similarity in the formulations of the 250 and 500 mg strengths.

Mesalamine (375 mg), Capsule (Extended Release), I (Basket), 100, Acid Stage: 0.1 N HCl Buffer stage: Phosphate Buffer, pH 6.8, Acid stage: 750 mL; Buffer stage: 1000 mL, Acid stage: 2 h; Buffer stage: 0.5, 1, 2, 4, 7, and 9 h, 06/10/2009.

Mesalamine (400 and 800 mg), Tablet (Delayed Release), Refer to USP, 11/05/2010.

Mesalamine Enema, Rectal Enema, II (Paddle), 50, Phosphate Buffer, pH 7.2, 900, 5, 10, 15, and 30, 06/18/2007.

Mesna, Tablet, II (Paddle), 50, 0.06 N HCl, 500, 5, 10, 15, 20, and 30, 02/09/2004.

Mestranol/Norethindrone, Tablet, Refer to USP, 03/25/2010.

Metaxalone, Tablet, II (Paddle), 100, 0.5% SLS in water, 900, 30, 60, 90, and 120, 02/06/2004, 400 mg based on (1) acceptable bioequivalence studies on the 800 mg strength, (2) proportional similarity of the formulations across all strengths, and (3) acceptable in vitro dissolution testing of all strengths.

Metformin HCl, Tablet, Refer to USP, 04/10/2008, 500 and 850 mg based on (1) acceptable bioequivalence studies on the 1000 mg strength, (2) acceptable dissolution testing across all strengths, and (3) proportional similarity in the formulations across all strengths.

Metformin HCl, Tablet (Extended Release), Refer to USP, 12/12/2008, 500 mg based on (1) acceptable bioequivalence studies on the 750 mg strength, (2) proportional similarity of the formulations across all strengths, and (3) acceptable in vitro dissolution testing of all strengths.

Metformin HCl/Pioglitazone HCl, Tablet, II (Paddle), 50, pH 2.5 McIlvaine buffer (0.1 M Citric acid adjusted to pH 2.5 with 0.2 M Na_2HPO_4), 900, 10, 20, 30, and 45, 01/03/2007, 1000 mg metformin hydrochloride/15 mg pioglitazone (base equiv) based on (1) acceptable bioequivalence studies on the 1000 mg metformin hydrochloride/30 mg pioglitazone (base equiv), (2) proportional similarity of all strengths, and (3) acceptable in vitro dissolution testing of all strengths.

(*Continued*)

Metformin HCl/Saxagliptin, Tablet (Extended Release), I (Basket), 100, Phosphate Buffer, pH 6.8, 1000, Metformin: 1, 2, 3, 4, 6, 8, 10, and 12 h. Saxagliptin: 5, 10, 15, 20, and 30 min, 01/26/2012, 1000 mg Metformin Hydrochloride; 2.5 mg Saxagliptin and 500 mg Metformin; 5 mg Saxagliptin based on acceptable (1) bioequivalence studies on the 1000 mg/5 mg strength, (2) proportional similarity of the formulations, and (3) acceptable in vitro dissolution testing across strengths.

Metformin HCl/Sitagliptin Phosphate, Tablet, II (Paddle), 75, 0.025 M NaCl, 900, 10, 15, 20, and 30, 10/06/2008, 50 mg/500 mg based on (1) acceptable bioequivalence studies on the 50 mg/1000 mg strength, (2) acceptable dissolution testing across all strengths, and (3) proportional similarity in the formulations across all strengths.

Metformin/Repaglinide, Tablet, II (Paddle), 50, Citric acid/phosphate buffer, pH 5.0, 900, 5, 10, 15, 20, and 30, 10/30/2009.

Methadone HCl, Tablet, Refer to USP, 07/14/2008, 5, 10, and 40 mg strengths based on acceptable in vitro dissolution testing under 21 CFR § 320.22(c).

Methenamine Hippurate, Tablet, Refer to USP, 07/31/2013.

Methimazole, Tablet, Refer to USP, 01/14/2008, 5 mg based on (1) acceptable bioequivalence studies on the 10 mg strength, (2) acceptable in vitro dissolution testing of all strengths, and (3) proportional similarity of formulations across all strengths.

Methocarbamol, Tablet, Refer to USP, 08/15/2013.

Methotrexate Sodium, Tablet, Refer to USP, 04/02/2009, 5, 7.5, and 10 mg based on (1) acceptable bioequivalence studies on the 15 mg strength, (2) proportional similarity of the formulations across all strengths, and (3) acceptable in vitro dissolution testing of all strengths.

Methoxsalen, Capsule, Refer to USP, 03/25/2010.

Methylphenidate, Capsule (Extended Release), II (Paddle), 50, Water, 500, 1, 2, 4, 6, 8, 12 h and until at least 80% released, 04/15/2008.

Methylphenidate, Tablet (Extended Release), VII (Reciprocating holder) with oral extended-release tablet holder (spring holder), 30 cycles/min, Water, pH 3.0, 250, 1, 2, 4, 6, 8, and 10 h, 04/15/2008.

Methylphenidate, Transdermal Patch, VI (Cylinder), 50, 0.01 N HCl at 32°C, 900, 0.5, 1.5, 3, 4 h and until at least 80% released, 04/15/2008.

Methylphenidate (BX), Capsule (Extended Release), I (Basket), 75, 0–2 h:0.01 N HCl. 2–10 h: Phosphate Buffer, pH 6.8, 0–2 h: 500. 2–10 h:500, 0.5, 1, 3, 6, 8, and 10 h, 07/25/2007.

Methylphenidate HCl, Tablet, Refer to USP, 07/14/2008, 5 and 10 mg strengths based on (1) acceptable bioequivalence studies on the 20 mg strength, (2) acceptable dissolution testing of all strengths, and (3) proportional similarity in the formulations across all strengths.

Methylphenidate HCl, Tablet (Chewable), I (Basket), 100, Water, 900, 15, 30, 45, and 60, 03/25/2010, 5 and 2.5 mg tablets, based on (1) acceptable bioequivalence studies on the 10 mg strength tablet, (2) proportional similarity of the formulations across the strengths, and (3) acceptable in vitro dissolution testing of all strengths.

Methylprednisolone, Tablet, Refer to USP, 01/29/2010, 4, 8, and 16 mg based on (1) acceptable bioequivalence studies on the 32 mg strength, (2) proportional similarity of the formulations across all strengths, and (3) acceptable in vitro dissolution testing of all strengths.

Methylprednisolone Acetate, Injectable Suspension, IV (Flow-Through Cell-Open system), 0.55% SDS, 15, 30, 45, 60, 90, and 120, 10/08/2009, 20 and 40 mg/mL (multidose vials) and 40 mg/mL (single-dose vial) based on (1) acceptable bioequivalence studies on the respective 80 mg/mL strength, (2) acceptable dissolution testing across all strengths, and (3) proportional similarity in the formulations across all strengths.

Methyltestosterone, Capsule, Refer to USP, 07/31/2013.

Methyltestosterone, Tablet, II (Paddle), 50, Water, 900, 10, 20, 30, 45, 60, and 75, 07/31/2013.

Metoclopramide HCl, Tablet, Refer to USP, 07/15/2009.

Metoclopramide HCl, Tablet (Orally Disintegrating), I (Basket), 50, Water, 900, 5, 10, 15, 20, 30, and 45, 04/08/2010, 5 mg based on (1) acceptable bioequivalence studies on the 10 mg strength, (2) acceptable in vitro dissolution testing of all strengths, and (3) proportional similarity of formulations across all strengths.

Metolazone, Tablet, II (Paddle), 75, 2% SLS in 0.05 M Sodium Phosphate Buffer, pH 7.5, 900, 30, 60, 90, 120, and 150, 02/10/2004, 2.5 mg strength based on (1) acceptable bioequivalence study on the 5 and 10 mg strengths, (2) acceptable in vitro dissolution testing of all strengths, and (3) proportional similarity of the formulations across all strengths.

Metoprolol Succinate, Tablet (Extended Release), Refer to USP, 07/25/2007, 25, 100 mg tablets, based on (1) acceptable bioequivalence studies on the 50 and 200 mg strengths, (2) proportionally similar across all strengths, and (3) acceptable in vitro dissolution testing of all strengths.

Metoprolol Tartrate, Tablet, Refer to USP, 07/25/2007.

(*Continued*)

Metronidazole, Tablet, Refer to USP, 08/05/2010, 250 mg based on (1) acceptable bioequivalence studies on the 500 mg strength, (2) acceptable in vitro dissolution testing of all strengths, and (3) proportional similarity of the formulations across all strengths.

Metronidazole, Capsule, I (Basket), 100, 0.1 N HCl, 900, 10, 20, 30, and 45, 02/09/2004.

Miconazole, Tablet (Buccal), I (Basket), 60, 0.5% SDS (Sodium dodecyl sulfate) in water pH adjusted to 6.5 ± 0.5, 1000, 1, 2, 4, 6, 8, 10, and 12 h, 10/28/2010.

Miconazole Nitrate, Suppository (Vaginal), I (Basket), 100, 0.45% SLS in water, 900, 15, 30, 45, and 60, 10/08/2009.

Midodrine HCl, Tablet, II (Paddle), 50, 0.1 N HCl, 900, 5, 10, 15, and 30, 02/06/2004, 2.5 and 10 mg based on (1) acceptable bioequivalence studies on the 5 mg strength, (2) acceptable dissolution testing across all strengths, and (3) proportional similarity in the formulations across all strengths.

Mifepristone, Tablet, II (Paddle), 75, 0.01 N HCl, 900, 5, 10, 15, 20, and 30, 01/14/2008.

Miglitol, Tablet, II (Paddle), 75, Water, 900, 10, 20, 30, and 45, 03/03/2011.

Miglustat, Capsule, I (Basket), 100, 0.1 N HCl, 1000, 10, 20, 30, and 45, 01/03/2007.

Milnacipran HCl, Tablet, II (Paddle), 50, 0.1 N HCl, 900, 10, 15, 30, 45, and 60, 08/05/2010, 12.5, 25, and 100 mg based on (1) acceptable bioequivalence studies on the 50 mg strength, (2) proportional similarity of the 12.5, 25, 50 and 100 mg strengths, and (3) acceptable in vitro dissolution testing of the 12.5, 25, 50, and 100 mg strengths.

Minocycline HCl, Tablet, Refer to USP, 07/25/2007, 50 and 75 mg based on (1) acceptable bioequivalence study on the 100 mg strength, (2) acceptable in vitro dissolution testing of all strengths, and (3) proportional similarity in the formulations across all strengths.

Minocycline HCl, ER Tablets, I (Basket), 100, 0.1 N HCl, 900, 1, 2, 4, 6 h and until 80% of drug released, 01/14/2008, 45, 55, 65, 80, 90, 105, and 115 mg based on (1) acceptable bioequivalence study on the 135 mg strength, (2) acceptable in vitro dissolution testing of all strengths, and (3) proportional similarity in the formulations across all strengths. Please refer to Mirtazapine Tablet Guidance for additional information regarding waiver of in vivo testing.

Minocycline HCl, Capsule, Refer to USP, 04/15/2008, 50 and 75 mg based on (1) acceptable bioequivalence studies on the 100 mg strength, (2) acceptable dissolution testing across all strengths, and (3) proportional similarity in the formulations across all strengths.

Minoxidil, Tablet, Refer to USP, 04/15/2008, 2.5 mg based on (1) acceptable bioequivalence studies on the 10 mg strength, (2) acceptable dissolution testing across all strengths, and (3) proportional similarity in the formulations across all strengths.

Mirabegron, Tablet (Extended Release), I (Basket), 100, Phosphate Buffer, pH 6.8, 900, 1, 3, 5, 7, 8.5, 10, and 12 h, 05/09/2013.

Mirtazapine, Tablet, II (Paddle), 50, 0.1 N HCl, 900, 5, 10, 15, and 30, 02/10/2004, 30 and 45 mg based on (1) acceptable bioequivalence studies on the 15 mg strength, (2) acceptable in vitro dissolution testing of all strengths, and (3) proportional similarity of the formulations across all strengths. Please see the succeeding text for additional information regarding waivers of in vivo testing.

Mirtazapine, Tablet (Orally Disintegrating [ODT]), II (Paddle), 50, 0.1 N HCl, 900, 5, 10, 15, 20, and 30, 03/04/2006, 30 and 45 mg based on (1) acceptable bioequivalence studies on the 15 mg strength, (2) proportional similarity of the formulations across all strengths, and (3) acceptable in vitro dissolution testing of all strengths.

Misoprostol, Tablet, II (Paddle), 50, Water (deaerated), 500, 5, 10, 20, and 30, 02/10/2004, 0.1 mg based on (1) acceptable bioequivalence studies on the 0.2 mg strength, (2) acceptable in vitro dissolution testing for all strengths, and (3) proportional similarity of the formulations across all strengths.

Mitotane, Tablet, Refer to USP, 06/10/2009.

Modafinil, Tablet, II (Paddle), 50, 0.1 N HCl, 900, 10, 20, 30, 45, and 60, 02/10/2004, 100 mg based on (1) acceptable bioequivalence studies on the 200 mg strength, (2) proportional similarity of the formulations across all strengths, and (3) acceptable in vitro dissolution testing of all strengths.

Moexipril HCl, Tablet, II (Paddle), 50, Water (deaerated), 900, 5, 10, 15, and 30, 02/10/2004.

Molindone HCl, Tablet, Refer to USP, 07/25/2007, 5 and 10 mg based on (1) an acceptable bioequivalence study on the 25 mg strength, (2) proportional similarity in the formulations of the 5, 10, and 25 mg strengths, and (3) acceptable in vitro dissolution testing of these strengths.

Montelukast, Granule, I (Basket) (100 mesh), 50, 0.5% w/v SDS in Water, 900, 5, 15, 20, and 30, 09/24/2008.

Montelukast Sodium, Tablet, II (Paddle), 50, 0.5% SDS in water, 900, 5, 10, 20, and 30, 04/09/2007.

Montelukast Sodium, Tablet (Chewable), II (Paddle), 50, 0.5% SDS in water, 900, 5, 10, 20, and 30, 03/04/2006, 4 mg based on (1) acceptable bioequivalence studies on the 5 mg strength, (2) proportional similarity of the formulations across all strengths, and (3) acceptable in vitro dissolution testing of all strengths.

(Continued)

Morphine Sulfate, Capsule (Extended Release), Refer to USP, 08/11/2008, 20, 30, 50, 60, and 80 mg strengths based on (1) acceptable bioequivalence studies on the 100 mg strength, (2) acceptable in vitro dissolution testing for the 20, 30, 50, 60, 80, and 100 mg strengths, and (3) proportional similarity of formulations across 20, 30, 50, 60, 80, and 100 mg strengths.

Morphine Sulfate, Tablet, II (Paddle), 50, Deionized Water, 900, 5, 15, 20, and 30, 01/15/2010, 15 mg strength based on (1) acceptable bioequivalence studies on the 30 mg strength, (2) acceptable in vitro dissolution testing for all strengths, and (3) proportional similarity in the formulations across all strengths.

Morphine Sulfate (AB), Tablet (Extended Release), I (Basket), 50, Water (deaerated), 900, 1, 2, 3, 6, 9, and 12 h, 12/23/2010, 15, 30, 60, and 200 mg based on (1) acceptable bioequivalence studies on the 100 mg strength, (2) acceptable in vitro dissolution testing of all strengths, and (3) proportional similarity of the formulations across all strengths. Please refer to the Mirtazapine Tablet Guidance for additional information regarding waivers of in vivo testing.

Morphine Sulfate (BC), Tablet (Extended Release), I (Basket), 100, Water, 500, 1, 2, 4, 6, 8, 10, and 12 h, 12/23/2010.

Morphine Sulfate/Naltrexone HCl, Capsule (Extended Release), II (Paddle), 50, Acid stage: 0.1 N HCl; Buffer stage: 0.05 Phosphate Buffer, pH 7.5, Acid stage: 500; Buffer stage: 500, Morphine Sulfate: Acid stage: 1 h; Buffer stage: 1, 3, 5, 8, and 10 h. Naltrexone HCl: Acid stage: 1 h; Buffer stage: 1, 12, 24, 48, 73, and 96 h, 01/26/2012, 20 mg/0.8 mg, 30 mg/1.2 mg, 50 mg/2 mg, 80 mg/3.2 mg and 100 mg/4 mg based on (1) acceptable bioequivalence studies on the 60 mg/2.4 mg strength, (2) acceptable in vitro dissolution testing of all strengths, and (3) proportional similarity in the formulations across all strengths. Please refer to the Mirtazapine Tablet Guidance for additional information regarding waivers of in vivo testing.

Moxifloxacin, Tablet, II (Paddle), 50, 0.1 N HCl, 900, 15, 30, 45, and 60, 06/18/2007.

Mycophenolate Mofetil, Suspension, II (Paddle), 40, 0.1 N HCl, 900, 5, 10, 20, and 30, 02/10/2004.

Mycophenolate Mofetil, Capsule, II (Paddle), 40, 0.1 N HCl, 900, 5, 10, 20, and 30, 02/10/2004.

Mycophenolate Mofetil, Tablet, II (Paddle), 50, 0.1 N HCl, 900, 5, 10, 15, and 30, 02/10/2004.

Mycophenolic acid, Tablet (Delayed Release), II (Paddle), 50, Acid Stage: 0.1 N HCl; Buffer Stage: Buffer Solution, pH 6.8 (After initial 120 min, 250 mL of 0.2 M sodium phosphate solution is added to acid stage medium. The pH of the mixture is adjusted to 6.8 using 0.2 M sodium phosphate, 2 N sodium hydroxide, or concentrated HCl acid solution if necessary.), 750 (Acid), 1000 (Buffer), 120 (Acid), 10, 20, 30, 45, and 60 (Buffer), 12/19/2008, 180 mg based on (1) acceptable bioequivalence study on the 360 mg strength, (2) acceptable in vitro dissolution testing of all strengths, and (3) proportional similarity of the formulations.

Nabilone, Capsule, II (Paddle), 50, 0.1% Tween 80 solution, 1000, 15, 30, 45, and 60, 07/14/2008.

Nabumetone, Tablet, Refer to USP, 07/25/2007, 500 mg based on (1) acceptable bioequivalence studies on the 750 mg strength, (2) proportional similarity of the formulations across all strengths, and (3) acceptable in vitro dissolution testing of all strengths.

Nadolol, Tablet, Refer to USP, 04/02/2009, 20 and 40 mg based on (1) acceptable bioequivalence studies on the 80 mg strength, (2) acceptable dissolution testing across all strengths, and (3) proportional similarity in the formulations across all strengths.

Naltrexone, Injectable Suspension, Phosphate buffered saline with 0.02% Tween 20 and 0.02% Sodium azide, pH 7.4 (final osmolality should be 270 ± 20 mOsm), or any other appropriate medium, at 37°C. Develop an in vitro release method using USP IV (Flow-Through Cell), and, if applicable, Apparatus II (Paddle) or any other appropriate method, for comparative evaluation by the agency, 09/01/2011.

Naltrexone HCl, Tablet, Refer to USP, 04/15/2008, 25 and 100 mg based on (1) acceptable bioequivalence studies on the 50 mg strength, (2) proportional similarity of the formulations across all strengths, and (3) acceptable in vitro dissolution testing of all strengths.

Naproxen, Tablet (Delayed Release), Refer to USP, 12/15/2009, 375 mg based on (1) acceptable bioequivalence studies on the 500 mg strength, (2) proportional similarity in the formulations of all strengths, and (3) acceptable in vitro dissolution testing of all strengths.

Naproxen, Tablet, Refer to USP, 07/25/2007.

Naproxen Sodium, Tablet, Refer to USP, 10/04/2012, 250 mg based on (1) acceptable bioequivalence studies on the 500 mg strength, (2) acceptable dissolution testing across all strengths, and (3) proportional similarity in the formulations across all strengths.

Naproxen Sodium, Tablet (Extended Release), II (Paddle), 50, Phosphate Buffer, pH 7.5, 900, 0.5, 1, 2, 3, 4, 6, 8, 10, 12, and 14 h, 04/08/2010, 375 and 500 mg based on (1) acceptable bioequivalence studies on the 750 mg strength, (2) proportional similarity in the formulations of all strengths, and (3) acceptable in vitro dissolution testing of all strengths.

Naproxen Sodium/Sumatriptan Succinate, Tablet, I (Basket), 75, Phosphate Buffer, pH 6.8, 900, 10, 15, 20, 30, and 45, 07/01/2010.

(Continued)

Naproxen/Esomeprazole Magnesium, Tablet (Delayed Release), II (Paddle) with sinkers, Naproxen::50 rpm Esomeprazole::75 rpm, Naproxen:: Acid Stage: 0.1 M HCl; Buffer Stage: 0.05 M Phosphate buffer, pH 6.8. Sampling for Acid stage: Transfer the undissolved tablet and sinker to the vessel containing the buffer stage medium. Add, 10 mL of 10 M NaOH to each vessel of the remaining acid stage medium. Continue rotation at 100 rpm for 30 min, withdraw aliquot, and analyze. Esomeprazole (second set of tablets)(without preexposure to acid stage):: 0.05 M Phosphate buffer, pH 7.4, Naproxen::Acid Stage: 1000; Buffer Stage: 1000; Esomeprazole::900, Naproxen:: Acid stage: 120; Buffer stage: 10, 20, 30, 45, 60, 75, and 90; Esomeprazole::10, 20, 30, 45, 60, 75, and 90, 06/06/2013, Eq 20 mg base/375 mg based on (1) acceptable bioequivalence studies on the Eq 20 mg base/500 mg strength, (2) acceptable in vitro dissolution testing of all strengths, and (3) proportional similarity in the formulations across all strengths. Please refer to the Mirtazapine Tablet Guidance for additional information regarding waivers of in vivo testing.

Naratriptan HCl, Tablet, Refer to USP, 07/25/2007, 1 mg based on (1) acceptable bioequivalence studies on the 2.5 mg strength, (2) acceptable dissolution testing across all strengths, and (3) proportional similarity in the formulations across all strengths.

Nateglinide, Tablet, II (Paddle), 50, 0.01 N HCl with 0.5% (w/v) SLS, 1000, 10, 20, 30, and 45, 01/03/2007, 60 mg, based on (1) acceptable bioequivalence studies on the 120 mg strength, (2) proportionally similar across all strengths, and (3) acceptable in vitro dissolution testing of all strengths.

Nebivolol HCl, Tablet, II (Paddle), 50, 0.01 N HCl, 900, 10, 20, 30, and 45, 01/15/2010, 2.5, 5, and 10 mg based on (1) acceptable bioequivalence studies on the 20 mg strength, (2) proportional similarity of the 2.5, 5, 10, and 20 mg strengths, and (3) acceptable in vitro dissolution testing of the 2.5, 5, 10, and 20 mg strengths.

Nefazodone HCl, Tablet, II (Paddle), 50, 0.1 N HCl, 900, 10, 20, 30, 45, and 60, 01/03/2007.

Nelfinavir Mesylate, Tablet, II (Paddle), 50, 0.1 N HCl, 900, 5, 10, 15, 20, 30, 45, 60, and 90, 01/03/2007.

Nelfinavir Mesylate, Powder for suspension, II (Paddle), 50, 0.1 N HCl, 900, 5, 10, 15, 20, 30, and 45, 09/13/2007.

Neomycin Sulfate, Tablet, II (Paddle), 50, 0.05 M Phosphate Buffer, pH 6.8, 900, 15, 30, 45, and 60, 01/14/2008.

Nevirapine, Tablet (Extended Release), I (Basket), 75, 0.04 M Sodium phosphate buffer pH 6.8 containing 2% sodium lauryl sulfate, 900, 1, 2, 3, 4, 5, 6, 8, 10, 12, 16, and 20 h, 01/31/2013.

Nevirapine, Suspension, II (Paddle), 25, 0.1 N HCl, 900, 10, 20, 30, 45, and 60, 02/11/2004.

Nevirapine, Tablet, Refer to USP, 09/13/2007.

Niacin, Tablet (Extended Release), I (Basket), 100, Water, 900, 1, 3, 6, 9, 12, 15, 20, and 24 h, 06/10/2009, 500 mg based on (1) acceptable bioequivalence studies on the 1000 mg strength, (2) proportionally similarity of the formulations across all strengths, and (3) acceptable in vitro dissolution testing of all strengths.

Niacin/Simvastatin, Tablet (Extended Release), Niacin: I (40 mesh rotating Basket); Simvastatin: I (10 mesh rotating Basket), 100, Niacin: Water; Simvastatin: 0.5% SDS in 0.01 M Sodium Phosphate, pH 7.0, 900, Niacin: 1, 3, 6, 9, 12, 15, 18, 20, and 24 h; Simvastatin: 10, 20, 30, 45, and 60, 01/15/2010, 500 mg/20 mg and 500 mg/40 mg based on (1) acceptable bioequivalence studies on the 1000 mg/40 mg strength, (2) acceptable in vitro dissolution testing of all strengths, and (3) proportional similarity of the formulations across all strengths.

Nicardipine HCl, Capsule (Extended Release), II (Paddle), 50, 0.1 N HCl, 1000, 0.5, 2, and 6 h, 07/14/2008, 45 and 30 mg based on (1) acceptable bioequivalence studies on the 60 mg strength, (2) acceptable dissolution testing across all strengths, and (3) proportional similarity in the formulations across all strengths.

Nicardipine HCl, Capsule, II (Paddle), 50, 0.033 M Citric Acid Buffer, pH 4.5, 900, 10, 20, 30, and 45, 02/11/2004.

Nicotine, Film, Transdermal (Extended Release), Refer to USP, 01/31/2013.

Nicotine Polacrilex, Lozenge, I (Basket), 100, Phosphate Buffer, pH 7.4, 900, 0.5, 1, 2, 3, 6, and 8 h, 12/23/2010, Eq. 2 mg based on (1) acceptable bioequivalence study on the Eq. 4 mg base strength, (2) acceptable in vitro dissolution testing on all strengths, and (3) proportional similarity of the formulation across all strengths. Please refer to the Mirtazapine Tablet Guidance for additional information regarding waivers of in vivo testing.

Lozenges with alternate flavors cannot be filed in the same ANDA as the mint flavored lozenge. For each flavor, a separate submission (ANDA) should be submitted.

Lozenges with an alternate flavor may be eligible for a waiver of the bioequivalence study requirements based on (1) an acceptable bioequivalence study on the 4 mg strength of the mint lozenge, (2) acceptable dissolution testing for the nicotine polacrilex lozenge with additional flavor versus the RLD, (3) proportional similarity in the formulations of the nicotine polacrilex lozenge with additional flavor and nicotine polacrilex lozenge with mint flavor, and (4) the additional flavor (the inactives) has been approved for the same route of administration.

Nicotine Polacrilex, Lozenge (Mini), I (Basket), 100, Phosphate Buffer, pH 7.4, 900, 30, 60, 90, 120, and 180, 12/23/2010.

(*Continued*)

Nicotine Polacrilex, Chewing Gum, Chewing Machine as described in European Pharmacopoeia (2.9.25), 60 cycles (chews) per minute, Phosphate Buffer, pH 7.4 or any other appropriate buffer and conditions, 20, 5, 10, 20, and 30, 01/05/2012.

Nifedipine, Capsule, Refer to USP, 03/03/2011, 20 mg based on (1) an acceptable bioequivalence studies on the 10 mg strength, (2) acceptable dissolution testing across all strengths, and (3) proportional similarity in the formulations across all strengths. Please refer to the Mirtazapine Tablet Guidance for additional information regarding waivers of in vivo testing.

Nifedipine, Tablet (Extended Release), Refer to USP, 07/25/2007, 30 and 60 mg based on (1) acceptable bioequivalence studies on the 90 mg strength, (2) acceptable dissolution testing across all strengths, and (3) proportional similarity in the formulations across all strengths.

Nilotinib Hydrochloride Monohydrate, Capsule, I (Basket), 100, 0.1 N HCl, 1000, 10, 15, 30, and 45, 10/30/2009.

Nilutamide, Tablet, Develop a dissolution method, 05/20/2009.

Nimodipine, Capsule, II (Paddle), 50, 0.5% SDS in water, 900, 10, 20, 30, and 45, 04/09/2007.

Nisoldipine, Tablet (Extended Release), II (Paddle) with option to use a sinker, 50, HCl with SLS (32.5 ± 0.1 g Sodium Lauryl Sulfate in 6489 mL of purified water containing 17.0 mL HCl, pH adjusted to 1.20 ± 0.05 with HCl), 900, 1, 4, 8, 12, 15, 18, and 24 h, 04/02/2009, 25.5 mg based on (1) acceptable bioequivalence studies on the 34 mg strength, (2) acceptable dissolution testing across 25.5 and 34 mg strengths, and (3) proportional similarity in the formulations across 25.5 and 34 mg strengths.

Nitazoxanide, Oral Suspension, II (Paddle), 100, Phosphate buffer at pH 7.5 with 6% hexadecyltrimethylammonium bromide, bath temperature at 25°C, 900, 10, 20, 30, 45, and 60, 10/21/2010.

Nitazoxanide, Tablet, II (Paddle), 75, Phosphate buffer at pH 7.5 with 6% hexadecyltrimethylammonium bromide, bath temperature at 25°C, 900, 10, 20, 30, 45, 60, 01/03/2007.

Nitrofurantoin, Capsule, Refer to USP, 04/02/2009.

Nitrofurantoin, Suspension, II (Paddle), 50, Phosphate Buffer, pH 7.2, 900, 15, 30, 60, 120, and 180, 04/02/2009.

Nitroglycerin, Film, Transdermal (Extended Release), Modified USP Type V (Paddle over disk), 100, Deionized Water at 32°C, 900, 10, 20, 30, 45, 60, 90, 120, and 180, 04/08/2010, 0.1, 0.2, 0.3, 0.6, and 0.8 mg/h based on (1) an acceptable bioequivalence study on the 0.4 mg/h strength, (2) proportion similarity across all strengths, and (3) acceptable in vitro dissolution testing of all strengths.

Nitroglycerin, Tablet (Sublingual), II (Paddle), 50, Phosphate Buffer, pH 6.5, 500, 1, 3, 5, 8, and 10, 01/15/2010, 0.3 and 0.4 mg based on (1) acceptable bioequivalence study on the 0.6 mg strength, (2) acceptable in vitro dissolution testing of all strengths, and (3) proportional similarity in the formulations across all strengths. Please refer to the Mirtazapine Tablet Guidance for additional information regarding waivers of in vivo testing.

Nizatidine, Capsule, Refer to USP, 01/14/2008.

Norethindrone, Tablet, Refer to USP, 10/08/2009.

Norethindrone Acetate, Tablet, Refer to USP, 08/27/2009.

Nystatin, Tablet, II (Paddle), 75, Water with 0.1% SLS, 900, 15, 30, 45, 60, and 90, 01/03/2007.

Nystatin, Oral Suspension, II (Paddle), 25, 50, and 75, 0.1% and 0.2% SLS in water, 900, 5, 10, 20, 30, 45, and 60, 10/28/2010, 100,000 units/mL pursuant to 21 CFR 320.22(c) provided that the in vitro dissolution profiles of the proposed product are comparable to those of the reference product.

Octreotide Injection, Injectable (Extended Release), Develop a dissolution method using USP IV (Flow-Through Cell), and, if applicable, Apparatus II (Paddle) or any other appropriate method, for comparative evaluation by the agency, 12/23/2010.

Ofloxacin, Tablet, I (Basket), 100, 0.1 N HCl, 900, 10, 20, 30, and 45, 02/12/2004.

Olanzapine, Tablet, II (Paddle), 50, 0.1 N HCl, 900, 5, 10, 20, and 30, 02/12/2004, 2.5, 7.5, 10, 15, and 20 mg based on (1) acceptable bioequivalence studies on the 5 mg strength, (2) acceptable dissolution testing across all strengths, and (3) proportional similarity in the formulations across all strengths.

Olanzapine (Orally disintegrating), Tablet (Orally Disintegrating), II (Paddle), 50, 0.1 N HCl, 900, 5, 10, 15, and 30, 02/12/2004, 10, 15, and 20 mg based on (1) acceptable bioequivalence studies on the 5 mg strength, (2) proportional similarity of the formulations across all strengths, and (3) acceptable in vitro dissolution testing of all strengths.

Olmesartan, Tablet, II (Paddle), 50, 0.05 M Phosphate Buffer, pH 6.8, 900, 10, 20, 30, and 45, 07/09/2007.

Olsalazine Sodium, Capsule, I (Basket), 100, Phosphate Buffer, pH 7.5, 900, 10, 20, 30, and 45, 02/12/2004.

Omeprazole, Capsule (Delayed Release), Refer to USP, 06/18/2007, 10 mg based on (1) acceptable bioequivalence studies on the 20 mg strength, (2) proportional similarity of the formulations on 10 and 20 mg strengths, and (3) acceptable in vitro dissolution testing of 10 and 20 mg strengths.

(*Continued*)

Omeprazole, OTC Tablet (Delayed Release), II (Paddle), 100, Tablets are preexposed to 750 mL of 0.1 M HCl for 2 h and then 250 mL of 0.2 M Na_3PO_4 is added to the medium to give 1000 mL with pH 6.8, Acid stage: 750; Buffer stage: 1000, Acid stage: 120; Buffer stage: 10, 20, 30, 45, and 60, 02/28/2013.

Omeprazole Magnesium, OTC Tablet (Delayed Release), II (Paddle), 100, Tablets are preexposed to 300 mL of 0.1 M HCl for 2 h and then 700 mL of 0.086 M Na_2HPO_4 is added to the medium containing the capsule to give 1000 mL with pH 6.8, 300 mL for the acid stage; 1000 mL for the buffer stage, Sampling started at the buffer stage 10, 20, 30, 45, and 60, 01/03/2007.

Omeprazole Sodium Bicarbonate, Capsule, II (Paddle), 75, Phosphate Buffer, pH 7.4, 900, 15, 30, 45, and 60, 07/14/2008, 20 mg/1.1 g based on (1) acceptable bioequivalence studies on the 40 mg/1.1 g strength, (2) acceptable dissolution testing across all strengths, and (3) proportional similarity in the formulations across all strengths.

Omeprazole Sodium Bicarbonate, Powder for suspension (Immediate Release), II (Paddle), 50, 0.25 mM Sodium Phosphate Buffer, pH 7.4, 900, 5, 10, 15, and 30, 06/20/2007, 20 mg/packet based on (1) acceptable bioequivalence study on the 40 mg strength, (2) proportional similarity of the formulations across all strengths, and (3) acceptable in vitro dissolution testing of all strengths.

Ondansetron, Tablet (Orally Disintegrating), Refer to USP, 06/18/2007, 4 mg based on (1) acceptable bioequivalence studies on the 8 mg strength, (2) proportional similarity of the formulations across all strengths, and (3) acceptable in vitro dissolution testing of all strengths.

Ondansetron, Film (Oral), V (Paddle over Disk) with a stainless steel disk (120 mesh screens), 50, 0.1 N HCl, 900, 5, 10, 15, 20, and 30, 01/26/2012, 4 mg based on (1) acceptable bioequivalence studies on the 8 mg strength, (2) acceptable in vitro dissolution testing of all strengths, and (3) proportional similarity of the formulations across all strengths.

Ondansetron HCl, Tablet, II (Paddle), 50, Water (deaerated), 500, 5, 10, 15, and 30, 02/12/2004, 4 and 8 mg based on (1) acceptable bioequivalence studies on the 24 mg strength, (2) proportional similarity of the formulations across all strengths, and (3) acceptable in vitro dissolution testing of all strengths.

Orlistat, Capsule, II (Paddle), 75, 3% SLS in 0.5% Sodium Chloride, pH 6.0, 900, 10, 20, 30, 45, and 60, 02/12/2004, 120 mg based on (1) acceptable bioequivalence studies on the 60 mg strength, (2) acceptable in vitro dissolution testing of all strengths, and (3) proportional similarity of the formulations across all strengths.

Orphenadrine Citrate, Tablet (Extended Release), Refer to USP, 08/27/2009.

Oseltamivir Phosphate, Capsule, II (Paddle), 50, 0.1 N HCl, 900, 5, 10, 20, and 30, 01/03/2007, 30 and 45 mg based on (1) acceptable bioequivalence studies on the 75 mg strength, (2) proportional similarity of the formulations across all strengths, and (3) acceptable in vitro dissolution testing of all strengths.

Oseltamivir Phosphate, Oral Suspension, II (Paddle), 25, 0.1 N HCl, 900, 5, 10, 15, 20, and 30, 07/15/2009.

Oxaprozin, Tablet, II (Paddle), 75, 0.05 M Phosphate Buffer, pH 7.4, 1000, 10, 20, 30, 45, and 60, 02/12/2004.

Oxcarbazepine, Suspension, II (Paddle), 75, 1% SDS in water, 900, 10, 20, 30, and 45, 02/12/2004.

Oxcarbazepine, Tablet (Extended Release), II (Paddle) with sinker, 75, 1.0% SDS in Deionized Water (degassed), 900, 1, 2, 4, 6, 8, and 10 h, 08/15/2013.

Oxcarbazepine (150 mg), Tablet, II (Paddle), 60, 0.3% SDS in water, 900, 10, 20, 30, 45, 60, and 90, 02/12/2004.

Oxcarbazepine (300 mg), Tablet, II (Paddle), 60, 0.6% SDS in water, 900, 10, 20, 30, 45, 60, and 90, 02/12/2004.

Oxcarbazepine (600 mg), Tablet, II (Paddle), 60, 1% SDS in water, 900, 10, 20, 30, 45, 60, and 90, 02/12/2004, 150 and 300 mg based on (1) acceptable bioequivalence studies on the 600 mg strength, (2) proportional similarity of the formulations across all strengths, and (3) acceptable in vitro dissolution testing of all strengths.

Oxybutynin, Transdermal, Paddle over Disk (Apparatus 5), 50, Phosphate Buffer, pH 4.5 at 32°C, 900, 1, 4, 24 h, 01/03/2007.

Oxybutynin Chloride, Tablet (Extended Release), Refer to USP, 5 and 10 mg based on (1) acceptable bioequivalence studies on the 15 mg strength, (2) acceptable dissolution testing across all strengths, and (3) proportional similarity in the formulations across all strengths.

Oxycodone HCl, Tablet, Refer to USP, 01/14/2008.

Oxycodone HCl, Tablet (Extended Release), Refer to USP, 07/01/2010, 10, 15, 20, 30, 60 and 80 mg based on (1) acceptable bioequivalence studies on the 40 mg strength, (2) acceptable in vitro dissolution testing of all strengths, and (3) proportional similarity of the formulations across all strengths.

Oxymorphone HCl, Tablets, II (Paddle), 50, 0.1 N HCl, 900, 10, 20, 30, and 45, 01/14/2008, 5 mg, based on acceptable (1) bioequivalence studies on the 10 mg strength, (2) proportional similarity of the formulations, and (3) acceptable in vitro dissolution testing of all strengths.

(Continued)

Oxymorphone HCl, Tablet (Extended Release), II (Paddle), 50, pH 4.5 phosphate buffer, 900, 1, 4, 6, 10, and 14 h, 12/03/2007, 5, 7.5, 10, 15, 20, and 30 mg based on (1) acceptable bioequivalence studies on the 40 mg strength, (2) acceptable dissolution testing across all strengths, and (3) proportional similarity in the formulations across all strengths.

If the 40 mg strength is not considered for manufacture, the highest manufactured strength should be used for the biostudies and approval should be sought on the lower strengths, based on the aforementioned criteria.

Paliperidone, Tablet (Extended Release), II (Paddle), 50, Modified SGF, pH 1.0 [NaCl (0.2% w/w) in 0.0825 N HCl], 500, 1, 2, 4, 6, 8, 12, 14, 18, and 24 h, 08/27/2009, 1.5, 3, and 9 mg based on (1) acceptable bioequivalence studies on the 6 mg strength, (2) acceptable in vitro dissolution testing of all strengths, and (3) proportional similarity in the formulations across all strengths. Please refer to the Mirtazapine Tablet Guidance for additional information regarding waivers of in vivo testing.

Paliperidone Palmitate, Injectable Suspension, II (paddle), 50, 0.001 M HCl containing 0.489% Polysorbate 20 at 25.0°C ± 0.5°C, 900, 1.5, 5, 8, 10, 15, 20, 30, and 45, 09/01/2011, 39 mg/0.25 mL, 78 mg/0.5 mL, 156 mg/mL, and 234 mg/1.5 mL based on (1) acceptable bioequivalence study on the 117 mg/0.75 mL strength, (2) acceptable in vitro dissolution testing of all strengths, and (3) proportional similarity of the formulations across all strengths.

Palonosetron HCl, Capsule, II (Paddle), 75, 0.1 N HCl, 500, 10, 15, 30, 45, and 60, 08/05/2010.

Pantoprazole Sodium, Delayed-Release Granules for Oral Suspension, II (Paddle), 100, Acid Stage: 0.1 N Hydrochloric Acid; Buffer Stage: 0.05 M Tribasic Sodium Phosphate (Add 250 mL of 0.2 mM Tribasic Sodium Phosphate after completion of acid stage); pH 6.8 (Method B), Acid stage: 750; Buffer stage: 1000, Acid stage: 60, 90, and 120; Buffer stage: 10, 20, 30, 45, and 60, 10/30/2009, 20 mg based on (1) acceptable bioequivalence studies on the 40 mg strength, (2) proportional similarity of the formulations across all strengths, and (3) acceptable in vitro dissolution testing of all strengths.

Pantoprazole Sodium, Tablet (Delayed Release), Refer to USP, 07/21/2009.

Paricalcitol, Capsule, I (Basket), 100, 4 mg/mL (0.4%) Lauryldimethylamine N-oxide (LDAO), 500, 20, 30, 45, 60, 06/18/2007, 2 or 1 mcg tablets, based on (1) acceptable bioequivalence studies of the 4 mcg strength, (2) proportionally similar across all strengths, and (3) acceptable in vitro dissolution testing of all strengths.

Paromomycin Sulfate, Capsule, I (Basket), 50, 0.05 M Phosphate Buffer, pH 6.8, 900, 5, 10, 15, 20, 30, and 45, 02/13/2004.

Paroxetine, Tablet (Extended Release), II (Paddle), 150, Step 1: 0.1 N HCl for 2 h; Step 2: Tris Buffer containing 50 mM Tris, pH 7.5, 750 (step 1); 1000 (step 2), 2 h (step 1), 1, 2, 4, and 6 h (step 2), 11/25/2008, 12.5, 25 mg based on (1) acceptable bioequivalence studies on the 37.5 mg strength, (2) proportionally similar across all strengths, and (3) acceptable in vitro dissolution testing of all strengths.

Paroxetine HCl, Tablet, Refer to USP, 01/14/2008, 10, 20, and 30 mg based on (1) acceptable bioequivalence studies on the 40 mg strength, (2) acceptable dissolution testing across all strengths, and (3) proportional similarity in the formulations across all strengths.

Paroxetine HCl, Suspension, II (Paddle), 100, SGF without enzyme, 900, 10, 20, 30, and 45, 02/13/2004.

Pazopanib HCl, Tablet, II (Paddle), 75, 50 mM Sodium Acetate buffer, pH 4.5, containing 0.75% SDS, 900, 10, 15, 30, 45, and 60, 08/05/2010.

Pemoline, Tablet, II (Paddle), 75, Water (deaerated), 900, 10, 20, 30, 45, 60, and 90, 02/13/2004.

Penbutolol Sulfate, Tablet, Refer to USP, 06/24/2010.

Penicillamine, Capsule, Refer to USP, 09/03/2008, 125 mg based on (1) acceptable bioequivalence studies on the 250 mg strength, (2) acceptable dissolution testing across all strengths, and (3) proportional similarity in the formulations across all strengths.

Penicillin V, Tablet, Refer to USP, 06/09/2011.

Penicillin V Potassium, Tablet, Refer to USP, 06/09/2011, Penicillin V Potassium Tablets, Eq 250 mg base, based on (1) acceptable bioequivalence studies on the 500 mg base strength, (2) acceptable in vitro dissolution testing of all strengths, and (3) proportional similarity of the formulations across all strengths. Please refer to the Mirtazapine Tablet Guidance for additional information regarding waivers of in vivo testing.

Pentosan Polysulfate Sodium, Capsule, I (Basket), 50, Water, 900, 5, 15, 30, 45, and 60, 04/15/2008.

Pentoxifylline, Tablet (Extended Release), Refer to USP, 06/09/2011.

Pergolide Mesylate, Tablet, II (Paddle), 50, Simulated gastric fluid TS with cysteine without enzymes, 500, 10, 20, 30, and 45, 03/04/2006.

Perindopril Erbumine, Tablet, II (Paddle), 50, 0.1 N HCl, 900, 10, 20, 30, and 45, 06/20/2007, 2 and 4 mg based on (1) acceptable bioequivalence studies on the 8 mg strength, (2) proportional similarity of the formulations across all strengths, and (3) acceptable in vitro dissolution testing of all strengths.

(*Continued*)

Perphenazine, Tablet, Refer to USP, 12/15/2009, 2, 4, and 8 mg based on (1) acceptable bioequivalence studies on the 16 mg strength, (2) proportional similarity of the formulations across all strengths, and (3) acceptable in vitro dissolution testing on all strengths.

Phendimetrazine Tartrate, Tablet, Refer to USP, 05/20/2009, 35 mg strength of the test product, based on acceptable formulation data and in vitro dissolution testing under 21 CFR § 320.22(c).

Phendimetrazine Tartrate, Capsule, Refer to USP, 06/10/2009.

Phendimetrazine Tartrate, Capsule (Extended Release), II (Paddle), 50, 1 h—SGF w/o Enzymes; after 1 h—SIF w/o Enzymes, 900, 1, 2, 4, 6, and 8 h, 06/10/2009.

Phenelzine Sulfate, Tablet, II (Paddle), 50, Simulated Gastric Fluid without enzymes, pH 1.2, 900, 10, 20, 30, and 45, 03/25/2010.

Phenoxybenzamine HCl, Capsule, Refer to USP, 04/10/2008.

Phentermine HCl, Tablet, II (Paddle), 50, Water, 900, 10, 20, 30, 45, and 60, 08/27/2009.

Phentermine HCl, Capsule, Refer to USP, 01/14/2008, 15, 30, and 37.5 mg strengths based on acceptable formulation data and in vitro dissolution testing under 21 CFR § 320.22(c).

Phentermine HCl, Tablet (Orally Disintegrating), II (Paddle), 50, Water, 500 mL (15 mg) or 900 mL (30 and 37.5 mg), 5, 10, 15, 20, and 30, 07/31/2013.

Phentermine HCl, Tablet, Refer to USP, 07/15/2009, 30 and 37.5 mg pursuant to 21 CFR 320.22(c) provided the in vitro dissolution profiles of your product are comparable to those of the reference product.

Phentermine HCl/Topiramate, Capsule (Extended Release), I (Basket), 100, Water (deionized and degassed), 750, Phentermine: 10, 15, 20, 30, and 45; Topiramate: 0.5, 1, 2, 3, 4, 6, and 8 h, 06/06/2013, Eq 3.75 mg/23 mg, Eq 7.5 mg base/46 mg, and Eq 11.25 mg base/69 mg based on (1) acceptable bioequivalence studies on the Eq 15 mg base/92 mg strength, (2) acceptable in vitro dissolution testing of all strengths, and (3) proportional similarity of the formulations across all strengths. Please refer to the Mirtazapine Tablet Guidance for additional information regarding waivers of in vivo testing.

Phenytoin, Tablet (Chewable), Refer to USP, 01/14/2008.

Phenytoin, Suspension, Refer to USP, 06/18/2007.

Phenytoin Sodium, Capsule, Refer to USP, 06/18/2007, 200 mg based on (1) acceptable bioequivalence studies on the 300 mg strength, (2) proportionally similar across all strengths, and (3) acceptable in vitro dissolution testing of all strengths.

Phytonadione, Tablet, Develop a dissolution method, 03/25/2010.

Pilocarpine HCl, Tablet, II (Paddle), 50, 0.1 N HCl, 500, 10, 20, 30, 45, and 60, 01/20/2004, 5 mg based on (1) acceptable bioequivalence studies on the 7.5 mg strength, (2) proportional similarity of the formulations across all strengths, and (3) acceptable in vitro dissolution testing of all strengths.

Pimozide, Tablet, Refer to USP, 02/19/2008, 1 mg based on (1) acceptable bioequivalence studies on the 2 mg strength, (2) proportional similarity of the formulations across all strengths, and (3) acceptable in vitro dissolution testing of all strengths.

Pioglitazone HCl, Tablet, II (Paddle), 75, HCl–0.3 M KCl Buffer, pH 2.0, 900, 5, 10, 15, and 30, 02/13/2004, Eq 15 mg base and Eq 30 mg base, based on (1) acceptable bioequivalence studies on the Eq 45 mg base strength, (2) acceptable in vitro dissolution testing of all strengths, and (3) proportional similarity of the formulations across all strengths.

Piroxicam, Capsule, Refer to USP, 10/04/2012, 10 mg based on (1) acceptable bioequivalence studies on the 20 mg strength, (2) acceptable in vitro dissolution testing of the 10 and 20 mg, and (3) proportional similarity in the formulations of the 10 and 20 mg strengths.

Pitavastatin Calcium, Tablet, I (Basket), 35, 0.05 M Phosphate Buffer, pH 6.8, 900, 5, 10, 15, 20, 30, and 45, 12/23/2010.

Posaconazole, Oral Suspension, II (Paddle), 25, 0.3% SLS, 900, 10, 20, 30, and 45, 12/03/2007.

Potassium Chloride, Tablet (Extended Release), Refer to USP, 07/25/2007, 10 and 15 mEq based on (1) acceptable bioequivalence study on the 20 mEq strength, (2) acceptable in vitro dissolution testing of all strengths, and (3) proportional similarity of the formulations across all strengths.

Potassium Chloride, Capsule (Extended Release), Refer to USP, 8 mEq based on (1) acceptable bioequivalence studies on the 10 mEq strength, (2) acceptable in vitro dissolution testing of all strengths, and (3) proportional similarity of the formulations across all strengths.

Potassium Citrate, Tablet, Refer to USP, 08/05/2010, 5 and 10 mEq based on (1) acceptable bioequivalence study on the 15 mEq strength, (2) acceptable in vitro dissolution testing of all strengths, and (3) proportional similarity of the formulations across all strengths.

(Continued)

Pramipexole Dihydrochloride, Tablet (Extended Release), I (Basket), 100, 0.05 M phosphate buffer, pH 6.8, 500, 1, 2, 4, 6, 9, 12, 16, 20, and 24 h, 09/02/2010, 0.75, 1.5, 3, and 4.5 mg based on (1) acceptable bioequivalence studies on the 0.375 mg strength, (2) acceptable in vitro dissolution testing of all strengths, and (3) proportional similarity of the formulations across all strengths.

Pramipexole Dihydrochloride, Tablet, II (Paddle), 50, 0.023 M Citrate/0.155 M Phosphate Buffer, pH 6.8, 500, 5, 10, 15, 30, and 45, 10/09/2007, 0.125, 0.5, 0.75, 1 and 1.5 mg based on (1) acceptable bioequivalence studies on the 0.25 mg strength, (2) acceptable dissolution testing across all strengths, and (3) proportional similarity in the formulations across all strengths.

Prasugrel HCl, Tablet, II (Paddle), 75, Citrate–Phosphate buffer (0.023 M Citric acid + 0.026 M Sodium Phosphate, Dibasic), pH 4.0, 900, 10, 15, 20, 30, and 45, 10/04/2012, 5 mg based on (1) acceptable bioequivalence studies on the 10 mg strength, (2) acceptable in vitro dissolution testing of both strengths, and (3) proportional similarity of the formulations across both strengths.

Pravastatin Sodium, Tablet, II (Paddle), 50, Water (deaerated), 900, 5, 10, 20, and 30, 02/13/2004, 10, 20, and 40 mg based on (1) acceptable bioequivalence studies on the 80 mg strength, (2) proportional similarity of the formulations across all strengths, and (3) acceptable in vitro dissolution testing of all strengths.

Prednisolone, Tablet, Refer to USP, 11/25/2008.

Prednisolone Sodium Phosphate, Tablet (Orally Disintegrating), II (Paddle), 50, 22 mM Sodium Acetate Buffer, pH 4.5, 500, 5, 15, 30, 45, and 60, 09/03/2008, 10 and 15 mg (base) based on (1) acceptable bioequivalence studies on the 30 mg strength, (2) acceptable dissolution testing across all strengths, and (3) proportional similarity in the formulations across all strengths.

Prednisone, Tablet (Delayed Release), II (Paddle) with sinker, 100, Water, 500, 1, 2, 3, 3.5, 4, 4.5, 5, 5.5, 6, 7, 8, and 10 h, 07/31/2013.

Pregabalin, Capsule, II (paddle), 50, 0.06 N HCl, 900, 10, 20, 30, and 45, 03/22/2006, 25, 50, 75, 100, 150, 200, 225 mg based on (1) acceptable bioequivalence studies on the 300 mg strength, (2) proportional similarity of the formulations across all strengths, and (3) acceptable in vitro dissolution testing of all strengths.

Primidone, Tablet, Refer to USP, 01/14/2008.

Procarbazine HCl, Capsules, II (Paddle), 50, Water, 900, 10, 20, 30, 45, and 60, 01/14/2008.

Prochlorperazine, Rectal Suppository, I (Suppository, dissolution baskets, Palmieri type), 100, 0.1 N HCl at 38°C, 900, 10, 20, 30, and 45, 08/17/2006.

Progesterone, Vaginal Insert, II (Paddle), 50, 0.25% sodium dodecyl sulfate (SDS) in DI water, 900, 5, 10, 15, 20, and 30, 10/04/2012.

Progesterone, Capsule, Develop a quantitative rupture test, 04/08/2010, 100 mg based on (1) acceptable bioequivalence studies on the 200 mg strength, (2) proportional similarity of the formulations across all strengths, and (3) acceptable in vitro dissolution testing of all strengths.

Promethazine HCl, Tablet, Refer to USP, 07/25/2007, 12.5 and 25 mg based on (1) acceptable bioequivalence studies on the 50 mg strength, (2) proportional similarity of the formulations across all strengths, and (3) acceptable in vitro dissolution testing of all strengths.

Propafenone HCl, Tablet, II (Paddle), 75, 0.1 N HCl, 900, 10, 20, 30, and 45, 02/13/2004, 150 and 225 mg based on (1) acceptable bioequivalence studies on the 300 mg strength, (2) acceptable dissolution testing across all strengths, and (3) proportional similarity in the formulations across all strengths.

Propafenone HCl, Capsule (Extended Release), II (Paddle), 50, 0–2 h: 0.08 N HCl 2–15 h: phosphate buffer, pH 6.8, 900, 1, 2, 4, 8, 10, 12, and 15 h, 03/11/2008, 225 and 325 mg based on (1) acceptable bioequivalence studies on the 425 mg strength, (2) proportional similarity of the formulations across all strengths, and (3) acceptable in vitro dissolution testing of all strengths.

Propranolol HCl, Capsule (Extended Release), Refer to USP, 07/25/2007.

Propranolol HCl, Tablet, Refer to USP, 03/03/2011, 10, 20, 40, and 60 mg based on (1) acceptable bioequivalence study on the 80 mg strength, (2) proportional similarity across all strengths, and (3) acceptable in vitro dissolution testing of all strengths. Please refer to the Mirtazapine Tablet Guidance for additional information regarding waivers of in vivo testing.

Propylthiouracil, Tablet, Refer to USP, 06/07/2012.

Protriptyline HCl, Tablet, Refer to USP, 01/14/2008, 5 mg based on (1) acceptable bioequivalence studies on the 10 mg strength, (2) proportional similarity of the 10 and 5 mg strengths, and (3) acceptable in vitro dissolution testing of both strengths. Please refer to the Mirtazapine Tablet Guidance for additional information regarding waivers of in vivo testing.

Pseudoephedrine HCl, Tablet (Extended Release), Refer to USP, 01/14/2008.

Pseudoephedrine HCl/Triprolidine HCl, Tablet, Refer to USP, 01/15/2010.

(Continued)

Pyridostigmine Bromide, Tablet, Refer to USP, 06/10/2009.

Pyridostigmine Bromide, Tablet (Extended Release), II (Paddle), 50, Water, 900, 1, 2, 4, 6, 8, and 12 h, 06/10/2009.

Quetiapine Fumarate, Tablet (Extended Release), I (Basket, with 20 mesh), 200, 0.05 M citric acid and 0.09 N NaOH (pH 4.8) [solution A]. At 5 h, pH adjusted to 6.6 by addition of 100 mL of 0.05 M dibasic sodium phosphate and 0.46 N NaOH [solution B], 900 [solution A]. 1000 [final], 1, 2, 4, 6, 8, 10, 12, 16, 20, and 24 h, 01/31/2013, 50, 150, 300, and 400 mg based on (1) acceptable bioequivalence studies on the 200 mg strength, (2) acceptable dissolution testing of all strengths, and (3) proportionally similar in the formulations across all strengths.

Quetiapine Fumarate, Tablet, II (Paddle), 50, Water (deaerated), 900, 10, 20, 30, and 45, 02/18/2004, 50, 100, 150, 200, and 400 based on (1) acceptable bioequivalence studies on the 25 and 300 mg strength, (2) proportional similarity of the formulations across all strengths, and (3) acceptable in vitro dissolution testing of all strengths.

Quinapril HCl, Tablet, Refer to USP, 07/25/2007, 5, 10, and 20 mg based on (1) acceptable bioequivalence studies on the 40 mg strength, (2) proportionally similar across all strengths, and (3) acceptable in vitro dissolution testing of all strengths.

Quinine Sulfate, Capsule, Refer to USP, 01/14/2008.

Rabeprazole Sodium, Tablet (Delayed Release), II (Paddle), 100, 700 mL 0.1 N HCl (Acid stage), after 2 h add 300 mL of 0.6 M Tris buffer; adjust to pH 8.0 (Buffer stage) with 2 N HCl or 2 N NaOH. Stabilize the samples with the addition of 0.5 N NaOH, Acid stage: 700; Buffer stage: 1000, Acid stage: 120; Buffer stage: 10, 20, 30, and 45, 09/22/2011.

Raloxifene HCl, Tablet, II (Paddle), 50, 0.1% Polysorbate 80 in water, 1000, 10, 20, 30, and 45, 02/18/2004.

Raltegravir Potassium, Tablet, II (Paddle) with option to use a sinker, 100, Water (deaerated), 900, 15, 30, 45, 60, and 120, 04/02/2009.

Ramelteon, Tablet, II (Paddle), 50, Water, 900, 10, 20, 30, and 45, 04/02/2009.

Ramipril, Tablet, II (Paddle), 50, 0.1 N HCl, 500, 5, 10, 15, and 30, 09/03/2008, 1.25, 2.5, and 5 mg based on (1) acceptable bioequivalence studies on the 10 mg strength, (2) acceptable dissolution testing across all strengths, and (3) proportional similarity in the formulations across all strengths.

Ramipril, Capsule, II (Paddle), 50, 0.1 N HCl, 500, 10, 20, 30, and 45, 02/18/2004, 1.25, 2.5, and 5 mg based on (1) acceptable bioequivalence studies on the 10 mg strength, (2) proportional similarity of the formulations across all strengths, and (3) acceptable in vitro dissolution testing of all strengths.

Ranitidine HCl, Capsule, II (Paddle), 50, Water (deaerated), 900, 10, 20, 30, and 45, 02/18/2004.

Ranitidine HCl, Tablet, Refer to USP, 07/25/2007, 75 mg regular and 150 mg cool mint tablets are eligible for a waiver of in vivo bioequivalence testing based on (1) acceptable bioequivalence studies on the 150 mg regular strength, (2) acceptable in vitro dissolution testing for all strengths, and (3) proportional similarity in the formulations across all strengths.

Ranitidine HCl, Tablet (Effervescent), Develop a dissolution method, 04/08/2010.

Ranolazine, Tablet (Extended Release), II (Paddle), 50, 0.1 N HCl, 900, 0.5, 2, 4, 8, 12, 20, and 24 h, 06/03/2008, 500 mg based on (1) acceptable bioequivalence studies on the 1000 mg strength, (2) acceptable in vitro dissolution testing of all strengths, and (3) proportional similarity of the formulations across all strengths.

Rasagiline Mesylate, Tablet, II (Paddle), 50, 0.1 N HCl, 500, 10, 15, 30, and 45, 01/29/2010, 0.5 mg tablet, based on (1) acceptable bioequivalence studies on the 1 mg strength tablet, (2) proportional similarity of the formulations across the strengths, and (3) acceptable in vitro dissolution testing of all strengths.

Repaglinide, Tablet, Refer to USP, 07/25/2007, 0.5 and 1 mg based on (1) acceptable bioequivalence studies on the 2 mg strength, (2) acceptable dissolution testing across all strengths, and (3) proportional similarity in the formulations across all strengths.

Ribavirin, Tablet, II (Paddle), 50, Water (deaerated), 900, 10, 20, 30, and 45, 02/18/2004, 200 and 400 mg strengths based on (1) acceptable bioequivalence studies on the 600 mg strength, (2) proportional similarity of the formulations across all strengths, and (3) acceptable in vitro dissolution testing of all strengths.

Ribavirin, Capsule, I (Basket), 100, Water (deaerated), 900, 10, 20, 30, and 45, 02/18/2004.

Rifabutin, Capsule, Refer to USP, 12/15/2009.

Rifampin, Capsule, Refer to USP, 06/18/2007, 150 mg based on (1) acceptable bioequivalence studies on the 300 mg strength, (2) proportional similarity of the formulations across all strengths, and (3) acceptable in vitro dissolution testing of all strengths.

Rifapentine, Tablet, II (Paddle), 50, 0.8% SLS in Phosphate Buffer, pH 7.0, 900, 10, 20, 30, 45, 60, and 90, 02/25/2004.

Rifaximin (200 mg), Tablet, II (Paddle), 75, 0.1 M sodium phosphate buffer pH 7.4 containing 0.45% Sodium Lauryl Sulfate, 1000, 10, 20, 30, 45, 60, 90, and 120, 07/21/2011.

Rifaximin (550 mg), Tablet, II (Paddle), 75, 0.1 M sodium phosphate buffer pH 7.4 containing 0.8% Sodium Lauryl Sulfate, 1000, 10, 20, 30, 45, 60, 90, and 120, 07/21/2011.

(Continued)

Rilpivirine HCl, Tablet, II (Paddle), 75, 0.5% Polysorbate 20 in 0.01 N HCl (pH = 2.0), 900, 10, 20, 30, 45, and 60, 08/15/2013.

Riluzole, Tablet, II (Paddle), 50, 0.1 N HCl, 900, 10, 20, 30, 45, and 60, 02/18/2004.

Rimantadine HCl, Tablet, II (Paddle), 50, Water, 900, 10, 20, 30, and 45, 01/03/2007.

Risedronate Sodium, Tablet, Refer to USP, 07/01/2010, 5, 30, and 75 mg based on (1) acceptable bioequivalence study on the 35 and 150 mg strength, (2) proportional similarity of the formulations across all strengths, and (3) acceptable in vitro dissolution testing of all strengths.

Risedronate Sodium, Tablet (Delayed Release), II (Paddle), 75, Acid stage: 0.1 N HCl; Buffer stage: Phosphate buffer, pH 6.8, Acid stage: 500; Buffer stage: 500, Acid stage: 120; Buffer Stage: 10, 15, 20, 30, and 45, 01/26/2012.

Risedronate Sodium/Calcium Carbonate, Tablet (Copackaged), For Risedronate Tablets: Refer to USP; For Calcium Carbonate Tablets: Refer to USP, 07/01/2010.

Risperidone, Injectable, Develop a dissolution method using USP IV (Flow-Through Cell), and, if applicable, Apparatus II (Paddle) or any other appropriate method, for comparative evaluation by the agency, 01/15/2010, 12.5, 37.5, and 50 mg/vial based on (1) acceptable bioequivalence study on the 25 mg/vial strength, (2) proportional similarity of the formulations across all strengths, and (3) acceptable in vitro dissolution testing of all strengths.

Risperidone, Tablet, II (Paddle), 50, 0.1 N HCl, 500, 10, 20, 30, 45, and 60, 03/04/2006, 0.25, 0.5, 2, 3, and 4 mg based on (1) acceptable bioequivalence studies on the 1 mg strength, (2) proportional similarity of the formulations across all strengths, and (3) acceptable in vitro dissolution testing of all strengths.

Risperidone, Tablet (Orally Disintegrating), II (Paddle), 50, 0.1 N HCl, 500, 5, 10, 15, 07/23/2004, 0.5, 2, 3, and 4 mg based on (1) acceptable bioequivalence studies on the 1 mg strength, (2) proportional similarity of the formulations for across all strengths, and (3) acceptable in vitro dissolution testing of all strengths.

Ritonavir, Capsule, II (Paddle), 50, 0.1 N HCl with 25 mM Polyoxyethylene 10 Laurylether (POE10LE), 900, 10, 20, 30, and 45, 02/18/2004.

Ritonavir, Tablet, II (Paddle), 75, 60 mM Polyoxyethylene 10 Laurylether (POE10LE), 900, 10, 20, 30, 45, 60, 90, 120, 150, and 180, 10/28/2010.

Rivastigmine, Film, Transdermal, Modified USP Type VI (cylinder), 50, 0.9% NaCl at 32°C, 500, 1, 2, 4, 7, 9, and 12 h, 06/10/2009, 4.6 mg/24 h based on (1) acceptable bioequivalence studies on the 9.5 mg/24 h strength, (2) acceptable in vitro dissolution testing of all strengths, and (3) proportional similarity in the formulations across all strengths.

Rivastigmine Tartrate, Capsule, II (Paddle), 50, Water (deaerated), 500, 10, 20, 30, and 45, 01/03/2007, 3, 4.5, and 6 mg based on (1) acceptable bioequivalence studies on the 1.5 mg strength, (2) proportional similarity of the formulations across all strengths, and (3) acceptable in vitro dissolution testing of all strengths.

Rizatriptan Benzoate, Tablet (Orally Disintegrating), II (Paddle), 50, Water (deaerated), 900, 5, 10, and 15, 02/18/2004, 5 mg based on (1) acceptable bioequivalence studies on the 10 mg strength, (2) acceptable dissolution testing across both strengths, and (3) proportional similarity in the formulations across both strengths. Please refer to the Mirtazapine Tablet Guidance for additional information regarding waivers of in vivo testing.

Rizatriptan Benzoate, Tablet, II (Paddle), 50, Water (deaerated), 900, 5, 10, 15, and 30, 02/18/2004, 5 mg, based on (1) acceptable bioequivalence studies on the 10 mg strength, (2) proportionally similar across all strengths, and (3) acceptable in vitro dissolution testing of all strengths.

Roflumilast, Tablet, II (Paddle), 50, 1.0% SDS (sodium dodecyl sulfate) in Phosphate Buffer, pH 6.8, 1000, 5, 10, 15, 20, 30, and 45, 08/15/2013.

Ropinirole HCl, Tablet, I (Basket), 50, Citrate Buffer, pH 4.0, 500, 5, 10, 15, and 30, 01/03/2007.

Ropinirole HCl, Tablet (Extended Release), II (Paddle), 100, pH 4.0 Citrate–THAM Buffer, 500, 1, 2, 4, 6, 8, 12, 16, 20, and 24 h, 08/27/2009, 4, 8, and 12 mg based on (1) acceptable bioequivalence studies on the 2 mg strength, (2) acceptable in vitro dissolution testing across all strengths, and (3) proportional similarity across all strengths.

Rosiglitazone Maleate, Tablet, II (Paddle), 50, 0.01 M Acetate Buffer, pH 4.0, 900, 10, 20, 30, and 45, 02/24/2004, 2 and 4 mg, based on (1) acceptable bioequivalence studies on the 8 mg strength, (2) proportionally similar across all strengths, and (3) acceptable in vitro dissolution testing of all strengths.

Rosuvastatin Calcium, Tablet, II (Paddle), 50, 0.05 M Sodium Citrate Buffer pH 6.6 ± 0.05, 900, 10, 20, 30, and 45, 11/10/2010, 5, 10, and 20 mg strengths based on (1) acceptable bioequivalence studies on the 40 mg strength, (2) proportional similarity across all strengths, and (3) acceptable dissolution testing of all strengths.

Rotigotine, Transdermal, Paddle over Disk (Apparatus 5), 50, Phosphate Buffer, pH 4.5 at 32°C, 900, 15, 30, 60, 90, 120, 150, and 180, 07/15/2009, 1 mg/24 h, 3 mg/24 h, 4 mg/24 h, 6 mg/24 h and 8 mg/24 h, based on (1) acceptable bioequivalence studies on the 2 mg/24 h strength, (2) proportional similarity across all strengths, and (3) acceptable in vitro dissolution testing of all strengths.

(*Continued*)

Rufinamide, Suspension, II (Paddle), 50, 2.0% SDS (sodium dodecyl sulfate) in water, 900, 5, 10, 15, 20, and 30, 08/15/2013.

Rufinamide, Tablet, Refer to USP, 08/15/2013, 200 mg based on (1) acceptable bioequivalence studies on the 400 mg strength, (2) acceptable in vitro dissolution testing of all strengths, and (3) proportional similarity of the formulations across all strengths. Please refer to the Mirtazapine Tablet Guidance for additional information regarding waivers of in vivo testing.

Sapropterin Dihydrochloride, Tablet, II (Paddle), 50, 0.1 N HCl, 900, 5, 10, 15, and 20, 10/06/2008.

Saquinavir Mesylate, Tablet, II (Paddle), 50, Citrate buffer (pH 3.0), 900, 10, 20, 30, and 45, 09/13/2007.

Saquinavir Mesylate, Capsule, Refer to USP, 09/13/2007.

Saxagliptin HCl, Tablet, II (Paddle), 50, 0.1 N HCl, 900, 5, 10, 15, 20, 30, and 45, 08/15/2013, 2.5 mg based on (1) acceptable bioequivalence studies on the 5 mg strength, (2) acceptable in vitro dissolution testing of all strengths, and (3) proportional similarity of the formulations across all strengths. Please refer to the Mirtazapine Tablet Guidance for additional information regarding waivers of in vivo testing.

Scopolamine, Transdermal, Reciprocating disk (Apparatus 7), Stroke depth: 2–3 cm; 30–60 cycles/min, Distilled Water, 25 × 150 mm test tubes containing 20 mL, 1, 2, 4, 6, 12, 18, 24, 36, 48, and 72 h, 07/15/2009.

Selegiline (20 mg/20 cm^2 and 30 mg/30 cm^2), Transdermal, Paddle over Disk (Apparatus 5), 50, 0.1 M Phosphate buffer, monobasic, pH 5 at 32°C, 500, 1, 2, 4, 8, 12, 16, 20, and 24 h, 07/15/2009.

Selegiline (40 mg/40 cm^2), Transdermal, Rotating Cylinder (Apparatus 6), 50, 0.1 M Phosphate buffer, monobasic, pH 5 at 32°C, 1000, 1, 2, 4, 8, 12, 16, 20, and 24 h, 07/15/2009.

Selegiline HCl, Tablet (Orally Disintegrating), I (Basket), 50, Water, 500, 5, 10, 15, and 20, 10/06/2008.

Sertraline HCl, Tablet, II (Paddle), 75, 0.05 M Sodium Acetate Buffer, pH 4.5, 900, 10, 20, 30, and 45, 02/20/2004, 25, 50, 150, and 200 mg based on (1) acceptable bioequivalence studies on the 100 mg strength, (2) proportional similarity of the formulations across all strengths, and (3) acceptable in vitro dissolution testing of all strengths.

Sevelamer Carbonate, Tablet, Disintegration Testing in 0.1 N HCl as per USP <701>, 10/06/2008, 800 mg/packet, based on (1) acceptable equilibrium and kinetic in vitro bioequivalence studies on the 2.4 g/packet strength and (2) proportional similarity in the formulations of the 800 mg/packet and 2.4 g/packet strengths.

Sevelamer HCl, Capsule, Disintegration Testing in 0.1 N HCl as per USP <701>, 04/09/2008.

Sevelamer HCl, Tablet, Disintegration Testing in 0.1 N HCl as per USP <701>, 04/09/2008, 400 mg, based on (1) acceptable equilibrium and kinetic in vitro bioequivalence studies on the 800 mg strength, (2) proportional similarity in the formulations of the 400 and 800 mg strengths, and (3) acceptable disintegration testing in 0.1 N HCl.

Sibutramine HCl, Capsule, II (Paddle), 50, 0.05 M Acetate Buffer, pH 4.0, 500, 10, 20, 30, 45, and 60, 02/25/2004.

Sildenafil Citrate, Tablet, I (Basket), 100, 0.01 N HCl, 900, 5, 10, 15, and 30, 03/04/2006, 20, 25, and 50 mg based on (1) acceptable bioequivalence studies on the 100 mg strength, (2) proportionally similar across all strengths, and (3) acceptable in vitro dissolution testing of all strengths.

Silodosin, Capsule, II (Paddle) with sinker, 50, 0.1 N HCl, 900, 5, 10, 15, 20, and 30, 06/07/2012, 8 mg based on (1) acceptable bioequivalence studies on the 4 mg strength, (2) proportional similarity of the formulations across all strengths, and (3) acceptable in vitro dissolution testing of all strengths.

Simvastatin, Tablet, Refer to USP, 06/18/2007, 5, 10, 20, and 40 mg based on (1) acceptable bioequivalence studies on the 80 mg strength, (2) proportional similarity of the formulations across all strengths, and (3) acceptable in vitro dissolution testing of all strengths.

Simvastatin, Tablet (Orally Disintegrating), II (Paddle), 75, 0.15% SDS Buffer, pH 6.8, 900, 5, 10, 15, and 30, 09/03/2008, 10, 20, and 40 mg based on (1) acceptable bioequivalence studies on the 80 mg strength, (2) acceptable dissolution testing across all strengths, and (3) proportional similarity in the formulations across all strengths.

Sirolimus, Tablet, Basket (20 mesh), 120, 0.4% SLS in water, 500, 10, 20, 30, 45, 60, and 120, 03/14/2007, 0.5 and 1 mg based on (1) acceptable bioequivalence studies on the 2 mg strength, (2) acceptable in vitro dissolution testing of all strengths, and (3) proportional similarity of the formulations across all strengths.

Sitagliptin Phosphate, Tablet, I (Basket), 100, Water, 900, 5, 10, 15, 20, and 30, 07/01/2010, 25 mg (base equiv.) and 50 mg (base equiv.) based on (1) acceptable bioequivalence studies on the 100 mg strength, (2) proportional similarity of the formulations across all strengths, and (3) acceptable in vitro dissolution testing of all strengths.

Sodium Iodide I-123, Capsule, I (Basket), 100, Water (deaerated), 500, 5, 10, 15, and 30, 07/14/2008, 100 and 200 μCi capsules of the test product, based on acceptable in vitro dissolution testing under 21 CFR § 320.22(c).

Sodium Phenylbutyrate, Powder for oral, II (Paddle), 75, Simulated Intestinal Fluid, 900, 15, 30, 45, 60, and 90, 04/02/2009.

(*Continued*)

Sodium Phosphate Dibasic Anhydrous/Sodium Phosphate Monobasic Monohydrate, Tablet, II (Paddle), 100, Water (deionized), 900, 20, 30, 45, 60, and 90, 01/15/2010.

Solifenacin Succinate, Tablet, II (Paddle), 50, Water, 900, 10, 15, 30, and 45, 02/19/2008, 5 mg based on (1) acceptable bioequivalence studies on the 10 mg strength, (2) proportional similarity of the formulations across all strengths, and (3) acceptable in vitro dissolution testing of all strengths.

Sorafenib Tosylate, Tablet, II (Paddle), 75, 0.1 M HCl with 1% SDS, 900, 5, 10, 15, 20, and 30, 06/10/2009.

Spironolactone, Tablet, Refer to USP, 04/15/2008, 25 and 50 mg based on (1) acceptable bioequivalence studies on the 100 mg strength, (2) proportional similarity of the formulations across all strengths, and (3) acceptable in vitro dissolution testing of all strengths.

Stavudine, Capsule, Refer to USP, 06/18/2007, 15, 20, and 30 mg based on (1) acceptable bioequivalence studies on the 40 mg strength, (2) acceptable dissolution testing of the 15, 20, 30, and 40 mg strengths, and (3) proportional similarity in the formulations of the 15, 20, 30, and 40 mg strengths.

Succimer, Capsule, II (Paddle), 50, 0.01 N Phosphoric Acid, 900, 10, 20, 30, 45, 60, and 90, 02/20/2004.

Sucralfate, Suspension, II (Paddle), 75, 0.1 N HCl/0.067 M KCl, pH 1.0, 900, 10, 20, 30, and 45, 03/04/2006.

Sucralfate, Tablet, II (Paddle), 75, 0.1 N HCl/0.067 M KCl, pH 1.0, 900, 15, 30, 45, 60, 180, 240, and 480, 04/02/2009.

Sulfadiazine, Tablet, Refer to USP, 07/14/2008.

Sulfamethoxazole/Trimethoprim, Tablet, Refer to USP, 01/14/2008, 400 mg/80 mg based on (1) acceptable bioequivalence studies on the 800 mg/160 mg strength, (2) acceptable dissolution testing across all strengths, and (3) proportional similarity in the formulations across all strengths.

Sulfamethoxazole/Trimethoprim, Suspension, II (Paddle), 50, 1 mL of 0.2 N HCl in water, 900, 10, 20, 30, 45, 60, and 90, 02/25/2004.

Sulfasalazine, Tablet, Refer to USP, 12/15/2009.

Sulfasalazine, Tablet (Delayed Release), Refer to USP, 12/15/2009.

Sulfisoxazole Acetyl, Oral Suspension (Pediatric), II (Paddle), 30, 1% SLS in 0.1 N HCl, 900, 15, 30, 45, 60, and 90, 08/17/2006.

Sumatriptan Succinate, Tablet, II (Paddle), 30, 0.01 M HCl, 900, 5, 10, 15, and 30, 03/04/2006, 25 and 50 mg, based on (1) acceptable bioequivalence studies on the 100 mg strength, (2) proportionally similar across all strengths, and (3) acceptable in vitro dissolution testing of all strengths.

Sunitinib Malate, Capsule, II (Paddle), 50, 0.1 N HCl, 900, 10, 15, 30, and 45, 10/30/2009, 12.5, 25, and 37.5 mg based on (1) acceptable bioequivalence studies on the 50 mg strength, (2) acceptable in vitro dissolution testing of all strengths, and (3) proportional similarity 12.5 and 25 mg formulations to the 50 mg strength.

Tacrolimus, Capsule, II (Paddle), 50, Hydroxypropyl Cellulose Solution (1 in 20,000). Adjust to pH 4.5 by Phosphoric Acid, 900, 30, 60, 90, and 120, 02/20/2004, 0.5 and 1 mg, based on (1) acceptable bioequivalence studies on the 5 mg strength, (2) proportional similarity of the formulations across all strengths, and (3) acceptable in vitro dissolution testing of all strengths.

Tadalafil, Tablet, II (Paddle), 50, 0.5% Sodium Lauryl Sulfate, 1000, 10, 20, 30, and 45, 01/26/2006, 2.5, 5, and 10 mg based on (1) acceptable bioequivalence studies on the 20 mg strength, (2) acceptable in vitro dissolution testing of all strengths, and (3) proportional similarity of the formulations across all strengths.

Tamoxifen Citrate, Tablet, Refer to USP, 04/02/2009, 10 mg based on (1) acceptable bioequivalence studies on the 20 mg strength, (2) acceptable dissolution testing across all strengths, and (3) proportional similarity in the formulations across all strengths.

Tamsulosin HCl, Capsule, II (Paddle), 100, 0–2 h: 0.003% polysorbate 80, pH 1.2 2–8 h: phosphate buffer, pH 7.2, 500, 1, 2, 3, 6, 8, and 10 h, 03/26/2007, 50, 100, 150, and 200 mg based on (1) acceptable bioequivalence studies on the 250 mg strength, (2) acceptable in vitro dissolution testing of all strengths, and (3) proportional similarity of the formulations across all strengths.

Tapentadol HCl, Tablet (Extended Release), II (Paddle) with sinker, 100, 0.05 M Phosphate Buffer of pH 6.8, Simulated intestinal fluid (without enzyme), 900, 0.5, 1, 2, 3, 5, 7, and 10 h, 10/04/2012.

Tapentadol HCl, Tablet, I (Basket), 75, 0.1 N HCl, 900, 10, 20, 30, 45, and 60, 10/28/2010, 50 and 75 mg based on (1) acceptable bioequivalence studies on the 100 mg strength, (2) acceptable in vitro dissolution testing across all strengths, and (3) proportional similarity in the formulations across all strengths.

Telaprevir, Tablet, II (Paddle), 50, 1% SLS in Water, 900, 5, 10, 15, 20, and 30, 05/09/2013.

Telbivudine, Tablet, II (Paddle), 50, 0.1 N HCl, 900, 15, 30, and 45, 04/02/2009.

(Continued)

Telithromycin, Tablet, II (Paddle), 50, 0.1 N HCl, 900, 10, 20, 30, and 45, 01/03/2007, 300 mg based on (1) acceptable bioequivalence studies on the 400 mg strength, (2) proportionally similar across both strengths, and (3) acceptable in vitro dissolution testing of both strengths.

Telmisartan, Tablet, Refer to USP, 01/05/2012, 20 and 40 mg based on (1) acceptable bioequivalence studies on the 80 mg strength, (2) acceptable in vitro dissolution testing of all strengths, and (3) proportional similarity of the formulations across all strengths.

Temazepam, Capsule, Refer to USP, 01/14/2008, 7.5, 15, and 22.5 mg based on (1) acceptable bioequivalence studies on the 30 mg strength, (2) proportional similarity of the formulations across all strengths, and (3) acceptable in vitro dissolution testing of all strengths.

Temozolomide, Capsule, I (Basket), 100, Water, 500 (for 5 mg); 900 mL (for other strengths), 10, 20, 30, and 45, 08/11/2008, 5, 20, 100, 140, and 180 mg based on (1) acceptable bioequivalence study on the 250 mg strength, (2) proportional similarity of the formulations across all strengths, and (3) acceptable in vitro dissolution testing of all strengths.

Tenofovir Disoproxil Fumarate, Tablet, II (Paddle), 50, 0.1 N HCl, 900, 10, 20, 30, and 45, 01/03/2007, 150, 200, and 250 mg based on (1) acceptable bioequivalence studies on the 300 mg strength, (2) acceptable in vitro dissolution testing of all strengths, and (3) proportional similarity of the formulations across all strengths.

Tenofovir Disoproxil Fumarate, Powder for Oral, II (Paddle), 100, 0.2% polysorbate 80 in 0.01 M HCl, 900, 10, 20, 30, 45, 60, and 75, 01/31/2013.

Terazosin HCl, Tablet, II (Paddle), 50, Water (deaerated), 900, 10, 20, 30, 45, and 60, 02/20/2004.

Terazosin HCl, Capsule, II (Paddle), 50, Water (deaerated), 900, 10, 20, 30, 45, 60, and 90, 02/20/2004, 1, 5, and 10 mg based on (1) acceptable bioequivalence studies on the 2 mg strength, (2) proportionally similar across all strengths, and (3) acceptable in vitro dissolution testing of all strengths.

Terbinafine HCl, Tablet, II (Paddle), 50, Citrate Buffer, pH 3.0 adjusted with HCl, 500, 10, 20, 30, and 45, 02/20/2004, 125 mg (base) based on (1) acceptable bioequivalence studies on the 187.5 mg (base) strength, (2) acceptable dissolution testing across all strengths, and (3) proportional similarity in the formulations across all strengths.

Terconazole, Suppository (Vaginal), I (with Palmieri type basket), 100, 0.12 N HCl with 1% SLS, 900, 15, 30, 45, 60, 90, 120, and 150, 10/08/2009.

Testosterone, Buccal Tablet (Extended Release), II (Paddle, may use sinker), 60, 1% sodium dodecyl sulfate in double distilled water, 1000, 1, 2, 4, 6, 10, 12, and 24 h, 01/03/2007.

Testosterone, Pellet Implant, Develop a dissolution method, 11/25/2008.

Testosterone, Film, Transdermal (Extended Release), V (Paddle over disk). Paddle 25 mm above the film on the disk, 50, 0.1 M sodium chloride containing 2.5% (v/v) of Tween 40 at 32°C ± 0.5°C. Delivery surface faces upwards toward the media, 900, 1, 3, 5, 7, 11, 16, 20, and 24 h, 06/30/2011.

Tetrabenazine, Tablet, II (paddle), 50, 0.1 N HCl, 900, 5, 10, 15, 30, and 45, 09/01/2011, 12.5 mg based on (1) acceptable bioequivalence studies on the 25 mg strength, (2) acceptable in vitro dissolution testing of all strengths, and (3) proportional similarity of the formulations across both strengths. Please refer to the Mirtazapine Tablet Guidance for additional information regarding waivers of in vivo testing.

Tetracycline HCl, Capsule, Refer to USP, 06/24/2010, 250 mg based on (1) acceptable bioequivalence studies on the 500 mg strength, (2) acceptable in vitro dissolution testing of all strengths, and (3) proportional similarity of the formulations across all strengths.

Tetracycline HCl, Tablet, Refer to USP, 01/29/2010.

Thalidomide, Capsule, II (Paddle), 100, 1.5% (w/v) SLS (pH 3.0, adj w/HCl), 900, 10, 20, 30, 60, and 90, 03/04/2006, 50, 100, and 150 mg based on (1) acceptable fasting and fed bioequivalence studies on the 200 mg strength, (2) proportional similarity of the formulations across all strengths, and (3) acceptable in vitro dissolution testing of all strengths.

Theophylline, Capsule (Extended Release), Refer to USP, 900, 10/06/2008, 100, 200, and 300 mg based on (1) acceptable bioequivalence studies on the 400 mg strength, (2) acceptable in vitro dissolution testing of all strengths, and (3) proportional similarity of the formulations across all strengths. Please refer to the Mirtazapine Tablet Guidance for additional information regarding waivers of in vivo testing.

Theophylline (100 and 200 mg), Tablet (Extended Release), II (Paddle), 50, SGF without enzyme, pH 1.2 during the first hour. Phosphate Buffer at pH 7.5 from end of hour 1 through the duration of testing, 900, 1, 4, 8, and 12 h, 10/06/2008.

Theophylline (300 and 450 mg), Tablet (Extended Release), II (Paddle), 50, SGF without enzyme, pH 1.2 during the first hour. Phosphate Buffer at pH 7.5 from end of hour 1 through the duration of testing, 900, 1, 4, 8, and 12 h, 10/06/2008, 300 mg based on (1) acceptable bioequivalence studies on the 450 mg strength, (2) proportional similarity of the formulations across all strengths, and (3) acceptable in vitro dissolution testing of all strengths.

(*Continued*)

Theophylline (600 and 400 mg), Tablet (Extended Release), I (Basket), 100, SGF without enzyme, pH 1.2 during the first hour. SIF without enzyme from end of hour 1 through the duration of the testing, 900, 1, 2, 4, 8, 12, and 24 h, 10/06/2008, 400 mg based on (1) acceptable bioequivalence studies on the 600 mg strength, (2) proportional similarity of the formulations across all strengths, and (3) acceptable in vitro dissolution testing of all strengths.

Thioguanine, Tablet, Refer to USP, 07/15/2009.

Tiagabine HCl, Tablet, II (Paddle), 50, Water, 900, 5, 10, 15, 20, and 30, 01/03/2007, 2, 6, 8, 10, 12, and 16 mg based on (1) acceptable bioequivalence studies on the 4 mg strength, (2) acceptable in vitro dissolution testing of all strengths, and (3) proportional similarity in the formulations across all strengths.

Ticlopidine HCl, Tablet, II (Paddle), 50, Water (deaerated), 900, 10, 20, 30, 45, and 60, 02/19/2004.

Timolol Maleate, Tablet, Refer to USP, 07/31/2013.

Tinidazole, Tablet, I (Basket), 100, Water (deaerated), 900, 10, 20, 30, and 45, 01/03/2007, 250 mg based on (1) acceptable bioequivalence studies of the 500 mg strength, (2) proportionally similar across all strengths, and (3) acceptable in vitro dissolution testing of all strengths.

Tipranavir, Capsule, II (Paddle), 50, 0.05 M phosphate buffer pH 6.8, 900, 15, 30, 45, and 60, 12/03/2007.

Tizanidine HCl, Capsule, II (Paddle), 50, 0.01 N HCl, 500, 5, 10, 15, and 30, 02/20/2004, 2 and 4 mg based on (1) acceptable bioequivalence studies on the 6 mg strength, (2) proportional similarity of the formulations across all strengths, and (3) acceptable in vitro dissolution testing of all strengths.

Tizanidine HCl, Tablet, I (Basket), 100, 0.1 N HCl, 500, 5, 10, 15, and 30, 02/20/2004, 2 mg based on (1) acceptable bioequivalence studies on the 4 mg strength, (2) proportional similarity of the formulations across all strengths, and (3) acceptable in vitro dissolution testing of all strengths.

Tolcapone, Tablet, Refer to USP, 05/09/2013.

Tolterodine Tartrate, Tablet, II (Paddle), 50, SGF without enzymes, pH 1.2, 900, 5, 10, 15, and 30, 02/20/2004, 1 mg based on (1) acceptable bioequivalence studies on the 2 mg strength, (2) proportional similarity of the formulations across all strengths, and (3) acceptable in vitro dissolution testing of all strengths.

Tolterodine Tartrate, Capsule (Extended Release), I (Basket), 100, Phosphate buffer (pH 6.8), 900, 1, 3, and 7 h, 06/18/2007, 2 mg, based on (1) acceptable bioequivalence studies on the 4 mg strength, (2) proportional similarity of the formulations across all strengths, and (3) acceptable in vitro dissolution testing of all strengths.

Topiramate, Capsule (Sprinkle), II (Paddle), 50, Water (deaerated), 900, 10, 20, 30, 45, and 60, 02/19/2004, 15 mg based on (1) acceptable bioequivalence studies on the 25 mg strength, (2) acceptable in vitro dissolution testing of all strengths, and (3) proportional similarity of the formulations across all strengths.

Topiramate, Tablet, II (Paddle), 50, Water (deaerated), 900, 5, 10, 20, and 30, 02/19/2004, 25, 50, and 200 mg based on (1) acceptable bioequivalence studies on the 100 mg strength, (2) acceptable in vitro dissolution testing of all strengths, and (3) proportional similarity of the formulations across all strengths. Please refer to the Mirtazapine Tablet Guidance for additional information regarding waivers of in vivo testing.

Topotecan HCl, Capsule, II (Paddle), 50, Acetate Buffer with 0.15% SDS, pH 4.5, 500, 5, 10, 20, 30, and 45, 04/27/2009, Topotecan capsule, 1 mg strength may be considered for a waiver of in vivo bioequivalence study based on (1) acceptable bioequivalence studies on the 0.25 mg strength, (2) acceptable dissolution testing on the 0.25 and 1.0 mg strengths, and (3) proportional similarity in the formulations of the 0.25 and 1 mg strengths.

Toremifene Citrate, Tablet, II (Paddle), 50, 0.02 N HCl, 1000, 10, 20, 30, and 45, 02/20/2004.

Torsemide, Tablet, II (Paddle), 50, 0.1 N HCl, 900, 5, 10, 15, and 30, 02/20/2004, 5, 10, and 100 mg, based on (1) acceptable bioequivalence studies on the 20 mg strength, (2) proportional similarity of the formulations across all strengths, and (3) acceptable in vitro dissolution testing of all strengths.

Tramadol HCl, Tablet, I (Basket), 100, 0.1 N HCl, 900, 10, 20, 30, and 45, 02/19/2004.

Tramadol HCl, Tablet (Extended Release), I (Basket), 75, 0.1 N HCl, 900, 2, 4, 8, 10, and 16 h, 01/03/2007.

Trandolapril, Tablet, II (Paddle), 50, Water (deaerated), 500, 10, 20, 30, 45, and 60, 02/20/2004, 1 and 2 mg based on (1) acceptable bioequivalence studies on the 4 mg strength, (2) proportional similarity of the formulations across all strengths, and (3) acceptable in vitro dissolution testing of all strengths.

Trandolapril/Verapamil HCl, Tablet (Extended Release), II (Paddle), 50, For Trandolapril: Water; For Verapamil: 0–1 h Gastric Fluid w/o Pepsin pH = 1.2, 1–8 h Intestinal Fluid w/o Pancreatin; For Trandolapril: 500; For Verapamil: 900; For Trandolapril: 5, 10, 20, 30, and 45; For Verapamil: 1, 2, 3.5, 5, and 8 h, 12/19/2008, 2 mg/180 mg, 1 mg/240 mg and 2 mg/240 mg based on (1) acceptable bioequivalence studies on the 4/240 mg strength, (2) proportional similarity of the formulations across all strengths, and (3) acceptable in vitro dissolution testing of all strengths.

Tranexamic Acid, Tablet, II (Paddle), 50, Water, 900, 15, 30, 45, 60, 90, and 120, 12/23/2010.

(*Continued*)

Trazodone HCl, Tablet, Refer to USP, 12/15/2009, 50, 150, and 300 mg based on (1) acceptable bioequivalence studies on the 100 mg strength, (2) proportional similarity of the formulations across all strengths, and (3) acceptable in vitro dissolution testing of all strengths.

Trazodone HCl, Tablet (Extended Release), II (Paddle), 50, Water, 900, 1, 2, 3, 5, 8, 10, 12, 16, 20, and 24 h, 06/30/2011, 300 mg strength based on (1) an acceptable bioequivalence study on the 150 mg strength, (2) acceptable in vitro dissolution testing of all strengths, and (3) proportional similarity in the formulations across all strengths. Please refer to the Mirtazapine Tablet Guidance for additional information regarding waivers of in vivo testing.

Tretinoin, Capsule, I (Basket), 100, 0.5% solid Lauryldimethylamine oxide (LDAO) in 0.05 M Phosphate Buffer, pH 7.8, 900, 10, 15, 20, 30, and 45, 08/05/2010.

Triamcinolone Acetonide, Injectable Suspension, Develop a dissolution method using USP IV (Flow-Through Cell), and, if applicable, Apparatus II (Paddle) or any other appropriate method, for comparative evaluation by the agency, 01/15/2010, 10 mg/mL based on (1) acceptable bioequivalence study on the 40 mg/mL strength, (2) acceptable dissolution testing across all strengths, and (3) proportional similarity in the formulations across all strengths, and (4) the formulation of the 10 mg/mL strength is qualitatively (Q1) and quantitatively (Q2) identical to the 10 mg/mL strength of the RLD.

Triamterene, Capsule, Refer to USP, 06/18/2007, 50 mg based on (1) acceptable bioequivalence study on the 100 mg strength, (2) proportional similarity of the formulations across all strengths, and (3) acceptable in vitro dissolution testing of all strengths.

Trientine HCl, Capsule, Refer to USP, 07/31/2013.

Trimethoprim, Tablet, Refer to USP, 01/29/2010, 100 mg based on (1) acceptable bioequivalence studies on the 200 mg strength, (2) proportional similarity across all formulations, and (3) acceptable in vitro dissolution testing of all strengths.

Trimipramine Maleate, Capsule, I (Basket), 100, Water (deaerated), 1000, 10, 20, 30, 45, 60, and 90, 03/04/2006.

Triptorelin Pamoate, Injectable Suspension, II (Paddle), 200, Water–Methanol (95:5); Reconstitute vial in 2 mL Water for Injection, add to 500 mL medium at 37°C, 500, 1, 6, 12, 24, 48, and 72 h, 07/14/2008, 3.75 mg base/vial based on (1) acceptable bioequivalence studies on the 11.25 mg base/vial strength, (2) acceptable dissolution testing across all strengths, and (3) proportional similarity in the formulations across all strengths.

Trospium Chloride, Tablet, II (Paddle), 50, 0.1 N HCl, 1000, 10, 20, 30, and 45, 12/03/2007.

Trospium Chloride, Capsule (Extended Release), II (Paddle) with sinker, 50, 0.1 N HCl, pH 1.1 for 2 h and then add 200 mL of 0.1 N NaOH in 200 mM Phosphate Buffer. Adjust pH to 7.5 using 2 N HCl and/or 2 N NaOH, 0–2 h: 750 mL, After 2 h: 950 mL, 2, 3, 4, 6, 8, 12, and 16 h, 07/15/2010.

Ulipristal Acetate, Tablet, II (Paddle), 50, 0.1 N HCl, 900, 5, 10, 15, 20, and 30, 01/31/2013.

Ursodiol, Capsules, Refer to USP, 07/21/2009.

Ursodiol, Tablet, Refer to USP, 04/15/2008, 250 mg based on (1) acceptable bioequivalence studies on the 500 mg strength, (2) proportionally similar across all strengths, and (3) acceptable in vitro dissolution testing of all strengths.

Valacyclovir Hydrochloride, Tablet, II (Paddle), 50, 0.1 N HCl, 900, 10, 20, 30, 45, and 60, 08/27/2009, 500 mg based on (1) acceptable bioequivalence studies on the 1000 mg strength, (2) proportional similarity of the formulations across all strengths, and (3) acceptable in vitro dissolution testing of all strengths.

Valganciclovir HCl, Tablet, II (Paddle), 50, 0.1 N HCl, 900, 10, 15, 30, 45, 60, 06/18/2007.

Valproic Acid, Capsule, Refer to USP, 12/15/2009.

Valsartan (Tab & Cap), Tablet, II (Paddle), 50, 0.067 M Phosphate Buffer, pH 6.8, 1000, 10, 20, 30, and 45, 12/13/2004, 40, 80, and 160 mg based on (1) acceptable bioequivalence studies on the 320 mg strength, (2) proportional similarity of the formulations across all strengths, and (3) acceptable in vitro dissolution testing of all strengths.

Vancomycin hydrochloride, Capsule, Refer to USP, 01/14/2008.

Vardenafil HCl, Tablet, II (Paddle), 50, 0.1 N HCl, 900, 5, 10, 15, and 30, 12/20/2005, 2.5, 5, and 10 mg, based on (1) acceptable bioequivalence studies on the 20-mg strength, (2) proportionally similar across all strengths, and (3) acceptable in vitro dissolution testing of all strengths.

Varenicline Tartrate, Tablet, I (Basket), 100, 0.01 N HCl, 500, 5, 10, 15, and 30, 12/03/2007, 0.5 mg based on (1) acceptable bioequivalence studies on the 1 mg strength, (2) proportional similarity of the 1 and 0.5 mg strengths, and (3) acceptable in vitro dissolution testing of both strengths.

Venlafaxine HCl, Tablet, II (Paddle), 50, Water (deaerated), 900, 5, 10, 15, and 30, 02/19/2004, 25, 37.5, 75, and 100 mg based on (1) acceptable bioequivalence studies on the 50 mg strength, (2) proportional similarity of the formulations across all strengths, and (3) acceptable in vitro dissolution testing of all strengths.

Venlafaxine HCl, Capsule (Extended Release), I (Basket), 100, Water, 900, 2, 4, 8, 12, and 20 h, 01/03/2007, 37.5 and 75 mg based on (1) acceptable bioequivalence studies on the 150 mg strength, (2) proportional similarity of the formulations across all strengths, and (3) acceptable in vitro dissolution testing of all strengths.

(Continued)

Verapamil HCl, Tablet, Refer to USP, 11/04/2008, 40 and 80 mg based on (1) acceptable bioequivalence studies on the 120 mg strength, (2) acceptable dissolution testing across all strengths, and (3) proportional similarity in the formulations across all strengths.

Verapamil HCl, Tablet (Extended Release), Refer to USP, 06/24/2010.

Verapamil HCl (100, 200, and 300 mg), Capsule (Extended Release), I (Basket), 75, Water, pH 3.0 (adjusted with 0.1 N or 2 N HCl), 1000, 1, 4, 8, 11, and 24 h, 01/03/2007, 100 and 200 mg based on (1) acceptable bioequivalence studies on the 300 mg strength, (2) proportionally similar across all strengths, and (3) acceptable in vitro dissolution testing of all strengths.

Verapamil HCl (120, 180, 240, and 360 mg), Capsule (Sustained Release), I (Basket), 75, Water, pH 3.0 (adjusted with 0.1 N or 2 N HCl), 1000, 1, 4, 8, 11, and 24 h, 01/03/2007, 120, 180, and 240 mg based on (1) acceptable bioequivalence studies on the 360 mg strength, (2) proportionally similar across all strengths, and (3) acceptable in vitro dissolution testing of all strengths.

Vilazodone HCl, Tablet, II (Paddle), 60, 0.1% v/v Acetic acid (pH 3.1), 1000, 10, 20, 30, and 45, 01/26/2012, 20 and 40 mg based on (1) acceptable bioequivalence studies on the 10 mg strength, (2) acceptable in vitro dissolution testing of all strengths, and (3) proportional similarity of the formulations across all strengths. Please refer to the Mirtazapine Tablet Guidance for additional information regarding waivers of in vivo testing.

Voriconazole, Suspension, II (Paddle), 50, 0.1 N HCl, 900, 10, 20, 30, and 45, 01/03/2007.

Voriconazole, Tablet, II (Paddle), 50, 0.1 N HCl, 900, 10, 20, 30, and 45, 11/25/2008, 50 mg based on (1) acceptable bioequivalence studies on the 200 mg strength, (2) proportional similarity of the formulations across all strengths, and (3) acceptable in vitro dissolution testing of all strengths.

Vorinostat, Capsule, II (Paddle) with sinker, 100, 2% Tween 80 in Water, 900, 5, 15, 30, 45, and 60, 09/03/2008.

Warfarin Sodium, Tablet, Refer to USP, 01/29/2010, 1, 2, 2.5, 3, 4, 5, and 6 mg are eligible for a waiver of in vivo bioequivalence testing based on (1) acceptable bioequivalence studies on the 10 mg strength, (2) acceptable in vitro dissolution testing of all strengths, and (3) proportional similarity of the formulations across all strengths.

Zafirlukast, Tablet, II (Paddle), 50, 1% w/v Aqueous Sodium Dodecyl Sulfate, 1000, 10. 30, 30, and 45, 10/09/2007, 10 mg based on acceptable (1) bioequivalence studies on the 20 mg tablet, (2) proportional similarity of the formulations, and (3) acceptable in vitro dissolution testing of all strengths.

Zalcitabine, Tablet, Refer to USP, 900, 02/19/2008, 0.375 mg based on (1) acceptable bioequivalence studies on the 0.75 mg strength, (2) proportional similarity of the formulations across all strengths, and (3) acceptable in vitro dissolution testing of all strengths.

Zaleplon, Capsule, II (Paddle), 75, Deionized Water, 900, 5, 10, 20, and 30, 01/03/2007.

Zidovudine, Tablet, Refer to USP, 07/25/2007.

Zidovudine, Capsule, Refer to USP, 06/18/2007.

Zileuton, Tablet, II (Paddle), 50, 0.05 M SLS in water, 900, 10, 20, 30, 45, and 60, 02/19/2004.

Zileuton, Tablet (Extended Release), II (Paddle) with sinker, 75, 0.1 M SDS (sodium dodecyl sulfate) in water, 900, 1, 2, 4, 6, 8, 10, and 12 h, 08/15/2013.

Zinc Acetate, Capsule, II (Paddle), 50, 0.1 N HCl, 900, 10, 20, 30, and 45, 02/19/2004.

Ziprasidone HCl, Capsule, II (Paddle), 75, Tier I: 0.05 M Na phosphate buffer, pH 7.5 + 2% SDS (w/w) Tier II: 0.05 M Na phosphate buffer, pH 7.5 (700 mL) + 1% pancreatin. After 15 min incubation, add 200 mL of phosphate buffer containing 9% SDS, 900, 10, 20, 30, 45, and 60, 03/04/2006, 40, 60, and 80 mg based on (1) acceptable bioequivalence studies on the 20 mg strength, (2) proportionally similar across all strengths, and (3) acceptable in vitro dissolution testing of all strengths.

Zolmitriptan, Tablet (Orally Disintegrating), II (Paddle), 50, 0.1 N HCl, 500, 5, 10, 15, 20, and 30, 06/18/2007, 2.5 mg based on acceptable (1) bioequivalence studies on the 5 mg strength, (2) proportional similarity of the formulations, and (3) acceptable in vitro dissolution testing of all strengths.

Zolmitriptan, Tablet, II (Paddle), 50, 0.1 N HCl, 500, 5, 10, 15, 20, and 30, 07/21/2009, 2.5 mg based on (1) acceptable bioequivalence studies on the 5 mg strength, (2) acceptable in vitro dissolution testing for all strengths, and (3) proportional similarity of the formulations across all strengths.

Zolpidem Tartrate, Tablet, II (Paddle), 50, 0.01 N HCl, pH 2.0, 900, 5, 10, 15, and 30, 02/19/2004.

Zolpidem Tartrate, Tablet (Sublingual), II (Paddle), 75, Phosphate buffer, pH 6.8, 900, 1, 3, 5, 7, 10, and 15, 09/02/2010, 1.75 mg based on (1) acceptable bioequivalence studies on the 3.5 mg strength, (2) formulation proportionally similar for both strengths, and (3) acceptable in vitro dissolution testing of all strengths.

Zolpidem Tartrate, Tablet (Extended Release), Refer to USP, 01/05/2012.

Zonisamide, Capsule, II (Paddle), 50, Water (deaerated), 900, 10, 20, 30, and 45, 01/03/200, 25 and 50 mg, based on acceptable (1) bioequivalence studies on the 100 mg capsule, (2) proportional similarity of the formulations, and (3) acceptable in vitro dissolution testing of all strengths.

Bibliography

PREFACE

Anton P, Silberglitt R, and Schneider R (2001) *The Global Technology Revolution: Bio/Nano/Materials Trends and Their Synergies with Information Technology by 2015*. Rand Corporation. Santa Monica, CA (http://www.rand.org/pubs/monograph_reports/MR1307/index.html).

Blumenthal M, Daniel E, Farnsworth N, and Riggins C (2003) *Botanical Medicine: Efficacy, Quality Assurance and Regulation*. Mary Ann Liebert.

Carpenter JF and Manning MC (eds.) (2003) *Rational Design of Stable Protein Formulations: Theory and Practice (Pharmaceutical Biotechnology)*. Plenum Press, New York.

Carstensen JT (1998) *Pharmaceutical Preformulation*. CRC Press, Boca Raton, FL.

Florence AT (ed.) (1984) *Materials Used in Pharmaceutical Formulation* (*Critical Reports on Applied Chemistry*, Vol 6). Blackwell Science.

Gibson M (2001) *Pharmaceutical Preformulation and Formulation: A Practical Guide from Candidate Drug Selection to Commercial Dosage Form*. CRC Press, Boca Raton, FL.

Hovgaard SF Jr. (1999) *Pharmaceutical Formulation Development of Peptides and Proteins*. CRC Press, Boca Raton, FL.

Mumoli N, Cei M, Luschi R, Carmignani G, and Camaiti A. Allergic reaction to Croscarmellose sodium used as excipient of a generic drug. *QJM*. 2011;104(8):709–710.

Niazi SK (2004) Pharmacokinetic and Pharmacodynamic Modeling in Early Drug Development in Charles G. Smith and James T. O'Donnell (eds.), *The Process of New Drug Discovery and Development* (2nd edn.). CRC Press, New York.

Niazi SK (2006) *Handbook of Preformulation: Drugs, Botanicals and Biological Pharmaceutical Products*. Informa, New York.

Niazi SK (2007) *Handbook of Bioequivalence Testing*. Informa, New York.

Niazi SK (2009) *Handbook of Biogeneric Therapeutic Proteins: Manufacturing, Regulatory, Testing and Patent Issues*. CRC Press, Boca Raton, FL.

Niazi SK (2009) *Handbook of Pharmaceutical Manufacturing Formulations*, 2nd edn., Volume 1: *Compressed Solids*. CRC Press, Boca Raton, FL.

Niazi SK (2009) *Handbook of Pharmaceutical Manufacturing Formulations*, 2nd edn., Vol. 2: *Uncompressed Solids*. CRC Press, Boca Raton, FL.

Niazi SK (2009) *Handbook of Pharmaceutical Manufacturing Formulations*, 2nd edn., Vol. 3: *Liquid Products*. CRC Press, Boca Raton, FL.

Niazi SK (2009) *Handbook of Pharmaceutical Manufacturing Formulations*, 2nd edn., Vol. 4: *Semisolid Products*, CRC Press, Boca Raton, FL.

Niazi SK (2009) *Handbook of Pharmaceutical Manufacturing Formulations*, 2nd edn., Vol. 5: *Over the Counter Products*. CRC Press, Boca Raton, FL.

Niazi SK (2009) *Handbook of Pharmaceutical Manufacturing Formulations*, 2nd edn., Vol. 6: *Sterile Products*. CRC Press, Boca Raton, FL.

Niazi SK (2010) *Textbook of Biopharmaceutics and Clinical Pharmacokinetics*. The Book Syndicate, Hyderabad, India.

Wells JI (1988) *Pharmaceutical Preformulation: The Physicochemical Properties of Drug Substances*. Ellis Horwood Ltd., West Sussex, U.K.

CHAPTER 1: HISTORICAL PERSPECTIVE ON GENERIC PHARMACEUTICALS

Baldwin FD. If it quacks like a duck.... *MedHunters*. Archived from the original on February 06, 2008. Retrieved October 13, 2007.

Barrett S. Quackwatch. Your guide to quackery, health fraud, and intelligent decisions (quackwatch). Retrieved October 13, 2007.

Benedetti F (2009) Placebo effects: Understanding the mechanisms in health and disease. Oxford University Press, Oxford, U.K.

Blanchard A, Helene D'Iorio, and Robert Ford (2010) *What You Need to Know to Succeed: Canada: Key Trends in Canada's Biotech Industry*. Insights, spring.

British Medical Association, Secret Remedies (1909) What they cost and what they contain.

Calo-Fernández B and Martínez-Hurtado J. Biosimilars: Company strategies to capture value from the biologics market. *Pharmaceuticals*. 2012 Dec;5(12):1393–1408. doi:10.3390/ph5121393.

Carson, G (1961) *One for a Man, Two for a Horse: A Pictorial History, Grave and Comic, of Patent Medicines*. Hardcover, Double Day & Co, New York.

Declerck PJ. Biosimilar monoclonal antibodies: A science-based regulatory challenge. *Expert Opin Biol Ther*. 2013 Feb;13(2):153–156.

Eisner DA (2000). *The Death of Psychotherapy: From Freud to Alien Abductions*. Praegner, Westport, CT.

EMEA Guideline on Similar Biological Medicinal Products, CHMP/437/04 London, U.K., October 30, 2005.

FDA. *Hearing: Assessing the Impact of a Safe and Equitable Biosimilar Policy in the United States*. Subcommittee on Health Wednesday, May 2, 2007.

FDA. Page on Biosimilars: http://www.fda.gov/Drugs/DevelopmentApprovalProcess/HowDrugsareDevelopedandApproved/ApprovalApplications/TherapeuticBiologicApplications/Biosimilars/.

Final report of the White House Commission on Complementary and Alternative Medicine (http://www.whccamp.hhs.gov).

Griffenhagen, George B. and James Harvey Young, Old English Patent Medicines in America, Contributions from the Museum of History and Technology (U.S. National Museum Bulletin 218, Smithsonian Institution, Washington, DC, 1959, pp. 155–183.

Heinze WF. Dead Patents Walking, IEEE Spectrum (2002). Consulted on November 26, 2009.

Hulda RC (1995) *The Cure For All Diseases*. New Century Press, San Diego, CA.

Janssen W (1993) The Gadgeteers. In S. Barrett and W. Jarvis (Eds.), *The Health Robbers. A Close Look at Quackery in America*. Prometheus Books, Buffalo, NY, pp. 321–335.

Jarvis WT. Quackery: A national scandal. *Clin Chem*. 1992;38(8B Pt 2):1574–1586.

Jarvis WT. Quackery: The National Council Against Health Fraud perspective. *Rheum Dis Clin North Am*. 1999;25(4):805–814.

Katona GP. The Myth of Submarine Patents, Pandab online newsletter, August 10, 1998. Consulted on March 28, 2010.

Ladimer I. The Health Advertising Program of the National Better Business Bureau. *Am J Public Nations Health*. 1965 Aug;55(8):1217–1227.

Nick C. The US biosimilars act: Challenges facing regulatory approval. *Pharm Med*. 2012;26(3):145–152.

Offit PA (2013) *Do You Believe in Magic? The Sense and Nonsense of Alternative Medicine*. Harper, New York. (also titled *Killing Us Softly: The Sense and Nonsense of Alternative Medicine*. London, U.K.).

Ozer J (2010–03–04) Ogg, MPEG LA, and Submarine Patents. *Streaming Media Magazine*. Retrieved 2011–07–26.

Paul R (2009–07–05) *Decoding the HTML 5 video codec debate*. Ars Technica. Retrieved 2011–07–26.

Quackery: How Should It Be Defined? Quackwatch.org. January 17, 2009.

Renckens CN. In the interest of all who value their purse and their health: A brief history of the Vereniging tegen de Kwakzalverij—Society Against Quackery—of the Netherlands. *Eval Health Prof*. 2009 Dec;32(4):343–348.

Rosenblum J. Paying for Patents, Legal Affairs, May/June 2005. Consulted on March 28, 2010.

Salla D (1999) *Mind Myths: Exploring Popular Assumptions about the Mind and Brain*. Wiley, New York.

Styles J (2000) Product innovation in early modern London. In: Past & Present 168, 124–169.

U.S. Senate Committee on the Judiciary. Testimony of Dr. Lester Crawford, Acting Commissioner, FDA, Silver Spring, MD, June 23, 2004.

Young JH (1961) *The Toadstool Millionaires: A Social History of Patent Medicines in America before Federal Regulation*. Princeton University Press, Princeton, NJ, 282pp.

CHAPTER 2: PHYSICOCHEMICAL PROPERTIES AFFECTING BIOEQUIVALENCE

Agoram B, Woltosz WS, and Bolger MB. Predicting the impact of physiological and biochemical processes on oral drug bioavailability. *Adv Drug Deliv Rev*. 2001 Oct 1;50(Suppl 1):S41–S67.

Al-Awqati Q. One hundred years of membrane permeability: Does Overton still rule? *Nat Cell Biol*. 1999 Dec;1(8):E201–E202.

Ansoborlo E, Henge-Napoli MH, Chazel V, Gibert R, Guilmette RA. Review and critical analysis of available in vitro dissolution tests. *Health Phys.* 1999 Dec;77(6):638–645.

Aungst BJ. Novel formulation strategies for improving oral bioavailability of drugs with poor membrane permeation or presystemic metabolism. *J Pharm Sci.* 1993 Oct;82(10):979–987.

Avdeef A. Physicochemical profiling (solubility, permeability and charge state). *Curr Top Med Chem.* 2001 Sep;1(4):277–351.

Avdeef A. The rise of PAMPA. *Expert Opin Drug Metab Toxicol.* 2005 Aug;1(2):325–342.

Avdeef A and Testa B. Physicochemical profiling in drug research: A brief survey of the state-of-the-art of experimental techniques. *Cell Mol Life Sci.* 2002 Oct;59(10):1681–1689.

Bachmann KA and Ghosh R. The use of in vitro methods to predict in vivo pharmacokinetics and drug interactions. *Curr Drug Metab.* 2001 Sep;2(3):299–314.

Balimane PV, Chong S, and Morrison RA. Current methodologies used for evaluation of intestinal permeability and absorption. *J Pharmacol Toxicol Methods.* 2000 Jul–Aug;44(1):301–312.

Bergstrom CA. Computational models to predict aqueous drug solubility, permeability and intestinal absorption. *Expert Opin Drug Metab Toxicol.* 2005 Dec;1(4):613–627.

Bergstrom CA. In silico predictions of drug solubility and permeability: Two rate-limiting barriers to oral drug absorption. *Basic Clin Pharmacol Toxicol.* 2005 Mar;96(3):156–161.

Bittner B and Mountfield RJ. Intravenous administration of poorly soluble new drug entities in early drug discovery: The potential impact of formulation on pharmacokinetic parameters. *Curr Opin Drug Discov Devel.* 2002 Jan;5(1):59–71.

Blake JF. Chemoinformatics—Predicting the physicochemical properties of "drug-like" molecules. *Curr Opin Biotechnol.* 2000 Feb;11(1):104–107.

Blume HH and Schug BS. The biopharmaceutics classification system (BCS): Class III drugs—Better candidates for BA/BE waiver? *Eur J Pharm Sci.* 1999 Dec;9(2):117–121.

Brockmeier D. Mean time concept and component analysis in pharmacokinetics. *Int J Clin Pharmacol Ther.* 1999 Nov;37(11):555–561.

Buchwald P and Bodor N. Octanol-water partition: Searching for predictive models. *Curr Med Chem.* 1998 Oct;5(5):353–380.

Buchwald P and Bodor N. A simple, predictive, structure-based skin permeability model. *J Pharm Pharmacol.* 2001 Aug;53(8):1087–1098.

Caldwell GW, Ritchie DM, Masucci JA, Hageman W, and Yan Z. The new pre-preclinical paradigm: Compound optimization in early and late phase drug discovery. *Curr Top Med Chem.* 2001 Nov;1(5):353–366.

Camenisch G, Folkers G, and van de Waterbeemd H. Review of theoretical passive drug absorption models: Historical background, recent developments and limitations. *Pharm Acta Helv.* 1996 Nov;71(5):309–327.

Charman WN, Porter CJ, Mithani S, and Dressman JB. Physiochemical and physiological mechanisms for the effects of food on drug absorption: The role of lipids and pH. *J Pharm Sci.* 1997 Mar;86(3):269–282.

Chiou WL. Determination of drug permeability in a flat or distended stirred intestine. Prediction of fraction dose absorbed in humans after oral administration. *Int J Clin Pharmacol Ther.* 1994 Sep;32(9):474–482.

Chiou WL. We may not measure the correct intestinal wall permeability coefficient of drugs: Alternative absorptive clearance concept. *J Pharmacokinet Biopharm.* 1995 Jun;23(3):323–331.

Constantinides PP. Lipid microemulsions for improving drug dissolution and oral absorption: Physical and biopharmaceutical aspects. *Pharm Res.* 1995 Nov;12(11):1561–1572.

Doherty MM and Pang KS. First-pass effect: Significance of the intestine for absorption and metabolism. *Drug Chem Toxicol.* 1997 Nov;20(4):329–344.

Dokoumetzidis A, Kosmidis K, Argyrakis P, and Macheras P. Modeling and Monte Carlo simulations in oral drug absorption. *Basic Clin Pharmacol Toxicol.* 2005 Mar;96(3):200–205.

Dressman JB, Amidon GL, Reppas C, and Shah VP. Dissolution testing as a prognostic tool for oral drug absorption: Immediate release dosage forms. *Pharm Res.* 1998 Jan;15(1):11–22.

Dressman JB and Reppas C. In vitro-in vivo correlations for lipophilic, poorly water-soluble drugs. *Eur J Pharm Sci.* 2000 Oct;11(Suppl 2):S73–S80.

Faassen F and Vromans H. Biowaivers for oral immediate-release products: Implications of linear pharmacokinetics. *Clin Pharmacokinet.* 2004;43(15):1117–1126.

Fenstermacher JD. The blood-brain barrier is not a "barrier" for many drugs. *NIDA Res Monogr.* 1992;120:108–120.

Fleisher D, Li C, Zhou Y, Pao LH, and Karim A. Drug, meal and formulation interactions influencing drug absorption after oral administration. Clinical implications. *Clin Pharmacokinet.* 1999 Mar;36(3):233–254.

Golub AL, Frost RW, Betlach CJ, and Gonzalez MA. Physiologic considerations in drug absorption from the gastrointestinal tract. *J Allergy Clin Immunol.* 1986 Oct;78(4 Pt 2):689–694.

Goodhart FW and Eichman ML. Pharmaceutical sciences—1975: Literature review of pharmaceutics II. *J Pharm Sci.* 1976 Aug;65(8):1101–1139.

Greenblatt DJ, Arendt RM, and Shader RI. Pharmacodynamics of benzodiazepines after single oral doses: Kinetic and physiochemical correlates. *Psychopharmacology Suppl.* 1984;1:92–97.

Habgood MD, Begley DJ, and Abbott NJ. Determinants of passive drug entry into the central nervous system. *Cell Mol Neurobiol.* 2000 Apr;20(2):231–253.

Halsey MJ. Physical chemistry applied to anaesthetic action. *Br J Anaesth.* 1974 Mar;46(3):172–180.

Hidalgo IJ. Assessing the absorption of new pharmaceuticals. *Curr Top Med Chem.* 2001 Nov;1(5):385–401.

Hofmann AF and Roda A. Physicochemical properties of bile acids and their relationship to biological properties: An overview of the problem. *J Lipid Res.* 1984 Dec 15;25(13):1477–1489.

Horter D and Dressman JB. Influence of physicochemical properties on dissolution of drugs in the gastrointestinal tract. *Adv Drug Deliv Rev.* 2001 Mar 1;46(1–3):75–87.

Hu J, Johnston KP, and Williams RO 3rd. Nanoparticle engineering processes for enhancing the dissolution rates of poorly water soluble drugs. *Drug Dev Ind Pharm.* 2004 Mar;30(3):233–245.

Idson B. Vehicle effects in percutaneous absorption. *Drug Metab Rev.* 1983;14(2):207–222.

Kalia YN and Guy RH. Modeling transdermal drug release. *Adv Drug Deliv Rev.* 2001 Jun 11;48(2–3):159–172.

Kaushal AM, Gupta P, and Bansal AK. Amorphous drug delivery systems: Molecular aspects, design, and performance. *Crit Rev Ther Drug Carrier Syst.* 2004;21(3):133–193.

Kerns EH. High throughput physicochemical profiling for drug discovery. *J Pharm Sci.* 2001 Nov;90(11):1838–1858.

Kerns EH and Di L. Utility of mass spectrometry for pharmaceutical profiling applications. *Curr Drug Metab.* 2006 Jun;7(5):457–466.

Khanvilkar K, Donovan MD, and Flanagan DR. Drug transfer through mucus. *Adv Drug Deliv Rev.* 2001 Jun 11;48(2–3):173–193.

Klopman G and Zhu H. Recent methodologies for the estimation of n-octanol/water partition coefficients and their use in the prediction of membrane transport properties of drugs. *Mini Rev Med Chem.* 2005 Feb;5(2):127–133.

Kubinyi H. Lipophilicity and biological acitivity. Drug transport and drug distribution in model systems and in biological systems. *Arzneimittelforschung.* 1979;29(8):1067–1080.

Lennernas H. Does fluid flow across the intestinal mucosa affect quantitative oral drug absorption? Is it time for a reevaluation? *Pharm Res.* 1995 Nov;12(11):1573–1582.

Lennernas H and Abrahamsson B. The use of biopharmaceutic classification of drugs in drug discovery and development: Current status and future extension. *J Pharm Pharmacol.* 2005 Mar;57(3):273–285.

Li S, He H, Parthiban LJ, Yin H, and Serajuddin AT. IV-IVC considerations in the development of immediate-release oral dosage form. *J Pharm Sci.* 2005 Jul;94(7):1396–1417.

Lien EJ. Structure-activity relationships and drug disposition. *Annu Rev Pharmacol Toxicol.* 1981;21:31–61.

Lindenberg M, Kopp S, and Dressman JB. Classification of orally administered drugs on the World Health Organization Model list of Essential Medicines according to the biopharmaceutics classification system. *Eur J Pharm Biopharm.* 2004 Sep;58(2):265–278.

Lipinski CA, Lombardo F, Dominy BW, Feeney PJ. Experimental and computational approaches to estimate solubility and permeability in drug discovery and development settings. *Adv Drug Deliv Rev.* 2001 Mar 1;46(1–3):3–26.

Lipinski CA. Drug-like properties and the causes of poor solubility and poor permeability. *J Pharmacol Toxicol Methods.* 2000 Jul–Aug;44(1):235–249.

Loftsson T, Konradsdottir F, and Masson M. Influence of aqueous diffusion layer on passive drug diffusion from aqueous cyclodextrin solutions through biological membranes. *Pharmazie.* 2006 Feb;61(2):83–89.

Macheras P and Argyrakis P. Gastrointestinal drug absorption: Is it time to consider heterogeneity as well as homogeneity? *Pharm Res.* 1997 Jul;14(7):842–847.

Malkia A, Murtomaki L, Urtti A, and Kontturi K. Drug permeation in biomembranes: In vitro and in silico prediction and influence of physicochemical properties. *Eur J Pharm Sci.* 2004 Sep;23(1):13–47.

Manganaro AM. Drug delivery via the transmucosal drug delivery. *Mil Med.* 1997 Jan;162(1):27–30.

Manners CN, Payling DW, and Smith DA. Distribution coefficient, a convenient term for the relation of predictable physico-chemical properties to metabolic processes. *Xenobiotica.* 1988 Mar;18(3):331–350.

Markuszewski MJ, Wiczling P, and Kaliszan R. High-throughput evaluation of lipophilicity and acidity by new gradient HPLC methods. *Comb Chem High Throughput Screen.* 2004 Jun;7(4):281–289.

Martinez M, Amidon G, Clarke L, Jones WW, Mitra A, and Riviere J. Applying the biopharmaceutics classification system to veterinary pharmaceutical products. Part II. Physiological considerations. *Adv Drug Deliv Rev*. 2002 Oct 4;54(6):825–850.

Martinez MN and Amidon GL. A mechanistic approach to understanding the factors affecting drug absorption: A review of fundamentals. *J Clin Pharmacol*. 2002 Jun;42(6):620–643.

Masimirembwa CM, Bredberg U, and Andersson TB. Metabolic stability for drug discovery and development: Pharmacokinetic and biochemical challenges. *Clin Pharmacokinet*. 2003;42(6):515–528.

McCarley KD and Bunge AL. Pharmacokinetic models of dermal absorption. *J Pharm Sci*. 2001 Nov;90(11):1699–1719.

Mochizuki M. Relationship between the transfer rate and the gap in partial pressure of gas molecules at a heterogeneous interface. *Jpn J Physiol*. 1988;38(5):591–605.

Morillon V, Debeaufort F, Blond G, Capelle M, and Voilley A. Factors affecting the moisture permeability of lipid-based edible films: A review. *Crit Rev Food Sci Nutr*. 2002 Jan;42(1):67–89.

Muranishi S. Characteristics of drug absorption via the rectal route. *Methods Find Exp Clin Pharmacol*. 1984 Dec;6(12):763–772.

Murthy KS and Ghebre-Sellassie I. Current perspectives on the dissolution stability of solid oral dosage forms. *J Pharm Sci*. 1993 Feb;82(2):113–126.

Pagliara A, Reist M, Geinoz S, Carrupt PA, and Testa B. Evaluation and prediction of drug permeation. *J Pharm Pharmacol*. 1999 Dec;51(12):1339–1357.

Parker PR and Parker WA. Pharmacokinetic considerations in the haemodialysis of drugs. *J Clin Hosp Pharm*. 1982 Jun;7(2):87–99.

Peppin SS and Elliott JA. Non-equilibrium thermodynamics of concentration polarization. *Adv Colloid Interface Sci*. 2001 Sep 3;92(1–3):1–72.

Potsch L, Skopp G, and Moeller MR. Biochemical approach on the conservation of drug molecules during hair fiber formation. *Forensic Sci Int*. 1997 Jan 17;84(1–3):25–35.

Pouton CW. Formulation of poorly water-soluble drugs for oral administration: Physicochemical and physiological issues and the lipid formulation classification system. *Eur J Pharm Sci*. 2006 Nov;29(3–4): 278–287. Epub 2006 May 16.

Prescott LF. Gastrointestinal absorption of drugs. *Med Clin North Am*. 1974 Sep;58(5):907–916.

Raunio H, Taavitsainen P, Honkakoski P, Juvonen R, and Pelkonen O. In vitro methods in the prediction of kinetics of drugs: Focus on drug metabolism. *Altern Lab Anim*. 2004 Oct;32(4):425–430.

Ritschel WA. Sorption promotors in biopharmaceutics. *Angew Chem Int Ed Engl*. 1969 Oct;8(10):699–710.

Sanders NN, De Smedt SC, and Demeester J. The physical properties of biogels and their permeability for macromolecular drugs and colloidal drug carriers. *J Pharm Sci*. 2000 Jul;89(7):835–849.

Scheuplein RJ. Properties of the skin as a membrane. *Adv Biol Skin*. 1972;12:125–152.

Segall MD, Beresford AP, Gola JM, Hawksley D, and Tarbit MH. Focus on success: Using a probabilistic approach to achieve an optimal balance of compound properties in drug discovery. *Expert Opin Drug Metab Toxicol*. 2006 Apr;2(2):325–337.

Sergeev PV and Shimanovskii NL. Physicochemical mechanisms of the organotropism of drugs. *Vestn Akad Med Nauk SSSR*. 1984;(11):85–88. Russian.

Seydel JK, Coats EA, Cordes HP, and Wiese M. Drug membrane interaction and the importance for drug transport, distribution, accumulation, efficacy and resistance. *Arch Pharm (Weinheim)*. 1994 Oct;327(10):601–610.

Shah AC and Herd AK. Pharmaceutical sciences—P 1972: Literature review of pharmaceutics. II. *J Pharm Sci*. 1973 Aug;62(8):1217–1252.

Siewert M. Perspectives of in vitro dissolution tests in establishing in vivo/in vitro correlations. *Eur J Drug Metab Pharmacokinet*. 1993 Jan–Mar;18(1):7–18.

Singh SS. Preclinical pharmacokinetics: An approach towards safer and efficacious drugs. *Curr Drug Metab*. 2006 Feb;7(2):165–182.

Smith D, Schmid E, and Jones B. Do drug metabolism and pharmacokinetic departments make any contribution to drug discovery? *Clin Pharmacokinet*. 2002;41(13):1005–1019.

Stenberg P, Bergstrom CA, Luthman K, and Artursson P. Theoretical predictions of drug absorption in drug discovery and development. *Clin Pharmacokinet*. 2002;41(11):877–899.

Stenberg P, Luthman K, and Artursson P. Virtual screening of intestinal drug permeability. *J Control Release*. 2000 Mar 1;65(1–2):231–243.

Subramanian K. truPK—Human pharmacokinetic models for quantitative ADME prediction. *Expert Opin Drug Metab Toxicol*. 2005 Oct;1(3):555–564.

Sun D, Yu LX, Hussain MA, Wall DA, Smith RL, and Amidon GL. In vitro testing of drug absorption for drug 'developability' assessment: Forming an interface between in vitro preclinical data and clinical outcome. *Curr Opin Drug Discov Devel.* 2004 Jan;7(1):75–85.

Thoma K and Albert K. Amphiphilic drugs. 2. Relation between colloidal properties and pharmaceutic-technological, pharmacokinetic and pharmacodynamic behavior. *Pharmazie.* 1983 Dec;38(12):807–817. German.

Touray JC and Baillif P. In vitro assessment of the biopersistence of vitreous fibers: State of the art from the physical-chemical point of view. *Environ Health Perspect.* 1994 Oct;102(Suppl 5):25–30.

van de Waterbeemd H. The fundamental variables of the biopharmaceutics classification system (BCS): A commentary. *Eur J Pharm Sci.* 1998 Dec;7(1):1–3.

van de Waterbeemd H, Smith DA, and Jones BC. Lipophilicity in PK design: Methyl, ethyl, futile. *J Comput Aided Mol Des.* 2001 Mar;15(3):273–286.

Varma MV, Khandavilli S, Ashokraj Y, Jain A, Dhanikula A, Sood A, Thomas NS et al. Biopharmaceutic classification system: A scientific framework for pharmacokinetic optimization in drug research. *Curr Drug Metab.* 2004 Oct;5(5):375–388.

Verkman AS. Water permeability measurement in living cells and complex tissues. *J Membr Biol.* 2000 Jan 15;173(2):73–87.

Vippagunta SR, Brittain HG, and Grant DJ. Crystalline solids. *Adv Drug Deliv Rev.* 2001 May 16;48(1):3–26.

Weisiger RA. When is a carrier not a membrane carrier? The cytoplasmic transport of amphipathic molecules. *Hepatology.* 1996 Nov;24(5):1288–1295.

Wu CY and Benet LZ. Predicting drug disposition via application of BCS: Transport/absorption/elimination interplay and development of a biopharmaceutics drug disposition classification system. *Pharm Res.* 2005 Jan;22(1):11–23.

Youdim KA, Avdeef A, and Abbott NJ. In vitro trans-monolayer permeability calculations: Often forgotten assumptions. *Drug Discov Today.* 2003 Nov 1;8(21):997–1003.

Yu LX, Lipka E, Crison JR, Amidon GL. Transport approaches to the biopharmaceutical design of oral drug delivery systems: Prediction of intestinal absorption. *Adv Drug Deliv Rev.* 1996 Jun 12;19(3):359–376.

CHAPTER 3: DRUG DELIVERY FACTORS AFFECTING BIOEQUIVALENCE

Abdul S and Poddar SS. A flexible technology for modified release of drugs: Multi layered tablets. *J Control Release.* 2004 Jul 7;97(3):393–405.

Abrams J. Glyceryl trinitrate (nitroglycerin) and the organic nitrates. Choosing the method of administration. *Drugs.* 1987 Sep;34(3):391–403.

Aldridge MA and Ito MK. Colesevelam hydrochloride: A novel bile acid-binding resin. *Ann Pharmacother.* 2001 Jul–Aug;35(7–8):898–907.

Atkins PJ. Dry powder inhalers: An overview. *Respir Care.* 2005 Oct;50(10):1304–1312;discussion 1312.

Bach M and Lippold BC. Percutaneous penetration enhancement and its quantification. *Eur J Pharm Biopharm.* 1998 Jul;46(1):1–13.

Baker MT and Naguib M. Propofol: The challenges of formulation. *Anesthesiology.* 2005 Oct;103(4):860–876.

Baldi F. Lansoprazole oro-dispersible tablet: Pharmacokinetics and therapeutic use in acid-related disorders. *Drugs.* 2005;65(10):1419–1426.

Baldi F and Malfertheiner P. Lansoprazole fast disintegrating tablet: A new formulation for an established proton pump inhibitor. *Digestion.* 2003;67(1–2):1–5.

Bang LM and Keating GM. Paroxetine controlled release. *CNS Drugs.* 2004;18(6):355–364;discussion 365–366.

Behar-Cohen F. Drug delivery systems to target the anterior segment of the eye: Fundamental bases and clinical applications. *J Fr Ophtalmol.* 2002 May;25(5):537–544.

Behrend M and Braun F. Enteric-coated mycophenolate sodium: Tolerability profile compared with mycophenolate mofetil. *Drugs.* 2005;65(8):1037–1050.

Bejan A and Turcu G. Liposomes: Presentation and actual applicative trends in medicine. *Rom J Intern Med.* 1995 Jul–Dec;33(3–4):141–149.

Benninger MS. Amoxicillin/clavulanate potassium extended release tablets: A new antimicrobial for the treatment of acute bacterial sinusitis and community-acquired pneumonia. *Expert Opin Pharmacother.* 2003 Oct;4(10):1839–1846.

Bernkop-Schnurch A, Kast CE, and Guggi D. Permeation enhancing polymers in oral delivery of hydrophilic macromolecules: Thiomer/GSH systems. *J Control Release.* 2003 Dec 5;93(2):95–103.

Bianchi Porro G and Parente F. Antacids for duodenal ulcer: Current role. *Scand J Gastroenterol Suppl.* 1990;174:48–53.

Bittner B and Mountfield RJ. Intravenous administration of poorly soluble new drug entities in early drug discovery: The potential impact of formulation on pharmacokinetic parameters. *Curr Opin Drug Discov Devel.* 2002 Jan;5(1):59–71.

Bourquin J, Schmidli H, van Hoogevest P, and Leuenberger H. Basic concepts of artificial neural networks (ANN) modeling in the application to pharmaceutical development. *Pharm Dev Technol.* 1997 May;2(2):95–109.

Brocklebank D, Ram F, Wright J, Barry P, Cates C, Davies L, Douglas G, Muers M, Smith D, and White J. Comparison of the effectiveness of inhaler devices in asthma and chronic obstructive airways disease: A systematic review of the literature. *Health Technol Assess.* 2001;5(26):1–149.

Budde K, Glander P, Diekmann F, Waiser J, Fritsche L, Dragun D, and Neumayer HH. Review of the immunosuppressant enteric-coated mycophenolate sodium. *Expert Opin Pharmacother.* 2004 Jun;5(6):1333–1345.

Bunker M, Davies M, and Roberts C. Towards screening of inhalation formulations: Measuring interactions with atomic force microscopy. *Expert Opin Drug Deliv.* 2005 Jul;2(4):613–624.

Butcher EC. Can cell systems biology rescue drug discovery? *Nat Rev Drug Discov.* 2005 Jun;4(6):461–467.

Cabrera J, Redondo P, Becerra A, Garrido C, Cabrera J Jr, Garcia-Olmedo MA, Sierra A, Lloret P, and Martinez-Gonzalez MA. Ultrasound-guided injection of polidocanol microfoam in the management of venous leg ulcers. *Arch Dermatol.* 2004 Jun;140(6):667–673.

Chang TM. Artificial cells with emphasis on bioencapsulation in biotechnology. *Biotechnol Annu Rev.* 1995;1:267–295.

Chaubal MV. Application of drug delivery technologies in lead candidate selection and optimization. *Drug Discov Today.* 2004 Jul 15;9(14):603–609.

Chen D, Maa YF, and Haynes JR. Needle-free epidermal powder immunization. *Expert Rev Vaccines.* 2002 Oct;1(3):265–276.

Chew NY and Chan HK. The role of particle properties in pharmaceutical powder inhalation formulations. *J Aerosol Med.* 2002 Fall;15(3):325–330.

Chourasia MK and Jain SK. Pharmaceutical approaches to colon targeted drug delivery systems. *J Pharm Pharm Sci.* 2003 Jan–Apr;6(1):33–66.

Cleland JL. Protein delivery from biodegradable microspheres. *Pharm Biotechnol.* 1997;10:1–43.

Cole P. Pharmacologic and clinical comparison of cefaclor in immediate-release capsule and extended-release tablet forms. *Clin Ther.* 1997 Jul–Aug;19(4):617–625;discussion 603.

Constantinides PP, Tustian A, and Kessler DR. Tocol emulsions for drug solubilization and parenteral delivery. *Adv Drug Deliv Rev.* 2004 May 7;56(9):1243–1255.

Courrier HM, Butz N, and Vandamme TF. Pulmonary drug delivery systems: Recent developments and prospects. *Crit Rev Ther Drug Carrier Syst.* 2002;19(4–5):425–498.

Dahlof B and Andersson OK. A felodipine-metoprolol extended-release tablet: Its properties and clinical development. *J Hum Hypertens.* 1995 Jul;9(Suppl 2):S43–S47.

Dando TM and Scott LJ. Abacavir plus lamivudine: A review of their combined use in the management of HIV infection. *Drugs.* 2005;65(2):285–302.

Darkes MJ and Perry CM. Clarithromycin extended-release tablet: A review of its use in the management of respiratory tract infections. *Am J Respir Med.* 2003;2(2):175–201.

Davis SS and Illum L. Absorption enhancers for nasal drug delivery. *Clin Pharmacokinet.* 2003;42(13): 1107–1128.

Davis SS. Coming of age of lipid-based drug delivery systems. *Adv Drug Deliv Rev.* 2004 May 7;56(9):1241–1242.

De Moor R, Verbeeck R, and Martens L. Evaluation of long-term release of fluoride by type II glass ionomer cements with a conventional hardening reaction. *Rev Belge Med Dent.* 1996;51(3):22–35.

Dempski RE, Scholtz EC, Oberholtzer ER, and Yeh KC. Pharmaceutical design and development of a Sinemet controlled-release formulation. *Neurology.* 1989 Nov;39(11 Suppl 2):20–24.

Devlin JW, Welage LS, and Olsen KM. Proton pump inhibitor formulary considerations in the acutely ill. Part 1: Pharmacology, pharmacodynamics, and available formulations. *Ann Pharmacother.* 2005 Oct;39(10):1667–1677. Epub 2005 Aug 23.

Dhaon NA. Amoxicillin tablets for oral suspension in the treatment of acute otitis media: A new formulation with improved convenience. *Adv Ther.* 2004 Mar–Apr;21(2):87–95.

Digenis GA, Gold TB, and Shah VP. Cross-linking of gelatin capsules and its relevance to their in vitro-in vivo performance. *J Pharm Sci.* 1994 Jul;83(7):915–921.

Doelker E. Recent advances in tableting science. *Boll Chim Farm.* 1988 Feb;127(2):37–49.

Doelker E and Massuelle D. Benefits of die-wall instrumentation for research and development in tabletting. *Eur J Pharm Biopharm.* 2004 Sep;58(2):427–444.

Dowson AJ and Almqvist P. Part III: The convenience of, and patient preference for, zolmitriptan orally disintegrating tablet. *Curr Med Res Opin*. 2005;21(Suppl 3):S13–S17.

Du Pont HL. Nonfluid therapy and selected chemoprophylaxis of acute diarrhea. *Am J Med*. 1985 Jun 28;78(6B):81–90.

Duchene D and Ponchel G. Principle and investigation of the bioadhesion mechanism of solid dosage forms. *Biomaterials*. 1992;13(10):709–714.

DuPont HL. Bismuth subsalicylate in the treatment and prevention of diarrheal disease. *Drug Intell Clin Pharm*. 1987 Sep;21(9):687–693.

Faassen F and Vromans H. Biowaivers for oral immediate-release products: Implications of linear pharmacokinetics. *Clin Pharmacokinet*. 2004;43(15):1117–1126.

Fedorak RN and Bistritz L. Targeted delivery, safety, and efficacy of oral enteric-coated formulations of budesonide. *Adv Drug Deliv Rev*. 2005 Jan 6;57(2):303–316.

Felt O, Buri P, and Gurny R. Chitosan: A unique polysaccharide for drug delivery. *Drug Dev Ind Pharm*. 1998 Nov;24(11):979–993.

Femia RA and Goyette RE. The science of megestrol acetate delivery: Potential to improve outcomes in cachexia. *BioDrugs*. 2005;19(3):179–187.

Figgitt DP, Plosker GL. Saquinavir soft-gel capsule: An updated review of its use in the management of HIV infection. *Drugs*. 2000 Aug;60(2):481–516.

Flores NA. Ezetimibe + simvastatin (Merck/Schering-Plough). *Curr Opin Investig Drugs*. 2004 Sep;5(9):984–992.

Frijlink HW and De Boer AH. Dry powder inhalers for pulmonary drug delivery. *Expert Opin Drug Deliv*. 2004 Nov;1(1):67–86.

Frishman WH, Sherman D, and Feinfeld DA. Innovative drug delivery systems in cardiovascular medicine: Nifedipine-GITS and clonidine-TTS. *Cardiol Clin*. 1987 Nov;5(4):703–716.

Fu Y, Yang S, Jeong SH, Kimura S, and Park K. Orally fast disintegrating tablets: Developments, technologies, taste-masking and clinical studies. *Crit Rev Ther Drug Carrier Syst*. 2004;21(6):433–476.

Fuseau E, Petricoul O, Moore KH, Barrow A, and Ibbotson T. Clinical pharmacokinetics of intranasal sumatriptan. *Clin Pharmacokinet*. 2002;41(11):801–811.

Gallen CC. Strategic challenges in neurotherapeutic pharmaceutical development. *NeuroRx*. 2004 Jan;1(1):165–180.

Garg S, Kandarapu R, Vermani K, Tambwekar KR, Garg A, Waller DP, and Zaneveld LJ. Development pharmaceutics of microbicide formulations. Part I: Preformulation considerations and challenges. *AIDS Patient Care STDS*. 2003 Jan;17(1):17–32.

Garg S, Tambwekar KR, Vermani K, Kandarapu R, Garg A, Waller DP, and Zaneveld LJ. Development pharmaceutics of microbicide formulations. Part II: Formulation, evaluation, and challenges. *AIDS Patient Care STDS*. 2003 Aug;17(8):377–399.

Gehl J. Electroporation: Theory and methods, perspectives for drug delivery, gene therapy and research. *Acta Physiol Scand*. 2003 Apr;177(4):437–447.

Gill J and Feinberg J. Saquinavir soft gelatin capsule: A comparative safety. *Drug Saf*. 2001;24(3):223–232.

Gillis JC, Benfield P, and Goa KL. Transnasal butorphanol. A review of its pharmacodynamic and pharmacokinetic properties, and therapeutic potential in acute pain management. *Drugs*. 1995 Jul;50(1):157–175.

Goldenberg MM. An extended-release formulation of oxybutynin chloride for the treatment of overactive urinary bladder. *Clin Ther*. 1999 Apr;21(4):634–642.

Goldenheim PD, Conrad EA, and Schein LK. Treatment of asthma by a controlled-release theophylline tablet formulation: A review of the North American experience with nocturnal dosing. *Chronobiol Int*. 1987;4(3):397–408.

Goldstein JL, Larson LR, and Yamashita BD. Prevention of nonsteroidal anti-inflammatory drug-induced gastropathy: Clinical and economic implications of a single-tablet formulation of diclofenac/misoprostol. *Am J Manag Care*. 1998 May;4(5):687–697.

Gooding OW. Process optimization using combinatorial design principles: Parallel synthesis and design of experiment methods. *Curr Opin Chem Biol*. 2004 Jun;8(3):297–304.

Goodnow RA. Current practices in generation of small molecule new leads. *J Cell Biochem Suppl*. 2001;(Suppl 37):13–21.

Gotfried MH. Clarithromycin (Biaxin) extended-release tablet: A therapeutic review. *Expert Rev Anti Infect Ther*. 2003 Jun;1(1):9–20.

Gregoriadis G. Liposomes as a drug delivery system: Optimization studies. *Adv Exp Med Biol*. 1988;238: 151–159. No abstract available.

Grosdidier J, Boissel P, Bresler L, and Vidrequin A. Stenosing and perforated ulcers of the small intestine related to potassium chloride in enteric-coated tablets. Apropos of 11 cases. *Chirurgie*. 1989;115(3): 163–169. French.

Grzeszczak W. Cardura XL—A unique drug formulation—Doxazosine administered in a slow-release form (doxazosine GITS). *Przegl Lek*. 2000;57(11):643–654.

Gudgin Dickson EF, Goyan RL, and Pottier RH. New directions in photodynamic therapy. *Cell Mol Biol* (Noisy-le-grand). 2002 Dec;48(8):939–954.

Guyton JR. Extended-release niacin for modifying the lipoprotein profile. *Expert Opin Pharmacother*. 2004 Jun;5(6):1385–1398.

Hadgraft J. Passive enhancement strategies in topical and transdermal drug delivery. *Int J Pharm*. 1999 Jul 5;184(1):1–6.

Haefner S, Knietsch A, Scholten E, Braun J, Lohscheidt M, and Zelder O. Biotechnological production and applications of phytases. *Appl Microbiol Biotechnol*. 2005 Sep;68(5):588–597. Epub 2005 Oct 26.

Harashima H, Ishida T, Kamiya H, and Kiwada H. Pharmacokinetics of targeting with liposomes. *Crit Rev Ther Drug Carrier Syst*. 2002;19(3):235–275.

Harashima H, Shinohara Y, and Kiwada H. Intracellular control of gene trafficking using liposomes as drug carriers. *Eur J Pharm Sci*. 2001 Apr;13(1):85–89.

Hardy IJ, Fitzpatrick S, and Booth SW. Rational design of powder formulations for tamp filling processes. *J Pharm Pharmacol*. 2003 Dec;55(12):1593–1599.

Harsch IA. Inhaled insulins: Their potential in the treatment of diabetes mellitus. *Treat Endocrinol*. 2005;4(3):131–138.

Hatefi A and Amsden B. Camptothecin delivery methods. *Pharm Res*. 2002 Oct;19(10):1389–1399.

Hausheer FH, Kochat H, Parker AR, Ding D, Yao S, Hamilton SE, Petluru PN, Leverett BD, Bain SH, and Saxe JD. New approaches to drug discovery and development: A mechanism-based approach to pharmaceutical research and its application to BNP7787, a novel chemoprotective agent. *Cancer Chemother Pharmacol*. 2003 Jul;52(Suppl 1):S3–S15. Epub 2003 Jun 18.

Heinig R. Clinical pharmacokinetics of nisoldipine coat-core. *Clin Pharmacokinet*. 1998 Sep;35(3):191–208. Erratum in: *Clin Pharmacokinet*. 1998 Nov;35(5):390.

Hiestand EN. Mechanical properties of compacts and particles that control tableting success. *J Pharm Sci*. 1997 Sep;86(9):985–990.

Hirata K and Iwanami H. The role of triptans and analgesics for primary headache treatment. *Nippon Rinsho*. 2005 Oct;63(10):1797–1801. Japanese.

Hirschberger R. Pharmacokinetics and -dynamics of retard formation of isradipine. Summary of studies. *Fortschr Med*. 1993 Oct 30;111(30):481–484.

Huang Y, Leobandung W, Foss A, and Peppas NA. Molecular aspects of muco- and bioadhesion: Tethered structures and site-specific surfaces. *J Control Release*. 2000 Mar 1;65(1–2):63–71.

Itkin YM and Trujillo TC. Intravenous immunoglobulin-associated acute renal failure: Case series and literature. *Pharmacotherapy*. 2005 Jun;25(6):886–892.

Jones DH, Partidos CD, Steward MW, and Farrar GH. Oral delivery of poly(lactide-co-glycolide) encapsulated vaccines. *Behring Inst Mitt*. 1997 Feb;(98):220–228.

Jovanovic N, Bouchard A, Hofland GW, Witkamp GJ, Crommelin DJ, and Jiskoot W. Stabilization of proteins in dry powder formulations using supercritical fluid technology. *Pharm Res*. 2004 Nov;21(11):1955–1969.

Kalia YN and Guy RH. Modeling transdermal drug release. *Adv Drug Deliv Rev*. 2001 Jun 11;48(2–3):159–172.

Kanjickal DG and Lopina ST. Modeling of drug release from polymeric delivery systems—A review. *Crit Rev Ther Drug Carrier Syst*. 2004;21(5):345–386.

Karlsson L, Torstensson A, and Taylor LT. The use of supercritical fluid extraction for sample preparation of pharmaceutical formulations. *J Pharm Biomed Anal*. 1997 Feb;15(5):601–611.

Kaur IP and Kanwar M. Ocular preparations: The formulation approach. *Drug Dev Ind Pharm*. 2002 May;28(5):473–493.

Kleinebudde P. Roll compaction/dry granulation: Pharmaceutical applications. *Eur J Pharm Biopharm*. 2004 Sep;58(2):317–326.

Knop K. Active ingredient release from solid drug forms—Test methods, evaluation, influencing parameters. *Pharm Unserer Zeit*. 1999 Nov;28(6):301–308. German.

Knox ED and Stimmel GL. Clinical review of a long-acting, injectable formulation of risperidone. *Clin Ther*. 2004 Dec;26(12):1994–2002.

Kostarelos K. Rational design and engineering of delivery systems for therapeutics: Biomedical exercises in colloid and surface science. *Adv Colloid Interface Sci*. 2003 Dec 1;106:147–168.

Kozubek A, Gubernator J, Przeworska E, and Stasiuk M. Liposomal drug delivery, a novel approach: PLARosomes. *Acta Biochim Pol.* 2000;47(3):639–649.

Kuhlmann J. Alternative strategies in drug development: Clinical pharmacological aspects. *Int J Clin Pharmacol Ther.* 1999 Dec;37(12):575–583.

Kulkarni SB, Betageri GV, and Singh M. Factors affecting microencapsulation of drugs in liposomes. *J Microencapsul.* 1995 May–Jun;12(3):229–246.

Labiris NR and Dolovich MB. Pulmonary drug delivery. Part II: The role of inhalant delivery devices and drug formulations in therapeutic effectiveness of aerosolized medications. *Br J Clin Pharmacol.* 2003 Dec;56(6):600–612.

Laube BL. The expanding role of aerosols in systemic drug delivery, gene therapy, and vaccination. *Respir Care.* 2005 Sep;50(9):1161–1176.

Leuenberger H. New trends in the production of pharmaceutical granules: Batch versus continuous processing. *Eur J Pharm Biopharm.* 2001 Nov;52(3):289–296.

Levy RA. Therapeutic inequivalence of pharmaceutical alternates. *Am Pharm.* 1985 Apr;NS25(4):28–39.

Lightman S. Somatuline Autogel: An extended release lanreotide formulation. *Hosp Med.* 2002 Mar;63(3):162–165.

Maa YF and Prestrelski SJ. Biopharmaceutical powders: Particle formation and formulation considerations. *Curr Pharm Biotechnol.* 2000 Nov;1(3):283–302.

Machida Y. Development of topical drug delivery systems utilizing polymeric materials. *Yakugaku Zasshi.* 1993 May;113(5):356–368. Japanese.

Mainardes RM and Silva LP. Drug delivery systems: Past, present, and future. *Curr Drug Targets.* 2004 Jul;5(5):449–455.

Man M and Rugo H. Paclitaxel poliglumex. Cell Therapeutics/Chugai Pharmaceutical. *IDrugs.* 2005 Sep;8(9):739–754.

Maroni A, Zema L, Cerea M, and Sangalli ME. Oral pulsatile drug delivery systems. *Expert Opin Drug Deliv.* 2005 Sep;2(5):855–871.

McKay B, Hoogenraad M, Damen EW, and Smith AA. Advances in multivariate analysis in pharmaceutical process development. *Curr Opin Drug Discov Devel.* 2003 Nov;6(6):966–977.

Mehnert W and Mader K. Solid lipid nanoparticles: Production, characterization and applications. *Adv Drug Deliv Rev.* 2001 Apr 25;47(2–3):165–196.

Melia CD and Davis SS. Review article: Mechanisms of drug release from tablets and capsules. 2. Dissolution. *Aliment Pharmacol Ther.* 1989 Dec;3(6):513–525.

Michele TM, Knorr B, Vadas EB, and Reiss TF. Safety of chewable tablets for children. *J Asthma.* 2002 Aug;39(5):391–403.

Mooradian AD. Towards single-tablet therapy for type 2 diabetes mellitus. Rationale and recent developments. *Treat Endocrinol.* 2004;3(5):279–287.

Moribe K and Maruyama K. Pharmaceutical design of the liposomal antimicrobial agents for infectious disease. *Curr Pharm Des.* 2002;8(6):441–454.

Morissette SL, Almarsson O, Peterson ML, Remenar JF, Read MJ, Lemmo AV, Ellis S, Cima MJ, and Gardner CR. High-throughput crystallization: Polymorphs, salts, co-crystals and solvates of pharmaceutical solids. *Adv Drug Deliv Rev.* 2004 Feb 23;56(3):275–300.

Nagai T. New drug development by innovative drug administration—"Change" in pharmaceutical field. *Yakugaku Zasshi.* 1997 Nov;117(10–11):963–971.

Nail SL, Jiang S, Chongprasert S, and Knopp SA. Fundamentals of freeze-drying. *Pharm Biotechnol.* 2002;14:281–360.

Najib J. Fenofibrate in the treatment of dyslipidemia: A review of the data as they relate to the new suprabioavailable tablet formulation. *Clin Ther.* 2002 Dec;24(12):2022–2250.

Nishida K. Development of drug delivery system based on a new administration route for targeting to the specific region in the liver. *Yakugaku Zasshi.* 2003 Aug;123(8):681–689. Japanese.

Norman TR and Olver JS. New formulations of existing antidepressants: Advantages in the management of depression. *CNS Drugs.* 2004;18(8):505–520.

Nuijen B, Bouma M, Schellens JH, and Beijnen JH. Progress in the development of alternative pharmaceutical formulations of taxanes. *Invest New Drugs.* 2001 May;19(2):143–153.

Nunn T and Williams J. Formulation of medicines for children. *Br J Clin Pharmacol.* 2005 Jun;59(6):674–676.

Odegard PS and Capoccia KL. Inhaled insulin: Exubera. *Ann Pharmacother.* 2005 May;39(5):843–853. Epub April 12, 2005.

O'Hagan DT and Singh M. Microparticles as vaccine adjuvants and delivery systems. *Expert Rev Vaccines.* 2003 Apr;2(2):269–283.

Okada N. Design and creation of cytomedicine for application to cell therapy. *Yakugaku Zasshi.* 2005 Aug;125(8):601–615. Japanese.

Ormrod D and Goa KL. Intranasal metoclopramide. *Drugs.* 1999 Aug;58(2):315–322;discussion 323–324.

Owens DR, Zinman B, and Bolli G. Alternative routes of insulin delivery. *Diabet Med.* 2003 Nov;20(11):886–898.

Panchagnula R, Agrawal S, Ashokraj Y, Varma M, Sateesh K, Bhardwaj V, Bedi S et al. Fixed dose combinations for tuberculosis: Lessons learned from clinical, formulation and regulatory perspective. *Methods Find Exp Clin Pharmacol.* 2004 Nov;26(9):703–721.

Pathak P, Meziani MJ, and Sun YP. Supercritical fluid technology for enhanced drug delivery. *Expert Opin Drug Deliv.* 2005 Jul;2(4):747–761.

Patro SY, Freund E, and Chang BS. Protein formulation and fill-finish operations. *Biotechnol Annu Rev.* 2002;8:55–84.

Pawar R, Ben-Ari A, and Domb AJ. Protein and peptide parenteral controlled delivery. *Expert Opin Biol Ther.* 2004 Aug;4(8):1203–1212.

Perkins AC and Frier M. Nuclear medicine techniques in the evaluation of pharmaceutical formulations. *Pharm World Sci.* 1996 Jun;18(3):97–104.

Perkins AC and Frier M. Radionuclide imaging in drug development. *Curr Pharm Des.* 2004;10(24):2907–2921.

Pettipher R and Cardon LR. The application of genetics to the discovery of better medicines. *Pharmacogenomics.* 2002 Mar;3(2):257–263.

Pfister WR and Hsieh DS. Permeation enhancers compatible with transdermal drug delivery systems: Part II: System design considerations. *Med Device Technol.* 1990 Nov–Dec;1(6):28–33.

Pfister WR and Hsieh DS. Permeation enhancers compatible with transdermal drug delivery systems. Part I: Selection and formulation considerations. *Med Device Technol.* 1990 Sep–Oct;1(5):48–55. Erratum in: *Med Device Technol.* 1990 Nov–Dec;1(6):33.

Rabiskova M, Vostalova L, Medvecka G, and Horackova D. Hydrophilic gel matrix tablets for oral administration of drugs. *Ceska Slov Farm.* 2003 Sep;52(5):211–217. Czech.

Ram CV and Featherston WE. Calcium antagonists in the treatment of hypertension. An overview. *Chest.* 1988 Jun;93(6):1251–1253.

Ramsay EC, Dos Santos N, Dragowska WH, Laskin JJ, and Bally MB. The formulation of lipid-based nanotechnologies for the delivery of fixed dose anticancer drug combinations. *Curr Drug Deliv.* 2005 Oct;2(4):341–351.

Ranade VV. Drug delivery systems 5B. Oral drug delivery. *J Clin Pharmacol.* 1991 Feb;31(2):98–115.

Reddy KR. Controlled-release, pegylation, liposomal formulations: New mechanisms in the delivery of injectable drugs. *Ann Pharmacother.* 2000 Jul–Aug;34(7–8):915–923.

Reeves RR, Wallace KD, and Rogers-Jones C. Orally disintegrating olanzapine: A possible alternative to injection of antipsychotic drugs. *J Psychosoc Nurs Ment Health Serv.* 2004 May;42(5):44–48.

Richard A and Margaritis A. Poly(glutamic acid) for biomedical applications. *Crit Rev Biotechnol.* 2001;21(4):219–232.

Ritschel WA. Microemulsion technology in the reformulation of cyclosporine: The reason behind the pharmacokinetic properties of Neoral. *Clin Transplant.* 1996 Aug;10(4):364–373.

Sable D and Murakawa GJ. Quinolones in dermatology. *Dis Mon.* 2004 Jul;50(7):381–394.

Sams-Dodd F. Target-based drug discovery: Is something wrong? *Drug Discov Today.* 2005 Jan 15;10(2):139–147.

Schmidt PC and Christin I. Effervescent tablets—A nearly forgotten drug form. *Pharmazie.* 1990 Feb;45(2):89–101.

Schneider G and Fechner U. Computer-based de novo design of drug-like molecules. *Nat Rev Drug Discov.* 2005 Aug;4(8):649–663.

Schumacher HR Jr. Ketoprofen extended-release capsules: A new formulation for the treatment of osteoarthritis and rheumatoid arthritis. *Clin Ther.* 1994 Mar–Apr;16(2):145–159.

Seager H. Drug-delivery products and the Zydis fast-dissolving dosage form. *J Pharm Pharmacol.* 1998 Apr;50(4):375–382.

Sezaki H. Drug delivery systems. *Gan To Kagaku Ryoho.* 1985 Nov;12(11):2077–2082.

Sharma VK. Comparison of 24-hour intragastric pH using four liquid formulations of lansoprazole and omeprazole. *Am J Health Syst Pharm.* 1999 Dec 1;56(23 Suppl 4):S18–S21. Erratum in: *Am J Health Syst Pharm.* 2000 Apr 1;57(7):699.

Sharpe M, Ormrod D, and Jarvis B. Micronized fenofibrate in dyslipidemia: A focus on plasma high-density lipoprotein cholesterol (HDL-C) levels. *Am J Cardiovasc Drugs.* 2002;2(2):125–132;discussion 133–134.

Shigeyama M. Preparation of a gel-forming ointment base applicable to the recovery stage of bedsore and clinical evaluation of a treatment method with different ointment bases suitable to each stage of bedsore. *Yakugaku Zasshi*. 2004 Feb;124(2):55–67. Japanese.

Siepmann J and Gopferich A. Mathematical modeling of bioerodible, polymeric drug delivery systems. *Adv Drug Deliv Rev*. 2001 Jun 11;48(2–3):229–247.

Singh B, Dahiya M, Saharan V, and Ahuja N. Optimizing drug delivery systems using systematic "design of experiments." Part II: Retrospect and prospects. *Crit Rev Ther Drug Carrier Syst*. 2005;22(3):215–294.

Singh B, Kumar R, and Ahuja N. Optimizing drug delivery systems using systematic "design of experiments." Part I: Fundamental aspects. *Crit Rev Ther Drug Carrier Syst*. 2005;22(1):27–105.

Singla AK and Chawla M. Chitosan: Some pharmaceutical and biological aspects—An update. *J Pharm Pharmacol*. 2001 Aug;53(8):1047–1067.

Singla AK, Garg A, and Aggarwal D. Paclitaxel and its formulations. *Int J Pharm*. 2002 Mar 20;235(1–2):179–192.

Sinha VR and Kumria R. Microbially triggered drug delivery to the colon. *Eur J Pharm Sci*. 2003 Jan;18(1):3–18.

Smart JD. Buccal drug delivery. *Expert Opin Drug Deliv*. 2005 May;2(3):507–517.

Song H, Guo T, Zhang R, Zheng C, Ma Y, Li X, Bi K, and Tang X. Preparation of the traditional Chinese medicine compound recipe heart-protecting musk pH-dependent gradient-release pellets. *Drug Dev Ind Pharm*. 2002 Nov;28(10):1261–1273.

Speiser PP. Nanoparticles and liposomes: A state of the art. *Methods Find Exp Clin Pharmacol*. 1991 Jun;13(5):337–342.

Strickley RG. Solubilizing excipients in oral and injectable formulations. *Pharm Res*. 2004 Feb;21(2):201–230.

Sun Y, Peng Y, Chen Y, and Shukla AJ. Application of artificial neural networks in the design of controlled release drug delivery systems. *Adv Drug Deliv Rev*. 2003 Sep 12;55(9):1201–1215.

Swainston Harrison T and Keating GM. Extended-release carbamazepine capsules: In bipolar I disorder. *CNS Drugs*. 2005;19(8):709–716.

Takada S and Ogawa Y. Design and development of controlled release of drugs from injectable microcapsules. *Nippon Rinsho*. 1998 Mar;56(3):675–679.

Takakura Y, Nishikawa M, Yamashita F, and Hashida M. Influence of physicochemical properties on pharmacokinetics of non-viral vectors for gene delivery. *J Drug Target*. 2002 Mar;10(2):99–104.

Takayama K, Fujikawa M, and Nagai T. Artificial neural network as a novel method to optimize pharmaceutical formulations. *Pharm Res*. 1999 Jan;16(1):1–6.

Takayama K, Fujikawa M, Obata Y, and Morishita M. Neural network based optimization of drug formulations. *Adv Drug Deliv Rev*. 2003 Sep 12;55(9):1217–1231.

Takeuchi H, Thongborisute J, Matsui Y, Sugihara H, Yamamoto H, and Kawashima Y. Novel mucoadhesion tests for polymers and polymer-coated particles to design optimal mucoadhesive drug delivery systems. *Adv Drug Deliv Rev*. 2005 Nov 3;57(11):1583–1594. Epub 2005 Sep 16.

Tobyn M, Staniforth JN, Morton D, Harmer Q, and Newton ME. Active and intelligent inhaler device development. *Int J Pharm*. 2004 Jun 11;277(1–2):31–37.

Todd PA and Faulds D. Felodipine. A review of the pharmacology and therapeutic use of the extended release formulation in cardiovascular disorders. *Drugs*. 1992 Aug;44(2):251–277.

Toguchi H, Ogawa Y, Okada H, and Yamamoto M. Once-a-month injectable microcapsules of leuprorelin acetate. *Yakugaku Zasshi*. 1991 Aug;111(8):397–409.

Turker S, Onur E, and Ozer Y. Nasal route and drug delivery systems. *Pharm World Sci*. 2004 Jun;26(3):137–142.

Tye H. Application of statistical 'design of experiments' methods in drug discovery. *Drug Discov Today*. 2004 Jun 1;9(11):485–491.

Valenta C and Auner BG. The use of polymers for dermal and transdermal delivery. *Eur J Pharm Biopharm*. 2004 Sep;58(2):279–289.

Wagstaff AJ and Figgitt DP. Extended-release metformin hydrochloride. Single-composition osmotic tablet formulation. *Treat Endocrinol*. 2004;3(5):327–332.

Wagstaff AJ and Goa KL. Once-weekly fluoxetine. *Drugs*. 2001;61(15):2221–2228; discussion 2229–2230.

Wassef NM, Alving CR, and Richards RL. Liposomes as carriers for vaccines. *Immunomethods*. 1994 Jun;4(3):217–222.

Wellington K. Rosiglitazone/Metformin. *Drugs*. 2005;65(11):1581–1592; discussion 1593–1594.

Wernsdorfer WH. Coartemether (artemether and lumefantrine): An oral antimalarial drug. *Expert Rev Anti Infect Ther*. 2004 Apr;2(2):181–196.

White NS and Errington RJ. Fluorescence techniques for drug delivery research: Theory and practice. *Adv Drug Deliv Rev*. 2005 Jan 2;57(1):17–42.

Willems L, van der Geest R, and de Beule K. Itraconazole oral solution and intravenous formulations: A review of pharmacokinetics and pharmacodynamics. *J Clin Pharm Ther*. 2001 Jun;26(3):159–169.

Wissing SA, Kayser O, and Muller RH. Solid lipid nanoparticles for parenteral drug delivery. *Adv Drug Deliv Rev*. 2004 May 7;56(9):1257–1272.

Wouters J and Ooms F. Small molecule crystallography in drug design. *Curr Pharm Des*. 2001 May;7(7):529–545.

Yalkowsky SH, Krzyzaniak JF, and Ward GH. Formulation-related problems associated with intravenous drug delivery. *J Pharm Sci*. 1998 Jul;87(7):787–796.

Yeo Y and Park K. Control of encapsulation efficiency and initial burst in polymeric microparticle systems. *Arch Pharm Res*. 2004 Jan;27(1):1–12.

Yilmaz E and Borchert HH. Effect of lipid-containing, positively charged nanoemulsions on skin hydration, elasticity and erythema—An in vivo study. *Int J Pharm*. 2006 Jan 13;307(2):232–238. Epub Nov 11, 2005.

Young SS, Lam RL, and Welch WJ. Initial compound selection for sequential screening. *Curr Opin Drug Discov Devel*. 2002 May;5(3):422–427.

Zargar A, Basit A, and Mahtab H. Sulphonylureas in the management of type 2 diabetes during the fasting month of Ramadan. *J Indian Med Assoc*. 2005 Aug;103(8):444–446.

Zhang GG, Law D, Schmitt EA, and Qiu Y. Phase transformation considerations during process development and manufacture of solid oral dosage forms. *Adv Drug Deliv Rev*. 2004 Feb 23;56(3):371–390.

Zimmermann U, Mimietz S, Zimmermann H, Hillgartner M, Schneider H, Ludwig J, Hasse C, Haase A, Rothmund M, and Fuhr G. Hydrogel-based non-autologous cell and tissue therapy. *Biotechniques*. 2000 Sep;29(3):564–572, 574, 576 *passim*.

Zore M, Harris A, Tobe LA, Siesky B, Januleviciene I, Behzadi J, Amireskandari A, Egan P, Garff K, and Wirostko B. Generic medications in ophthalmology. *Br J Ophthalmol*. 2013 Mar;97(3):253–257.

CHAPTER 4: PHARMACOKINETIC-PHARMACODYNAMIC MODELING

Aarons L. Population pharmacokinetics: Theory and practice. *Br J Clin Pharmacol*. 1991;32:669–670.

Bischoff KB and Brown RG. Drug distribution in mammals. *Chem. Eng. Prog. Symp*. 1966;62:33–45.

Campbell DB. The use of kinetic-dynamic interactions in the evaluation of drugs. *Psychopharmacology*. 1990;100(4):433–450.

Carithers RL et al. Methylprednisolone therapy in patients with severe alcoholic hepatitis. *Ann Intern Med*. 1989;110:685–690.

Figg WD et al. Comparison of quantitative methods to assess hepatic function: Pugh's classification, indocyanine green, antipyrine, and dextromethorphan. *Pharmacotherapy*. 1995;15:693–700.

Fuseau E and Sheiner LB. Simultaneous modeling of pharmacokinetics and pharmacodynamics with a nonparametric pharmacodynamic model. *Clin Pharmacol Ther*. 1984;35(6):733–741.

Hammarlund-Udenaes M and Benet LZ. Furosemide pharmacokinetics and pharmacodynamics in health and disease: An update. *J Pharmacokinet Biopharm*. 1989;17(1):1–46.

Holford NH and Sheiner LB. Pharmacokinetic and pharmacodynamic modeling in vivo. *Crit Rev Bioeng*. 1981;5(4):273–322.

Holford NH and Sheiner LB. Kinetics of pharmacologic response. *Pharmacol Ther*. 1982;16(2):143–166.

Karlsson MO et al. A general model for time–dissociated pharmacokinetic–pharmacodynamic relationships exemplified by paclitaxel myelosuppression. *Clin Pharmacol Ther*. 1998;63(1):11–25.

Landoni MF and Lees P. Pharmacokinetic/pharmacodynamic modeling of non-steroidal anti-inflammatory drugs. *J Vet Pharmacol Ther*. 1997;20(suppl 1):118–120.

Machado SG, Miller R, and Hu C. A regulatory perspective on Pharmacokinetic/Pharmacodynamic modeling. *Stat. Methods Med. Res*. 1999;8(3):217–245.

Maddrey WC et al. Corticosteroid therapy of alcoholic hepatitis. *Gastroenterology*. 1978;75:193–199.

Mager DE and Jusko WJ. Pharmacodynamic modeling of time-dependent transduction systems. *Clin Pharmacol Ther*. 2001;70(3):210–216.

Mattie H. Antibiotic efficacy in vivo predicted by in vitro activity. *Int J Antimicrob Agents*. 2000;14(2):91–98.

Niazi S. Volume of distribution and tissue level errors in instantaneous intravenous input assumptions. *J Pharm Sci*. 1976;65(10):1539–1540.

Peck CC et al. Opportunities for integration of pharmacokinetics, pharmacodynamics, and toxicokinetics in rational drug development. *Int J Pharm*. 1992;82:9–19.

Pugh RNH et al. Transection of the oesophagus for bleeding oesophageal varices. *Brit J Surg*. 1973;60:646–649.

Sheiner LB and Steimer JL. Pharmacokinetic/Pharmacodynamic modeling in drug development. *Ann Rev Pharmacol Toxicol*. 2000;40:67–95.

Tang H-S and Hu OY-P. Assessment of liver function using a novel galactose single point method. *Digestion*. 1992;52:222–231.

Teorell T. Kinetics of distribution of substances administered to the body. I: The extravascular modes of administration. *Arch Intern Pharmacodyn.* 1937;57:205–225.

Teorell T. Kinetics of distribution of substances administered to the body. II: The intravascular mode of administration. *Arch Intern Pharmacodyn.* 1937;57:226–240.

Testa R et al. Monoethylglycinexylidide formation measurement as a hepatic function test to assess severity of chronic liver disease. *Am J Gastroenterol.* 1997;92:2268–2273.

Toutain PL. Pharmacokinetic/Pharmacodynamic integration in drug development and dosage-regimen optimization for veterinary medicine. *AAPS Pharm Sci.* 2002;4(4):1–29.

Wiesner RH et al. Primary sclerosing cholangitis: Natural history, prognostic factors and survival analysis. *Hepatology.* 1989;10:430–436.

Zakim D and Boyer TD (1996) *Hepatology: A Textbook of Liver Disease.* WB Saunders Co., Philadelphia, PA.

CHAPTER 5: BIOEQUIVALENCE TESTING RATIONALE AND PRINCIPLES

Allec A, Chatelus A, and Wagner N. Skin distribution and pharmaceutical aspects of adapalene gel. *J Am Acad Dermatol.* 1997;36:S119–S125.

Bhattycharyya L, Dabbah R, Hauck W, Sheinin E, Yeoman L, and Williams R. Equivalence studies for complex active ingredients and dosage forms. *AAPS J.* 2005 17 Nov;7(4):E786–E812.

Byron PR and Notari RE. Critical analysis of "Flip-flop" phenomenon in two-compartment pharmacokinetic model. *J Pharm Sci.* 1976;65:1140–1144.

Caron D, Queille-Roussel C, Shah VP, and Schaefer H. Correlation between the drug penetration and the blanching effect of topically applied hydrocortisone creams in human beings. *J Am Acad Dermatol.* 1990;23:458–462.

Chen ML, Shah VP, Patnaik R, Adams W, Hussain A, Conner D, Mehta M et al. Bioavailability and bioequivalence: An FDA regulatory overview. *Pharm Res.* 2001 Dec;18(12):1645–1650.

Chen ML, Shah VP, Crommelin DJ, Shargel L, Bashaw D, Bhatti M, Blume H et al. Harmonization of regulatory approaches for evaluating therapeutic equivalence and interchangeability of multisource drug products: Workshop summary report. *Eur J Pharm Sci.* 2011b Nov 20;44(4):506–513.

Cifani C, Costantino S, Massi M, and Berrino L. Commercially available lipid formulations of amphotericin b: Are they bioequivalent and therapeutically equivalent? *Acta Biomed.* 2012 Aug;83(2):154–163.

D'Argenio DA. Optimal sampling times for pharmacokinetic experiments. *J. Pharm Biopharm.* 1981;9:739–756.

Devine JW, Cline RR, and Farley JF. Follow-on biologics: Competition in the biopharmaceutical marketplace. *J Am Pharm Assoc.* (Washington, DC). 2006 Mar–Apr;46(2):193–201;quiz 202–204.

Endrenyi L and Tothfalusi L. Metrics for the evaluation of bioequivalence of modified-release formulations. *AAPS J.* 2012 Dec;14(4):813–819.

Fareed J, Leong W, Hoppensteadt DA, Jeske WP, Walenga J, Bick RL. Development of generic low molecular weight heparins: A perspective. *Hematol Oncol Clin North Am.* 2005 Feb;19(1):53–68, v–vi.

Fareed J, Leong WL, Hoppensteadt DA, Jeske WP, Walenga J, Wahi R, and Bick RL. Generic low-molecular-weight heparins: Some practical considerations. *Semin Thromb Hemost.* 2004 Dec;30(6):703–713.

FDA. *Topical Dermatological Corticosteroids: In Vivo Bioequivalence*, CDER, June 1995.

FDA. *SUPAC-SS Nonsterile Semisolid Dosage Forms, Scale-Up and Post Approval Changes: Chemistry, Manufacturing, and Controls; In Vitro Release Testing and In Vivo Bioequivalence Documentation*, Center for Drug Evaluation and Research (CDER), May 1997.

FDA. *Guidance for Industry: Waiver of in vivo Bioavailability and Bioequivalence Studies for IR Solid Oral Dosage Forms Based on a Biopharmaceutics Classification System*, Federal Drug and Food Administration, Rockville, MD, 2002.

Forbes A, Cartwright A, Marchant S, McIntyre P, and Newton M. Review article: Oral, modified-release mesalazine formulations—proprietary versus generic. *Aliment Pharmacol Ther.* 2003 May 15;17(10):1207–1214.

Graffner C. Regulatory aspects of drug dissolution from a European perspective. *Eur J Pharm Sci.* 2006 Nov;29(3–4):288–293. Epub May 16, 2006.

Hauschke D, Steinijans VW, Diletti E, and Burke M. Sample size determinationfor bioequivalence assessment using a multiplicative model. *J Pharm Biopharm.* 1992;20:557–561.

Hueber F, Schaefer H, and Wepierre J. Role of transepidermal and transfollicular routes in percutaneous absorption of steroids: In vitro studies on human skin. *Skin Pharmacol.* 1994;7:237–244.

Hung HM, Wang SJ, and O'Neill R. A regulatory perspective on choice of margin and statistical inference issue in non-inferiority trials. *Biom J.* 2005 Feb;47(1):28–36;discussion 99–107.

Illel B, Schaefer H, Wepierre J, and Doucet O. Follicles play an important role in percutaneous absorption. *J Pharm Sci.* 1991;80:424–427.

Jackson AJ, Robbie G, and Marroum P. Metabolites and bioequivalence: Past and present. *Clin Pharmacokinet.* 2004;43(10):655–672.

Joshi L and Lopez LC. Bioprospecting in plants for engineered proteins. *Curr Opin Plant Biol.* 2005 Apr;8(2):223–236.

Kaminski L, Schepers U, and Wätzig H. Analytical method transfer using equivalence tests with reasonable acceptance criteria and appropriate effort: Extension of the ISPE concept. *J Pharm Biomed Anal.* 2010 Dec 15;53(5):1124–1129.

Karalis V and Macheras P. Current regulatory approaches of bioequivalence testing. *Expert Opin Drug Metab Toxicol.* 2012 Aug;8(8):929–942.

Karalis V, Symillides M, and Macheras P. Novel methods to assess bioequivalence. *Expert Opin Drug Metab Toxicol.* 2011 Jan;7(1):79–88.

Keiding N and Budtz-Jorgensen E. The Precautionary Principle and statistical approaches to uncertainty. *Int J Occup Med Environ Health.* 2004;17(1):147–151.

Koren G, Nordeng H, and MacLeod S. Gender differences in drug bioequivalence: Time to rethink practices. *Clin Pharmacol Ther.* 2013 Mar;93(3):260–262.

Lathers CM. Challenges and opportunities in animal drug development: A regulatory perspective. *Nat Rev Drug Discov.* 2003 Nov;2(11):915–918.

Lees P, Hunter RP, Reeves PT, and Toutain PL. Pharmacokinetics and pharmacodynamics of stereoisomeric drugs with particular reference to bioequivalence determination. *J Vet Pharmacol Ther.* 2012 Apr;35 Suppl 1:17–29.

Lennernas H and Abrahamsson B. The use of biopharmaceutic classification of drugs in drug discovery and development: Current status and future extension. *J Pharm Pharmacol.* 2005 Mar;57(3):273–285.

Lovering E, McGilveray I, McMillan I, and Tostowaryk W. Comparative bioavailabilies from truncated blood level curves. *J Pharm Sci.* 1975;64:1521–1524.

Maibach HI (ed.) (1996) *Dermatologic Research Techniques.* CRC Press Inc., Boca Raton, FL.

Martinez MN and Hunter RP. Current challenges facing the determination of product bioequivalence in veterinary medicine. *J Vet Pharmacol Ther.* 2010 Oct;33(5):418–433.

Martinez MN and Jackson AJ. Suitability of various noninfinity area under the plasma concentration-time curve (AUC) estimates for use in bioequivalence determinations: Relationship to AUC from zero to time infinity (AUC(0-INF)). *Pharm Res.* 1991;8:512–517.

Martinez MN and Riviere JE. Review of the 1993 veterinary drug bioequivalence workshop. *J Vet Pharm Therap.* 1994;17:85–119.

Mathias NR and Crison J. The use of modeling tools to drive efficient oral product design. *AAPS J.* 2012 Sep;14(3):591–600.

Mehta CR, Patel NR, and Tsiatis AA. Exact significance testing to establish treatment equivalence with ordered categorical data. *Biometrics.* 1984 Sep;40:819–825.

Midha KK, Rawson MJ, and Hubbard JW. The bioequivalence of highly variable drugs and drug products. *Int J Clin Pharmacol Ther.* 2005 Oct;43(10):485–498.

Mitra A and Wu Y. Challenges and opportunities in achieving bioequivalence for fixed-dose combination products. *AAPS J.* 2012 Sep;14(3):646–655.

Parry GE, Dunn P, Shah VP, and Pershing LK, Acyclovir bioavailability in human skin. *J Invest Dermatol* 1992;98:856–863.

Pershing L, Silver BS, Krueger GG et al. Feasibility of measuring the bioavailability of topical betamethasone dipropionate in commercial formulations using drug content in skin and a skin blanching bioassay. *Pharm Res.* 1992;9:45–51.

Pershing LK, Corlett J, and Jorgensen C. In vivo pharmacokinetics and pharmacodynamics of topical ketoconazole and miconazole in human stratum corneum. *Antimicrob Agents Chemother.* 1994;38:90–95.

Pershing LK, Lambert L, Wright ED et al. Topical 0.05% betamethasone dipropionate: Pharmacokinetic and pharmacodynamic dose-response studies in humans. *Arch Dermatol.* 1994;130:740–747.

Pershing LK, Lambert LD, Shah VP, and Lam SY. Variability and correlation of chromameter and tape-stripping methods with the visual skin blanching assay in quantitative assessment of topical 0.05% betamethasone dipropionate bioavailability in humans. *Int J Pharmaceutics.* 1992;86:201–210.

Price D, Summers M, and Zanen P. Could interchangeable use of dry powder inhalers affect patients? *Int J Clin Pract Suppl.* 2005 Dec;(149):3–6.

Raw AS, Furness MS, Gill DS, Adams RC, Holcombe FO, and Yu LX. Regulatory considerations of pharmaceutical solid polymorphism in Abbreviated New Drug Applications (ANDAs). *Adv Drug Deliv Rev.* 2004 Feb 23;56(3):397–414.

Ronfeld RA and Benet LZ. Interpretation of plasma concentration-time curves after oral dosing. *J Pharm Sci* 1977;66:178–180.

Rougier A, Dupuis D, Lotte C et al. In vivo correlation between stratum corneum reservoir function and percutaneous absorption. *J Inv Derm.* 1983;81:275–278.

Rougier A, Dupuis D, Lotte C et al. Regional variation in percutaneous absorption in man: Measurement by the stripping method. *Arch Dermatol Res.* 1986;278:465–469.

Rougier A, Rallis M, Kiren P and Lotte C. In vivo percutaneous absorption: A key role for stratum corneum/vehicle partitioning. *Arch Dermatol Res.* 1990;282:498–505.

Rudolph MW, Klein S, Beckert TE, Petereit H, and Dressman JB. A new 5-aminosalicylic acid multi-unit dosage form for the therapy of ulcerative colitis. *Eur J Pharm Biopharm.* 2001 May;51(3):183–190.

Sampaio C, Costa J, and Ferreira JJ. Clinical comparability of marketed formulations of botulinum toxin. *Mov Disord.* 2004 Mar;19(Suppl 8):S129–S136.

Schaefer H and Redelmeir TE (eds.) (1996) *Skin Barrier, Principles of Percutaneous Absorption.* Karger Publishers, New York.

Schellekens H. Follow-on biologics: Challenges of the "next generation". *Nephrol Dial Transplant.* 2005 May;20(Suppl 4):iv31–iv36.

Schuirmann DJ. A comparison of the two one-sided tests procedure and the power approach for assessing the equivalence of average bioavailability. *J Pharm Biopharm.* 1987;15:657–680.

Shah VP and Maibach HI (eds.) (1993) *Topical Drug Bioavailability, Bioequivalence and Penetration.* Plenum Press, New York.

Shah VP, Clynn GL, Yacobi A et al. Bioequivalence of topical dosage forms—Methods of evaluation of bioequivalence. Workshop report. *Pharm Res.* 1998;15:167–171, 1998.

Shah VP, Midha KK, Dighe S et al. Analytical methods validation: Bioavailability, bioequivalence and pharmacokinetic studies. Workshop report. *Pharm Res.* 1992;9:588–592.

Srinivas NR. Considerations for metabolite pharmacokinetic data in bioavailability/bioequivalence assessments. Overview of the recent trends. *Arzneimittelforschung.* 2009;59(4):155–165.

Steinijans VW, Hauck WW, Diletti E, Hauschke D, and Anderson S. Effect of changing the bioequivalence range from (0.80, 1.20) to (0.80, 1.25) on the power and sample size. *J Clin Pharm Therap Toxicol.* 1992;30:571–575.

Surber C, Wilhelm K-P, Bermann D, and Maibach HI. In vivo skin penetration of acitretin in volunteers using three sampling techniques. *Pharm Res.* 1993;9:1291–1294.

Verbeeck RK, Kanfer I, and Walker RB. Generic substitution: The use of medicinal products containing different salts and implications for safety and efficacy. *Eur J Pharm Sci.* 2006 May;28(1–2):1–6. Epub January 18, 2006.

Ward LS. Levothyroxine and the problem of interchangeability of drugs with narrow therapeutic index. *Arq Bras Endocrinol Metabol.* 2011 Oct;55(7):429–434.

Westlake WJ (1988) Bioavailability and bioequivalence of pharmaceutical formulations. In K. Peace (Ed.), *Biopharmaceutical Statistics for Drug Development*, Marcel Dekker, Inc., New York, pp. 329–352.

Wilding I. Bioequivalence testing for locally acting gastrointestinal products: What role for gamma scintigraphy? *J Clin Pharmacol.* 2002 Nov;42(11):1200–1210.

Penner N, Xu L, and Prakash C. Radiolabeled absorption, distribution, metabolism, and excretion studies in drug development: Why, when, and how? *Chem Res Toxicol.* 2012 Mar 19;25(3):513–531.

CHAPTER 6: BIOEQUIVALENCE WAIVERS

Aamir MN, Ahmad M, Akhtar N et al. Development and in vitro-in vivo relationship of controlled-release microparticles loaded with tramadol hydrochloride. *Int J Pharm.* 2011;407:38–43.

Abdou HM (1989) *Dissolution, Bioavailability and Bioequivalence.* Mack Printing, Easton, PA.

Alkhalidi BA, Ghazawi MA, AlKhatib HS et al. Development of a predictive in vitro dissolution for clarithromycin granular suspension based on in vitro-in vivo correlations. *Pharmaceut Develop Technol.* 2010;15:286–295.

Amidon GL, Lennernas H, Shah VP, and Crison JR. A theoretical basis for a biopharmaceutic drug classification: The correlation of in vitro drug product dissolution and in vivo bioavailabilty. *Pharm Res* 1995;12:413–419.

Amidon GL, Robinson JR, and Williams RL (1997) *Scientific Foundations for Regulating Drug Product Quality.* American Association of Pharmaceutical Scientists, AAPS Press, Alexandria, VA, pp. 99–113.

Amidon GL, Lennernäs H, Shah VP, and Crison JR, A theoretical basis for a biopharmaceutics drug classification: The correlation of in vitro drug product dissolution and in vivo bioavailability. *Pharm Res.* 1995;12:413–420.

Araújo LU, Albuquerque KT, Kato KC, Silveira GS, Maciel NR, Spósito PÁ, Barcellos NM, Souza Jd, Bueno M, and Storpirtis S. Generic drugs in Brazil: Historical overview and legislation. *Rev Panam Salud Publica.* 2010 Dec;28(6):480–492.

Arnal J, Gonzalez-Alvarez I, Bermejo M, Amidon GL, Junginger HE, Kopp S, Midha KK et al. Biowaiver monographs for immediate release solid oral dosage forms: Aciclovir. *J Pharm Sci.* 2008 Dec;97(12):5061–5073.

Avramoff A and Domb AJ. In-vitro and in-vivo characteristics of a modified-release double-pulse formulation for a water soluble drug. *Int J Clin Pharmacol Ther.* 2010;48:250–258.

Bansal T, Singh M, Mishra G, Talegaonkar S, Khar RK, Jaggi M, and Mukherjee R. Concurrent determination of topotecan and model permeability markers (atenolol, antipyrine, propranolol and furosemide) by reversed phase liquid chromatography: Utility in Caco-2 intestinal absorption studies. *J Chromatogr B Analyt Technol Biomed Life Sci.* 2007 Nov 15;859(2):261–266.

Baynes R, Riviere J, Franz T, Monteiro-Riviere N, Lehman P, Peyrou M, Toutain. Challenges obtaining a biowaiver for topical veterinary dosage forms. *J Veterinary Pharmacology and Therapeutics.* 2012;35:103–114.

Beal SL and Sheiner LB (1992) NONMEM User's Guides, NONMEM Project Group, University of California, San Francisco, CA.

Becker C, Dressman JB, Amidon GL, Junginger HE, Kopp S, Midha KK, Shah VP, Stavchansky S, and Barends DM. Biowaiver monographs for immediate release solid oral dosage forms: Isoniazid. *J Pharm Sci.* 2006 Nov 20.

Becker C, Dressman JB, Amidon GL, Junginger HE, Kopp S, Midha KK, Shah VP, Stavchansky S, and Barends DM. Biowaiver monographs for immediate release solid oral dosage forms: Pyrazinamide. *J Pharm Sci.* 2008a Sep;97(9):3709–3720.

Becker C, Dressman JB, Amidon GL, Junginger HE, Kopp S, Midha KK, Shah VP, Stavchansky S, and Barends DM. Biowaiver monographs for immediate release solid oral dosage forms: Ethambutol dihydrochloride. *J Pharm Sci.* 2008b Apr;97(4):1350–1360.

Becker C, Dressman JB, Amidon GL, Junginger HE, Kopp S, Midha KK, Shah VP, Stavchansky S, Barends DM, and International Pharmaceutical Federation, Groupe BCS. Biowaiver monographs for immediate release solid oral dosage forms: Isoniazid. *J Pharm Sci.* 2007 Mar;96(3):522–531.

Becker C, Dressman JB, Junginger HE, Kopp S, Midha KK, Shah VP, Stavchansky S, and Barends DM. Biowaiver monographs for immediate release solid oral dosage forms: Rifampicin. *J Pharm Sci.* 2009 Jul;98(7):2252–2267.

Benet LZ and Larregieu CA. The FDA should eliminate the ambiguities in the current BCS biowaiver guidance and make public the drugs for which BCS biowaivers have been granted. *Clin Pharmacol Ther.* 2010 Sep;88(3):405–407.

Benet LZ et al. The use of BDDCS in classifying the permeability of marketed drugs. *Pharm Res.* 2008;25:483–488.

Blume H, Schug B, Tautz J, and Erb K. New guidelines for the assessment of bioavailability and bioequivalence. *Bundesgesundheitsblatt Gesundheitsforschung Gesundheitsschutz.* 2005 May;48(5):548–555.

Brahmankar DM and Jaiswal BS. *Biopharmaceutics and Pharmacokinetics –a Treatise.* Vallabh Prakashan, New Delhi, India, Vol. 2, pp. 335–336.

Breda SA, Jimenez-Kairuz AF, Manzo RH, and Olivera ME. Solubility behavior and biopharmaceutical classification of novel high-solubility ciprofloxacin and norfloxacin pharmaceutical derivatives. *Int J Pharm.* 2009 Apr 17;371(1–2):106–113.

Brockmeier D, Voegele D, and von Hattingberg HM. In vitro-in vivo correlation, a time scaling problem? Basic techniques for testing equivalence. *Arzneimittelforschung.* 1983;33:598–601.

Buchwald P. Direct, differential-equation-based in-vitro-in-vivo correlation (IVIVC) method. *J Pharm Pharmacol.* 2003;55:495–504.

Buice RG et al. Bioequivalence of a highly variable drug: An experience with nadolol. *Pharm Res.* 1996;13:1109–1115.

Canada. Health Canada's Guideline on Preparation of DIN Submissions (February 22, 1995) http:/www.hc-s.gc.ca/hpb-dgps/therapeut/htmleng/guidemain.html#PrepDIN.

Cardot JM, Beyssac E, and Alric M. In vitro–in vivo correlation: Importance of dissolution in IVIVC. *Dissolution Technol.* 2007;14:15–19.

Charkoftaki G, Dokoumetzidis A, Valsami G, and Macheras P. Elucidating the role of dose in the biopharmaceutics classification of drugs: The concepts of critical dose, effective in vivo solubility, and dose-dependent BCS. *Pharm Res.* 2012 Nov;29(11):3188–3198.

Cheng CL, Yu LX, Lee HL, Yang CY, Lue CS, and Chou CH. Biowaiver extension potential to BCS Class III high solubility-low permeability drugs: Bridging evidence for metformin immediate-release tablet. *Eur J Pharm Sci*. 2004 Jul;22(4):297–304.

Chilukuri DM and Sunkara G. IVIVC: An important tool in the development of drug delivery systems. *Drug Deliv Technol*. 2003;3:4.

Chilukuri DM, Sunkara G, and Young D. In vitro-in vivo correlation: Transdermal drug delivery systems. Pharmaceutical product development: In vitro-in vivo correlation. *Informa Healthcare USA*. 2007;165:153–176.

Chowdhury AK and Islam S. In vitro–in vivo correlation as a surrogate for bioequivalence testing: The current state of play. *Asian J Pharm Sci*. 2011;6(3–4):176–190.

Chuasuwan B, Binjesoh V, Polli JE, Zhang H, Amidon GL, Junginger HE, Midha KK et al. Biowaiver monographs for immediate release solid oral dosage forms: Diclofenac sodium and diclofenac potassium. *J Pharm Sci*. 2009 Apr;98(4):1206–1219.

Cook JA and Bockbrader HN. An industrial implementation of the Biopharmaceutics Classification System. *Dissolution Technol*. May 2002. http://www.dissolutiontech.com/D Tresour/0502art/DTMay02_art1.htm.

Corrigan OI. The biopharmaceutic drug classification and drugs administered in extended release (ER) formulations. *Adv Exp Med Biol*. 1997;423:111–128.

Crison JR, Timmins P, Keung A, Upreti VV, Boulton DW, and Scheer BJ. Biowaiver approach for biopharmaceutics classification system class 3 compound metformin hydrochloride using in silico modeling. *J Pharm Sci*. 2012 May;101(5):1773–1782.

Cristofoletti R, Chiann C, Dressman JB, and Storpirtis S. A comparative analysis of biopharmaceutics classification system and biopharmaceutics drug disposition classification system: A cross-sectional survey with 500 bioequivalence studies. *J Pharm Sci*. 2013 Sep;102(9):3136–3144.

Cristofoletti R, Nair A, Abrahamsson B, Groot DW, Kopp S, Langguth P, Polli JE, Shah VP, and Dressman JB. Biowaiver monographs for immediate release solid oral dosage forms: Efavirenz. *J Pharm Sci*. 2013 Feb;102(2):318–329.

Csoka I, Csanyi E, and Zapantis G. In vitro and in vivo percutaneous absorption of topical dosage forms: Case studies. *Int J Pharm*. 2005;291:11–19.

Cutler DJ. Linear system analysis in pharmacokinetics. *Pharmacokinet Biopharm* 1978;6:265–282.

Dahan A, Miller JM, and Amidon GL. Prediction of solubility and permeability class membership: Provisional BCS classification of the world's top oral drugs. *AAPS J*. 2009 Dec;11(4):740–746.

Dash AK, Haney PW, and Garavalia MJ. Development of an in vitro dissolution method using microdialysis sampling technique for implantable drug delivery systems. *J Pharm Sci*. 1998;8:1036–1040.

de Campos DR, Klein S, Zoller T, Vieria NR, Barros FA, Meurer EC, Coelho EC, Marchioretto MA, and Pedrazzoli J. Evaluation of pantoprazole formulations in different dissolution apparatus using biorelevant medium. *Arzneimittelforschung*. 2010;60(1):42–47.

Demirturk E and Oner L. In vitro-in vivo correlations. *J Pharm Sci*. 2003;28:215–224.

Devane J. Oral drug delivery technology: Addressing the solubility/permeability paradigm. *Pharmaceut Technol*. 1998;22:68–80.

Devane J and Butler J. The impact of in vitro-in vivo relationships on product development. *Pharm Tech*. 1997;21:146–159.

Dezani AB, Pereira TM, Caffaro AM, Reis JM, and Serra CH. Determination of lamivudine and zidovudine permeability using a different ex vivo method in Franz cells. *J Pharmacol Toxicol Methods*. 2013 May–Jun;67(3):194–202.

Dressman J, Butler J, Hempenstall J, and Reppas C. The BCS: Where do we go from here? *Pharmaceut Technol*. 2001;25:68–76.

Dressman JB, Amidon GL, Reppas C et al. Dissolution testing as a prognostic tool for oral drug absorption: Immediate release dosage forms. *Pharm Res*. 1999;15:11–22.

Dressman JB, Nair A, Abrahamsson B, Barends DM, Groot DW, Kopp S, Langguth P, Polli JE, Shah VP, and Zimmer M. Biowaiver monograph for immediate-release solid oral dosage forms: Acetylsalicylic acid. *J Pharm Sci*. 2012 Aug;101(8):2653–2667.

Dubey R. Bioequivalence challenges in development of fixed-dose combination products: Looking beyond reformulation. *Expert Opin Drug Deliv*. 2012 Mar;9(3):325–332.

Dunne A (2007) Approaches to developing IVIVC models, Chapter 5. In Chilukuri, Sunkara and Young (Eds.), *Pharmaceutical Product Development: In Vitro—In Vivo Correlation*. Taylor & Francis, New York.

Emami J. Comparative in vitro and in vivo evaluation of three tablet formulations of amiodarone in healthy subjects. *DARU J Pharm Sci*. 2010;18:193–199.

Emami J. In vitro-in vivo correlation: From theory to applications. *J Pharm Pharmaceut Sci*. 2006;9:169–189.

EMEA. Committee for Medicinal Products for Human Use. Concept Paper on BCS-based Biowaiver. May 2007. http://www.emea.europa.eu/pdfs/human/ewp/21303507en.pdf.

EMEA. Committee for Proprietary Medicinal Products. Note for Guidance on the Investigation of Bioavailability and Bioequivalence. July 2001. http://www.emea.europa.eu/pdfs/human/ewp/140198en.pdf.

Emmanuel S. Predictive in vitro dissolution tools: Application during formulation development, PhD thesis Pharmaceutical Sciences, University Clermont-Ferrand, Faculty of Pharmacy, Clermont-Ferrand, France, 2010.

FDA. *Guidance Waiver of In Vivo Bioavailability and Bioequivalence Studies for Immediate-Release Solid Oral Dosage Forms Based on a Biopharmaceutics Classification System*. http://www.fda.gov/cder/guidance/3618fnl.htm.

FDA. *Interim Policy on Exceptions to the Batch-Size and Production Condition Requirements for Non-Antibiotic, Solid, Oral-Dosage Form Drug Products Supporting Proposed ANDA's. Policy and Procedure* Guide #22–90. Office of Generic Drugs, CDER, September 13, 1990.

FDA. *Guidance for Dissolution Testing of Immediate Release Solid Oral Products*, 1997.

FDA. *Guidance for the Development, Evaluation and Application of In Vitro/In Vivo Correlations for Extended Release Solid Oral Dosage Forms*, 1997.

FDA. *Stability Testing of New Drug Substances and Products; ICH Guideline, Federal Register*, Vol. 59, No. 183, pp. 48754–48759, September 1994.

FDA. *Guidance for Industry: Immediate Release Solid Oral Dosage Forms: Scale Up and Post Approval Changes*, Center for Drug Evaluation, USFDA, 1995.

FDA. *Dissolution Bioavailability and Bioequivalence*, Abdou, HM (ed.), Mack Printing, Easton, PA, 1989.

FDA. *Guidance for Industry: Extended Release Oral Dosage Forms: Development, Evaluation and Application of In Vitro/In Vivo Correlations*, U.S. Department of Health and Human Services, FDA, Center for Drug Evaluation and Research (CDER), 1997.

FDA. *U.S. Government Printing Office. Code of Federal Regulations Title 21—Food and Drugs. Part 320—Bioavailability and Bio-equivalence Requirements*. http://www.access.gpo.gov/nara/cfr/waisidx_03/21cfr320_03.html.

FDA. *Content and Format of an Abbreviated Drug Application—Establishes the Requirement for BE in ANDAs*, (21 CFR 314. 94(a) (7)).

FDA. *Approved Drug Products with Therapeutic Equivalence Evaluations*, 28th edn., 2013. http://www.fda.gov/cder/orange/obannual.pdf.

FDA. *Draft Guidance for Industry: Bioequivalence Recommendations for Specific Products*, May 2007. http://www. fda.gov/cder/guidance/6772dft.pdf.

FDA. *Guidance for Industry: Bioavailability and Bioequivalence Studies for Orally Administered Drug Products—General Considerations,* March 2003. http://www.fda.gov/cder/guidance/5356fnl.pdf.

FDA. *Guidance for Industry: Waiver of in vivo Bioavailability and Bioequivalence Studies for Immediate-Release Solid Oral Dosage Forms Based on a Biopharmaceutics Classification System*, August 2000. http://www.fda.gov/cder/guidance/3618fnl.htm.

FDA. *Guidance for Industry: Extended Release Oral Dosage Forms: Development, Evaluation, and Application of In Vitro/In Vivo Correlations*, September 1997. http://www.fda.gov/cder/guidance/1306fnl.pdf.

FDA. *Guidance for Industry: Immediate-Release Solid Oral Dosage Forms: Scale-Up and Post-Approval Changes: Chemistry, Manufacturing and Controls, in vitro Dissolution Testing, and in vivo Bioequivalence Documentation*, November 1995. http://www.fda.gov/cder/guidance/cmc5.pdf.

FDA. *Individual Product Bioequivalence Recommendations*. http://www.fda.gov/cder/guidance/bioequivalence/#Ind_Rec.

FDA. *SUPAC-IR: Immediate-Release Solid Oral Dosage Forms: Scale-Up and Post-Approval Changes: Chemistry, Manufacturing and Controls, In Vitro Dissolution Testing, and In Vivo Bioequivalence Documentation*. http://www.fda.gov/cder/guidance/cmc5.pdf.

FDA. *SUPAC-IR/MR: Immediate Release and Modified Release Solid Oral Dosage Forms Manufacturing Equipment Addendum*. http://www.fda.gov/cder/guidance/1721fnl.pdf.

FDA. *SUPAC-MR: Modified Release Solid Oral Dosage Forms Scale-Up and Post approval Changes: Chemistry, Manufacturing, and Controls; In Vitro Dissolution Testing and In Vivo Bioequivalence Documentation*. http://www.fda.gov/cder/guidance/1214fnl.pdf.

FDA. *United States Code of Federal Regulations*, title 21 (21 CFR 314 and 320).

FDA. *Waiver of In Vivo Bioavailability and Bioequivalence Studies for Immediate-Release Solid Oral Dosage Forms Based on a Biopharmaceutics Classification System*. http://www.fda.gov/cder/guidance/3618fnl.htm.

Fernández-Teruel C, Gonzalez-Alvarez I, Navarro-Fontestad C, Arieta AG, Bermejo M, and Casabó VG. Computer simulations of bioequivalence trials: Selection of design and analyte in BCS drugs with first-pass hepatic metabolism: Part II. Non-linear kinetics. *Eur J Pharm Sci.* 2009 Jan 31;36(1):147–156.

Fong SY, Liu M, Wei H, Löbenberg R, Kanfer I, Lee VH, Amidon GL, and Zuo Z. Establishing the pharmaceutical quality of Chinese herbal medicine: A provisional BCS classification. *Mol Pharm.* 2013 May 6;10(5):1623–1643.

Frosta AB, Larsenb F, Ostergaard J et al. On the search for in vitro in vivo correlations in the field of intra-articular drug delivery: Administration of sodium diatrizoate to the horse. *Eur J Pharmaceut Sci.* 2010, 41:10–15.

Galia E, Nicolaides E, Horter D, Lobenberg R, Reppas C et al. (1998) Evaluation of various dissolution media for predicting in vivo performance of class I and class II drugs. *Pharm Res.* 15:698–705.

Gaynor C, Dunne A, and Davis J (2008) A differential equations approach to in vitro—In vivo correlation modelling. *International Biometric Conference*, Dublin, Ireland.

Gaynor C, Dunne A, and Davis J. A comparison of the prediction accuracy of two IVIVC modelling techniques. *J Pharm Sci.* 2008;97:3422–3432.

Ghosh A, Choudhury GK. In vitro-in vivo correlation (IVIVC): A review. *J Pharm Res.* 2009, 2:1255–1260.

Gleiter CH et al. When are bioavailability studies required? A German proposal. *J Clin Pharmacol.* 1998;38:904–911.

Gordon, J. Joint UNICEF/UNFPA/WHO Meeting with Manufacturers Copenhagen, September 23–25, 2013.

Granero GE, Longhi MR, Becker C, Junginger HE, Kopp S, Midha KK, Shah VP, Stavchansky S, Dressman JB, and Barends DM. Biowaiver monographs for immediate release solid oral dosage forms: Acetazolamide. *J Pharm Sci.* 2008 Sep;97(9):3691–3699.

Granero GE, Longhi MR, Mora MJ, Junginger HE, Midha KK, Shah VP, Stavchansky S, Dressman JB, and Barends DM. Biowaiver monographs for immediate release solid oral dosage forms: Furosemide. *J Pharm Sci.* 2010 Jun;99(6):2544–2556.

Grube S, Langguth P, Junginger HE, Kopp S, Midha KK, Shah VP, Stavchansky S, Dressman JB, and Barends DM. Biowaiver monographs for immediate release solid oral dosage forms: Quinidine sulfate. *J Pharm Sci.* 2009 Jul;98(7):2238–2251.

Gupta E et al. Review of global regulations concerning biowaivers for immediate release solid oral dosage forms. *Eur J Pharm Sci.* 2006;29:315–324.

Heisler M et al. The health effects of restricting prescription medication use because of cost. 2004;*Med Care* 42:626–634.

Helga M (2002) The Biopharmaceutical Classification System (BCS) and its usage. *Drugs Made Ger.* 45:63–65.

Homsek I, Parojcić J, Dacević M, Petrović L, and Jovanović D. Justification of metformin hydrochloride biowaiver criteria based on bioequivalence study. *Arzneimittelforschung.* 2010;60(9):553–559.

Huhna E, Buchholzb HG, Shazlya G et al. Predicting the in vivo release from a liposomal formulation by IVIVC and non-invasive positron emission tomography imaging. *Eur J Pharm Sci.* 2010, 41:71–77.

Huic M et al. How safe are bioequivalence studies in healthy volunteers? *Therapie* 1996;51:410–413.

Hwang SS, Bayne W, and Theeuwes F. In vivo evaluation of controlled-release products. *J Pharm Sci.* 1993;82:1145–1150.

Hwang SS, Gorsline JJ, Louie J, Dye D, Guinta D et al. In vitro and in vivo evaluation of a once-daily controlled release pseudoephedrine product. *J Clin Pharmacol.* 1995;35:259–267.

Jantratid E, Prakongpan S, Amidon GL, and Dressman JB. Feasibility of biowaiver extension to biopharmaceutics classification system class III drug products: Cimetidine. *Clin Pharmacokinet.* 2006;45(4):385–399.

Jantratid E, Prakongpan S, Dressman JB, Amidon GL, Junginger HE, Midha KK, and Barends DM. Biowaiver monographs for immediate release solid oral dosage forms: Cimetidine. *J Pharm Sci.* 2006 May;95(5):974–984.

Jantratid E, Strauch S, Becker C, Dressman JB, Amidon GL, Junginger HE, Kopp S et al. Biowaiver monographs for immediate release solid oral dosage forms: Doxycycline hyclate. *J Pharm Sci.* 2010 Apr;99(4):1639–1653.

Jayaprakasam B, Seeram NP, and Nairs MG (2003) Anticancer and antiinflammatory activities of cucurbitacins from *Cucurbita andreana*. *Cancer Lett.* 189:11–16.

Kalantzi L, Reppas C, Dressman JB, Amidon GL, Junginger HE, Midha KK, Shah VP, Stavchansky SA, and Barends DM. Biowaiver monographs for immediate release solid oral dosage forms: Acetaminophen (paracetamol). *J Pharm Sci.* 2006 Jan;95(1):4–14.

Kannan K, Manavalan R, Karar PK. In vitro in vivo correlation of sustained release capsules of metoprolol. *J Pharm Sci Res.* 2010;2:562–566.

Khaled AA, Pervaiz K, Karim S, Farzana K, and Murtaza G. Development of in vitro-in vivo correlation for encapsulated metoprolol tartrate. *Acta Pol Pharm.* 2013 Jul–Aug;70(4):743–747.

Khan MZ, Rausl D, Radosevic S, Filic D, Danilovski A, Dumic M, and Knezevic Z. Classification of torasemide based on the Biopharmaceutics Classification System and evaluation of the FDA biowaiver provision for generic products of CLASS I drugs. *J Pharm Pharmacol.* 2006 Nov;58(11):1475–1482.

Khan SA, Ahmad M, Murtaza G, Aamir MN, Kousar R, Rasool F, and Shahiq-u-Zaman. In vitro-in vivo correlation study on nimesulide loaded hydroxypropylmethylcellulose microparticles. *Yao Xue Xue Bao.* 2010 Jun;45(6):772–777.

Khandelwal A et al. Computational models to assign biophar-maceutics drug disposition classification from molecular structure. *Pharm Res.* 2007;24:2249–2262.

Kim JS, Mitchell S, Kijek P, Tsume Y, Hilfinger J, and Amidon GL. The suitability of an in situ perfusion model for permeability determinations: Utility for BCS class I biowaiver requests. *Mol Pharm.* 2006 Nov–Dec;3(6):686–694.

Kleina S, Rudolphb MW, Skalsky B et al. Use of the BioDis to generate a physiologically relevant IVIVC. *J Control Release.* 2008, 130:216–219.

Koeppe MO, Cristofoletti R, Fernandes EF, Storpirtis S, Junginger HE, Kopp S, Midha KK et al. Biowaiver monographs for immediate release solid oral dosage forms: Levofloxacin. *J Pharm Sci.* 2011 May;100(5):1628–1636.

Kortejärvi H, Shawahna R, Koski A, Malkki J, Ojala K, and Yliperttula M. Very rapid dissolution is not needed to guarantee bioequivalence for biopharmaceutics classification system (BCS) I drugs. *J Pharm Sci.* 2010 Feb;99(2):621–625.

Kortejärvi H, Urtti A, and Yliperttula M. Pharmacokinetic simulation of biowaiver criteria: The effects of gastric emptying, dissolution, absorption and elimination rates. *Eur J Pharm Sci.* 2007 Feb;30(2):155–166.

Kortejarvi H, Yliperttula M, Dressman JB, Junginger HE, Midha KK, Shah VP, and Barends DM. Biowaiver monographs for immediate release solid oral dosage forms: Ranitidine hydrochloride. *J Pharm Sci.* 2005 Aug;94(8):1617–1625.

Kovacević I, Parojcić J, Homsek I, Tubić-Grozdanis M, and Langguth P. Justification of biowaiver for carbamazepine, a low soluble high permeable compound, in solid dosage forms based on IVIVC and gastrointestinal simulation. *Mol Pharm.* 2009 Jan–Feb;6(1):40–47.

Kovacevi I, Parojci J, Tubi-Grozdanis M, and Langguth P. An investigation into the importance of "very rapid dissolution" criteria for drug bioequivalence demonstration using gastrointestinal simulation technology. *AAPS J.* 2009 Jun;11(2):381–384.

Kubbinga M, Langguth P, and Barends D. Risk analysis in bioequivalence and biowaiver decisions. *Biopharm Drug Dispos.* 2013 Jul;34(5):254–261.

Leeson LJ (1995) In vitro/in vivo correlations. *Drug Info J.* 29:903–915.

Lehto P (February 2010) Mechanistic studies of drug dissolution testing. Academic Dissertation, Division of Pharmaceutical Technology, Faculty of Pharmacy, University of Helsinki, Helsinki, Finland.

Limberg J and Potthast H. Regulatory status on the role of in vitro dissolution testing in quality control and biopharmaceutics in Europe. *Biopharm Drug Dispos.* 2013 Jul;34(5):247–253.

Lindenberg M, Kopp S, and Dressman JB. Classification of orally administered drugs on the World Health Organization Model list of Essential Medicines according to the biopharmaceutics classification system. *Eur J Pharm Biopharm.* 2004 Sep;58(2):265–278.

Loo JC and Riegelman S (1968) New method for calculating the intrinsic absorption rate of drugs. *J Pharm Sci.* 57:918–928.

Lootvoeta G, Beyssaca E, Shiub GK et al. Study on the release of indomethacin from suppositories: In vitro-in vivo correlation. *Int J Pharmaceut.* 1992;85:113–120.

Lue BM, Nielsen FS, Magnussen T et al. Using biorelevant dissolution to obtain IVIVC of solid dosage forms containing a poorly-soluble model compound. *Eur J Pharm Biopharm.* 2008;69:648–657.

Mandal U, Ray KK, Gowda V et al. In-vitro and in-vivo correlation for two gliclazide extended-release tablets. *J Pharm Pharmacol.* 2007;59:971–976.

Manzo RH, Olivera ME, Amidon GL, Shah VP, Dressman JB, and Barends DM. Biowaiver monographs for immediate release solid oral dosage forms: Amitriptyline hydrochloride. *J Pharm Sci.* 2006 May;95(5):966–973.

Martinez Y, Rodriguez O, Vazquez M et al. Correlation of in vivo-in vitro potency assays for the cuban Hepatitis B vaccine. *Biotecnologia Aplicada.* 2005;22:34–36.

Mehta MU et al. Comparison of clinical pharmacology (CP) and biopharmaceutics (BP) studies submitted in NDAs during 1995 and 1997. 1998 ASCPT annual meeting abstract.

Mehta MU. Presentation (September 25, 2002) Classification of New Drugs: NDA 1995–2001 Survey at AAPS workshop biopharmaceutics classification system: Implementation challenges and extension opportunities, Arlington, VA.

Mehta MU Presentation (May 21, 2007) FDA Regulatory Use of BCS. AAPS workshop on bioequivalence, biopharmaceutics classification system, and beyond, North Bethesda, MD.

Meyer MC, Straughn AB, Mhatre RM, Shah VP, and Williams RL (1998) Lack of in vitro/in vivo correlations of 50 and 250 mg primidone tablets. *Pharm Res.* 1998;15:1085–1089.

Mishra V, Gupta U, and Jain NK. Biowaiver: An alternative to in vivo pharmacokinetic bioequivalence studies. *Pharmazie.* 2010 Mar;65(3):155–161.

Modi NB, Lam A, Lindemulder E, Wang B, and Gupta SK (2000) Application of in vitro-in vivo correlation (IVIVC) in setting formulation release specifications. *Biopharm Drug Dispos.* 21:321–326.

Moore JW and Flanner HH. Mathematical comparison of dissolution profiles. *Pharm Technol.* 1996;6:64–74.

Morais JA and Lobato Mdo R. The new European Medicines Agency guideline on the investigation of bioequivalence. *Basic Clin Pharmacol Toxicol.* 2010 Mar;106(3):221–225.

Nair A, Abrahamsson B, Barends DM, Groot DW, Kopp S, Polli JE, Shah VP, and Dressman JB. Biowaiver monographs for immediate-release solid oral dosage forms: Primaquine phosphate. *J Pharm Sci.* 2012 Mar;101(3):936–945.

Nair A, Abrahamsson B, Barends DM, Groot DW, Kopp S, Polli JE, Shah VP, and Dressman JB. Biowaiver monographs for immediate release solid oral dosage forms: Amodiaquine hydrochloride. *J Pharm Sci.* 2012 Dec;101(12):4390–4401.

Nair AK, Anand O, Chun N, Conner DP, Mehta MU, Nhu DT, Polli JE, Yu LX, and Davit BM. Statistics on BCS classification of generic drug products approved between 2000 and 2011 in the USA. *AAPS J.* 2012 Dec;14(4):664–666.

O'Hara T, Hayes S, Davis J, Devane J, Smart T et al. (2001) In vivo-in vitro correlation (ivivc) modeling incorporating a convolution step. *J Pharmacokinet Pharmacodyn.* 2001;28:277–298.

Ochoaa L, Igartuaa M, Hernandez RM et al. In vivo evaluation of two new sustained release formulations elaborated by one-step melt granulation: Level A in vitro–in vivo correlation. *Eur J Pharm Biopharm.* 2010;75:232–237.

Okumu A, DiMaso M, and Löbenberg R. Computer simulations using GastroPlus to justify a biowaiver for etoricoxib solid oral drug products. *Eur J Pharm Biopharm.* 2009 May;72(1):91–98.

Olivera ME, Manzo RH, Junginger HE, Midha KK, Shah VP, Stavchansky S, Dressman JB, and Barends DM. Biowaiver monographs for immediate release solid oral dosage forms: Ciprofloxacin hydrochloride. *J Pharm Sci.* 2011 Jan;100(1):22–33.

Ostrowski M, Wilkowska E, and Baczek T. IVIVC for amoxicillin trihydrate 1000 mg dispersible tablet. *Drug Dev Industrial Pharm.* 2009;35:981–985.

Patel JR and Barve KH. Intestinal permeability of Lamivudine using single pass intestinal perfusion. *Indian J Pharm Sci.* 2012 Sep;74(5):478–481.

Peternel L, Kristan K, Petruševska M, Rižner TL, and Legen I. Suitability of isolated rat jejunum model for demonstration of complete absorption in humans for BCS-based biowaiver request. *J Pharm Sci.* 2012 Apr;101(4):1436–1449.

Ping H. Bridging in vitro dissolution tests to in vivo dissolution for poorly soluble acidic drugs. A dissertation submitted in partial fulfillment of the requirements for the degree of Doctor of Philosophy (Pharmaceutical Sciences), The University of Michigan, Ann Arbor, MI, 2010.

Polli JE et al. "Pavlovian" food effect on the enterohepatic recirculation of piroxicam. *Biopharm Drug Dispos.* 1996;17:635–641.

Polli JE et al. Summary workshop report: Biopharmaceutics classification system—Implementation challenges and extension opportunities. *J Pharm Sci.* 2004;93:1375–1381.

Polli JE. In vitro studies are sometimes better than conventional human pharmacokinetic in vivo studies in assessing bioequivalence of immediate-release solid oral dosage forms. *AAPS J.* 2008 Jun;10(2).

Polli JE and Ginski MJ. Human drug absorption kinetics and comparison to Caco-2 monolayer permeabilities. *Pharm Res.* 1998;15:47–52.

Potthast H, Dressman JB, Junginger HE, Midha KK, Oeser H, Shah VP, Vogelpoel H, and Barends DM. Biowaiver monographs for immediate release solid oral dosage forms: Ibuprofen. *J Pharm Sci.* 2005 Oct;94(10):2121–2131.

Qureshi SA. In vitro-in vivo correlation (IVIVC) and determining drug concentrations in blood from dissolution testing—A simple and practical approach. *Open Drug Delivery J.* 2010;4:38–47.

Rediguieri CF, Porta V, G Nunes DS, Nunes TM, Junginger HE, Kopp S, Midha KK et al. Biowaiver monographs for immediate release solid oral dosage forms: Metronidazole. *J Pharm Sci.* 2011 May;100(5):1618–1627.

Rettig H and Mysicka J. IVIVC: Methods and applications in modified-release product Development. *Dissolution Technol.* 2008;15:6–8.

Safran DG, Neuman P, Schoen C, Kitchman MS et al. Prescription drug coverage and seniors: Findings from a 2003 national survey. *Health Aff.* 0.1377/hlthaff.w5.152 (2005).

Sakore S and Chakroborty B. In vitro–in vivo correlation (IVIVC): A strategic tool in drug development. *J Bioequiv* Availab, 2011, S3: http://dx.doi.org/10.4172/jbb.S3–001.

Sakuma S, Tachiki H, Uchiyama H, Fukui Y, Takeuchi N, Kumamoto K, Satoh T et al. A perspective for biowaivers of human bioequivalence studies on the basis of the combination of the ratio of AUC to the dose and the biopharmaceutics classification system. *Mol Pharm.* 2011 Aug 1;8(4):1113–1119.

Schliecker G, Schmidt C, and Fuchs S. in vitro and in vivo correlation of buserelin release from biodegradable implants using statistical moment analysis. *J Control Release.* 2004, 94:25–37.

Shah VP, Siewert M, Dressman J et al. Dissolution/in vitro release testing of special dosage forms. *Dissolution Technol.* 2002;9:1–5.

Shargel L and Yu ABC (1999) Biopharmaceutic considerations in drug product design. *Applied Biopharmaceutics & Pharmacokinetics*, 4th edn., Appleton and Lange, Stamford, CT, pp. 129–167.

Shohin IE, Kulinich JI, Ramenskaya GV, Abrahamsson B, Kopp S, Langguth P, Polli JE et al. Biowaiver monographs for immediate-release solid oral dosage forms: Ketoprofen. *J Pharm Sci.* 2012 Oct;101(10):3593–3603.

Shohin IE, Kulinich JI, Vasilenko GF, and Ramenskaya GV. Interchangeability evaluation of multisource Ibuprofen drug products using biowaiver procedure. *Indian J Pharm Sci.* 2011 Jul;73(4):443–446.

Silva AL, Cristofoletti R, Storpirtis S, Sousa VD, Junginger HE, Shah VP, Stavchansky S, Dressman JB, and Barends DM. Biowaiver monographs for immediate-release solid oral dosage forms: Stavudine. *J Pharm Sci.* 2012 Jan;101(1):10–16.

Sirisuth N and Eddington ND. In vitro in vivo correlations, systemic methods for the development and validation of an IVIVC metoprolol and naproxen drug examples. *Int J Gen Drugs* 2002;3:250–258.

Sirisuth N and Eddington ND. In vitro-in vivo correlation definitions and regulatory guidance. *Int J Gen Drugs.* 2002, Part 2:1–11.

Skelly JP et al. Workshop report: Scaleup of oral extended-release dosage forms. *Pharm Res.* 1993; 10(12):1800–1805.

Smetanová L, Stětinová V, Svoboda Z, and Kvetina J. Caco-2 cells, biopharmaceutics classification system (BCS) and biowaiver. *Acta Medica (Hradec Kralove).* 2011;54(1):3–8.

Soares KC, Rediguieri CF, Souza J, Serra CH, Abrahamsson B, Groot DW, Kopp S et al. Biowaiver monographs for immediate-release solid oral dosage forms: Zidovudine (azidothymidine). *J Pharm Sci.* 2013 Aug;102(8):2409–2423.

Soumerai S. et al. Cost-related medication nonadherence among elderly and disabled medicare beneficiaries: A national survey 1 year before the medicare drug benefit. *Arch Intern Med.* 2006;166:1829–1835.

Spiegeleer BD, Voorena LV, Voorspoelsb J et al. Dissolution stability and IVIVC investigation of a buccal tablet. *Anal Chim Acta.* 2001, 446:343–349.

Stein. CM. Managing risk in healthy subjects participating in clinical research. *J Clin Pharm Ther.* 2003;74:511–512.

Stosik AG, Junginger HE, Kopp S, Midha KK, Shah VP, Stavchansky S, Dressman JB, and Barends DM. Biowaiver monographs for immediate release solid oral dosage forms: Metoclopramide hydrochloride. *J Pharm Sci.* 2008 Sep;97(9):3700–3708.

Strauch S, Dressman JB, Shah VP, Kopp S, Polli JE, Barends DM. Biowaiver monographs for immediate-release solid oral dosage forms: Quinine sulfate. *J Pharm Sci.* 2012 Feb;101(2):499–508.

Strauch S, Jantratid E, Dressman JB, Junginger HE, Kopp S, Midha KK, Shah VP, Stavchansky S, Barends DM. Biowaiver monographs for immediate release solid oral dosage forms: Mefloquine hydrochloride. *J Pharm Sci.* 2011 Jan;100(1):11–21.

Strauch S, Jantratid E, Dressman JB, Junginger HE, Kopp S, Midha KK, Shah VP, Stavchansky S, and Barends DM. Biowaiver monographs for immediate release solid oral dosage forms: Lamivudine. *J Pharm Sci.* 2011 Jun;100(6):2054–2063.

Strauch S, Jantratid E, and Dressman JB. Comparison of WHO and US FDA biowaiver dissolution test conditions using bioequivalent doxycycline hyclate drug products. *J Pharm Pharmacol.* 2009 Mar;61(3):331–337.

Strauch S, Jantratid E, Stahl M, Rägo L, and Dressman JB. The biowaiver procedure: Its application to antituberculosis products in the WHO prequalification programme. *J Pharm Sci.* 2011 Mar;100(3):822–830.

Strickley RG. Parenteral formulations of small molecules therapeutics marketed in the United States Part I, PDA. *J Pharm Sci Tech* 1999;53:324–349.

Sunkara G, Chilukuri DM. IVIVC applications. *Drug Deliv Technol.* 2003, 3(4).

Takagi T et al. A Provisional Biopharmaceutical Classification of the Top 200 Oral Drug Products in the United States, Great Britain, Spain, and Japan. *Mol. Pharmaceutics* 3:631–643 (2006).

Tanguay M et al. When will a drug formulation pass or fail bioequivalence criteria? Experience From 1200 Studies. *AAPS Pharm Sci.* 2002;4(4):Abstract R6193.

Tiwari G, Tiwari R, Pandey S et al. In vitro-in vivo correlation and biopharmaceutical classification system (BCS): A review. *Der Pharm Chem.* 2010;2:129–140.

Tothfalusi L et al. Evaluation of the bioequivalence of highly-variable drugs and drug products. *Pharm Res.* 2001;18:728–733.

Tsume Y and Amidon GL. The biowaiver extension for BCS class III drugs: The effect of dissolution rate on the bioequivalence of BCS class III immediate-release drugs predicted by computer simulation. *Mol Pharm.* 2010 Aug 2;7(4):1235–1243.

Tsume Y, Langguth P, Arieta AG, and Amidon GL. In silico prediction of drug dissolution and absorption with variation in intestinal pH for BCS class II weak acid drugs: Ibuprofen and ketoprofen. *Biopharm Drug Dispos.* 2012 Oct;33(7):366–377.

United States Pharmacopoeia. In vitro and in vivo Evaluations of Dosage Forms, 27th edn, Revision, Mack Publishing Co., Easton, PA, 2004.

Uppoor VRS. Regulatory perspectives on in vitro (dissolution)/in vivo (bioavailability) correlations. *J Control Release.* 2001;72:127–132.

Venkatesh S and Lipper RA (2000) Role of the development scientist in compound lead selection and optimisation. *J Pharm Sci* 89:145–154.

Verbeeck RK, Junginger HE, Midha KK, Shah VP, and Barends DM. Biowaiver monographs for immediate release solid oral dosage forms based on biopharmaceutics classification system (BCS) literature data: Chloroquine phosphate, chloroquine sulfate, and chloroquine hydrochloride. *J Pharm Sci.* 2005 Jul;94(7):1389–1395.

Vogelpoel H, Welink J, Amidon GL, Junginger HE, Midha KK, Moller H, Olling M, Shah VP, and Barends DM. Biowaiver monographs for immediate release solid oral dosage forms based on biopharmaceutics classification system (BCS) literature data: Verapamil hydrochloride, propranolol hydrochloride, and atenolol. *J Pharm Sci.* 2004 Aug;93(8):1945–1956.

Vogt M, Derendorf H, Kramer J, Junginger HE, Midha KK, Shah VP, Stavchansky S, Dressman JB, and Barends DM. Biowaiver monographs for immediate release solid oral dosage forms: Prednisolone. *J Pharm Sci.* 2007 Jan;96(1):27–37.

Vogt M, Derendorf H, Krämer J, Junginger HE, Midha KK, Shah VP, Stavchansky S, Dressman JB, and Barends DM. Biowaiver monographs for immediate release solid oral dosage forms: Prednisone. *J Pharm Sci.* 2007 Jun;96(6):1480–1489.

Wagh MP and Patel JS. Biopharmaceutical classification system: Scientific basis for biowaiver extensions. *Int J Pharm Pharmceut Sci.* 2010;2:12–19.

Wagner JG and Nelson E. Kinetic analysis of blood levels and urinary excretion in the absorptive phase after single doses of drug. *J Pharm Sci.* 1964;53:1392–1403.

WHO document (1999) entitled Marketing Authorization of Pharmaceutical Products with Special Reference to Multisource (Generic) Products: A Manual for Drug Regulatory Authorities.

Wilding I (1999) Evolution of the Biopharmaceutics Classification System (BCS) to Modified Release (MR) formulations: What do we need to consider? *Eur J Pharm Sci* 1999;8:157–159.

Woo BH, Kostanski W, Gebrekidan S et al. Preparation, characterization and in vivo evaluation of 120-day poly (D, L-lactide) leuprolide microspheres. *J Control Release.* 2001;75:307–315.

Wu CY and Benet LZ, Predicting drug disposition via application of BCS: Transport/absorption/elimination interplay and development of a biopharmaceutics drug disposition classification system. *Pharm Res.* 2005;22:11–23.

Xu L, Li S, and Sunada H. Preparation and evaluation in vitro and in vivo of captopril elementary osmotic pump tablets. *Asian J Pharm Sci.* 2006;1:236–245.

Yang SG. Biowaiver extension potential and IVIVC for BCS Class II drugs by formulation design: Case study for cyclosporine self-microemulsifying formulation. *Arch Pharm Res.* 2010 Nov;33(11):1835–1842.

Yang Y, Shah RB, Yu LX, and Khan MA. In vitro bioequivalence approach for a locally acting gastrointestinal drug: Lanthanum carbonate. *Mol Pharm.* 2013 Feb 4;10(2):544–550.

Young D, Chilukuri D, Becker R, Bigora S, Farrell C et al. Approaches to developing a Level-A IVIVC for injectable dosage forms. *AAPS Pharm Sci* 2002;4:M1357.

Young D, Devane JG, and Butler J (1997) *In Vitro-In Vivo Correlations.* Plenum Press, New York.

Yu H and Joves R. The STATs of cancer—New molecular targets come of age. *Nat Rev Cancer* 2004 4:97–105.

Yu LX, Amidon GL, Polli JE, Zhao H, Mehta MU, Conner DP, Shah VP et al. Biopharmaceutics classification system: The scientific basis for biowaiver extensions. *Pharm Res.* 2002 Jul;19(7):921–925.

Yu LX et al. Influence of drug release properties of conventional solid dosage forms on the systemic exposure of highly soluble drugs. *AAPS Pharm Sci.* 2001;3(3):Article 24.

Žakelj S, Berginc K, Roškar R, Kraljič B, and Kristl A. Do the recommended standards for in vitro biopharmaceutic classification of drug permeability meet the "passive transport" criterion for biowaivers? *Curr Drug Metab.* 2013 Jan;14(1):21–27.

CHAPTER 7: STATISTICAL EVALUATION OF BIOEQUIVALENCE DATA

Anderson S and Hauck WW. Considerations of individual bioequivalence. *J Pharmacokin Biopharm.* 1993;18:259–273.

Anderson S. Individual Bioequivalence: A Problem of Switchability (with discussion). *Biopharm Rep.* 1993;2(2):1–11.

Anderson S. Current issues of individual bioequivalence. *Drug Inf. J.* 1995;29:961–964.

Anderson S and Hauck WW. Consideration of individual bioequivalence. *J Pharmacokin Biopharm.* 1990;18:259–273.

Barnard GA. Comparing the means of two independent samples. *J Stat.* 1984;33(3):266–271.

Boddy AW, Snikeris FC, Kringle RO, Wei GCG, Opperman JA, and Midha KK. An approach for widening the bioequivalence acceptance limits in the case of highly variable drugs. *Pharm Res.* 1995;12:1865–1868.

Boomer: http://www.boomer.org/pkin/soft.html.

Chen ML, Davit B, Lionberger R, Wahba Z, Ahn HY, and Yu LX. Using partial area for evaluation of bioavailability and bioequivalence. *Pharm Res.* 2011 Aug;28(8):1939–1947.

Chen ML. Individual bioequivalence—A regulatory update (with discussion). *J Biopharm Stat.* 1997;7:5–111.

Chen ML, Patnaik R, Hauck WW, Schuirmann DJ, Hyslop T, Williams RL, and the FDA Population and Individual Bioequivalence Working Group. An individual bioequivalence criterion: Regulatory considerations. *Stat Med.* 2000;19:2821–2842.

Chen ML, Lee SC, Ng M-J, Schuirmann DJ, Lesko LJ, and Williams RL. Pharmacokinetic analysis of bioequivalence trials: Implications for sex-related issues in clinical pharmacology and biopharmaceutics. *Clin Pharmacol Ther.* 2000;68(5):510–521.

Chen ML. Individual bioequivalence—A regulatory update. *J Biopharm Stat.* 1997;7:5–11.

Chinchilli VM. The assessment of individual and population bioequivalence. *J Biopharm Stat.* 1996;6:1–14.

Chinchilli VM and Esinhart JD. Design and analysis of intra-subject variability in cross-over experiments. *Stat Med.* 1996;15:1619–1634.

Chow SC and Shao J. Statistical methods for two-sequence three-period cross-over designs with incomplete data. *Stat Med.* 1997 May 15;16(9):1031–1039.

Chow SC and Liu JP. On assessment of bioequivalence under a higher-order crossover design. *J Biopharm Stat.* 1992;2(2):239–256.

Chow SC. Individual bioequivalence—A review of the FDA draft guidance. *Drug Inf J* 1999;33:435–444.

Davit BM, Chen ML, Conner DP, Haidar SH, Kim S, Lee CH, Lionberger RA et al. Implementation of a reference-scaled average bioequivalence approach for highly variable generic drug products by the US Food and Drug Administration. *AAPS J.* 2012 Dec;14(4):915–924.

Diletti E, Hauschke D, and Steinijans VW. Sample size determination for bioequivalence assessment by means of confidence intervals. *Int J Clin Pharmacol Ther Toxicol.* 1991a Jan;29:1–8.

Diletti E, Hauschke D, and Steinijans VW. Sample size determination for bioequivalence assessment by means of confidence intervals. *Int J Pharm Ther Toxicol.* 1991b;29:1–8.

Efron B. Better bootstrap confidence intervals (with discussion). *J Amer Stat Assoc.* 1987;82:171–201.

Efron B and Tibshirani RJ (1993) *An Introduction to the Bootstrap*, Chapman and Hall, New York, Chapter 14.

Ekbohm G and Melander H. The subject-by-formulation interaction as a criterion of interchangeability of drugs. *Biometrics* 1989;45:1249–1254. (Furosemide—Lasix and Furix)

Ekbohm G. and Melander H. (1990) On variation, bioequivalence and interchangeability. Report 14, Department of Statistics, Swedish University of Agricultural Sciences, Uppsala, Sweden.

Endrenyi L (1993) A procedure for the assessment of individual bioequivalence. In H.H. Blume and K.K. Midha (Eds.), *Bio-International: Bioavailability, Bioequivalence and Pharmacokinetics*. Medpharm Publications, pp. 141–146.

Endrenyi L. A method for the evaluation of individual bioequivalence. *Int J Clin Pharmacol Therap.* 1994;32:497–508.

Endrenyi L. A simple approach for the evaluation of individual bioequivalence. *Drug Inf J.* 1995;29:847–855.

Endrenyi L, Amidon GL, Midha KK, and Skelly JP. Individual bioequivalence: Attractive in principle, difficult in practice. *Pharm Res*. 1998;15:1321–1325.

Endrenyi L and Hao Y. Asymmetry of the mean-variability tradeoff raises questions about the model in investigations of individual bioequivalence. *Int J Clin Pharmacol Ther*. 1998;36:450–457.

Endrenyi L and Midha KK. Individual bioequivalence—Has its time come? *Eur J Pharm Sci*. 1998;6:271–278.

Endrenyi L and Tothfalusi L. Subject-by-formulation interaction in determinations of individual bioequivalence: Bias and prevalence. *Pharm Res*. 1999 Feb;16(2):186–190.

Endrenyi L and Tothfalusi L. Regulatory conditions for the determination of bioequivalence of highly variable drugs. *J Pharm Pharm Sci*. 2009;12:138–149.

Endrenyi L and Schulz M. Individual variation and the acceptance of average bioequivalence. *Drug Inf J*. 1993;27:195–201.

Esinhart JD and Chinchilli VM. Extension to the use of tolerance intervals for the assessment of individual bioequivalence. *J Biopharm Stat*. 1994;4(1):39–52. (Verapamil).

Esinhart JD and Chinchilli VM. Sample size considerations for assessing individual bioequivalence based on the method of tolerance interval. *Int J Clin Pharmacol Ther*. 1994;32(1):26–32.

European Medicines Agency. Guideline on the investigation of bioequivalence. London, U.K., January 20, 2010. http://www.ema.europa.eu/docs/en_GB/document_li brary/Scientific_guideline/2010/01/WC500070039.pdf.

FDA. http://www.fda.gov/downloads/Drugs/GuidanceComplianceRegulatoryInformation/Guidances/UCM2092 94.pdf.

FDA. *Draft Guidance for Industry: In vivo Bioequivalence Studies Based on Population and Individual Bioequivalence Approaches*, Food and Drug Administration, Center for Drug Evaluation and Research (CDER), Rockville, MD, October, 1997.

FDA. *Guidance Statistical For Bioequivalence Studies Using A Standard Two-Treatment Crossover Design, Division of Bioequivalence Office of Generic Drugs (OGD)*, FDA, 1992.

Fleiss JL. A critique of recent research on the two-treatment crossover design. *Control Clin Trials*. 1989;10:237–243.

Fluchler H, Hirtz J, and Moser A. An aid to decision-making in bioequivalence assessment. *J Pharmacokinet Biopharm*. 1981 April;9(2):235–243.

Freeman PR. The role of P-values in analysing trial results. *Stat Med*. 1993;12:1443–1452.

Graybill F and Wang CM. Confidence intervals on nonnegative linear combinations of variances. *J Am Stat Assoc*. 1980;75:869–873.

Guilbaud O. Exact comparisons of meanss annnnd within-subject variances in 2×2 crossover trials. *Drug Inf J*. 1995;29:857–870.

Haidar S, Davit HB, Chen ML, Conner D, Lee LM, Li QH, Lionberger R et al. Bioequivalence approaches for highly variable drugs and drug products. *Pharm Res*. 2008a;15:237–241.

Haidar WH, Makhlouf F, Schuirmann DJ, Hyslop T, Davit B, Conner D, and Yu LX. Evaluation of a scaling approach for the bioequivalence of highly variable drugs. *AAPS J*. 2008;10:450–454.

Hauck WW and Anderson S. A new statistical procedure for testing equivalence in two-group comparative bioavailability trials. *J Pharmacokin Biopharm*. 1984;12:83–91.

Hauck WW and Anderson S. Types of bioequivalence and related statistical considerations. *Int J Clin Pharmacol Therap*.1992;30:181–187.

Hauck, WW and Anderson S. Measuring switchability and prescribability: When is average bioequivalence sufficient? *J Pharmacokin Biopharm*. 1994;22:551–564.

Hauck WW, Chen M-L, Hyslop T, Patnaik R, Schuirmann D, and Williams RL. for the FDA Population and Individual Bioequivalence Working Group. Mean difference vs. variability reduction: Tradeoffs in aggregate measures for individual bioequivalence. *Int J Clin Pharmacol Therap*. 1996;34:535–541.

Hauschke D, Steinijans V, and Pigeot I (2007) *Bioequivalence Studies in Drug Development: Methods and Applications*. Wiley, Chichester, U.K.

Hauschke D, Steinijans VW, Diletti E, and Burke M. Sample size determination for bioequivalence assessment using a multiplicative model. *J Pharmacokin Biopharm*. 1992;20:557–561.

Holder DJ and Hsuan F. Moment-based criteria for determining bioequivalence. *Biometrika*. 1993;80:835–846.

Holder DJ and Hsuan F. A moment-based method for determining individual bioequivalence. *Drug Inf J*. 1995;29:965–979.

Howe WG. Approximate confidence limits on the mean of X+Y where X and Y are two tabled independent random variables. *J Amer Stat Assoc*. 1994;69:789–794.

Hsu JC, Hwang JTG, Liu H-K, and Ruberg SJ. Confidence intervals associated with tests for bioequivalence. *Biometrika*. 1994;81:103–114.

Hwang S, Huber PB, Hesney M, and Kwan KC. Bioequivalence and Interchangeability. *J Pharm Sci.* 1978;67:IV "Open Forum."

Hyslop T, Hsuan F, and Holder DJ. A small-sample confidence interval approach to assess individual bioequivalence. *Stat Med.* 2000 Oct 30;19:2885–2897.

Jones B and Doney AN. Modelling and design of cross-over trials. *Stat Med.* 1996;15:1435–1446.

Keiding N and Budtz-Jorgensen E. The precautionary principle and statistical approaches to uncertainty. *Int J Occup Med Environ Health.* 2004;17(1):147–151.

Kimanani EK and Potvin D. Parametric confidence interval for a moment-based scaled criterion for individual bioequivalence. *J Pharm Biopharm.* 1998;25:595–614.

Laird N. MN., J. Skinner. M. Keward. An analysis of two-period cross-over designs with carry-over effects. *Stat Med.* 1992;11:1967–1979.

Littell RC, Henry PR, and Ammerman CB. Statistical analysis of repeated measures data using SAS procedures. *J Anim Sci.* 1998 Apr;76(4):1216–1231.

Littell RC, Henry PR, Lewis AJ, and Ammerman CB. Estimation of relative bioavailability of nutrients using SAS procedures. *J Anim Sci.* 1997 Oct;75(10):2672–2683.

Littell RC, Pendergast J, and Natarajan R. Modelling covariance structure in the analysis of repeated measures data. *Stat Med.* 2000 Jul 15;19(13):1793–1819.

Liu J-P. Use of the repeated crossover designs in assessing bioequivalence. *Stat Med.* 1995;14:1067–1078.

Liu JP and Chow SC. Sample size determination for the two one-sided test procedure in bioequivalence. *J Pharmacokin Biopharm.* 1992;20:101–104.

Locke CS. Pharmacometrics. An exact confidence interval from untrasformed data for the ratio of two formulation means. *J Pharmacokinet Biopharm.* 1984 Dec;12(6):649–655.

Mandallaz D and Mau J. Comparison of different methods for decision making in bioequivalence assessment. *Biometrics.* 1981;37:213–222.

McCarthy WF. http://biostats.bepress.com/cgi/viewcontent.cgi?article = 1057&context = cobra.

McLachlan GJ. On the EM algorithm for overdispersed count data. *Stat Methods Med Res.* 1997 Mar;6(1):76–98.

Pabst G and Jaeger H. Review of methods and criteria for the evaluation of bioequivalence studies. *Eur J Clin Pharmacol.* 1990;38(1):5–10. Comment in *Eur J Clin Pharmacol.* 1991;40(2):201–203.

Pack SE. Hypothesis testing for proportions with overdispersion. *Biometrics.* 1986;42:967–972.

Patnaik RN, Lesko LJ, Chen M-L, and Williams R.L. Individual bioequivalence—New concepts in the statistical assessment of bioequivalence metrics. *Clin Pharmacokin.* 1997;33:1–6.

Patnaik RN, Lesko LJ, Chen ML, Williams RL, and the FDA Population and Individual Bioequivalence Working Group. Individual bioequivalence: New concepts in the statistical assessment of bioequivalence metrics. *Clin Pharmacokin.* 1997;33:1–6.

Phillips KF. Power of the two one-sided tests procedure in bioequivalence. *J Pharmacokin Biopharm.* 1990;18:137–144.

Powers J and Powers T. Statistical analysis of pharmacokinetic data with special applications to bioequivalence studies. *Ann Rech Vet.* 1990;21(Suppl 1):87S–92S.

Quesenberry CP, Whiteker TB, and Dickens JW. On testing normality using several samples. *Biometrics.* 1976;32:753–759.

Rocke DM. On testing bioequivalence. *Biometrics.* 1984;40:225–230.

Rom DM and Hwang E. Testing for individual and population equivalence base on the proportion of similar responses. *Stat Med.* 1996;15:1489–1505.

Ryde M, Huitfeldt B, and Pettersson R. Relative bioavailability of Olsalazine from tablets and capsules: A drug targeted for local effect in the colon. *Biopharm Drug Dispos.* 1991;12:233–246. Also, in Chow and Liu, p. 280. (N-acetyl-5-aminosalicyclic acid).

Schall R. Assessment of individual and population bioequivalence using the probability that bioavailabilities are similar. *Biometrics.* 1995;51:615–626.

Schall R (1995) Unified view of individual, population and average bioequivalence. In H.H. Blume and K.K. Midha (Eds.), *Bio-International 2: Bioavailability, Bioequivalence and Pharmacokinetic Studies.* Medpharm Scientific Publishers, pp. 91–105.

Schall R and Luus HE. On population and individual bioequivalence. *Stat Med.* 1993;12:1109–1124.

Schall R and Williams RL for the FDA Individual Bioequivalence Working Group. Towards a practical strategy for assessing individual bioequivalence. *J. Pharmacokin Biopharm.* 1996 Feb;24(1):133–149.

Schuirmann DJ. A comparison of the two one-sided tests procedure and the power approach for assessing the equivalence of average bioavailability. *J Pharmacokin Biopharm.* 1987;15:657–680.

Schuirmann DJ (1989) Treatment of bioequivalence data: Log transformation, In *Proceedings of Bio-International '89—Issues in the Evaluation of Bioavailability Data*, Toronto, Ontario, Canada, October 1–4, pp. 159–161.

Schumi J and Wittes JT. Through the looking glass: Understanding non-inferiority. *Trials*. 2011 May 3;12:106.

Selwyn MR, Denpster AP, and Hall NR. A bayesian approach to bioequivalence for the 2╫2 changeover design. *Biometrics*. 1981;37:11–21.

Senn S and Lambrou D. Robust and realistic approaches to carry-over. *Stat Med*. 1998;17:2849–2864.

Sheiner LB. Bioequivalence revisited. *Stat Med*. 1992;11:1777–1788.

Steinijans VW and Hauschke D. Update on the statistical analysis of bioequivalence studies. *Int J Clin Pharmacol Ther Toxicol*. 1990;28:105–110.

Tempelman RJ. Experimental design and statistical methods for classical and bioequivalence hypothesis testing with an application to dairy nutrition studies. *J Anim Sci*. 2004;82(E-Suppl):E162–E172.

Ting N, Burdick RK, Graybill FA, Jeyaratnam S, and Lu TFC. Confidence intervals on linear combinations of variance components that are unrestricted in sign. *J Stat Comp Sim*. 1990;35:135–143.

Tothfalusi L and Endrenyi L. Sample sizes for designing bioequivalence studies for highly variable drugs. *J Pharm Pharmaceut Sci*. 2012;15(1):73–84.

Tothfalusi L and Endrenyi. Limits for the scaled average bioequivalence of highly variable drugs and drug products. *Pharm Res*. 2003;20:382–389.

Tothfalusi L, Endrenyi L, and Arieta AG. Evaluation of bioequivalence for highly variable drugs with scaled average bioequivalence. *Clin Pharmacokinet*. 2009;48(11):725–743.

Tothfalusi L, Endrenyi L, Midha KK, Rawson MJ, and Hubbard JW. Evaluation of the bioequivalence of highly-variable drugs and drug products. *Pharm Res*. 2001;18:728–733.

USP 23 - NF - 18, (1090) (1995) *In Vivo Bioequivalence Guidances*. Twinbrook Parkway, Rockville, MD, pp. 1929–1932.

Van Peer A. Variability and impact on design of bioequivalence studies. *Basic Clin Pharmacol Toxicol*. 2010 Mar;106(3):146–153.

Vuorinen J and Turunen J. A simple three-step procedure for parametric and nonparametric assessment of bio-equivalence. *Drug Inf J*. 1997;31:167–180.

Vuorinen J and Turunen J. A three-step procedure for assesing bioequivalence in the general mixed model framework. *Stat Med*. 1996;15:2635–2655.

Wallenstein S and Fisher AC. An analysis of the two-period repeated measurements crossover design with application to clinical trials. *Biometrics*. 1977;33:261–269.

Wang WW, Mehrotra DV, Chan IS, and Heyse JF. Statistical considerations for noninferiority/equivalence trials in vaccine development. *J Biopharm Stat*. 2006;16(4):429–441.

Westlake WJ (1973) The design and analysis of comparative blood-level trials, In J. Swarbrick (Ed.), *Current Concepts in the Pharmaceutical Sciences, Dosage Form Design and Bioavailability*. Lea and Febiger, Philadelphia, PA. pp. 149–179.

Westlake WJ. Statistical aspects of comparative bioavailability trials. *Biometrics*. 1979;35:273–280.

Westlake WJ. Response to Kirkwood, TBL: Bioequivalence testing—A need to rethink. *Biometrics*. 1981;37:589–594.

Westlake WJ (1988) Bioavailability and bioequivalence of pharmaceutical formulations, In K.E. Peace (Ed.), *Biopharmaceutical Statistics for Drug Development*. Marcel Dekker, Inc., New York. pp. 329–352.

Yadav M and Shrivastav PS. Incurred sample reanalysis (ISR): A decisive tool in bioanalytical research. *Bioanalysis*. 2011 May;3(9):1007–1024.

Yee KF. The calculation of probabilities in rejecting bioequivalence. *Biometrics*. 1986;42:961–965.

CHAPTER 8: REGULATORY INSPECTION PROCESS

21 CFR 11, Electronic Records. Electronic Signatures Regulation effective August 1997.

21 CFR part 11, Electronic Records; Electronic Signatures, part 50, Protection of Human Subjects, part 56, Institutional Review Boards, part 200.10, "Contract Facilities (Including Consulting Laboratories) Utilized as Extramural Facilities by Pharmaceutical Manufacturers, part 207, Registration of Producers of Drugs, part 312, Investigational New Drug Application, part 314, Applications for FDA Approval to Market a New Drug or An Antibiotic Drug, part 314.125, Refusal to Approve an Application or Abbreviated Antibiotic Application, part 320, Bioavailability and Bioequivalence Requirements, part 361.1, Radioactive Drugs for Certain Research Uses.

21 CFR part 54, Financial Disclosure by Clinical Investigators.

21 CFR 58.1–58.219, Good Laboratory Practice Regulations effective June 1979, and amended effective October 1987.
21 CFR part 314, Applications for FDA Approval to Market a New Drug.
21 CFR part 320, Bioavailability and Bioequivalence Requirements.
Compliance Program Guidance Manual (CPGM), CPGM 7348.811, Clinical Investigators.
FD & C Act section 301 (e), 505 and 510.
FDA. *Compliance Program Guidance Manual (CPGM), Compliance Program 7348.001 – Bioresearch Monitoring – in vivo Bioequivalence.*
Good Laboratory Practice Regulations, Management Briefings, Post Conference Report, August 1979.
Good Laboratory Practice Regulations, Questions and Answers, June 1981.
Guide to Inspection of Computerized Systems in Drug Processing, February 1983.
Guide For Detecting Fraud in Bioresearch Monitoring Inspections, April 1993.
Software Development Activities, Technical Report, July 1987.
The Federal Food, Drug, and Cosmetic Act, section 505(k)(2)

CHAPTER 9: FED BIOEQUIVALENCE STUDIES

Ahmed T, Sajid M, Singh T, Saini GS, Monif T, Saha N, and Pillai KK. Influence of grape juice and orange juice on the pharmacokinetics and pharmacodynamics of diltiazem in healthy human male subjects. *Int J Clin Pharmacol Ther.* 2008 Oct;46(10):511–518.

Alkhalidi BA, Tamimi JJ, Salem II, Ibrahim H, and Sallam AA. Assessment of the bioequivalence of two formulations of clarithromycin extended-release 500-mg tablets under fasting and fed conditions: A single-dose, randomized, open-label, two-period, two-way crossover study in healthy Jordanian male volunteers. *Clin Ther.* 2008 Oct;30(10):1831–1843.

Álvarez C, Gómez E, Simón M, Govantes C, Guerra P, Frías J, and Arieta AG. Differences in lercanidipine systemic exposure when administered according to labelling: In fasting state and 15 minutes before food intake. *Eur J Clin Pharmacol.* 2012 Jul;68(7):1043–1047. doi: 10.1007/s00228–012–1215–8.

Amsden GW, Whitaker AM, and Johnson PW. Lack of bioequivalence of levofloxacin when coadministered with a mineral-fortified breakfast of juice and cereal. *J Clin Pharmacol.* 2003 Sep;43(9):990–995.

Anschütz M, Wonnemann M, Schug B, Toal C, Donath F, Pontius A, Pauli K, Brendel E, and Blume H. Differences in bioavailability between 60 mg of nifedipine osmotic push-pull systems after fasted and fed administration. *Int J Clin Pharmacol Ther.* 2010 Feb;48(2):158–170.

Araujo MV, Ifa DR, Ribeiro W, Moraes ME, Moraes MO, and de Nucci G. Determination of minocycline in human plasma by high-performance liquid chromatography coupled to tandem mass spectrometry: Application to bioequivalence study. *J Chromatogr B Biomed Sci Appl.* 2001 May 5;755(1–2):1–7.

Bae SK, Kim SH, Lee HW, Seong SJ, Shin SY, Lee SH, Lim MS, Yoon YR, and Lee HJ. Pharmacokinetics of a new once-daily controlled-release formulation of aceclofenac in Korean healthy subjects compared with immediate-release aceclofenac and the effect of food: A randomized, open-label, three-period, crossover, single-centre study. *Clin Drug Investig.* 2012 Feb 1;32(2):111–119.

Bass A, Stark JG, Pixton GC, Sommerville KW, Zamora CA, Leibowitz M, and Rolleri R. Dose proportionality and the effects of food on bioavailability of an immediate-release oxycodone hydrochloride tablet designed to discourage tampering and its relative bioavailability compared with a marketed oxycodone tablet under fed conditions: A single-dose, randomized, open-label, 5-way crossover study in healthy volunteers. *Clin Ther.* 2012 Jul;34(7):1601–1612.

Becker C, Dressman JB, Junginger HE, Kopp S, Midha KK, Shah VP, Stavchansky S, and Barends DM. Biowaiver monographs for immediate release solid oral dosage forms: Rifampicin. *J Pharm Sci.* 2009 Jul;98(7):2252–2267.

Bello CL, Sherman L, Zhou J, Verkh L, Smeraglia J, Mount J, and Klamerus KJ. Effect of food on the pharmacokinetics of sunitinib malate (SU11248), a multi-targeted receptor tyrosine kinase inhibitor: Results from a phase I study in healthy subjects. *Anticancer Drugs.* 2006 Mar;17(3):353–358.

Bhargava V, Lenfant B, Perret C, Pascual MH, Sultan E, and Montay G. Lack of effect of food on the bioavailability of a new ketolide antibacterial, telithromycin. *Scand J Infect Dis.* 2002;34(11):823–826.

Boulton DW, Smith CH, Li L, Huang J, Tang A, and LaCreta FP. Bioequivalence of saxagliptin/metformin extended-release (XR) fixed-dose combination tablets and single-component saxagliptin and metformin XR tablets in healthy adult subjects. *Clin Drug Investig.* 2011;31(9):619–630.

Buice RG, Subramanian VS, Duchin KL, and Uko-Nne S. Bioequivalence of a highly variable drug: An experience with nadolol. *Pharm Res.* 1996 Jul;13(7):1109–1115.

Buice RG, Subramanian VS, and Lane E. Bioequivalence of two orally administered nicardipine products. *Biopharm Drug Dispos*. 1996 Aug;17(6):471–480.

Chandra KP, Shiwalkar A, Kotecha J, Thakkar P, Srivastava A, Chauthaiwale V, Sharma SK, Cross MR, and Dutt C. Phase I clinical studies of the advanced glycation end-product (AGE)-breaker TRC418 safety, tolerability and pharmacokinetics in healthy subjects. *Clin Drug Investig*. 2009;29(9):559–575.

Chen ML, Lee SC, Ng MJ, Schuirmann DJ, Lesko LJ, and Williams RL. Pharmacokinetic analysis of bioequivalence trials: Implications for sex-related issues in clinical pharmacology and biopharmaceutics. *Clin Pharmacol Ther*. 2000 Nov;68(5):510–521.

Chen ML and Lesko LJ. Individual bioequivalence revisited. *Clin Pharmacokinet*. 2001;40(10):701–706. Review.

Chen ML, Patnaik R, Hauck WW, Schuirmann DJ, Hyslop T, and Williams R. An individual bioequivalence criterion: Regulatory considerations. *Stat Med*. 2000 Oct 30;19(20):2821–2842.

Chen ML, Straughn AB, Sadrieh N, Meyer M, Faustino PJ, Ciavarella AB, Meibohm B, Yates CR, and Hussain AS. A modern view of excipient effects on bioequivalence: Case study of sorbitol. *Pharm Res*. 2007 Jan;24(1):73–80. Epub Oct 18 2006.

Chen YX, Cabana B, Kivel N, and Michaelis A. Effect of food on the pharmacokinetics of rifalazil, a novel antibacterial, in healthy male volunteers. *J Clin Pharmacol*. 2007 Jul;47(7):841–849.

Chung M, Calcagni A, Glue P, and Bramson C. Effect of food on the bioavailability of amlodipine besylate/atorvastatin calcium combination tablet. *J Clin Pharmacol*. 2006 Oct;46(10):1212–1216.

Chung M, Vashi V, Puente J, Sweeney M, and Meredith P. Clinical pharmacokinetics of doxazosin in a controlled-release gastrointestinal therapeutic system (GITS) formulation. *Br J Clin Pharmacol*. 1999 Nov;48(5):678–687.

D'Angelo L, De Ponti F, Crema F, Caravaggi M, and Crema A. Effect of food on the bioavailability of pidotimod in healthy volunteers. *Arzneimittelforschung*. 1994 Dec;44(12A):1473–1475.

De Guchtenaere A, Van Herzeele C, Raes A, Dehoorne J, Hoebeke P, Van Laecke E, and Vande Walle J. Oral lyophylizate formulation of desmopressin: Superior pharmacodynamics compared to tablet due to low food interaction. *J Urol*. 2011 Jun;185(6):2308–1213. doi: 10.1016/j.juro.2011.02.039. Epub Apr 21, 2011.

De Mey C. Hemodynamic surrogate end-points in phase I studies? *Int J Clin Pharmacol Ther*. 1997 Apr;35(4):175–179. Review.

Digenis GA, Sandefer EP, Page RC, Doll WJ, Gold TB, and Darwazeh NB. Bioequivalence study of stressed and nonstressed hard gelatin capsules using amoxicillin as a drug marker and gamma scintigraphy to confirm time and GI location of in vivo capsule rupture. *Pharm Res*. 2000 May;17(5):572–582.

Disanto AR and Golden G. Effect of food on the pharmacokinetics of clozapine orally disintegrating tablet 12.5 mg: A randomized, open-label, crossover study in healthy male subjects. *Clin Drug Investig*. 2009;29(8):539–549.

Drabant S, Klebovich I, Gachályi B, Renczes G, and Farsang C. Role of food interaction pharmacokinetic studies in drug development. Food interaction studies of theophylline and nifedipine retard and buspirone tablets. *Acta Pharm Hung*. 1998 Sep;68(5):294–306.

Drake J, Kirkpatrick CT, Aliyar CA, Crawford FE, Gibson P, and Horth CE. Effect of food on the comparative pharmacokinetics of modified-release morphine tablet formulations: Oramorph SR and MST Continus. *Br J Clin Pharmacol*. 1996 May;41(5):417–420.

Endrenyi L and Midha KK. Individual bioequivalence—Has its time come? *Eur J Pharm Sci*. 1998 Oct;6(4):271–278. Review.

Endrenyi L, Taback N, and Tothfalusi L. Properties of the estimated variance component for subject-by-formulation interaction in studies of individual bioequivalence. *Stat Med*. 2000 Oct 30;19(20):2867–2878.

Endrenyi L and Tothfalusi L. Subject-by-formulation interaction in determinations of individual bioequivalence: Bias and prevalence. *Pharm Res*. 1999 Feb;16(2):186–190.

Falcoz C, Jenkins JM, Bye C, Hardman TC, Kenney KB, Studenberg S, Fuder H, and Prince WT. Pharmacokinetics of GW433908, a prodrug of amprenavir, in healthy male volunteers. *J Clin Pharmacol*. 2002 Aug;42(8):887–898.

Fitzpatrick SC, Brynes SD, and Guest GB. Dietary intake estimates as a means to the harmonization of maximum residue levels for veterinary drugs. I. Concept. *J Vet Pharmacol Ther*. 1995 Oct;18(5):325–327.

Fiumara K and Goldhaber SZ. Cardiology patient pages. A patient's guide to taking coumadin/warfarin. *Circulation*. 2009 Mar 3;119(8):e220–e222.

Flesch G, Tudor D, Souppart C, D'Souza J, and Hossain M. Oxcarbazepine final market image tablet formulation bioequivalence study after single administration and at steady state in healthy subjects. *Int J Clin Pharmacol Ther*. 2002 Nov;40(11):524–532.

Fontes-Ribeiro C, Macedo T, Nunes T, Neta C, Vasconcelos T, Cerdeira R, Lima R et al. Dosage form proportionality and food effect of the final tablet formulation of eslicarbazepine acetate: Randomized, open-label, crossover, single-centre study in healthy volunteers. *Drugs R D*. 2008;9(6):447–454.

Fuchs WS, Hens C, and von Nieciecki A. Requirements for product quality of theophylline sustained-release preparations. *Arzneimittelforschung*. 1998 May;48(5A):556–561. German.

Fuhr U, Müller-Peltzer H, Kern R, Lopez-Rojas P, Jünemann M, Harder S, and Staib AH. Effects of grapefruit juice and smoking on verapamil concentrations in steady state. *Eur J Clin Pharmacol*. 2002 Apr;58(1):45–53.

Gai MN, Costa E, and Arancibia A. Bioavailability of a controlled-release cyclobenzaprine tablet and influence of a high fat meal on bioavailability. *Int J Clin Pharmacol Ther*. 2009 Apr;47(4):269–274.

Galanello R, Piga A, Cappellini MD, Forni GL, Zappu A, Origa R, Dutreix C et al. Effect of food, type of food, and time of food intake on deferasirox bioavailability: Recommendations for an optimal deferasirox administration regimen. *J Clin Pharmacol*. 2008 Apr;48(4):428–435.

Godfrey AR, Digiacinto J, and Davis MW. Single-dose bioequivalence of 105-mg fenofibric acid tablets versus 145-mg fenofibrate tablets under fasting and fed conditions: A report of two phase I, open-label, single-dose, randomized, crossover clinical trials. *Clin Ther*. 2011 Jun;33(6):766–775.

Golovenko NIa and Borisiuk IIu. Biopharmaceutical classification system—Experimental model of the prediction of drug bioavailability. *Biomed Khim*. 2008 Jul–Aug;54(4):392–407.

González MA and Straughan AB. Effect of meals and dosage-form modification on theophylline bioavailability from a 24-hour sustained-release delivery system. *Clin Ther*. 1994 Sep–Oct;16(5):804–814.

Gourlay GK. Sustained relief of chronic pain. Pharmacokinetics of sustained release morphine. *Clin Pharmacokinet*. 1998 Sep;35(3):173–190.

Haessler F, Tracik F, Dietrich H, Stammer H, and Klatt J. A pharmacokinetic study of two modified-release methylphenidate formulations under different food conditions in healthy volunteers. *Int J Clin Pharmacol Ther*. 2008 Sep;46(9):466–476.

Hauck WW, Hyslop T, Chen ML, Patnaik R, and Williams RL. Subject-by-formulation interaction in bioequivalence: Conceptual and statistical issues. FDA Population/Individual Bioequivalence Working Group. Food and Drug Administration. *Pharm Res*. 2000 Apr;17(4):375–380.

Heinig R and Sachse R. The effect of food and time of administration on the pharmacokinetic and pharmacodynamic profile of metrifonate. *Int J Clin Pharmacol Ther*. 1999 Sep;37(9):456–464.

Henney HR 3rd, Fitzpatrick A, Stewart J, and Runyan JD. Relative bioavailability of tizanidine hydrochloride capsule formulation compared with capsule contents administered in applesauce: A single-dose, open-label, randomized, two-way, crossover study in fasted healthy adult subjects. *Clin Ther*. 2008 Dec;30(12):2263–2271.

Henney HR 3rd and Shah J. Relative bioavailability of tizanidine 4-mg capsule and tablet formulations after a standardized high-fat meal: A single-dose, randomized, open-label, crossover study in healthy subjects. *Clin Ther*. 2007 Apr;29(4):661–669.

Hoch M, Hoever P, Haschke M, Krähenbühl S, and Dingemanse J. Food effect and biocomparison of two formulations of the dual orexin receptor antagonist almorexant in healthy male subjects. *J Clin Pharmacol*. 2011 Jul;51(7):1116–1121.

Holdich T and Sawyer J. Influence of food on the pharmacokinetics of apricitabine, a novel deoxycytidine analogue reverse transcriptase inhibitor. *Expert Opin Pharmacother*. 2008 Aug;9(12):2021–2025.

Houin G. Bioequivalence studies: A new EMA guideline. *Arzneimittelforschung*. 2010;60(4):169–170.

Huang Y, Shi R, Gee W, and Bonderud R. Regulated drug bioanalysis for human pharmacokinetic studies and therapeutic drug management. *Bioanalysis*. 2012 Aug;4(15):1919–1931. doi: 10.4155/bio.12.157. Review.

Hwang SS, Gorsline J, Louie J, Dye D, Guinta D, and Hamel L. In vitro and in vivo evaluation of a once-daily controlled-release pseudoephedrine product. *J Clin Pharmacol*. 1995 Mar;35(3):259–267.

Ibarra M, Fagiolino P, Vázquez M, Ruiz S, Vega M, Bellocq B, Pérez M, González B, and Goyret A. Impact of food administration on lopinavir-ritonavir bioequivalence studies. *Eur J Pharm Sci*. 2012 Aug 15;46(5):516–521.

Jansat JM, Martinez-Tobed A, Garcia E, Cabarrocas X, and Costa J. Effect of food intake on the bioavailability of almotriptan, an antimigraine compound, in healthy volunteers: An open, randomized, crossover, single-dose clinical trial. *Int J Clin Pharmacol Ther*. 2006 Apr;44(4):185–190.

Jenner P, Könen-Bergmann M, Schepers C, Haertter S. Pharmacokinetics of a once-daily extended-release formulation of pramipexole in healthy male volunteers: Three studies. *Clin Ther*. 2009 Nov;31(11):2698–2711.

Jiang W, Kim S, Zhang X, Lionberger RA, Davit BM, Conner DP, and Yu LX. The role of predictive biopharmaceutical modeling and simulation in drug development and regulatory evaluation. *Int J Pharm*. 2011 Oct 14;418(2):151–160.

Jiménez Torres NV, Romero Crespo I, Ballester Solaz M, Albert Marí A, and Jiménez Arenas V. Antineoplastic oral agents and drug-nutrient interactions: A sistematic review. *Nutr Hosp*. 2009 May–Jun;24(3):260–272.

Johnston A, Keown PA, and Holt DW. Simple bioequivalence criteria: Are they relevant to critical dose drugs? Experience gained from cyclosporine. *Ther Drug Monit*. 1997 Aug;19(4):375–381.

Kalantzi L, Persson E, Polentarutti B, Abrahamsson B, Goumas K, Dressman JB, and Reppas C. Canine intestinal contents vs. simulated media for the assessment of solubility of two weak bases in the human small intestinal contents. *Pharm Res*. 2006 Jun;23(6):1373–1381.

Kapil R, Nolting A, Roy P, Fiske W, Benedek I, and Abramowitz W. Pharmacokinetic properties of combination oxycodone plus racemic ibuprofen: Two randomized, open-label, crossover studies in healthy adult volunteers. *Clin Ther*. 2004 Dec;26(12):2015–2025.

Karalis V, Macheras P, Van Peer A, and Shah VP. Bioavailability and bioequivalence: Focus on physiological factors and variability. *Pharm Res*. 2008 Aug;25(8):1956–1962.

Kaul S, Ji P, Lu M, Nguyen KL, Shangguan T, and Grasela D. Bioavailability in healthy adults of efavirenz capsule contents mixed with a small amount of food. *Am J Health Syst Pharm*. 2010 Feb 1;67(3):217–222.

Kees F, Bucher M, Schweda F, Gschaidmeier H, Faerber L, and Seifert R. Neoimmun versus Neoral: A bioequivalence study in healthy volunteers and influence of a fat-rich meal on the bioavailability of Neoimmun. *Naunyn Schmiedebergs Arch Pharmacol*. 2007 Aug;375(6):393–399.

Kim Y, Morikawa M, Ohsawa H, Kou M, Ishida E, Igarashi J, Kajimoto T et al. Effects of foods on the pharmacokinetics and clinical efficacy of quazepam. *Nihon Shinkei Seishin Yakurigaku Zasshi*. 2003 Oct;23(5):205–210.

Kimanani EK, Lavigne J, and Potvin D. Numerical methods for the evaluation of individual bioequivalence criteria. *Stat Med*. 2000 Oct 30;19(20):2775–2795.

Klebovich I. Review of bioanalytic methods in pharmacokinetics. *Acta Pharm Hung*. 1998 Jul;68(4):234–248. Review. Hungarian.

Klein S, Butler J, Hempenstall JM, Reppas C, and Dressman JB. Media to simulate the postprandial stomach I. Matching the physicochemical characteristics of standard breakfasts. *J Pharm Pharmacol*. 2004 May;56(5):605–610.

Klueglich M, Ring A, Scheuerer S, Trommeshauser D, Schuijt C, Liepold B, and Berndl G. Ibuprofen extrudate, a novel, rapidly dissolving ibuprofen formulation: Relative bioavailability compared to ibuprofen lysinate and regular ibuprofen, and food effect on all formulations. *J Clin Pharmacol*. 2005 Sep;45(9):1055–1061.

Kunz K, Lorkowski G, Petersen G, Samcova E, Schaffler K, and Wauschkuhn CH. Bioavailability of escin after administration of two oral formulations containing aesculus extract. *Arzneimittelforschung*. 1998 Aug;48(8):822–825.

la Porte C, Verweij-van Wissen C, van Ewijk N, Aarnoutse R, Koopmans P, Reiss P, Stek M Jr, Hekster Y, and Burger D. Pharmacokinetic interaction study of indinavir/ritonavir and the enteric-coated capsule formulation of didanosine in healthy volunteers. *J Clin Pharmacol*. 2005 Feb;45(2):211–218.

Lainesse A, Hussain S, Monif T, Reyar S, Tippabhotla SK, Madan A, and Thudi NR. Bioequivalence studies of tacrolimus capsule under fasting and fed conditions in healthy male and female subjects. *Arzneimittelforschung*. 2008;58(5):242–247.

Lee D, Lim LA, Jang SB, Lee YJ, Chung JY, Choi JR, Kim K et al. Pharmacokinetic comparison of sustained- and immediate-release oral formulations of cilostazol in healthy Korean subjects: A randomized, open-label, 3-part, sequential, 2-period, crossover, single-dose, food-effect, and multiple-dose study. *Clin Ther*. 2011 Dec;33(12):2038–2053.

Lee J, Zhang W, Moy S, Kowalski D, Kerbusch V, van Gelderen M, Sawamoto T, Grunenberg N, and Keirns J. Effects of food intake on the pharmacokinetic properties of mirabegron oral controlled-absorption system: A single-dose, randomized, crossover study in healthy adults. *Clin Ther*. 2013 Mar;35(3):333–341.

Lennernäs H and Abrahamsson B. The use of biopharmaceutic classification of drugs in drug discovery and development: Current status and future extension. *J Pharm Pharmacol*. 2005 Mar;57(3):273–285.

Lincoln J, Stewart ME, and Preskorn SH. How sequential studies inform drug development: Evaluating the effect of food intake on optimal bioavailability of ziprasidone. *J Psychiatr Pract*. 2010 Mar;16(2):103–114.

Lippert C, Keung A, Arumugham T, Eller M, Hahne W, and Weir S. The effect of food on the bioavailability of dolasetron mesylate tablets. *Biopharm Drug Dispos*. 1998 Jan;19(1):17–19.

Mahmood I. Development of a limited sampling approach in pharmacokinetic studies: Experience with the antiepilepsy drug tiagabine. *J Clin Pharmacol*. 1998 Apr;38(4):324–330.

Maia J, Vaz-da-Silva M, Almeida L, Falcão A, Silveira P, Guimarães S, Graziela P, and Soares-da-Silva P. Effect of food on the pharmacokinetic profile of eslicarbazepine acetate (BIA 2–093). *Drugs R D*. 2005;6(4):201–206.

Manfio JL, dos Santos MB, Favreto WA, Weich A, Pugens AM, and Donaduzzi CM. Bioequivalence study of two formulations of 100 mg capsule of itraconazole. Quantification by tandem mass spectrometry. *Arzneimittelforschung*. 2010;60(3):157–161.

Marathe PH, Arnold ME, Meeker J, Greene DS, and Barbhaiya RH. Pharmacokinetics and bioavailability of a metformin/glyburide tablet administered alone and with food. *J Clin Pharmacol*. 2000 Dec;40(12 Pt 2):1494–1502.

Marzo A. Open questions on bioequivalence: Some problems and some solutions. *Pharmacol Res*. 1999 Oct;40(4):357–368. Review.

Marzo A. Open questions on bioequivalence: An updated reappraisal. *Curr Clin Pharmacol*. 2007 May;2(2):179–189.

Marzo A and Fontana E. Critical considerations into the new EMA guideline on bioequivalence. *Arzneimittelforschung*. 2011;61(4):207–220.

McLean A, Browne S, Zhang Y, Slaughter E, Halstenson C, and Couch R. The influence of food on the bioavailability of a twice-daily controlled release carbamazepine formulation. *J Clin Pharmacol*. 2001 Feb;41(2):183–186.

Mehvar R and Jamali F. Bioequivalence of chiral drugs. Stereospecific versus non-stereospecific methods. *Clin Pharmacokinet*. 1997 Aug;33(2):122–141.

Meyer MC, Straughn AB, Jarvi EJ, Patrick KS, Pelsor FR, Williams RL, Patnaik R, Chen ML, and Shah VP. Bioequivalence of methylphenidate immediate-release tablets using a replicated study design to characterize intrasubject variability. *Pharm Res*. 2000 Apr;17(4):381–384.

Midha KK, Rawson MJ, and Hubbard JW. The bioequivalence of highly variable drugs and drug products. *Int J Clin Pharmacol Ther*. 2005 Oct;43(10):485–498.

Moore KH, Shaw S, Laurent AL, Lloyd P, Duncan B, Morris DM, O'Mara MJ, and Pakes GE. Lamivudine/zidovudine as a combined formulation tablet: Bioequivalence compared with lamivudine and zidovudine administered concurrently and the effect of food on absorption. *J Clin Pharmacol*. 1999 Jun;39(6):593–605.

Moore KT, St-Fleur D, Marricco NC, Ariyawansa J, Pagé V, Natarajan J, Morelli G, and Richarz U. A randomized study of the effects of food on the pharmacokinetics of once-daily extended-release hydromorphone in healthy volunteers. *J Clin Pharmacol*. 2011 Nov;51(11):1571–1579.

Neuhofel AL, Wilton JH, Victory JM, Hejmanowsk LG, and Amsden GW. Lack of bioequivalence of ciprofloxacin when administered with calcium-fortified orange juice: A new twist on an old interaction. *J Clin Pharmacol*. 2002 Apr;42(4):461–466.

Oostendorp RL, Loftiss J, Goel S, Smith DA, Dar MM, Witteveen PO, Cohen RB et al. Bioequivalence study of a new oral topotecan formulation, relative to the current topotecan formulation, in patients with advanced solid tumors. *Int J Clin Pharmacol Ther*. 2009 Mar;47(3):195–206.

Ozbay L, Unal DO, and Erol D. Food effect on bioavailability of modified-release trimetazidine tablets. *J Clin Pharmacol*. 2012 Oct;52(10):1535–1539.

Peloquin CA, Zhu M, Adam RD, Singleton MD, and Nix DE. Pharmacokinetics of para-aminosalicylic acid granules under four dosing conditions. *Ann Pharmacother*. 2001 Nov;35(11):1332–1338.

Raber M, Schulz HU, Schürer M, Bias-Imhoff U, and Momberger H. Pharmacokinetic properties of tramadol sustained release capsules. 2nd communication: Investigation of relative bioavailability and food interaction. *Arzneimittelforschung*. 1999 Jul;49(7):588–593.

Rackley RJ. Re: Acknowledgement of human bioequivalency of matrix's lopinavir-ritonavir formulation to Kaletra, comment to article on Bioavailability of generic ritonavir and lopinavir/ritonavir tablet products in a dog model, by Garren et al. *J Pharm Sci*. 2010 Feb;99(2):568–571;author reply 574.

Retzow A, Schürer M, and Schulz HU. Influence of food on the bioavailability of a carbamazepine slow-release formulation. *Int J Clin Pharmacol Ther*. 1997 Dec;35(12):557–560.

Retzow A, Vens-Cappell B, and Wangemann M. Influence of food on the pharmacokinetics of a new multiple unit sustained release sodium valproate formulation. *Arzneimittelforschung*. 1997 Dec;47(12):1347–1350.

Ricart AD, Sarantopoulos J, Calvo E, Chu QS, Greene D, Nathan FE, Petrone ME, Tolcher AW, and Papadopoulos KP. Satraplatin, an oral platinum, administered on a five-day every-five-week schedule: A pharmacokinetic and food effect study. *Clin Cancer Res*. 2009 Jun 1;15(11):3866–3871.

Rostami-Hodjegan A, Shiran MR, Ayesh R, Grattan TJ, Burnett I, Darby-Dowman A, and Tucker GT. A new rapidly absorbed paracetamol tablet containing sodium bicarbonate. I. A four-way crossover study to compare the concentration-time profile of paracetamol from the new paracetamol/sodium bicarbonate tablet and a conventional paracetamol tablet in fed and fasted volunteers. *Drug Dev Ind Pharm*. 2002 May;28(5):523–531.

Rubino CM, Bhavnani SM, Ambrose PG, Forrest A, and Loutit JS. Effect of food and antacids on the pharmacokinetics of pirfenidone in older healthy adults. *Pulm Pharmacol Ther.* 2009 Aug;22(4):279–285.

Saffar F, Aiache JM, and Andre P. Influence of food on the disposition of the antidiabetic drug metformin in diabetic patients at steady-state. *Methods Find Exp Clin Pharmacol.* 1995 Sep;17(7):483–487.

Sathyan G, Xu E, Thipphawong J, and Gupta SK. Pharmacokinetic profile of a 24-hour controlled-release OROS formulation of hydromorphone in the presence and absence of food. *BMC Clin Pharmacol.* 2007 Feb 2;7:2.

Sauron R, Wilkins M, Jessent V, Dubois A, Maillot C, and Weil A. Absence of a food effect with a 145 mg nanoparticle fenofibrate tablet formulation. *Int J Clin Pharmacol Ther.* 2006 Feb;44(2):64–70.

Schmitt C, Charoin-Pannier A, McIntyre C, Zandt H, Ciorciaro C, Zweigler L, Winters K, and Pepper T. Effect of food on the pharmacokinetics and pharmacodynamics of R1663, an oral factor Xa inhibitor, in healthy male volunteers. *Int J Clin Pharmacol Ther.* 2012 Aug;50(8):566–572.

Schug BS, Brendel E, Chantraine E, Wolf D, Martin W, Schall R, and Blume HH. The effect of food on the pharmacokinetics of nifedipine in two slow release formulations: Pronounced lag-time after a high fat breakfast. *Br J Clin Pharmacol.* 2002 Jun;53(6):582–588.

Shah A, Liu MC, Vaughan D, and Heller AH. Oral bioequivalence of three ciprofloxacin formulations following single-dose administration: 500 mg tablet compared with 500 mg/10 mL or 500 mg/5 mL suspension and the effect of food on the absorption of ciprofloxacin oral suspension. *J Antimicrob Chemother.* 1999 Mar;43(Suppl A):49–54.

Shepard DR, Mani S, Kastrissios H, Learned-Coughlin S, Smith D, Ertel P, Magnum S et al. Estimation of the effect of food on the disposition of oral 5-fluorouracil in combination with eniluracil. *Cancer Chemother Pharmacol.* 2002 May;49(5):398–402.

Smith S, Collaku A, Heaslip L, Yue Y, Starkey YY, Clarke G, and Kronfeld N. A new rapidly absorbed paediatric paracetamol suspension. A six-way crossover pharmacokinetic study comparingthe rate and extent of paracetamol absorption from a new paracetamol suspension with two marketed paediatric formulations. *Drug Dev Ind Pharm.* 2012 Mar;38(3):372–379. doi: 10.3109/03639045.2011.605141.

Solans A, Carbó ML, Peña J, Nadal T, Izquierdo I, and Merlos M. Influence of food on the oral bioavailability of rupatadine tablets in healthy volunteers: A single-dose, randomized, open-label, two-way crossover study. *Clin Ther.* 2007 May;29(5):900–908.

Sostek MB, Chen Y, and Andersson T. Effect of timing of dosing in relation to food intake on the pharmacokinetics of esomeprazole. *Br J Clin Pharmacol.* 2007 Sep;64(3):386–390. Epub Apr 10, 2007.

Souliman S, Blanquet S, Beyssac E, and Cardot JM. A level A in vitro/in vivo correlation in fasted and fed states using different methods: Applied to solid immediate release oral dosage form. *Eur J Pharm Sci.* 2006 Jan;27(1):72–79.

Spénard J, Aumais C, Massicotte J, Brunet JS, Tremblay C, Grace M, and Lefebvre M. Effects of food and formulation on the relative bioavailability of bismuth biskalcitrate, metronidazole, and tetracycline given for Helicobacter pylori eradication. *Br J Clin Pharmacol.* 2005 Oct;60(4):374–377.

Steinijans VW, Sauter R, Böhm A, Dietrich R, and Benedikt G. Chronotherapy concept and the pharmacokinetic validation of a theophylline retard preparation for once-nightly administration (Euphylong). *Pneumologie.* 1991 Nov;45(Suppl 4):827–833. German.

Teo SK, Scheffler MR, Wu A, Stirling DI, Thomas SD, Stypinski D, and Khetani VD. A single-dose, two-way crossover, bioequivalence study of dexmethylphenidate HCl with and without food in healthy subjects. *J Clin Pharmacol.* 2004 Feb;44(2):173–178.

Tulloch SJ, Zhang Y, McLean A, and Wolf KN. SLI381 (Adderall XR), a two-component, extended-release formulation of mixed amphetamine salts: Bioavailability of three test formulations and comparison of fasted, fed, and sprinkled administration. *Pharmacotherapy.* 2002 Nov;22(11):1405–1415.

Van Peer A. Variability and impact on design of bioequivalence studies. *Basic Clin Pharmacol Toxicol.* 2010 Mar;106(3):146–153.

Vanover KE, Robbins-Weilert D, Wilbraham DG, Mant TG, van Kammen DP, Davis RE, and Weiner DM. The effects of food on the pharmacokinetics of a formulated ACP-103 tablet in healthy volunteers. *J Clin Pharmacol.* 2007 Jul;47(7):915–919.

Vaz-da-Silva M, Nunes T, Rocha JF, Falcão A, Almeida L, and Soares-da-Silva P. Effect of food on the pharmacokinetic profile of etamicastat (BIA 5–453). *Drugs R D.* 2011;11(2):127–136.

Waldman SA and Morganroth J. Effects of food on the bioequivalence of different verapamil sustained-release formulations. *J Clin Pharmacol.* 1995 Feb;35(2):163–169.

Wallace AW, Victory JM, and Amsden GW. Lack of bioequivalence of gatifloxacin when coadministered with calcium-fortified orange juice in healthy volunteers. *J Clin Pharmacol.* 2003 Jan;43(1):92–96.

Wallace AW, Victory JM, and Amsden GW. Lack of bioequivalence when levofloxacin and calcium-fortified orange juice are coadministered to healthy volunteers. *J Clin Pharmacol.* 2003 May;43(5):539–544.

Wan H and Chow SC. On statistical power for average bioequivalence testing under replicated crossover designs. *J Biopharm Stat.* 2002 Aug;12(3):295–309.

Wigal SB, Gupta S, Heverin E, and Starr HL. Pharmacokinetics and therapeutic effect of OROS methylphenidate under different breakfast conditions in children with attention-deficit/hyperactivity disorder. *J Child Adolesc Psychopharmacol.* 2011 Jun;21(3):255–263.

Wilder BJ, Leppik I, Hietpas TJ, Cloyd JC, Randinitis EJ, and Cook J. Effect of food on absorption of Dilantin Kapseals and Mylan extended phenytoin sodium capsules. *Neurology.* 2001 Aug 28;57(4):582–589.

Yacobi A, Masson E, Moros D, Ganes D, Lapointe C, Abolfathi Z, LeBel M et al. Who needs individual bioequivalence studies for narrow therapeutic index drugs? A case for warfarin. *J Clin Pharmacol.* 2000 Aug;40(8):826–835.

Yerino GA, Halabe EK, Zini E, and Feleder EC. Bioequivalence study of two oral tablet formulations containing tenofovir disoproxil fumarate in healthy volunteers. *Arzneimittelforschung.* 2011;61(1):55–60.

Yin OQ, Rudoltz M, Galetic I, Filian J, Krishna A, Zhou W, Custodio J, Golor G, and Schran H. Effects of yogurt and applesauce on the oral bioavailability of nilotinib in healthy volunteers. *J Clin Pharmacol.* 2011 Nov;51(11):1580–1586. doi: 10.1177/0091270010384116.

Zariffa NM and Patterson SD. Population and individual bioequivalence: Lessons from real data and simulation studies. *J Clin Pharmacol.* 2001 Aug;41(8):811–822.

Zhang X, Lionberger RA, Davit BM, and Yu LX. Utility of physiologically based absorption modeling in implementing Quality by Design in drug development. *AAPS J.* 2011 Mar;13(1):59–71. doi: 10.1208/s12248-010-9250-9.

CHAPTER 10: BIOEQUIVALENCE OF TOPICAL DRUGS

Boix-Montanes A. Relevance of equivalence assessment of topical products based on the dermatopharmacokinetics approach. *Eur J Pharm Sci.* 2011 Feb 14;42(3):173–179.

Farahmand S and Maibach HI. Transdermal drug pharmacokinetics in man: Interindividual variability and partial prediction. *Int J Pharm.* 2009 Feb 9;367(1–2):1–15.

Narkar Y. Bioequivalence for topical products—An update. *Pharm Res.* 2010 Dec;27(12):2590–2601.

Payette M and Grant-Kels JM. Generic drugs in dermatology: Part I. *J Am Acad Dermatol.* 2012 Mar;66(3):343.

Payette M and Grant-Kels JM. Generic drugs in dermatology: Part II. *J Am Acad Dermatol.* 2012 Mar;66(3):353.

Puig L. Biosimilars or follow-on biologics in dermatology. *Actas Dermosifiliogr.* 2010 Jan-Feb;101(1):4–6.

Ranjan N, Mahajan VK, and Misra M. Biosimilars: The 'future' of biologic therapy? *J Dermatolog Treat.* 2011 Dec;22(6):319–322.

Smit P, Neumann HA, and Thio HB. The skin-blanching assay. *J Eur Acad Dermatol Venereol.* 2012 Oct;26(10):1197–1202.

Strober BE, Armour K, Romiti R, Smith C, Tebbey PW, Menter A, and Leonardi C. Biopharmaceuticals and biosimilars in psoriasis: What the dermatologist needs to know. *J Am Acad Dermatol.* 2012 Feb;66(2):317–322.

Topical Holmgaard R, Nielsen JB, and Benfeldt E. Microdialysis sampling for investigations of bioavailability and bioequivalence of topically administered drugs: Current state and future perspectives. *Skin Pharmacol Physiol.* 2010;23(5):225–243.

CHAPTER 11: BIOEQUIVALENCE OF NASAL PRODUCTS

Aerosols, metered-dose inhalers, and dry powder inhalers. *Pharm Forum.* 1998;24:6936–6971.

Allec A, Chatelus A, and Wagner N. Skin distribution and pharmaceutical aspects of adapalene gel. *J Am Acad Dermatol.* 1997;36:S119–S125.

Apiou-Sbirlea G, Newman S, Fleming J, Siekmeier R, Ehrmann S, Scheuch G, Hochhaus G, and Hickey A. Bioequivalence of inhaled drugs: Fundamentals, challenges and perspectives. *Ther Deliv.* 2013 Mar;4(3):343–367.

Byron PR (ed.) (1990) Aerosol formulation, generation, and delivery using metered systems, Chapter 7. *Respiratory Drug Delivery.* CRC Press, Boca Raton, FL.

Caron D, Queille-Roussel C, Shah VP, and Schaefer H. Correlation between the drug penetration and the blanching effect of topically applied hydrocortisone creams in human beings. *J Am Acad Dermatol.* 1990;23:458–462.

Daley-Yates PT and Parkins DA. Establishing bioequivalence for inhaled drugs; weighing the evidence. *Expert Opin Drug Deliv*. 2011 Oct;8(10):1297–1308.

FDA. *Topical Dermatological Corticosteroids: In Vivo Bioequivalence*, CDER, June 1995.

FDA. *Draft Guidance for Industry: Metered Dose Inhaler (MDI) and Dry Powder Inhaler (DPI) Drug Products—Chemistry, Manufacturing, and Controls Documentation*, October 1998.

FDA. *Guidance for industry: Points to Consider: Clinical Development Programs for MDI and DPI Drug Products*, September 19, 1994.

FDA. *Draft Guidance on Specifications: Test Procedures and Acceptance Criteria for New Drug Substances and New Drug Products: Chemical Substances*, International Conference on Harmonisation (ICH) Q6A, November 1997.

FDA. *SUPAC-SS Nonsterile Semisolid Dosage Forms, Scale-Up and Post Approval Changes: Chemistry, Manufacturing, and Controls; In Vitro Release Testing and* in vivo *Bioequivalence Documentation*, Center for Drug Evaluation and Research (CDER), May 1997.

Fuglsang A. The US and EU regulatory landscapes for locally acting generic/hybrid inhalation products intended for treatment of asthma and COPD. *J Aerosol Med Pulm Drug Deliv*. 2012 Aug;25(4):243–247.

Heykants J, Van Peer A, Van de Velde V, Snoeck E et al. The pharmacokinetic properties of topical levocabastine: A review. *Clin Pharmacokinet*. 1995;29:221–230.

Hueber F, Schaefer H, and Wepierre J. Role of transepidermal and transfollicular routes in percutaneous absorption of steroids: In vitro studies on human skin. *Skin Pharmacol*. 1994;7:237–244.

Hyslop T, Hsuan F, and Holder DJ. A small-sample confidence interval approach to assess individual bioequivalence. *Stat Med*. 2000 Oct 30;19(20):2885–2897.

Illel B, Schaefer H, Wepierre J, and Doucet O. Follicles play an important role in percutaneous absorption. *J Pharm Sci*. 1991;80:424–427.

Kublic H and Vidgren MT. Nasal delivery systems and their effect on deposition and absorption. *Adv Drug Deliv Rev*. 1998;29:157–177.

Lee SL, Adams WP, Li BV, Conner DP, Chowdhury BA, and Yu LX. In vitro considerations to support bioequivalence of locally acting drugs in dry powder inhalers for lung diseases. *AAPS J*. 2009 Sep;11(3):414–423.

Maibach HI (ed.) (1996) *Dermatologic Research Techniques*, CRC Press Inc., Boca Raton, FL.

Parry GE, Dunn P, Shah VP, and Pershing LK. Acyclovir bioavailability in human skin. *J Invest Dermatol*. 1992;98:856–863.

Pershing LK, Corlett J, and Jorgensen C. In vivo pharmacokinetics and pharmacodynamics of topical ketoconazole and miconazole in human stratum corneum. *Antimicrob Agents Chemother*. 1994;38:90–95.

Pershing LK, Lambert LD, Shah VP, and Lam SY. Variability and correlation of chromameter and tape-stripping methods with the visual skin blanching assay in quantitative assessment of topical 0.05% betamethasone dipropionate bioavailability in humans. *Int J Pharm*. 1992;86:201–210.

Pershing LK, Lambert LD, Wright ED et al. Topical 0.05% betamethasone dipropionate: Pharmacokinetic and pharmacodynamic dose-response studies in humans. *Arch Dermatol*. 1994;130:740–747.

Pershing L, Silver BS, Krueger GG et al. Feasibility of measuring the bioavailability of topical betamethasone dipropionate in commercial formulations using drug content in skin and a skin blanching bioassay. *Pharm Res*. 1992;9:45–51.

Rougier A, Dupuis D, Lotte C et al. In vivo correlation between stratum corneum reservoir function and percutaneous absorption. *J Inv Derm*. 1983;81:275–278.

Rougier A, Dupuis D, Lotte C et al. Regional variation in percutaneous absorption in man: Measurement by the stripping method. *Arch Dermatol Res*. 1986;278:465–469.

Rougier A, Rallis M, Kiren P, and Lotte C. In vivo percutaneous absorption: A key role for stratum corneum/vehicle partitioning. *Arch Dermatol Res*. 1990;282:498–505.

Schaefer H and Redelmeir TE (eds.) (1996) *Skin Barrier, Principles of Percutaneous Absorption*. Karger Publishers, New York.

Schultz RK. Drug delivery characteristics of metered-dose inhalers. *J Allergy Clin Immunol*. 1995;96:284–287.

Sciarra JJ and Cutie AJ (1989) Aerosol suspensions and emulsions, Vol. 2. In H.A. Lieberman, M.M. Rieger, and G.S. Banker (Eds.), *Pharmaceutical Dosage Forms: Disperse Systems*. Marcel Dekker, New York, pp. 455–458.

Shah VP and Maibach HI (eds.) (1993) *Topical Drug Bioavailability, Bioequivalence and Penetration*. Plenum Press, New York.

Shah VP, Flynn GL, Yacobi A et al. Bioequivalence of topical dermatological dosage forms—Methods of evaluation of bioequivalence. *Pharm Res*. 1998;15:167–171.

Shah VP, Midha KK, Dighe S et al. Analytical methods validation: Bioavailability, bioequivalence and pharmacokinetic studies. Workshop report. *Pharm Res*. 1992;9, 588–592.

Stegemann S, Kopp S, Borchard G, Shah VP, Senel S, Dubey R, Urbanetz N et al. Developing and advancing dry powder inhalation towards enhanced therapeutics. *Eur J Pharm Sci.* 2013 Jan 23;48(1–2):181–194.

Surber C, Wilhelm K-P, Bermann D, and Maibach HI. In vivo skin penetration of acitretin in volunteers using three sampling techniques. *Pharm Res.* 1993;9:1291–1294.

Task group on lung dynamics: Deposition and retention models for internal dosimetry of the human respiratory tract. *Health Phys.* 1966;12:173–207.

Wilson AM, Sims EJ, McFarlane LC, and Lipworth BJ. Effects of intranasal corticosteroids on adrenal, bone, and blood markers of systemic activity in allergic rhinitis. *J Allergy Clin Immunol.* 1998;102:598–604.

CHAPTER 12: BIOEQUIVALENCE OF COMPLEMENTARY AND ALTERNATIVE MEDICINES

American Association of Medical Colleges. *Contemporary Issues in Medicine: Communication in Medicine. Medical Schools Objectives Project Report III (MSOP III).* American Association of Medical Colleges, Washington, DC, 1999.

Astin JA, Marie A, Pelletier KR, Hansen E et al. A review of the incorporation of complementary and alternative medicine by mainstream physicians. *Arch. Intern Med.* 1998;158(21):2303–2310.

Astin JA. Why patients use alternative medicine: Results of a national study. *JAMA* 1998;279:1548–1553.

Berliner HS and Salmon JW. The holistic alternative to scientific medicine: History and analysis. *Int J Health Serv.* 1980;10:133–147.

Boschma G. The meaning of holism in nursing: Historical shifts in holistic nursing ideas. *Public Health Nurs* 1994;11(5):324–330.

Caspi O, Bell IR, Rychener D, Gaudet TW et al. The Tower of Babel: Communication and medicine: An essay on medical education and complementary-alternative medicine. *Arch Intern Med.* 2000;160(21):3193–3195.

Cassileth BR. Complementary therapies: Overview and state of the art. *Cancer Nurs.* 1999 Feb;22(1):85–90.

Clinical practice guidelines in complementary and alternative medicine. An analysis of opportunities and obstacles. Practice and Policy Guidelines Panel, National Institutes of Health Office of Alternative Medicine. *Arch Fam Med.* 1997;6(2):149–154.

DeFriese GH, Woomert A, Guild PA, Steckler AB et al. From activated patient to pacified activist: A study of the self-care movement in the United States. *Soc Sci Med.* 1989;29(2):195–204.

Edelson PJ. Adopting Osler's principles: Medical textbooks in American medical schools, 1891–1906. *Bull Hist Med.* 1994;68:67–84.

Eisenberg DM, Davis RB, Ettner SL, Appel S et al. Trends in alternative medicine use in the United States. *J Am Med Assoc.* 1998;280:1569–1675.

Eisenberg DM, Kessler RC, Foster C et al. Unconventional medicine in the United States: Prevalence, costs, and patterns of use. *N Engl J Med.* 1993;328:246–252.

Elder NC, Gillerist A, and Mina R. Use of alternative health care by family practice patients. *Arch Fam Med.* 1997;6:1131–1134.

Eliason BC, Huebner J, and Marchand L. What physicians can learn from consumers of dietary supplements. *J Fam Pract.* 1999;48(6):459–463.

Ernst E and Fugh-Berman A. Complementary and alternative medicine needs an evidence base before regulation. *West J Med.* 1999;171(3):149–150.

Furnham A and Smith C. Choosing alternative medicine: A comparison of the beliefs of patients visiting a general practitioner and a homoeopath. *Soc Sci Med.* 1988;26(7):685–689.

Goldstein MS (2000) The culture of fitness and the growth of CAM, In: M. Kelner, K. Wellman, B. Pescosolido and M. Saks (Eds.), *Complementary and Alternative Medicine: Challenge and Change.* Harwood Academic Publishers, Canada;2000:27–38.

Gordon RJ, Nienstedt BC, and Gesler WM (eds) (1998) *Alternative Therapies: Expanding Options in Health Care.* Springer Publishing Company, New York.

Joyce CR. Placebo and complementary medicine. *Lancet.* 1994;344(8932):1279–1281.

Kaptchuk TJ. Intentional ignorance: The history of blind assessment and placebo controls in medicine. *Bull Hist Med.* 1998;72:389–433.

Kaufman M (1971) *American Medical Education: The Formative Years, 1765–1910.* Greenwood Press, Westport, CT.

Kickbusch I. Self-care in health promotion. *Soc Sci Med.* 1989;29(2):125–130.

Kim C and Kwok VS. Navajo use of native healers. *Arch Intern Med.* 1998;158:2245–2249.

King LS. The Flexnor report of 1910. *J Am Med Assoc.* 1984;251(8):1079–1086.

Lowenberg JS (1989) *Caring and Responsibility: The Crossroads between Holistic Practice and Traditional Medicine*. University of Pennsylvania Press, Philadelphia, PA.

Marbella AM, Harris MC, Diehr S, and Ignace C. Use of Native American healers among Native American patients in an urban Native American health center. *Arch Fam Med*. 1998:7:182–185.

Moore SR.125 years of public health in the USA. *J R Soc Health* 2001;121(4):262–267.

Murray J and Shepherd S. Alternative or additional medicine? A new dilemma for the doctor. *J R Coll Gen Pract*. 1988;38(316):511–514.

Muscat, M. Beth Israel's Center for Health and Healing: Realizing the Goal of Fully Integrative Care. *Altern Ther Health Med*. 2000;6(5):100–101.

Reston J. Now, let me tell you about my appendectomy in Peking. *The New York Times*, July 26, 1971.

Starr C, Benjamin S, Berman B, and Jacobs J. Exploring complementary therapies in conventional practice. *J Am Assoc Phys Assist*. 1999;12(3):18–20, 23–26, 29–30.

Starr P. *The Social Transformation of American Medicine*. Basic Books, Inc., New York, 1982.

Wagner PJ, Jester D, LeClair B, Taylor AT et al. Taking the edge off: Why patients choose St. John's Wort. *J Fam Pract* 1999 Aug;48(8):615–619.

Whorton JC (1999). The history of complementary and alternative medicine. In W.B. Jonas and J.S. Levin JS (Eds.), *The Essentials of Complementary and Alternative Medicine*. Lippincott Williams & Wilkins, Philadelphia, PA.

Zaldivar A and Smolowitz J. Perceptions of the importance placed on religion and folk medicine by non-Mexican-American Hispanic adults with diabetes. *Diab. Educ*. 1994;20:303–306.

CHAPTER 13: BIOEQUIVALENCE OF BIOSIMILAR PRODUCTS

Arieta AG and Blázquez A. Regulatory considerations for generic or biosimilar low molecular weight heparins. *Curr Drug Discov Technol*. 2012 Jun 1;9(2):137–142.

Biological Gecse KB, Khanna R, van den Brink GR, Ponsioen CY, Löwenberg M, Jairath V, Travis SP, Sandborn WJ, Feagan BG, and D'Haens GR. Biosimilars in IBD: Hope or expectation? *Gut*. 2013 Jun;62(6):803–807.

Biosimilar Aapro MS. What do prescribers think of biosimilars? *Target Oncol*. 2012 Mar;7 Suppl 1:S51–S55.

Chow SC, Lu Q, Tse SK, and Chi E. Statistical methods for assessment of biosimilarity using biomarker data. *J Biopharm Stat*. 2010 Jan;20(1):90–105.

Chirino AJ and Mire-Sluis A. Characterizing biological products and assessing comparability following manufacturing changes. *Nat Biotechnol*. 2004;22:1383–1391.

Chow SC, Endrenyi L, Lachenbruch PA, Yang LY, and Chi E. Scientific factors for assessing biosimilarity and drug Interchangeability of follow-on biologics. *Biosimilars* 2011;1:1–14.

Chow SC, Hsieh TC, Chi E, and Yang J. A comparison of moment-based and probability-based criteria for assessment of follow-on biologics. *J Biopharm Stat*. 2010;20:31–45.

Chow SC and Liu JP (2008) *Design and Analysis of Bioavailability and Bioequivalence Studies*, 3rd edn., Chapman Hall/CRC Press, New York/Taylor & Francis, New York.

Chow SC and Liu JP. Statistical assessment of biosimilar products. *J Biopharm Stat*. 2010;20:10–30.

Chow SC, Shao J, and Li L. Assessing bioequivalence using genomic data. *J Biopharm Stat*. 2004;14:869–880.

Chow SC, Shao J, and Wang H. Individual bioequivalence testing under 2×3 crossover designs. *Stat Med*. 2002;21:629–648.

Crommelin D, Bermejo T, Bissig M, Damianns J, Kramer I et al. Biosimilars, generic versions of the first generation of therapeutic proteins: Do they exist? *Contrib Nephrol*. 2005;149:287–294.

Del Bono R, Martini G, and Volpi R. Update on low molecular weight heparins at the beginning of third millennium. Focus on reviparin. *Eur Rev Med Pharmacol Sci*. 2011 Aug;15(8):950–959.

Dissanayake S. Assessing the bioequivalence of analogues of endogenous substances ('endogenous drugs'): Considerations to optimize study design. *Br J Clin Pharmacol*. 2010 Mar;69(3):238–244.

Dranitsaris G, Amir E, and Dorward K. Biosimilars of biological drug therapies: Regulatory, clinical and commercial considerations. *Drugs*. 2011 Aug 20;71(12):1527–1536.

Drouet L. Low molecular weight heparin biosimilars: How much similarity for how much clinical benefit? *Target Oncol*. 2012 Mar;7(Suppl 1):S35–S42.

EMEA (2001) Note for Guidance on the Investigation of Bioavailability and Bioequivalence. The European Medicines Agency Evaluation of Medicines for Human Use. EMEA/EWP/QWP/1401/98, London, U.K.

EMEA (2003a) Note for Guidance on Comparability of Medicinal Products Containing Biotechnology-derived Proteins as Drug Substance—Non Clinical and Clinical Issues. The European Medicines Agency Evaluation of Medicines for Human Use. EMEA/CHMP/3097/02, London, U.K.

EMEA (2003b) Rev. 1 Guideline on Comparability of Medicinal Products Containing Biotechnology-derived Proteins as Drug Substance—Quality Issues. The European Medicines Agency Evaluation of Medicines for Human Use. EMEA/CHMP/BWP/3207/00/Rev 1, London, U.K.

EMEA (2005a) Guideline on Similar Biological Medicinal Products. The European Medicines Agency Evaluation of Medicines for Human Use. EMEA/CHMP/437/04, London, U.K.

EMEA (2005b) Draft Guideline on Similar Biological Medicinal Products Containing Biotechnology-derived Proteins as Drug Substance: Quality Issues. The European Medicines Agency Evaluation of Medicines for Human Use. EMEA/CHMP/49348/05, London, U.K.

EMEA (2005c) Draft Annex Guideline on Similar Biological Medicinal Products Containing Biotechnology-derived Proteins as Drug Substance—Non Clinical and Clinical Issues—Guidance on Biosimilar Medicinal Products containing Recombinant Erythropoietins. The European Medicines Agency Evaluation of Medicines for Human Use. EMEA/CHMP/94526/05, London, U.K.

EMEA (2005d) Draft Annex Guideline on Similar Biological Medicinal Products Containing Biotechnology-derived Proteins as Drug Substance—Non Clinical and Clinical Issues—Guidance on Biosimilar Medicinal Products containing Recombinant Granulocyte-Colony Stimulating Factor. The European Medicines Agency Evaluation of Medicines for Human Use. EMEA/CHMP/31329/05, London, U.K.

EMEA (2005e) Draft Annex Guideline on Similar Biological Medicinal Products Containing Biotechnology-derived Proteins as Drug Substance—Non-Clinical and Clinical Issues—Guidance on Biosimilar Medicinal Products containing Somatropin. The European Medicines Agency Evaluation of Medicines for Human Use. EMEA/CHMP/94528/05, London, U.K.

EMEA (2005f) Draft Annex Guideline on Similar Biological Medicinal Products Containing Biotechnology-derived Proteins as Drug Substance—Non Clinical and Clinical Issues–Guidance on Biosimilar Medicinal Products containing Recombinant Human Insulin. The European Medicines Agency Evaluation of Medicines for Human Use. EMEA/CHMP/32775/05, London, U.K.

EMEA (2005g) Guideline on the Clinical Investigating of the Pharmacokinetics of Therapeutic proteins. The European Medicines Agency Evaluation of Medicines for Human Use. EMEA/CHMP/89249/04, London, U.K.

FDA. *Guidance on Statistical Approaches to Establishing Bioequivalence*, Center for Drug Evaluation and Research, the U.S. Food and Drug Administration, Rockville, MD, 2001.

FDA. *Guidance on Bioavailability and Bioequivalence Studies for Orally Administrated Drug Products—General Considerations*, Center for Drug Evaluation and Research, the U.S. Food and Drug Administration, Rockville, MD, 2003a.

FDA. *Guidance on Bioavailability and Bioequivalence Studies for Nasal Aerosols and Nasal Sprays for Local Action*, Center for Drug Evaluation and Research, the U.S. Food and Drug Administration, Rockville, MD, 2003b.

Gascon P. Presently available biosimilars in hematology-oncology: G-CSF. *Target Oncol.* 2012 Mar;7(Suppl 1):S29–S34.

Kálmán-Szekeres Z, Olajos M, and Ganzler K. Analytical aspects of biosimilarity issues of protein drugs. *J Pharm Biomed Anal.* 2012 Oct;69:185–195.

Lee JF, Litten JB, and Grampp G. Comparability and biosimilarity: Considerations for the healthcare provider. *Curr Med Res Opin.* 2012 Jun;28(6):1053–1058.

Haidar SH, Davit B, Chen ML, Conner D, Lee L et al. Bioequivalence approaches for highly variable drugs and drug products; *Pharm Res.* 2008 Jan;25(1):237–241.

Hsieh TC, Chow SC, Liu JP, Hsiao CF, and Chi E. Statistical test for evaluation of biosimilarity in variability of follow-on biologics. *J Biopharm Stat.* 2010 Jan;20(1):75–89.

Hsieh TC, Chow SC, Yang LY, and Chi E. (2011) Statistical test for evaluation of biosimilarity based on reproducibility probability. Submitted for publication consideration.

ICH (1996) Q5C Guideline on Quality of Biotechnological Products: Stability Testing of Biotechnological/Biological Products. Center for Drug Evaluation and Research, Center for Biologics Evaluation and Research, the U.S. Food and Drug Administration, Rockville, MD.

ICH (1999) Q6B Guideline on Test Procedures and Acceptance Criteria for Biotechnological/Biological Products. Center for Drug Evaluation and Research, Center for Biologics Evaluation and Research, the U.S. Food and Drug Administration, Rockville, MD.

ICH (2005) Q5E Guideline on Comparability of Biotechnological/Biological Products Subject to Changes in Their Manufacturing Process. Center for Drug Evaluation and Research, Center for Biologics Evaluation and Research, the U.S. Food and Drug Administration, Rockville, MD.

Owens DR, Landgraf W, Schmidt A, Bretzel RG, and Kuhlmann MK. The emergence of biosimilar insulin preparations—A cause for concern? *Diabetes Technol Ther.* 2012 Nov;14(11):989–996.

Roger SD. Biosimilars: How similar or dissimilar are they? *Nephrology*. 2006;11:341–346.

Roger SD and Mikhail A. Biosimilars: Opportunity or cause for concern? *J Pharm Sci*. 2007;10:405–410.

Schuirmann DJ. A comparison of the two one-sided tests procedure and the power approach for assessing the equivalence of average bioavailability. *J Pharmacokinet Biopharm*. 1987;15:657–680.

Schellekens H. Assessing the bioequivalence of biosimilars The Retacrit case. *Drug Discov Today*. 2009 May;14(9–10):495–499.

Schellekens H. How similar do 'biosimilar' need to be? *Nat Biotechnol*. 2004;22:1357–1359.

Shao J and Chow SC. Reproducibility probability in clinical trials. *Stat Med*. 2002;21:1727–1742.

Tamilvanan S, Raja NL, Sa B, and Basu SK. Clinical concerns of immunogenicity produced at cellular levels by biopharmaceuticals following their parenteral administration into human body. *J Drug Target*. 2010 Aug;18(7):489–498.

Vulto AG, Crow SA. Risk management of biosimilars in oncology: Each medicine is a work in progress. *Target Oncol*. 2012 Mar;7(Suppl 1):S43–S49.

Walker E, Nowacki AS. Understanding equivalence and noninferiority testing. *J Gen Intern Med*. 2011 Feb;26(2):192–196.

Wangge G, Klungel OH, Roes KC, de Boer A, Hoes AW, and Knol MJ. Room for improvement in conducting and reporting non-inferiority randomized controlled trials on drugs: A systematic review. *PLoS One*. 2010 Oct 27;5(10):e13550.

WHO (2005) World Health Organization Draft Revision on Multisource (Generic) Pharmaceutical Products: Guidelines on Registration Requirements to Establish Interchangeability, Geneva, Switzerland.

Wiatr C. US biosimilar pathway unlikely to be used: Developers will opt for a traditional BLA filing. *BioDrugs*. 2011 Feb 1;25(1):63–67.

CHAPTER 14: BIOEQUIVALENCE TESTING—THE U.S. PERSPECTIVE

Benet LZ. Understanding bioequivalence testing. *Transplant Proc*. 1999 May;31(3A Suppl):7S–9S. Review.

Bristol DR. Clinical equivalence. *J Biopharm Stat*. 1999 Nov;9(4):549–561.

Chang RK, Raw A, Lionberger R, and Yu L. Generic development of topical dermatologic products: Formulation development, process development, and testing of topical dermatologic products. *AAPS J*. 2013 Jan;15(1):41–52. doi: 10.1208/s12248-012-9411-0. Epub Oct 9, 2012. Review.

Christians U, Klawitter J, and Clavijo CF. Bioequivalence testing of immunosuppressants: Concepts and misconceptions. *Kidney Int Suppl*. 2010 Mar;(115):S1–S7.

Daley-Yates PT and Parkins DA. Establishing bioequivalence for inhaled drugs; weighing the evidence. *Expert Opin Drug Deliv*. 2011 Oct;8(10):1297–1308.

Davit B and Conner D (2010). Reference-scaled average bioequivalence approach. In: I. Kanfer, L. Shargel, (Eds.), *Generic Drug Product Development—International Regulatory Requirements for Bioequivalence*. Informa Healthcare, New York. pp. 271–272.

Davit BM, Chen ML, Conner DP, Haidar SH, Kim S, Lee CH, Lionberger RA et al. Implementation of a reference-scaled average bioequivalence approach for highly variable generic drug products by the U.S. Food and Drug Administration. *AAPS J*. 2012 Dec;14(4):915–924. doi: 10.1208/s12248-012-9406-x.

Ferner U and Neumann N. Active control equivalence trials: Some methodological aspects. *Psychopharmacology (Berl)*. 1992;106 Suppl):S93–S95. Review.

Hsieh TC, Chow SC, Liu JP, Hsiao CF, and Chi E. Statistical test for evaluation of biosimilarity in variability of follow-on biologics. *J Biopharm Stat*. 2010 Jan;20(1):75–89.

Karalis V and Macheras P. Current regulatory approaches of bioequivalence testing. *Expert Opin Drug Metab Toxicol*. 2012 Aug;8(8):929–942.

Keiding N and Budtz-Jørgensen E. The precautionary principle and statistical approaches to uncertainty. *Int J Occup Med Environ Health*. 2004;17(1):147–151.

Meredith PA. Generic drugs. Therapeutic equivalence. *Drug Saf*. 1996 Oct;15(4):233–242.

Midha KK, Rawson MJ, and Hubbard JW. The bioequivalence of highly variable drugs and drug products. *Int J Clin Pharmacol Ther*. 2005 Oct;43(10):485–498.

Peters JR, Hixon DR, Conner DP, Davit BM, Catterson DM, and Parise CM. Generic drugs—Safe, effective, and affordable. *Dermatol Ther*. 2009 May–Jun;22(3):229–240.

Pidgen AW. Statistical aspects of bioequivalence—A review. *Xenobiotica*. 1992 Jul;22(7):881–893.

Tubert-Bitter P, Manfredi R, Lellouch J, and Bégaud B. Sample size calculations for risk equivalence testing in pharmacoepidemiology. *J Clin Epidemiol*. 2000 Dec;53(12):1268–1274.

Walker E and Nowacki AS. Understanding equivalence and noninferiority testing. *J Gen Intern Med.* 2011 Feb;26(2):192–196.

Warren JB. Generics, chemisimilars and biosimilars: Is clinical testing fit for purpose? *Br J Clin Pharmacol.* 2013 Jan;75(1):7–14.

Wellek S. A comment on so-called individual criteria of bioequivalence. *J Biopharm Stat.* 1997 Mar;7(1):17–21.

WHO Expert Committee on specifications for pharmaceutical preparations. *World Health Organ Tech Rep Ser.* 1996;86:1–194.

Yomota C. Trends in the quality evaluation of generic products and bioequivalence guidelines. *Kokuritsu Iyakuhin Shokuhin Eisei Kenkyusho Hokoku.* 2012;(130):1–12.

CHAPTER 15: BIOEQUIVALENCE TESTING— THE EUROPEAN PERSPECTIVE

Amidon GL, Lennernas H, Shah VP, Crison JR. A theoretical basis for a biopharmaceutic drug classification: The correlation of in vitro drug product dissolution and in vivo bioavailability. *Pharm Res.* 1995;12:413–420.

Arieta AG and Gordon J. Bioequivalence requirements in the European Union: Critical discussion. *AAPS J.* 2012 Dec;14(4):738–748.

Belgian Federal Agency for Medicines and Health Products: International Non-Proprietary Name (INN) Prescription, December 09, 2011 (http://www.fagg-afmps.be/en/).

Benet LZ. Relevance of pharmacokinetics in narrow therapeutic index drugs. *Transplant Proc.* 1999;31:1642–1644.

Benet LZ and Goyan JE. Bioequivalence and narrow therapeutic index drugs. *Pharmacotherapy.* 1995;52. 15:433–440.

Bialer M and Midha KK. Generic products of anti-epileptic drugs: A perspective on bioequivalence and interchangeability. *Epilepsia.* 2010;51:941–950.

Blume H and Midha KK. Bio-international'92, conference on bioavailability, bioequivalence and pharmacokinetic studies: Bad Homburg, Germany, 20–22, May 1992. *Eur J Pharm Sci.* 1993;1:165–171.

Blume HH, Schug BS. The biopharmaceutics classification system (BCS): Class III drugs—Better candidates for BA/bioequivalence waiver? *Eur J Pharm Sci.* 1999;9:117–121.

Boddy WO, Snikeris FC, Kringle RO, Wei GCG, Oppermann JA, and Midha KK. An approach for widening the bioequivalence acceptance limits in the case of highly variable drugs. *Pharm Res.* 1995;12:1865–1868.

Common Technical Document for the Registration of Pharmaceuticals for Human Use: Organisation of Common Technical Document, CPMP/ICH/2887/99, London, U.K., February 2004.

CPMP Note for Guidance on the Investigation of Bioavailability and Bioequivalence, Brussels, Belgium, December 1991.

Danish Health and Medicines Authority: Bioequivalence and labelling of medicinal Products with regard to generic substitution, February 07, 2012 (http://laegemiddelstyrelsen.dk/en/topics/authorisa tion-and-supervision/licensing-of-medicines/marketing-authorisation/application for-marketing-authorisation/bioequivalence-and-labelling-of-medicine-substitution).

Danish Health and Medicines Authority: Generic substitution terminated for oral medicines containing cyclosporine or tacrolimus, July 13, 2011(http://laegemiddelstyrelsen.dk/en/topics/auth orisation-and-supervision/licensing-of-medicines/news/generic-substitution-terminated for-oral—tacrolimus).

Davit BM, Conner DP, Fabian-Fritsch B, Haidar S, Jiang X, Patel DT, Seo PRH, Suh K, Thompson CL, and Yu LX. Highly variable drugs: Observations from bioequivalence data submitted to the FDA for new generic drug applications. *AAPS J.* 2008;.10:148–156.

Di Girolamo G, Czerniuk P, Bertuola R, and Keller GA. Bioequivalence of two tablet formulations of clopidogrel in healthy Argentinian volunteers: A single-dose, randomized-sequence, open-label crossover study. *Clin Ther.* 2010;32:161–170.

El Ahmady O, Ibrahim M, Hussain AM, and Bustami RT. Bioequivalence of two oral formulations of clopidogrel tablets in healthy male volunteers. *Int J Clin Pharmacol Ther.* 2009;47:780–784.

EMA Questions & Answers: Positions on specific questions addressed to the Pharmacokinetics Working Party, EMA/618604/2008 Rev. 3, February 16, 2012.

EMEA Concept Paper on BCS-Biowaiver, EMEA/CHMP/EWP/213035/2007, May 20, 2007 (http://www.emea.europa.eu/pdfs/human/ewp/213 03507en.pdf).

EMEA Concept Paper on the need for revision of the note for guidance on modified release oral and transdermal dosage forms: Section II (pharmacokinetic and clinical evaluation), EMA/CHMP/EWP/1303/2010,May 20, 2010 (http://www.ema.europa.eu/docs/en_GB/document_library/Scientific_guideline/2010/06/WC500091 662.pdf).

EMEA Guideline on Clinical Development of Fixed Combination Medicinal Products, CHMP/EWP/240/95 Rev. 1, February 19, 2009 http://www.ema.europa.eu/docs/en_GB/document _library/Scientific_guideline/2009/09/WC500003 686.pdf.

EMEA Guideline on the Investigation of Bioequivalence, CPMP/EWP/QWP/1401/98 Rev. 1/Corr **, London, U.K., January 20, 2010.

EMEA Note for Guidance on Modified Release Oral and Transdermal Dosage Forms: Section II (Pharmacokinetic and Clinical Evaluation), CPMP/EWP/280/96Corr * London, U.K., July 28,1999, (http://www.ema.europa.eu/docs/en_GB/documen t_library/Scientific_guideline/2009/09/WC500003 126.pdf). (Accessed June 5, 2014).

EMEA Note for Guidance on the Investigation of Bioavailability and Bioequivalence, London, U.K., July 26, 2001, CPMP/EWP/QWP/EMEA (http://www.emea.europa.eu/pdfs/human/qwp/140 198enfin.pdf).

EMEA Recommendation on the Need for Revision of CHMP. Note for Guidance on the Investigation of Bioavailability and Bioequivalence. London, U.K., May 2007, EMEA/CHMP/EWP/200943/2007.

FDA. *ACPS-CP Meeting: Bioequivalence and Quality Standards for Narrow Therapeutic Index Drug Products*, July 26, 2011. http://www.fda.gov/downloads/AdvisoryCommittees/CommitteesMeetingMaterials/Drugs/Advisory CommitteeforPharmaceuticalScienceandClinicalP harmacology/UCM263465.pdf.

FDA. *Draft Guidance on Progesterone*, February 2011. http://www.fda.gov/downloads/drugs/Guidance ComplianceRegulatoryInformation/Guidances/UCM209294.pdf.

FDA. *Guidance for Industry: Bioavailability and Bioequivalence Studies for Orally Administered Drug Products—General Considerations*, CDER/FDA, Washington, March 2003. http://www.fda.gov/downloads/Drugs/GuidanceComplianceRegulatoryInformation/Guidances/ucm0 70124.pdf.

FDA. *Guidance for Industry: Bioequivalence Recommendations for Specific Products*, CDER/FDA, Washington, June 2010. http://www.fda.gov/downloads/Drugs/GuidanceComplianceRegulatoryInformation/Guidances/ucm0 72872.pdf.

FDA. *Guidance for Industry: Waiver of in vivo Bioavailability and Bioequivalence Studies for Immediate Release Solid Oral Dosage Forms Based on a Biopharmaceutics Classification System*, CDER/FDA, Washington, August 2000. http://www.fda.gov/cder/guidance/index.htm.

FDA. *Meeting of the Advisory Committee for Pharmaceutical Science and Clinical Pharmacology; Topic 1: Revising the Bioequivalence (BE) Approaches for Critical Dose Drugs*, April 13, 2010. http://www.fda.gov/downloads/AdvisoryCommittees/CommitteesMeetingMaterials/Drugs/Advisory CommitteeforPharmaceutical-ScienceandClinicalPharmacology/UCM207955.pdf.

Gagne JJ, Avorn J, Shrank WH, and Schneeweiss S. Refilling and switching of antiepileptic drugs and seizure-related events. *Cin Pharmacol Ther*. 2010;88:347–353.

Guidance on Losartan Potassium, FDA, May 2008 (http://www.fda.gov/Downloads/Drugs/Guidance ComplianceRegulatoryInformation/Guidances/uc m088645.pdf).

Haidar SH, Davit B, Chen M-L, Conner D, Lee LM, Li QH, Lionberger R, Makhlouf F, Patel D, Schuirmann DJ, and Yu LX. Bioequivalence approaches for highly variable drugs and drug products. *Pharm Res*. 2008;25:237–241.

Hauschke D, Steinijans VW, Diletti E, and Burke M. Sample size determination for bioequivalence assessment using a multiplicative model. *J Pharmacokinet Biopharm*. 1992;20:557–561.

HPFB Guidance for Industry: Bioequivalence requirements: Critical dose drugs, Ottawa, Ontario, Canada, May 31, 2006 (http://www.hc-sc.gc.ca/dhp¬mps/alt_formats/pdf/prodpharma/applic-demande/guide-ld/bio/critical_dose_critique¬eng.pdf).

HPFB Guidance for Industry: Conduct and Analysis of Bioavailability and Bioequivalence Studies—Part A: Oral Dosage Formulations, not in Modified used for Systemic Effects, Ottawa, Ontario, Canada, 1992 (http://www.hc-sc.gc.ca/dhp¬mps/alt_formats/hpfb-dgpsa/pdf/prodpharma/bio¬a-eng.pdf).

Jackson AJ, Robbie G, and Marroum P. Metabolites and bioequivalence: Past en present. *Clin Pharmacokinet*. 2004;43:655–672.

Jantratid E, Prakongpan S, Amidon GL, and Dressman JB. Feasibility of biowaiver extension to biopharmaceutics classification system III drug products-cimetidine. *Clin Pharmacokinet*. 2006;75. 45:385–399.

Karalis V, Symillides M, and Macheras P. Bioequivalence of highly variable drugs: A comparison of the newly proposed regulatory approaches by FDA and EMA. *Pharm Res*. 2012;50. 29:1066–1077.

Kesselheim AS, Misono AS, Lee JL, Stedman MR, Brookhart MA, Choudhry NK, and Shrank WH. Clinical equivalence of generic and brand-name drugs used in cardiovascular disease. *J Am Med Assoc*. 2008;300:2514–2526.

Lenz TL and Wilson AF. Clinical pharmacokinetics of antiplatelet agents used in the secondary prevention of stroke. *Clin Pharmacokinet*. 2003;42:909–920.

Levy G. What are narrow therapeutic index drugs? *Clin Pharmacol Ther*. 1998;63:501–505.

Marzo A and Fontana E. Critical considerations into the new EMA guideline on bioequivalence. *Arzneimittelforschung*. 2011;61(4):207–220.

Midha KK and McKay G. Bioequivalence: Its history, practice and future. *AAPS J*. 2009;11:664–670.

Midha KK, Rawson MJ, and Hubbard JW. The bioequivalence of highly variable drugs and drug products. *Int J Clin Pharmacol Ther*. 2005;43:485–498.

Midha KK, Rawson MJ, and Hubbard JW. The role of metabolites in bioequivalence. *Pharm Res*. 2004;21:1331–1344.

Montague TH, Potvin D, DiLiberti CE, Hauck WW, Parr AF, and Schuirmann DJ. Additional results for Sequential design approaches for bioequivalence studies with crossover designs. *Pharmaceut Stat*. 10, published on-line February 10, 2011.

Moore N, Berdaï D, and Bégaud B. Are generic drugs really inferior medicines? *Clin Pharmacol Ther*. 2010;88:302–304.

Morais JA and Lobato Mdo R. The new European Medicines Agency guideline on the investigation of bioequivalence. *Basic Clin Pharmacol Toxicol*. 2010 Mar;106(3):221–225.

Nation RL and Sansom LN. Bioequivalence requirements for generic products. *Pharmac Ther*. 1994;62:41–55.

NIHS Japan Guideline for Bioequivalence Studies of Generic Products, December 2006 http://www.nihs.go.jp/drug/be-guide(e)/be2006e.pdf.

Nirogi RV, Kandikere VN, Shukla M, Mudigonda K, Maurya S, and Boosi R. Quantification of clopidogrel in human plasma by sensitive liquid chromatography/tandem mass spectrometry. *Rapid Commun Mass Spectrom*. 20:1695–1700.

Pharmacokinetic studies in man, 1987 (http://www.emea.europa.eu/pdfs/human/ewp/3cc 3aen.pdf).

Plosker GL and Lyseng-Williamson KA. Clopidogrel a review of its use in the revention of thrombosis. *Drugs*. 2007;67:613–646.

Pocock SJ. Group sequential methods in the design and analysis of clinical trials. *Biometrika*. 1977;64:191–199.

Polli JE, Yu LX, Cook JA et al. Summary workshop report: Biopharmaceutics Classification System—Implementation challenges and extension opportunities. *J Pharm Sci*. 2004;93:1375–1381.

Potvin D, DiLiberti CE, Hauck WW, Parr AF, Schuirmann DJ, and Smith RA. Sequential design approaches for bioequivalence studies with crossover designs. *Pharmaceut Stat*. 2008;7:245–262.

Richter W, Erenmemisoglu A, Van der Meer MJ, Emritte N, Tuncay E, and Koytchev R. Bioequivalence study of two different clopidogrel bisulfite film-coated tablets. *Arzneim-Forsch/Drug Res*. 2009;59:297–302.

Schuirmann DJ. A comparison of the two one-sided tests procedure and the power approach for assessing the equivalence of average bioavailability. *J Pharmacokinet Biopharm*. 1987 Dec;15(6):657–680.

Shah VP, Yacobi A, Barr WH, Benet LZ, Breimer D, Dobrinska MR, Endrenyi L et al. Evaluation of orally administered highly variable drugs and drug formulations. *Pharm Res*. 1996;13:1590–1594.

Silvestro L, Gheorghe MC, Tarcomnicu J, Savu S, Savu SR, Iordachescu A, and Dulea C. Development and validation of an HPLC-MS/MS method to determine clopidogrel in human plasma: Use of incurred samples to test back-conversion. *J Chromatogr B*. 2010;878:3134–3142.

Staatz CE and Tett SE. Clinical pharmacokinetics and pharmacodynamics of mycophenolate in solid organ transplant recipients. *Clin Pharmacokinet*. 2007;46:13–58.

Tothfalusi L and Endrenyi L. Sample sizes for designing bioequivalence studies for highly variable drugs. *J Pharm Pharmaceut Sci*. 2012;15:73–84.

Tothfalusi L, Endrenyi L, and Arieta AG. Evaluation of bioequivalence for highly variable drugs with scaled average bioequivalence. *Clin Pharmacokinet*. 2009;48:725–743.

Tothfalusi L, Endrenyi L, Midha KK, Rawson MJ, and Hubbard JW. Evaluation of bioequivalence of highly-variable drugs and drug products. *Pharm Res*. 2001;18:728–733.

Tsang YC, Pop R, Gordon P, Hems J, and Spino M. High variability in drug pharmacokinetics complicates determination of bioequivalence: Experience with verapamil. *Pharm Res*. 1996;13:846–850.

Verbeeck RK. Bioequivalence, therapeutic equivalence and generic drugs. *Acta Clin Belg*. 2009;64:379–383.

Verbeeck RK and Warlin J (2010) The European Union, In I. Kanfer and L. Shargel(Eds.), *Generic Drug Product Development: International Regulatory Requirements for Bioequivalence*. Informa Healthcare, New York. pp. 95–113.

WHO Expert Committee on Specifications for Pharmaceutical Preparations, Annex 7: Multisource (generic) pharmaceutical products guidelines on registration requirements to establish interchangeability. WHO Technical Report Series No. 937, 2006 (http://www.who.int/medicines/publications/pharmprep/PDF_TRS953_WEB.pdf).

WHO Expert Committee on Specifications for Pharmaceutical Preparations, Annex 8: Proposal to waive in vivo bioequivalence requirements for WHO Model List of Essential Medicines immediate-release, solid oral dosage forms. WHO Technical Report Series No. 937, 2006 (http://www.who.int/medicines/publications/pharmprep/PDF_TRS953_WEB.pdf).

Williams RL. Therapeutic equivalence of generic drugs—Response to National Association of Boards of Pharmacy, Center for Drug Evaluation and Research, FDA, 1997 (www.fda.gov/cder/news.ntiletter.htm).

Wu C-Y and Benet LZ. Predicting drug disposition via application of BCS: Transport/absorption/elimination interplay and development of a biopharmaceutics drug disposition classification system. *Pharm Res.* 2005;22:11–23.

Yu LX, Amidon GL, Polli JE, Zhao H, Mehta MU, Conner DP, Shah VP et al. Biopharmaceutics classification system: The scientific basis for biowaiver extensions. *Pharm Res.* 2002;19:921–925.

CHAPTER 16: BIOEQUIVALENCE TESTING— THE ROW PERSPECTIVE

Amidon GL, Lennernas H, Shah VP, and Crison JR. A theoretical basis for a biopharmaceutics drug classification: The correlation of in vitro drug product dissolution and in vivo bioavailability. *Pharm Res.* 1995;12:413–420.

Asean Association of Southeast Asian Nations Consultative Committee for Standards and Quality (http://www.aahsa.org.sg/asean-regulation/asean-consultative-committee-on-standards-and-quality-accsq/). (Accessed June 5, 2014).

Asean Guidelines for the Conduct of Bioavailability and Bioequivalence Studies. http://www.hsa.gov.sg/publish/etc/medialib/hsa_library/health_products_regulation/ western_medicines/files_guidelines.Par.59188.File.dat/ACTR_GuidelinesforConductofBioavailabilityandBioequivalenceStudies_Nov05.pdf. Accessed July 21, 2004.

Beers DO (1999) *Generic and Innovator Drugs: A Guide to FDA Approval Requirements*, 5th edn. Aspen Publishers, Frederick, MD. 1999;3–24.

Birkett DJ. Generics—Equal or not? *Aust Prescr.* 2003, 26(4):85–87.

Brazil: Health Minister. National Health Surveillance Agency. MS/GM Ordinance no. 3916 of October 30, 1998. Approval of the National Drug Policy. Brazil. Bras´ılia, DF, 1998. http://e-legis.bvs/leisref/public/showAct.php?id = 751.

Brazil. Health Minister. National Health Surveillance Agency. Resolution N. 391 of August 9, 1999. Technical regulation for generic drugs. Bras´ılia, DF; 1999. http://e-legis.bvs/leisref/public/showAct.php?id = 251.

Brazil. Law No. 9782 of January 26, 1999. Definition of the National Health Surveillance System, establishment of the National Health Surveillance Agency, among other provisions. Bras´ılia, DF; 1999. http://e-legis.bvs/leisref/public/showAct.php?id = 182.

Brazil. Law No. 9787 of February 10, 1999. Amendment of Act No 6360 deals with health surveillance and provides for the use of generic names in pharma¬ceutical and other provisions. Bras´ılia, DF, 1999. http://e-legis.bvs/leisref/public/showAct.php?id = 245.

British Medical Association and the Royal Pharmaceutical Society of Great Britain (2012) *British National Formulary*, 63rd edn. BMJ Publishing Group, London, U.K.

Bulsara C, McKenzie A, Sanfilippo F, Holman CDJ, and Emery JE. "Not the full Monty": A qualitative study of seniors' perceptions of generic medicines in Western Australia. *Aust J Prim Health.* 2010;16(3):240–245.

Canada: Expert Advisory Committee on Bioavailability, Health Protection Branch, December 1992. Report C: Report on bioavailability of oral dosage formulations, not in modified-release form, of drugs used for systemic effects, having complicated or variable pharmacokinetics. Health Canada, 1992. http://www.hc-sc.gc.ca/dhp-mps/prodpharma/applic-demande/guide-ld/bio/biorepc biorapc-eng.php.

Canada: Guidance for Industry. Conduct and Analysis of Bioavailability and Bioequivalence Studies—Part A: Oral Dosage Formulations Used for Systemic Effects. Health Canada 1992. http://www.hc-sc.gc.ca/dhp-mps/alt formats/hpfb-dgpsa/pdf/prodpharma/bio-a e.pdf.

Canada: Guidance for Industry. Conduct and Analysis of Bioavailability and Bioequivalence Studies—Part B: Oral Modified Release Formulations, Health Canada 1996 http://www.hc-sc.gc.ca/dhp-mps/prodpharma/applic-demande/guide-ld/bio/bio-b-.

Canadian Health Protection Branch (HPB), Health Canada. Guidance for Industry Conduct and Analysis of Bioavailability and Bioequivalence Studies—Part A: Oral Dosage Formulations Used for Systemic Effects. Published by authority of the Minister of Health. 1992. Health Products and Food Branch Guidance Document. http://www.hc-sc. gc.ca/dhp-mps/alt_formats/hpfb-dgpsa/pdf/prodpharma/bio-a-eng.pdf.

Chen ML, Shah V, Patnaik R, Adams W, Hussain A, and Conner D et al. Bioavailability and Bioequivalence: An FDA Regulatory Overview. *Pharm Res*. 2001;18:1645–1650.

Chinese Pharmacopoeia (2000) 2nd edn. Drug bioavailability and bioequivalence testing guidance Principle, Appendix 193–197.

CPMP Note for Guidance on the Investigation of Bioavailability and Bioequivalence, Brussels, Belgium, December 1991.

Davit BM, Conner DP, Fabian-Fritsch B et al. Highly variable drugs: Observations from bioequivalence data submitted to the FDA for new generic drug applications. *AAPS J*. 2008;10(1):148–156.

Davit BM, Nwakama PE, Buehler GJ, Conner DP, Haidar SH, Patel DT, Yang Y, Yu LX, and Woodcock J. Comparing generic and innovator drugs: A review of 12 years of bioequivalence data from the United States Food and Drug Administration. *Ann Pharmacother*. 2009;43(10):1583–1597.

Duerden MG and Hughes DA. Generic and therapeutic substitutions in the UK: Are they a good thing? *Br J Clin Pharmacol*. 2010;70(3):335–341.

EC Notice to Applicants (November 2005) Marketing Authorisation, Chapter 1, Vol. 2A. In *Procedures for Marketing Authorisation. The Rules Governing Medicinal Products in the European Union*. November 2005. http://ec.europa.eu/health/files/eudralex/vol-2/a/vol2a_chap1_2005-11_en.pdf.

EMA European Medicines Agency. Committee for MEDICINAL Products for Human Use. Guideline on the Investigation of ioequivalence. London, U.K., January 20, 2010. http://www.ema.europa.eu/docs/en_GB/document_ library/Scientific_guideline/2010/01/WC500070039.pdf.

EMA European Medicines Agency. Doc. Ref. EMEA/CHMP/EWP/200943/2007. Committee for Medicinal Products for Human Use (CHMP). Recommendation on the Need for Revision of (CHMP).Note For Guidance on the Investigation of Bioavailability and Bioequivalence. CPMP/EWP/QWP/1401/98. London, U.K., May 24, 2007. http://www.ema. europa.eu/docs/en_GB/document_library/Scientific_guideline/2009/09/WC500003009.pdf.

EMA European Medicines Agency. Pre-Authorisation Evaluation of Medicines for Human Use. Doc. Ref. CPMP/EWP/QWP/1401/98 Rev. 1. Committee for Medicinal Products for Human Use (CHMP). Draft. Guideline on the Investigation of Bioequivalence. London, U.K., July 24, 2008. http://www.ema.europa.eu/docs/en_GB/document_library/Scientific_ guideline/2009/09/WC500003011.pdf.

EMA Q&A on Generic Medicine. http://www.ema.europa.eu/docs/en_GB/document_library/Medicine_QA/2009/11/WC500012382.pdf.

EMA: European Agency for the Evaluation of Medicinal Products (EMEA) (http://www.emea.europa.eu/).

EMEA (GCPs, GLPs, new draft guidance on BE, see URL http://www.emea.europa.eu/pdfs/human/qwp/140198enrev1.pdf).

EMEA Circular 14/95, Data Required as Evidence of Efficacy, Annexure 13, Medicines Control Council, Department of Health, October, 2, 1995.

EMEA Guideline on the Investigation of Bioequivalence, EMEA, London, U.K., July 24, 2008, CPMP/EWP/QWP/1401/98 Rev. 1. http://www.emea.europa.eu/pdfs/human/qwp/140198enrev1.pdf.

EMEA: Committee for Proprietary Medicinal Products (http://www.emea.europa.eu/pdfs/human/ewp/20094307en.pdf).

EMEA: European Union (European Commission and EMEA) (http://www.emea.europa.eu/htms/human/ich/ichquality.htm).

EU Authorization Procedures. http://ec.europa.eu/health/authorizationprocedures_en.htm.

EU Committee For Medicinal Products For Human Use: Guideline on the Investigation of Bioequivalence.In European Medicines Agency website, European Medicines Agency; 2010.

EU Council Directive 65/65/EEC of January 26, 1965 on the approximation of provisions laid down by Law, Regulation or Administrative Action relating to proprietary medicinal products, Official Journal of the European Union: Eur-Lex: The Council of The European Communities; 1965.

EU Council Directive 75/318/EEC of May 20, 1975 on the approximation of the laws of Member States relating to analytical, pharmaco-toxicological and clinical standards and protocols in respect of the testing of proprietary medicinal products, Official Journal of the European Union: Eur-Lex: The Council of The European Communities; 1975.

EU Council Directive 93/41/EEC of June 14, 1993 repealing Directive 87/22/EEC on the approximation of national measures relating to the placing on the market of high-technology medicinal products, particularly those derived from biotechnology, Official Journal of the European Union.: Eur-Lex: The Council of The European Communities; 1993.

EU Directive 2001/83/EC of the European Parliament and of the Council of 6 November 2001 on the Community code relating to medicinal products for human use. http://eur-lex.europa.eu/LexUriServ/LexUriServ.do?uri = CELEX:32001L0083:EN:HTML.

EU Directive 2011/62/EU of The European Parliament and of the Council of 8 June 2011 Amending Directive 2001/83/EC On The Community Code Relating To Medicinal Products For Human Use, As Regards The Prevention Of The Entry Into The Legal Supply Chain Of Falsified Medicinal Products. http://ec.europa.eu/health/files/eudralex/vol-1/dir_2011_62/dir_2011_62_en.pdf.

EU European Medicines Agency. Committee for MEDICINAL Products for Human Use. Guideline on the Investigation of ioequivalence. London, U.K., January 20, 2010. http://www.ema.europa.eu/docs/en_GB/document_ library/Scientific_guideline/2010/01/WC500070039.pdf.

EU Falsified medicines. http://ec.europa.eu/health/human-use/falsified_medicines/index_en.htm.

EU New visual identity, web/e-mail addresses and organisation chart of the European Medicines Agency. http://www.epha.org/IMG/pdf/.

EU Notice to Applicants Volume 2A Procedures for marketing authorization. http://ec.europa.eu/health/files/eudralex/vol-2/a/vol2a_chap1_2005–11_en.pdf.

EU Pharmaceuticals in The European Union. http://ec.europa.eu/enterprise/newsroom/cf/_getdocument.cfm?doc_id = 1684.

EU Questions and Answers on biosimilar medicines (similar biological medicinal products). http://www.ema.europa.eu/docs/en_GB/document_library/Medicine_QA/2009/12/WC500020062.pdf.

EU Second Council Directive 75/319/EEC of May 20, 1975 on the approximation of provisions laid down by Law, Regulation or Administrative Action relating to proprietary medicinal products, *Official Journal of the European Union*: Eur-Lex: The Council of the European Communities; 1975.

EU The Centralised Procedure. http://ec.europa.eu/health/authorization¬procedures-centralised_en.htm.

EU The Decentralised Procedure. http://ec.europa.eu/health/authorization¬procedures-decentralised_en.htm.

EU The Mutual Recognition Procedure. http://ec.europa.eu/health/authorization¬procedures-mutual-recognition_en.htm.

European Generic Medicines Association—Healthcare Economics. http://198.170.119.137/gen-economics.htm.

European Medicines Agency. London, U.K., 24 May 2007, Doc. Ref. EMEA/CHMP/EWP/213035/2007.

European Generic Medicines Associations, FAQ on Biosimilar Medicines. http://www.egagenerics.com/index.php/biosimilar-medicines/faq-on-biosimilars.

FDA. *Approved Drug Products with Therapeutics Equivalence Evaluations, Electronic Orange Book. Approved Drug Products with Therapeutic Equivalence Evaluations*, Current through April 2009, Department of Health and Human Services, Public Health Service, Food and drug Administration, Center for Drug Evaluation and Research, Office of Information Technology, Division of Data Management and Services. http://www.accessdata.fda.gov/scripts/cder/ob/default.cfm.

FDA. *Generic Drugs: Overview of ANDA Review Process*. http://www.fda.gov/downloads/Drugs/NewsEvents/UCM167310.pdf.

FDA. *Guidance for Industry. Food-Effect Bioavailability and Fed Bioequivalence Studies*, US Department of Health and Human Services Food and Drug Administration Center for Drug Evaluation and Research (CDER), Issued date. December 2002, BP. http://www.fda. gov/downloads/Drugs/GuidanceComplianceRegulatoryInformation/Guidances/ucm070241.pdf.

FDA. *Guidance for Industry Bioavailability and Bioequivalence Studies for Orally Administered Drug Products—General Considerations*, US Department of Health and Human Services, Food and Drug Administration Center for Drug Evaluation and Research (CDER), March 2003. http://www.fda.gov/downloads/Drugs/GuidanceComp lianceRegulatoryInformation/Guidances/ucm070124.pdf.

Fusier I, Tollier C, and Husson MC. Medicines containing pharmaceutical excipients with known effects: A French review. *Pharm World Sci*. 2003;25(4):152–155.

Generic Substitution: Medicines Control Council, Department of Health, December 2003. http://www.mccza.com/showdocument.asp?Cat = 17&Desc = Guidelines%20-%20Human%20Medicines.

Global Harmonization Task Force (http://www.ghtf.org/).

Guidance for medication interchangeability of Mexican Official Journal (1999) section I: 45–68.

Guidance on the establishment of new INN stems. http://www.who.int/medicines/services/inn/StemBook_2011_Final.pdf.

Haidar SH, Davit B, Chen ML, Conner D, Lee LM, Li QH et al. Bioequivalence approaches to highly variable drugs and drug products. *Pharm Res*. 2008;25:237–241.

Haidar SH, Makhlouf F, Schuirmann DJ et al. Evaluation of a scaling approach for the bioequivalence of highly variable drugs. *AAPS J*. 2008;10(3):450–454.

Hassali MAA, Shafie AA, Jamshed S, Ibrahim MIM, and Awaisu A. Consumers' views on generic medicines: A review of the literature. *Int J Pharm Pract*. 2009;17(2):79–88.

India: Global Regulatory GMP Effort by ISPE (http://www.kppub.com/articles/india-Q7A-symposium-2006/ISPE-india-Q7A-symposium-2006.html).

India: Guidelines for Bioavailability and Bioequivalence Studies. Central Drugs Standard Control Organization, Directorate General of Health Services, Ministry of Health and Family Welfare, Government of India, New Delhi, India, March 2005. http://cdsco.nic.in/html/be%20guidelines%20draft%20ver10%20march%2016,%2005.pdf.

India: The Patents (Amendment) Act 2005, The Gazette of India: Extraordinary, Part II-Sec. I, 1–18, 2005.

Japan: Global GMP Harmonization by Japan (http://www.nihs.go.jp/drug/section3/hiyama070518–3.pdf).

Japan: Guideline for Bioequivalence Studies of Generic Products (December, 2006). National Institute of Health Sciences. Japan NIHS. http://www. nihs.go.jp/drug/be-guide(e)/be2006e.pdf.

Japan Ministry of Health, Labor and Welfare. Guideline for Bioequivalence Studies for Additional Dosage Forms of Oral Solid Dosage Forms, 2001.

Japan Ministry of Health, Labour and Welfare. Guideline for Bioequivalence Studies for Different Strengths of Oral Solid Dosage Forms, 2000b. http://www.nihs.go.jp/drug/be-guide(e)/strength/strength.html.

Japan Ministry of Health, Labour and Welfare. Guideline for Bioequivalence Studies of Generic Products, 1997. http://www.nihs.go.jp/drug/be-guide(e)/Generic/be97E.html.

Japan Ministry of Health, Labor and Welfare. Guideline for Bioequivalence Studies of Generic Products for Topical Dermal Application, 2003. http://www.nihs.go.jp/drug/DrugDiv-E.html.

Japan Ministry of Health, Labor and Welfare. Guideline for Formulation Changes of Oral Solid Dosage Forms, 2000a. http://www.nihs.go.jp/drug/DrugDiv-E.html.

Johnston A, Asmar R, Dahlof B, Hill K, Jones DA, Jordan J, Livingston M et al. Generic and therapeutic substitution: A viewpoint on achieving best practice in Europe. *Br J Clin Pharmacol.* 2011;72(5):727–730.

Kingdom of Saudi Arabia Saudi Food and Drug Authority Drug Sector. Bioequivalence Requirements Guidelines. http://www.sfda.gov.sa/NR/rdonlyres/6A114B70–4201–46EF-B4C7–127FD66D3314/0/BioequivalenceRequirementGuidelines.pdf.

Korea: Guidance Document for Bioequivalence Study of Korea Food and Drug Administration Notification #2008–22 (May 07, 2008). http://betest.kfda.go.kr/country/GUIDANCE%20FOR%20INDUSTRY%20(KFDA_2005).pdf.

Lainesse C. International veterinary bioequivalence guideline similarities and differences between Australia, Canada, Europe, Japan, New Zealand and the United States. *AAPS J.* 2012 Dec;14(4):792–798.

Levitt GM (2002) The drugs/biologics approval process, 2nd edn. In K.R. Pina and W.L. Pines (Eds.), *A Practical Guide to Food and Drug Law Regulation.* Food and Drug Law Institute, Washington, DC.

Medicines Control Council, Department of Health, 2.06 Biostudies, June 2007, v2, 2007. http://www.mccza.com/showdocument.asp?Cat = 17&Desc = Guidelines%20-%20Human%20Medicines.

Medicines Control Council, Department of Health, 2.07 Dissolution, June 2007, v2, 2007. http://www.mccza.com/showdocument.asp?Cat = 17&Desc = Guidelines%20-%20Human%20Medicines.

Medicines Control Council. Biostudies. June 2007. http://www.mccza. com/genericDocuments/2.06_Biostudies_Jun07_v2.zip.

Medicines and Related Substances Control Act, 1965 (Act No. 101 of 1965) as amended by Act No. 90 of 1997 and Act No. 59 of 2002. http://www.mccza.com.

Middle East Regulatory Conference (MERC)—United Arab Emirates (http://www.blnz.com/news/2008/05/13/Middle_East_Regulatory_Conference_MERC_6132.html).

Midhal KK and McKay G. Bioequivalence: Its history, practice, and future. *AAPS J.* 2009;11:664–670.

New Zealand: Medicines Act 1981 (Reprint 2006) (published under the Authority of the New Zealand Government).

New Zealand Regulatory Guidelines for Medicines Vol. 1. Guidance notes for applicants for consent to distribute new and changed medicines and related products. 5th edn. October 2001. http://www.medsafe. govt.nz/downloads/Vol1.doc.

OECD Health Data 2012 -Frequently Requested Data. http://www.oecd.org/health/healthpoliciesanddata/oecdhealthdata2012-frequentlyrequesteddata.htm.

Poh J and Tam KT. Registration of similar biological products—Singapore's approach. *Biologicals.* 2011 Sep;39(5):343–345.

Quintal C and Mendes P. Underuse of generic medicines in Portugal: An empirical study on the perceptions and attitudes of patients and pharmacists. *Health Policy.* 2012;104(1):61–68.

Schuirmann DJ. A comparison of the two one-sided tests procedure and the power approach for assessing the equivalence of average bioavailability. *Pharmacokinet Biopharm.* 1987;15:657–680.

Simoens S. Generic medicine pricing in Europe: Current issues and future perspective. *J Med Econ.* 2008;11(1):171–175.

Simoens S. International comparison of generic medicine prices. *Curr Med Res Opin.* 2007;23(11):2647–2654.

Sinclair U (1906) *The Jungle.* Doubleday, Jabber & Company, New York.

Tamboli AM, Todkar P, Zope P, and Sayyad FJ. An overview on bioequivalence: Regulatory consideration for generic drug products. *J Bioequiv Availab*. 2010;2:86–92.

TGA Note for Guidance on the investigation of bioavailability and bioequivalence (CPMP/EWP/QWP/1401/98). CPMP Guidance—As adopted by the TGA—With amendment. April 10, 2002. http://www.tga.gov.

TGA Therapeutic Goods Act 1989. Australian Government Publishing Service, Canberra, Australian Capital Territory, Australia.

TGA Therapeutic Goods Administration. CPMP Guideline—As adapted in Australia by the TGA—With Amendment—Note for Guidance on the Investigation of Bioavailability and Bioequivalence (CPMP/EWP/QWP/1401/98). Effective April 10, 2002. http://www.tga.gov.au/docs/pdf/euguide/ewp/140198entga.pdf.

TGA: Therapeutic Good Administration—Australia (http://www.tga.gov.au/).

The Drugs and Cosmetics Act 1940 and The Drugs and Cosmetics Rules 1945, Ministry of Health and Family Welfare, Government of India, 2005, pp. 425–459.

Turkey Regulation on Licensing for Medicinal Products for Human Use, Official Gazetta No: 25725/19 January 2005.

Turkey Regulation on the Evaluation of Bioavailability and Bioequivalence of Medicinal Products Official Gazetta of Turkey No: 21942. http://rega.basbakanlik.gov.tr/.

UK: Items Unsuitable for Generic Prescribing. July 2008; http://www.nhssb.n-i.nhs.uk/prescribing/documents/Regional List Generic Exceptions Jul08.pdf.

U.S. Food and Dr ug Administration. Orange Book: Approved Drug Products with Therapeutic Equivalence Evaluations. http://www.accessdata.fda.gov/scripts/cder/ob/default.cfm.

Westlake WJ. Use of confidence intervals in analysis of comparative bioavailability trials. *J Pharm Sci*. 1972;61:1340–1341.

WHO Generic Drugs. http://www.who.int/trade/glossary/story034/en/index.html.

WHO Multisource (Generic) Pharmaceutical Products: Guidelines on registration requirements to establish interchangeability. In: WHO Expert Committee on Specifications for Pharmaceutical Preparations, Fortieth Report. World Health Organization, Geneva, Swtzerland, 2006 (WHO Technical Report Series, No. 937, Annex 7) pp. 347–390.

WHO Multisource Pharmaceutical Products: WHO Guideline on Registration Requirements.

CHAPTER 17: BIOEQUIVALENCE TESTING PROTOCOLS

Abbas M, Khan AM, Amin S, Riffat S, Ashraf M, and Waheed N. Bioequivalence study of montelukast tablets in healthy Pakistani volunteers. *Pak J Pharm Sci*. 2013 Mar;26(2):255–259.

Agarwal S, Das A, Ghosh D, Sarkar AK, Chattaraj TK, and Pal TK. Comparative bioequivalence study of leflunomide tablets in Indian healthy volunteers. *Arzneimittelforschung*. 2012 Mar;62(3):145–148. doi: 10.1055/s-0031-1298024. Epub Jan 25 2012.

Ahmed MU, Islam MS, Shohag H, Karim R, Mustafa AG, Bhuiyan NH, Rahim M, and Hasnat A. Comparative pharmacokinetic and bioequivalence study of azithromycin 500 mg tablet in healthy Bangladeshi volunteers. *Int J Clin Pharmacol Ther*. 2012 Jun;50(6):452–458.

Al-Ghazawi M, Alzoubi M, and Faidi B. Pharmacokinetic comparison of two 4 mg tablet formulations of tizanidine. *Int J Clin Pharmacol Ther*. 2013 Mar;51(3):255–262.

Al-Talla ZA, Akrawi SH, Tolley LT, Sioud SH, Zaater MF, and Emwas AH. Bioequivalence assessment of two formulations of ibuprofen. *Drug Des Devel Ther*. 2011;5:427–433.

Almeida S, Pedroso P, Filipe A, Neves RI, Tanguay M, and Torns A. Bioequivalence of two formulations of escitalopram. *Arzneimittelforschung*. 2012 Jul;62(7):307–312.

Álvarez C, Gómez E, Simón M, Govantes C, Guerra P, Frías J, and Arieta AG. Differences in lercanidipine systemic exposure when administered according to labelling: In fasting state and 15 minutes before food intake. *Eur J Clin Pharmacol*. 2012 Jul;68(7):1043–1047.

Basmenji S, Valizadeh H, and Zakeri-Milani P. Comparative in vitro dissolution and in vivo bioequivalence of two diclofenac enteric coated formulations. *Arzneimittelforschung*. 2011;61(10):566–570. doi: 10.1055/s-0031-1300554.

Bass A, Stark JG, Pixton GC, Sommerville KW, Zamora CA, Leibowitz M, and Rolleri R. Dose proportionality and the effects of food on bioavailability of an immediate-release oxycodone hydrochloride tablet designed to discourage tampering and its relative bioavailability compared with a marketed oxycodone tablet under fed conditions: A single-dose, randomized, open-label, 5-way crossover study in healthy volunteers. *Clin Ther*. 2012 Jul;34(7):1601–1612.

Batolar LS, Iqbal M, Monif T, Khuroo A, and Sharma PL. Bioequivalence and pharmacokinetic comparison of 3 metformin extended/sustained release tablets in healthy Indian male volunteers. *Arzneimittelforschung*. 2012 Jan;62(1):22–26.

Benedek IH, Jobes J, Xiang Q, and Fiske WD. Bioequivalence of oxymorphone extended release and crush-resistant oxymorphone extended release. *Drug Des Devel Ther*. 2011;5:455–463.

Bitto A, Burnett BP, Polito F, Russo S, D'Anna R, Pillai L, Squadrito F, Altavilla D, and Levy RM. The steady-state serum concentration of genistein aglycone is affected by formulation: A bioequivalence study of bone products. *Biomed Res Int*. 2013;2013:273498.

Bockbrader HN, Alvey CW, Corrigan BW, and Radulovic LL. Bioequivalence assessment of a pregabalin capsule and oral solution in fasted healthy volunteers: A randomized, crossover study. *Int J Clin Pharmacol Ther*. 2013 Mar;51(3):244–248.

Bus-Kwasnik K, Ksycinska H, Les A, Serafin-Byczak K, Rudzki PJ, Raszek J, Lazowski T, Bielak A, and Wybraniec A. Bioequivalence and pharmacokinetics of two 10-mg bisoprolol formulations as film-coated tablets in healthy white volunteers: A randomized, crossover, open-label, 2-period, single-dose, fasting study. *Int J Clin Pharmacol Ther*. 2012 Dec;50(12):909–919.

Cai W, Wang ZT, Li J, Hu JJ, and Zhong J. Bioequivalence study of two esomeprazole enteric coated formulations in healthy Chinese volunteers. *Arzneimittelforschung*. 2011;61(9):502–505.

Cánovas M, Arcabell M, Martínez G, Canals M, and Cabré F. Bioequivalence studies of film-coated tablet and chewable tablet generic formulations of montelukast in healthy volunteers. *Arzneimittelforschung*. 2011;61(11):610–616.

Cánovas M, Canals M, Polonio F, and Cabré F. Bioequivalence study of 2 orodispersible formulations of zolmitriptan 5 mg in healthy volunteers. *Arzneimittelforschung*. 2012 Oct;62(10):482–486.

Cánovas M, Rios J, Domenech G, Cebrecos J, Pelagio P, Canals M, Polonio F, and Cabré F. Bioequivalence study of 2 orodispersible formulations of ondansetron 8 mg in healthy volunteers. *Arzneimittelforschung*. 2012 Feb;62(2):59–62.

Cánovas M, Torres F, Domenech G, Cebrecos J, Pelagio P, Martínez G, Polonio F, and Cabré F. Bioequivalence evaluation of two oral formulations of quetiapine fumarate in healthy volunteers. *Arzneimittelforschung*. 2011;61(9):489–493.

Cao GY, Li KX, Jin PF, Yue XY, Yang C, Hu X. Comparative bioavailability of ferrous succinate tablet formulations without correction for baseline circadian changes in iron concentration in healthy Chinese male subjects: A single-dose, randomized, 2-period crossover study. *Clin Ther*. 2011 Dec;33(12):2054–2059.

Cawello W, Bonn R, and Boekens H. Bioequivalence of intravenous and oral formulations of the antiepileptic drug lacosamide. *Pharmacology*. 2012;90(1–2):40–46.

Chang MJ and Shin WG. Comparative pharmacokinetics and bioequivalence of two 50 mg atenolol tablet formulations in healthy Korean male volunteers. *Arzneimittelforschung*. 2012 Sep;62(9):410–413. doi: 10.1055/s-0032–1314853.

Chatsiricharoenkul S, Thangboonjit W, Pongnarin P, Konhan K, Sathirakul K, and Kongpatanakul S. Bioequivalence study of 10 mg ramipril tablets in healthy Thai volunteers. *J Med Assoc Thai*. 2011 Oct;94(10):1260–1266.

Chen J, Jiang B, Lou H, Yu L, and Ruan Z. Bioequivalence evaluation of cefdinir in healthy fasting subjects. *Arzneimittelforschung*. 2012 Jan;62(1):9–13.

Chen J, Jiang B, Lou H, Yu L, and Ruan Z. Bioequivalence studies of 2 oral cefaclor capsule formulations in chinese healthy subjects. *Arzneimittelforschung*. 2012 Mar;62(3):134–137.

Chen Q, Zhang MQ, Liu Y, Liu YM, Li SJ, Lu C, Liu GY, Qi YL, Yu C, and Jia JY. Pharmacokinetics and bioequivalence of 2 tablet formulations of olanzapine in healthy Chinese volunteers: A randomized, open-label, single-dose study. *Arzneimittelforschung*. 2012 Nov;62(11):508–512.

Chinsangaram J, Honeychurch KM, Tyavanagimatt SR, Bolken TC, Jordan R, Jones KF, Marbury T et al. Pharmacokinetic comparison of a single oral dose of polymorph form i versus form V capsules of the antiorthopoxvirus compound ST-246 in human volunteers. *Antimicrob Agents Chemother*. 2012 Jul;56(7):3582–3586.

Choi HY, Noh YH, Jin SJ, Kim YH, Kim MJ, Sung H, Jang SB, Lee SJ, Bae KS, and Lim HS. Bioavailability and tolerability of combination treatment with revaprazan 200 mg + itopride 150 mg: A randomized crossover study in healthy male Korean volunteers. *Clin Ther*. 2012 Sep;34(9):1999–2010.

Day DG, Walters TR, Schwartz GF, Mundorf TK, Liu C, Schiffman RM, and Bejanian M. Bimatoprost 0.03% preservative-free ophthalmic solution versus bimatoprost 0.03% ophthalmic solution (Lumigan) for glaucoma or ocular hypertension: A 12-week, randomised, double-masked trial. *Br J Ophthalmol*. 2013 Aug;97(8):989–993. doi: 10.1136/bjophthalmol-2012–303040.

de Freitas Silva M, Schramm SG, Kano EK, Koono EE, Manfio JL, Porta V, and dos Reis Serra CH. Metronidazole immediate release formulations: A fasting randomized open-label crossover bioequivalence study in healthy volunteers. *Arzneimittelforschung*. 2012 Oct;62(10):490–495.

de Mey C, Dimitrova V, Lennartz P, and Wangemann M. Bioequivalence of a novel minitablet formulation of levetiracetam. *Arzneimittelforschung*. 2012 Feb;62(2):94–98. doi: 10.1055/s-0031–1297965.

Dickinson PA, Abu Rmaileh R, Ashworth L, Barker RA, Burke WM, Patterson CM, Stainforth N, and Yasin M. An investigation into the utility of a multi-compartmental, dynamic, system of the upper gastrointestinal tract to support formulation development and establish bioequivalence of poorly soluble drugs. *AAPS J*. 2012 Jun;14(2):196–205.

do Carmo Borges NC, Astigarraga RB, Sverdloff CE, Galvinas PR, Borges BC, and Moreno RA. Comparative bioavailability of two oral formulations of clozapine in steady state administered in schizophrenic volunteers under individualized dose regime. *Curr Clin Pharmacol*. 2012 Nov;7(4):241–253.

Edwards ES, Gunn R, Simons ER, Carr K, Chinchilli VM, Painter G, and Goldwater R. Bioavailability of epinephrine from Auvi-Q compared with EpiPen. *Ann Allergy Asthma Immunol*. 2013 Aug;111(2):132–137.

Ehinger KH, Hansson MJ, Sjövall F, and Elmér E. Bioequivalence and tolerability assessment of a novel intravenous ciclosporin lipid emulsion compared to branded ciclosporin in Cremophor® EL. *Clin Drug Investig*. 2013 Jan;33(1):25–34.

Esseku F, Joshi A, Oyegbile Y, Edowhorhu G, Gbadero D, and Adeyeye M. A randomized pahse I bioequivalence clinincal trial of a paediatric fixed-dose combination antiretroviral reconstitutable suspension in healthy adult volunteers. *Antivir Ther*. 2013;18(2):205–212. Review.

FDA. http://www.fda.gov/Drugs/GuidanceComplianceRegulatoryInformation/Guidances/ucm075207.html.

Galli C, Maggi FM, Risé P, and Sirtori CR. Bioequivalence of two omega-3 fatty acid ethyl ester formulations: A case of clinical pharmacology of dietary supplements. *Br J Clin Pharmacol*. 2012 Jul;74(1):60–65.

Glue P, Gale C, Menkes DB, and Hung N. Evaluation of bioequivalence between clozapine suspension and tablet formulations: A multiple-dose, fed and fasted study. *Clin Drug Investig*. 2012 Nov;32(11):723–727.

Gu N, Yi S, Kim TE, Kim J, Shin SG, Jang IJ, and Yu KS. Comparative pharmacokinetics and tolerability of branded etanercept (25 mg) and its biosimilar (25 mg): A randomized, open-label, single-dose, two-sequence, crossover study in healthy Korean male volunteers. *Clin Ther*. 2011 Dec;33(12):2029–2037.

Gutierrez MM, Nicolas LB, Donazzolo Y, and Dingemanse J. Relative bioavailability of a newly developed pediatric formulation of bosentan vs. the adult formulation. *Int J Clin Pharmacol Ther*. 2013 Jun;51(6):529–536.

Gwaza L, Gordon J, Welink J, Potthast H, Hansson H, Stahl M, and Arieta AG. Statistical approaches to indirectly compare bioequivalence between generics: A comparison of methodologies employing artemether/lumefantrine 20/120 mg tablets as prequalified by WHO. *Eur J Clin Pharmacol*. 2012 Dec;68(12):1611–1618.

Helmy SA and El Bedaiwy HM. Comparative in vitro dissolution and in vivo bioavailability of diflunisal/naproxen fixed-dose combination tablets and concomitant administration of diflunisal and naproxen in healthy adult subjects. *Drug Res (Stuttg)*. 2013a Mar;63(3):150–158.

Helmy SA and El Bedaiwy HM. Effect of the formulation on the bioequivalence of meloxicam: Tablet and suspension. *Drug Res (Stuttg)*. 2013b Jul;63(7):331–337. doi: 10.1055/s-0033–1337979.

Huang J, Chen R, Li R, Wei CM, Yuan GY, Liu XY, Wang BJ, and Guo RC. Bioequivalence of two misoprostol tablets in healthy Chinese female volunteers: A single-dose, two-period, double crossover study. *Arzneimittelforschung*. 2012 Jan;62(1):35–39.

Huang J, Chen R, Wei C, Li R, Yuan G, Liu X, Wang B, and Guo R. Pharmacokinetics and bioequivalence evaluation of two acipimox tablets: A single-dose, randomized-sequence, two-way crossover study in healthy Chinese male volunteers. *Drug Res (Stuttg)*. 2013 Feb;63(2):79–83.

Huang M, Shen-Tu J, Hu X, Chen J, Liu J, and Wu L. Comparative fasting bioavailability of dispersible and conventional tablets of risperidone: A single-dose, randomized-sequence, open-label, two-period crossover study in healthy male Chinese volunteers. *Clin Ther*. 2012 Jun;34(6):1432–1439.

Ibarra M, Fagiolino P, Vázquez M, Ruiz S, Vega M, Bellocq B, Pérez M, González B, and Goyret A. Impact of food administration on lopinavir-ritonavir bioequivalence studies. *Eur J Pharm Sci*. 2012 Aug 15;46(5):516–521.

Innes S, Norman J, Smith P, Smuts M, Capparelli E, Rosenkranz B, and Cotton M. Bioequivalence of dispersed stavudine: Opened versus closed capsule dosing. *Antivir Ther*. 2011;16(7):1131–1134.

Islam MS, Trini AB, Shohag H, Ahmed MU, Maruf AA, and Hasnat A. Bioavailability of omeprazole 20 mg capsules containing omeprazole 22.5% enteric coated pellets versus a reference product in healthy Bangladeshi male subjects: An open-label, single-dose, randomized-sequence, two-way crossover study. *Int J Clin Pharmacol Ther*. 2011 Dec;49(12):778–786.

Jang JW, Seo JH, Jo MH, Lee YJ, Cho YW, Yim SV, and Lee KT. Relative bioavailability of levodropropizine 60 mg capsule and syrup formulations in healthy male Korean volunteers: A singledose, randomized-sequence, open-label, two-way crossover study. *Int J Clin Pharmacol Ther*. 2013 Feb;51(2):152–160.

Jiang T, Rong Z, Xu Y, Chen B, Xie Y, Chen C, Lu Y et al. Pharmacokinetics and bioavailability comparison of generic and branded citalopram 20 mg tablets: An open-label, randomized-sequence, two-period crossover study in healthy Chinese CYP2C19 extensive metabolizers. *Clin Drug Investig*. 2013 Jan;33(1):1–9.

Jin J, Liu J, Chen J, Zhao L, Ma Z, Chen X, Huang M, and Zhong G. Bioequivalence evaluation of 2 tablet formulations of entecavir in healthy chinese volunteers: A single-dose, randomized-sequence, open-label crossover study. *Arzneimittelforschung*. 2012 Mar;62(3):113–116.

Kanjanawart S, Gaysonsiri D, Phunikom K, Simasathiansopon S, Tangsucharit P, Vannaprasaht S, Kaewkamson T, and Tassaneeyakul W. Comparative bioavailability of two moxifloxacin tablet products after single dose administration under fasting conditions in a balanced, randomized and cross-over study in healthy volunteers. *Int J Clin Pharmacol Ther*. 2013 Mar;51(3):249–254.

Kaza M, Leś A, Serafin-Byczak K, Ksycińska H, Rudzki PJ, Gutkowskpi P, Drewniak T et al. Bioequivalence study of 500 mg cefuroxime axetil film-coated tablets in healthy volunteers. *Acta Pol Pharm*. 2012 Nov–Dec;69(6):1356–1363.

Khandave SS, Sawant SV, Sahane RV, Murthi V, Dhanure SS, and Surve PG. Bioequivalence study of two losartan tablet formulations with special emphasis on cardiac safety. *Int J Clin Pharmacol Ther*. 2012 May;50(5):349–359.

Kim HT, Song YK, Lee SD, Park Y, and Kim CK. Relative bioavailability of two 5-mg montelukast sodium chewable tablets: A single dose, randomized, open-label, 2-period crossover comparison in healthy korean adult male volunteers. *Arzneimittelforschung*. 2012 Mar;62(3):123–127.

Kim SH, Park YS, Kim JM, Park HJ, Lee MH, Park HK, Kim YJ, Cho SH, Shaw LM, and Kang JS. Pharmacokinetic and bioavailability studies of 5 mg mosapride tablets in healthy Korean volunteers. *Int J Clin Pharmacol Ther*. 2012 Jul;50(7):524–531.

King J, McCall M, Cannella A, Markiewicz MA, James A, Hood CB, and Acosta EP. A randomized crossover study to determine relative bioequivalence of tenofovir, emtricitabine, and efavirenz (Atripla) fixed-dose combination tablet compared with a compounded oral liquid formulation derived from the tablet. *J Acquir Immune Defic Syndr*. 2011 Apr 15;56(5):e130–e132.

Kocic I, Homsek I, Dacevic M, Grbic S, Parojcic J, Vucicevic K, Prostran M, and Miljkovic B. A case study on the in silico absorption simulations of levothyroxine sodium immediate-release tablets. *Biopharm Drug Dispos*. 2012 Apr;33(3):146–159.

Lee D, Lim LA, Jang SB, Lee YJ, Chung JY, Choi JR, Kim K et al. Pharmacokinetic comparison of sustained- and immediate-release oral formulations of cilostazol in healthy Korean subjects: A randomized, open-label, 3-part, sequential, 2-period, crossover, single-dose, food-effect, and multiple-dose study. *Clin Ther*. 2011 Dec;33(12):2038–2053.

Lee J, Zhang W, Moy S, Kowalski D, Kerbusch V, van Gelderen M, Sawamoto T, Grunenberg N, and Keirns J. Effects of food intake on the pharmacokinetic properties of mirabegron oral controlled-absorption system: A single-dose, randomized, crossover study in healthy adults. *Clin Ther*. 2013 Mar;35(3):333–341.

Lee S, Chung JY, Hong KS, Yang SH, Byun SY, Lim HS, Shin SG, Jang IJ, and Yu KS. Pharmacodynamic comparison of two formulations of Acarbose 100-mg tablets. *J Clin Pharm Ther*. 2012 Oct;37(5):553–557.

Lehman PA and Franz TJ. Assessing the bioequivalence of topical retinoid products by pharmacodynamic assay. *Skin Pharmacol Physiol*. 2012;25(5):269–280.

Li G, Zhang X, Tian Y, Zhang Z, Rui J, and Chu X. Pharmacokinetics and bioequivalence study of two indapamide formulations after single-dose administration in healthy Chinese male volunteers. *Drug Res (Stuttg)*. 2013 Jan;63(1):13–18.

Liu Y, Lu C, Chen Q, Wang W, Liu GY, Lu XP, Zhang MQ, Yu C, and Jia JY. Bioequivalence and pharmacokinetic evaluation of two tablet formulations of carvedilol 25-mg: A single-dose, randomized-sequence, open-label, two-way crossover study in healthy Chinese male volunteers. *Drug Res (Stuttg)*. 2013 Feb;63(2):74–78.

Liu YM, Chen Q, Zhang MQ, Liu GY, Jia JY, Pu HH, Liu Y et al. A parallel design study to assess the bioequivalence of generic and branded hydroxychloroquine sulfate tablets in healthy volunteers. *Arzneimittelforschung*. 2012 Dec;62(12):644–649.

Liu YM, Zhang KE, Liu Y, Zhang HC, Song YX, Pu HH, Lu C et al. Pharmacokinetic properties and bioequivalence of two sulfadoxine/pyrimethamine fixed-dose combination tablets: A parallel-design study in healthy Chinese male volunteers. *Clin Ther*. 2012 Nov;34(11):2212–2220.

Lu C, Jia Y, Yang J, Song Y, Liu W, Ding Y, Sun X, and Wen A. Relative bioavailability study of a novel prodrug of tenofovir, tenofovir dipivoxil fumarate, in healthy male fasted volunteers. *Clin Drug Investig*. 2012 May 1;32(5):333–338.

Lu HM and Ye M. The relative bioavailability study and fasting and fed states pharmacokinetics of bicalutamide 50-mg tablets in healthy Chinese volunteers. *Arzneimittelforschung*. 2012 Jan;62(1):18–21.

Macha S, Mattheus M, Pinnetti S, Woerle HJ, and Broedl UC. Effect of empagliflozin on the steady-state pharmacokinetics of ethinylestradiol and levonorgestrel in healthy female volunteers. *Clin Drug Investig*. 2013 May;33(5):351–357.

Marcelín-Jiménez G, Angeles-Moreno AP, Contreras-Zavala L, García-González A, and Ramírez-San Juan E. Comparison of fasting bioavailability among 100-mg commercial, 100-mg generic, and 50-mg chewable generic sildenafil tablets in healthy male Mexican volunteers: A single-dose, 3-period, crossover study. *Clin Ther*. 2012 Mar;34(3):689–698.

Müller C, Lötsch J, Giessmann T, Franke G, Walter R, Zschiesche M, and Siegmund W. Relative bioavailability and pharmacodynamic effects of methantheline compared with atropine in healthy subjects. *Eur J Clin Pharmacol*. 2012 Nov;68(11):1473–1481.

Nave R, Connolly SM, Popper L, Lahu G, and Schmitt H. Single-dose and multi-dose delivery systems for intranasal fentanyl spray are bioequivalent as demonstrated in a replicate pharmacokinetic study. *Int J Clin Pharmacol Ther*. 2012 Oct;50(10):751–759. doi: 10.5414/CP201729.

Nedogoda SV, Lediaeva AA, Mazina GG, Chumachek EV, Salasiuk AS, and Tsoma VV. Comparative efficacy of original and generic valsartan arterial hypertension. *Kardiologiia*. 2011;51(9):22–28.

Nicolas LB, Gutierrez MM, and Dingemanse J. Comparative pharmacokinetic, pharmacodynamic, safety, and tolerability profiles of 3 different formulations of epoprostenol sodium for injection in healthy men. *Clin Ther*. 2013 Apr;35(4):440–449.

Noh YH, Ko YJ, Cho SH, Ghim JL, Choe S, Jung JA, Kim UJ et al. Pharmacokinetic comparison of 2 formulations of anastrozole (1 mg) in healthy Korean male volunteers: A randomized, single-dose, 2-period, 2-sequence, crossover study. *Clin Ther*. 2012 Feb;34(2):305–313.

Ohls RK, Christensen RD, Kamath-Rayne BD, Rosenberg A, Wiedmeier SE, Roohi M, Lacy CB et al. A randomized, masked, placebo-controlled study of darbepoetin alfa in preterm infants. *Pediatrics*. 2013 Jul;132(1):e119–e127.

Ozbay L, Unal DO, and Erol D. Food effect on bioavailability of modified-release trimetazidine tablets. *J Clin Pharmacol*. 2012 Oct;52(10):1535–1539. Epub 2011 Dec 12.

Parikh N, Goskonda V, Chavan A, and Dillaha L. Single-dose pharmacokinetics of fentanyl sublingual spray and oral transmucosal fentanyl citrate in healthy volunteers: A randomized crossover study. *Clin Ther*. 2013 Mar;35(3):236–243.

Perez-Lloret S, Olmos L, de Mena F, Pieczanski P, and Rodriguez Moncalvo JJ. Bioequivalence of lamotrigine 50-mg tablets in healthy male volunteers: A randomized, single-dose, 2-period, 2-sequence crossover study. *Arzneimittelforschung*. 2012 Oct;62(10):470–476.

Queckenberg C, Wachall B, Erlinghagen V, Di Gion P, Tomalik-Scharte D, Tawab M, Gerbeth K, and Fuhr U. Pharmacokinetics, pharmacodynamics, and comparative bioavailability of single, oral 2-mg doses of dexamethasone liquid and tablet formulations: A randomized, controlled, crossover study in healthy adult volunteers. *Clin Ther*. 2011 Nov;33(11):1831–1841.

Radicioni M, Connolly S, Stroppolo F, Granata G, Loprete L, and Leuratti C. Bioequivalence study of a novel orodispersible tablet of meloxicam in a porous matrix after single-dose administration in healthy volunteers. *Int J Clin Pharmacol Ther*. 2013 Mar;51(3):234–243.

Roh H, Son H, Lee D, Yeon KJ, Kim HS, Kim H, and Park K. Pharmacokinetic comparison of an orally disintegrating film formulation with a film-coated tablet formulation of sildenafil in healthy Korean subjects: A randomized, open-label, single-dose, 2-period crossover study. *Clin Ther*. 2013 Mar;35(3): 205–214.

Saavedra SI, Sasso AJ, Quiñones SL, Saavedra BM, Gaete GL, Boza TI, Carvajal HC and Soto LJ. Relative bioavailability study of two oral formulations of mycophenolate mofetil in healthy volunteers. *Rev Med Chil*. 2011 Jul;139(7):902–908.

Sanki UK, Mandal BK, and Chandrakala V. Comparative pharmacokinetics study of two different clindamycin capsule formulations: A randomized, two-period, two-sequence, two-way crossover clinical trial in healthy volunteers. *Arzneimittelforschung*. 2011;61(9):538–543.

Sasso J, Carmona P, Quiñones L, Ortiz M, Tamayo E, Varela N, Cáceres D, and Saavedra I. Bioequivalence of acenocoumarol in chilean volunteers: An open, randomized, double-blind, single-dose, 2-period, and 2-sequence crossover study for 2 oral formulations. *Arzneimittelforschung*. 2012 Aug;62(8): 395–399.

Schmitt C, Charoin-Pannier A, McIntyre C, Zandt H, Ciorciaro C, Zweigler L, Winters K, and Pepper T. Effect of food on the pharmacokinetics and pharmacodynamics of R1663, an oral factor Xa inhibitor, in healthy male volunteers. *Int J Clin Pharmacol Ther.* 2012 Aug;50(8):566–572.

Schug BS, Donath F, and Blume HH. Bioavailability and pharmacodynamics of two 10-mg estradiol valerate depot formulations following IM single dose administration in healthy postmenopausal volunteers. *Int J Clin Pharmacol Ther.* 2012 Feb;50(2):100–117.

Segal D, Tupy D, Distiller L. The Biosulin equivalence in standard therapy (BEST) study - a multicentre, open-label, non-randomised, interventional, observational study in subjects using Biosulin 30/70 for the treatment of insulin-dependent type 1 and type 2 diabetes mellitus. *S Afr Med J.* 2013 Apr 2;103(7):458–460.

Setiawati E, Yunaidi DA, Handayani LR, Santoso ID, Setiawati A, and Tjandrawinata RR. Bioequivalence study of two clopidogrel film-coated tablet formulations in healthy volunteers. *Arzneimittelforschung.* 2011;61(12):681–684.

Shen Y, Zhang YF, Chen XY, Guo LX, and Zhong DF. Pharmacokinetics and bioequivalence of atorvastatin calcium tablets in healthy male Chinese volunteers. *Zhonghua Xin Xue Guan Bing Za Zhi.* 2012 Mar;40(3):243–247.

Shohag MH, Islam MS, Ahmed MU, Joti JJ, Islam MS, Hasanuzzaman M, and Hasnat A. Pharmacokinetic and bioequivalence study of etoricoxib tablet in healthy Bangladeshi volunteers. *Arzneimittelforschung.* 2011;61(11):617–621.

Shram MJ, Quinn AM, Chen N, Faulknor J, Luong D, Sellers EM, and Endrenyi L. Differences in the in vitro and in vivo pharmacokinetic profiles of once-daily modified-release methylphenidate formulations in Canada: Examination of current bioequivalence criteria. *Clin Ther.* 2012 May;34(5):1170–1181.

Sultana TA, Islam MS, Bhuiyan NH, Shohag MH, Ahmed MU, Naznin SR, Al Maruf A, Huq SM, and Hasnat A. Comparative pharmacokinetic and relative bioavailability study of coated and uncoated azithromycin powder for suspension in healthy Bangladeshi male volunteers. *Arzneimittelforschung.* 2011;61(10):594–598.

Thyssen A, Solanki B, and Treem W. Randomized, open-label, single-dose, crossover, relative bioavailability study in healthy adults, comparing the pharmacokinetics of rabeprazole granules administered using soft food or infant formula as dosing vehicle versus suspension. *Clin Ther.* 2012 Jul;34(7):1636–1645.

Tippabhotla SK, Betha MR, Gadiko C, Battula R, Nakkawar M, Cheerla R, Khan SM, Yergude S, Thota S, and Vobalaboina V. Bioequivalence of fixed dose combination of atorvastatin 10 mg and aspirin 150 mg capsules: A randomized, open-label, single-dose, two-way crossover study in healthy human subjects. *Drug Res (Stuttg).* 2013 May;63(5):250–257.

Tjandrawinata RR, Setiawati E, Yunaidi DA, Santoso ID, Setiawati A, and Susanto LW. Bioequivalence study of two formulations of bisoprolol fumarate film-coated tablets in healthy subjects. *Drug Des Devel Ther.* 2012;6:311–316.

Tjandrawinata RR, Setiawati E, Yunaidi DA, Santoso ID, Setiawati A, and Susanto LW. Bioequivalence study of 2 formulations of film-coated tablets containing a fixed dose combination of bisoprolol fumarate 5 mg and hydrochlorothiazide 6.25 mg in healthy subjects. *Drug Res (Stuttg).* 2013 May;63(5):243–249. doi: 10.1055/s-0033–1334922.

Trabelsi F, Bartůnek A, Vlavonou R, Navrátilová L, Dubé C, Tanguay M, and Hauser T. Single-dose, 2-way crossover, bioequivalence study of two rosuvastatin formulations in normal healthy subjects under fasting conditions. *Int J Clin Pharmacol Ther.* 2012 Oct;50(10):741–750.

Trellu M, Fau JB, Cortez P, Cheng S, Paty I, Boëlle E, Donat F, and Sanderink GJ. Bioequipotency of idraparinux and idrabiotaparinux after once weekly dosing in healthy volunteers and patients treated for acute deep vein thrombosis. *Br J Clin Pharmacol.* 2013 May;75(5):1255–1264.

Upreti VV, Keung CF, Boulton DW, Chang M, Li L, Tang A, Hsiang BC, Quamina-Edghill D, Frevert EU, and Lacreta FP. Bioequivalence of saxagliptin/metformin immediate release (IR) fixed-dose combination tablets and single-component saxagliptin and metformin IR tablets in healthy adult subjects. *Clin Drug Investig.* 2013 May;33(5):365–374.

Valizadeh H, Hamishehkar H, Ghanbarzadeh S, Zabihian N, and Zakeri-Milani P. Pharmacokinetics and bioequivalence evaluation of two brands of ciprofloxacin 500 mg tablets in Iranian healthy volunteers. *Arzneimittelforschung.* 2012 Dec;62(12):566–570.

Vilenchik R, Berkovitch M, Jossifoff A, Ben-Zvi Z, and Kozer E. Oral versus rectal ibuprofen in healthy volunteers. *J Popul Ther Clin Pharmacol.* 2012;19(2):e179–e186.

Wiesinger H, Eydeler U, Richard F, Trummer D, Blode H, Rohde B, and Diefenbach K. Bioequivalence evaluation of a folate-supplemented oral contraceptive containing ethinylestradiol/drospirenone/levomefolate calcium versus ethinylestradiol/drospirenone and levomefolate calcium alone. *Clin Drug Investig.* 2012 Oct 1;32(10):673–684.

Wu H, Liu M, Wang S, Zhao H, Yao W, Feng W, Yan M, Tang Y, and Wei M. Comparative fasting bioavailability and pharmacokinetic properties of 2 formulations of glucosamine hydrochloride in healthy Chinese adult male volunteers. *Arzneimittelforschung*. 2012 Aug;62(8):367–371.

Wu Y, Mao M, Wang L, and Jiang X. Bioequivalence research of cyclosporin soft capsules. *Sheng Wu Yi Xue Gong Cheng Xue Za Zhi*. 2012 Apr;29(2):311–314, 331.

Xu J, Jin H, Zhu H, Zheng M, Wang B, Liu C, Chen M et al. Oral bioavailability of rifampicin, isoniazid, ethambutol, and pyrazinamide in a 4-drug fixed-dose combination compared with the separate formulations in healthy Chinese male volunteers. *Clin Ther*. 2013 Feb;35(2):161–168.

Yan X, Lowe PJ, Fink M, Berghout A, Balser S, Krzyzanski W. Population pharmacokinetic and pharmacodynamic model-based comparability assessment of a recombinant human Epoetin Alfa and the Biosimilar HX575. *J Clin Pharmacol*. 2012 Nov;52(11):1624–1644.

Yi S, Kim SE, Park MK, Yoon SH, Cho JY, Lim KS, Shin SG, Jang IJ, and Yu KS. Comparative pharmacokinetics of HD203, a biosimilar of etanercept, with marketed etanercept (Enbrel®): A double-blind, single-dose, crossover study in healthy volunteers. *BioDrugs*. 2012 Jun 1;26(3):177–184.

Zaid AN, Natour S, Qaddomi A, and Abu Ghoush A. Formulation and in vitro and in vivo evaluation of film-coated montelukast sodium tablets using Opadry® yellow 20A82938 on an industrial scale. *Drug Des Devel Ther*. 2013;7:83–91.

Zaid AN, Qaddomi A, and Khammash S. Formulation and comparative bioavailability of 2 ciprofloxacin sustained release tablets. *Arzneimittelforschung*. 2012 Jul;62(7):319–323.

Zakeri-Milani P, Ghanbarzadeh S, and Valizadeh H. Comparative in vitro dissolution and in vivo bioequivalence of 2 pentoxifylline sustained release formulations. *Arzneimittelforschung*. 2012 Jul;62(7):335–339.

Zakeri-Milani P, Islambulchilar Z, Ghanbarzadeh S, and Valizadeh H. Single dose bioequivalence study of two brands of olanzapine 10 mg tablets in Iranian healthy volunteers. *Drug Res (Stuttg)*. 2013 Jul;63(7):346–350. doi: 10.1055/s-0033-1341427.

Zhai H, Zheng Y, Matravers P, Hicks DA, Wiener S, and Maibach HI. Bioequivalence in keratolytic activity of formulations vs its vehicle and comparator formulation: Randomized, double-blind clinical trial. *Skinmed*. 2013 Jan–Feb;11(1):21–25.

Zhang M, Yang J, Tao L, Li L, Ma P, and Fawcett JP. Acarbose bioequivalence: Exploration of new pharmacodynamic parameters. *AAPS J*. 2012 Jun;14(2):345–351.

Zhang R, Yuan G, Li R, Liu X, Wei C, Wang B, Gao H, and Guo R. Pharmacokinetic and bioequivalence studies of trospium chloride after a single-dose administration in healthy Chinese volunteers. *Arzneimittelforschung*. 2012 May;62(5):247–251.

Zhang S, Kan Q, Wen JG, Zhao J, Sheng Y, Li Y, Sun S et al. Pilot and pivotal study to evaluate the bioequivalence of two paroxetine 40 mg tablet formulations in healthy Chinese subjects. *Int J Clin Pharmacol Ther*. 2012 Jul;50(7):514–523.

Zhang XY, Tian Y, Zhang ZJ, Rui JZ, and Cao XM. Pharmacokinetics and bioequivalence study of two digoxin formulations after single-dose administration in healthy Chinese male volunteers. *Arzneimittelforschung*. 2011;61(11):601–604.

Zheng L, Yu Q, Miao J, Xiang J, and Xu N. Bioequivalence study of two mirtazapine oral tablet formulations in healthy Chinese male volunteers. *Int J Clin Pharmacol Ther*. 2012 May;50(5):368–374.

Zhu Y, Zhang Q, Yu C, Chen J, Hu Y, Zou J, Yuan L, and Ma J. Relative fasting bioavailability of two formulations of nateglinide 60 mg in healthy male Chinese volunteers: An open-label, randomized-sequence, single-dose, two-way crossover study. Clin Ther. 2012 Jul;34(7):1505–1510.

CHAPTER 18: BIOEQUIVALENCE DOCUMENTATION

CDER Guidance for Industry: BE Recommendations for Specific Products, http://www.fda.gov/Drugs/GuidanceComplianceRegulatoryInformation/Guidances/ucm075207.htm.

CDER Guidance for Industry: Submission of Summary Bioequivalence Data for ANDAs, http://www.fda.gov/downloads/Drugs/GuidanceComplianceRegulatoryInformation/Guidances/UCM134846.pdf.

http://www.fda.gov/downloads/Drugs/…/Guidances/UCM134846.pdf.

http://www.fda.gov/downloads/Drugs/DevelopmentApprovalProcess/HowDrugsareDevelopedandApproved/ApprovalApplications/AbbreviatedNewDrugApplicationANDAGenerics/UCM292669.pdf.

http://www.accessdata.fda.gov/scripts/cdrh/cfdocs/cfCFR/CFRSearch.cfm?fr=320.21.

http://www.fda.gov/Drugs/DevelopmentApprovalProcess/HowDrugsareDevelopedandApproved/ApprovalApplications/InvestigationalNewDrugINDApplication/ucm226358.htm.

http://www.ema.europa.eu/docs/en_GB/document_library/Scientific_guideline/2010/01/WC500070039.pdf.

http://www.ema.europa.eu/docs/en_GB/document_library/Scientific_guideline/2011/11/WC500117887.pdf.
http://www.fda.gov/downloads/Drugs/DevelopmentApprovalProcess/HowDrugsareDevelopedandApproved/ApprovalApplications/AbbreviatedNewDrugApplicationANDAGenerics/UCM120957.pdf.

CHAPTER 19: GOOD LABORATORY PRACTICES

Belgium: http://www.glp.be.
FDA. http://www.accessdata.fda.gov/scripts/cdrh/cfdocs/cfcfr/CFRSearch.cfm?CFRPart=58.htm.
FDA. http://www.fda.gov/ICECI/EnforcementActions/BioresearchMonitoring/ucm135197.htm.
http://www.cgalp.com.
http://www.noormd.com.
http://www1.bipm.org/en/committees/jc/jctlm/.
ISO: http://www.iso.org/iso/home/store/catalogue_ics/catalogue_detail_ics.htm?csnumber=56115.
Jump up ^ Staff, World Health Organization (2009) Handbook: Good Laboratory Practice (GLP).
Klimisch HJ, Andreae M, and Tillmann U. A systematic approach for evaluating the quality of experimental toxicological and eco-toxicological data. *Regul Toxicol Pharmacol.* 1997;25(1):1–5. http://dx.doi.org/10.1006/rtph.1996.1076.
MHRA: http://www.mhra.gov.uk/Howweregulate/Medicines/Inspectionandstandards/GoodLaboratoryPractice/Structure/index.htm.
OECD: http://www.oecd.org/chemicalsafety/testing/oecdseriesonprinciplesofgoodlaboratorypracticeglpandcompliancemonitoring.htm.
OECD: http://www.oecd.org/env/ehs/testing/goodlaboratorypracticeglp.htm.
Schneider, K. Faking it: The case against Industrial Bio-Test Laboratories. *Amicus J* (Natural Resources Defence Council) 1983; 14–26.
Webster, Gregory K et al.; Kott, L;Maloney, T. JALA tutorial: Considerations when implementing automated methods into GxP laboratories. *J AssocLab Autom (Elsevier).* 2005;10(3):182–191.

CHAPTER 20: BIOANALYTICAL METHOD VALIDATION

Baumann A. Early development of therapeutic biologics—Pharmacokinetics. *Curr Drug Metab.* 2006 Jan;7(1):15–21.
Belanger BA, Davidian M, and Giltinan DM. The effect of variance function estimation on nonlinear calibration inference in immunoassay data. *Biometrics.* 1996;52:158–175.
Boulanger B, Hubert P, Chiap P, and Dewé W (2000a) Objectives of pre-study validation and decision rules. Presented at AAPS APQ Open forum, Indianapolis, IN.
Boulanger B, Hubert P, Chiap P, Dewé W, and Crommen J (2000b). Analyse statistique des résultats de validation de méthodes chromatographiques. Presented at Journées du GMP, Bordeaux, France.
Colburn WA. Biomarkers in drug discovery and development: From target identification through drug marketing. *J Clin Pharmacol.* 2003 Apr;43(4):329–341.
Colburn WA and Lee JW. Biomarkers, validation and pharmacokinetic-pharmacodynamic modelling. *Clin Pharmacokinet.* 2003;42(12):997–1022.
Dadgar D and Burnett PE. Issues in evaluation of bioanalytical method selectivity and drug stability. *J Pharm Biomed Anal.* 1995 Dec;14(1–2):23–31.
Dadgar D, Burnett PE, Choc MG, Gallicano K, and Hooper JW. Application issues in bioanalytical method validation, sample analysis and data reporting. *J Pharm Biomed Anal.* 1995 Feb;13(2):89–97.
DeSilva B, Smith W, Weiner R, Kelley M, Smolec J, Lee B, Khan M, Tacey R, Hill H, and Celniker A. Recommandations for bioanalytical method validation of ligand-binding assays to support pharmacokinetic assessments of macromolecules. *Pharm Res.* 2003;20, 1885–1900.
Fast D, Workshop Report. *AAPS J.* 2009;11:238–241.
FDA. *Guidance for industry: Bioanalytical methods validation*, 2001.
Findlay JWA, Smith WC, Lee JW, Nordblom GD, Das I, Desilva BS, Khan MN, and Bowsher RR. Validation of immunoassays for bioanalysis: A pharmaceutical industry perspective. *J Pharm Biomed Anal.* 2001;21, 1249–1273.
Finney D (1978). *Statistical Methods in Biological Assays.* Charles Griffin, London, U.K.
Hartmann C, Smeyers-Verbeke J, Massart DL, and McDowall RD. Validation of bioanalytical chromatographic methods. *J Pharm Biomed Anal.* 1998 Jun;17(2):193–218.

Hoffman D and Kringle R. Two-sided tolerance intervals for balanced and unbalanced random effects models. *J Biopharm Stat*. 2005;15, 283–293.

Hubert P, Nguyen-Huu J-J, Boulanger B, Chapuzet E, Chiap P, Cohen N, Compagnon P-A et al. Harmonization of strategies for the validation of quantitative analytical procedures. A SFSTP proposal Part I. *J Pharm Biomed Anal*. 2004;36, 579–586.

Hubert P, Chiap P, Crommen J, Boulanger B, Chapuzet E, Mercier N, Bervoas-Martin S et al. The SFSTP guide on the validation of chromatographic methods for drug bioanalysis: From the Washington Conference to the laboratory. *Anal Chim Acta*. 1999;391, 135–148.

ICH (International Conference on Harmonization). (1995). Guideline on validation of analytical procedures: Definitions and terminology.

ICH (International Conference on Harmonization). (1997). Guideline on validation of analytical procedures: Methodology.

International Organization for Standardization. ISO 5725–1:1994 *Accuracy (Trueness and Precision) of Measurement Methods and Results—Part 2: Basic Method for the Determination of Repeatability and Reproducibility of a Standard Measurement Method (Edition 1)*.

Karnes HT, Shiu G, and Shah VP. Validation of bioanalytical methods. *Pharm Res*. 1991 Apr;8(4):421–426.

Klebovich I. Review of bioanalytic methods in pharmacokinetics. *Acta Pharm Hung*. 1998 Jul;68(4):234–248.

Lee JW, Nordblom GD, Smith WC, and Bowsher RR (2003). Validation of bioanalytical assays for novel biomarkers: Practical recommendations for clinical investigation of new drug entities. In J. Bloom and R.A. Dean (Eds.), *Biomarkers in Clinical Drug Development*. Marcel Dekker, New York, pp. 119–148.

Mee RW. β-expectation and β-content tolerance limits for balanced one-way ANOVA pandom model. *Technometrics*. 1984;26, 251–254.

Mire-Sluis AR, Barrett YC, Devanarayan V, Koren E, Liu H, Maia M, Parish T et al. Recommendations for the design of immunoassays used in the detection of antibodies against biological products. *J Immunol Methods*. 2004;289, 1–16.

O'Connell MA, Belanger BA, and Haaland PD. Calibration and assay development using the four-parameter logistic model. *Chemometr Intell Lab Syst*. 1993;20:97–114.

SAS Institute Inc. *SAS/STAT User's Guide*. SAS Institute Inc., Cary, NC.

Shah VP et al. Workshop Report. *Pharm Res*. 1992;9:588–592.

Shah VP et al. Workshop Report. *Pharm Res*. 2000;17:1551–1557.

Smith WC and Sittampalam SG. Conceptual and statistical issues in the validation of analytical dilution assays for pharmaceutical applications. *J Biopharm Stat*. 1998;8:509–532.

Smolec J, DeSilva B, Smith W, Weiner R, Kelly M, Lee B, Khan M et al. Bioanalytical method validation for macromolecules in support of pharmacokinetic studies. *Pharm Res*. 2005 Sep;22(9):1425–1431. Epub Aug 24, 2005.

Srinivas NR. Applicability of bioanalysis of multiple analytes in drug discovery and development: Review of select case studies including assay development considerations. *Biomed Chromatogr*. 2006 May;20(5):383–414.

Taylor PJ. Matrix effects: The Achilles heel of quantitative high-performance liquid chromatography-electrospray-tandem mass spectrometry. *Clin Biochem*. 2005 Apr;38(4):328–334.

Timmerman PM, de Vries R, and Ingelse BA. Tailoring bioanalysis for PK studies supporting drug discovery. *Curr Top Med Chem*. 2001 Nov;1(5):443–462.

Viswanathan CT. Workshop report. *Pharm Res*. 2007;24:1962–1967.

CHAPTER 21: GOOD CLINICAL PRACTICE

Alfaro V. Clinical trials, good publication practice and legal regulations. *Med Clin (Barc)*. 2004 Jun 19;123(3):100–103.

Anhalt E. The development of good clinical practice in the EEC and in Germany. *Methods Find Exp Clin Pharmacol*. 1993 May;15(4):217–222.

Arcangeli G, Overgaard J, Gonzalez Gonzalez D, and Shrivastava PN. International clinical trials in radiation oncology. Hyperthermia trials. *Int J Radiat Oncol Biol Phys*. 1988;14(Suppl 1):S93–S109.

Atherton PJ and Sloan JA. Rising importance of patient-reported outcomes. *Lancet Oncol*. 2006 Nov;7(11):883–884.

Belorgey C, Pletan Y, and Goehrs JM. Round Table No. 5, Giens XIX. Adaptation of the clinical trials directive: Recommendations on the contents of a dossier for the request for authorisation of the first trials in human subjects. *Therapie*. 2004 May–Jun;59(3):329–347. English, French.

Bertoye PH, Courcier-Duplantier S, and Best N. Adaptation of the application of good clinical practice depending on the features of specific research projects. *Therapie*. 2006 Jul–Aug;61(4):279–285, 271–277. English, French.

Bhagat K, Kurashe J, and Nyazema NZ. Ethical issues in clinical and industry-sponsored research. *Cent Afr J Med*. 2000 Apr;46(4):108–111.

Boll M, Binder J, Siegel A, and Grundmann R. Quality assurance in clinical studies: A necessity. *Z Exp Chir Transplant Kunstliche Organe*. 1990;23(2):65–72. German.

Bothner U, Seeling W, Schwilk B, Pfenninger E, and Georgieff M. Clinical randomized controlled studies in anesthesiology according to quality guidelines of good clinical practice.

Brunetti MM. Critical aspects in the application of the principles of good laboratory practice (GLP). *Ann Ist Super Sanita*. 2002;38(1):41–45.

Cartwright JC, Hickman SE, Bevan L, and Shupert CL. Navigating federalwide assurance requirements when conducting research in community-based care settings. *J Am Geriatr Soc*. 2004 Sep;52(9):1567–1571.

Cavero I and Crumb W. ICH S7B draft guideline on the non-clinical strategy for testing delayed cardiac repolarisation risk of drugs: A critical analysis. *Expert Opin Drug Saf*. 2005 May;4(3):509–530.

Collins JF and Sather MR. Regulatory issues for clinical trials in humans. *Epidemiol Rev*. 2002;24(1):59–66.

Commens CA. Truth in clinical research trials involving pharmaceutical sponsorship. *Med J Aust*. 2001 Jun 18;174(12):648–649.

Dejgaard A and Thomsen MK. The roles and responsibilities of the pharmaceutical industry. *Ugeskr Laeger*. 2003 Apr 14;165(16):1676–1679.

Dent NJ and Sweatman WJ. Audit Working Party of the European Forum for Good Clinical Practice. Can non-regulators audit Independent Ethic Committees (IEC), and if so, how? *Qual Assur*. 2001 Jan-2002 Mar;9(1):43–54.

Dirach J. Good clinical practice. Requirements for clinical documentation on the introduction of new drugs. *Ugeskr Laeger*. 1990 Apr 2;152(14):992–994.

Donato BJ and Gibson TR. Does your clinical investigator understand the consequences of non-compliance? *Qual Assur*. 1999 Jul–Sep;7(3):135–145.

Dubois MY. Conflicts of interest with the health industry. *Pain Med*. 2006 Sep–Oct;7(5):463–465.

Englev E, Petersen KP. ICH-GCP Guideline: Quality assurance of clinical trials. Status and perspectives. *Ugeskr Laeger*. 2003 Apr 14;165(16):1659–1662.

Fischer TW and Elsner P. Good clinical practice. Significance for dermatological research. *Hautarzt*. 2000 Sep;51(9):704–711;quiz 711–712.

Fukushima M. Quality control in clinical trials. *Gan To Kagaku Ryoho*. 1996 Jan;23(2):172–182.

Gassman JJ, Owen WW, Kuntz TE, Martin JP, and Amoroso WP. Data quality assurance, monitoring, and reporting. *Control Clin Trials*. 1995 Apr;16(2 Suppl):104S–136S.

Giacinti L, Lopez M, and Giordano A. Clinical trials. *Front Biosci*. 2006 Sep 1;11:2918–2923.

Gordon BG, Kessinger A, Mann SL, and Prentice ED. The impact of escalating regulatory requirements on the conduct of clinical research. Cytotherapy. 2003;5(4):309–313.

Hartford CG, Petchel KS, Mickail H, Perez-Gutthann S, McHale M, Grana JM, and Marquez P. Pharmacovigilance during the pre-approval phases: An evolving pharmaceutical industry model in response to ICH E2E, CIOMS VI, FDA and EMEA/CHMP risk-management guidelines. *Drug Saf*. 2006;29(8):657–673.

Helft PR, Ratain MJ, Epstein RA, and Siegler M. Inside information: Financial conflicts of interest for research subjects in early phase clinical trials. *J Natl Cancer Inst*. 2004 May 5;96(9):656–661.

Herman ZS. Progress and dilemma of contemporary clinical pharmacology. *Int J Clin Pharmacol Ther*. 2005 Jan;43(1):43–50.

Hippe E, Westin J, Wisloff F. Nordic Myeloma Study Group, the first 15 years: Scientific collaboration and improvement of patient care. *Eur J Haematol*. 2005 Mar;74(3):185–193.

Hurst SA. Ethics consultation: What does it serve? *Rev Med Suisse*. 2006 Sep 27;2(80):2195, 2197–2199.

Hvidberg EF, Pedersen PA. Therapeutic trials in general practice, an unmet need. *Dan Med Bull*. 1990 Feb;37(1):84–86.

Johann-Liang R, James AN, Behr VL, Struble K, Birnkrant DB. Reporting of deaths during pre-approval clinical trials for advanced HIV-infected populations. *Drug Saf*. 2005;28(7):559–564.

Jones TC. Call for a new approach to the process of clinical trials and drug registration. *BMJ*. 2001 Apr 14;322(7291):920–923.

Jongen PJ. Data handling in clinical trials: An ongoing debate. *Mult Scler*. 1995;1 Suppl 1:S60–S63.

Keller K. Special therapeutic guidelines from the viewpoint of BfArM (Federal Institute for Drug and Medical Products) *Z Arztl Fortbild Qualitatssich*. 1997 Nov;91(7):669–674.

Kern P. Medical treatment of echinococcosis under the guidance of Good Clinical Practice (GCP/ICH). *Parasitol Int.* 2006;55 Suppl:S273–S282. Epub Dec 9, 2005.

Kusche J. Good clinical practice and phytotherapy. *Methods Find Exp Clin Pharmacol.* 1993 May;15(4):241–247.

Legler UF. Experiences with GLP/GCP from the pharmaceutical industry's viewpoint. *Methods Find Exp Clin Pharmacol.* 1993 May;15(4):233–236.

Links M. Analogies between reading of medical and religious texts. *BMJ.* 2006 Nov 18;333(7577):1068–1070.

Mabunda G. Ethical issues in HIV research in poor countries. *J Nurs Scholarsh.* 2001;33(2):111–114.

Madelaine-Chambrin I. Characteristics of clinical trials in oncology: Pharmacist's implication. *Ann Pharm Fr.* 2006 Jan;64(1):36–42. French.

Massari C. A general guide for conducting in-process inspections. *Qual Assur.* 1998 Apr–Jun;6(2):97–105.

Matsuura C. Roles of nurses in clinical trials of anticancer drug development. *Gan To Kagaku Ryoho.* 1999 Jan;26(2 Suppl):231–234. Japanese.

Mizuno K. [Clinical studies conducted in the U.S.] Gan To Kagaku Ryoho. 1999 Jan;26(2 Suppl):217–224. Japanese.

Moazam F. Research and developing countries: Hopes and hypes. *East Mediterr Health J.* 2006;12 Suppl 1:S30–S36.

Nagata R, Fukase H, and Rafizadeh-Kabe JD. East-West development: Understanding the usability and acceptance of foreign data in Japan. *Int J Clin Pharmacol Ther.* 2000 Feb;38(2):87–92.

Nakano S. Improving the infrastructure for clinical trials in Japan. *Gan To Kagaku Ryoho.* 1999 Jan;26(2 Suppl):225–230. Japanese.

Nelson DF, Gonzalez DG, Bleehen N. International Clinical Trials in Radiation Oncology. Brain sites. *Int J Radiat Oncol Biol Phys.* 1988;14 Suppl 1:S135–S145.

Ohno R. Development of anti-leukemic drugs and clinical studies in Japan. *Rinsho Ketsueki.* 1999 Mar;40(3):168–171.

Garty N. Clinical trials in Israel—The need for good clinical practice guidelines. *Harefuah.* 1996 Feb 15;130(4):253–255.

Ono S and Kodama Y. Clinical trials and the new good clinical practice guideline in Japan. An economic perspective. *Pharmacoeconomics.* 2000 Aug;18(2):125–141.

Otte A, Maier-Lenz H, and Dierckx RA. Good clinical practice: Historical background and key aspects. *Nucl Med Commun.* 2005 Jul;26(7):563–574.

Osswald W. Ethical controls of drug trials. *Rev Port Cardiol.* 1999 Sep;18(9):833–837.

Pandya DP and Dave J. Protection of human subjects in clinical research: The pitfalls in clinical research. *Compr Ther.* 2005 Spring;31(1):72–77.

Popp RJ, Smith SC, Adams RJ, Antman EM, Kavey RE, DeMaria AN, Ohman EM et al.; American College of Cardiology Foundation; American Heart Association. ACCF/AHA consensus conference report on professionalism and ethics. *Circulation.* 2004 Oct 19;110(16):2506–2549.

Regnier B. Good clinical practice. *Eur J Clin Microbiol Infect Dis.* 1990 Jul;9(7):519–522.

Rodriguez-Villanueva J, Alsar MJ, Avendano C, Gomez-Piqueras C, and Garcia-Alonso F. Pharmacogenetic studies: Evaluation guidelines for research ethics committees. Scientific background and legal framework (I). *Med Clin (Barc).* 2003 Jan 25;120(2):63–67.

Salzberg M and Muller E. Quality requirements in clinical studies: A necessary burden? *Swiss Med Wkly.* 2003 Aug 9;133(31–32):429–432.

Samanta A and Samanta J. Research governance: Panacea or problem? *Clin Med.* 2005 May–Jun;5(3):235–239.

Sandstrom B. Quality criteria in human experimental nutrition research. *Eur J Clin Nutr.* 1995 May;49(5):315–322.

Shah RR. Drugs, QTc interval prolongation and final ICH E14 guideline: An important milestone with challenges ahead. *Drug Saf.* 2005;28(11):1009–1028.

Shimoyama M. Need to establish academic research organization for cancer clinical trial and clinical research system having clinical investigators and study coordinators, with special reference to promotion of medical and life science research policy. *Gan To Kagaku Ryoho.* 1999 Jan;26(2 Suppl):235–246. Japanese.

Suzuki-Nishimura T, Toyoshima S, Uyama Y, Yamada H, Hosoki R, Fujimori K, and Nagao T. Contribution of pharmacology for new drug application in Japan. *Nippon Yakurigaku Zasshi.* 2002 Sep;120(3):187–194. Japanese.

Sweatman J. Good clinical practice: A nuisance, a help or a necessity for clinical pharmacology? *Br J Clin Pharmacol.* 2003 Jan;55(1):1–5.

Takano T and Saijo N. Doctor-initiated clinical trials and the revised pharmaceutical affairs law. *Gan To Kagaku Ryoho.* 2003 Oct;30(10):1391–1397.

van Dongen AJ. Good clinical practice, a transparent way of life. A review. *Comput Med Imaging Graph.* 2001 Mar–Apr;25(2):213–216.

Venulet J and ten Ham M. Methods for monitoring and documenting adverse drug reactions. *Int J Clin Pharmacol Ther.* 1996 Mar;34(3):112–129.

Wegener T, Schneider B and Working Party of the German Society of Phytotherapy. Proposals to enhance the quality of observational cohort studies. *Phytomedicine.* 2003 Nov;10(8):700–707.

CHAPTER 22: COMPUTER AND SOFTWARE VALIDATION

Adrion WR, Branstad MA, and Cherniavsky JC. *75 Validation, Verification, and Testing of Computer Software*, Computer Sciences and Technology; 500(75), National bureau of standards. Special publication.

Alan K (1997) Software validation, *Current Issues in Medical Device Quality Systems.* Association for the Advancement of Medical Instrumentation, Arlington, VA.

ANSI/ANS-10.4–1987 (1987) *Guidelines for the Verification and Validation of Scientific and Engineering Computer Programs for the Nuclear Industry*, American National Standards Institute. http://www.ans.org/store/i_240150

ANSI/ASQC Standard D1160–1995 (1995) *Formal Design Reviews*, American Society for Quality Control.

ANSI/UL 1998:1998 (1998) *Standard for Safety for Software in Programmable Components.* Underwriters Laboratories, Inc.

AS 3563.1–1991, *Software Quality Management System, Part 1: Requirements.* Published by Standards Australia (Standards Association of Australia), Homebush, New South Wales, Australia.

AS 3563.2–1991, *Software Quality Management System, Part 2: Implementation Guide.* Published by Standards Australia (Standards Association of Australia), Homebush, New South Wales, Australia.

Beizer B (1990) *Software Testing Techniques*, 2nd edn., Van Nostrand Reinhold.

Beizer B (1995) *Black Box Testing, Techniques for Functional Testing of Software and Systems.* John Wiley & Sons.

Beizer B (1996) *Software System Testing and Quality Assurance.* International Thomson Computer Press.

Bender R (1996) *Writing Testable Requirements, Version 1.0.* Bender & Associates, Inc., Larkspur, CA.

Branstad MA, Cherniavsky JC, and Adrion WR, *NBS Special Publication 500–56, Validation, Verification, and Testing for the Individual Programmer*, Center for Programming Science and Technology, Institute for Computer Sciences and Technology, National Bureau of Standards, U.S. Department of Commerce, February 1980.

Bryant JL and Wilburn NP (1987) *Handbook of Software Quality Assurance Techniques Applicable to the Nuclear Industry*, NUREG/CR-4640, U.S. Nuclear Regulatory Commission.

Castano S et.al. (1995) *Database Security*, ACM Press, New York/Addison-Wesley Publishing Company, Workingham, U.K.

Cem K, Jack F, Nguyen HC (1993) *Testing Computer Software*, 2nd edn., Vsn Nostrand Reinhold.

Computerized Data Systems for Nonclinical Safety Assessment, Current Concepts and Quality Assurance, Drug Information Association, Maple Glen, PA, September 1988.

Design Control Guidance for Medical Device Manufacturers, Center for Devices and Radiological Health, Food and Drug Administration, March 1997.

Deutsch MS (1982) *Software Verification and Validation, Realistic Project Approaches.* Prentice Hall.

Do It by Design, An Introduction to Human Factors in Medical Devices, Center for Devices and Radiological Health, Food and Drug Administration, March 1997.

Dunn RH and Ullman RS (1994) *TQM for Computer Software*, 2nd edn. McGraw-Hill, Inc.

Ebenau RG and Strauss SH (1994) *Software Inspection Process.* McGraw-Hill.

Electronic Records; Electronic Signatures Final Rule (March 20, 1997) 62 Federal Register 13430.

Elfriede Dustin, Jeff Rashka, and John Paul (1999). *Automated Software Testing—Introduction, Management and Performance,* Addison Wesley Longman, Inc., Reading, MA.

Fairley RE (1985) *Software Engineering Concepts.* McGraw-Hill Publishing Company.

FDA. *21 CFR part 11*, Electronic Records, Electronic Signatures, Final Rule. *Federal Register* Vol. 62, No. 54, 13429, March 20, 1997.

FDA. *Compliance Program Guidance Manual,* Compliance Program 7348.810—Sponsors, Contract Research Organizations and Monitors, October 30, 1998.

FDA. *Compliance Program Guidance Manual*, Compliance Program 7348.811—Bioresearch Monitoring—Clinical Investigators, September 2, 1998.

FDA. *General Principles of Software Validation; Guidance for Industry and FDA Staff*, 2002.

FDA. *Glossary of Computerized System and Software Development Terminology*, 1995.

FDA. *Good Clinical Practice VICH GL9*, 2001.

FDA. *Guideline for the Monitoring of Clinical Investigations*, 1988.

FDA. *Information Sheets for Institutional Review Boards and Clinical Investigators*, 1998.

FDA. *Electronic Records; Electronic Signatures—Scope and Application*, 2003.

FDA. *Software Development Activities*, part 11, 1987.

Frederick P. Brooks (1995) *The Mythical Man-Month, Essays on Software Engineering*, Addison-Wesley Longman, Anniversary Edition.

Friedman MA and Voas JM (1995) *Software Assessment—Reliability, Safety, Testability*, Wiley-Interscience, John Wiley & Sons Inc.

GAMP Guide For Validation of Automated Systems in Pharmaceutical Manufacture,Version V3.0, Good Automated Manufacturing Practice (GAMP) Forum, March 1998: *Volume 1, Part 1: User Guide; Part 2: Supplier Guid; Volume 2: Best Practice for User and Suppliers.*

Grigonis GJ, Subak EJ, and Wyrick M. Validation key practices for computer systems used in regulated operations. *Pharm Technol*. 1997 Jun;21(6):74.

Myers GJ (1979) *The Art of Software Testing*, John Wiley & Sons.

Glossary of Computerized System and Software Development Terminology, Division of Field Investigations, Office of Regional Operations, Office of Regulatory Affairs, Food and Drug Administration, August 1995.

Grady RB (1992) *Practical Software Metrics for Project Management and Process Improvement*. PTR Prentice-Hall Inc.

Guidance for Industry, FDA Reviewers and Compliance on Off-the-Shelf Software Use in Medical Devices, Office of Device Evaluation, Center for Devices and Radiological Health, Food and Drug Administration, September 1999.

Guide to Inspection of Computerized Systems in Drug Processing, Reference Materials and Training Aids for Investigators, Division of Drug Quality Compliance, Associate Director for Compliance, Office of Drugs, National Center for Drugs and Biologics, & Division of Field Investigations, Associate Director for Field Support, Executive Director of Regional Operations, Food and Drug Administration, February 1983.

Guideline on General Principles of Process Validation, Center for Drugs and Biologics, & Center For Devices and Radiological Health, Food and Drug Administration, May 1987.

Halvorsen JV. A software requirements specification document model for the medical device industry, *Proceedings IEEE SOUTHEASTCON '93, Banking on Technology*. April 4th–7th, 1993, Charlotte, NC.

Hatton L (1994) *Safer C: Developing Software for High-Integrity and Safety-Critical Systems*. McGraw-Hill Book Company.

Herrmann DS (1999) *Software Safety and Reliability: Techniques, Approaches and Standards of Key Industrial Sectors,* IEEE Computer Society.

Hecht H et.al., *Review Guidelines on Software Languages for Use in Nuclear Power Plant Safety Systems, Final Report*. NUREG/CR-6463. Prepared for U.S. Nuclear Regulatory Commission, 1996.

Hecht H et.al., *Verification and Validation Guidelines for High Integrity Systems*. NUREG/CR-6293. Prepared for U.S. Nuclear Regulatory Commission, 1995.

Hetzel B (1988) *The Complete Guide to Software Testing*, 2nd edn., A Wiley-QED Publication, John Wiley & Sons, Inc.

Ian Sommerville (1989) *Software Engineering*, 3rd edn., Addison Wesley Publishing Co., 1989.

IEC 60601–1–4:1996, *Medical electrical equipment, Part 1: General requirements for safety, 4. Collateral Standard: Programmable electrical medical systems*. International Electrotechnical Commission, 1996.

IEC 61506:1997, *Industrial process measurement and control—Documentation of application software*. International Electrotechnical Commission, 1997.

IEC 61508:1998, *Functional safety of electrical/electronic/programmable electronic safety-related systems*. International Electrotechnical Commission, 1998.

IEEE Standards Collection, Software Engineering, Institute of Electrical and Electronics Engineers, Inc., 1994.

IEEE Std 1012–1986, *Software Verification and Validation Plans*, Institute for Electrical and Electronics Engineers, 1986.

International Conference on Harmonisation, E6 Good Clinical Practice: Consolidated Guideline, *Federal Register*, Vol. 62, No. 90, 25711, May 9, 1997.

ISO 13485:1996, *Quality systems—Medical devices—Particular requirements for the application of ISO 9001*. International Organization for Standardization, 1996.

ISO 14971–1:1998, *Medical Devices—Risk Management—Part 1: Application of Risk Analysis.* International Organization for Standardization, 1998.

ISO 8402:1994, *Quality management and quality assurance—Vocabulary.* International Organization for Standardization, 1994.

ISO 9000–3:1997, *Quality management and quality assurance standards - Part 3: Guidelines for the application of ISO 9001:1994 to the development, supply, installation and maintenance of computer software.* International Organization for Standardization, 1997.

ISO 9001:1994, *Quality systems—Model for quality assurance in design, development, production, installation, and servicing.* International Organization for Standardization, 1994.

ISO/IEC 12119:1994, *Information technology—Software packages—Quality requirements and testing,* Joint Technical Committee ISO/IEC JTC 1, International Organization for Standardization and International Electrotechnical Commission, 1994.

ISO/IEC 12207:1995, *Information technology—Software life cycle processes,* Joint Technical Committee ISO/IEC JTC 1, Subcommittee SC 7, International Organization for Standardization and International Electrotechnical Commission, 1995.

ISO/IEC 14598:1999, *Information technology—Software product evaluation,* Joint Technical Committee ISO/IEC JTC 1, Subcommittee SC 7, International Organization for Standardization and International Electrotechnical Commission, 1999.

Jones C (1997) *Software Quality, Analysis and Guidelines for Success.* International Thomson Computer Press.

Juran JM and Gryna FM (1993) *Quality Planning and Analysis*, 3rd edn., McGraw-Hill.

Kaplan C, Clark R, and Tang V (1995) *Secrets of Software Quality, 40 Innovations from IBM.* McGraw-Hill.

Kit E (1995) *Software Testing in the Real World.* Addison-Wesley Longman, Reading, MA.

Lawrence JD and Persons WL (1994) *Survey of Industry Methods for Producing Highly Reliable Software*, NUREG/CR-6278, U.S. Nuclear Regulatory Commission.

Lawrence JD and Preckshot GG (1994) *Design Factors for Safety-Critical Software*, NUREG/CR-6294, U.S. Nuclear Regulatory Commission.

Leveson NG (1995) *Safeware, System Safety and Computers.* Addison-Wesley Publishing Company.

Lyu MR (ed) (1996) *Handbook of Software Reliability Engineering.* IEEE Computer Society Press, McGraw-Hill.

Mallory SR (1994) *Software Development and Quality Assurance for the Healthcare Manufacturing Industries.* Interpharm Press Inc., Englewood, CO.

Marick B (1995) *The Craft of Software Testing.* Prentice Hall PTR.

McConnell S (1996) *Rapid Development.* Microsoft Press, Redmond, WA.

Neumann PG (1995) *Computer Related Risks.* ACM Press/Addison-Wesley Publishing Co., 1995.

Olivier DP (1994) *Conducting Software Audits, Auditing Software for Conformance to FDA Requirements*, Computer Application Specialists, San Diego, CA.

Olivier DP. Validating Process Software. *FDA Investigator Course: Medical Device Process Validation*, Food and Drug Administration.

Patricia B. Powell, Editor. *NBS Special Publication 500–93, Software Validation, Verification, and Testing Technique and Tool Reference Guide*, Center for Programming Science and Technology, Institute for Computer Sciences and Technology, National Bureau of Standards, U.S. Department of Commerce, September 1982.

Powell PB (ed.) *NBS Special Publication 500–98, Planning for Software Validation, Verification, and Testing*, Center for Programming Science and Technology, Institute for Computer Sciences and Technology, National Bureau of Standards, U.S. Department of Commerce, November 1982.

Pressman RS (1992) *Software Engineering, A Practitioner's Approach*, 3rd edn., McGraw-Hill Inc.

Pressman RS (1993) *A Manager's Guide to Software Engineering.* McGraw-Hill Inc.

Sage AP and Palmer JD (1990) *Software Systems Engineering.* John Wiley & Sons, New York.

Sanders J and Curran E (1994) *Software Quality.* Addison-Wesley Publishing Co.

Shumate K and Keller M (1992) *Software Specification and Design, A Disciplined Approach for Real-Time Systems.* John Wiley & Sons.

Smith DD (1999) *Designing Maintainable Software.* Springer-Verlag.

Software Considerations in Airborne Systems and Equipment Certification. Special Committee 167 of RTCA. RTCA Inc., Washington, DC. Tel: 202-833-9339. Document No. RTCA/DO-178B, December 1992.

Stephen H. Kan, *Metrics and Models in Software Quality Engineering.* Addison-Wesley Publishing Company, 1995.

Student Manual 1, Course INV545, Computer System Validation, Division of Human Resource Development, Office of Regulatory Affairs, Food and Drug Administration, 1997.

Technical Report No. 18, Validation of Computer-Related Systems. PDA committee on validation of computer-related systems. *J Pharm Sci Technol*. 1995 Jan–Feb;49(1), Supplement.

Technical Report, Software Development Activities, Division of Field Investigations, Office of Regional Operations, Office of Regulatory Affairs, Food and Drug Administration, July 1987.

The Application of the Principles of GLP to Computerized Systems, Environmental Monograph #116, Organization for Economic Cooperation and Development (OECD), 1995.

Tom G and Dorothy G (1993) *Software Inspection*, Addison-Wesley Publishing Company, Workingham, U.K.

Validation Compliance Annual 1995, International Validation Forum, Inc.

Wallace DR (ed) (August 1996) *NIST Special Publication 500–235, Structured Testing: A Testing Methodology Using the Cyclomatic Complexity Metric*. Computer Systems Laboratory, National Institute of Standards and Technology, U.S. Department of Commerce.

Wallace DR et.al. (March 1996) *NIST Special Publication 500–234, Reference Information for the Software Verification and Validation Process*. Computer Systems Laboratory, National Institute of Standards and Technology, U.S. Department of Commerce.

Wallace DR and Fujii RU (September 1995) *NIST Special Publication 500–165, Software Verification and Validation: Its Role in Computer Assurance and Its Relationship with Software Project Management Standards*, National Computer Systems Laboratory, National Institute of Standards and Technology, U.S. Department of Commerce.

Wallace DR, Ippolito LM, and Kuhn DR (September 1992) *NIST Special Publication 500–204, High Integrity Software, Standards and Guidelines*, Computer Systems Laboratory, National Institute of Standards and Technology, U.S. Department of Commerce.

Watts SH (1989) *Managing the Software Process*. Addison-Wesley Publishing Company, Reading, MA.

Watts SH (1995) *A Discipline for Software Engineering*. Addison-Wesley Longman, Reading, MA.

Wiegers KE (April 1995) *Software Inspection, Improving Quality with Software Inspections*. Software Development, pp. 55–64.

Wiegers KE (1996) *Creating a Software Engineering Culture*, Dorset House Publishing.

Wiegers KE (1999) *Software Requirements*. Microsoft Press.

William EP and Randall WR (1997) *Surviving the Top Ten Challenges of Software Testing*. Dorset House Publishing, New York.

William P (1995). *Effective Methods for Software Testing*. John Wiley & Sons, Inc., New York.

CHAPTER 23: OUTSOURCING AND MONITORING OF BIOEQUIVALENCE STUDIES

ADAMON study (http://ctj.sagepub.com/content/6/6/585.full.pdf+html).

Baigent et al. Ensuring trial validity by data quality assurance and diversification of monitoring methods. *Clin Trials*. 2008;5:49–55.

Bakobaki et al. The potential for central monitoring techniques to replace on-site monitoring: Findings from an international multi-centre clinical trial. *Clin Trials*. 2012;9:257–264.

Brosteanu et al. Risk analysis and risk adapted on-site monitoring in noncommercial clinical trials. *Clin Trials*. 2009;6:585–595.

Buyse et al. The role of biostatistics in the prevention, detection and treatment of fraud in clinical trials. *Stat Med*. 1999;18:3435–3451.

Califf et al. Developing Systems for Cost-Effective Auditing of Clinical Trials. *Control Clin Trials*. 18:651–660 (1997).

CBER. http://www.fda.gov/BiologicsBloodVaccines/GuidanceComplianceRegulatoryInformation/Guidances/default.htm.

CBER. http://www.fda.gov/BiologicsBloodVaccines/GuidanceComplianceRegulatoryInformation/Guidances/default.htm. FDA.http://www.fda.gov/MedicalDevices/DeviceRegulationandGuidance/GuidanceDocuments/default.htm.

CDRH. http://www.fda.gov/MedicalDevices/DeviceRegulationandGuidance/GuidanceDocuments/default.htm.

CLINICAL TRIALS https://www.ctti-clinicaltrials.org/websiteadministration/documents/COLLINS%20FDA%20trial%20quality%200811%20FINAL_no%20animation.pdf/view.

Collins, Rory. (2010, October) Quality Design of Clinical Trials. Presentation at CTTI work stream 3 expert meeting. Available at https://www.ctti-clinicaltrials.org/website- administration/documents/COLLINS%20FDA%20trial%20quality%200811%20FINAL_no%20animation.pdf/view.

CPGM 7348.810: Sponsors, contract research organizations and monitors (March 11, 2011), available at: http://www.fda.gov/ICECI/EnforcementActions/BioresearchMonitoring/ucm133777.htm.

CPGM 7348.811: Clinical investigators and sponsor-investigators (December 8, 2008), available at: http://www.fda.gov/ICECI/EnforcementActions/BioresearchMonitoring/ucm133562.htm.

FDA. *Concept Paper: Quality in FDA-Regulated Clinical Research; Background to HSP/BIMO Workshop* 5/10- 5/11/07, (4/26/07).

FDA. *Guidance for Industry. Oversight of clinical investigations—A risk-based approach to monitoring*. http://www.fda.gov/Drugs/GuidanceComplianceRegulatoryInformation/Guidances/default.htm.

Glickman et al. Ethical and scientific implications of the globalization of clinical research. *N Engl J Med.* 2009;360:816–823.

Gruppo Italiano per lo Studio della Sopravvivenza nell'Infarto Miocardico-Italian group for the study of the survival of myocardial infarction.

Guidance for Industry, E6 Good Clinical Practice: Consolidated Guidance, 1996, section 5.18.3.

Guidance for industry, Q9 Quality Risk Management, June 2006.

ICH E6, section 5.19 and ISO 14155:2011, section 6.11.

IDE regulations (21 CFR 812.25(e)) require that written monitoring procedures be submitted as part of the IDE application.

ISO 14155:2011, Clinical investigation of medical devices for human subjects—Good clinical practice, sections 5.7 and 6.3.

ISO 31010:2009 Risk management—Risk assessment techniques.

Morrison et al. Monitoring the quality of conduct of clinical trials: A survey of current practices. *Clin Trials.* 2011;8:342–349.

OPTIMON study https://ssl2.isped.u-bordeaux2.fr/optimon/Default.aspx.

Tantsyura et al. Risk-based source data verification approaches: Pros and cons. *Drug Inf J.* 2010;44:745–756.

Temple, R. Policy developments in regulatory approval. *Stat Med.* 2002;21:2939–2948.

Usher, R. PhRMA bioresearch monitoring committee perspective on acceptable approaches for clinical trial monitoring. *Drug Inf J.* 2010;44:477–483.

Venet et al. A statistical approach to central monitoring of data quality in clinical trials. *Clin Trials.* 2012;0:1–9.

CHAPTER 24: EPILOG: FUTURE OF BIOEQUIVALENCE TESTING

See also Chapter 3 and 6

AAPS PharmSciTech. 2010 Dec;11(4):1726–1729.

Akerlof G. Dielectric constants of some organic solvent-water mixtures at various temperatures. *J Am Chem Soc.* 1932;54:4125–4139.

Amidon GL, Lennernäs H, Shah VP, and Crison JR. A theoretical basis for a biopharmaceutic drug classification: The correlation of in vitro drug product dissolution and in vivo bioavailability. *Pharm Res.* 1995 Mar;12(3):413–420.

Beerbower A, Wu PL, and Martin A. Expanded solubility parameter approach 1. Naphthalene and benzoic acid in individual solvents. *J Pharm Sci.* 1984;73:179–188.

Biopharmaeutics Classification System (BCS). Therapeutic System Research Laboratories (TSRL inc.). http://tsrlinc.com/resources/services/. Accessed June 5, 2014.

Blume HH and Schug BS. The biopharmaceutics classification system (BCS): Class III drugs-better candidate for BA/BE waiver? *Eur J Pharm Sci.* 1999;9:117–121.

Born M. Volumen und Hydratationswärme der Ionen. *Z Phys.* 1920;1:45–48.

Cheng CL, Yu LX, Lee HL, Yang CY, Lue CS, and Chou CH. Biowaiver extension potential to BCS Class III high solubility-low permeability drugs: Bridging evidence for metformin immediate-release tablet. *Eur J Pharm Sci.* 2004;22:297–304.

EMEA/CPMP/EWP/1401/98. Note for guidance on the investigation of bioavailability and bioequivalence. EMEA, London, U.K., July 26, 2001.

EMEA/CPMP/EWP/QWP/1401/98 Rev.1/Corr*. Guideline on the investigation of bioequivalence. London, U.K., January 20, 2010. Available at: http://www.ema.europa.eu/docs/en_GB/document_library/Scientific_guideline/2010/01/WC500070039.pdf.

Fakhree MA et al. The Importance of Dielectric Constant for Drug Solubility Prediction in Binary Solvent Mixtures: Electrolytes and Zwitterions in Water+Ethanol. *AAPS PharmSciTech*. 2010;11(4):1726–1729.

FDA. *Guidance for Industry: Waiver of In Vivo Bioavailability and Bioequivalence Studies for Immediate Release Solid Oral Drug Products Based on a Biopharmaceutics Classification System*, Food and Drug Administration, Rockville, MD, 2000. http://www.fda.gov/RegulatoryInformation/Guidances/default.htm.

FDA. *Orange Book Preface, Statistical Criteria for Bioequivalence. Approved Drug Products with Therapeutic Equivalente Evaluations*. 29th edn., U.S. Food and Drug Administration. Center for Drug Evaluation and Research. 2009–06–18. 2009. http://www.fda.gov/drugs/developmentapprovalprocess/ucm079068.htm.

Folkers G, Waterbeemd, Hvan de, Lennernäs H, Per Artursson; Raimund Mannhold; Hugo Kubinyi (2003). *Drug Bioavailability: Estimation of Solubility, Permeability, Absorption and Bioavailability (Methods and Principles in Medicinal Chemistry)*. Wiley-VCH, Weinheim, Germany.

Hauschke D, Steinijans VW, Diletti E, and Burke M. Sample size determination for bioequivalence assessment using a multiplicative model. *J Pharmacokinet Biopharm*. 1992;20:557–561.

Jouyban A. Review of the cosolvency models for predicting solubility of drugs in water-cosolvent mixtures. *J Pharm Pharm Sci*. 2008;11:32–58.

Jouyban A (2009) *Handbook of Solubility Data for Pharmaceuticals*. CRC Press, Boca Raton, FL.

Jouyban A and Acree WE., Jr. In silico prediction of drug solubility in water–ethanol mixtures using Jouyban-Acree model. *J Pharm Pharm Sci*. 2006;9:262–269.

Jouyban A, Soltanpour Sh, and Acree WE, Jr. Improved prediction of drug solubility in ethanol + water mixtures at various temperatures. *Biomed Int*. 2010;1:19–24.

Jouyban A, Soltanpour Sh, and Chan H-K. A simple relationship between dielectric constant of mixed solvents with solvent composition and temperature. *Int J Pharm*. 2004;269:353–360.

Julious SA. Tutorial in biostatistics. *Statistic Med*. 2004;23:1921–1986.

Larson RG and Hunt H. Molecular forces and solvent power. *J Phys Chem*. 1939;43:417–423.

Pan American Health Organization. Regional Office of the World Health Organization. Science based criteria for bioequivalence in vivo and in vitro, bio-waivers, and strategic framework for implementation. http://www.paho.org/english/ad/ths/ev/be-doct-draft-eng.pdf.

Paruta AN, Sciarrone BJ, and Lordi NG. Correlation between solubility parameters and dielectric constants. *J Pharm Sci*. 1962;51:704–705.

Polli JE, Yu LX, Cook JA, Amidon GL, Borchardt RT, Burnside BA, Burton PS et al. Summary workshop report: Biopharmaceutics classification system-implementation challenges and extension opportunities. *J Pharm Sci*. 2004;93:1375–1381.

Ramirez E et al. Acceptability and characteristics of 124 human bioequivalence studies with active substances classified according to the Biopharmaceutic Classification System. *Br J Clin Phamacol*. 2010;70(5):694–702.

Ramirez E, Guerra P, Laosa O, Duque B, Tabares B, Lei AS, Carcas AJ, and Frias J. The importance of sample size, log-mean ratios, and intra-subject variability in the acceptance criteria of bioequivalence studies. *Eur J Clin Pharmacol*. 2008;64:783–793.

Reillo A, Cordoba M, Escalera B, Selles E, and Cordoba M Jr. Prediction of sulfamethiazole solubility in dioxane-water mixtures. *Pharmazie*. 1995;50:472–475.

Takagi T, Ramadrasekharan C, Bermejo M, Yamashita S, Yu LX, and Amidon LG. A provisional Biopharmaceutical Classification of the Top 200 oral drugs products in the Unites States, Great Britain, Spain and Japan. *Mol Pharm*. 2006;3:631–643.

WHO: Proposal to waive in vivo bioequivalence requirements for WHO Model List of Essential Medicines immediate-release, E. Ramirez et al. solid oral drug products. Technical Report Series, No 937, 40th Report, Annex 8 of WHO Expert committee on specifications for pharmaceutical preparations. http://whqlibdoc.who.int/trs/WHO_TRS_937_eng.pdf.

Widenski DJ, Abbas A, and Romagnoli JA. Effect of the solubility model on antisolvent crystallization predicted volume mean size. *Chem Eng Trans*. 2009;17:639–644.

Wu CY and Benet LZ. Predicting drug disposition via application of BCS: Transport/absorption/elimination interplay and development of a biopharmaceutics drug disposition classification system. *Pharm Res*. 2005;22:11–23.

Yu LX, Amidon GL, Polli GL, Zhao H, Mehta MU, Conner DL, Shah VP et al. Biopharmaceutics classification system: The scientific basis for biowaiver extensions. *Pharm Res*. 2002;19:921–925.

Index

A

Abacavir sulfate, 523–524
Abbreviated new animal drug application (ANADA), 95, 111, 142–143, 683
Abbreviated new drug applications (ANDAs), 95, 245–249, 264–265, 293, 296–298, 305–306, 312, 683, 755
  fed bioequivalence studies, 278–282
  IVIVC, 151
  methods validation, 249–252
  submission requirements, 394–395
ABE, *see* Average bioequivalence (ABE)
ABE approach with expanding limits (ABEL), 230, 234–236, 429–430
Absorption, distribution, metabolism, and excretion (ADME) assays, 186–187, 787, 789
Acamprosate calcium, 524
Acetaminophen, 582
Acitretin, 524
Active pharmaceutical ingredients (APIs), 755, 757
Acyclovir, 524
ADME assays, *see* Absorption, distribution, metabolism, and excretion (ADME) assays
Agencia Nacional de Vigilancia Sanitaria (ANVISA), 464
Albuterol metered-dose inhalers, 133
Alendronate sodium, 524
Alfuzosin hydrochloride, 525
Almotriptan malate, 525
Alosetron hydrochloride, 525
Alprazolam, 525
American Council of Research Organizations (ACRO), 757
Amlodipine besylate, 526
Amorphous forms, 28, 49, 189
Amoxicillin, 526–527
Amprenavir, 527
Anagrelide hydrochloride, 527
Analysis of variance (ANOVA), 427–428
Analytical deconvolution, 93
Analytical laboratory inspection
  antibiotic analyses, 266
  data handling and storage, 267
  equipment, 264
  methods validation, 264–265
  pre-study analysis, 263
  protocol acceptance, 264
  radiometric analyses, 267
  sample analysis, 265–266
Anastrozole, 527
ANDAs, *see* Abbreviated new drug applications (ANDAs)
Animal drug bioequivalence testing
  reference product, 142
  species selection, 142
  subject characteristics, 143
  tissue residue depletion study, 142–144
  waivers, 142
Animal model testing, 189
Antibiotic analyses, 266
APIs, *see* Active pharmaceutical ingredients (APIs)
Aprepitant, 527–528
Area under the curve (AUC)
  computer applications, 119
  integration method, 118
  multiple-dose study, 117–118
  single-dose study, 117
  trapezoidal rule, 118
Aripiprazole, 528–529
Armodafinil, 529
Atazanavir sulfate, 529
Atomoxetine hydrochloride, 529
Atorvastatin calcium, 530
Atovaquone, 530
AUC, *see* Area under the curve (AUC)
Average bioequivalence (ABE), 195
  data analysis, 212–214
  HVDs/HVDPs, 428
  sample size determination, 204–205
  scaling, 199
  standards, 203
Azithromycin, 530

B

Balaam design, 227
Balanced incomplete block design, 197
Balsalazide disodium, 530–531
Bayesian inference using Gibbs sampling (BUGS) project, 735
BCS, *see* Biopharmaceutical classification system (BCS)
Benazepril hydrochloride, 526
Benzonatate, 531
Benzphetamine hydrochloride, 531
Bicalutamide, 531
Bioanalytical method validation
  acceptance criteria, 665
  accuracy, 668, 672–673
  application domain, 665
  back-calculated quantities/inverse predictions, 671
  biomarkers, 696
  calibration, 664, 668
  chromatographic methods
    accuracy, precision, and recovery, 686
    calibration curve, 686–687
    reference standards, 685
    reproducibility, 688
    selectivity, 685–686
    sensitivity, 688
    stability, 688
    use, data analysis, and reporting, 689–690
  computational aspects, 669
  cross validation, 667
  decision rule, 674
  definitive quantitative assay, 662

diagnostic kits, 696–697
documentation of
bioanalytical report, 699–701
method establishment, 681
routine drug analysis, 681–682
summary information, 681, 698
system suitability/equilibration, 698
dried blood spot, 697
endogenous compounds, 695
guidance, 683
history, 683–685
incurred sample reanalysis, 695
ligand-binding assays
accuracy, precision, and recovery, 691
calibration curve, 691–693
quantification issues, 678–679
reagents, 690
reproducibility, 693
selectivity, 677–278, 691
sensitivity, 693
stability, 693
use, data analysis, and reporting, 694
limit of detection, 671–672
linear and polynomial models, 669
linearity, 672
LLOQ, 671
microbiological assay, 677–679
nonlinear model
PROC NLIN, 669–670
PROC NLMIXED, 670
partial validation, 666–667
precision method, 668, 670, 673
pre-study validation, 663, 665–666
principles of, 676
qualitative methods, 661
quantitative methods, 661–662
quasi-quantitative assay, 663
recovery, 668
reference standard, 667
relative quantitative assay, 663
requirements, 661–662
selectivity, 668
specific instructions, 676–677
specificity–selectivity, 672
specific laboratory investigations, 665
stability, 675
total error/measurement error, 673–674
ULOQ, 671
use in routine analysis, 679–680
validation samples, 664
Bioequivalence documentation
ANDA, 587
CRO monitoring, 587–591
drug product formulation
complex dosage forms, 597
IR formulations, 594–595
nonrelease controlling excipients, 595–596
release controlling excipients, 596
semisolid dosage forms, 597
international submissions
appendices, 600–601
assay method and validation, 600
clinical study, 599
in vitro testing, 599
in vivo study, 599
pharmacokinetic parameters, 600
report contents, 598
screening record (*see* Screening record)
statistical analysis, 600
title page, 598
in vivo and in vitro testing, 591
LANDSCAPE orientation
bioanalytical method validation, 618
demographic profile, 618
dropout information, 620
formulation data, 624
incidence of, 619
in vitro dissolution study, 623
meal composition, 621
product information, 620
protocol deviations, 620
sample reanalysis, 624
SOP, 621
standard curve and QC data, 621
statistical summary, BA data, 622–623
study information, 619
PORTRAIT orientation, 617
standard SOPs, 592–593
summary report, 593
Bioequivalence testing software, 241–243
Bioequivalent drug products, 105, 113, 397, 405–406, 426, 491, 515, 522
Biological license applications (BLAs), 683
Biologics Price Competition and Innovation Act (BPCI Act), 8, 329–330
Biopharmaceutical classification system (BCS), 278, 785
ANDAs, 178
applications, 191–192
biowaivers, EMA
complete drug absorption, 456
drug substance characteristics, 455
excipients, 457–458
vs. FDA, 432
fixed combinations, 458
immediate-release drug product, 455
in vitro dissolution, 456–457
pH–solubility profile, 456
strength requirements, 441–442
WHO reports and guidelines, 433
complete absorption, 193
dimensionless numbers, 162
dissolution profiles
noncompendia and compendia medias, 166
quality by design, 166
similarity factor, 166, 172
drugs, classification of, 161–162
excipients, 172
expected IVIVC
extended-release drug products, 162–163
immediate-release drug products, 162–163
FDA definition, 164
immediate-release drug products, 192
INDs/NDAs, 177–178
in vitro dissolution experiments, 193–194
manufacturing changes, 178–180
narrow therapeutic range drug products, 173

permeability class boundary, 165, 167–168
pharmacokinetic studies
absolute bioavailability studies, 168–171
mass balance, 168
pH-solubility profile, 167, 192–193
postapproval changes, 178
prodrugs, 172
products absorbed in oral cavity, 173
solubility class boundary, 165
stereoisomerism, 172–173
supporting data
high permeability, 183
high solubility, 183
rapid and similar dissolution, 184
transporter efflux factors, 163–165
Biorelevant dissolution medium (BDM), 175
Biosimilar products
biosimilarity index
advantages, 349
Bayesian approach, 350
characteristics, 349–350
CI approach, 349
finite sample size performance, 351
reproducibility probability curves, 350
R–R study, 349
degree of similarity, 343, 346–347
doses
clinical efficacy, 369–370
clinical studies, 370–371
pharmacodynamics, 369
PK parameters, 368
routes of administration, 368
sampling times, 368
timing, PK evaluation, 368–369
erythropoietins
clinical safety, 375
clinical studies, 372–373
pharmacodynamics studies, 372
routes of administration, 373–374
toxicological studies, 372
European guidance
clinical safety, 356–357
clinical studies, 353–354
efficacy endpoints, 355–356
efficacy trials, 355
in vitro studies, 352
in vivo studies, 352–353
nonclinical study, 351
pharmacodynamic studies, 354–355
pharmacokinetic studies, 354
safety profile, 351
study designs, 355
hGH
clinical efficacy studies, 384–385
clinical safety, 385
nonclinical studies, 383–384
pharmacodynamics studies, 384
pharmacodynamic studies, 384
pharmacokinetic studies, 384
toxicological studies, 384
human follicle-stimulating hormone
clinical efficacy, 359
clinical safety, 359–360
pharmacodynamic studies, 359
pharmacokinetic studies, 358
regulating reproductive function, 357
r-hFSH, 357
toxicological studies, 358
human G-CSF
clinical efficacy studies, 382–383
clinical safety, 383
nonclinical studies, 381
pharmacodynamic studies, 382
pharmacokinetic studies, 382
toxicological studies, 382
human insulin
clinical efficacy studies, 387
clinical safety, 387
nonclinical studies, 386
pharmacodynamic studies, 386–387
pharmacokinetic studies, 386
toxicological studies, 386
IFN-α (2a/2b)
clinical studies, 380
efficacy, 380
extrapolation, 381
immunogenicity, 381
nonclinical studies, 379
pharmacodynamics studies, 379
primary and secondary treatment, 380
safety, 380–381
toxicological studies, 379
IFN-β
clinical efficacy, 361–362
clinical safety, 363
clinical studies, 361
pharmacodynamics, 361
pharmacokinetics, 361
interchangeability, 347–348
limitations, 346
LMWH
anti-FXa and anti-FIIa activity, 375
biological origin and manufacturing process, 375
clinical studies, 377–378
pharmacodynamic studies, 376
toxicological studies, 376
monoclonal antibodies
clinical studies, 366
in vivo study, 365–366
nonclinical studies, 364
pharmacokinetics, 366
postauthorization safety studies, 364
study design, 366–367
practical issues, 343–344
quantitative evaluation, 344
statistical methods, 345–347
study design, 345, 347
U.S. regulations
analytical methodology, 334–335
animal data, 335–336
animal immunogenicity assessment, 336
animal PK and PD measures, 336
animal study, 331
biosimilarity, 329–331
clinical data extrapolation, 342
clinical immunogenicity assessment, 338–339
clinical safety and effectiveness data, 339–340
clinical study, 331, 336–337, 340–342

functional assays, 335
human pharmacology data, 337–338
manufacturing processes, 332–333
mechanism of action, 333
postmarketing safety monitoring, 342–343
product specificity, 334
scientific basis, 332
stepwise approach, 333
totality-of-the-evidence approach, 333–334
Biowaivers, *see* Waivers
Bisoprolol fumarate, 531
Bolus-response function, 91–92
BPCI Act, *see* Biologics Price Competition and Innovation Act (BPCI Act)
B-spline function, 92
Bupropion hydrochloride, 531–532
Bureau of Pharmaceutical Affairs (BPA), 465

C

Calcium carbonate, 574
CAM, *see* Complementary and alternative medicines (CAM)
Canadian Society of Pharmaceutical Sciences (CSPS), 462
Candesartan cilexetil, 532
Capsules, 55–57, 62, 413–414
Carbamazepine, 532–533
Carbidopa, 533
Carr's index, 51
Carvedilol, 534
Case report forms (CRFs), 712
informed consent, 775
monitor and study site staff, 775
records maintenance, 775–776
verification and corroboration, 776
Cefdinir, 534
Cefditoren pivoxil, 534
Cefixime, 534
Celecoxib, 534
Center for Drug Evaluation and Research (CDER), 195, 522
Center for Veterinary Medicine (CVM), 110–111, 142–144
Central Drugs Standard Control Organization (CDSCO), 463
Centralized procedure (CP), 424
Certificate of pharmaceutical products (CPP), 597
Certified copy, 737–738
Cetirizine hydrochloride, 534–535
Cevimeline hydrochloride, 535
Chamber of Drug Market Regulation (CMED), 464
Change control, 742
Chemical drug products, 792
Chemical modifications, 19
Chemistry, manufacturing, and control (CMC) tests, 95
Chewable tablets, 416, 452
Cilostazol, 535
Cinacalcet hydrochloride, 535
Ciprofloxacin, 536
Ciprofloxacin hydrochloride, 535–536
Clarithromycin, 536
Clavulanate potassium, 526–527
Clearance, 89–90
Clinical facilities, inspection
abbreviated report, 262–263
human subjects, consent of, 261
inspection procedures, 259
IRB, 261
protocol modifications, 259–260
records retention, 262
sponsor, 261
study responsibility, 259
subjects' records, 260–261
test article accountability, 261–262
Clinical quality assurance (CQA), 756
ClinicalTrials.gov, 515–516
Clinical Trials Transformation Initiative (CTTI), 771
Clonidine hydrochloride, 536–537
Clopidogrel bisulfate, 537
Coated particles, 58
Code of Federal Regulations (CFR), 389–390
Combinatorial chemistry, 21–22
COMIS database
controlled correspondence, 248
dissolution data (DIS), 247
dissolution waiver (DIW), 247
DSI, 247–248
other (OTH), 247–248
protocols, 248
study amendment (STA), 247
study types, 247
waiver (WAI), 247
Commissioner of Food and Drugs, 406–407
Committee for Medicinal Products for Human Use (CHMP), 228, 462
Committee for Proprietary Medicinal Products (CPMP), 423, 461, 790
Comparator product, 459–461, 466, 667
Compartment pharmacokinetic modeling
AUC function, 89
clearance, 89–90
distribution volume, 88
effective concentration, 88
rate of elimination, 87
two-stage approach, 90
Complementary and alternative medicines (CAM)
aging population, 317
characteristics, 315
chronic illness and health-care costs, 317
clinical experience, 315
counterculture movement, 317
definition, 315
FDA guidance
act's food additive provisions, 321
bioequivalence testing, 321
dietary supplementation, 321
FDC act, 318
HACCP system, 319
history, 318
NCCAM (*see* National Center for Complementary and Alternative Medicine (NCCAM))
PHC act, 318
traditional Chinese medicine and Ayurvedic medicine, 323
health-care systems, 315–316
history, 315
holistic health-care movement, 317
onset of illness, risk of, 317
prevalence, 317–318
revolution, 316
therapy and products, 315–316

Complex bioequivalence, 31, 791
- albuterol metered-dose inhalers, 133
- digoxin, 131
- levothyroxine sodium, 131–132
- warfarin sodium, 132–133

Compliance Program audit
- analytical testing noncompliance, 256
- clinical/analytical facility, 253, 256–257
- clinical testing noncompliance, 256
- directed data audit, 254
- GBIB assignment memos, 254
- inspectional operations, 254–255
- routine data audit, 254

Compliance program guidance manuals (CPGMs), 772

Computer validation
- access controls, 738
- audit trails, 738–739
- data retrieval and record retention, 740
- date/time stamps, 739
- direct entry of data, 739
- electronic records copies, 743
- electronic signature certification, 743
- legacy systems, 741
- OTS software, 741–742
- security measures, 738–739
- system dependability, 741
- systems control, 742–743
- system security, 740
- training, 743

Contract research organizations (CROs), 245–246, 463, 714, 755–756
Control article, 273–274, 629–631, 633–634, 637, 641, 651, 654–655
Controlled-release dosage forms, 61–62
Convection model, 40–41
Convolution model, 155–157
Coumadin, 132–133
CPMP, *see* Committee for Proprietary Medicinal Products (CPMP)
CRFs, *see* Case report forms (CRFs)
CROs, *see* Contract research organizations (CROs)
Crystal morphology, 23–24
CVM, *see* Center for Veterinary Medicine (CVM)
Cube root model, 41
CYP3A4-transfected Caco-2 cells, 189

D

Danazol, 537
Danish Health and Medicines Authority, 430
Dantrolene sodium, 537
Darifenacin hydrobromide, 537
Darunavir ethanolate, 537
DBE, *see* Division of bioequivalence (DBE)

Debarment
- debarment act, 758
- Debarment Task Force, 763
- Division of Scientific Investigations, 764
- FDA list, 759–762
- mandatory debarment, 763
- Office of Enforcement, 762
- permissive debarment, 763

Decentralized procedure (DCP), 424
Deconvolution technique, 157
Deferasirox, 537–538
Definitive quantitative assay, 662
Delavirdine mesylate, 538
Delayed-release drug products, 415

Dermatopharmacokinetics (DPK)
- analytical methods, 286–287
- application, 286
- crossover design and subject selection, 288
- drug uptake/elimination, 290
- metrics and statistics, 291
- sample collection, 289
- sample pilot study, 287–288
- sites and duration, 289
- skin-stripping methodology, 287, 290
- stratum corneum concentration–time curves, 286
- test and reference products, application and removal of, 289

DESI program, *see* Drug Efficacy Study Implementation (DESI) program
Desloratadine, 538
Dexmethylphenidate, 538
Dextromethorphan polistirex, 538–539
Diclofenac sodium, 539
Dicloxacillin sodium, 539
Didanosine, 539
Differential equation–based model, 157–158
Diffusion model, 40
Digoxin, 539–540
Diltiazem hydrochloride, 540–541
Dipyridamole, 541
Directed inspection, 249, 268–269, 636, 644
Disintegration test, 64–65

Dissolution testing, 45–46, 65–66
- concentration gradient, 44
- constant, 44–45
- in vitro bioequivalence system, 134–135
- models, 40–44
- profiles, 794

Distribution coefficient, 16–19
Distribution volume, 88
Divalproex sodium, 541–542
Division of bioequivalence (DBE), 245–248

Division of Scientific Investigations (DSI)
- debarment act, 764
- inspection reports, 247–249

Dofetilide, 542
Donepezil hydrochloride, 542–543

Dosage forms
- attributes, 47
- biosimilar products
  - clinical efficacy, 369–370
  - clinical studies, 370–371
  - pharmacodynamics, 369
  - PK parameters, 368
  - routes of administration, 368
  - sampling times, 368
  - timing, PK evaluation, 368–369
- capsules, 55–57
- characteristics, 47
- chemical content evaluations, 63
- coated particles, 58
- contaminants, 64
- content uniformity, 64
- controlled-release forms, 61–62

disintegration test, 64–65
dissolution rate test, 65–66
emulsion formulation, 60
formulation factors, 48
inhalers, 56
modified-release products, 56
powder, 52–56
solid form (*see* Solid dosage forms)
solutions, 58–59
suspensions, 60–61
tablets, 58
therapeutic systems, 62–63
Doxazosin mesylate, 543
Doxycycline, 543
DPK, *see* Dermatopharmacokinetics (DPK)
Drospirenone, 543
Drug bioavailability
food effects, 74–75
gastrointestinal drug biotransformation, 73–74
pathophysiological disorders, 75
Drug delivery systems
absorption, 66–77
bioavailability variations, 47
dosage forms (*see* Dosage forms)
evaluation of, 63–66
first-pass hepatic biotransformation, 77–78
intra-arterial routes, 79
intramuscular routes, 79–80
intravenous administration, 78–79
nasal delivery, 85–86
ophthalmic routes, 84–85
percutaneous routes, 81–82
pulmonary system, 82–84
rectal route, 78
subcutaneous administration, 80–81
sublingual/buccal administration, 78
Drug Efficacy Study Implementation (DESI) program, 6, 101, 786
Drug Price Competition and Patent Term Restoration Act of 1984, *see* Hatch–Waxman Act (HWA)
Drug regulatory authority, 96, 137–138
Drug Safety and Evaluation Branch (DSEB), 461
Drugs and Cosmetics Act 1988, 462
Dry sieving, 49
DSI, *see* Division of Scientific Investigations (DSI)
Duloxetine hydrochloride, 544
Dutasteride, 544
Dutch Society Against Quackery, 4
Dynamic vapor sorption (DVS), 49

E

Efavirenz, 544
EIR, *see* Establishment inspection report (EIR)
Electronic records/signatures, 258
EMA, *see* European Medicines Agency (EMA)
Emtricitabine, 544
Emulsion formulation, 60, 453
Entacapone, 533, 544
Entecavir, 544–545
Environmental exposure unit (EEU), 304–305
Eplerenone, 545
Eprosartan mesylate, 545
Equivalence documentation
fixed combination products, 97
immediate-release oral pharmaceutical products, 97
non-oral and nonparenteral pharmaceutical products, 97
nonsolution pharmaceutical products, 97
oral drug products, 96
test methods, 96
Erlotinib hydrochloride, 545
Erythropoietins
clinical safety, 375
clinical studies, 372–373
pharmacodynamics studies, 372
routes of administration, 373–374
toxicological studies, 372
Escitalopram oxalate, 545
Esomeprazole magnesium, 545–546
Establishment inspection report (EIR), 259, 263, 269, 271–273, 276, 626
equipment, 639–640
facilities, 639
nonclinical laboratory study, 641–642
organization and personnel, 637–638
QAU, 638–639
reagents and solutions, 641
records and reports, 642–643
test and control articles, 641
testing facility operations, 640
Esterified estrogens, 546
Estradiol, 543
Eszopiclone, 546
Ethambutol hydrochloride, 546
Ethinyl estradiol, 546–547
Etidronate disodium, 547
European Medicines Agency (EMA)
acceptance interval, 444–445
analysis of variance, 427–428, 445
BCS-based biowaivers, 432–433, 441–442, 454–458
bioequivalence testing guidance, 425–427
biosimilar drugs, 424
bracketing approach, 442–443
carryover effects, 445
centralized procedure, 424
data presentation, 445–446
decentralized procedure, 424
dosage strength(s), 432
enantiomers, 440
endogenous substances, 441
Falsified Medicines Directive, 423
fasting and fed conditions, 426, 439
fixed combinations (FCs), 443
generic medicine applications, 424–425, 433–434
good laboratory practice, 443
HVDs/HVDPs, 428–430, 446–447
immediate-release formulation, 425–426, 434
in vitro dissolution test, 447
legal basis, 434–435
linear pharmacokinetics, 442
mutual recognition procedure, 424
narrow therapeutic index drugs, 430–432, 446
national procedures (NPs), 424
nonlinear pharmacokinetics, 442
number of subjects, 437
packaging process, 437

parent compound *vs.* metabolites, 427, 440–441
pharmaceutical directives, 423, 434–435
pharmacokinetic parameters, 427, 439–440, 442, 449
reason for exclusion, 444
reference product, 436
sampling schedule, 439
scope, 434
statistical analysis, 445
strength evaluation, 441
study conduct standardization, 438
study design, 435–436
study report, 448–449
subject accountability, 443
subject population, 437–438
test product, 436–437
"two-stage"/"add-on" design, 426, 445
urinary excretion data, 440–441
variation applications, 448–449
European Medicines Evaluation Agency (EMEA), 790
European pharmaceutical directive
65/65/EEC, 423
75/318/EEC, 423
75/319/EEC, 423
93/41/EEC, 423
Falsified Medicines Directive [2011/62/EU], 423–424
Everted intestinal sac method, 135
Excipients, 172, 457, 595
Exemestane, 547
Extended-release drug products
nonrelease controlling excipients, 595–596
release controlling excipients, 596
U.S. FDA, 415

F

Falsified Medicines Directive, 423
Famotidine, 547–548
FDA, *see* Food and Drug Administration (FDA)
FDCA, *see* Federal Food, Drug, and Cosmetic Act (FDCA)
Fed bioequivalence studies, 145
administration, 148
fasted treatments, 280
fed treatments, 280–281
data analysis and labeling, 148, 281–282
dosage strength, 147, 280
general design, 147
immediate-release dosage forms, 146
modified-release drug products, 146
ANDAs, 279
INDs/NDAs, 279
sample collection, 148, 281
special vehicles, 149
sprinkles, 149
study design, 147, 279
subject selection, 147, 280
test meal, 147, 280
U.S. FDA, 412, 420
Federal Agency for Medicines and Health Products (FAMHP), Belgium, 431
Federal Food, Drug, and Cosmetic Act (FDCA), 5, 95, 318, 323, 587
Felbamate, 548
Fenofibrate, 548–549
Fexofenadine hydrochloride, 549
Finished dosage forms (FDFs), 755, 757
First-pass hepatic biotransformation, 77–78
Fixed combinations (FCs), 443, 452
Flavoxate hydrochloride, 549
Fluconazole, 549
Fluoxetine hydrochloride, 549–550
Fluvastatin sodium, 550
Follow-up BA/BE safety report, 405
Food and Drug Administration (FDA), 95, 245, 285, 291
capsules, 57
debarment (*see* Debarment)
IVIVC definition, 152
powder dosage forms, 56
Food and Drug Administration Amendments Act of 2007 (FDAAA), 515–516
Food and Drug Administration Modernization Act of 1997 (FDAMA), 515
Food-effect bioavailability studies, 145
administration
fasted treatments, 280
fed treatments, 280–281
data analysis and labeling, 281–282
dosage strength, 280
drug product and substances, 277–278
immediate release drug products
ANDAs, 278–279
INDs/NDAs, 278
mechanisms, 277
modified-release drug products
ANDAs, 279
INDs/NDAs, 279
sample collection, 281
study designs, 279
subject selection, 280
test meal, 280
Fosamprenavir calcium, 550
Fosinopril sodium, 550–551
Four-period, two-sequence, two-formulation BE design, 203–204
Four-sequence, four-period design, 226
Freedom of Information Act, 626

G

Gabapentin, 551
Galantamine hydrobromide, 551–552
Ganciclovir, 552
GBIB, *see* Good Laboratory Practice and Bioequivalence Investigations Branch (GBIB)
GCP, *see* Good clinical practice (GCP)
Gemifloxacin mesylate, 552
General Directorate of Drug and Pharmacy (GDDP), 465
Generic Drug Enforcement Act (GDEA), 758, 762–763
Generic drugs
vs. branded pharmaceuticals, 2–4, 10
FDA's approval process, 8–9
legal challenges, 9
life expectancy, critical analysis of, 1
quack medicines, 3–5
submarine patenting, 9–10
Generic Drug User Fee Act of 2012 (GDUFA), 757
Generic Pharmaceutical Association (GPhA), 6
Glimepiride, 552–553

Glipizide, 553
Global regulatory agencies and organizations, 469–471, 785
  bioequivalence assessment
    clinical endpoint studies/comparative clinical trials, 473
    in vitro endpoint studies, 473
    PD endpoint studies, 472
    PK endpoint studies, 471–472
GLP, *see* Good laboratory practice (GLP)
Glyburide, 553
GMP, *see* Good manufacturing practice (GMP)
Good clinical practice (GCP)
  clinical/non clinical biomedical research, 703–704
  clinical trial and amendments protocol
    amendments, 723
    background information, 723
    data handling and record keeping, 725
    efficacy assessment, 724
    ethics, 725
    financing and insurance, 725
    general information, 723
    objectives and purpose, 723
    publication policy, 725
    quality control and assurance, 725
    safety assessment, 724
    source data/documents, 725
    statistics, 724–725
    subject selection and withdrawal, 724
    supplements, 725
    treatments, 724
    trial design, 723–724
  essential documents
    after completion/termination, 732–733
    during clinical file, 731–732
    clinical phase, before, 730–731
    master files, 730
  IB
    confidentiality statement, 726
    data and guidance, 729
    humans effects, 728–729
    nonclinical studies, 727–728
    pharmaceutical properties and formulation, 727
    physical and chemical properties, 727
    table of contents, 726
    title page, 726
  ICH principles, 704–705
  investigator reports
    adequate resources, 707
    final reports, 713
    informed consent, 709–712
    investigational product(s), 709
    IRB/IEC communication, 708
    premature termination/suspension, 713
    progress reports, 713
    protocol compliance, 708
    qualifications and agreements, 707
    randomization procedures and unblinding, 709
    records and reports, 712
    safety reports, 713
    trial subjects, medical care, 708
  IRB/IEC
    adequate resources, 704
    composition, functions, and operations, 706
    procedures, 706–707
    records, 707
    responsibilities, 705–706
  sponsor
    adverse drug reaction reports, 719
    allocation of responsibilities, 716
    audits, 721
    clinical trial/study reports, 722
    coding investigational products, 717–718
    compensation, 716
    CRO, 714
    electronic trial data handling, 714–715
    financial aspects, 716
    investigational products, 717
    investigator selection, 716
    IRB/IEC conditions, 717
    manufacturing, packaging, labeling products, 717–718
    medical expertise, 714
    monitors, 719–721
    multicenter trials, 722
    noncompliance, 722
    notification/submission, 716–717
    premature termination/suspension, 722
    protocol design, 714
    quality assurance and quality control, 714
    record access, 718
    record retention, 714–715
    regulatory requirements, 716
    safety information, 718–719
    supplying and handling investigational products, 718
    trial management, 714–715
Good laboratory practice (GLP), 443
  acting/deputy study director, 626–627
  agency investigators, 626
  conforming amendment statement, 659
  equipment, 632, 652
  facilities, audit of, 631–632
    administrative and personnel facilities, 652
    animal care and supply facilities, 651
    computerized systems, 644
    data audit, 643–644
    directed inspections, 636
    EIR (*see* Establishment inspection report (EIR))
    handling test and control articles, 651
    laboratory operation areas, 652
    master schedule sheet, 636
    organization chart, 636
    personnel, 644–645
    specimen and data storage facilities, 652
    study identification, 636
    surveillance inspections, 636
  FDA-483 lists, 625–626
  laboratory's computer specialists, 628
  nonclinical laboratory (*see* Nonclinical laboratory study)
  organization and personnel
    QAU, 630–631, 650–651
    study director, 630, 650
    testing facility management, 630, 650
  quality assurance unit functions, 627
  records and reports, 634–635, 657–659
  regulations
    computerized operations, 270–271
    data audit, 275
    directed inspections, 268–269

EIR documentation and reporting, 271–273
equipment, 272
facility floor-plan diagram, 269, 271–272
final report *vs.* raw data, 275
inspectional observations, 276
key laboratory and management personnel, 270
management involvement, 269–270
master schedule sheet, 269
nonclinical laboratory study, 274
objective, 268
organization and personnel, 269
protocol, 274
protocol *vs.* final report, 275
QAU operations, 271, 275
reagents and solutions, 273
records and reports, 274
sample collection, 276
SOPs, 272–273
storage and retrieval data, 274–275
study director, 270
surveillance inspections, 268–269
test and control articles, 273–274
requirements, 628
test and control articles, 625, 633, 654–655
testing facilities operation
animal care, 653–654
reagents and solutions, 633, 653
SOP, 627–628, 632, 652–653
Good Laboratory Practice and Bioequivalence Investigations Branch (GBIB), 249, 254–255, 263
Good manufacturing practice (GMP), 705
Granisetron hydrochloride, 553

H

Hatch–Waxman Act (HWA), 6–7
Hatch–Waxman law, 785
Hazard analysis and critical control point (HACCP), 319
Health Products and Food Branch (HPFB), Canada, 431, 462
Henderson–Hasselbalch equation, 14–15
Highly variable drugs/drug products (HVDs/HVDPs), 428–430, 446–447, 490–491
HIV protease inhibitors, 27
Human follicle-stimulating hormone (hFSH)
clinical efficacy, 359
clinical safety, 359–360
pharmacodynamic studies, 359
pharmacokinetic studies, 358
regulating reproductive function, 357
r-hFSH, 357
toxicological studies, 358
Human granulocyte-colony-stimulating factor (Human G-CSF)
clinical efficacy studies, 382–383
clinical safety, 383
nonclinical studies, 381
pharmacodynamic studies, 382
pharmacokinetic studies, 382
toxicological studies, 382
Human growth hormone (hGH)
clinical efficacy studies, 384–385
clinical safety, 385
nonclinical studies, 383–384
pharmacodynamic studies, 384
pharmacokinetic studies, 384
toxicological studies, 384
Human insulin
clinical efficacy studies, 387
clinical safety, 387
nonclinical studies, 386
pharmacodynamic studies, 386–387
pharmacokinetic studies, 386
toxicological studies, 386
HVDs/HVDPs, *see* Highly variable drugs/drug products (HVDs/HVDPs)
Hydrates, 29–31
Hydrochlorothiazide, 531–532, 545, 551, 553–554, 558–559
Hygroscopicity, 33

I

Ibandronate sodium, 554–555
Ibuprofen, 555
Immediate release drug products, 594–595
food-effect bioavailability studies
ANDAs, 278–279
INDs/NDAs, 278
U.S. FDA, 413–414, 418–419
Incurred sample reanalysis (ISR), 695
Independent data-monitoring committee (IDMC), 715
Indinavir sulfate, 555
Individual bioequivalence
confidence intervals
constrained REML, 222–224
method of moments, 223–224
data analysis
linearized criteria, 219–220
statistical model, 218–219
95% upper confidence bound, 220–222
variance estimation, 222
mixed-scaling approach, 200–201
sample sizes, 206
standards, 202–203
Individual difference ratio (IDR), 202
INDs, *see* Investigational new drugs (INDs)
Informed consent, 709–711, 775
Inhalers, 56
In situ bioequivalence system, 38, 136
Institutional Review Board (IRB), 255, 259–261
Institutional Review Board/Independent Ethics Committee (IRB/IEC), 717
adequate resources, 704
composition, functions, and operations, 706
investigator reports, 708
procedures, 706–707
records, 707
responsibilities, 705–706
Interchangeability generic product, 466–468
Interchangeable pharmaceutical products, 791–792
definition, 347–348
study design, 348
Interferon-α (2a/2b)
clinical studies, 380
efficacy, 380
extrapolation, 381
immunogenicity, 381
nonclinical studies, 379
pharmacodynamics studies, 379

primary and secondary treatment, 380
safety, 380–381
toxicological studies, 379
Interferon beta (IFN-β)
clinical efficacy, 361–362
clinical safety, 363
clinical studies, 361
pharmacodynamics, 361
pharmacokinetics, 361
International normalized ratio (INR), 132
Intra-arterial drug administration, 79
Intramuscular drug administration, 79–80
Intravenous drug administration, 78–79
Investigational new drugs (INDs), 683, 766
BCS, 177–178
food-effect bioavailability studies
immediate-release drug products, 278
modified-release dosage forms, 279
in vivo bioavailability/bioequivalence studies, 389, 404–405
nasal product documentation, 296
systemic exposure studies, 294, 305
Investigators brochure (IB)
confidentiality statement, 726
data and guidance, 729
humans effects, 728–729
nonclinical studies, 727–728
pharmaceutical properties and formulation, 727
physical and chemical properties, 727
table of contents, 726
title page, 726
In vitro bioavailability/bioequivalence studies
dissolution testing, 134–135
dosage forms, disintegration of, 133–134
everted intestinal sac method, 135
isolated perfused liver, 135
nasal products
ANDAs, 298
dose/spray content uniformity, 299
droplet and drug particle size distribution, 299–301
INDs/NDAs, 297–298
plume geometry, 302
priming and repriming, 302
profile comparison, 313–314
spray pattern, 301
statistical analyses, 307–311
tail-off profile, 302–303
In vitro dissolution testing, 107, 416, 418, 447
bioequivalence surrogate inference, 450
product quality, 449
profile similarity, 450–451
In vitro–in vivo correlation (IVIVC), 413
BDM, 175
biodegradable parenteral delivery systems, 175
buccal tablet, 174
colonic drug delivery system, 175
definitions, 152
development, 159–161
driving forces, 151
failure reasons, 173
levels, 153–154
models
convolution model, 155–157
deconvolution technique, 157
differential equation–based model, 157–158
external predictability, 158–159
internal predictability, 158
modified-release dosage forms
dissolution specifications, 176
drug candidate selection, 176
in vitro dissolution, 176
mapping process, 176–177
multiple-unit enteric-coated granules, 174
nasal drug delivery systems, 174
parenteral administration, 174, 177
purpose, 152–153
suppositories, 174
transdermal drug delivery systems, 174
In vivo bioavailability/bioequivalence studies
absorption profiling, 137
acute pharmacological effect, 403
analytical methods, 403–404
accuracy, 112
analyte stability, 112
analytical system stability, 113
concentration range and linearity, 111
CVM reviewer, 111
limit of detection, 112
limit of quantitation, 112
precision, 112
prior review, 113–114
quality control samples, 113
record maintenance, 114–115
replicate and repeat analyses, 113
specificity, 112
animal species, choice of, 136
AUC
computer applications, 119
integration method, 118
multiple-dose study, 117–118
single-dose study, 117
trapezoidal rule, 118
blood level study, 105
clinical endpoint study, 105, 110–111
Commissioner of Food and Drugs, 113–114
design criteria, 399
drugs, 103
evidence, 398–399, 406–407
extended-release formulation, 400
fasted states study, 109–110
FDA, 97, 101–102
fed study, 109–110
guidelines, 399–401
multiple-dose design, 402–403
single-dose design, 401–402
health risk concept, 122–125
hepatobiliary cannulation, 136
IND application, 389, 404–405
inquiries and protocol, 404
in vivo and in vitro approaches, 106
LD50 comparisons, 136
measuring methods, 397–399
multiple-dose study, 108–109
pharmacological endpoint study, 105, 110
pharmacotherapeutic evidence, 103
physicochemical evidence, 103–104
plasma concentration profiles, 104–105
product's rate and extent of absorption, 105

rate of absorption, 119
single-dose study, 108–109
statistical analysis
log-transformed data, 140–141
untransformed data, 139–140
study design
blood level pivotal parameters, 117
one-period parallel design, 116–117
protein binding, 116
sample size, 116
sampling times, 115–116
two-period crossover design, 116
washout period, 117
submission of evidence, 395–397
supplemental application, 97
thiry–vella loop, 136
waivers, 142
Ionization, 14–15
IRB, *see* Institutional Review Board (IRB)
IRB/IEC, *see* Institutional Review Board/Independent Ethics Committee (IRB/IEC)
Irbcsartan, 553–555
IS, *see* Internal standard (IS)
Isolated perfused liver, 135
Isosorbide mononitrate, 555
ISR, *see* Incurred sample reanalysis (ISR)
Isradipine, 556
Itraconazole, 556
IVIVC, *see* In vitro–in vivo correlation (IVIVC)

J

Japanese bioequivalence study guidance
acceptance criteria, 500, 509
acidic drugs, 501
assessment, 508
bioequivalence range, 500
clinical study, 501, 508
coated products, 502
dissolution equivalence, 509–511
drug administration, 495, 499
enteric-coated products, 503–506
lag time, 512
neutral/basic drugs, 502
non-oral dosage form, 511–512
number of subjects, 494–495
parallel design, 494
parameters, 499–500
pharmacodynamic study, 500–501, 508
reference product, 494, 507–508
samples, measurement, 499
soluble drugs, 502
statistical analysis, 500
subject selection, 495–498
terminology, 493
test condition, 508–509
testing time, 508
test product, 507–508
test results, 506–507, 511
time point testing, 512
washout periods, 499
Just another Gibbs sampler (JAGS), 735

K

Kefauver–Harris Drug Amendments (KHDA), 5–6

L

Lamivudine, 523–524, 556–557
Lamotrigine, 557
Lanoxin, 131
Latin square design, 197
Leflunomide, 557
Levodopa, 533
Levonorgestrel, 546–547, 557
Levothroid, 131
Levothyroxine sodium tablets, 131–132
Lidocaine, 557–558
Limit of detection (LOD), 671–672
Linezolid, 558
Liothyronine sodium, 558
Lipophilizing modifications, 19–20
Liposomal dosage forms, 453
Lisinopril, 558–559
LLOQ, *see* Lower limit of quantification (LLOQ)
LMWH, *see* Low molecular weight heparins (LMWH)
Local delivery bioavailability
clinical studies
batches, 303
EEU, 304–305
endpoints, 303
outdoor allergen study, 304
traditional treatment, 304
solution formulations, 294
suspension formulation products, 294
Locally acting gastrointestinal drugs, 144–145
Locally acting nasal aerosols and sprays, *see* Nasal products
Locally applied, locally acting medicinal products, 454
LOD, *see* Limit of detection (LOD)
Lopinavir, 559
Loratadine, 559
Losartan potassium, 554, 559
Lower limit of quantification (LLOQ), 671
Low molecular weight heparins (LMWH)
anti-FXa and anti-FIIa activity, 375
biological origin and manufacturing process, 375
clinical studies, 377–378
pharmacodynamic studies, 376
toxicological studies, 376

M

Magnesium hydroxide, 567
Manual of Policies and Procedures (MAPP), 245
Markov chain Monte Carlo (MCMC) methods, 735
Medicines and Related Substances Control Act (MRSCA), 463
Medicines Control Council (MCC), 463
Mefloquine hydrochloride, 559
Meloxicam, 559–560
Mercaptopurine, 560
Mesalamine, 560–561
Metaxalone, 561
Metered dose inhaler (MDI) formulation, 51, 56, 83
Metformin hydrochloride, 553, 561

Metoprolol succinate, 562
Micellar dosage forms, IV administration, 453
Micronization, 49
Miglustat, 562
Millipore MultiScreen Caco-2 assay system, 188
Minister of Health, Labour and Welfare (MHLW), 463
Ministry of Health (MoH), 461, 465, 493, 599, 755
Minocycline hydrochloride, 562
Mirtazapine, 562
Misoprostol, 539
Mixed-scaling approach
  discontinuity, 228
  individual bioequivalence, 200–201
  population bioequivalence, 199–200
Modafinil, 563
Modified-release drug products, 56, 454
  food-effect bioavailability studies
    ANDAs, 279
    INDs/NDAs, 279
  U.S. FDA, 415–416, 419
Moexipril hydrochloride, 563
MoH, *see* Ministry of Health (MoH)
Molecular size, 38–40
Monitoring methods, CRO
  centralized monitoring, 774
  clinical investigation, 770–771
  clinical trial, 782
  CPGMs, 772
  CRFs, 775–776
  critical data and processes, 776–777
  CTTI, 771
  definition, 771
  developmental factors, 778
  documentation, 781
  electronic data capture, 779
  factors, 778
  ISO standards, 772
  on-site monitoring, 773–774
  planning scheme
    communicating monitoring result, 780
    monitoring approach, 779–780
    noncompliance management, 780
    plan amendment, 781
    quality monitoring, 780–781
  quality risk management approach, 771
  rationale, 772–773
  risk assessment, 777–778
  selection, 784
  study quality
    clinical investigator training and communication, 782
    IND study delegation, 782
    protocol and CRF design, 781
    site selection and initiation activities, 782
  study-specific monitoring plan, 773
Montelukast sodium, 563
Morphine sulfate, 563–564
Multiple-dose in vivo bioavailability study, 402–403
Multiple dosing levels, 790
Multisource drug products, 466
Mutual recognition procedure (MRP), 424
Mycophenolate mofetil, 564
Mycophenolate mofetil hydrochloride, 564

N

Nabumetone, 564
Narrow therapeutic index drugs (NTIDs), 103, 430–432, 446, 491–492
Nasal products
  bioequivalence studies, decision tree for, 295
  CMC tests, 294–295
  container and closure system, 295–296
  container sizes, 312
  documentation
    ANDAs, 296–297
    INDs/NDAs, 296
    postapproval change, 297
  drug delivery factors, 85–86
  formulation, 295
  in vitro studies
    ANDAs, 298
    dose/spray content uniformity, 299
    droplet and drug particle size distribution, 299–301
    INDs/NDAs, 297–298
    plume geometry, 302
    priming and repriming, 302
    profile comparison, 313–314
    spray pattern, 301
    statistical analyses, 307–311
    tail-off profile, 302–303
  local delivery bioavailability, 293–294, 303–305
  solution formulation nasal sprays, 311–312
  suspension formulation nasal sprays, 312
  systemic exposure and absorption, 294
    clinical studies, 305–307
    PK study, 305
Nateglinide, 564–565
National Center for Complementary and Alternative Medicine (NCCAM)
  botanical products, 319, 324
  functional food, 319
  manipulative and body-based practices, 320, 323–325
  mind-body medicine, 320, 323–325
  probiotics, 319, 324
  veritable and putative energy fields, 320, 324–325
  whole medical systems, 320–321
National Formulary (NF) revision committees, 95
National Policy for Drug Products (1998), Brazil, 461
NCCAM, *see* National Center for Complementary and Alternative Medicine (NCCAM)
NDAs, *see* New drug applications (NDAs)
Nelfinavir mesylate, 565
Nevirapine, 565–566
New animal drug application (NADA), 95
New drug applications (NDAs), 95, 293, 302–303, 306, 312, 389, 683
  BCS, 177–178
  food-effect bioavailability studies
    immediate-release drug products, 278
    modified-release dosage forms, 279
    special vehicles, 283
    sprinkles, 282
    test meal, 280
  in vitro bioavailability/bioequivalence studies, 297–298
  nasal products documentation, 296
  PK studies, 294, 305

Nonclinical laboratory study
calibration, frequency of, 627
conduct of, 634, 657
master schedule sheet, 627
protocol, 633–634, 656–657
raw data, 628
reagents and chemicals, 628
reporting of, 634–635, 657–658
study director, 630
testing facility management, 630
NONMEM software package, 155
Non-oral immediate-release dosage forms, 452–453
Nonreplicated crossover design, 203
average bioequivalence, 212–213
population bioequivalence, 218
unequal carryover effects, 225–226
Notice of Opportunity for Hearing (NOOH), 101
Noyes–Whitney model, 42–44, 50
NTIDs, *see* Narrow therapeutic index drugs (NTIDs)

O

Office of Generic Drugs (OGD), 245–246, 249–252, 522
Off-the-shelf (OTS) software, 741–744, 746, 748, 752–753
Olanzapine, 549–550, 566
Olmesartan medoxomil, 554, 566
Olsalazine sodium, 566
Omeprazole, 566–567
Omeprazole magnesium, 567
Ondansetron, 567
Ondansetron hydrochloride, 567
Onetime testing, 791
Ophthalmic drug administration, 84–85
Oral powders, 55
Oral solutions, 413, 452
Orodispersible tablets (ODTs), 451–452
Orphan Drug Act (ODA), 6
Outlier data, 227–228
Outsourcing
active pharmaceutical ingredients, 755, 757
clinical quality assurance, 756
contract research organizations, 755–756
debarment (*see* Debarment)
efficiency and cost, 755
finished dosage forms, 755, 757
GDEA, 758
GDUFA, 757
inhalant products, 769–770
multiple studies and shipments retention, 765
reserve samples, 765–766
sampling technique, 764–765
SMOs (*see* Site management organizations (SMOs))
study settings, 766
testing facility, 766
Ovarian hyperstimulation syndrome (OHSS), 357
Oxcarbazepine, 567–568
Oxymorphone hydrochloride, 568–569

P

Paliperidone, 569
Pantoprazole sodium, 569
Parallel artificial membrane permeability analysis (PAMPA), 186–187
Parallel group design, 196
average bioequivalence, 214
population bioequivalence, 218
Parenteral dosage forms, 142, 177, 453
Paricalcitol, 569
Partial validation, 666, 684
Particle size, 33–35, 39, 48–49, 55, 65, 83, 300
Partition coefficient, 15–16, 20
Patent medicines
advertisement of, 3
biosimilar products, 329
definition, 2
quack, 3–5
PDx-IVIVC software program, 156
Percutaneous drug administration, 81–82
Perindopril erbumine, 570
Permeability assay, 185–187
Pharmaceutical alternatives, 449
Pharmaceutical and Medical Devices Agency (PMDA), 463
Pharmaceutical equivalence, 449
Pharmacodynamic (PD) endpoint studies, 472
Pharmacokinetic/pharmacodynamic (PK/PD) modeling
assumptions, 87–88
bioavailability
dose–response curve, 93
PD drug responses, 94
compartment modeling
AUC function, 89
clearance, 89–90
distribution volume, 88
effective concentration, 88
rate of elimination, 87
two-stage approach, 90
deconvolution techniques, 92–93
physiological modeling, 90–91
selection process, 87
systemic exposure models, 91–92
Pharmacokinetic (PK) studies
global regulatory agencies, 471–472
systemic exposure studies
ANDA, 305
INDs/NDAs, 294, 305
suspension formulations, 297
topical dermatological drug products, 292
U.S. FDA
bioanalytical method validation, 411
CI values, 421
data deletion, 421
fed BE studies, 412, 420
pilot studies, 410
pivotal BE studies, 410
predose plasma concentration, 420
rate and extent of absorption, 411–412
replicate crossover study designs, 410
sampling technique, 420
single-dose studies, 411
specific beverages, 412
sprinkle BE studies, 412

statistical information, 421
steady-state studies, 411
study conduct, 419–420
study population, 410–411
Phenytoin, 570–571
Phenytoin sodium, 570–571
Physicochemical basis
amorphous forms, 28
chemical modifications, 19
complexation, 31
crystal morphology, 23–24
dissolution
concentration gradient, 44
constant, 44–45
models, 40–44
testing, 45–46
distribution coefficient, 16–19
hydrates, 29–31
hygroscopicity, 33
ionization, 14–15
lipophilizing modifications, 19–20
molecular size, 38–40
particle size, 33–35
partition coefficient, 15–16
polymorphism, 24–28
salt forms, 21–22
solubility, 35–38
solvates, 29
surfactants, 31–33
Physiological pharmacokinetic modeling, 89–91
Pilocarpine hydrochloride, 571
Pilot scale study, 107–108, 115–117, 159, 248, 287–288, 410, 489, 756
Pimozide, 571–572
Pioglitazone hydrochloride, 561
Polymorphism, 24–28, 54, 495
Population bioequivalence
confidence intervals, 223
data analysis
linearized criteria, 215–216
statistical model, 214–215
95% upper confidence bound, 216–218
mixed-scaling approach, 199–200
nonreplicated crossover design, 218
parallel group design, 218
replicated crossover designs, 218
sample sizes, 205
standards, 201–203
Population difference ratio (PDR), 202
Posaconazole, 572
Postauthorization safety studies (PASS), 364
Powder dosage forms
caking, 54
dispersion problems, 52
electrostaticity, 53
FDA classification, 56
flow properties, 53
milling, 54–55
mixing mechanisms, 52–53
oral, 55
particle size, 55
polymorphism, 54
relative humidity, 54
PROC GLM procedure, 238–241
Product failure, 228
Product-specific BE recommendations
abacavir sulfate, 523–524
acamprosate calcium, 524
acetaminophen, 582
acitretin, 524
acyclovir, 524
alendronate sodium, 524
alfuzosin hydrochloride, 525
almotriptan malate, 525
alosetron hydrochloride, 525
alprazolam, 525
amlodipine besylate, 526
amoxicillin, 526–527
amprenavir, 527
anagrelide hydrochloride, 527
anastrozole, 527
aprepitant, 527–528
aripiprazole, 528–529
armodafinil, 529
atazanavir sulfate, 529
atomoxetine hydrochloride, 529
atorvastatin calcium, 530
atovaquone, 530
azithromycin, 530
balsalazide disodium, 530–531
benazepril hydrochloride, 526
benzonatate, 531
benzphetamine hydrochloride, 531
bicalutamide, 531
bisoprolol fumarate, 531
bupropion hydrochloride, 531–532
calcium carbonate, 574
candesartan cilexetil, 532
carbamazepine, 532–533
carbidopa, 533
carvedilol, 534
cefdinir, 534
cefditoren pivoxil, 534
cefixime, 534
celecoxib, 534
cetirizine hydrochloride, 534–535
cevimeline hydrochloride, 535
cilostazol, 535
cinacalcet hydrochloride, 535
ciprofloxacin, 536
ciprofloxacin hydrochloride, 535–536
clarithromycin, 536
clavulanate potassium, 526–527
clonidine hydrochloride, 536–537
clopidogrel bisulfate, 537
danazol, 537
dantrolene sodium, 537
darifenacin hydrobromide, 537
darunavir ethanolate, 537
deferasirox, 537–538
delavirdine mesylate, 538
desloratadine, 538
dexmethylphenidate, 538
dextromethorphan polistirex, 538–539
diclofenac sodium, 539
dicloxacillin sodium, 539
didanosine, 539

digoxin, 539–540
diltiazem hydrochloride, 540–541
dipyridamole, 541
divalproex sodium, 541–542
dofetilide, 542
donepezil hydrochloride, 542–543
doxazosin mesylate, 543
doxycycline, 543
drospirenone, 543
duloxetine hydrochloride, 544
dutasteride, 544
efavirenz, 544
emtricitabine, 544
entacapone, 533, 544
entecavir, 544–545
eplerenone, 545
eprosartan mesylate, 545
erlotinib hydrochloride, 545
escitalopram oxalate, 545
esomeprazole magnesium, 545–546
esterified estrogens, 546
estradiol, 543
eszopiclone, 546
ethambutol hydrochloride, 546
ethinyl estradiol, 546–547
etidronate disodium, 547
exemestane, 547
famotidine, 547–548
felbamate, 548
fenofibrate, 548–549
fexofenadine hydrochloride, 549
flavoxate hydrochloride, 549
fluconazole, 549
fluoxetine hydrochloride, 549–550
fluvastatin sodium, 550
fosamprenavir calcium, 550
fosinopril sodium, 550–551
gabapentin, 551
galantamine hydrobromide, 551–552
ganciclovir, 552
gemifloxacin mesylate, 552
glimepiride, 552–553
glipizide, 553
glyburide, 553
granisetron hydrochloride, 553
hydrochlorothiazide, 531–532, 545, 551, 553–554, 558–559
ibandronate sodium, 554–555
ibuprofen, 555
indinavir sulfate, 555
irbesartan, 553–555
isosorbide mononitrate, 555
isradipine, 556
itraconazole, 556
lamivudine, 523–524, 556–557
lamotrigine, 557
leflunomide, 557
levodopa, 533
levonorgestrel, 546–547, 557
lidocaine, 557–558
linezolid, 558
liothyronine sodium, 558
lisinopril, 558–559
lopinavir, 559
loratadine, 559
losartan potassium, 554, 559
magnesium hydroxide, 567
mefloquine hydrochloride, 559
meloxicam, 559–560
mercaptopurine, 560
mesalamine, 560–561
metaxalone, 561
metformin hydrochloride, 553, 561
metoprolol succinate, 562
miglustat, 562
minocycline hydrochloride, 562
mirtazapine, 562
misoprostol, 539
modafinil, 563
moexipril hydrochloride, 563
montelukast sodium, 563
morphine sulfate, 563–564
mycophenolate mofetil, 564
mycophenolate mofetil hydrochloride, 564
nabumetone, 564
nateglinide, 564–565
nelfinavir mesylate, 565
nevirapine, 565–566
olanzapine, 549–550, 566
olmesartan medoxomil, 554, 566
olsalazine sodium, 566
omeprazole, 566–567
omeprazole magnesium, 567
ondansetron, 567
ondansetron hydrochloride, 567
oxcarbazepine, 567–568
oxymorphone hydrochloride, 568–569
paliperidone, 569
pantoprazole sodium, 569
paricalcitol, 569
perindopril erbumine, 570
phenytoin, 570–571
phenytoin sodium, 570–571
pilocarpine hydrochloride, 571
pimozide, 571–572
pioglitazone hydrochloride, 561
posaconazole, 572
pseudoephedrine hydrochloride, 535, 555
quetiapine fumarate, 572
quinapril hydrochloride, 572–573
quinine sulfate, 573
raloxifene hydrochloride, 573
ramipril, 573
ribavirin, 573–574
rifampin, 574
riluzole, 574
risedronate sodium, 574
risperidone, 574
ritonavir, 559, 574
rizatriptan benzoate, 575
rosiglitazone maleate, 552–553, 575
rosuvastatin calcium, 575
saquinavir mesylate, 575
sertraline hydrochloride, 575
sibutramine hydrochloride, 575–576
sildenafil citrate, 576
simvastatin, 576
sirolimus, 576

sodium bicarbonate, 567
solifenacin succinate, 576–577
stavudine, 577
sulfamethoxazole, 577
sumatriptan succinate, 577
tacrolimus, 577
tadalafil, 577
tamsulosin hydrochloride, 577
telithromycin, 578
telmisartan, 578
tenofovir disoproxil fumarate, 578
terazosin hydrochloride, 578
terbinafine hydrochloride, 578
testosterone, 578–579
ticlopidine hydrochloride, 579
tinidazole, 579
tipranavir, 579
tizanidine hydrochloride, 579–580
tolterodine tartrate, 580
topiramate, 580–581
topiramate sprinkle, 580
torsemide, 581
tramadol, 581
tramadol hydrochloride, 582
trandolapril, 582
triamterene, 582
trimethoprim, 577
trospium chloride, 582
valacyclovir hydrochloride, 582–583
valsartan, 554, 583
vardenafil hydrochloride, 583
varenicline tartrate, 583
venlafaxine hydrochloride, 583–584
verapamil hydrochloride, 584
voriconazole, 584
zafirlukast, 585
zalcitabine, 585
zaleplon, 585
zidovudine, 523–524, 556–557, 585
zileuton, 585
ziprasidone, 585–586
zolmitriptan, 586
zolpidem, 586
zonisamide, 586
Proprietary medicines, *see* Patent medicines
Pseudoephedrine, 555
Pseudoephedrine hydrochloride, 535
Public Health Service (PHS) Act, 5, 318, 323
Pulmonary drug administration, 82–84
Pure Drug and Cosmetic Act (PDCA), 5

Q

Quack medicines, 3–5
Quality assurance unit (QAU)
establishment inspections, 638–639
organization and personnel, 630–631
Quasi-quantitative assay, 663
Quetiapine fumarate, 572
Quinapril hydrochloride, 572–573
Quinine sulfate, 573

R

Radiometric analyses, 267
Raloxifene hydrochloride, 573
Ramipril, 573
Rectal drug administration, 68, 78
Reference-listed drug (RLD), 95, 394, 409–410, 412–416, 420
Reference member state (RMS), 462
Reference product, 138, 298
Regulatory acceptance criteria
bioanalysis
analyte selection, 478
components, 474
definition, 474
drug products, 485, 488
enantiomers *vs.* racemate, 485
parent drug *vs.* metabolite(s), 478, 485
bioequivalence study design
Add-On Criteria, 473, 485
demographics, 473, 478
fasting and fed study requirements, 473, 483
fluid intake, and posture and physical activity, 473, 484
immediate and modified-release formulations, 473, 481–482
linear and nonlinear pharmacokinetics, 473, 476–477
sample size, 473, 479–480
specific information agency, 473–474
HVD, 490–491
logarithm transformed data, 490
NTID, 491–492
statistical approaches
90% confidence interval, 489
75/75 rule, 489
study power, 489
Regulatory inspection process
analytical data and operations (*see* Analytical laboratory inspection)
clinical data and operations (*see* Clinical facilities)
COMIS database
controlled correspondence, 248
dissolution data (DIS), 247
dissolution waiver (DIW), 247
DSI, 247–248
other (OTH), 247–248
protocols, 248
study amendment (STA), 247
study types, 247
waiver (WAI), 247
Compliance Program audit
analytical testing noncompliance, 256
clinical/analytical facility, 253, 256–257
clinical testing noncompliance, 256
directed data audit, 254
GBIB assignment memos, 254
inspectional operations, 254–255
routine data audit, 254
electronic records/signatures, 258
GLP regulations (*see* Good laboratory practice (GLP))

MAPP procedures, 245–246, 249
methods validation
ANDA approval, 249–250
policy, 250–251
procedures, 251–252
reviewer assignment, 248–249
Relative quantitative assay, 663
Release-controlling excipients, 596–597
Replicated crossover designs
average bioequivalence, 213–214
four-period, two-sequence, two-formulation design, 203–204
population bioequivalence, 218
three-period design, 203–204
unequal carryover effects, 225–226
Ribavirin, 573–574
Rifampin, 574
Riluzole, 574
Risedronate sodium, 574
Risk-based bioequivalence, 122–123
Risperidone, 574
Ritonavir, 559, 574
Rizatriptan benzoate, 575
RLD, *see* Reference-listed drug (RLD)
Rosiglitazone maleate, 552–553, 575
Rosin–Rammler–Sperling–Weibull equation, 42
Rosuvastatin calcium, 575
Routine inspection, 249, 632

S

Salt forms, 21–23, 34, 60, 82
Saquinavir mesylate, 575
SAS GLM procedure, 237–241
Scaled average bioequivalence (SABE), 428
Scale-up and postapproval changes (SUPAC), 135, 151, 153, 177, 785
Screening record
adverse events, 611, 614
alcohol screen, 604
blood glucose strip, 609, 613
blood sampling form, 610, 613
clinical record, 608
drug administration form, 609, 612
drugs of abuse, 605
ECG examination, 606
follow-up record, 616
inclusion and exclusion criteria, 606–607
medical history, 603–604
physical examination, 605
subject evaluation form, 615
subject's identification data, 601–602
vital signs measurement form, 608, 612
Sequential design, 225
Sertraline hydrochloride, 575
SFI, *see* Subject-by-formulation interaction (SFI)
Sibutramine hydrochloride, 575–576
Side effect, 19, 80, 103, 327–328, 357, 495
Sieving, 49, 53
Sildenafil citrate, 576
Simvastatin, 576
Single-dose in vivo bioavailability study, 401–402
Sirolimus, 576
Site management organizations (SMOs)
in vitro studies, 769
in vivo studies, 768–769
pharmacodynamic/clinical endpoints, 767–768
study sponsor/drug manufacturer, 767
Site of action requirement, 13, 47, 90, 96, 144, 286, 471, 786
Skin-stripping methodology, 287, 290
SMOs, *see* Site management organizations (SMOs)
Sodium bicarbonate, 567
Software validation
automated process equipment
OTS software, 752–753
quality system software, 750–752
construction/coding, 749
design, 749
IQ/OQ/PQ, 745–746
principles of, 746–748
quality planning, 748
quality system regulations, 744
requirements, 749
software developer testing, 749
user site testing, 749–750
verification and validation, 744–745
Solid dosage forms
bulk density, 51
capsules, 55–57
coated particles, 58
particle size, 48–49
porosity, 50
powder (*see* Powder dosage forms)
real density, 51
sieve analysis, 49
size distribution, 49
solutions, 58–59
surface area, 50
suspensions, 60–61
tablets, 58
tapped density, 51
true density, 51
Solifenacin succinate, 576–577
Solubility, 35–38
Solution formulation nasal sprays, 311–312
Solutions, 58–59
Solvates, 29
Sprinkle bioequivalence studies, 282, 412
Standard operating procedures (SOPs), 627–628, 632, 737–738, 740–742
Statistical evaluation
average bioequivalence (*see* Average bioequivalence (ABE))
constant scaling, 199
EMA, 445
highly variable drugs, 228–237
individual bioequivalence (*see* Individual bioequivalence)
logarithmic transformation, 210–212
more than two formulations, comparisons of
balanced incomplete block design, 197
Latin square design, 197
nasal products
nonprofile analyses, 307–309
profile analyses, 310–311
supportive nonprofile and profile analyses, 309

nonreplicated design, 203
number of subjects, 197–198
population bioequivalence (*see* Population bioequivalence)
reference scaling, 199
regulatory model, 198–199
replicated crossover designs
four-period, two-sequence, two-formulation design, 203–204
three-period design, 203–204
two formulations, comparisons of
crossover designs, 196–197
parallel group design, 196
Statistical modeling errors, 786–788
Statistical Procedures for Bioequivalence Studies Using a Standard Two-Treatment Crossover Design, 195
Stavudine, 577
StudySize 3.0, 204
Subcutaneous drug administration, 62, 80–81, 354, 454
Subject-by-formulation interaction (SFI), 195, 198, 204, 218, 228
Subject identification code, 715, 732–733
Sublingual/buccal drug administration, 78
Submarine patenting, 9–10
Sulfamethoxazole, 577
Sumatriptan succinate, 577
SUPAC, *see* Scale-up and postapproval changes (SUPAC)
Surface reaction model, 41
Surfactants, 31–33
Suspension formulation nasal sprays, 312
Synthroid, 131
Systemic absorption, 294
clinical studies, nasal products
batches, 306
crossover design, 306–307
endpoints, 306
parallel design, 306
PK study, 305
Systemic exposure models, 91–92

T

Tablet dissolution model, 42
Tablets, 58, 64
Tacrolimus, 577
Tadalafil, 577
Tamsulosin hydrochloride, 577
Telithromycin, 578
Telmisartan, 578
Tenofovir disoproxil fumarate, 578
Terazosin hydrochloride, 578
Terbinafine hydrochloride, 578
TEST™, 800
Testosterone, 578–579
Therapeutic Goods Act 1989, 461
Therapeutic Goods Administration (TGA), 461
Therapeutic systems, 62–63; *see also* Dosage forms
Thermodynamic potential
bipolarity, 799
comprehensive approach, 794
dielectric properties, 796–797, 801
dissolution profiles, 794
duration of testing, 800
electrical field, 799
lipophilicity, 799
osmolality, 799
pH, 797–799
physical stress, 800
sink condition, 800
surfactants, 797
temperature, 796
TEST™, 800
thermodynamic equivalence surrogate test, 794–795
Three-period replicate BE design, 203–204
Ticlopidine hydrochloride, 579
Tinidazole, 579
Tipranavir, 579
Tizanidine hydrochloride, 579–580
Tolterodine tartrate, 580
Topical dermatological drug products
comparative clinical trials, 285
DPK approach
analytical methods, 286–287
application, 286
crossover design and subject selection, 288
drug uptake/elimination, 290
metrics and statistics, 291
sample collection, 289
sample pilot study, 287–288
sites and duration, 289
skin-stripping methodology, 287, 290
stratum corneum concentration–time curves, 286
test and reference products, 289
inactive ingredients, safety of, 285
in vivo BE, 285
pharmacodynamic approaches
formulation and manufacturing changes, 292
in vitro release rate, 291–292
systemic exposure studies, 292
Topiramate, 580–581
Topiramate sprinkle, 580
Torsemide, 581
Total nasal symptom scores (TNSS), 303–304
Tramadol, 581
Tramadol hydrochloride, 582
Trandolapril, 582
Triamterene, 582
Trimethoprim, 577
Trospium chloride, 582
Turlington's Balsam of Life, 4
Two one-sided test (TOST) procedure, 195
Two-period replicated crossover designs, 227
Two-sequence, three-period design, 227
Two-way crossover power analysis study, 204–210

U

ULOQ, *see* Upper limit of quantification (ULOQ)
Unexpected adverse drug reaction, 177, 381
United States Pharmacopeia (USP), 28, 95, 152
Upper limit of quantification (ULOQ), 671
U.S. FDA
abbreviated new drug application, 394–395
alcoholic beverages, 417
amendment procedure, 405
BA/BE safety report, 405
batch testing and certification, 407

biopharmaceutics guidelines, 389–390
biosimilar products
analytical methodology, 334–335
animal and clinical study, 331
animal data, 335–336
animal immunogenicity assessment, 336
animal PK and PD measures, 336
biosimilarity, 329–331
clinical data extrapolation, 342
clinical immunogenicity assessment, 338–339
clinical safety and effectiveness data, 339–340
clinical study, 336–337, 340–342
functional assays, 335
human pharmacology data, 337–338
manufacturing processes, 332–333
mechanism of action, 333
postmarketing safety monitoring, 342–343
product specificity, 334
scientific basis, 332
stepwise approach, 333
totality-of-the-evidence approach, 333–334
chewable tablets, 416
Code of Federal Regulations, 389–390
complex drug substances, 417
dissolution testing, 418–419
enantiomers *vs.* racemates, 416–417
endogenous compounds, 418
first point C_{max}, 417
immediate-release products, 413–414
in vitro studies, 407, 413
in vivo studies (*see* In vivo bioavailability/bioequivalence studies)
IVIVC studies, 413
legislative events, 389, 391–394
long half-life drugs, 417
new drug application, 389
oral solutions, 413
parent drug *vs.* metabolites, 416
pharmacodynamic studies, 413
PK studies (*see* Pharmacokinetic (PK) studies)
record maintenance, 407
reserve samples, 407–408
submission of evidence, 395–397
supplemental application, 395
suspensions, 415–416
therapeutic equivalence code classifications, 98–101
USP, *see* United States Pharmacopeia (USP)

V

Valacyclovir hydrochloride, 582–583
Valsartan, 554, 583
Vardenafil hydrochloride, 583
Varenicline tartrate, 583
Venlafaxine hydrochloride, 583–584
Verapamil hydrochloride, 584
Voriconazole, 584

W

Waivers
BCS, 788–789 (*see also* Biopharmaceutical classification system (BCS))
European perspective
bracketing approach, 191
fixed combinations, 191
linear pharmacokinetics, 190
nonlinear pharmacokinetics, 191
strengths, 190
intestinal drug transport
animal model testing, 189
Caco-2 assay, 187–189
log *P* factor, 185
PAMPA assay, 186–187
permeability assay, 185–186
IVIVC (*see* In vitro–in vivo correlation (IVIVC))
risks, 180, 182
supporting data
high permeability, 183
high solubility, 183
rapid and similar dissolution, 184
worldwide requirements, 181
Wald's method, 670
Warfarin sodium tablets, 132–133
96-Well permeability testing method, 185
World Health Organization (WHO), 465, 492–493

Z

Zafirlukast, 585
Zalcitabine, 585
Zaleplon, 585
Zidovudine, 523–524, 556–557, 585
Zileuton, 585
Ziprasidone, 585–586
Zolmitriptan, 586
Zolpidem, 586
Zonisamide, 586